Study Guide and Solutions Manual to Accompany

ORGANIC CHEMISTRY

Stanislaw Skonieczny

University of Toronto

George H. Schmid

University of Toronto

St. Louis Baltimore Boston Carlsbad Chicago Naples New York Philadelphia Portland
London Madrid Mexico City Singapore Sydney Tokyo Toronto Wiesbaden

Mosby
Dedicated to Publishing Excellence

A Times Mirror
Company

Copyright © 1996 by Mosby-Year Book, Inc

All rights reserved. No part of this publication may be reproduced, stored in a retrieval system, or transmitted, in any form or by any means, electronic, mechanical, photocopying, recording, or otherwise, without prior written permission from the publisher.

Permission to photocopy or reproduce solely for internal or personal use is permitted for libraries or other users registered with the Copyright Clearance Center, provided that the base fee of $4.00 per chapter plus $.10 per page is paid directly to the Copyright Clearance Center, 27 Congress Street, Salem, MA 01970. This consent does not extend to other kinds of copying, such as copying for general distribution, for advertising or promotional purposes, for creating new collected works, or for resale.

Printed in the United States of America

Mosby-Year Book, Inc
11830 Westline Industrial Drive
St. Louis, Missouri 63146

International Standard Book Number 0-80167-491-3

96 97 98 99/ 9 8 7 6 5 4 3 2 1

Contents

Chapter 1 1
Chapter 2 27
Chapter 3 51
Chapter 4 71
Chapter 5 Spectroscopy 99
Chapter 6 Stereochemistry 115
Chapter 7 Alkenes 135
Chapter 8 More Addition Reactions of Alkenes 157

Examination 1 187

Chapter 9 Alkynes 191
Chapter 10 Nuclear Magnetic Resonance Spectroscopy 221
Chapter 11 Alcohols 249
Chapter 12 Nucleophilic Substitution and Elimination Reactions 285
Chapter 13 Ethers and Epoxides 315
Chapter 14 Aldehydes and Ketones 343

Examination 2 391

Chapter 15 Carboxylic Acids 397
Chapter 16 Acyl Transfer Reactions 433
Chapter 17 Enols and Enolate Anions 479
Chapter 18 Free Radical Reactions 521
Chapter 19 π Electron Delocalization in Acyclic Compounds and Intermediates 557
Chapter 20 Aromaticity 581
Chapter 21 Chemistry of Benzene and Its Derivatives 601
Chapter 22 Amines 633

Examination 3 677

Chapter 23 Halobenzenes, Phenols, and Quinones 681
Chapter 24 Chemistry of Difunctional Compounds 713
Chapter 25 Carbohydrates 753
Chapter 26 Amino Acids, Polypeptides, Proteins, and Enzymes 787
Chapter 27 Nucleic Acids 819

Examination 4 831

Appendix A Synthesis of Important Functional Groups 835
Appendix B Reactions of Important Functional Groups 859
Appendix C Uses of Important Reagents 887

Answers to Examination 1 891
Answers to Examination 2 893
Answers to Examination 3 899
Answers to Examination 4 903

TO THE STUDENT

This Study Guide and Solutions Manual is a supplement to your textbook that has been prepared to help you study and learn organic chemistry. Each chapter of the Study Guide and Solutions Manual starts with a review of the concepts in the corresponding chapter of the textbook. This is followed by detailed explanations of how the answers to the in-text Exercises and end-of-chapter Additional Exercises are obtained.

Four tests are included in the Study Guide and Solutions Manual to help you evaluate your knowledge of the material in the textbook. The first is found after chapter 8, the second after chapter 14, the third after chapter 22, and the fourth after chapter 27. Each test is designed to be answered in a specific time period, either one or two hours, just like the tests that you will have in your course. To obtain the maximum benefits from these tests, try to simulate the conditions in a real test center. Find a quiet place that will be free of distractions for the length of the test, keep your textbook closed, and stop writing after the prescribed time has expired. Then check your answers with the answers to each of the tests that are found at the end of the book. Remember that the ability to answer the test questions in the time allotted is also a test of your knowledge of the subject material.

The Appendix of the Study Guide and Solutions Manual contains three summaries that are useful for review. They are Synthesis of Important Functional Groups, Reactions of Important Functional Groups, and Uses of Important Reagents.

Stanislaw Skonieczny
George H. Schmid

CHAPTER 1

Concepts

1. **Organic chemistry** is the study of the chemistry of compounds that contain carbon atoms in combination with atoms of other elements such as hydrogen, oxygen, nitrogen, and the halogens (F, Cl, Br, I).
2. The **electron configuration of atoms** is designated by the energies of the electrons associated with a given atom, their location relative to the nucleus, and their spin.
3. Electrons are located around the nuclei of atoms in defined regions of space called **atomic orbitals.** Each orbital can hold zero, one, or two electrons (the **Pauli Exclusion Principle**).
4. **Electronegativity** is a measure of the ability of an atom to attract electrons from other atoms to which it is bonded. The electronegativity difference between two atoms that are bonded together is used to assess the degree of polarity in the bond. On the Pauling scale of electronegativity, a large number signifies a greater affinity for electrons. Values of electronegativity range from 4.0 for fluorine, the most electronegative element, to values below 1 for the alkali metals, the most electropositive elements.
5. The shape of many simple molecules can be predicted using the **valence-shell electron-pair repulsion theory** (written **VSEPR** and pronounced "vesper"). Electron pairs repel one another, whether they are in chemical bonds (bonding pairs) or unshared (lone pairs). Electron pairs assume orientation about an atom to *minimize* repulsions. Thus, sp^3-hybrid carbon atoms assume a tetrahedral geometry with bond angles of 109.5°; sp^2-hybrid carbon atoms are trigonal planar with bond angles of 120°; and sp-hybrid carbon atoms have linear geometry with bond angles of 180°.
6. Electrons in the valence shells of atoms form the **chemical bonds** in molecules. Atoms gain, lose, or share valence electrons to attain electron configurations that are **isoelectronic** with the nearest noble gas (**noble gas configuration**). A hydrogen atom, for example, must gain one or share two electrons to attain the electron configuration of helium, the nearest noble gas. Carbon, nitrogen, oxygen, and fluorine must share eight electrons (the **octet rule**) to attain the configuration of neon.
7. **Ionic bonds** are formed between elements of very different electronegativities. The more electropositive atom transfers one or more electrons to the other atom creating negatively and positively charged ions that attract each other.
8. **Covalent bonds** are formed between atoms by sharing electrons. When the electronegativity between the atoms forming a covalent bond is only slightly different (about 1.8) the bond is polarized. Electrons are attracted to the more electronegative atom forming a partial negative charge. The less electronegative (or more electropositive) atom has a partial positive charge. Polarized bonds have **dipole moments** (bond dipole moments). Vectorial addition of bond dipole moments results in the molecular dipole moment.

9. A **Lewis structure** is a representation of the structure of a compound in which electrons that form a single bond between two atoms are represented by a pair of dots or a single dash. Nonbonded or a lone pair of electrons in a molecule are represented by a pair of dots.

Solutions to the Problems

1.1

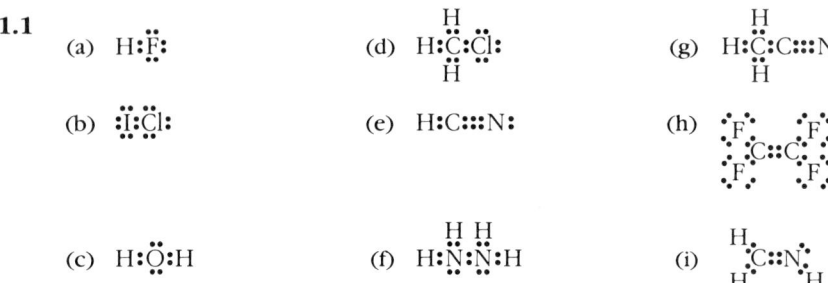

1.2 The structure **A** is an incorrect Lewis structure for CO_2 because the central atom of carbon does not have eight electrons around it. It has only four electrons around it and, thus, the *octet rule* is not fulfilled.

structure **A** (incorrect)   structure **B** (correct)

1.3 (a) Nitric acid:

$$H-\overset{\delta}{\underset{..}{\overset{..}{O}}}-N\overset{\overset{\alpha}{\overset{..}{O}:}}{\underset{\underset{..}{\overset{..}{O}:\beta}}{}}$$

Number of valence electrons at α O:
 (a) in Lewis structure:
  from N—O double bond 2
  from lone pairs   4
          total 6
 (b) in isolated O atom: 6
Difference: (b) - (a) = 6 - 6 = 0
Hence, formal charge = 0

Number of valence electrons at β O:
 (a) in Lewis structure:
  from N—O bond   1
  from lone pairs   6
          total 7
 (b) in isolated O atom: 6
Difference: (b) - (a) = 6 - 7 = -1
Hence, formal charge = -1

Number of valence electrons at N:
   (a) in Lewis structure:
                               from three N-O bonds   4
                                   from lone pairs $\underline{0}$
                                             total   4
   (b) in isolated N atom: 5
Difference: (b) − (a) = 5 − 4 = +1
Hence, formal charge = +1

Number of valence electrons at δ O:
   (a) in Lewis structure:
                                   from N—O bond   1
                                   from H—O bond   1
                                   from lone pairs $\underline{4}$
                                             total   6
   (b) in isolated O atom: 6
Difference: (b) − (a) = 6 − 6 = 0
Hence, formal charge = 0

Net charge on molecule:
                                       Charge on β O   −1
                                         Charge on N   +1
                          Net charge on molecule   0

The Lewis structure of nitric acid is written as:

$$\text{H} - \overset{..}{\underset{..}{\text{O}}} - \overset{+}{\text{N}} \begin{array}{c} \nearrow \overset{..}{\text{O}}: \\ \searrow \underset{..}{\overset{..}{\text{O}}}:^{-} \end{array}$$

(b) Methoxide ion:

$$\text{H} - \underset{\underset{\text{H}}{|}}{\overset{\overset{\text{H}}{|}}{\text{C}}} - \overset{..}{\underset{..}{\text{O}}}:$$

Number of valence electrons at C:
   (a) in Lewis structure:
                                  from three C—H bonds   3
                                  from C—O bond $\underline{1}$
                                          total   4
   (b) in isolated C atom: 4
Difference: (b) − (a) = 4 − 4 = 0
Hence, formal charge = 0

Number of valence electrons at O:
   (a) in Lewis structure:
                                     from C—O bond   1
                                     from lone pairs $\underline{6}$
                                          total   7
   (b) in isolated O atom: 6
Difference: (b) − (a) = 6 − 7 = −1
Hence, formal charge = −1

Net charge on molecule = -1
The Lewis structure of methoxide ion is written as:

$$H-\underset{\underset{H}{|}}{\overset{\overset{H}{|}}{C}}-\ddot{\underset{\cdot\cdot}{O}}\mathbf{:}^-$$

(c) Borohydride ion:

$$H-\underset{\underset{H}{|}}{\overset{\overset{H}{|}}{B}}-H$$

Number of valence electrons at each H:
   (a) in Lewis structure:
                           from B—H bond $\underline{1}$
                                      total   1
   (b) in isolated H atom: 1
Difference: (b) - (a) = 1 - 1 = 0
Hence, formal charge = 0

Number of valence electrons at B:
   (a) in Lewis structure:
                        from four B—H bonds =   4
                             from lone pairs = $\underline{0}$
                                     total =   4
   (b) in isolated B atom: 3
Difference: (b) - (a) = 3 - 4 = -1
Hence, formal charge = -1

Net charge on molecule = -1
The Lewis structure of borohydride ion is written as:

$$H-\underset{\underset{H}{|}}{\overset{\overset{H}{|}}{B}}^{-}-H$$

(d) Diazomethane:

$$\underset{H}{\overset{H}{\diagdown}}C=\overset{\gamma}{N}=\overset{\delta}{N}\overset{\cdot\cdot}{\underset{\cdot\cdot}{:}}$$

Number of valence electrons at each H:
   (a) in Lewis structure:
                            from C—H bond $\underline{1}$
                                      total   1
   (b) in isolated H atom: 1
Difference: (b) - (a) = 1 - 1 = 0
Hence, formal charge = 0

Number of valence electrons at C:
   (a) in Lewis structure:

from two C—H bonds	2
from C—N bond	2
total	4

   (b) in isolated C atom: 4
Difference: (b) - (a) = 4 - 4 = 0
Hence, formal charge = 0

Number of valence electrons at γ N:
   (a) in Lewis structure:

from C—N bonds	2
from N—N bonds	2
from lone pairs	0
total	4

   (b) in isolated N atom: 5
Difference: (b) - (a) = 5 - 4 = +1
Hence, formal charge = +1

Number of valence electrons at δ N:
   (a) in Lewis structure:

from N—N bonds	2
from lone pairs	4
total	6

   (b) in isolated N atom: 5
Difference: (b) - (a) = 5 - 6 = -1
Hence, formal charge = -1

Net charge of molecule = 0
Lewis structure is written as:

$$\begin{array}{c}H\\ \diagdown\\ C=\overset{+}{N}=\overset{..}{\underset{..}{N}}{}^{-}\\ \diagup\\ H\end{array}$$

(e) Methyl nitrite:

$$H-\underset{\underset{H}{|}}{\overset{\overset{H}{|}}{C}}-\overset{\gamma}{\underset{..}{\overset{..}{O}}}-\overset{..}{N}=\overset{\varepsilon}{\underset{..}{\overset{..}{O}}}$$

Number of valence electrons at each H:
   (a) in Lewis structure:

from C—H bond	1
total	1

   (b) in isolated H atom: 1
Difference: (b) - (a) = 1 - 1 = 0
Hence, formal charge = 0

Number of valence electrons at C:
   (a) in Lewis structure:

from three C—H bonds	3
from C—O bond	1
total	4

   (b) in isolated C atom: 4
Difference: (b) − (a) = 4 − 4 = 0
Hence, formal charge = 0

Number of valence electrons at γ O:
   (a) in Lewis structure:

from C—O bond	1
from N—O bond	1
from lone pairs	4
total	6

   (b) in isolated O atom: 6
Difference: (b) − (a) = 6 − 6 = 0
Hence, formal charge = 0

Number of valence electrons at N:
   (a) in Lewis structure:

from N—O bond	1
from N=O bond	2
from lone pairs	2
total	5

   (b) in isolated N atom: 5
Difference: (b) − (a) = 5 − 5 = 0
Hence, formal charge = 0

Number of valence electrons at ε O:
   (a) in Lewis structure:

from N—O bond	2
from lone pairs	4
total	6

   (b) in isolated O atom: 6
Difference: (b) − (a) = 6 − 6 = 0
Hence, formal charge = 0

Net charge on molecule = 0
Lewis structure of methyl nitrite is written as:

$$H-\underset{\underset{H}{|}}{\overset{\overset{H}{|}}{C}}-\ddot{\underset{..}{O}}-\ddot{N}=\ddot{\underset{..}{O}}:$$

1.4 Thionyl chloride:

$$\begin{array}{c}:\ddot{C}l: \\ \diagdown \\ \phantom{:\ddot{C}l:}S-\ddot{\underset{..}{O}}: \\ \diagup \\ :\ddot{C}l:\end{array}$$

Number of valence electrons at each Cl:
(a) in Lewis structure:

from Cl—S bond	1
from lone pairs	6
total	7

(b) in isolated Cl atom: 7
Difference: (b) - (a) = 7 - 7 = 0
Hence, formal charge = 0

Number of valence electrons at S:
(a) in Lewis structure:

from two Cl—S bonds	2
from S—O bond	1
from lone pairs	2
total	5

(b) in isolated S atom: 6
Difference: (b) - (a) = 6 - 5 = +1
Hence, formal charge = +1

Number of valence electrons at O:
(a) in Lewis structure:

from S—O bond	1
from lone pairs	6
total	7

(b) in isolated O atom: 6
Difference: (b) - (a) = 6 - 7 = -1
Hence, formal charge = -1

1.5 The *valence-shell electron-pair repulsion (VSEPR)* theory is a method of explaining molecular shapes based on the idea that electron pairs, both bonded and nonbonded, repel each other and will orient themselves as far away from each other as possible around the central atom in a molecule. The electron pairs in multiple bonds are treated together as one. The geometry of an atom will depend on the number of electron pairs around the nucleus.

Thus, the *VSEPR theory* predicts that:

(a) An atom having two pairs of electrons will be linear and the bond angle between the two electron pairs will be 180°.
(b) An atom having three pairs of electrons will be trigonal planar and the bond angles will be 120°.
(c) An atom with four electron pairs will be tetrahedral with the bond angles of 109.5°.

The *VSEPR theory* is in most cases successful in predicting the shape of molecules. However, the *shape* of a molecule is determined by the positions of its atoms and not its electron pairs. For instance, ammonia is a trigonal pyramid and water is angular, although the electron pairs in both molecules exist in a tetrahedral geometry.

(a)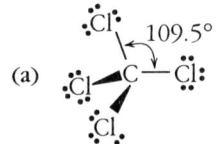
Tetrahedral

(b)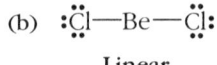
Linear

(c)
$$\text{H} \quad | \quad \text{H} \blacktriangleleft \text{N}^+ \blacktriangleright \text{H} \quad | \quad \text{H}$$
Tetrahedral

(d)
$$\text{H} \blacktriangleleft \text{O}^+ \blacktriangleright \text{H} \quad | \quad \text{H}$$
Trigonal pyramid

(e)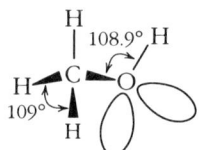
Angular

(f)
H 108.9° H
H—C—O
109° |
H

The geometry about the C atom is predicted to be tetrahedral while that about the O atom is predicted to be angular as observed.

(g)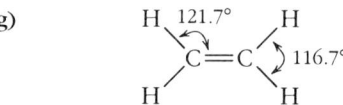

The geometry about each C atom is predicted to be trigonal planar, which is close to the actual bond angles.

(h)
$$\text{H}-\text{C}\equiv\text{C}-\text{H} \quad 180°$$

The geometry about each C atom is predicted to be linear as observed.

1.6 Each of the two carbon atoms in dimethyl ether is surrounded by four bonding electron pairs (three C—H bonds and one C—O bond). Thus, the geometry is tetrahedral and the bond angles around carbon atoms will be close to 109.5°. The oxygen atom is also surrounded by four pairs of electrons (two C—O single bonds and two lone electron pairs). The geometry is again tetrahedral even though the methyl groups will try to take up a little more space than the lone electron pairs. Thus, the bond angle (C—O—C) should be slightly larger than 109.5° (reported bond angle is 111.7°).

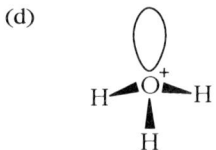

1.7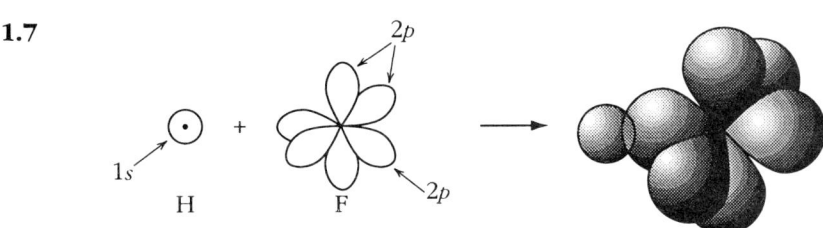

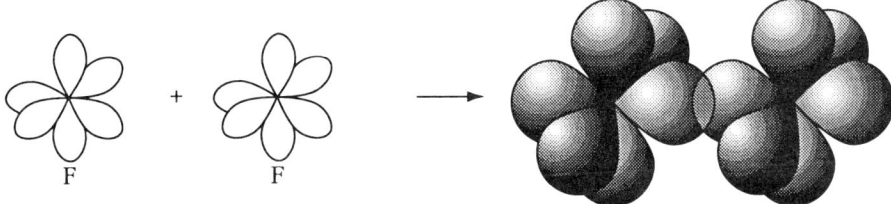

1.8 This is a theoretical problem because we know that the BeH$_2$ molecule is linear. This geometry can be obtained by overlapping the two sp hybrid orbitals of Be with 1s orbitals of two hydrogen atoms. However, for the sake of an argument we could imagine what would happen if we did not use an sp hybrid Be atom. Then we would have one σ bond arising from an overlap of the 2p orbital of Be with one 1s hydrogen orbital. This bond would point in a specific direction. The other σ bond arises from the overlap of the 2s orbital of Be with the 1s orbital of the other hydrogen atom. This bond has no direction because the 2s orbital is spherically symmetrical. As a result, we would have one bond with a specific direction in space and the other one without direction so we cannot predict the geometry of the BeH$_2$ molecule using an unhybridized Be atom.

1.9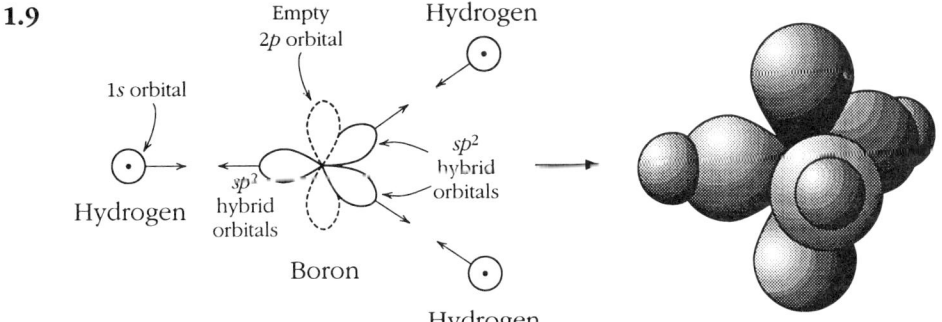

1.10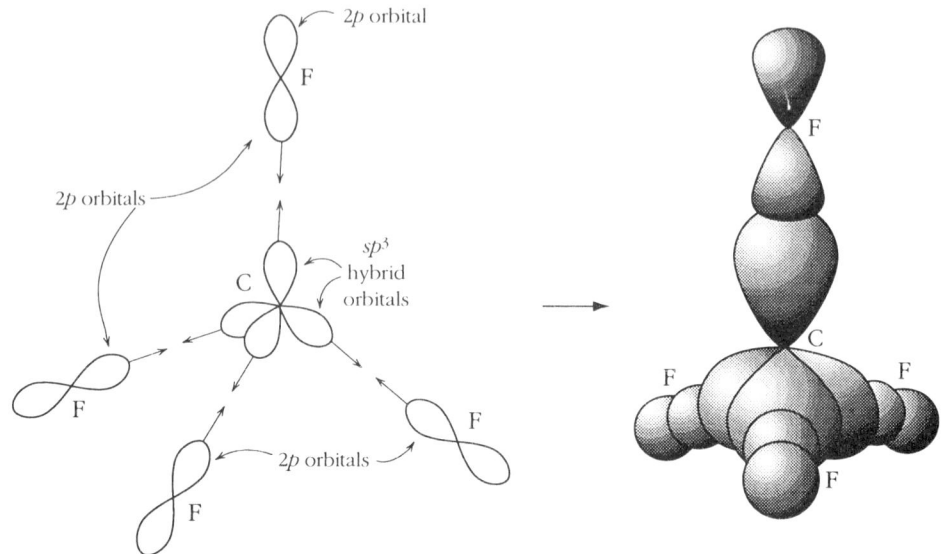

1.11

(a) $CH_3-C\overset{\overset{\ddot{\ddot{O}}:}{\|}}{\underset{\underset{:\ddot{O}:^-}{}}{}} \longleftrightarrow CH_3-C\overset{\overset{:\ddot{O}:^-}{|}}{\underset{\underset{:\ddot{O}:}{\|}}{}}$

(b) $:\ddot{\ddot{O}}-\overset{+}{N}\overset{\overset{\ddot{\ddot{O}}:}{\|}}{\underset{\underset{:\ddot{O}:^-}{}}{}} \longleftrightarrow :O=\overset{+}{N}\overset{\overset{:\ddot{O}:^-}{|}}{\underset{\underset{:\ddot{O}:^-}{}}{}} \longleftrightarrow {}^-:\ddot{\ddot{O}}-\overset{+}{N}\overset{\overset{:\ddot{O}:^-}{|}}{\underset{\underset{:\ddot{O}:}{\|}}{}}$

(c) $CH_2=CH-\overset{+}{C}H_2 \longleftrightarrow \overset{+}{C}H_2-CH=CH_2$

1.12

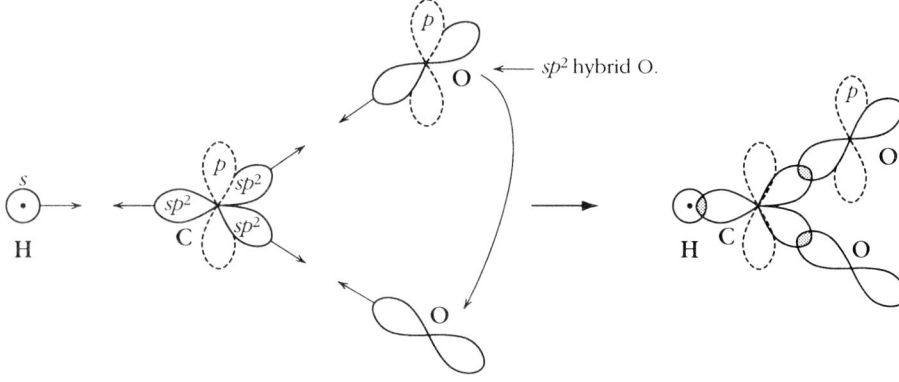

1.13

He$_2$ is not a stable molecule because two of its electrons are located in a bonding molecular orbital and its other two electrons are in an antibonding molecular orbital. Thus, it has the same number of electrons in both bonding and antibonding orbitals. If a molecular species is to be stable, it must have *more* electrons in bonding than in antibonding orbitals.

1.14 For diatomic molecules (which contain two atoms), the bond dipole moment is the molecular dipole moment, μ. Bond polarity is due to the difference in the **electronegativity** of the two atoms. An arrow (+⟶) is used to indicate the direction of polarity. By convention, the cross-base arrow points to the atom that attracts electrons more strongly.

(a) H—H
nonpolar bond

(b) $\overset{\delta+}{H}—\overset{\delta-}{Cl}$
polar bond
+⟶

(c) $\overset{\delta+}{I}—\overset{\delta-}{Cl}$
polar bond
+⟶

(d) I—I
nonpolar bond

(e) $\overset{\delta+}{Br}—\overset{\delta-}{Cl}$
polar bond
+⟶

1.15 (a)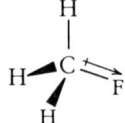

The C—F bond moment is responsible for the molecular dipole moment of fluoromethane.

(b)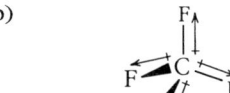

Tetrafluoromethane has polar bonds but the C—F bond moments cancel each other due to the tetrahedral geometry of the molecule; tetrafluoromethane, therefore, is nonpolar.

(c)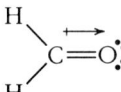

Methanal (formaldehyde) is a planar molecule. The C=O bond moment is responsible for the large molecular dipole moment.

(d)

H_2S has a molecular dipole moment because the H—S bond moments do not cancel each other.

(e)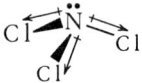

NCl_3 has a molecular dipole moment because the N—Cl bond moments do not cancel each other.

1.16 (a) Hydrogen bonding is the intermolecular attraction in liquid HF. The H—F bond is strongly polarized so the hydrogen atom of one molecule is attracted to a nonbonding pair of electrons on the fluorine atom of a neighboring H—F molecule.
(b) London forces are the intermolecular attractions between helium atoms.
(c) Dipole-dipole interactions are the intermolecular attractions between polar IBr molecules.

1.17 (a) **Isotopes** are atoms which contain the same number of protons and a different number of neutrons in their nuclei. This means that isotopes of an element have the same atomic number but different mass numbers. The following are isotopes that are most frequently encountered in organic compounds: ^{12}C, ^{13}C, and ^{14}C; ^{1}H and ^{2}H (D); ^{14}N and ^{15}N; ^{16}O and ^{18}O; ^{35}Cl and ^{37}Cl; ^{79}Br and ^{81}Br.
(b) **Valence electrons** are the outermost electrons of an atom. Valence electrons on the average are farthest from the nucleus. In terms of quantum numbers, valence electrons have the highest principal quantum number. A carbon atom has four outermost electrons. These electrons all have a principal quantum number of 2.

(c) **Lewis structures** represent the structure of compounds by joining the symbols of the elements present in the compound by lines or pairs of dots. Each line or pair of dots represents a bond made up of a pair of electrons.

(d) A **sigma bond** (σ bond) is a bond formed by the head-on overlap of simple or hybrid atomic orbitals along the straight line joining the nuclei of the bonded atoms. In this way, electrons are shared between the bonded atoms. In the Lewis structures of organic compounds, sigma bonds are represented by a single dash.

(e) A **pi bond** (π bond) is a bond formed by the sideways overlap of two p orbitals, producing a high electron density above and below a plane joining the bonded atoms. Double bonds usually contain both a pi and a sigma bond. A double bond is represented in a Lewis structure by two lines (=).

(f) Three pairs of electrons shared between two bonded atoms is a **triple bond**. It is represented in a Lewis structure by three lines ($\equiv$). Triple bonds usually consist of one sigma (σ) bond formed by the overlap of sp hybrid orbitals (one from each bonded atom) and two pi (π) bonds. Each π bond is formed by the overlap of two unhybridized $2p$ orbitals (one from each bonded atom).

(g) **Formal charge** is the number of outer-shell (valence) electrons in an isolated atom minus the number of electrons assigned to that atom in a Lewis structure.

(h) **Nonbonding electrons** are electrons in the valence shell of an atom in a molecule which are not involved in bond formation. Nonbonding electrons are also called lone-pair electrons. Examples include the nonbonding electrons on the nitrogen atom of ammonia or the two lone pairs of electrons on the oxygen atom of water.

(i) An **atomic orbital** is a region around the nucleus of an atom where the electron density (or the probability of finding an electron) is high. There are several kinds of atomic orbitals (s, p, d, f) which differ from one another in the shapes of the regions of high electron density.

(j) A **localized valence bond orbital** is an orbital constructed by the overlap of the valence orbitals of the two atoms forming the bond.

(k) An **ionic bond** is formed when positive and negative ions are held together by electrostatic forces of attraction. Positive and negative ions are formed when an electron (or electrons) is (are) transferred between atoms of significantly different electronegativity (by more than 1.8 on the electronegativity scale).

(l) A **bonding molecular orbital** is an orbital surrounding two or more nuclei that is lower in energy than any of the atomic orbitals that combine to form the molecular orbital.

(m) A **polar covalent bond** is a bond in which a pair of electrons is shared unevenly between two atoms. The pair of electrons is shared unequally because of a difference in electronegativity between the two atoms. The shared electron pair is drawn more closely to the more electronegative atom (indicated by the symbol $\delta-$) while the other atom develops a partial positive charge ($\delta+$). Examples of the polar covalent bonds in organic chemistry are carbon-halogen, carbon-oxygen, and carbon-nitrogen bonds.

(n) A **nonpolar covalent bond** is formed between two atoms of the same electronegativity (usually atoms of the same element). The distribution of electron density between such two atoms is even.

(o) The **bond dipole moment** is a measure of polarity of a polar covalent bond and it is a product of charge separation between two bonded atoms of different electronegativity.

(p) **Molecular dipole moment**, μ, is the vector sum of the individual bond dipole moments of a molecule.

(q) A **polar molecule** is a molecule with a resultant molecular dipole moment greater than zero.

(r) **Resonance structures** are two or more correct Lewis structures that can be written for a molecule or ion. The true structure of the molecule or ion is a composite (or hybrid) of these different contributing resonance structures.

(s) **Resonance hybrid** is a description of the electron structure of a molecule for which there is no single Lewis structure that is a correct description of the bonding but there are two or more Lewis structures that satisfy the octet rule. The true electron structure of such a molecule is regarded as a hybrid (or composite) of these contributing Lewis or resonance structures.

(t) The *sp^3* **hybrid carbon atom** is a carbon atom whose four valence electrons occupy four sp^3 hybrid orbitals. The sp^3 hybrid orbitals (also called sp^3 hybrids) arise from a combination of one $2s$ orbital and three $2p$ orbitals of a carbon atom. Thus, four equivalent sp^3 orbitals are formed and they are spatially oriented toward the four corners of a tetrahedron (bond angles are 109.5°). A typical example of a pure tetrahedron sp^3 hybrid atom is the carbon atom in methane.

(u) A **hydrogen bond** is an intermolecular force of attraction in which a positively polarized H atom, covalently bonded to one atom, is attracted simultaneously to a negatively polarized atom of the same or a nearby molecule. Hydrogen bonds typically involve the hydrogen of O—H or N—H bonds and an atom with a lone pair of electrons, like another O or N. The strength of a hydrogen bond involving an oxygen, fluorine, or nitrogen atom ranges from 3 to 10 kcal/mol, making the hydrogen bond the strongest known type of intermolecular interaction.

(v) **Dipole-dipole interactions** are defined as the interactions of permanent dipoles in different molecules. Dipole-dipole interactions are weaker than hydrogen bonds and their energy ranges from 1 to 3 kcal/mol.

(w) The *sp^2* **hybrid carbon atom** is a carbon atom whose four valence electrons occupy one $2p$ atomic orbital and three sp^2 hybrid orbitals. The sp^2-hybrid orbitals are created when two of the $2p$ orbitals of the carbon atom are combined with the $2s$ orbital of the same carbon atom to form three trigonal planar sp^2 orbitals. The third $2p$ orbital of an sp^2 hybridized carbon is unhybridized and retains its shape as an atomic orbital that is perpendicular to the plane defined by the three sp^2 hybrid orbitals.

(x) **Ion-dipole interactions** are interactions between ions and polar molecules. Attractive interaction occurs between ions and the opposite charged end of a dipolar molecule. Repulsive interaction occurs between ions and the similarly charged end of a dipolar molecule.

(y) The *sp* **hybrid carbon atom** is a carbon atom whose four valence electrons occupy two $2p$ atomic orbitals and two sp hybrid orbitals. Two sp hybrid orbitals are formed by a combination of one $2s$ and one $2p$ orbital of a carbon atom. The two sp hybrid orbitals point away from each other along a straight line so that the bond angle equals 180°. The remaining two

2p atomic orbitals remain unhybridized and at right angles to each other as well as to the two *sp* hybrid orbitals.

(z) An **antibonding molecular orbital** is an orbital surrounding two or more nuclei with an energy higher than any of the atomic orbitals that combine to form the molecular orbital.

1.18 (a) The electron configuration of ground state chlorine is $1s^22s^22p^63s^23p^5$. The valence electrons of an element have the highest principal quantum number so chlorine has a valence of 7 because it has seven electrons in orbitals of principal quantum number 3.
(b) 5
(c) 5
(d) 6
(e) 4

1.19 (a) N
(b) F
(c) C

1.20 (a) $1s^22s^1$
(b) $1s^22s^22p^63s^23p^4$
(c) $1s^22s^22p^63s^23p^3$
(d) $1s^22s^22p^63s^23p^5$
(e) $1s^22s^22p^63s^23p^2$

1.21 (a) 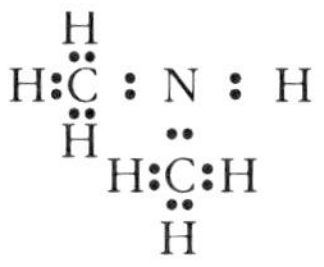 methylamine (b) 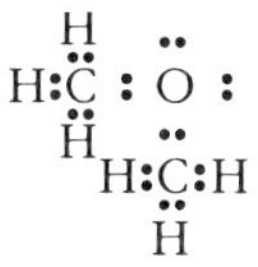 methanol (c) 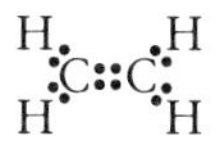 ethene

(d)  formaldehyde (e)  methyl fluoride (f) 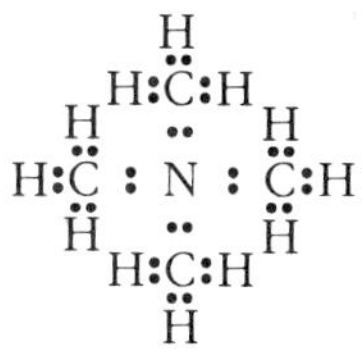 trimethylamine

(g) 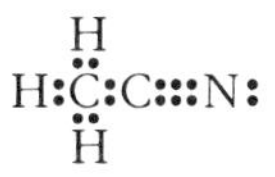 acetonitrile (h)  carbon tetrachloride

1.22 (a) 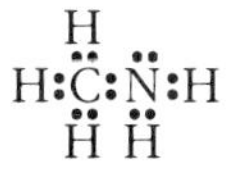 (b) 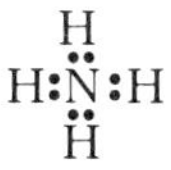 (c)

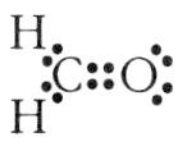

(d)  SiF₄ (e) HC≡C: (f) :C≡N:

1.23 Formal charges are *apparent* charges associated with atoms in a Lewis structure. They arise when atoms have not made equal contributions of electrons to the covalent bond joining them. The sum of formal charges of the atoms in a Lewis structure must equal *zero* for a neutral molecule and must equal the ionic charge for a polyatomic ion.

(a)

$$\text{H}-\underset{\underset{\text{H}}{|}}{\overset{\overset{\text{H}}{|}}{\text{C}_\alpha}}-\text{C}_\beta\underset{\ddot{\underset{..}{\text{O}}}:^\delta}{\overset{\ddot{\text{O}}:^\gamma}{\diagup}}$$

	H	C_α	C_β	O_γ	O_δ
Valence electrons	1	4	4	6	6
Electrons assigned	1	4	4	6	7
Formal charge	**0**	**0**	**0**	**0**	**−1**

Net charge for the molecule = 0 + 0 + 0 + 0 + (−1) = −1

(b)

$$\text{H}_\alpha-\underset{\underset{\text{H}_\alpha}{|}}{\overset{\overset{\text{H}_\alpha}{|}}{\text{C}}}-\underset{\underset{\text{H}_\beta}{|}}{\overset{\overset{\text{H}_\beta}{|}}{\text{N}}}-\ddot{\text{O}}:$$

	H_α	C	H_β	N	O
Valence electrons	1	4	1	5	6
Electrons assigned	1	4	1	4	7
Formal charge	**0**	**0**	**0**	**+1**	**−1**

Net charge for the molecule = 0 + 0 + 0 + 1 + (−1) = 0

(c)

$$\text{H}-\underset{\underset{\text{H}}{|}}{\overset{\overset{\text{H}}{|}}{\text{C}}}:$$

	H	C
Valence electrons	1	4
Electrons assigned	1	5
Formal charge	**0**	**−1**

Net charge for the molecule = 0 + (−1) = −1

(d)

$$\underset{\text{H}}{\overset{\text{H}}{\diagdown}}\text{C}:$$

	H	C
Valence electrons	1	4
Electrons assigned	1	4
Formal charge	**0**	**0**

Net charge for the molecule = 0 + 0 = 0

1.24

$$:\ddot{O}:$$
$$|$$
$$:\ddot{O}-Cl-\ddot{O}:$$
$$|$$
$$:\ddot{O}:$$

	O	Cl
Valence electrons	6	7
Electrons assigned	7	4
Formal charge	**−1**	**+3**

Net charge for the molecule = 4 × (−1) + 3 = −4 + 3 = −1

$$\overset{\alpha}{\ddot{O}}$$
$$\parallel$$
$$^\beta:\ddot{O}-S-\ddot{O}:^\beta$$
$$\parallel$$
$$\underset{\alpha}{\ddot{O}}$$

	O_α	O_α	O_β	O_β	S
Valence electrons	6	6	6	6	6
Electrons assigned	6	6	7	7	6
Formal charge	**0**	**0**	**−1**	**−1**	**0**

Net charge for the molecule = 0 + 0 + (−1) + (−1) + 0 = −2

1.25 (a) There are three bonds to the carbon atom of CH_3^+ so its structure is predicted to be trigonal planar.
(b) There are three bonds and one lone pair of electrons around the carbon atom of CH_3^- so its structure is predicted to be trigonal pyramidal.
(c) There are three bonds and one lone pair of electrons around the phosphorus atom of PCl_3 so its structure is predicted to be trigonal pyramidal.
(d) There are three bonds and one lone pair of electrons around the oxygen atom so it should be trigonal pyramidal.
(e) Trigonal planar.
(f) Tetrahedral.
(g) There are three bonds and two nonbonding pairs of electrons around the iodine atom of ICl_3. Five pairs of electrons about an atom arrange themselves into a trigonal bipyramid. The nonbonding electron pairs always occupy the equatorial positions in a trigonal bipyramid so the geometry of ICl_3 is T shaped.

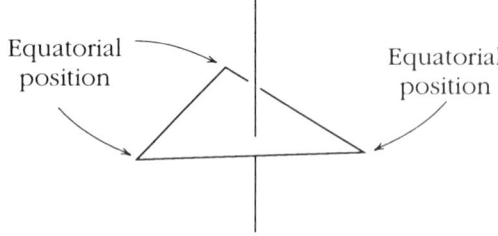

Trigonal bipyramid

T-shaped ICl₃ molecule

(h) Trigonal pyramid; the same shape as ammonia because this molecule can be considered as a molecule of ammonia in which one hydrogen atom is replaced by a CH₃ group. Such a replacement does not greatly affect the geometry around the nitrogen atom.

1.26

:Cl:N⋯O: α
       ⋯O: β

	Cl	N	O_α	O_β
Valence electrons	7	5	6	6
Electrons assigned	7	4	6	7
Formal charge	0	+1	0	–1

Net charge for the molecule = 0 + 1 + 0 + (–1) = 0

There are three bonds about the nitrogen atom so the bond angles around the nitrogen atom will be about 120°.

1.27 (a) The carbon atom is bonded to four chlorine atoms so an sp^3 hybrid carbon atom will account for the tetrahedral geometry.
(b) The carbon atom is part of a triple bond. An sp hybrid carbon atom will account for the linear geometry of the triple bond.
(c) sp^2 Hybrid carbon atom.
(d) sp^2 Hybrid carbon atom.

1.28 (a) H:N̈:C:::N:
          H

(b) Linear

(c)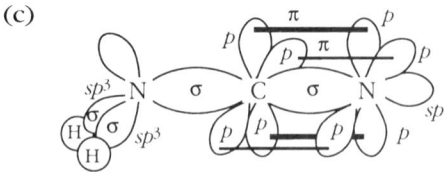

(d) There are four sigma bonds and two pi bonds in the cyanamide molecule.

1.29 The VSEPR theory predicts that the two hydrogen atoms and the two non-bonding electron pairs of water should point to the corners of a tetrahedron. This predicts a H—O—H bond angle of 109.5°, which is close but not exactly equal to the experimentally determined value of 104.5°. The hybrid that comes the closest to this geometry is an *sp³* hybrid formed by combining the 2*s* and three 2*p* atomic orbitals of an oxygen atom.

1.30 (a) The three hydrogen atoms and the nonbonding electron pair of ammonia point to the corners of a tetrahedron. This geometry can be obtained by use of an *sp³* hybrid nitrogen atom.
(b) The linear geometry around the nitrogen atoms of the triple bond can be obtained by using an *sp* hybrid nitrogen atom.
(c) The trigonal planar geometry of the nitrogen atom can be obtained by using an *sp²* hybrid nitrogen atom.

1.31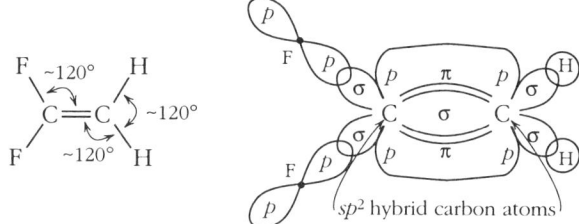

1.32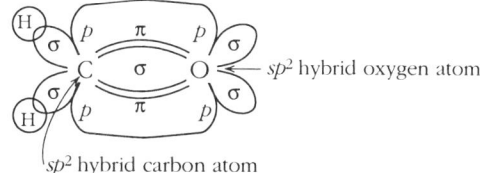

1.33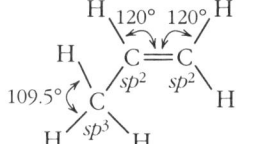

Propene   Acetonitrile

1.34 :N̈=N⁺=Ö: ⟷ :N≡N⁺—Ö:⁻

1.35

(a) Two resonance structures of O$_3$ with a central O$^+$ double-bonded to one O and single-bonded to O$^-$, shown in both directions.

(b) Two resonance structures of SO$_2$ with analogous arrangement, S$^+$ center.

(c) Four resonance structures of CrO$_4^{2-}$ showing Cr$^+$ with various combinations of single- and double-bonded oxygens bearing negative charges.

1.36 No, the two structures are not a resonance hybrid since the way the atoms in these two molecules are joined is different. Acceptable contributing structures to a resonance hybrid must all have the same skeleton structure; they can differ only in how electrons are distributed within the structure.

1.37 The bonding molecular orbitals are lower in energy than the antibonding molecular orbitals of simple diatomic molecules. The electron distribution is also different. Considerable electron density is located between the nuclei in bonding molecular orbitals, which stabilizes the molecule because the electrons are simultaneously associated with both nuclei and because the electrons diminish nuclear-nuclear repulsion. In contrast, little electron density is located between the nuclei of nonbonding molecular orbitals, which destabilizes diatomic molecules.

1.38 (a) The atomic and molecular orbital energy diagram of molecular hydrogen is the same one as shown in Figure 1.22 (p. 33). A molecule of H$_2^+$ has only one electron, which is placed in σ_{1s}, the lowest energy molecular orbital. H$_2^+$ is a stable molecule because the molecule with one electron in σ_{1s} is more stable (lower in energy) than an electron in an isolated hydrogen atom.

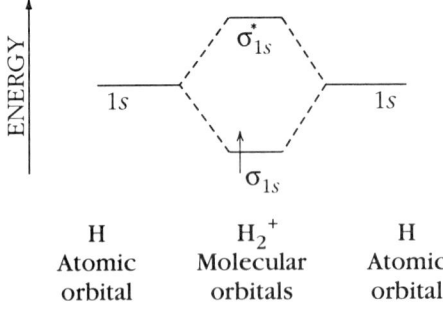

(b) To determine if Li$_2$ is a stable molecule only the valence electrons of a lithium atom need be considered. Combining the 2s atomic orbitals of two lithium atoms forms two molecular orbitals, one bonding σ_{2s} and one antibonding σ^*_{2s}. The two electrons of Li$_2$ are placed in σ_{2s}, the lowest energy molecular orbital. Li$_2$ is a stable molecule because the molecule with two electrons in σ_{2s} is more stable than one electron in each of two isolated lithium atoms.

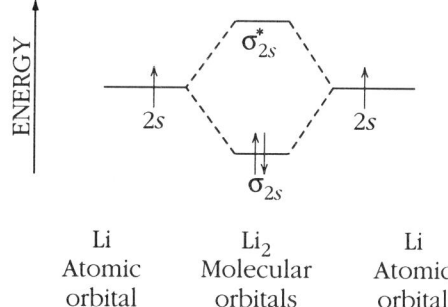

(c) H_2^- is a stable molecule.

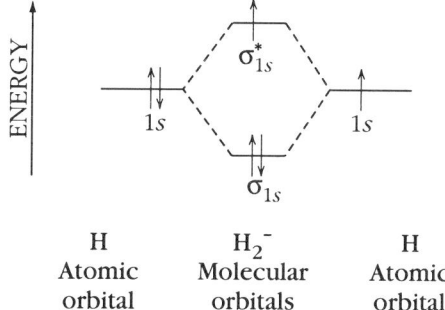

(d) He_2^+ is a stable molecule.

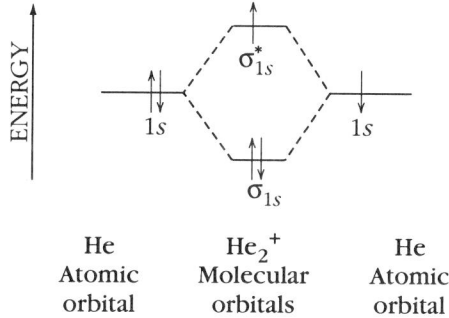

(e) Be_2 is not a stable molecule because the molecule with two electrons in σ_{2s} and two electrons in σ^*_{2s} is not any more stable than two electrons in each of two isolated Be atoms.

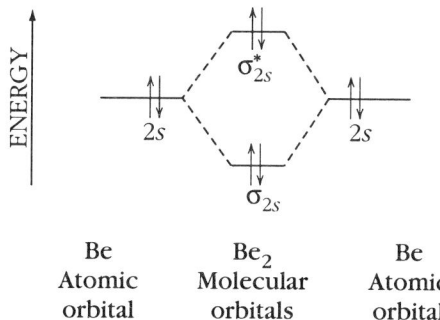

1.39 (a) Fluorine is more electronegative than silicon so electrons are attracted to the fluorine atom of the bond. The bond dipole is:

$$\text{Si} \longrightarrow \text{F}$$
$$\longmapsto$$

(b) Nitrogen is the more electronegative atom so the bond dipole is:

$$\text{C} \longrightarrow \text{N}$$
$$\longmapsto$$

(c) Carbon is more electronegative than lithium so the bond dipole is:

$$\text{C} \longrightarrow \text{Li}$$
$$\longleftarrow$$

(d) $\overset{\longmapsto}{\text{C} \longrightarrow \text{O}}$

(e) $\overset{\longmapsto}{\text{N} \longrightarrow \text{F}}$

(f) $\text{C} \longrightarrow \text{F}$
$\longmapsto$

1.40 (a) NaF   Electronegativity of sodium is 0.9 (from Figure 1.26)
Electronegativity of fluorine is 4.0
Difference: 4.0 - 0.9 = 3.1. A difference of 3.1 is larger than 1.8 so the bond is ionic.

(b) CCl$_4$ Electronegativity of carbon is 2.5, electronegativity of chlorine is 2.8.
Difference: 2.8 - 2.5 = 0.3. A difference of 0.3 is much smaller than 1.8 so the bond is covalent. However, there is a difference in electronegativity between C and Cl, so the carbon-chlorine bond is a polar covalent bond.

(c) MgF$_2$ The two Mg-F bonds are ionic.

(d) IBr   The I-Br bond is a polar covalent bond.

(e) I$_2$ There is no difference of electronegativity between two atoms of the same element so the bond must be nonpolar covalent.

1.41 (a) BCl$_3$ Boron trichloride is a trigonal planar molecule.
Chlorine is more electronegative than boron, so the bond moments are not zero. However, the three B—Cl bonds are symmetrically arranged around the boron atom so the B—Cl bond dipole moments cancel. Hence, the molecule is nonpolar.

(b) The ammonia molecule, NH$_3$, has three bonds and one lone pair tetrahedrally arranged around the central nitrogen atom.
The bond dipole moments do not cancel so NH$_3$ is polar.
Because nitrogen is more electronegative than hydrogen, the N atom is the negative end of the dipole. The dipole moment of ammonia is $\mu = 1.46$ D.

(c) The bond dipole moments of the C—O and the O—H bonds do not cancel because the C—O—H angle is 108.5°. Therefore, CH$_3$OH is a polar molecule. The dipole moment of methanol is $\mu = 1.7$ D.

(d) The S—O bond has a bond dipole moment because oxygen is more electronegative than sulfur. The S—O bond dipole moments do not cancel because SO_2 is a bent molecule. Therefore, SO_2 is a polar molecule. The dipole moment of SO_2 is $\mu = 1.60$ D.

(e) The Si—F bonds have bond dipole moments but due to the tetrahedral symmetry of the molecule, these bond moments cancel and the resultant dipole moment is zero. SiF_4 is a nonpolar molecule.

(f) The C—O bond has a bond dipole moment. The C—O—C bond angle is 112° so the bond dipole moments do not cancel, which makes CH_3OCH_3 a polar molecule. The dipole moment of CH_3OCH_3 is $\mu = 1.30$ D.

1.42 The vectorial sum of the bond dipole moments of the C—F bonds of CF_4 is zero because of the symmetrical tetrahedral shape of the molecule. Consequently CF_4 is nonpolar.

The value of the bond dipole moment of the C—F bond is different from that of the bond dipole moment of the C—Cl bond so even though a molecule of CCl_2F_2 is tetrahedral, the vectorial sum of the bond dipole moments is not zero. The dipole moment of CCl_2F_2 is $\mu = 0.51$ D.

1.43 *cis*-1,2-Dichloroethene and *trans*-1,2-dichloroethene have the same molecular formula ($C_2H_2Cl_2$), the same number and type of bonds, but different molecular structures.

cis-1,2-Dichloroethene
$\mu = 1.89$ D

trans-1,2-Dichloroethene
$\mu = 0$

This pair of compounds is an excellent illustration of the importance of molecular shape. In *trans*-1,2-dichloroethene the strong C—Cl bond dipole moments point in opposite directions and cancel each other so the dipole moment of the molecule is zero. In *cis*-1,2-dichloroethene, on the other hand, the vector sum of the C—Cl bond moments gives a resultant dipole moment of $\mu = 1.89$ D. Thus, if the sample in hand has a dipole moment, it must be the *cis* isomer; if the sample has *no* dipole moment, it must be the *trans* compound.

1.44 A hydrogen bond is an intermolecular force of attraction between a positively polarized H atom, covalently bonded to one atom, and a negatively polarized atom of the same or a nearby molecule. Hydrogen bonds typically involve the hydrogen of O—H or N—H bonds (so called *hydrogen bond donors*) and an atom with a lone pair of electrons, like another O or N (*hydrogen bond acceptors*).

(a) Methanol (CH_3OH) is both a donor and an acceptor of hydrogen bonds. The hydrogen atom of the O—H bond is the hydrogen bond donor and the oxygen atom of another molecule of methanol is the hydrogen bond acceptor.

(b) The fluorine atom of fluoromethane, which has three lone electron pairs, can act as a hydrogen bond acceptor. However, there is *no* hydrogen bond donor in fluoromethane so hydrogen bonding between molecules of fluoromethane cannot take place.
(c) Ethane has *neither* hydrogen bond donors *nor* acceptors, so *no* hydrogen bonding can take place.
(d) Acetonitrile ($CH_3C\equiv N{:}$) has a lone electron pair on the nitrogen atom that acts as a hydrogen bond acceptor. However, there are *no* acidic hydrogen atoms that can be hydrogen bond *donors*, so hydrogen bonding cannot occur.
(e) Methanamine (CH_3NH_2) has both donors and acceptors of hydrogen bonds. The hydrogen atom of the N—H bond is the hydrogen bond donor and the nitrogen atom is the hydrogen bond acceptor so hydrogen bonding occurs between amine molecules.

(f) Trichloromethane (chloroform, $CHCl_3$) has chlorine atoms with three lone electron pairs each, which can act as hydrogen bond acceptors. However, there is *no* hydrogen bond *donor* so hydrogen bonding cannot take place between molecules of chloroform.

1.45 **Vaporization** is the change of a solid or a liquid into its vapor. In order to vaporize, molecules of a liquid must have enough energy to overcome the intermolecular interactions and, in addition, have enough kinetic energy to overcome the atmospheric pressure. The boiling point of a substance, therefore, is a function of molecular weight and intermolecular interactions.
Intermolecular interactions are the most important factor in determining boiling

points so that is why particular attention is paid to the type of attractive forces present when discussing the relationship between the structure of a compound and its boiling point.

There are three types of attractive forces between neutral molecules: dipole-dipole forces, London (or dispersion) forces, and hydrogen bonding forces. The term **van der Waals forces** is *a general term for those intermolecular forces that include dipole-dipole and London forces.*

(a) Water. Studies of the structure of water in its different physical states show that water contains many hydrogen bonds. The hydrogen atom of one water molecule is attracted to the electron pair of the oxygen atom of another water molecule. Many H_2O molecules can be linked this way to form clusters of hydrogen bonded molecules.

(b) Ethanol. The reason for the high boiling points of alcohols is that alcohols, like water, can form clusters of hydrogen bonded molecules. However, one molecule of ethanol has only one hydrogen bond donor (the hydrogen atom of the O—H bond) while a water molecule has two. As a result, less energy is needed to break the hydrogen bonds in ethanol than in water. As a result the boiling point of ethanol (b.p. 78.5 °C) is lower than that of water (b.p. 100.0 °C).

(c) Ethane is a symmetrical nonpolar molecule. London (or dispersion) forces are the only type of intermolecular interactions between ethane molecules. London forces involve the weak attractive forces between molecules resulting from small, instantaneous dipoles that occur because of the varying positions of electrons during their motion about the nuclei. London forces tend to increase with the size of the molecule since the number of electrons (and polarizability) increases.

However, in the case of ethane, which is a relatively small molecule, the attractive force between ethane molecules is small so it has a low boiling point (-88.6 °C).

(d) Dimethyl ether (methoxymethane) is a polar molecule ($\mu = 1.3$ D) that has a hydrogen bond acceptor (the nonbonding electron pairs of the oxygen atom) but no hydrogen bond donor. Consequently no hydrogen bonds are formed. The predominant intermolecular interaction is dipole-dipole interaction. Dipole-dipole attractions are weaker than hydrogen bonds but stronger than London forces so the boiling point of dimethyl ether is higher (-25 °C) than that of ethane but much lower than that of ethanol (b.p. 78.5 °C) or water (b.p. 100 °C).

CHAPTER 2

Concepts

1. **Hydrocarbons** are compounds that contain only carbon and hydrogen. They can be divided into two main classes: **aliphatic hydrocarbons** and **aromatic hydrocarbons.** Aliphatic hydrocarbons can be further subdivided into two major groups: saturated and unsaturated. **Saturated hydrocarbons,** or **alkanes,** contain only single bonds. **Cycloalkanes** are alkanes in which carbon atoms form a ring. **Unsaturated hydrocarbons** contain one or more double bonds or triple bonds.

2. **Functional groups** are structural units responsible for the characteristic reactions of a molecule. Different compounds containing the same functional groups undergo the same types of reactions.

3. Alkanes and cycloalkanes do not have functional groups and they are nonpolar and insoluble in water. The only forces of attraction between nonpolar molecules are relatively weak **induced dipole-induced dipole attractions.** These forces are referred to as **van der Waals attractions, London forces,** or **dispersion forces.** Branched alkanes have lower boiling points than their unbranched isomers because their smaller surface area affords fewer points of contact between molecules. There is a limit to how closely two molecules can approach one another, which is given by the sum of the van der Waals radii of the proximate atoms.

4. **Nomenclature.** Alkanes may contain straight (unbranched) chains, branched chains, or rings. Alkanes are named systematically according to the rules of the IUPAC. The name of a compound is based on its parent chain, which, for an alkane, is the longest continuous carbon chain in the molecule.

5. **Alkyl groups** are structural units that lack one of the hydrogen atoms of an alkane. Unbranched alkyl groups in which the point of attachment is at the end of the chain are named in systematic nomenclature by replacing the *-ane* ending by *-yl.* Branched alkyl groups are named by using the longest continuous chain that begins at the point of attachment as the parent name.

6. The "R" notation is used as a general abbreviation for alkyl groups; Ph is the abbreviation for a phenyl group, and Ar is the abbreviation for an aryl group (unspecified aromatic group). Thus, RH is an alkane, PhH is benzene, ArH is an aromatic compound.

Solutions to the Exercises

2.1

(a)

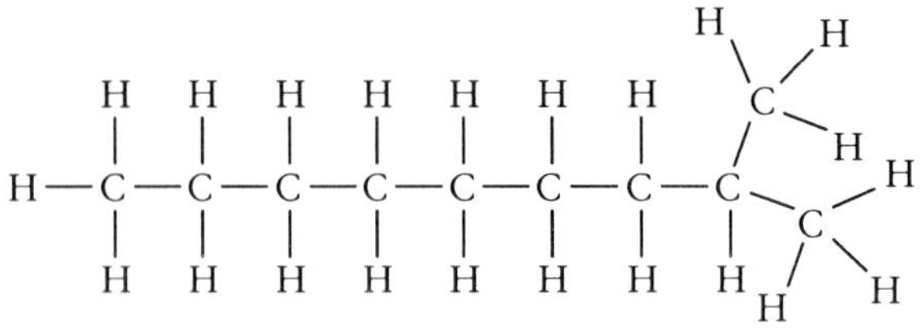

(b)

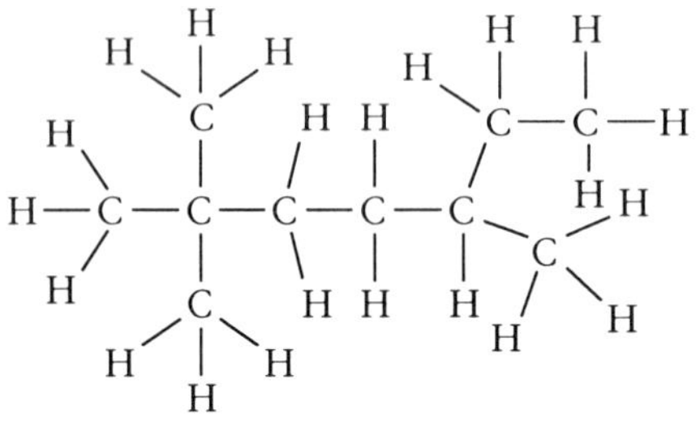

(c)

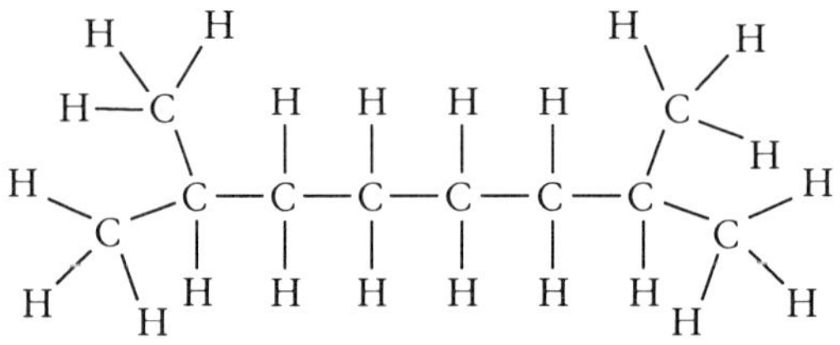

(d)

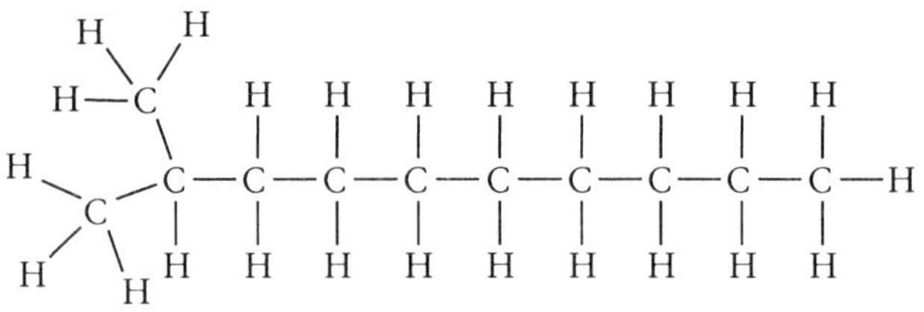

(e)

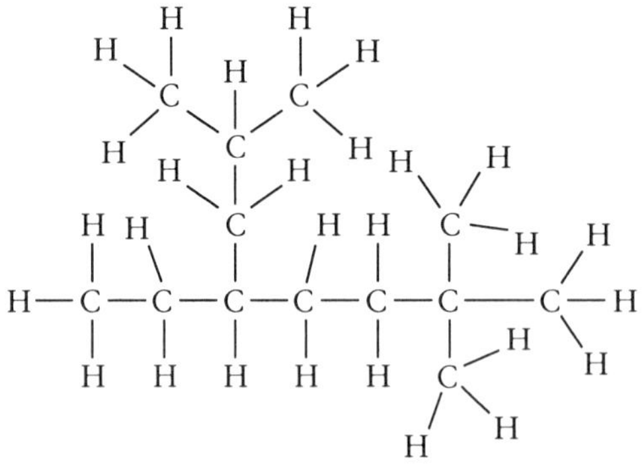

(f) [structure]

(g) [structure]

(h) [structure]

(i) [structure]

2.2 Steps in Writing Isomers of Alkanes

The use of line structures to write the structures of possible isomers is convenient because the structures can be written quickly and the hydrogen atoms bonded to the various carbon atoms do not have to be counted. We must be careful, however, that each carbon atom has no more than *four* bonds.

Let us use line structures to write the structures of the isomers in this exercise:

(a) Step 1. Begin with the unbranched isomer of C_5H_{12}:

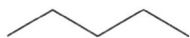

Step 2. Next, remove one carbon atom from the end of the continuous chain of five carbon atoms which, in this case, gives a four carbon atom chain. Place the remaining one carbon atom (methyl group) at the carbon atom next to the end of the four carbon chain. *Caution! We cannot attach the methyl group to a terminal carbon in the chain because this would give us back our original straight-chain structure.*

In this way we obtain the following branched isomer of C_5H_{12}:

Step 3. We can try to move the methyl group further along the four carbon atom chain but this new structure is simply a different representation of the branched isomer obtained in Step 2. Thus, it is not a new isomer.

Step 4. The length of the five carbon chain can be further reduced by removing two carbon atoms (methyl groups) and adding them to the internal carbon atoms of the remaining three carbon atom chain. Both methyl groups must be attached to the same carbon atom because a three carbon chain has only one internal carbon. In this way we obtain another branched isomer of C_5H_{12}.

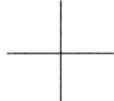

There are no other possible isomers of C_5H_{12} so the three constitutional isomers of C_5H_{12} are:

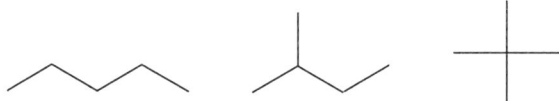

Now, let us apply the same method to solve the next exercise.

(b) C_6H_{14}:

Step 1. The first isomer is the straight-chain alkane containing six carbon atoms.

Step 2. The continuous carbon chain is shortened by one carbon atom and this carbon atom is attached to the carbon next to the end of the five carbon atom chain.

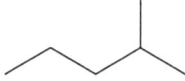

Step 3. Moving the methyl group further along the chain gives another branched isomer of C_6H_{14}.

This isomer is not identical to the isomer obtained in Step 2. However, moving the methyl group still further along the carbon chain would result in a different representation of the isomer written in Step 2.

Step 4. The six carbon atom chain is reduced by two carbon atoms and the two methyl groups are attached to the internal carbon atoms. There are two internal carbon atoms so the two methyl groups can be attached in two different ways:

(i) One methyl group can be attached to each internal carbon atom:

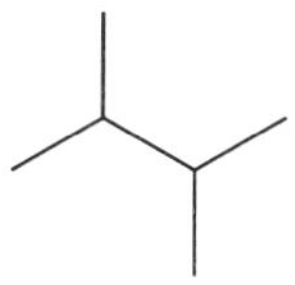

(ii) Both methyl groups can be attached to one of the internal carbon atoms:

Note! It does not matter which of the two internal carbon atoms ends up with the two methyl groups because both structures obtained in this way are equivalent.

Finally we write the line structures of all isomers of C_6H_{14}.

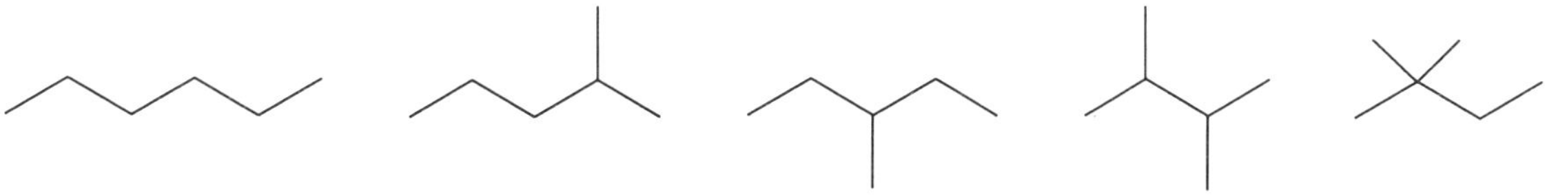

(c) The nine isomers of C_7H_{16} are:

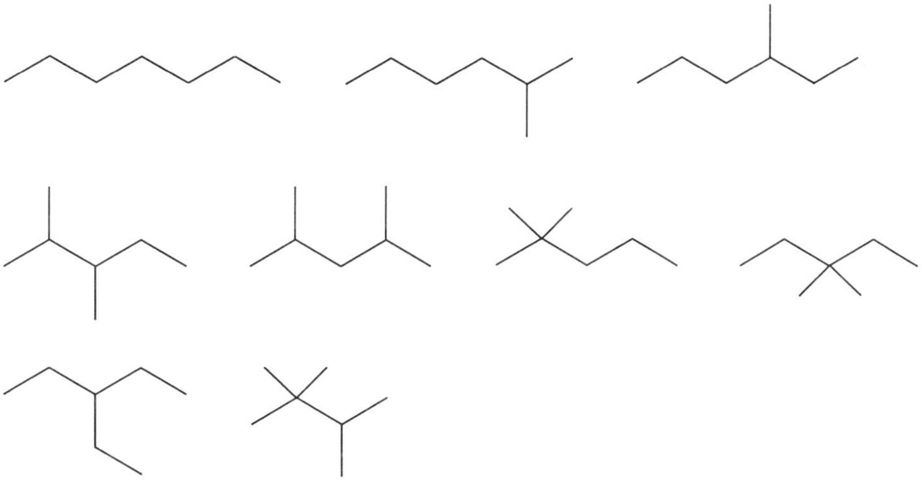

2.3 An **alkyl group** is named according to the following pattern:
*Replace the -***ane** *ending of the parent alkane with an -***yl** *ending.* For instance,

Number of carbon atoms	Name of the parent alkane	Name of the alkyl group
1	meth*ane*	meth*yl*
2	eth*ane*	eth*yl*
3	prop*ane*	prop*yl*

Thus,

(a) CH$_3$CH$_2$CH$_2$CH$_2$-
(b) CH$_3$(CH$_2$)$_6$CH$_2$-
(c) CH$_3$(CH$_2$)$_9$CH$_2$-
(d) CH$_3$(CH$_2$)$_{18}$CH$_2$-
(e) CH$_3$(CH$_2$)$_{11}$CH$_2$-

2.4 (a)

$$\text{CH}_3\text{CHCH}_2\text{CH}_3$$
$$|$$
$$\text{CH}_2\text{CH}_3$$

IUPAC RULE 1. Find the longest continuous carbon chain in the molecule and name it. In order to locate the longest carbon chain in this case we have to *turn corners*. We find that the longest chain, or **stem chain,** contains five carbon atoms. Thus, the compound will be named as a derivative of **pentane.**

$$\text{H}_3\text{C} \text{—} \text{CHCH}_2\text{CH}_3$$
$$|$$
$$\text{CH}_2\text{CH}_3$$

IUPAC RULE 2. Name all groups attached to the longest chain as alkyl substituents. In this case we have one methyl group.

IUPAC RULE 3. Number the carbons of the longest chain beginning with the end that is closest to a substituent.

(i) Begin the numbering at the end *nearer the first substituent*.

$$\overset{3\ \ 4\ \ 5}{\text{H}_3\text{C} \text{—} \text{CHCH}_2\text{CH}_3}$$
$$\overset{2|\ \ \ 1}{|}$$
$$\text{CH}_2\text{CH}_3$$

In this case it does not matter from which end we begin since the first substituent is found at C3 in both numberings.

(ii) If there is a substituent an equal distance away from both ends of the parent chain, begin numbering at the end nearer the *second substituent*. We do not have a second substituent, so either numbering is correct.

Thus, we have a methyl group attached to C3. So we write: **3-methyl.**

IUPAC RULE 4. Write the name of the compound. The name is a single word and we use hyphens to separate different prefixes (*di-, tri-, tetra-,* and so forth) and we use commas to separate numbers.

In this case we do not have any prefix so the name will be **3-methylpentane.**

(b) Step 1. Identify the longest carbon chain:

$$\text{CH}_3$$
$$|$$
$$\text{CH}_3\text{CH}_2 \text{—} \text{C} \text{—} \text{CH}_2\text{CH} \text{—} \text{CH}_2\text{CH}_3$$
$$|\ \ \ \ \ \ \ \ \ |$$
$$\text{CH}_3\ \ \ \ \text{CH}_2\text{CH}_2\text{CH}_3$$

It has *eight* carbon atoms so the parent hydrocarbon is **octane.**

Step 2. Number the carbon atoms in the main chain beginning from the left since in this way we find a first substituent at C3 (and not at C4 as we would have found by beginning from the right).

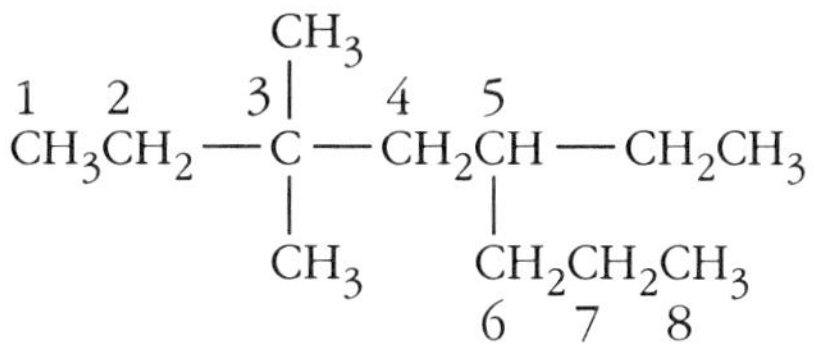

$$\underset{1}{CH_3}\underset{2}{CH_2}-\underset{\underset{\underset{\underset{}{CH_3}}{|}}{\overset{\overset{CH_3}{|}}{C}}}{\overset{3}{}}-\underset{4}{CH_2}\underset{\underset{\underset{\underset{\underset{8}{CH_3}}{}}{\underset{7}{CH_2}}}{\underset{6}{CH_2}}}{\overset{5}{CH}}-\underset{}{CH_2CH_3}$$

Step 3. Identify the substituents and assign their positions. On carbon C3 we find two methyl groups. We write them twice and both methyl groups get the same number. On carbon C5 we find an ethyl group. Thus, we write the substituents:

3-methyl
3-methyl
5-ethyl

Step 4. Write the name of the compound.
(i) Arrange the substituents alphabetically:

5-ethyl
3-methyl
3-methyl

(ii) Since we have two identical substituents we use a prefix (*di-*) and the number (locant) is repeated twice. However, the prefixes are not used for alphabetizing purposes. Thus, we alphabetize *dimethyl* as *methyl* and the name of the compound is:

5-ethyl-3,3-dimethyloctane.

(c)

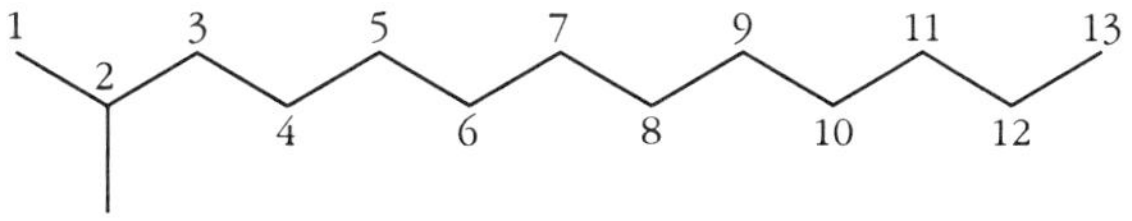

Name: **2-methyltridecane.**

(d) In this exercise, there is a choice in the designation of the longest continuous carbon chain. We have a choice between three chains, each containing *seven* carbon atoms (so the parent chain is a **heptane**).

Step 1 (ii): If two or more different chains of equal length are present, choose the one with the largest *number* of substituents. Thus we choose the following parent chain:

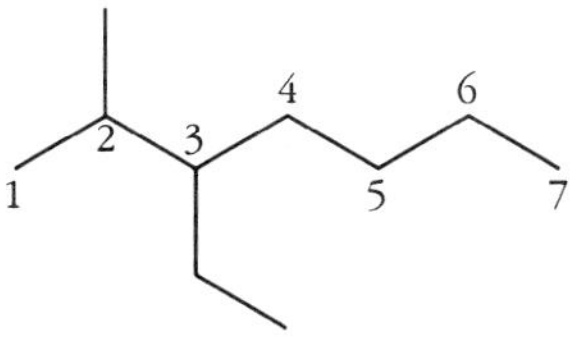

Name: **3-ethyl-2-methylheptane.**

(e)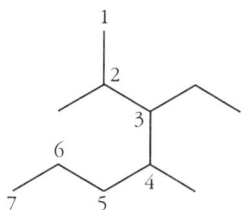

Name: **4-ethyl-3,6-dimethylnonane** (*Remember:* we alphabetize *dimethyl* as *methyl*).

(f)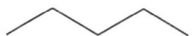

Name: **3-ethyl-2,4-dimethylheptane** (*Remember:* we alphabetize *dimethyl* as *methyl*).

2.5 To write the structure of a compound from its IUPAC name, first identify its parent chain and write its line structure. Then number the chain and place the substituents on the appropriate carbon atom of the chain. Let's apply this to writing the line structure of 2-methylpentane.

(a) The parent chain in this molecule is pentane, a five carbon atom chain:

Next number the carbon atoms. Numbering from left to right or vice versa makes no difference.

Finally we place the methyl group on carbon atom 2 of the chain to give the correct structure of 2-methylpentane.

(a) 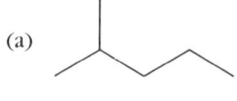 (b)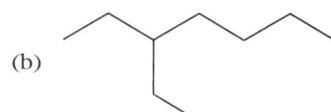

(c)  (d)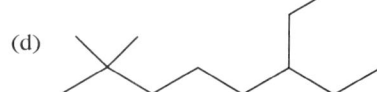

(e)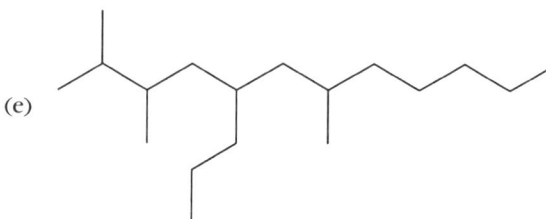

2.6 The structure of 1-ethylhexane is

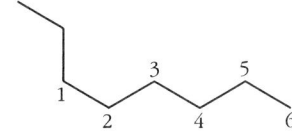

because the name *1-ethylhexane* means that an ethyl group is attached to the carbon number 1 of a six-carbon chain. This compound is incorrectly named because the parent chain contains eight carbon atoms not six. Thus, the correct IUPAC name for the compound is **octane.** *Remember* that it is incorrect to have an alkyl group on carbon atom 1 of the parent chain of a hydrocarbon.

2.7 (a) IUPAC name: 2,2,6,6-tetramethyl-8-(1,1-dimethylethyl)undecane. The alternative name, which is accepted by IUPAC, is 8-*tert*-butyl-2,2,6,6-tetramethylundecane

Note: (i) *tetra-, di-,* and *tert-* are not considered while alphabetizing substituents, (ii) *methyl* is located before *methylethyl* while alphabetizing substituents.

(b) IUPAC name: 4-methyl-5-(1-methylethyl)decane.
The alternative name, which is accepted by IUPAC, is 5-isopropyl-4-methyldecane.
Please note: *isopropyl* is one word and all the letters (including *iso*) are considered while alphabetizing substituents.

(c) IUPAC name: 2,3,10-trimethyl-6-(2-methylpropyl)-7-(1-methylethyl)dodecane. The alternative name, which is accepted by IUPAC, is 6-isobutyl-7-isopropyl-2,3,10-trimethyldodecane.

(d) 10-(1-Methylethyl)-5-(2-methylpropyl)-7-methyltetradecane.
The alternative name, which is accepted by IUPAC, is 5-isobutyl-10-isopropyl-7-methyltetradecane.

(e) 4,10-Diethyl-2,14-dimethyl-6-(1-methylpropyl)pentadecane.
The alternative name, which is accepted by IUPAC, is 6-*sec*-butyl-4,10-diethyl-2,14-dimethylpentadecane.

Note: *sec-* is not considered while alphabetizing the substituents.

2.8

(a)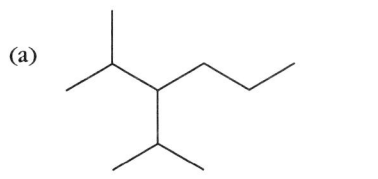

(b)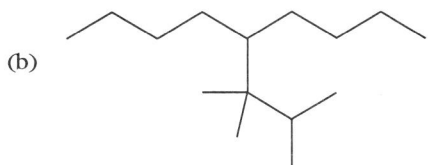

(c)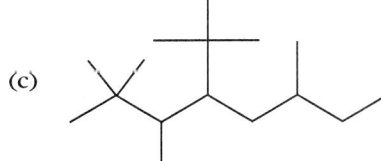

(d)

(e)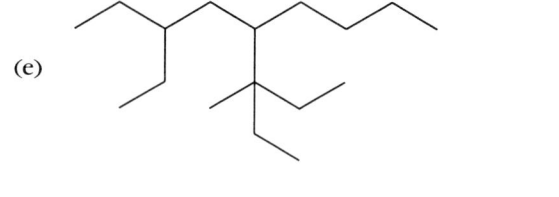

2.9

A: 1-ethylpropyl
(*correct*)

B: 1-ethylpropane
(*incorrect!*) The name of this alkane is *pentane*

The name 1-ethylpropyl is correct because it is the name of an alkyl group. Recall that alkyl groups are named choosing the longest carbon chain of the substituent as the parent name. The parent chain is then numbered beginning at the point of attachment. In 1-ethylpropyl, an *ethyl* is the substituent attached to carbon atom 1 of a propyl group (the same carbon from which a hydrogen atom was removed).

The name 1-ethylpropane is incorrect because it is the name of an alkane. An alkyl group cannot be attached to carbon atom 1 of an alkane because it extends the parent chain. That is why the correct IUPAC name of compound **B** is *pentane*.

2.10 (a) (1-Methylpropyl)cycloheptane
The alternative name, which is accepted by IUPAC, is *sec*-butylcycloheptane.
(b) 1,2,3,4-Tetramethylcyclobutane
(c) 1-Cyclobutyl-2,3-dimethylhexane
(d) 4-Ethyl-1-methyl-2-(1,1-dimethylethyl)cyclooctane
The alternative name, which is accepted by IUPAC, is 2-*tert*-butyl-4-ethyl-1-methylcyclooctane
(e) 1-Methyl-3-(1-methylethyl)cyclodecane
The alternative name, which is accepted by IUPAC, is 3-isopropyl-1-methylcyclodecane.

2.11

(a)   (b)

(c)   (d)

2.12 (a) 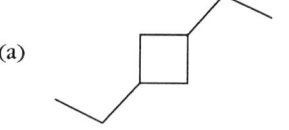 (b) 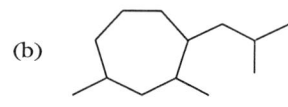 (c)

2.13 2,2-Dimethylpropane (neopentane) is the most nearly spherical, pentane is linear (least spherical).

2.14 $C_nH_m + O_2 \rightarrow CO_2 + H_2O$
% Carbon: (mass of CO_2/mass of sample) × (12.01 g C/44.01 g CO_2) × 100% = (26.53/9.76) × 27.29% = 74.18%
% Hydrogen: (mass of H_2O/mass of sample) × (2 × 1.008 g H/18.02 g H_2O) × 100% = (21.56/9.76) × 11.19% = 24.72%
Assuming a 100 g sample,
Moles C = (74.18 g) (1 mol/12.01 g) × 100% = 6.18 mol C
Moles H = (24.72 g) (1 mol/1.008 g) × 100% = 24.52 mol H

The empirical formula is $C_{6.18}H_{24.52}$. Dividing the subscripts by 6.18 and rounding up gives CH_4 for the empirical formula. Thus, the empirical formula molar mass is 16, and because the molecular weight of the gas is also 16, the molecular formula of the gas is CH_4.

We divide the numbers in the empirical formula by 6.18 in order to get the smallest ratio of integers for the relative number of atoms. Usually we divide by the smallest number of moles (here 6.18 is smaller than 24.52) in order to get number 1 as the smallest integer.

2.15 (a) 2-Iodo-2-methyloctane,
(b) 1-Bromo-2-chloro-3-fluorocyclopropane,
(c) 1-Bromo-3-(1-methylethyl)cyclohexane,
The alternative name, which is accepted by IUPAC, is 1-bromo-3-isopropylcyclohexane.

2.16

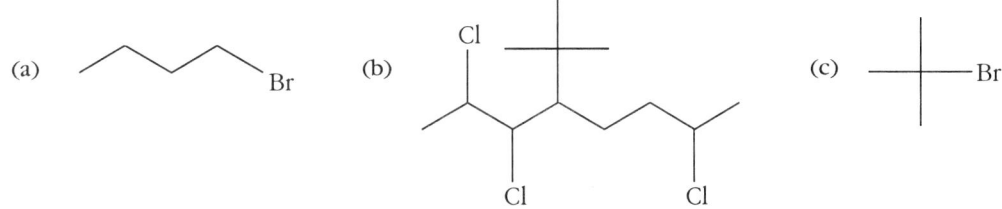

2.17

2.18 (a) Carbon-carbon double bond (alkene)
(b) Aromatic ring and carboxyl group (two functional groups)
(c) 2° (secondary) amine
(d) Ketone
(e) 1° (primary) amine
(f) Alcohol (hydroxyl group)
(g) Triple bond (alkyne)
(h) Halogen (chlorine)
(i) Aldehyde

2.19

(a) 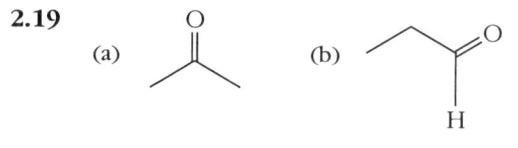 (b)

2.20

(a) 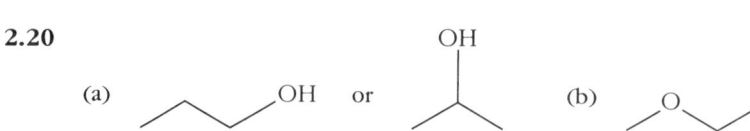 (b)

2.21 (a) An **alkyl group** is a molecular fragment derived from an alkane by the removal of one hydrogen atom. The simplest alkyl group is a methyl group (CH_3—) which is obtained by the removal of one hydrogen atom of methane. The general symbol for an alkyl group is R—.

(b) An **unsaturated hydrocarbon** is a hydrocarbon that contains at least one carbon-carbon multiple bond: a double bond (alkene) or a triple bond (alkyne). The name "unsaturated" stems from the fact that these compounds have fewer hydrogen atoms per carbon than related alkanes.

(c) A **molecular formula** specifies the atomic composition of a compound. In the molecular formulas of organic compounds, carbon is cited first, hydrogen second, and the remaining elements in alphabetical order. Subscripts are used to indicate the number of atoms of each element in a molecule.

(d) A **branched chain alkane** is an alkane in which at least one of the main chain carbons bears an alkyl group in the place of a hydrogen. Such an alkane has at least one 3° (tertiary) or 4° (quaternary) carbon.

(e) An **acyclic compound** (or noncyclic compound) is a molecule in which carbon atoms are *not* arranged in a ring. Sometimes such compounds are called open-chain compounds.

(f) **Ethers** are a class of compounds with an oxygen atom bonded to two carbon atoms that are part of alkyl or aryl groups.

(g) **Ketones** are a class of compounds in which the carbon atom of the carbonyl group ($\diagdown$C=O) is bonded to two alkyl or aryl groups.

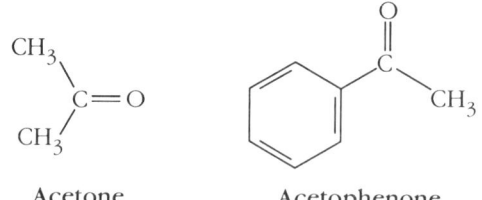

Acetone   Acetophenone

(h) A **cycloalkane** is an alkane in which some of the carbon atoms form a ring. The smallest cycloalkane is cyclopropane, where the three carbon atoms form a three-membered ring.

(i) A **hydrocarbon** is an organic compound composed of only hydrogen and carbon.

(j) A **functional group** is the group of atoms and their associated bonds that defines the structure and chemical properties of a particular family of compounds.

(k) **Empirical formula** is the simplest formula that gives the relative numbers of different atoms in a compound. In some cases the empirical and molecular formulas are identical; in other cases the molecular formula is a "multiple" of the empirical formula. For instance, the empirical formula for acetic acid is (CH$_2$O) while the molecular formula for acetic acid is C$_2$H$_4$O$_2$. Please note that the empirical formula for glucose is also CH$_2$O, while the molecular formula is C$_6$H$_{12}$O$_6$.

(l) **Amides** are a class of compounds in which an NH$_2$ group is attached to the carbonyl carbon atom. Examples include acetamide and *N*-methylbenzamide.

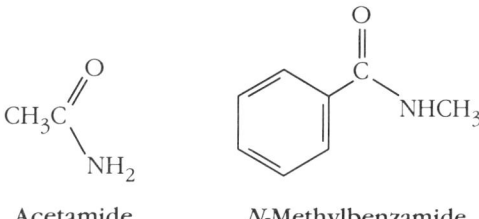

Acetamide   *N*-Methylbenzamide

(m) A **2° hydrogen atom** is a hydrogen atom attached to a secondary (2°) carbon atom. A *secondary carbon atom* is attached to two other carbon atoms. This classification applies only to saturated carbon atoms and hydrogen atoms attached to them.

(n) **Hydrophobic** is a term meaning "water-hating" and it applies to substances (or parts of molecules) that are not soluble in water. Hydrophobic molecules (or hydrophobic parts of compounds) are soluble in nonpolar, hydrocarbonlike media.

(o) **Line structures (bond-line formulas)** are simplified Lewis structures in which neither carbon nor hydrogen atoms are shown explicitly. A carbon atom is assumed to be at the end of each line and at the intersection of any two, three, or four lines. The correct number of hydrogen atoms is assumed to be bonded to each carbon atom. The following are line structures of pentane and hexane.

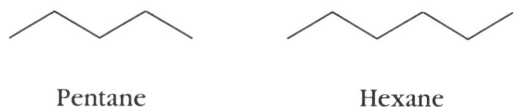

Pentane   Hexane

(p) **Amines** are a class of organic compounds that contain a nitrogen atom bonded to one, two, or three organic groups. Amines can be primary, 1° (RN̈H$_2$), secondary, 2° (RN̈HR), or tertiary, 3° (:NR$_3$).

(q) **Aldehydes** are a class of organic compounds in which the carbon atom of the carbonyl group (C=O) is bonded to a hydrogen atom and an alkyl or aryl group, R—CHO or Ar—CHO. Examples include acetaldehyde and benzaldehyde.

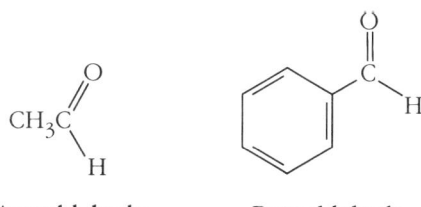

Acetaldehyde   Benzaldehyde

(r) **Saturated hydrocarbons** are hydrocarbons in which there are only carbon-carbon single bonds.

(s) **Constitutional isomers** are compounds having the same molecular formula but their atoms are connected in different orders.

(t) A **1° (primary) carbon atom** is a carbon atom bonded to only one other carbon atom.

(u) **Aromatic compounds** are a class of compounds that contain an aromatic ring. Benzene is a common example of an aromatic compound.

Benzene

(v) **Haloalkanes** (also called *alkyl halides*) are a class of compounds in which at least one hydrogen atom of an alkane is replaced by a halogen. The structure of *haloalkanes* is abbreviated as R—X, in which X corresponds to a halogen atom (F, Cl, Br, or I).

(w) **Carboxylic acids** are a class of compounds which have a carboxyl group (—COOH) attached to a hydrogen, alkyl, or aryl group. Examples of carboxylic acids include *methanoic* (common name: *formic*) *acid* and *ethanoic acid* (common name: *acetic acid*).

Formic acid   Acetic acid

(x) **Alcohols** are a class of compounds that contain a *hydroxyl* (—OH) group. The structure of alcohols is abbreviated as ROH and examples include *methanol* (CH_3OH) and *ethanol* (CH_3CH_2OH).

(y) **Esters** are a class of compounds that have the general formula R—COOR'. The following are examples of esters:

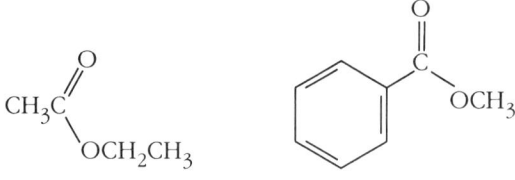

Ethyl acetate   Methyl benzoate

(z) A **straight chain alkane** is a hydrocarbon in which carbon atoms are arranged in a straight chain. Sometimes such compounds are called *unbranched alkanes*. Examples include methane, ethane, propane, etc.

2.22 One way to solve such exercises is to name each of the compounds to be compared. Structural formulas that represent the same compound must have the same name.

(a) 2-Methylbutane   Pentane   2-Methylbutane
(common name: *isopentane*)   (common name: *isopentane*)
Thus, the first and third structures represent the same compound (2-methylbutane) but the middle one represents a different alkane (pentane).

(b) 1-Chloro-2-methylpropane   1-Chloro-2-methylpropane
1-Chloro-2-methylpropane
Thus, all three structures represent the same compound.

(c) 3-Methylhexane   3-Methylhexane   3-Methylhexane
Thus, all three structures represent the same compound.

2.23 (a) 2,3-Difluoro-2-methylbutane
Note: (i) Halides and alkyl groups are treated equally for the purpose of alphabetizing.
(ii) Prefixes (*di-, tri-, hexa-, tert-,* etc.) are *not* considered for the purpose of alphabetizing.
(b) Hexachloroethane
(c) 2-Chloro-5-cyclohexyl-4-methylheptane
(d) 2,3-Dimethyl-1,4-diphenylbutane
(e) 3,4,6,7,8,9-Hexamethyldodecane
(f) 1,1-Dibromocyclopropane
(g) 11-Ethyl-9-methyl-4-(1-methylethyl)pentadecane
The alternative name, which is accepted by IUPAC, is
11-ethyl-4-isopropyl-9-methylpentadecane.
(h) 2-Fluoro-1-iodo-4-methylcyclohexane
(i) 1-Cyclopropyl-4-(1,1-dimethylethyl)cyclooctane or 1-cyclopropyl-4-*tert*-butylcyclooctane

2.24

2.25 (a) 2-*tert*-Butylpentane

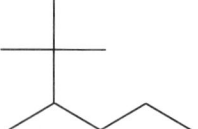

IUPAC RULE 1 was not correctly applied because *the longest chain in the molecule* was *not* properly chosen. It should include the two carbon atoms of the *tert-butyl* group.
The correct IUPAC name is **2,2,3-trimethylhexane.**

(b) 4-Methylpentane

IUPAC RULE 3 was not correctly applied because the carbon atoms were *not* numbered properly. The numbering should *begin at the end that is closest to a substituent.* If we begin numbering with the other end, we get the methyl group at carbon atom number 2.
Thus, the correct IUPAC name is **2-methylpentane.**

(c) 2-Chloro-2-butylpentane

IUPAC RULE 1 was not correctly applied because *the longest chain in the molecule* was *not* properly chosen. The longest chain should include the carbon atoms of the *butyl* group.
Thus, the correct IUPAC name is **4-chloro-4-methyloctane.**

(d) 1,5-Dimethylpentane

IUPAC RULE 1 was not correctly applied because *the longest chain in the molecule* was *not* properly chosen. The longest chain should include the carbon atoms of the *1-methyl* and *5-methyl* groups.
Thus, the correct IUPAC name is **heptane.**

(e) 3-Cyclohexylbutane

(i) **IUPAC RULE 3** was not properly applied because the carbon atoms in *butane* were *not* numbered properly. One should *begin at the end that is closest to a substituent.* If we begin numbering with the other end, we get the cyclohexyl group at C-2.

(ii) The *butane* chain is incorrectly chosen as the parent structure. *If the cyclic portion of the compound contains the same or higher number than the acyclic portion, the ring becomes the parent structure.*

Thus, the proper IUPAC name is **(1-methylpropyl)cyclohexane.**

(f) 2-Dimethylpentane

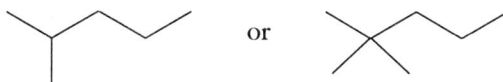

The prefix *di-* is used to indicate that the following substituent appears twice in the parent structure. In such a case the second *locant* should be used to indicate the position of the second methyl group. Since the second *locant* is missing we cannot be sure if the prefix *di-* is used unnecessarily or if the second *locant* is simply missing.

Thus, the proper IUPAC name might be **2-methylpentane** or **2,2-dimethylpentane.**

2.26

(a), (b), (c), (d), (e), (f), (g)

2.27

(a), (b), (c), (d)

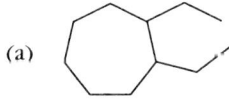

(e) 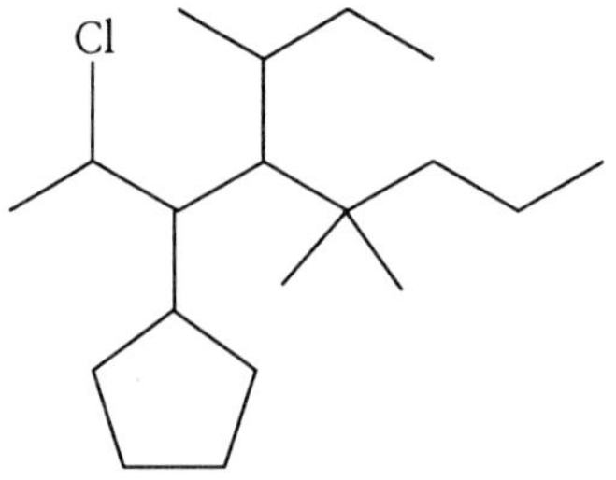

(f) 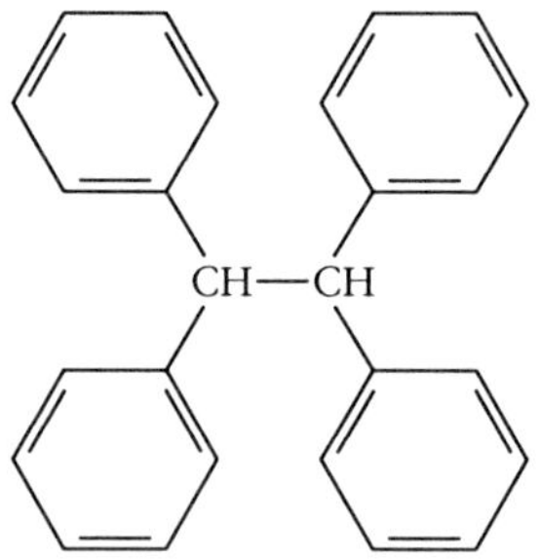

(g) 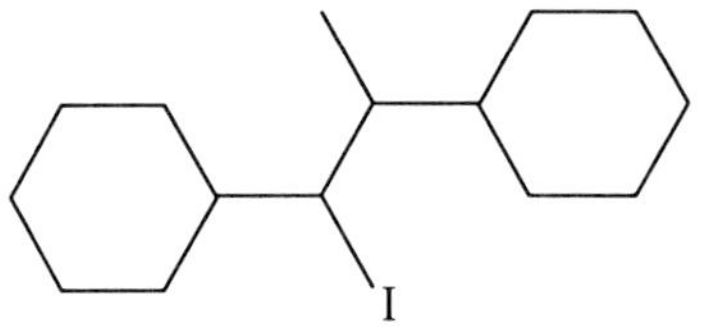

(h) 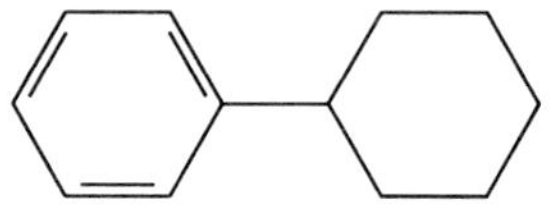

(i) 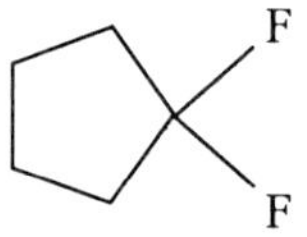

(j) 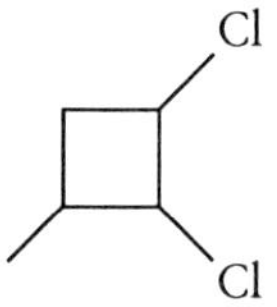

2.28

(a) 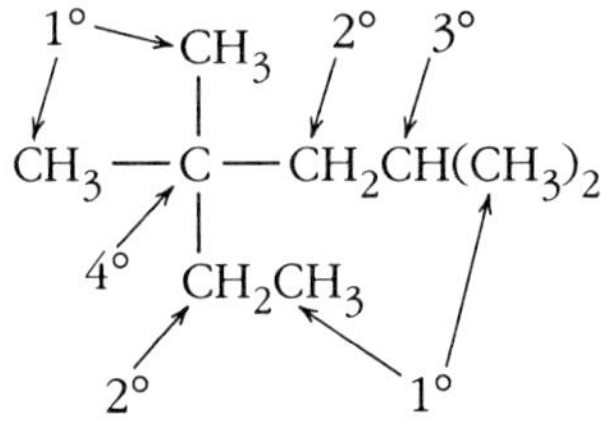

(b) 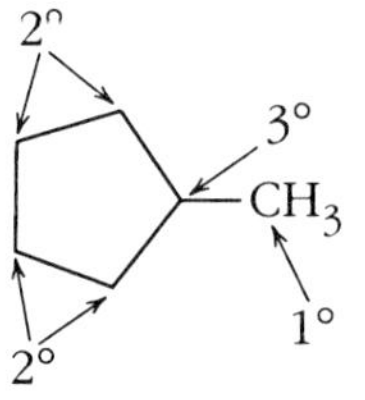

(c) 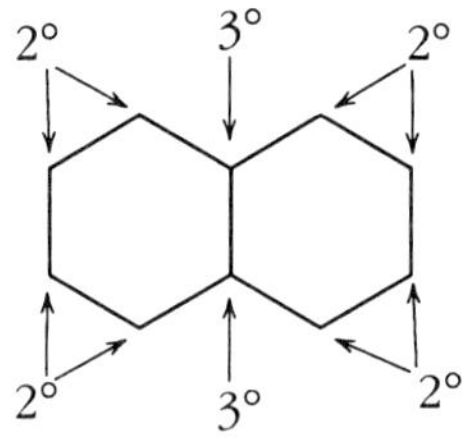

(d) 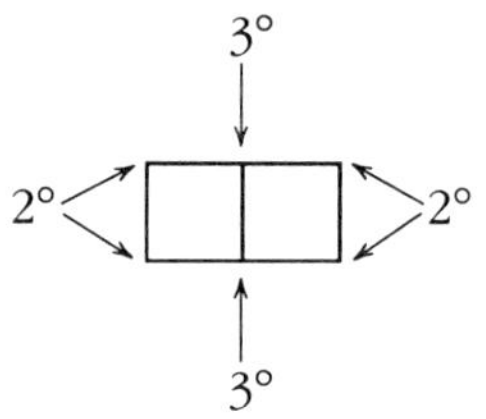

2.29

(a) 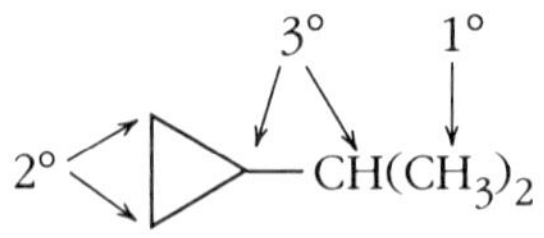

(b) 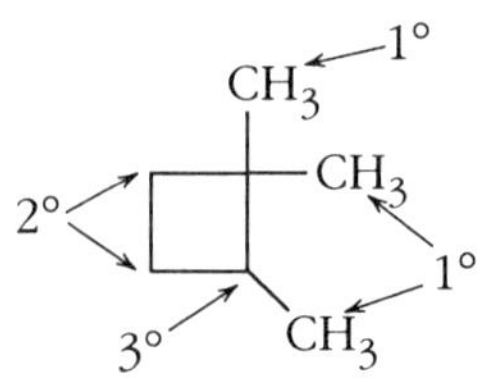

(c) 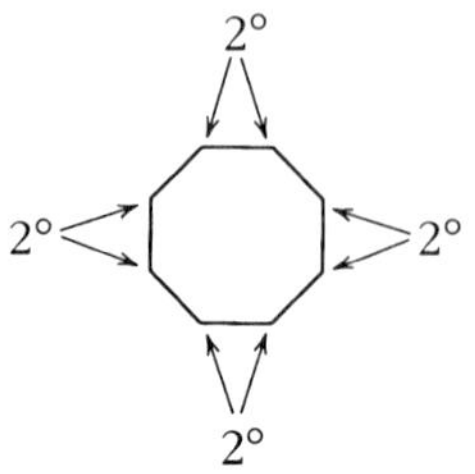

(d) 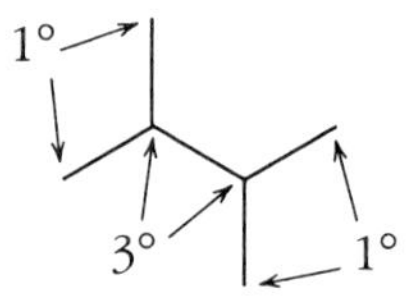

2.30 (a) There are many possible answers. The simplest is 2,2-dimethylpropane, the second simplest is 2,2,3,3-tetramethylbutane, etc.

(b) First, let us write the partial structures available:
CH CH$_2$ CH$_2$ CH$_3$ CH$_3$ CH$_3$
and then arrange them keeping in mind that carbon is tetravalent:

$$\begin{array}{c} H_3C \\ \diagdown \\ H_3C \diagup \end{array} CH-CH_2-CH_2-CH_3 \quad \text{2-Methylpentane}$$

(c) First, let us write the partial structures available:
—CH$_3$—CH$_3$—CH$_2$—
and then arrange them as follows:
CH$_3$—CH$_2$—CH$_3$ (propane)

(d) Primary (1°) hydrogen can be found only in a methyl group. To have 12 primary hydrogen atoms we must have (12:3 = 4) 4 methyl groups. Secondary (2°) hydrogens are found in methylene groups (—CH$_2$—). To have 8 secondary hydrogens we must have (8:2 = 4) 4 methylene groups. Thus, we have (4 + 4 = 8) 8 carbon atoms and all the necessary hydrogen atoms. We should have 9 carbon atoms, so one carbon atom must bear no hydrogen atoms (must be quaternary, 4°).

Now, let us list the partial structures available:
CH$_2$ CH$_2$ CH$_2$ CH$_2$ CH$_3$ CH$_3$ CH$_3$ CH$_3$ C
and then arrange them as follows:

$$\begin{array}{c} CH_3CH_2 \diagdown \diagup CH_2-CH_3 \\ C \\ CH_3CH_2 \diagup \diagdown CH_2CH_3 \end{array} \quad \text{3,3-Diethylpentane}$$

2.31

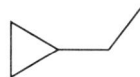   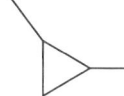

Ethylcyclopropane   1,1-Dimethylcyclopropane   1,2-Dimethylcyclopropane

There are three isomers of 1,2-dimethylcyclopropane, as we will learn in Chapter 4.

2.32 (a) Octane; propane; dodecane
All of these compounds are straight-chain alkanes. The intermolecular forces present are only London forces. Since molecules with relatively large surfaces experience greater London attractions it follows that propane, having the smallest molecule, should have the lowest boiling point. The molecular weight of propane is lower than any of the other alkanes listed which also contributes to the lower boiling point of propane.

(b) Heptane; 2-methylhexane; 3,3-dimethylpentane
All of the above compounds are alkanes with the same molecular weight. Only their structures are different.
  Branched alkanes have smaller surface areas than their straight-chain isomers. As a result, they experience smaller London attractions and are unable to pack as well. The weaker attractions result in lower boiling points, so 3,3-dimethylpentane should have the lowest boiling point.

(c) Methanol; ethane; cyclooctane
Methanol forms strong hydrogen bonds in the liquid phase. Ethane's molecule is roughly the same size as that of methanol, but London forces are the only intermolecular interactions in alkanes so the boiling point of ethane will be much lower than that of methanol. Cyclooctane is also an alkane but with a much larger molecule leading to greater London attractions and thus a much higher boiling point. As a result, ethane should have the lowest boiling point.

2.33 The straight-chain five-carbon alkane (*pentane*) has three different types of hydrogen atoms. Therefore, we have three different alkyl groups resulting from removal of one of the hydrogen atoms.

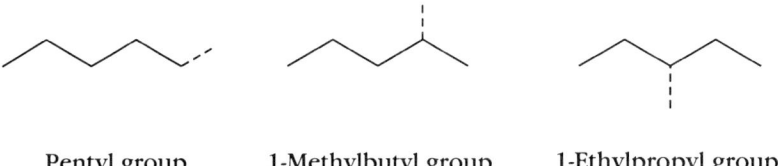

Pentyl group   1-Methylbutyl group   1-Ethylpropyl group

Another possible arrangement of five carbon atoms is that in 2-methylbutane. In this case we have four different types of hydrogen atoms, and thus four different alkyl groups

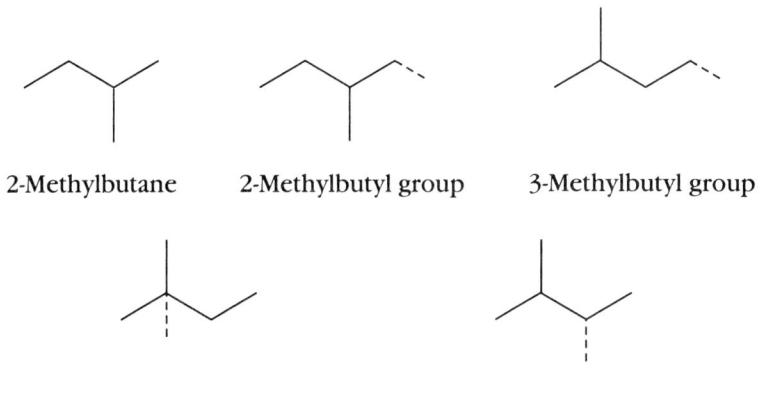

2-Methylbutane   2-Methylbutyl group   3-Methylbutyl group

1,1-Dimethylpropyl group   1,2-Dimethylpropyl group

The third possible arrangement of five carbon atoms is that in 2,2-dimethylpropane. Here we have only one type of hydrogen atom and, as a result, we have only one alkyl group originating from this structure.

2,2-Dimethylpropane   2,2-Dimethylpropyl group

2.34 (a)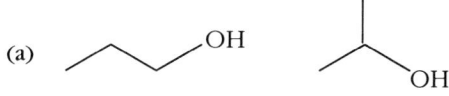

(b) [structures: diethyl ether; methyl propyl ether; methyl isopropyl ether]

(c) Any two of the following:

[structures: ethyl acetate; methyl propanoate; propyl formate; isopropyl formate]

(d) [structures: pentanal; 3-methylbutanal; 2-methylbutanal; 2,2-dimethylpropanal]

(e) Any five of the following:

[structures: cyclopentanone; 2-methylcyclobutanone; 3-methylcyclobutanone; 2-methylcyclopropanone with methyl; 2-ethylcyclopropanone; 2,2-dimethylcyclopropanone]

2.35
CH₃CH₂CH₂NH₂   (CH₃)₂CHNH₂
1° (primary)   1° (primary)
1-Propanamine   1-Methylethanamine
                                (common name: isopropylamine)

CH₃CH₂NHCH₃   N(CH₃)₃
2° (secondary)   3° (tertiary)
N-Methylethanamine   *N,N*-Dimethylmethanamine

2.36 Bromine constitutes 94.85% of the molecule of MW 253. Thus, we have (94.85 × 253) : 100 = 239.97 g of bromine, which means that we have 239.97:79.9 = 3 bromine atoms in the molecule.

    Hydrogen constitutes 0.40% of the molecule of MW 253. Thus, we have (0.40 × 253) : 100 = 1.01 g of hydrogen, which means that we have 1.01 : 1.01 = 1 hydrogen atom in the molecule.

    Carbon constitutes 4.75% of the molecule of MW 253. Thus, we have (4.75 × 253) : 100 = 12.02 g of carbon, which means that we have 12.02 : 12.01 = 1 carbon atom in the molecule.

    As a result, we get the molecular formula of bromoform: CHBr₃

2.37 $C_lH_mO_n + x\, O_2 \rightarrow l\, CO_2 + m/2\, H_2O$

10.53 mg sample gives 9.78 mg of water.
1 mole of water (MW = 18.016 g) contains 2 atoms of hydrogen (from the molecular formula).

9.78 mg of water contains:

(9.78 × 2.016) : 18.016 = 1.09 mg of hydrogen

Thus, hydrogen constitutes (1.09 : 10.53) × 100% = 10.35% of the sample.
1 mole of CO_2 (MW = 44 g) contains 12.01 g of carbon (from the formula) 23.92 mg of carbon dioxide contains:

(23.92 × 12.01) : 44 = 6.53 mg of carbon.

Thus, carbon constitutes (6.53:10.53) × 100% = 62.01%

Oxygen constitutes the rest:

100% − (10.35% + 62.01%) = 100% − 72.36% = 27.64%

Molecular weight of the molecule (MW) = 116; so we have 10.35% of 116 g of hydrogen in one molecule; this is (116 × 10.35) : 100 = 12.006 g of hydrogen (or 12 atoms).

62.01% of 116 g of carbon; which is (116 × 62.01) : 100 = 71.93 g of carbon (or 71.93 : 12.01 = 6 atoms).

27.64% of 116 g of oxygen in one molecule; this is (116 × 27.64) : 100 = 32.06 g of oxygen (or 32.06 : 16 = 2 atoms).

Thus, the molecular formula is $C_6H_{12}O_2$.

2.38 There are more correct answers than the question asks for. Any appropriate combination of the following answers is acceptable.

(a) CH_3CH_2COOH CH_3COOCH_3 $HCOOCH_2CH_3$

(b)

(c)

2.39

(a) $HOCH_2CH_2\overset{\overset{OH}{|}}{\underset{\underset{CH_3}{|}}{C}}CH_2COOH$

(b)

carboxyl
hydroxyl (2)

carbonyl (ketone), hydroxyl,
carbon-carbon double bond (alkene)

(c)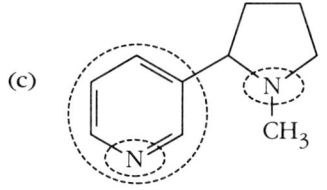

aromatic ring
amine (tertiary) (2)

(d)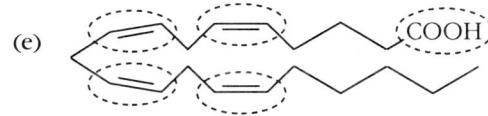

carbon-carbon double bond (5)
hydroxyl

(e)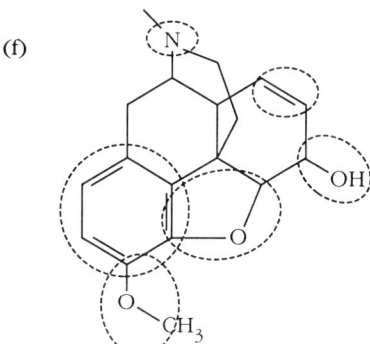

carbon-carbon double bond (4)
carboxyl

(f)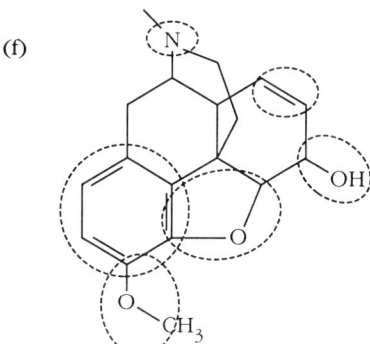

amine (tertiary), aromatic ring, carbon-carbon double bond, ether (2), hydroxyl

CHAPTER 3

Concepts

1. A compound is a **Brønsted acid** when it reacts by donating a proton. A **Brønsted base** reacts by accepting a proton. A pair of species that can be interconverted by the loss and gain of a proton is called a **conjugate acid-base pair**. The conjugate acid of any base will have an additional hydrogen and an increase in positive charge (or a decrease in negative charge). The conjugate base of an acid will have one hydrogen fewer and will have an increase in negative charge (or a decrease in positive charge). Brønsted acids are also called **protic acids** because they react via the transfer of a proton.

2. The strength of a Brønsted acid is indicated by the magnitude of its dissociation constant, K_a. Because some K_a values are very small and some are large, they are best expressed as logarithms. Thus, pK_a is defined as $-\log K_a$. Strong acids have negative values of pK_a while weak acids have positive values of pK_a. *The weaker the acid, the stronger its conjugate base. The weaker the acid, the more positive its pK_a value.* The strength of a base is expressed by the pK_a of its conjugate acid.

3. The **Lewis theory of acids and bases** defines an acid as an electron-pair acceptor and a base as an electron-pair donor. Thus, a proton is only one of a large number of species that may function as a Lewis acid. Lewis acids can also be *aprotic acids,* which are compounds that react with bases by accepting pairs of electrons, not by donating protons.

4. The **equilibrium constant, K_{eq}** for a reaction is related to the standard free-energy difference, $\Delta G°$, between products and reactants by the relationship $\Delta G° = 2.3\ RT \log K_{eq}$. Reactions with $K_{eq} < 1$ have positive $\Delta G°$ values and favor reactants at equilibrium. Reactions with $K_{eq} > 1$ have negative $\Delta G°$ values and favor products at equilibrium.

5. The **curved-arrow formalism** is a symbolism used to depict the flow of electrons in chemical reactions. A curved arrow is drawn from the site of electron density, such as a pair of bonding or nonbonding electrons, to the site of electron deficiency, such as an atom with a positive charge or a partial positive charge. The curved arrows can also be used to derive resonance structures that are related by the movement of one or more electron pairs.

Solutions to the Exercises

3.1 The structure of the conjugate acid of a base is obtained by adding a proton to the structure of the base.

(a) $HO^- + H^+ \longrightarrow HOH$

                                  Conjugate
                                  acid of HO^-

(b) $Br^- + H^+ \longrightarrow HBr$

                                    Conjugate
                                  acid of Br^-

(c) HNO_3
(d) H_2CO_3
(e) NH_4^+
(f) HSO_4^-

3.2 The structure of a conjugate base of an acid is that of the acid minus its acidic proton.

(a) $HClO_4 - H^+ \longrightarrow ClO_4^-$

        Acid   Conjugate
                                base of $HClO_4$

(b) I^-
(c) HO^-
(d) CO_3^{2-}
(e) NH_3
(f) $H_2PO_4^-$

3.3 (a) $HBr + H_2O \longrightarrow Br^- + H_3O^+$

        Conjugate   Conjugate   Conjugate   Conjugate
        acid of Br^- base of H_3O^+ base of HBr   acid of H_2O

(b) $H_2SO_4 + H_2O \longrightarrow HSO_4^- + H_3O^+$

        Conjugate   Conjugate   Conjugate   Conjugate
        acid of HSO_4^- base of H_3O^+ base of H_2SO_4 acid of H_2O

(c) $NH_2^- + H_2O \longrightarrow NH_3 + HO^-$

        Conjugate   Conjugate   Conjugate   Conjugate
        base of NH_3 acid of HO^- acid of NH_2^- base of H_2O

3.4 (a) $K_a = 10^{+5}$
$pK_a = -\log K_a = -\log 10^{+5} = -(+5) = -5$
(b) $K_a = 10^{-6}$
$pK_a = -\log K_a = -\log 10^{-6} = -(-6) = 6$
(c) $K_a = 1.8 \times 10^{-16}$
$pK_a = -\log K_a = -\log(1.8 \times 10^{-16}) = -(0.255 - 16) = -(-15.745) = 15.745$
(d) $K_a = 1.5 \times 10^{+8}$
$pK_a = -\log K_a = -\log(1.5 \times 10^{+8}) = -(0.176 + 8) = -(+8.176) = -8.176$

CHAPTER 3   53

3.5 (a) $pK_a = 10$
$K_a = $ –INVERT LOG $(pK_a) = 10^{-pK_a}$
$K_a = 1 \times 10^{-10}$

(b) $pK_a = -10$
$K_a = $ –INVERT LOG $(pK_a) = 10^{-pK_a}$
$K_a = 1 \times 10^{-(-10)} = 1 \times 10^{10}$

(c) $pK_a = 3.5$
$K_a = $ –INVERT LOG $(pK_a) = 10^{-pK_a}$
$K_a = 10^{-3.5} = 10^{+0.5-4} = 10^{0.5} \times 10^{-4} = 3.16 \times 10^{-4}$

(d) $pK_a = -1.5$
$K_a = $ –INVERT LOG $(pK_a) = 10^{-pK_a}$
$K_a = 10^{-(-3.5)} = 10^{+3.5} = 10^{0.5+3} = 10^{0.5} \times 10^{3} = 3.16 \times 10^{3}$

(e) $pK_a = 0.5$
$K_a = $ –INVERT LOG $(pK_a) = 10^{-pK_a}$
$K_a = 10^{-0.5} = 10^{+0.5-1} = 10^{0.5} \times 10^{-1} = 3.16 \times 10^{-1} = 0.316$

3.6 The acidity constant for any generalized acid, HA, is the equilibrium constant:

$$HA + H_2O \rightleftharpoons A^- + H_3O^+$$

$$K_a = \frac{[H_3O^+][A^-]}{[HA]}$$

Stronger acids have their equilibria toward the right and thus have larger acidity constants, K_a, whereas weaker acids have their equilibria toward the left and have smaller acidity constants.

Acid strengths are usually expressed as pK_a values, where pK_a is equal to the negative logarithm of the acidity constant:

$$pK_a = -\log K_a$$

A stronger acid (larger acidity constant, K_a) has a *less positive* pK_a, and a weaker acid (smaller K_a) has a *more positive* pK_a.

(a) Let us compare the pK_a of methanesulfonic acid (-1.8) with that of formic acid (3.7). Methanesulfonic acid has a *more negative* pK_a value (-1.8 is *more negative* than 3.7) so **methanesulfonic acid** is the stronger acid.

(b) Acetic acid (pK_a 4.7) has a *less positive* pK_a value than that of phenol (pK_a 10.0) (4.7 is *less positive* than 10.0) so **acetic acid** is the stronger acid.

(c) Glucose (pK_a 12.3) has a *less positive* pK_a value than that of ethanol (pK_a 15.9) (12.3 is *less positive* than 15.9) so **glucose** is the stronger acid.

3.7 There is an inverse relationship between the acid strength of an acid, pK_a, and the base strength of its conjugate base. The conjugate base of a *stronger* acid is a *weaker* base because it has less affinity for a proton. Similarly, the conjugate base of a *weaker* acid is a *stronger* base.

(a) To find out which of the two bases is stronger, let us compare the pK_a values of their conjugate acids: pK_a for hydrofluoric acid is 3.2; pK_a for carbonic acid equals 3.75. Hydrofluoric acid is a stronger acid (less positive pK_a) and carbonic acid is a weaker acid; that is why a conjugate base of carbonic acid (hydrogen carbonate) is the stronger base.

(b) The pK_a for hydronium ion is -1.7; pK_a for ammonium ion equals 9.24. Hydronium ion is a stronger acid (less positive pK_a) and ammonium ion is the weaker acid; that is why a conjugate base of ammonium ion (ammonia) is the stronger base.

(c) The pK_a for hydrocyanic acid is 9.2; pK_a for sulfuric acid equals ≈ −9. Sulfuric acid is a stronger acid (less positive pK_a) and hydrocyanic acid is a weaker acid; that is why a conjugate base of hydrocyanic acid (cyanide ion) is the stronger base.

3.8

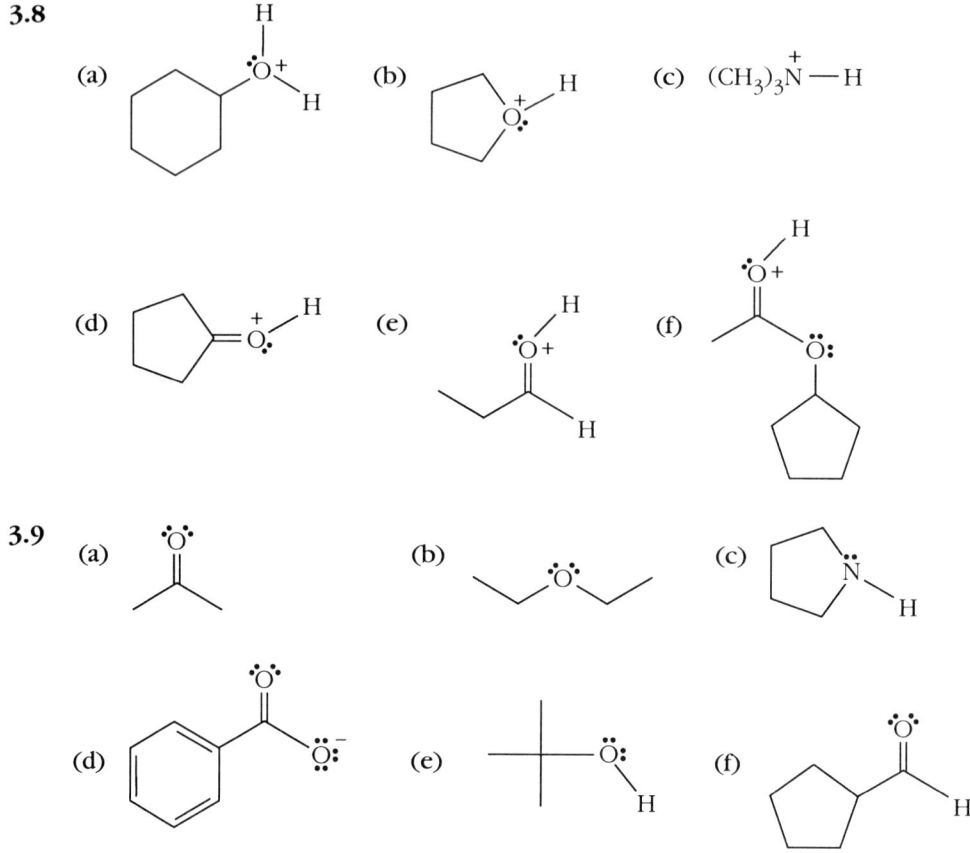

3.9

3.10 (a) First, we list the pK_a for H_2CO_3 and the conjugate acid of CH_3O^-. Thus, the pK_a of H_2CO_3 is 6.4 and pK_a of CH_3OH is 15.2. Then we compare the pK_a values to decide if the proton transfer takes place.
Note: An acid will donate a proton to the conjugate base of any acid weaker than itself (that is, one whose pK_a is more positive). In this case, H_2CO_3 is a stronger acid than methanol (less positive pK_a) and will donate a proton to CH_3O^-.

(b) The pK_a of H_2CO_3 is 6.4 and the pK_a of CH_3COOH is 4.76. H_2CO_3 is a weaker acid than acetic acid (more positive pK_a) so it will *not* donate a proton to CH_3COO^-.

(c) The pK_a of H_2CO_3 is 6.4 and the pK_a of H_3O^+ is −1.7. H_2CO_3 is a weaker acid than protonated water (more positive pK_a) so it will *not* donate a proton to H_2O.

(d) The pK_a of H_2CO_3 is 6.4 and the pK_a of RNH_3^+ is 11. H_2CO_3 is a stronger acid than RNH_3^+, (less positive pK_a) so it will donate a proton to RNH_2.

(e) The pK_a of H_2CO_3 is 6.4 and the pK_a of H_2SO_4 is ≈−9. H_2CO_3 is a weaker acid than sulfuric acid (more positive pK_a) so it will *not* donate a proton to HSO_4^-.

3.11 (a) First, we list the pK_a for the conjugate acids of NH_3 and CH_3OH. Thus, pK_a for NH_4^+ is 9.2 and that of CH_3OH is 15.2. Then we compare the pK_a values and decide if the proton transfer takes place.
Note: An acid will donate a proton to NH_3 if the acid is stronger than the conjugate acid, NH_4^+ (that is, if pK_a of the acid is lower than 9.2). In this case, CH_3OH is a weaker acid than the ammonium cation (more positive pK_a) so it will *not* donate a proton to NH_3.

(b) The pK_a for NH_4^+ is 9.2 and that of CH_3COOH is 4.76. In this case, CH_3COOH is a stronger acid than the ammonium cation (less positive pK_a) so it will transfer a proton to NH_3.

(c) The pK_a for NH_4^+ is 9.2 and that of H_2O is 15.7. In this case, H_2O is a weaker acid than the ammonium cation (more positive pK_a) so it will *not* donate a proton to NH_3.

(d) The pK_a for NH_4^+ is 9.2 and that for H_2SO_4 is -9. In this case, H_2SO_4 is a stronger acid than the ammonium cation (less positive pK_a) so it will transfer a proton to NH_3.

(e) The pK_a for NH_4^+ is 9.2 and that of HF is 3.2. In this case, HF is a stronger acid than the ammonium cation (less positive pK_a) so it will transfer a proton to NH_3.

3.12 (a) First, we write the equation and identify the pairs of conjugate acids and bases in the reaction.

$$RCOOH + R_2NH \rightleftharpoons RCOO^- + R_2NH_2^+$$

Conjugate acid of $RCOO^-$, p$K_a \approx 5$

Conjugate base of $R_2NH_2^+$

Conjugate base of RCOOH

Conjugate acid of R_2NH, p$K_a \approx 10.5$

Second, we list the acids: RCOOH and $R_2NH_2^+$.
Third, we find pK_a values for the acids: 5 for RCOOH and 10.5 for $R_2NH_2^+$.
Fourth, we compare the strength of acids and bases on both sides and decide where the equilibrium lies. In this case, we have a stronger acid and a stronger base on the left side of the equation, so the equilibrium lies to the right.

(b) $$H_3O^+ + ROR \rightleftharpoons H_2O + ROR\text{–}H^+$$

Conjugate acid of H_2O, p$K_a = -1.7$

Conjugate base of $ROR\text{–}H^+$

Conjugate base of H_3O^+

Conjugate acid of ROR, p$K_a \approx -4$

In this case, we have a stronger acid and a stronger base on the right side of the equation, so the equilibrium lies to the left.

3.13 The following equilibrium constants, K_{eq}, were calculated for the *forward* reactions as specified in Exercise 3.12. If K_{eq} is higher than 1, the equilibrium lies to the right. However, if the K_{eq} is lower than 1, the reverse reaction prevails and the equilibrium lies to the left.

(a) pK_{eq} = pK_a(RCOOH) - pK_a($R_2NH_2^+$) = 5 - 10.5 = -5.5
$K_{eq} = 10^{-pK_{eq}} = 10^{-(-5.5)} = 10^{5.5} = 3.16 \times 10^5$

(b) pK_{eq} = pK_a(H_3O^+) - pK_a(R_2OH^+) = -1.7 - (-4) = 2.3
$K_{eq} = 10^{-pK_{eq}} = 10^{-(2.3)} = 10^{-2.3} = 5.0 \times 10^{-3}$

3.14 We solve these problems in the following way. Write the equation, identify acids, find their pK_as, label the weaker acid (based on the difference in the pK_a: *the less positive the pK_a value, the stronger the acid*), and label the weaker base (Note: *the weaker the acid, the stronger its conjugate base*).

(a) $HCl(g)$ + $CH_3OH(l)$ ⇌ $CH_3OH_2^+$ + Cl^-
p$K_a \approx -7$ p$K_a = -2$
                                               Weaker acid   Weaker base

(b) $NaHCO_3(s)$ + $H_2SO_4(l)$ ⇌ $NaHSO_4$ + H_2CO_3
                   p$K_a \approx -9$ p$K_a = 6.4$
                                               Weaker base   Weaker acid

(c) $CH_3NH_2(g)$ + $p\text{-}CH_3C_6H_4SO_3H(l)$ ⇌ $CH_3NH_3^+$ + $p\text{-}CH_3C_6H_4SO_3^-$
                    p$K_a = -0.6$ p$K_a = 11$
                                                     Weaker acid   Weaker base

(d) $NaOH(s)$ + $(CH_3)_3COH(l)$ ⇌ $(CH_3)_3CO^-$ + Na^+ + H_2O
                  p$K_a = 17$ p$K_a = 15.7$
Weaker base   Weaker acid

(e) $NH_3(g)$ + $CH_3CH_2CH_2CH_2Li(l)$ ⇌ $CH_3CH_2CH_2CH_3$ + Li^+ + NH_2^-
p$K_a \approx 38$ p$K_a \approx 50$
                                              Weaker acid   Weaker base

(f) $NaOCH_3(s)$ + $H_2O(l)$ ⇌ CH_3OH + Na^+ + HO^-
                    p$K_a = 15.7$ p$K_a = 15.2$
Weaker base   Weaker acid

(g) $NaNH_2(s)$ + $CH_3CH_2OH(l)$ ⇌ NH_3 + $CH_3CH_2O^-$ + Na^+
                   p$K_a = 16$ p$K_a \approx 38$
                                             Weaker acid   Weaker base

3.15 Please note: The following equilibrium constants, K_{eq}, were calculated for the *forward* reactions as written in Exercise 3.14. If K_{eq} is higher than 1, the equilibrium lies to the right. However, if the K_{eq} is lower than 1, the reverse reaction prevails and the equilibrium lies to the left.

(a) pK_{eq} = pK_a(HCl) − pK_a(CH$_3$OH$_2^+$) = −7 − (−2) = −5
    $K_{eq} = 10^{-pK_{eq}} = 10^{-(-5)} = 10^5$

(b) pK_{eq} = pK_a(H$_2$SO$_4$) − pK_a(H$_2$CO$_3$) = −9 − 6.4 = −15.4
    $K_{eq} = 10^{-pK_{eq}} = 10^{-(-15.4)} = 10^{15.4} = 2.5 \times 10^{15}$

(c) pK_{eq} = pK_a(p-CH$_3$C$_6$H$_4$SO$_3$H) − pK_a(CH$_3$NH$_3^+$) = −0.6 − (11) = −22
    $K_{eq} = 10^{-pK_{eq}} = 10^{-(11.6)} = 10^{11.6}$

(d) pK_{eq} = pK_a((CH$_3$)$_3$COH) − pK_a(H$_2$O) = 17 − 15.7 = 1.3
    $K_{eq} = 10^{-pK_{eq}} = 10^{-1.3} = 5 \times 10^{-2}$

(e) pK_{eq} = pK_a(NH$_3$) − pK_a(CH$_3$CH$_2$CH$_2$CH$_3$) = 38 − 50 = −12
    $K_{eq} = 10^{-pK_{eq}} = 10^{-(-12)} = 10^{12}$

(f) pK_{eq} = pK_a(H$_2$O) − pK_a(CH$_3$OH) = 15.7 − 15.2 = 0.5
    $K_{eq} = 10^{-pK_{eq}} = 10^{-0.5} = 3.16 \times 10^{-1} = 0.316$

(g) pK_{eq} = pK_a(CH$_3$CH$_2$OH) − pK_a(NH$_3$) = 16 − 38 = −22
    $K_{eq} = 10^{-pK_{eq}} = 10^{-(-22)} = 10^{22}$

3.16 (a) NH_3 + H_3O^+ ⇌ NH_4^+ + H_2O

Lewis base — Lewis acid

(b) BF_3 + F^- ⇌ BF_4^-

Lewis acid — Lewis base

(c) $AlCl_3$ + Cl^- ⇌ $AlCl_4^-$

Lewis acid — Lewis base

(d) CH_3COCH_3 + BF_3 ⇌ $(CH_3)_2CO^{+-}BF_3$

Lewis base — Lewis acid

(e) NH_3 + BCl_3 ⇌ $H_3N^{+-}BCl_3$

Lewis base — Lewis acid

3.17 (a) Step 1: Rewrite the equation to show all nonbonding electrons:

$$H-\overset{+}{\underset{H}{\overset{H}{N}}}-H \ + \ {}^-{:}\ddot{\underset{..}{O}}-H \ \rightleftharpoons \ H\cdots\overset{\cdot\cdot}{N}(H)(H) \ + \ H-\overset{\cdot\cdot}{\underset{\cdot\cdot}{O}}-H$$

Step 2: Identify the electron donor and electron acceptor.

One of the lone pairs of electrons of HO⁻ becomes the new bond between oxygen and hydrogen. Therefore, the oxygen of the hydroxide ion is the electron donor and the nitrogen is the electron acceptor.

Step 3: Draw an arrow from the source of electrons, one lone pair on the oxygen, to their destination, the hydrogen atom of H_4N^+.

$$H-\overset{+}{\underset{H}{\overset{H}{N}}}-H \ \curvearrowleft \ {}^-{:}\ddot{\underset{..}{O}}-H \ \rightleftharpoons \ H\cdots\overset{\cdot\cdot}{N}(H)(H) \ + \ H-\overset{\cdot\cdot}{\underset{\cdot\cdot}{O}}-H$$

Step 4: Add a second arrow indicating the electron pair displacement by the first electron movement.

$$H-\overset{+}{\underset{H}{\overset{H}{N}}}\curvearrowright H \ \curvearrowleft \ {}^-{:}\ddot{\underset{..}{O}}-H \ \rightleftharpoons \ H\cdots\overset{\cdot\cdot}{N}(H)(H) \ + \ H-\overset{\cdot\cdot}{\underset{\cdot\cdot}{O}}-H$$

The arrows show that a nonbonding electron pair from oxygen forms a new O–H bond in water. The bonding pair between nitrogen and hydrogen becomes the nonbonding electron pair on the nitrogen atom.

(b) Step 1: Rewrite the equation to show all nonbonding electrons:

$$(CH_3)_3C^+ \;+\; :\!\ddot{\underset{..}{Cl}}\!:^- \;\rightleftarrows\; (CH_3)_3C-\ddot{\underset{..}{Cl}}:$$

Step 2: Identify the electron donor and electron acceptor.
One of the lone pairs of electrons of Cl⁻ becomes the new bond between chlorine and carbon. Therefore, chloride ion is the electron donor and the carbocation is the electron acceptor.

Step 3: Draw an arrow from the source of electrons, one lone pair on the chlorine, to their destination, the carbon atom of $(CH_3)_3C^+$.

$$(CH_3)_3\overset{\frown}{C^+} \;+\; :\!\ddot{\underset{..}{Cl}}\!:^- \;\rightleftarrows\; (CH_3)_3C-\ddot{\underset{..}{Cl}}:$$

(c) Step 1: Rewrite the equation to show all nonbonding electrons:

[Structure showing acetone + H–Cl ⇌ protonated acetone + Cl⁻]

Step 2: Identify the electron donor and electron acceptor.
One of the two lone pairs of electrons of the oxygen becomes the new bond between oxygen and hydrogen. The hydrogen-chlorine bonding electron pair becomes a lone electron pair on the chlorine. Therefore, the oxygen atom is the electron donor and the chlorine atom is the electron acceptor.

Step 3: Draw an arrow from the source of electrons, one lone pair on the oxygen, to their destination, the hydrogen atom of HCl.

[Structure with curved arrow from O lone pair to H of HCl]

Step 4: Add a second arrow indicating the electron pair displacement by the first electron movement. The H—Cl bonding electron pair becomes a lone electron pair on the chlorine.

[Structure with both curved arrows]

(d)

$$H-\underset{\underset{H}{|}}{\overset{\overset{H}{|}}{B}}-H \;+\; :\!\ddot{\underset{..}{O}}\!\overset{H}{\diagdown}_H \;\rightleftarrows\; \underset{H\;\;\;H}{\overset{\overset{H}{|}}{B}} \;+\; H_2 \;+\; H-\ddot{\underset{..}{O}}:^-$$

3.18 A **curved arrow** is drawn from the site of electron density, such as a pair of bonding or nonbonding electrons, to the site of electron deficiency, such as an atom with a positive charge or a partial positive charge.

(a) The chloride anion should have either eight electrons around it or the sign of the negative charge (Cl⁻) but not both. If the eight valence electrons are shown, the arrow should begin at one pair of electrons.

Another problem that we have here is the structure of the product. It has 10 electrons around the nitrogen atom, which is a violation of the octet rule for a second period atom. The arrow should point to one of the hydrogen atoms of the ammonium ion and result in a proton removal from the nitrogen atom. The N—H bonding electrons would then become a lone electron pair on the nitrogen.

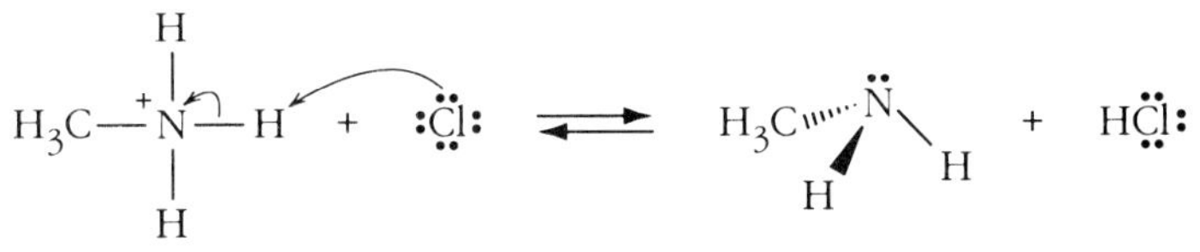

(b) The arrow shows the movement of an atom (hydrogen atom) and it should be used to denote the *movement of an electron pair.* The net result of the reaction is the transfer of a proton from sulfuric acid to ammonia. However, the arrow should begin at the lone pair of electrons on the nitrogen and point to the hydrogen atom (and not its bonding electrons!). Then, the second arrow should be used to indicate the movement of the previously H—O bonding electrons to the oxygen atom. The correct mechanism should be drawn in the following manner:

(c) The arrow should show the movement of an electron pair. It should begin at the lone pair of electrons on the oxygen and should point to the boron atom:

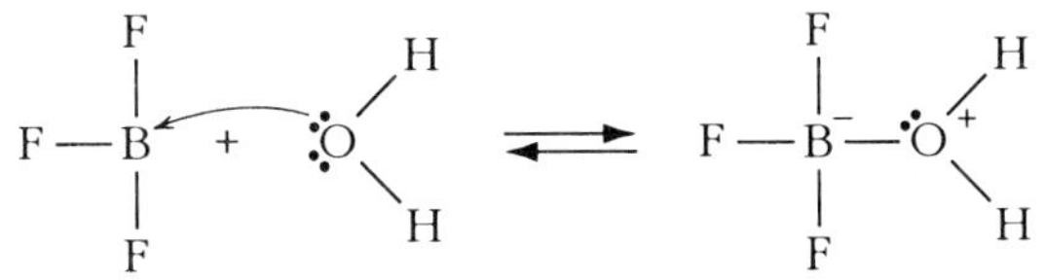

3.19 The quantitative relationship between the standard free energy change of a reaction ($\Delta G°$) and its equilibrium constant (K_{eq}) is the following:

$$\Delta G° = -2.30\, RT\, [\log K_{eq}]$$

where T is the absolute temperature in kelvin (K) and R is the gas constant (8.315×10^{-3} kJ/mol·K or 1.987×10^{-3} kcal/mol·K).

In order to calculate the equilibrium constant, the above equation has to be transformed. We transform it by dividing both sides of the equation by $(-2.30\ RT)$:

$$\frac{\Delta G°}{-2.3\ RT} = \log K_{eq}\ \text{or}\ K_{eq} = 10^{-\Delta G°/2.3\ RT}$$

(a) The standard Gibbs free energy is defined for the reactions at 25 °C (298.2 K). Since $\Delta G°$ is expressed in kcal/mol, R must be given in the same units:
1.987×10^{-3} kcal/mol·K.
Thus,
$K_{eq} = 10^{-\Delta G°/2.3\ RT} = 10^{-(-21.2)/(2.3 \times 1.987 \times 0.001 \times 298.2)}$
$K_{eq} = 10^{21.2/1.363} = 10^{15.554} = \mathbf{3.58 \times 10^{15}}$

(b) Since $\Delta G°$ is expressed in kJ/mol, R must be given in the same units: 8.315×10^{-3} kJ/mol·K.
Thus,
$K_{eq} = 10^{-\Delta G°/2.3\ RT} = 10^{-(-41.8)/(2.3 \times 8.315 \times 0.001 \times 298.2)}$
$K_{eq} = 10^{41.8/5.703} = 10^{7.329} = \mathbf{2.13 \times 10^7}$

3.20 The quantitative relationship between the standard free energy change of a reaction ($\Delta G°$) and its equilibrium constant (K_{eq}) is the following:

$$\Delta G° = 2.30\ RT\ [\log K_{eq}]$$

where T is the absolute temperature in kelvin (K) and R is the gas constant (8.315×10^{-3} kJ/mol·K or 1.987×10^{-3} kcal/mol·K). The standard Gibbs free energy is defined for the reactions at 25 °C (298.2 K).

(a) $\Delta G° = -2.3\ RT\ [\log K_{eq}]$
$\Delta G° = -2.3 \times 8.315 \times 10^{-3}$ kJ/mol·K̸ $\times 298.2$ K̸ $\times (\log 1.8 \times 10^{-5})$
$= -5.703 \times (-4.745) = \mathbf{27.1\ kJ/mol}$
or
$\Delta G° = -2.3 \times 1.987 \times 10^{-3}$ kcal/mol·K̸ $\times 298.2$ K̸ $\times (\log 1.8 \times 10^{-5})$
$= -1.363 \times (-4.745) = \mathbf{6.47\ kcal/mol}$

(b) $\Delta G° = -2.3\ RT\ [\log K_{eq}]$
$\Delta G° = -2.3 \times 8.315 \times 10^{-3}$ kJ/mol·K̸ $\times 298.2$ K̸ $\times (\log 1.3 \times 10^6)$
$= -5.703 \times (6.114) = \mathbf{-34.9\ kJ/mol}$
or
$\Delta G° = -2.3 \times 1.987 \times 10^{-3}$ kJ/mol·K̸ $\times 298.2$ K̸ $\times (\log 1.3 \times 10^6)$
$= -1.363 \times (6.114) = \mathbf{-8.33\ kcal/mol}$

3.21 Initial rate of reaction = $k\ [NO]_0\ [O_3]_0$

(a) If the concentration of NO is decreased by half, we have the following equilibrium for the new reaction rate:
New rate = $k\ [NO]_n\ [O_3]_n$, where $[NO]_n = 1/2\ [NO]_0$ and $[O_3]_n = [O_3]_0$

$$\frac{\text{new rate}}{\text{initial rate}} = \frac{k\ 1/2\ [NO]_0[O_3]_0}{k\ [NO]_0\ [O_3]_0} = \frac{1}{2}$$

Hence, the new rate is going to be a half of the initial rate.

(b) If the concentration of O_3 is decreased by half, the new reaction rate will be expressed by:
New rate = $k\,[NO]_n\,[O_3]_n$, where $[NO]_n = [NO]_0$ and $[O_3]_n = 1/2\,[O_3]_0$

$$\frac{\text{new rate}}{\text{initial rate}} = \frac{k\,[NO]_0\,1/2\,[O_3]_0}{k\,[NO]_0\,[O_3]_0} = \frac{1}{2}$$

Hence, the new rate is going to be a half of the initial rate.

(c) If the concentrations of both NO and O_3 are decreased by half, the new reaction rate is expressed by:
New rate = $k\,[NO]_n\,[O_3]_n$, where $[NO]_n = 1/2\,[NO]_0$ and $[O_3]_n = 1/2\,[O_3]_0$

$$\frac{\text{new rate}}{\text{initial rate}} = \frac{k\,1/2\,[NO]_0\,1/2\,[O_3]_0}{k\,[NO]_0\,[O_3]_0} = \frac{1}{4}$$

Hence, the new rate is going to be a quarter of the initial rate.

3.22 (a) The reaction is first order with respect to cyclopropane and first order overall.

(b) The reaction is first order with respect to 2-chloro-2-methylpropane, zero order with respect to water, and first order overall.

(c) The reaction is first order with respect to ethene, second order with respect to bromine, and third order overall.

(d) The reaction is first order with respect to chloromethane, first order with respect to sodium hydroxide, and second order overall.

3.23 (a) $CH_3OCH_2CH_2OCH_3$
1,2-Dimethoxyethane (ε 7.2) (like diethyl ether, ε 4.3) is a nonpolar, aprotic solvent with oxygen atoms acting as Lewis bases (electron-donor solvent).

(b) Toluene (ε 2.38) (like benzene, ε 2.23) is a nonpolar and aprotic solvent.

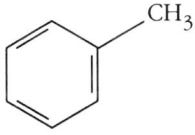

(c) $CH_3COOCH_2CH_3$
Ethyl acetate (ε 6.02) is a polar and aprotic solvent with oxygen atoms acting as Lewis bases (electron-donor solvent).

(d) CH_3NO_2
Nitromethane (ε 35.9) is a polar and aprotic solvent with oxygen atoms acting as Lewis bases (electron-donor solvent).

3.24 The free energy of activation is the difference between the free energy of the transition state and the free energy of the reactants. In this case, we need to compare only the difference between the free energy of the transition state in solvent **A** to that in solvent **C** because the free energy of the reactants is about the same in both solvents. The free energy of the transition state is lower in solvent **C**, so the reaction will be faster in solvent **C**.

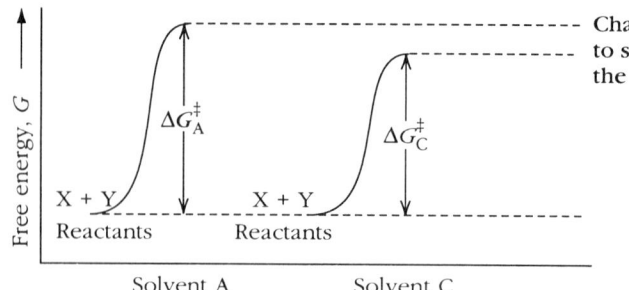

3.25 (a) A **rate law** is an equation that describes the dependence of the reaction rate on concentrations of reactants (and sometimes even products). For example, for a reaction A + 2B ⟶ C the rate law is of the form rate = k [A] [B]^2, where k is a proportionality constant called a **rate constant**.

The examples like the one above lead to the idea that stoichiometry and order are related. Be careful, however, because there is no relationship between the orders of the reactants and their coefficients in the balanced chemical equation.

(b) A **Brønsted-Lowry base** is a species that can *accept a proton* from an acid. A Brønsted-Lowry base has a pair of electrons that is capable of binding the proton. It may be an anion (such as HO$^-$, Cl$^-$, CH$_3$O$^-$) or a neutral molecule with at least one lone pair of electrons (such as H$_3$N, H$_2$O).

(c) **Activation energy** is a certain minimum amount of energy that colliding molecules must have in order to reach the transition state (so that the reaction can take place). Usually only a small portion of molecules will have this energy at any given time.

(d) A **conjugate acid-base pair** is a pair of Brønsted acids and bases whose formulas differ by a single proton. Some examples of such pairs are H$_3$O$^+$ and H$_2$O, H$_2$O and HO$^-$, HO$^-$ and O^{2-} (rarely), NH$_3$ and H$_2$N$^-$, etc.

(e) A **Lewis acid** is a substance that accepts electron pairs. Examples are H$^+$, BF$_3$, CO$_2$, ZnCl$_2$, AlCl$_3$.

(f) K_a, the **acid ionization constant,** is the equilibrium constant for the ionization of an acid in water. For example, acetic acid ionizes in water:

$$CH_3COOH(aq) \rightleftharpoons H^+(aq) + CH_3COO^-(aq)$$

and its ionization constant is expressed by the following formula:

$$K_a = \frac{[H^+][CH_3COO^-]}{[CH_3COOH]}$$

At 25 °C K_a is 1.76×10^{-5}.

(g) A **Brønsted-Lowry acid** is a substance that can donate a proton to a base. Examples of such acids are: H$_3$O$^+$, H$_2$O, (but also HO$^-$ since it can donate its proton to become O^{2-}), CH$_3$COOH, H$_2$SO$_4$, HSO$_4^-$, etc.

(h) An **endergonic reaction** is a reaction with a positive value of ΔG. Such a reaction is not spontaneous in the direction written (that is from left to right). However, the reverse reaction will have a negative value of ΔG and will proceed spontaneously.

(i) A **catalyst** is a substance that increases the rate of a reaction without being consumed in the reaction. In effect, a catalyst provides an alternative reaction mechanism with a lower activation energy than the uncatalyzed reaction. Thus, in *homogeneous catalysis* a catalyst first combines with a reactant to form an activated complex whose potential energy is less than that of the complex formed in the absence of the catalyst. In *heterogeneous catalysis,* a contact catalyst provides a surface on which the reaction takes place.

(j) An **aprotic solvent** is a solvent that does not have easily exchangeable protons. Examples of aprotic solvents are hexane, ethoxyethane (diethyl ether), tetrachloromethane (carbon tetrachloride), dichloromethane (methylene chloride), and benzene, etc.

(k) A **Lewis base** is a substance that donates an electron pair. Examples include both ionic (HO$^-$, $^-$CN, Cl$^-$, HC≡C$^-$) and neutral molecules (H$_2$O, NH$_3$, CH$_3$CH$_2$Br).

(l) A **spontaneous reaction** is a reaction that occurs by itself without external intervention. It takes place when the system *is not at equilibrium* and there is a *net driving force* in the forward direction. At equilibrium, the rates of the forward and reverse reactions are equal, so the net driving force is zero.

(m) An **exergonic reaction** is a reaction with a negative value of ΔG. Such a reaction takes place *spontaneously* at constant temperature and pressure in the direction written (that is, from left to right).

(n) **pK_a** is the negative logarithm of the equilibrium constant for the ionization of an acid in water (or an **acid ionization constant, K_a**):

$$pK_a = -\log K_a$$

For example, if K_a for acetic acid at 25 °C is 1.76×10^{-5}, then the pK_a of acetic acid at 25 °C is

$$pK_a = -\log(1.76 \times 10^{-5}) = 4.75$$

(o) A **transition state** is the highest point on the energy reaction path diagram for an elementary step of the reaction. The terms *transition state* and *activated complex* are often used interchangeably.

(p) A **protic solvent** is a solvent that has easily exchangeable protons. Water and alcohols are common examples of protic solvents used in organic chemistry.

3.26 $\quad$ CH$_3$COOH + HCO$_3^-$ $\longrightarrow$ CH$_3$COO$^-$ + H$_2$CO$_3$

$\quad\quad$ Acid $\quad\quad$ Base $\quad\quad\quad\quad$ Conjugate base $\quad$ Conjugate acid
$\quad\quad\quad\quad\quad\quad\quad\quad\quad\quad\quad\quad\quad$ of CH$_3$COOH $\quad\quad$ of HCO$_3^-$

3.27 (a) $\quad$ (CH$_3$)$_3\overset{+}{\text{N}}$H $\quad\quad\quad\quad$ (CH$_3$)$_3$N:

$\quad\quad\quad\quad$ Conjugate acid $\quad\quad\quad\quad$ Conjugate base

(b)

Conjugate acid $\quad\quad\quad\quad$ Conjugate base

(c)

$$\begin{array}{cc} \overset{\overset{H}{|}}{\underset{H_3C}{O^+}}\diagdown_{CH_3} & \overset{\overset{..}{\underset{..}{O}}}{\underset{H_3C}{}}\diagdown_{CH_3} \\ \text{Conjugate acid} & \text{Conjugate base} \end{array}$$

(d) $\quad CH_4 \qquad\qquad H_3C^-$
    Conjugate acid   Conjugate base

(e) $\quad CH_3CH=CHCOOH \qquad CH_3CH=CHCOO^-$
    Conjugate acid   Conjugate base

(f) $\quad (CH_3)_2NH \qquad\qquad (CH_3)_2N^-$
    Conjugate acid   Conjugate base

3.28 (a) [structure: methoxy-decalin with O: — Conjugate base] [structure: protonated methoxy-decalin O+–H — Conjugate acid]

(b) [benzaldehyde with C=O: — Conjugate base] [protonated benzaldehyde C=O+–H — Conjugate acid]

(c) [N-methylpiperidine with N: — Conjugate base] [N-methylpiperidinium N+–H — Conjugate acid]

(d) [benzophenone with C=O: — Conjugate base] [protonated benzophenone C=O+–H — Conjugate acid]

(e)

    Conjugate base   Conjugate acid

(f)

    Conjugate base   Conjugate acid

3.29 (a) $CH_3CH_2OH < H_2O < CH_3COOH$
       pK_a 16   15.7   4.76
       (b) $H_2O < HCN < NH_4^+$
       pK_a 15.7   9.1   9.25
       (c) $CH_3OH < H_2CO_3 < HBr$
       pK_a 15.2   3.58 ≈ -9
       (d) $CH_3NH_3^+ < CH_3COOH < H_3O^+$
       pK_a 11   4.76   -1.7
       (e) $CH_3\overset{+}{N}H_3 < CH_3CH_2\overset{+}{O}H_2 < HCl$
       pK_a 9 ≈ -2 ≈ -6

3.30 $HA + (CH_3)_3N: \longrightarrow (CH_3)_3NH^+ + A^-$

We have to find pK_a values for the "HA" acid and $(CH_3)_3NH^+$. When the pK_a of the "HA" acid is less positive than that of $(CH_3)_3NH^+$, then the protonation will occur. If "HA" is a weaker acid (more positive pK_a) then the proton transfer will not take place.

(a) $(CH_3)_3N: + H_3O^+ \longrightarrow (CH_3)_3NH^+ + H_2O$
     pK_a -1.7   9.81

The hydronium ion is a much stronger acid than the conjugate acid of N,N-dimethylmethanamine so the equilibrium will be shifted to the right. Thus, H_3O^+ is strong enough to react with an aqueous solution of $(CH_3)_3N$.

(b) $(CH_3)_3N: + HCO_3^- \rightleftharpoons (CH_3)_3NH^+ + CO_3^{2-}$
     pK_a 10.25   9.81

The hydrogen carbonate is a little weaker acid than the conjugate acid of N,N-dimethylmethanamine so the equilibrium will be shifted to the left. Thus, HCO_3^- is not strong enough to react with an aqueous solution of $(CH_3)_3N$.

(c) $(CH_3)_3N: + CH_3COOH \longrightarrow (CH_3)_3NH^+ + CH_3COO^-$
     pK_a 4.76   9.81

Acetic acid is a stronger acid than the conjugate acid of N,N-dimethylmethanamine so the equilibrium will be shifted to the right. Thus, CH_3COOH is strong enough to react with an aqueous solution of $(CH_3)_3N$.

3.31 (a) $Na^+NH_2^- + CH_3(CH_2)_2CH_3 \rightleftharpoons CH_3(CH_2)_2CH_2^-Na^+ + NH_3$
pK_a ≈ 50 ≈ 38

$pK_{eq} = pK_a(CH_3(CH_2)_2CH_3) - pK_a(NH_3) = 50 - 38 \approx 12$

$K_{eq} = 10^{-pK_{eq}} \approx \mathbf{10^{-12}}$

(b) $NaH + (CH_3)_3COH \rightleftharpoons (CH_3)_3CO^-Na^+ + H_2$
pK_a 16 ≈ 35

$pK_{eq} = pK_a((CH_3)_3COH) - pK_a(H_2) = 16 - 35 \approx -19$

$K_{eq} = 10^{-pK_{eq}} = 10^{-(-19)} \approx \mathbf{10^{19}}$

(c) $CH_3(CH_2)_2CH_2^-Li^+ + H_2O \rightleftharpoons LiOH + CH_3(CH_2)_2CH_3$
pK_a 15.7 ≈ 50

$pK_{eq} = pK_a(H_2O) - pK_a(CH_3(CH_2)_2CH_3) = 15.7 - 50 \approx -34.3$

$K_{eq} = 10^{-pK_{eq}} = 10^{-(-34.3)} = 10^{34.3} \approx \mathbf{2 \times 10^{34}}$

(d) $HCl(aq) + CH_3CH_2OH \rightleftharpoons CH_3CH_2\overset{+}{O}H_2 + Cl^-$
pK_a -1.7 ≈ -2

$pK_{eq} = pK_a(HCl) - pK_a(CH_3CH_2\overset{+}{O}H_2) = (-1.7) - (-2) \approx 0.3$

$K_{eq} = 10^{-pK_{eq}} = 10^{-0.3} \approx \mathbf{0.5}$

(e) $H_2SO_4 + CH_3\overset{\overset{\ddot{O}\cdot}{\|}}{C}CH_3 \rightleftharpoons CH_3\overset{\overset{\overset{+}{\ddot{O}}-H}{\|}}{C}CH_3 + HSO_4^-$
pK_a ≈ -9 ≈ -7

$pK_{eq} = pK_a(H_2SO_4) - pK_a(CH_3\overset{\overset{\overset{+}{O}-H}{\|}}{C}CH_3) \approx (-9) - (-7) \approx -2$

$K_{eq} = 10^{-pK_{eq}} = 10^{-(-2)} \approx \mathbf{10^2}$

3.32 The quantitative relationship between the standard free energy change of a reaction ($\Delta G°$) and its equilibrium constant (K_{eq}) is the following:

$$\Delta G° = -2.30\ RT\ [\log K_{eq}]$$

where T is the absolute temperature in kelvin (K) and R is the gas constant (8.315×10^{-3} kJ/mol · K or 1.987×10^{-3} kcal/mol · K). The standard Gibbs free energy is defined for the reactions at 25 °C (298.2 K).
Thus,
$\Delta G° = -2.30\ RT\ [\log K_{eq}] = -2.3 \times 1.987 \times 10^{-3}$ kcal/mol · K $\times 298.2$ K $\times (\log K_{eq})$
$= -1.363$ kcal/mol $\times \log K_{eq}$

(a) $\Delta G° = -1.363 \times \log K_{eq}$ kcal/mol $= -1.363 \times \log(10^{-12})$ kcal/mol
$= \mathbf{16.3\ kcal/mol\ (or\ 68.4\ kJ/mol)}$

(b) $\Delta G° = -1.363 \times \log K_{eq}$ kcal/mol $= -1.363 \times \log(10^{19})$ kcal/mol
$= \mathbf{-25.9\ kcal/mol\ (or\ -108.4\ kJ/mol)}$

(c) $\Delta G° = 1.363 \times \log K_{eq}$ kcal/mol = $-1.363 \times \log (2 \times 10^{34})$ kcal/mol =
= **–46.7 kcal/mol** (or **–195.6 kJ/mol**)
(d) $\Delta G° = 1.363 \times \log K_{eq}$ kcal/mol = $-1.363 \times \log (0.5)$ kcal/mol =
= **–0.95 kcal/mol** (or **–3.99 kJ/mol**)
(e) $\Delta G° = 1.363 \times \log K_{eq}$ kcal/mol = $-1.363 \times \log (10^2)$ kcal/mol =
= **–2.72 kcal/mol** (or **–11.4 kJ/mol**)

3.33 When hydrochloric acid is dissolved in water, its dissociation produces hydronium and chloride ions. The hydronium ion is the strongest acid that can exist in water. Then, hydronium ion establishes an equilibrium with acetone:

$$HCl + H_2O \longrightarrow H_3O^+ + Cl^-$$

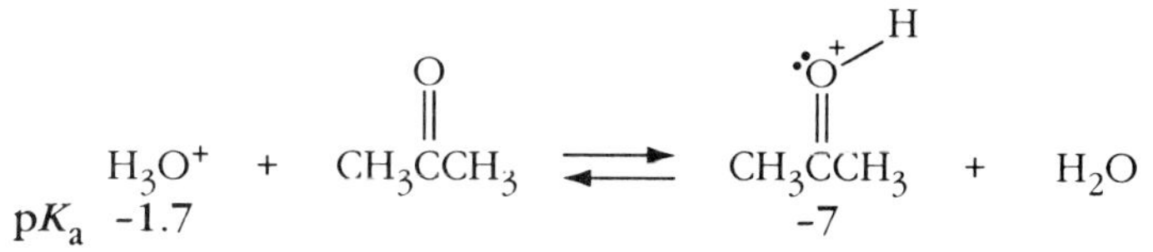

$$\underset{pK_a\ -1.7}{H_3O^+} + CH_3\overset{O}{\underset{\|}{C}}CH_3 \rightleftharpoons \underset{-7}{CH_3\overset{\overset{+}{O}H}{\underset{\|}{C}}CH_3} + H_2O$$

Protonated acetone is a stronger acid (*lower* pK_a) so the equilibrium will be shifted to the left. Let us confirm this conclusion by calculating the equilibrium constant, K_{eq}.

$$pK_{eq} = pK_a(H_3O^+) - pK_a(CH_3\overset{\overset{+}{O}H}{\underset{\|}{C}}CH_3) = -1.7 - (-7) \approx 5.3$$
$$K_{eq} = 10^{-pK_{eq}} = 10^{-5.3} \approx 5 \times 10^{-6}$$

K_{eq} is less than 1, so the equilibrium lies to the left.

3.34 If the reaction is carried out in hexane, hexane is such a weak base that it does not react with HCl and the acid in this solution is HCl.

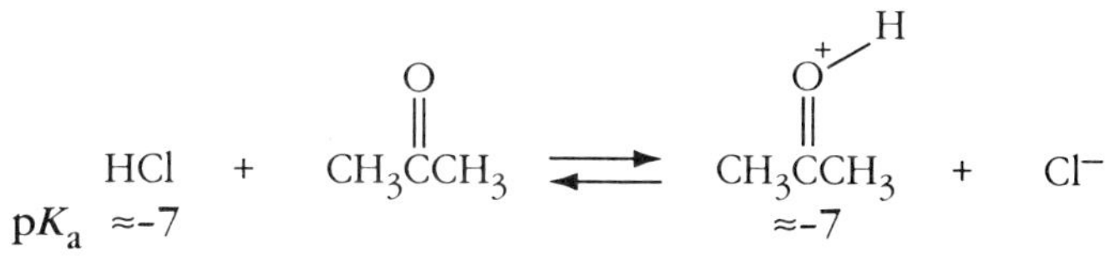

$$\underset{pK_a\ \approx -7}{HCl} + CH_3\overset{O}{\underset{\|}{C}}CH_3 \rightleftharpoons \underset{\approx -7}{CH_3\overset{\overset{+}{O}H}{\underset{\|}{C}}CH_3} + Cl^-$$

The pK_a values for both above acids are about the same, so $K_{eq} \approx 1$. Thus, at equilibrium about equal amounts of HCl and $CH_3\overset{\overset{+}{O}H}{\underset{\|}{C}}CH_3$ are present in solution so proton transfer from HCl to acetone does occur.

3.35 $$\underset{pK_a\ \approx 25}{HC \equiv C-H} + CH_3O^- \rightleftharpoons \underset{\text{Stronger base}}{HC \equiv C^-} + \underset{\underset{\text{Stronger acid}}{15.5}}{CH_3OH}$$

The above reaction will not occur because the reverse reaction will predominate.

3.36 $$\underset{pK_a}{NaOH} + \underset{\approx 16}{CH_3OH} \rightleftharpoons CH_3O^-Na^+ + \underset{15.7}{H_2O}$$

Water and methanol have very similar pK_a values, so at equilibrium about equal amounts of HO⁻ and CH_3O^- are present in solution.

$$NaH + CH_3OH \rightleftharpoons CH_3O^-Na^+ + H_2$$
$$pK_a \quad\quad\quad 16 \quad\quad\quad\quad\quad\quad\quad\quad \approx 35$$

Methanol is a much stronger acid than hydrogen, so the reaction occurs as written to form almost 100% of sodium methoxide.

Thus, sodium hydride (NaH) will react with methanol to form methoxide ion in high concentration.

3.37 When hydrochloric acid is dissolved in water, its dissociation produces hydronium and chloride ions. The hydronium ion is the strongest acid that can exist in water. Then, hydronium ion establishes an equilibrium with ethanol:

$$HCl + H_2O \longrightarrow H_3O^+ + Cl^-$$
$$H_3O^+ + CH_3CH_2OH \rightleftharpoons CH_3CH_2O^+H_2 + H_2O$$
$$pK_a \quad -1.7 \quad\quad\quad\quad\quad\quad\quad\quad \approx -2$$
$$pK_{eq} = pK_a(H_3O^+) - pK_a(CH_3CH_2O^+H_2) = -1.7 - (-2) = 0.3$$
$$K_{eq} = 10^{-pK_{eq}} = 10^{-0.3} = 5 \times 10^{-1} = 0.5$$

When HCl is dissolved in tetrahydrofuran (THF), a proton is transferred from HCl to THF as follows:

HCl + [THF] ⇌ [THF-H⁺] + Cl⁻

Protonated THF has a pK_a of about -4.

[THF-H⁺] + CH_3CH_2OH ⇌ $CH_3CH_2\overset{+}{O}H_2$ + [THF]
$pK_a \quad \approx -4 \quad\quad\quad\quad\quad\quad\quad\quad \approx -2$

$$pK_{eq} = pK_a(THF-H^+) - pK_a(CH_3CH_2\overset{+}{O}H_2) = -4 - (-2) = -2$$
$$K_{eq} = 10^{-pK_{eq}} = 10^{-(-2)} = 10^2 = 100$$

Therefore a higher concentration of protonated ethanol is formed when tetrahydrofuran is used as a solvent.

3.38 The **Lewis theory** defines a **Lewis acid** as an *electron-pair acceptor* and a **Lewis base** as an *electron-pair donor.*

(a)

[THF] + H_2SO_4 ⇌ [THF-H⁺] + HSO_4^-

Lewis base \quad\quad Lewis acid

(b)

$H_3C-\overset{\overset{\cdot\cdot}{\overset{\cdot\cdot}{O}}}{C}-\overset{\cdot\cdot}{\underset{\cdot\cdot}{O}}-CH_3$ + BF_3 ⇌ $H_3C-\overset{\overset{\overset{-BF_3}{\underset{\cdot\cdot}{O^+}}}{\parallel}}{C}-\overset{\cdot\cdot}{\underset{\cdot\cdot}{O}}-CH_3$

Lewis base   Lewis acid

(c) Br_2 + $FeBr_3$ ⇌ $FeBr_4^-\ Br^+$
Lewis base   Lewis acid

3.39 Organic chemists have developed a symbolic device for keeping track of electron pairs in chemical reactions; this device is called the **curved-arrow formalism.** According to this formalism, the formation of a chemical bond is described by a "flow" of electrons from *the electron donor* (Lewis base) to *the electron acceptor* (Lewis acid). This "electron flow" is indicated by a curved arrow *from the electron source to the electron acceptor.*

(a) [tetrahydrofuran with O lone pair arrow] + H—OSO$_3$H ⇌ [protonated tetrahydrofuran with O$^+$—H] + HSO$_4^-$

(b) $H_3C-\overset{\overset{\cdot\cdot}{\overset{\cdot\cdot}{O}}}{C}-\overset{\cdot\cdot}{\underset{\cdot\cdot}{O}}-CH_3$ + BF_3 ⇌ $H_3C-\overset{\overset{\overset{-BF_3}{\underset{\cdot\cdot}{O^+}}}{\parallel}}{C}-\overset{\cdot\cdot}{\underset{\cdot\cdot}{O}}-CH_3$

(c) $:\!\overset{\cdot\cdot}{\underset{\cdot\cdot}{Br}}\!-\!\overset{\cdot\cdot}{\underset{\cdot\cdot}{Br}}\!:$ + $FeBr_3$ ⇌ $FeBr_4^-\ Br^+$

3.40

(a) $H_3\overset{+}{N}\!-\!H$ + $:\!\overset{-}{N}H_2$ ⇌ $\overset{\cdot\cdot}{N}H_3$ + $\overset{\cdot\cdot}{N}H_3$

(b) [tetrahydrofuran] + H—OSO$_3$H ⇌ [protonated THF, O$^+$—H] + HSO$_4^-$

(c) [cyclohexanone protonated, O$^+$—H] + $H-\overset{\cdot\cdot}{\underset{\cdot\cdot}{O}}-H$ ⇌ [cyclohexanone] + $H-\overset{+}{\underset{H}{O}}-H$ (H on top)

(d) H_2N^- + $H-C\equiv CH$ ⇌ NH_3 + $^-C\equiv CH$

3.41 (a) A **Lewis acid** is a species that acts as an electron-pair acceptor:

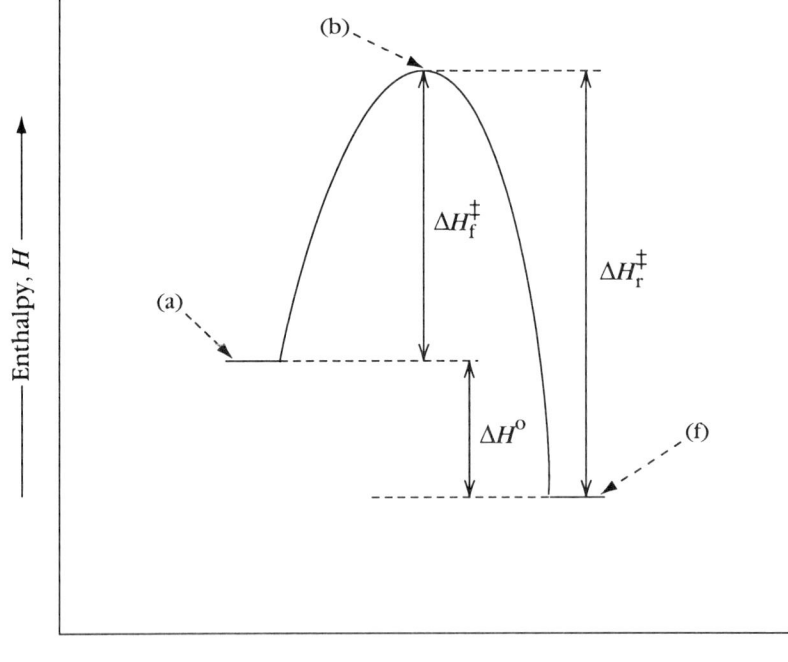

(b) A **Brønsted-Lowry** acid is a species that is a proton donor:

3.42 The reactants are represented by *(a)*; the transition state by *(b)*; the activation energy E_a for the forward reaction by $\Delta H_f^{\ddagger}$; the activation energy E_a for the reverse reaction by $\Delta H_r^{\ddagger}$; the enthalpy change for the reaction by $\Delta H°$; and products by *(f)*.

3.43 (a) $K_{eq} = \dfrac{\text{[1-methylethylidenecyclohexane]}}{\text{[1-methylethylcyclohexene]}} = \dfrac{30}{70} = 0.43$

(b) $\Delta G° = -2.3\ RT\ (\log K_{eq}) = -1.363 \times (-0.368) =$ **0.50 kcal/mol** or **2.1 kJ/mol**

CHAPTER 4

Concepts

1. Rotation occurs about carbon-carbon single bonds so alkanes exist as a large number of rapidly interchanging conformations. **Staggered conformations** of alkanes are more stable than **eclipsed conformations** because of **torsional strain,** which is due to the eclipsing of bonds on adjacent atoms. The most stable conformation of straight chain alkanes is the **anti conformation** in which two alkyl groups are as far away as possible.

Staggered conformation	Eclipsed conformation	Anti conformation

2. Cycloalkanes have varying degrees of ring strain. Cyclopropane has the highest ring strain followed by cyclobutane. Cyclohexane is relatively strain free.

   Cycloalkanes adopt their minimum-energy conformations for a combination of the following three reasons:

   Angle strain, the strain due to expansion or compression of bond angles.
   Torsional strain, the strain due to eclipsing of neighboring bonds.
   Steric strain, the strain due to repulsive interaction of atoms approaching each other too closely.

3. The cyclohexane ring is relatively strain free because it can adopt a **chair conformation** in which all bond angles are close to 109° and all C—H bonds on adjacent carbon atoms are in staggered conformations. There are two kinds of C—H bonds in the chair conformation of cyclohexane, six **axial** and six **equatorial**. The chair conformation is conformationally mobile so axial and equatorial bonds are converted by **ring flipping**.

   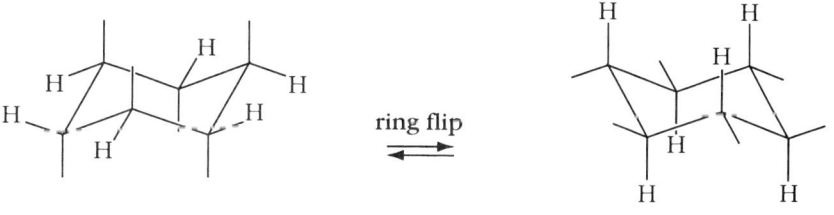

   Equatorial hydrogens   become   *axial hydrogens* after ring flip

4. Disubstituted cyclohexanes exist as stereoisomers that differ only in the arrangement of their atoms in space. Stereoisomers of cyclohexanes *cannot* be interconverted by ring flipping.

Solutions to the Exercises

4.1 A **Newman projection** is a view of a molecule down the axis of a carbon-carbon bond. The carbon atom closest to the viewer is represented by a dot and the carbon atom towards the rear is represented by a circle. The atom or groups on the carbon atoms are shown as being bonded to the dot or the circle.

Let us make the model of the molecule shown on the left (Exercise 4.1).

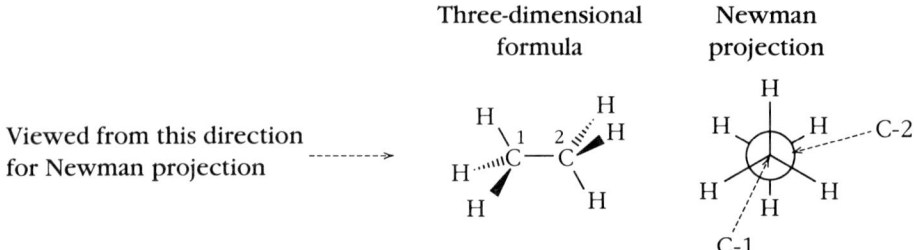

Staggered conformation of ethane

Now we repeat the same operation with the structure shown on the right:

Eclipsed conformation of ethane

4.2

4.3 Totally eclipsed (θ = 0°)

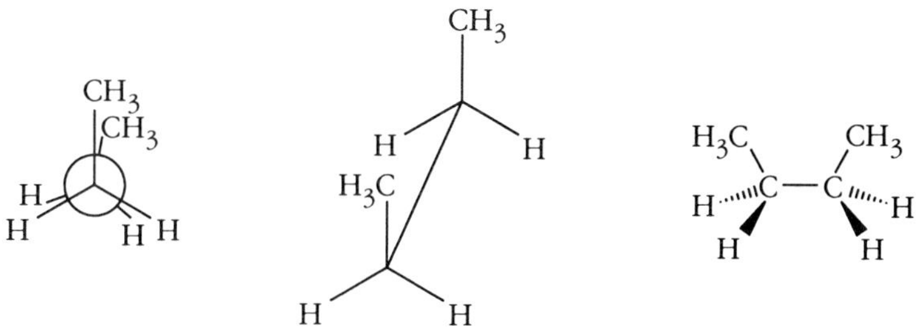

Gauche (θ = 60°)

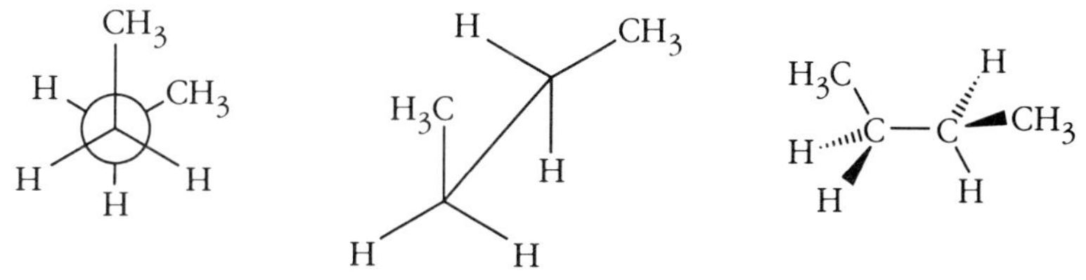

Eclipsed (θ = 120°)

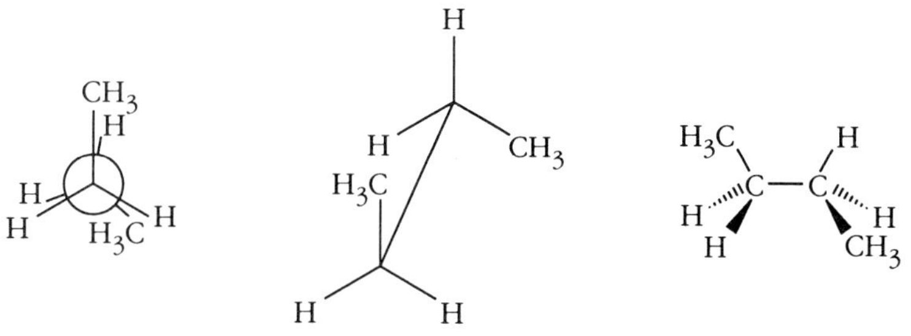

Anti (θ = 180°)

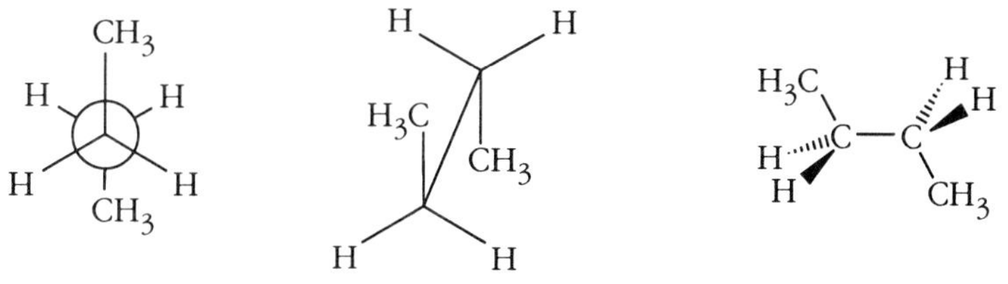

Eclipsed (θ = 240°)

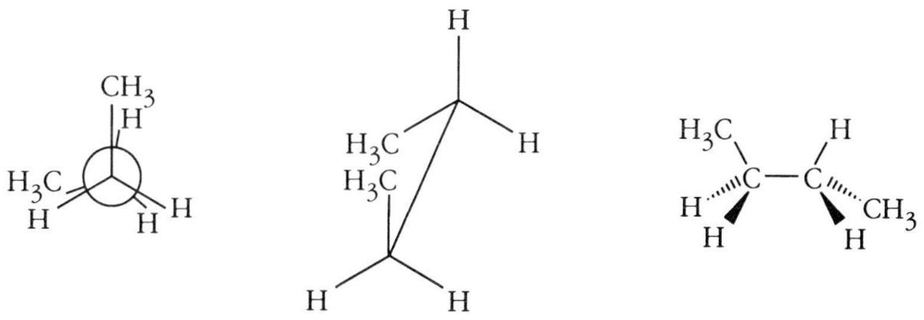

Gauche (θ = 300°)

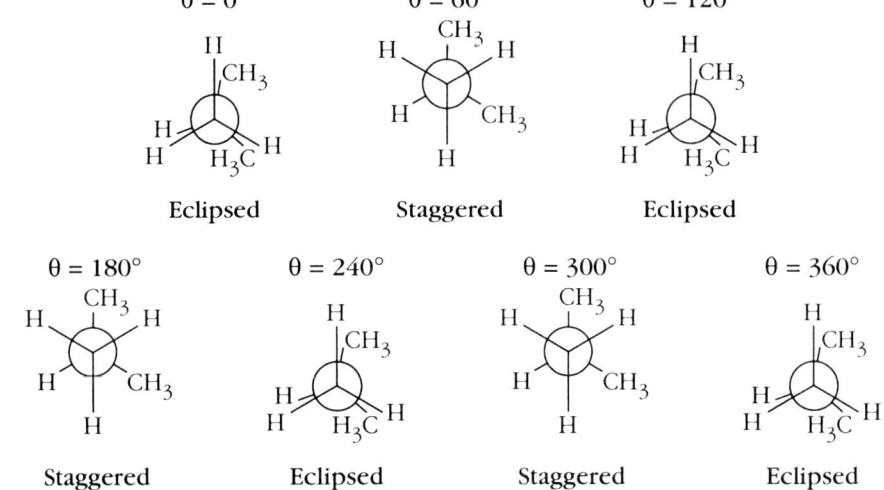

Totally eclipsed (θ = 360°)

4.4 (a) Rotating about front carbon atom gives the following staggered and eclipsed conformations:

θ = 0°   Eclipsed
θ = 60° Staggered
θ = 120° Eclipsed
θ = 180° Staggered
θ = 240° Eclipsed
θ = 300° Staggered
θ = 360° Eclipsed

(b) There are three equivalent staggered conformations of 2-methylpropane: θ = 60°, 180°, 300°.

(c) There are three equivalent eclipsed conformations of 2-methylpropane: θ = 0°, 120°, 240°.

(d) The eclipsed conformation of 2-methylpropane has one pair of eclipsed C—H bonds (1.0 kcal/mol) and two pairs of C—H and C—C eclipsed bonds (1.4 kcal/mol each). Thus, the total difference between the eclipsed and the staggered conformation of 2-methylpropane equals: 1.0 + 1.4 + 1.4 = 3.8 kcal/mol (or 16 kJ/mol).

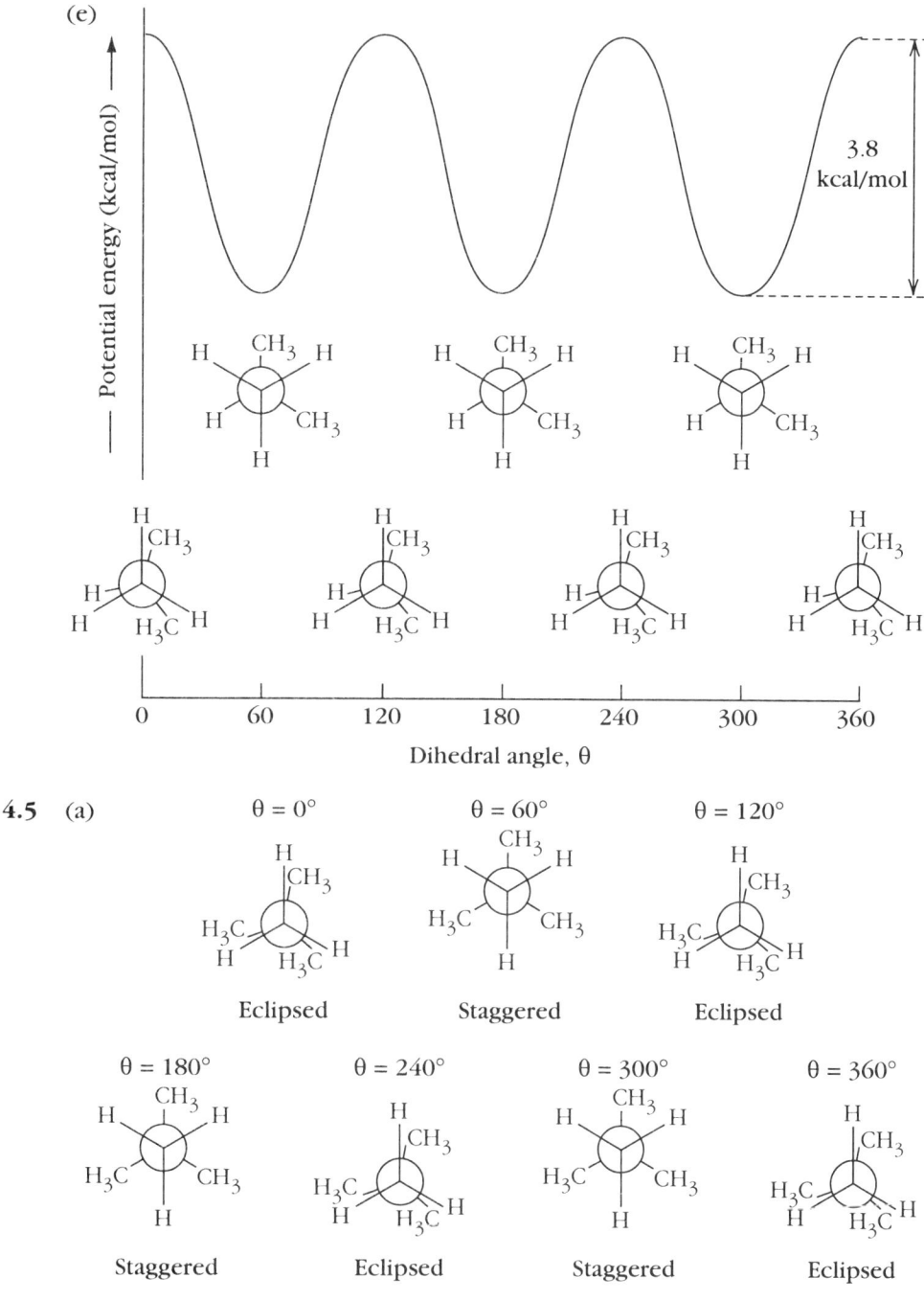

(b) There are three equivalent staggered conformations of 2,2-dimethylpropane.
(c) There are three equivalent eclipsed conformations of 2,2-dimethylpropane.
(d) The eclipsed conformation of 2,2-dimethylpropane has three pairs of C—H and C—C eclipsed bonds (1.4 kcal/mol each). Thus, the total difference between the eclipsed and the staggered conformation of 2,2-dimethylpropane equals:

$$1.4 + 1.4 + 1.4 = 4.2 \text{ kcal/mol (or 18 kJ/mol)}$$

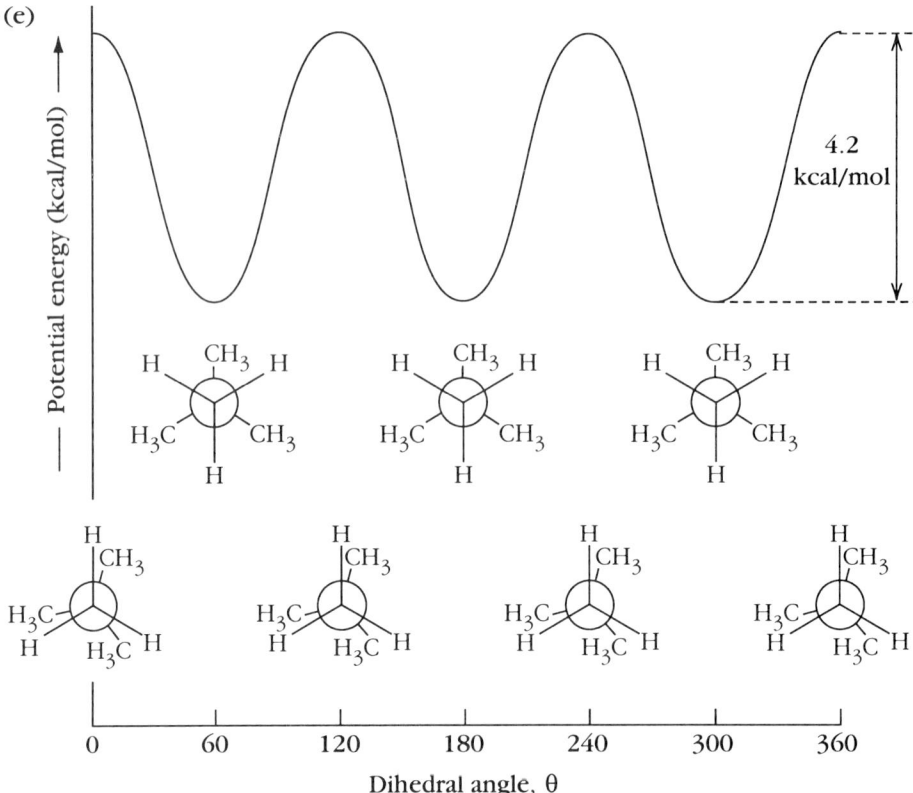

4.6 Cyclobutane is not planar. One of the atoms in the ring is bent out of the plane of the other three by about 25°. This causes the expected internal bond angles of 90° to be reduced to 88° but it also minimizes the eclipsing of the hydrogen atoms on adjacent carbon atoms.

Bent cyclobutane:

Flat cyclobutane:

4.7 There are five pairs of C—H bonds in a planar cyclopentane. If each pair contributes 1 kcal/mol (4 kJ/mol) to the torsional strain, then the total torsional strain is 5 × 1 kcal/mol = 5 kcal/mol (21 kJ/mol).

The reported ring strain is 6.5 kcal/mol (27 kJ/mol). Therefore, the total ring strain is greater than the torsional strain by 1.5 kcal/mol (7 kJ/mol).

There are only four pairs of eclipsed C—H bonds in an envelope conformation of cyclopentane. Thus, the envelope conformation relieves 1 kcal/mol (4 kJ/mol) in torsional strain.

4.8 (a) According to the data in Table 4.3, the free energy difference between the axial and equatorial conformers for ethylcyclohexane is 1.9 kcal/mol (8.0 kJ/mol). The equilibrium constant is related to the standard free energy:

$$\Delta G° = -2.3\, RT\, (\log K_{eq})$$

−1.9 kcal/mol = −2.3 (1.987 cal/mol K)(298.2 K) log K_{eq}
−1.9 kcal/mol = −1,363 cal/mol log K_{eq} = −1.36 kcal/mol log K_{eq}
log K_{eq} = 1.4
K_{eq} = 25.3
Let x = percent of equatorial conformer

$$K_{eq} = \frac{[\%\ \text{equatorial}]}{[\%\ \text{axial}]} = \frac{x}{100 - x}$$

25.3 = x/(100 − x)
25.3 (100 − x) = x
2530 − 25.3 x = x
2530 = 26.3 x
x = 96.2

Conclusion: Ethylcyclohexane consists of 96.2% of the molecules in the equatorial conformation and 3.8% of the molecules in the axial conformation.

(b) According to the data in Table 4.3, the free energy difference between the axial and equatorial conformers for cyclohexanol is 1.0 kcal/mol. The equilibrium constant is related to the standard free energy:

$$\Delta G° = -2.3\, RT\, (\log K_{eq})$$

−1.0 kcal/mol = −2.3 (1.987 cal/mol K)(298.2 K) log K_{eq}
−1.0 = −1.36 log K_{eq}
log K_{eq} = 0.735
K_{eq} = 5.44
Let x = percent of equatorial conformer

$$K_{eq} = \frac{[\%\ \text{equatorial}]}{[\%\ \text{axial}]} = \frac{x}{100 - x}$$

5.44 = x/(100 − x)
x = 84.5

Conclusion: Cyclohexanol consists of 84.5% of the molecules in the equatorial conformation and 15.5% in the axial conformation.

(c)

Cyclohexanecarboxylic acid

According to the data in Table 4.3, the free energy difference between the axial and equatorial conformers for cyclohexanecarboxylic acid is 1.4 kcal/mol. The equilibrium constant is related to the standard free energy:

$$\Delta G° = -2.3 \ RT \ (\log K_{eq})$$

-1.4 kcal/mol $= -2.3$ (1.987 cal/mol K)(298.2 K) log K_{eq}
$K_{eq} = 10.8$
Let x = percent of equatorial conformer

$$K_{eq} = \frac{[\% \text{ equatorial}]}{[\% \text{ axial}]} = \frac{x}{100 - x}$$

x = 91.5
Conclusion: Cyclohexanecarboxylic acid consists of 91.5% of the molecules in the equatorial conformation and 8.5% in the axial conformation.

(d)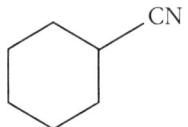

Cyclohexane carbonitrile

According to the data in Table 4.3, the free energy difference between the axial and equatorial conformers for cyclohexanecarbonitrile is 0.2 kcal/mol. The equilibrium constant is related to the standard free energy:

$$\Delta G° = -2.3 \ RT \ (\log K_{eq})$$

-0.2 kcal/mol $= -2.3$ (1.987 cal/mol K)(298.2 K) log K_{eq}
$K_{eq} = 1.38$
Let x = percent of equatorial conformer

$$K_{eq} = \frac{[\% \text{ equatorial}]}{[\% \text{ axial}]} = \frac{x}{100 - x}$$

x = 58
Conclusion: Cyclohexanecarbonitrile consists of 58% of the molecules in the equatorial conformation and 42% in the axial conformation.

4.9 (a) 1. The parent name is cyclobutane.
2. There are two substituents: a methyl group and an ethyl group. One of the carbon atoms is C1, the other C2. *Note*: We assign the lower number to the substituent that comes first in alphabetical order.
3. The two substituents are on opposite sides of the plane of the ring, or *trans* to each other.
4. The name is *trans*-1-ethyl-2-methylcyclobutane.

(b) 1. The parent name is cyclopentane.
2. There are two substituents: a chlorine atom and a bromine atom. One of the carbon atoms will be C1 and the other C2. *Note*: We assign the lower number to the substituent that comes first in alphabetical order.
3. The two substituents are on the same side of the plane of the ring, or *cis* to each other.
4. The name is *cis*-1-bromo-2-chlorocyclopentane.

(c) 1. The parent name is cyclopentane.
2. There are two substituents: a 1-methylethyl group and a chlorine atom. One of the carbon atoms will be C1, the other C2. *Note*: We assign the lower number to the substituent that comes first in alphabetical order.

3. The two substituents are on opposite sides of the plane of the ring, or *trans* to each other.
4. The name is *trans*-1-chloro-2-(1-methylethyl)cyclopentane.

(d) 1. The parent name is cyclopropane.
2. There are two identical substituents: 1,1-dimethylethyl groups (trivial name: *tert*-butyl).
3. The two substituents are on opposite sides of the plane of the ring, or *trans* to each other.
4. The name is *trans*-1,2-*bis*(1,1-dimethylethyl)cyclopropane (or *trans*-1,2-di-*tert*-butylcyclopropane). *Note:* We use *bis*- instead of *di*- when there are two complex substituents.

(e) 1. The parent name is cyclopentane.
2. There are two substituents: a 1,1-dimethylethyl group (trivial name: *tert*-butyl group) and a fluorine atom. One of the carbon atoms is C1, the other is C3. *Note:* We assign the lower number to the substituent that comes first in alphabetical order.
3. The two substituents are on opposite sides of the plane of the ring, or *trans* to each other.
4. The name is *trans*-1-fluoro-3-(1,1-dimethylethyl)cyclopentane (or *trans*-1-*tert*-butyl-3-fluorocyclopentane). *Note:* Substituents are cited alphabetically. When the numbering could begin with either of two groups (as in this case), we must begin with the one that is alphabetically first. That is why we have different numbering depending on the choice of IUPAC or trivial name of the alkyl group.

(f) 1. The parent name is cyclooctane.
2. There are two substituents: a methyl group and a bromine atom. One of the carbon atoms is C1, the other is C4. *Note:* We assign the lower number to the substituent that comes first in alphabetical order.
3. The two substituents are on the same side of the plane of the ring, or *cis* to each other.
4. The name is *cis*-1-bromo-4-methylcyclooctane.

4.10 (a)

 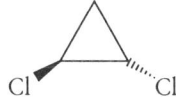

cis-1,2-Dichlorocyclopropane   *trans*-1,2-Dichlorocyclopropane

(b) *Note:* We assign a lower number to the substituent that comes first in alphabetical order. That is why the numbering is different depending if we assign IUPAC or trivial name.

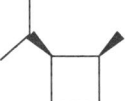 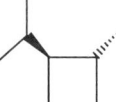

cis-1-Methyl-2-(methylethyl)cyclobutane   *trans*-1-Methyl-2-(methylethyl)cyclobutane
or *cis*-1-isopropyl-2-methylcyclobutane   or *trans*-1-isopropyl-2-methylcyclobutane

(c)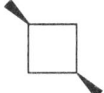

cis-1,3-Dimethylcyclobutane   *trans*-1,3-Dimethylcyclobutane

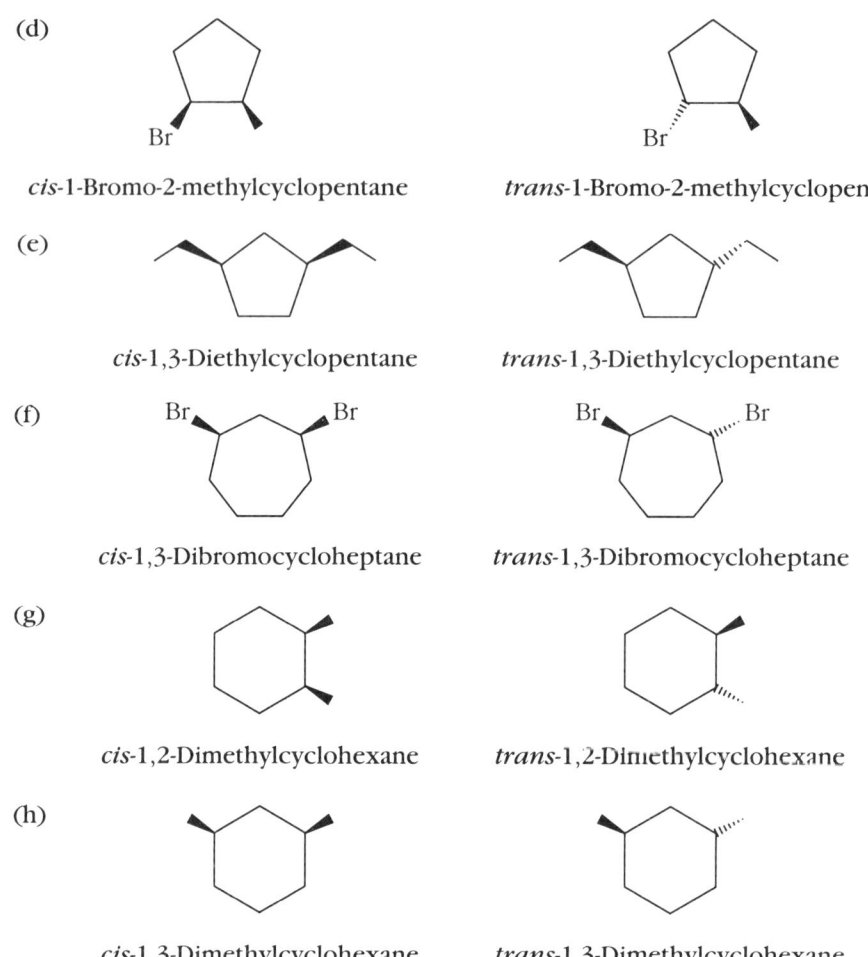

(d) cis-1-Bromo-2-methylcyclopentane / trans-1-Bromo-2-methylcyclopentane

(e) cis-1,3-Diethylcyclopentane / trans-1,3-Diethylcyclopentane

(f) cis-1,3-Dibromocycloheptane / trans-1,3-Dibromocycloheptane

(g) cis-1,2-Dimethylcyclohexane / trans-1,2-Dimethylcyclohexane

(h) cis-1,3-Dimethylcyclohexane / trans-1,3-Dimethylcyclohexane

4.11 *trans*-1,4-Dimethylcyclohexane exists as an equilibrium mixture of diequatorial and diaxial chair conformations.

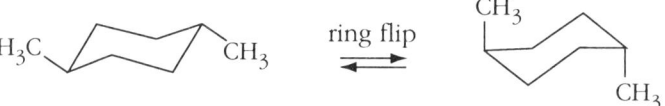

The diequatorial conformation has *no strain energy* due to 1,3–diaxial interactions. The diaxial conformation contains *two* pairs of 1,3–diaxial interactions, so the difference in energy between the two conformations equals 2 × 1.7 kcal/mol = 3.4 kcal/mol (or 14.2 kJ/mol). Therefore, *trans*-1,4-dimethylcyclohexane exists predominantly in the diequatorial conformation.

The two chair conformations of *cis*-1,4-dimethylcyclohexane have one axial and one equatorial methyl group. Thus, each has *one* pair of 1,3-diaxial interactions so its strain energy is 1.7 kcal/mol (7.1 kJ/mol). As a result, the diequatorial conformation of *trans*-1,4-dimethylcyclohexane, which has no strain energy, is more stable than *cis*-1,4-dimethylcyclohexane by 1.7 kcal/mol (or 7.1 kJ/mol), close to the experimental value of 1.6 kcal/mol (6.7 kJ/mol).

4.12 (a) *trans*-1-Ethyl-4-isopropylcyclohexane (or *trans*-1-ethyl-4-(methylethyl)cyclohexane):

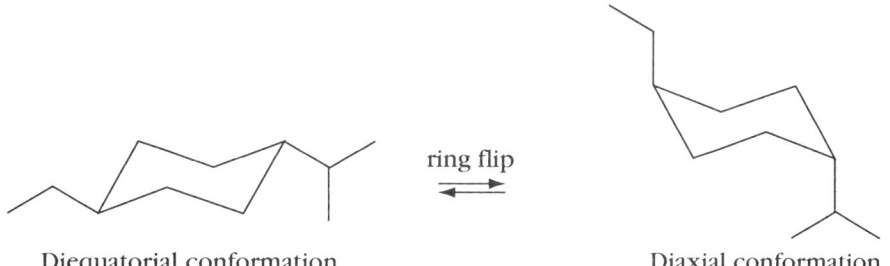

Diequatorial conformation   Diaxial conformation

The diequatorial conformation has *no 1,3-diaxial interactions* (both substituents are in equatorial positions). However, both substituents experience 1,3-diaxial interactions in the diaxial conformation. Thus, the total strain energy equals the sum of individual 1,3-diaxial interactions. According to data from Table 4.3, an ethyl group in an axial position experiences *strain energy* of 1.9 kcal/mol (8.0 kJ/mol) and an isopropyl group experiences 2.2 kcal/mol (9.2 kJ/mol). As a result, the diequatorial conformation is more stable by 1.9 + 2.2 = 4.1 kcal/mol (17.1 kJ/mol).

(b) *trans*-1-Ethyl-3-methylcyclohexane:

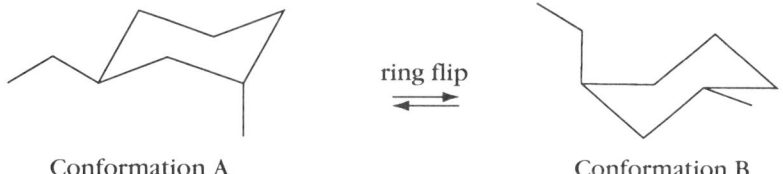

Conformation A   Conformation B

Both conformation A and B have one of the alkyl substituents in an equatorial position and the other in an axial position. Conformation B has an ethyl group in the axial position (strain energy = 1.9 kcal/mol or 7.9 kJ/mol); conformation A has a methyl group (strain energy = 1.7 kcal/mol or 7.1 kJ/mol). Thus, conformation A has the smaller methyl group in the axial position and is slightly more stable (by 0.2 kcal/mol or 0.84 kJ/mol).

(c) *cis*-1-Bromo-2-isopropylcyclohexane or *cis*-1-Bromo-2-(methylethyl)cyclohexane:

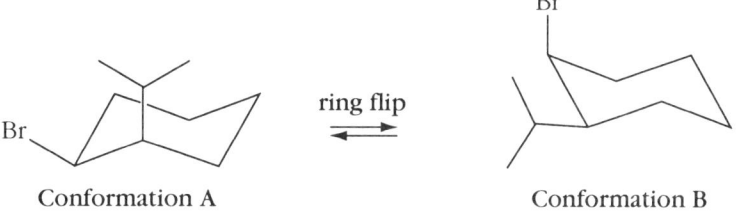

Conformation A   Conformation B

Both conformation A and B have one of the substituents in an equatorial position and the other in an axial position. Conformation B has a bromine atom in the axial position; conformation A has an isopropyl group. Because the isopropyl group is bulkier and exerts larger 1,3-diaxial interactions, conformation A is less stable.

According to data in Table 4.3, an isopropyl group has the strain energy equal to 2.2 kcal/mol (9.2 kJ/mol); the bromine atom has the strain energy equal to 0.7 kcal/mol (2.9 kJ/mol). Thus, conformation B is more stable by 2.2 − 0.7 = 1.5 kcal/mol (6.3 kJ/mol).

4.13 When a cycloalkane ring bears two substituents on different carbon atoms, these substituents may be on the same side or on opposite sides of the ring. When the substituents are on the same side, we say they are *cis* to each other; when substituents are on opposite sides, we say they are *trans* to each other. It is *not* important whether the substituents are in axial or equatorial positions; it is important whether the substituents are above or below the plane of the ring.

There are six *axial* positions: three pointing up and three pointing down.

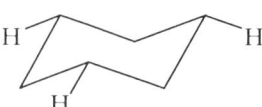

There are also six *equatorial* positions in the chair conformation of cyclohexane. Atoms or groups in equatorial positions lie in a ring outside of the carbon skeleton and roughly in the plane of the molecule. A closer examination demonstrates that the equatorial bonds point slightly up or slightly down (three in each direction).

Three equatorial bonds pointing slightly *up*

Three equatorial bonds pointing slightly *down*

(a) The methyl group *(axial)* is pointing up and the bromine *(equatorial)* is pointing down so it is a *trans*-stereoisomer. Name: *trans*-1-bromo-3-methylcyclohexane.

(b) The methyl group is pointing up and the hydroxyl group is pointing down so it is a *trans*-stereoisomer. Name: *trans*-2-methylcyclohexanol.

(c) The bromine atom *(equatorial)* is pointing up and the 1,1-dimethylethyl group (or *tert*-butyl group) *(equatorial)* is also pointing up so it is a *cis*-stereoisomer. Name: *cis*-1-bromo-3-(1,1-dimethylethyl)cyclohexane or *cis*-1-bromo-3-*tert*-butylcyclohexane.

(d) The bromine atom *(equatorial)* is pointing down and the chlorine atom *(equatorial)* is pointing up so it is a *trans*-stereoisomer. Name: *trans*-1-bromo-2-chlorocyclohexane.

(e) The bromine atom *(equatorial)* is pointing down and the chlorine atom *(axial)* is pointing up so it is a *trans*-stereoisomer. Name: *trans*-1-bromo-3-chlorocyclohexane.

(f) The chlorine atom *(equatorial)* is pointing down and the other chlorine atom (also *equatorial*) is pointing up so it is a *trans*-stereoisomer. Name: *trans*-1,4-dichlorocyclohexane.

4.14 (a)

(b) No, it is not possible to carry out a chair-chair interconversion on both rings of *trans*-decalin. To form the second ring of *trans*-decalin, the two carbon atoms would have been in axial positions and the remaining two methylene groups could not form sufficiently long links to close the ring. As a result, ring flipping is blocked.

Note: Only two carbon atoms are needed to close the ring and provide a chain long enough to reach between the *axial* and *equatorial* carbon atoms of *cis*-decalin. As a result, chair-chair ring flipping can, and indeed does, take place quite readily (Exercise 4.14 a).

(c)

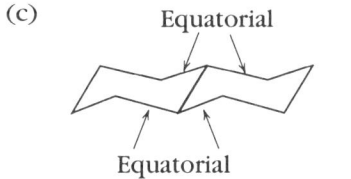

trans-Decalin

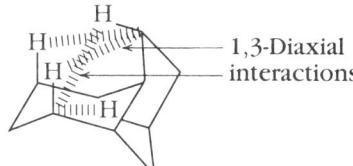

cis-Decalin

(d) In *trans*-decalin, both of the methylene groups on the ring junction are equatorial and there are *no* 1,3-diaxial interactions.

In *cis*-decalin, one of the methylene groups on the ring junction is axial in each case. Thus, there are two methyl-like substituents in axial position on each "cyclohexane" ring. Each methyl group in an axial position exerts two 1,3-diaxial interactions that result in a strain of 1.7 kcal/mol (Table 4.3). Here also there are two 1,3-diaxial interactions per axial carbon, except that one is common to both carbons. Hence, there are a total of three 1,3-diaxial interactions and the free energy difference between *trans*- and *cis*-decalin is 1.7 + 0.85 = 2.55 kcal/mol (10.7 kJ/mol).

4.15 (a) The **eclipsed conformation** is a conformation in which bonds on adjacent atoms are aligned parallel to each other. For example, the chlorine and hydrogen atoms are "pointing in the same direction". The C–H bond and C–Cl bond are eclipsed.

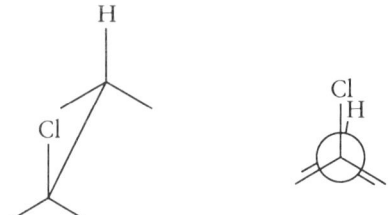

(b) **Torsional energy** is the amount of energy by which the stability of a molecule is decreased as a result of the eclipsing of bonds.
(c) **Steric strain** (also known as van der Waals strain or van der Waals repulsion) is the destabilization of a molecule that results when two atoms or groups approach one another too closely.
(d) **Standard heat of combustion** is the amount of energy ($-\Delta H°$) released on complete combustion (oxidation) of a compound under standard conditions. This number provides the same information on the thermodynamic stability of a molecule as the heat of formation, but the heat of combustion is obtained experimentally while heats of formation are calculated using the known heats of combustion of the elements.
(e) **Ring strain** is the destabilization of a molecule that results when its atoms form a ring. Total ring strain is the sum of three strains: (i) *angle strain* (the strain due to expansion or compression of bond angles, (ii) *torsional strain* (the strain due to eclipsing of neighboring bonds), and (iii) *steric strain* (the strain due to repulsive interactions when atoms approach each other too closely).
(f) **Angle strain** is the amount of strain a molecule possesses when its bond angles are distorted from their normal values.
(g) The **chair conformation of cyclohexane** is usually the most stable conformation of a cyclohexane ring. In the chair conformation, all the carbon-carbon bonds have staggered conformations and each carbon is free to adopt its preferred tetrahedral geometry.
(h) An **equatorial bond** is one of the two possible orientations that substituents can adopt in the most stable "chair" conformation of cyclohexane. The equatorial bonds point out horizontally but also slightly *up* or slightly *down* relative to the ring.

(i) A **fused polycyclic molecule** is a molecule in which the rings share two adjacent atoms and the bond between them.
(j) The **Newman projection formula** is a method for depicting conformations by which one sights down a carbon-carbon bond and represents the front carbon by a point and the back carbon by a circle.

Three-dimensional formula   Newman projection

Viewed from this direction for Newman projection ------>

(k) The **gauche conformation** is a staggered conformation in which the dihedral angle between bonds of two substituents on adjacent carbon atoms is 60°. Thus, the term *gauche* describes the orientation in space of groups or atoms on adjacent carbon atoms. For example, —H and —CH₃ are gauche to one another:

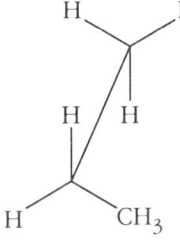

(l) The **staggered conformation** is a conformation of a molecule in which the bonds on adjacent carbon atoms are as far away from one another as possible:

(m) **Torsional strain** is the amount of energy by which the stability of a molecule is decreased as a result of eclipsing of the bonds.

(n) The **anti conformation** is a staggered conformation in which the angle between bonds of two substituents on adjacent carbon atoms is 180°. Thus, the term *anti* describes the orientation in space of groups or atoms on adjacent carbon atoms that are as far apart as possible. For example, the —H and —Cl atoms are *anti* to one another in the following conformation:

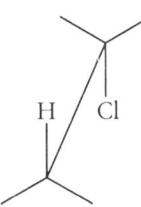

(o) The **boat conformation of cyclohexane** is shown below. The boat conformation of cyclohexane is less stable than the chair conformation of cyclohexane by 6.9 kcal/mol (29 kJ/mol).

(p) An **axial bond** is one of two possible orientations that substituents can adopt in the most stable "chair" conformation of cyclohexane. The axial bonds are perpendicular to the plane of the ring. Three of them are on each side of the general plane of the ring.

(q) A **1,3-diaxial interaction** is a repulsive force between axial substituents on the same side of a cyclohexane ring:

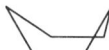1,3-Diaxial interactions

(r) **Stereoisomers** are isomers that have the same constitution but which differ in respect to the arrangement of their atoms in space. *cis*-1,2-Dimethylcyclohexane and *trans*-1,2-dimethylcyclohexane are examples of stereoisomers.

4.16

| Anti conformation | Gauche conformation | Least stable eclipsed conformation | Eclipsed conformation |

(a) Newman projections shown for each conformation.

(b) Sawhorse projections shown for each conformation.

4.17 (a) Staggered conformations:

Eclipsed conformations:

(b) Staggered conformations:

Eclipsed conformations:

(c) Staggered conformations:

Eclipsed conformations:

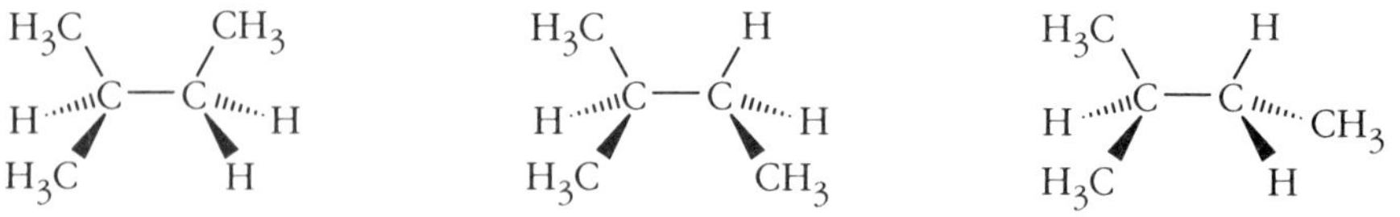

4.18 (a)

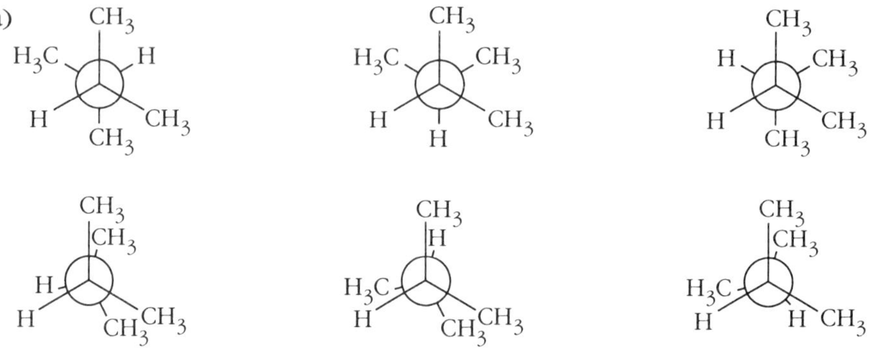

(b)

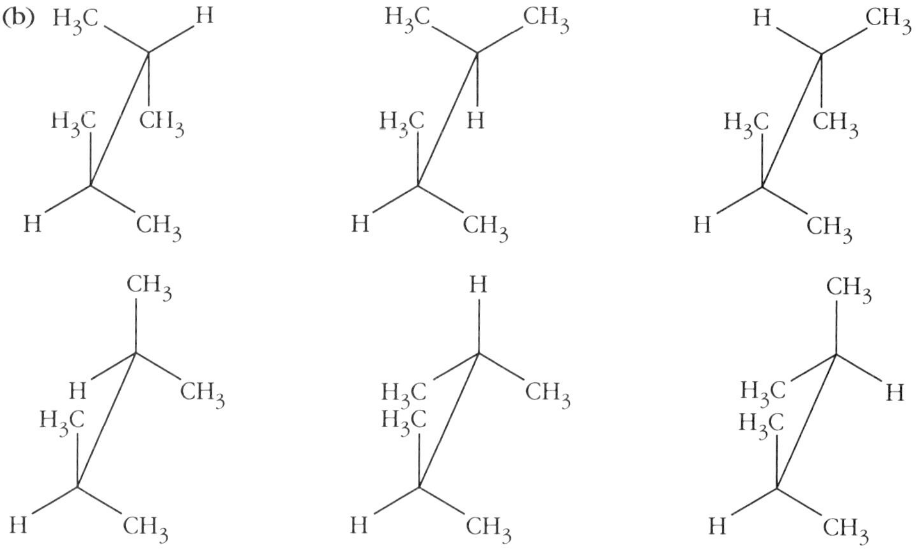

(c)

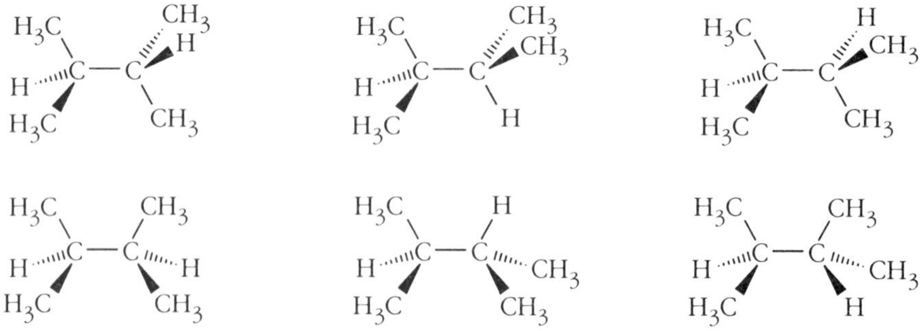

4.19 (a) We begin the **conformational analysis** (the study of the energetics of different conformations) of 2-methylbutane with the **totally eclipsed** conformation. It has two eclipsed methyl groups (2.5 kcal/mol), eclipsed hydrogen and a methyl group (1.4 kcal/mol) and two eclipsed hydrogen atoms (1.0 kcal/mol). The contributions of various interactions are taken from Table 4.1. The total strain energy of this conformation is 2.5 + 1.4 + 1.0 = 4.9 kcal/mol (20.5 kJ/mol).

As rotation around the C2—C3 bond occurs, a **gauche** conformation is reached at θ = 60°. It is a staggered conformation and even though it has no eclipsing interactions, it has two gauche butane interactions (0.77 kcal/mol each). The total strain energy of this conformation is 0.77 + 0.77 = 1.54 kcal/mol (6.44 kJ/mol).

As rotation around the C2—C3 bond continues, another **totally eclipsed** conformation is reached at θ = 120°. It has two eclipsed methyl groups (2.5 kcal/mol), eclipsed hydrogen and a methyl group (1.4 kcal/mol) and two eclipsed hydrogen atoms (1.0 kcal/mol). The total strain energy of this conformation is 2.5 + 1.4 + 1.0 = 4.9 kcal/mol (20.5 kJ/mol).

At θ = 180°, a staggered conformation is reached. This time we have only one gauche butane interaction and the total strain energy will have a value of 0.77 kcal/mol (3.22 kJ/mol).

At θ = 240°, an eclipsed conformation is reached. This time we do not have eclipsed methyl groups but only three hydrogen–methyl group interactions (each 1.4 kcal/mol). The total strain energy equals 1.4 + 1.4 + 1.4 = 4.2 kcal/mol (17.6 kJ/mol).

At θ = 300°, we have a similar staggered conformation as occurs at θ = 180°. The total strain energy is 0.77 kcal/mol (3.22 kJ/mol).

And finally, at θ = 360°, we return to the **totally eclipsed** conformation we discussed for θ = 0°; total strain energy 4.9 kcal/mol (20.5 kJ/mol).

Now we can construct a plot of potential energy *versus* dihedral angle, θ.

(b) Again, we begin the **conformational analysis** of 2,3-dimethylbutane with **totally eclipsed** conformation. It has two pairs of eclipsed methyl groups

(2.5 kcal/mol each), and two eclipsed hydrogen atoms (1.0 kcal/mol) (Table 4.1). The total strain energy of this conformation is 2.5 + 2.5 + 1.0 = 6.0 kcal/mol (25.1 kJ/mol).

As rotation around the C2–C3 bond occurs, a **gauche** conformation is reached at θ = 60°. It is a staggered conformation and even though it has no eclipsing interactions, it has three gauche butane interactions (0.77 kcal/mol each). The total strain energy of this conformation is 0.77 + 0.77 + 0.77 = 2.31 kcal/mol (9.66 kJ/mol).

As rotation around the C2–C3 bond continues, another **totally eclipsed** conformation is reached at θ = 120°. Two eclipsed interactions of a hydrogen and a methyl group (1.4 kcal/mol each) and one pair of eclipsed methyl groups (2.5 kcal/mol) contribute to the total strain energy of this conformation of 5.3 kcal/mol (22.2 kJ/mol).

At θ = 180°, a staggered conformation is reached. This time we have only two gauche butane interactions (0.77 kcal/mol) so the total strain energy will be 1.54 kcal/mol (6.44 kJ/mol).

At θ = 240°, an eclipsed conformation is reached. The two hydrogen-methyl group interactions (each 1.4 kcal/mol) plus one methyl-methyl interaction (2.5 kcal/mol) will give the total strain energy of 1.4 + 1.4 + 2.5 = 5.3 kcal/mol (22.2 kJ/mol).

At θ = 300°, we have a similar staggered conformation as occurs at θ = 60°. The total strain energy is 2.31 kcal/mol (9.66 kJ/mol).

And finally, at θ = 360°, we return to the **totally eclipsed** conformation we discussed for θ = 0°; total strain energy 6.0 kcal/mol (25.1 kJ/mol).

Now we can construct a plot of potential energy versus dihedral angle, θ.

4.20 (a) *cis*-1,3-Dimethylcyclobutane exists in the two following conformations:

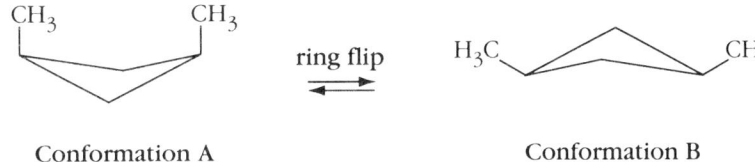

Conformation A   Conformation B

(b) The two methyl groups in Conformation A of *cis*-1,3-dimethylcyclobutane are positioned similarly to the two methyl groups in *axial* positions of *cis*-1, 3-dimethylcyclohexane. For this reason we should expect interactions between those two methyl groups to be similar to 1,3-diaxial interactions in the chair conformation of cyclohexane. Conformation B, on the other hand, is free of such interactions and the two methyl groups resemble the two methyl groups in *equatorial* positions of *cis*-1,3-dimethylcyclohexane. Hence, conformation B would be expected to be the more stable of the two.

4.21 (a) Because cyclopropane releases more heat on combustion ($\Delta H° = -469.6$ kcal/mol or -1964.8 kJ/mol) than propene ($\Delta H° = -461.8$ kcal/mol or -1932.2 kJ/mol), cyclopropane contains more energy and is less stable than propene.

(b) Because 2-methylbutane releases less heat on combustion ($\Delta H° = -782.4$ kcal/mol or -3273.6 kJ/mol) than pentane ($\Delta H° = -784.2$ kcal/mol or -3281.1 kJ/mol), 2-methylbutane contains less energy and is more stable.

(c) Because 2,2-dimethylpropane releases less heat on combustion ($\Delta H° = -779.0$ kcal/mol or -3259.3 kJ/mol) than pentane ($\Delta H° = -784.2$ kcal/mol or -3281.1 kJ/mol), 2,2-dimethylpropane contains less energy and is more stable.

4.22 (a) The methyl group is in the axial position (perpendicular to the general plane of the ring).

(b) The chlorine atom is in the equatorial position ("almost" in the plane of the ring).

(c) The carboxyl group is in the equatorial position ("almost" in the plane of the ring).

(d) The bromine atom is in the axial position (perpendicular to the general plane of the ring).

4.23 (a) *cis*-3-Methylcyclohexanol exists in the following two conformations:

Diaxial   Diequatorial

cis-3-Methylcyclohexanol is more stable when both substituents are in the equatorial positions.

(b) *trans*-1-Chloro-2-isopropylcyclohexane exists in the following two conformations:

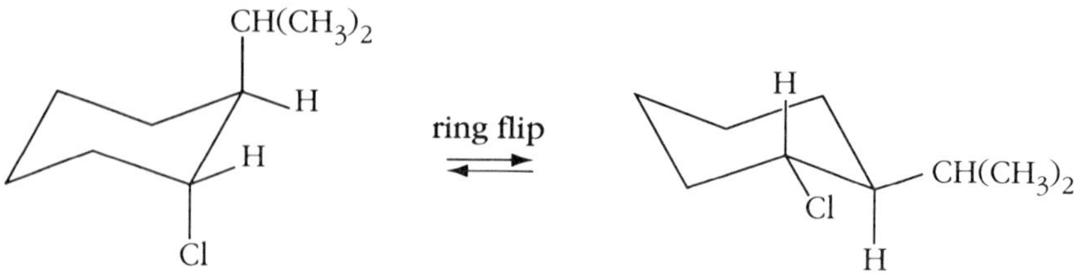

Diaxial ⇌ (ring flip) Diequatorial

trans-1-Chloro-2-isopropylcyclohexane is more stable when both substituents are in equatorial positions.

(c) *cis*-1,2-Diethylcyclohexane exists in the following two conformations:

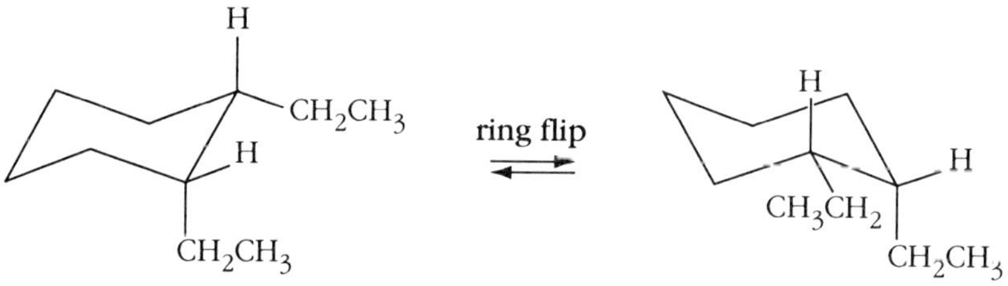

One ethyl group is in an axial position and the other is in an equatorial position in each conformation. Both conformations are equally stable so they will exist in equal amounts.

(d) *trans*-1-Chloro-4-ethylcyclohexane exists in the following two conformations:

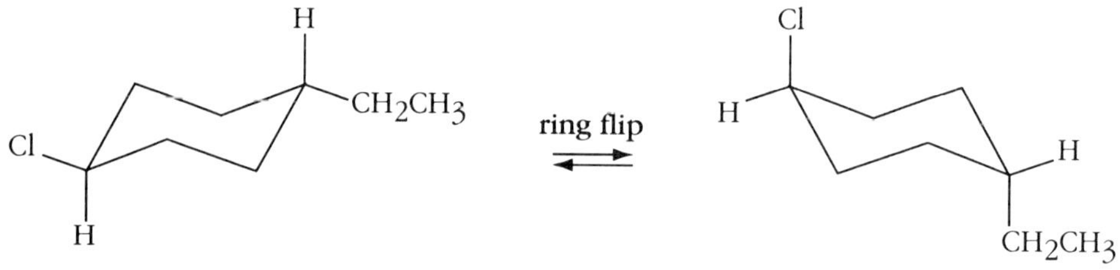

Diequatorial ⇌ (ring flip) Diaxial

trans-1-Chloro-4-ethylcyclohexane is more stable when both substituents are in equatorial positions.

(e) *cis*-4-(1,1-Dimethylethyl)cyclohexanol (or *cis*-4-*tert*-butylcyclohexanol) exists in the following two conformations:

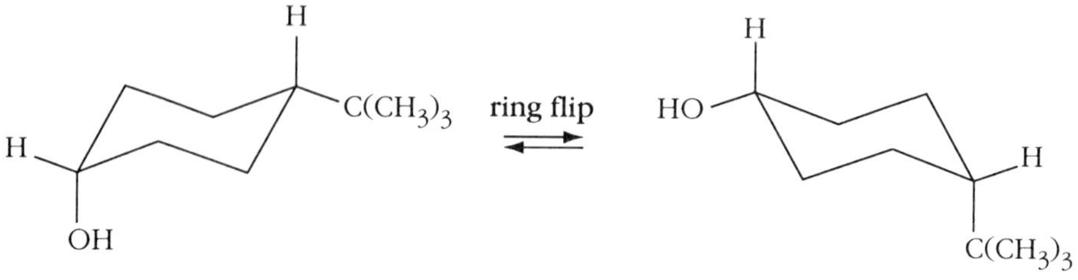

The hydroxyl group (—OH) is in the axial position and the 1,1-dimethylethyl (*tert*-butyl) group is in the equatorial position in the conformation on the left. The —OH group is in the equatorial and the *tert*-butyl group is in the axial position in the conformation on the right. Because the *tert*-butyl group is much bulkier than the hydroxyl group, the conformation with the bulky group in the equatorial position will be much more stable. Thus, the conformation on the left should be more stable.

(f) *cis*-2-Ethylcyclohexanol exists in the following two conformations:

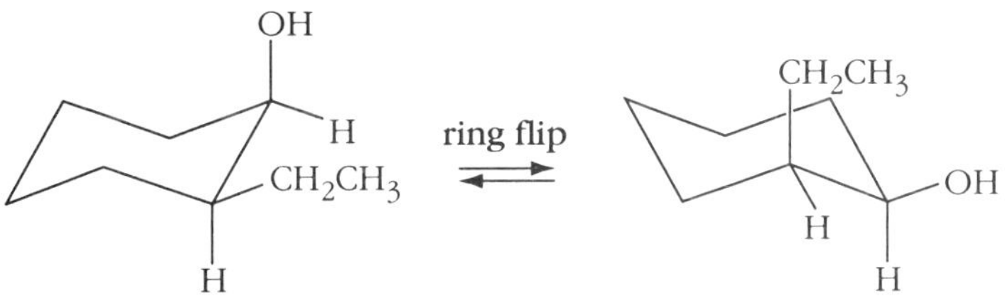

The hydroxyl group (—OH) is in the axial position and the ethyl group is in the equatorial position in the conformation on the left. The —OH group is in equatorial and the ethyl group is in axial position in the conformation on the right. Because the ethyl group is bulkier than the hydroxyl group (Table 4.3), the conformation with the bulky group in the equatorial position will be more stable. Thus, the conformation on the left should be the more stable.

4.24 All of the following examples have one substituent in an axial position and one in an equatorial position. When the cyclohexane ring flips, the axial substituent becomes equatorial and the equatorial substituent becomes axial. Therefore, the conformation with a smaller substituent in the axial position and a larger substituent in the equatorial position will be preferred (Table 4.3).

(a) 1-Methyl-1-(methylethyl)cyclohexane or 1-isopropyl-1-methylcyclohexane:

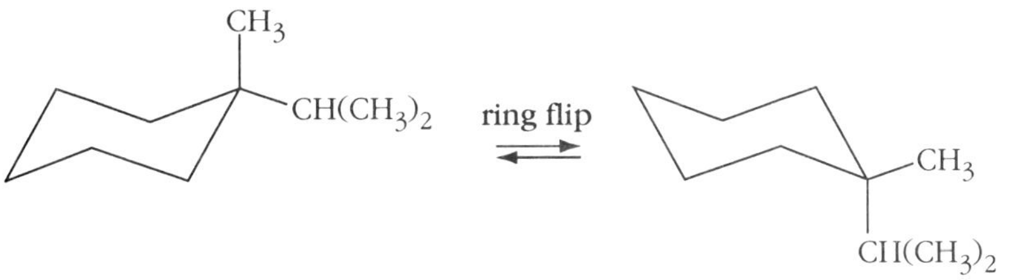

According to the data in Table 4.3, an isopropyl (methylethyl) group has a greater preference for the equatorial position than a methyl group. As a result, the conformation on the left is preferred.

(b) *cis*-1-Bromo-2-(methylethyl)cyclohexane or
cis-1-bromo-2-isopropylcyclohexane:

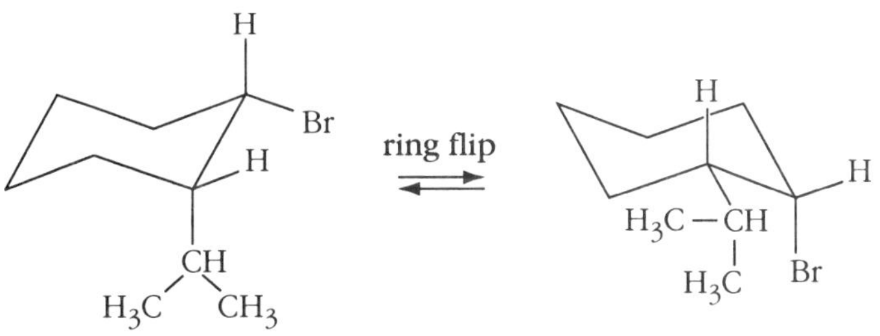

According to the data in Table 4.3, an isopropyl (methylethyl) group has a greater preference for the equatorial position than bromine. As a result, the conformation on the right is preferred.

(c) *trans*-1-Methyl-3-(methylethyl)cyclohexane or *trans*-1-isopropyl-3-methylcyclohexane:

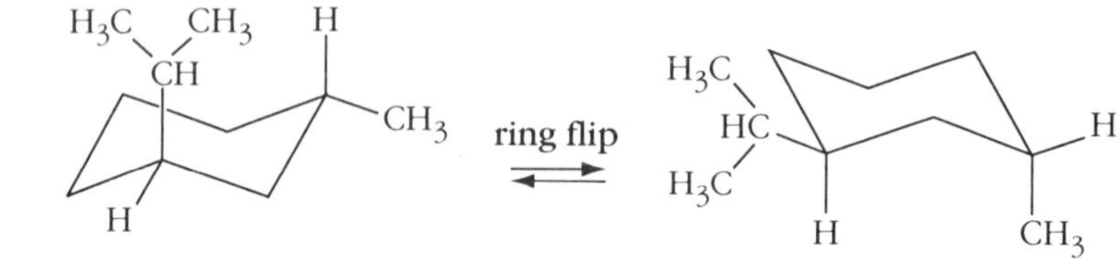

According to the data in Table 4.3, an isopropyl (methylethyl) group has a greater preference for the equatorial position than a methyl group. As a result, the conformation on the right is preferred.

(d) *cis*-1-Chloro-4-(methylethyl)cyclohexane or *cis*-1-chloro-4-isopropylcyclohexane:

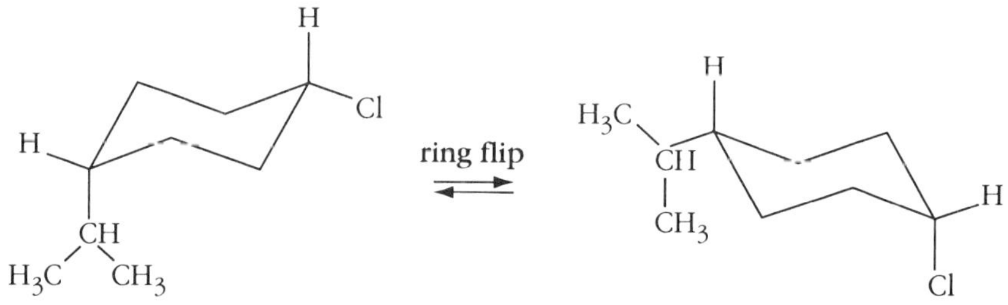

According to the data in Table 4.3, an isopropyl (methylethyl) group has a greater preference for the equatorial position than chlorine. As a result, the conformation on the right is preferred.

4.25 (a) *cis* isomer (carboxyl group and methyl group are pointing *up*)
(b) *cis* isomer (bromine and —CN group are pointing *up*)
(c) *cis* isomer (methyl group and hydroxyl group are pointing *up*)
(d) *trans* isomer (chlorine is pointing *down*; hydroxyl group is pointing *up*)
(e) *trans* isomer (bromine is pointing *up*; *tert*-butyl group is pointing *down*)
(f) *trans* isomer (one carboxyl group is pointing *up*, the other is pointing *down*)

4.26 All *trans*-1,2,3,4,5,6-hexachlorocyclohexane exists in the following conformations. The one on the left has all six chlorine atoms in equatorial positions while the conformation on the right has all six chlorine atoms in axial positions. The conformation on the left is the preferred conformation.

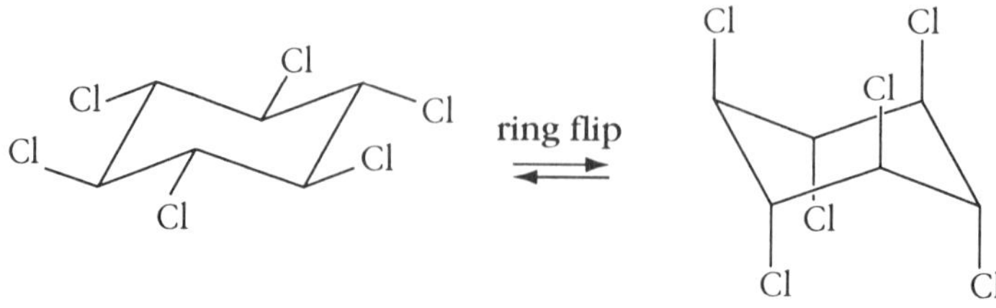

4.27 (a) Constitutional isomers (1-chloropropane and 2-chloropropane)
(b) Conformers (two conformations of *trans*-1-chloro-4-methylcyclohexane)
(c) Constitutional isomers (2-bromo-2-methylpropane and 2-bromobutane)
(d) Stereoisomers (*cis*-1,3-dichlorocyclopentane and *trans*-1,3-dichlorocyclopentane)
(e) Stereoisomers because they differ only in the arrangement of atoms or groups around C2
(f) Conformers (two conformations of *trans*-3-bromocyclohexanecarboxylic acid)

4.28 The two stereoisomers of 1,2-dichlorocyclohexane are *cis*- and *trans*-1,2-dichlorocyclohexane. Each of the stereoisomers can exist in two conformations:

cis-1,2-Dichlorocyclohexane:

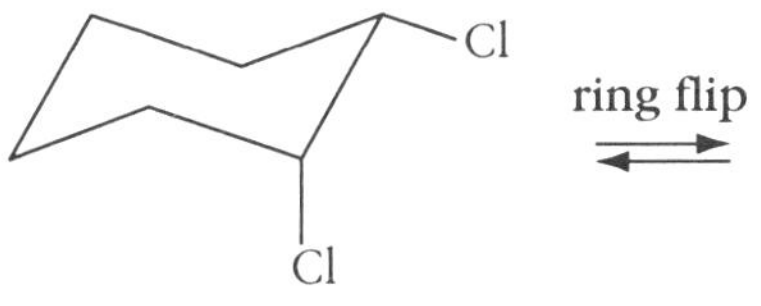

trans-1,2-Dichlorocyclohexane:

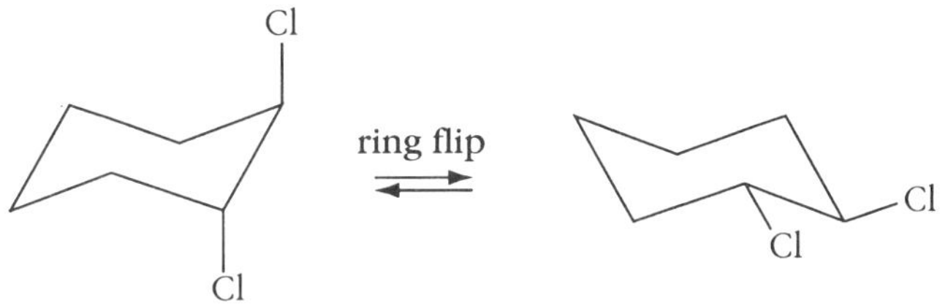

(a) The two chair conformations of *cis*-1,2-dichlorocyclohexane can be interconverted by ring flipping as can the two chair conformations of *trans*-1,2-dichlorocyclohexane.
(b) Neither chair conformation of *cis*-1,2-dichlorocyclohexane can be interconverted by ring flipping into a chair conformation of *trans*-1,2-dichlorocyclohexane. Confirm this with your molecular model. Make a model of *cis*-1,2-dichlorocyclohexane and another of *trans*-1, 2-dichlorocyclohexane. Ring flipping of one model does not convert it into the other.

4.29 One way to approach such problems is first to check if the structures represent the same compound and not constitutional isomers. Are the substituents in the same relative position to each other (e.g., 1,2 or 1,3) ? If they are in the same relative position to each other, we check to see if the structures are conformers by converting one structure into the other by ring flipping and/or rotating the molecule.

(a) The chlorine atom has a 1,3 relationship to the methyl group in both structures so these structures both represent 1-chloro-3-methylcyclohexane. The chlorine is in an axial position in the structure on the left and in an equatorial position in the structure on the right. The methyl group is in an equatorial position in the structure on the left and in an axial position in the structure on the right. Ring flipping interconverts the positions of both substituents:

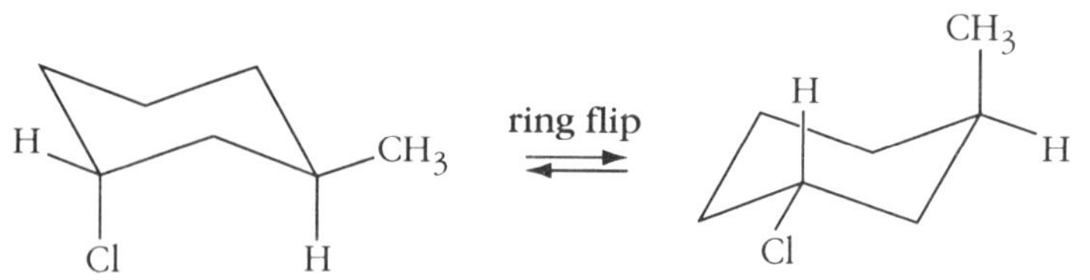

After flipping the ring we rotate the structure around axis AB and we obtain the structure identical to the structure drawn in the problem on the right-hand side.

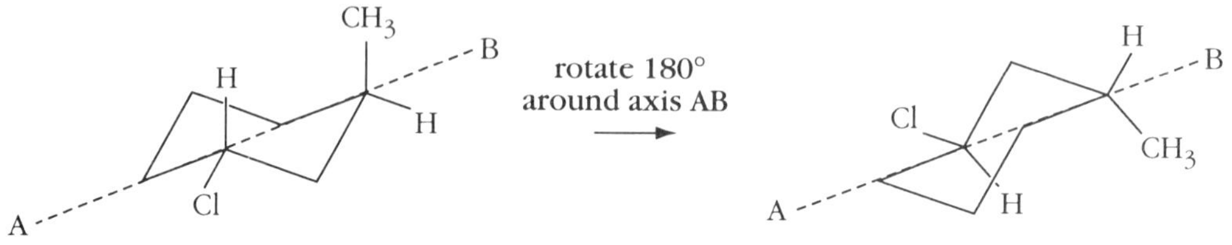

Thus, the two chair forms are conformers.

(b) The chlorine atom has a 1,2 relationship to the methyl group in both structures. The chlorine is in an equatorial position in the structure on the left and in an axial position in the structure on the right. The methyl group is in the equatorial position in the structure on the left and in the equatorial position in the structure on the right. If we flip the ring, the equatorial chlorine atom will become axial (what we observe) and the equatorial methyl group becomes axial (what we do not observe). Thus, the two structures are not conformers but stereoisomers. The structure on the left is *trans*-1-chloro-2-methylcyclohexane and the other is *cis*-1-chloro-2-methylcyclohexane.

(c) The chlorine atom has a 1,4 relationship to the methyl group in both structures. The chlorine is in an equatorial position in the structure on the left and in an axial position in the structure on the right. The methyl group is in an axial position in the structure on the left and in an equatorial position in the structure on the right. Thus, ring flipping will interconvert the positions of both substituents.

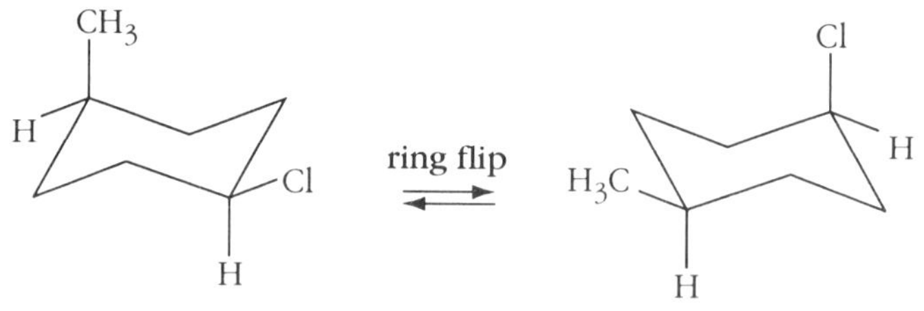

After further rotation of the ring we obtain the second structure:

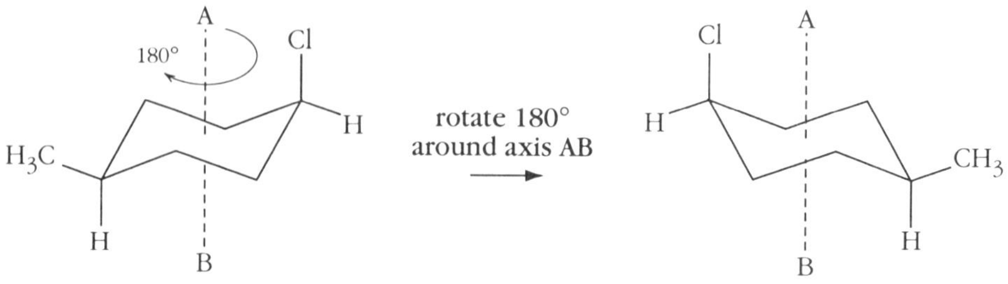

Thus, the two structures are conformers.

4.30 First look for structures that represent the same compound. All the structures are substituted cyclohexanes. To be the same compound they must have the same substituents in the same relative position to each other (e.g., 1,2 or 1,3). If they are in the same relative position to each other, we check to see if the structures are conformers by converting one structure to the other by ring flipping and/or rotating the molecule.

- (a) Structure A has a carboxyl group and chlorine in 1,4 relationship. The only other structure that has the same groups is E. Structures A and E are converted one into the other by the ring flip so they are conformational isomers. Hence, the first pair is A and E.
- (b) Structure B is 1,3,5-trimethylcyclohexane with two methyl groups in the equatorial positions and one methyl group in the axial position. The only other structure with three methyl groups in 1,3,5 relationship is G. After ring flipping of B, one gets G with two methyl groups in the axial positions and one in the equatorial position. Thus, B and G are conformational isomers.
- (c) Structure C is 1,2,4-trimethylcyclohexane. We have three other structures of 1,2,4-trimethylcyclohexane, so we have to look at the positions of the methyl groups in structure C: all three methyl groups are in axial positions. After the ring flip all methyl groups will become equatorial, as in structure D. We have to rotate structure C clockwise by 60° to get structure D. Thus, the third pair of conformational isomers is C and D.
- (d) Structure F has one methyl group (at C1) in the equatorial position and two methyl groups (at C2 and C4) in the axial positions. After the ring flip the methyl group at C1 becomes axial and the two methyl groups at C2 and C4 become equatorial (like those in structure H). The fourth pair of conformational isomers is F and H.

4.31 The methyl group in *trans*-9-methyldecalin has 1,3-diaxial interactions with four hydrogen atoms (at C2, C4, C5, and C7). The methyl group in *cis*-9-methyldecalin has 1,3-diaxial interactions only with two hydrogen atoms (at C2 and C4). Those hydrogen atoms are shown below. Thus, the methyl group in *cis*-9-methyldecalin will *cause* less instability than that in *trans*-9-methyldecalin by 1.7 kcal/mol (7.1 kJ/mol) and the net difference will be smaller by that amount. Thus, *trans*-9-methyldecalin will be more stable than *cis*-9-methyldecalin by 2.55 −1.7 = 0.85 kcal/mol (or 3.55 kJ/mol).

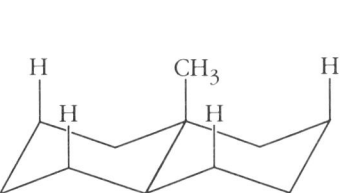

trans-9-Methyldecalin

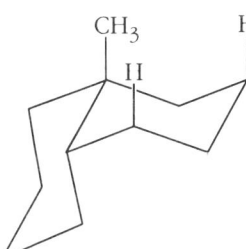

cis-9-Methyldecalin

CHAPTER 5

SPECTROSCOPY

Spectroscopy is the study of the interactions of electromagnetic radiation with matter. Electromagnetic radiation can be described like a wave, in terms of its wavelength and frequency. **Wavelength** is the distance between any two identical points on a wave and is given the symbol λ (lambda). Wavelength is expressed in units of distance, so it is usually expressed in meters (m). **Frequency** is the number of full cycles of a wave that pass a given point in a fixed period of time. Frequency is given the symbol ν (nu) and is usually expressed in hertz (1 Hz = s^{-1}). 1 Hz corresponds to one cycle per second and 2 MHz corresponds to 2×10^6 cycles per second. Specific wavelengths within the infrared (IR) region are usually expressed in micrometers (1 μm = 10^{-4} cm), and frequencies are expressed in **wavenumbers** rather than in hertz.

Infrared spectroscopy provides information about functional groups. The main use of infrared spectroscopy is to determine the presence or absence of certain functional groups in a molecule. A molecule absorbs electromagnetic radiation that corresponds to transitions between fixed energy levels. Transitions occur between vibrational energy levels in a molecule upon absorption of infrared radiation, usually at a frequency from 4000 to 600 cm^{-1}. Chemical bonds undergo various stretching and bending vibrations. Vibrations that give rise to a change in the dipole moment result in absorption of radiation that can be monitored by a spectrophotometer. A spectrum is a record of the change in the absorption of energy by the compound plotted against the wavelength or the frequency of the radiation being used.

The frequencies at which a molecule absorbs depend on the types of bonds present. The stretching frequency of a bond is related to the masses of the two atoms involved in the bond and to the strength of the bond. An infrared spectrum is generated by the molecule as a whole, but certain absorption bands are typical of particular groups of atoms, such as a carbon-carbon double bond, a carbonyl group, and a hydroxyl group.

Nuclear magnetic resonance (NMR) is a method for structure determination based on the effect of molecular environment on the energy required to promote a given nucleus from a lower-energy spin state to a higher-energy state. The nuclei of atoms such as hydrogen and carbon-13 behave like small magnets. When a sample of an organic compound is placed in a strong magnetic field, slightly more than half of the nuclei of atoms such as ^1H and ^{13}C align themselves with the external field. In this state, the nuclei are in the lower of the two energy levels available to them. They absorb energy in the radio frequency range in going to a higher-energy level in which they are aligned against the external magnetic field. The exact amount of energy necessary to cause the nucleus to undergo a transition from a lower energy state to a higher one depends on the strength of the magnetic field used and on the environment of the atom in the molecule.

Like IR spectroscopy, NMR can be used with a very small sample, and does not harm the sample. The NMR spectrum provides a great deal of information about the structure of the compound, and some structures can be determined using only the NMR spectrum. More commonly, however, the NMR spectrum is used in conjunction with other forms of spectroscopy and chemical analysis to determine the structures of complicated organic molecules.

Solutions to the Exercises

5.1 Wavelength and frequency are inversely related by the following equation:

$\lambda = c/\nu$, where c is the speed of light (2.99×10^8 m/s),
λ is the wavelength in meters, and
ν is the frequency in hertz (1 Hz = s^{-1}).

(a) $\nu = 4.41 \times 10^{14}$ Hz,
$\lambda = 2.99 \times 10^8$ m $s^{-1}/4.41 \times 10^{14}$ s^{-1} = **6.78×10^{-7} m** = 0.678 μm.

(b) $\nu = 102.7$ MHz = 102.7×10^6 s^{-1} = 1.027×10^8 s^{-1}
$\lambda = 2.99 \times 10^8$ m $s^{-1}/1.027 \times 10^8$ s^{-1} = **2.91 m.**

(c) $\nu = 60$ s^{-1},
$\lambda = 2.99 \times 10^8$ m $s^{-1}/60$ s^{-1} = **5×10^6 m.**

5.2 The relationship between the frequency of electromagnetic radiation and its energy is expressed by the following equation:

Energy = h ν = h c/λ, where h is Planck's constant (6.63×10^{-37} kJ s mol^{-1}).
c is the speed of light = 2.99×10^8 m/s,
ν is the frequency of the electromagnetic radiation
(1 Hz = s^{-1}), and λ is its wavelength.

(a) $\lambda = 2.3 \times 10^{-3}$ cm = 2.3×10^{-5} m,
$\nu = c/\lambda = 2.99 \times 10^8$ m $s^{-1}/2.3 \times 10^{-5}$ m = 1.3×10^{13} s^{-1},
E = h ν = 6.63×10^{-37} kJ s $mol^{-1} \times 1.3 \times 10^{13}$ s^{-1} = **8.6×10^{-24} kJ/mol (2.1×10^{-24} kcal/mol).**

(b) $\nu = 10^{17}$ s^{-1},
E = h ν = 6.63×10^{-37} kJ s $mol^{-1} \times 10^{17}$ s^{-1} = **6.63×10^{-20} kJ/mol (1.6×10^{-20} kcal/mol).**

(c) $\nu = 6.5 \times 10^{15}$ s^{-1},
E = h ν = 6.63×10^{-37} kJ s $mol^{-1} \times 6.5 \times 10^{15}$ s^{-1} = **4.3×10^{-21} kJ/mol (1.03×10^{-21} kcal/mol).**

(d) $\lambda = 450$ nm = 450×10^{-9} m = 4.5×10^{-7} m,
$\nu = c/\lambda = 2.99 \times 10^8$ m $s^{-1}/4.5 \times 10^{-7}$ m = 6.64×10^{14} s^{-1},
E = h ν = 6.63×10^{-37} kJ s $mol^{-1} \times 6.64 \times 10^{14}$ s^{-1} = **4.40×10^{-22} kJ/mol (1.05×10^{-22} kcal/mol).**

(e) $\lambda = 1$ m,
$\nu = c/\lambda = 2.99 \times 10^8$ m $s^{-1}/1$ m = 2.99×10^8 s^{-1},
E = h ν = 6.63×10^{-37} kJ s $mol^{-1} \times 2.99 \times 10^8$ s^{-1} = **1.98×10^{-28} kJ/mol (4.7×10^{-29} kcal/mol).**

5.3 Increasing energy means shorter wavelength (λ) or higher frequency (ν).

(a) Radio waves (λ = 1-1000 m), X-ray (λ = 1 nm), visible light (400-800 nm); thus: radio waves < visible light < X-ray.

(b) Infrared light (800-1200 nm), ultraviolet light (200-400 nm), visible light (400-800 nm); thus: infrared light < visible light < ultraviolet light

(c) $\nu = 10^5$ Hz < $\nu = 10^{14}$ Hz < $\nu = 10^{26}$ Hz

(d) $\lambda = 10^6$ m < $\lambda = 10$ m < $\lambda = 10$ nm = 10^{-8} m

(e) First, we have to convert either wavelengths to frequencies or frequencies to wavelengths. There is only one radiation characterized by frequency, so let us convert it to wavelength.
    The relationship between wavelength and frequency is the following:
$\lambda = c/\nu$; thus $\lambda = 2.99 \times 10^8$ m $s^{-1}/10^{26}$ s^{-1} = 2.99×10^{-18} m

Hence: $\lambda = 10^{-3}$ m < $\lambda = 10^{-9}$ m < $\lambda = 2.99 \times 10^{-18}$ m

5.4 Alkanes contain absorption bands at 2960 to 2850 cm^{-1} and 1500 to 1350 cm^{-1}. The absorption band of the carbonyl group of ketones is found in the region of 1730 to 1700 cm^{-1}. Thus, the IR spectrum of a ketone can be obtained in mineral oil since the alkane absorption does not mask the absorption of the carbonyl group.

5.5 (a) The carboxyl group (—COOH) has two characteristic absorption bands in the IR spectrum: the absorption due to the carbonyl group at about 1700 cm^{-1} and the broad absorption due to the O—H bond, which begins at about 3400 cm^{-1} and extends to about 2450 cm^{-1}.

(b) The hydroxyl group (—OH) has a broad, strong absorption band centered at about 3300 cm^{-1}.

(c) Carbon-carbon triple bond (C≡C) and halogen (Cl). The IR spectra of alkynes contain a characteristic absorption band at 2260 to 2100 cm^{-1} due to the carbon-carbon triple bond. Absorption bands due to the carbon-chlorine bond appear in the fingerprint region from 1350 to 670 cm^{-1}.

(d) Carbonyl group (C═O) bonded to one hydrogen (aldehyde). The IR spectrum contains an intense absorption band at 1725 cm^{-1} due to the carbonyl group and the carbon-hydrogen stretching frequency at 2715 cm^{-1} for the carbon-hydrogen bond on the carbonyl group.

(e) Ether: oxygen atom (—O—) bonded to two carbon atoms. The IR spectrum contains an intense absorption band in the region 1250 to 1050 cm^{-1} due to the C—O bond.

(f) Carbon-carbon double bond (C═C). The IR spectrum contains an absorption band in the 1680 to 1620 cm^{-1} region due to the carbon-carbon double bond (if the alkene is symmetrical that absorption is not observed) and two bands slightly above 3000 cm^{-1} due to sp^2 C—H bonds.

(g) Carbonyl group (C═O) bonded to two alkyl groups (ketone). The absorption band of the carbonyl group is found in the region 1715 to 1700 cm^{-1}.

(h) Carbon-carbon double bond (C═C) and carbonyl group (ketone). The carbon-carbon double bond is symmetrically substituted with no permanent dipole, so only a weak or no C═C stretching should be seen in the 1680 to 1620 cm^{-1} region. There will be two bands slightly above 3000 cm^{-1} due to sp^2 C—H bonds as well as an intense absorption band at 1715 to 1700 cm^{-1} due to the carbonyl group.

(i) The carboxyl group (—COOH) has two characteristic absorption bands in the IR spectrum: the absorption due to the carbonyl group at about 1700 cm^{-1} and the broad absorption due to the O—H bond, which begins at about 3400 cm^{-1} and extends to about 2450 cm^{-1}.

5.6 (a) The presence of an absorption band at ≈ 1700 cm^{-1} means that the compound contains a carbonyl group. A broad absorption that begins at about 3600 cm^{-1} and extends to about 2350 cm^{-1} indicates the presence of an O—H group bonded to the carbonyl carbon atom of a carboxyl group. Therefore, we conclude that the compound is a carboxylic acid.

(b) The presence of an absorption band at 3080 cm^{-1} means that the compound contains sp^2 C—H bonds (an alkene). An absorption band at around 1650 cm^{-1} confirms the presence of the C═C bond. The presence of an absorption band at 3000 to 2800 cm^{-1} indicates an alkane C—H absorption. Therefore, we conclude that the compound is an alkene also containing an alkane part.

(c) The presence of a broad, strong absorption band at 3500 to 3100 cm^{-1} is due to the —O—H bond. The presence of an absorption band at 3000 to

2800 cm^{-1} indicates the alkane C—H absorption. Therefore, we conclude that the compound is an aliphatic alcohol.

(d) The presence of absorption bands at 3000 to 2800 cm^{-1}, 1500 cm^{-1}, and 1350 cm^{-1} indicates an alkane C—H absorption. The lack of other absorption bands characteristic of other functional groups indicates that the compound is an alkane.

5.7 (a) One compound is a carboxylic acid; the other an aldehyde. The carbonyl group absorption occurs at 1700 cm^{-1} in both compounds. However, the carboxylic acid O—H absorbs as a broad band at 3500 to 2500 cm^{-1} while an aldehyde shows a sharp medium absorption band at around 2725 cm^{-1} due to an sp^2 C—H bond (aldehyde group). Consequently, the presence of one of the preceding absorption bands distinguishes a carboxylic acid from an aldehyde.

(b) One compound is an aldehyde; the other an alcohol. The aldehyde shows the characteristic strong absorption band at 1700 cm^{-1} due to the presence of a carbonyl group and a sharp medium absorption band at around 2800 cm^{-1} due to a C—H bond involving sp^2-hybridized carbon. The alcohol shows neither of the above but a broad, strong band at 3500 to 3100 cm^{-1} due to the O—H bond. Consequently, the presence or the lack of these characteristic features distinguishes an aldehyde from an alcohol.

(c) Both compounds are alkenes. One is a terminal alkene while the other is an internal alkene. The terminal alkene shows two absorption bands slightly above 3000 cm^{-1} due to the terminal methylene group (=CH$_2$). The internal alkene shows only one absorption band for the vinyl hydrogens. Thus, it is possible to distinguish between a terminal alkene and an internal alkene.

5.8 (a) The presence of an absorption band at 3080 cm^{-1} means that the compound contains sp^2 C—H bonds (an alkene). The presence of an absorption band at around 1650 cm^{-1} is due to the C=C bond. On the other hand, a terminal alkyne would show an absorption band at 3300 cm^{-1} due to an sp C—H bond and a sharp absorption band at about 2100 cm^{-1}. There is no evidence of the sp C—H or C≡C absorption, and there is evidence of sp^2 C—H and C=C absorption so we identify the compound as H$_2$C=CH(CH$_2$)$_3$CH$_3$.

(b) The presence of the broad, strong absorption band stretching from 3500 to 2500 cm^{-1} indicates the presence of an O—H group attached to the carbonyl group. Moreover, we observe a strong absorption at ≈ 1700 cm^{-1} due to the presence of the carbonyl group. An alcohol would show just a strong, broad absorption band at 3500 to 3100 cm^{-1} and no absorption at ≈ 1700 cm^{-1}. Thus, we conclude that the IR spectrum matches (CH$_3$)$_2$CHCH$_2$COOH.

(c) The presence of the broad, strong absorption band at 3500 to 3100 cm^{-1} indicates the presence of the —O—H group. An alkene would show a sharp absorption at ≈ 1650 cm^{-1} due to the C=C bond and another absorption at ≈ 3080 cm^{-1} due to sp^2 C—H bonds. There is no evidence of the latter, and the presence of a strong band centered at 3300 cm^{-1} indicates that the compound is CH$_3$CH$_2$CH$_2$CH$_2$OH.

5.9 We solve the problems in the same manner as Sample Problem 5.6 was solved. We just change the frequency of the observed position of the signal in each case and we get the following values of δ:

(a) $\delta = \dfrac{(1550 - 0) \text{ Hz}}{90 \text{ MHz}} = 17.2 \text{ ppm}$

(b) $\delta = \dfrac{(2150 - 0) \text{ Hz}}{90 \text{ MHz}} = 23.9$ ppm

(c) $\delta = \dfrac{(2790 - 0) \text{ Hz}}{90 \text{ MHz}} = 31$ ppm

5.10 (a) 2-Chloropropane ($CH_3CHClCH_3$) has two groups of nonequivalent carbon atoms: one is the terminal methyl (—CH_3) group (two equivalent methyl groups) and the other is the central chloromethylene (—CHCl—) group.

(b) 1-Chloropropane ($CH_3CH_2CH_2Cl$) has three groups of nonequivalent carbon atoms: the terminal methyl (—CH_3) group, the central methylene (—CH_2—) group, and the carbon atom with two hydrogen atoms and one chlorine (—CH_2Cl).

(c) 3-Methylhexane ($CH_3CH_2CH(CH_3)CH_2CH_2CH_3$) has seven nonequivalent carbon atoms (each of them is different).

(d) Methylpropane ($CH_3CH(CH_3)_2$) has two groups of nonequivalent carbon atoms: a methyl (—CH_3) group (three of them) and the central C2 carbon atom.

(e) 2,2-Dimethylpropane ($CH_3C(CH_3)_3$) has two groups of nonequivalent carbon atoms: a methyl (—CH_3) group (four of them) and the central C2 carbon atom.

5.11 As is evident from Exercises **5.10** (a) and **5.10** (b), 1-chloropropane has three nonequivalent carbon atoms, so there are three signals in its ^{13}C NMR spectrum. On the other hand, 2-chloropropane has two nonequivalent carbon atoms, so there are two signals in its ^{13}C NMR spectrum. Thus, it is possible to distinguish 1-chloropropane from 2-chloropropane by the number of signals in the ^{13}C NMR spectrum.

5.12 Yes, it is possible to distinguish the five isomeric compounds by using ^{13}C NMR spectroscopy. Cyclobutane has only *one* type of carbon atom, so it has only *one* signal in the ^{13}C NMR spectrum; 1-butene has *four* different types of carbon atoms and would show *four* signals; 2-butene has *two* types of carbon atoms and would show *two* signals. Methylcyclopropane and 2-methylpropene both have three nonequivalent carbon atoms, but 2-methylpropene can be distinguished by the chemical shift (δ 100–150 ppm) of the carbon atoms of the double bond.

5.13 Step 1: Determine from the number of carbon atoms (n_C), halogens (n_X), and nitrogen atoms (n_N) in the molecular formula the number of H atoms required for the molecule to be saturated (H_{sat}):

$$H_{sat} = 2\,n_C + 2 - n_X + n_N$$

The above formula is derived from the alkane formula C_nH_{2n+2}. The formula has been modified for compounds containing elements other than just carbon and hydrogen.

Organohalogen compounds. Because a halogen substituent is simply a replacement for hydrogen in an organic molecule (both are monovalent), we can *add* the number of halogens and hydrogens to arrive at an equivalent hydrocarbon formula.

Organooxygen compounds. Because oxygen is divalent, it does not affect the formula of an equivalent hydrocarbon and can be ignored when calculating the degree of unsaturation.

Organonitrogen compounds. Because nitrogen is trivalent, an organonitrogen compound has one more hydrogen than an equivalent hydrocarbon has, and we therefore *subtract* the number of nitrogens from the number of hydrogens to arrive at the equivalent hydrocarbon formula.

Step 2: Compare H_{sat} with the actual number of H atoms in the molecular formula, H_{act}, to determine the degree of unsaturation: Degree of unsaturation = $(H_{sat} - H_{act})/2$

(a) $H_{sat} = 2 \times 6 + 2 = 14$; $H_{act} = 12$;
Degree of unsaturation = $(14 - 12)/2 = 1$.

(b) $H_{sat} = 2 \times 6 + 2 = 14$; $H_{act} = 6$;
Degree of unsaturation = $(14 - 6)/2 = 4$.

(c) $H_{sat} = 2 \times 8 + 2 = 18$; $H_{act} = 14$;
Degree of unsaturation = $(18 - 14)/2 = 2$.

(d) $H_{sat} = 2 \times 14 + 2 = 30$; $H_{act} = 28$;
Degree of unsaturation = $(30 - 28)/2 = 1$.

(e) $H_{sat} = 2 \times 20 + 2 = 42$; $H_{act} = 34$;
Degree of unsaturation = $(42 - 34)/2 = 4$.

5.14 A saturated hydrocarbon with 4 carbon atoms would have $H_{sat} = 2 \times 4 + 2 = 10$ hydrogen atoms. The given compound has only 6 H atoms. Thus, the degree of unsaturation = $(H_{sat} - H_{act})/2 = (10 - 6)/2 = 2$.
Examples:

(a) A triple bond:

$$CH_3CH_2C \equiv CH \quad \text{or} \quad CH_3C \equiv CCH_3$$

(b) Two double bonds:

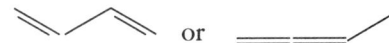

(c) A ring and a double bond:

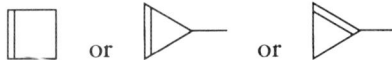

(d) Two rings that share a side:

5.15 Step 1: Determine from the number of carbon atoms (n_C), halogens (n_X), and nitrogen atoms (n_N) in the molecular formula the number of H atoms required for the molecule to be saturated (H_{sat}):

$H_{sat} = 2\,n_C + 2 - n_X + n_N$ (Disregard oxygen and sulphur atoms).

Step 2: Compare H_{sat} with the actual number of H atoms in the molecular formula, H_{act}, to determine the degree of unsaturation: Degree of unsaturation = $(H_{sat} - H_{act})/2$

(a) C_3H_4ClBr
$H_{sat} = 2 \times 3 + 2 - 2 = 6$
Degree of unsaturation = $(6 - 4)/2 = 1$
Many different examples are possible. Here are just some possibilities:

(b) $C_6H_{10}O$
$H_{sat} = 2 \times 6 + 2 = 14$
Degree of unsaturation = (14 - 10)/2 = 2
Many isomers of the following compounds are possible:

(c) C_4H_5OCl
$H_{sat} = 2 \times 4 + 2 - 1 = 9$
Degree of unsaturation = (9 - 5)/2 = 2
Many isomers of the following compounds are possible:

(d) C_5H_9N
$H_{sat} = 2 \times 5 + 2 + 1 = 13$
Degree of unsaturation = (13 - 9)/2 = 2
Many isomers of the following compounds are possible:

(e) C_5H_8NOBr
$H_{sat} = 2 \times 5 + 2 + 1 - 1 = 12$
Degree of unsaturation = $(12 - 8)/2 = 2$
Many isomers of the following compounds are possible:

(f) $C_3H_6O_2$
$H_{sat} = 2 \times 3 + 2 = 8$
Degree of unsaturation = $(8 - 6)/2 = 1$
Many isomers of the following compounds are possible:

5.16 (a) C_2H_6O. $H_{sat} = 2 \times 2 + 2 = 6$; $H_{act} = 6$; Degree of unsaturation = 0.

There are two possible *functional group isomers* for a saturated organooxygen compound: CH_3—O—CH_3 (ether) or CH_3CH_2OH (alcohol). The presence of the broad, strong absorption at 3500 to 3100 cm^{-1} indicates the O—H functional group. The ^{13}C NMR spectrum shows two signals, which is consistent with ethanol (alcohol). Methoxymethane (dimethyl ether) would have one signal in the ^{13}C NMR spectrum. Hence, the structure is CH_3CH_2OH (ethanol).

(b) $C_5H_{10}O$. $H_{sat} = 2 \times 5 + 2 = 12$; $H_{act} = 10$; Degree of unsaturation = $2/2 = 1$. (Possibilities: a ring, a carbon-oxygen double bond, or a carbon-carbon double bond.)

The presence of the strong absorption at 1710 cm^{-1} indicates the C=O functional group. The ^{13}C NMR spectrum shows one signal at 211 ppm (carbon of the carbonyl group) and two signals in the alkane region (8 and 34 ppm, respectively). We know two functional groups with the carbon-oxygen double bond: an aldehyde and a ketone. The IR spectrum lacks a signal at 2715 cm^{-1} due to the hydrogen bonded to the carbon of the carbonyl group and that eliminates the possibility of an aldehyde. There are two ketones with molecular formula $C_5H_{10}O$:

O   O
‖   ‖
CH₃—C—CH₂CH₂CH₃   CH₃CH₂—C—CH₂CH₃

2-Pentanone   3-Pentanone

The ^{13}C NMR spectrum of 2-pentanone would show five signals due to five nonequivalent carbon atoms while the ^{13}C NMR spectrum of 3-pentanone would show three signals due to three groups of nonequivalent carbon atoms. We observe three signals in the ^{13}C NMR spectrum, and we conclude that the unknown compound is **3-pentanone**.

(c) $C_5H_{10}O_2$. $H_{sat} = 2 \times 5 + 2 = 12$; $H_{act} = 10$; Degree of unsaturation = $2/2 = 1$.

The presence of the strong absorption at 1700 cm^{-1} indicates the C=O functional group. A strong and broad absorption over a wide range 3400 to 2600 cm^{-1} is characteristic of a carboxyl group. The ^{13}C NMR spectrum shows a signal at δ 186 ppm (carbon of the carboxyl group) and two signals in the alkane region (38 and 27 ppm, respectively). We may have the following carboxylic acids with the formula $C_5H_{10}O_2$:

                              CH₃
                              |
CH₃CH₂CH₂CH₂COOH   CH₃CHCH₂COOH

        A   B

        CH₃   CH₃
        | |
CH₃CH₂CHCOOH   CH₃CCOOH
                              |
                              CH₃
        C   D

Compound A has five nonequivalent carbon atoms and would show five signals in the ^{13}C NMR spectrum; compound B would show four signals; compound C, five signals; and compound D, three signals. Thus, only structure D can be the unknown compound.

5.17 (a) A **functional group** is a structural unit consisting of an atom or a group of atoms that serve as a site of chemical reactivity in a molecule.

(b) **Spectroscopy** is the study of the interaction of electromagnetic radiation with matter. The results of this interaction provide information about the structure of compounds. There are three types of spectroscopy that are most often used in organic chemistry: 1) infrared (IR) spectroscopy, 2) ultraviolet (UV) spectroscopy, 3) nuclear magnetic resonance (NMR) spectroscopy. Interaction of matter with IR radiation provides information about the functional group or groups in a molecule. Interaction of matter with UV radiation provides information about the π-electron structure of a molecule. Interaction of matter in a magnetic field with radiofrequency

(c) **Distillation** is a procedure for separating a liquid from a mixture by vaporizing the liquid and condensing the vapor. In distillation the components of a liquid solution are separated from each other on the basis of their different volatilities. When boiling points of liquid components of a mixture differ by 30° or more, a *simple distillation* can be used to separate them. A *fractional distillation* is used to separate liquids with a difference in boiling points as small as a couple of degrees.

(d) An **IR spectrum** is a chart on which the amount of IR radiation absorbed by a compound is plotted as a function of the wavelength. The IR radiation (λ = 800 to 2000 nm) causes vibrational excitation of a compound's bonds and, therefore, the IR spectrum provides information about functional groups present in the molecule.

(e) The **degree of unsaturation** (or **index of hydrogen deficiency**) is the number of rings and/or multiple bonds present in a compound. It is one half of the difference between the number of hydrogen atoms present in a given compound and the number of hydrogen atoms in a straight-chain alkane with the same number of carbon atoms.

(f) **TMS** (tetramethylsilane, $(CH_3)_4Si$) is a compound used to calibrate ^{13}C NMR spectra (and ^1H NMR spectra as we will learn in Chapter 10). The hydrogen atoms and carbon atoms of TMS are more shielded than those of almost all organic compounds. Peak positions are measured in frequency units (Hz) downfield from the TMS peak.

(g) The **electromagnetic spectrum** is the range of electromagnetic radiation. Usually the whole range of electromagnetic radiation is divided into smaller regions characterized by different energy ranges. Thus, we distinguish infrared (energy range from 0.1 to 30 kcal/mol), visible radiation (energy range from 30 to 60 kcal/mol), and ultraviolet radiation (energy range from 60 to 150 kcal/mol).

(h) **Wavelength** is the distance between successive crests in a wave. In other words, it is the length of a wave from peak to peak. Thus, the unit of wavelength is the unit of distance: m, cm, nm, etc.

(i) **Chromatography** is a method of separation and analysis of mixtures based on the different rates at which different compounds are removed from a stationary phase by a moving phase.

(j) The **wavenumber** (cm^{-1}) of light is the number of its wavelengths in a centimeter and is related to its wavelength (in centimeters) as follows:

$$\text{Wavenumber } (\bar{\nu}) = 1/\lambda \text{ (in cm)}$$

(k) The **fingerprint region** is the region in the infrared (from 400 to about 1100 cm^{-1}) that usually exhibits a series of complex, low-energy bands that are characteristic of a specific molecule (rather than a functional group).

(l) **Chemical shift** is the position of an absorption in an NMR spectrum relative to the position of the absorption of TMS.

(m) A **^{13}C NMR spectrum** is a pattern of absorption peaks (signals) of nonequivalent carbon atoms present in a molecule. Different magnetic environments of different carbon atoms cause spinning ^{13}C nuclei to absorb at different combinations of magnetic field and radiofrequency.

(n) **Frequency** is the number of wave crests of a wave that pass a fixed point in a unit of time (usually 1 second). The unit of frequency is thus 1/s or s^{-1} = 1 Hz (hertz).

(o) **Crystallization** is a technique for the purification of solid organic compounds. It involves dissolving the impure solid in a minimum volume of hot solvent, filtering to remove insoluble impurities, cooling the solution slowly whereupon crystals of pure compound are formed, separating the crystals by filtration, and drying the crystals. Crystallization is the most important method for purification of solid organic compounds.

5.18 IR absorption bands are described by either wavelength of the absorbed light, λ (expressed in μm; 1 μm = 10^{-6} m), or wavenumber, $\bar{v}$ (in units of cm^{-1}). Wavenumber is reciprocal to wavelength: $\bar{v} = 1/\lambda$. Thus,

(a) $\lambda = 3.05$ μm $= 3.05 \times 10^{-4}$ cm
$\bar{v} = 1/\lambda = 1/(3.05 \times 10^{-4}$ cm$) = 3280$ cm^{-1}

(b) $\lambda = 3.35$ μm $= 3.35 \times 10^{-4}$ cm
$\bar{v} = 1/\lambda = 1/(3.35 \times 10^{-4}$ cm$) = 2985$ cm^{-1}

(c) $\lambda = 4.71$ μm $= 4.71 \times 10^{-4}$ cm
$\bar{v} = 1/\lambda = 1/(4.71 \times 10^{-4}$ cm$) = 2123$ cm^{-1}

(d) $\lambda = 5.81$ μm $= 5.81 \times 10^{-4}$ cm
$\bar{v} = 1/\lambda = 1/(5.81 \times 10^{-4}$ cm$) = 1721$ cm^{-1}

(e) $\lambda = 6.06$ μm $= 6.06 \times 10^{-4}$ cm
$\bar{v} = 1/\lambda = 1/(6.06 \times 10^{-4}$ cm$) = 1650$ cm^{-1}

5.19 Absorption band   Structural feature

(a) 3279 cm^{-1} C—H bond of a terminal alkyne (R—C≡C—H)
(b) 2985 cm^{-1} C—H bond of an alkane (R_3C—H)
(c) 2123 cm^{-1} C≡C bond of an alkyne (R—C≡C—R')
(d) 1721 cm^{-1} carbonyl group (\C=O)
(e) 1650 cm^{-1} C=C bond of an alkene (\C=C/)

5.20 One can attempt this problem by analyzing either the functional groups in the structures shown and looking for the absorption bands in the spectra or by identifying the functional groups from the spectra and then looking for structures that contain that functional group. Either method is good and should give the same result. Let us identify the functional groups in each structure and look for spectra that contain the absorption bands of those functional groups.

(a) This compound contains a carboxyl group (—COOH), which has two characteristic absorption bands in the IR spectrum: the absorption due to the carbonyl group at about 1700 cm^{-1} and the broad absorption due to the O—H bond, which begins at about 3400 cm^{-1} and extends to about 2450 cm^{-1}. Only Spectrum 1 shows all those absorption bands.

(b) This compound contains a terminal —C≡C—H group. The IR spectrum of a terminal alkyne contains a characteristic absorption band at 2260 to 2100 cm^{-1} due to the carbon-carbon triple bond and the sharp band at 3300 cm^{-1} due to the stretching of the terminal *sp* C—H bond. Only Spectrum 4 exhibits those absorption bands.

(c) This compound is a ketone. The absorption band of the carbonyl group is found in the region 1715 to 1700 cm^{-1}. Spectrum 3 shows such an absorption but so does Spectrum 5. Spectrum 5 also shows a sharp absorption band at 2715 cm^{-1}, however, which should be absent from the spectrum of a ketone because we do not have any hydrogens bonded to the carbonyl group. Thus, only Spectrum 3 matches the expected absorption band for this ketone.

(d) This compound is an aldehyde. The IR spectrum of an aldehyde contains an intense absorption band at 1725 cm^{-1} due to the carbonyl group and an absorption at 2715 cm^{-1} due to the hydrogen bonded to the carbon of the carbonyl group. Only Spectrum 5 shows both those absorption bands.

(e) This compound is a secondary amine, which would give one absorption peak in the region of 3500 to 3200 cm^{-1} due to stretching of the N—H bond and an absorption band at 1220 to 1020 cm^{-1} due to stretching of the C—N bond. Spectrum 2 seems to show such absorption bands. However, the absorption band at 3400 cm^{-1} is too strong and too broad for an amine. It is rather characteristic of an alcohol. Hence, we do not have any spectrum matching the expected absorption bands.

(f) This compound is an alkene. The IR spectrum of an alkene contains an absorption band in the 1680 to 1620 cm^{-1} region due to the carbon-carbon double bond and two bands slightly above 3000 cm^{-1} due to sp^2 C—H bonds. There is not any spectrum exhibiting such absorption bands.

(g) This compound has a hydroxyl functional group (—OH), which shows a broad, strong absorption band centered at about 3300 cm^{-1} due to stretching of the O—H bond and an absorption band at 1300 to 1000 cm^{-1} due to stretching of the C—O bond. Spectrum 2 matches the expected absorption pattern.

5.21 One compound is an alcohol and the other a primary amine. An alcohol shows a broad, strong band at 3500 to 3100 cm^{-1} due to stretching of the O—H bond and an absorption band at 1300 to 1000 cm^{-1} due to stretching of the C—O bond. Spectrum B shows such absorption bands.

A primary amine shows two peaks in the region 3500 to 3300 cm^{-1} due to stretching of the N—H bond and an absorption band at 1220 to 1020 cm^{-1} due to stretching of the C—N bond. Spectrum A shows such absorption bands.

5.22 The structure contains three functional groups: the hydroxyl group (—OH), carbon-carbon double bond (C=C), and the ester group.

The hydroxyl functional group (—OH) is expected to show a broad, strong absorption band centered at about 3300 cm^{-1} due to stretching of the O—H bond and an absorption band at 1300 to 1000 cm^{-1} due to stretching of the C—O bond. The O—H absorption is clearly seen, but the C—O absorption is lost in the region of many strong absorption bands in the region 1250 to 1000 cm^{-1}.

The IR spectrum of an alkene is expected to contain an absorption band in the 1680 to 1620 cm^{-1} region due to stretching of the carbon-carbon double bond and two bands slightly above 3000 cm^{-1} due to sp^2 C—H bonds. The stretching of the C=C bond is observed at 1650 cm^{-1} and the stretching of the C—H bond is observed at 3020 cm^{-1} (small peak).

The carbonyl group of an ester is expected to show a strong absorption at 1730 cm^{-1} and the C—O bond shows absorption (C—O stretching) at 1300 to 1000 cm^{-1}. The carbonyl absorption is clearly visible but the C—O bond absorption is obscured by many strong absorption bands in the region 1250 to 1000 cm^{-1}.

5.23 Step 1: Determine from the number of carbon atoms (n_C), halogens (n_X), and nitrogen atoms (n_N) in the molecular formula the number of H atoms required for the molecule to be saturated (H_{sat}):
$H_{sat} = 2\,n_C + 2 - n_X + n_N$ (Disregard oxygen and sulphur atoms.)

Step 2: Compare H_{sat} with the actual number of H atoms in the molecular formula, H_{act}, to determine the degree of unsaturation:
Degree of unsaturation = $(H_{sat} - H_{act})/2$

(a) $C_{10}H_{18}$
$H_{sat} = 2 \times 10 + 2 = 22 \qquad H_{act} = 18$
Degree of unsaturation = $(22 - 18)/2 = 2$

(b) $C_{14}H_{22}O$
$H_{sat} = 2 \times 14 + 2 = 30 \qquad H_{act} = 22$
Degree of unsaturation = $(30 - 22)/2 = 4$

(c) C_3H_9N
$H_{sat} = 2 \times 3 + 2 + 1 = 9 \qquad H_{act} = 9$
Degree of unsaturation = $(9 - 9)/2 = 0$

(d) C_5H_5N
$H_{sat} = 2 \times 5 + 2 + 1 = 13 \qquad H_{act} = 5$
Degree of unsaturation = $(13 - 5)/2 = 4$

(e) $C_8H_{12}O_3$
$H_{sat} = 2 \times 8 + 2 = 18 \qquad H_{act} = 12$
Degree of unsaturation = $(18 - 12)/2 = 3$

(f) C_6H_5NO
$H_{sat} = 2 \times 6 + 2 + 1 = 15 \qquad H_{act} = 5$
Degree of unsaturation = $(15 - 5)/2 = 5$

5.24 The following problems can be solved in two different ways. One method would be to analyze the structures and simply count the units of unsaturation. The other method would be to calculate the degree of unsaturation using the following formula:
Degree of unsaturation = $(H_{sat} - H_{act})/2$, where $H_{sat} = 2\,n_C + 2 - n_X + n_N$

(a) Carbon-oxygen double bond = 1 unit of unsaturation, or
C_4H_8O; $\quad H_{sat} = 2 \times 4 + 2 = 10 \qquad H_{act} = 8$
Degree of unsaturation = $(10 - 8)/2 = 1$

(b) Ring = 1 unit of unsaturation, or
$C_5H_{11}N$; $\quad H_{sat} = 2 \times 5 + 2 + 1 = 13 \quad H_{act} = 11$
Degree of unsaturation = $(13 - 11)/2 = 1$

(c) Carbon-oxygen double bond and a ring = 2 units of unsaturation, or
$C_6H_{10}O$; $\quad H_{sat} = 2 \times 6 + 2 = 14 \qquad H_{act} = 10$
Degree of unsaturation = $(14 - 10)/2 = 2$

(d) Two carbon-carbon double bonds and three rings = 5 units of unsaturation, or
C_8H_8; $\quad H_{sat} = 2 \times 8 + 2 = 18 \qquad H_{act} = 8$
Degree of unsaturation = $(18 - 8)/2 = 5$

5.25 Equivalent atoms have the same relationship to the rest of the molecule.

(a)

Carbon atom C1 is the only carbon atom bonded to the oxygen atom (double bond) and it constitutes one group of carbon atoms.
Carbon atoms C2 and C6 are both α to the carbonyl group. Each has two hydrogen atoms and identical hydrocarbon chain so they are equivalent (second group of carbon atoms).

Carbon atoms C3 and C5 are both β to the carbonyl group. Each has two hydrogen atoms and both are bonded to the same methylene group so they are equivalent (third group of carbon atoms).

Carbon atom C4 is γ to the carbonyl group. It is nonequivalent with any other carbon atom, and it constitutes a fourth group of carbon atoms.

Thus, we have four different groups of carbon atoms in cyclohexanone.

(b)

$$\text{CH}_3\text{CHCH}_2-\underset{\underset{\text{CH}_3}{|}}{\overset{\overset{\text{CH}_3}{|}}{\text{C}}}-\text{CH}_3$$
$$\overset{\text{OH}}{|}$$

(with OH on the CHCH₂ carbon)

The circled methyl groups are equivalent. All other carbon atoms are nonequivalent. Thus, we have five nonequivalent carbon atoms in this molecule (4,4-dimethyl-2-pentanol).

(c)

$$\underset{\text{H}_3\text{C}}{\overset{\text{H}_3\text{C}}{\diagdown}}\text{CH}-\text{CH}\underset{\diagdown\text{CH}_3}{\overset{\diagup\text{CH}_3}{}}$$

All four methyl groups are equivalent. Both $\diagdown\text{CH}-\diagup$ groups are also equivalent.

Thus, we have two different groups of carbon atoms in 2,3-dimethylbutane.

5.26 (a) All six carbon atoms of cyclohexane are equivalent so its ^{13}C NMR spectrum would show one signal. The other structure (2,3-dimethyl-2-butene) has two sets of carbon atoms so its ^{13}C NMR spectrum would show two signals.

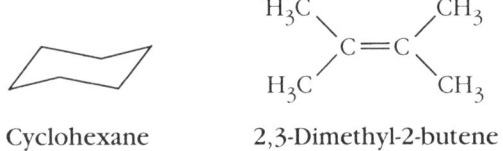

Cyclohexane   2,3-Dimethyl-2-butene

The spectrum shown has two signals so it must be 2,3-dimethyl-2-butene.

(b)

$$\text{H}_3\text{C}-\underset{\underset{\text{CH}_3}{|}}{\overset{\overset{\text{CH}_3}{|}}{\text{C}}}-\text{OH} \qquad \underset{\text{H}_3\text{C}}{\overset{\text{H}_3\text{C}}{\diagdown}}\text{CHCH}_2\text{OH}$$

1,1-Dimethylethanol   2-Methylpropanol
(*tert*-Butyl alcohol)   (Isobutyl alcohol)

tert-Butyl alcohol has two groups of carbon atoms (all three methyl groups are equivalent), so its ^{13}C NMR spectrum would show two signals. Isobutyl alcohol has three sets of carbon atoms (the methyl groups are equivalent), so its ^{13}C NMR spectrum would show three signals. The spectrum shown has two signals so it is that of *tert*-butyl alcohol (1,1-dimethylethanol).

(c)

CH₃CH₂CH₂C(=O)H

Butanal
(Butyraldehyde)

Oxolane
(Tetrahydrofuran)

Butanal has four different (nonequivalent) carbon atoms, so its ^{13}C NMR spectrum would show four signals. Oxolane (tetrahydrofuran) has two sets of carbon atoms, so its ^{13}C NMR spectrum would show two signals. The spectrum shown has two signals so it is that of tetrahydrofuran.

5.27 (a) Compound $C_3H_6Br_2$ is saturated (zero degrees of unsaturation). Two signals in its ^{13}C NMR spectrum suggest that the compound has symmetry. Two bromine atoms can be bonded either to end carbon atoms of propane or to the central carbon atom.

BrCH₂CH₂CH₂Br   CH₃—C(Br)(Br)—CH₃

1,3-Dibromopropane   2,2-Dibromopropane

We will learn how to use ^{1}H NMR and ^{13}C NMR spectroscopy to distinguish between these isomers in Chapter 10.

(b) Compound C_5H_8O has two degrees of unsaturation. Three signals in its ^{13}C NMR spectrum suggest that the compound has symmetry. One unit of unsaturation is a carbonyl group (1730 cm^{-1} in the IR spectrum; 219 ppm in ^{13}C NMR spectrum), and the other must be a ring because no bands appear in the region of the IR spectrum where carbon-carbon double bonds or sp^2 C—H bonds absorb. The compound is cyclopentanone:

(c) Compound $C_5H_{11}N$ has one degree of unsaturation. It must be a ring because no bands appear in the region of the IR spectrum where carbon-carbon double bonds or sp^2 C—H bonds absorb. There is one peak in the region of 3300 cm^{-1} indicating a secondary amine. Three signals in its ^{13}C NMR spectrum suggest that the compound has symmetry. The compound is piperidine:

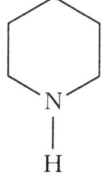

(d) Compound $C_4H_8O_2$ has one degree of unsaturation. It must be a ring because no bands appear in the region of the IR spectrum where carbon-carbon double bonds or sp^2 C—H bonds absorb or at 1700 cm^{-1} where the carbonyl group absorption is observed. There is no significant absorption in the region of 3300 cm^{-1} indicating the lack of the hydroxyl group(s). It must be an ether. One signal in its ^{13}C NMR spectrum suggests that the compound has symmetry. The compound is 1,4-dioxane:

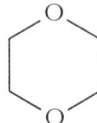

(e) Compound $C_5H_{10}O$ has one degree of unsaturation. It must be a ring because no bands appear in the region of the IR spectrum where carbon-carbon double bonds or sp^2 C—H bonds absorb or at 1700 cm^{-1} where the carbonyl group absorption is observed. There is a strong absorption band in the region of 3300 cm^{-1} indicating an alcohol. Three signals in its ^{13}C NMR spectrum suggest that the compound has symmetry. The compound is cyclopentanol:

CHAPTER 6

STEREOCHEMISTRY
Concepts

Stereochemistry deals with the spatial properties of compounds and chemical reactions. Compounds that differ from each other only in how their atoms are arranged in space are known as **stereoisomers.** Stereoisomers can be classified as **conformational isomers,** which are interconvertible by rotations around single bonds at room temperature, and as configurational isomers. **Configurational isomers** can be either **enantiomers,** which are related to each other as nonsuperposable mirror images, or **diastereomers,** which include all other stereoisomers.

An important structural feature of many stereoisomers is the **stereocenter** (an atom in a molecule at which interchange of two atoms or groups bonded to that atom produces a different stereoisomer). A carbon atom with four different groups bonded to it is a **tetrahedral stereocenter.** The presence of a stereocenter is a sufficient condition for **chirality,** but not a necessary condition.

The *configuration* at any stereocenter can be designated by the **Cahn-Ingold-Prelog convention,** known also as the ***R-S* convention.** To apply this convention, (1) each atom or group of atoms bonded to the sterocenter is assigned a priority, (2) each atom or group of atoms is numbered from highest priority to lowest priority, (3) the molecule is oriented in space so that the group of lowest priority is directed away from the observer, and (4) the remaining three groups are read in order from highest priority to lowest priority. If reading of groups is *clockwise,* the configuration is ***R*** (Latin: *rectus*, right). If reading of groups is *counterclockwise,* the configuration is ***S*** (Latin: *sinister,* left).

Light that vibrates in only one plane is said to be **plane-polarized.** A **polarimeter** is an instrument used to detect and measure the magnitude of optical activity. **Observed rotation** is the number of degrees the plane of polarized light has been rotated. If the analyzing prism must be turned clockwise to restore the zero point, the compound is **dextrorotatory** (+). If the analyzing prism must be turned counterclockwise to restore the zero point, the compound is **levorotatory** (–).

Enantiomers have identical physical properties (including IR and NMR spectra) except that they rotate a plane of polarized light in opposite directions. Each enantiomer rotates the plane-polarized light an equal number of degrees. Any compound that rotates the plane of polarized light is said to be **optically active.** **Specific rotation** is the observed rotation measured in a cell of light path length of 1 dm (10 cm) and at a concentration of 1 g/mL. A **racemic mixture** is a mixture of equal amounts of two enantiomers and has a specific rotation of zero. If one enantiomer of a pair is present to a greater extent, the mixture will show rotation corresponding to the percentage of the species that is present in excess. The percentage of the enantiomer that is present in excess is known as the **enantiomeric excess.**

Diastereomers have different physical properties and may be separated from each other by ordinary physical means. Diastereomers have optical activity if they contain stereocenters; however, if they have a plane of symmetry, they are **meso** compounds and are not optically active. Diastereomers such as *cis* and *trans* isomers of alkenes do not have optical activity because they do not contain a stereocenter.

For a molecule with n stereocenters, the maximum number of stereoisomers possible is 2^n. Certain molecules have special symmetry properties (for instance a plane of symmetry in a *meso* compound) that reduce the number of stereoisomers to fewer than that predicted by the 2^n rule.

Solutions to the Exercises

6.1 (a) A rake—*achiral,* (b) a baseball glove—**chiral,** (c) a baseball bat—*achiral,* (d) a screw—**chiral,** (e) a screwdriver—*achiral,* (f) a sweatshirt with the word **TORONTO** on the front—**chiral.**

6.2 (a) 1-Chloroethane *(achiral)*

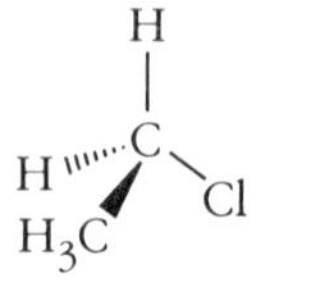

(b) 1-Bromo-1-chloroethane **(chiral)**

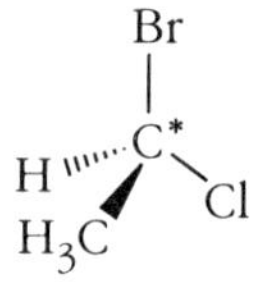

(c) 3-Chloropentane *(achiral)*

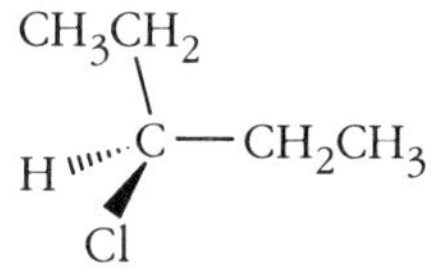

(d) 3-Bromohexane **(chiral)**

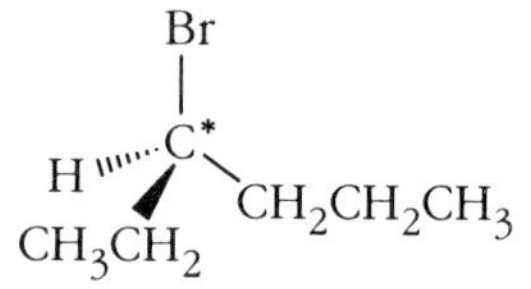

(e) 2-Chloro-1,5-diiodohexane **(chiral)**

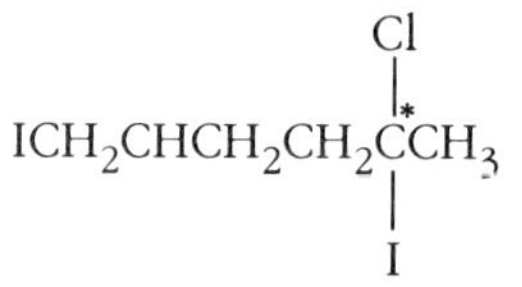

(f) 3,4,5-Trichloro-3,5-diethylheptane *(achiral)*

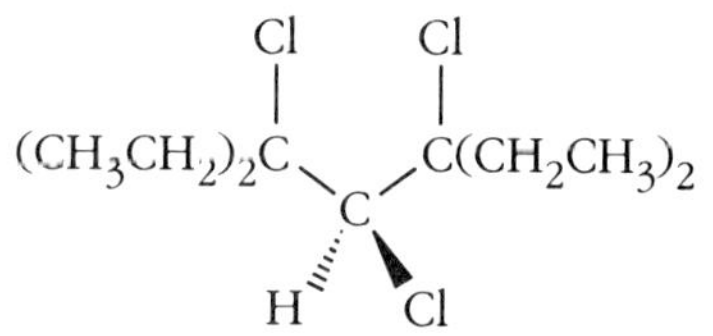

6.3

1-Bromo-1-chloroethane

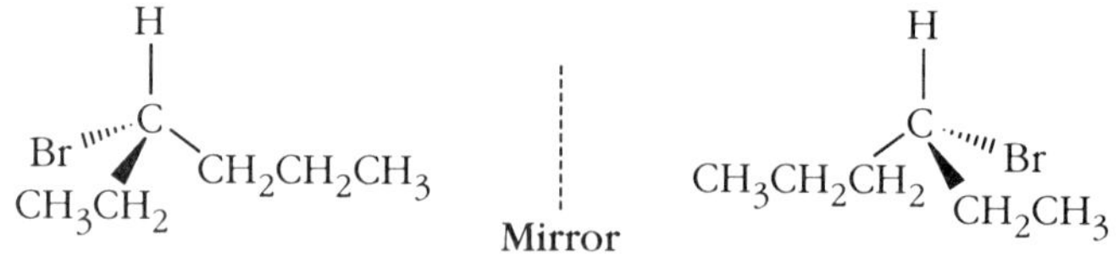

3-Bromohexane

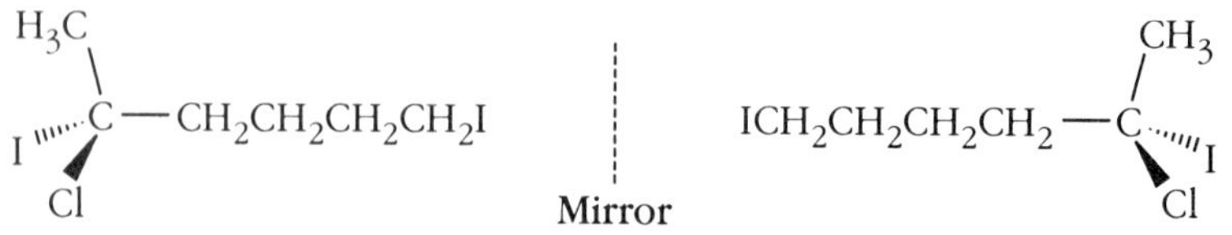

5-Chloro-1,5-diiodohexane

6.4 (a) The word "DAD" has no symmetry so it can be placed anywhere and the cup will no longer have a plane of symmetry. The word "MOM" has symmetry so it must be placed in such a way that its symmetry line does not coincide with the symmetry line of the cup.

(b) The word "DAD" has no symmetry so it cannot be placed anywhere to retain the symmetry of the cup. The word "MOM" has symmetry but it must be placed in such a way that its symmetry line *does* coincide with the symmetry line of the cup. It means that the word "MOM" must be placed opposite to the handle or (rather impractical) under the handle.

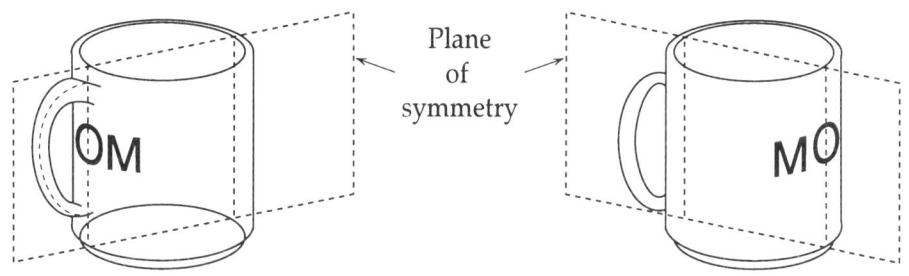

6.5 (a) Chiral (b) Chiral

(c) *Achiral* (plane of symmetry) (d) Chiral

(e) *Achiral,* (plane of symmetry)

6.6

(*R*)-Bromochlorofluoromethane
(as given in Sample Problem 6.1)

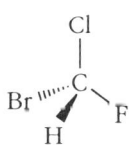

Positions of Cl and F changed

(*S*)-Bromochlorofluoromethane

6.7 There are many ways to determine the relationship between the structures. We must do a series of rotations of one of the structures so they both are represented in the same way. Here are some examples of such rotations, but many other series of rotations are possible. However, each sequence of rotations should give the same result.

Let us rotate the first structure in space to make it similar to the other one:

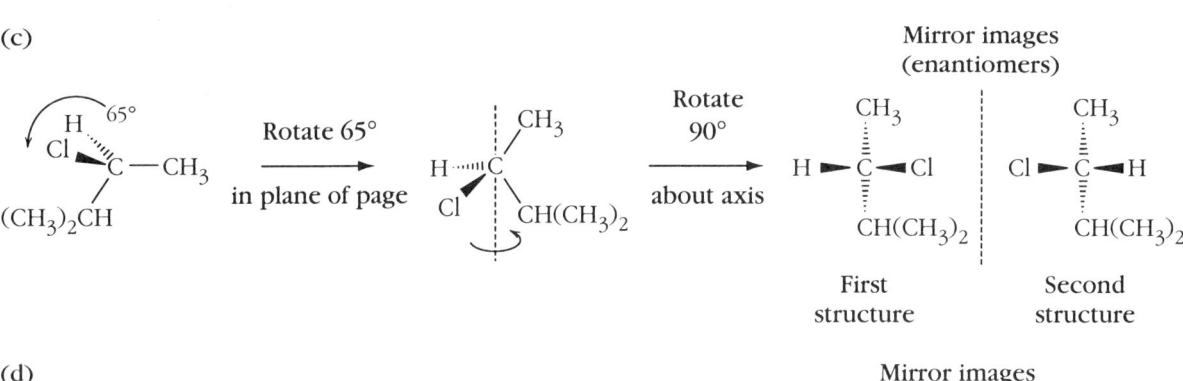

6.8 The priority of each atom or group attached to the stereocenter is determined by the following rules.

(1) Priority is based first on atomic number. The higher the number of the atom bonded directly to the stereocenter, the higher the priority of the substituent.

(2) For isotopes, the higher the atomic weight, the higher the priority.

(3) If a priority based on atomic number cannot be assigned to the atoms directly bonded to the stereocenter, then look at the next set of atoms and continue until the *first point of difference is found.*

(4) Atoms of double or triple bonds are considered as if they are bonded to an equivalent number of similar atoms by single bonds (not applicable in Exercise **6.8**).

The following substituents are listed in the order of increasing priority (the lowest priority substituent is listed first and the highest priority substituent is listed last).

(a) —H, —CH$_3$, —Cl, —I
(b) —CH$_3$, —CH$_2$CH$_3$, —CH(CH$_3$)$_2$, —C(CH$_3$)$_3$,
(c) —CH$_3$, —CH$_2$OH, —CH$_2$Cl, —OH
(d) —H, —CH$_3$, —CH$_2$D, —CH$_2$CH$_3$.

6.9 (a) Priority 1: —Cl
Priority 2: —CH(CH$_3$)$_2$
Priority 3: —CH$_3$
Priority 4: —H

Counterclockwise: therefore (*S*)
Name: (*S*)-2-chloro-3-methylbutane

(b) Priority 1: —Br
Priority 2: —F
Priority 3: —CH$_2$CH$_3$
Priority 4: —H

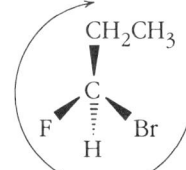

Clockwise: therefore (*R*)
Name: (*R*)-1-bromo-1-fluoropropane

(c) Priority 1: —Cl
Priority 2: —CH$_2$Cl
Priority 3: —CH$_3$
Priority 4: —H

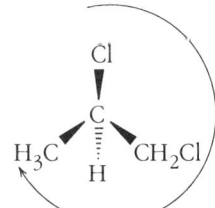

Clockwise: therefore (*R*)
Name: (*R*)-1,2-dichloropropane

(d) Priority 1: —Cl
Priority 2: —CH$_2$CH$_2$CH$_3$
Priority 3: —CH$_2$CH$_3$
Priority 4: —H

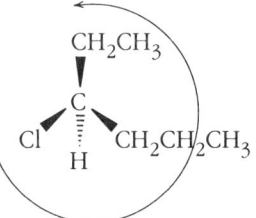

Counterclockwise: therefore (*S*)
Name: (*S*)-3-chlorohexane

6.10 (a) Name: (*S*)-1-chloro-1-fluoroethane. Priorities of substituents bonded to the stereocenter:
Priority 1: —Cl
Priority 2: —F
Priority 3: —CH₃
Priority 4: —H
The configuration is *S*, so the substituent of the lowest priority (4) is directed away from the viewer and the three substituents of higher priority, (1) – (3), must be projected *counterclockwise*:

(b) Name: (*R*)-3-iodo-2-methylpentane. Priorities of substituents bonded to the stereocenter:
Priority 1: —I
Priority 2: —CH(CH₃)₂
Priority 3: —CH₂CH₃
Priority 4: —H
The configuration is *R*, so the substituent of the lowest priority (4) is directed away from the viewer and the three substituents of higher priority, (1) – (3), must be projected *clockwise*:

(c) Name: (*R*)-3-chloro-2,2,3-trimethylpentane. Priorities of substituents bonded to the stereocenter:
Priority 1: —Cl
Priority 2: —C(CH₃)₃
Priority 3: —CH₂CH₃
Priority 4: —CH₃
The configuration is *R*, so the substituent of the lowest priority (4) is directed away from the viewer and the three substituents of higher priority, (1) – (3), must be projected *clockwise*:

(d) Name: (*S*)-1-chloro-1-deuterioethane. Priorities of substituents bonded to the stereocenter:
Priority 1: —Cl
Priority 2: —CH₃
Priority 3: —D
Priority 4: —H
The configuration is *S*, so the substituent of the lowest priority (4) is directed away from the viewer and the three substituents of higher priority, (1) – (3), must be projected *counterclockwise*:

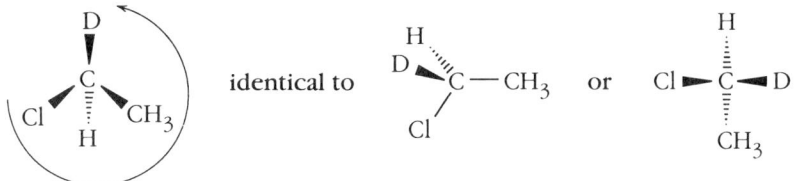

6.11 The following substituents are listed in the order of increasing priority (the lowest priority substituent is listed first and the highest priority substituent is listed last).

(a)   —CH₃, —CH₂CH₂Cl, —CH(CH₃)₂, —OH.
(b)   —CONH₂, —COOH, —COOCH₃, —COCl.
(c)   —C(CH₃)=CH₂, —CH₂NH₂, —C≡N, —CH₂Cl.

6.12 (a) Priority 1: —CHO
Priority 2: —CH=CH₂
Priority 3: —CH₃
Priority 4: —H

Counterclockwise: therefore (*S*)

(b) Priority 1: —NH₂
Priority 2: —COOH
Priority 3: —CH₃
Priority 4: —H

Counterclockwise: therefore (*S*)

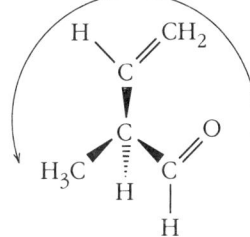

(c) Priority 1: —C≡CH
Priority 2: —CH=CH₂
Priority 3: —CH(CH₃)₂
Priority 4: —CH₃

Counterclockwise: therefore (*S*)

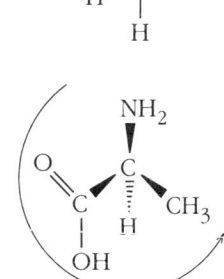

(d) Priority 1: —CH$_2$Br
Priority 2: —CH=CH$_2$
Priority 3: —CH$_2$CH$_3$
Priority 4: —H

Clockwise: therefore (R)

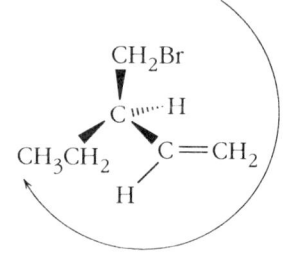

6.13 $[\alpha]_D = \dfrac{\alpha}{l \times C}$

$\alpha = +3.835°; l = 10$ cm $= 1$ dm; $C = 2.350$ g/100 mL $= 0.0235$ g/mL
$[\alpha]_D^{20} = +3.835/(1 \times 0.0235) = +163.2°$

6.14 $[\alpha]_D = \dfrac{\alpha}{l \times C}$

$\alpha = +18.88°; l = 10$ cm $= 1$ dm; $C = 28.35$ g/100 mL $= 0.2835$ g/mL
$[\alpha]_D = +18.88/(1 \times 0.2835) = +66.6°$

6.15 $[\alpha]_D = \dfrac{\alpha}{l \times C} \rightarrow C = \dfrac{\alpha}{l \times [\alpha]_D}$

$\alpha = +5.38°; l = 10$ cm $= 1$ dm; $[\alpha]_D = +66.6°$
$C = +5.38/(1 \times 66.6) = 0.0808$ g/mL

6.16 $[\alpha]_D = \dfrac{\alpha}{l \times C} \rightarrow \alpha = [\alpha]_D(l \times C)$

$[\alpha]_D = +66.6°$

(a) If we now add another 100 mL of water, concentration of the solution is determined as follows:
$C_1 \times V_1 = C_2 \times V_2$ (C_1—concentration before dilution, V_1—volume before dilution (100 mL); C_2—concentration after dilution (V_2—volume after dilution (100 + 100 = 200 mL))

$C_2 = C_1 \dfrac{V_1}{V_2} = 0.0404$ g/mL

$l = 10$ cm $= 1$ dm
$\alpha = +66.6 (1 \times 0.0404) = +2.69°$

(b) $[\alpha]_D = \dfrac{\alpha}{l \times C} \rightarrow \alpha = [\alpha]_D(l \times C)$

$l = 20$ cm $= 2$ dm; $[\alpha]_D = +66.6°$; $C = 0.0808$ g/mL
$\alpha = +66.6 (2 \times 0.0808) = +10.76°$

6.17 **Enantiomeric excess** is the percentage of the enantiomer that is present in excess. Thus, the total mixture contains 90% of the levorotatory 2-chlorobutane and (100 − 90 = 10) 10% of a racemic mixture containing equal amounts of levorotatory and dextrorotatory 2-chlorobutane (10%/2 = 5%), 5% of each enantiomer. Therefore, there is 95% of levorotatory and 5% of dextrorotatory 2-chlorobutane in the mixture.

6.18 This reaction occurs without breaking or forming any of the bonds to the stereocenter. Thus, the configuration of the stereocenter has not changed in this reaction; this means that (S)-2-methyl-1-butanol gives (S)-2-methylbutanoic acid on oxidation.

Notice that both compounds have the (S)-configuration, yet one is levorotatory and the other is dextrorotatory. We cannot, therefore, predict the sign of the specific rotation of the product because *there is no correlation between the sign of the specific rotation of a compound and the absolute configuration of its stereocenter.*

6.19 (a)

CH₃ H—C—OH Br—C—H CH₃ (2S,3S) **A**	CH₃ HO—C—H H—C—Br CH₃ (2R,3R) **B**	CH₃ HO—C—H Br—C—H CH₃ (2R,3S) **C**	CH₃ H—C—OH H—C—Br CH₃ (2S,3R) **D**

Structures (**A** and **B**) and (**C** and **D**) are pairs of enantiomers, respectively. The following pairs: (**A** and **C**), (**A** and **D**), (**B** and **C**), and (**B** and **D**) are diastereomers.

(b)

CH₃ H—C—OH HO—C—H CH₂CH₃ (2S,3S) **A**	CH₃ HO—C—H H—C—OH CH₂CH₃ (2R,3R) **B**	CH₃ HO—C—H HO—C—H CH₂CH₃ (2R,3S) **C**	CH₃ H—C—OH H—C—OH CH₂CH₃ (2S,3R) **D**

Structures (**A** and **B**) and (**C** and **D**) are pairs of enantiomers, respectively. The following pairs: (**A** and **C**), (**A** and **D**), (**B** and **C**), and (**B** and **D**) are diastereomers.

6.20 (a)

CH₂OH H—C—Br Br—C—H CH₂OH (2S,3S) **A**	CH₂OH Br—C—H H—C—Br CH₂OH (2R,3R) **B**	CH₂OH Br—C—H Br—C—H CH₂OH (2R,3S) **C**	≡	CH₂OH H—C—Br H—C—Br CH₂OH (2S,3R) **D**

Structure **C** is a *meso* compound and it is identical to **D**. Structures **A** and **B** are enantiomers. The following pairs of structures (**A** and **C**) and (**B** and **C**) are diastereomers.

(b) First, let us identify all stereocenters. This compound contains only two stereocenters, so there are a maximum of four stereoisomers.

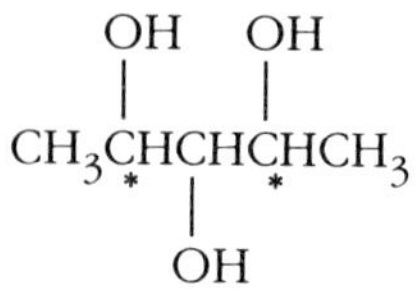

Now, let us draw all 2^2 structures:

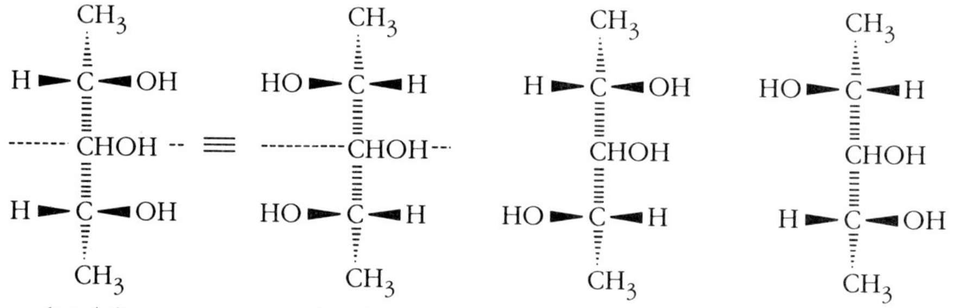

(2*S*,4*R*) (2*R*,4*S*) (2*S*,4*S*) (2*R*,4*R*)

A **B** **C** **D**

Structure **A** is a *meso* compound and it is identical to **B**. Compounds **C** and **D** are chiral and are enantiomers. Thus, there are only three stereoisomers. The following pairs: (**A** and **C**) and (**A** and **D**) are diastereomers.

(c)

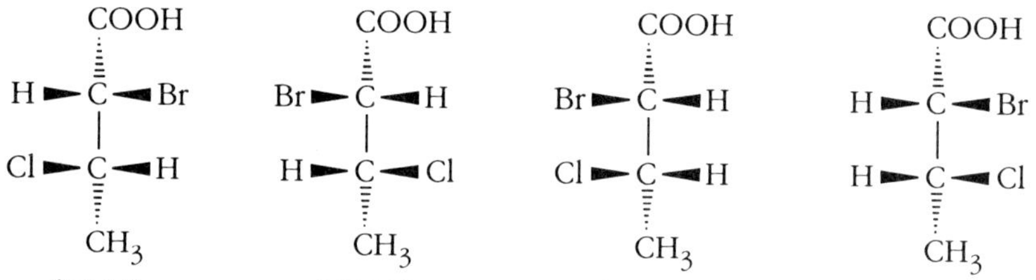

(2*S*,3*S*) (2*R*,3*R*) (2*R*,3*S*) (2*S*,3*R*)

A **B** **C** **D**

We have two pairs of enantiomers: (**A** and **B**) and (**C** and **D**). We have four pairs of diastereomers: (**A** and **C**), (**A** and **D**), (**B** and **C**), and (**B** and **D**).

6.21 1,2-Dichlorocyclopropane:

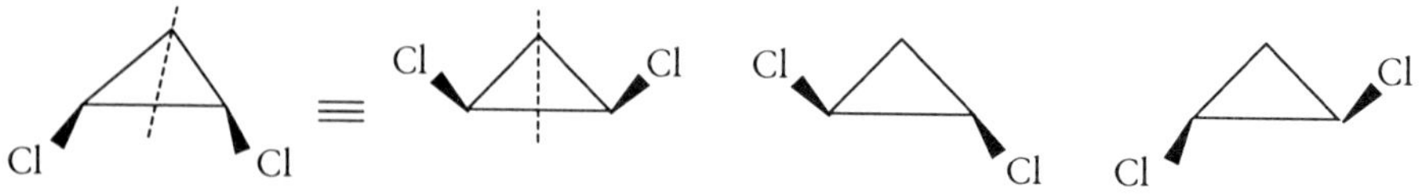

(1*R*,2*S*) (1*S*,2*R*) (1*S*,2*S*) (1*R*,2*R*)

A **B** **C** **D**

Structure **A** is a *meso* compound and it is identical to **B**. Structures **C** and **D** are enantiomers. The following pairs of structures (**A** and **C**) and (**A** and **D**) are diastereomers.

1-Bromo-2-chlorocyclopropane:

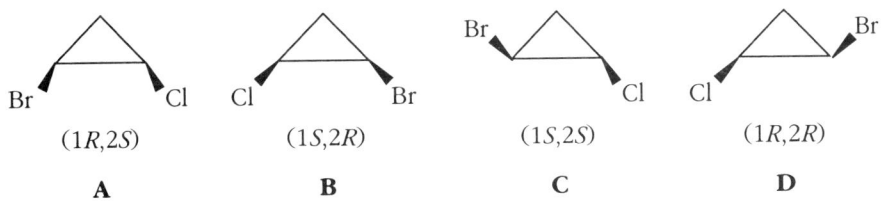

We have two pairs of enantiomers: (**A** and **B**) and (**C** and **D**). We have four pairs of diastereomers: (**A** and **C**), (**A** and **D**), (**B** and **C**), and (**B** and **D**).

6.22 Chirality of bromochlorocyclohexanes:

Position of substituents	Diastereomer	Chirality
1, 2	cis	enantiomers
1, 2	trans	enantiomers
1, 3	cis	enantiomers
1, 3	trans	enantiomers
1, 4	cis	achiral
1, 4	trans	achiral

6.23 If the priority of H_a is made higher than that of H_b, then H_b becomes the lowest priority and we position the molecule so that the lowest priority substituent is placed as far from the observer as possible. The progression from the ethyl group (highest priority substituent) to the methyl group (second highest priority) to H_a (third highest priority) is clockwise and the configuration at the stereocenter is R. Therefore, H_a is H_R.

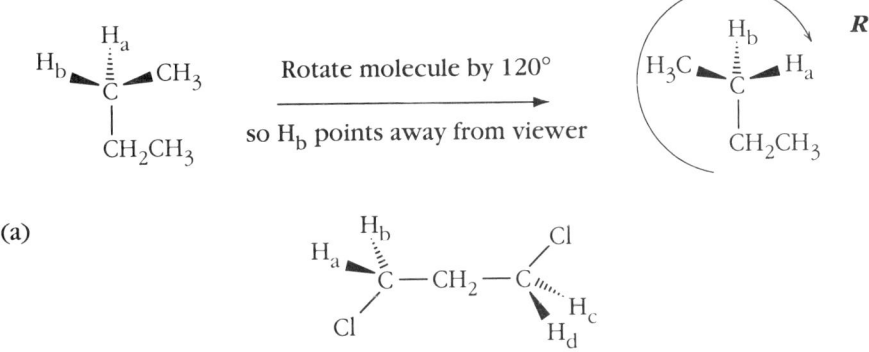

6.24 (a)

The terminal methylene hydrogens are enantiotopic because their replacement by deuterium forms enantiomers.

The central methylene hydrogens are *not* enantiotopic because the same compound is formed if any of the hydrogen atoms is replaced by deuterium (*meso* forms).

(b)

The methyl hydrogens of propanoic acid are not enantiotopic because the same compound is formed by the successive replacement of the three hydrogen atoms by deuterium.

The methylene hydrogens are enantiotopic because their successive replacement by deuterium forms enantiomers.

(c)

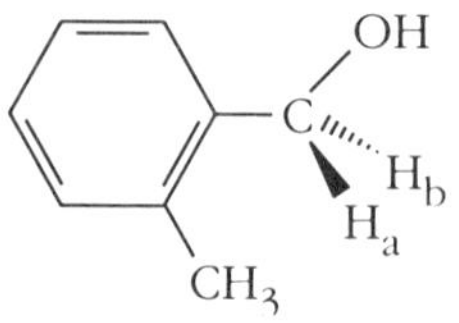

The methyl hydrogens in the above structure are not enantiotopic because the same compound is formed by the successive replacement of the three hydrogen atoms by deuterium.

The methylene hydrogens are enantiotopic because their successive replacement by deuterium forms enantiomers.

6.25 The enantiotopic hydrogen atoms in Exercise 6.24 were indicated as H_a, H_b, H_c, and H_d. In order to designate them as pro-R or pro-S we arbitrarily give priority to one over the other and we determine the configuration of the prochiral center in each case. If the configuration is R, the hydrogen atom is designated as pro-R; if the configuration is S, the hydrogen atom is designated as pro-S.

(a) (b) 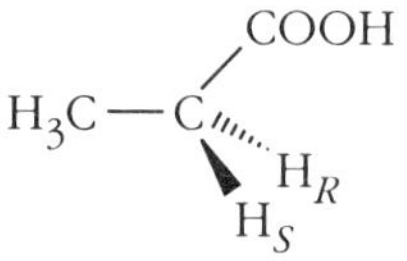 (c)

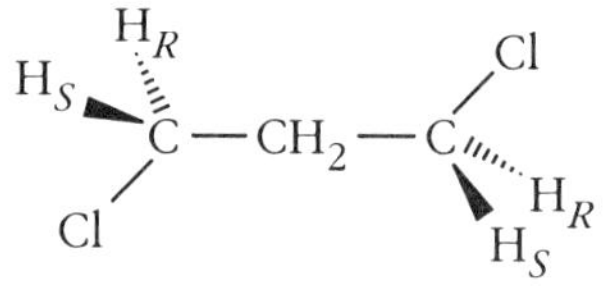

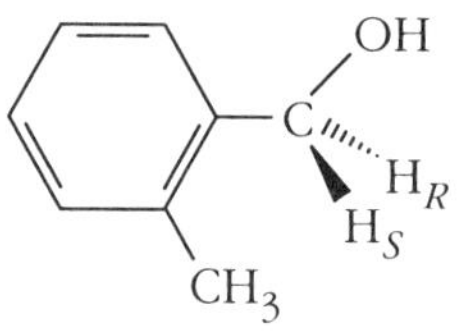

6.26 (a) After flipping a Fischer projection we get its mirror image. The compound on the right is (R)-lactic acid, an enantiomer of (S)-lactic acid. Flipping a Fischer projection therefore inverts the configuration of the stereocenter. Conclusion: *DO NOT flip a Fischer projection because the configuration of the stereocenter in the compound represented by the flipped Fischer projection is different from the original Fischer projection.*

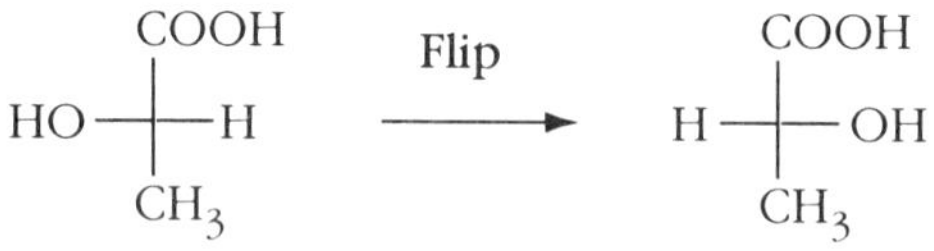

(b) Rotating a Fischer projection also changes the configuration of the stereocenter. The compound on the right is (R)-lactic acid, the enantiomer of (S)-lactic acid. Conclusion: *DO NOT rotate a Fischer projection because the configuration of the stereocenter in the compound represented by the rotated Fischer projection is different from the original Fischer projection.*

$$\begin{array}{c} \text{COOH} \\ \text{HO}\!\!-\!\!\!\!\underset{\text{CH}_3}{\overset{|}{\text{C}}}\!\!-\!\!\text{H} \\ \text{A} \end{array} \longrightarrow \begin{array}{c} \text{OH} \\ \text{H}_3\text{C}\!\!-\!\!\!\!\underset{\text{H}}{\overset{|}{\text{C}}}\!\!-\!\!\text{COOH} \\ \text{C} \end{array} \quad \text{same as} \quad \begin{array}{c} \text{COOH} \\ \text{H}\!\!-\!\!\!\!\underset{\text{CH}_3}{\overset{|}{\text{C}}}\!\!-\!\!\text{OH} \\ \text{B} \end{array}$$

6.27 (a) A **stereocenter** (synonymous with *stereogenic center*) is an atom in a molecule at which interchange of two atoms or groups of atoms bonded to that atom produces a different stereoisomer. The type of stereocenter encountered most often in organic chemistry is a tetrahedral carbon atom that bears four different substituents. At various times it has been called a *chiral center*, a *chiral carbon atom*, an *asymmetric center*, or an *asymmetric carbon atom*. Example: (*R*)-2-butanol:

(Four different substituents at C2)

(b) An **enantiomer** is one of two molecules that exist as nonsuperposable mirror images of each other. Each of the enantiomers is optically active and rotates plane-polarized light by equal amounts but in opposite directions.

(*R*)-(−)2-Butanol $[\alpha]_D^{25} = -13.52°$

(*S*)-(+)-2-Butanol $[\alpha]_D^{25} = +13.52°$

mirror

(c) A **racemic mixture** is a mixture containing equal quantities of enantiomers. Since the enantiomers rotate plane-polarized light by equal amounts but in opposite directions, the specific rotation of a 1:1 mixture of (+) and (−) enantiomers is 0°. Alternatively, we say that a racemic mixture is *optically inactive*. Example: (±)-2-Butanol [or (*R,S*) 2-butanol] indicates a 1:1 mixture of the above enantiomers of 2-butanol.

(d) A **chiral molecule** is a molecule that is not superposable on its mirror image. Example: Each enantiomer of 2-butanol is a chiral molecule, because neither of them is superposable on its mirror image (its mirror image is another enantiomer).

(e) A **prochiral center** is an atom that is not a stereocenter but becomes one upon substitution of one of the atoms or groups bonded to it. If the replacement of a given group or atom gives an *R* center, that group or atom replaced is called pro-*R*; if replacement gives an *S* center, the group or atom is called pro-*S*. For example, the C2 carbon in propanoic acid is a prochiral center:

(f) An **enantiomeric excess** is a difference between the percentage of the major enantiomer present in a mixture and the percentage of its mirror image. An optically pure material has an enantiomeric excess of 100%. A racemic mixture has an enantiomeric excess of zero.

(g) A **plane of symmetry** (also called a **mirror plane**) is a plane that bisects an object, such as a molecule, into two mirror-image halves. When a line is drawn from any element in the object perpendicular to such a plane and extended an equal distance in the opposite direction, a duplicate of the element is encountered. The dotted line in *cis*-1,3-cyclopentanediol represents a plane of symmetry (or a mirror plane):

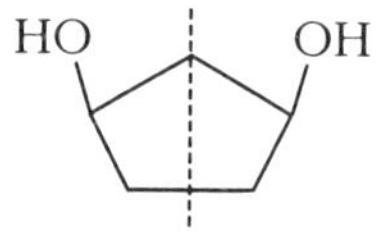

cis-1,3-Cyclopentanediol

(h) A **diastereomer** is a stereoisomer that is neither a mirror image, nor superposable. It means that compounds that are stereoisomers of one another, but are not enantiomers, are called *diastereomers*.

(2*R*,3*R*)- and (2*R*,3*S*)-3-bromo-2-butanols are neither mirror images nor superposable, so they are diastereomers.

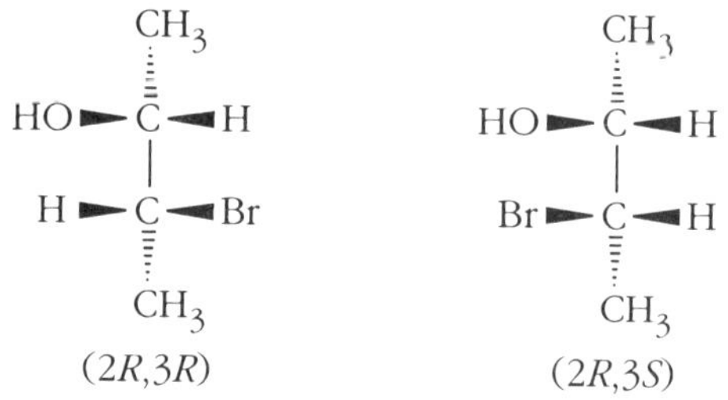

3-Bromo-2-butanol

(i) A ***meso*** **compound** is an achiral molecule that has stereocenters. The most common kind of *meso* compound is a molecule with two or more stereocenters and a plane of symmetry. (2*R*,3*S*)-2,3-Dibromobutane is an example of a *meso* compound:

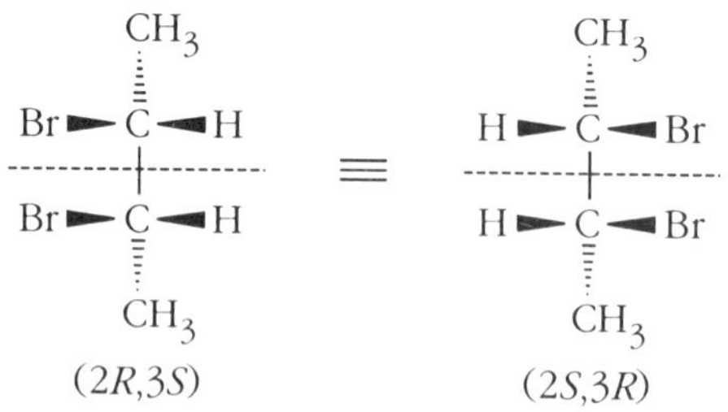

(j) **Superposable**—to place one object or structure in the space occupied by another so that the two coincide throughout their whole extent. The two structures shown in Exercise **6.27(i)** are superposable.

(k) Optical activity is the ability of a substance to rotate the plane of polarized light. In order to be **optically active**, a substance must be chiral, and one enantiomer must be present in excess of the other.

(l) **Absolute configuration** is the actual three-dimensional arrangement of atoms or groups at a stereocenter. The configuration of (+)-glyceraldehyde is:

(m) **Specific rotation**, $[\alpha]_D^{25}$, of a chiral compound is a physical constant that is defined by the equation:

$$[\alpha]_D^{25} = \frac{\alpha}{l \times C}$$

where α is the observed rotation in degrees, l is the path length in decimeters (dm), and C is the concentration in g/mL.

(n) **Relative configuration** is a stereochemical configuration of two compounds established on a comparative rather than an absolute basis, even though the absolute configuration of either may be unknown. For example, the configuration of (R)-(+)-glyceraldehyde can be easily related to that of (R)-(-)-glyceric acid because the spatial arrangement of the substituents in the two molecules remains the same during the oxidation reaction.

(R)-(+)-Glyceraldehyde → Mild oxidation → (R)-(-)-Glyceric acid

(o) **Enantiotopic hydrogens** are hydrogen atoms that lie on opposite sides of the plane of symmetry of an achiral molecule.

(p) An **achiral molecule** is a molecule that is superposable on its mirror image. Compounds that have no stereocenter and those that have one or more stereocenters but also have a plane or center of symmetry are *achiral*.

6.28

(R)-2-Chloro-3-methylbutane   (S)-2-Chloro-3-methylbutane

6.29 An alkane contains only sp^3-hybridized carbon atoms and hydrogen atoms. A chiral carbon atom contains four different substituents. Thus we take a carbon atom and attach to it a hydrogen atom and the three smallest different alkyl groups: methyl, ethyl, and propyl. There are two enantiomers of 3-methylhexane:

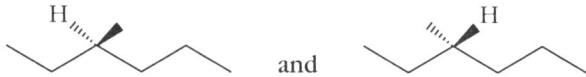
and

6.30 (a) 2-Methyl butane does not contain a stereocenter; (b) 1,2-dichloropropane contains a stereocenter (C2); (c) 3-methyl-1-hexene contains a stereocenter (C3).

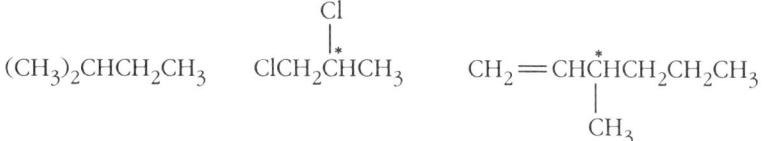

6.31 (a) The central carbon atom is bonded to two identical substituents (carboxyl groups) so this is an achiral molecule.
(b) The structure represents a stereoisomer that is expected to be optically active because the central carbon atom is bonded to four different substituents: carboxyl group, amino group, methyl, and isopropyl (methylethyl).
(c) The structure represents an achiral molecule because it has a plane of symmetry defined by the hydrogen, carbon, and chlorine atoms.

6.32 (a) (S)-2-Bromo-1-chloro-3-iodopropane,
(b) (S)-2,2,3-Trimethylpentane,
(c) (S)-4-Bromo-1,1-dichloro-2-methylbutane.

6.33 (a) R (b) S (c) S

6.34
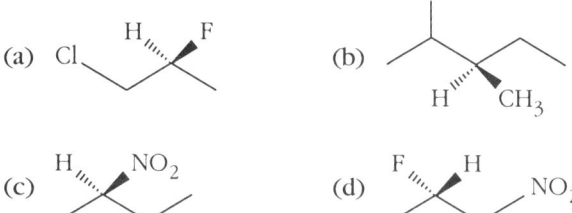

6.35 $[\alpha]_D = \dfrac{\alpha}{l \times C}$

$\alpha = +17.3°$ for $l = 10$ cm.

(a) Because $[\alpha]_D$ is constant for a given compound and C is not changed, halving the value of l (from 10 cm to 5 cm) means the value of α will also be halved:

$$\alpha = +17.3/2 = +8.65°$$

(b) Because $[\alpha]_D$ is constant for a given compound and l is not changed, decreasing C to half its original value means α will also be halved:

$$\alpha = +17.3/2 = +8.65°$$

6.36 In a mixture of enantiomers, the (+) and (−) enantiomers have identical magnitudes of specific rotation but opposite signs. If we let x = mole fraction of (+)-amphetamine and y = mole fraction of (−)-amphetamine, then $x + y = 1$ and $y = 1 - x$. The observed rotation is the sum of the contributions of the two enantiomers.

Since $[\alpha]_D(+) = +40.1°$, $[\alpha]_D(-) = -40.1°$

$-10.3° = +40.1° \, x + (-40.1°)(1-x) = 80.2° \, x - 40.1°$

$+29.8° = 80.2° \, x$

$x = 0.37$

$y = 1 - x = 0.63$

Answer: The mixture should contain 37% of (+)-amphetamine and 63% of (−)-amphetamine.

6.37 The configuration of the stereocenter of the product remains R due to the following two facts: (a) none of the bonds to the chiral center is either formed or broken and (b) priorities of substituents bonded to the chiral center did not change (highest priority —CH=CH$_2$ group was transformed into the highest priority carboxyl group, —COOH).

6.38 The central carbon atom was originally bonded to four different substituents: —H, —CH$_2$CH$_3$, —CH=CH$_2$, and —COOH. During the reaction, however, the —CH=CH$_2$ group was transformed into the carboxyl group, —COOH. Thus, there are two identical carboxyl groups bonded to the central carbon atom after the reaction and the product is achiral. As a result, the optical activity was lost.

6.39 (a) Superposable, (b) Enantiomers, (c) Superposable (*meso* compound), (d) Enantiomers, (e) Superposable, (f) Enantiomers, (g) Superposable.

6.40 The stereocenters are identified by the asterisks. There are three stereocenters, and thus $2^3 = 8$ stereoisomers.

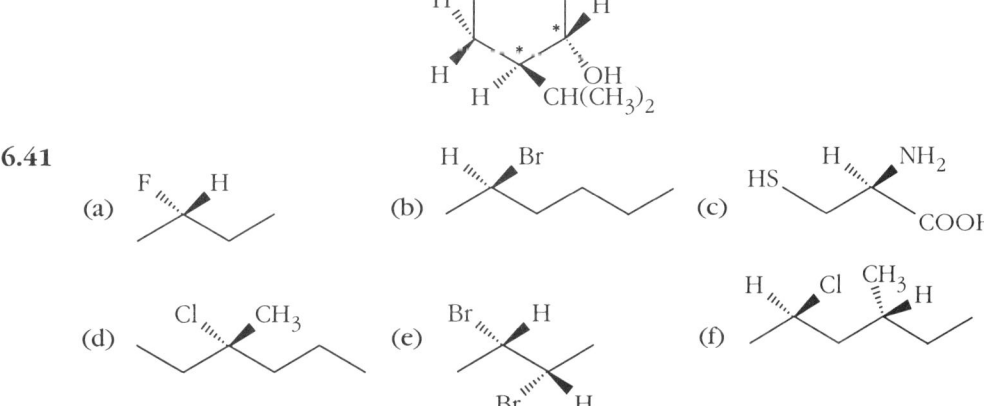

6.42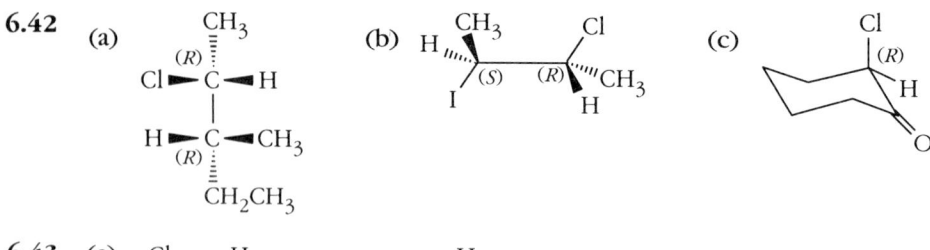

6.43 (a)

Cl,,,,   H   H,,,,   Cl
    \\ /   \\ /
     C   or   C
    / \\   / \\

(b) Many different answers are possible. The following are some examples (all have a plane of symmetry):

(c) Many different answers are possible. The following are just two pairs of enantiomers:

6.44 (a) B & C
 (b) A & B, A & C, A & D, B & E, C & E, D & E
 (c) A & E
 (d) B & D, C & D
 (e) A

6.45 (a) Enantiomers
 (b) Diastereomers
 (c) Diastereomers
 (d) Diastereomers

6.46 There are two stereocenters and $2^2 = 4$ stereoisomers.

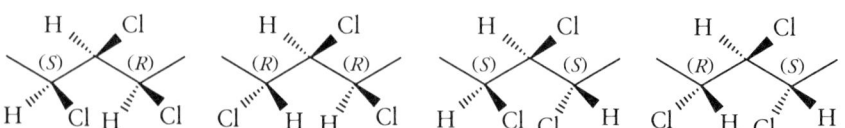

6.47 If the chlorine atom on C3 of 2,3,4-trichloropentane is replaced by a bromine atom, the number of stereoisomers will not change because C3 is still not a stereocenter.

If the chlorine atom on C2 of 2,3,4-trichloropentane is replaced by a bromine atom, a new stereocenter at C3 is created and the number of stereoisomers increases to eight ($2^3 = 8$).

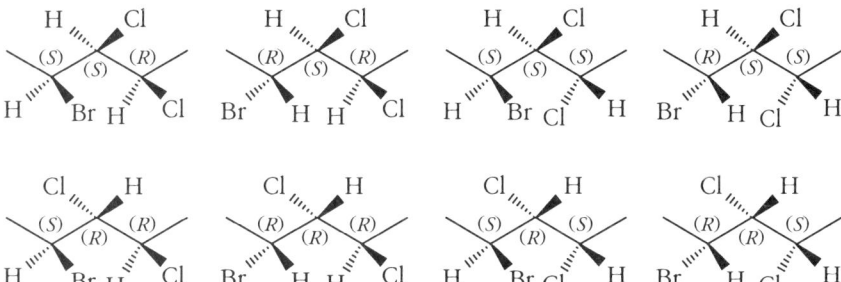

CHAPTER 7

ALKENES

Concepts

Alkenes are hydrocarbons that contain carbon-carbon double bonds. The double bond is a functional group in alkenes. The carbon atoms of the double bond and the four atoms bonded to them lie in a plane, with bond angles of approximately 120° around the two carbon atoms. This geometry is obtained by using sp^2 hybrid carbon atoms to construct the localized valence bond model of the double bond.

The double bond in an alkene consists of a σ bond resulting from the overlap of sp^2 hybrid orbitals on two carbon atoms and a π bond resulting from the side-to-side overlap of the remaining p orbital on each carbon atom.

A **terminal alkene,** in which the double bond is at the end of a chain, is less stable than an **internal alkene,** which has the double bond somewhere in the middle of the chain. A *trans* alkene, in which the large substituents on the double bond are farther apart from each other, is more stable than the corresponding *cis* alkene in which the larger substituents are closer together. The stability of an alkene increases with the number of alkyl groups that are substituted on the carbon atoms of the double bond.

So far, we know only one reaction leading to the synthesis of alkenes, namely the *acid-catalyzed dehydration of secondary or tertiary alcohols:*

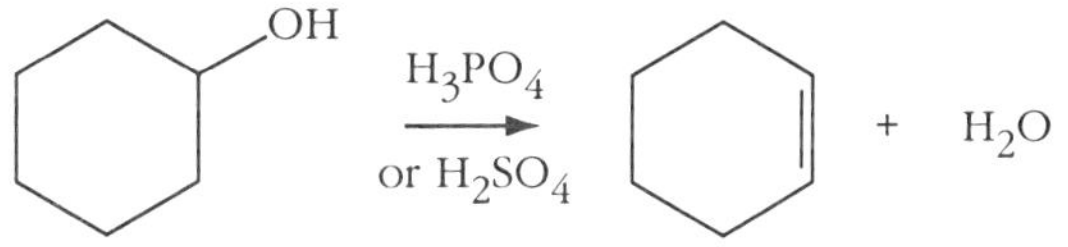

Addition to the double bond is the predominant reaction of alkenes. In this chapter, the addition of Brønsted acids to the double bond is discussed. The general mechanism of the addition of Brønsted acids occurs in two steps.

Step 1 is an attack of the π bond on the electrophile (frequently a proton) to form a carbocation intermediate:

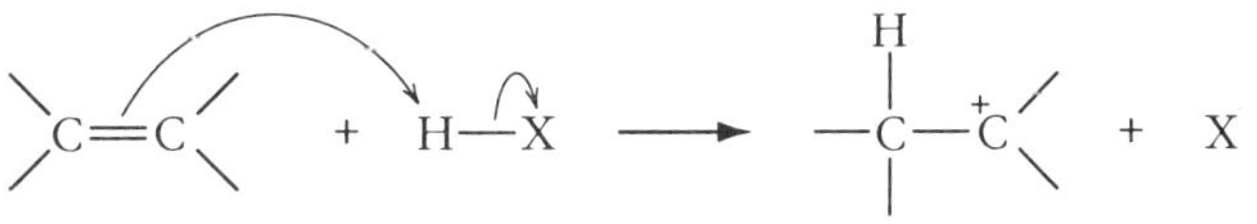

Step 2 is the reaction of the carbocation intermediate with a nucleophile:

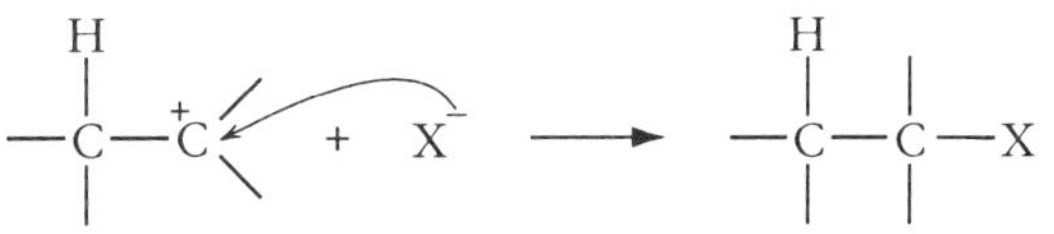

Solutions to the Exercises

7.1 (a) 2-Butene exists as a pair of diastereomers: *cis* and *trans*.

cis-2-Butene   *trans*-2-Butene

(b) 1-Bromo-1-fluoropropene exists as a pair of diastereomers: (*Z*) and (*E*)

(*Z*)-1-Bromo-1-fluoropropene   (*E*)-1-Bromo-1-fluoropropene

(c) 1-Chloro-2-methylpropene does not exist as a pair of diastereomers because the C2 carbon of the double bond is bonded to two identical groups (methyl groups).

(d) 1,1-Dichloro-2,2-difluoroethene does not exist as a pair of diastereomers because both carbon atoms of the double bond are bonded to two identical atoms (C1 carbon to Cl atoms, C2 carbon to F atoms).

7.2 (a) Step 1: The longest continuous chain containing both carbons of the double bond is seven atoms long.
Step 2: The name of the straight chain alkane with seven carbon atoms is *heptane*. There is one double bond in the molecule so the *-ane* ending of *heptane* is changed to *-ene* making *heptene* the parent name of this alkene.
Step 3: Numbering the chain from left to right gives the lowest number, number 2, to the first carbon of the double bond.

Step 4: There are two methyl groups (on C2 and C5) and bromine (on C4). Listing bromine alphabetically before the methyl groups, we have:
4-bromo-2,5-dimethyl-2-heptene

(b) 3-Ethyl-2-methyl-2-pentene
(c) 3,3-Dichloropropene
(d) 4-Isopropyl-2,7-dimethyl-2,6-octadiene or 2,7-dimethyl-4-(1-methylethyl)-2,6-octadiene
(e) 2-Isopropyl-5,6-dimethyl-1,5-heptadiene or 5,6-dimethyl-2-(1-methylethyl)-1,5-heptadiene

CHAPTER 7

7.3

(a) (CH₃)₂C=CCH₂CH₃
                |
               CH₃

(b) CH₃CHCH=CHCHCH₃ with Br on each CHCH₃

(c) CH₂=C(CH₃)−C(CH₃)=C(CH₃)−C=CH₂ (as drawn)

(d) CH₃CH₂CHCH=CHCH₂C(CH₃)₃
              |
             CH₃

7.4
(a) 5-Bromo-1,3-cyclopentadiene
(b) 4-isopropyl-3-methyl-1,5-cyclooctadiene or 3-methyl-4-(1-methylethyl)-1,5-cyclooctadiene
(c) 1,2-dimethylcyclobutene

7.5
(a) 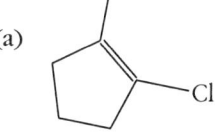   (b) 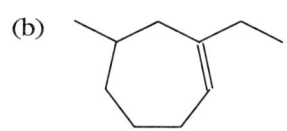   (c)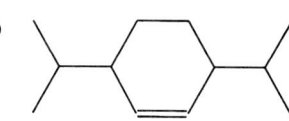

7.6

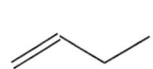   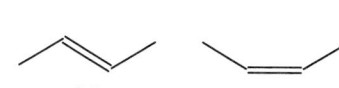   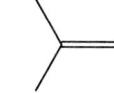

1-Butene   trans-2-Butene   cis-2-Butene   2-Methylpropene

7.7

	Higher priority	Lower priority
(a)	I	Cl
(b)	CH₃CH₂	CH₃
(c)	Cl	(CH₃)₂CH
(d)	Cl	HO
(e)	HO	CH₃
(f)	CH₂OH	CH₃

7.8
(a) (Z)-1,2-Dichloroethene
(b) (E)-3,4-Dimethylpentene
(c) (Z)-2-Bromo-1-chloro-1-fluoropropene
(d) (E)-1-Chloro-2-methyl-2-pentene

7.9

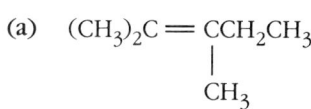

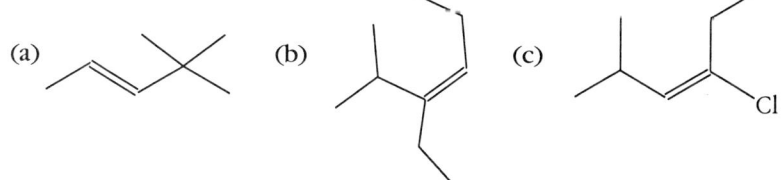

7.10 (2*E*,4*Z*)-3-Methyl-2,4-heptadiene

7.11

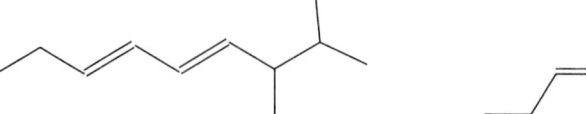
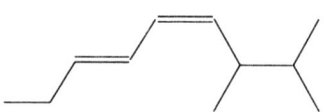

(3*E*,5*E*)-7,8-Dimethyl-3,5-nonadiene   (3*E*,5*Z*)-7,8-Dimethyl-3,5-nonadiene

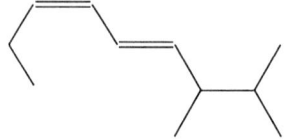
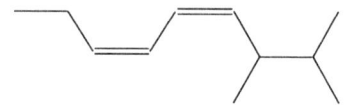

(3*Z*,5*E*)-7,8-Dimethyl-3,5-nonadiene   (3*Z*,5*Z*)-7,8-Dimethyl-3,5-nonadiene

7.12 The relative stabilities of alkenes increase as the degree of substitution of the double bond increases.

(a) 2-Pentene is a disubstituted alkene, 2-methyl-2-butene is a trisubstituted alkene, and 1-pentene is a monosubstituted alkene. Thus, the order of *increasing* stability is 1-pentene, 2-pentene, and 2-methyl-2-butene.

(b) (*Z*)-3-Methyl-3-hexane and (*E*)-3-methyl-3-hexene are trisubstituted alkenes, whereas 2-methyl-1-hexene is a disubstituted alkene. (*E*) isomers are usually more stable than the corresponding (*Z*) isomers because the substituents are farther apart than they are in the (*Z*) isomers. Thus, the order of *increasing* stability is 2-methyl-1-hexene, (*Z*)-3-methyl-3-hexene, and (*E*)-3-methyl-3-hexene.

(c) 1-Heptene is a monosubstituted alkene, 2,3-dimethyl-2-pentene is a tetra-substituted alkene, and 2-methyl-1-hexene is a disubstituted alkene. Thus, the order of *increasing* stability is 1-heptene, 2-methyl-1-hexene, and 2,3-dimethyl-2-pentene.

7.13 The molecular formula, C_5H_8, indicates two units of unsaturation; it may be an alkyne, an alkadiene, or a cycloalkene. There is no C≡C stretch observed around 2200 to 2100 cm^{-1}. The IR spectrum shows a sharp peak at 3060 cm^{-1} that indicates the sp^2-hybridized carbon-hydrogen bond stretching vibration. The IR spectrum suggests an alkene. The absence of the carbon-carbon double bond stretching vibration in the region 1670 to 1645 cm^{-1} indicates a symmetrically substituted alkene. The same conclusion can be drawn from the ^{13}C NMR spectrum because only one signal of the sp^2-hybridized carbon is observed. Its chemical shift, δ 130 ppm, indicates it is a monosubstituted alkene, =CHR. Thus it cannot be a pentadiene because a pentadiene cannot have four equivalent sp^2-hybridized carbon atoms. The only option left, then is cyclopentene. The ^{13}C NMR spectrum is consistent with cyclopentene because we observe two peaks of sp^3-hybridized carbon atoms: C3 and C5 at δ 32 ppm and C4 at δ 22 ppm.

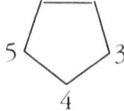

7.14 (a) $CH_2=CH_2 + HCl \longrightarrow CH_3CH_2Cl$

(b) $CH_2=CH_2 + CF_3COOH \longrightarrow CF_3C(=O)OCH_2CH_3$

(c) $CH_2=CH_2 + KI \xrightarrow{H_3PO_4} CH_3CH_2I$

(d) $CH_2=CH_2 + CH_3OH \xrightarrow{H_2SO_4} CH_3CH_2OCH_3$

(e) $CH_2=CH_2 + H_2O \xrightarrow{H_2SO_4} CH_3CH_2OH$

7.15

(a) 1,2-dimethylcyclopentene + HCl → two stereoisomeric products (Cl and CH$_3$ on one carbon; H and CH$_3$ on adjacent carbon)

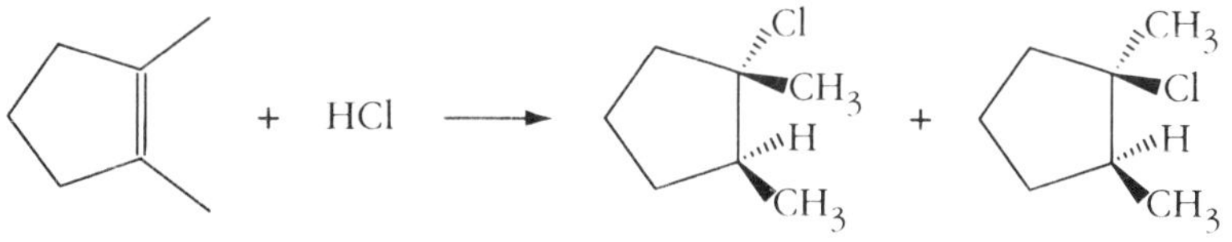

(b) 1,2-dimethylcyclopentene + CF$_3$COOH → two stereoisomeric products with OOCCF$_3$ group

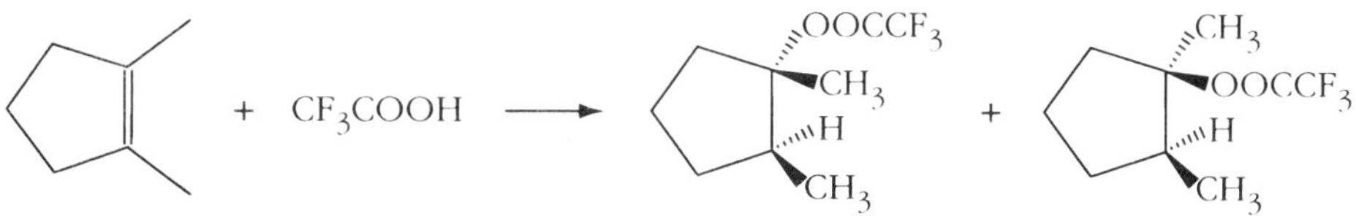

(c) 1,2-dimethylcyclopentene + KI, H$_3$PO$_4$ → two stereoisomeric products with I

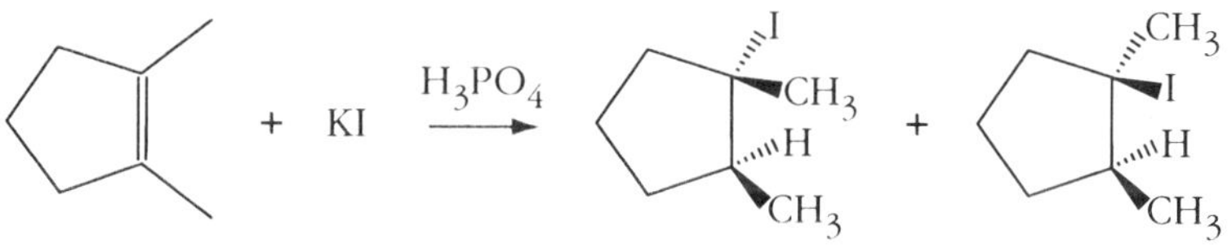

(d) 1,2-dimethylcyclopentene + CH$_3$OH, H$_2$SO$_4$ → two stereoisomeric products with OCH$_3$

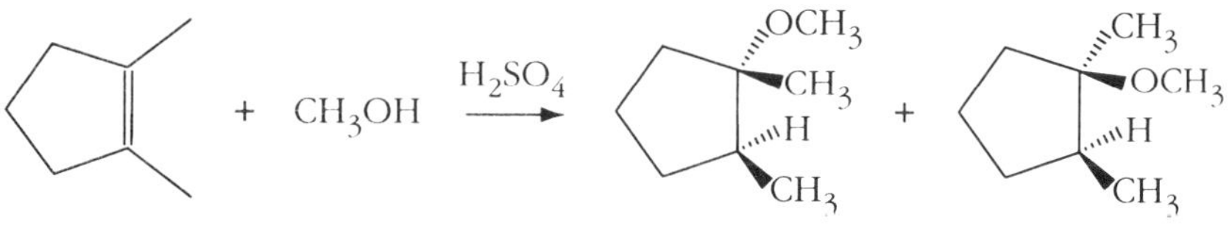

(e) 1,2-dimethylcyclopentene + H$_2$O, H$_2$SO$_4$ → two stereoisomeric products with OH

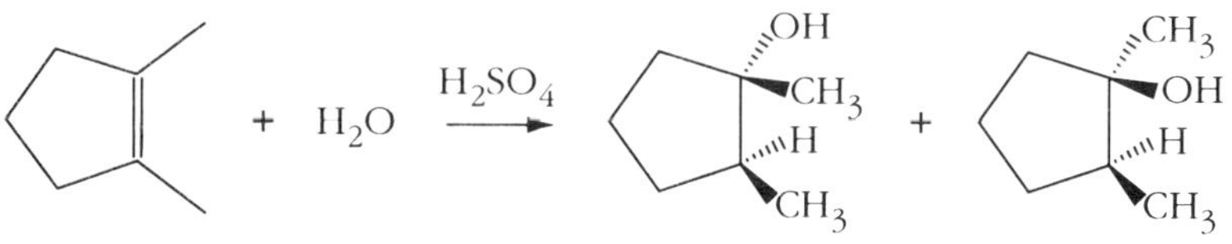

7.16 Step 1 of the mechanism shown in Figure 7.11 is protonation of the π bond. The alkene acts as a nucleophile in this step as it provides its pair of π electrons to form a new carbon-hydrogen bond.

Step 2 of the mechanism is a reaction of the chloride ion with the carbocation intermediate. The carbocation acts as an electrophile because it accepts a pair of electrons from the chloride ion to form a new carbon-chlorine bond.

7.17 Step 1 is an acid-base reaction in which the alkene is protonated to form a carbocation:

$H_2C=CH_2$ + H—Ö:—C(=Ö:)CF$_3$ ⟶ H_3C—$\overset{+}{C}H_2$ + CF$_3$C(=Ö:)(—:Ö:⁻)

In Step 2 the carbocation formed in Step 1 accepts an electron pair from a carboxylate anion to form the product:

H_3C—$\overset{+}{C}H_2$ + :Ö:⁻—C(=:Ö:)CF$_3$ ⟶ CH$_3$CH$_2$Ö:—C(=:Ö:)CF$_3$

7.18 (a) 2°   (b) 1°   (c) 3°   (d) 3°

7.19 Increasing stability ⟶

(c) < (b) < (a) < (d)   because   methyl < 1° < 2° < 3°

7.20 The addition of HCl to the following alkenes follows the Markownikoff Rule: *In the addition of protic acids to alkenes, the proton adds to the carbon atom of the double bond that has the greater number of hydrogen atoms.*

(a), (b), (c), (d) [reactions of alkenes with HCl showing Markovnikov addition products]

7.21

(*E*)-3-Methyl-3-hexene on addition of proton forms

(*Z*)-3-Methyl-3-hexene on addition of proton forms

Yes, the two carbocations are superposable.

7.22 In the case of methylpropene, two carbocations can be formed during Step 1 of the addition of HCl. The two carbocations are a tertiary carbocation and a primary carbocation. The tertiary carbocation is much more stable than the primary carbocation. The Hammond Postulate allows us to use the relative stabilities of carbocations as a guide to the relative stabilities of transition states. The more stable the carbocation, the lower is the energy of the transition state for its formation, and the faster it is formed. Thus, the tertiary carbocation will be formed much faster. That is why 2-chloro-2-methylpropane is the only observed product.

7.23 The acid-catalyzed addition of an alcohol to an alkene is a three-step process:
Step 1: The acid protonates methanol and protonated methanol (being a stronger acid) protonates the alkene to give a tertiary carbocation.

Step 2: The carbocation reacts with the methanol nucleophile to form the conjugate acid of the ether (oxonium ion).

Step 3: The conjugate acid of the ether is deprotonated.

7.24 The stabilization by an amino group is due to the availability of the free electron pair on nitrogen:

7.25 Step 1: Reaction of 3-methyl-1-butene with HCl forms a 2° carbocation.

Step 2a: Migration of a hydrogen atom with its bonding electrons (a 1,2-hydride shift) from an adjacent carbon atom gives a more stable carbocation.

Step 3a: The more stable carbocation reacts with a nucleophile to give the rearrangement product.

Step 2b: If, in Step 2, a methyl group shifts from an adjacent carbon atom instead of the hydrogen atom then the same carbocation is formed. Thus, such a rearrangement is not productive because it leads to the same 2° carbocation.

Step 3b: The 2° carbocation reacts with the chloride nucleophile to form 2-chloro-3-methylbutane.

Rearrangements of the sort just shown are a common feature of carbocation chemistry. We will see at numerous places in subsequent chapters that their occurrence in a reaction provides strong mechanistic evidence of carbocation intermediates.

7.26

(a) [structure: H₃C—C(CH₃)(CH₃)—⁺CH₂ → (H₃C)(H₃C)⁺C—CH₂CH₃]

(b) [structure: H₃C—C(CH₃)(CH₃)—⁺CH(H)(CH₃) → (H₃C)(H₃C)⁺C—CH₂CH₂CH₃ ... shown as (H₃C)₂⁺C—CH₂CH₃ with rearrangement]

(c) [structure: cyclopentyl—⁺CH₂ with H migration → cyclopentyl cation with —CH₃]

7.27
(a) An **electron-withdrawing substituent** (an electron-withdrawing group) is an atom or group of atoms that withdraws electrons from a neighboring center by an inductive or resonance effect. Such a substituent always decreases the electron density on adjacent carbon atoms either by an inductive effect or by a resonance effect or by both. Examples of some electron-withdrawing groups are: —$\overset{+}{N}R_3$, —NO_2, —CN, —SO_3H, —CHO, —COOR, —Cl, etc.

(b) An **electrophile** ("electron seeker") is a species (ion or compound) that can act as a Lewis acid or electron pair acceptor. For example, carbocations are electrophiles.

(c) A **carbocation** is a species that contains a positively charged carbon atom. The planar geometry about the positively charged carbon atom can be described in terms of an sp^2 hybrid carbon atom bonded to three other atoms (or groups). Carbocations are classified as methyl, primary (1°), secondary (2°), or tertiary (3°) according to the number of carbons that are directly attached to the positively charged carbon atom.

(d) An **electrophilic addition reaction** is a reaction in which the species that reacts with a π bond is an electrophile.

(e) A **nucleophile** is an atom or a group of atoms that has an unshared electron pair that can be donated to another atom to form a bond.

(f) A **rate-determining step** in a multistep reaction mechanism is an elementary process that is the slowest step in the mechanism. The rate of the overall reaction cannot occur faster than the rate-determining step.

(g) The **Markownikoff rule** states that an unsymmetrical reagent adds to an unsymmetrical double bond in the direction that places the positive part of the reagent on the carbon of the double bond that has the greater number of hydrogen atoms. In mechanistic terms, an electrophile adds to a π bond to form the more stable carbocation.

(h) An **electron-donating substituent** is an atom or a group of atoms that donates electrons to a neighboring center by an inductive or resonance effect. Such a substituent always increases the electron density on the adjacent carbon atom either by an inductive effect or by a resonance effect or by both. Examples of some electron-donating groups are: —NH$_2$, —OH, —OR, and —R.

(i) An **enzyme-substrate complex** is an association of the starting material with an enzyme. The word *enzyme* means "in yeast." An enzyme fits itself around the substrate (the molecule to be acted upon) to form an enzyme-substrate complex. The bonds of the substrate may be strained by attractions between itself and the enzyme. Strained bonds are of higher energy and are more easily broken. Therefore, the desired reaction proceeds easily and yields an enzyme-product complex.

7.28 (a) 3-Methyl-1-butene (b) 1,1-Dibromo-2-methylpropene
(c) 2,4-Dimethyl-1-pentene (d) 4-Methylcyclopentene
(e) *cis*-4,4-Dimethyl-2-pentene, (f) 5,7-Dimethyl-2,5-nonadiene
(g) (7*R*,3*Z*)-3-Chloro-7-fluoro-4-methyl-3-octene
(h) 3-Bromocyclohexene (i) 3,3-Dichlorocyclopropene

7.29

7.30 (a) The numbering of the carbon chain is incorrect. The chain should be numbered from the end closest to the double bond. In this case, we begin the numbering from right to left because then the double bond starts on C1. The correct name is **1-butene**.

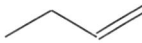
1-Butene

(b) The numbering of the chain carbon atoms is incorrect. The chain should be numbered from the end closest to the double bond. In this case, numbering from both ends gives the double bond number 4 (4-octene). If the double bond is an equal distance away from both ends of the parent chain, begin numbering at the end nearer the first branch point. Again, the numbering should start from the right, so the methyl group will be placed on C4.

Moreover, the name is incomplete since we may have the (E) or (Z) configuration.

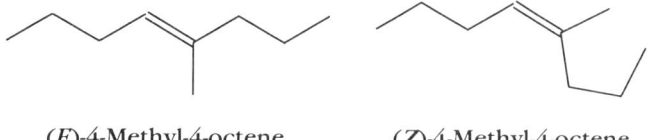

(E)-4-Methyl-4-octene   (Z)-4-Methyl-4-octene

(c) The longest continuous carbon chain was not correctly found. If you continue the chain along the isopropyl substituent, you will find a longer chain (5 carbon atoms). Thus, the parent alkene is 2-pentene. Moreover, the configuration of the double bond is not specified. It could be (E) or (Z):

(Z)-3,4-Dimethyl-2-pentene   (E)-3,4-Dimethyl-2-pentene

7.31 The double bond is assigned a Z (for *zusammen*, the German word for *together*) configuration if the two groups of higher priority at each end of the double bond are on the same side of the molecule. If the two groups of higher priority are on the opposite sides of the double bond, the configuration is denoted by an E (for *entgegen*, the German word for *opposite*).

(a) Chlorine is higher priority than the methyl group; the carboxyl group is higher priority than the methyl group. Groups of higher priority are on opposite sides of the double bond; therefore the compound has the (E)-configuration.

(b) The methyl group is higher priority than hydrogen; the methylene groups (—CH₂—) are identical so we go further down the chain. Further down the chain, chlorine is higher priority than the hydroxyl group. Groups of higher priority are on the same side of the double bond; therefore the compound has the (Z)-configuration.

(c) Fluorine is higher priority than hydrogen atom; the carbonyl groups are identical so we go further down the chain. Further down the chain, chlorine is higher priority than hydrogen. Groups of higher priority are on the same side of the double bond; therefore the compound has the (Z)-configuration.

7.32 (a) 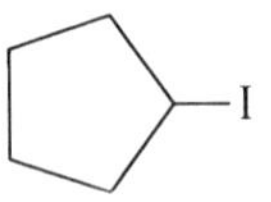 cyclopentyl-I (b) 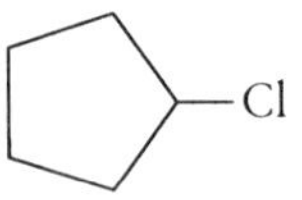 cyclopentyl-Cl (c) 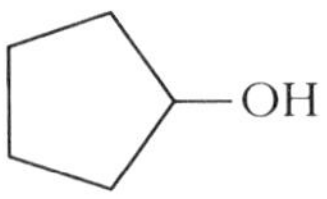 cyclopentyl-OH

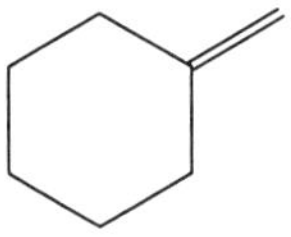

(d) 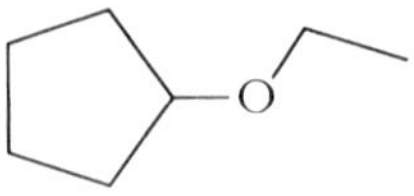 cyclopentyl-O-CH₂CH₃ (e) 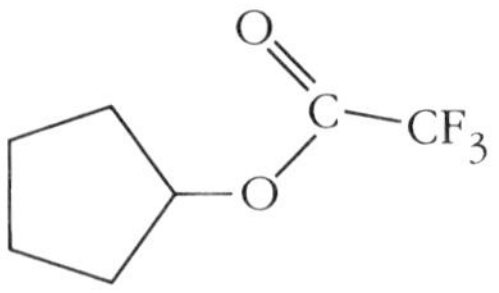 cyclopentyl-O-C(=O)-CF₃

(f) 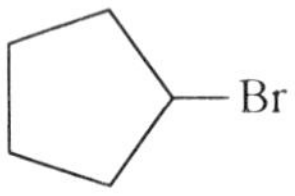 cyclopentyl-Br

7.33 (a) 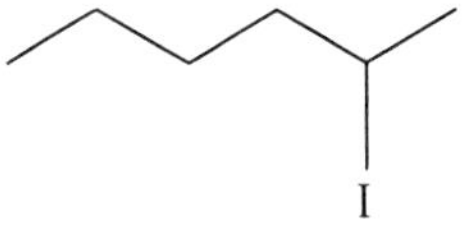 5-carbon chain with methyl and I on C2 (b) 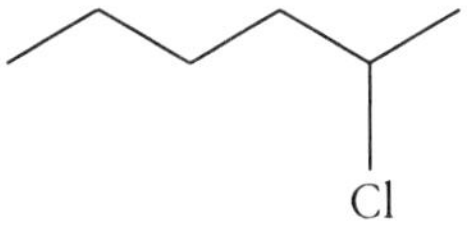 same with Cl

(c) 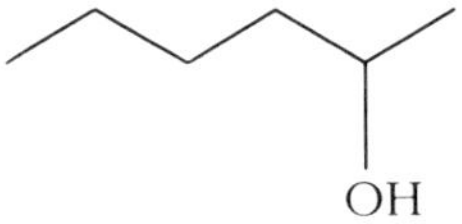 same with OH (d) 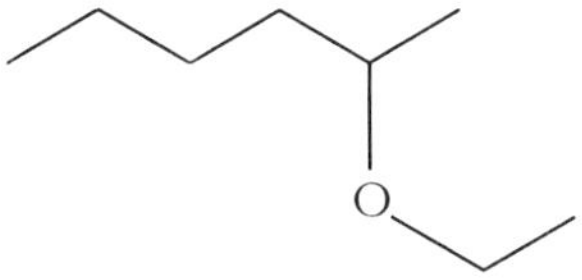 same with -OCH₂CH₃

(e) 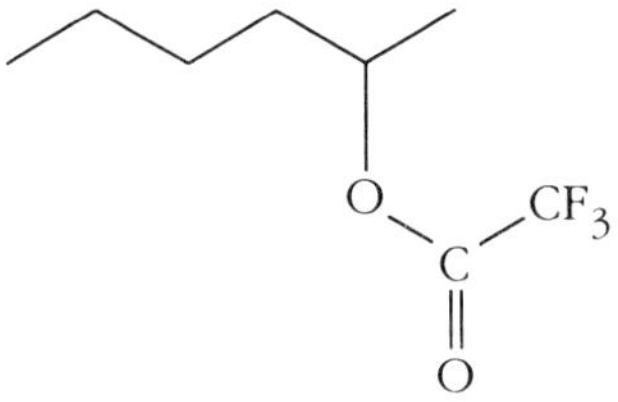 same with -O-C(=O)-CF₃ (f) 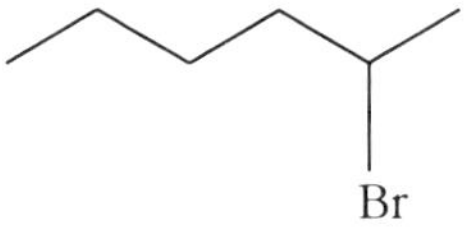 same with Br

7.34 (a) 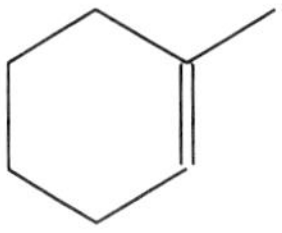 1-methylcyclohexene or methylenecyclohexane (b) CH₃OH, cat. H₂SO₄

(c) 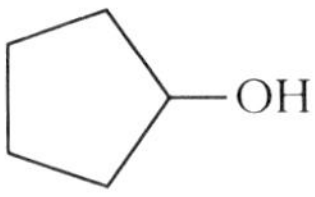 cyclopentanol (d) KI, H₃PO₄

7.35 The *regioselectivity* of addition of HCl to an alkene is predicted by the Markownikoff Rule: *In the addition of protic acids to alkenes, the proton adds to the carbon atom of the double bond that has the greater number of hydrogen atoms.* In mechanistic terms this means that *an electrophile reacts with the double bond of an alkene to form the more stable carbocation.*

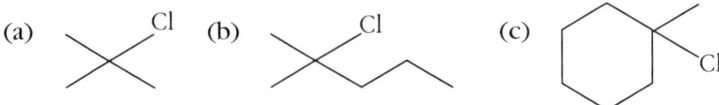

(d) The addition of HCl to 3-methyl-1-butene is accompanied by the rearrangement of the secondary carbocation (a 1,2-hydride shift) to a more stable tertiary carbocation. Thus, two products are observed: 2-chloro-3-methylbutane (40%) and 2-chloro-2-methylbutane (60%).

$$CH_2=CHCH(CH_3)_2 + H-Cl \longrightarrow CH_3-\overset{+}{C}HCH(CH_3)_2 \xrightarrow{Cl^-}$$

$$\underset{\underset{CH_3CHCH(CH_3)_2}{|}}{Cl}$$

Non-rearranged product

$$CH_3-\overset{+}{C}H-\underset{\underset{H}{|}}{C}(CH_3)_2 \longrightarrow CH_3CH_2-\overset{+}{C}(CH_3)_2 \xrightarrow{Cl^-}$$

$$\underset{\underset{CH_3CH_2C(CH_3)_2}{|}}{Cl}$$

(e)   Only one product, 1-bromo-1-chloroethane is observed:

$$H_2C=CHBr + H-Cl \longrightarrow \begin{array}{c} H \\ | \\ H-C-\overset{+}{C} \\ | \diagdown \\ H Br \end{array} \xrightarrow{Cl^-} \begin{array}{cc} H & H \\ | & | \\ H-C-C-Cl \\ | & | \\ H & Br \end{array}$$

$$\updownarrow$$

$$\begin{array}{c} H H \\ | \diagup \\ H-C-C \\ | \diagdown \\ H Br^+ \end{array}$$

(f)   The rearrangement product predominates:

$$CH_2=\underset{\underset{CH_3}{|}}{\overset{\overset{CH_3}{|}}{C}}CCH_2CH_3 + H-Cl \longrightarrow CH_3-\underset{\underset{CH_3}{|}}{\overset{\overset{CH_3}{|}}{\overset{+}{C}}}HCCH_2CH_3 \xrightarrow{Cl^-}$$

$$\underset{\underset{CH_3}{|}}{\overset{Cl CH_3}{\underset{||}{CH_3CHCCH_2CH_3}}}$$

Non-rearranged product

$$CH_3-\overset{+}{\underset{\underset{CH_3}{|}}{C}}HCCH_2CH_3 \quad \longrightarrow \quad CH_3-\underset{\underset{CH_3}{|}}{\overset{\overset{CH_3}{|}}{C}}H\overset{+}{C}CH_2CH_3 \quad \xrightarrow{Cl^-} \quad (CH_3)_2CH\underset{\underset{Cl}{|}}{\overset{\overset{CH_3}{|}}{C}}CH_2CH_3$$

3-Chloro-2,3-dimethylpentane
(Major product)

(g) Only one carbocation is formed and only one product is expected: 2-chloro-2-methylbutane.

7.36 The alkene hydration is an acid-catalyzed reaction and involves formation of a carbocation intermediate. Any substituent that stabilizes the carbocation will facilitate the rate-determining step of the process and, thus, will increase the reaction rate.

(a) A 3° carbocation is more stable than a 2° one so 2-methylpropene undergoes acid-catalyzed hydration faster than propene.

(b) 2-Methylpropene forms a 3° carbocation upon protonation. Chloroethene forms the following carbocation, which is stabilized by resonance but, at the same time, it is destabilized by the inductive, electron-withdrawing chlorine atom.

$$H_2C=CHCl \;+\; H-B \;\longrightarrow\; H-\underset{H}{\overset{H}{\underset{|}{C}}}-\overset{+}{C}\overset{H}{\underset{Cl}{\diagdown}} \;\longleftrightarrow\; H-\underset{H}{\overset{H}{\underset{|}{C}}}-C\overset{H}{\underset{Cl^+}{\diagdown}}$$

Therefore, the relative stabilities of the two carbocations are $CH_3\overset{+}{C}HCl < (CH_3)_3C^+$. As a result, 2-methylpropene undergoes acid-catalyzed hydration faster than chloroethene.

(c) 2-Methyl-1-butene forms a 3° carbocation and 1-pentene forms a 2° carbocation. A 3° carbocation is more stable and, thus, 2-methyl-1-butene reacts faster.

(d) Propene and 3,3,3-trifluoropropene form carbocations. However, there is a strong electron-withdrawing trifluoromethyl group in 3,3,3-trifluoropropene, which destabilizes the carbocation. Thus, propene reacts faster.

More stable   Less stable

(e) 2-Methoxy-2-butene will form a resonance stabilized 3° carbocation, which will be additionally stabilized by the strongly electron-donating methoxy group. 2-Butene will form a 2° carbocation. The carbocation formed by 2-methoxy-2-butene will be much more stable and 2-methoxy-2-butene will react faster.

7.37 The steps in the mechanism of the dehydration of an alcohol are as follows.
Step 1: Protonation of the alcohol.

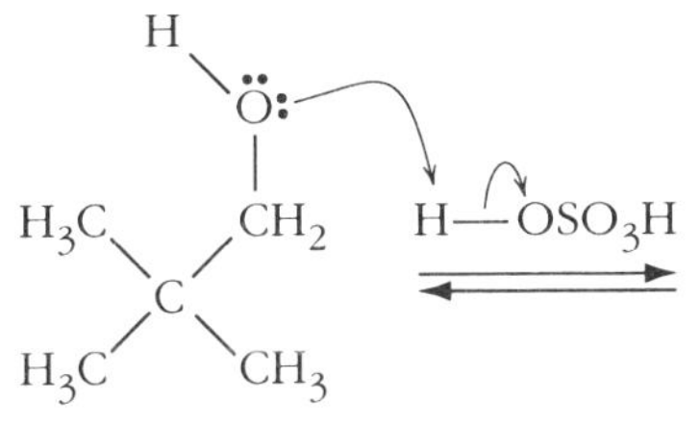

Step 2: Loss of water to form a carbocation.

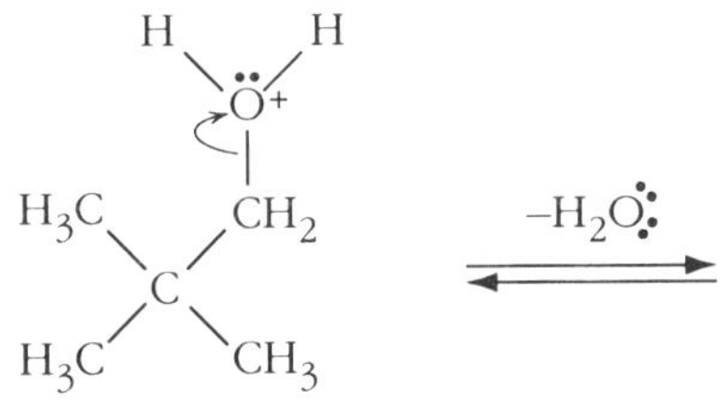

Step 3: Rearrangement of the 1° carbocation to a more stable 3° carbocation.

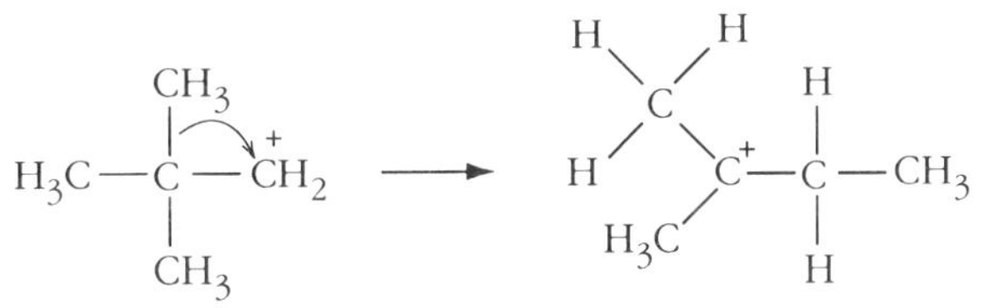

Step 4: Removal of a β-hydrogen.

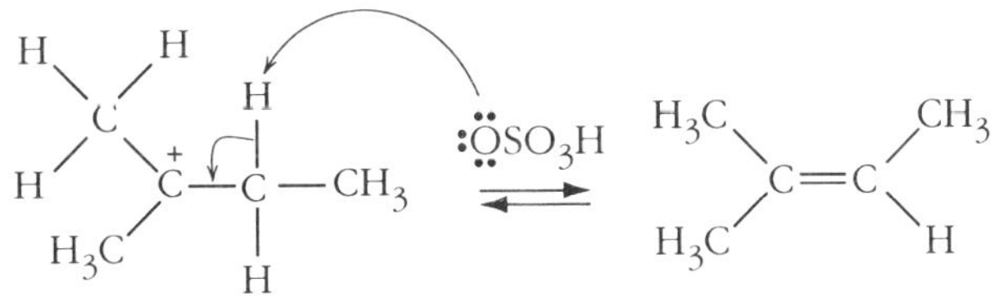

2-Methyl-2-butene

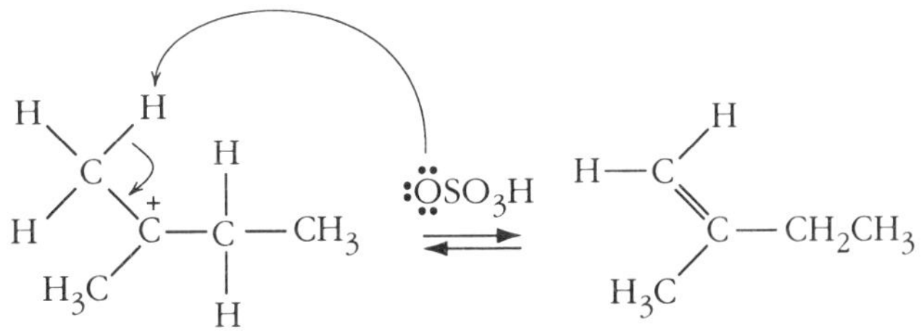

2-Methyl-1-butene

There are two types of β-hydrogens and that is why a mixture of two products is expected. The most stable alkene (the one with the greatest number of branches at the double bond) is formed in greatest amount. Thus, 2-methyl-2-butene is formed in 85% yield and 2-methyl-1-butene in 15% yield.

7.38 Step 1: Protonation of the double bond to form a carbocation.

Step 2: Reaction of the carbocation with acetic acid to form the conjugate acid of the product. Reaction occurs equally well from either side to form a racemic mixture.

Step 3: Removal of a proton to form the product.

7.39 Step 1: Protonation of the alkene to form a carbocation.

Step 2: Reaction of the carbocation with chloride ion. The attack of the chloride ion may take place equally well from either side to form a mixture of stereoisomers (diastereomers).

7.40 (a) Step 1: Protonation of the double bond to form a carbocation.

Step 2: Rearrangement (a 1,2-methylene shift) to form a more stable carbocation.

Step 3: Stabilization of the carbocation by elimination of a proton.

(b) Step 1: Protonation of the double bond to form a carbocation. There are two double bonds in the molecule. The one that forms a 3° carbocation reacts faster.

Step 2: Intramolecular attack of the double bond on the carbocation:

Step 3: Reaction of the carbocation with the formate ion.

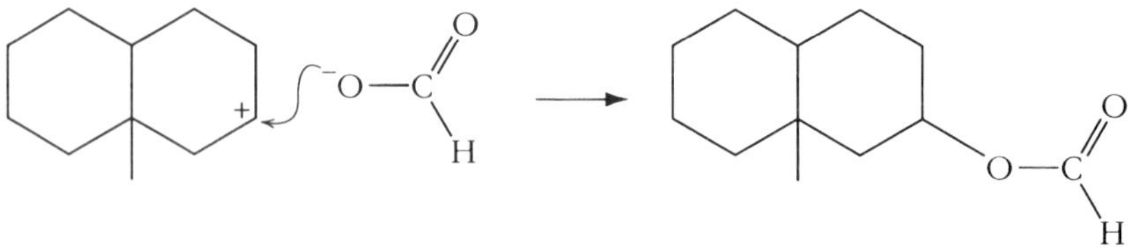

7.41 The IR spectrum shows a strong absorption band at 3500 to 3200 cm^{-1} (—O—H stretching vibrations) so the product is an alcohol. The alcohol is probably **2-methyl-2-pentanol** formed by hydration of 2-methyl-1-pentene by the following mechanism:

Step 1: The reaction begins with the protonation of the carbon-carbon double bond in the direction that leads to the more stable carbocation.

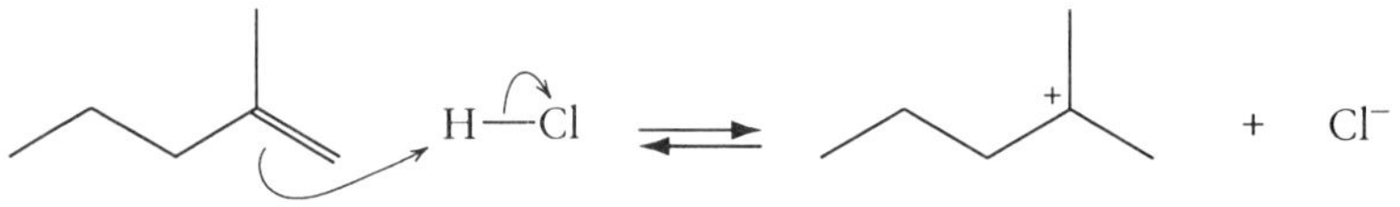

Step 2a: The carbocation can react with a chloride anion to give the expected product, 2-chloro-2-methylpentane.

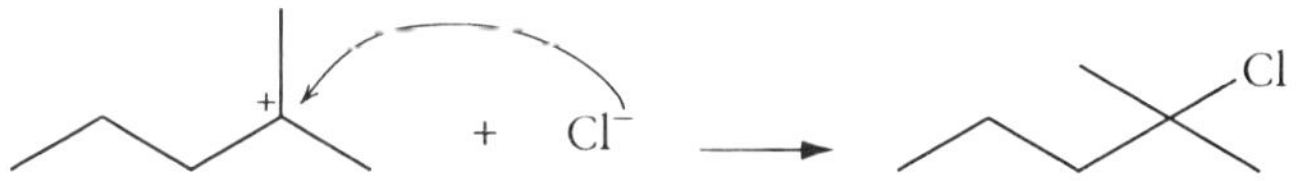

Step 2b: However, the carbocation can also be attacked by a water molecule to give a different product, a protonated alcohol.

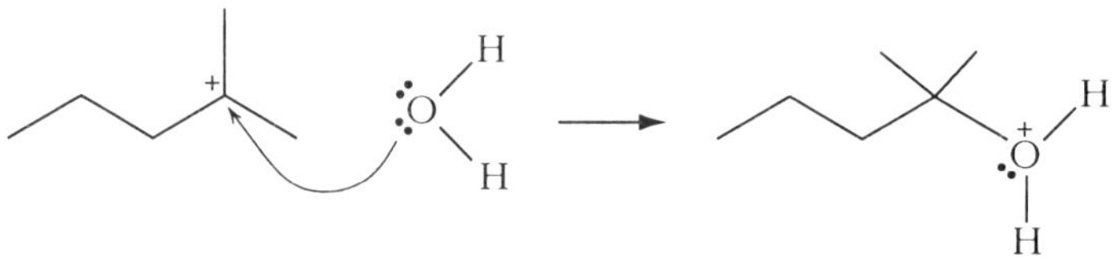

Step 3: Deprotonation gives 2-methyl-2-pentanol as the other product.

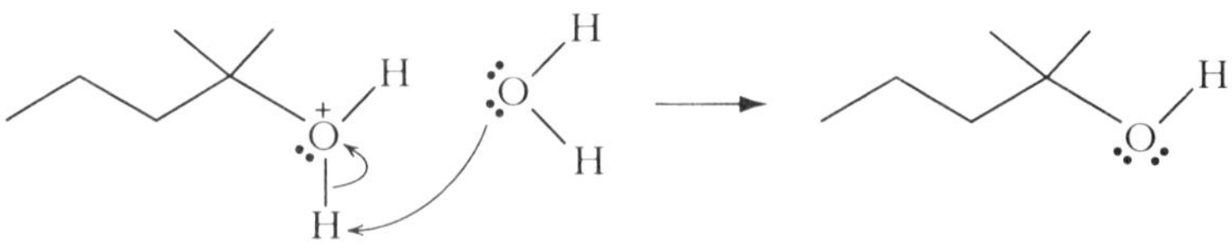

7.42 (a)

Compound	IR important absorption bands	^{13}C NMR approximate chemical shifts, δ:
$CH_3CH{=}CHOCH_2CH_3$	3100-3000 cm^{-1} (sp^2 C—H), 1680-1620 cm^{-1} (C=C), 1200-1000 cm^{-1} (C—O)	143 (=CHO), 133 (R—CH=), 60 (O—CH_2—), 20 (CH_3—sp^2C) 16 (sp^3C—CH_3)
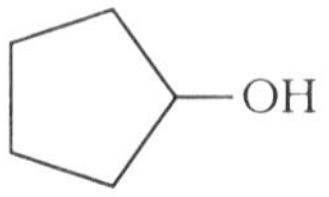	3500-3200 cm^{-1} (O—H)	60 (CH—OH), 32 (C2, C5), 23 (C3 and C4).

The IR spectrum of the alcohol will have a strong absorption at 3500 to 3200 cm^{-1} that is absent in the spectrum of the ether. The ether will have five signals in its ^{13}C NMR spectrum while the ^{13}C NMR spectrum of the alcohol will have only three signals.

(b)

CH$_3$CH$_2$CH$_2$COOH

3200-2500 cm^{-1} (—COO—H), 1700 cm^{-1} (C=O), 1200-1000 cm^{-1} (C—O)

180 (—COOH), 30 (—CH$_2$—, C2) 20 (—CH$_2$—, C3), 16 (—CH$_3$)

CH$_3$CH=CHCH$_2$OH

3500-3200 cm^{-1} (O—H), 3100-3000 cm^{-1} (sp^2 C—H), 1680-1620 cm^{-1} (C=C), 1200-1000 cm^{-1} (C—O)

135 (=CH—, C2), 130 (=CH—, C3), 70 (=CH—CH$_2$—OH, C1), 20 (sp^2C—CH$_3$)

The IR spectrum of the carboxylic acid has a strong and broad absorption at 3200 to 2500 cm^{-1} due to the O—H bond of the carboxyl group and another strong absorption at 1700 cm^{-1} due to the carbonyl group. The IR spectrum of the alcohol shows a strong absorption at 3500 to 3200 cm^{-1} due to the O—H group. The shape of the —COOH absorption band, which is characteristic of a carboxylic acid, should distinguish between these two compounds.

Both compounds have four signals in their ^{13}C NMR spectra but the carboxylic acid has one at δ 180 characteristic of a carbonyl carbon atom while the alcohol has two signals in the alkene region (δ 130-140).

(c)

[structure: H$_2$C=CH—C(=O)—CH$_3$]

3100-3000 cm^{-1} (sp^2C—H), 1700 cm^{-1} (C=O), 1680-1620 cm^{-1} (C=C)

220 (C=O), 143 (=CH—, C3), 118 (=CH$_2$), 30 (—CH$_3$)

H$_2$C=CHCH$_2$C(=O)OH

3200-2500 cm^{-1} (—COO—H), 3100-3000 cm^{-1} (sp^2—C—H)(hidden), 1700 cm^{-1} (C=O), 1680-1620 cm^{-1} (C=C)

180 (—COOH, C1), 130 (=CH—, C3), 118 (=CH$_2$, C4), 30 (—CH$_2$—, C2)

IR: The characteristic strong absorption at 3200 to 2500 cm^{-1} of the carboxylic acid group serves to distinguish between these compounds.

^{13}C NMR: Both compounds show four signals; the only difference is the chemical shift of the carbonyl carbon, which is around δ 180 in a carboxylic acid and above δ 210 in a ketone.

7.43 (a) The IR spectrum is not very informative because both compounds are expected to show sp^2 C—H stretching vibrations at 3100 to 3000 cm^{-1} and C—C stretching vibrations at 1650 cm^{-1}.

The ^{13}C NMR is decisive, however. Cycloheptene shows one signal at δ 130 for two sp^2-hybridized carbon atoms (equivalent) and three signals in the alkane region (δ 20-40), 1,6-heptadiene should show two signals in the alkene region (δ 120-140) and two signals in the alkane region (δ 20-40). The ^{13}C NMR spectrum shown is consistent with the expected spectrum for **cycloheptene.**

(b) In this case, the ^{13}C NMR spectrum is not very informative because both compounds are expected to have three signals (two in the alkene region and one around δ 60).

The IR spectrum is decisive, however. An ether should not show a strong —O—H stretching absorption band at 3500 to 3200 cm^{-1} expected of an alcohol. The spectrum shown does have a strong absorption at 3500 to 3200 cm^{-1}. The unknown is **2-propen-1-ol:**

$$CH_2=CHCH_2OH$$

7.44 1-Bromo-1-propene exists as a pair of stereoisomers: (E)- and (Z)-1-bromo-1-propene. Because the magnetic environments of all carbon atoms in both diastereomers are different, we observe different chemical shifts and, as a result, six peaks in total, three from each isomer.

7.45 (a) Molecular formula: $C_2H_2Cl_2$ (one unit of unsaturation). Possible structures:

The IR indicates a carbon-carbon double bond absorption (unsymmetrical alkene); the ^{13}C NMR shows two different alkene carbon atoms, one of which is characteristic of the terminal —CH_2 (δ 115). These data are consistent with the first structure: **1,1-dichloroethene.**

(b) Molecular formula: $C_4H_6Cl_2$ (one unit of unsaturation). There are many possible structures. The IR spectrum shows a weak absorption of carbon–carbon double bond (alkene). The absorption is weak, so the molecule is symmetrical (it lacks a dipole moment). The symmetrical structure is confirmed by its ^{13}C NMR spectrum because only two signals are observed: one signal of the alkene carbon at δ 130 (monosubstituted) and one signal in the alkane region at δ 38.9 (R—CH_2—Cl). Thus, the unknown is *cis*- or *trans*-**1,4-dichloro-2-butene:**

trans-1,4-Dichloro-2-butene   *cis*-1,4-Dichloro-2-butene

(c) Molecular formula: C_6H_{10} (two units of unsaturation). There are many possible structures. The IR spectrum shows an absorption of a carbon–carbon double bond (alkene). The ^{13}C NMR spectrum shows two signals for the alkene carbon atoms at δ 143.4 (two substituents on the carbon atom, —CR_2), and at δ 113 (no alkyl substituents, terminal —CH_2) and one signal in the alkane region, at δ 20.6 (—CH_3). Thus, the unknown is **2,3-dimethyl-1,3-butadiene:**

(d) Molecular formula: C_3H_3N (three units of unsaturation). The IR shows a very characteristic absorption band at 2220 cm^{-1} due to the C≡N of a nitrile. This information is confirmed by the ^{13}C NMR spectrum: δ 137 (carbon of the nitrile functional group). Furthermore, the IR spectrum shows a carbon-carbon double bond (1605 cm^{-1}). This is also confirmed by the ^{13}C NMR because there are two peaks at δ 117 and 107. Thus, the unknown is (IUPAC name: **propenenitrile**; trivial name: acrylonitrile):

$$H_2C=CH-C\equiv N$$

CHAPTER 8

MORE ADDITION REACTIONS OF ALKENES
Concepts

In addition reactions of carbon-carbon double bonds, the π bond is broken and two new σ bonds are formed. The formation of the two sigma bonds on the same side or face of the double bond is called **syn addition.** Hydroboration, catalytic hydrogenation, hydroxylation, and epoxidation of alkenes occur by syn addition.

The formation of the σ bonds on the opposite side or face of the double bond is called **anti addition.** Chlorination and bromination of aliphatic alkenes occur by anti addition. When addition of a reagent to either face or side of the plane of a double bond is equally probable, the product formed is a racemic mixture or an achiral compound. When addition to one face is preferred, the product is an unequal mixture of stereoisomers.

Electrophilic addition is the most important reaction of alkenes. HCl, HBr, and HI add to carbon-carbon double bonds by a two-step mechanism involving initial reaction of the nucleophilic double bond with H$^+$ to form a carbocation intermediate, followed by attack of halide ion nucleophile on the carbocation intermediate. Bromine and chlorine add to alkenes via three-membered-ring **halonium ion** intermediates to give addition products having anti stereochemistry. If water is present during halogen addition reactions, a **halohydrin** is formed.

Addition to the alkene double bond occurs by a variety of mechanisms:

1. Mechanisms involving formation of carbocation intermediates (addition of hydrogen halides, hydration).
2. Mechanisms involving formation of cyclic ion intermediates (oxymercuration, halogenation).
3. Concerted mechanisms (hydroboration, formation of 1,2-diols, ozonolysis).
4. Mechanisms involving free-radical intermediates (free-radical addition of HBr).

Some useful transformations of alkenes involve additions followed by other transformations. These include oxymercuration-reduction and hydroboration-oxidation (alcohols are formed in both cases); ozonolysis followed by treatment with a reducing agent [(CH$_3$)$_2$S or Zn/H$_2$O] to give aldehydes and/or ketones; ozonolysis followed by treatment with an oxidizing agent (H$_2$O$_2$) to give carboxylic acids and/or ketones; addition of OsO$_4$ or KMnO$_4$ followed by hydrolysis to give 1,2-diols.

Solutions to the Exercises

8.1 The following reactions exemplify the Markownikoff addition to an alkene.

(a) The hydration reaction: the alkene reacts with hydronium ion to give a tertiary carbocation. The nucleophiles present in the reaction mixture are water and hydrogen sulphate ion. Some of the carbocations combine with water to form 1,1-dimethylethanol (*tert*-butyl alcohol); some combine with hydrogen sulphate to form an ester. The sulphate ester is decomposed by heating with dilute aqueous acid, and the alcohol is isolated as the product.

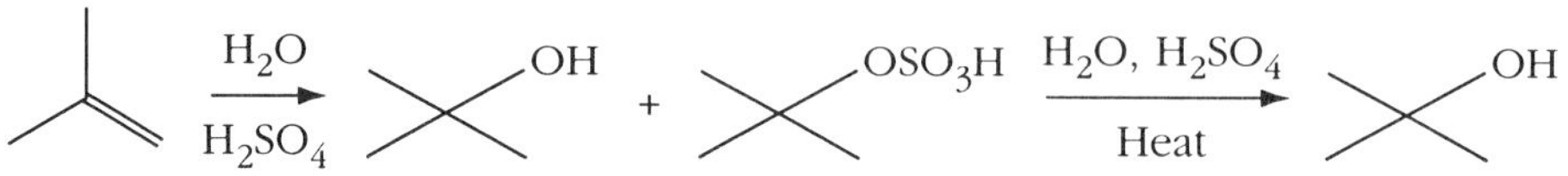

Hydroboration is the addition of borane (BH_3) to an alkene. Because boron is the electrophilic part of borane, it adds mainly to the *less substituted* carbon atom of the double bond. Better control of the regiochemistry is observed when 9-BBN is used because 9-BBN is bulkier than BH_3.

(b)

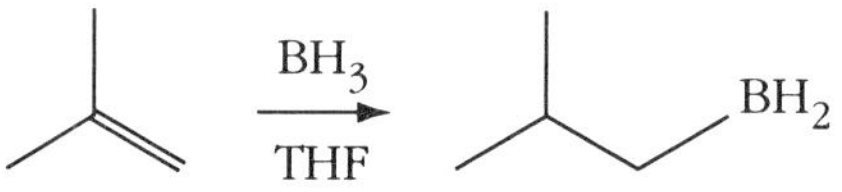

(c)

8.2 The bulkiness of substituents on the boron atom gives the reagent high selectivity in its reactions with substituted alkenes:

(a)

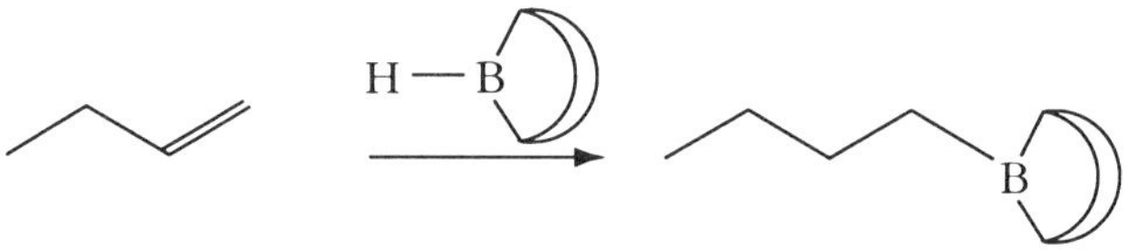

(b)

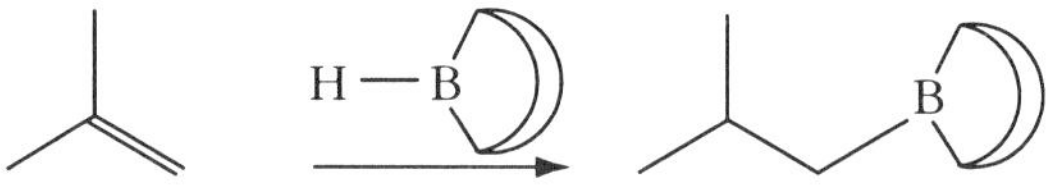

(c)

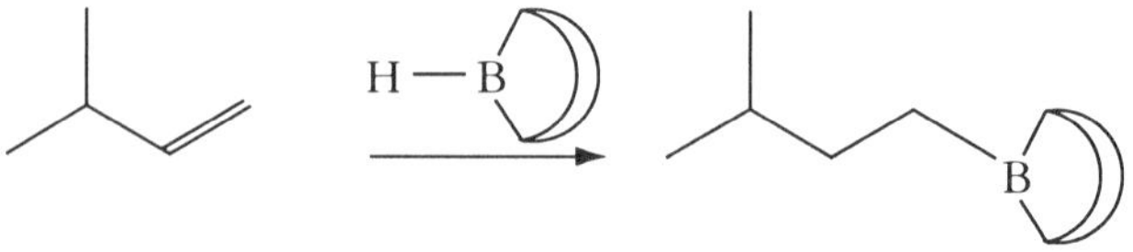

8.3 In the hydroboration reaction, syn addition of boron and hydrogen to the double bond occurs. A syn addition is an addition reaction in which the incoming groups are added to the same side of the molecule.

8.4 Organoboranes undergo reactions in which the boron atom is replaced by other atoms or functional groups. *Protonolysis* is the reaction of an organoborane with propanoic acid in which the boron atom is replaced by a hydrogen atom to form an alkane (Exercise 8.4(a)) or by deuterium to form a deuteroalkane (Exercise 8.4(d)). *Bromination* of organoboranes replaces the boron atom by a bromine atom to form a bromoalkane (Exercise 8.4(c)). *Oxidation* of organoboranes with an alkaline solution of hydrogen peroxide replaces the boron atom by a hydroxyl group and forms alcohols (Exercise 8.4(b)).

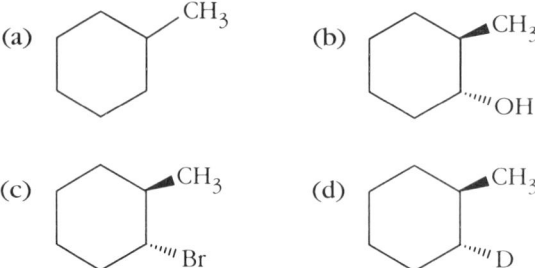

8.5 The reaction of osmium tetraoxide with an alkene leads to addition of two hydroxyl groups to the double bond and is referred to as *hydroxylation*. Both oxygens of the diol come from osmium tetraoxide via a cyclic osmate ester intermediate. The reaction of OsO$_4$ with the alkene is a syn addition, and the conversion of the cyclic osmate to the diol involves cleavage of the bonds between oxygen and osmium. Thus, both hydroxyl groups of the diol become attached to the same face of the double bond so syn hydroxylation of the alkene is observed.

(a) [structure: 1-heptene + OsO$_4$/H$_2$O$_2$/H$_2$O → two enantiomeric diols]

(b) [structure: cycloheptane-1,2-diol, both OH wedge]

(c) [structure: CH$_3$CH(OH)CH(OH)CH$_2$CH$_3$ with H's shown]

8.6 Hydroxylation of alkenes with OsO$_4$/H$_2$O$_2$/H$_2$O is a syn addition of two hydroxyl groups to the carbon atoms of the double bond. The addition may occur from either side of the plane. Thus, the hydroxylation of (*E*)-2-butene gives a mixture of two enantiomers (2*S*,3*S*)-2,3-butanediol and (2*R*,3*R*)-2,3-butanediol (*racemic* mixture). The hydroxylation of (*Z*)-2-butene gives (2*R*,3*S*)-2,3-butanediol, which is not optically active because it is a *meso* compound.

[reaction scheme: (*E*)-2-Butene + OsO$_4$/H$_2$O$_2$/H$_2$O → (2*S*,3*S*)-2,3-Butanediol + (2*R*,3*R*)-2,3-Butanediol; A pair of enantiomers (racemic mixture)]

[reaction scheme: (*Z*)-2-Butene + OsO$_4$/H$_2$O$_2$/H$_2$O → (2*R*,3*S*)-2,3-Butanediol + (2*S*,3*R*)-2,3-Butanediol; Identical (a *meso* compound)]

8.7 The structure of the alkene is obtained by removing both —OH groups and placing a double bond between the two carbon atoms. Then, we write the reaction of the alkene with OsO_4 just to confirm that the desired diol is obtained.

(a) [alkene] $\xrightarrow{OsO_4, H_2O_2/H_2O}$ [diol product 1] + [diol product 2]

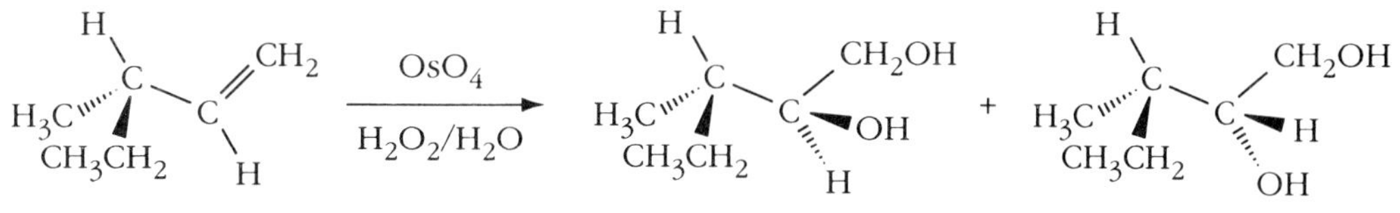

A pair of diastereomers

(b) [alkene] $\xrightarrow{OsO_4, H_2O_2/H_2O}$ [diol product 1] + [diol product 2]

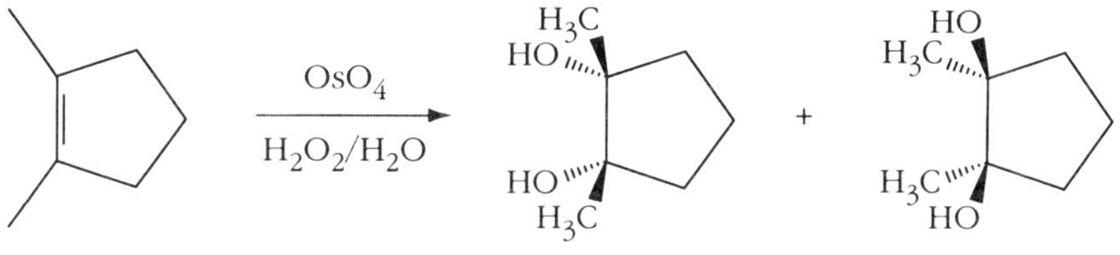

Identical (a *meso* compound)

8.8 The epoxidation is characterized by syn addition of an oxygen to the double bond. Substituents that are *cis* to each other in the alkene are also *cis* in the oxirane; those that are *trans* in the alkene are also *trans* in the oxirane.

(a)

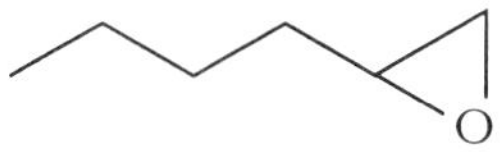

(b)

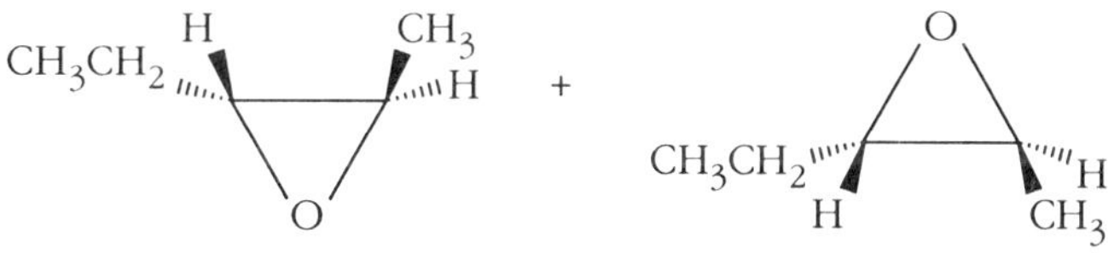

(c) 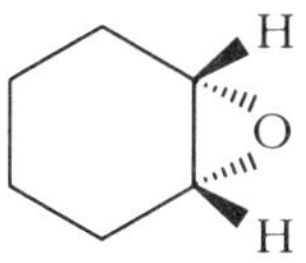   (d)

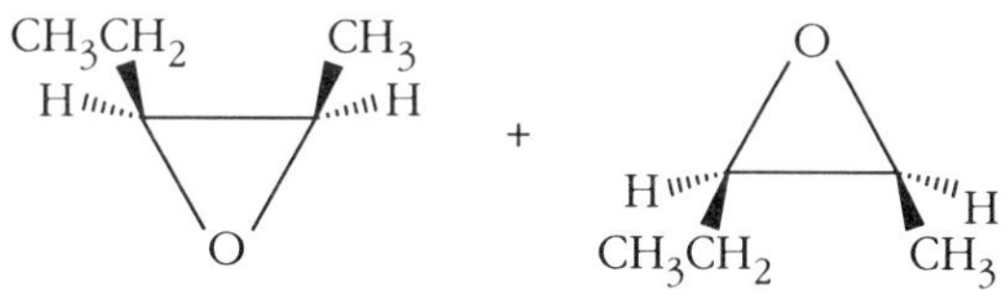

In Exercise 8.8 (e) we may epoxidize selectively one of two differently substituted double bonds. The more highly substituted double bond will react more quickly with one equivalent of peroxycarboxylic acid.

(e)

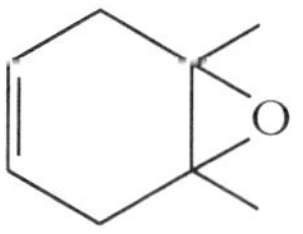

8.9 The reaction of chloroform with potassium *t*-butoxide in *t*-butanol produces dichlorocarbene, which adds to the double bond to form a dichlorocyclopropane derivative. The addition of dichlorocarbene is stereospecific, which means that only a single stereoisomer is formed as product. Thus, starting from the *trans* alkene, only *trans*-disubstituted cyclopropane is produced.

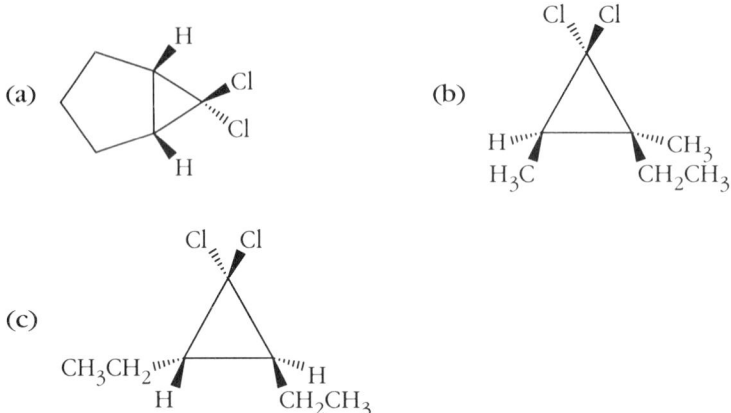

8.10 The treatment of metallic zinc with a salt of copper gives a zinc-copper couple that reacts with diiodomethane to give the active species usually formulated as CH_2ZnI_2. The exact structure of the reagent is not known, but it adds a methylene group to an alkene to form a cyclopropane ring. The addition is stereospecific, which means that only a single stereoisomer is formed as product. Thus, starting from the *cis* alkene, only *cis*-disubstituted cyclopropane is produced.

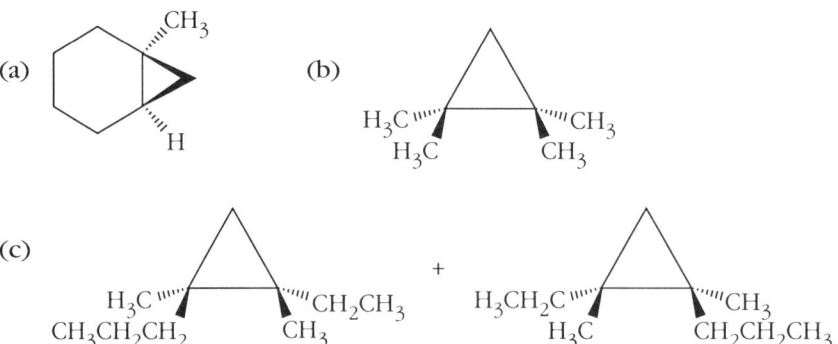

8.11 The addition of either chlorine or bromine to an alkene is stereospecific and occurs by anti addition.

(a) A pair of enantiomers is formed:

(b) One product is observed:

$(CH_3)_2C=CH_2 + Br_2 \longrightarrow BrCH_2C(CH_3)_2Br$

$(CH_3)_2C=CH_2 + Cl_2 \longrightarrow ClCH_2C(CH_3)_2Cl$

(c) A pair of enantiomers is formed:

cyclohexene + Br$_2$ ⟶ trans-1,2-dibromocyclohexane (pair of enantiomers)

cyclohexene + Cl$_2$ ⟶ trans-1,2-dichlorocyclohexane (pair of enantiomers)

(d) A pair of enantiomers is formed:

1-methylcyclohexene + Br$_2$ ⟶ trans-1,2-dibromo-1-methylcyclohexane (pair of enantiomers)

1-methylcyclohexene + Cl$_2$ ⟶ trans-1,2-dichloro-1-methylcyclohexane (pair of enantiomers)

8.12 Step 1: Formation of chloronium ion

$H_2C=CH_2 + Cl-Cl \longrightarrow$ chloronium ion + Cl^-

Step 2: Opening of the chloronium ion

chloronium ion + $Cl^- \longrightarrow ClCH_2CH_2Cl$

8.13 Step 1: Formation of chloronium ion

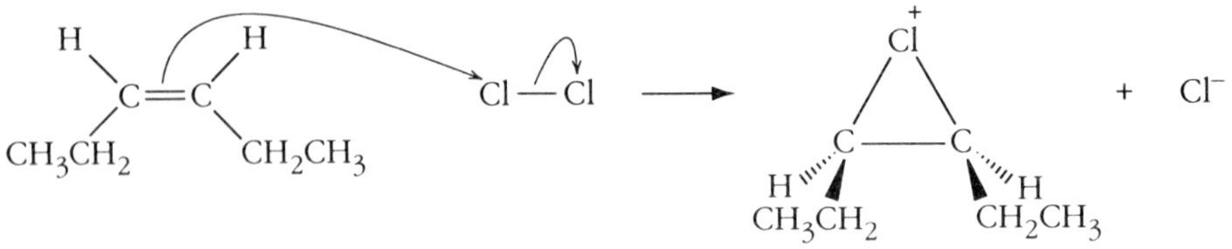

Step 2: Opening of the chloronium ion

(3R,4R)-3,4-Dichlorohexane

(3S,4S)-3,4-Dichlorohexane

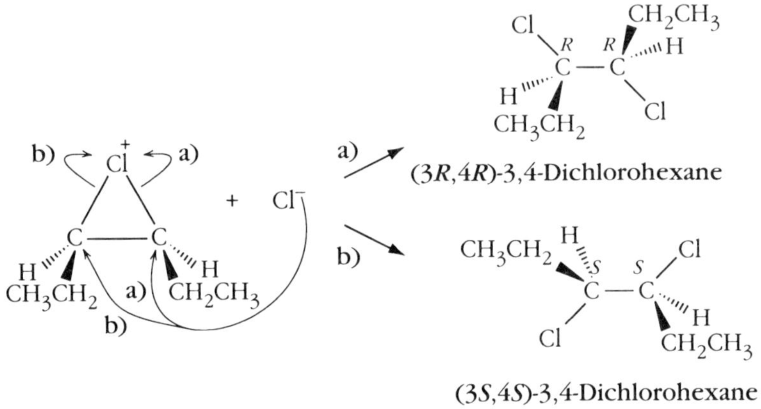

The two products are (3R,4R)-3,4-dichlorohexane and (3S,4S)-3,4-dichlorohexane. These two products are enantiomers and they are formed in equal amounts because the nucleophile can attack either carbon atom of the chloronium ion. As a result, a racemic mixture is formed.

8.14 Mechanism for the reaction of chlorine with (E)-3-hexene:

Step 1: Formation of the chloronium ion

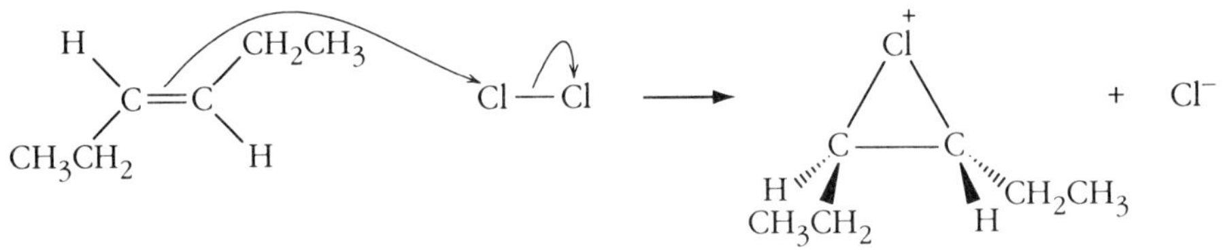

Step 2: Opening of the chloronium ion

(3R,4S)-3,4-Dichlorohexane

(3S,4R)-3,4-Dichlorohexane

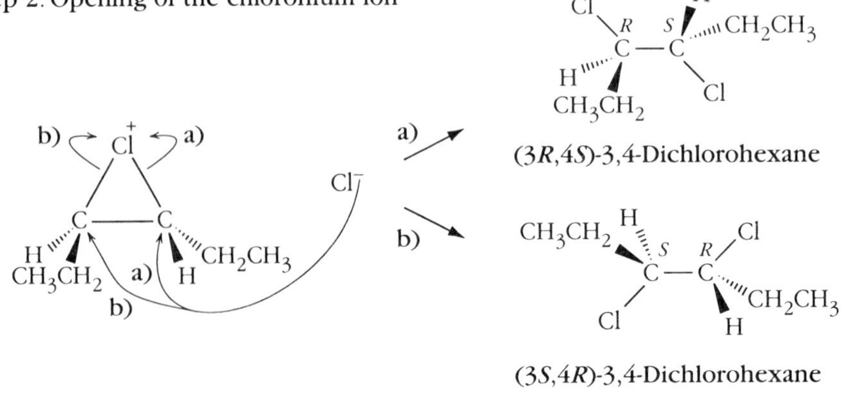

The two structures are identical. The product is *meso*-3,4-dichlorohexane.

8.15 When addition of a halogen to an alkene is carried out in water as solvent, halohydrins are formed. Halogenation of an unsymmetrical alkene is a regioselective reaction; that is, the halogen atom is always added to the less substituted carbon atom of the original double bond.

(a) [cyclohexene] + Cl_2/H_2O → trans-2-chlorocyclohexanol (two enantiomers shown)

Racemic mixture

(b) [2-methylpropene] + Cl_2/H_2O → 1-chloro-2-methyl-2-propanol (OH on more substituted C, Cl on less substituted C)

(c) [1-methylcyclohexene] + Cl_2/H_2O → 1-methyl-2-chloro-cyclohexanol (two enantiomers shown with Cl on less substituted carbon and OH on more substituted carbon bearing CH₃)

Racemic mixture

8.16 Halogenation of an unsymmetrical alkene in the presence of alcohol is a regioselective reaction; that is, the halogen atom is bonded to the less substituted carbon atom. The intermediate chloronium ion is unsymmetrical with considerable cationic character at the more highly substituted carbon atom and that is why the product obtained has the methoxy group on the carbon atom that could support more positive charge in the intermediate.

Step 1: Formation of the chloronium ion

Step 2: Attack of a nucleophile takes place at the more substituted carbon atom

Step 3: Deprotonation of the protonated ether

8.17 The oxymercuration-demercuration (also known as oxymercuration-reduction) sequence produces alcohols with a regioselectivity identical to that of acid-catalyzed hydration. Hydrogen is introduced at the carbon that has the greater number of hydrogen substituents, and hydroxyl is introduced at the carbon that has the fewer number of hydrogens, so the addition of water to the double bond follows the Markownikoff rule. Rearrangements of the carbon skeleton do not occur.

(a) Reaction 1:

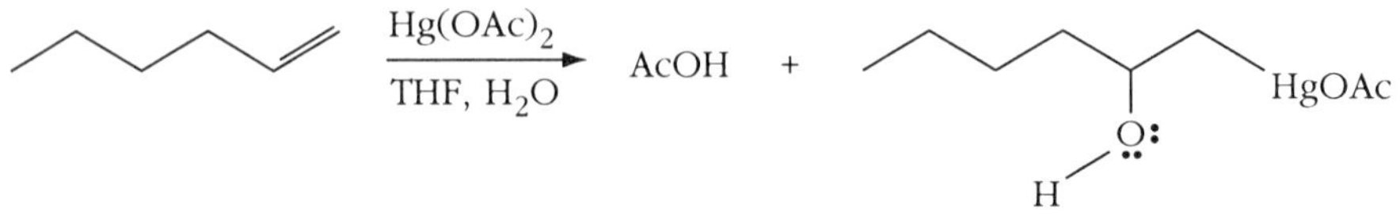

[structure: 1-hexene] $\xrightarrow{\text{Hg(OAc)}_2, \text{THF, H}_2\text{O}}$ AcOH + [structure: 2-(acetoxymercurio)hexan-2-ol with OH and HgOAc]

Reaction 2:

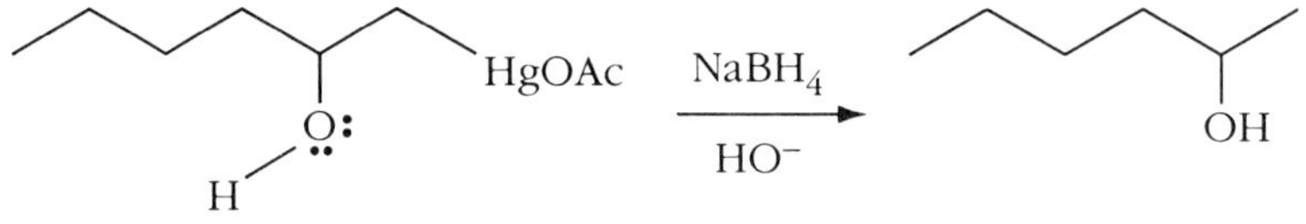

[structure from Reaction 1] $\xrightarrow[\text{HO}^-]{\text{HgOAc, NaBH}_4}$ [structure: 2-hexanol]

(b) Reaction 1:

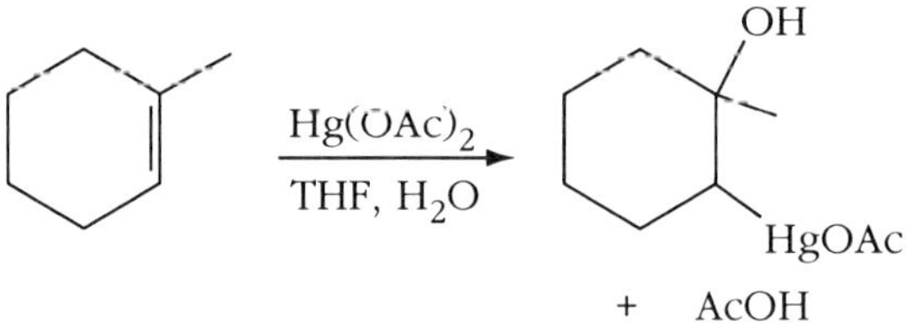

[1-methylcyclohexene] $\xrightarrow{\text{Hg(OAc)}_2, \text{THF, H}_2\text{O}}$ [1-methyl-2-(acetoxymercurio)cyclohexan-1-ol] + AcOH

Reaction 2:

[structure from (b) Reaction 1] $\xrightarrow[\text{HO}^-]{\text{NaBH}_4}$ [1-methylcyclohexanol]

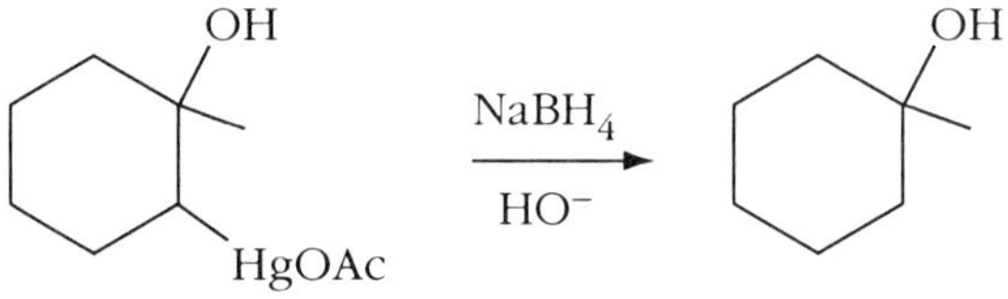

(c) The overall reaction:

[2-methyl-1-pentene] $\xrightarrow[\text{(2) NaBH}_4, \text{HO}^-]{\text{(1) Hg(OAc)}_2, \text{THF, H}_2\text{O}}$ [2-methyl-2-pentanol]

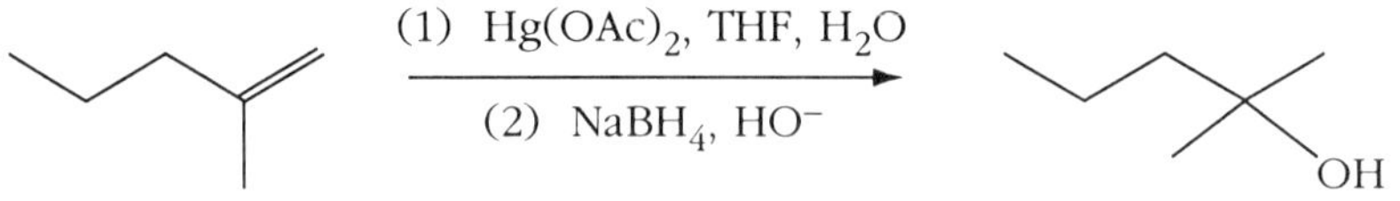

8.18 The alcohol is formed by addition of a hydrogen to one carbon and a hydroxyl to the other carbon of a double bond. In order to find a suitable alkene we must do the opposite operation. We detach the hydroxyl group and a hydrogen from an adjacent carbon atom. If the alcohol is symmetrical, it does not matter from which adjacent carbon atom we remove the hydrogen. For example, in Exercise 8.18 (a) cyclohexene is the alkene. However, we can get two different alkenes in Exercise 8.18 (c): 1-pentene or 2-pentene. We choose the alkene that gives the highest yield of the desired alcohol. 2-Pentene would give two alcohols, 2-pentanol and 3-pentanol, on mercuration-demercuration. 1-Pentene, on the other hand, gives exclusively 2-pentanol. As a result, we choose 1-pentene as the starting alkene.

(a) cyclohexene $\xrightarrow[(2)\ NaBH_4,\ HO^-]{(1)\ Hg(OAc)_2,\ THF,\ H_2O}$ cyclohexanol

(b) 2,3-dimethyl-2-butene $\xrightarrow[(2)\ NaBH_4,\ HO^-]{(1)\ Hg(OAc)_2,\ THF,\ H_2O}$ 2,3-dimethyl-2-butanol

(c) 1-pentene $\xrightarrow[(2)\ NaBH_4,\ HO^-]{(1)\ Hg(OAc)_2,\ THF,\ H_2O}$ 2-pentanol

8.19 The acid-catalyzed hydration of 3,3-dimethyl-1-butene is accompanied by a rearrangement of the carbon skeleton according to the following mechanism:

Step 1: Protonation of the carbon-carbon double bond in the direction that leads to the more stable carbocation

Step 2: Rearrangement of the carbocation. The driving force of this rearrangement is the higher stability of a tertiary carbocation than a secondary carbocation.

Step 3: Nucleophilic attack of water

Step 4: Deprotonation of the oxonium ion

Thus, 2,3-dimethyl-2-butanol is the major product.

There is no rearrangement in the oxymercuration-reduction sequence because there is no carbocation intermediate. The expected 3,3-dimethyl-2-butanol is the product.

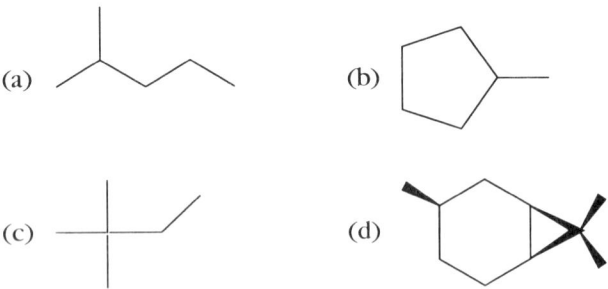

8.20 Metal-catalyzed addition of hydrogen to an alkene is an example of syn addition. Both hydrogen atoms are added to the same face of the double bond.

A second stereochemical aspect of alkene hydrogenation concerns its stereoselectivity. A reaction in which a single starting material can give two or more stereoisomeric products but yields one of them is said to be *stereoselective*. Exercise 8.20 (d) illustrates this principle. The bottom face of the double bond is more exposed, and hydrogen is transferred from the catalyst to that face.

(a) (b)

(c) (d)

8.21 Catalytic hydrogenation of alkenes occurs by syn addition. Thus, a symmetrical alkene such as (Z)-4,5-dimethyl-4-octene gives only one product, (4S,5R)-4,5-dimethyloctane. The (E)-isomer, however, gives two enantiomers, each of which results from the addition of hydrogen to a different side of the alkene.

8.22 If we have heats of hydrogenation of isomeric alkenes and if they give the same hydrogenation product, then the heats of hydrogenation reflect the differences in stabilities of the alkenes. Thus, the alkene with the highest amount of heat evolved during its hydrogenation is the least stable and the one with the least heat evolved is the most stable. The order of the increasing stability is E < A < B < D < C.

8.23 (a)

(b) 3-Methyl-1-butene is the least stable because it is a terminal alkene with only one alkyl group on carbon atoms of the double bond. 2-Methyl-1-butene has two alkyl groups on carbon atoms of the double bond and it is more stable. The most stable of the three alkenes is 2-methyl-2-butene with three alkyl groups on carbon atoms of the double bond.

8.24 (a) **Anti addition** is an addition reaction in which the two portions of the adding reagent add to opposite faces (sides) of the multiple bond.

(b) **Acid-catalyzed hydration** is an acid-catalyzed addition of the elements of water (H, OH) to a multiple bond.

(c) **Hydroboration** is an unsymmetrical addition of hydrogen and boron or a derivative of borane to a multiple carbon-carbon bond. The reaction is a syn addition and shows high regioselectivity. Hydrogen adds to the more substituted carbon atom and the boron-containing group adds to the less highly substituted carbon. The intermediate in the reaction of borane (BH_3) with an alkene, R—BH_2, reacts with other alkene molecules until all hydrogen atoms have been replaced by alkyl groups.

(d) **Carbene addition** is an addition of the carbene (a reactive neutral molecule in which one of the carbon atoms has two single bonds and two nonbonding electrons around its nucleus) to an alkene to form a cyclopropane ring. The reaction is stereospecific. A carbene, such as dibromomethylene (Br_2C:), formed by the reaction of $CHBr_3$ and $(CH_3)_3CO^-K^+$, reacts with (Z)-2-butene to give cis-1,1-dibromo-2,3-dimethylcyclopropane and with (E)-2-butene to give the trans isomer.

(Z)-2-Butene + $CHBr_3$ + $(CH_3)_3CO^-K^+$ ⟶ cis-1,1-Dibromo-2,3-dimethylcyclopropane + $(CH_3)_3COH$ + KBr

(E)-2-Butene + $CHBr_3$ + $(CH_3)_3CO^-K^+$ ⟶ trans-1,1-Dibromo-2,3-dimethylcyclopropane + $(CH_3)_3COH$ + KBr

(e) **Hydroxylation** is the syn addition of two hydroxyl groups (—OH) to an alkene to form a vicinal diol (1,2-dialcohol also called *glycol*). Because oxygen is added to the alkene during the reaction, we call this an oxidation. Common reagents are potassium permanganate ($KMnO_4$) and osmium tetraoxide (OsO_4).

Cyclohexene   A cyclic manganate intermediate   cis-1,2-Cyclohexanediol

- (f) **Stereospecific addition** is an addition reaction in which one of the possible stereoisomeric products predominates.
- (g) **Syn addition** is an addition reaction in which the two portions of the adding reagent add to the same side (or face) of the multiple bond.
- (h) **Epoxidation** is a conversion of an alkene to an epoxide (three-membered cyclic ether known also as an *oxirane*) by treatment with a peroxyacid (e.g., *m*-chloroperbenzoic acid).
- (i) **Halogenation** is a general term given to the reaction of a halogen with an organic compound. There are two types of halogenation: (i) a replacement of a hydrogen atom by a halogen atom and (ii) an addition of Cl_2 and/or Br_2 to a multiple bond (in alkenes or alkynes).
- (j) **Carbenoid** ("like a carbene"), also known as the *Simmons-Smith reagent,* is the active species $CH_2(ZnI_2)$, which is obtained from zinc-copper alloy and diiodomethane. The carbenoid reacts with an alkene by syn addition to yield a cyclopropane derivative.
- (k) **Catalytic hydrogenation** is the syn addition of hydrogen gas to an alkene or alkyne. The addition is catalyzed by a metal such as Pt, Pd, Ni, or Rh.
- (l) **Oxymercuration-demercuration** is a synthetic method for converting alkenes to alcohols. An alkene is treated with a solution of mercury(II) salt (usually acetate) in aqueous THF (tetrahydrofuran), followed by reduction of the adduct with sodium borohydride. Hydration of the double bond occurs with a regiochemistry consistent with Markownikoff's rule. Unlike the reaction of an alkene and H_3O^+, there is *no* rearrangement.
- (m) A **halohydrin** is a compound that contains both a halogen and a hydroxyl group. The term is most often used for compounds in which the halogen and the hydroxyl group are on adjacent sp^3 hybrid carbon atoms (vicinal halohydrins). The most commonly encountered halohydrins are chlorohydrins (obtained when chlorine is added to an alkene in the presence of water) and bromohydrins (when bromine is added to an alkene in the presence of water).
- (n) A **vicinal diol** (trivial name: "vicinal glycol") is a compound containing two hydroxyl groups (—OH) on adjacent sp^3 hybrid carbon atoms.
- (o) **Hydroboration-oxidation** is a two-step reaction sequence used to convert alkenes to alcohols. The first stage is an addition of a hydroborating agent to the alkene to give an organoborane. The organoborane is oxidized with hydrogen peroxide to give an alcohol in the second stage. The hydroboration-oxidation is a syn hydration of an alkene, which gives a product with anti-Markownikoff orientation.
- (p) The **heat of hydrogenation** is the amount of heat evolved during hydrogenation of a substance. It is the change in enthalpy, ΔH, for that reaction.
- (q) **Hydroboration-protonolysis** is a sequence of two reactions used to convert an alkene to an alkane. First, the alkene reacts with a hydroborating agent to form an organoborane. Alkanes are formed by

reacting the organoborane with propanoic acid, which replaces the boron atom by a hydrogen atom.

(r) An **organoborane** is a compound containing a carbon-boron bond. Typical examples of organoboranes are R—BH$_2$, R$_2$—BH, and R$_3$—B, which are formed by the reaction of alkenes and a hydroborating reagent, such as borane or 9-BBN, in a reaction known as *hydroboration*.

(s) **Hydroboration-bromination** is a sequence of two reactions used to convert an alkene to a bromoalkane. First, the alkene reacts with a hydroborating reagent to form an organoborane, which reacts with bromine to form a bromoalkane by replacing the boron atom by a bromine atom.

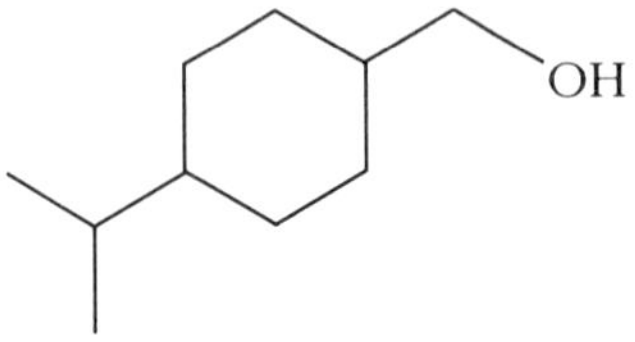

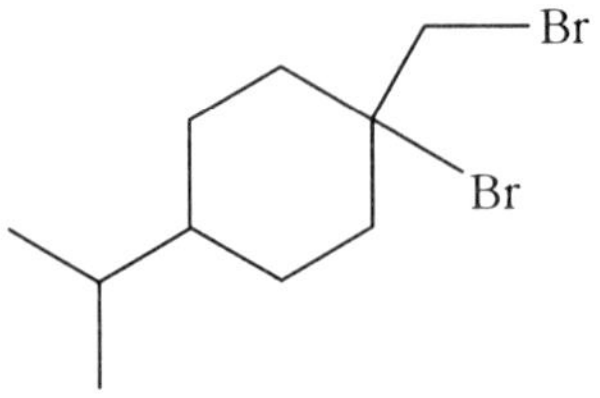

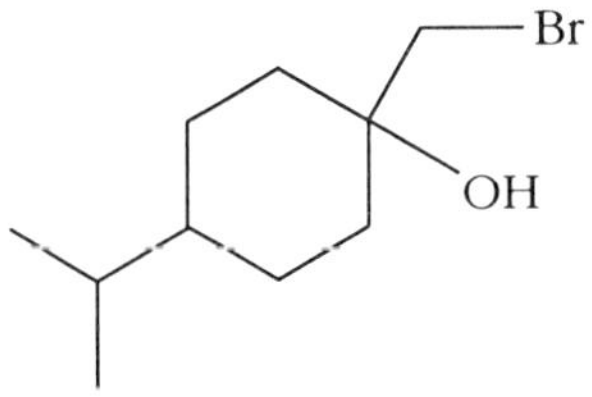

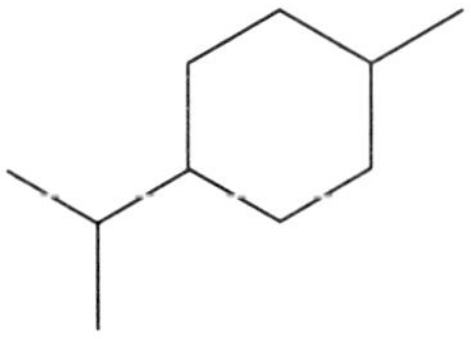

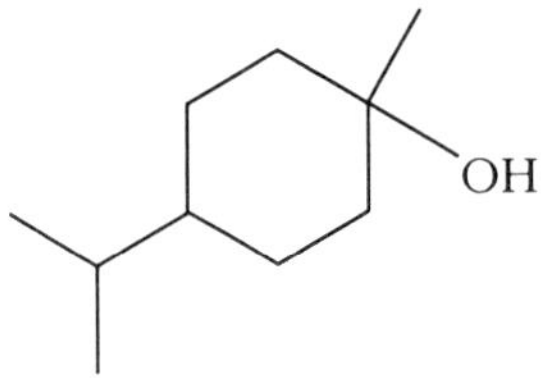

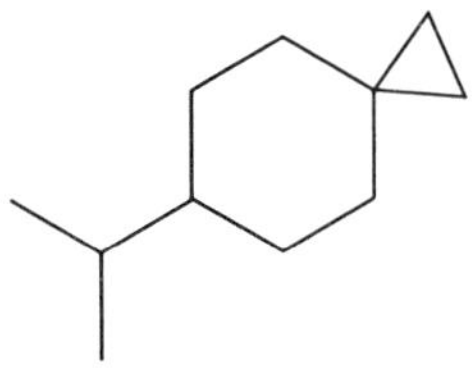

8.25

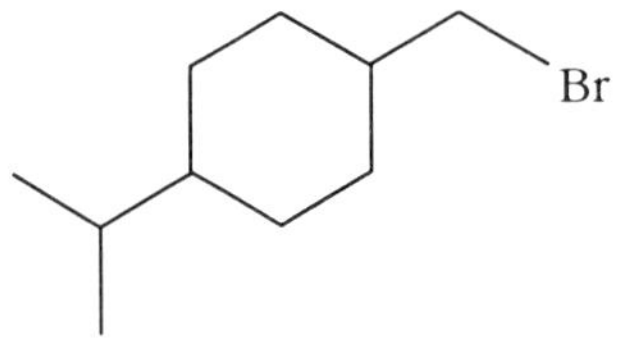

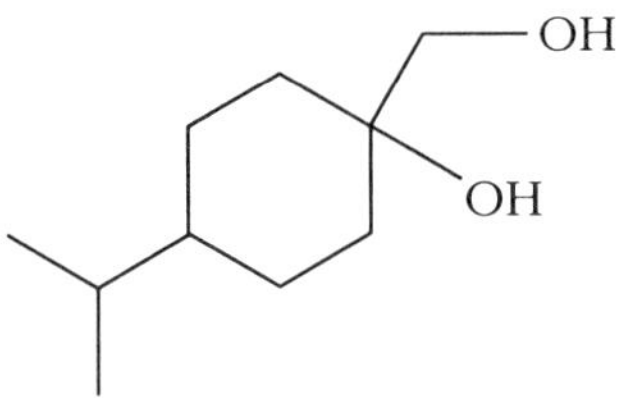

8.26

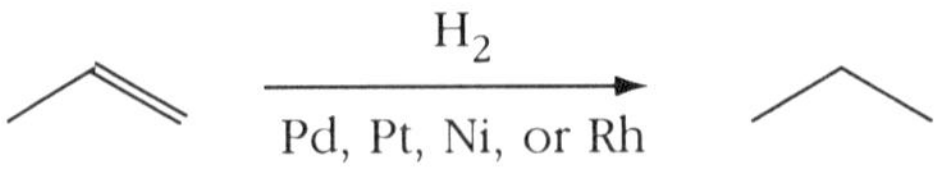

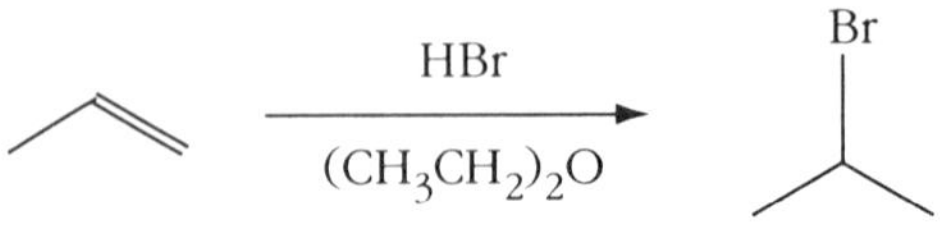

(c) CH₂=CHCH₃ —[Cl₂ / CCl₄]→ CH₃CHClCH₂Cl

(d) CH₂=CHCH₃ —[(1) 9-BBN; (2) Br₂/NaOCH₃]→ CH₃CH₂CH₂Br

(e) CH₂=CHCH₃ —[Cl₂ / H₂O]→ CH₃CH(OH)CH₂Cl

(f) CH₂=CHCH₃ —[H₃O⁺ or (1) Hg(OAc)₂, H₂O (2) NaBH₄, HO⁻]→ (CH₃)₂CHOH

(g) CH₂=CHCH₃ —[OsO₄ / H₂O₂/H₂O]→ CH₃CH(OH)CH₂OH

(h) CH₂=CHCH₃ —[RCOOOH]→ methyloxirane (epoxide)

8.27 The starting material in the following Exercises is 1-ethylcyclohexene. Identify the functional groups in each product and review the reactions of alkenes leading to the corresponding functional groups. Then, suggest the reagents needed for each conversion.

(a) The product shown has two hydroxyl groups on two adjacent carbon atoms (1,2-dialcohol or vicinal diol). A carbon-carbon double bond can be converted to a *cis* 1,2-diol by hydroxylation, which can be achieved by a reaction with OsO₄, H₂O₂/H₂O or cold KMnO₄/H₂O.

(b) The product is ethylcyclohexane (an alkane). Alkenes are converted to alkanes by catalytic hydrogenation. The reagent is hydrogen gas and a metal catalyst: H₂/Pd, Pt, Ni, or Rh.

(c) The product contains bromine and ethoxy group. Their relationship is *trans* so anti addition would form the product with the correct stereochemistry. Addition of bromine to an alkene occurs via a bromonium ion. When the bromonium ion is attacked by ethanol, the expected product is formed. The reagents are Br₂ in CH₃CH₂OH.

(d) The product contains a cyclopropane ring (carbenoid addition to the alkene). The reagents needed: CH₂I₂+ Zn(Cu).

(e) The product contains a three-membered cyclic ether, also called an *epoxide*, or *oxirane*. Peroxycarboxylic acids react with alkenes to give good yields of epoxides.

(f) The product contains a hydroxyl group on the adjacent carbon to the ethyl group (—OH and —H were added to the alkene so the reaction is a hydration). Both elements of water (H and OH) were added from the same side (face) of the carbon-carbon double bond (syn addition). The hydroxyl group was added to the less substituted carbon atom. Only hydroboration-oxidation of the alkene would form such a product. Reagents: (1) 9-BBN or BH_3, (2) H_2O_2/HO^-.

8.28

(a) 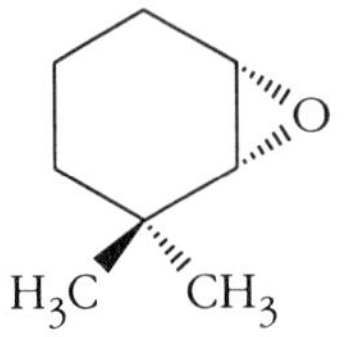
O
H3C
CH3

(b) 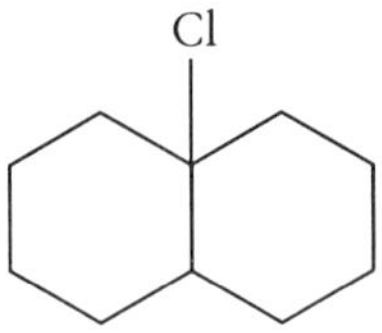
Cl

(c) 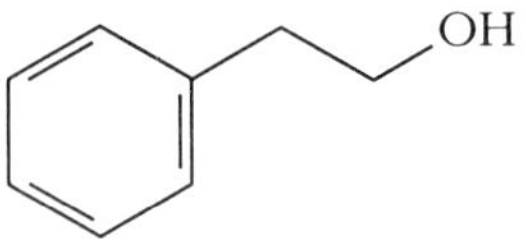
OH

(d) 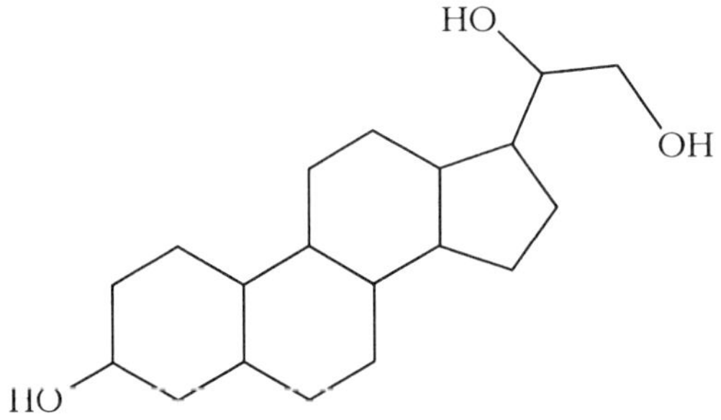
HO
OH
HO

(e) 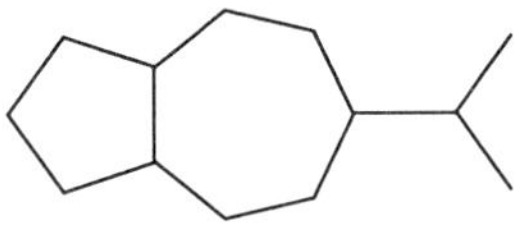

(f) The anti addition of chlorine to *trans*-cyclooctene gives *cis*- 1,2-dichlorocyclooctane.

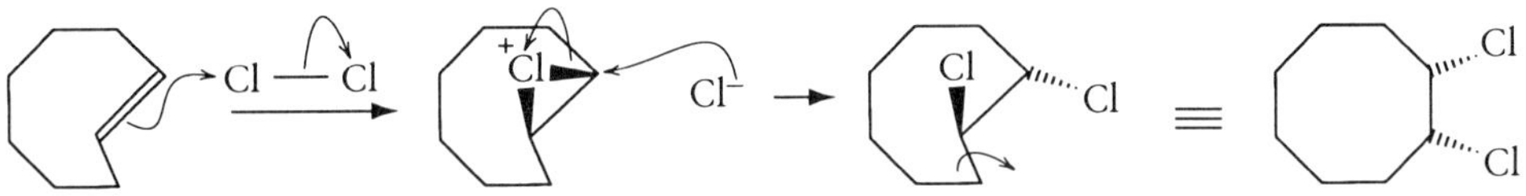
Cl — Cl
Cl
Cl⁻
Cl
Cl
≡
Cl
Cl

8.29 The oxymercuration-demercuration reaction with an alkene forms an alcohol by the addition of the elements of water according to the Markownikoff rule. Hydroboration-oxidation, in contrast, forms an alcohol by addition of water contrary to the Markownikoff rule. The same product is observed only when either of these reagents reacts with a symmetrical alkene; for example:

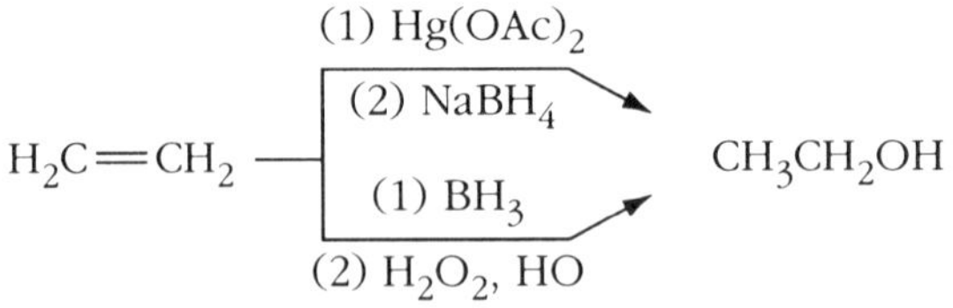
(1) Hg(OAc)2
(2) NaBH4
H2C=CH2
(1) BH3
(2) H2O2, HO
CH3CH2OH

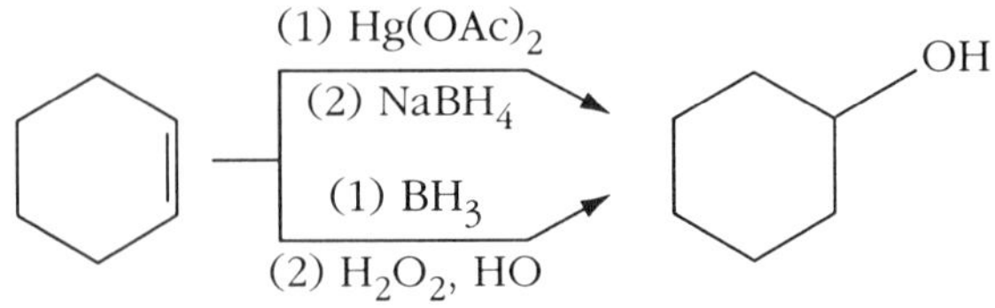
(1) Hg(OAc)2
(2) NaBH4
(1) BH3
(2) H2O2, HO
OH

8.30 1,4-Cyclohexadiene has two peaks in its ^{13}C NMR spectrum while 1,3-cyclohexadiene has three peaks in its ^{13}C NMR spectrum.

8.31 Epoxidation is characterized by syn addition of an oxygen atom to the double bond. Attack can occur from above and below the plane of the double bond to form oxiranes (epoxides) that are diastereomers:

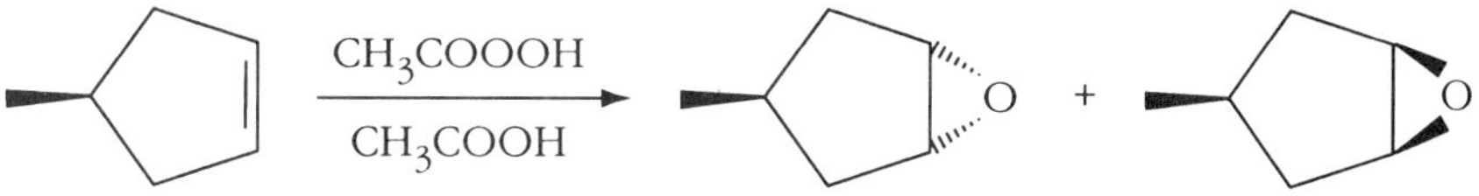

8.32 Review the functional groups and stereochemistry of the products and the starting materials. Suggest missing structures to complete the reactions.

(a) This reaction is the addition of water to the carbon-carbon double bond in a manner contrary to the Markownikoff rule (H is added to the less substituted carbon atom of the double bond). This can be done by the hydroboration-oxidation sequence of reactions.

$$(CH_3)_2C=CHCH_3 \xrightarrow[(2)\ H_2O_2/HO^-,\ H_2O]{(1)\ 9\text{-BBN/THF}} (CH_3)_2CH-\underset{\underset{OH}{|}}{C}HCH_3$$

(b) The product contains two hydroxyl groups on two adjacent carbon atoms (1,2-dialcohol or vicinal diol). A carbon-carbon double bond can be converted to a *cis* 1,2-diol by hydroxylation, which can be achieved by a reaction with $OsO_4, H_2O_2/H_2O$ or cold $KMnO_4/H_2O$. We can try to write such a reaction with *trans*-2-pentene and with *cis*-2-pentene and we find that the *trans* stereoisomer affords the required product.

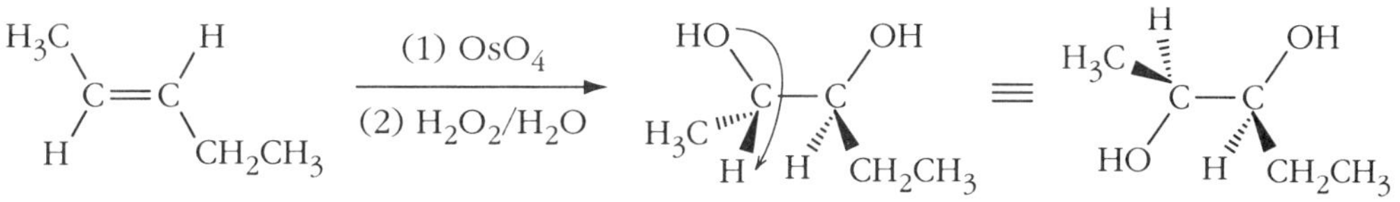

(c) The starting material contains a carbon-carbon double bond and a methyl group attached to the sp^2 hybrid carbon. The product contains a hydroxyl group on the adjacent carbon to the methyl group (—OH and —H were added to the alkene: hydration). Both elements of water (H and OH) were added from the same side (face) of the carbon-carbon double bond (syn addition). The hydroxyl group was added to the less substituted carbon atom. Only hydroboration-oxidation of the alkene would form such a product. Possible reagents: (1) 9-BBN or BH_3, (2) H_2O_2/HO^-. The reaction gives a racemic mixture.

(d) Bromine adds to an alkene by the bromonium ion mechanism. In the presence of methanol, methanol opens the bromonium ion to complete the anti addition. There is an equal chance that the bromonium ion can be formed from either side (face) of the ring and a racemic mixture of products will result.

(e) Only the product, *meso*-2,3-dichlorobutane, is given. We draw the structure of the product and analyze how it can be formed. We know that a halogen addition to an alkene results in anti orientation of halogen atoms. Working backward, we arrive at the structure of a possible starting material.

Now we can write the complete reaction.

(f) The product contains the 1,1-dichlorocyclopropane ring. It can be formed in a reaction of dichlorocarbene with an alkene. The products of cyclopropanation retain any *cis* or *trans* stereochemistry of the reactant. The product contains two *cis* alkyl groups so our substrate must be *cis*-2-pentene.

8.33 The ^{13}C NMR spectrum of compound **D** shows three peaks. There are three possible constitutional isomers of alkane **D**, C_5H_{12}: pentane (**D**$_1$), 2-methylbutane (**D**$_2$) and 2,2-dimethylpropane (**D**$_3$). Pentane exhibits three peaks in the

^{13}C NMR spectrum, 2-methylbutane exhibits four peaks, and 2,2-dimethylpropane exhibits two peaks. Thus, only pentane matches the spectral data of the hydrogenation product **D**.

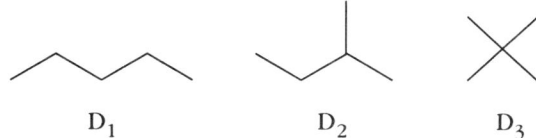

Now let us consider possible structures for the alkene **A**. It can be either 1-pentene or 2-pentene. However, 2-pentene would give a mixture of alcohols in both hydroboration-oxidation and oxymercuration-demercuration reactions. Thus, only 1-pentene is consistent with the experimental observations so 1-pentene is alkene **A**.

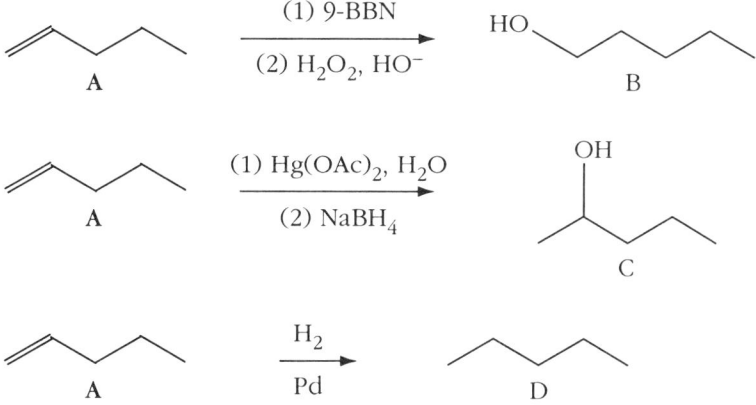

Hence, A = 1-pentene, B = 1-pentanol, C = 2-pentanol, D = pentane.

8.34 To account for all the products observed, we must consider a mechanism analogous to that proposed for the formation of halohydrins. Such a mechanism involves the following three steps:

Step 1 is the reaction of the halogen with the π bond of the alkene to form a bridged halonium ion intermediate. The attack may take place at each side of the double bond.

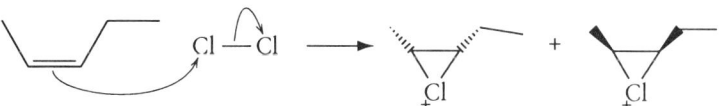

Step 2 is a nucleophilic attack at one of the carbon atoms of the halonium intermediate. If the nucleophile is a halide, we observe a dihaloalkane product.

Reaction of the intermediate with a solvent molecule (acetic acid) in Step 2 results in an addition of acetic acid

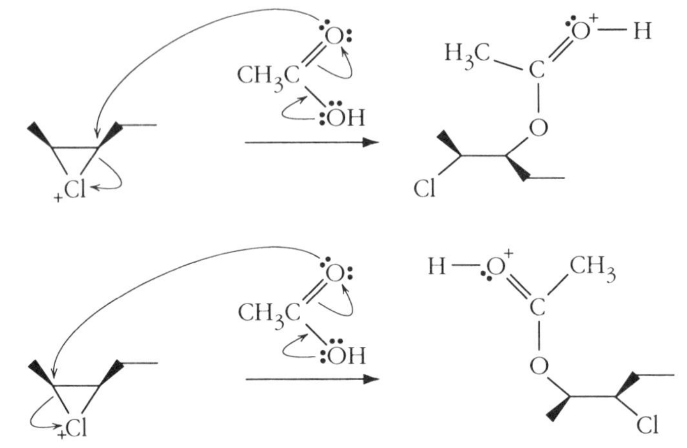

Step 3: Loss of a proton.

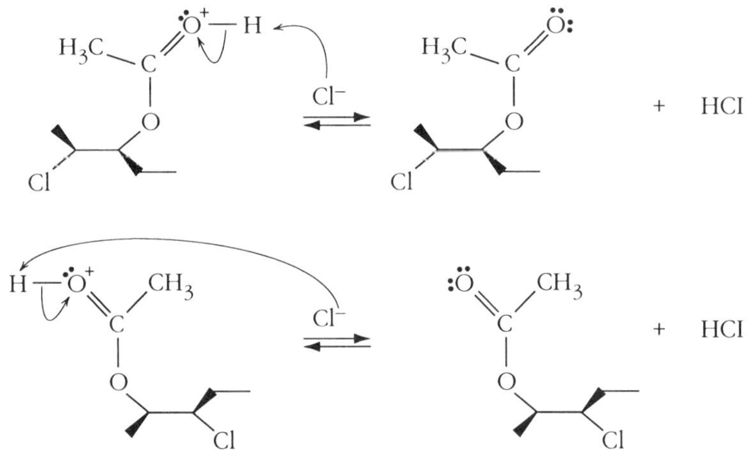

8.35

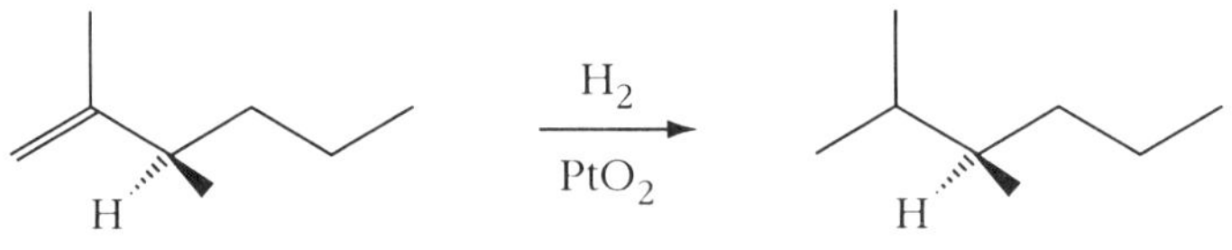

(R)-2,3-Dimethyl-1-hexene → (R)-2,3-Dimethylhexane

(H₂, PtO₂)

The product is optically active because the stereocenter on the C3 carbon is not disturbed during hydrogenation, so its configuration remains unchanged. The priorities of substituents also did not change, so the product is (R)-2,3-dimethylhexane.

8.36 Metal-catalyzed addition of hydrogen to an alkene occurs by syn addition. When one side (or face) of the double bond is hindered, addition occurs preferentially to the other face. In such cases, alkene hydrogenation occurs stereoselectively.

A stereoselective reaction is a reaction in which a single starting material can give two or more stereoisomeric products but yields preferentially one of them. Exercise 8.36 (d) and 8.36 (e) illustrate this principle. The bottom face of the double bond is more exposed, and deuterium is transferred from the catalyst to that face.

(a) [cyclohexane with two D substituents]

(b) CH_3CH_2—C—C—CH_2CH_3 with D, H, D, H substituents ≡ CH_3CH_2—C(H)(D)—C(D)(H)—CH_2CH_3

(c) CH_3CH_2—C(H)(D)—C(D)(H)—CH_2CH_3

(d) [bicyclic structure with two D substituents]

(e) [bicyclic structure with two D substituents]

8.37 The mechanism of the addition of hydrogen halides to alkenes occurs in two successive steps:

Step 1: A carbon of the π bond is protonated.

Step 2: A halide ion attacks the resulting carbocation. Sometimes a carbocation intermediate can be stabilized through a rearrangement, which is what is observed in this example.

Step 1: The reaction begins when the double bond is protonated to yield the carbocation with the greater number of alkyl branches at the electron-deficient carbon (secondary).

[spiro compound] + H—Cl ⇌ [protonated spiro cation]$^+$ + Cl$^-$

However, the carbocation undergoes further stabilization through a rearrangement. The methylene group moves with its pair of bonding electrons from the carbon adjacent to the electron-deficient carbon.

[spiro cation] → [methylcyclohexyl cation]

The carbon from which this group departs becomes electron deficient and positively charged. As a result, a tertiary carbocation is formed.

Step 2: Reaction of the carbocation with Cl⁻ yields the final product.

8.38 First we must determine the stereochemistry of the addition reaction. We do this by analyzing the products observed and checking if the addition is syn or anti. The following example demonstrates this kind of analysis.

We take the first product of addition of BrCl to fumaric acid and we rotate about the central carbon-carbon bond so Br and Cl are in the anti conformation. Next, we operate in the reverse order to the addition process; that is, we remove Br and Cl and we add a second bond between the central carbon atoms to arrive at fumaric acid.

The preceding operations strongly suggest that the addition results in an anti orientation of the two parts of the adding reagent. Let us check to see if we get the starting material by removing Br and Cl from *cis* positions. We take the same product and rotate about the C—C bond so Br and Cl are eclipsed, and then remove Br and Cl and add a second bond between carbon atoms. We arrive at maleic acid. This means that the reverse sequence of operations (syn addition of Br and Cl to maleic acid) does *not* produce the observed product.

Similar analysis of the other products leads to the conclusion that addition of BrCl to fumaric acid and maleic acid occurs by anti addition.

Anti addition suggests a mechanism involving a halonium ion intermediate. In general, the nucleophilic double bond attacks the electrophilic nucleus of a halogen. In this case we have an unsymmetrical electrophile with a more electronegative chlorine and a more electropositive bromine. Thus, BrCl is polarized ($\overset{\delta+}{Br}-\overset{\delta-}{Cl}$). The double bond, therefore, reacts with the bromine atom of BrCl to form a bromonium ion and a chloride ion in Step 1 of the mechanism. The chloride ion attacks the bromonium ion from the back side in Step 2 to form the product. In this case, the bromonium ion is symmetrical and we observe a mixture of enantiomers.

Mechanism of addition of BrCl to fumaric acid:

Step 1: Formation of the bromonium ion

Step 2: Opening of the bromonium ion by chloride

Mechanism of addition of BrCl to maleic acid:

Step 1: Formation of the bromonium ion

Step 2: Opening of the bromonium ion by chloride

8.39 The nucleophilic double bond reacts with the bromine atom (the more electrophilic part of the BrCl molecule) to form the bromonium ion as an intermediate. The more substituted carbon of the bromonium ion is more positive because it better supports a partial positive charge. Therefore, attack by the chloride ion is directed toward the more highly substituted carbon to give the Markownikoff product. Hence, the result is in accord with the mechanism described in Exercise 8.38.

Step 1: Formation of the bromonium ion

Step 2: Chloride ion acts as a nucleophile and opens the bromonium ion ring

8.40 The catalytic addition of hydrogen to an alkene may occur according to syn addition or anti addition. Let us add hydrogen atoms syn to the alkene from above the plane of the page and from below the page and write structures of the products obtained in both cases.

The experimental results indicate that the syn addition of hydrogen occurred in both cases.

8.41 Step 1: Formation of an intermediate iodonium ion

The iodonium ion may be opened by either of the two nucleophiles, namely iodide ion or the oxygen atom of the carboxyl group. Although the iodide ion is a very good nucleophile, the intramolecular reactions are usually faster because the nucleophilic oxygen is very close to the electrophilic carbon all the time. The carboxyl group, therefore, reacts with the iodonium ion in Step 2.

Step 2: Nucleophilic attack of the carboxyl group

Step 3: Removal of a proton:

[Structure with I, O, and :O-H being deprotonated by HCO₃⁻] → [Deprotonated structure with :O:] + H₂CO₃

8.42 Borane (BH₃) adds regioselectively to alkenes so that the boron becomes bonded to the less branched carbon of the double bond. In this case, however, the alkene is symmetrical and both carbon atoms of the double bond have one alkyl group attached. Despite the proximity of the isopropyl group, the steric effects seem to be negligible.

In the case of 9-BBN {bora(9)-bicyclo[3.3.1]nonane} the steric effects are appreciable and, as a result, this reagent becomes more regiospecific.

8.43

	Compound	IR (important absorption bands)	¹³C NMR (approximate chemical shifts)
(a)	$\underset{\underset{CH_3CHCH_3}{\mid}}{Cl}$	1430–600 cm⁻¹ (C—Cl)	Two signals: δ36 (\C—Cl) δ20 (—CH₃)
	CH₃CH₂CH₂Cl	1430–600 cm⁻¹ (C—Cl)	Three signals: δ36 (—CH₂—Cl) δ30 (—CH₂—) δ19 (—CH₃)

The IR spectrum is not very informative because both compounds show C—Cl stretching vibrations in the fingerprint region. The ¹³C NMR should be decisive, however, because 2-chloropropane shows two signals while 1-chloropropane shows three signals in the alkane region.

	Compound	IR	¹³C NMR
(b)	Epoxycyclohexane	1200–1000 cm⁻¹ (C—O)	Three signals: δ68 (—C—O—) δ35 (—CH₂—C—O) δ27 (—CH₂—)
	Cyclohexanone	1710 cm⁻¹ (C=O) 1200–1000 cm⁻¹ (C—O)	Four signals: δ210 (C=O) (C1) δ41 (—CH₂—C=O)(C2, C6) δ29 (C3 and C5) δ27 (C4)

The IR spectrum distinguishes clearly between the two because epoxycyclohexane has no absorption in the 1700 cm⁻¹ region, while cyclohexanone has a very strong absorption band at 1710 cm⁻¹.

The ¹³C NMR is also helpful because epoxycyclohexane shows three signals while cyclohexanone shows four signals including that of the carbonyl carbon at 210 ppm.

	Compound	IR	¹³C NMR
(c)	$\underset{\underset{CH_3CHCH_2Br}{\mid}}{OH}$ 1-Bromo-2-propanol	3500–3200 cm⁻¹ (O—H)	Three signals: δ60 (—CH—OH) δ42 (—CH₂Br) δ23 (—CH₃)
	$\underset{\underset{CH_3CHCH_2Br}{\mid}}{Br}$ 1,2-Dibromopropane	1430–600 cm⁻¹ (C—Br)	Three signals: δ39 (\CH—Br) δ36 (—CH₂Br) δ20 (—CH₃)

In this case, the ¹³C NMR spectrum is not very informative because both compounds are expected to have three signals.

The IR spectrum is decisive, however, because 1,2-dibromopropane does *not* show a strong absorption band at 3500–3200 cm⁻¹ due to —OH stretching while 1-bromo-2-propanol does.

8.44 (a) Molecular formula: $C_2H_4Br_2$; degree of unsaturation: 0, so it is a saturated compound. The ^{13}C NMR spectrum shows one signal, which means that both carbon atoms have the same substituents. Hence, the only solution is **1,2-dibromoethane.**

$$BrCH_2CH_2Br$$

(b) Molecular formula: $C_6H_{14}O_2$; degree of unsaturation: 0, so it is a saturated compound. The ^{13}C NMR spectrum shows two signals, which means that the compound has a high degree of symmetry. The IR spectrum indicates the presence of a hydroxyl group (or groups) so the compound has one or two hydroxyl groups attached to carbon atoms. To account for all the carbon atoms we place two hydroxyl groups and four methyl groups on two carbon atoms.
2,3-Dimethyl-2,3-butanediol is consistent with both the IR and ^{13}C NMR spectra.

2,3-Dimethyl-2,3-butanediol

8.45 (a) Molecular formula: C_2H_5ClO; degree of unsaturation: 0, so it is a saturated compound. The ^{13}C NMR spectrum shows two signals, one at δ 61 (carbon bonded to oxygen) and the other at δ 43 (carbon bonded to a halide). The IR spectrum shows a very strong absorption band at 3300 cm^{-1}, which indicates the presence of the hydroxyl group. If we put the Cl on one carbon atom and the —OH on the other we obtain **2-chloroethanol** as the structure of the compound.

$$ClCH_2CH_2OH$$

2-Chloroethanol

1-Chloroethanol, $CH_3\overset{\underset{\mid}{OH}}{C}HCl$, might seem to be a viable alternate structure.

The ^{13}C NMR peak of the CH_3— group of 1-chloroethanol, however, would be expected at δ 0 to 30. The absence of a peak in this region of the spectrum means the compound must be 2-chloroethanol, not 1-chloroethanol.

(b) Molecular formula: C_4H_8O; degree of unsaturation: 1, so it is an unsaturated compound. The possible sources of unsaturation are a ring, a carbon-carbon double bond, or a carbon-oxygen double bond. The ^{13}C NMR spectrum shows four signals, two in the alkane region [one at δ 62 (carbon bonded to oxygen) and one at δ 15 (terminal methyl group)] and two in the alkene region (at 151 ppm and 86 ppm). Thus, the ^{13}C NMR spectrum does not show a signal at δ 200 to 215 characteristic of a carbonyl carbon. Also, the IR spectrum does not show any absorption peak at 1700 cm^{-1} characteristic of a carbonyl group. However, the IR spectrum shows a band at 1615 cm^{-1}, which is characteristic of carbon-carbon double bond stretching vibrations. There is also a sharp absorption band at 3000 cm^{-1}, which is characteristic of the vinylic C—H stretching. The lack of the absorption in the 3500 to 3200 cm^{-1} region eliminates the possibility of an alcohol. The only functional group left with an oxygen atom is an ether. The strong absorption at 1200 cm^{-1} confirms the presence of a carbon-oxygen bond. Hence, the unknown is **ethyl vinyl ether** (IUPAC name: **ethoxyethene**).

$$\underset{H}{\overset{H}{\diagdown}}C=C\underset{O-CH_2CH_3}{\overset{H}{\diagup}}$$

Ethoxyethene

(c) Molecular formula: $C_6H_{12}O_2$; degree of unsaturation: 1; unsaturated compound; possible sources of unsaturation: ring, carbon-carbon double bond, or carbon-oxygen double bond. The ^{13}C NMR spectrum shows three signals, all three of which are in the alkane region [one at δ 76 (carbon bonded to oxygen), one at δ 33, and one at δ 23]. Thus, the ^{13}C NMR spectrum does not show a signal at δ 200 to 215 characteristic of a carbonyl carbon. Also, the IR spectrum does not show any absorption peak at 1700 cm^{-1} characteristic of a carbonyl group. The IR spectrum does not show a band at 1615 cm^{-1} characteristic of carbon-carbon double bond stretching vibrations. The presence of the absorption in the 3500 to 3000 cm^{-1} region indicates the presence of an alcohol. The strong absorption in the range of 1200 to 1000 cm^{-1} confirms the presence of a carbon-oxygen bond. Thus, there is a strong indication that the unknown contains two —OH groups attached to a cyclohexane ring. There are four possible diols: 1,1-cyclohexanediol, 1,2-cyclohexanediol, 1,3-cyclohexanediol, and 1,4-cyclohexanediol.

1,1-Cyclohexanediol   1,2-Cyclohexanediol   1,3-Cyclohexanediol   1,4-Cyclohexanediol

1,1-Cyclohexanediol and 1,3-cyclohexanediol would have four signals in the ^{13}C NMR spectrum, 1,4-cyclohexanediol would have only two signals and 1,2-cyclohexanediol has three signals in the ^{13}C NMR spectrum. Hence, the unknown is **1,2-cyclohexanediol.**

EXAMINATION 1

To get the maximum benefit from the following one-hour exam, choose a quiet location and spend exactly one hour answering the questions. Check your answers against those at the end of the Study Guide to evaluate your knowledge of the material.

Chapters 1–8

Time allowed: 1 hour

1. *(12%)* Draw structures of the following compounds. Use the partial structures suggested.

    (a) *(S)*-2-Chlorobutane

    (b) *(Z)*-3-Methyl-2-hexene

    (c) *cis*-1-Methyl-3-(1-methylethyl) cyclohexane or *cis*-1-isopropyl-3-methylcyclohexane (*most stable conformation!*)

    (d) *meso*-2,4-Dibromopentane

2. *(12%)* Give the IUPAC name for each of the following compounds. Use *cis*, *trans*, *(R)*, *(S)*, *(E)*, *(Z)* where applicable.

    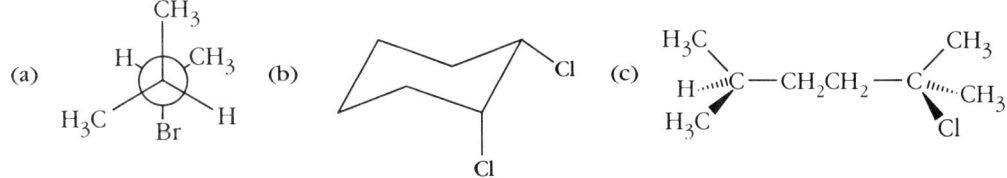

3. *(8%)* Arrange the following compounds in order of increasing boiling points.

    CH₃CH₂CH₂OH   CH₃CH₂CH₂CH₂CH₃   CH₃CHCH₂CH₃ (with Cl on C2)

    1-Propanol   Pentane   2-Chlorobutane

4. Consider the following Newman projection of butane:

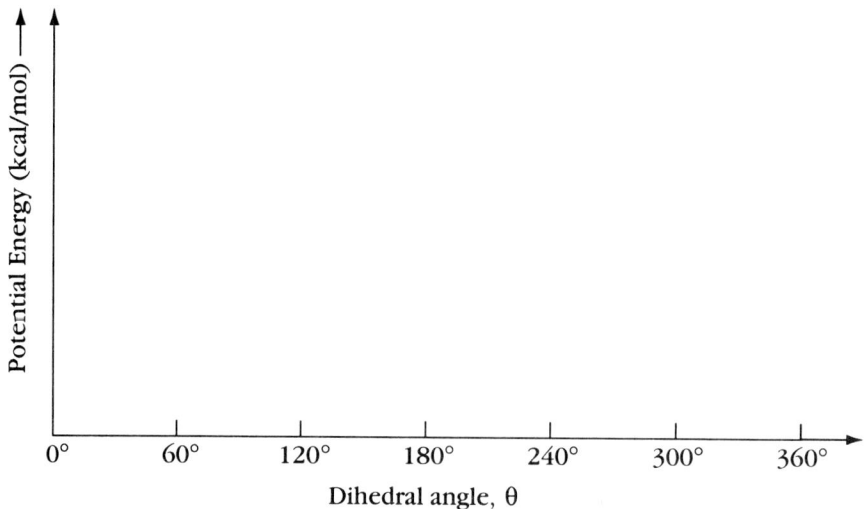

The Newman projections representing some conformations obtained by internal rotation around the central carbon-carbon bond in butane are shown on the following page.

(a) *(10%)* Draw a potential energy diagram for the clockwise rotation of the front carbon.

(b) *(10%)* The following are six structures numbered 1 to 6, which represent conformations of butane. On the energy diagram constructed above, indicate by appropriate numbers (1-6) the points that represent the following conformations.

(c) *(8%)* Name each conformation (e.g., staggered, gauche, etc.) below each Newman projection.

5. **(40%)** Write structures of missing products or give necessary organic and inorganic reagents for the following transformations. Indicate if more than one step is needed.

(a) cyclopentyl-Cl ⟶ cyclopentene

(b) 1-methylcyclopentene ⟶ methylcyclopentane

(c) bicyclic alkene $\xrightarrow{\text{OsO}_4, \text{H}_2\text{O}_2/\text{H}_2\text{O}}$

(d) bicyclic alkene $\xrightarrow{\text{Cl}_2, \text{H}_2\text{O}}$

(e) 1-methylcyclopentene ⟶ (product with OH and CH₃ shown with stereochemistry)

(f) 2-methyl-1-pentene ⟶ 2-methylpentan-1-ol (with CH₂OH)

(g) cyclohexene $\xrightarrow{\text{Br}_2, \text{CCl}_4}$ [Draw the chair conformation of the product]

(h) 1-methylcyclopentene $\xrightarrow{\text{RCOOOH (1 equivalent)}}$

CHAPTER 9

ALKYNES

Concepts

Alkynes are hydrocarbons that contain carbon-carbon triple bonds. Alkynes are also called **acetylenes**, after the simplest alkyne—acetylene (H—C≡C—H). As a consequence of the triple bond, an alkyne has four fewer hydrogens than the corresponding alkane. Its molecular formula for a noncyclic alkyne is like that of a molecule with two double bonds: C_nH_{2n-2}. Therefore, the triple bond contributes two elements of unsaturation.

The IUPAC nomenclature for alkynes is similar to that for alkenes. We find the longest continuous chain of carbon atoms that includes the triple bond and change the *-ane* ending of the parent alkane to *-yne*. The chain is numbered from the end closest to the triple bond, and the position of the triple bond is designated by its lower-number carbon atom.

$$H-C\equiv C-H \qquad CH_3\underset{\underset{CH_3}{|}}{C}HC\equiv CH \qquad CH_3\underset{\underset{CH_3}{|}}{C}HC\equiv CCH_2CH_2Cl$$

Ethyne   3-Methyl-1-butyne   1-Chloro-5-methyl-3-hexyne
(acetylene)

When double bonds and triple bonds are present in the same molecule, the numerical precedence of a double or triple bond is given to the bond that gives the lowest number in the name of the compound. If a triple bond and a double bond are equal distance from either end, the double bond takes precedence in numbering. The double bond is always cited first in the name by dropping the terminal *-e* from the *-ene* suffix, as in the following examples:

$$\overset{1}{HC}\equiv\overset{2}{C}-\overset{3}{C}H=\overset{4}{C}H\overset{5}{C}H_3 \qquad \overset{1}{C}H_2=\overset{2}{C}H-\overset{3}{C}\equiv\overset{4}{C}-\overset{5}{C}H_3$$

3-Penten-1-yne   1-Penten-3-yne

$$HC\equiv C-CH_2-CH_2-\underset{\underset{OH}{|}}{C}H-CH_3$$

5-Hexyn-2-ol

Many of the chemical properties of an alkyne depend on whether there is an acetylenic hydrogen (—C≡C—H) or, in other words, if the triple bond comes at the end of a carbon chain. Such an alkyne is called a **terminal alkyne**. Alkynes with triple bonds located elsewhere are called **internal alkynes**.

Alkynes undergo many of the same reactions as alkenes. Alkynes, however, have two π bonds so they can react with two equivalents of electrophilic reagents. Addition of the second equivalent is usually more difficult so the reaction can be stopped after the addition of only one equivalent. Thus, hydroboration, hydrogenation, and halogenation of alkenes occur to form products of either mono- or di-addition. Sometimes there are differences between the chemistry of alkenes and alkynes. For example, the products of hydration of alkynes, unlike those of alkenes, are carbonyl-containing compounds.

Terminal alkynes are much stronger acids than alkenes or alkanes. They react with strong bases, such as NaNH$_2$, to form **acetylide ions**. Acetylide ions react with methyl or primary haloalkanes to form internal alkynes.

$$CH_3-C\equiv C-H \xrightarrow[\text{liq. NH}_3]{\text{NaNH}_2} CH_3-C\equiv C^-\ Na^+ + R-CH_2X$$

$$\longrightarrow CH_3-C\equiv C-CH_2R$$

Solutions to the Exercises

9.1 The IUPAC names of alkynes are obtained by a modification of the usual rules for hydrocarbons: (i) the longest continuous carbon chain that contains the triple bond determines the *parent name*; (ii) the suffix *-ane* is replaced by *-yne*; (iii) the triple bond is given the lowest possible number.

(a) 3,3-Dimethyl-1-pentyne (b) 6-Bromo-3,8-dimethyl-4-nonyne
(c) 2,2-Dimethyl-4-decyne (d) 3-Cyclopentyl-7-ethyl-9-methyl-4-decyne
(e) Cyclopentylethyne (f) (Z)-3-Nonen-6-yne

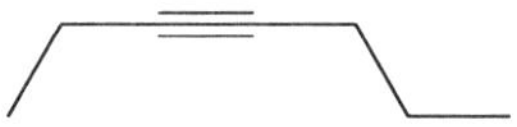

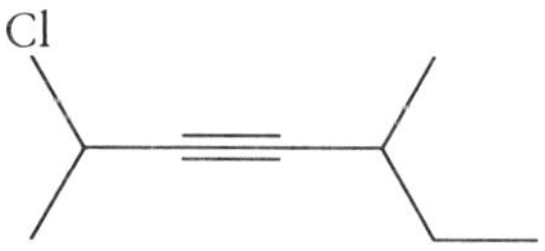

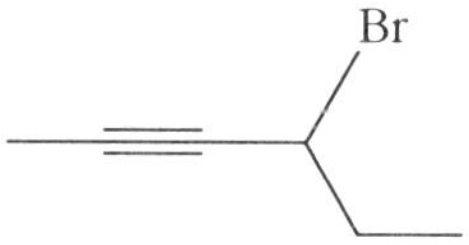

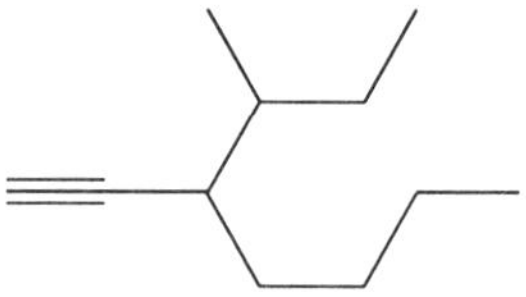

9.2

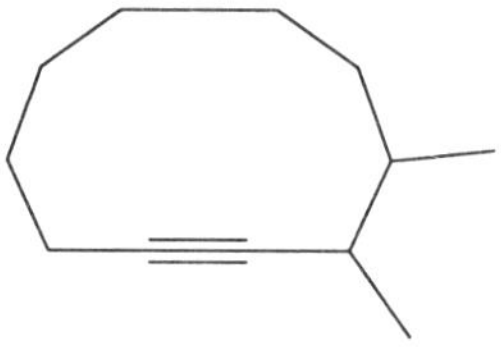

(g) CH$_2$=CHCH=CHC≡CCH$_2$CH$_3$

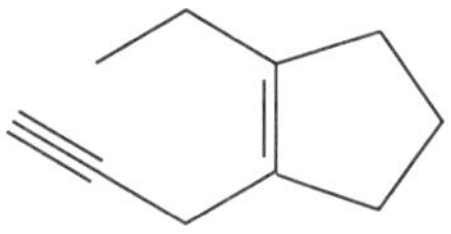

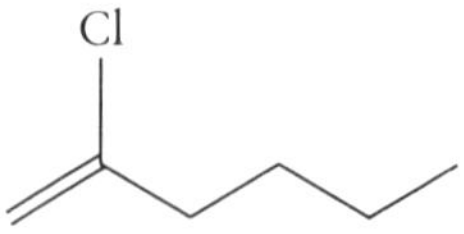

9.3

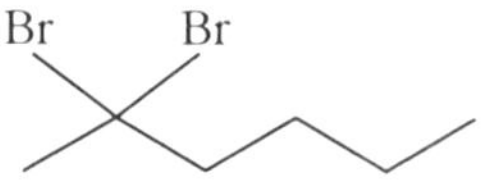

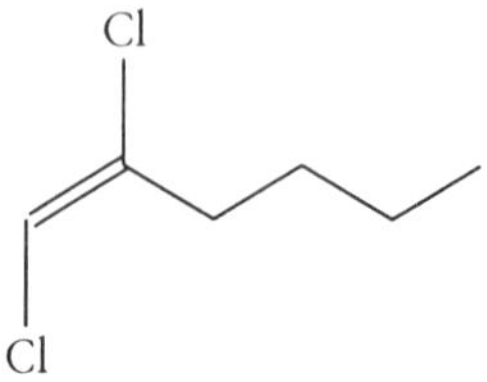

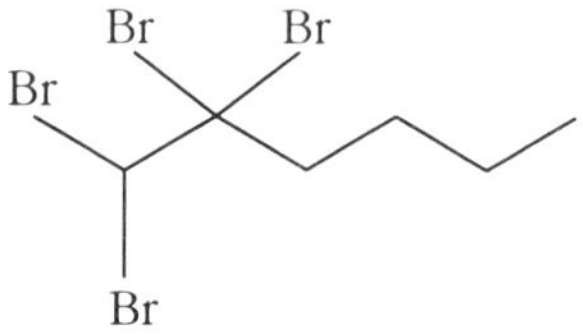

9.4

(a) 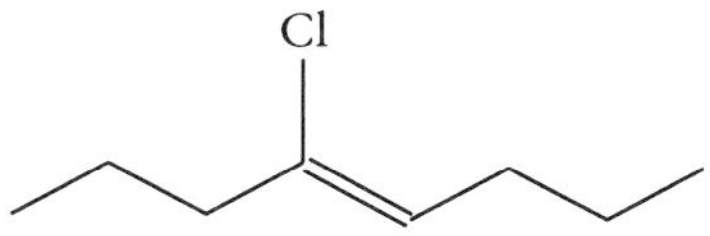 (b) 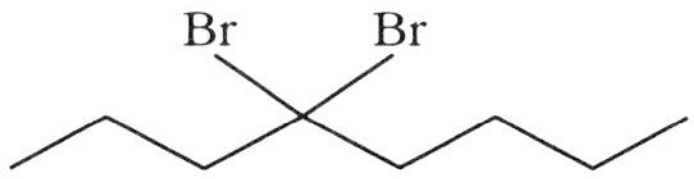

(c) 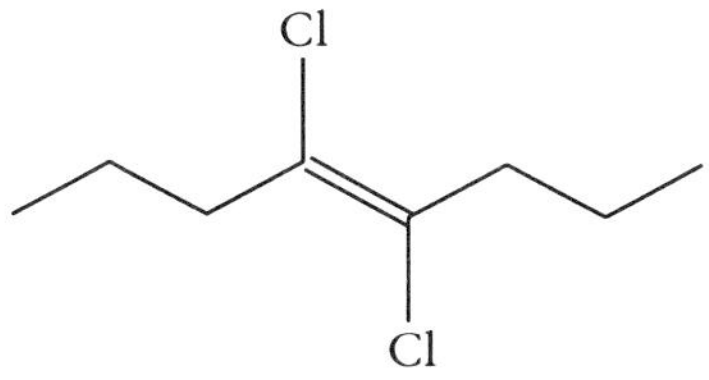 (d) 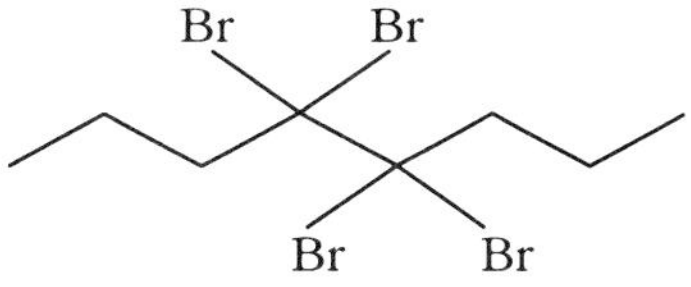

9.5

(a) 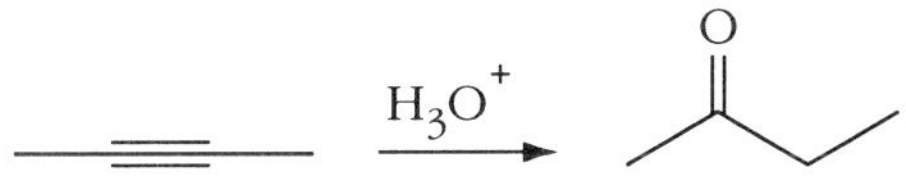

(b) 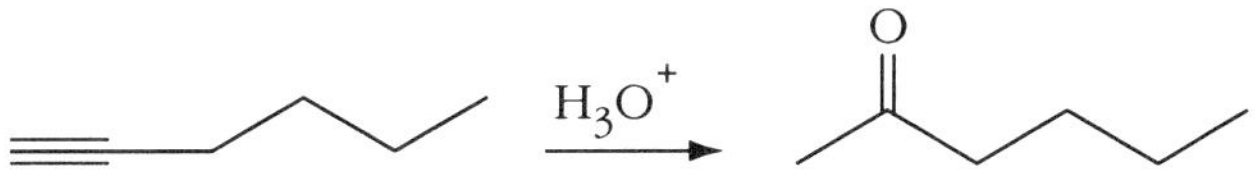

(c) 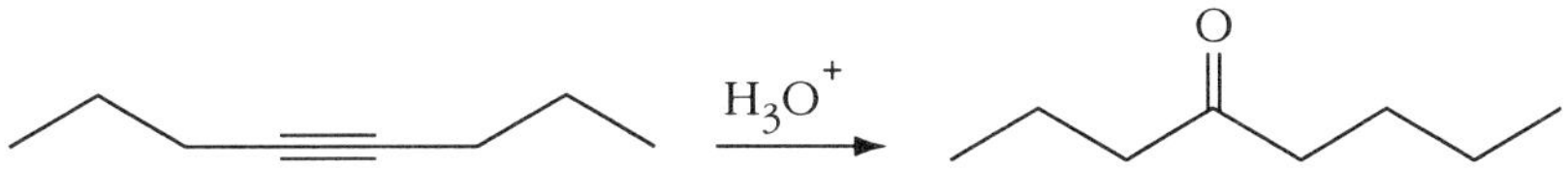

(d) 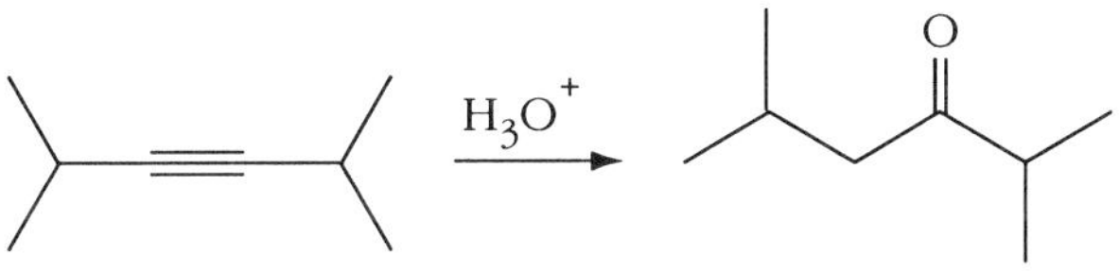

9.6 The process by which enols are converted to aldehydes or ketones is called **keto-enol isomerism** (or **keto-enol tautomerism**). In this process, the hydrogen atom of the —OH group of the enol is lost and a hydrogen atom is added to the carbon atom of the double bond that is not bonded to the —OH group and the double bond is moved between carbon and oxygen atoms.

(a) 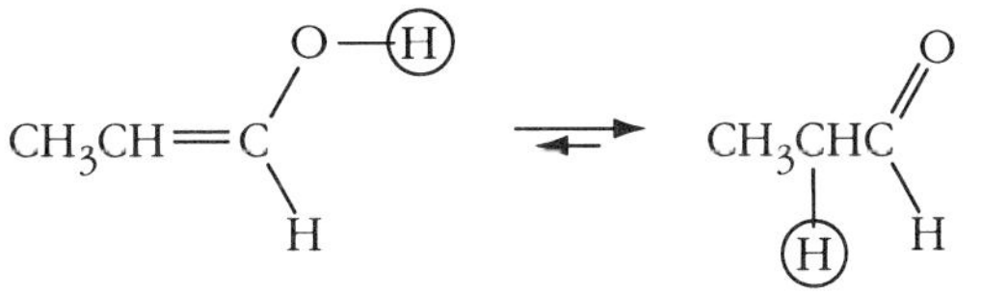

(b) 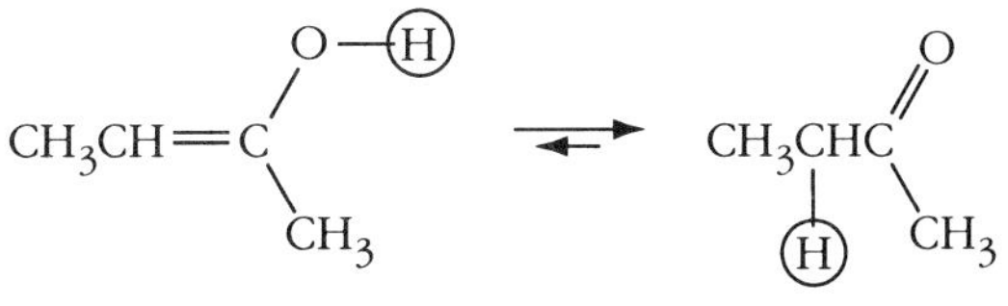

(c) 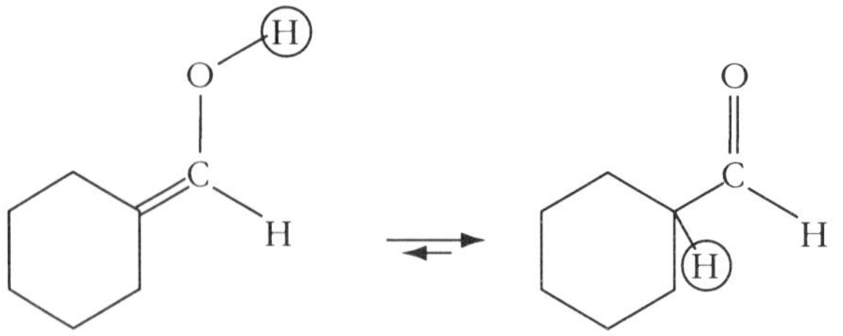 [structure]

9.7 Because of the regioselectivity of acid-catalyzed alkyne hydration, acetylene is the only alkyne structurally capable of yielding an aldehyde. Terminal alkynes yield methyl-substituted ketones.

(a) 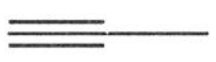 [structure]   (b) 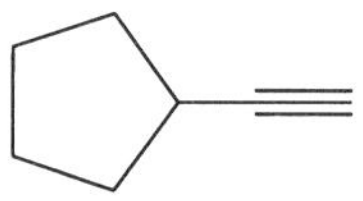 [structure]

In the case of internal alkynes, only symmetrical alkynes form a single product upon hydration. Thus, in order to obtain 4-octanone we must use 4-octyne as a substrate.

(c) 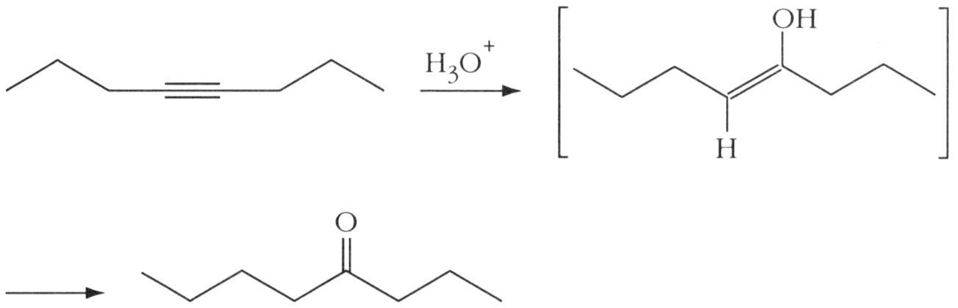 [reaction scheme]

Note: Hydration of 3-octyne will result in a mixture of two products, 3-octanone and 4-octanone:

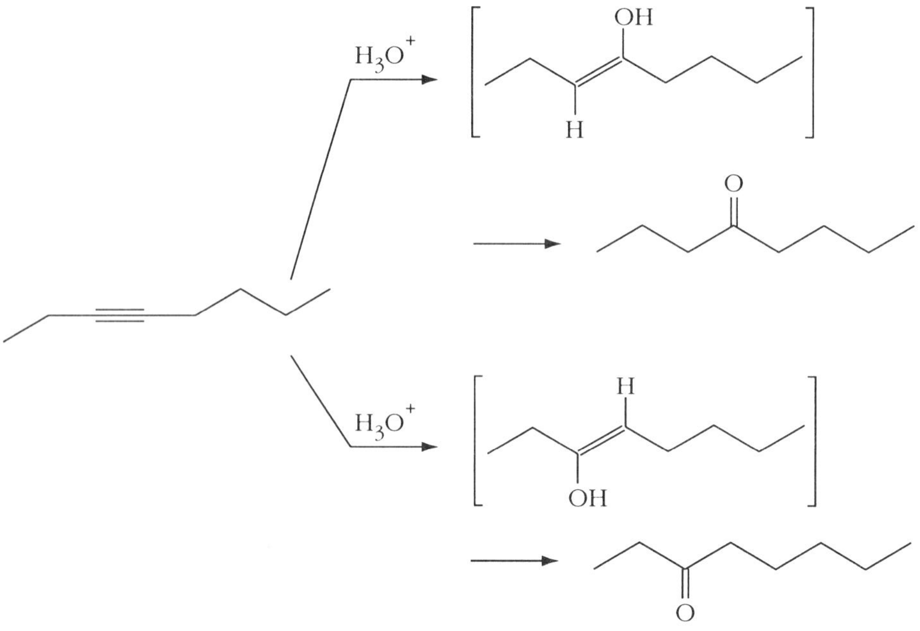

[reaction scheme]

9.8 Step 1: Hydration of an alkyne is initiated by protonation of the triple bond. In the case of 2-hexyne, two vinyl cations are formed.

Step 2: Each vinyl cation reacts with a nucleophile (water).

Step 3: The products of the preceding reactions are deprotonated by water to yield enols, which rapidly are converted to much more stable keto forms.

9.9 The reaction of terminal alkynes with 9-BBN followed by aqueous alkaline hydrogen peroxide (so-called hydroboration-oxidation) yields aldehydes, whereas acid-catalyzed hydration forms methyl ketones.

In the case of internal alkynes, both acid-catalyzed hydration and hydroboration-oxidation give the same product(s).

9.10 Because each carbon of the triple bond in 2-pentyne has the same degree of substitution, very little regiospecificity occurs during reaction with 9-BBN. Two enols are formed, and from them isomeric ketones are formed.

9.11

(a)

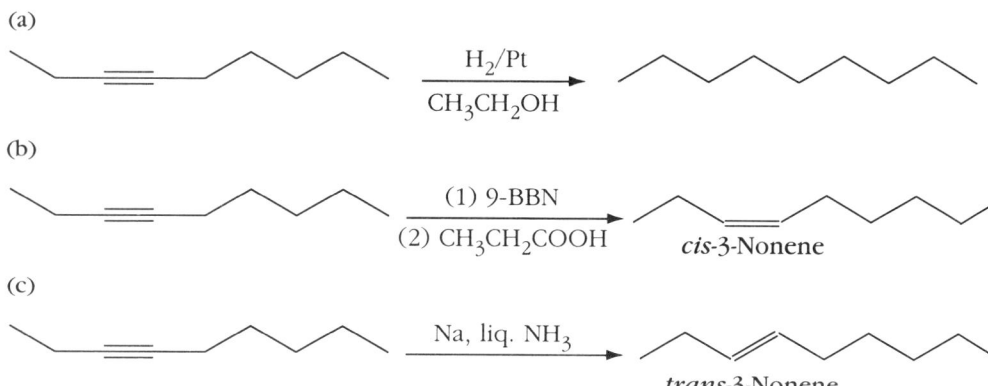

(b) [reaction giving cis-3-Nonene with (1) 9-BBN, (2) CH₃CH₂COOH]

(c) [reaction giving trans-3-Nonene with Na, liq. NH₃]

9.12 (i) We know that reaction of an internal alkyne with a solution of an alkali metal (usually sodium or lithium) in liquid ammonia gives a *trans*-alkene. That is why we should choose 2-hexyne as a starting material. (ii) One of the best methods to prepare *cis*-alkenes is hydrogenation of alkynes using Lindlar's catalyst. Hydroboration-protonation also produces a *cis*-alkene. Either of the two methods may be used in Problem 9.12 (b), although catalytic hydrogenation with deuterium gas seems to be a more practical choice for Problem 9.12 (c).

(a) [reaction with Na, liq. NH₃]

(b) [reaction with (1) 9-BBN, (2) CH₃CH₂COOH]

(c) [reaction with D₂, Lindlar's catalyst]

9.13 The ^{13}C NMR spectrum shown is not that of (*E*)-4-octene because it does not show peaks of sp^2-hybrid carbon atoms (δ 100–140). The spectrum shows only *four* signals in the region where sp^3-hybrid carbon atoms are found. Thus, the spectrum is that of octane.

[reaction of 4-octyne with H₂, Ni + Pd/C → octane]

9.14 Reaction of 1-hexyne with sodium hydride produces the acetylide ion. The reaction takes place because 1-hexyne is a much stronger acid than hydrogen.

[Reaction scheme: 1-hexyne ($pK_a \approx 25$) + NaH in Hexane → sodium hexynide + H_2 ($pK_a \approx 35$)]

[Reaction scheme: sodium hexynide + $CH_3(CH_2)_4CH_2$—H ($pK_a \approx 49$) ⇌ 1-hexyne ($pK_a \approx 25$) + $CH_3(CH_2)_4CH_2^-Na^+$]

Hexane can be used as the solvent because it is a much weaker acid than 1-hexyne and hexane does not react with acetylide ions.

9.15

[Reaction scheme: sodium hexynide + H_2O ($pK_a \approx 15.7$) ⇌ 1-hexyne ($pK_a \approx 25$) + NaOH]

Water is a stronger acid than 1-hexyne so the reaction occurs to form 1-hexyne, NaOH, and H_2 as the products.

9.16 The anions of the terminal alkynes are very strong nucleophiles and they easily substitute a halide on a primary carbon atom. This method is used to convert terminal alkynes into larger internal alkynes.

(a) [structure]   (b) [structure]

9.17 3-Heptyne is an unsymmetrical alkyne and it can be prepared by the alkylation of an acetylide ion in two different ways. We must use a terminal alkyne as the starting material so we detach the bond between the sp carbon and sp^3 carbon at each end of the product to see what haloalkane and acetylide ion pieces we need. The process of reasoning backward from the target molecule to suitable starting materials is known as **retrosynthesis**. A symbol used to indicate a retrosynthetic step is an open arrow written from product to suitable precursors or fragments of those precursors:

Target molecule ⟹ precursors

Thus,

[Retrosynthesis 1: 3-heptyne ⟹ 1-butynyl sodium + 1-bromopropane]

[Retrosynthesis 2: 3-heptyne ⟹ 1-bromobutane + sodium propynide]

Now we can write the complete chemical equation to show the synthesis of the 3-heptyne:

(a) [1-butynyl sodium + 1-bromopropane → 3-heptyne]

(b)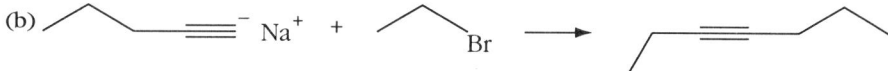

9.18 To design a synthesis, a chemist has to know what reactions can be used to give different types of functional groups. The reactions that create carbon-carbon bonds are among the most important. The design of a synthesis is quite complex and one should examine the structures of starting materials allowed and products. In the case of a multistep synthesis, it is best to work backwards from the target molecule to suitable starting materials. The name for this process is retrosynthesis and it is described on page 391 of the text. Let us follow those guidelines for our Exercises:

(a) Step 1: What are the functional groups and their positions on the carbon skeleton in the product?

In our case, the double bond starts at the first carbon of the chain and bromine is bonded to the second carbon.

What functional groups are there in the starting material? A carbon-carbon triple bond.

Step 2: What reaction(s) can we use to accomplish the conversion? Here we should review the changes that were made to convert the substrate into the product.

The alkyl group was added to one end of the alkyne. The triple bond was then converted to a double bond and bromine was added to the second carbon atom.

Is it possible to dissect the structures of the starting material and product to see which bonds must be broken and which formed?

$$H-C\equiv C \;\vdots\; H \qquad \underset{CH_3CH_2}{\overset{Br}{\diagdown}}C=C\overset{H}{\diagup}_{H}$$

Step 3: Choose the general chemical reaction. New bonds are created when an electrophile reacts with a nucleophile. Do we recognize any part of the product molecule as coming from a good nucleophile or an electrophile?

(i) The carbon-carbon bonds in the alkyne must be the result of nucleophilic substitution reactions. Acetylene can be converted into a good nucleophile. The alkyl group, attached to leaving group, must supply the electrophile. The alkyne can be put together from these reagents:

$$H-C\equiv C-H \qquad CH_3CH_2Br \qquad NaNH_2$$

(ii) How do we convert the alkyne into the product? An addition of one mole of hydrogen bromide to a terminal alkyne, 1-butyne, will give 2-bromo-1-butene.

Step 4: Write the chemical equation with the specific compounds that undergo the reaction to form the desired product.

$$HC\equiv CH \xrightarrow[\text{liq. NH}_3]{\text{NaNH}_2} HC\equiv C^-Na^+ \xrightarrow{CH_3CH_2Br} CH_3CH_2C\equiv CH \xrightarrow{HBr}$$

$$\underset{\overset{|}{Br}}{CH_3CH_2C}=CH_2$$

(b) Step 1: 2-Butanone is a methyl ketone.

Step 2: Methyl ketones are easily prepared from terminal alkynes. Thus, 1-butyne will be a good precursor to the product, 2-butanone.

Step 3: Acid-catalyzed or mercury-salt-catalyzed hydration of a terminal alkyne will form a methyl alkyl ketone. 1-Butyne, in turn, is synthesized from acetylene and bromoethane.

Step 4:

$$HC\equiv CH \xrightarrow[\text{liq. NH}_3]{\text{NaNH}_2} HC\equiv C^-Na^+ \xrightarrow{CH_3CH_2Br} CH_3CH_2C\equiv CH \xrightarrow[\text{HgSO}_4]{H_3O^+}$$

$$CH_3CH_2\overset{\overset{O}{\|}}{C}CH_3$$

(c) Step 1: The final product, 1-butanol, is a saturated primary alcohol.

Step 2: There is no one-step conversion of an alkyne that gives a saturated alcohol. As a result, we must think of another precursor that can be used. There is more than one good precursor. Examples include an aldehyde (butanal) or an alkene (1-butene).

Step 3: We know the reaction that converts 1-butene to 1-butanol. 1-Butene, in turn, can be obtained from 1-butyne by partial hydrogenation, and 1-butyne can be easily obtained from acetylene (ethyne).

Step 4: The following is a good method to synthesize 1-butanol:

$$HC\equiv CH \xrightarrow[\text{liq. NH}_3]{\text{NaNH}_2} HC\equiv C^-Na^+ \xrightarrow{CH_3CH_2Br} CH_3CH_2C\equiv CH$$

$$\downarrow \text{H}_2/\text{Lindlar's catalyst or Li or Na in liq. NH}_3$$

$$CH_3CH_2CH_2CH_2OH \xleftarrow[\text{(2) H}_2\text{O}_2, \text{HO}^-]{\text{(1) 9-BBN}} CH_3CH_2CH=CH_2$$

9.19 The yield of the final product in a multistep synthesis is the product of the yields of each individual step.

(Yield of 2-butyne from propyne) × (Yield of (Z)-2-butene from 2-butyne) = (Yield of (Z)-2-butene from propyne)

(0.97)(0.96) = 0.93 or **93%**

Answer: The yield of (Z)-2-butene from propyne is **93%**.

9.20 (a) An **alkyne** is a hydrocarbon containing a carbon-carbon triple bond (—C≡C—). A *terminal* alkyne has a triple bond at the end of a chain, with an acetylenic hydrogen. An *internal* alkyne has the triple bond somewhere other than at the end of the chain.

Acetylenic hydrogen

$CH_3CH_2CH_2C\equiv C-\boxed{H}$

1-Pentyne

No acetylenic hydrogen

$CH_3C\equiv CCH_2CH_3$

2-Pentyne

(b) An **enol** is an alkene containing an —OH group bonded to an sp^2 hybridized carbon atom. Most enols are unstable, spontaneously isomerizing to their carbonyl tautomer, called the *keto* form of the compound.

$$\underset{}{\overset{}{}}C=C\overset{OH}{}$$

(c) **Tautomerism** is a special type of isomerism in which two interconvertible constitutional isomers exist in equilibrium with each other and differ only in the location of a hydrogen and a double bond. Most typical examples of tautomers are keto-enol tautomers.

Enol form $\xrightarrow{H_3O^+ \text{ or } HO^-}$ Keto form

(d) A **vinyl cation** is a carbocation in which the positive charge is located at an sp^2 hybrid carbon atom.

A vinyl cation

(e) **Hydroboration-oxidation** is a sequence of two reactions. In the first reaction (hydroboration), a hydroborating reagent such as borane or 9-BBN reacts with an alkene or alkyne to form an organoborane. In the second reaction (oxidation), the organoborane is oxidized with hydrogen peroxide. The result of these two reactions is hydration (addition of —H and —OH) of carbon-carbon double bond or triple bond in an anti-Markownikoff manner.

$CH_3-C\equiv C-H \xrightarrow[(2)\ H_2O_2,\ NaOH]{(1)\ Sia_2BH\ THF}$ [Vinyl alcohol (Unstable)] ⟶ CH_3-CH_2-CHO (Propanal)

Sia = $CH_3\underset{CH_3}{\overset{H_3C}{CHCH}}-$

(f) An **acetylide ion** is an anion derived from a terminal alkyne by removal of a hydrogen atom from the *sp* hybrid carbon atom by a strong base such as sodium amide.

$$R-C\equiv C^-$$

(g) **Alkylation of the acetylide ion** is a reaction in which the acetylide ion reacts with a primary or secondary haloalkane to form alkynes with longer carbon chains. A new carbon-carbon bond is formed in this reaction.

$$R-C\equiv C-H \xrightarrow[\text{liq. NH}_3]{\text{NaNH}_2} R-C\equiv C^- \text{Na}^+ + R'-CH_2-X \longrightarrow R-C\equiv C-CH_2-R'$$

(h) **Hydroboration-protonolysis of an alkyne** is a sequence of two reactions that converts an internal alkyne into the (Z)-isomer of an alkene. In the first reaction, an alkyne reacts with a hydroborating reagent to form an organoborane. In the second reaction, the organoborane reacts with propionic acid to form an alkene.

9.21

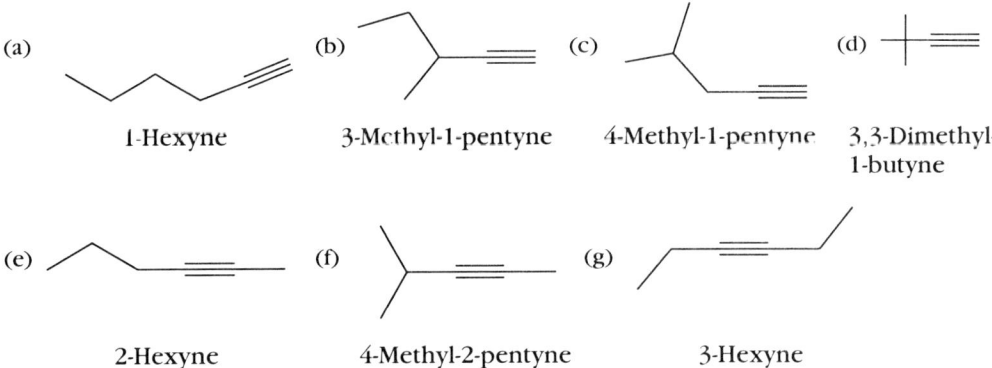

(a) 1-Hexyne (b) 3-Methyl-1-pentyne (c) 4-Methyl-1-pentyne (d) 3,3-Dimethyl-1-butyne

(e) 2-Hexyne (f) 4-Methyl-2-pentyne (g) 3-Hexyne

Compounds (a), (b), (c), and (d) are all terminal alkynes that will react with NaNH$_2$ in liquid ammonia.

Only compound (b) exists as a pair of enantiomers:

(R)-3-Methyl-1-pentyne   (S)-3-Methyl-1-pentyne

9.22 (a) 1,7-Nonadiyne
(b) 2,3,4-Trimethyl-1-hepten-6-yne
(c) 6-Phenyl-2,4-hexadiyne
(d) 4-Chloro-2-ethynyl-1-methylcyclohexane
(e) (3E,6Z)-3,4-Dimethyl-3,6-tetradecadien-1,10-diyne
(f) 1,7-Cyclododecadiyne

9.23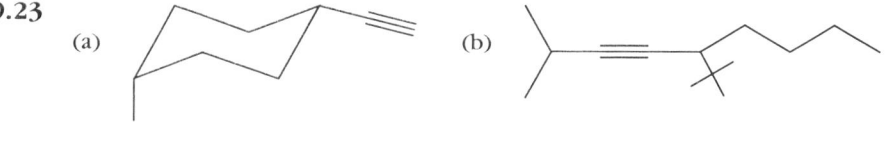

9.24 Alkynes react with electrophiles in electrophilic addition reactions. Exercises 9.24(a), (c), (d), (e), (g), and (l) involve the addition of an unsymmetrical reagent, such as an acid or 9-BBN, to 1-pentyne with different substituents on the two carbons of the triple bond. Such reactions show high regioselectivity, which is determined by the formation of the most stable intermediate. As a result, we observe that only one product is formed.

(g) [structure with two Cl on same carbon] (h) [structure with Na⁺] (i) [alkyne structure]

(j) [structure with Br groups] (k) [alkene structure] (l) [structure with OOCCF₃]

9.25 As in Exercise 9.24, some of the following reactions are electrophilic addition reactions. Exercises 9.25(a), (c), (d), (e), (g), and (l) involve the addition of an unsymmetrical reagent, such as an acid or 9-BBN, to 2-pentyne. However, 2-pentyne is an internal alkyne and each of the two carbons of the triple bond has an alkyl group. Such substrates do not show a regioselectivity, because both intermediates have very similar stability. 2-Pentyne is an unsymmetrical compound and we observe formation of two products in those cases.

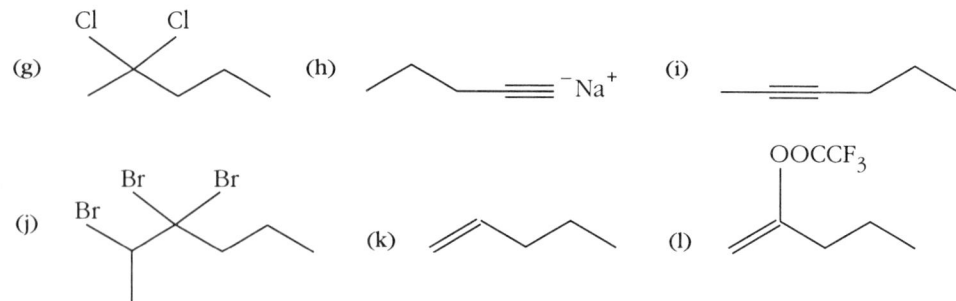

Both products are observed because the stability of both carbocations is about the same.

(a) [two chloroalkene products] (b) [pentane]

(c) [two ketone products] (d) [two ketone products]

(e) [alkene] (f) [dibromoalkene] (g) [two dichloro products]

(h) No reaction (i) No reaction (j) [structure: C(Cl)(Cl)-C(Cl)(Cl) with ethyl] (k) [trans-2-pentene]

(l) [2-methyl-2-butenyl trifluoroacetate] + [trifluoroacetate-substituted alkene] (m) [cis-2-butene]

9.26 The following Exercises are useful for testing your knowledge of organic reactions. You should analyze the structures given as well as the reagents and reaction conditions and it should become apparent what is missing.

(a) The product contains three functional groups: a carbon-carbon double bond, a carboxyl group, and a carbonyl group. The reaction is hydrogenation in the presence of Lindlar's catalyst. We know that Lindlar's catalyst catalyzes hydrogenation of an alkyne to a *cis*-alkene. We have a *cis*-alkene fragment in the product, so the starting material should contain a carbon-carbon triple bond in its place.

[Reaction scheme: alkyne with cyclopropyl ketone and COOH → H₂/Lindlar's catalyst → cis-alkene product with cyclopropyl ketone and COOH]

(b) In the reaction of a terminal alkyne with 9-BBN, boron adds to the terminal carbon atom of the triple bond, and hydrogen adds to the other carbon atom. After the oxidation step a vinyl alcohol is formed, which tautomerizes into the aldehyde.

[Reaction scheme: bicyclic terminal alkyne + H–B (9-BBN) → vinyl borane; then H₂O₂/HO⁻ → aldehyde product]

(c) Acetylenic anions undergo S_N2 reactions with haloalkanes to form internal alkynes.

(d) This reaction converts a carbon-carbon triple bond into a *cis* carbon-carbon double bond. We know that such a reaction takes place when Lindlar's catalyst is used to catalyze an addition of hydrogen gas to an alkyne.

9.27 The starting material, 1-butyne, is given along with a target molecule in each problem. We begin with an examination of the target compound and then examine possible intermediates and synthetic routes.

(a) The desired product is a vicinal dibromide (a compound containing two bromine atoms bonded to adjacent carbon atoms). We know that vicinal dibromides can be prepared by adding bromine to alkenes so 1,2-dibromobutane can be prepared by adding Br_2 to 1-butene:

$$CH_2=CHCH_2CH_3 + Br_2 \xrightarrow{CCl_4} BrCH_2CHBrCH_2CH_3$$

1-Butene can be prepared by catalytic hydrogenation of 1-butyne. Lindlar's catalyst can be used because it reduces an alkyne only to an alkene and the stereochemistry of the alkene is not important in this case:

$$HC\equiv CCH_2CH_3 \xrightarrow{H_2/\text{Lindlar's catalyst}} CH_2=CHCH_2CH_3$$

The complete synthesis is the following:

$$HC\equiv CCH_2CH_3 \xrightarrow{H_2/\text{Lindlar's catalyst}} CH_2=CHCH_2CH_3 + Br_2 \xrightarrow{CCl_4}$$

$$BrCH_2CHBrCH_2CH_3$$

(b) The product is a bromohydrin, which can be prepared by addition of an aqueous bromine solution to an alkene:

$$CH_2\!=\!CHCH_2CH_3 \xrightarrow[H_2O]{Br_2} BrCH_2\overset{\underset{\mid}{OH}}{C}HCH_2CH_3$$

1-Butene can be prepared by partial hydrogenation of 1-butyne as in Exercise 9.27(a), so the complete synthesis is the following:

$$HC\!\equiv\!CCH_2CH_3 \xrightarrow{H_2/\text{Lindlar's catalyst}} CH_2\!=\!CHCH_2CH_3 \xrightarrow[H_2O]{Br_2} BrCH_2\overset{\underset{\mid}{OH}}{C}HCH_2CH_3$$

(c) The desired product is a chloroalkane. Addition of HCl to an alkene is one way of preparing a chloroalkane. 2-Chloroalkane can be prepared in either of the following addition reactions:

$$CH_2\!=\!CHCH_2CH_3 \xrightarrow[(CH_3CH_2)_2O]{HCl} CH_3\overset{\underset{\mid}{Cl}}{C}HCH_2CH_3$$

$$CH_3CH\!=\!CHCH_3 \xrightarrow[(CH_3CH_2)_2O]{HCl} CH_3\overset{\underset{\mid}{Cl}}{C}HCH_2CH_3$$

In this case, addition of HCl to 1-butene is the preferred synthesis because 1-butene can easily be prepared from 1-butyne and it is not so easy to prepare 2-butene from 1-butyne. The complete synthesis is the following:

$$HC\!\equiv\!CCH_2CH_3 \xrightarrow{H_2/\text{Lindlar's catalyst}} CH_2\!=\!CHCH_2CH_3 \xrightarrow[(CH_3CH_2)_2O]{HCl} CH_3\overset{\underset{\mid}{Cl}}{C}HCH_2CH_3$$

(d) The desired product is 1-butene. It can be easily prepared from 1-butyne by a reduction using hydrogen and Lindlar's catalyst:

$$HC\!\equiv\!CCH_2CH_3 \xrightarrow{H_2/\text{Lindlar's catalyst}} CH_2\!=\!CHCH_2CH_3$$

(e) The target molecule is butanal. Aldehydes can be synthesized from terminal alkynes by hydroboration-oxidation. 9-BBN will show much better regioselectivity than borane. Moreover, the bulky alkyl part of 9-BBN prevents a second addition to the boron-substituted alkene. The synthesis is the following:

$$\text{HC}\!\equiv\!\text{CCH}_2\text{CH}_3 \xrightarrow[(2)\ H_2O_2,\ HO^-]{(1)\ 9\text{-BBN}} \text{CH}_3\text{CH}_2\text{CH}_2\text{CHO}$$

(f) The desired product is 2-butanone, a methyl ketone. Acid-catalyzed addition of water to an alkyne gives an enol, which immediately rearranges to a ketone. Terminal alkynes are less reactive toward addition of water.

The mercuric ion acts as a catalyst to increase the rate of the addition reaction. The synthesis is the following:

(g) The desired product is 3-heptyne, an internal alkyne. It contains three more carbon atoms than the starting material, 1-butyne. One reaction that results in the formation of a new carbon-carbon bond is the reaction of an acetylide anion with an alkyl halide. We should just choose the appropriate size of the haloalkane (in this case bromo- or iodopropane). The synthesis is the following:

9.28 The starting material, 1-hexyne, is given along with a target molecule in each problem. Begin with an examination of the target compound and then examine possible intermediates and synthetic routes.

(a) Hydrogenation of 1-hexyne in the presence of a metal catalyst such as platinum, palladium, or nickel accomplishes this synthesis in one step.

(b) Addition of 1 equivalent of HCl will give 2-chloro-1-hexene because addition of hydrogen halide occurs preferentially to the alkyne rather than to the halogen-substituted alkene.

(c) The desired product is 2-hexanone, a methyl ketone. Acid-catalyzed addition of water to an alkyne gives an enol, which immediately rearranges to a ketone. Terminal alkynes are less reactive toward addition of water. The mercuric ion acts as a catalyst to increase the rate of the addition reaction. The synthesis is the following:

(d) The desired product is 2-heptyne, an internal alkyne. It contains one more carbon atom than the starting material, 1-hexyne. The acetylide anion of 1-hexyne can react with bromo- or iodomethane to form 2-heptyne. The synthesis is the following:

$$\text{1-hexyne} \xrightarrow[\text{liq. NH}_3]{\text{NaNH}_2} \text{hexynyl}^- \text{Na}^+ \xrightarrow{\text{CH}_3\text{Br}} \text{2-heptyne}$$

(e) The target molecule is hexanal. Aldehydes can be synthesized from terminal alkynes by hydroboration-oxidation:

$$\text{1-hexyne} \xrightarrow[\text{(2) H}_2\text{O}_2,\ \text{HO}^-]{\text{(1) 9-BBN}} \text{hexanal}$$

9.29 Exercises 9.27 and 9.28 dealt with synthetic problems involving, in most cases, one-step reactions. The following Exercises require more than one reaction to synthesize desired products from the starting materials given. Chemists find that it is easier to design a multistep synthesis by working backward from products to reactants. This process is known as retrosynthesis, and it is described in Exercises 9.17 and 9.18. We will use retrosynthesis to solve the following Exercises.

(a) 2-Chlorooctane can be obtained by addition of HCl to 1-octene. 1-Octene can be prepared by a reaction of the acetylide anion with 1-halohexane. Those observations can be expressed in the following way by retrosynthesis:

$$\text{2-chlorooctane} \Rightarrow \text{HCl} + \text{1-octene}$$

$$\Rightarrow \text{HC}\equiv\text{C---hexyl} \Rightarrow \text{HC}\equiv\text{C}^-\text{Na}^+ + \text{1-bromohexane}$$

$$\text{HC}\equiv\text{CH} \xrightarrow[\text{liq. NH}_3]{\text{NaNH}_2} \text{HC}\equiv\text{C}^-\text{Na}^+ + \text{hexyl-Br} \longrightarrow \text{1-octyne}$$

$$\downarrow \text{H}_2/\text{Lindlar's catalyst}$$

$$\text{2-chlorooctane} \xleftarrow{\text{HCl}} \text{1-octene}$$

The proposed synthesis is the following:

The proposed synthesis is not the only possible one. Synthetic chemists simultaneously consider time, cost, and yield. A well-designed synthesis should require as few steps as possible and those steps should each involve a reaction that is easy to do. In the case of converting 1-octyne to 2-chlorooctene, one may suggest adding HCl to 1-octyne to form a vinyl chloride (2-chloro-1-octene), which can be hydrogenated in the presence of a metal catalyst. The latter reaction, however, involves the cleavage of a carbon-halide bond to yield octane, an undesired side-product.

(b) The product is *meso*-2,3-dibromobutane, a vicinal dibromide (a compound containing two bromine atoms bonded to adjacent carbon atoms, C2 and C3). We know that vicinal dibromides can be prepared by adding bromine to alkenes so 2,3-dibromobutane can be prepared by adding Br_2 to 2-butene. There are two 2-butenes: *cis*- and *trans*-2-butene. Let us check which one will give *meso*-2,3-dibromobutane. We know that addition of halogens to the double bond proceeds with *anti* stereochemistry.

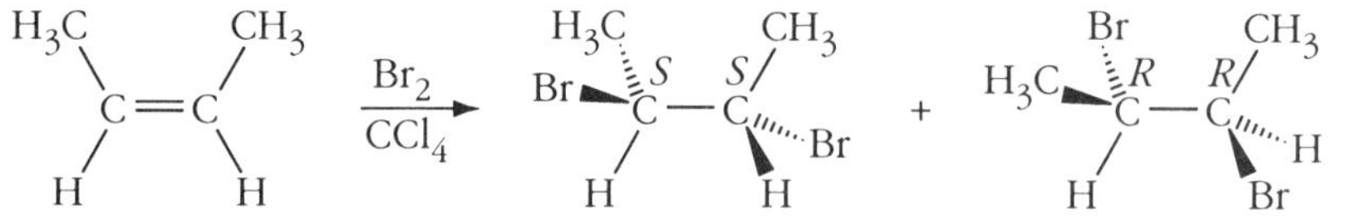

A mixture of enantiomers

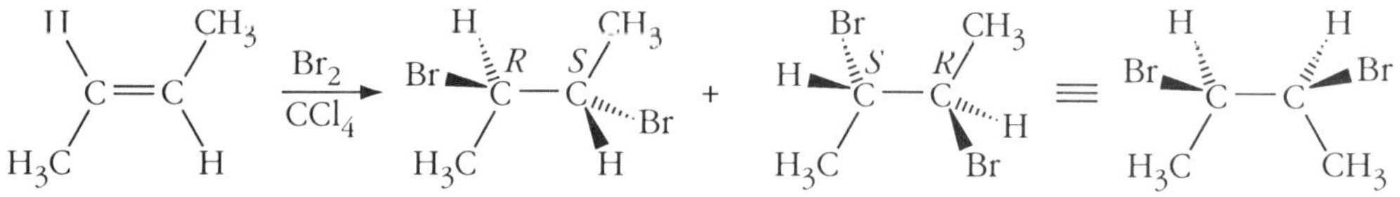

meso compound

Conclusion: *trans*-2-Butene will give *meso*-2,3-dibromobutane, so we need to figure out how to prepare *trans*-2-butene. *trans*-2-Butene can be prepared from 2-butyne by reduction with sodium or lithium in liquid ammonia and 2-butyne can be prepared by methylation of the acetylide anion. The complete retrosynthesis is the following:

CH_3I + $HC\equiv C^- Na^+$ ⇐ $CH_3C\equiv C^- Na^+$ + CH_3I

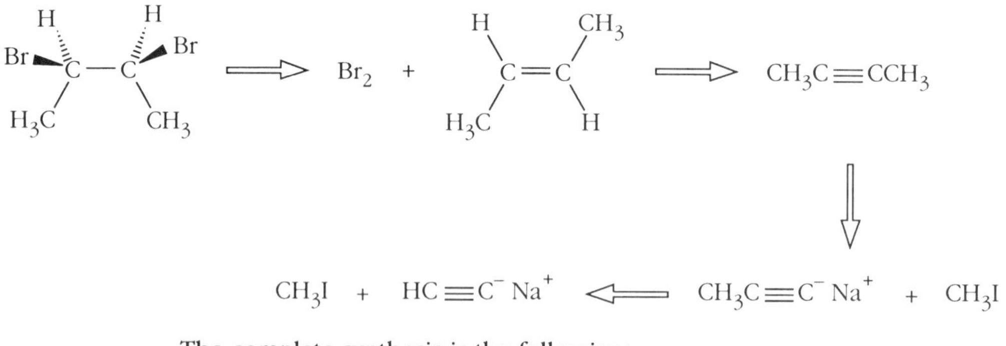

The complete synthesis is the following:

$$\equiv \xrightarrow[\text{liq. NH}_3]{\text{NaNH}_2} \equiv^- \text{Na}^+ \xrightarrow{\text{CH}_3\text{I}} -\!\!\!\equiv\!\!\!- \xrightarrow[\text{liq. NH}_3]{\text{NaNH}_2} -\!\!\!\equiv^- \text{Na}^+$$

$$\downarrow \text{CH}_3\text{I}$$

[structure: meso-2,3-dibromobutane with H₃C, CH₃, H, H, Br, Br] ≡ [2,3-dibromobutane with wedge/dash bonds] ←—Br₂— [2-butene] ←——Na, liq. NH₃—— [2-butyne] ←—— [from CH₃I step above]

(c) 1-Pentanal can be easily prepared from 1-pentene, and 1-pentene from 1-pentyne. The retrosynthesis is the following:

pentanal ⟹ 1-pentene ⟹ ≡⁻Na⁺ + CH₃CH₂CH₂—I

The proposed synthesis is the following:

$$\equiv \xrightarrow[\text{liq. NH}_3]{\text{NaNH}_2} \equiv^- \text{Na}^+ \;+\; \diagup\!\!\!\diagdown\!\!\!\diagup\!\!\text{I} \longrightarrow \text{1-pentyne}$$

$$\xrightarrow[\text{(2) H}_2\text{O}_2,\,\text{HO}^-]{\text{(1) 9-BBN}} \text{pentanal}$$

(d) 1-Pentanol can be prepared by a hydroboration-oxidation of 1-pentene; 1-pentene can be obtained by reduction of 1-pentyne; 1-pentyne is formed when the acetylide anion is alkylated with a halopropane. The retrosynthesis is the following:

1-pentanol ⟹ 1-pentene ⟹ 1-pentyne ⟹ ≡⁻Na⁺ + CH₃CH₂CH₂—Br

The proposed synthesis is the following:

$$\equiv \xrightarrow[\text{liq. NH}_3]{\text{NaNH}_2} \equiv^- \text{Na}^+ \xrightarrow{\text{CH}_3\text{CH}_2\text{CH}_2\text{Br}} \text{1-pentyne} \xrightarrow[\text{Lindlar's catalyst}]{\text{H}_2} \text{1-pentene}$$

$$\xrightarrow[\text{(2) H}_2\text{O}_2,\,\text{HO}^-]{\text{(1) 9-BBN}} \text{1-pentanol}$$

(e) 1-Bromohexane can be prepared by the addition of HBr to 1-hexene in the presence of radicals; 1-hexene can be prepared from 1-hexyne, etc. The retrosynthesis is the following:

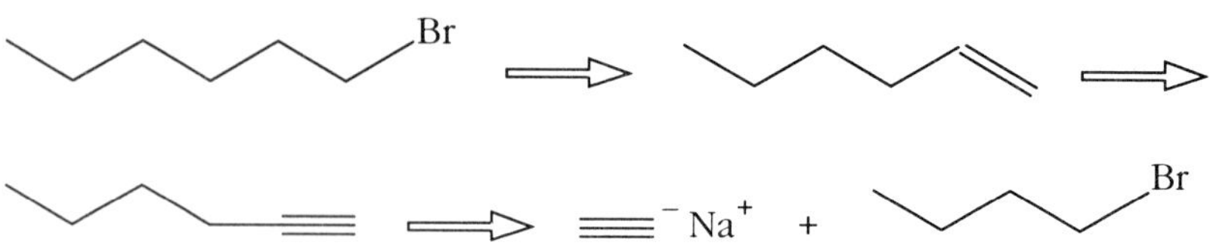

[retrosynthesis scheme: 1-bromohexane ⇒ 1-hexene ⇒ 1-hexyne ⇒ sodium acetylide + 1-bromobutane]

The synthesis is the following:

$$\text{HC≡CH} \xrightarrow[\text{liq. NH}_3]{\text{NaNH}_2} \text{HC≡C}^- \text{Na}^+ \xrightarrow{\text{CH}_3\text{CH}_2\text{CH}_2\text{CH}_2\text{Br}} \text{1-hexyne} \xrightarrow[\text{Lindlar's catalyst}]{\text{H}_2} \text{1-hexene}$$

(1) BH$_3$/THF
(2) BR$_2$, NaOCH$_3$/CH$_3$OH

↓

[1-bromohexane]

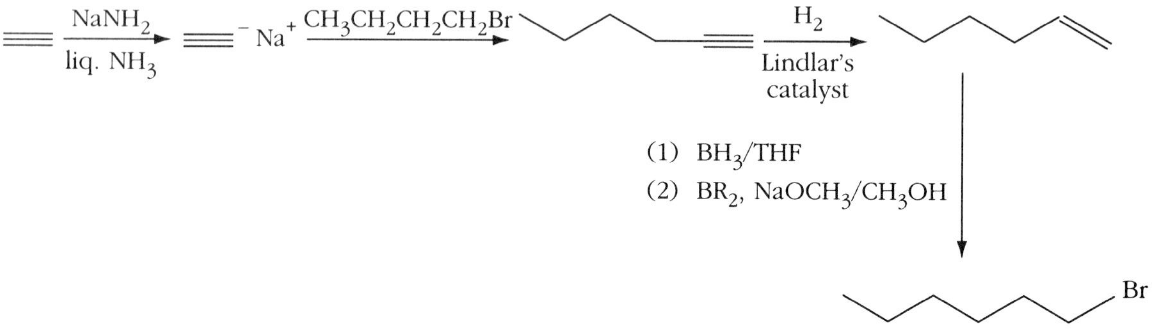

(f) Racemic 2,3-pentanediol is a vicinal diol (or 1,2-diol), which can be prepared from 2-pentene by hydroxylation (reaction with OsO$_4$). The retrosynthesis is the following:

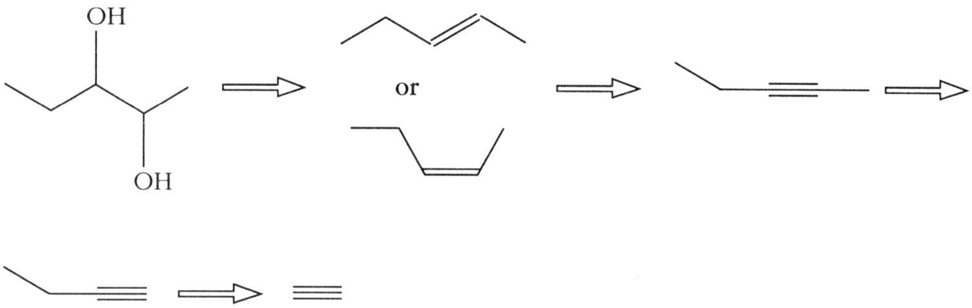

[retrosynthesis: 2,3-pentanediol ⇒ trans- or cis-2-pentene ⇒ 2-pentyne ⇒ propyne + ...]

The synthesis is the following:

(g) 3-Hexanone can be prepared by hydration of 3-hexyne because it is a symmetrical alkyne, so it will give only one product on hydration. The retrosynthesis is the following:

The synthesis is the following:

$$\equiv \xrightarrow[\text{liq. NH}_3]{\text{NaNH}_2} \equiv^-\text{Na}^+ \xrightarrow{\text{CH}_3\text{CH}_2\text{Br}} \diagup\!\!\!\equiv \xrightarrow[\text{liq. NH}_3]{\text{NaNH}_2} \diagup\!\!\!\equiv^-\text{Na}^+$$

$$\Bigg\downarrow \text{CH}_3\text{CH}_2\text{I}$$

$$\text{(3-hexanone)} \xleftarrow[\text{HgSO}_4]{\text{H}_3\text{O}^+} \diagup\!\!\!\equiv\!\!\!\diagdown$$

(h) 2-Hexanone is a methyl ketone and it is best prepared by the mercuric ion-catalyzed hydration of a terminal alkyne, which in this case is 1-hexyne.

(2-hexanone) ⇒ (1-hexyne) ⇒ (acetylene)

The synthesis is the following:

$$\equiv \xrightarrow[\text{liq. NH}_3]{\text{NaNH}_2} \equiv^-\text{Na}^+ \xrightarrow{\text{CH}_3\text{CH}_2\text{CH}_2\text{CH}_2\text{Br}} \text{1-hexyne}$$

$$\xrightarrow[\text{HgSO}_4]{\text{H}_3\text{O}^+} \text{2-hexanone}$$

(i) 1-Deutero-1-heptyne can be readily prepared from the acetylide anion of 1-heptyne by hydrolysis with D$_2$O. Acetylide anions are very strong bases because they are conjugate bases of weak acids (pK_a ≈ 25) and they hydrolyze instantly in the presence of water. The retrosynthsis is the following:

(1-deutero-1-heptyne)—D ⇒ (1-heptyne) ⇒ ≡

The synthesis is the following:

$$\equiv \xrightarrow[\text{liq. NH}_3]{\text{NaNH}_2} \equiv^-\text{Na}^+ \xrightarrow{\text{CH}_3(\text{CH}_2)_4\text{Br}} \text{1-heptyne}$$

$$\Bigg\downarrow \text{NaNH}_2, \text{ liq. NH}_3$$

$$\text{1-deutero-1-heptyne}—D \xleftarrow{\text{D}_2\text{O}} \text{heptynyl}^-\text{Na}^+$$

(j) (Z)-2,3-Dideutero-5-methyl-2-hexene can be prepared by reduction of 5-methyl-2-hexyne with D_2 in the presence of Lindlar's catalyst. The retrosynthesis is the following:

The synthesis:

(k) (E)-2,3-Dideutero-2-undecene can be prepared by the metal reduction of 2-undecyne. Either lithium or sodium metal can be used and the deuterium must be provided by the solvent. Both liquid ammonia and liquid amines are successfully used for that purpose. It is much more convenient to prepare and use a deuterated amine rather than deuterated ammonia. Any deuterated amine can be used. The retrosynthesis is the following:

The synthesis is the following:

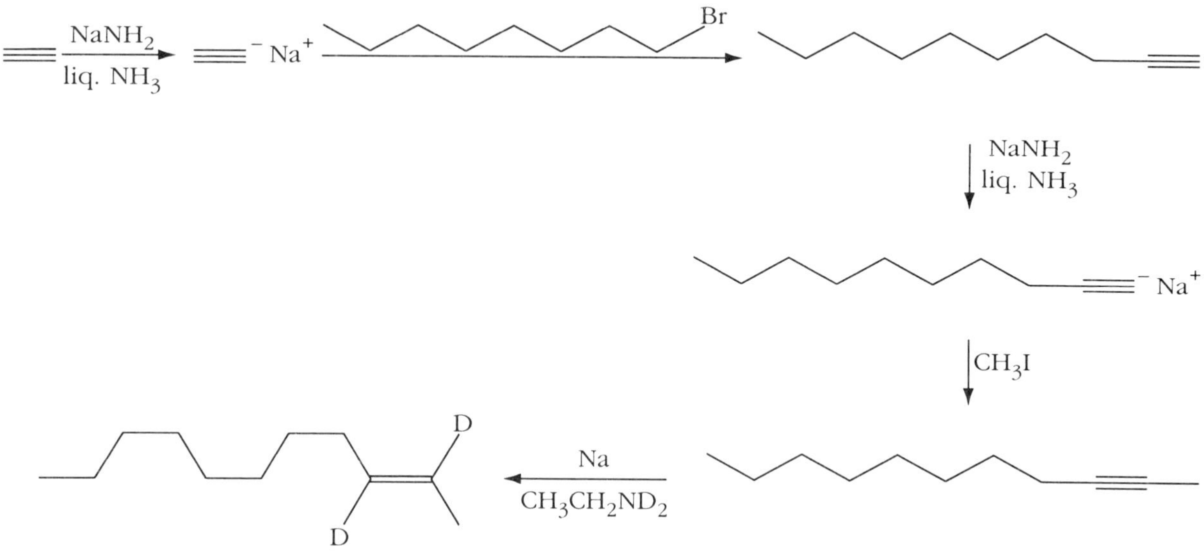

$\equiv\!\!\equiv \xrightarrow[\text{liq. NH}_3]{\text{NaNH}_2} \equiv\!\!\equiv^- \text{Na}^+ \xrightarrow{\text{CH}_3(\text{CH}_2)_n\text{Br}}$ [alkyl-terminal alkyne]

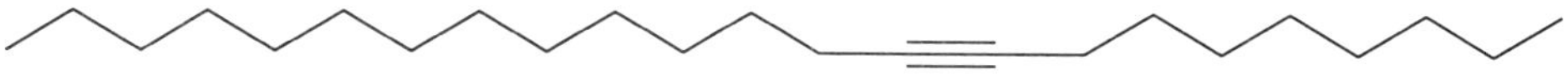

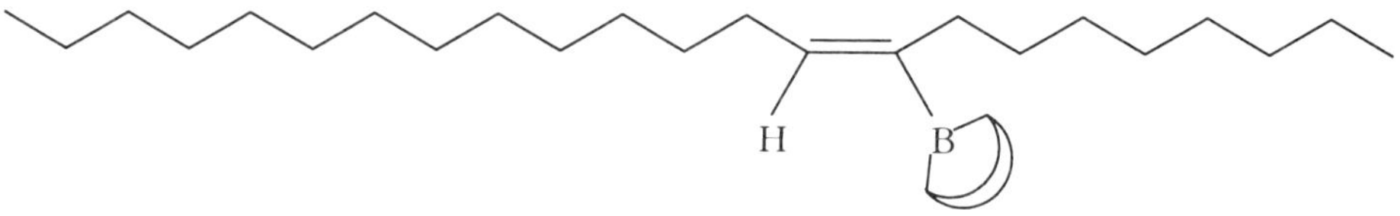

$\downarrow \text{NaNH}_2 \;/\; \text{liq. NH}_3$

[alkynide] $\equiv\!\!\equiv^- \text{Na}^+$

$\downarrow \text{CH}_3\text{I}$

[internal alkyne]

[alkene with D, D] $\xleftarrow[\text{CH}_3\text{CH}_2\text{ND}_2]{\text{Na}}$

9.30

A = [long chain alkynide]$\equiv\!\!\equiv^-\text{Na}^+$

B = [long chain internal alkyne]

C = [long chain alkene with H on one side and B (partial bond) on the other]

Muscalure = [long chain cis alkene with H, H]

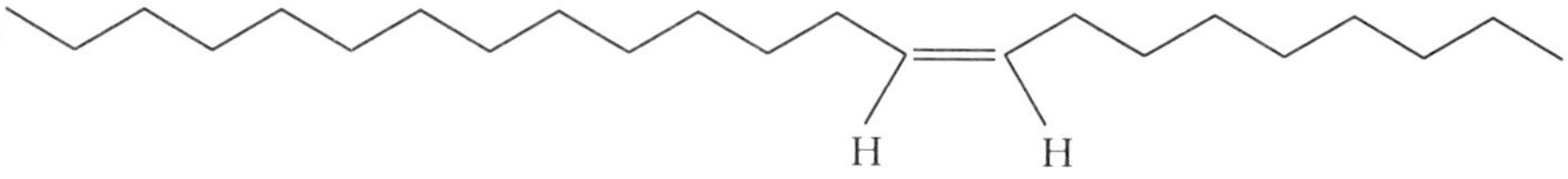

9.31

$-\text{C}\!\equiv\!\text{C}-$ 2260–2100 cm^{-1} $\text{H}-\text{C}\!\equiv\!\text{C}-$ 3300–3260 cm^{-1}

$\diagdown\!\!\text{C}\!=\!\text{C}\!\diagup$ 1680–1640 cm^{-1} $\diagdown\!\!\text{C}\!=\!\text{C}\!\diagup$ (with H)   3100–3040 cm^{-1}

There are significant differences between the IR spectra of alkynes and alkenes. Unsymmetrical alkynes have a characteristic carbon-carbon absorption in the 2260–2100 cm^{-1} region whereas alkenes show no absorption in this region. Unsymmetrical alkenes, on the other hand, have a characteristic carbon-carbon absorption in the 1680–1640 cm^{-1} region whereas alkynes show no absorption in this region. The hydrogen-carbon absorption can be used to decide if we have an alkyne or an alkene.

To differentiate between an internal or terminal alkyne we may also use the carbon-hydrogen bond stretching absorptions:

an internal alkyne has no $\equiv$C—H band at 3300–3260 cm^{-1};
the terminal alkyne shows absorption at 3300–3260 cm^{-1}.

9.32 The ^{13}C NMR spectrum of an alkyne would have signals in the region of δ 65 to 70 due to the carbon atoms of the triple bond. The ^{13}C NMR spectrum of an alkene or a diene would have signals in the region of δ 120 to 128 due to the carbon atoms of the double bond. The ^{13}C NMR spectrum shows peaks at δ 68 and 84 and no signals at δ 120 to 128. Thus, the compound is not a diene but an alkyne. It could be 1-hexyne, 2-hexyne, or 3-hexyne. 3-Hexyne is a symmetrical molecule and it has three signals in its ^{13}C NMR spectrum. There are six signals in the spectrum so the unknown cannot be 3-hexyne.

Moreover, it is a terminal alkyne. Hence, the compound must be 1-hexyne.

9.33 The amount of energy that a bond absorbs depends on the change in the bond moment during the absorption of a large amount of energy. Symmetrically substituted unsaturated bonds do not absorb infrared radiation because there is no change in bond moment when the bonds oscillate. 3-Hexyne is a symmetrical molecule so it will not show absorption at 2100 to 2250 cm^{-1}, which is characteristic of bonds between sp hybrid carbon atoms (—C≡C—). Because it is an internal alkyne, we also do not observe a characteristic sp C—H absorption at 3300 to 3260 cm^{-1}. So, although we do not observe any absorption bands typical of an alkyne, the chemist does *not* have to be concerned.

9.34

Compound	IR (important absorption bands)	^{13}C NMR (approximate chemical shifts):
(a) H$_2$C=CHCH$_2$CH$_2$CH=CH$_2$	3100–3040 cm^{-1} (sp^2 C—H) 1680–1640 cm^{-1} (C=C)	Three signals: δ 125 (—CH=) δ 118 (=CH$_2$) δ 30 (—CH$_2$—)
(CH$_3$)$_2$CHCH$_2$C≡CH	3300–3260 cm^{-1} (sp C—H) 2260–2100 cm^{-1} (C≡C)	Five signals: δ 76 (terminal sp C) δ 67 (internal sp C) δ 34 (>CH—) δ 28 (—CH$_2$—) δ 19 (—CH$_3$)

Both the IR and ^{13}C NMR spectra are very informative and each can be used to distinguish between the diene and the alkyne. Thus, in the IR spectrum, one should look for an absorption at 3300 to 3260 cm^{-1} (terminal alkyne, if present). If not present, check for a band at around 3080 cm^{-1} (alkene). Furthermore, the multiple carbon-carbon bonds show characteristic absorption bands: carbon-carbon triple bond at 2260 to 2100 cm^{-1}; carbon-carbon double bond at 1680 to 1640 cm^{-1}. Both the alkene and alkyne are unsymmetrical and therefore have permanent dipoles, so the stretching of multiple carbon-carbon bonds will have a prominent absorption band.

The ^{13}C NMR spectra also differ substantially. The diene is symmetrical and will have only three signals, including two in the δ 120 to 130 ppm region. The alkyne is unsymmetrical and will show five signals in its spectrum in the δ 20 to 80 ppm region.

(b) CH$_3$C≡CCH$_3$ Two signals:
                                                                                                δ 65 (internal sp C)
                                                                                                δ 25 (—CH$_3$)

H$_2$C=CHCH=CH$_2$ 3100-3040 cm^{-1} (sp^2 C—H)   Two signals:
                                     1680-1640 cm^{-1} (C=C)   δ 125 (—CH=)
                                                                                               δ 118 (=CH$_2$)

The IR spectrum is very informative. 2-Butyne is a symmetrical internal alkyne, so there is no significant dipole moment associated with the C≡C bond, and no large change in dipole moment associated with bond stretching. Thus we do not observe any signals in the IR spectrum typical of an alkyne. 1,3-Butadiene, on the other hand, will show strong absorption of both C=C stretching and sp^2 C—H stretching (at 1680 to 1640 and 3100 to 3040 cm^{-1}, respectively).

The ^{13}C NMR spectrum of each compound will show only two peaks. One may be tempted to distinguish 2-butyne from 1,3-butadiene based on expected difference in chemical shifts of carbon atoms: sp^2-hybridized carbon atoms appear at δ 120 to 130 ppm.

(c)
            O
            ‖
HC≡CCH$_2$CCH$_3$ 3300-3260 cm^{-1} (sp C—H)   Five signals:
1-Pentyn-4-one   2260-2100 cm^{-1} (C≡C)   δ 210 (C=O)
                            1700 cm^{-1} (C=O)   δ 76 (terminal sp C)
                                                                      δ 65 (internal sp C)
                                                                      δ 35 (—CH$_2$—)
                                                                      δ 19 (—CH$_3$)

HC≡CCH=CHCH$_2$OH   3500-3200 cm^{-1} (O—H)   Five signals:
2-Penten-4-yn-1-ol   3300-3260 cm^{-1} (sp C—H)   δ 125 (—CH=)
                                     3100-3040 cm^{-1} (sp^2 C—H)   δ 118 (=CH—)
                                     2260-2100 cm^{-1} (C≡C)   δ 76 (terminal sp C)
                                     1680-1640 cm^{-1} (C=C)   δ 70 (—CH—OH)
                                                                          δ 65 (internal sp C)

In this case, the ^{13}C NMR spectrum of 1-pentyn-4-one is distinct from that of 2-penten-4-yn-1-ol by the presence of a peak at δ 210 due to the carbonyl carbon atom.

In the IR spectrum, 1-pentyn-4-one does *not* show a strong absorption band at 3500 to 3200 cm^{-1} while 2-penten-4-yn-1-ol *does* (—O—H stretching). Furthermore, 1-pentyn-4-one shows a strong absorption band at about 1700 cm^{-1} (carbonyl group) while 2-penten-4-yn-1-ol does *not* show absorption in that region. Moreover, one should see a clear absorption of C=C stretching at 1680 to 1640 cm^{-1} (2-penten-4-yn-1-ol) and the lack of such absorption in the spectrum of 1-pentyn-4-one.

9.35 (a) Molecular formula: C$_3$H$_4$O; degree of unsaturation: 2; unsaturated compound (alkyne?); ^{13}C NMR spectrum shows three signals, two in the range δ 70 to 90 (alkyne) and one at δ 50 (carbon bonded to an oxygen atom).

The IR spectrum is even more informative because it shows a very strong absorption at 3500 to 3100 cm^{-1} (an OH group) and a sharp absorption at 2170 cm^{-1} (C≡C stretch). Hence, the unknown compound is **2-propyn-1-ol**.

HC≡CCH$_2$OH

(b) Molecular formula: C_5H_8; degree of unsaturation: 2; unsaturated compound (alkyne, diene, or cycloalkene).

The ^{13}C NMR spectrum shows five signals, two in the range δ 73 to 81 (internal alkyne); there are no signals at δ 120 to 130 (no alkene). The IR spectrum lacks the presence of absorption bands characteristic of terminal alkynes. Even the C≡C stretch is not observed, which confirms that there are two alkyl groups attached to the sp-hybridized carbon atoms, making the molecule quite symmetrical. Hence, the unknown compound is **2-pentyne**.

$$CH_3C{\equiv}CCH_2CH_3$$

(c) Molecular formula: C_8H_6; degree of unsaturation: 6; unsaturated compound. The ^{13}C NMR spectrum shows six signals, two in the range δ 77 to 84 (alkyne) and four signals in the δ 120 to 135 range. The IR spectrum shows the presence of absorption characteristic of a terminal alkyne: 3300 cm^{-1} (sp C—H) and 2100 cm^{-1} (C≡C stretch). There are also carbon-hydrogen stretching vibrations with frequency above 3000 cm^{-1} (aromatic ring). The presence of a monosubstituted benzene ring is supported by a series of overtones in the 1900 to 1700 cm^{-1} region. The presence of a monosubstituted ring is also consistent with the ^{13}C NMR spectrum (four signals). Hence, the unknown is **phenylethyne** (phenylacetylene).

Ph—C≡CH

9.36 (a) Molecular formula: C_5H_8; degree of unsaturation: 2; unsaturated hydrocarbon: alkyne, diene, or cycloalkene.

The ^{13}C NMR spectrum shows five signals, two signals in the range δ 68 to 84 (alkyne) and no signals in the δ 120 to 135 range (no alkene). The IR spectrum shows the presence of absorption characteristic of a terminal alkyne: 3300 cm^{-1} (sp C—H) and 2120 cm^{-1} (C≡C stretch).
Hence, the structure is **1-pentyne**.

$$CH_3CH_2CH_2C{\equiv}CH$$

(b) Molecular formula: C_5H_7Cl; degree of unsaturation: 2; unsaturated compound: alkyne, diene, or cycloalkene.

The ^{13}C NMR spectrum shows four signals: two signals in the range δ 70 to 87 (alkyne) and no signals in the δ 120 to 135 range (no alkene). The chemical shift of δ 56.9 is consistent with one expected for a carbon atom bonded to chlorine. The signal at δ 34.5 could be the one next to the carbon bonded to chlorine. We observe four signals for five carbon atoms, which means that there is some symmetry in the molecule and one of the signals represents two carbon atoms. It cannot be either of the two representing sp-hybridized carbon atoms (δ 71.8 and δ 86.5). It cannot be the one with chlorine because there is only one chlorine atom in the molecular formula. Therefore, by the process of elimination, it must be the signal at δ 34.5 that represents two carbon atoms.

The IR spectrum shows the presence of absorption characteristic of a terminal alkyne: 3290 cm^{-1} (sp C—H) and 2100 cm^{-1} (C≡C stretch). Hence, the structure is **3-chloro-3-methyl-1-butyne**.

$$H_3C-\underset{\underset{CH_3}{|}}{\overset{\overset{Cl}{|}}{C}}-C\equiv CH$$

(c) Molecular formula: $C_3H_2O_2$; degree of unsaturation: 3; unsaturated compound: alkyne, alkene, carbonyl group, or ring.

The ^{13}C NMR spectrum shows three signals, two signals in the range δ 70 to 80 (alkyne) and no signals in the δ 120 to 135 range (no alkene). The signal at δ 157.3 is characteristic of the carboxyl carbon atom. *Note:* The carboxyl carbon atom is shielded by the neighboring carbon-carbon triple bond because it usually appears at δ 170 to 185.

The IR spectrum shows a very strong and broad absorption band from 3500 to 2600 cm^{-1}, which is characteristic of the oxygen-hydrogen single bond of —COOH. This band most likely overlaps with the band at 3300 cm^{-1}, which is the stretching frequency for the terminal *sp*-hybridized carbon-hydrogen bond. Another characteristic absorption of terminal alkynes is the C≡C stretch appearing at 2100 cm^{-1}.

Hence, the structure is **propynoic acid**.

$$HC\equiv C-COOH$$

(d) Molecular formula: C_4H_7N; degree of unsaturation: 2; unsaturated compound: alkyne, alkene, nitrile, or ring.

The ^{13}C NMR spectrum shows four signals: three signals in the range δ 13 to 20 (alkane) and one signal at δ 119.8 (carbon-nitrogen triple bond). The presence of a nitrile functional group is confirmed by a characteristic infrared absorption band at 2240 cm^{-1}.

Butanenitrile or 2-methylpropanenitrile are possible structures. However, 2-methylpropanenitrile has two equivalent methyl groups and will show only three signals in the ^{13}C NMR spectrum. Hence, the structure is **butanenitrile**.

$$CH_3CH_2CH_2C\equiv N$$

CHAPTER 10

NUCLEAR MAGNETIC RESONANCE SPECTROSCOPY
Concepts

When spinning nuclei of ^{1}H or ^{13}C are placed in an applied magnetic field, their spins orient either with or against the field. On irradiation with radio frequency radiation, energy is absorbed and the nuclei "spin-flip" from the lower energy state to the higher energy state. This absorption of energy is detected, amplified, and displayed as a **nuclear magnetic resonance (NMR) spectrum**.

The experimental conditions required to cause nuclei to resonate are affected by the local chemical and magnetic environment. Electrons also have spin and create local magnetic fields that **shield** nuclei from the applied field. Any factors that increase the exposure of nuclei to an applied field are said to **deshield** them.

The resonance signals in ^{1}H NMR spectra are reported by how far they are shifted from the resonance signal of hydrogen atoms of **tetramethylsilane (TMS)**. The resonance signals in ^{13}C NMR spectra are reported by how far they are shifted from the resonance signal of the carbon atoms of TMS. The NMR chart is calibrated in **delta (δ) units**, where 1 δ = 1 part per million (ppm).

NMR spectra are displayed on charts that show the applied field strength increasing from left to right. Thus, the left part of the chart is the low-field side (**downfield**), and the right part is the high-field side (**upfield**).

Both ^{1}H and ^{13}C NMR spectra display four general features:

1. **Number of signals.** Each nonequivalent ^{1}H or ^{13}C nucleus in a molecule can give rise to a unique signal.

2. **Chemical shift.** The exact position of each signal is its chemical shift. By correlating chemical shifts with chemical and magnetic environments, we can learn about the chemical nature of each nucleus.

3. **Integration.** All modern spectrometers can electronically integrate the area under each peak. The area under each peak is proportional to the relative number of hydrogens giving rise to that peak.

4. **Spin-spin splitting.** The tiny magnetic field of one nucleus affects the magnetic field felt by a neighboring nucleus. This phenomenon is due to the fact that the nuclear spin of an atom interacts, or **couples**, with the nuclear spin of a nearby atom.

Empirically, the degree of spin-spin splitting in ^{1}H NMR spectra can be predicted on the basis of the **(n + 1) rule**, which predicts that a signal of a hydrogen having n nonequivalent neighboring hydrogens (on the same or adjacent atoms) is split into (n + 1) peaks. A **coupling constant (J)** is the distance between adjacent peaks in a multiplet. The magnitude of a coupling constant is expressed in hertz (Hz) and is measured on the same scale as the chemical shift. Because the value of J depends only on internal forces within a molecule it has the same value regardless of the value of the applied magnetic field.

^{13}C—^{13}C coupling is not observed in ^{13}C NMR because only 1% of the carbon atoms in a sample are ^{13}C and are magnetic, so there is a small probability (0.01 × 0.01 = 0.0001, or 0.01%) that an observed ^{13}C nucleus is adjacent to another ^{13}C nucleus. Therefore, carbon-carbon splitting can be ignored. Carbon-hydrogen coupling is quite common and can be observed for carbon atoms bonded to hydrogen atoms or sufficiently close to hydrogen atoms. **Off-resonance decoupling (proton-coupled ^{13}C NMR** spectra) simplifies the spectrum but retains the splitting information of signals of carbons coupled with the protons directly bonded to them. To further simplify ^{13}C NMR spectra, they are commonly recorded using **proton spin decoupling** (broadband spectra). As a result, each carbon signal appears as a single, unsplit peak because any carbon-hydrogen splitting has been eliminated.

Interpreting ^{1}H NMR Spectra

Learning to obtain the structure of a compound from its NMR spectrum requires practice with a large number of examples and problems. This section is intended to provide some hints that can help make spectral analysis a little easier.

When you first look at a spectrum, consider the major features before getting to the minor details. The following are a few major characteristics you might watch for:

1. If the molecular formula is known, use it to determine the number of elements of unsaturation. The elements of unsaturation suggest rings, double bonds, or triple bonds. Matching the integrated peak areas with the number of protons in the formula gives the numbers of protons represented by the individual peaks.

2. Any broadened singlets in the spectrum might be due to hydrogen atoms of O—H or N—H groups. If the broad singlet is deshielded past 10 ppm, a hydrogen atom of a carboxy group, —COOH, is likely.

3. An absorption in the range of 0.8 to around 2.0 ppm indicates the presence of hydrogen atoms of alkyl groups. Learn to recognize typical fragments by their characteristic splitting patterns: ethyl groups (a quartet and a triplet, relative areas 2:3), propyl groups (a triplet, a multiplet (sextet), and a triplet, relative areas 2:2:3), and an isopropyl group (multiplet (septet) and a doublet, relative areas 1:6).

4. Absorption around 2.1 to 2.5 ppm may suggest hydrogen atoms on carbon atoms adjacent to a carbonyl group or next to an aromatic ring.

   A singlet at δ 2.1 often results from a methyl group bonded to a carbonyl group.

5. A sharp peak around δ 2.5 suggests a terminal alkyne: R—C≡C—H

6. An absorption around 3 to 4 ppm suggests protons on a carbon atom bearing an electronegative element such as oxygen or a halogen. Protons that are more distant from the electronegative atom will be less strongly deshielded.

7. Absorptions around 5 to 6 ppm suggest vinyl hydrogen atoms. Coupling constants can differentiate *cis* and *trans* hydrogen atoms.

8. Absorptions around 7 to 8 ppm suggest the presence of hydrogen atoms attached to an aromatic ring. If some of the aromatic protons absorb farther downfield than 7.2 ppm, an electron-withdrawing substituent may be attached.

9. A signal in the range 9 to 10 ppm (singlet or triplet, 1H) suggests the hydrogen atom bonded to the carbonyl group of an aldehyde.

The above hints can be used to identify partial structures (fragments) of the unknown molecule. The next step is to put together all the clues and to arrive at a proposed structure. The structure should be rechecked to make sure it is consistent with the molecular formula, the proton ratios given by the integrals, the chemical shifts of the signals, and the spin-spin coupling.

Solutions to the Exercises

10.1 Signals on the left of the spectrum (higher δ and ν values) are deshielded and are located downfield while those on the right are shielded (lower δ and ν values) and are located upfield (higher field). Thus,

(a) (lowest shielding) δ 9.56, δ 5.34, δ 2.74 (highest shielding) (^{1}H NMR)
(b) (lowest shielding) δ 205.7, δ 105.4, δ 38.5 (highest shielding) (^{13}C NMR)
(c) (lowest shielding)   750 Hz, 450 Hz, 150 Hz   (highest shielding)

10.2 The missing values in the following table are calculated from the formula:

$$\delta = \frac{\text{distance from TMS, [Hz]}}{\text{frequency of NMR, [MHz]}}$$

Frequency of spectrometer (MHz)	Distance of signal downfield from TMS, (Hz)	Chemical shift of signal (δ)
200	453	2.27
60	156	2.60
301	1054	3.50
400	1000	2.50
200	500	2.50

10.3 The difference of δ 0.02 is equal to $(60 \times 0.02) = 1.2$ Hz for a 60 MHz spectrometer (the same formula as in the preceding Exercise). For a 400 MHz spectrometer, these two signals would be separated by $(400 \times 0.02) = 8$ Hz.

10.4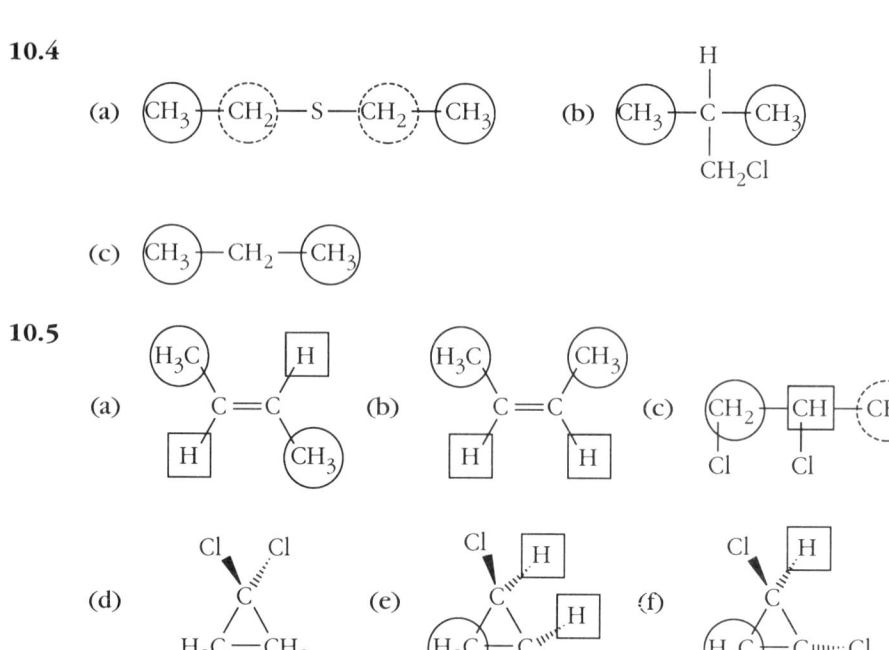

10.5

(d) All hydrogen atoms are equivalent

10.6 (a) Two types of hydrogen atoms: (i) hydrogen atoms of the methylene group and (ii) hydrogen atoms of the methyl groups.
The ratio of the actual numbers of hydrogen atoms is 2 : 6.
Integration of the ^{1}H NMR spectrum will reveal the ratio of 1 : 3.

(b) Three types of hydrogen atoms: (i) on aromatic ring, (ii) in the methylene groups, and (iii) in the methyl groups.
The ratio of the actual numbers of hydrogen atoms is 4 : 4 : 6.
Integration of the ^{1}H NMR spectrum will reveal the ratio of 2 : 2 : 3.

(c) Two types of hydrogen atoms: (i) in the methylene group and (ii) in the methyl groups.
The ratio of the actual numbers of hydrogen atoms is 2 : 9.
Integration of the ^{1}H NMR spectrum will reveal the ratio of 2 : 9.

10.7 The following are actual values of chemical shifts of the indicated hydrogen atoms:
(a) δ 3.3
(b) δ 3.7
(c) δ 1.7
(d) δ 9.8
(e) δ 2.2

10.8 The following are actual values of the chemical shifts of the carbon atoms:
(a) δ 58.5
(b) δ 43.5
(c) δ 20.4 ($-CH_3$), δ 123.4 (sp^2 hybrid carbon atoms)
(d) δ 199.7
(e) δ 30.8 ($-CH_3$), δ 206.6 (C=O)

10.9 (a) The molecular formula, $C_4H_8O_2$, indicates that there is one unit of unsaturation in the compound (double bond or ring). The ^{1}H NMR spectrum shows only one signal, at δ 3.7 ppm. The chemical shift indicates that the hydrogen atoms are bonded to a carbon attached to an electronegative element (halide, oxygen). Because the molecular formula contains oxygen, it must be bonded to the oxygen atom. The total number of hydrogen atoms is eight. The alkyl groups are all equivalent, so each must contain two hydrogen atoms. (This is the only way you can divide eight without a remainder. No alkyl group can contain four hydrogen atoms so the alkyl groups are methylene groups, —CH$_2$—.) Each —CH$_2$— group is bonded to the oxygen atom (—CH$_2$—O—CH$_2$—). Joining these two parts together gives the structure of the compound called **1,4-dioxane**.

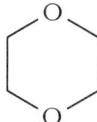

(b) The molecular formula, C$_8$H$_{10}$, indicates that there are four units of unsaturation in the compound (a combination of double bonds and/or rings). The ^{1}H NMR spectrum shows two singlets, one at δ 7.0 and the other at δ 2.3. The chemical shift of the signal at δ 7.0 is in the region of hydrogen atoms bonded to an aromatic ring. The integration data show a 2:3 ratio. The total number of hydrogens in the molecule is 10, so the actual ratio must be 4:6. The four equivalent aromatic hydrogens indicate a symmetrically substituted benzene derivative.

The rest of the molecular formula, C$_2$H$_6$, represents two methyl groups. Joining together these partial structures gives **1,4-dimethylbenzene** (also known as *p*-xylene) as the unknown:

$$\underset{\text{CH}_3}{\underset{|}{\bigcirc}}\text{—CH}_3$$

(c) The molecular formula, C$_6$H$_{12}$O, indicates that there is one unit of unsaturation in the compound (a double bond or ring). The ^{1}H NMR spectrum shows two signals, one at δ 2.1 and the other at δ 1.1. The chemical shift of the singlet at δ 2.1 indicates that the hydrogen atoms are bonded to a carbon attached to a carbonyl group. The integration data indicate a ratio of 1:3. The total number of hydrogens in the molecule is 12, so the actual numbers of hydrogens are three and nine, respectively. The singlet at δ 1.1 of area nine is typical of a *tert*-butyl group. Joining these partial structures gives **3,3-dimethyl-2-butanone** as the unknown:

$$\text{CH}_3\text{C}(=\text{O})\text{C}(\text{CH}_3)_3$$

10.10 There are two ways of solving an Exercise like this. We can examine each structure and look for characteristic patterns in the ^{1}H NMR spectrum, then look for a spectrum that matches such a pattern. If it is not possible to pick up distinguishing features immediately, it may be necessary to sketch an approximate spectrum of each compound for comparison.

The other approach involves looking at each spectrum and assigning the signals observed to fragments of the structures. Both methods should give the same results.

Let us use the former method to solve this Exercise.

(a) There are two groups of nonequivalent hydrogens in Cl_2CHCH_3. They should show the following splitting patterns: a quartet (δ 5.5–6.0, 1H) and a doublet (δ 2.0, 3H). Spectrum **B** matches this pattern.

(b) There is only one kind of hydrogen atoms in $Cl_3CCH_2CCl_3$. The singlet of the methylene group will appear at approximately δ 3.4. Neither spectrum shows such a pattern.

(c) The spectrum of $CH_3COOCH_2CH_3$ should show a singlet at δ 2.0 due to the methyl group adjacent to the carbonyl group and a familiar pattern of an ethyl group, namely a quartet (δ 4.1, 2H) and a triplet (1.2, 3H). Spectrum **A** matches that pattern.

10.11 1-Bromopropane and 2-bromopropane can be distinguished by symmetry. 1-Bromopropane lacks the symmetry of 2-bromopropane, and thus there are three types of nonequivalent hydrogen atoms in 1-bromopropane and only two different types in 2-bromopropane.

The hydrogen atoms on C1 in 1-bromopropane are shifted downfield by the electronegative bromine atom and appear as a triplet at δ 3.4, split by the two hydrogen atoms on C2. Signals for the two hydrogen atoms of C2 are farther upfield and appear as a sextet at δ 1.9. The hydrogen atoms on C3 are farther upfield at δ 1.05, split into a triplet by the coupling to the C2 hydrogen atoms. These three signals integrate in the ratio 2:2:3. Spectrum **A** matches such a pattern.

In contrast with the somewhat complex spectrum for 1-bromopropane, the spectrum of 2-bromopropane is much simpler, owing to its symmetry. Thus, the hydrogen atoms on C1 and C3 are equivalent, split into a doublet by the hydrogen atom on C2. The signal of the hydrogen atom on C2 is shifted downfield by the electronegative bromine and split into a septet by the six equivalent hydrogen atoms on C1 and C3. The two signals integrate in a ratio of 6:1. Spectrum **B** matches such a pattern.

10.12 Let us consider the first structure:

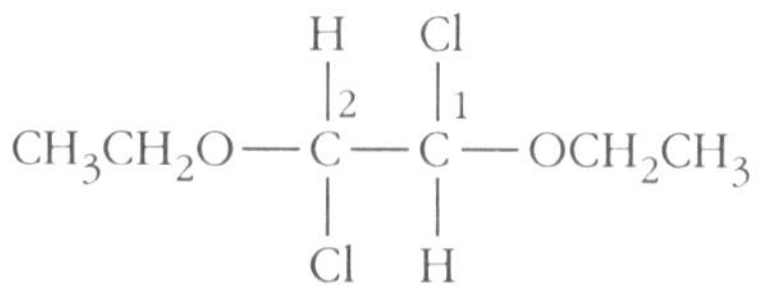

$$CH_3CH_2O-\underset{\underset{Cl}{|}}{\overset{\overset{H}{|}}{C}}-\underset{\underset{H}{|}}{\overset{\overset{Cl}{|}}{C}}-OCH_2CH_3$$

The broadband decoupled ^{13}C NMR spectrum on page 429 shows four signals for ^{13}C atoms. The structure has three types of carbon atoms due to the symmetry of the molecule.

The proton-coupled ^{13}C NMR spectrum shows two doublets for the internal carbon atoms. The spectrum of the above structure would have a single doublet for the equivalent —CHCl— groups.

The ^{1}H NMR spectrum shows two doublets for the hydrogen atoms on C1 and C2. The above structure has two equivalent —CHCl— groups, so only a singlet would be observed.

Thus, the above structure is *not* consistent with the spectra.

The second structure is:

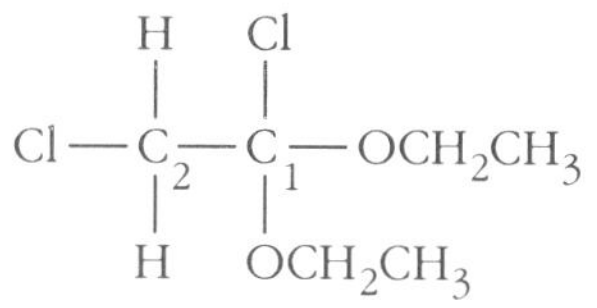

$$\text{Cl}-\underset{\underset{\text{H}}{|}}{\overset{\overset{\text{H}}{|}}{\underset{2}{C}}}-\underset{\underset{\text{OCH}_2\text{CH}_3}{|}}{\overset{\overset{\text{Cl}}{|}}{\underset{1}{C}}}-\text{OCH}_2\text{CH}_3$$

Let us consider the spectra. The broadband decoupled ^{13}C NMR spectrum on page 429 shows four signals for ^{13}C atoms. The above structure has four nonequivalent carbon atoms. Thus, the first spectrum is consistent with this structure.

The proton-coupled ^{13}C NMR spectrum shows two doublets for the internal carbon atoms. The spectrum of the above structure would have a singlet for C1 and a triplet for C2.

The ^{1}H NMR spectrum shows two doublets for the hydrogen atoms on C1 and C2, indicating that there is one hydrogen on each carbon and the carbon atoms are not equivalent. The above structure would have only one singlet due to the hydrogen atoms on the —CH$_2$Cl— group.

Thus, the above structure is *not* consistent with the spectra.

10.13 The molecular formula, C$_3$H$_6$Br$_2$, indicates zero units of unsaturation.

The ^{1}H NMR spectrum shows only two types of hydrogen atoms in the molecule. A triplet at δ 3.55 indicates that the carbon bearing those hydrogen atoms is also bonded to an electronegative atom. The molecular formula shows bromine, so it must be —CH$_2$Br or —CHBr—. Another signal is a quintet at δ 2.35. The integration shows the ratio 2:1. There are six hydrogen atoms in total, so the triplet must represent four hydrogens and the quintet must represent two hydrogen atoms. Out of four possible isomers of dibromopropane only **1,3-dibromopropane** is consistent with the ^{1}H NMR spectrum.

The proton-decoupled ^{13}C NMR spectrum shows two signals. This confirms the symmetry of 1,3-dibromopropane, which contains two types of nonequivalent carbon atoms.

The proton-coupled ^{13}C NMR spectrum shows a multiplet. Each carbon atom bears two hydrogen atoms so it should be split to a triplet. The triplets partially overlap because the signals for the carbon atoms in the proton-decoupled ^{13}C NMR spectrum are close to each other.

Thus, all the spectra are consistent with the structure of **1,3-dibromopropane**.

$$\text{BrCH}_2\text{CH}_2\text{CH}_2\text{Br}$$

10.14 (a) The structure shows two types of hydrogen atoms, namely methyl groups and the methylene group. The two types of hydrogen atoms are not coupled because they are separated by a carbon atom without a hydrogen. Thus, we would expect two singlets. The singlet representing the —CH$_2$Br group is strongly deshielded by bromine so it will appear at approximately δ 3.4. The other signal should be expected at δ 1.0.

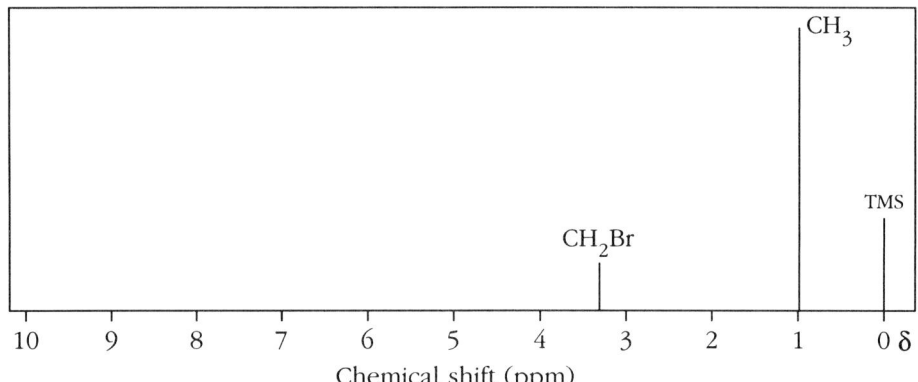

(b) There are three types of hydrogen atoms, namely a methyl group and two nonequivalent methylene groups. The methyl group should give rise to a singlet at δ 2.1. The signals of both methylene groups will be split to triplets. The chlorine atom is very electronegative and should deshield the hydrogen atoms, so the triplet for the terminal methylene group should appear at δ 3.7. The triplet for the internal methylene group should appear at δ 2.8.

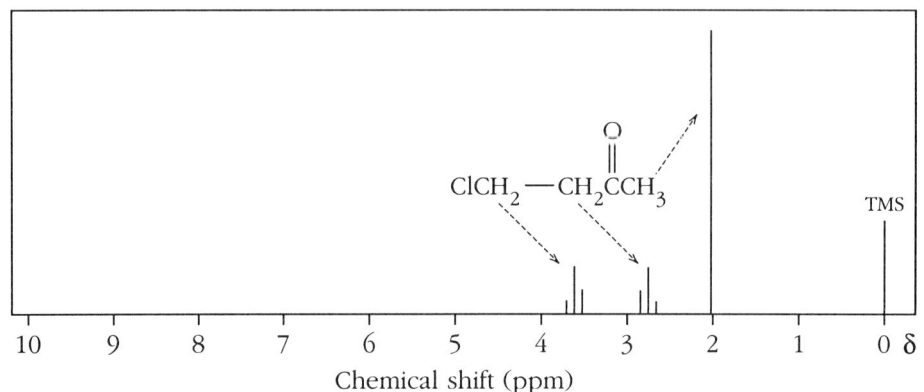

(c) The structure contains two alkyl groups bonded to an oxygen atom. The ethyl group shows a quartet-triplet pattern while the isopropyl group shows septet-doublet. The —CH$_2$— and —CH— groups are bonded directly to the oxygen atom, so their signals are found downfield in the region of δ 3.6 to 3.7. The signal for —CH$_3$ of the ethyl group (triplet) overlaps with the signal for —CH$_3$ of the isopropyl group (doublet) in the δ1.0 to 1.2 region. Similarly, signals of hydrogen atoms of the methylene group (quartet) and those of —CH— overlap in the δ 3.4 to 3.8 region.

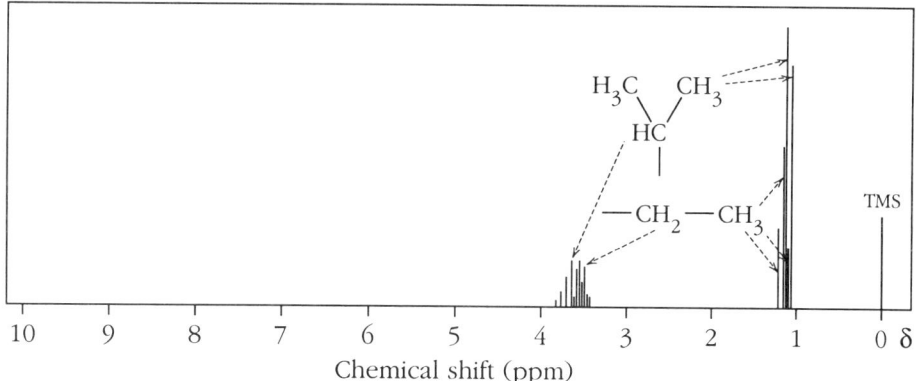

(d) The structure shows two equivalent ethyl groups bonded to oxygen atoms, so the spectrum should have a quartet at δ 4.25 and a triplet at δ 1.3. The two vinyl hydrogen atoms are also equivalent and they will show a singlet at δ 6.25.

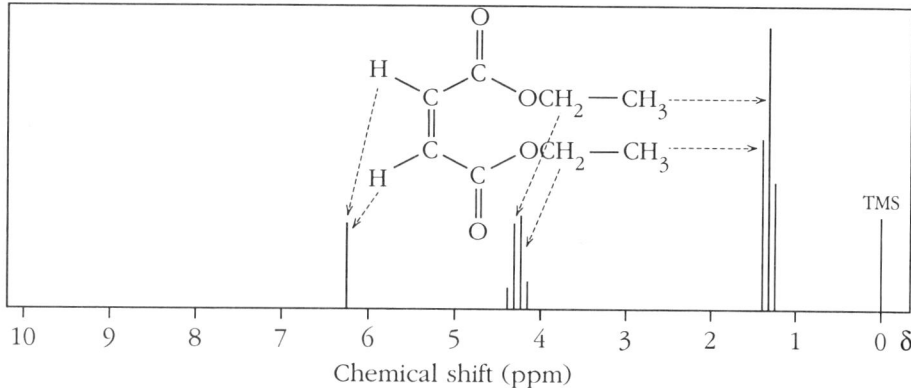

10.15 The four possible spin combinations are the following:

The chemical shift of a hydrogen, H_a, is influenced by whether the spin of a hydrogen on an adjacent carbon atom, H_b, is aligned with the applied field or aligned against it. If the spin of the hydrogen atom on an adjacent carbon is aligned with the applied field and adds to it, then H_a absorbs at a lower applied field. If, on the other hand, the spin of H_b is aligned against the applied field and subtracts from it, then H_a absorbs at a higher applied field.

When we have three neighboring hydrogen atoms, their spins can be oriented in the four possible ways, namely all three with the external field, two with and one against, one with and two against, and all three against the applied field. The probability that any one of these situations will occur is 1:3:3:1. All eight possible spin orientations are shown above.

A spin-spin splitting diagram, also called a **tree diagram,** is a convenient method for the analysis of splitting patterns.

If H_a undergoes no spin-spin coupling, its signal is not split. Each neighboring proton splits the signal into two halves. When we have three neighboring protons, the original signal is split three times. The 1:3:3:1 ratio of the areas arises from the fact that all the hydrogen atoms (being equivalent) have the same coupling constant and, consequently, the protons have superimposed absorption positions.

10.16 The splitting patterns for vinyl hydrogen atoms are more complex than those for alkyl protons. The complexity arises from the lack of rotation around the double bond.

In our example, all three vinyl hydrogen atoms (H_a, H_b, and H_c) are nonequivalent. Therefore, they exhibit different chemical shifts and give rise to three separate signals. In addition, the coupling constants between any two of the hydrogen atoms (J_{ab}, J_{ac}, and J_{bc}) are different.

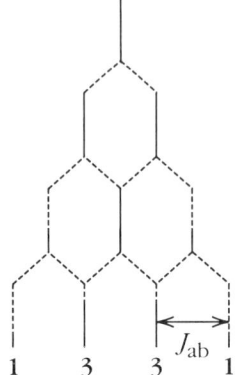

The tree diagram for the splitting patterns of H_b into a doublet of doublets by coupling with H_a and H_c is shown below. First, the signal is split into a doublet by coupling with H_c. Then, each signal is further split into a doublet by H_a. J_{bc} is usually in the range of 6 to 14 Hz, while J_{ab} varies from 0 to 3 Hz.

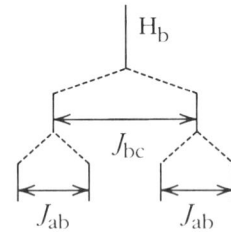

10.17

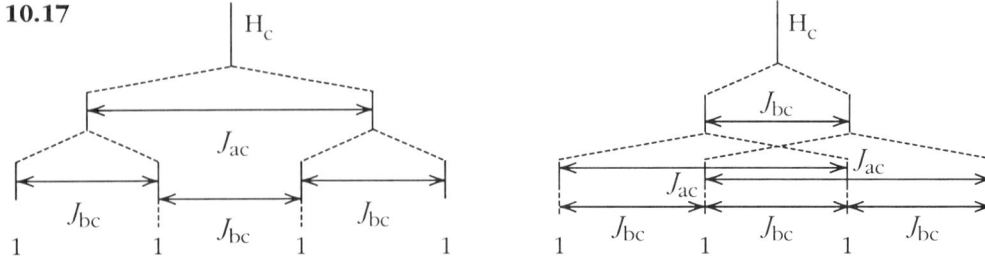

10.18

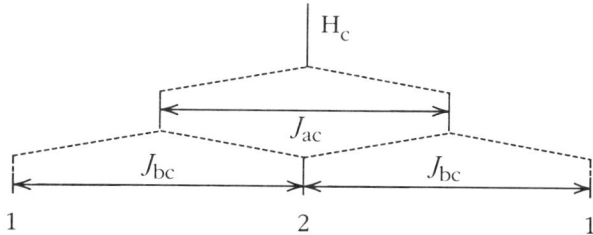

10.19 The ^{1}H NMR spectrum of vinyl bromide (IUPAC name: bromoethene) is complex for two reasons. First, the three alkene hydrogen atoms are chemically nonequivalent. Hence, the absorptions of all three hydrogen atoms occur at different chemical shifts. Second, each hydrogen atom is split by the other two with different coupling constants.

Let us now construct a splitting diagram for each hydrogen atom, assuming the following coupling constants: $J_{ab} = 1.5$ Hz, $J_{ac} = 14$ Hz, and $J_{bc} = 7$ Hz. The values of the coupling constants are average values observed for vinyl hydrogen atoms.

The resonance position of each hydrogen atom (given in Exercise 10.19) is shown along the top of the figure, and the successive applications of splittings give the observed spectrum, shown along the bottom of the figure.

Notice that the sum of the intensities of all lines in a splitting pattern equals the intensity of the corresponding unsplit line.

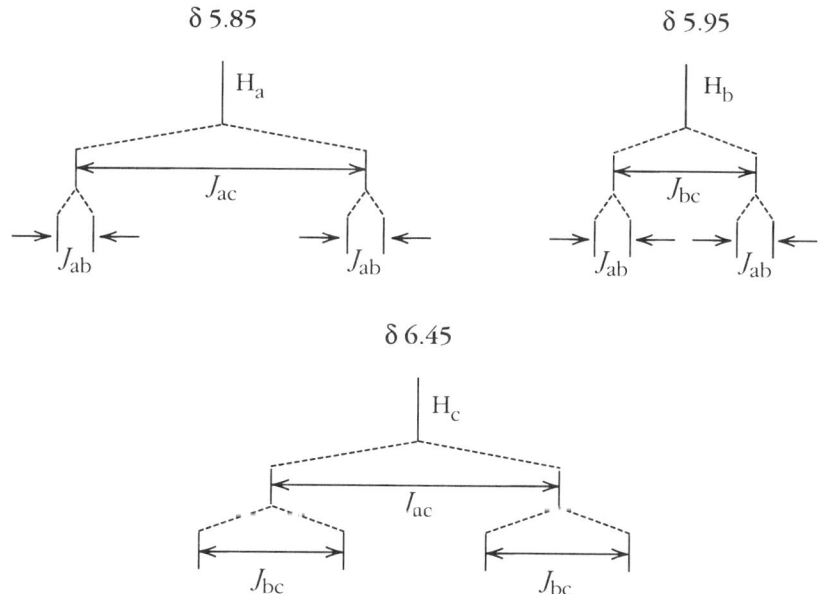

Now we can trace a spectrum of vinyl bromide. Note that some signals of H_a and H_b overlap.

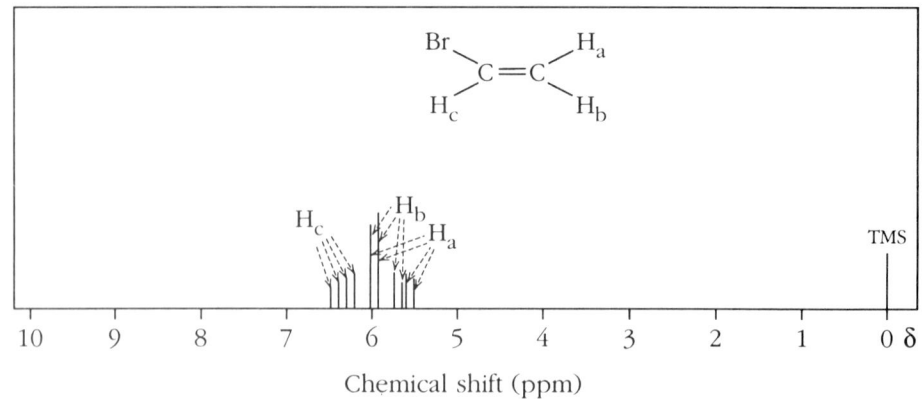

10.20 The relationship between δ, distance of a signal from TMS, and operating frequency of a spectrometer is described by the following formula:

$$\delta = \frac{\text{Distance from TMS, [Hz]}}{\text{Frequency of NMR, [MHz]}}$$

(a) 100 Hz
(b) 300 Hz
(c) 500 Hz

10.21 Methylcyclohexane exists in the chair form. The methyl group is in either an equatorial position or an axial position. Partial rotation around the carbon-carbon bonds in the cyclohexane ring results in the conversion of one chair form to another. Thus, the methyl group flips from equatorial to axial position and *vice versa*. The rapid flipping of the ring at room temperature results in a single peak at δ 23.1. However, at -90 °C the flipping between the two forms is much slower and two signals are observed, one for the methyl group in the equatorial position and the other for the methyl group in the axial position.

10.22

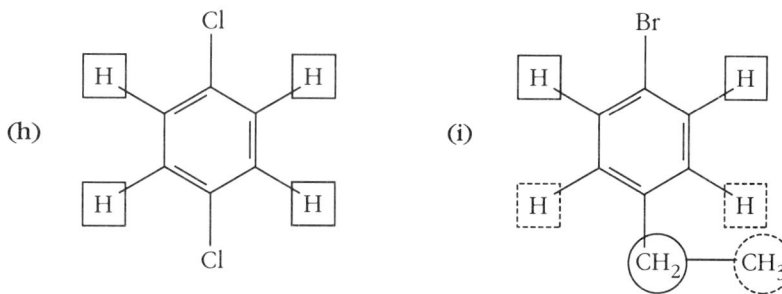

10.23 (a) 3 signals
(b) 3 signals
(c) 3 signals
(d) 1 signal
(e) 1 signal
(f) 4 signals
(g) 6 signals
(h) 2 signals
(i) 6 signals

10.24 NMR spectra are displayed on charts that show the applied field strength increasing from left to right. Thus, the left part of the chart is the low-field (or downfield) side, and the right part is the high-field (or upfield) side.

All nuclei in molecules are surrounded by electron clouds. When an external magnetic field is applied, the circulating electron clouds induce local magnetic fields. These local magnetic fields act in opposition to the applied field so that the effective field actually felt by the nucleus is a bit smaller than the applied field. This phenomenon is known as shielding. If a hydrogen atom is bonded to a carbon bearing an electronegative atom or group, the electron density around such a hydrogen will decrease and the shielding effect will be smaller (signals will be moved downfield). The more electronegative the atom or group is, the more deshielded the neighboring hydrogen atoms are. An electron-donating substituent will increase electron density around a hydrogen atom and it will shield the nucleus of hydrogen (as a result, its signal is moved upfield).

(a) Chlorine is more electronegative than bromine so chlorine will deshield hydrogen atoms slightly more than bromine. Thus, hydrogen atoms of chloroethane will be found farther downfield (at δ 3.5, compared to δ 3.4 in bromoethane).

(b) Alkyl groups shield hydrogen atoms. A carbonyl group deshields slightly. Thus, hydrogen atoms in acetone will be found farther downfield than those in propane.

(c) The carbon-carbon double bonds deshield the vinyl hydrogens. However, the hydrogen in chloroethene is additionally deshielded by the electronegative chlorine. Thus, the hydrogen atom in chloroethene will be found farther downfield than those in ethene.

(d) The carbonyl group *strongly* deshields a hydrogen atom directly bonded to a carbonyl carbon. This effect is much larger than that in alkenes. Thus, the hydrogen atom of the aldehyde group in acetaldehyde (ethanal) will be found farther downfield than the hydrogen atoms in ethene.

(e) The π electrons in benzene induce a magnetic field that deshields the aromatic hydrogen atoms. Thus, hydrogen atoms in benzene will be found farther downfield than those in cyclohexane.

10.25 The following are the actual chemical shifts of indicated hydrogen atoms (δ ± 0.1):

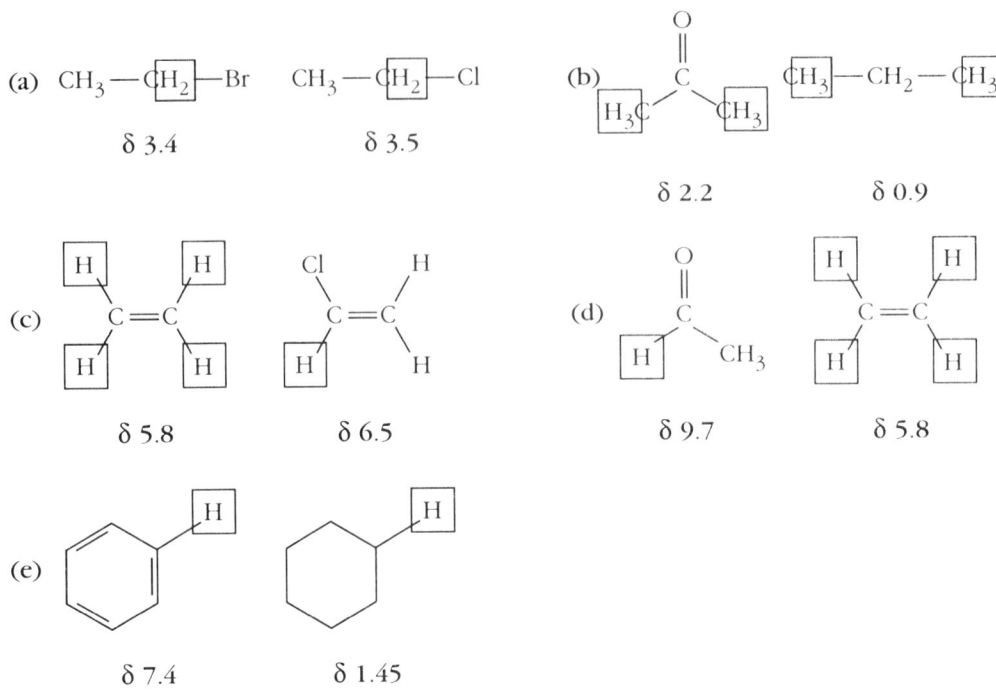

(a) CH₃—C̄H₂—Br   CH₃—C̄H₂—Cl
   δ 3.4   δ 3.5

(b) structure with H₃C—C(=O)—CH₃, δ 2.2;   C̄H₃—CH₂—C̄H₃, δ 0.9

(c) H₂C=CH₂ (boxed H's), δ 5.8;   ClHC=CH₂ (boxed H on CHCl side), δ 6.5

(d) H—C(=O)—CH₃, δ 9.7;   H₂C=CH₂ (boxed H's), δ 5.8

(e) benzene with H, δ 7.4;   cyclohexane with H, δ 1.45

10.26 First focus on the structural features that differentiate the two compounds, and then choose a spectroscopic method that can distinguish this difference.

(a) Ethanol has an —OH group while CH₃CH₂OCH₂CH₃ does not. Therefore ethanol can be distinguished from diethyl ether by IR spectroscopy. The infrared spectrum of ethanol shows a strong O—H stretch at 3500 to 3200 cm⁻¹, which is absent from the IR spectrum of diethyl ether.

(b) Ethyl formate (HCOOCH₂CH₃) has a hydrogen atom bonded to the carbonyl carbon atom while diethyl carbonate does not. Diethyl carbonate, therefore, can be distinguished from ethyl formate by ¹H NMR spectroscopy. The ¹H NMR spectrum of diethyl carbonate contains a quartet at δ 4.2 (2H) and a triplet at δ 1.3 (3H). The ¹H NMR spectrum of ethyl formate contains, in addition to two signals of the ethyl group (the same chemical shifts and coupling pattern as those for diethyl carbonate), a sharp singlet of the hydrogen bonded to the *sp*² hybrid carbon at δ 8.05. This singlet is absent in the spectrum of diethyl carbonate.

(c) Ethyl 2,2-dimethylpropionate [(CH₃)₃CCOOCH₂CH₃] can be distinguished from ethyl formate (HCOOCH₂CH₃) by ¹H or ¹³C NMR. The ¹H NMR spectrum of ethyl 2,2-dimethylpropionate will show a singlet (9H) at around δ 1.2, which is absent from the spectrum of ethyl formate. Ethyl formate has a sharp singlet of the hydrogen bonded to the *sp*² hybrid carbon at δ 8.05, which is absent from the ¹H NMR spectrum of ethyl 2,2-dimethylpropionate. The ¹³C NMR spectrum of ethyl 2,2-dimethyl-propionate will show two extra signals at around δ 19, which are absent from the spectrum of ethyl formate.

(d) One compound is a terminal alkyne while the other is an internal alkene. All three spectroscopic techniques (IR, ¹³C NMR, and ¹H NMR) can distinguish between them. The IR spectrum of 1-butyne will show a strong absorption for the *sp* C—H stretch at 3300 cm^{-1} and medium absorption for the C≡C stretch at 2200 cm^{-1}. These absorption bands are absent from the IR spectrum of 2-butene. 2-Butene will show a medium absorption band for the sp^2 C—H stretch at 3100 to 3020 cm^{-1}, which is absent from the IR spectrum of 1-butyne.

The ^{1}H NMR spectrum of 1-butyne will show a pair of quartets (δ 2.15, 2H), a triplet (δ 1.5, 3H), and a triplet (δ 1.95, 1H). The ^{1}H NMR spectrum of 2-butene will show a doublet (δ 1.6, 6H) and a multiplet of vinyl hydrogens at δ 5.4 (2H).

1-Butyne can be best distinguished from 2-butene by ^{13}C NMR. 1-Butyne will show four signals (δ 68.2, 84.5, 14.4, 12.4) while 2-butene will show only two signals (δ 120–133 and 14).

(e) 1,3-Dimethylcyclohexane and 1,4-dimethylcyclohexane differ in symmetry and can be best distinguished by ^{13}C NMR spectroscopy. 1,4-Dimethylcyclohexane has three types of carbon atoms and will show only three signals in its ^{13}C NMR spectrum. In 1,3-dimethylcyclohexane, a symmetry plane passes through C2 and C5, causing five signals to be observable in its ^{13}C NMR spectrum.

(f) An analysis similar to that for Exercise 10.26 (e) applies to 1,4-dioxane and 1,3-dioxane. They differ in symmetry and can be best distinguished by ^{13}C NMR spectroscopy. 1,4-Dioxane has only one type of carbon atom and will show only *one* signal in its ^{13}C NMR spectrum. In 1,3-dioxane, a symmetry plane passes through C2 and C5, so it has three types of nonequivalent carbon atoms and it will show three signals in its ^{13}C NMR spectrum.

The difference in symmetry also causes a difference in the splitting pattern observed in their ^{1}H NMR spectra. 1,4-Dioxane will show only one singlet at δ 3.7, while 1,3-dioxane will show a singlet at δ 4.75, a triplet at δ 3.75, and a multiplet at δ 1.5.

10.27 To solve a problem of matching the structures with spectra, it is best to examine each structure and look for characteristic patterns in the ^{13}C NMR spectrum. Then look for a spectrum that matches such a pattern. If it is not possible to pick up distinguishing features immediately, it may be necessary to estimate chemical shifts for each type of carbon or to sketch an approximate spectrum of each compound for comparison.

(a) The ^{13}C NMR spectrum of 2-methylpropanol should show the presence of three nonequivalent carbon atoms, namely two in the alkyl region of δ 15 to 50 and the one bonded to the oxygen atom in the region δ 50 to 80. Spectrum **B** shows such a pattern.

(b) The ^{13}C NMR spectrum of acetophenone will show *six* signals for six types of carbon atoms, namely the carbonyl carbon (δ 190 to 210), four aromatic carbon atoms in the region δ 110 to 150, and one of the methyl group bonded to a carbonyl carbon at δ 20 to 30. Spectrum **D** matches this pattern.

(c) The ^{13}C NMR spectrum of cyclohexene should show the presence of three unique carbons, namely two in the alkyl region (δ 15–50) and one in the alkene region (δ 100 to 150). Spectrum **E** matches this pattern.

(d) The ^{13}C NMR spectrum of 1,3,5-trimethylbenzene will show *three* signals for three types of carbon atoms, namely two types of aromatic carbons in the region δ 110 to 150 and one for the methyl groups bonded to the aromatic ring (δ 20 to 30). Spectrum **A** matches this pattern.

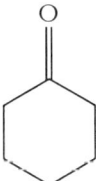

(e) The ^{13}C NMR spectrum of cyclohexanone will show *four* signals for four types of carbon atoms, namely the carbonyl carbon (δ 190 to 210) and three alkyl carbons in the region δ 20 to 50. Spectrum **C** matches this pattern.

10.28 The spin-spin splitting is caused by the presence of vicinal, or neighboring hydrogen atoms (hydrogen atoms on an adjacent carbon) that are nonequivalent to the hydrogen atom in question. The $n + 1$ rule predicts the number of spin-spin splitting peaks.

(a) The methylene hydrogen atoms of 1,1-dichloro-2-iodoethane see *one* neighboring, nonequivalent hydrogen atom. Their NMR signal is split into (1 + 1 = 2 peaks) a doublet with relative areas 1:1.

(b) The methylene hydrogen atoms of 2-butanone see *three* neighboring, nonequivalent hydrogen atoms of the methyl group. Their NMR signal is split into (3 + 1 = 4 peaks) a quartet of relative areas 1:3:3:1.

(c) The methylene hydrogen atoms of 1-bromo-2-chloroethane see *two* neighboring, nonequivalent hydrogen atoms. Their NMR signal is split into (2 + 1 = 3 peaks) a triplet of relative areas 1:2:1.

(d) The hydrogen atom on C2 of 2-methylpropanal sees *six* neighboring, nonequivalent hydrogen atoms of the methyl group and one nonequivalent hydrogen atom of the aldehyde group. The NMR signal of the CH hydrogen atom is split into (6 + 1 = 7 peaks) a septet, each of which is further split into a doublet by the hydrogen atom of the aldehyde group. The relative area of the hydrogen atom on C2 is so small (1H) compared to the other peaks in the spectrum that it is unlikely that this splitting will be visible. The signal will most likely appear as a broad multiplet.

(e) The aldehydic proton of acetaldehyde sees three neighboring hydrogen atoms of the methyl group. Its NMR signal is split into (3 + 1 = 4) a quartet of relative areas 1:3:3:1.

10.29 (a)

Benzaldehyde   Acetophenone

Benzaldehyde exhibits a characteristic downfield shift for the aldehydic proton (δ 9.5) that is absent in the ketone. The methyl group of acetophenone gives rise to a singlet at δ 2.1.

(b) $CH_3-C\equiv C-CH_3$ $H-C\equiv C-CH_2-CH_3$
    2-Butyne   1-Butyne

2-Butyne and 1-butyne differ in symmetry. All hydrogen atoms in 2-butyne are equivalent and they give rise to singlet at δ 1.75. The spectrum of 1-butyne contains a pair of quartets at δ 2.15, a triplet at δ 1.5 and a triplet of the terminal alkyne hydrogen at δ 1.95.

(c)

Acetone   Methyl acetate

Acetone and methyl acetate differ by one oxygen atom. The chemical shift of the hydrogen atoms of the methyl group attached to the oxygen (δ 3.6 ppm) differs from that of the methyl group attached to the carbonyl group (δ 2.1 ppm).

(d)

1,1-Dibromoethene   1,2-Dibromoethene

Both 1,1-dibromoethene and 1,2-dibromoethene have only equivalent hydrogen atoms. The ^1H NMR spectrum of each compound will consist only of one singlet. The only difference is the distance between electron-withdrawing bromine and hydrogen atoms. The chemical shift of the singlet in the spectrum of 1,1-dibromoethene is δ 5.4, while that of 1,2-dibromoethene is δ 6.7.

10.30 (a) The molecular formula, C_4H_8O, indicates one unit of unsaturation (double bond or a ring). The ^1H NMR spectrum shows three signals, namely a quartet (δ 2.47), a singlet (δ 2.1), and a triplet (δ 1.01). The integration shows the ratio of those signals to be 2:3:3, respectively. The singlet at δ 2.1, integrating to three hydrogen atoms, best matches a methyl group attached to a carbonyl group. The other two signals, a quartet (2H) and a triplet (3H), indicate an ethyl group attached to a carbonyl group. Thus, the structure is **2-butanone**.

$$\underset{H_3C}{\overset{1}{}}\overset{\overset{O}{\|}}{\underset{2}{C}}\underset{CH_2}{\overset{3}{}}\underset{CH_3}{\overset{4}{-}}$$

(b) The hydrogen atoms of the methyl group on C4 give rise to the triplet at δ 1.01 (coupling with two hydrogen atoms of the methylene group on C3 according to the $n + 1$ rule). The hydrogen atoms of the other methyl group, on C1, give rise to a singlet at δ 2.1 (no coupling because there are no neighboring hydrogens on C2). The hydrogen atoms of the methylene group on C3 give rise to a quartet at δ 2.47 (coupling with three hydrogens of the methyl group on adjacent C4).

(c) [splitting tree diagrams showing $J_{3,4} = 2.4$ Hz with intensities 1:2:1 and 1:3:3:1]

10.31 (a) 2,2-Dimethylbutane has three types of hydrogen atoms, namely three methyl groups attached to C2, a methylene group on C3, and a methyl group on C4.

$$\overset{4}{CH_3}-\overset{3}{CH_2}-\overset{\overset{CH_3}{|}}{\underset{\underset{CH_3}{|}}{\overset{2}{C}}}-CH_3$$

The three methyl groups attached to C2 carbon atom contain nine equivalent hydrogen atoms that do not have any neighboring, nonequivalent hydrogen atoms because there are no hydrogens on C2 carbon. Thus, the nine hydrogens will give rise to a singlet at around δ 0.9.

The methylene group on C3 contains two hydrogen atoms that have three neighboring, nonequivalent hydrogens on C4. Thus, the signal of those two hydrogens will be split into a quartet centered around δ 1.15.

The methyl group on C4 contains three hydrogen atoms that have two neighboring, nonequivalent hydrogens on C3. Thus, the signal of those three hydrogens will be split into a triplet centered around δ 0.9.

Thus, the ^{1}H NMR spectrum of 2,2-dimethylbutane will show overlapping of the triplet and singlet because both are expected to appear at the same chemical shift (δ 0.9).

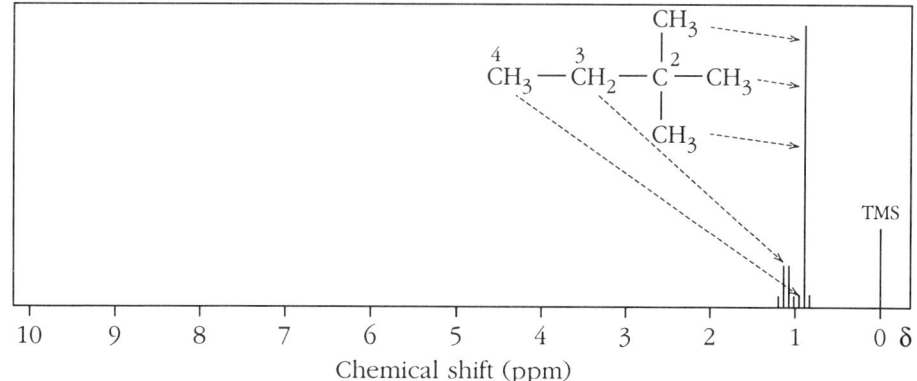

(b) 2-Methoxypropane (trivial name: isopropyl methyl ether) has three types of hydrogen atoms, namely a methyl group bonded to oxygen, the hydrogen atom attached to C2 of the isopropyl group, and the two methyl groups (C1 and C3 of the isopropyl group).

$$H_3C-\overset{..}{\underset{..}{O}}-\underset{\underset{^3CH_3}{|}}{\overset{2}{C}H}-{}^1CH_3$$

The methyl group bonded to oxygen contains three hydrogen atoms that do not have any neighboring, nonequivalent hydrogen atoms because the carbon to which they are bonded is attached to an oxygen (without hydrogens). Their signal will appear as a singlet. The neighboring oxygen deshields the hydrogens, moreover, so their singlet will be shifted downfield to around δ 3.3.

The hydrogen atom attached to C2 sees *six* equivalent neighboring hydrogen atoms on C1 and C3. Thus, its signal will be split into a septet and will be shifted downfield to approximately δ 3.6 because the C2 carbon is bonded to the oxygen atom.

The third group of hydrogens contains six hydrogen atoms that have only *one* neighboring hydrogen on C2, so their signal will be split into a doublet. The doublet is expected to appear at about δ 1.1.

The chemical shifts of the three types of hydrogen atoms are sufficiently far apart that none of the signals of the spectrum should overlap.

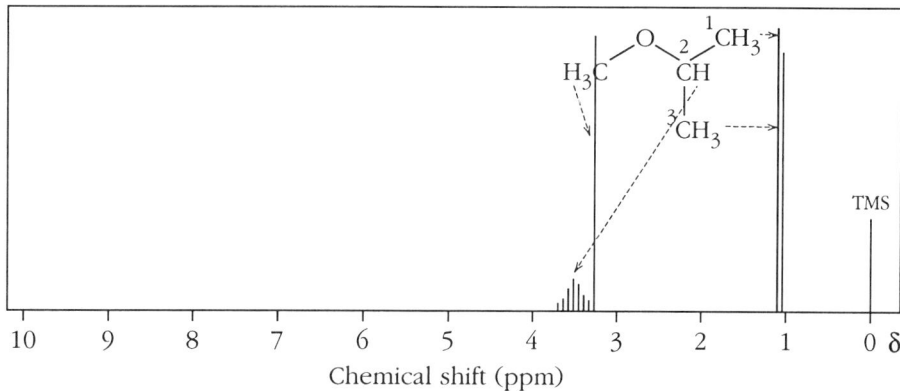

(c) 3-Buten-2-one has four different groups of hydrogen atoms, namely the methyl group (C1) and each of the three hydrogen atoms of the vinyl group.

The methyl group (C1) has three equivalent hydrogens and will give rise to a singlet (no neighboring hydrogen) at δ 2.3. The position of the peak is moved slightly downfield due to the anisotropic effect of the vinyl group.

3-Buten-2-one

A complicated splitting pattern results when a hydrogen atom is coupled unequally to two or more nonequivalent hydrogen atoms. This is what we observe in case of the vinyl group. Each of the three hydrogen atoms (H_a, H_b, and H_c) is different. Thus, each signal is split into a doublet by one neighbor, and the peaks of each doublet are split again into another doublet by the other neighbor. This pattern is called a doublet of doublets. None of these groupings of four peaks is a quartet because the four peaks are not evenly spaced and do not have relative intensities of 1:3:3:1.

10.32 (a) The formula, $C_9H_{18}O$, shows one unit of unsaturation (a ring or a double bond). The IR spectrum indicates a carbonyl group. The ^1H NMR spectrum shows only one signal at δ 0.95 ppm, which suggests methyl groups with no neighboring hydrogen atoms and no immediate proximity of the carbonyl group. Thus, we must have (18 : 3 = 6) six equivalent methyl groups in the molecule. What remains from the molecular formula is C_3O. When we take C=O for the carbonyl group, only two carbon atoms are left. The only solution seems to be **2,2,4,4-tetramethyl-3-pentanone**.

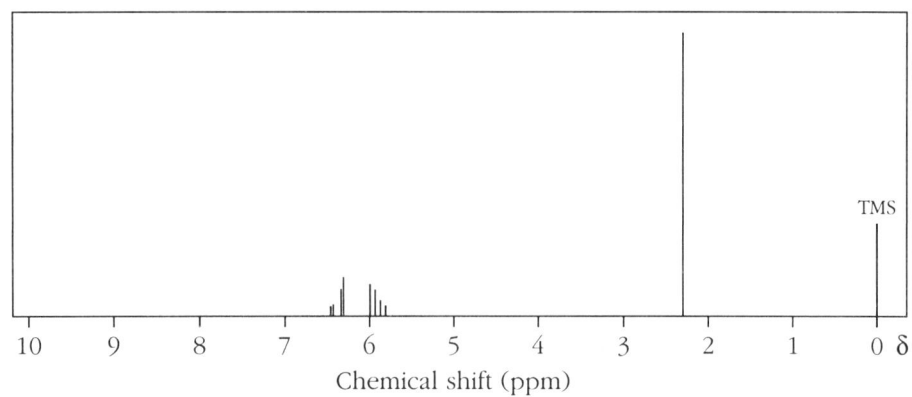

(b) The molecular formula, C_4H_7NO, shows two units of unsaturation (multiple bonds or rings). The IR spectrum shows the presence of a nitrile or unsymmetrical alkyne. The ^1H NMR spectrum shows a singlet at δ 3.30 (3H), which suggests a methyl group attached to an electron-withdrawing element such as oxygen. The other two signals are triplets (2H each). This suggests two methylene groups bonded together, —CH_2—CH_2—. When we add the methoxy group, we get C_3H_7O. What is left from the molecular formula is CN (nitrile group). The unknown is **3-methoxy-propanenitrile**.

$$CH_3OCH_2CH_2C≡N$$

(c) The molecular formula, C_5H_{12}, shows no units of unsaturation. Thus, the unknown is a saturated hydrocarbon. The ^{13}C NMR spectrum shows only two signals. Let us consider possible isomers of C_5H_{12} and examine which of the isomers will have only two types of carbon atoms.
The three isomers are:

$$CH_3CH_2CH_2CH_2CH_3 \quad CH_3\underset{|}{\overset{CH_3}{C}}HCH_2CH_3 \quad CH_3-\underset{\underset{CH_3}{|}}{\overset{\overset{CH_3}{|}}{C}}-CH_3$$

Pentane   2-Methylbutane   2,2-Dimethylpropane
               (Isopentane)   (Neopentane)

Pentane has three types of carbon atoms (C1 is equivalent with C5; C2 is equivalent with C4). 2-Methylbutane has four types of carbon atoms (C1 is equivalent with carbon of the methyl group bonded to C2). 2,2-Dimethylpropane has two types of carbon atoms, namely the C2 carbon and four equivalent methyl groups. Thus, the unknown is **2,2-dimethylpropane**.

(d) The formula, $C_5H_{10}O$, shows one unit of unsaturation (a ring or a double bond). The IR spectrum indicates a carbonyl group. The ^{13}C NMR spectrum shows three signals, namely one at δ 210.7 (a carbonyl carbon in a ketone) and two in the alkyl region. Two equivalent ethyl groups attached to the carbonyl group matches that requirement. Thus, the unknown is **3-pentanone**.

$$CH_3CH_2\overset{\overset{O}{\|}}{C}CH_2CH_3$$

(e) The molecular formula, C_4H_7N, shows two units of unsaturation (multiple bonds or rings). The IR spectrum shows the presence of a nitrile or unsymmetrical alkyne. The molecular formula contains nitrogen and the ^{13}C NMR spectrum shows a carbon signal at δ 123.7 (the carbon of the nitrile group). The ^{13}C NMR spectrum also shows two other carbon signals in the alkane region.

The ^1H NMR spectrum shows a characteristic pattern of the isopropyl group, namely a septet (1H) and a doublet (6H). This is consistent with two peaks in the alkane region of the ^{13}C NMR spectrum. Thus, the unknown is **2-methylpropanenitrile**.

$$\text{H}_3\text{C} \diagdown \atop \text{H}_3\text{C} \diagup \text{HC} - \text{C} \equiv \text{N}$$

10.33 (a) The molecular formula, C_4H_9Br, shows no units of unsaturation. Thus, the unknown is a saturated bromohydrocarbon.

The ^{1}H NMR spectrum shows some symmetry in the compound. There are three peaks in the spectrum, namely a doublet at δ 3.3 (2H), a multiplet at δ 1.95 (1H), and a doublet at δ 0.97 (6H). The doublet at δ 3.3 is shifted downfield because of attachment to an electronegative atom. The molecular formula contains bromine so we can suggest a —CH$_2$Br group. The signal of those two protons is split into a doublet, which means that there is one neighboring hydrogen, —CH— . The remaining two signals, a multiplet (1H) and a doublet (6H), are a characteristic feature of the isopropyl group. Thus, the isopropyl group is attached to —CH$_2$Br and the unknown is **1-bromo-2-methylpropane.**

$$\text{BrCH}_2 - \text{CH} \diagup \text{CH}_3 \diagdown \text{CH}_3$$

(b) The two hydrogen atoms on C1 are split into a doublet (δ 3.3, 2H) by the hydrogen atom on C2; the hydrogen on C2 sees eight neighboring hydrogens so its signal is split into a nonet. However, even the expended multiplet shows only seven peaks. The explanation is that relative intensities of peaks in a nonet are 1:8:28:56:70:56:28:8:1. The intensities of the most outside peaks is eight times lower than those of second last peaks. The total integration of all peaks is one hydrogen. Thus, we usually see only the seven (or even five) peaks of highest intensity.

The six equivalent hydrogens of the two methyl groups bonded to C2 have one neighboring hydrogen on C2 and they are split into a doublet at δ 0.97.

(c) The signal of the —CH$_2$Br group, as well as that of the two methyl groups bonded to C2, appear as doublets because they are split by the one hydrogen on C2:

Assume $J_{1,2} = J_{2,3}$

10.34 (a) The molecular formula, C_6H_8, shows three units of unsaturation (multiple bonds or rings). From the ^{1}H NMR spectrum, we can tell that the molecule is symmetrical with only two types of hydrogen atoms, namely vinyl protons (signal at δ 5.7) and alkyl protons (δ 2.7). For three units of unsaturation, we could have three double bonds in a six-carbon chain or two double bonds in a cyclohexane ring. 1,3,5-Hexatriene has all six sp^2 hybrid carbon atoms and all hydrogen atoms will appear in the alkene range of the ^{1}H NMR spectrum (δ 5 to 6). That is not what we see. We may have 1,3- or 1,4- cyclohexadiene. 1,3-Cyclohexadiene has two types of vinyl hydrogen atoms and 1,4-cyclohexadiene has one type of vinyl hydrogen atom and one type of hydrogen atom bonded to an sp^3 hybridized carbon. The ^{13}C NMR spectrum shows two signals of carbon atoms at δ 26.0 (sp^3 hybrid carbon atom) and at δ 124.5 (sp^2 hybrid carbon atoms). 1,3-Cyclohexadiene has two nonequivalent sp^2 hybrid carbon atoms and one type of sp^3 hybrid carbon atoms and its ^{13}C NMR spectrum would show three signals. **1,4-Cyclohexadiene** has one type of sp^2 hybrid carbon atoms and one type of sp^3 hybrid carbon atoms and meets all the requirements of the spectrum.

(b) The molecular formula, $C_{10}H_{18}O$, shows two units of unsaturation (multiple bonds or rings). From the ^{1}H NMR spectrum, we can tell that the molecule is symmetrical with only three types of hydrogen atoms, namely three singlets at δ 2.15 (4H), δ 1.6 (2H), and δ 1.0 (12H). The position of the signal at δ 2.15 is characteristic of hydrogen atoms adjacent to a carbonyl group. The IR spectrum shows an absorption band at 1720 cm^{-1} and, thus, confirms the presence of the carbonyl group.

The signal at δ 1.0 is typical of a methyl group. Twelve equivalent hydrogen atoms can be found on four methyl groups. There are no signals in the alkene region, so the other unsaturation must come from a ring. To maintain the symmetry of the molecule, the methyl groups are placed on carbon atoms C3 and C5. The unknown is **3,3,5,5-tetramethylcyclohexanone**.

(c) The molecular formula, $C_7H_{14}O$, shows one unit of unsaturation (multiple bond or a ring). The IR spectrum shows an absorption band at 1695 cm^{-1}, which indicates the presence of a carbonyl group. From the ^{1}H NMR spectrum, we can tell that the molecule is symmetrical with only two types of hydrogen atoms, namely a septet at δ 2.79 (2H) and a doublet at δ 1.08 (12H). This is a characteristic splitting pattern of an isopropyl group, so it appears as if we have two equivalent isopropyl groups. Two isopropyl groups contribute C_6H_{14} to the molecular formula. What is left is CO (a carbonyl group). The unknown is then **2,4-dimethyl-3-pentanone**.

$$\begin{array}{c}\overset{1}{H_3C}\overset{O}{\underset{\|}{}}\overset{5}{CH_3}\\ \diagdown\underset{2}{C}\diagup\underset{3}{C}\diagdown\underset{4}{C}\diagup\\ HCCH\\ ||\\ CH_3CH_3\end{array}$$

(d) The molecular formula, C_6H_{10}, shows two units of unsaturation (multiple bonds or rings). There are four types of hydrogen signals, namely a multiplet at δ 2.05 to 2.11 (2H), a triplet at δ 1.96 (1H), a septet at δ 1.82 (1H), and a doublet at δ 0.99 (6H). The ^{1}H NMR spectrum shows no hydrogen atoms bonded to sp^2 hybrid carbon atoms. The septet-doublet pattern is characteristic of an isopropyl group. The IR spectrum shows absorption bands at 3250 cm^{-1} (terminal alkyne, —C≡C—H) and at 2150 cm^{-1} (carbon-carbon triple bond, —C≡C—). What is left from the molecular formula is —CH$_2$—. Joining the partial structures together gives the structure of the unknown as **4-methyl-1-pentyne**.

$$\begin{array}{c}H_3C\\ \diagdown\underset{4}{}\\ HC-\overset{5}{CH_3}\\ \underset{1}{HC}\equiv\underset{2}{C}-\overset{3}{\underset{}{CH_2}}\diagup\end{array}$$

(e) The molecular formula, $C_5H_{11}Cl$, shows no units of unsaturation (saturated chloroalkane). The ^{1}H NMR spectrum shows three signals of hydrogen atoms in the alkane region. The three signals are a quartet (δ 1.77, 2H), a singlet (δ 1.5, 6H), and a triplet (δ 1.03, 3H). However, there is no signal around δ 3.5, which would be expected for a —CHCl— partial structure. Thus, chlorine is attached to a carbon without hydrogen atoms, $\diagdown$CCl—.
$\diagup$

A quartet (2H) and a triplet (3H) are the characteristic pattern of an ethyl group, —CH$_2$CH$_3$. The singlet at δ 1.5 can result from two methyl groups. Joining these partial structures gives the structure of the unknown as **2-chloro-2-methylbutane**.

$$\begin{array}{c}\overset{1}{H_3C}\overset{3}{CH_2}\overset{4}{CH_3}\\ \diagdown\underset{2}{C}\diagup\\ H_3C\diagup\diagdown Cl\end{array}$$

(f) The molecular formula, C_5H_8, shows two units of unsaturation (multiple bonds or rings). The IR absorption band at 2220 cm^{-1} suggests the presence of a carbon-carbon triple bond, —C≡C—. There are three types of hydrogen signals in the ^{1}H NMR spectrum, namely a multiplet at δ 2.05 to 2.19 (2H), a triplet at δ 1.75 (3H), and a triplet at δ 1.1 (3H). The multiplet at δ 2.05 to 2.19 shows characteristic "long-range" coupling through a triple bond. The chemical shift of that signal also indicates a methylene group adjacent to a triple bond. The triplet at δ 1.1 indicates a methyl group bonded to that methylene group. The triplet at δ 1.75 indicates another methyl group attached to an sp-hybridized carbon atom. The unknown is **2-pentyne**.

$$H_3\overset{1}{C}-\overset{2}{C}\equiv\overset{3}{C}-\overset{4}{C}H_2\overset{5}{C}H_3$$

10.35 (a) The coupling of the H_b hydrogens with five (3 + 2) neighboring hydrogen atoms (all having the same coupling constant, $J_{ab} = J_{bc} = 7$ Hz) results in a regular sextet with relative intensities of peaks of 1:5:10:10:5:1.

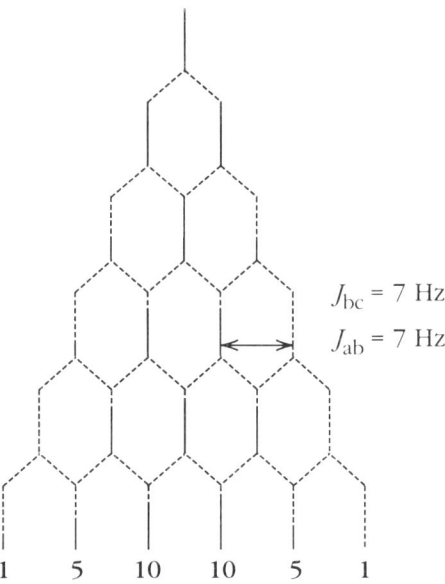

(b) The coupling of the H_b hydrogens with five (3 + 2) neighboring hydrogen atoms (each group having a different coupling constant, $J_{ab} = 7$ Hz, $J_{bc} = 14$ Hz) results in a symmetrical but irregular octet with relative intensities of peaks of 1:3:5:7:7:5:3:1. Note that it is not important if the signal of the H_b hydrogens is first coupled with H_a and then with H_c or first with H_c.

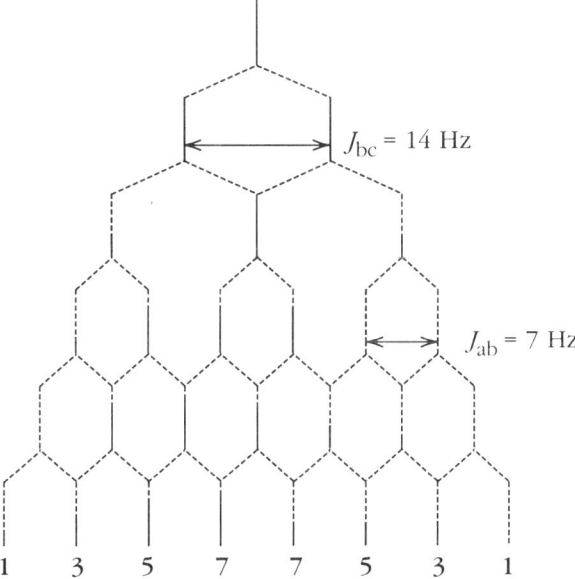

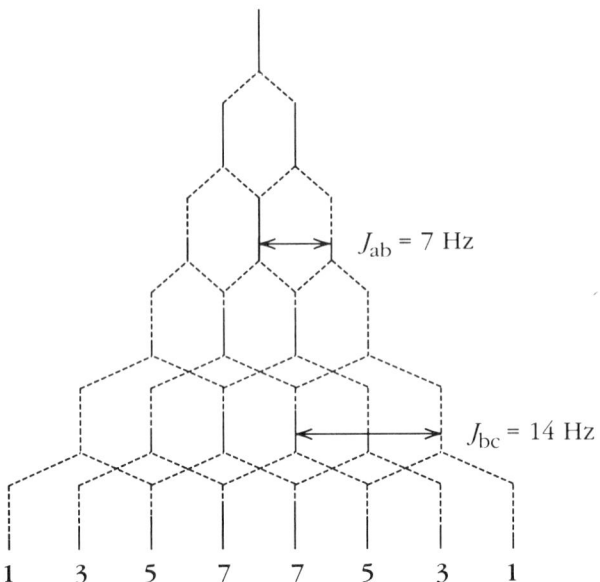

1 3 5 7 7 5 3 1

(c) The coupling of the H_b hydrogens with five (3 + 2) neighboring hydrogen atoms (each group having a different coupling constant, J_{ab} = 14 Hz, J_{bc} = 8 Hz) results in a symmetrical but very irregular multiplet with relative intensities of peaks of 1:2:3:1:6:3:3:6:1:3:2:1. Note that it is not important if the signal of the H_b hydrogens is first coupled with H_a and then with H_c or first with H_c.

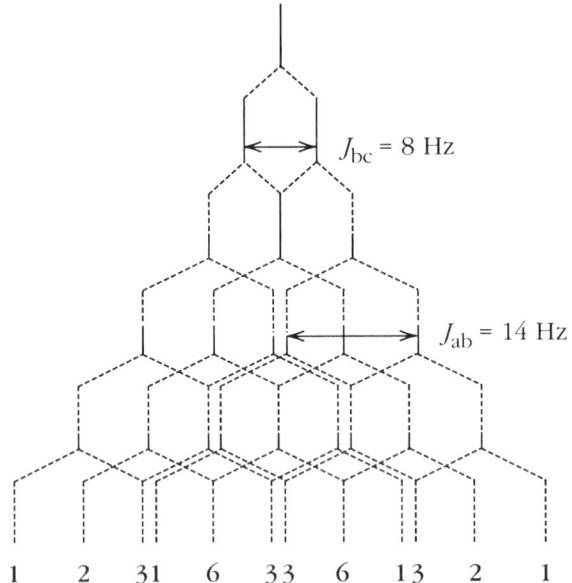

1 2 3 1 6 3 3 6 1 3 2 1

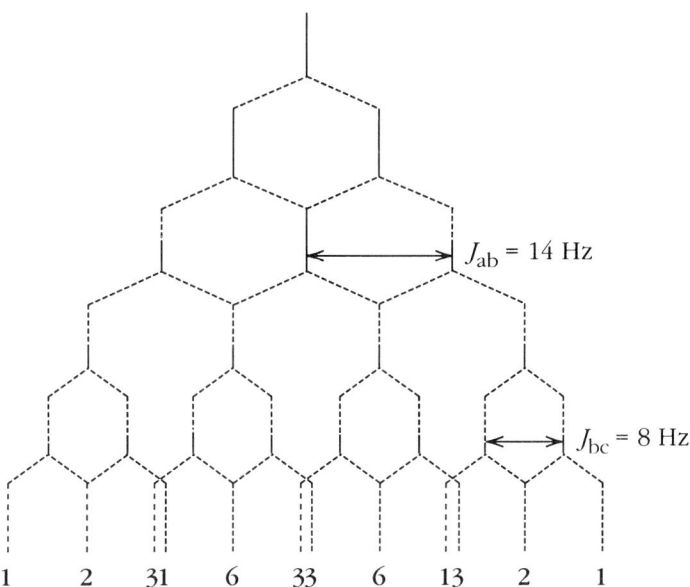

CHAPTER 11

ALCOHOLS

Concepts

Alcohols are compounds that contain a hydroxyl group (—OH) bonded to an sp^3 hybrid carbon atom. There are both IUPAC and radicofunctional (trivial) nomenclature systems for alcohols and for ethers. With alcohols, all but the simplest are named by the IUPAC method, which involves an important concept of nomenclature called the **principal group.** The principal group is the chemical group on which the name is based, and is *always cited as a suffix in the name*. For example, in a simple alcohol, the —OH group is the principal group, and its suffix is *-ol*. The name of an alcohol is constructed by dropping the final "*e*" from the name of the parent alkane and adding this suffix. If there are several nonequivalent carbon atoms in the molecule, then a number is written immediately before the name to indicate the position of the hydroxyl group. The name of the alcohol is based on the longest carbon chain *that contains the —OH group*, numbered so that the carbon atom bearing this group has the lowest possible number. Some examples of IUPAC names are given here.

CH₃OH	CH₃CH₂OH	CH₃CH₂CH₂OH	CH₃CHCH₃ with OH	CH₃CHCH₂OH with CH₃
Methanol	Ethanol	1-Propanol	2-Propanol	2-Methyl-1-propanol

When more than one hydroxyl group is present in the compound, the hydrocarbon name is retained, the positions of the hydroxyls are numbered, and a suffix denoting the number of hydroxyls (e.g., *-diol*, *-triol*, etc.) is added to the name of the appropriate alkane *without* dropping the final *e*.

1,3-Propanediol   (2R,4R)-2,4-Hexanediol   (Z)-4-Chloro-4-hepten-2-ol

A diol (especially a 1,2-diol) is often referred to as a **glycol.** The trivial name for a 1,2-diol is that of the corresponding *alkane* followed by the word *glycol*.

Classification of Alcohols

Alcohols, like haloalkanes, are classified as **methyl, primary, secondary,** or **tertiary.**

CH₃OH	CH₃CHCH₂OH with CH₃	CH₃CHCH₃ with OH	HO—C(CH₂CH₃)(CH₃)—CH₂CH₂CH₃
Methanol	2-Methyl-1-propanol	2-Propanol	3-Methyl-3-hexanol
Methyl	*Primary (1°)*	*Secondary (2°)*	*Tertiary (3°)*

Alcohols are an important class of compounds because the —OH group can be converted into a number of other functional groups.

(a) Primary and secondary alcohols undergo *oxidation* to give various products:
   (i) *primary* alcohols are oxidized by most oxidizing agents to carboxylic acids,
   (ii) oxidation of *primary* alcohols can be stopped at the aldehyde stage by the use of specific reagents, such as pyridinium chlorochromate (PCC) in anhydrous solvents,
   (iii) *secondary* alcohols are oxidized to ketones.

(b) Alcohols can be converted to haloalkanes. Alcohols react with hydrogen halides (HCl, HBr, or HI) to form haloalkanes, with thionyl chloride ($SOCl_2$) to form chloroalkanes, and with phosphorus tribromide (PBr_3) to form bromoalkanes.

(c) Alcohols react with acids to form esters and water. The acid can be a carboxylic acid or one of a number of other acids such as arenesulfonic acid, sulfuric acid, nitric acid, or phosphoric acid.

Alcohols are prepared in a number of ways: (a) from haloalkanes by hydrolysis, (b) from Grignard reagents by reaction with carbonyl compounds, (c) from alkenes by hydration, (d) from carbonyl compounds by reduction with $NaBH_4$ or $LiAlH_4$.

The reduction of aldehydes and/or ketones can be *stereospecific*. When addition of a hydride ion occurs equally well to either side of the plane of the carbonyl group of a prochiral aldehyde or ketone, the product is a racemic mixture because both enantiomers are formed in equal amounts. When addition to one side of the plane of the carbonyl group of a prochiral aldehyde or ketone is preferred, the product is optically active because one enantiomer is formed in excess.

Solutions to the Exercises

11.1 Alcohols are classified according to the type of carbon atom to which the —OH group is bonded. If this carbon atom is primary (bonded to one other carbon atom), the compound is a **primary alcohol.** A **secondary alcohol** has the —OH group attached to a secondary carbon atom; a **tertiary alcohol** has the —OH group bonded to a tertiary carbon atom.

(a) Secondary (2°) (b) Primary (1°) (c) Tertiary (3°)
(d) Tertiary (3°) (e) Primary (1°) (f) Secondary (2°)

11.2 (a) Methyl alcohol (b) Pentyl alcohol
(c) Cyclobutyl alcohol (d) Isobutyl alcohol

11.3 (a) Methanol (b) Pentanol (c) Cyclobutanol
(d) 2-Methyl-1-propanol

11.4

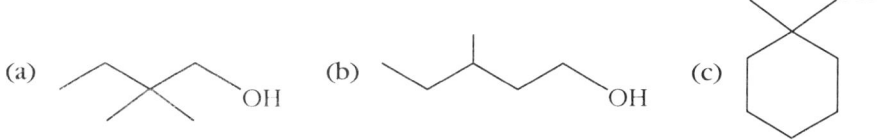

11.5 (a) 2-Chloroethanol (b) 4-Methyl-1-pentanol
(c) 5,5-Dibromo-2-methyl-1-hexanol (d) *cis*-3-Chlorocyclobutanol
(e) (*R*)-3-Ethyl-5-iodo-3-hexanol (f) 4-Ethyl-1,2,5-pentanetriol

11.6 Compound A:
Molecular formula: $C_5H_{12}O$; zero units of unsaturation (saturated alcohol).

The ¹H NMR spectrum shows a triplet at δ 0.87 (3H), a singlet at δ 1.15 (6H), and a quartet at δ 1.45 (2H). There is no signal in the range of 3.5 to 4.2 ppm, which indicates that there is no hydrogen atom attached to the carbon atom to which the —OH group is bonded. It means that we have a *tertiary alcohol* (that is, the carbon atom with the —OH group is bonded to three alkyl groups). One such group is an ethyl group (quartet and triplet). The other two must be two equivalent methyl groups (singlet in the spectrum). Thus, the compound A is **2-methyl-2-butanol**.

$$CH_3CH_2-\underset{\underset{CH_3}{|}}{\overset{\overset{CH_3}{|}}{C}}-OH$$

Compound B:
Molecular formula: $C_4H_{10}O$; zero units of unsaturation (saturated alcohol).

The ¹H NMR spectrum shows a doublet at δ 0.85 (6H), a septet at δ 1.7 (1H), a multiplet at δ 2.67 (1H), and a doublet at δ 3.33 (2H). The signals at δ2.67 disappear when the sample is shaken with D_2O. The doublet at 3.33 ppm corresponds to the hydrogen atoms attached to the carbon atom to which the —OH group is bonded. The integration tells us that there are two hydrogen atoms bonded to the carbon atom to which the —OH group is attached. That means we have a *primary alcohol*. The only alkyl group is an isopropyl group [septet (1H) and doublet (6H)]. Thus compound B is **2-methyl-1-propanol**.

$$(CH_3)_2CHCH_2-OH$$

11.7 The hydroboration-oxidation of alkenes gives alcohols with anti-Markownikoff regiochemistry (the —OH group is added to the less branched carbon of the double bond). Oxymercuration-demercuration of alkenes gives alcohols with Markownikoff regiochemistry (the —OH group is bonded to the more branched carbon of the double bond).

(a) [alkene] $\xrightarrow{\text{(1) 9-BBN} \quad \text{(2) } H_2O_2, H_2O/HO^-}$ [alcohol with OH on less branched carbon]

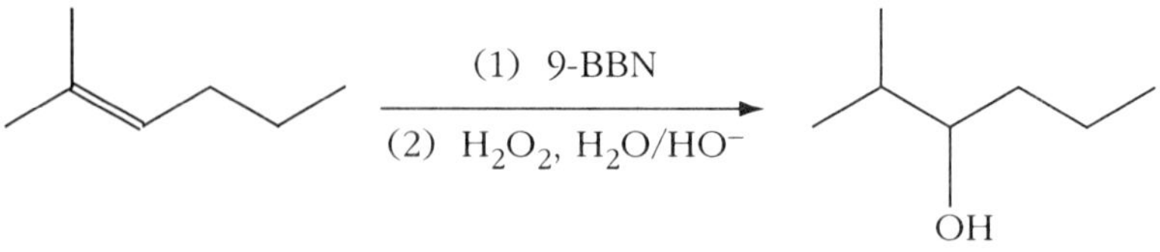

(b) [alkene] $\xrightarrow{\text{(1) Hg(OOCCH}_3)_2 \quad \text{(2) NaBH}_4, HO^-}$ [alcohol with OH on more branched carbon]

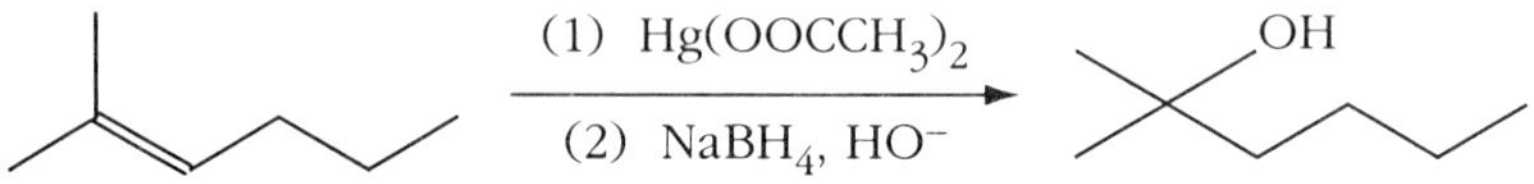

11.8 The reaction of an organolithium reagent with a carbonyl group occurs in two steps. In the first step, the organolithium reagent adds to the carbonyl group (the nucleophilic carbon atom of the organolithium reagent attacks the electrophilic carbon of the carbonyl group). Reaction of the lithium salt with aqueous acid in the second step forms an alcohol.

Step 1:

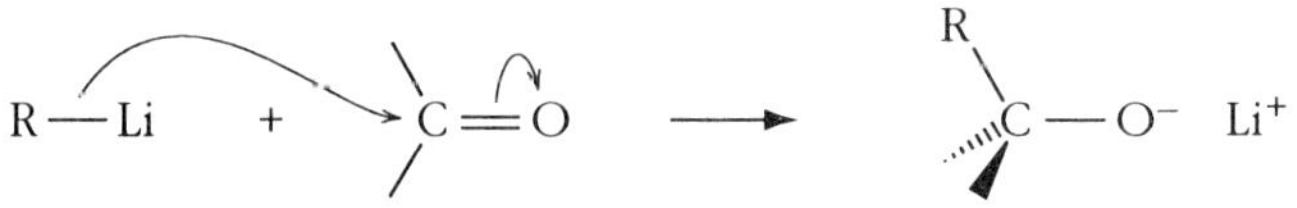

R—Li + C=O ⟶ R—C—O⁻ Li⁺

Step 2:

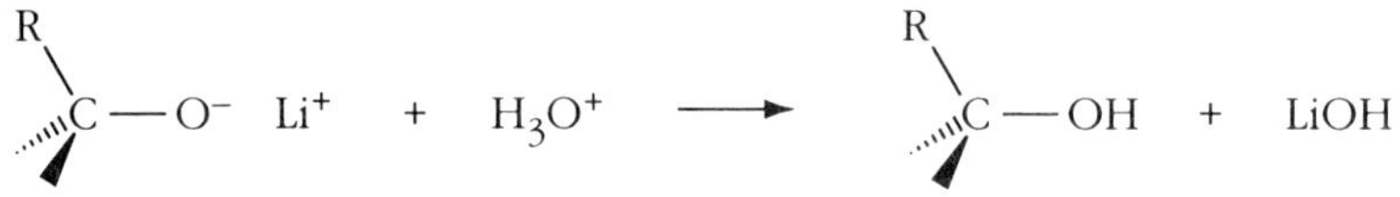

R—C—O⁻ Li⁺ + H_3O^+ ⟶ R—C—OH + LiOH

11.9

(a) [t-BuBr] + 2 Li ⟶ [t-BuLi] + LiBr

(b) [t-BuLi] + HCHO ⟶ [alkoxide O⁻ Li⁺]

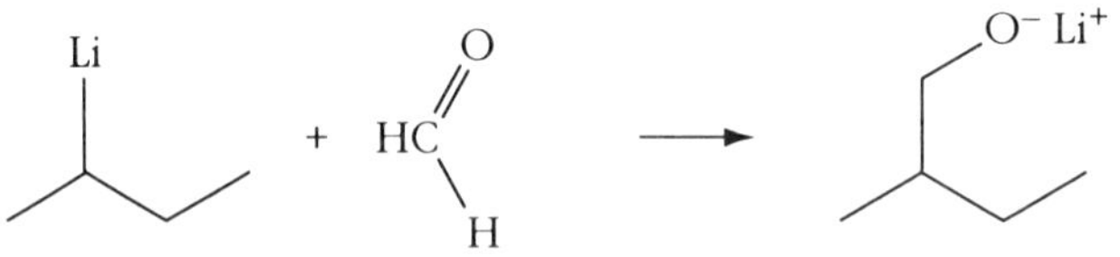

(c) [alkoxide O⁻ Li⁺] + H_3O^+ ⟶ [alcohol OH] + LiOH

(d) [sec-butyllithium] $\xrightarrow[(2)\ H_3O^+]{(1)\ CH_3CH_2CHO}$ [3-methyl-2-pentanol structure with OH]

(e) [sec-butyllithium] $\xrightarrow[(2)\ H_3O^+]{(1)\ \text{3-pentanone}}$ [tertiary alcohol product]

11.10 The reaction of Grignard or organolithium reagents with a carbonyl group occurs in two steps. First, the carbon group of the Grignard or organolithium reagent attacks the carbonyl carbon (the nucleophilic carbon atom of the Grignard or organolithium reagent attacks the electrophilic carbon of the carbonyl group). Addition of dilute acid to the reaction mixture in the second step forms an alcohol.

(a) $(CH_3)_2CHI \xrightarrow[(CH_3CH_2)_2O]{Mg} (CH_3)_2CHMgI$

(b) [cyclopentanone] $\xrightarrow[\text{THF}]{\text{butyl}-Li}$ [1-butylcyclopentanolate $O^-\ Li^+$] $\xrightarrow{H_3O^+}$ [1-butylcyclopentanol OH]

(c) [acetophenone: PhC(=O)CH₃] $\xrightarrow[(CH_3CH_2)_2O]{CH_3CH_2CH_2MgBr}$ [Ph-C(CH₃)(CH₂CH₂CH₃)-O⁻ $^+$MgBr]

$\xrightarrow{H_3O^+}$ [Ph-C(CH₃)(OH)(CH₂CH₂CH₃)]

(d) [2-methyl-2-butenyl Li] $\xrightarrow[\text{THF}]{H_2C=O}$ [allylic $O^-\ Li^+$] $\xrightarrow{H_3O^+}$ [allylic OH]

(e) [PhMgBr] + CH₃CH₂CHO —THF→ [Ph-CH(O⁻MgBr⁺)-CH₂CH₃] —H₃O⁺→ [Ph-CH(OH)-CH₂CH₃]

11.11 When you are asked to prepare an alcohol from an organometallic reagent and a carbonyl-containing compound, you should apply a *retrosynthetic analysis* (the technique of reasoning backward from a target molecule to suitable starting materials). The target molecule is an alcohol. In the Grignard synthesis of an alcohol, the carbon atom bearing the —OH group in the product is the carbonyl carbon atom of the starting material. Keeping this in mind, we can devise the following syntheses.

(a) Because 2-hexanol is a secondary alcohol, disconnection of bonds to the hydroxyl-bearing carbon generates two pairs of structural fragments. One route involves the addition of a methyl Grignard reagent to a five-carbon aldehyde, while the other route requires addition of a butylmagnesium halide to a two-carbon aldehyde:

Retrosynthesis:

2-hexanol ⟹ CH₃MgBr + pentanal

2-hexanol ⟹ CH₃CHO + CH₃CH₂CH₂CH₂MgCl

Synthesis:

pentanal —CH₃MgBr, Dry ether→ 2-hexanol magnesium bromide alkoxide —H₃O⁺→ 2-hexanol

(b) 1-Hexanol is a primary alcohol. It can be prepared from formaldehyde and a five-carbon Grignard reagent:

Retrosynthesis:

[Structure: CH₃CH₂CH₂CH₂CH₂CH₂OH with disconnection ⇒ H₂C=O + CH₃CH₂CH₂CH₂CH₂MgCl]

Synthesis:

[CH₃CH₂CH₂CH₂CH₂MgCl + H₂C=O, Dry ether → CH₃CH₂CH₂CH₂CH₂CH₂O⁻ ⁺MgCl]

[H₃O⁺ → CH₃CH₂CH₂CH₂CH₂CH₂OH]

(c) 1-Methylcyclopentanol may be conveniently prepared from cyclopentanone and a methyl Grignard reagent. The alternative disconnection leads to an impractical starting material containing both a carbonyl group and a Grignard reagent.

Retrosynthesis:

[cyclopentane with CH₃ and OH, disconnection of CH₃ ⇒ CH₃MgBr + cyclopentanone]

[cyclopentane with CH₃ and OH, disconnection of ring ⇏ ClMg-CH₂CH₂CH₂CH₂-C(=O)CH₃]

Synthesis:

[cyclopentanone + CH₃MgBr, Dry ether → 1-methylcyclopentyl-O⁻ ⁺MgBr, H₃O⁺ → 1-methylcyclopentanol]

(d) The target alcohol, 3-ethyl-3-pentanol, is tertiary and so is prepared by addition of a Grignard reagent to a ketone. Since all three substituents on the hydroxyl-bearing carbon are ethyl groups, there is only *one*, not three, distinct way of preparing 3-ethyl-3-pentanol by a Grignard reaction.

Retrosynthesis:

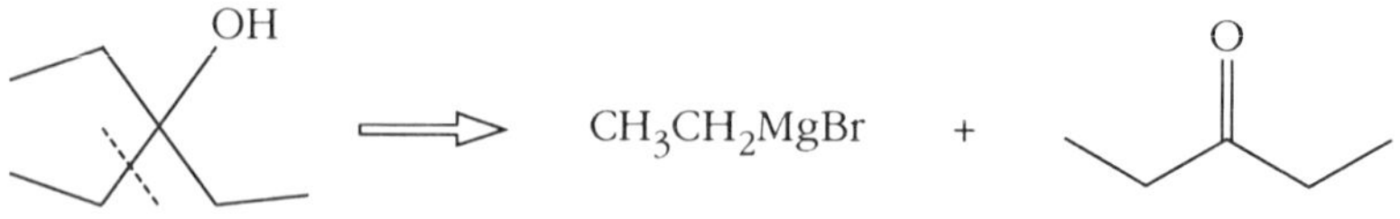

[Structure: 3-methyl-3-pentanol with bond cleavage] ⟹ CH₃CH₂MgBr + [Structure: CH₃CH₂C(O)CH₃... showing 2-butanone]

Synthesis:

[2-butanone] $\xrightarrow[\text{Dry ether}]{\text{CH}_3\text{CH}_2\text{MgBr}}$ [alkoxide-MgBr intermediate] $\xrightarrow{\text{H}_3\text{O}^+}$ [3-methyl-3-pentanol]

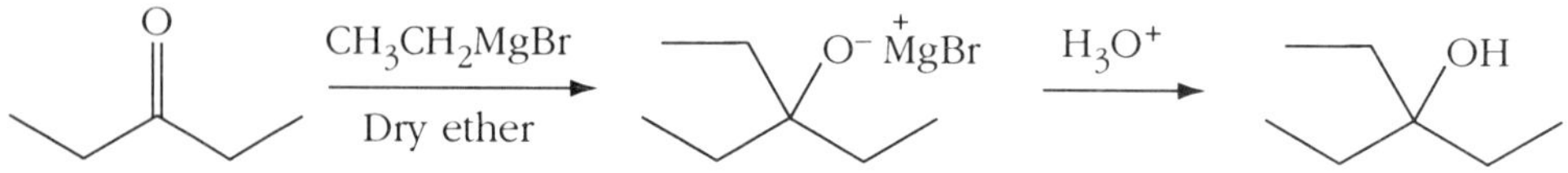

(e) The target alcohol is secondary and it can be prepared by the reaction of a Grignard reagent with an aldehyde. There are two retrosynthetic transformations and two plausible syntheses:

Retrosynthesis:

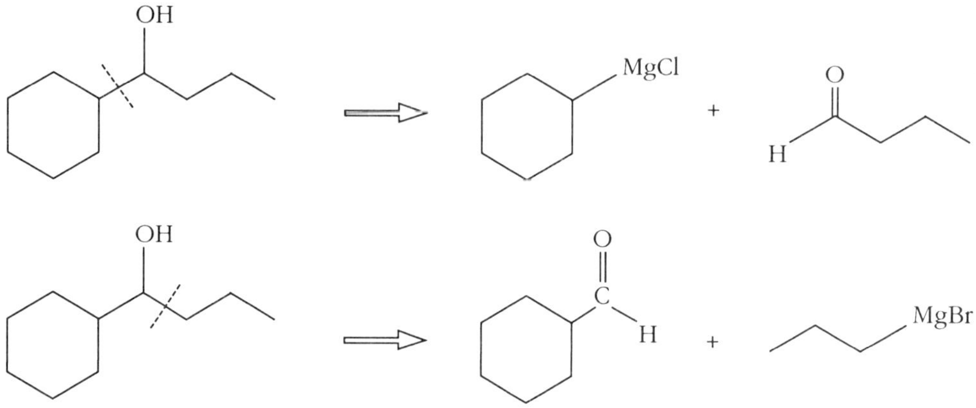

[1-cyclohexyl-1-butanol with cleavage at C–cyclohexyl bond] ⟹ [cyclohexyl-MgCl] + [butanal]

[1-cyclohexyl-1-butanol with cleavage at C–propyl bond] ⟹ [cyclohexanecarbaldehyde] + [propyl-MgBr]

Synthesis:

[butanal] $\xrightarrow[\text{Dry ether}]{\text{cyclohexyl-MgCl}}$ [alkoxide intermediate with O⁻ MgCl⁺]

$\xrightarrow{\text{H}_3\text{O}^+}$ [1-cyclohexyl-1-butanol]

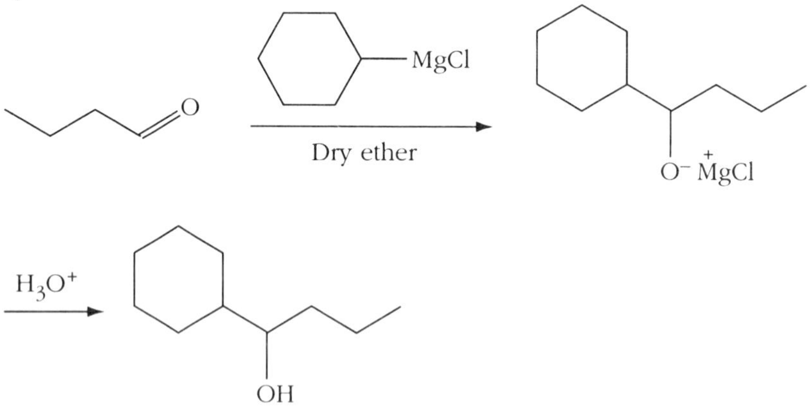

[Reaction scheme: cyclohexanecarbaldehyde + propylMgBr / dry ether → alkoxide intermediate; then H₃O⁺ → 1-cyclohexyl-1-butanol with OH]

11.12 Organolithium reagents react with carbonyl compounds in the same way that Grignard reagents do. The only difference is that in their reactions with aldehydes and ketones, organolithium reagents are somewhat more reactive than Grignard reagents. The retrosynthetic analysis employed here is identical to the one shown in Exercise 11.11.

(a) The target molecule is 2-methyl-2-pentanol, a tertiary alcohol. Disconnection of bonds to the hydroxyl-bearing carbon generates two pairs of structural fragments. One route involves the addition of methyllithium to 2-pentanone, while the other route requires addition of propyllithium to acetone.

Retrosynthesis:

[Structure of 2-methyl-2-pentanol ⟹ CH₃Li + 2-pentanone]

[Structure of 2-methyl-2-pentanol ⟹ propyllithium + acetone]

Synthesis:

[2-pentanone + CH₃Li / dry ether → lithium alkoxide; then H₃O⁺ → 2-methyl-2-pentanol]

[acetone + propyllithium / dry ether → lithium alkoxide; then H₃O⁺ → 2-methyl-2-pentanol]

(b) (CH₃)₂CH—Li + H₂C=O —Dry ether→ (CH₃)₂CHCH₂O⁻Li⁺ —H₃O⁺→ (CH₃)₂CHCH₂OH

(c) CH₃C(=O)H + CH₂=CH—Li —Dry ether→ CH₂=CH—CH(CH₃)—O⁻Li⁺ —H₃O⁺→ CH₂=CH—CH(CH₃)—OH

CH₂=CH—CHO + CH₃Li —Dry ether→ CH₂=CH—CH(CH₃)—O⁻Li⁺ —H₃O⁺→ CH₂=CH—CH(CH₃)—OH

(d) PhLi + CH₃CH₂C(=O)CH₃ —Dry ether→ Ph—C(CH₃)(CH₂CH₃)—O⁻Li⁺ —H₃O⁺→ Ph—C(CH₃)(CH₂CH₃)—OH

PhC(=O)CH₃ + CH₃CH₂Li —Dry ether→ Ph—C(CH₃)(CH₂CH₃)—O⁻Li⁺ —H₃O⁺→ Ph—C(CH₃)(CH₂CH₃)—OH

PhC(=O)CH₂CH₃ + CH₃Li —Dry ether→ Ph—C(CH₃)(CH₂CH₃)—O⁻Li⁺ —H₃O⁺→ Ph—C(CH₃)(CH₂CH₃)—OH

11.13 In these exercises, the principles of retrosynthetic analysis are applied. The only restriction is the size of an alkyl group (four or fewer carbon atoms). This restriction leaves only one possibility in each case. We can use organolithium or Grignard reagents.

(a) The target compound is 1-isopropylcyclobutanol, a tertiary alcohol. The retrosynthetic analysis suggests two possible ways of disconnection. The second transformation, however, leads to starting material with more than four carbon atoms (disallowed!). So, at least one more step in the synthesis would be needed to prepare the starting material. The starting material would also be impractical because it contains a carbonyl group and lithium reagent that react with each other intramolecularly and intermolecularly leading to a mixture of products.

Retrosynthesis:

[Retrosynthesis scheme showing cyclobutane with isopropyl and OH → cyclobutanone + isopropyllithium]

[Alternative disconnection crossed out: cyclobutane with isopropyl and OH ⇏ Li-CH₂CH₂CH₂-C(=O)-CH(CH₃)₂]

The only successful synthesis:

[Cyclobutanone + isopropyllithium, Dry ether → alkoxide intermediate, H₃O⁺ → 1-isopropylcyclobutanol]

(b) The restriction of the size of an alkyl group to four or fewer carbon atoms leads to the only one-step synthesis of the desired product:

[Isobutyl MgCl + CH₃CH₂CH₂C(=O)H, Dry ether → alkoxide (O⁻ ⁺MgCl), H₃O⁺ → alcohol product]

11.14 **Oxidation** corresponds to an *increase* in the number of bonds between carbon and oxygen and/or a *decrease* in the number of carbon-hydrogen bonds. Conversely, **reduction** corresponds to an increase in the number of carbon-hydrogen bonds and/or a *decrease* in the number of carbon-oxygen bonds.

Adding or removing any element *more electronegative* than carbon will have the same effect on the oxidation state of carbon as adding or removing oxygen. Adding or removing any element *less electronegative* than carbon will have the same effect on the oxidation state of carbon as adding or removing hydrogen.

(a) Oxidation
(b) Neither oxidation nor reduction (both H and Cl atoms are added)
(c) Reduction
(d) Oxidation
(e) Neither oxidation nor reduction (atom of O is replaced by Cl)

11.15 (a) Reducing agent (H_2 and a metal catalyst)
(b) Reducing agent ($LiAlH_4$ in dry ether)
(c) Oxidizing agent (O_2)
(d) Reducing agent (H_2 and a catalyst)

11.16 Lithium aluminum hydride reduces ketones to secondary alcohols and aldehydes to primary alcohols. Thus,

(a) cyclopentanone $\xrightarrow{(1)\ LiAlH_4,\ (2)\ H_3O^+}$ cyclopentanol

(b) $(CH_3)_2CHCH_2CCH_3$ (with C=O) $\xrightarrow{(1)\ LiAlH_4,\ (2)\ H_3O^+}$ $(CH_3)_2CHCH_2CHCH_3$ (with OH)

(c) $H_2C=CHCH_2CH_2CHO$ $\xrightarrow{(1)\ LiAlH_4,\ (2)\ H_3O^+}$ $H_2C=CHCH_2CH_2CH_2OH$

Please note that the carbon-carbon double bond is not reduced with a metal hydride.

11.17 Reduction of aldehydes yields primary alcohols, whereas reduction of ketones gives secondary alcohols. Catalytic hydrogenation is not selective and hydrogen is added to both carbon-carbon double bonds and carbon-oxygen double bonds.

(a) cyclobutanol with OH

(b) 1-cyclohexylethanol (cyclohexyl-CH(OH)-CH3)

(c) benzyl alcohol (PhCH2OH)

(d) pentan-1-ol (HOCH2CH2CH2CH2CH3)

11.18 (a) 2-Pentanol can be prepared by the reduction of the carbonyl group of 2-pentanone.
(b) 1,1-Dimethyl-1-propanol *cannot* be prepared by the reduction of a carbonyl group because such a substrate would have had to have a carbonyl group and three alkyl groups at the same carbon atom. Thus, tertiary alcohols cannot be prepared by the reduction of carbonyl compounds.
(c) 3,3-Dimethyl-1-butanol can be prepared by the reduction of 3,3-dimethylbutanal.

11.19 The reduction of the carbonyl group would result in a hydrogen atom and a hydroxyl group attached to the "original" carbonyl carbon atom. A carbon stereocenter must have four different substituents and that is why a prochiral carbonyl compound must have two different alkyl substituents attached to the carbonyl group. Hence, only acetophenone (d) is prochiral.

(a) The reduction product is ethanol, which does not contain a stereocenter, so ethanal is not prochiral.

(b) The reduction product is 3-pentanol, which does not contain a stereocenter, so 3-pentanone is not prochiral.

(c) The reduction product is methanol, which does not contain a stereocenter, so methanal is not prochiral.

(d) The reduction product is 1-phenylethanol, which contains a stereocenter, so acetophenone *is* prochiral.

11.20 The order of priority of the groups bonded to the carbonyl carbon is determined using the Cahn-Ingold-Prelog rules. If the direction from highest to lowest priority of the groups is clockwise, that side is called the Re side. If the direction of the groups from highest to lowest priority is counterclockwise, that side is called the Si side.

(a) The order of increasing priority is H < CH_3 < O. The direction from highest to lowest priority of the groups is clockwise, so this is the Re face.

(b) The order of increasing priority is CH_3 < C_6H_5 < O. The direction from highest to lowest priority of the groups is counterclockwise, so this is the Si face.

(c) Re face

(d) Re face

11.21 L-Lactate dehydrogenase is stereospecific and it catalyzes the reduction of the keto group in α-keto acids to (S)-2-hydroxyacids. Thus, the expected product is (S)-2-hydroxybutanoic acid:

11.22

(a) (CH₃)₂CHOH (pK_a 18.0) + CH₃MgI ⟶ (CH₃)₂CH–O⁻ ⁺MgI + CH₄ (pK_a ≈ 50)

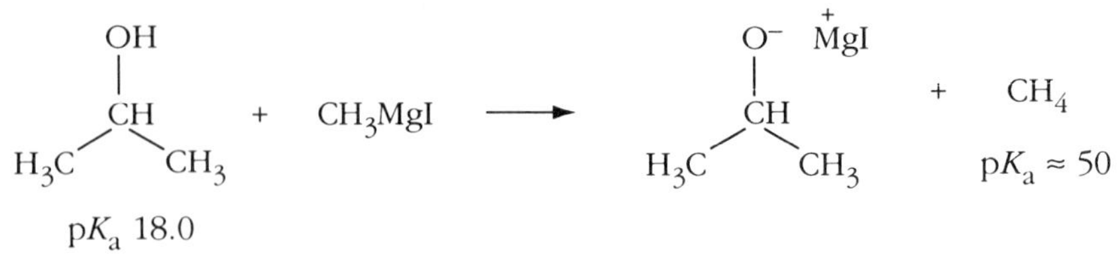

(b) (CH₃)₂CHOH (pK_a 18.0) + CH₃C≡C⁻ Na⁺ ⟶ (CH₃)₂CH–O⁻ Na⁺ + CH₃C≡CH (pK_a ≈ 25)

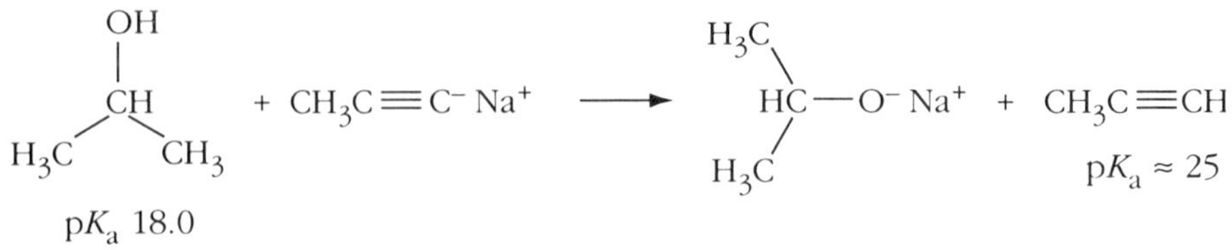

(c) Water is a stronger acid than 2-propanol and the equilibrium lies to the left.
Conclusion: Sodium hydroxide does not convert an alcohol to an alkoxide to a great extent.

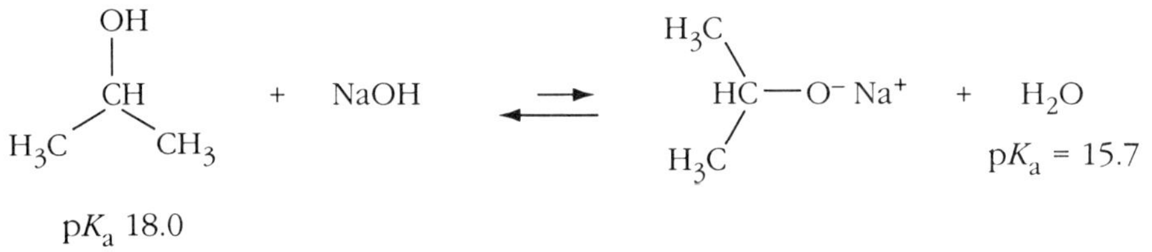

(CH₃)₂CHOH (pK_a 18.0) + NaOH ⇌ (CH₃)₂CH–O⁻ Na⁺ + H₂O (pK_a = 15.7)

(d) (CH₃)₂CHOH (pK_a 18.0) + CH₃CH₂CH₂Li ⟶ (CH₃)₂CH–O⁻ Li⁺ + CH₃CH₂CH₃ (pK_a ≈ 50)

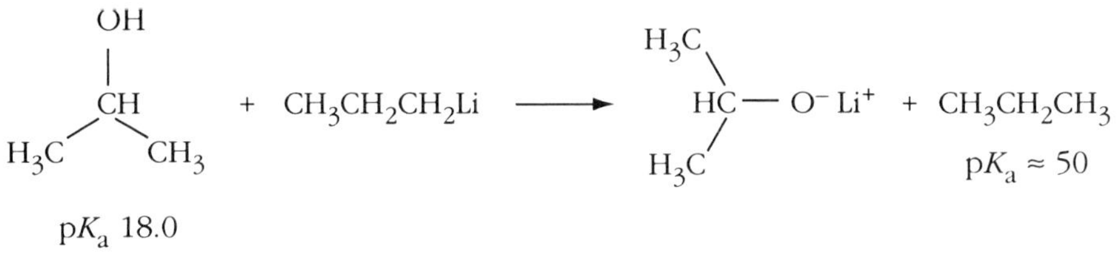

11.23

(a) CH₃CH₂CH₂CH₂Br (b) 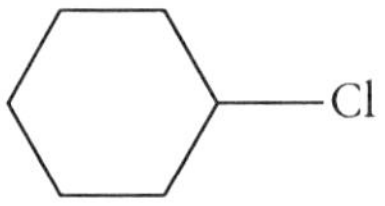 cyclohexyl–Cl

(c) 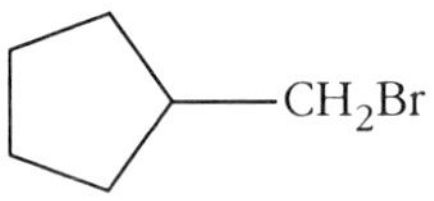 cyclopentyl–CH₂Br (d) 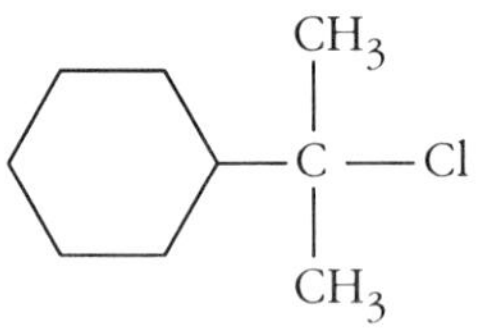 cyclohexyl–C(CH₃)₂–Cl

11.24 (a) KMnO₄ or H₂CrO₄ (CrO₃/H₂SO₄),
(b) Collins' reagent or PCC in dry CH₂Cl₂,
(c) Any oxidizing agent, e.g., Collins' reagent, PCC, or H₂CrO₄.

11.25 (a) A **carbonyl group** is a functional group that contains a carbon-oxygen double bond ($\mathrm{C}=\mathrm{O}$) and two other atoms or groups bonded to the carbonyl carbon. The simplest compounds in which there is a carbonyl group are aldehydes and ketones. However, the carbonyl group may also be a part of a more complex functional group such as the carboxyl group, ester, amide, or anhydride.

(b) A **Grignard reagent** is an organomagnesium compound (RMgX) formed by the reaction of magnesium with an alkyl, vinyl, or aryl halide (R—X).

(c) A **1° alcohol** is a compound of the type RCH$_2$OH; thus, the hydroxyl group is attached to a *primary* (1°) carbon atom (a carbon atom bonded only to *one* other carbon atom).

(d) **Oxidation** is a chemical reaction resulting in a decrease in the number of electrons associated with an atom. In organic chemistry, oxidation of a carbon atom occurs when a bond between the carbon atom and an atom that is less electronegative than carbon is replaced by a bond to an atom that is more electronegative than carbon.

(e) A **thiol** (or an *alkanethiol*) is an organic compound containing the —SH functional group (*sulfhydryl group* or *thiol group*); R—SH or Ar—SH.

(f) An **organolithium reagent** is an organometallic compound that contains a carbon-lithium bond; R—Li or Ar—Li.

(g) The **Re face of a carbonyl group** is the side of the carbonyl group on which the order of decreasing priority of the atoms and groups bonded to the carbonyl carbon atom is clockwise (rectus).

(h) A **2° alcohol** is a compound of the type RR'CHOH; thus, the hydroxyl group is attached to a *secondary* (2°) carbon atom (a carbon atom bonded to *two* other carbon atoms).

(i) The **Si face of a carbonyl group** is the side of the carbonyl group on which the order of decreasing priority of the atoms and groups bonded to the carbonyl carbon atom is counterclockwise (sinister).

(j) **Reduction** is a chemical reaction resulting in an increase in the number of electrons associated with an atom. In organic chemistry, reduction of a carbon atom occurs when a bond between the carbon atom and an atom that is more electronegative than carbon is replaced by a bond to an atom that is less electronegative than carbon.

(k) A **3° alcohol** is a compound of the type R$_3$COH; thus, the hydroxyl group is attached to a *tertiary* (3°) carbon atom (a carbon atom bonded to *three* other carbon atoms).

(l) A **disulfide** is a compound containing the —S—S— linkage; R—S—S—R, R—S—S—R', etc.

11.26 (a) (1*S*,2*S*)-2-(1,1-Dimethylethyl)cyclohexanol or (1*S*,2*S*)-2-*tert*-butylcyclohexanol
(b) (*S*)-2,5-Dimethyl-3-hexanol
(c) 4-Cyclobutyl-2-methyl-2-butanol
(d) *trans*-(*S*)-3-Methyl-5-hepten-3-ol or (*E*)-(*S*)-3-methyl-5-hepten-3-ol
(e) (*E*)-2,3-Dimethyl-3-nonen-2-ol
(f) 4-Methyl-2-heptanethiol

11.27

(a) (b)

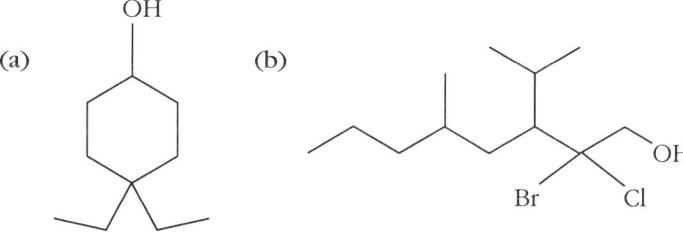

(c) [structure: bromoethyl-cyclohexanol with H stereochemistry] (d) [structure: cyclopentane-1,2-diol, (1R,2S)]

(e) HO—CH₂—C(H)(OH)—C(H)(OH)—CH₃ (with wedge/dash stereochemistry) (f) HS—CH₂CH₂—SH

11.28 Hydroboration-oxidation is an indirect hydration of an alkene giving an alcohol with a regioselectivity opposite to that of Markownikoff's rule (a so-called anti-Markownikoff product).

Oxymercuration-demercuration is another method for converting alkenes to alcohols but, in this case, we obtain alcohols with Markownikoff orientation.

(a) (CH₃)₂C=CH₂ $\xrightarrow{\text{(1) BH}_3 \quad \text{(2) H}_2\text{O}_2, \text{H}_2\text{O/HO}^-}$ (CH₃)₂CH—CH₂OH

(b) (CH₃)₂C=CH₂ $\xrightarrow{\text{(1) Hg(OAc)}_2, \text{H}_2\text{O} \quad \text{(2) NaBH}_4}$ (CH₃)₃C—OH

11.29

(a) cyclopentanone $\xrightarrow{\text{NaBH}_4, \text{CH}_3\text{OH}}$ cyclopentanol

(b) cyclopentanone $\xrightarrow{\text{(CH}_3)_3\text{CLi}, \text{(CH}_3\text{CH}_2)_2\text{O}}$ Li⁺O⁻–C(cyclopentyl)(C(CH₃)₃) $\xrightarrow{\text{H}_3\text{O}^+}$ HO–C(cyclopentyl)(C(CH₃)₃)

(c) cyclopentanone $\xrightarrow{\text{(CH}_3)_2\text{CHCH}_2\text{MgBr}, \text{(CH}_3\text{CH}_2)_2\text{O}}$ BrMg⁺O⁻–C(cyclopentyl)(CH₂CH(CH₃)₂) $\xrightarrow{\text{H}_3\text{O}^+}$ HO–C(cyclopentyl)(CH₂CH(CH₃)₂)

11.30 (a) The target alcohol, 2-hexanol, is a secondary alcohol. Secondary alcohols can be prepared by reduction of ketones. Thus, 2-hexanone will be the starting material. Aldehydes and ketones are reduced to alcohols with either lithium aluminum hydride (LiAlH₄) or sodium borohydride (NaBH₄). Sodium borohydride is usually the reagent of choice for reduction of a simple aldehyde or ketone, such as 2-hexanone, because of its safety and ease of handling. Alternatively, catalytic hydrogenation (H$_2$/Pt) can also be used.

(b) Because 2-hexanol is a secondary alcohol, disconnection of bonds to the hydroxyl-bearing carbon generates two pairs of structural fragments. One route involves the addition of a methyl Grignard reagent to a five-carbon aldehyde, while the other route requires addition of a butylmagnesium halide to a two-carbon aldehyde:

(c) Organolithium reagents react with carbonyl compounds in the same way that Grignard reagents do. The only difference is that in their reactions with aldehydes and ketones, organolithium reagents are somewhat more reactive than Grignard reagents. The retrosynthetic analysis employed here is identical to the one shown in Exercise 11.30 b. Thus, the carbonyl containing materials are also identical:

11.31 (a)

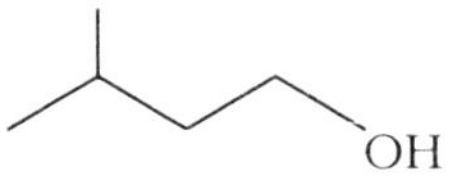

(b) **A** 1-Pentanol **B** (S)-2-Pentanol **C** (R)-2-Pentanol
D 3-Pentanol **E** (S)-2-Methyl-1-butanol **F** (R)-2-Methyl-1-butanol
G 3-Methyl-1-butanol **H** 2,2-Dimethyl-1-propanol
I 2-Methyl-2-butanol **J** (R)-3-Methyl-2-butanol
K (S)-3-Methyl-2-butanol

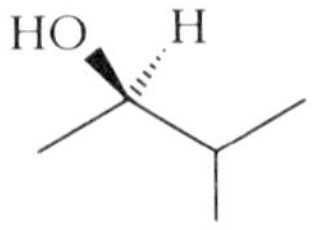

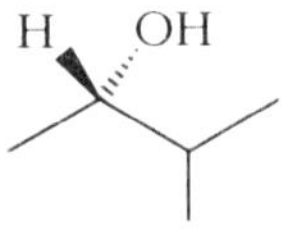

(c) **A** - 1° **B** - 2° **C** - 2° **D** - 2° **E** - 1° **F** - 1°
G - 1° **H** - 1° **I** - 3° **J** - 2° **K** - 2°

(d) **B, C; E, F; J, K**

(e) We will apply retrosynthetic analysis (the technique of reasoning backward from a target molecule to suitable starting materials). We have to review all reactions leading to the expected alcohol and, then, think of starting materials suitable for each synthetic route.

1-Pentanol is a primary alcohol. Primary alcohols can be prepared via anti-Markownikoff hydration of an alkene (borohydration-oxidation) or the reaction of an organometallic reagent with formaldehyde or oxirane.

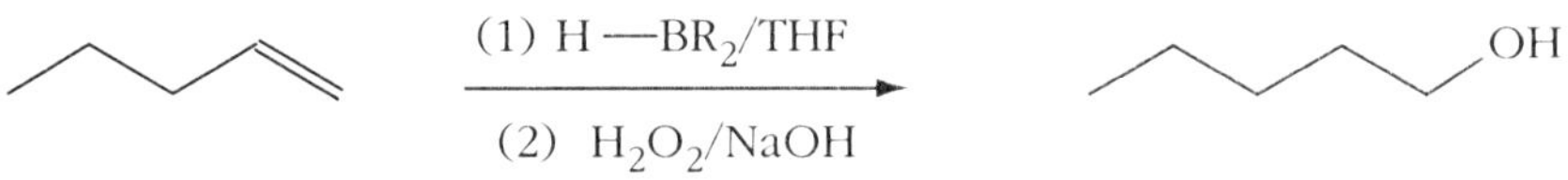

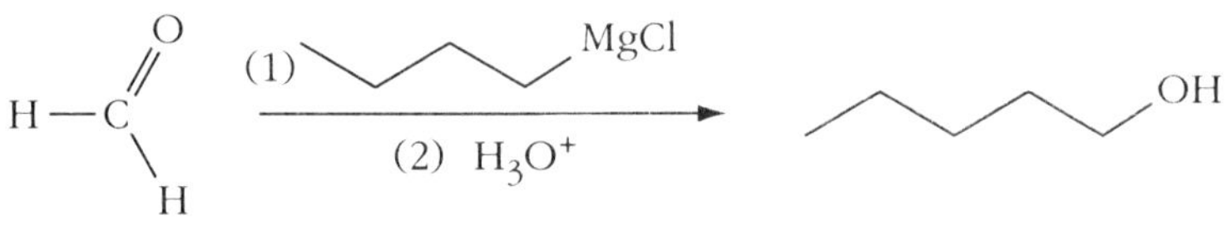

Compounds **B** and **C** are enantiomers. A racemic mixture of **B** and **C** can be prepared by reduction of 2-pentanone:

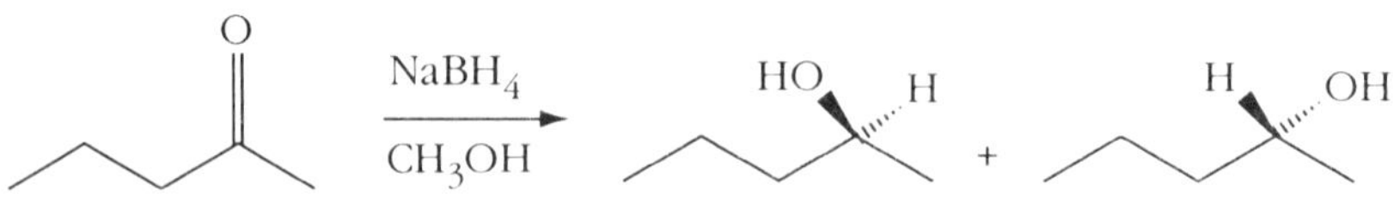

The racemic mixture can be resolved by the method described in Section 6.13 (page 262). Reaction of an optically active carboxylic acid with the racemic mixture forms diastereomeric esters. After separation and hydrolysis, compounds **B** and **C** are obtained.

Diastereomers

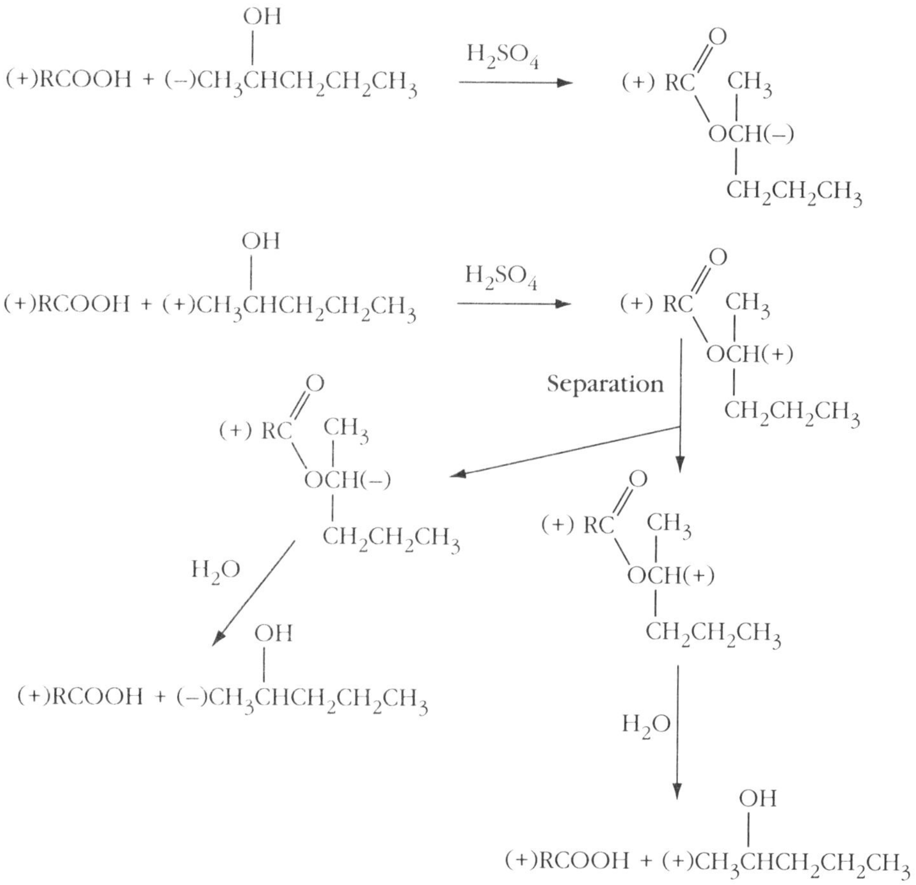

$$(+)RCOOH + (-)CH_3CHCH_2CH_2CH_3 \xrightarrow{H_2SO_4} (+)RC(=O)O-CH(-)(CH_3)(CH_2CH_2CH_3)$$

$$(+)RCOOH + (+)CH_3CHCH_2CH_2CH_3 \xrightarrow{H_2SO_4} (+)RC(=O)O-CH(+)(CH_3)(CH_2CH_2CH_3)$$

Separation →

(+) RC(=O)O-CH(−)(CH₃)(CH₂CH₂CH₃)   (+) RC(=O)O-CH(+)(CH₃)(CH₂CH₂CH₃)

↓ H₂O   ↓ H₂O

$(+)RCOOH + (-)CH_3CHCH_2CH_2CH_3$ (with OH) $(+)RCOOH + (+)CH_3CHCH_2CH_2CH_3$ (with OH)

Compound **D**, 3-pentanol, is a secondary alcohol. It may be easily prepared by reduction of 3-pentanone:

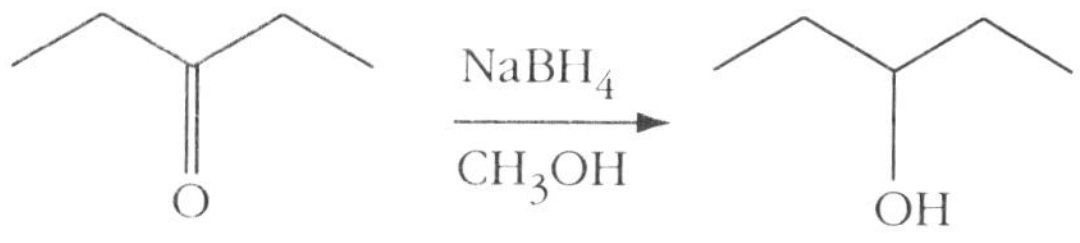

$$\text{3-pentanone} \xrightarrow[\text{CH}_3\text{OH}]{\text{NaBH}_4} \text{3-pentanol}$$

Compounds **E** and **F** are enantiomers. A racemic mixture of **E** and **F** can be prepared in a number of ways, including by the reaction of a Grignard reagent with formaldehyde. The racemic mixture can be resolved by the same method used to resolve compounds **B** and **C**.

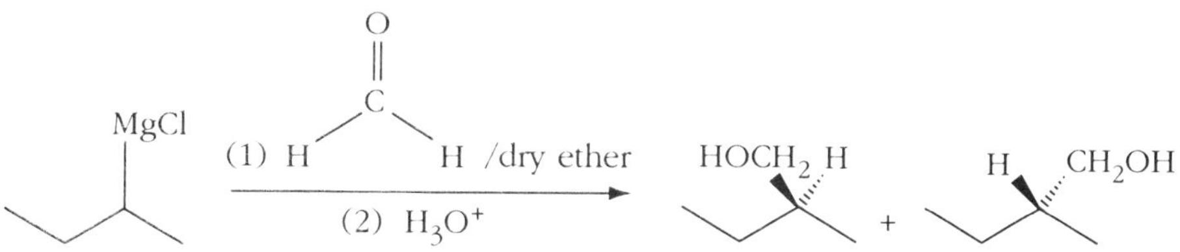

sec-butyl-MgCl + H-C(=O)-H /dry ether $\xrightarrow{(1),\ (2)\ H_3O^+}$ HOCH₂-C*(H)(Et)(Me) + H-C*(Et)(Me)-CH₂OH

Compound **G**, 3-methyl-1-butanol, is a primary alcohol and it can be prepared via nucleophilic substitution on a haloalkane or tosylate, anti-Markownikoff hydration of an alkene (borohydration-oxidation), or reaction of an organometallic reagent with formaldehyde or oxirane. We are already familiar with the following methods of synthesis:

$$\text{(CH}_3)_2\text{CHCH=CH}_2 \xrightarrow[\text{(2) H}_2\text{O}_2/\text{NaOH/H}_2\text{O}]{\text{(1) H—BR}_2} (\text{CH}_3)_2\text{CHCH}_2\text{CH}_2\text{OH}$$

$$(\text{CH}_3)_2\text{CHCH}_2\text{MgBr} \xrightarrow[\text{(2) H}_3\text{O}^+]{\text{(1) HCHO /dry ether}} (\text{CH}_3)_2\text{CHCH}_2\text{CH}_2\text{OH}$$

Compound **H**, 2,2-dimethyl-1-propanol, can be conveniently prepared in the following ways:

$$(\text{CH}_3)_3\text{CCHO} \xrightarrow[{[\text{H}] = \text{H}_2/\text{Pt, NaBH}_4, \text{LiAlH}_4}]{[\text{H}]} (\text{CH}_3)_3\text{CCH}_2\text{OH}$$

$$(\text{CH}_3)_3\text{CLi} \xrightarrow[\text{(2) H}_3\text{O}^+]{\text{(1) HCHO}} (\text{CH}_3)_3\text{CCH}_2\text{OH}$$

Compound **I**, 2-methyl-2-butanol, is a tertiary alcohol and the best method to synthesize it involves reaction of a ketone (or an ester as we will learn later) with Grignard (or organolithium) reagent:

$$\text{CH}_3\text{CH}_2\text{COCH}_3 \xrightarrow[\text{(2) H}_3\text{O}^+]{\text{(1) CH}_3\text{MgBr}} \text{CH}_3\text{CH}_2\text{C(OH)(CH}_3)_2$$

$$(\text{CH}_3)_2\text{C=O} \xrightarrow[\text{(2) H}_3\text{O}^+]{\text{(1) CH}_3\text{CH}_2\text{MgBr}} \text{CH}_3\text{CH}_2\text{C(OH)(CH}_3)_2$$

$$\text{CH}_3\text{CH}_2\text{COOR} \xrightarrow[\text{(2) H}_3\text{O}^+]{\text{(1) CH}_3\text{MgBr (2 moles)}} \text{CH}_3\text{CH}_2\text{C(OH)(CH}_3)_2$$

Compounds **J** and **K** are enantiomers. A racemic mixture of **J** and **K** can be formed by reduction of the appropriate ketone:

The racemic mixture can be resolved by the same method used to resolve compounds **B** and **C**.

11.32 The first four compounds to be synthesized are alcohols. Retrosynthetic analysis (the technique of reasoning backward from a target molecule to suitable starting materials) is used to disconnect the bonds to the hydroxyl-bearing carbon atom so that at least one of the structural fragments contains not more than two carbon atoms. This fragment is obtained from iodoethane, one starting material.

(a) One route involves the addition of an ethyl Grignard reagent to an aldehyde. We can go back one step further to synthesize the Grignard reagent (or organolithium reagent) from iodoethane. Thus,

The second synthesis is better because it involves two steps from the starting material specified:

(b) Retrosynthesis:

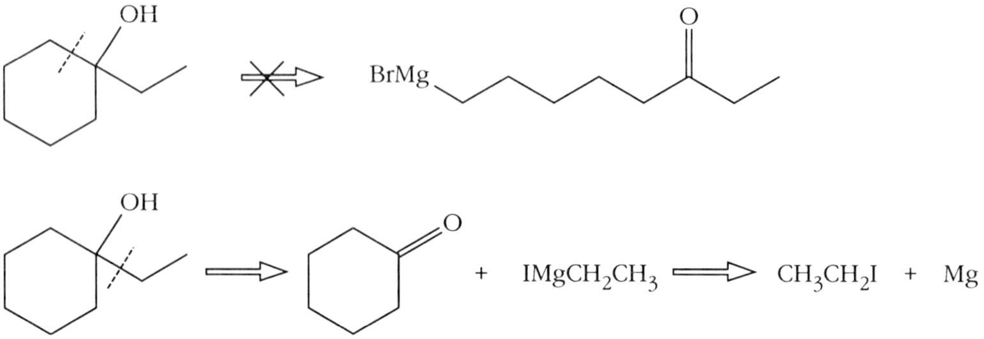

The first disconnection leads to an impractical intermediate having a Grignard reagent and a carbonyl group in the same molecule. The second method will work, though, so the complete synthesis is the following:

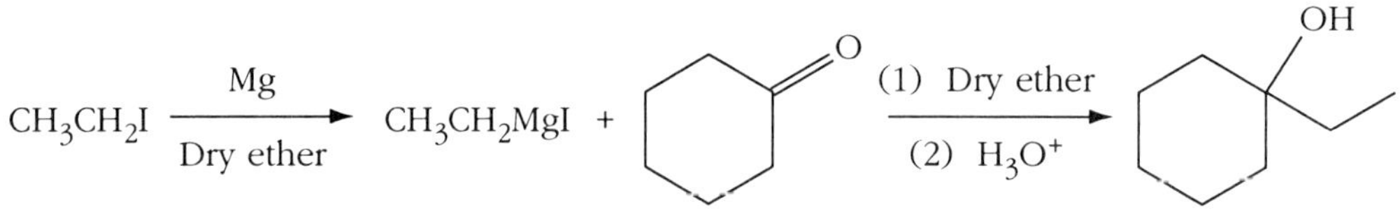

$$CH_3CH_2I \xrightarrow[\text{Dry ether}]{Mg} CH_3CH_2MgI + \text{cyclohexanone} \xrightarrow[(2)\ H_3O^+]{(1)\ \text{Dry ether}} \text{1-ethylcyclohexanol}$$

(c) Only one disconnection is possible and it leads to the desired starting material:

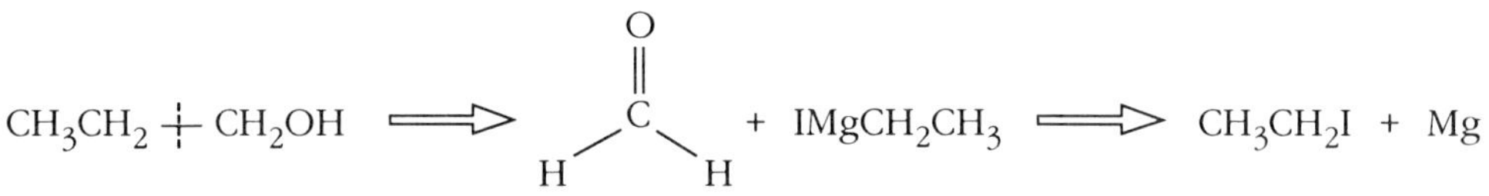

$$CH_3CH_2 \!+\! CH_2OH \Longrightarrow HCHO + IMgCH_2CH_3 \Longrightarrow CH_3CH_2I + Mg$$

The complete synthesis:

$$CH_3CH_2I \xrightarrow[\text{Dry ether}]{Mg} CH_3CH_2MgI + HCHO \xrightarrow[(2)\ H_3O^+]{(1)\ \text{Dry ether}} CH_3CH_2CH_2OH$$

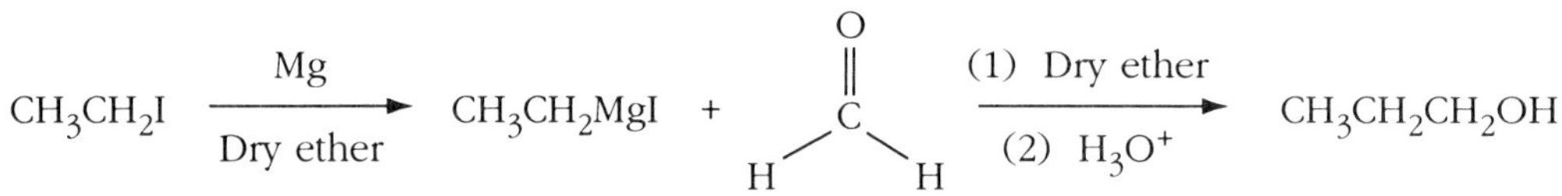

(d) Retrosynthesis:

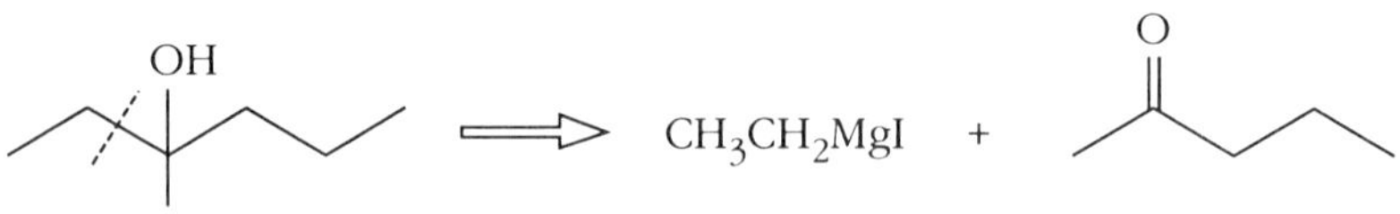

Only this disconnection leads to a short synthesis starting with iodoethane.

The complete synthesis:

$$CH_3CH_2I \xrightarrow[\text{Dry ether}]{Mg} CH_3CH_2MgI + CH_3CCH_2CH_2CH_3\ (\text{C=O})$$

$$\xrightarrow[\text{(2) }H_3O^+]{\text{(1) Dry ether}} CH_3CH_2\underset{\underset{CH_3}{|}}{\overset{\overset{OH}{|}}{C}}CH_2CH_2CH_3$$

(e) The most useful synthetic method for converting an alkyl halide into an alkane involves a Grignard reagent intermediate. Thus:

$$CH_3CH_2 \text{—} D \implies D_2O + CH_3CH_2MgI \implies CH_3CH_2I + Mg$$

The complete synthesis:

$$CH_3CH_2I \xrightarrow[\text{(2) }D_2O]{\text{(1) Mg/dry ether}} CH_3CH_2D$$

11.33 We know that a carbonyl-containing compound (aldehyde or ketone) undergoes reduction with $NaBH_4$ and, after workup with an aqueous acid solution, an alcohol is isolated as a product.

$$\underset{R'}{\overset{R}{>}}C=O \xrightarrow[\text{(2) }H_3O^+]{\text{(1) }NaBH_4} \underset{R'}{\overset{R}{>}}CH\text{—}OH$$

The ^{1}H NMR spectrum shows a signal at δ 1.68 (1H) which is very characteristic of that of —OH. Furthermore, we have a doublet at δ 0.92 (12H), a multiplet at δ 1.75 (2H) and a triplet at δ 3.0 (1H). The doublet-septet pattern is characteristic of an isopropyl group. Integration indicates that there are two equivalent isopropyl groups. The signal at δ 3.0 is in the region of a hydrogen atom of a H—C—OH group. Hence, the structure of the product is **2,4-dimethyl-3-pentanol** and that of the starting material is **2,4-dimethylpentanone**.

$$\text{2,4-Dimethylpentanone} \xrightarrow[\text{(2) }H_3O^+]{\text{(1) }NaBH_4} \text{2,4-dimethyl-3-pentanol}$$

2,4-Dimethylpentanone

11.34 (a) The formula C_3H_8O has no units of unsaturation so the compound must be a saturated alcohol. There are only two possibilities: 1-propanol or 2-propanol. The ^{1}H NMR spectrum shows a triplet at δ 0.93 (3H), a sextet at δ 1.62 (2H), and a triplet at δ 3.57 (2H). This pattern is characteristic for a propyl group because an isopropyl group would contain a doublet (6H) and a septet (1H). Notice that a peak for the hydrogen atom of the OH group is missing. Thus, the compound is 1-propanol.

$$CH_3CH_2CH_2OH$$

(b) The molecular formula, $C_9H_{10}O$, has five units of unsaturation (combination of rings and multiple bonds). The ^1H NMR spectrum shows a doublet at δ 4.3 (2H), a pair of triplets at δ 6.35 (1H), a doublet at δ 6.61 (1H), and a multiplet at δ 7.3-7.5 (5H). The multiplet at δ 7.3-7.5 (5H) indicates a monosubstituted benzene ring. The doublet at δ 6.61 indicates a vinyl hydrogen atom conjugated to one hydrogen. The pair of triplets at δ 6.35 indicates a vinyl hydrogen atom conjugated to the other vinyl hydrogen (the same coupling constants) and to two other hydrogen atoms (a methylene group). The doublet at δ 4.3 indicates a methylene group coupled with one hydrogen. The coupling constant is the same as that in the triplets at δ 6.35 so the methylene group must be bonded to the carbon-carbon double bond, —CH=CH—CH$_2$—, C_3H_4. When we add this fragment to the monosubstituted benzene ring, C_6H_5, we have only OH left from the molecular formula. Notice that a peak for the hydrogen atom of the OH group is missing. Thus the hydroxyl group is the last fragment of the structure. Joining the fragments together we get two possible structures (**A** or **B**). Structure **B**, however, is an unstable enol that tautomerizes immediately to a keto form **C**, so the only possible structure is **A**.

(c) The formula $C_5H_{10}O$ shows one unit of unsaturation (a ring or a double bond). The ^1H NMR spectrum shows a pair of singlets at δ 1.68 (3H) and δ 1.75 (3H), a singlet at δ 2.1 (1H), a doublet at δ 4.14 (2H), and a triplet at δ 5.4 (1H). The signal at δ 5.4 is characteristic of a vinyl hydrogen (bonded to an alkene carbon). There is only *one* such hydrogen, which indicates a trisubstituted alkene. *Note:* The two signals in the δ 1.65-1.75 region are a pair of singlets and *not* a doublet because (i) the distance between the two signals does not match any coupling constant of any other signal, and (ii) the "leaning" of the doublet points away from any other signal. *Note:* A multiplet often *leans* toward the hydrogen atoms that are causing the splitting. Thus, we have the following fragments:

$$-CH_3 \quad -CH_3 \quad -CH_2- \quad \underset{/}{\overset{\backslash}{C}}=\underset{\backslash}{\overset{/}{C}}\overset{H}{\underset{}{}} \quad -OH$$

The signal of the vinyl hydrogen is split into a triplet, which indicates a methylene group ($-CH_2-$) attached to the same carbon atom to which the hydrogen is bonded. The signal of the methylene group would be split to a doublet. We observe such a doublet at δ 4.14 (2H). The methylene group must be bonded to the hydroxyl group, $-OH$, because its signal is so far downfield. What remains now are two methyl groups and two positions on the alkene carbon. Thus, the structure of the unknown alcohol is the following:

$$\underset{H_3C}{\overset{H_3C}{\backslash}}C=C\underset{CH_2-OH}{\overset{H}{/}}$$

Name: 3-Methyl-2-buten-1-ol

11.35 (a) 3-Hexanol lacks symmetry and each of the six carbon atoms is unique. We will observe six signals in the ^{13}C NMR spectrum.

(b) 6 signals: [(4 in cyclohexane ring: C1, C2 and C6, C3 and C5, C4) and two signals of the ethyl group],

(c) 4 signals (two methyl groups are equivalent and they will show the same chemical shift),

(d) 5 signals (the methyl groups will show two signals because they are not magnetically equivalent—different distance from the $-OH$ group),

(e) 2 signals [(all methylene groups, $-CH_2-$, are equivalent and they will show one signal) and the other signal will arise from the central carbon atom].

11.36 (a) This synthesis requires the conversion of an alkene into an alcohol. This can be done in one step by hydration of the double bond. The addition must occur according to Markownikoff's rule, so mercuration-demercuration is the best method.

$$\text{methylcycloheptene} \xrightarrow[\text{(2) NaBH}_4/\text{NaOH}]{\text{(1) Hg(OAc)}_2/\text{H}_2\text{O}} \text{1-methylcycloheptanol}$$

(b) This synthesis can be carried out by the anti-Markownikoff addition of water to the starting alkene. The best method to do this is by the hydroboration-oxidation process:

$$\text{cycloheptene} \xrightarrow[\text{(2) } H_2O_2/NaOH]{\text{(1) } H-BR_2/THF} \text{trans-cycloheptanol (one enantiomer)} + \text{trans-cycloheptanol (other enantiomer)}$$

(c) No single reaction presented so far can convert a ketone directly into a bromoalkane. 2-Bromohexane, however, can be prepared by the reaction of 2-hexanol and PBr_3, and 2-hexanol is formed by reduction of the ketone given as a starting material.

2-bromohexane ⇒ 2-hexanol ⇒ 2-hexanone

The complete synthesis consists of the following two steps:

$$\text{2-hexanone} \xrightarrow[\text{(2) } H_3O^+]{\text{(1) } NaBH_4} \text{2-hexanol} \xrightarrow{PBr_3 \text{ or } HBr} \text{2-bromohexane}$$

(d) A Grignard reagent (or organolithium reagent) can accomplish the following conversion:

$$\text{cyclohexanone} + \text{CH}_3\text{CH=CH(MgBr)} \xrightarrow[\text{(2) } H_3O^+]{\text{(1) } (CH_3CH_2)_2O} \text{1-(propenyl)cyclohexanol}$$

11.37 (a)

(i) $CH_3(CH_2)_5 - \overset{OH}{\underset{CH_3}{\overset{|}{C}}} \cdots H$

(ii) $(CH_3)_3C - \overset{OH}{\underset{D}{\overset{|}{C}}} \cdots H$

(iii) $\overset{H}{\underset{H_3C}{}} \overset{OH}{\underset{CH_2CH_2OH}{\overset{|}{C}}}$

(iv) $\overset{H}{\underset{H_3C}{}} \overset{OH}{\overset{|}{C}} - \overset{CH_3}{\underset{OH}{\overset{|}{C}}} - H$

(b) The enzyme transfers a hydride ion to the Re face of each carbonyl group. For instance:

$$\underset{\text{Re face}}{\overset{O}{\underset{H_3C}{\|}}C-CH_2CH_2OH} \xrightarrow[\text{to Re face}]{\text{Add hydrogen atom}} \underset{(S)\text{-1,3-Butanediol}}{\overset{H\;\;\;\;OH}{\underset{H_3C}{\;}C\;CH_2CH_2OH}}$$

11.38 The first equivalent of Grignard reagent reacts with the hydroxyl group (—OH) according to the first equation (below). This reaction is extremely fast and no other reaction can take place until the last —OH group is present. Once all hydrogen atoms are removed from all —OH groups by adding one equivalent of Grignard reagent, the second equivalent of Grignard reagent can react with the carbonyl group (second reaction below).

(1) [3-(hydroxymethyl)cyclohexanone] + CH$_3$MgI ⟶ [3-(CH$_2$O$^-$MgI$^+$)cyclohexanone] + CH$_4$

(2) [3-(CH$_2$O$^-$MgI$^+$)cyclohexanone] + CH$_3$MgI ⟶ [1-CH$_3$, 1-O$^-$MgI$^+$, 3-CH$_2$O$^-$MgI$^+$ cyclohexane] $\xrightarrow{H_3O^+}$ [1-CH$_3$, 1-OH, 3-CH$_2$OH cyclohexane]

11.39
(a) pentan-1-ol $\xrightarrow[\text{Pentane}]{\text{NaH}}$ pentyl-O$^-$ Na$^+$ + H$_2$

(b) pentan-1-ol $\xrightarrow{\text{conc. aq. HBr}}$ 1-bromopentane

(c) pentan-1-ol $\xrightarrow{\text{SOCl}_2}$ 1-chloropentane

(d) pentan-1-ol + H$_3$C—C$_6$H$_4$—SO$_2$Cl ⟶ H$_3$C—C$_6$H$_4$—SO$_2$O—pentyl

(e) CH₃CH₂CH₂CH₂CH₂OH →[Na₂Cr₂O₇ / aq. H₂SO₄] CH₃CH₂CH₂CH₂COOH

(f) CH₃CH₂CH₂CH₂CH₂OH →[PCC / CH₂Cl₂] CH₃CH₂CH₂CH₂CHO

(g) CH₃CH₂CH₂CH₂CH₂OH →[PBr₃] CH₃CH₂CH₂CH₂CH₂Br

11.40

(a) pentan-2-ol →[NaNH₂ / NH₃] sodium pentan-2-olate (O⁻ Na⁺)

(b) pentan-2-ol →[CH₃MgI / (CH₃CH₂)₂O] pentan-2-olate ⁺MgI salt

(c) pentan-2-ol + H₃C–C₆H₄–SO₂Cl → H₃C–C₆H₄–SO₂O–CH(CH₃)CH₂CH₂CH₃ (tosylate ester)

(d) pentan-2-ol →[CrO₃ / aq. H₂SO₄] pentan-2-one

(e) pentan-2-ol →[KMnO₄ / aq. H₂SO₄] pentan-2-one

(f) pentan-2-ol →[Na metal] sodium pentan-2-olate (O⁻ Na⁺) + H₂

(g) pentan-2-ol →[NaI / aq. H₂SO₄] 2-iodopentane

11.41

(a) (CH₃)₃C–OH + CH₃Li → (CH₃)₃C–O⁻ Li⁺ + CH₄

(b) (CH₃)₃C–OH + KMnO₄ / aq. H₂SO₄ → No reaction*

(c) (CH₃)₃C–OH + conc. aq. HCl → (CH₃)₃C–Cl + H₂O

(d) (CH₃)₃C–OH + PCC / CH₂Cl₂ → No reaction

*Note: In acid solution, dehydration of the tertiary alcohol may occur, e.g.,

11.42 In some cases, the choice of the reagent(s) is very obvious. In other cases, we can use various reagents. For instance, 1-butanol can be oxidized to butanoic acid (Exercise 11.42b) with potassium permanganate or chromic acid.

(a) CH₃CH₂CH₂CH₂OH + SOCl₂ → CH₃CH₂CH₂CH₂Cl

(b) CH₃CH₂CH₂CH₂OH + KMnO₄ / aq. H₂SO₄ → CH₃CH₂CH₂C(=O)OH

(c) CH₃CH₂CH₂CH₂OH + PCC / CH₂Cl₂ → CH₃CH₂CH₂CHO

(d) CH₃CH₂CH₂CH₂OH + PBr₃ → CH₃CH₂CH₂CH₂Br + HC≡C⁻ Na⁺ → CH₃CH₂CH₂CH₂C≡CH

(e) The retrosynthetic analysis indicates two possible disconnections of the carbon-carbon bonds:

[Retrosynthetic scheme: 4-octanol disconnects to butanal + butylmagnesium bromide, both derived from 1-butanol (via 1-bromobutane for the Grignard). An alternative disconnection gives propylmagnesium bromide + pentanal.]

The top disconnection is much better because both reactants can be readily obtained from 1-butanol. Thus, we synthesize first butanal and the butyl Grignard reagent, and then we synthesize 4-octanol, the desired product.

[Scheme: 1-butanol + PCC/CH$_2$Cl$_2$ → butanal]

[Scheme: 1-butanol + PBr$_3$ → 1-bromobutane + Mg/Dry ether → butylmagnesium bromide]

[Scheme: butanal + butyl MgBr, (1) Dry ether, (2) H$_3$O$^+$ → 4-octanol]

(f) Many different syntheses are possible. One may involve reactions of 4-octanol obtained as in Exercise **11.42 e**. For instance:

[Scheme: 4-octanol + H$_3$O$^+$/Heat → octene + H$_2$/Pd → octane]

4-Octanol can also be oxidized to 4-octanone (e.g., with PCC/CH$_2$Cl$_2$) and 4-octanone can be reduced to octane (later we will learn the Clemmensen reaction and the Wolf-Kishner reaction).

We already know all the reactions in the following sequence that also leads to octanone:

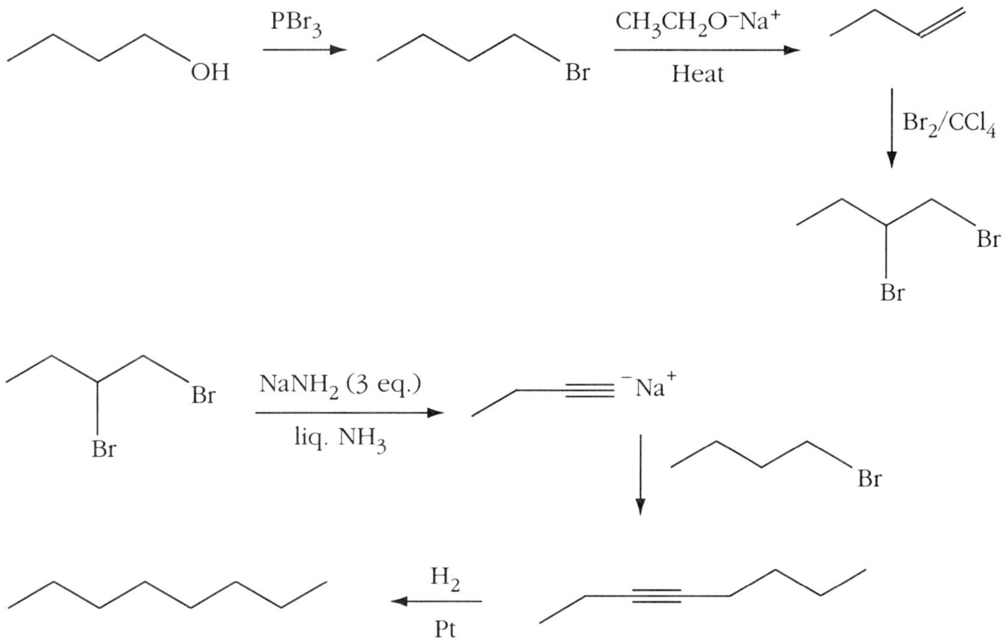

$CH_3CH_2CH_2CH_2OH \xrightarrow{PBr_3} CH_3CH_2CH_2CH_2Br \xrightarrow[\text{Heat}]{CH_3CH_2O^-Na^+} CH_2=CHCH_2CH_3 \xrightarrow{Br_2/CCl_4}$ 2,3-dibromobutane

$\xrightarrow[\text{liq. NH}_3]{NaNH_2 \ (3 \text{ eq.})} CH_3C\equiv C^-Na^+ \xrightarrow{CH_3CH_2CH_2CH_2Br} CH_3C\equiv CCH_2CH_2CH_2CH_3 \xrightarrow[Pt]{H_2}$ octane

11.43

(a) $^{13}CH_3OH \xrightarrow[CH_2Cl_2]{PCC} ^{13}CH(=O)H \xrightarrow[(2) \ H_3O^+]{(1) \ CH_3MgI} CH_3{}^{13}CH_2OH \xrightarrow[CH_2Cl_2]{PCC} CH_3{}^{13}CH(=O)H$

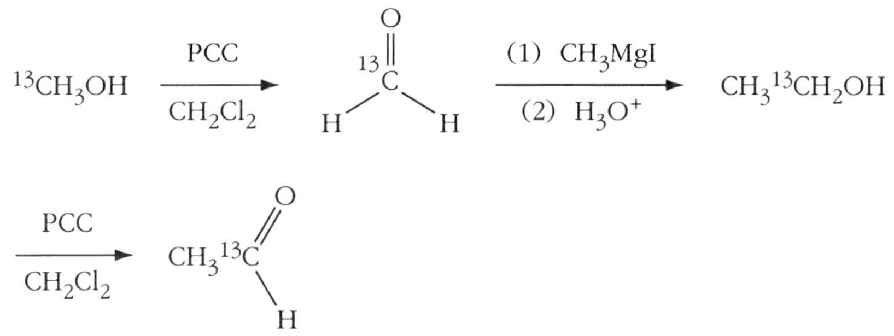

(b) $^{13}CH_3OH \xrightarrow{PBr_3} {}^{13}CH_3Br \xrightarrow[\text{Dry ether}]{Mg} {}^{13}CH_3MgBr$

$\xrightarrow[(2) \ H_3O^+]{(1) \ HCHO} {}^{13}CH_3CH_2OH \xrightarrow[CH_2Cl_2]{PCC} {}^{13}CH_3CH(=O)H$

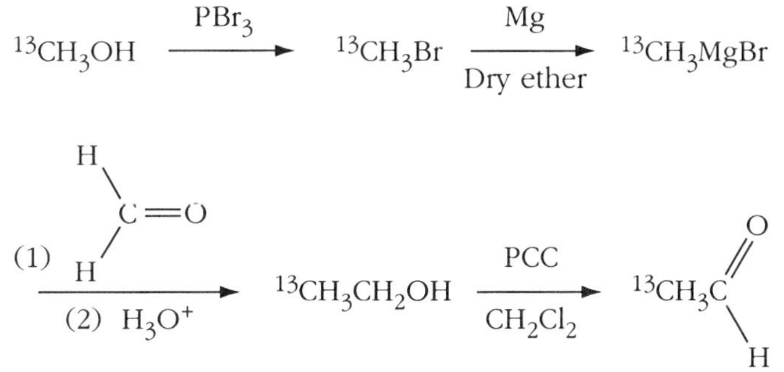

(c) $^{13}CH_3CHO \xrightarrow[(2) \ H_3O^+]{(1) \ CH_3MgI} {}^{13}CH_3CH(OH)CH_3 \xrightarrow[CH_2Cl_2]{PCC} {}^{13}CH_3CCH_3(=O)$

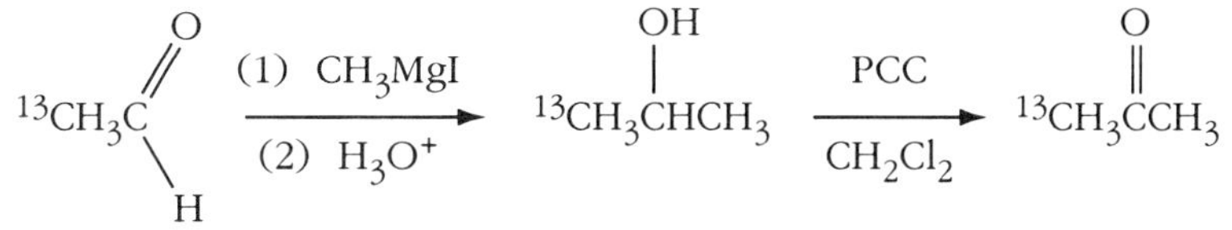

(d) 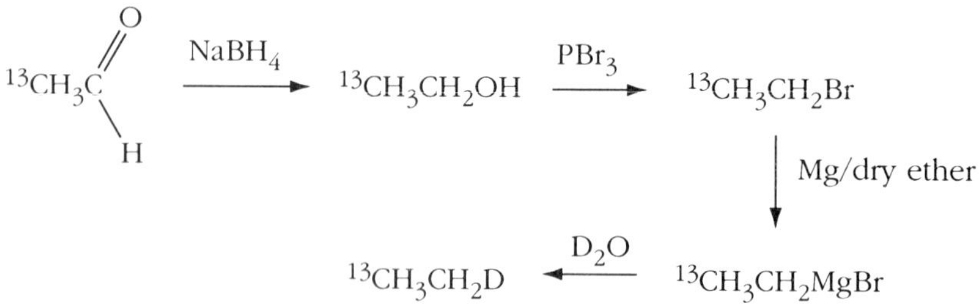 $^{13}CH_3\underset{H}{\overset{O}{\overset{\|}{C}}}$ $\xrightarrow{NaBH_4}$ $^{13}CH_3CH_2OH$ $\xrightarrow{PBr_3}$ $^{13}CH_3CH_2Br$

$\downarrow$ Mg/dry ether

$^{13}CH_3CH_2D$ $\xleftarrow{D_2O}$ $^{13}CH_3CH_2MgBr$

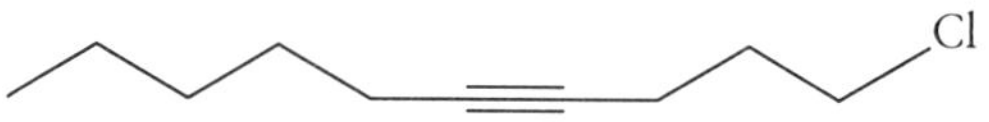

(e) 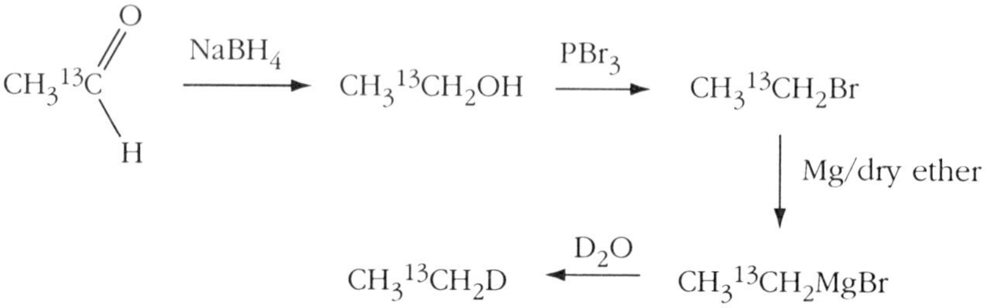 $CH_3{}^{13}\underset{H}{\overset{O}{\overset{\|}{C}}}$ $\xrightarrow{NaBH_4}$ $CH_3{}^{13}CH_2OH$ $\xrightarrow{PBr_3}$ $CH_3{}^{13}CH_2Br$

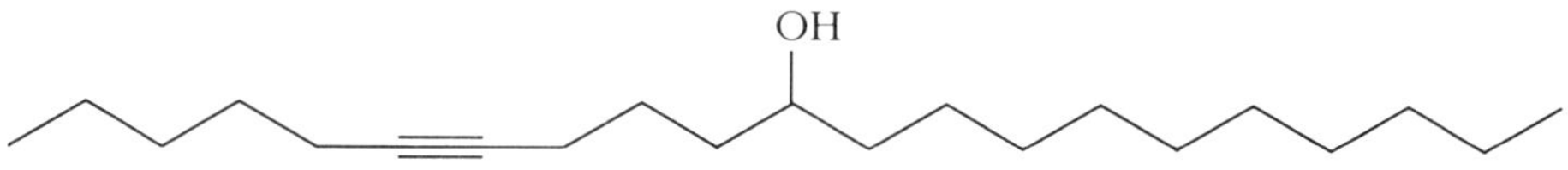

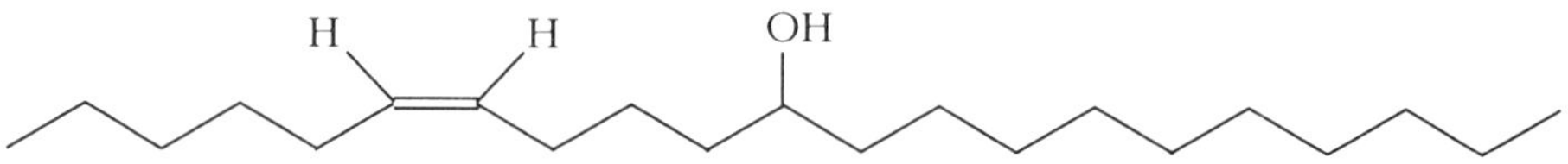

$\downarrow$ Mg/dry ether

$CH_3{}^{13}CH_2D$ $\xleftarrow{D_2O}$ $CH_3{}^{13}CH_2MgBr$

(f) 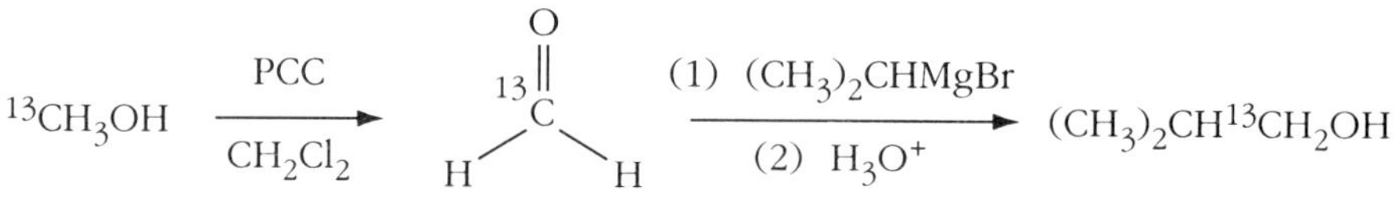 $^{13}CH_3OH$ $\xrightarrow[CH_2Cl_2]{PCC}$ $H\underset{H}{\overset{O}{\overset{\|}{{}^{13}C}}}$ $\xrightarrow[(2)\ H_3O^+]{(1)\ (CH_3)_2CHMgBr}$ $(CH_3)_2CH^{13}CH_2OH$

11.44

A = [structure: hex-5-ynyl sodium acetylide]⁻Na⁺

B = [structure: chloroalkyne with terminal Cl]

C = [structure: alkyne with terminal MgCl]

D = [structure: long-chain alkyne with OH]

E = [structure: long-chain cis-alkene with OH]

Sex attractant = [structure: long-chain cis-alkene with ketone]

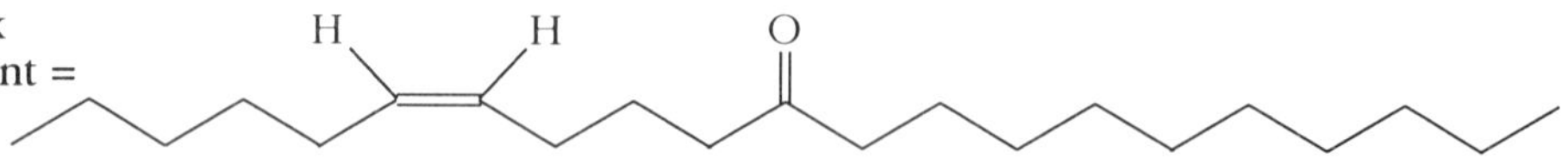

11.45

$CH_3-\overset{OH}{\underset{D}{\overset{|}{C}}}\cdots H$ $CH_3-\overset{OH}{\underset{H}{\overset{|}{C}}}\cdots D$

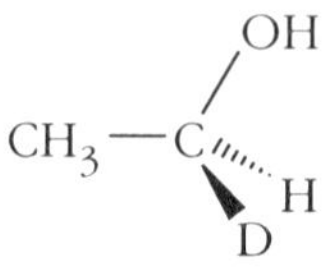

(*S*)-1-Deuterioethanol 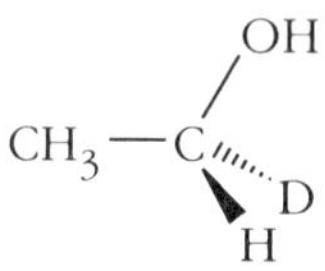   (*R*)-1-Deuterioethanol

The oxidation with PCC in CH_2Cl_2 results in dehydrogenation (literally, the removal of two hydrogen atoms, one from the oxygen and one from the carbon atom). Oxidation with PCC is not stereospecific. Instead, a hydrogen or deuterium may be removed at random, which is why we obtain a mixture of CH_3CDO and CH_3CHO.

Alcohol dehydrogenase is an enzyme that is stereospecific. NAD^+ is an oxidizing agent that removes a hydride ion from the Re face of the would-be carbonyl compound. In the case of (S)-1-deuterioethanol, only ^1H will be removed while ^2D will remain.

11.46

F = [vinyl-methylcyclohexane structure]

G = [1-(1-methylcyclohexyl)ethanol structure]

H = [1-(1-methylcyclohexyl)ethanone structure]

I = [2-(1-methylcyclohexyl)ethanol structure]

J = [(1-methylcyclohexyl)acetaldehyde structure]

K = [2-(1-methylcyclohexyl)-2-propanol structure]

L = [1-chloro-1-methyl-(isopropyl)cyclohexane structure]

11.47 One way to solve this exercise is to look at the structure and list the expected chemical shifts and spin–spin splitting patterns for each group of hydrogen atoms. Then look at which ^1H NMR data given on the right matches the expected pattern.

Note: The integration data gives the relative ratio of hydrogen atoms and not the actual number, which can be a multiple of that ratio.

(a) 2,3-Dibromobutane is expected to show a doublet for the methyl groups (δ 1.8, 3H) and a quartet for the —CHBr— groups (δ 4.2, 1H). The set of data number (4) matches the expected ^1H NMR spectrum of 2,3-dibromobutane.

(b) 1,4-Dibromobutane is expected to show a triplet for the terminal —CH_2Br groups (δ 3.4, 2H) and a triplet for the internal methylene groups, —CH_2—, (δ 2.0, 2H). The set of data number (1) matches the expected ^1H NMR spectrum of 1,4-dibromobutane.

(c) 1,2-Dibromo-2-methylpropane is expected to show two singlets, one for the hydrogen atoms on C1 (—CH$_2$Br) at δ 3.9 (2H), and the other for the two equivalent methyl groups on C2 at δ 1.9 (6H). The set of data number (2) matches the expected ^{1}H NMR spectrum of 1,2-dibromo-2-methylpropane.

(d) 1,3-Dibromobutane is expected to show a triplet for the hydrogen atoms on C1 (δ 3.5, 2H), a quartet for the internal methylene group on C2, —CH$_2$—, (δ 2.3, 2H), a sextet for the hydrogen on C3 (δ 4.3, 1H), and a doublet for the terminal methyl group on C4 (δ 1.8, 3H). The set of data number (3) matches the expected ^{1}H NMR spectrum of 1,3-dibromobutane.

11.48 (a) The molecular formula, C$_4$H$_{10}$O$_2$, has no units of unsaturation so the compound must be a saturated alcohol or ether. The IR spectrum shows an absorption band at 3350 cm^{-1} indicating at least one hydroxyl group. The ^{13}C NMR spectrum shows only two signals, which indicates the symmetry of the molecule. The signal at δ 70.9 indicates a carbon atom bonded to an oxygen atom that deshields the carbon nucleus. There are two oxygen atoms in the molecule. Only one signal of a carbon bearing an oxygen in the ^{13}C NMR spectrum indicates either two equivalent carbon atoms with —OH groups or two —OH groups attached to the same carbon atom. It is not possible to maintain the symmetry of the molecule with two —OH groups bonded to the same carbon atom. Thus the two —OH groups must be bonded to two carbon atoms in such a way that the carbon atoms are equivalent. There are only two possibilities: 1,4-butanediol or 2,3-butanediol.

$$HOCH_2CH_2CH_2CH_2OH \qquad \begin{array}{c} CH_3CH-CHCH_3 \\ |\ \ \ \ \ \ | \\ OH\ \ \ OH \end{array}$$

The ^{1}H NMR spectrum shows a doublet at δ 1.15 (3H), and a quartet at δ 3.49 (1H), and a broad singlet at δ 4.22 (2H). The singlet at δ 4.22 is a signal of —OH groups. The integration data confirm that there are two hydroxyl groups.

1,4-Butanediol would show two triplets: at about δ 1.5 (4H) and at about δ 3.5 (4H). 2,3-Butanediol would show a doublet for the terminal methyl groups (δ 1.1, 6H) and a doublet for the hydrogen atoms on C2 and C3 (δ 3.4, 2H). The ^{1}H NMR spectrum given is that of 2,3-butanediol.

(b) The molecular formula, C$_4$H$_6$O, has two units of unsaturation (a combination of a ring and multiple bonds). The IR spectrum shows an absorption band at 3295 cm^{-1} and at 2120 cm^{-1}. The absorption at 3295 cm^{-1} can be due to a hydroxyl group (there is an oxygen in the molecular formula) or a terminal alkyne (or both). The absorption at 2120 cm^{-1} is due to a carbon–carbon triple bond stretching.

The ^{13}C NMR spectrum shows four signals indicating a lack of symmetry in the molecule. It also confirms the lack of carbon-carbon double bonds (no signals in the 110–150 ppm region).

The ^{1}H NMR spectrum shows three signals: a triplet at δ 1.86 (3H), a singlet at δ 3.1 (1H), and a quartet at δ 4.23 (2H). The quartet at δ 4.23 is due to a methylene group bonded to an oxygen that deshields hydrogen atoms. So we have a first fragment: —CH$_2$OH. The singlet at δ 3.1 confirms the possibility of the —OH group. Three equivalent hydrogen atoms can be found in a methyl group, so we can assign the triplet at δ 1.86 to the —CH$_3$ group.

We have the following fragments: —C≡C—, —CH$_2$OH, and —CH$_3$. Joining those fragments together we get the structure of the unknown: **2-butyn-1-ol**.

$$CH_3C{\equiv}CCH_2OH$$

(c) The molecular formula, C$_6$H$_{12}$, has one unit of unsaturation (a carbon-carbon double bond or a ring). The IR spectrum shows an absorption at 1640 cm^{-1} indicating a carbon-carbon double bond.

The presence of C=C is confirmed by the ^{13}C NMR spectrum (signals at δ 108.8 and 149.8). The ^{13}C NMR spectrum shows four signals. It means that we have some equivalent carbon atoms.

The ^{1}H NMR spectrum shows three groups of signals. There is a singlet at δ 1.0 (9H), a pair of doublets at δ 4.80–4.95 (2H), and a pair of doublets at δ 5.78–5.93 (1H). The singlet at δ 1.0 must represent three equivalent methyl groups—most likely a *tert*-butyl group [—C(CH$_3$)$_3$]. What is left from the molecular formula is C$_2$H$_3$ or —CH=CH$_2$. Joining those two fragments together we get a structure that is consistent with all the spectral data. The unknown is **3,3-dimethyl-1-butene**.

$$\underset{H}{\overset{H}{\diagdown}}C=C\underset{H}{\overset{C(CH_3)_3}{\diagup}}$$

CHAPTER 12

NUCLEOPHILIC SUBSTITUTION AND ELIMINATION REACTIONS
Concepts

Nucleophilic substitution reactions and elimination reactions are two reaction types that are common in organic chemistry.

Nucleophilic substitution reactions are reactions in which an atom or group of atoms called the **leaving group** is substituted by a **nucleophile**.

$$\text{Nu}:^- + \text{C}-\text{LG} \longrightarrow \text{Nu}-\text{C} + \text{LG}^-$$

Nucleophile   Reaction site   Leaving group

Nucleophilic substitution reactions occur by two mechanisms, an S_N1 or an S_N2. A reaction that occurs by an S_N2 mechanism follows a second order rate law. The reaction is first order in nucleophile and first order in substrate and inversion of configuration occurs at the reaction site. The S_N2 mechanism was proposed to explain these observations. An **S_N2 mechanism** is a one-step process in which bond making and bond breaking to the reactive site occur at the same time. The designation S_N2 emphasizes the important features of this mechanism: the letter S stands for *substitution*, the letter N stands for *nucleophilic*, and the number 2 stands for *bimolecular*, meaning that two molecules or ions are involved in the rate-determining transition state.

$$\text{Nu}^- + \text{C}-\text{LG} \longrightarrow [\text{Nu}\cdots\text{C}\cdots\text{LG}]^\ddagger \longrightarrow \text{Nu}-\text{C} + \text{LG}^-$$

Transition state

The rate of reactions that occur by an S_N2 mechanism is very sensitive to steric hindrance about the reactive site. The more steric hindrance about the reactive site, the slower the rate of the nucleophilic substitution reaction. As a result, the rate of reactions that occur by an S_N2 mechanism decreases in the order methyl > primary > secondary starting materials. Tertiary compounds do not undergo nucleophilic substitution reactions by an S_N2 mechanism.

A reaction that occurs by an S_N1 mechanism follows a first order rate law. The reaction is first order in the substrate and independent of the concentration of the nucleophile. The products of nucleophilic substitution reactions of optically active tertiary and some secondary derivatives are mixtures of enantiomers. The S_N1 mechanism was proposed to explain these observations. An **S_N1 mechanism** is a two-step process in which the first slow step is breaking the bond to the leaving group to form a carbocation. The carbocation reacts rapidly with a nucleophile to form the product in the second step. The designation S_N1

emphasizes the important features of this mechanism. The letter *S* stands for *substitution,* the letter *N* stands for *nucleophilic,* and the number *1* stands for *unimolecular* because the rate-determining transition state contains only one reactant.

$$\text{C—LG} \xrightarrow[\text{First step}]{\text{Rate-determining}} \left[\text{C}^+ \cdots \text{LG}^- \right] \xrightarrow{\text{Nu}^-} \text{Nu—C} + \text{C—Nu}$$

Carbocation intermediate

Nucleophilic substitution reactions at a methyl or primary carbon atom occur exclusively by an S_N2 mechanism while reaction at a tertiary carbon atom occurs exclusively by an S_N1 mechanism. No general conclusions about the mechanism, S_N1 or S_N2, can be made for nucleophilic substitution reactions at a secondary carbon atom because both mechanisms are possible and each case must be evaluated separately.

Elimination reactions are reactions in which a π bond is formed by the loss of atoms or groups of atoms from adjacent atoms.

$$-\underset{H}{\overset{}{\text{C}}}-\underset{}{\overset{X}{\text{C}}}- \longrightarrow \text{C=C} + \text{HX}$$

Elimination reactions occur to a greater or lesser extent along with substitution reactions because nucleophiles are also bases. Elimination reactions occur by two mechanisms, E1 or E2.

An **E1 mechanism** is a two-step process. The first step of both the S_N1 mechanism of nucleophilic substitution reactions and the E1 step of elimination reactions is a slow rate-determining ionization of the substrate to form a carbocation. The elimination product is formed by transfer of a proton from the carbocation to a base. The designation *E1* emphasizes the important features of this mechanism. The letter *E* stands for *elimination* and the number *1* stands for *unimolecular* and indicates that the rate–determining transition state contains only one reactant.

$$\text{H—C—C—X} \xrightarrow{\text{Slow}} \text{H—C—C}^+ + \text{X}^- \xrightarrow[\text{Fast}]{:\text{Base}} \text{C=C} + \text{H:Base}^+$$

An **E2 mechanism** is a single-step mechanism. In the transition state of an E2 mechanism three things occur simultaneously. First, the electrons of the base begin to abstract a β hydrogen atom with the result that the β-carbon-hydrogen bond begins to break. At about the same time, a new carbon-carbon double bond begins to form as the leaving group begins to leave with its pair of electrons. In the designation E2, the letter *E* stands for *elimination* and the number *2* indicates that the rate-determining state contains two reactants, the base and the substrate.

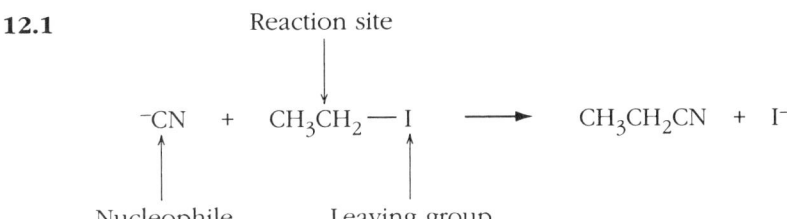

An E2 mechanism usually occurs when the β hydrogen atom and the leaving group are in a staggered conformation and are *anti* to each other.

Solutions to the Exercises

12.1

$^-$CN (Nucleophile) + CH$_3$CH$_2$—I (Reaction site, Leaving group) ⟶ CH$_3$CH$_2$CN + I$^-$

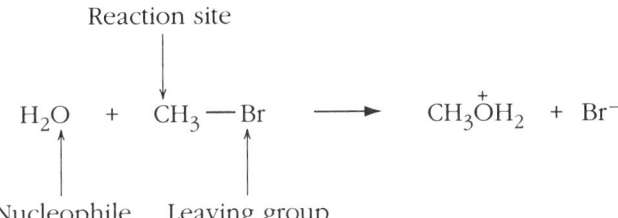

H$_2$O (Nucleophile) + CH$_3$—Br (Reaction site, Leaving group) ⟶ CH$_3$$\overset{+}{O}H_2$ + Br$^-$

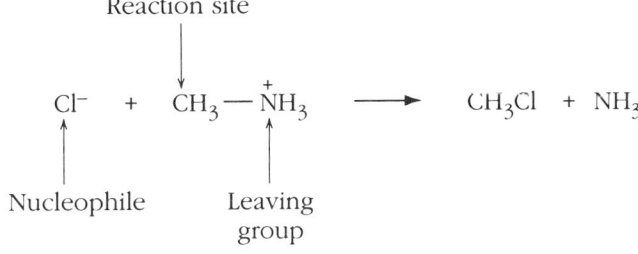

12.2 (a) CH$_3$CH$_2$CH$_2$CH$_2$OTs 1° (The other is also 1° but, like the 2,2-dimethylpropyl group, it is much more crowded about the reaction site so butyl *p*-toluenesulfonate will react faster.)

(b)   CH$_3$CH$_2$CH$_2$Cl   1°   (The other compound is a 2° chloroalkane.)

(c)   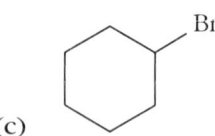   2°   (The other compound is a 3° bromoalkane.)

12.3 The reaction of iodomethane and methanol is an example of a *solvolysis* reaction. In this and related solvolyses, nucleophilic substitution is the first step and is rate-determining. The product formed in this step is an oxonium ion.

Step 1:

$$CH_3\ddot{\underset{H}{O}}: \quad CH_3-I \longrightarrow CH_3\underset{H}{\overset{+}{\underset{|}{O}}}\!\!-\!\!CH_3 \;+\; I^-$$

Dimethyloxonium ion

In order to form the observed product (dimethyl ether), the oxonium ion must lose a proton to the solvent in the second step. This occurs readily because dimethyloxonium ion is a stronger acid than methyloxonium ion.

Step 2:

$$CH_3\overset{+}{\underset{H}{O}}\!-\!CH_3 \;+\; :\ddot{O}CH_3 \rightleftharpoons CH_3\ddot{O}CH_3 \;+\; CH_3\overset{+}{\underset{H}{O}}\!-\!H$$

Dimethyloxonium ion   Methyloxonium ion
p$K_a \approx -3.6$ p$K_a \approx -2.4$

12.4 The effect of changing the leaving group on the rate of reactions that occur by an S$_N$2 mechanism is given in Table 12.3. The data in this Table can be used to predict the relative abilities of groups to be displaced by a nucleophile.

(a)   The starting material, 1-bromo-6-chlorohexane, contains two different halogens, bromine and chlorine. When we use only one mole of sodium 1-butynide, only one of the terminal halogens can be substituted. Both halogens are bonded to terminal (1°) carbon atoms so only their ability as leaving group should be considered. Bromine is a much better leaving group than chlorine (relative reaction rates are 1.0 and 0.016, respectively). Thus, we should expect 10-chloro-3-decyne to be the product of the reaction:

ClCH$_2$(CH$_2$)$_4$CH$_2$Br   +   CH$_3$CH$_2$C≡C$^-$ Na$^+$ ⟶ ClCH$_2$(CH$_2$)$_4$CH$_2$C≡CCH$_2$CH$_3$

(b)   When two moles of sodium 1-butynide is used, both halogens will be substituted:

ClCH$_2$(CH$_2$)$_4$CH$_2$Br   +   2 CH$_3$CH$_2$C≡C$^-$ Na$^+$

⟶ CH$_3$CH$_2$C≡CCH$_2$(CH$_2$)$_4$CH$_2$C≡CCH$_2$CH$_3$

12.5 Sodium methoxide will react with the compounds by an S$_N$2 mechanism because they are methyl derivatives. When all factors are kept constant, the data in

Table 12.3 indicate the following order of reactivity. Methyl tosylate is the most reactive and methyl acetate is the least reactive of the four compounds given:

$$CH_3OTs \quad > \quad CH_3I \quad > \quad CH_3Br \quad > \quad CH_3COOCH_3$$

Methyl tosylate   Methyl acetate

12.6 The difference in free energy between reactants and the transition state is defined as the **free energy of activation**. The reaction rate is related to the activation energy; the lower the energy of activation, the higher the reaction rate.

If both solvent **A** and **B** solvate the reactants to the same amount and solvent **A** solvates the transition state more than solvent **B** does, then the energy of activation must be lower in solvent **A**. Thus, the reaction occurs faster in solvent **A**.

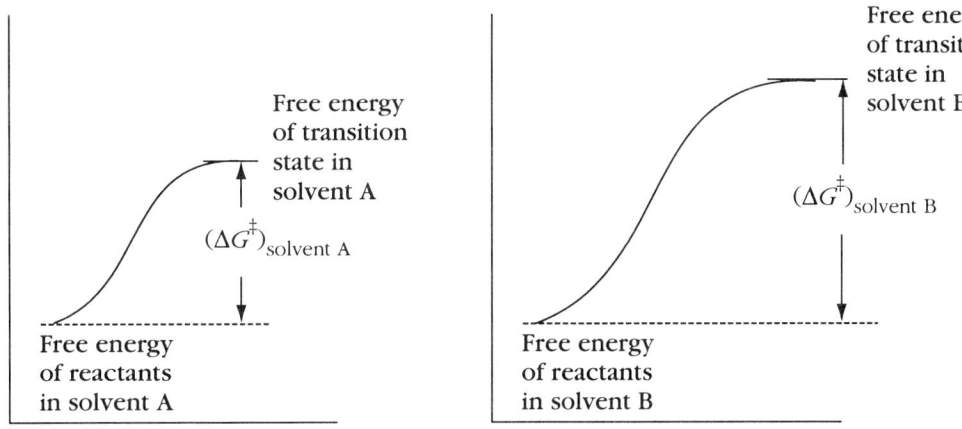

Reaction path in solvent A   Reaction path in solvent B

$$(\Delta G^{\ddagger})_{\text{solvent B}} > (\Delta G^{\ddagger})_{\text{solvent A}}$$
$$\text{so } k_{\text{solvent B}} < k_{\text{solvent A}}$$

and reaction occurs faster in solvent A.

12.7 Solvolysis of tertiary haloalkanes occur by an S_N1 mechanism. An S_N1 mechanism proceeds in two steps:

Step 1: The rate-determining step in which ionization of the haloalkane occurs to form a carbocation intermediate.

Step 2: The carbocation intermediate reacts with methanol. The carbocation is planar so the attack of a nucleophile can come from the either side.

Step 3: Loss of a proton to the solvent.

12.8 (a) The rate law of the reaction is second order overall, first order in both the nucleophile and the haloalkane, and the reaction occurs with inversion of configuration at the reaction site. The observations are consistent with a reaction that occurs by an S_N2 mechanism.

(b) The rate law of the reaction is first order, proportional to only the concentration of the alkyl tosylate and a mixture of enantiomers is formed. The observations are consistent with a reaction that occurs by an S_N1 mechanism.

Step 1: The ionization of the tosylate.

Step 2: Reaction of the carbocation intermediate with formic acid. The carbocation is planar so the attack of the nucleophile can come from either side.

Step 3: Removal of a proton by one of possible bases, B:, found in the reaction mixture.

12.9 The reaction of a tertiary alcohol with a hydrogen halide occurs by a three-step mechanism.
Step 1: An acid-base reaction in which the alcohol molecule is protonated to give an oxonium ion. Note: In water, HCl is completely dissociated to H_3O^+ and chloride ion.

Step 2: Dissociation of the alkyloxonium ion to a molecule of water and a carbocation.

Step 3: The carbocation reacts with chloride ion to yield *tert*-butyl chloride.

Very frequently we join all those steps and write a mechanism as a sequence of consecutive transformations. The above three steps can be written in the following manner:

12.10 The mechanism that describes the formation of all three products involves the formation of a carbocation intermediate by ionization of 2-bromo-2-methylbutane.

Like the dissociation of an oxonium ion into a carbocation, this step is rate-determining. The carbocation may react with ethanol as a nucleophile to form the ether product after deprotonation.

Ethanol may also act as a base to remove a proton from the carbocation to give the alkene product in an elimination step. This reaction gives rise to a mixture of alkenes in which the more highly substituted alkene is formed in larger amount.

12.11 Both these reactions occur by an E2 mechanism. Elimination by an E2 mechanism occurs when the hydrogen atom to be eliminated and the leaving group are in axial positions because only in this conformation can the hydrogen atom and the leaving group acquire the required *anti* relationship. Thus, the following mechanism can be written. There is only one hydrogen atom in the *anti* position in *trans*-2-methyl-bromocyclohexane and that is why we observe only one product, 3-methylcyclohexene.

On the other hand, there are two hydrogen atoms *anti* to bromine in *cis*-2-methyl-bromocyclohexane and, consequently, we observe two products.

12.12 2,3-Dimethyl-2-butene is a tetrasubstituted alkene, so it is more stable than 2,3-dimethyl-1-butene, which is a disubstituted alkene.

2,3-Dimethyl-2-butene, therefore, should be formed in the greater amount.

12.13 The mechanism for the acid-catalyzed dehydration of primary alcohols involves a proton transfer from the acid catalyst (H—A) to the hydroxyl group of the alcohol to generate an oxonium ion in the first step. Notice that the acid catalyst converts a poor leaving group, the —OH group, into a good leaving group, —OH$_2$.

Step 1: Protonation of the alcohol forms an oxonium ion.

Step 2: The oxonium ion eliminates a molecule of water by an E2 mechanism to form a terminal alkene.

$$CH_3CH_2-\underset{\underset{H}{|}}{\overset{\overset{H}{|}}{C}}-CH_2-\overset{+}{O}\overset{H}{\underset{H}{\diagdown}} \longrightarrow CH_3CH_2\underset{\underset{H}{|}}{C}=CH_2 \;+\; HA \;+\; H_2O$$

A^-

Step 3: Protonation of the terminal alkene forms a 2° carbocation.

$$CH_3CH_2\underset{\underset{H}{|}}{C}=CH_2 \quad \xrightarrow{H-A} \quad CH_3CH_2-\underset{\underset{CH_3}{|}}{\overset{\overset{H}{|}}{C}}+ \;+\; A^-$$

Step 4: The 2° carbocation undergoes elimination reactions to form a mixture of alkenes.

$$CH_3CH_2-\underset{H_2C}{\overset{H}{\diagup}}C^+\overset{A^-}{\diagdown H} \longrightarrow CH_3CH_2CH=CH_2 \;+\; H-A$$

$$\underset{H_3C}{\overset{H}{\diagdown}}\overset{H}{C}-\underset{CH_3}{\overset{H}{C^+}} \overset{A^-}{\longrightarrow} \underset{H_3C}{\overset{H}{\diagdown}}C=C\underset{CH_3}{\overset{H}{\diagup}} \;+\; H-A$$

$$\underset{H_3C}{\overset{H}{\diagdown}}\overset{H}{C}-\underset{H}{\overset{CH_3}{C^+}} \overset{A^-}{\longrightarrow} \underset{H_3C}{\overset{H}{\diagdown}}C=C\underset{H}{\overset{CH_3}{\diagup}} \;+\; H-A$$

12.14 I Straight chain aliphatic primary haloalkanes react by a bimolecular mechanism (S_N2 or E2). The substitution reaction usually predominates except in cases where highly hindered base is used. For example, when *t*-butoxide is used as the base, the major product is formed by an E2 mechanism.

[structure: 1-bromopentane] $\xrightarrow[\text{Acetone}]{Na^+\;I^-}$ [structure: 1-iodopentane]

[structure: 1-bromopentane] $\xrightarrow{CH_3OH}$ Exceedingly slow substitution reaction to form [structure: pentyl methyl ether, OCH₃]

12.14 II Secondary haloalkanes, such as 2-bromobutane, have three competing routes of reactions with bases or nucleophiles. They are (a) substitution of the leaving group by an S_N2 mechanism, (b) elimination by an E2 mechanism, and (c) substitution by an S_N1 mechanism or elimination by an E1 mechanism.

12.14 III Steric hindrance at the tertiary carbon atom of tertiary haloalkanes prevents substitution by an S_N2 mechanism. As a result, elimination is the major reaction with bases such as HO⁻ or RO⁻. Very weak nucleophiles, such as H_2O or ROH, react with tertiary haloalkanes to form substitution products by an S_N1 mechanism and elimination products by an E1 mechanism.

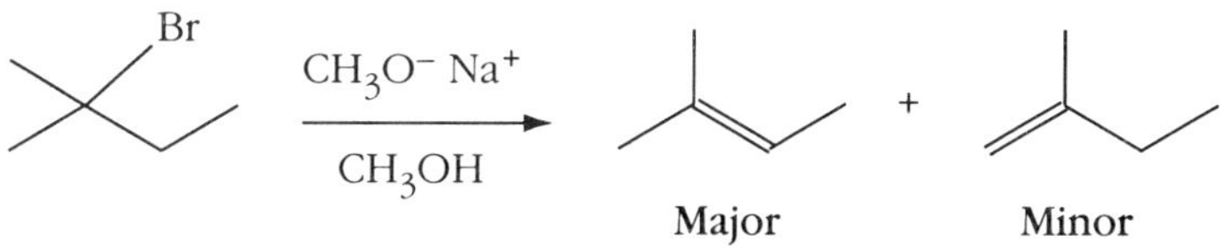

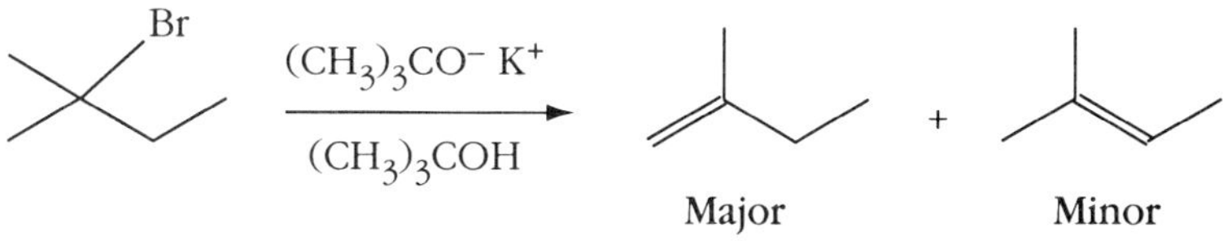

[Reaction scheme 1: 2-bromo-2,3-dimethylbutane + CH₃O⁻ Na⁺ / CH₃OH → 2,3-dimethyl-2-butene (Major) + 2,3-dimethyl-1-butene (Minor)]

[Reaction scheme 2: 2-bromo-2,3-dimethylbutane + (CH₃)₃CO⁻ K⁺ / (CH₃)₃COH → 2,3-dimethyl-1-butene (Major) + 2,3-dimethyl-2-butene (Minor)]

12.15 (a) A **nucleophile** (or *nucleophilic reagent*) is any species that contains a pair of electrons that it can donate to make a bond. Typical nucleophiles are RS^-, CN^-, I^-, HO^-, Br^-, ROH, H_2O, etc.

(b) The S_N1 **mechanism** (*S*ubstitution, *N*ucleophilic, *U*nimolecular) is a description of a substitution reaction that involves only one molecule in the rate-determining step. This rate-determining step usually involves ionization of the carbon-leaving group bond, which is followed by a fast attack by the nucleophile. The S_N1 mechanism usually results in a loss of configuration at the carbon atom (racemization). The rate depends on carbocation stability (3° > 2° > 1°).

(c) The **Saytzeff (Zaitsev) rule** predicts that in elimination reactions the more stable alkene, the one with the greatest number of alkyl groups on the double bonded carbon atoms, predominates in the product mixture.

(d) A **nucleophilic substitution reaction** is a replacement of an atom or a group of atoms on a carbon atom by a nucleophile. Such a reaction results in the substitution of one group for another bonded to a carbon atom. There are two types of mechanisms that operate in these reactions: S_N1 and S_N2.

(e) A **leaving group** is the atom or group of atoms that departs in a substitution or an elimination reaction. In general, the less basic the departing species is, the better a leaving group it is. Typical good leaving groups are halogens, $-OSO_2-C_6H_5$, $-OH_2$, $-NR_3$, $-SR_2$.

(f) The **E2 mechanism** (*E*limination, *B*imolecular) is the most common mechanism for an elimination reaction. It involves simultaneous removal of the proton by a base, formation of the π bond, and departure of the leaving group. All the orbitals must be coplanar, and this is usually accomplished in an *anti* relationship:

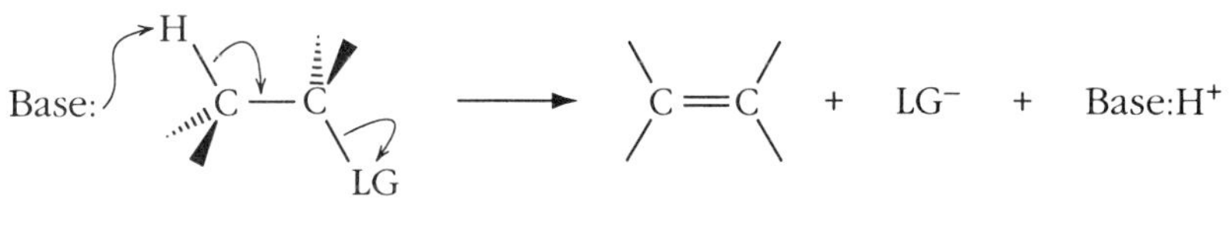

[Mechanism diagram: Base abstracts H from C–C with LG, giving C=C + LG⁻ + Base:H⁺]

(g) The S_N2 **mechanism** (*S*ubstitution, *N*ucleophilic, *B*imolecular) is a description of a displacement reaction that involves two molecules in the rate-determining step. These reactions usually occur with inversion of configuration at the carbon undergoing attack. The rate depends on the strength of the nucleophile as well as on steric hindrance, and is slowed either by α-branching or β-branching (e.g., "neopentyl" systems are *very slow*).

$$\text{Nu}^- \quad \overset{\frown}{\underset{\blacktriangle}{\text{C}}}-\text{LG} \longrightarrow \left[\text{Nu}\cdots\underset{\blacktriangle}{\text{C}}\cdots\text{LG}\right]^{\ddagger} \longrightarrow \text{Nu}-\underset{\blacktriangle}{\text{C}} \quad + \quad \text{LG}^-$$

(h) An **elimination reaction** is a chemical reaction in which a molecule eliminates the elements of a small molecule (like HX or H_2O) with the formation of the π bond, most frequently an alkene. The presence of a leaving group on a carbon atom makes possible two different kinds of reactions: substitution and elimination. Thus, elimination reactions always compete with substitution reactions and elimination reactions are promoted by strong bases and high temperatures. The order of reactivity in elimination reactions is 3° > 2° >> 1° (for both E1 and E2 mechanisms).

(i) The **Hofmann rule** predicts that the less-substituted alkene predominates as product when a bulky base such as *tert*-BuO$^-$ is used or when there is considerable steric hindrance about the leaving group. Thus, the less-substituted alkene, instead of the thermodynamically more stable alkene, is formed.

(j) The **E1 mechanism** (*E*limination, *U*nimolecular) is a two-step mechanism in which a carbocation is formed in Step 1 by ionization of the substrate. In Step 2, the carbocation loses a proton to a base to form an alkene. Notice that the first step in an E1 mechanism is identical to the first step in an S_N1 mechanism (ionization of the substrate).

$$-\underset{|}{\overset{H}{\underset{|}{C}}}-\underset{|}{\overset{|}{C}}-X \xrightarrow{\text{Slow}} -\underset{|}{\overset{|}{C}}-\overset{|}{C^+} + X^- \xrightarrow{\text{Fast}} C=C + B:H^+ + X^-$$

(k) A **carbocation** is a high-energy, unstable intermediate in which a carbon atom is only trivalent and has only six *valence* electrons. This trigonal geometry can be represented by an sp^2 hybrid carbon atom bonding to the three other atoms via the three sp^2 orbitals and leaving the remaining p orbital vacant. The four atoms (central carbon atom and the three others to which it is bonded) lie in a plane.

$$\overset{120°}{\underset{|}{C^+}}$$

(l) A **second order rate law** is an equation in which the reaction rate depends on the concentration of two components or on the square of the concentration of one component (total sum of the exponents of all the concentrations equals two). A typical example:

Rate = k [A][B] (where k is a rate constant)

12.16 (a) $(CH_3)_3P > (CH_3)_3N$ For nucleophiles from a given group in the periodic table, nucleophilicity increases going down the group. Thus, phosphorus is a better nucleophile than nitrogen.

(b) $H_2N^- > NH_3$ Anions are more powerful nucleophiles than their uncharged conjugate acids.

(c) CH₃OH > CH₃Cl
(d) CH₃COO⁻ > CH₃COOH   Anions are more powerful nucleophiles than their uncharged conjugate acids.

12.17

(a) 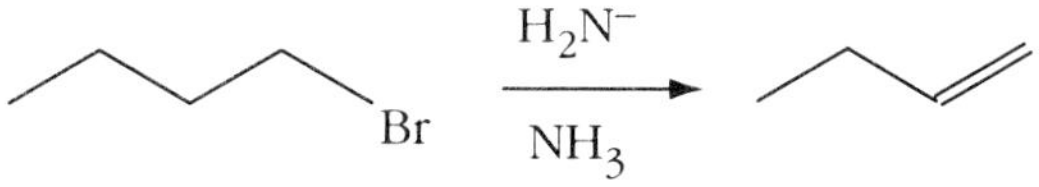 CH₃CH₂CH₂CH₂Br $\xrightarrow{\text{H}_2\text{N}^-,\ \text{NH}_3}$ CH₃CH₂CH=CH₂

(b) CH₃CH₂CH₂CH₂Br $\xrightarrow{\text{NaI}}$ CH₃CH₂CH₂CH₂I

(c) CH₃CH₂CH₂CH₂Br $\xrightarrow{\text{H}_2\text{O}}$ CH₃CH₂CH₂CH₂OH

(d) CH₃CH₂CH₂CH₂Br $\xrightarrow{\text{NaOH},\ \text{H}_2\text{O}}$ CH₃CH₂CH₂CH₂OH (Major) + CH₃CH₂CH=CH₂ (Minor)

(e) CH₃CH₂CH₂CH₂Br $\xrightarrow{\text{NaC}\equiv\text{N},\ \text{CH}_3\text{OH}}$ CH₃CH₂CH₂CH₂C≡N

(f) CH₃CH₂CH₂CH₂Br $\xrightarrow{\text{CH}_3\text{COONa},\ \text{CH}_3\text{COOH}}$ CH₃CH₂CH₂CH₂OOCCH₃

(g) CH₃CH₂CH₂CH₂Br $\xrightarrow{\text{CH}_3\text{C}\equiv\text{C}^-\ \text{Na}^+}$ CH₃CH₂CH₂CH₂C≡CCH₃

(h) CH₃CH₂CH₂CH₂Br $\xrightarrow{\text{CH}_3\text{NH}_2,\ (\text{CH}_3\text{CH}_2)_2\text{O}}$ CH₃CH₂CH₂CH₂N⁺H₂CH₃

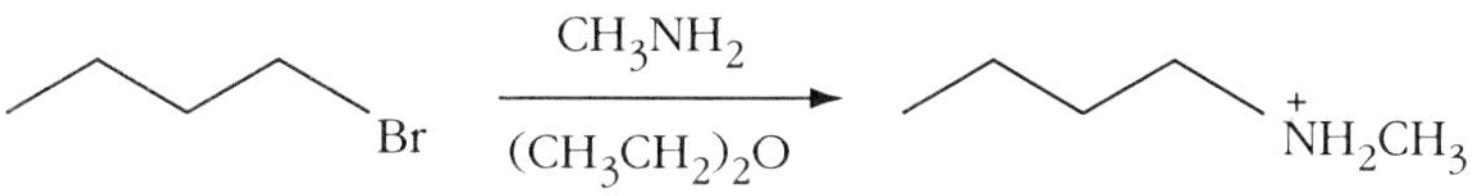

12.18

(a) 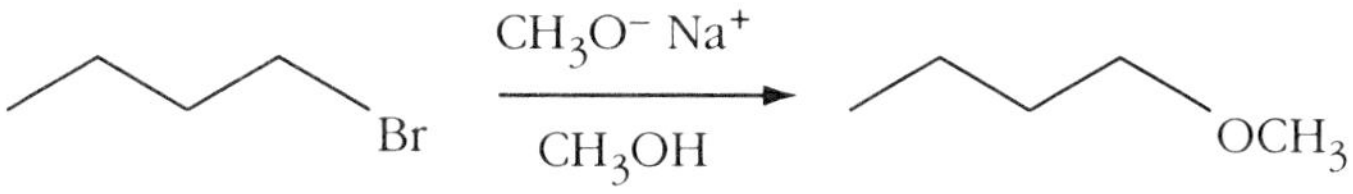 CH₃CH₂CH₂CH₂Br $\xrightarrow{\text{CH}_3\text{O}^-\ \text{Na}^+,\ \text{CH}_3\text{OH}}$ CH₃CH₂CH₂CH₂OCH₃

(b) 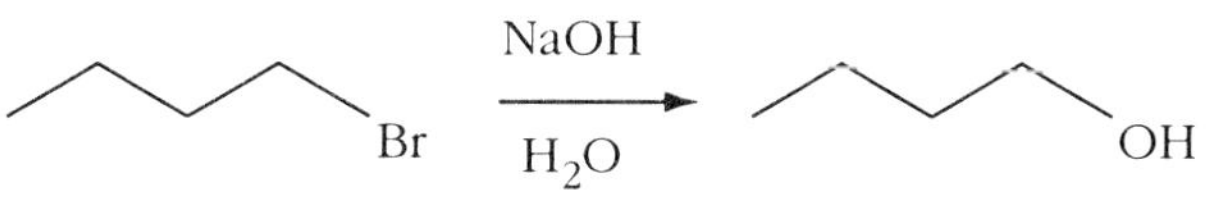 (CH₃)₂CHCH₂Br $\xrightarrow{\text{NaOH},\ \text{H}_2\text{O}}$ (CH₃)₂CHCH₂OH

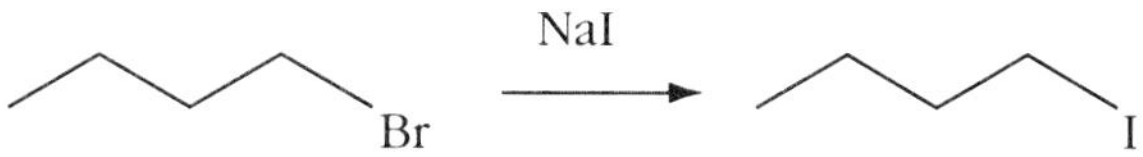

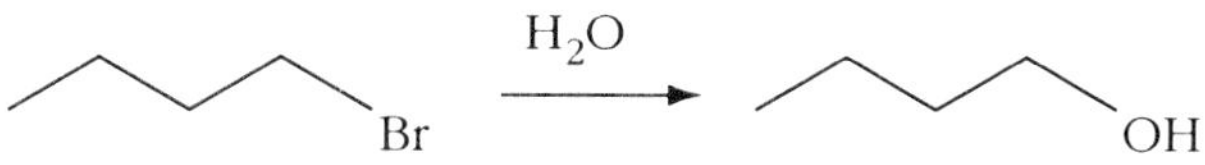

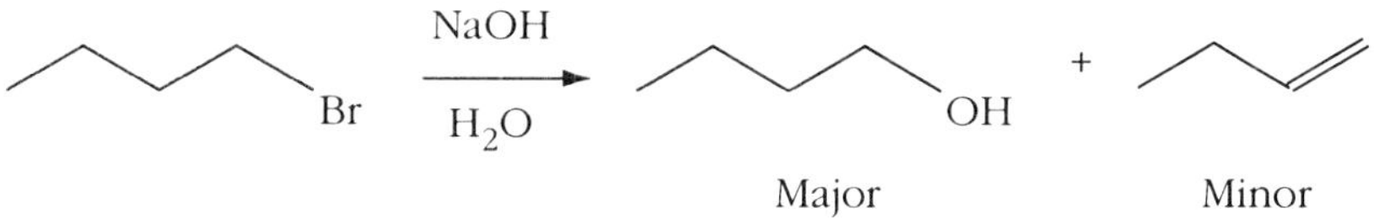

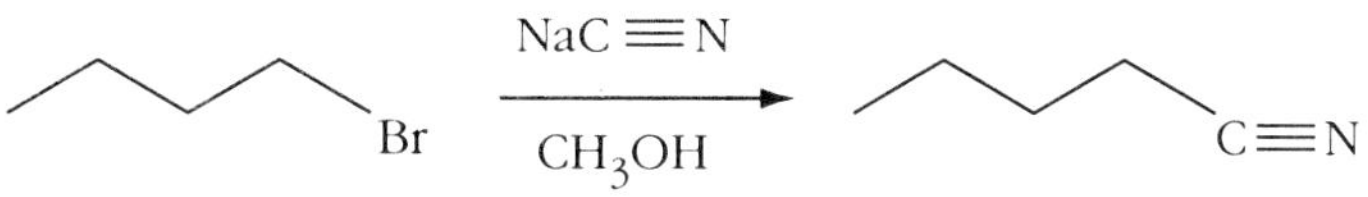

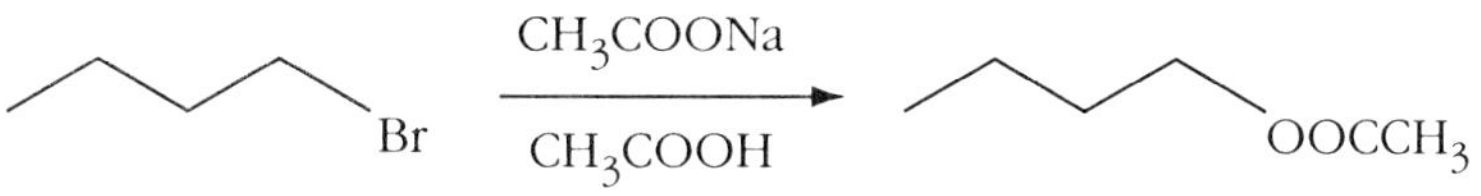

(c) 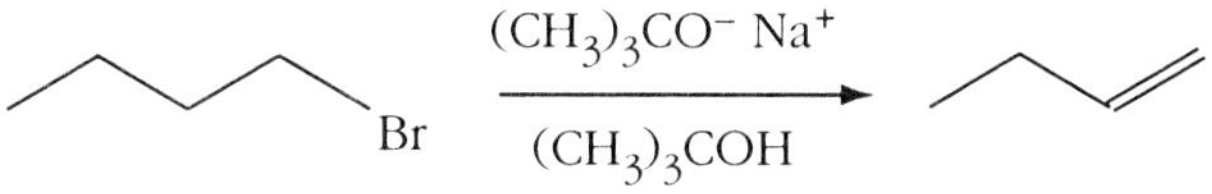 CH₃CH₂CH₂CH₂Br $\xrightarrow{(\text{CH}_3)_3\text{CO}^-\ \text{Na}^+,\ (\text{CH}_3)_3\text{COH}}$ CH₃CH₂CH=CH₂

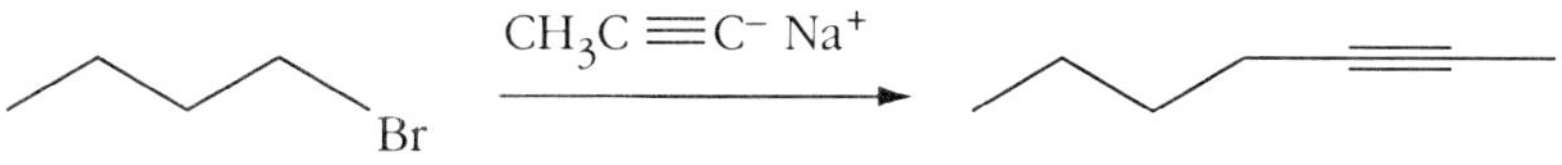

(d) [CH₃CH₂CH₂CH₂Br] —NaCN/CH₃CH₂OH→ [CH₃CH₂CH₂CH₂CN]

(e) [CH₃CH₂CH₂CH₂Br] —(1) Mg/dry ether; (2) CH₃CCH₃ (‖ O); (3) H₃O⁺→ [CH₃CH₂CH₂C(CH₃)₂OH]

12.19 In order to predict the product of a substitution and/or elimination reaction, we have to know the mechanism of that reaction. Predictions about the mechanism of a particular reaction must be based on considerations of the structure of the starting material, the reagent(s), the temperature, the nature of the solvent, and so on. Such analysis is shown for the first five reactions.

(a) Bromide ion is a good leaving group and it is on a secondary carbon. The sulfide ion is a very strong nucleophile. Acetone is a polar solvent and it is a good medium in which to carry out S_N2 reactions but a poor medium in which to carry out S_N1 reactions. We conclude that the product is formed by substitution and the reaction mechanism is S_N2.

$$(CH_3)_2CHBr + NaSCH_3 \longrightarrow (CH_3)_2CHSCH_3$$

(b) 2-Methyl-2-propanol is a tertiary alcohol that is protonated by concentrated HCl in a rapid, reversible proton–transfer reaction to form an oxonium ion. Oxonium ion formation is followed by loss of a molecule of water to give a tertiary carbocation. Once formed, the tertiary carbocation reacts with chloride ion to give 2-chloro-2-methylpropane. The reaction occurs by an S_N1 mechanism to form a substitution product.

$$(CH_3)_3COH \xrightarrow{conc.\ HCl} (CH_3)_3CCl$$

(c) 2-Chloro-2-methylpentane is a tertiary halide and potassium *tert*-butoxide is a strong base. The major reaction of tertiary haloalkanes with strong bases is elimination by an E2 mechanism. Because of steric hindrance of the *tert*-butyl group, the less stable product would be formed preferentially according to the Hofmann rule.

$$(CH_3)_2CClCH_2CH_2CH_3 \xrightarrow[(CH_3)_3COH]{(CH_3)_3CO^-\ K^+} H_2C=C(CH_3)(CH_2CH_2CH_3)$$

(d) Bromide ion is a good leaving group and it is on a secondary carbon. The phenoxide ion is a moderately strong nucleophile and ethanol is a moderately ionizing solvent. All favor formation of the substitution product by an S_N2 mechanism.

(e) The substrate is a secondary alcohol and the reaction is an acid-catalyzed dehydration. Step 1 of the mechanism of acid-catalyzed dehydration of alcohols involves protonation of the hydroxyl group and formation of an oxonium ion. In Step 2, the C—OH$_2^+$ bond breaks to give a carbocation and a molecule of water. Migration of an adjacent methylene group with its pair of electrons in Step 3 to the positively charged carbon atom gives a more stable tertiary carbocation. The tertiary carbocation then loses a proton in Step 4 to form the final product, 1,2-dimethyl-cyclopentene.

(g)

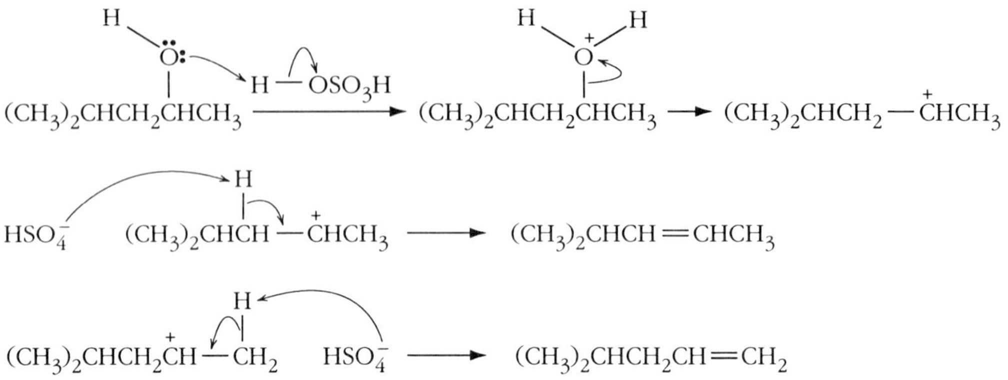

$$\text{(CH}_3\text{)}_2\text{CHCH}_2\text{CHCH}_3 \xrightarrow{\overset{H}{\underset{|}{\ddot{O}:}} \quad H-OSO_3H} \text{(CH}_3\text{)}_2\text{CHCH}_2\text{CHCH}_3 \text{ (with } \overset{+}{O}H_2) \longrightarrow \text{(CH}_3\text{)}_2\text{CHCH}_2-\overset{+}{\text{C}}\text{HCH}_3$$

$$HSO_4^- \quad \text{(CH}_3\text{)}_2\text{CHCH}-\overset{+}{\text{C}}\text{HCH}_3 \text{ (with H)} \longrightarrow \text{(CH}_3\text{)}_2\text{CHCH}=\text{CHCH}_3$$

$$\text{(CH}_3\text{)}_2\text{CHCH}_2\overset{+}{\text{C}}\text{H}-\text{CH}_2 \text{ (with H)} \quad HSO_4^- \longrightarrow \text{(CH}_3\text{)}_2\text{CHCH}_2\text{CH}=\text{CH}_2$$

(h)

[Newman projection and E2 elimination of 1,2-dibromo-1,2-diphenylethane with CH₃CH₂O⁻ giving (E)-1,2-dibromo-stilbene type alkene: BrC(C₆H₅)=C(C₆H₅)H with trans arrangement]

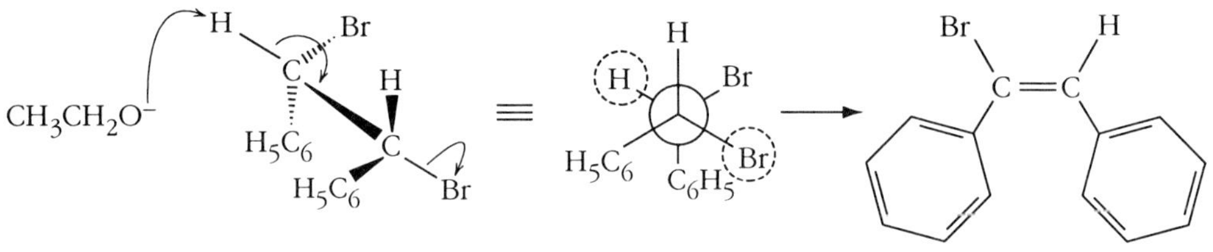

(i)

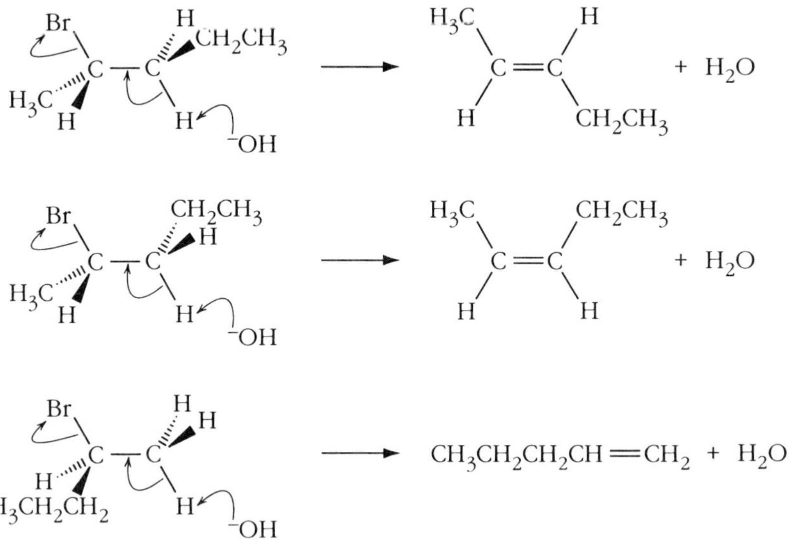

[Three E2 elimination diagrams from 2-bromo-3-methylpentane-type substrate with ⁻OH giving:
- H₃C/H C=C CH₂CH₃/H + H₂O
- H₃C/H C=C CH₂CH₃/H (other stereochemistry) + H₂O
- CH₃CH₂CH₂CH=CH₂ + H₂O]

12.20 (a)

[Br–(CH₂)₄–CHBr–... with ⁻OH → Br–(CH₂)₄–CH₂OH + Br⁻]

[Br–(CH₂)₄–CH₂–O–H with ⁻OH → Br–(CH₂)₄–CH₂–O⁻ + H₂O]

[Cyclic intermediate with Br and O → tetrahydrofuran + Br⁻]

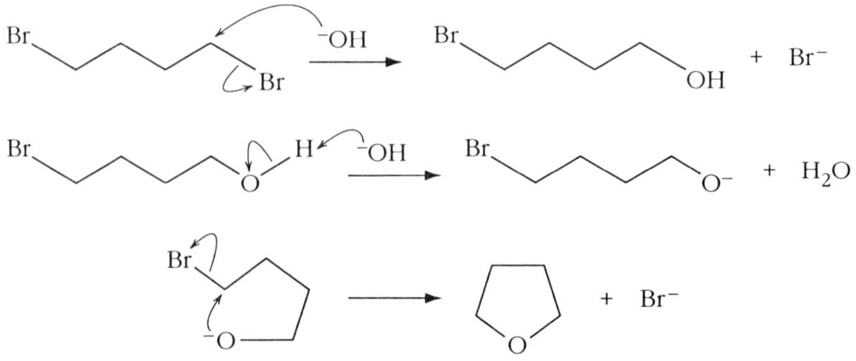

(b)

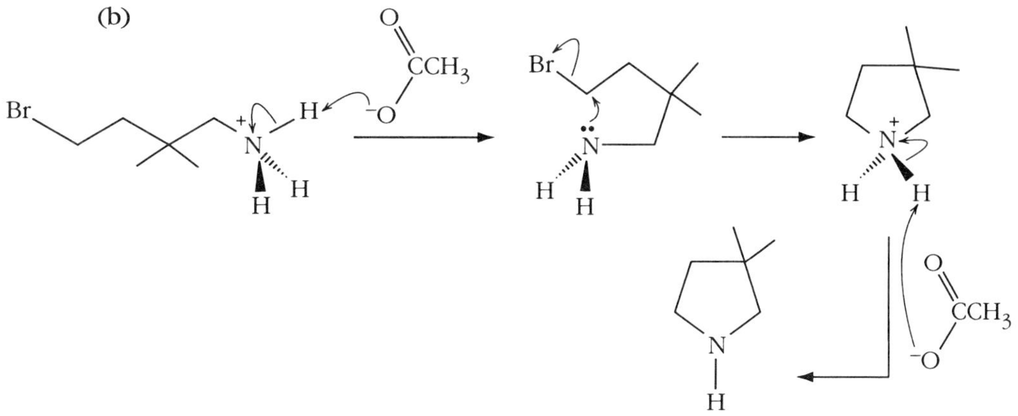

[Mechanism scheme showing intramolecular cyclization of a bromoamine with acetate acting as base, forming a pyrrolidinium intermediate and then 3,3-dimethylpyrrolidine.]

(c)

[Mechanism scheme showing intramolecular substitution of a trans-2-chlorocyclohexanethiol, with hydroxide deprotonating SH, then internal displacement of Cl to form a thiirane/episulfide via chair conformation.]

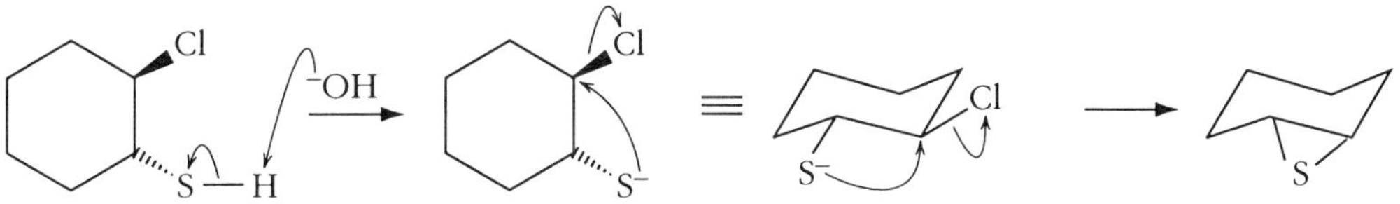

12.21 (*R*)-1-Deuterioethanol is unprotonated in pure water. However, addition of a small amount of sulfuric acid catalyzes the following racemization reaction.

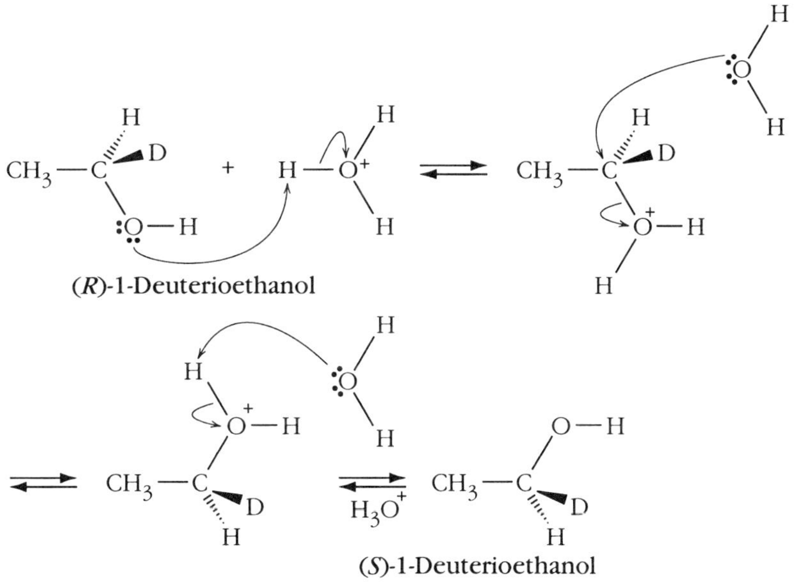

[Mechanism showing (*R*)-1-deuterioethanol being protonated by H_3O^+, water loss to form carbocation, re-addition of water from opposite face, and deprotonation to give (*S*)-1-deuterioethanol.]

12.22 The solvolysis of 2-bromo-2-methylbutane occurs by an S_N1 mechanism because the substrate is a tertiary bromoalkane and methanol is a polar solvent but a poor nucleophile, which favors the S_N1 mechanism.

Step 1: Spontaneous dissociation of 2-bromo-2-methylbutane yields an intermediate carbocation in a slow, rate-determining step.

[Reaction scheme: (CH3)2C(Br)(CH2CH3) ⇌ Br⁻ + (CH3)2C⁺(CH2CH3)]

Step 2a: The carbocation intermediate transfers a proton to a base (E1 mechanism) to give an alkene. There are two nonequivalent types of β hydrogens in the carbocation and, therefore, two alkenes are formed (a less stable **A** and a more stable **B**).

[Mechanism showing formation of alkene A with CH2=C(CH3)(CH2CH3) structure + HBr]

[Mechanism showing formation of alkene B, (CH3)2C=CHCH3 + HBr]

Step 2b: The carbocation intermediate reacts with methanol to afford ether **C**, which is the major product.

[Mechanism showing carbocation + CH3OH → oxocarbenium intermediate → ether C: (CH3)2C(OCH3)(CH2CH3) + CH3OH2⁺]

C

(a) Sodium methoxide is a strong base and the major product formed under these conditions comes from an elimination reaction that occurs by an E2 mechanism. Elimination reactions using CH_3O^- give the Saytzeff product (**B**) as a major product.

(b) The addition of water to methanol will increase the dielectric constant of the solution (about threefold). We know that a very polar solvent (such as water) encourages the S_N1 mechanism by helping stabilize the carbocation through solvation.

Apart from increasing the overall reaction rate, the addition of water will lead to the formation of a new compound, 2-methyl-2-butanol, which will become the chief product in 10% methanol.

(c) Potassium *tert*-butoxide is a strong and sterically hindered base. Thus, elimination by an E2 mechanism is most probable under these conditions, with the major product being alkene **A** (Hofmann product).

12.23 Step 1: Protonation of ethanol by sulfuric acid.

$$CH_3-CH_2-\overset{..}{\underset{..}{O}}-H \;+\; H-OSO_2OH \;\rightleftharpoons\; CH_3-CH_2-\overset{+}{O}(H)-H \;+\; {}^-OSO_3H$$

Step 2a: Ethanol or sulfate anion can act as nucleophiles in a nucleophilic substitution. Note that water is a good leaving group.

$$CH_3-CH_2-\overset{+}{O}(H)-H \;+\; {}^-OSO_3H \longrightarrow CH_3CH_2OSO_3H \;+\; H_2O$$

$$CH_3-CH_2-\overset{+}{O}(H)-H \;+\; H\overset{..}{\underset{..}{O}}CH_2CH_3 \rightleftharpoons CH_3CH_2\overset{+}{O}(H)CH_2CH_3 + H_2O \underset{H_2SO_4}{\overset{{}^-OSO_3H}{\rightleftharpoons}} CH_3CH_2OCH_2CH_3$$

Step 2b: The sulfate anion can also act as a base to remove a proton and to complete the dehydration of ethanol by an E2 mechanism.

$$HOSO_2O^- \;\; H-CH_2-CH_2-\overset{+}{O}(H)-H \longrightarrow H_2SO_4 \;+\; CH_2{=}CH_2 \;+\; H_2O$$

12.24 This reaction is the conversion of a 3° alcohol to a 3° bromoalkane in an acid solution. First, the alcohol is protonated in aqueous acid solution:

The conjugate acid of the alcohol forms a carbocation:

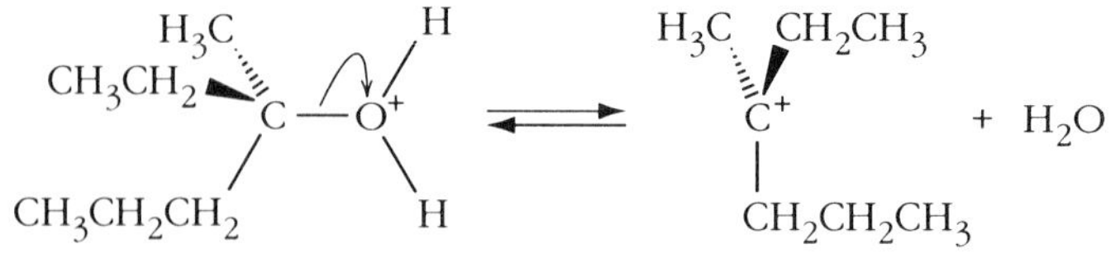

The bromide nucleophile may attack from either side, so a racemic mixture of 3-bromo-3-methylhexane is formed:

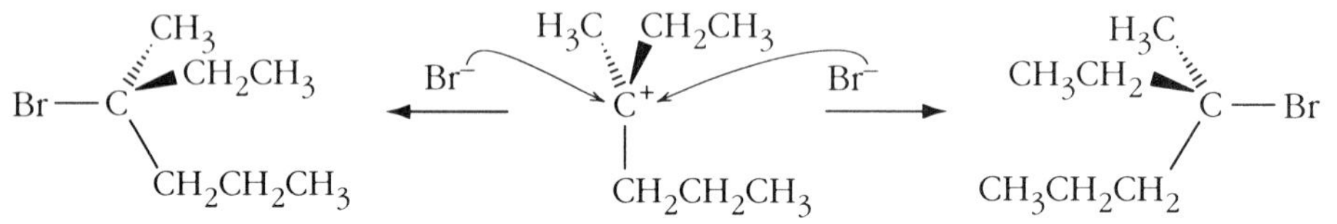

12.25

(a)

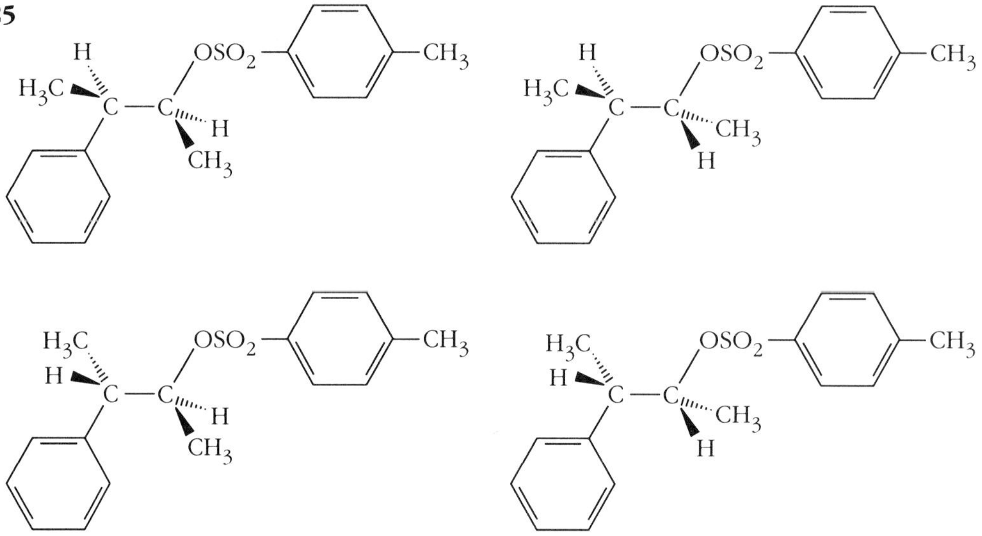

(b)

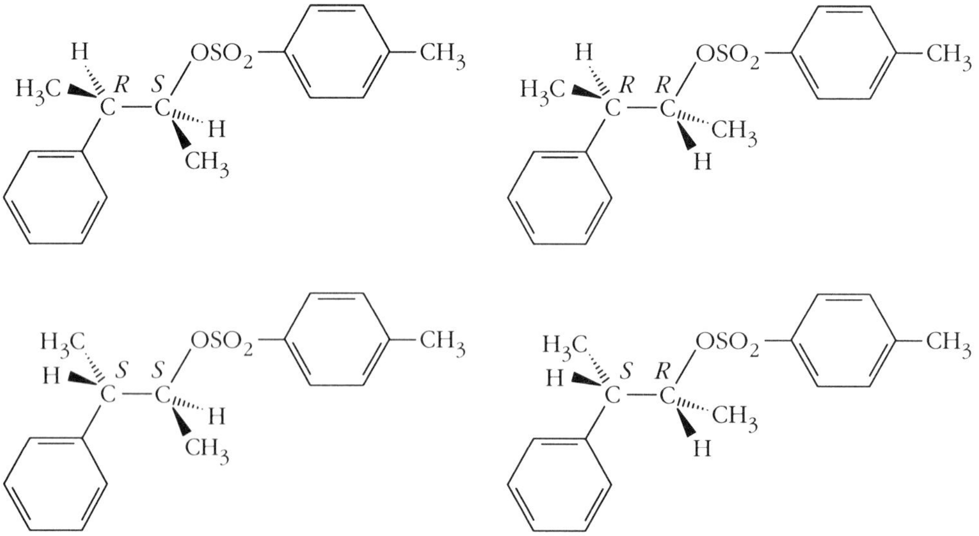

(c)

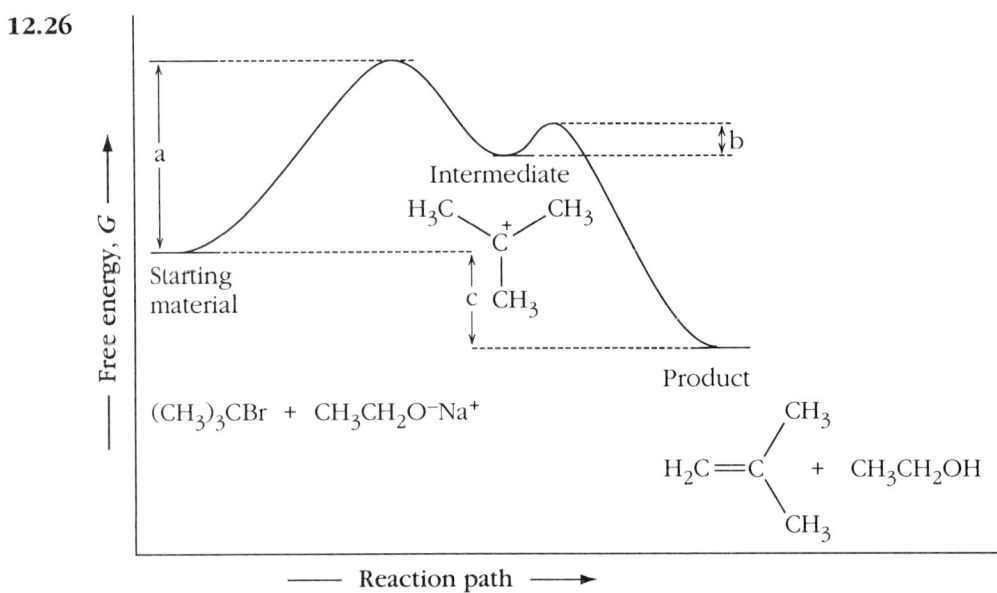

(a) Energy of activation of the first step (dissociation of *t*-butyl bromide),
(b) Energy of activation of the second step (combining of the *t*-butyl carbocation with $CH_3CH_2O^-$),
(c) Standard Gibbs free-energy change ($\Delta G°$)—the total amount of energy change during the reaction.

12.27 (a) Nucleophilicity increases while going down a column of the periodic table. Thus, CH_3S^- is more nucleophilic than CH_3O^-, and CH_3Se^- is more nucleophilic than CH_3S^-. The change from CH_3ONa to CH_3SeNa will significantly increase the reaction rate with iodomethane.
(b) Stable anions (weak bases) make good leaving groups. F^- is a much poorer leaving group than I^- and the change from CH_3I to CH_3F will significantly lower the reaction rate with sodium methoxide.
(c) Tosylate is a better leaving group than I^-. Thus, the change from CH_3I to CH_3OTs will increase the reaction rate with sodium methoxide.
(d) The most important variable in the S_N2 mechanism is steric hindrance about the reactive site. Alkyl branching next to the leaving group slows the reaction greatly. Thus, the change from CH_3I (methyl halide) to $(CH_3)_2CHI$

(2° haloalkane) will significantly slow the reaction rate of substitution with methoxide, as well as give some elimination product.

(e) Polar aprotic solvents favor S_N2 reactions. Thus, the change of solvent from methanol to DMF (dimethylformamide) will increase the reaction rate of iodomethane with sodium methoxide.

12.28 1-Bromo-2,2-dimethylpropane (neopentyl bromide) reacts by an S_N2 mechanism very slowly. Although it is a primary haloalkane, the *tert*-butyl group attached to the α carbon atom provides significant steric hindrance about the reactive site. To account for the observed results, we may propose that the reaction occurs by an S_N1 mechanism. The carbocation formed in the first step rapidly rearranges by a methyl shift to form a more stable 3° carbocation.

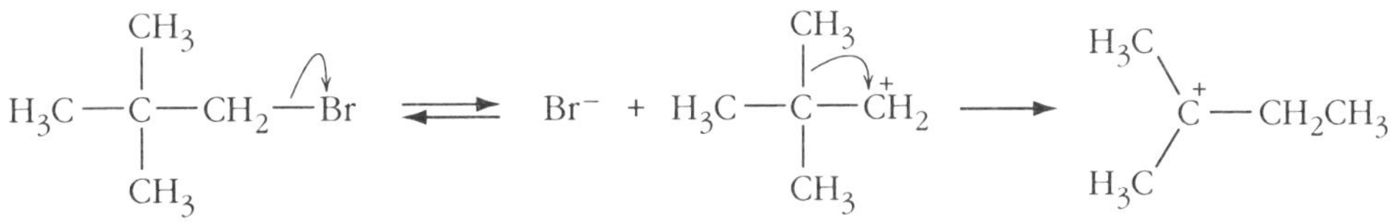

The carbocation reacts with either water or ethanol as a nucleophile:

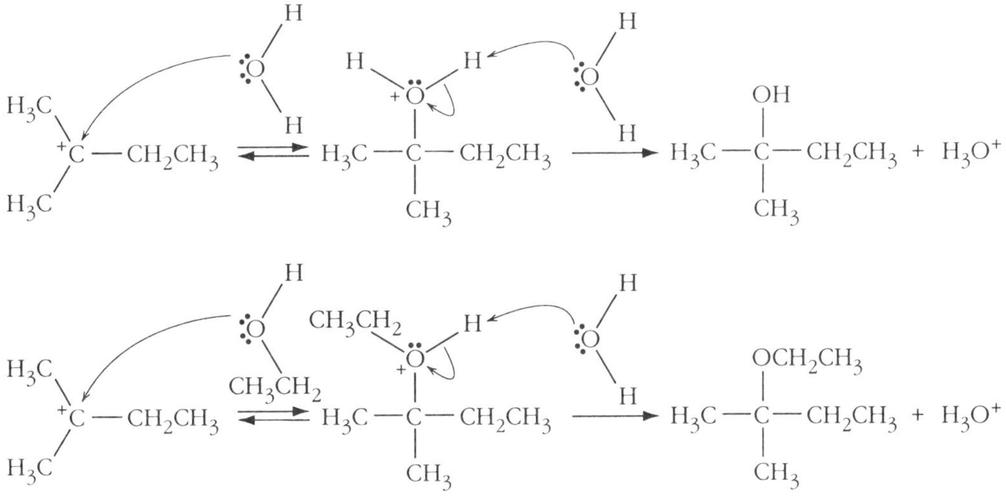

The carbocation may also react with a base and undergo elimination by a loss of a proton:

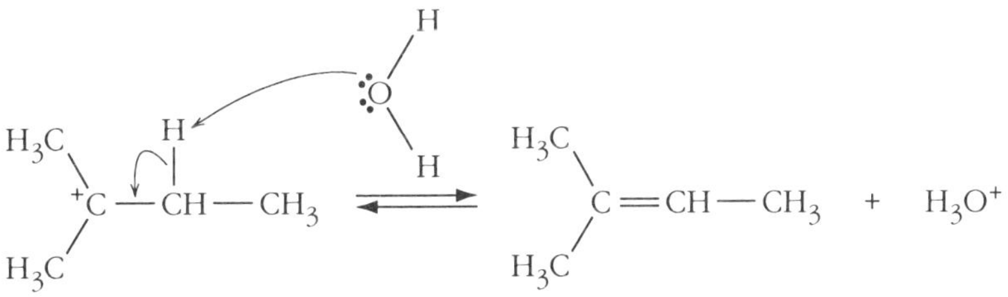

12.29 *p*-Toluenesulfonate is a better leaving group than methoxide ion so the reaction site is the carbon atom bonded to the —OTs group. Methoxide is a moderately strong nucleophile so the reaction occurs by an S_N2 mechanism. The nucleophilic attack from the side opposite the bond to the leaving group in an S_N2 mechanism results in the inversion of configuration ("back-side displacement"). This forms a *meso* compound, which is optically inactive, as product.

12.31 The elimination reaction of menthyl chloride with sodium ethoxide (strong base) at an elevated temperature occurs by an E2 mechanism. It is a requirement of an E2 reaction that hydrogen atom and halide are *anti* and *coplanar*. There is only one hydrogen atom in menthyl chloride that fulfils such a requirement. So only one product is observed.

[Reaction scheme showing chair conformation of chloro-methyl-isopropyl cyclohexane converting via E2 with ethanol to alkene product]

In the second case, both ethanol and water are very weak bases and the elimination reaction may occur by an E1 mechanism, or a combination of E1 and E2 mechanisms. As a result, two isomeric alkenes are formed as products.

[Reaction scheme showing E1 mechanism with carbocation intermediates leading to two isomeric alkene products]

12.32 Compounds **A** and **B** are both tertiary bromoalkanes so their solvolysis reactions will occur by an S_N1 mechanism. In order to react by an S_N1 mechanism the substrate must be able to form a planar carbocation. This is not a problem for compound **B**. In fact, the steric strain about the reactive site in compound **B** is relieved by forming a planar carbocation.

[Structure of Compound B: (CH₃CH₂)₃CBr ionizing to triethyl carbocation (planar) + Br⁻]

Compound B

Compound **A**, on the other hand, cannot form a planar carbocation without distorting the bicyclic ring system. Distorting the bicyclic ring system to form a planar carbocation requires additional energy in the first, rate-determining step of the solvolysis of compound **A**.

As a result, the free energy of activation needed to form a planar carbocation is much greater for compound **A** than compound **B** so compound **B** reacts faster than compound **A**.

12.33 The molecular formula of the product, C_7H_{12}, indicates two units of unsaturation, which is consistent with the formation of an alkyne as the product of a nucleophilic substitution reaction. The ^1H NMR spectrum of the product shows a doublet at δ 0.95 (6H), a triplet at δ 1.8 overlapping a septet at δ 1.7 (total relative area 4H), and a multiplet around δ 2 (2H).

The doublet at δ 0.95 and the septet at δ 1.7 is the characteristic pattern of an isopropyl group. The coupling constant of the doublet is the same as the coupling constant of the septet, which is consistent with an isopropyl group. This gives us two partial structures accounting for five of the seven carbon atoms in the compound: —C≡C— and —CH(CH$_3$)$_2$.

The triplet at δ 1.8 (3H) suggests a methyl group split by a methylene (—CH$_2$—) group. We can join these partial structures in two ways:

$$CH_3CH_2C≡CCH(CH_3)_2 \quad \text{or} \quad CH_3C≡CCH_2CH(CH_3)_2$$

Compound A Compound B

In both of these structures, the hydrogen atoms (H$_a$) of the methyl group will be split into a triplet by the hydrogen atoms of the methylene group (H$_b$). Also in both compounds, the hydrogen atoms of the methylene group (H$_b$) will be split by the hydrogen atoms of the methyl group (H$_a$) and by the methine hydrogen atom (H$_c$) of the isopropyl group. We must rely on the values of the coupling constants (J_{ab} and J_{bc}) and the resulting splitting patterns to distinguish between these two isomeric structures.

Long range coupling constants transmitted across the triple bond are smaller than the typical alkane coupling constants. Consequently, in compound **A**, $J_{ab} > J_{bc}$ so the splitting pattern should be as follows:

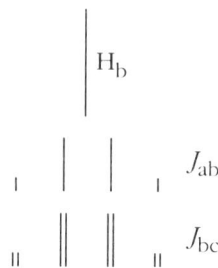

In contrast, in compound B, $J_{bc} > J_{ab}$ so the splitting pattern should be as follows:

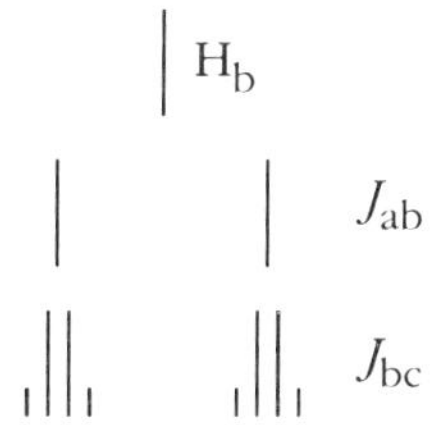

The pattern of the multiplet at about δ 2 resembles the predicted splitting pattern for compound **B**, so the product is $(CH_3)_2CHCH_2C\equiv CCH_3$.

The product is an internal alkyne, so there are two possible ways it can be synthesized from sodium acetylide and a bromoalkane. The reaction equations are as follows:

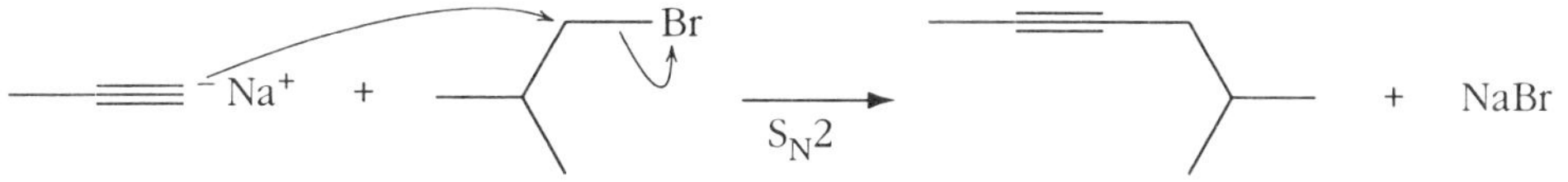

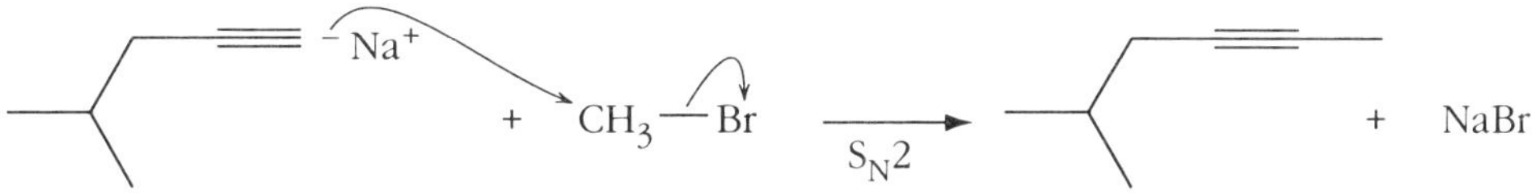

12.34 The molecular formula of the starting material, $C_6H_{14}O$, indicates no degrees of unsaturation. The reaction of the starting material with concentrated sulfuric acid to give an alkene as a product indicates that the starting material is a saturated alcohol. The starting material shows a singlet at δ 0.85 (9H), a doublet at δ 1.1 (3H), a singlet at 2.05 (1H) and a quartet at δ 3.45 (1H). The singlet at δ 0.85 represents three methyl groups bonded to a quaternary carbon characteristic of a *tert*-butyl group (no coupling). The chemical shift of the quartet at δ 3.45 indicates that the hydrogen atom giving rise to that signal is bonded to a secondary carbon atom to which the hydroxyl group is also attached. The signal at δ 3.45 is split into a quartet, which indicates that the adjacent carbon(s) must have a total of three hydrogen atoms. This is consistent with a methyl group being attached because its signal at δ 1.1 is a doublet (coupling with the —CH(OH)— unit). Putting all those pieces together allows us to suggest that the structure of the alcohol is that of **3,3-dimethyl-2-butanol**:

$$(CH_3)_3C\overset{\overset{\displaystyle OH}{|}}{C}HCH_3$$

CHAPTER 13

ETHERS AND EPOXIDES
Concepts

The **ether** functional group is composed of an oxygen atom bonded to two carbon atoms that are part of alkyl, aryl, or vinyl groups. Ethers can act as Lewis bases because they have two lone pairs of electrons on their oxygen atoms.

$$\underset{\text{Water}}{H \overset{\ddot{O}}{\underset{105°}{\frown}} H} \qquad \underset{\text{Methanol}}{H \overset{\ddot{O}}{\underset{108.5°}{\frown}} CH_3} \qquad \underset{\substack{\text{Methoxymethane} \\ \text{(dimethyl ether)}}}{H_3C \overset{\ddot{O}}{\underset{112°}{\frown}} CH_3} \qquad \underset{\text{Oxirane}}{\overset{1\ \ddot{O}}{H_2C \underset{2\ \ \ \ 3}{—} CH_2}}$$

Ethers are prepared in the laboratory by either an alkoxymercuration-demercuration sequence of reactions or the Williamson ether synthesis.

Ethers are inert to most reagents. The major exceptions are HI and HBr, which react with ethers to form cleavage products.

The cleavage reaction occurs by an S_N2 mechanism when the alkyl groups bonded to oxygen are primary or secondary and by an S_N1 mechanism if the alkyl groups bonded to oxygen are tertiary.

Epoxides (the IUPAC name: **oxiranes**) are compounds containing an oxygen atom in a three-membered ring.

$$H_3C \overset{1\ \ddot{O}}{\underset{\underset{H}{C} \underset{3}{—} \underset{2}{C} \underset{H}{}}{}} CH_2CH_3$$

IUPAC name: 2-Ethyl-3-methyloxirane
Trivial name: 2,3-Epoxypentane

Epoxides are readily prepared by the epoxidation of alkenes. The epoxide ring opens readily under acidic or basic conditions. Epoxide ring opening under basic conditions occurs by an S_N2 mechanism at the least sterically hindered ring carbon atom. Ring opening under acidic conditions occurs preferentially at the most substituted carbon atom. In both reactions, the configuration of the reactive site is inverted.

Solutions to the Exercises

13.1 (a) 2-Ethoxypropane (b) Methoxyethene
(c) 2-Isopropoxypropane or 2-(1-Methylethoxy)propane
(d) 1,2-Dimethoxyethane
(e) (1R, 3S)-1-Ethoxy-3-(1,1-dimethylethoxy)cyclohexane, or *cis*-1-ethoxy-3-(1,1-dimethylethoxy)cyclohexane, or *cis*-1-*tert*-butoxy-3-ethoxycyclohexane

13.2

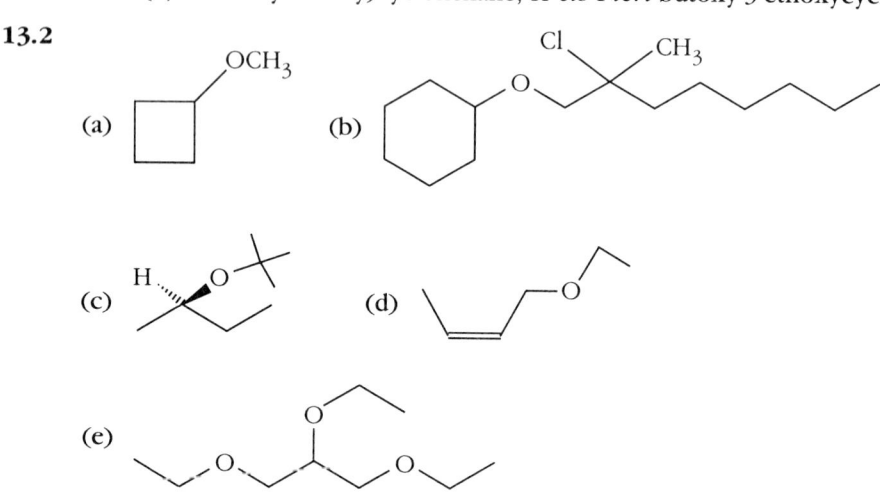

13.3 (a) (2R, 3S)-3-Methoxy-2-methyloxane
(b) 2-Methyl-2-pentyloxirane or 1,2-epoxy-2-methylheptane
(c) 3-Chloromethyloxolane

13.4

13.5 (a) (3R,4R,6S)-3,4-Epoxy-6-methyloctane
(b) 1,2-Epoxy-4,4-dimethylcyclohexane
(c) (1R,2S,4S)-4-Chloro-1,2-epoxycyclohexane or *cis*-4-chloro-1,2-epoxycyclohexane

13.6 (a), 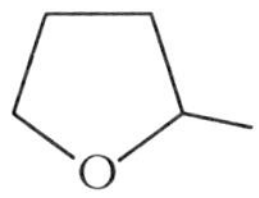 (b), 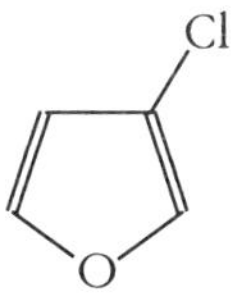 (c), 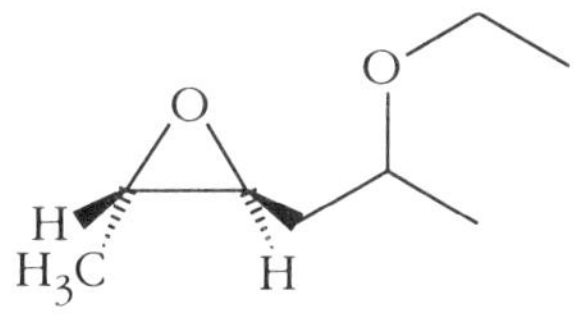 (d) 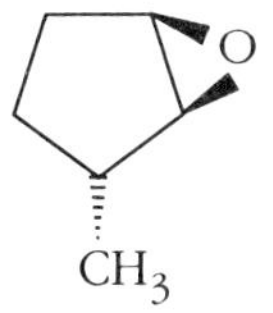 [structures]

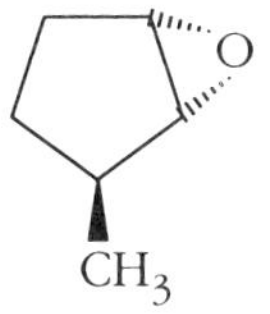

13.7 [mechanism scheme]

⟶ H—B⁺ + CH₂=CH₂ + H₂O

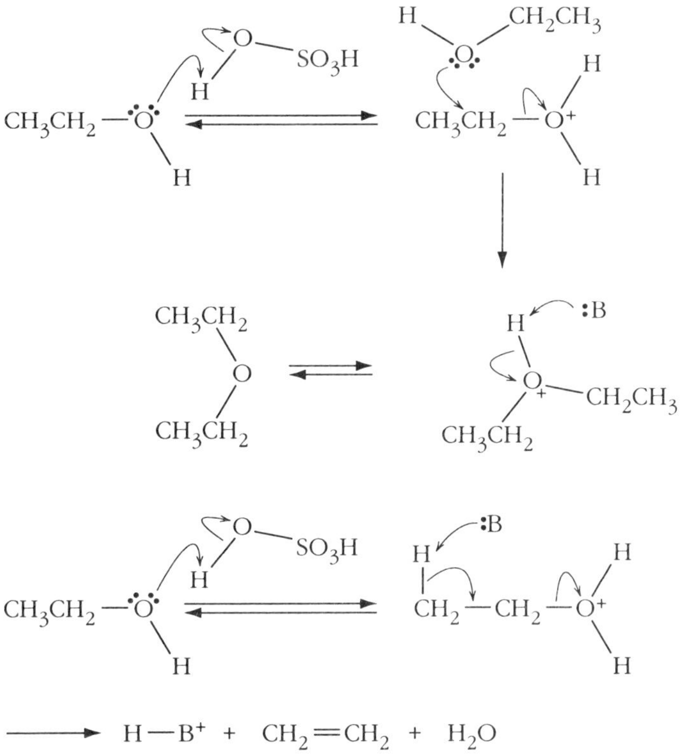

The first step (protonation of ethanol by sulfuric acid) and the first intermediate (protonated ethanol) are the same for both reactions. This is, however, where the similarity ends.

In the first reaction, the protonated ethanol reacts with another molecule of unprotonated ethanol acting as the nucleophile in an S_N2 reaction to form a protonated ether as product.

In the second reaction, the protonated ethanol reacts with a base (HSO_3^- or ethanol) to afford ethene, the product of elimination by an E2 mechanism.

13.8 1,2-Ethanediol must be used as the starting alcohol. There are three steps of the synthesis:

Step 1: Formation of diethylene glycol as an intermediate. The mechanism for the formation of diethylene glycol begins with protonation of one of the hydroxyl groups of 1,2-ethanediol followed by a nucleophilic attack by another molecule of 1,2-ethanediol on the oxonium ion to give diethylene glycol.

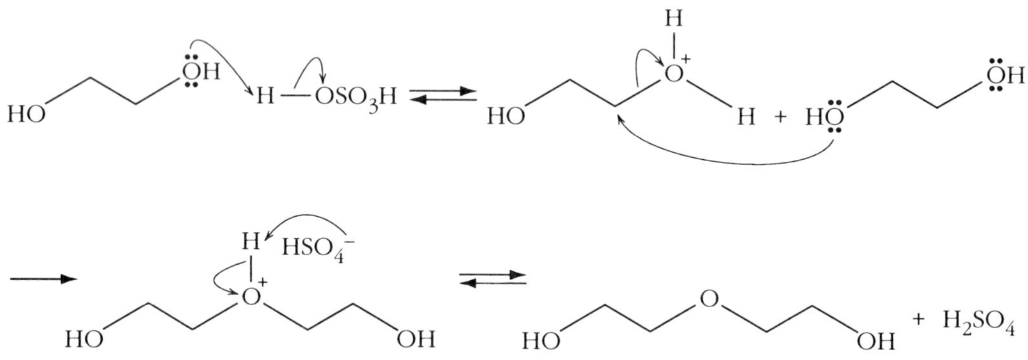

Step 2: Formation of 1,4-dioxane from diethylene glycol. Protonation of one of the hydroxyl groups of diethylene glycol followed by a nucleophilic attack by an unprotonated hydroxyl group forms a cyclic product that loses a proton and forms 1,4-dioxane.

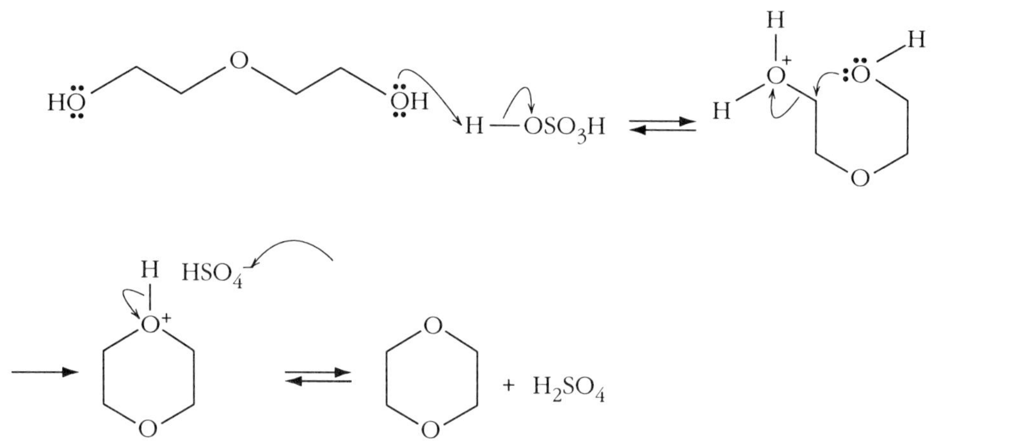

Step 3: Formation of side products. Diethylene glycol protonated on a hydroxyl group can also react with either 1,2-ethanediol or diethylene glycol as nucleophile to form linear polyethers as side products.

13.9

(a) cyclopentene $\xrightarrow{\text{(1) Hg(OAc)}_2/\text{CH}_3\text{CH}_2\text{OH}}_{\text{(2) NaBH}_4/\text{NaOH, H}_2\text{O}}$ cyclopentyl ethyl ether

(b) 1-methylcyclohexene $\xrightarrow{\text{(1) Hg(OAc)}_2/\text{CH}_3\text{CH}_2\text{CH}_2\text{CH}_2\text{OH}}_{\text{(2) NaBH}_4/\text{NaOH, H}_2\text{O}}$ 1-methylcyclohexyl butyl ether

(c) 2-methyl-2-butene $\xrightarrow{\text{(1) Hg(OAc)}_2/\text{CH}_3\text{CH}_2\text{OH}}_{\text{(2) NaBH}_4/\text{NaOH, H}_2\text{O}}$ 2-ethoxy-2-methylbutane

(d) [alkene] $\xrightarrow{\text{(1) Hg(OAc)}_2/\text{CH}_3\text{CH}_2\text{CH}_2\text{OH}}_{\text{(2) NaBH}_4/\text{NaOH, H}_2\text{O}}$ [ether product]

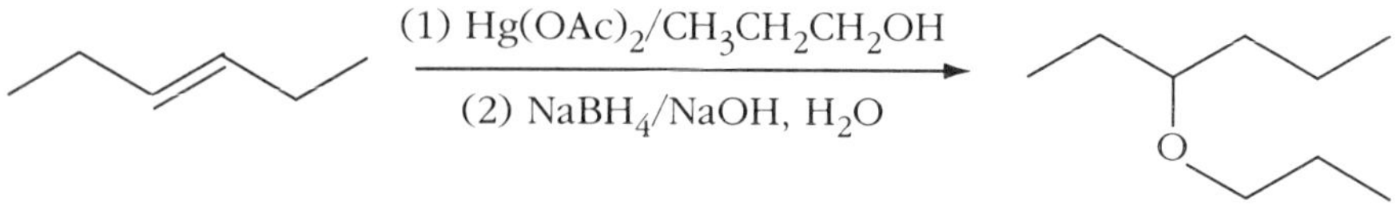

13.10 The student will obtain a mixture of products (2-ethoxyhexane and 2-hexanol) because both oxymercuration and methoxymercuration reactions will occur.

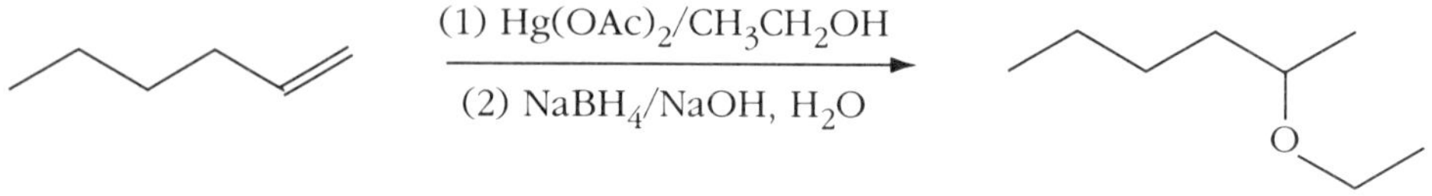

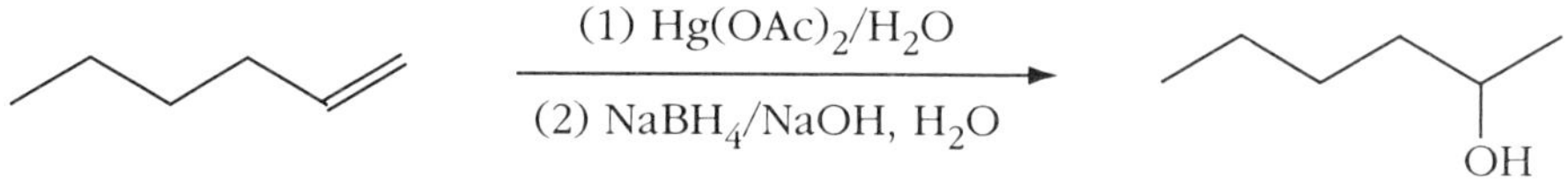

[1-hexene] $\xrightarrow{\text{(1) Hg(OAc)}_2/\text{CH}_3\text{CH}_2\text{OH}}_{\text{(2) NaBH}_4/\text{NaOH, H}_2\text{O}}$ [2-ethoxyhexane]

[1-hexene] $\xrightarrow{\text{(1) Hg(OAc)}_2/\text{H}_2\text{O}}_{\text{(2) NaBH}_4/\text{NaOH, H}_2\text{O}}$ [2-hexanol]

13.11 All the ethers to be synthesized (except Exercise 13.11 f) are unsymmetrical so there are two possible ways of preparing each. One part of an ether molecule must be derived from the alkoxide ion and the other from the haloalkane. Preparation of ethers by the Williamson ether synthesis is most successful when the haloalkane is the one that gives the highest yield of substitution reaction that occurs by an S_N2 mechanism. Methyl halides and primary haloalkanes, therefore, are the best substrates. Secondary and tertiary haloalkanes are poor substrates because they tend to react with alkoxide bases to form elimination products rather than substitution products. Both possible pairs of starting materials are shown below to illustrate the expected outcome in each case.

(a) [n-butanol] $\xrightarrow{\text{Na}}$ [n-BuO⁻ Na⁺] $\xrightarrow{\text{H}_3\text{C—I}}$ [butyl methyl ether]

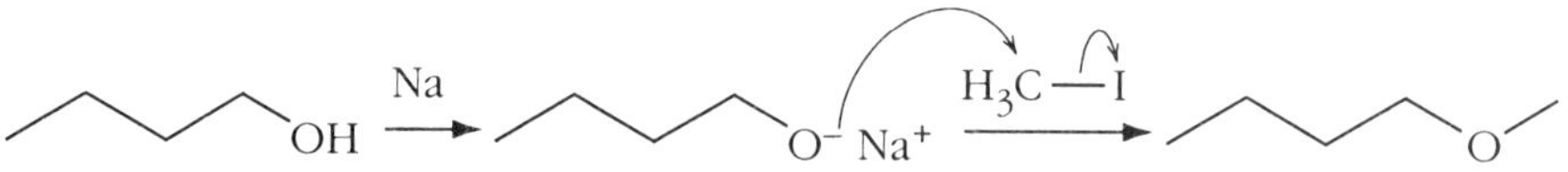

Very good

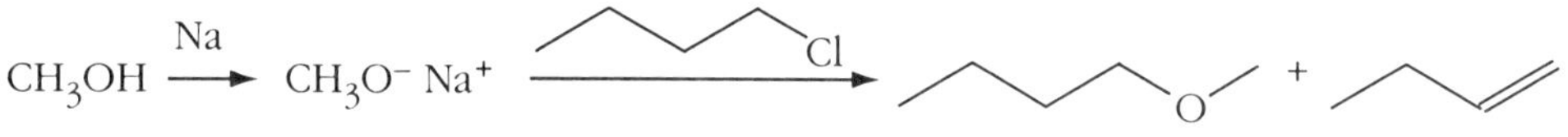

$\text{CH}_3\text{OH} \xrightarrow{\text{Na}} \text{CH}_3\text{O}^-\text{Na}^+ \xrightarrow{\text{[BuCl]}}$ [butyl methyl ether] + [butene]

Poor

(b) HO—[ ]—OH $\xrightarrow{\text{2 Na}}$ Na⁺ ⁻O—[ ]—O⁻ Na⁺ $\xrightarrow{\text{2 CH}_3\text{I}}$ H₃CO—[ ]—OCH₃

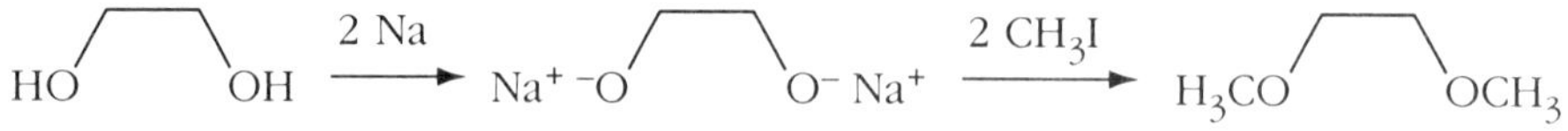

Very good

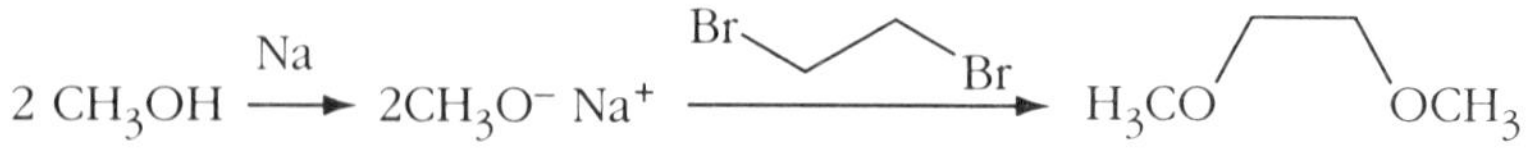

$2\,\text{CH}_3\text{OH} \xrightarrow{\text{Na}} 2\text{CH}_3\text{O}^-\text{Na}^+ \xrightarrow{\text{Br—[ ]—Br}}$ H₃CO—[ ]—OCH₃   Poor

(c)

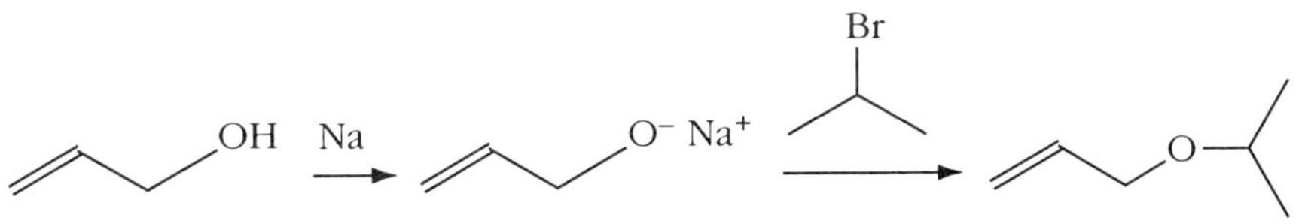

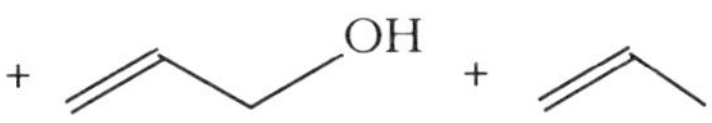

Very poor

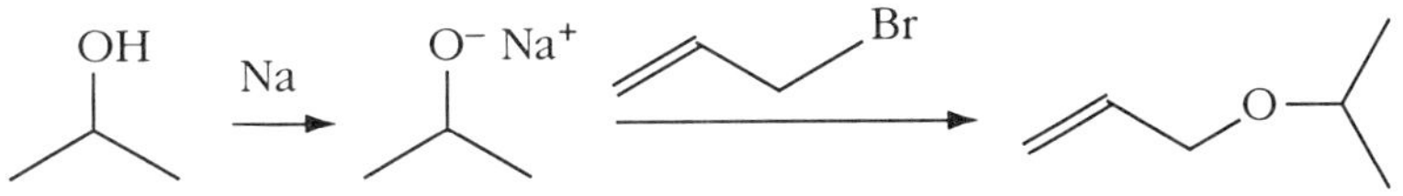

Very good

(d)

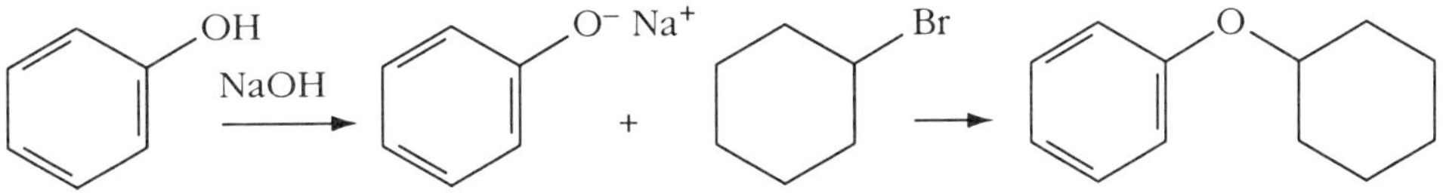

Good

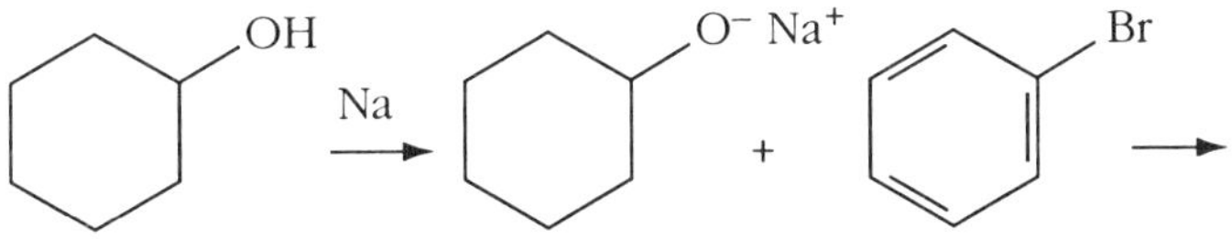

No reaction

(e)

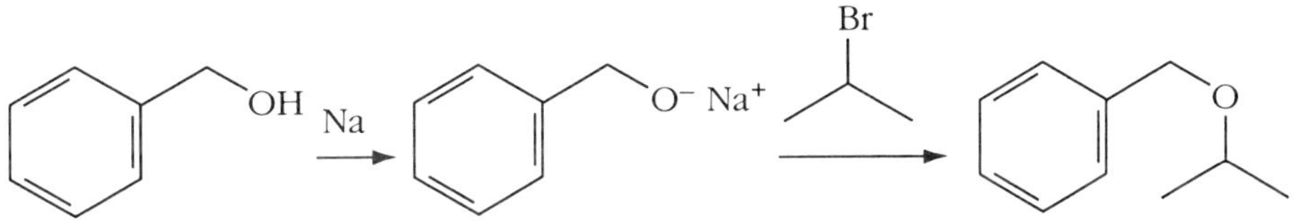

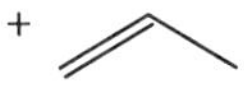

Poor

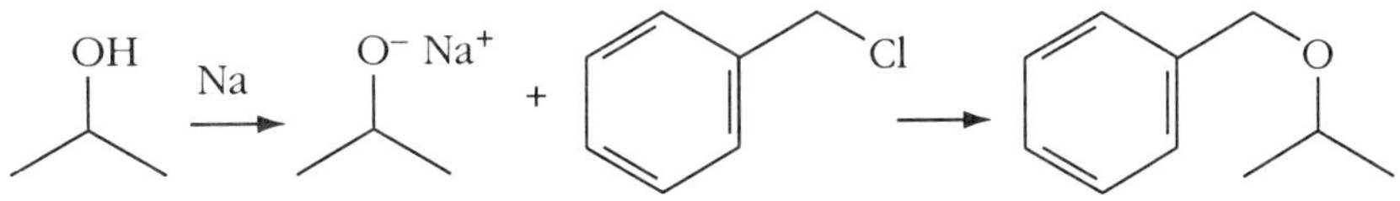

Very good

(f)

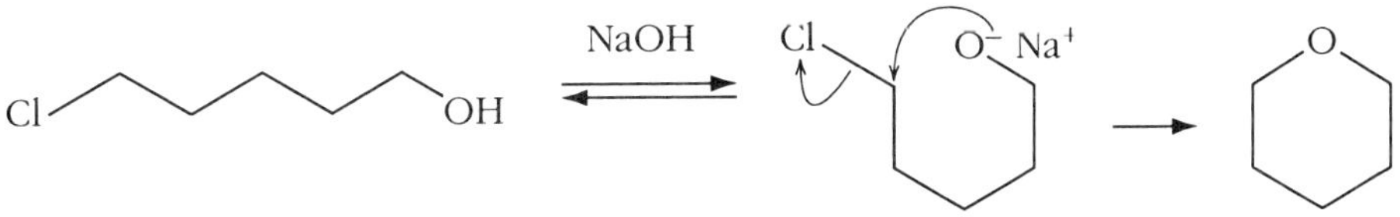

The above is the best method to synthesize tetrahydropyran (oxane). The alternative method, which is not a Williamson synthesis, is the intramolecular acid-catalyzed dehydration of 1,5-pentanediol.

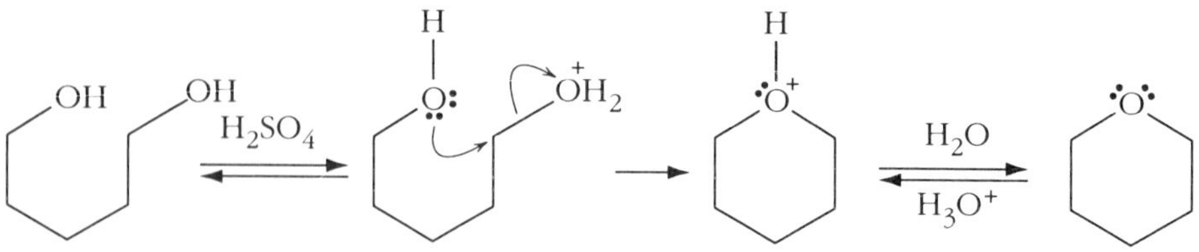

With the latter method, however, a competitive intermolecular reaction gives rise to side-products (see Exercise **13.8**, *Step 3*).

13.12 CH_3CH_2OH + Na^+ ^-OH ⇌ $CH_3CH_2O^-$ Na^+ + H_2O

pK_a 17 pK_a 15.7

The equilibrium of this reaction lies towards ethanol (pK_a 17) because water is a stronger acid (pK_a 15.7). As a result, not much nucleophile (ethoxide) is present in the reaction mixture. Another problem is that under the reaction conditions iodomethane undergoes hydrolysis to produce methanol.

CH_3I + Na^+ ^-OH ⟶ CH_3OH + NaI

In the reaction mixture, therefore, there is very little ethoxide ion and iodomethane is being converted to methanol so it is unlikely that any methoxyethane will be formed.

13.13

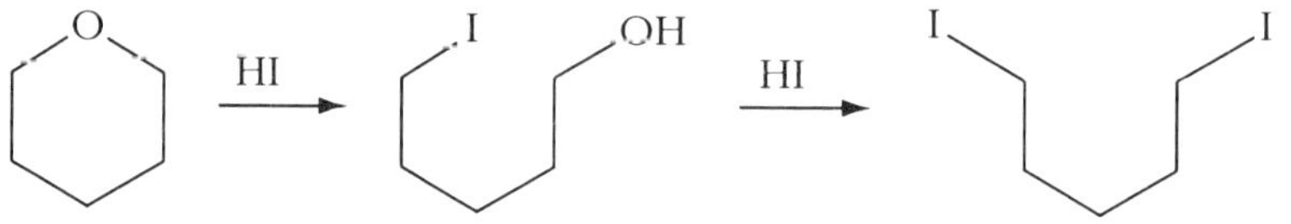

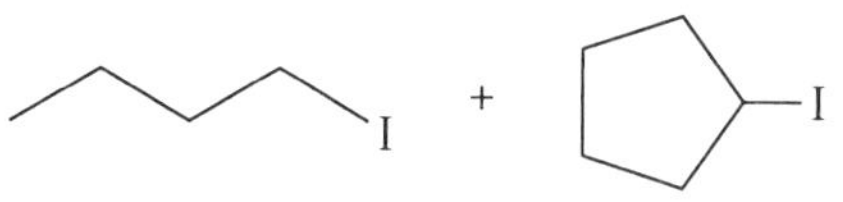

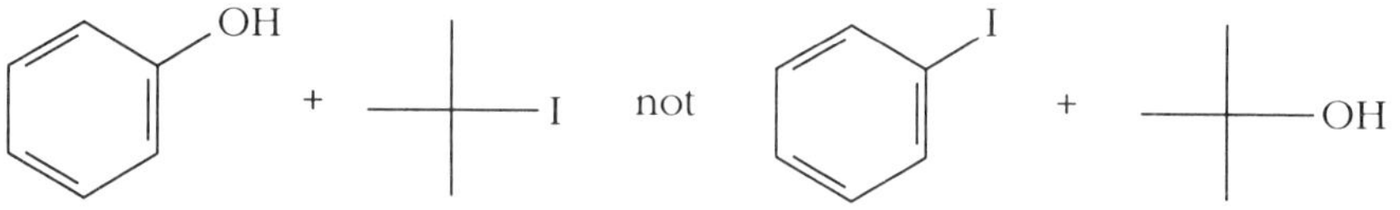

13.14

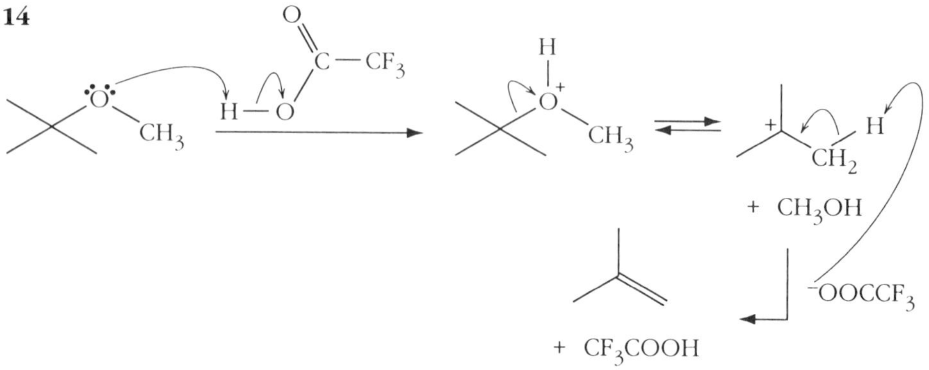

13.15 The first step in the reaction of both these ethers with HCl is a proton transfer to the ether oxygen atom to form an oxonium ion. Protonated diethyl ether is cleaved by a nucleophilic substitution reaction that occurs by an S_N2 mechanism.

Protonated *tert*-butyl ether reacts by a different mechanism. It ionizes by an S_N1 mechanism to form a stable tertiary carbocation as the intermediate.

Diethyl ether reacts very slowly because it is cleaved by a nucleophilic attack of chloride ion onto the oxonium ion of the ether, and we know that chloride ion is a poor nucleophile. *tert*-Butyl ether reacts by a different mechanism (S_N1). A tertiary carbocation is formed as the intermediate, and it is a quite stable carbocation.

13.16

The peroxyacid may react with the double bond from either side giving rise to two enantiomers. There is an equal chance of reaction with either side, so an equal amount of each enantiomer is obtained.

13.17 Epoxidation is a *syn*-addition of an oxygen atom to the double bond of an alkene. The substituents that are *cis* in the alkene are also *cis* in the oxirane. The epoxidation of (Z)-2-butene, therefore, forms *cis*-1,2-dimethyloxirane.

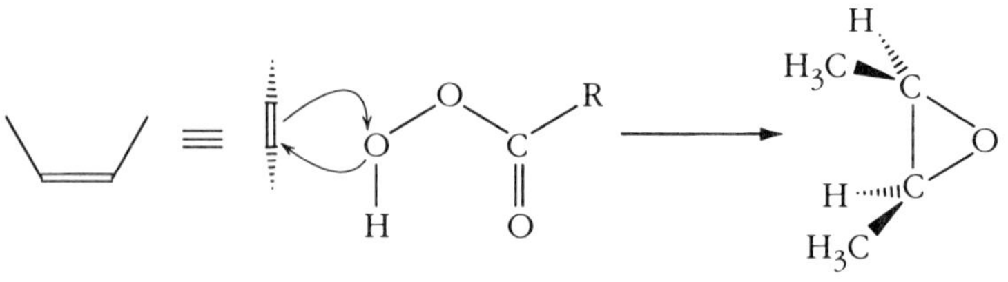

Reaction of an aqueous chlorine solution and an alkene forms a chlorohydrin in which the chlorine atoms and —OH groups are added by *anti*-addition.

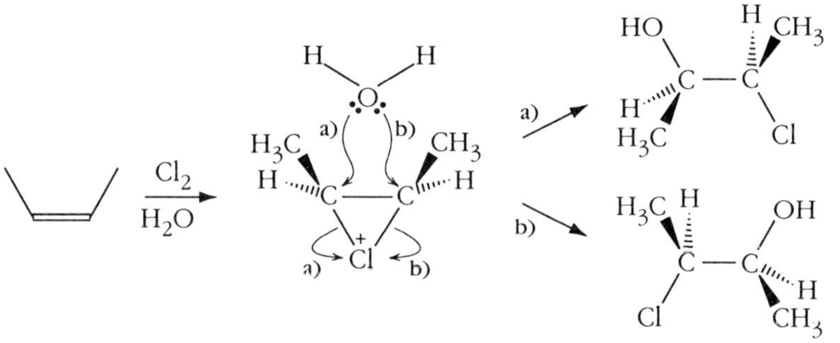

The chlorine atoms and —OH groups are correctly placed to undergo an internal Williamson ether synthesis to form *cis*-1,2-dimethyloxirane, which is the same product as the epoxidation reaction.

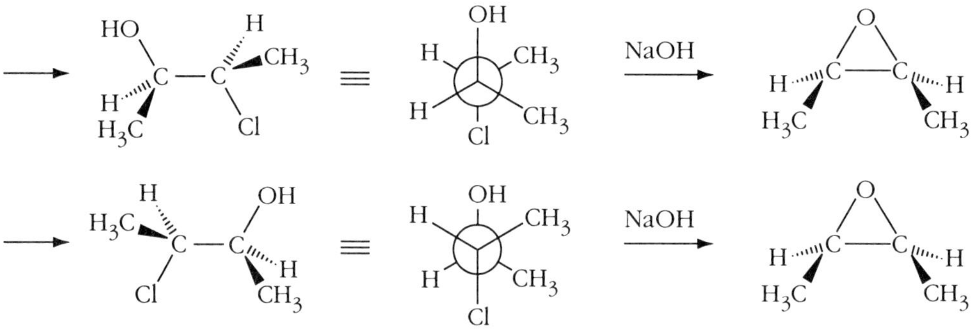

13.18

(a)

(b)

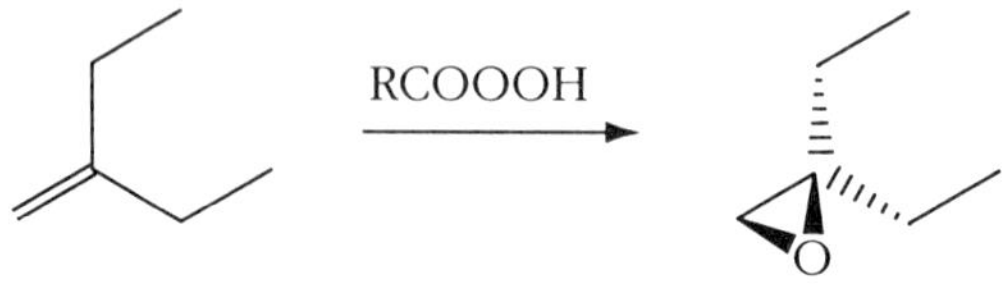

(c) 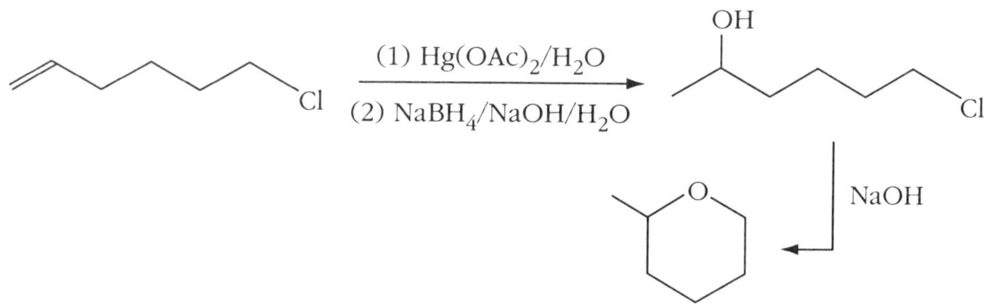 ![hexenyl chloride] $\xrightarrow[(2)\ NaBH_4/NaOH/H_2O]{(1)\ Hg(OAc)_2/H_2O}$ 6-chloro-2-hexanol $\xrightarrow{NaOH}$ 2-methyltetrahydropyran

(d) 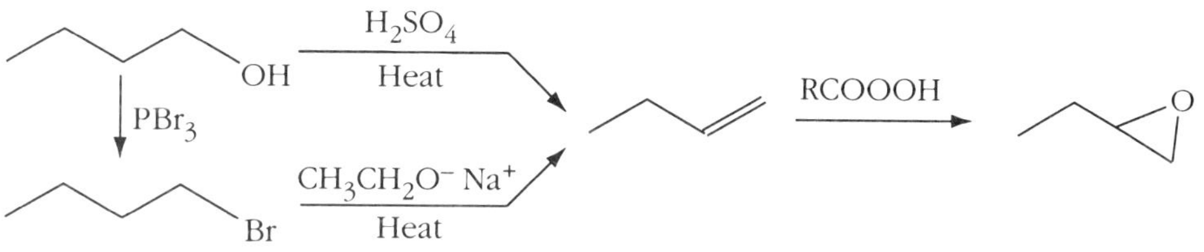 CH₃CH₂CH(OH)CH₂ ... $\xrightarrow{H_2SO_4,\ Heat}$ allyl-type alkene $\xrightarrow{RCOOOH}$ epoxide

with PBr₃ → alkyl bromide $\xrightarrow{CH_3CH_2O^-\ Na^+,\ Heat}$ same alkene

13.19

(a) 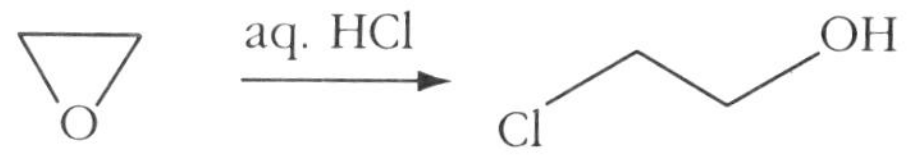 epoxide $\xrightarrow{aq.\ HCl}$ ClCH₂CH₂OH

(b) 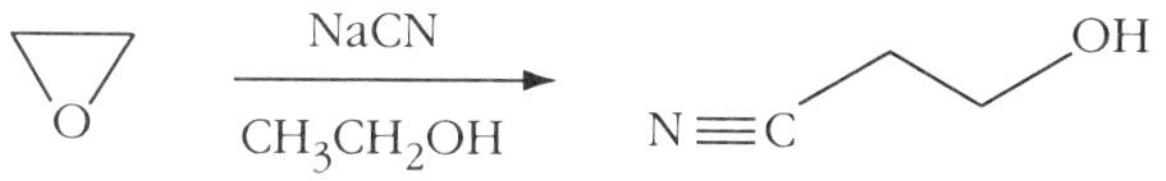 epoxide $\xrightarrow[CH_3CH_2OH]{NaCN}$ N≡C–CH₂CH₂OH

(c) epoxide $\xrightarrow[H_2O]{NaOH}$ HOCH₂CH₂OH

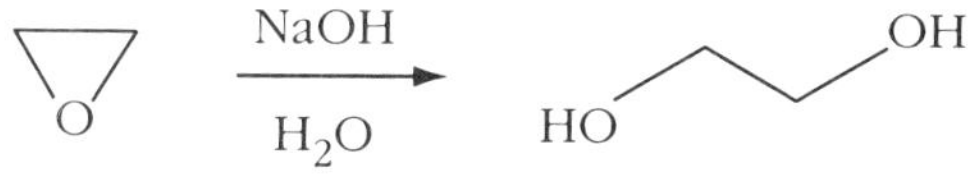

(d) 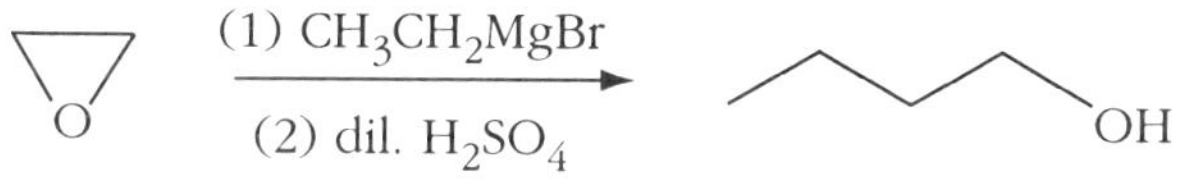 epoxide $\xrightarrow[(2)\ dil.\ H_2SO_4]{(1)\ CH_3CH_2MgBr}$ CH₃CH₂CH₂CH₂OH

(e) 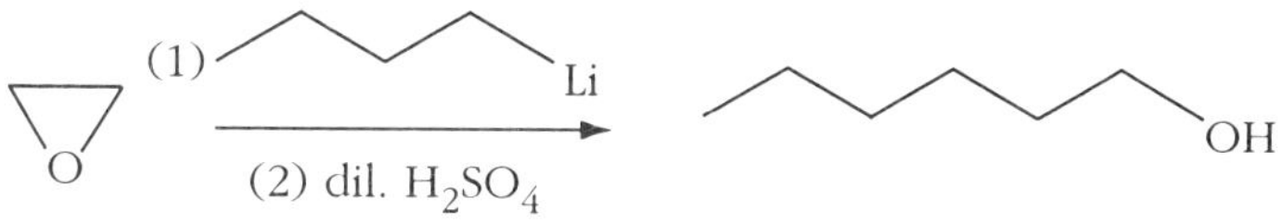 epoxide $\xrightarrow[(2)\ dil.\ H_2SO_4]{(1)\ CH_3CH_2CH_2CH_2Li}$ CH₃(CH₂)₅OH

13.20

(a) 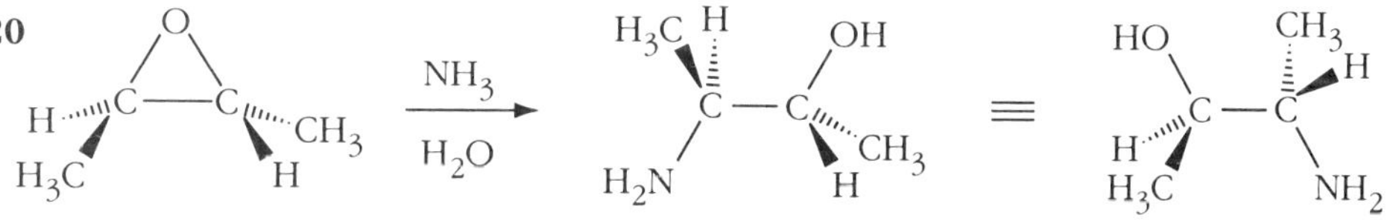 (2R,3S)-2,3-dimethyloxirane $\xrightarrow[H_2O]{NH_3}$ amino alcohol ≡ rearranged structure

(b) 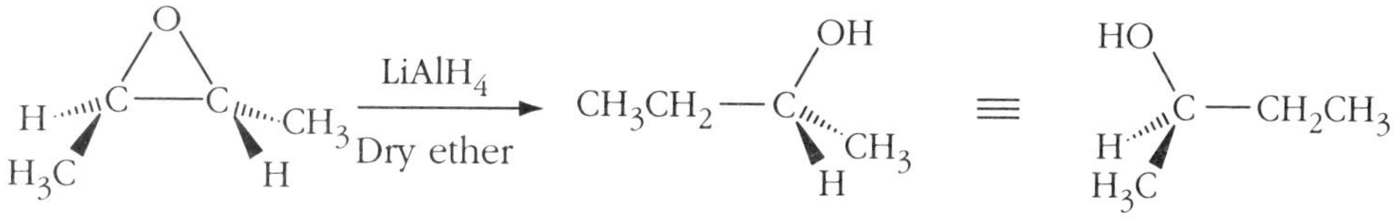 (2R,3S)-2,3-dimethyloxirane $\xrightarrow[Dry\ ether]{LiAlH_4}$ CH₃CH₂–C(OH)(CH₃)H ≡ rearranged structure

(c) 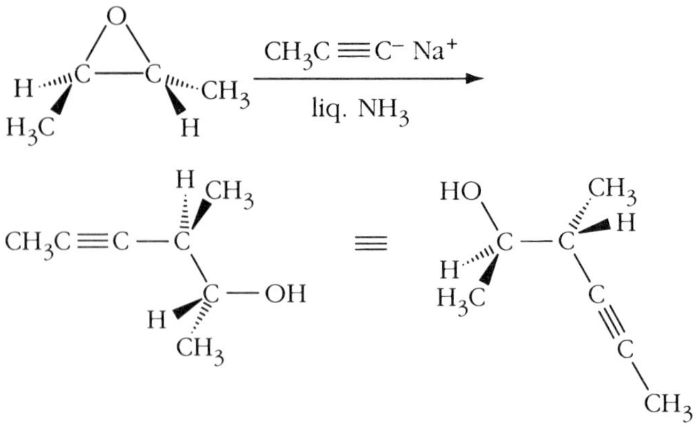

13.21 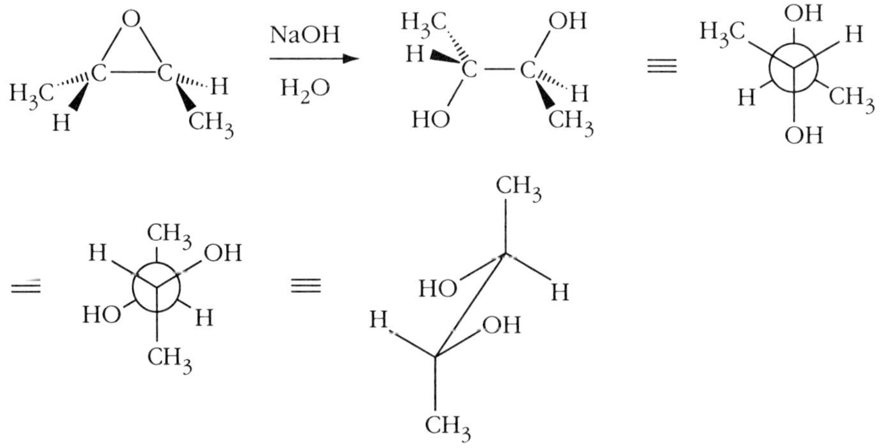

Compound **B** is identical with the product of the reaction of (2*S*,3*S*)-*trans*-2,3-dimethyloxirane with an aqueous NaOH solution.

13.22

(a) 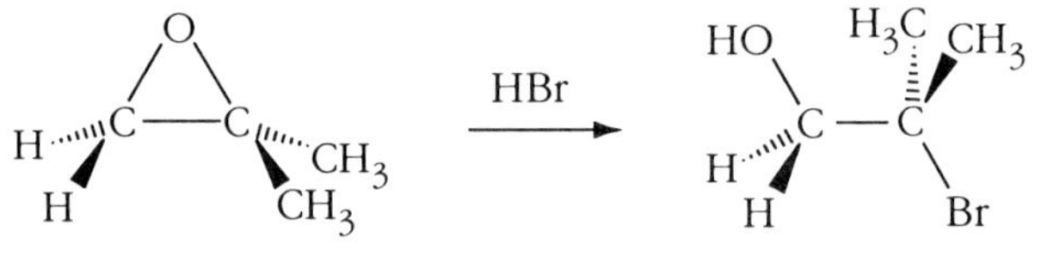

(b) 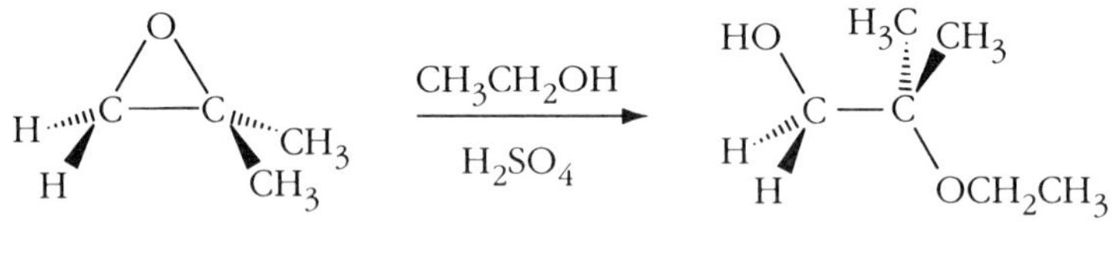

(c) 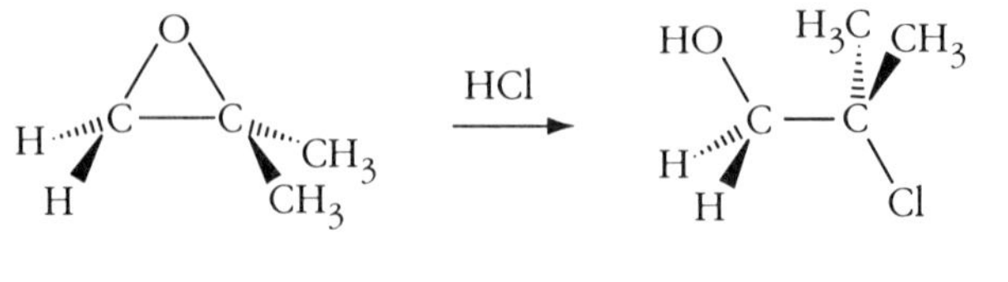

13.23 (a) The compound to be prepared has both an internal carbon-carbon triple bond and a hydroxyl group as functional groups. Internal alkynes can be prepared from terminal alkynes. Retrosynthetic analysis provides the following synthesis starting with an epoxide:

$$CH_3CH_2C \equiv C \dashv CH_2CH_2OH \implies CH_3CH_2C \equiv C^- + H_2C\underset{O}{-}CH_2$$

This reaction of the acetylide ion with oxirane introduces both the triple bond and the hydroxy functional groups in the molecule at the same time.
The complete synthesis is:

$$\text{(epoxide)} \xrightarrow[\text{(2) } H_3O^+]{\text{(1) } CH_3CH_2C \equiv C^-\ Na^+/\text{liq. } NH_3} CH_3CH_2C \equiv C-CH_2CH_2OH$$

(b) The product contains a chlorine atom and a hydroxyl group on adjacent carbon atoms as functional groups. The best way to introduce these two functional groups is by the addition of HCl to an epoxide. In the product, the chlorine atom is bonded to the more substituted carbon atom, which is the product expected by the addition of HCl to 2,2-dimethyloxirane. The one-step synthesis is:

$$\text{2,2-dimethyloxirane} \xrightarrow{HCl} \text{HO-CH}_2\text{-C(CH}_3)_2\text{Cl}$$

(c) A primary alcohol can be formed by reaction of oxirane with a Grignard or lithium reagent, so the one-step synthesis is:

$$\text{oxirane} \xrightarrow[\text{(2) } H_3O^+]{\text{(1) } CH_3(CH_2)_6MgBr} CH_3(CH_2)_7CH_2OH$$

(d) Hydrolysis of an epoxide under either acidic or basic condition opens the epoxide ring of *trans*-2,3-diethyloxirane to form the desired product:

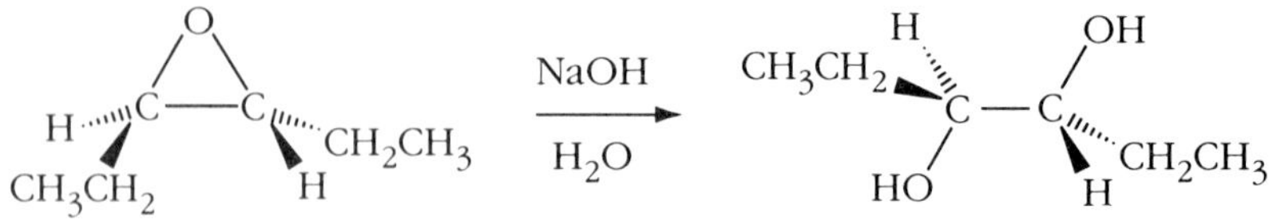

13.24

(a) 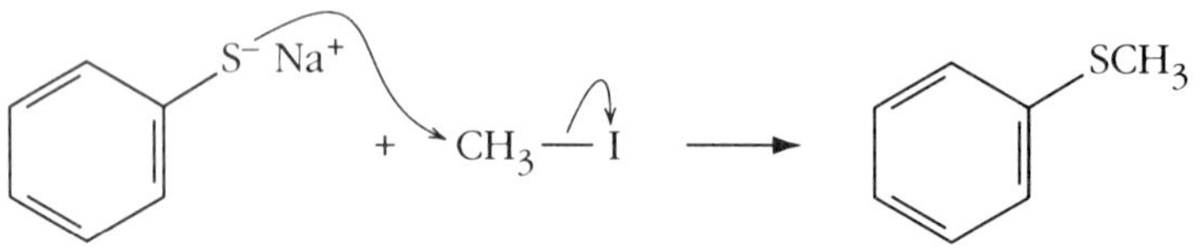 [PhS⁻Na⁺ + CH₃—I → PhSCH₃]

(b) 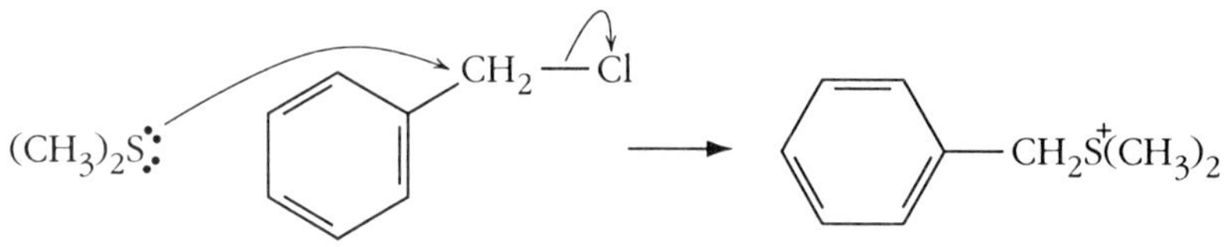 (CH₃)₂S: + PhCH₂—Cl → PhCH₂S⁺(CH₃)₂

(c) 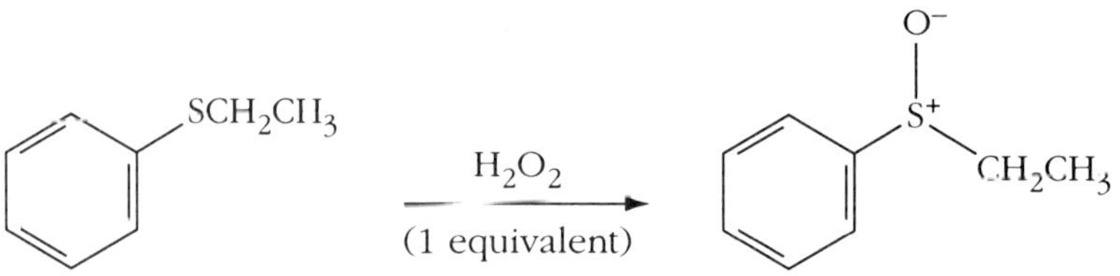 PhSCH₂CH₃ + H₂O₂ (1 equivalent) → PhS(=O)CH₂CH₃ (sulfoxide)

(d) (CH₃)₂S⁺CH₂C(CH₃)₃ I⁻ + :NH₃ ⟶ CH₃—S—CH₂C(CH₃)₃ + CH₃N⁺H₃ I⁻

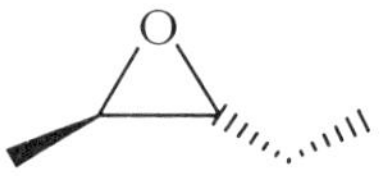

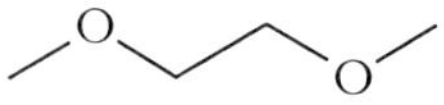

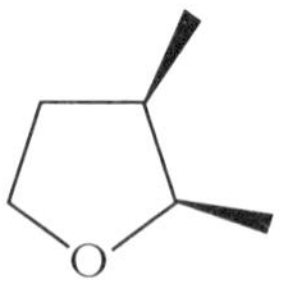

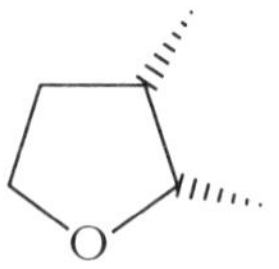

13.25

(a), 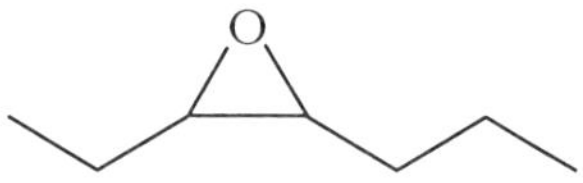 (b), 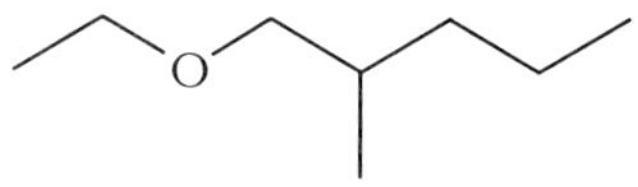 (c), 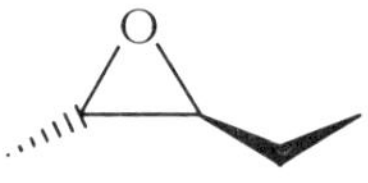 (d), (e), (f) 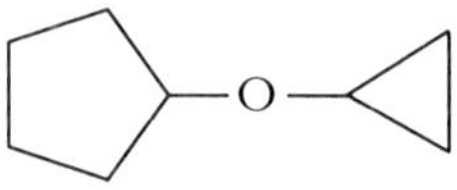 structures shown

13.26
(a) *cis*-1-Isopropoxy-4-*tert*-butoxycyclohexane
(b) (2S,4R)-4-Ethyl-2-methyltetrahydropyran [or (2S,4R)-4-ethyl-2-methyloxane]
(c) (R)-2-Ethoxy-1-methoxy-2-methylbutane
(d) (R)-2,3-Epoxy-2-methyldecane
(e) 3-Ethoxyoxetane

13.27

(a) epoxide + H₂O / Heat → HOCH₂CH₂OH

(b) epoxide + liq. NH₃ → H₂N-CH₂CH₂-OH

(c) epoxide + NaCN / CH₃CH₂OH → N≡C-CH₂CH₂-OH

(d) epoxide + NaN₃ / CH₃CH₂OH → N₃-CH₂CH₂-OH

(e) epoxide + (1) CH₃C≡C⁻ Na⁺ (2) dil. H₂SO₄ → CH₃C≡C-CH₂CH₂-OH

(f) epoxide + (1) CH₃CH₂MgI (2) dil. H₂SO₄ → CH₃CH₂CH₂CH₂OH

(g) epoxide + (1) CH₃S⁻ Na⁺/CH₃CH₂OH (2) dil. H₂SO₄ → CH₃S-CH₂CH₂-OH

13.28

(a) 2,2-dimethyloxirane + (1) LiAlH₄/dry ether (2) H₃O⁺ → (CH₃)₂C(OH)CH₃

(b) 2,2-dimethyloxirane + (1) 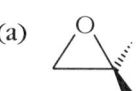MgBr/dry ether (2) H₃O⁺ → isobutyl-C(CH₃)₂OH

(c) 2,2-dimethyloxirane + (1) NaNH₂/liq. NH₃ (2) H₃O⁺ → H₂N-CH₂-C(CH₃)₂-OH

(d) 2,2-dimethyloxirane + 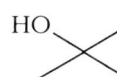S⁻ Na⁺ / CH₃CH₂OH → CH₃CH₂-S-CH₂-C(CH₃)₂-OH

(e) 2,2-dimethyloxirane + H₃O⁺ → HO-CH₂-C(CH₃)₂-OH

(f) 2,2-dimethyloxirane + (1) HC≡C⁻ Na⁺ (2) H₃O⁺ → HC≡C-CH₂-C(CH₃)₂-OH

(g) 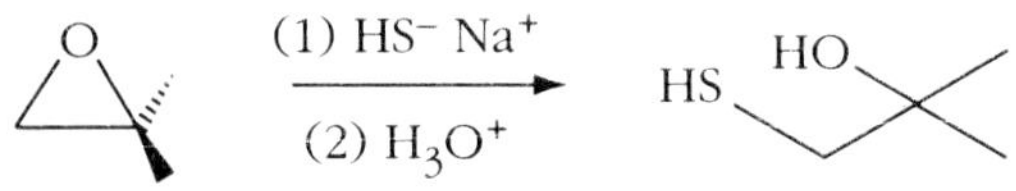 [epoxide with CH₃] $\xrightarrow[(2)\ H_3O^+]{(1)\ HS^-\ Na^+}$ HS—C(CH₃)₂—OH (shown as HS-CH₂-C(CH₃)₂-OH)

13.29

(a) 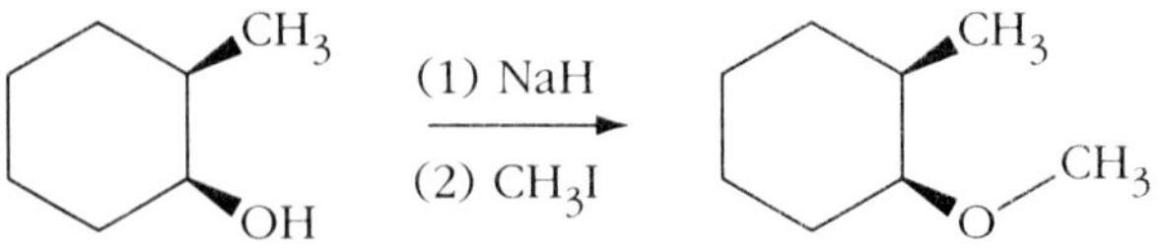 [cis-2-methylcyclohexanol] $\xrightarrow[(2)\ CH_3I]{(1)\ NaH}$ [cis-2-methylcyclohexyl methyl ether]

The reaction occurs with retention of configuration because the carbon-oxygen bond is not broken in the reaction.

(b) 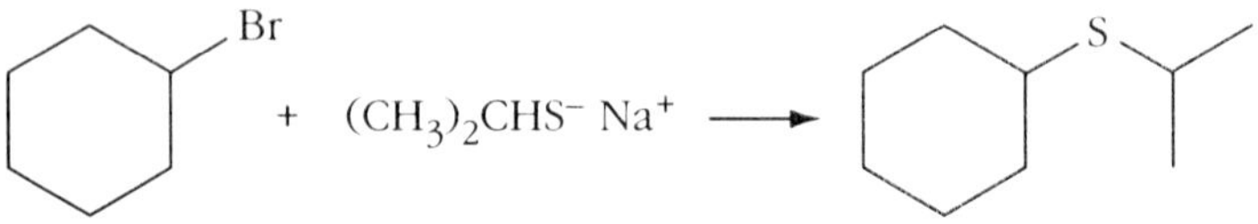 [cyclohexyl bromide] + $(CH_3)_2CHS^-\ Na^+$ ⟶ [cyclohexyl isopropyl sulfide]

(c) 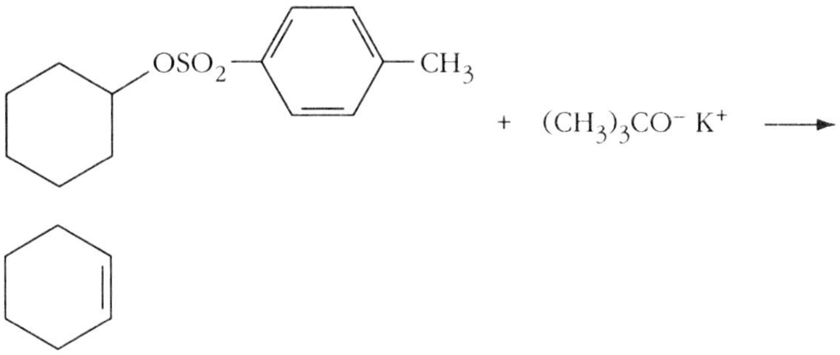 [cyclohexyl tosylate] + $(CH_3)_3CO^-\ K^+$ ⟶ [cyclohexene]

(d) 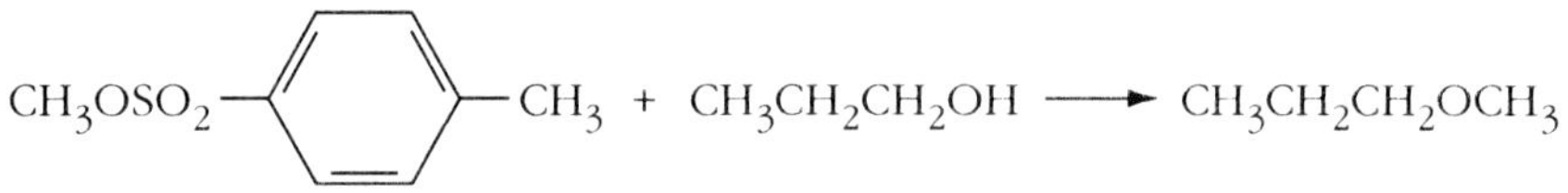 CH_3OSO_2—C₆H₄—CH_3 + $CH_3CH_2CH_2OH$ ⟶ $CH_3CH_2CH_2OCH_3$

(e) 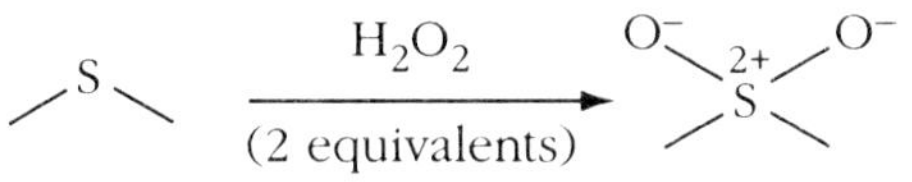 CH_3–S–CH_3 $\xrightarrow[\text{(2 equivalents)}]{H_2O_2}$ $(CH_3)_2S(O^-)_2^{2+}$

(f) $[(CH_3)_3CS(CH_3)_2]^+\ I^-$ + CH_3CH_2OH ⟶ $CH_3CH_2OCH_3$

13.30

(a) 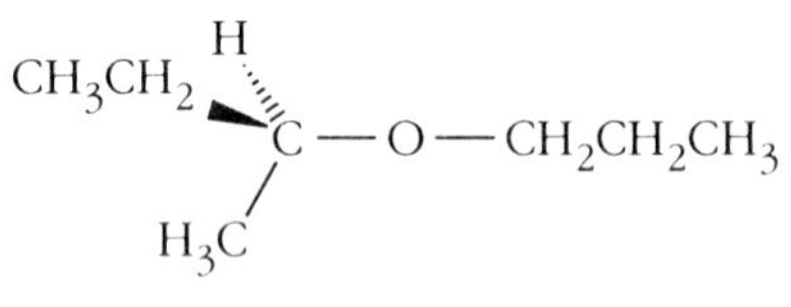 CH_3CH_2–C(H)(CH₃)–O–$CH_2CH_2CH_3$

(b) H_3C–C(H)(C≡N)–C(OH)(CH₃)–CH_3

(c) CH_3CH_2CH—CH_2 (epoxide with O)

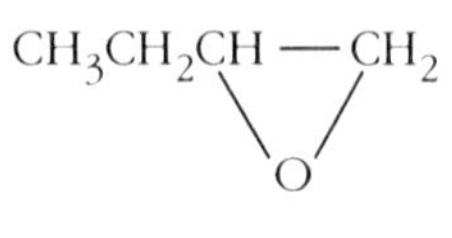

(d) 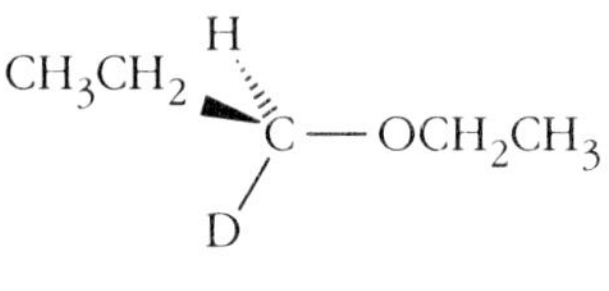 CH_3CH_2–C(H)(D)–OCH_2CH_3

(e) 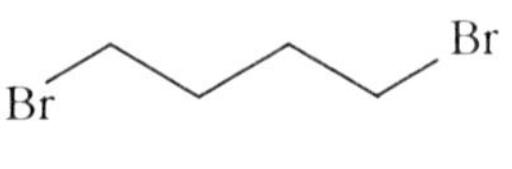 Br—CH₂CH₂CH₂CH₂—Br

(f) 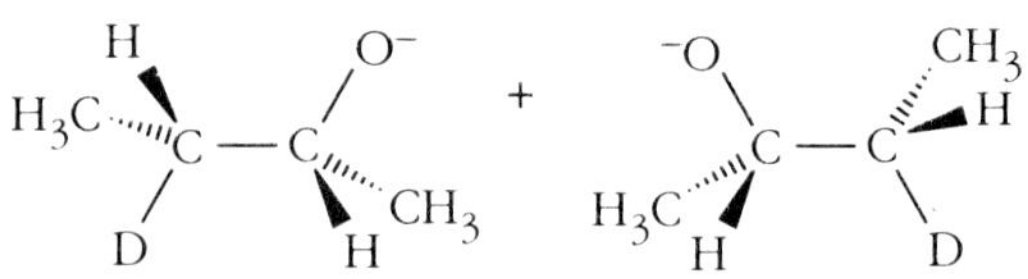 H_3C–C(H)(D)–C(O⁻)(CH₃)(H) + ⁻O–C(CH₃)(H)–C(H₃C)(H)(D)

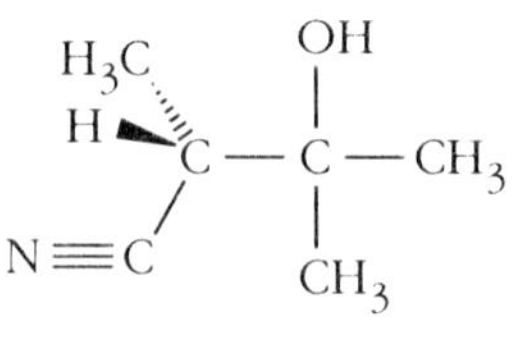

13.31

(a) cyclohexene → (1) OsO₄ (2) H₂O, NaHSO₃ → cis-cyclohexane-1,2-diol (both OH on wedges)

(b) cyclohexene → (1) Hg(OAc)₂, CH₃OH (2) NaBH₄/NaOH/H₂O → methoxycyclohexane

The alternative synthesis is also possible but it is not as good as the preceding one because methoxide ion will form some elimination product.

cyclohexene → (1) 9-BBN (2) H₂O₂/NaOH/H₂O → cyclohexanol → (1) NaH (2) CH₃I → methoxycyclohexane

(c) cyclohexene → m-chloroperoxybenzoic acid (Cl-C₆H₄-COOOH) → cyclohexene oxide → CH₃O⁻ K⁺ / CH₃OH →

trans-2-methoxycyclohexanol (OH wedge, OCH₃ dash) + trans-2-methoxycyclohexanol (OCH₃ wedge, OH dash)

(d) cyclohexene → Br₂ / H₂O → trans-2-bromocyclohexanol (Br wedge, OH dash) + trans-2-bromocyclohexanol (Br wedge, OH dash — enantiomer)

(e) cyclohexene → HBr → bromocyclohexane → CH₃S⁻ Na⁺ → (methylthio)cyclohexane

(f) cyclohexene → m-chloroperoxybenzoic acid (Cl-C₆H₄-COOOH) → cyclohexene oxide → H₃O⁺ → trans-cyclohexane-1,2-diol (OH wedge, OH dash)

13.32

$$\underset{H_3C}{\overset{H}{>}}C=C\underset{H}{\overset{CH_3}{<}} \xrightarrow{\text{Cl}\underset{}{\bigcirc}\text{COOOH}} \underset{H_3C}{\overset{H}{\cdots}}C\underset{}{\overset{CH_3}{\underset{}{\diagdown}}}\underset{}{\overset{}{\diagup}}O$$

$$\underset{H_3C}{\overset{H}{>}}C=C\underset{H}{\overset{CH_3}{<}} \xrightarrow[\text{H}_2\text{O}]{\text{Cl}_2} \underset{H_3C\ \ \ \text{OH}}{\overset{\text{Cl}\ \ \ \ \text{CH}_3}{\underset{H\cdots C}{\overset{C\blacktriangleleft H}{|}}}} \xrightarrow[\text{H}_2\text{O}]{\text{NaOH}} \underset{H_3C\ \ \ \text{O}^-}{\overset{\text{Cl}\ \underset{\curvearrowleft}{\ }\ \text{CH}_3}{\underset{H\cdots C}{\overset{C\blacktriangleleft H}{|}}}} \longrightarrow \underset{H_3C}{\overset{H\ \ \ \ \text{CH}_3}{\underset{}{\diagdown\diagup}}}\underset{}{\overset{}{\diagup}}O$$

13.33 When you are asked to propose a synthesis that requires two or more steps and a reaction sequence is not immediately evident, it is best to reason backward from the target molecule to suitable starting materials. This procedure is known as *retrosynthetic analysis*. A symbol used to indicate a retrosynthetic step is an open arrow written from product to suitable precursors or fragments of those precursors. The answer is then written forwards with all reaction conditions, solvents, etc.

(a) First we have to look for the sequence of as many atoms as possible of the starting material in the product. Once we have found this sequence, we must then decide what reactions or series of reactions the starting material must undergo to form the desired product.

Retrosynthesis:

$$CH_3 \dashv O-CH_2\underset{\underset{CH_3}{|}}{\overset{\overset{CH_3}{|}}{C}}-O\dashv CH_3 \Longrightarrow CH_3OCH_2\underset{\underset{CH_3}{|}}{\overset{\overset{CH_3}{|}}{C}}-O^- + CH_3I$$

$$CH_3\dashv O-CH_2\underset{\underset{CH_3}{|}}{\overset{\overset{CH_3}{|}}{C}}-O^- \Longrightarrow H_2C\overset{O}{\overset{\diagdown\diagup}{-}}C(CH_3)_2 + CH_3O^-$$

The complete synthesis is:

$$\overset{O}{\underset{\triangle}{\ }} \xrightarrow[\text{CH}_3\text{OH}]{\text{CH}_3\text{O}^-\ \text{Na}^+} \underset{}{\overset{}{\diagup O\diagdown\diagup}}\text{OH} \xrightarrow[(2)\ \text{CH}_3\text{I}]{(1)\ \text{NaH}} \underset{}{\overset{}{\diagup O\diagdown\diagup}}\text{OCH}_3$$

(b) Retrosynthesis:

$$(CH_3)_2CH\dashv CH_2\underset{\underset{CH_3}{|}}{\overset{\overset{CH_3}{|}}{C}}-O\overset{\times}{\underset{H}{\ }} \Longrightarrow (CH_3)_2CH^-\ Li^+ + H_2C\overset{O}{\overset{\diagdown\diagup}{-}}C(CH_3)_2$$

The complete synthesis is:

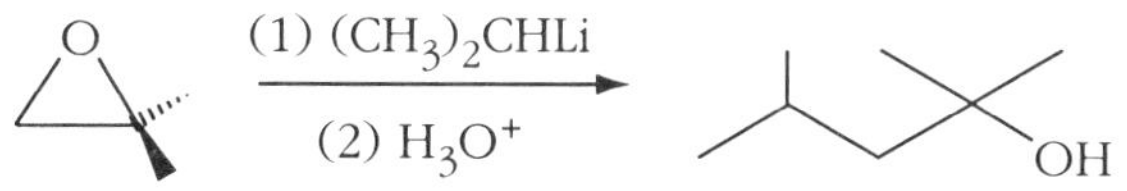

$$\text{(epoxide with CH}_3\text{)} \xrightarrow[\text{(2) H}_3\text{O}^+]{\text{(1) (CH}_3)_2\text{CHLi}} \text{(CH}_3)_2\text{CHCH}_2\text{C(CH}_3)_2\text{OH}$$

(c) Retrosynthesis:

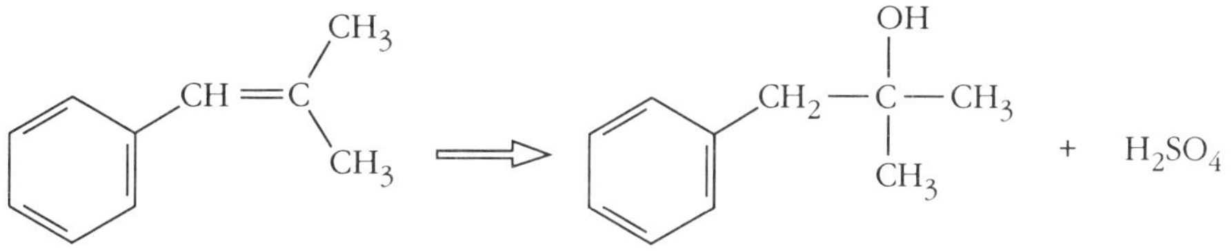

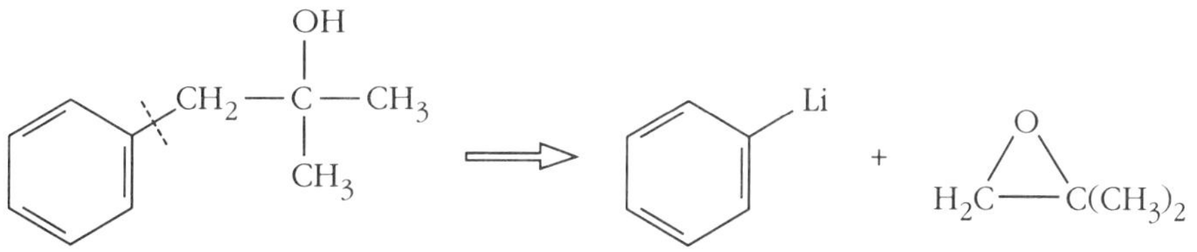

PhCH=C(CH₃)₂ ⟹ PhCH₂—C(OH)(CH₃)—CH₃ + H₂SO₄

PhCH₂⫶—C(OH)(CH₃)—CH₃ ⟹ PhLi + H₂C—C(CH₃)₂ epoxide

The complete synthesis:

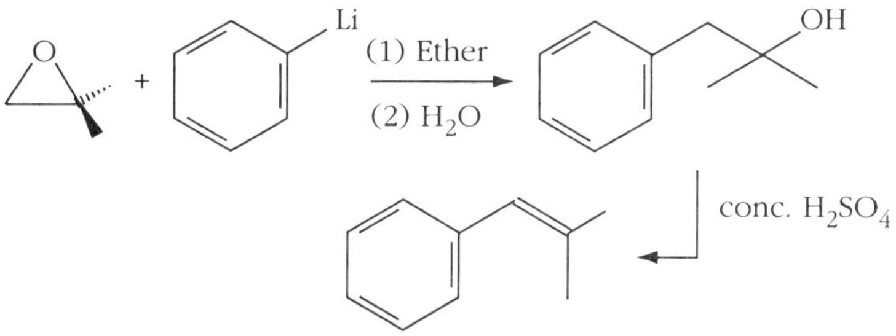

epoxide + PhLi $\xrightarrow[\text{(2) H}_2\text{O}]{\text{(1) Ether}}$ PhCH₂C(OH)(CH₃)₂ $\xrightarrow{\text{conc. H}_2\text{SO}_4}$ PhCH=C(CH₃)₂

(d) 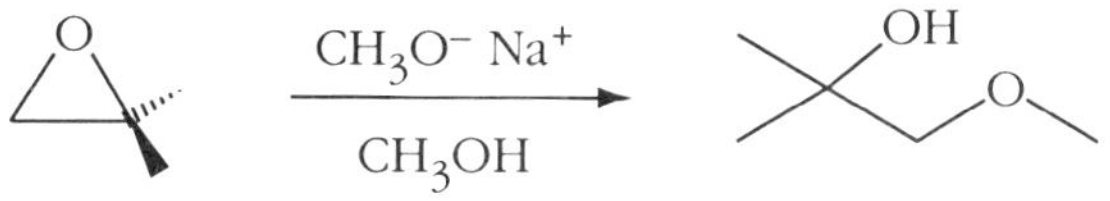 epoxide $\xrightarrow[\text{CH}_3\text{OH}]{\text{CH}_3\text{O}^-\text{Na}^+}$ (CH₃)₂C(OH)CH₂OCH₃

(e) 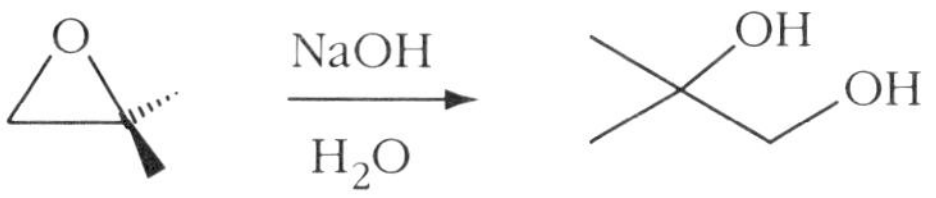 epoxide $\xrightarrow[\text{H}_2\text{O}]{\text{NaOH}}$ (CH₃)₂C(OH)CH₂OH

(f) 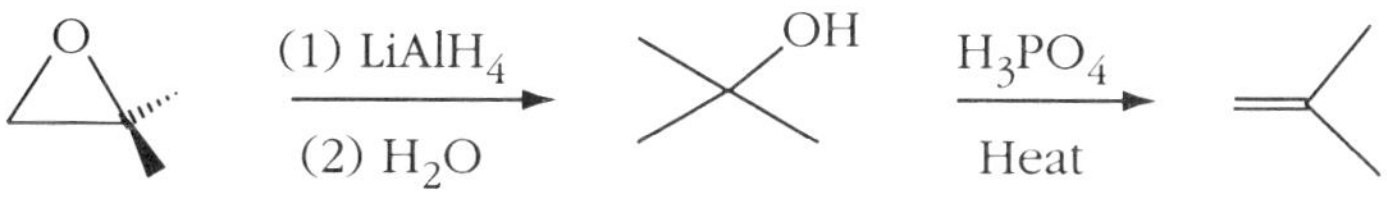 epoxide $\xrightarrow[\text{(2) H}_2\text{O}]{\text{(1) LiAlH}_4}$ (CH₃)₃COH $\xrightarrow[\text{Heat}]{\text{H}_3\text{PO}_4}$ (CH₃)₂C=CH₂

13.34

[Reaction mechanism showing 4-bromo-1-butanol formation and cyclization to tetrahydrofuran via Na⁺ ⁻OH / H₂O, proceeding through alkoxide intermediate to form the cyclic ether.]

13.35

[Mechanism showing (S)-2-methoxybutane reacting with H—Br to form protonated ether intermediate, which then gives BrCH₃ + (S)-2-butanol.]

The configuration of the alcohol [(S)-2-butanol] is the same as that of the ether because the (C2)- carbon-oxygen bond is not broken during the reaction. Bromide ion is a good nucleophile and the substitution is going to be much faster on the methyl carbon as compared with the secondary carbon of the 2-butyl group.

13.36 When you are asked to propose a synthesis that requires two or more steps and a reaction sequence is not immediately evident, it is best to reason backward from the target molecule to suitable starting materials. This procedure is known as *retrosynthetic analysis*. A symbol used to indicate a retrosynthetic step is an open arrow written from product to suitable precursors or fragments of those precursors. The answer is then written forwards with all reaction conditions, solvents, etc.

(a) Retrosynthesis: We want to break the target molecule into four-carbon (or smaller) pieces in such a way that it will require as few synthetic steps as possible. In this case, we can break the molecule of (Z)-3-penten-1-ol into two pieces in two ways:

[Retrosynthetic scheme: (Z)-3-penten-1-ol ⇒ propyne + ethylene oxide (via propynyl sodium + oxirane); alternatively (Z)-3-penten-1-ol ⇒ (Z)-1-chloromagnesio-2-butene + formaldehyde (H₂C=O).]

Complete syntheses:

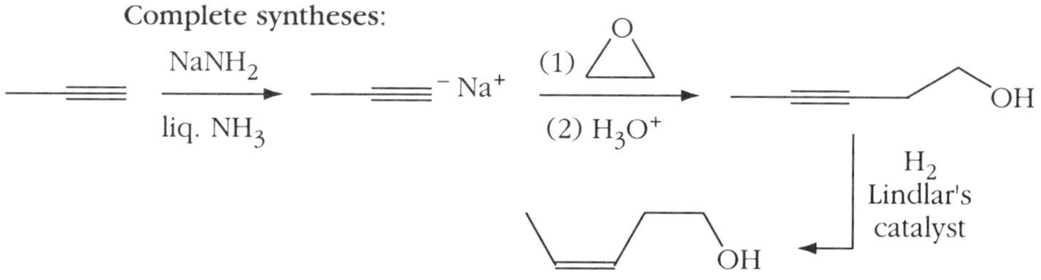

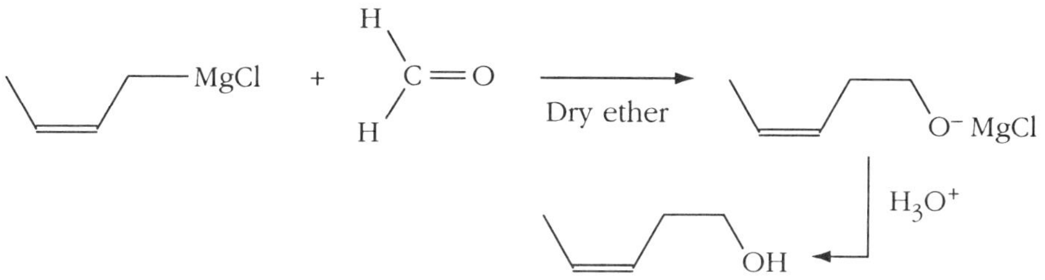

$$\text{HC}\equiv\text{CH} \xrightarrow[\text{liq. NH}_3]{\text{NaNH}_2} \text{HC}\equiv\text{C}^-\text{Na}^+ \xrightarrow[\text{(2) H}_3\text{O}^+]{\text{(1) ethylene oxide}} \text{HC}\equiv\text{C-CH}_2\text{CH}_2\text{OH}$$

$$\xrightarrow[\text{Lindlar's catalyst}]{\text{H}_2} \text{cis-CH}_2=\text{CH-CH}_2\text{CH}_2\text{OH}$$

$$\text{CH}_2=\text{CH-CH}_2-\text{MgCl} + \text{HCHO} \xrightarrow{\text{Dry ether}} \text{CH}_2=\text{CH-CH}_2\text{CH}_2\text{O}^-\text{MgCl}$$

$$\xrightarrow{\text{H}_3\text{O}^+} \text{CH}_2=\text{CH-CH}_2\text{CH}_2\text{OH}$$

The second synthetic pathway is much more efficient because it requires fewer steps.

(b) Retrosynthesis:

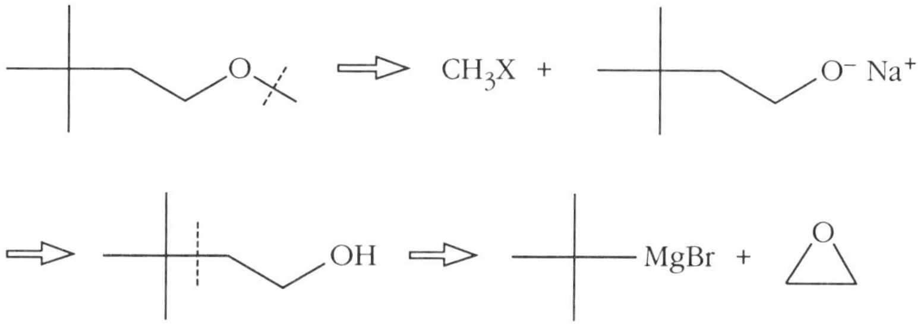

[retrosynthesis diagram: t-Bu-CH2CH2-O-CH3 ⟹ CH3X + t-Bu-CH2CH2-O⁻ Na⁺ ⟹ t-Bu-CH2CH2-OH ⟹ t-Bu-MgBr + ethylene oxide]

Complete synthesis:

$$\text{(CH}_3)_3\text{C-Br} \xrightarrow[\text{Dry ether}]{\text{Mg}} \text{(CH}_3)_3\text{C-MgBr} \xrightarrow{\text{ethylene oxide}}$$

$$\text{(CH}_3)_3\text{C-CH}_2\text{CH}_2\text{OMgBr} \xrightarrow{\text{CH}_3\text{I}} \text{(CH}_3)_3\text{C-CH}_2\text{CH}_2\text{OCH}_3$$

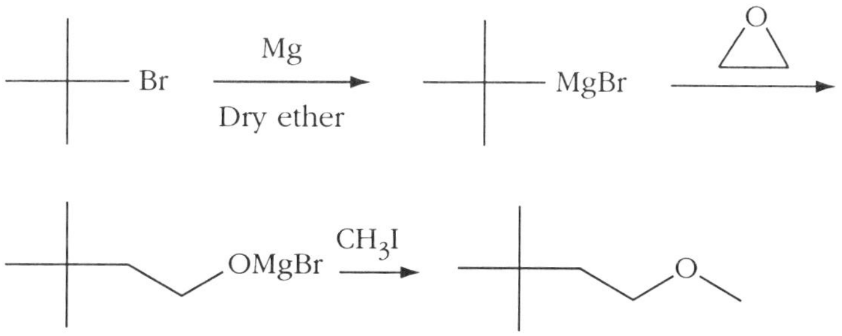

(c) *cis*-2,3-Epoxyheptane has seven carbon atoms, so it can be disconnected between C4 and C5 or between C3 and C4. The disconnection between C4 and C5 does not look promising because we do not know any reaction that could connect the two pieces again. The other direction looks better.

Retrosynthesis:

$$\underset{H_3C}{\overset{H\cdots}{}}\!\!C\!\!-\!\!\overset{O}{\underset{CH_2-X}{\overset{\cdots H}{C}}} \quad + \quad Y-CH_2CH_2CH_3$$

⟰̸

$$\underset{H_3C}{\overset{H\cdots}{}}\!\!C\!\!-\!\!\overset{O}{\underset{CH_2 \dotplus CH_2CH_2CH_3}{\overset{\cdots H}{C}}}$$

⟹

$$\underset{H_3C}{\overset{H}{}}C\!=\!C\underset{CH_2CH_2CH_2CH_3}{\overset{H}{}} \quad \Longrightarrow \quad CH_3C\!\equiv\!C\dotplus CH_2CH_2CH_2CH_3$$

⟹

$$CH_3C\equiv CH \;+\; BrCH_2CH_2CH_2CH_3$$

The complete synthesis:

$$\equiv \quad \xrightarrow[\text{liq. NH}_3]{\text{NaNH}_2} \quad \equiv^- \text{Na}^+ \quad \xrightarrow[\text{(2) H}_3\text{O}^+]{\text{(1) Br}\frown\frown} \quad \equiv\frown\frown$$

$$\xrightarrow{\text{H}_2 \text{ Lindlar's catalyst}} \quad \frown=\frown\frown$$

$$\xrightarrow{\text{Cl-C}_6\text{H}_4\text{-COOOH}} \quad \underset{H_3C}{\overset{H\cdots}{}}\!\!C\!\!-\!\!\overset{O}{\underset{CH_2CH_2CH_2CH_3}{\overset{\cdots H}{C}}}$$

(d) **Retrosynthesis:**

$$\underset{H_3C}{\overset{H_3C\dotplus O}{\underset{H\cdots}{}C\!-\!CH_2CH_2CH_3}} \Longrightarrow \underset{H_3C}{\overset{HO}{\underset{H\cdots}{}C\dotplus CH_2CH_2CH_3}} \Longrightarrow CH_3\overset{O}{\underset{H}{C}}$$

$$+ \;\; BrMgCH_2CH_2CH_3$$

The synthesis of 2-pentanol from ethanal and a propyl Grignard reagent forms a racemic mixture that would form a racemic mixture of 2-methoxypentane. Therefore, we must resolve the mixture of 2-pentanols to obtain pure (R)-2-pentanol before forming the ether. The resolution is easy if we react the racemic mixture with an optically active carboxylic acid to form a mixture of diastereomeric esters; such esters are easily separated due to a difference in physical properties. Separated esters are hydrolysed to give individual enantiomeric alcohols.

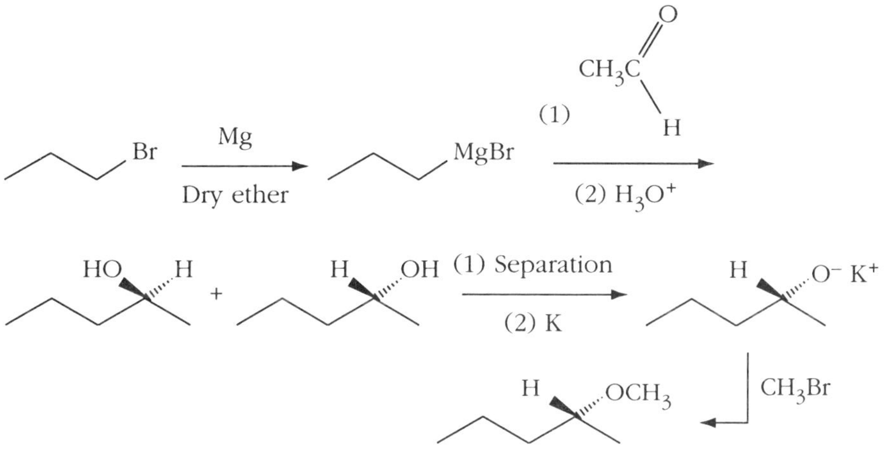

(e) Retrosynthesis:

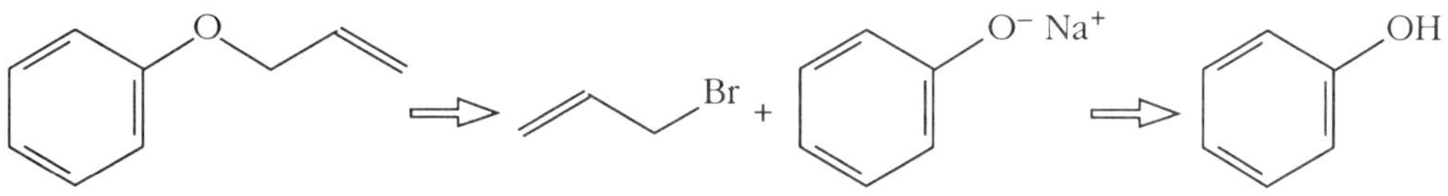

Synthesis:

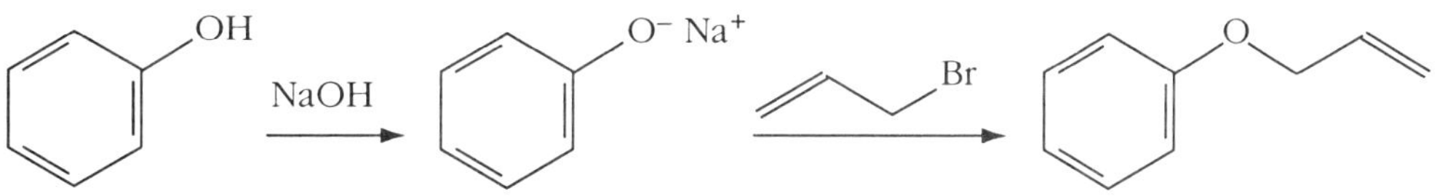

13.37 The results of the experiment suggest that 2-bromocyclohexanol is a mixture of *cis* and *trans* stereoisomers. The *trans* isomer reacts in the expected manner with epoxycyclohexane as the major product. The reaction occurs easily because in one conformation of the molecule the nucleophilic alkoxide is in position to displace bromide ion by an intramolecular S_N2 mechanism (route *a*). In this conformation the alkoxide ion and the leaving group are *anti* to each other as shown in the Newman projection for that portion of the cyclohexane ring. The elimination is also possible (route *b*) but the elimination product will be a minor product.

The *cis* isomer cannot react in the same way because it cannot achieve an *anti* orientation between the nucleophile and the leaving group. The only reaction *cis*-2-bromocyclohexanol can undergo is the elimination of HBr to give 2-cyclohexenol.

Thus, the results of the experiment strongly suggest that the student had a mixture of *cis*- and *trans*-2-bromocyclohexanol.

13.38 The reaction of (*R*)-2-octanol with NaH forms an alkoxide, which reacts with bromoethane to form an ether. The (C2)-oxygen bond of (*R*)-2-octanol is not broken in any reaction of the sequence, so the configuration of the C2 carbon atom remains unchanged. As a result, the product is (*R*)-2-ethoxyoctane.

The reaction pathway for the other method to synthesize 2-ethoxyoctane is shown on the following page. The reaction of ethoxide ion and (*R*)-2-bromooctane occurs by an S_N2 mechanism so the configuration of C2 atom is inverted. As a result, the product is (*S*)-2-ethoxyoctane.

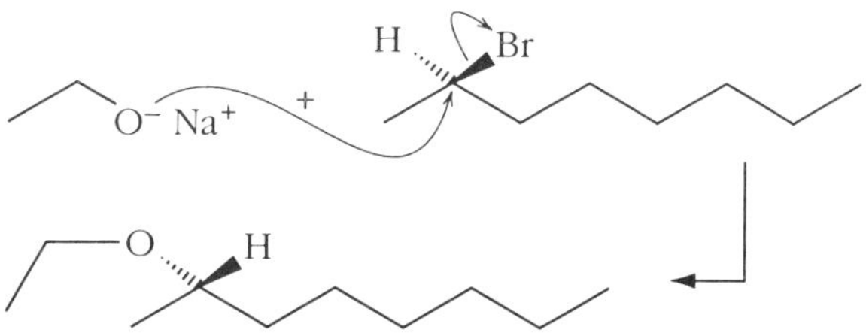

The above two methods of synthesis of 2-ethoxyoctane give us two different stereoisomers (enantiomers). The optical rotation for two enantiomers is known to have the same degree of rotation but in the opposite direction. This is what we observe (within the experimental error) in this example.

13.39

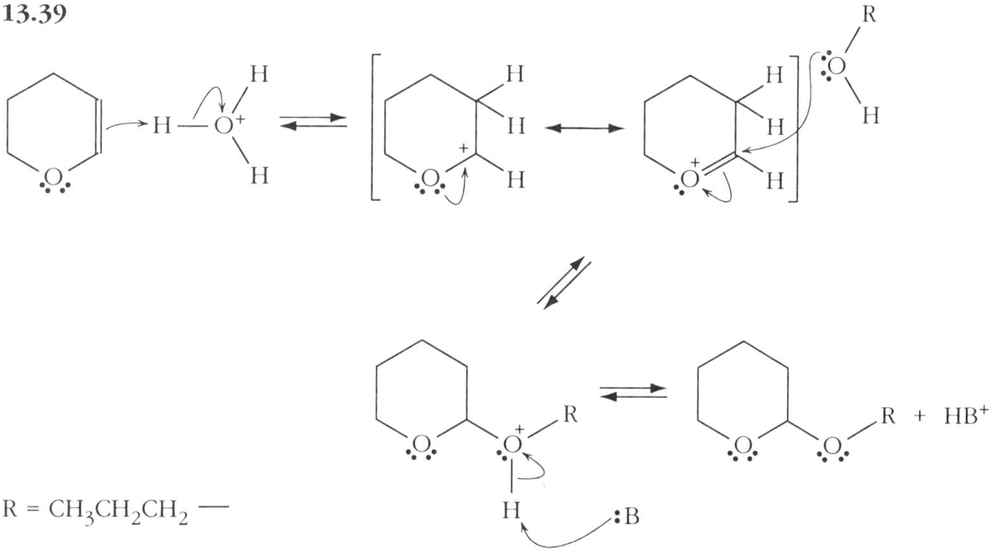

R = CH$_3$CH$_2$CH$_2$—

13.40 We have learned three spectroscopic techniques: ^{1}H NMR, ^{13}C NMR and IR. Let us compare the spectra of each pair of compounds and then decide which is the best method to distinguish between the two.

	^{13}C NMR	^{1}H NMR	IR
(a) 1,2-Dimethoxyethane	two signals	two singlets due to nonequivalent hydrogen atoms (relative areas, 3:2)	No absorption at 3300 cm^{-1}
2-Methoxyethanol	three signals	one singlet (methyl group) and two triplets (—CH$_2$—groups)	Strong peak at 3300 cm^{-1}

Any of the three spectroscopic techniques can distinguish between the two compounds. However, the best seems to be IR because the presence of strong absorption at 3300 cm^{-1} due to the —OH absorption identifies 2-methoxyethanol while the lack of that absorption identifies 1,2-dimethoxyethane.

		¹³C NMR	¹H NMR	IR
(b)	Diethyl ether	two signals	triplet (3H), methyl group quartet (2H), methylene group	1050–1260 cm⁻¹ (C—O)
	Tetrahydrofuran	two signals	two triplets (1:1 ratio), two methylene groups	1050–1260 cm⁻¹ (C—O)

IR and ¹³C NMR will not distinguish between these two compounds. However, ¹H NMR can differentiate them. The chemical shifts of the signals, their splitting, and their relative areas distinguish the two compounds.

		¹³C NMR	¹H NMR	IR
(c)	Diethyl ether	two signals	triplet (3H), methyl group quartet (2H), methylene group	1050–1260 cm⁻¹ (C—O)
	bis-2-Chloroethyl ether	two signals	two triplets (1:1 ratio), two methylene groups	1050–1260 cm⁻¹ (C—O)

IR and ¹³C NMR will not distinguish between these two compounds. However, ¹H NMR can differentiate them. The chemical shifts of the signals, their splitting, and their relative areas distinguish the two compounds.

13.41 Compound **A**: Molecular formula $C_6H_{14}O_2$ shows zero units of unsaturation. Three types of protons appear in the ¹H NMR spectrum: a triplet at δ 1.14, a quartet at δ 3.46, and a singlet at δ 3.51.

Note: The singlet overlaps with the most downfield signal of the quartet. It obscures the symmetry of the quartet and integration data. We know it is a quartet because we see three signals in the ratio of 1:3:3 and we know that a fourth one of ratio 1 should be further downfield at δ 3.52.

The absorptions at δ 1.14 and 3.46 resemble an ethyl group (a triplet and a quartet). The ethyl group is probably bonded to an electronegative element, since its methylene protons (—CH₂—) absorb at δ 3.46. Because the molecular formula contains oxygen, an ethoxy group is suggested. The compound must have symmetry since only three signals account for 14 protons. There are two oxygen atoms, so let us assume there are two ethoxy groups. The two ethoxy groups would have a partial molecular formula of $C_2H_{10}O_2$. What remains is C_2H_4 and it is represented as a singlet in the NMR spectrum. The position of the singlet suggests that the hydrogen atoms are bonded to the same carbon atom to which oxygen is attached. Thus, we have two methylene groups bonded to two ethoxy groups. Compound **A** is **1,2-diethoxyethane**.

Compound **B**: Molecular formula C_3H_5ClO implies one unit of unsaturation (a double bond or a ring). There are no signals of vinyl protons in the range of δ 5 to 6. The ¹H NMR spectrum shows four types of hydrogens. The integration data indicates the ratio of 1:1:1:2. The doublet (2H) must represent a methylene group bonded to an electronegative element because the signal is shifted downfield to δ 3.59. Because the molecular formula contains chlorine, a chloromethyl group (—CH₂Cl) is suggested. The other part of the molecule, C_2H_3O, must contain a ring. The only possibility is an oxirane ring. Thus, compound **B** is **1-chloro-2,3-epoxypropane**.

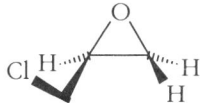

13.42 (a) The molecular formula, C$_2$H$_4$Cl$_2$O, shows no units of unsaturation (saturated compound). The IR spectrum exhibits an absorption band at 1200 cm^{-1} indicating the presence of carbon-oxygen bonding. The ^{1}H NMR spectrum shows two signals:

δ 3.68 (3H)—a methyl group attached to an electron-withdrawing atom (the molecular formula contains oxygen and chlorine; addition of chlorine would make chloromethane—not good; addition of oxygen makes a methoxy group—can be),
δ 7.33 (1H)—a hydrogen strongly deshielded by electron-withdrawing atoms (the two chlorine atoms) suggest a dichloromethyl group, —CHCl$_2$. Joining these two fragments gives the structure of the unknown:

CH$_3$OCHCl$_2$
Dichloromethoxymethane

(b) The molecular formula, C$_3$H$_8$O$_2$, shows no units of unsaturation (a saturated compound). The IR spectrum has an absorption band at 1200 cm^{-1}, which indicates the presence of a carbon-oxygen bond and an absorption band at 3350 cm^{-1} indicating the presence of an —OH group. The ^{13}C NMR spectrum shows three signals of carbon atoms in the range δ 58 to 74 (each carbon atom bonded to an oxygen atom). The ^{1}H NMR spectrum shows four signals:

δ 3.40 (s, 3H)—a methyl group bonded to an oxygen atom (methoxy group,—OCH$_3$),
δ 3.51 (t, 2H)—a methylene group
δ 3.72 (m, 2H)—a methylene group bonded to another methylene group and an oxygen atom

(—O—CH$_2$—CH$_2$—),

δ 3.10 (m, 1H)—may be a hydroxy group.
The spectra helped us to identify the following fragments:

—OH (IR), —OCH$_3$, and —O—CH$_2$—CH$_2$—

There is only one way that these fragments can be joined:

CH$_3$OCH$_2$CH$_2$OH
2-Methoxyethanol

(c) The molecular formula, C$_4$H$_8$O, shows one unit of unsaturation (a ring, a carbon-carbon double bond, or a carbon-oxygen double bond). The ^{13}C NMR shows only two signals, which indicates symmetry in the molecule. The ^{1}H NMR spectrum shows also only two types of hydrogen atoms:

δ 1.27 (d, 3H)—a methyl group bonded to —CH⟨ ,

δ 3.07 (q, 1H)—a hydrogen atom seeing a neighboring methyl group.

Thus, we have the following fragment:

$$\underset{/}{\overset{\backslash}{C}}H - CH_3$$

The fragment has only four hydrogen atoms, yet the molecule has eight hydrogen atoms. This implies two identical fragments. Two fragments contribute C_4H_8 to the molecular formula. What is left is one oxygen atom. Joining the two fragments and —O— we get the structure of the unknown:

$$H_3CCH \overset{O}{\underset{\triangle}{-}} CHCH_3$$

2,3-Epoxybutane or 2,3-dimethyloxirane

There could be two 2,3-dimethyloxiranes: *cis*- and *trans*-2,3-dimethyloxirane. However, there are no data to distinguish between them.

(d) The molecular formula, $C_4H_8O_2$, shows one unit of unsaturation (a ring, a carbon-carbon double bond, or a carbon-oxygen double bond). The ^{13}C NMR shows three signals, which indicates some symmetry in the molecule. The ^1H NMR spectrum shows also only three signals for the eight hydrogen atoms:

δ 1.78 (m, 1H)—?
δ 3.91 (t, 2H)—a methylene group bonded to an oxygen atom and another methylene group, —O—CH_2—CH_2—
δ 4.84 (s, 1H)—?

The total number of hydrogen atoms calculated from integration data is four, while the molecular formula shows eight hydrogen atoms. This means that we should multiply the number of each type of hydrogen atom by two.

δ 1.78 (m, 2H)—a methylene group bonded to carbon atoms with hydrogen atoms,
δ 3.91 (t, 4H)—two methylene groups, each bonded to an oxygen atom and another methylene group,
δ 4.84 (s, 2H)—a methylene group bonded to two oxygen atoms, —O—CH_2—O—.

Thus, we have three fragments:

—O—CH_2—O—, —O—CH_2—, —O—CH_2—, and —CH_2—

Joining these fragments we get the structure of the unknown:

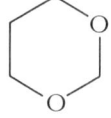

1,3-Dioxane

CHAPTER 14

ALDEHYDES AND KETONES
Concepts

The characteristic feature of **aldehydes** and **ketones** is the **carbonyl group**, C=O. Two of the more notable aspects of the carbonyl group are its *geometry* and its *polarity*. The carbonyl group and the atoms attached to it all lie in the same plane.

The bond angles involving the carbonyl group are close to 120° and do not vary much among simple aldehydes and ketones.

Because of the high electronegativity of oxygen, the electron density in both the σ and π components of the carbon-oxygen double bond is displaced toward oxygen. The carbonyl group is polarized so that carbon is partially positive and oxygen is partially negative.

The polarity of the carbonyl group makes low-molecular-weight aldehydes and ketones miscible with water. The oxygen acts as an acceptor of hydrogen bonds, and the carbonyl compound dissolves in water as long as the hydrocarbon portion of the molecule does not contain more than four or five carbon atoms.

The polarity of the carbonyl group makes molecules containing such a group reactive towards both acidic and basic reagents. The protonation of the nonbonding electrons on the oxygen atom of the carbonyl group is an important first step in the reactions under acidic conditions.

The most common reaction of ketones and aldehydes is the nucleophilic addition reaction. A nucleophile attacks the electrophilic carbon of the polar carbonyl group from a direction approximately perpendicular to the plane of the carbonyl sp^2 orbitals. Rehybridization of the carbonyl carbon from sp^2 to sp^3 then occurs, and a tetrahedral alkoxide ion intermediate is formed.

The attacking nucleophile can be either negatively charged (Nu:⁻ = HO⁻, H⁻, R₃C⁻, RO⁻, N≡C⁻) or neutral (H₂O, ROH, H₃N, RNH₂). Nucleophilic additions to ketones and aldehydes have two variations, as shown below. (1) The tetrahedral intermediate can be protonated by water or acid to give an alcohol, or (2) the carbonyl oxygen atom can be expelled as H₂O to give a new carbon-nucleophile double bond.

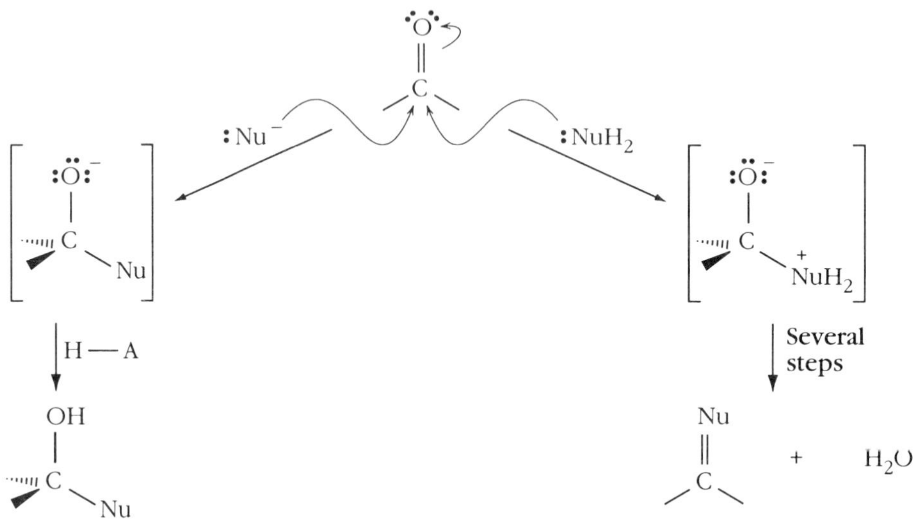

Nomenclature. The IUPAC nomenclature for aldehydes and ketones follows the familiar pattern of selecting as the parent alkane the longest chain of carbon atoms that contains the functional group.

The **aldehyde group** is shown by changing the suffix *-e* of the parent alkane to *-al*. The carbonyl group of an aldehyde can only appear at the end of a parent chain and numbering starts with it as carbon 1. Because its position is unambiguous, there is no need to use a number to locate it. For unsaturated aldehydes, the presence of carbon-carbon double or triple bond is indicated by the infix *-en-* or *-yn-* as shown below.

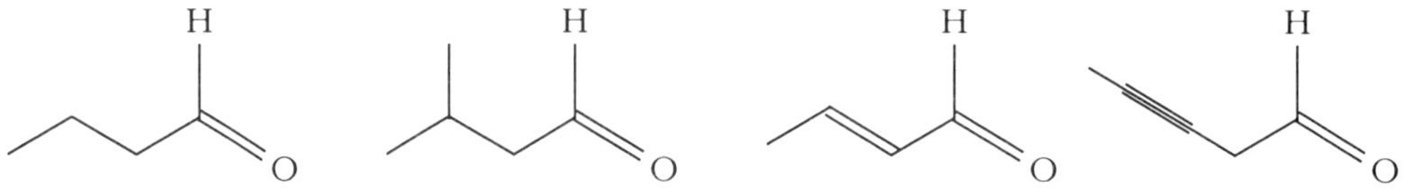

Butanal   3-Methylbutanal   2-Butenal   3-Pentynal

For cycloalkanes in which —CHO is attached directly to the ring, the molecule is named by adding the suffix *-carbaldehyde* to the name of the ring. The atom of the ring to which the aldehyde group is attached is numbered 1. Among the aldehydes for which the IUPAC system retains common names are benzaldehyde and cinnamaldehyde.

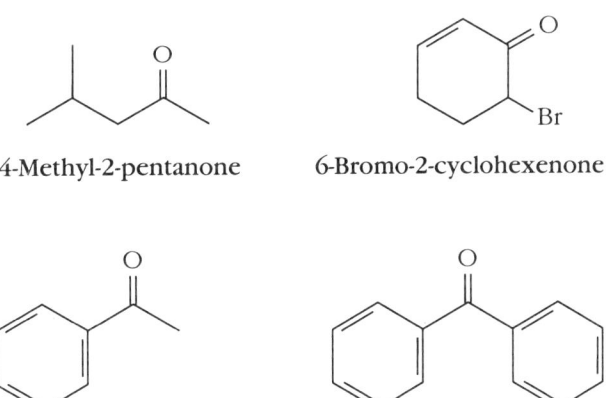

Cyclohexanecarbaldehyde   2-Cyclopentenecarbaldehyde

Benzaldehyde   Cinnamaldehyde

In the IUPAC system, ketones are named by selecting as the parent alkane the longest chain that contains the carbonyl group and then indicating the presence of the carbonyl group by changing the suffix from -e to -one. The parent chain is numbered from the direction that gives the carbonyl carbon the lowest number. The IUPAC system retains the common names acetone, acetophenone, and benzophenone.

4-Methyl-2-pentanone   6-Bromo-2-cyclohexenone

Acetophenone   Benzophenone

Solutions to the Exercises

14.1 (a) Hexanal
(b) 3-Methylbutanal
(c) *cis*-2-Methylcyclobutanecarbaldehyde or
(1*S*,2*R*)-2- methylcyclobutanecarbaldehyde
(d) Pentanedial

14.2

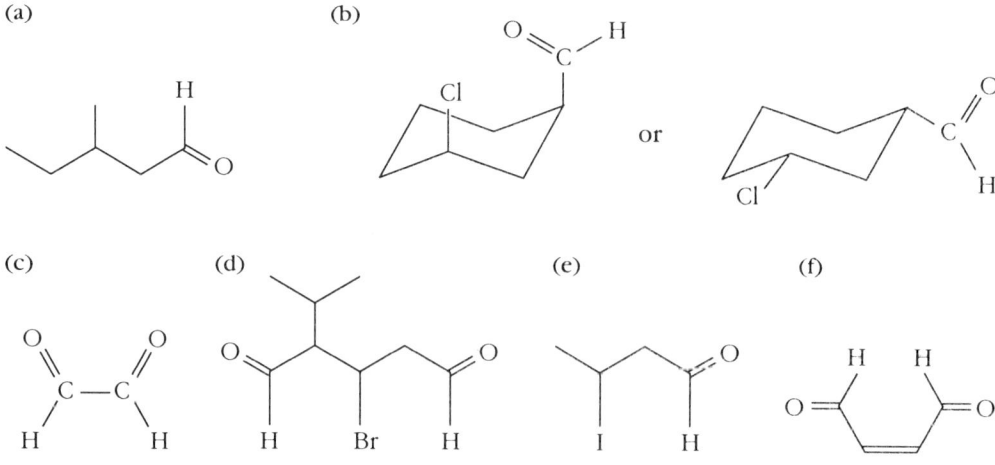

14.3 (a) Cycloheptanone
(b) 1,3-Cyclooctanedione
(c) 3-Buten-2-one
(d) 2-Methyl-3-pentanone
(e) 2,4-Cyclopentadienone
(f) 3-*tert*-Butyl-5- trifluoromethyl-2-hexanone or 5-trifluoromethyl-3-(1,1-dimethylethyl)-2-hexanone

14.4

14.5 Butane (−0.6 °C) < diethyl ether (36 °C) < 2-butanone (80 °C) < 1-butanol (118 °C)

1-Butanol forms hydrogen bonds, which are the strongest type of intermolecular interactions. 2-Butanone has a large dipole moment ($\mu = 2.9$ D) so it has strong dipole-dipole interactions between molecules. Diethyl ether has a smaller dipole moment (μ = 1.3 D) and the molecules experience only weak attractive interactions. Butane is nonpolar with a dipole moment of μ = 0 D, so only intermolecular forces that attract molecules, van der Waals interactions, are present. As a result, butane has the lowest boiling point of all four compounds.

14.6 The carbonyl group ($\ce{\backslash C=O}$) is a functional group that is very easy to detect by IR spectroscopy because it shows a strong absorption in the region of 1710 to 1750 cm^{-1}. 2-Hexanone, for example, shows an absorption at 1715 cm^{-1}. 2-Hexanol shows no absorption in this region so the absence of an absorption at 1715 cm^{-1} in the IR spectrum of a small sample of the reaction mixture indicates that all the starting material has reacted.

14.7 (a) 2-Butanone shows a strong absorption of the carbonyl group at 1715 cm^{-1} while butanal shows C=O bond stretching in the region of 1720 to 1725 cm^{-1}. The most important difference, however, will be found at 2700 cm^{-1}. Butanal shows a sharp peak as a result of the stretching of the carbonyl carbon-hydrogen bond while 2-butanone lacks such a hydrogen bonded to its carbonyl carbon atom and lacks absorption at 2700 cm^{-1}.

(b) 3-Hexanone shows the strong absorption of the carbonyl group at 1715 cm^{-1} and no absorption above 3000 cm^{-1}. Acetic acid exhibits typical C=O absorption at 1705 cm^{-1} and also shows a very distinctive O—H band, which spans from about 3330 cm^{-1} to 2500 cm^{-1}. The presence of such absorption indicates a carboxylic acid; the absence of such absorption indicates a ketone in this case.

(c) 1-Pentanol exhibits a distinctive O—H stretching absorption at 3500 to 3100 cm^{-1} and no absorption around 1700 cm^{-1}. 3-Pentanone, on the other hand, shows a strong stretching absorption at 1715 cm^{-1} and no absorption above 3000 cm^{-1}.

14.8 (a) The ^{1}H NMR spectrum of pentane shows the methylene groups (—CH$_2$—) as a single, broad peak at δ 1.25 and a triplet of two terminal methyl groups (—CH$_3$) at δ 0.85. 3-Pentanone shows a quartet at δ 2.40 and a triplet at δ 1.05.
The ^{13}C NMR spectra of both compounds show three signals. However, 3-pentanone shows a characteristic signal of the carbonyl carbon at δ 211 and two signals in the alkane region at δ 35 and δ 8, while pentane shows all three signals in the alkane region, at δ 14, δ 22, and δ 34.

(b) The ^{1}H NMR spectrum of 1-butene shows two doublets of the terminal =CH$_2$ group at δ 4.7 to 5.1 and a multiplet of the nonterminal alkenyl proton at δ 5.4 to 6.1 as well as a multiplet for the methylene group at δ 2.0 and triplet for the methyl group at δ 1.0.
2-Butanone shows a singlet for the α methyl group at δ 2.1, a quartet for the methylene group at δ 2.40 and a triplet for the β methyl group at δ 1.05.
^{13}C NMR spectra of both compounds show four signals. However, 1-butene shows two signals in the alkene region (δ 114 and δ 139) and two signals in the alkane region (δ 21 and δ 35) while 2-butanone shows one signal at δ 209 and three in the alkane region (δ 35, δ 27, and δ 7).

(c) The ^{1}H NMR spectrum of cyclopentanone exhibits only two overlapping multiplets at δ 2.0 to 2.2. Pentanal shows a singlet at δ 9.8 (—CHO), a triplet at δ 2.45 (—CH$_2$—CHO), a multiplet at δ 1.4 to 1.9, and a triplet at δ 1.0 (terminal —CH$_3$). The difference is significant: two multiplets for cyclopentanone at δ 2.0 to 2.2 and a very characteristic aldehyde signal at δ 9.8 for pentanal.

The ^{13}C NMR spectra are also different. The ^{13}C NMR spectrum of cyclopentanone has three signals, including the carbonyl carbon signal at δ 211, while the spectrum of pentanal shows five signals, including the one for the carbonyl carbon at δ 201 (the others are at δ 14, δ 22, δ 24, and δ 44).

(d) The ^{1}H NMR spectrum of cyclohexanone exhibits two multiplets: δ 2.2 to 2.5

$$(-CH_2-\overset{\overset{O}{\|}}{C}-CH_2-)$$

and δ 1.7 to 1.9 (other methylene groups). 2-Cyclohexenone shows a multiplet for the hydrogen attached to C3 at δ 6.9 to 7.2, a doublet for the hydrogen bonded to C2 at δ 6.0, a triplet for the C6 methylene group at δ 2.35 and a multiplet for the C4 and C5 methylene groups at δ 1.9 to 2.5.

The ^{13}C NMR spectra of the two compounds are distinctly different. The ^{13}C NMR spectrum of cyclohexanone has four signals while that of 2-cyclohexenone has six signals.

14.9 The molecular formula, C$_5$H$_{10}$O, indicates one unit of unsaturation. The IR spectrum indicates that this unsaturation is the carbon-oxygen double bond (characteristic absorption of the carbonyl group). That means that the rest of each molecule is made up of saturated hydrocarbon groups.

Compound I: The ^{1}H NMR spectrum shows a triplet at δ 1.06 and a quartet at δ 2.44 (ratio of 3:2). This kind of splitting pattern is characteristic of an ethyl group. Because we do not see any other signals, we can assume that there are two ethyl groups. The chemical shift of the quartet (δ 2.44) is characteristic of hydrogen atoms bonded to a carbon bonded to a carbonyl group or an aromatic ring. There is a carbonyl group in the molecule, so we can propose that the two ethyl groups are attached to it. Thus, the structure of compound I is **3-pentanone**.

Compound II: The ^{1}H NMR spectrum shows a triplet at δ 0.89, a multiplet (sextet) at δ 1.33, a multiplet (quintet) at δ 1.59, a triplet at δ 2.4 and a triplet at δ 9.74 (ratio of 3:2:2:2:1, respectively). The signal at δ 9.74 (1H) is characteristic of an aldehyde. The split triplet at δ 2.4 (2H) is that of a methylene group bonded directly to the carbonyl group. The triplet at δ 0.89 (3H) arises from a terminal methyl group and the two multiplets at δ 1.3 to 1.6 are the remaining two methylene groups. The sextet at δ 1.33 can be assigned to the methylene group next to the terminal methyl group (the $n + 1$ rule indicates five hydrogen atoms on neighboring carbon atoms). The quintet at δ 1.59 can be assigned to the second to last methylene group (C3). Thus, the structure of compound II is **pentanal**.

$$CH_3CH_2CH_2CH_2C\overset{\diagup\!\!\!\!\diagup O}{\diagdown H}$$

Compound III: The ^{1}H NMR spectrum shows two singlets: δ 1.1 (9H) and δ 9.5 (1H). The signal at δ 9.5 (1H) is characteristic of an aldehyde functional group. The signal at δ 1.1 (9H) is characteristic of three equivalent methyl groups of a *tert*-butyl group. Thus, the structure of compound III is **2,2-dimethylpropanal**.

14.10 (a) $(CH_3)_3CCH_2C\equiv CH \xrightarrow[(2)\ H_2O_2/NaOH/H_2O]{(1)\ 9\text{-BBN}} (CH_3)_3CCH_2CH_2CHO$

(b) Cyclohexyl-CH$_2$OH $\xrightarrow[CH_2Cl_2]{PCC}$ cyclohexyl-CHO

(c) $(CH_3)_2CHCOCl \xrightarrow[(2)\ H_3O^+]{(1)\ LiAl[OC(CH_3)_3]_3H} (CH_3)_2CHCHO$

14.11 (a) To convert a 1° alcohol to an aldehyde we must use the special oxidizing reagents that oxidize alcohols but not aldehydes, such as PCC/CH$_2$Cl$_2$ or Collin's reagent.

(b) To convert a terminal alkyne to an aldehyde, we must add water in an anti-Markownikoff manner. This can be done by hydroboration-oxidation: 1. 9-BBN, 2. H$_2$O$_2$/NaOH.

(c) The partial reduction of an acid chloride to an aldehyde can be effectively carried out with lithium tri-*tert*-butoxyaluminum hydride: 1. LiAlH(O-*t*-Bu)$_3$, 2. H$_3$O$^+$.

14.12 (a) Hydration of a symmetrical alkyne can be accomplished by either hydroboration-oxidation (1. 9-BBN, 2. H$_2$O$_2$/NaOH) or mercuric ion-catalyzed hydration (HgSO$_4$, H$_2$O, H$_2$SO$_4$).

(b) Friedel-Crafts acylation of benzene: CH$_3$CH$_2$COCl/AlCl$_3$

(c) (CH$_3$CH$_2$)$_2$CuLi

(d) Hydration in an anti-Markownikoff manner by 9-BBN followed by an oxidation of the alcohol to a ketone: 1. 9-BBN, 2. H$_2$O$_2$/NaOH, 3. H$_2$CrO$_4$ (Na$_2$Cr$_2$O$_7$ + H$_2$SO$_4$/H$_2$O).

14.13

14.14 (a)

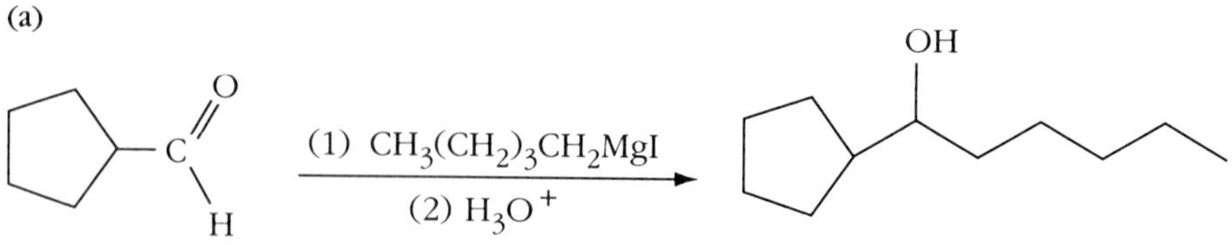
O
C
H
(1) $CH_3(CH_2)_3CH_2MgI$
(2) H_3O^+
OH

(b)

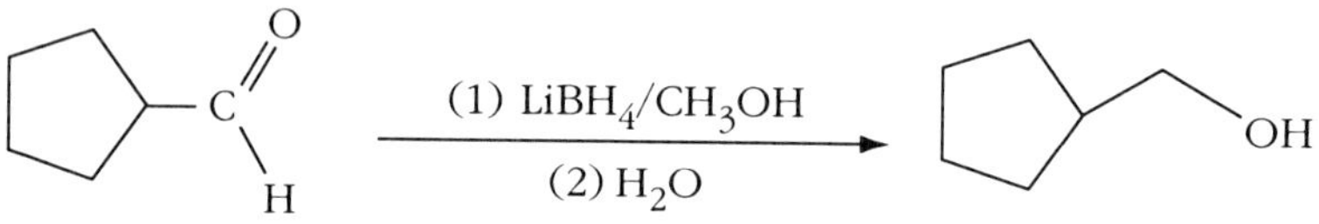
O
C
H
(1) $LiBH_4/CH_3OH$
(2) H_2O
OH

(c)

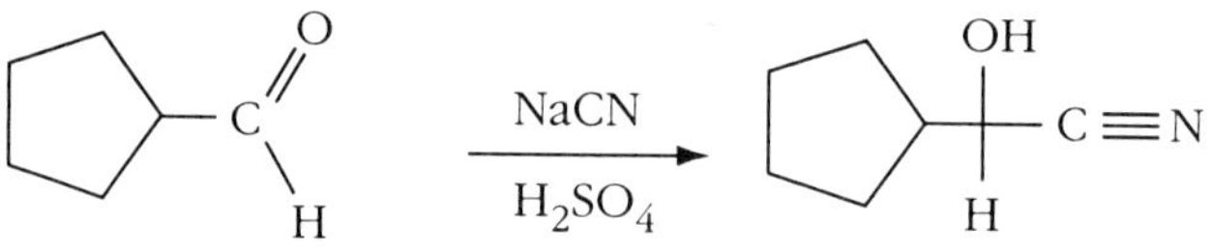
O
C
H
NaCN
H_2SO_4
OH
C≡N
H

14.15

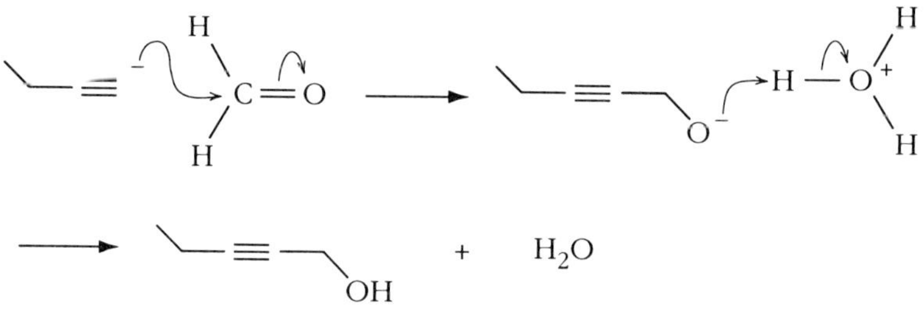
H
C=O
H
H
O
+
H
H
OH
+
H_2O

14.16

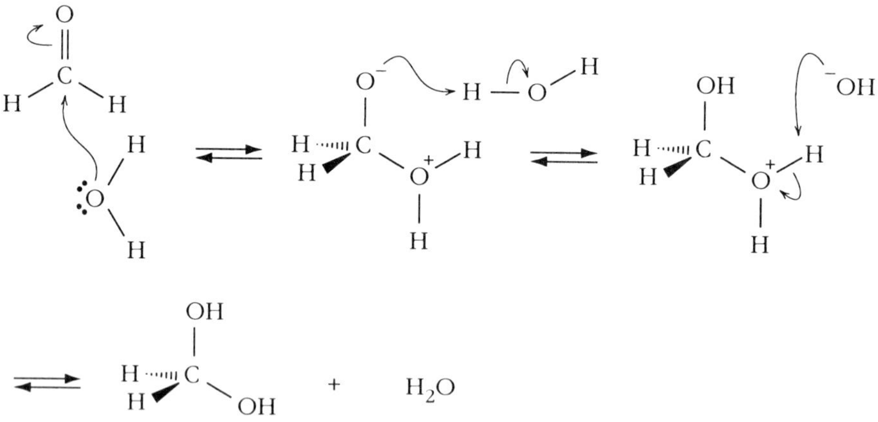
O
C
H
H
H
O
H
H
C
H
O
+
H
H
OH
OH
+
H_2O

14.17

(a) CH₃CH₂CHO $\xrightleftharpoons{CH_3CH_2OH, HCl}$ CH₃CH₂−C(OH)(H)(OCH₂CH₃) $\xrightleftharpoons{CH_3CH_2OH, HCl}$ CH₃CH₂−C(H)(OCH₂CH₃)(OCH₂CH₃)

(b) CH₃CH₂COCH₃ $\xrightleftharpoons{CH_3CH_2OH, HCl}$ CH₃CH₂−C(OH)(CH₃)(OCH₂CH₃) $\xrightleftharpoons{CH_3CH_2OH, HCl}$ CH₃CH₂−C(CH₃)(OCH₂CH₃)(OCH₂CH₃)

(c) C₆H₅CHO $\xrightleftharpoons{CH_3CH_2OH, HCl}$ C₆H₅−C(OH)(H)(OCH₂CH₃) $\xrightleftharpoons{CH_3CH_2OH, HCl}$ C₆H₅−C(H)(OCH₂CH₃)(OCH₂CH₃)

While only one enantiomer of the hemiacetal is written in these answers, in fact both enantiomers are formed in the reaction mixture.

14.18 Step 1: Protonation of the carbonyl group

CH₃−C(=O:)(H) + H−B ⇌ CH₃−C(=O⁺−H)(H) + :B⁻

Step 2: Nucleophilic attack on the carbonyl group

Step 3: Deprotonation

14.19

14.20

(a)
$$CH_3\underset{H}{\overset{O}{\underset{\|}{C}}} + HOCH_2CH_2OH \xrightleftharpoons{\text{Acid catalyst}} CH_3CH\begin{matrix}O-CH_2\\ \\O-CH_2\end{matrix} + H_2O$$

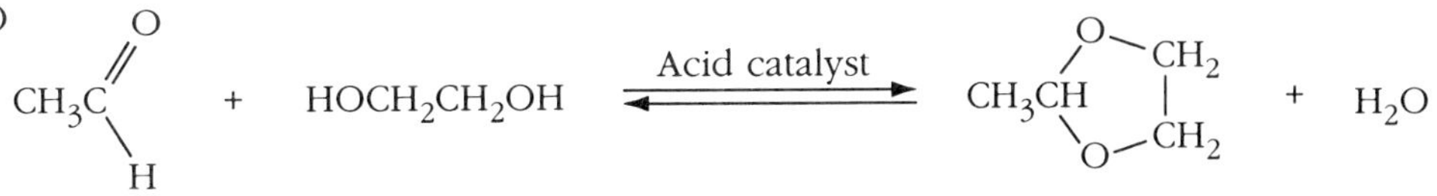

(b)
$$C_6H_5\underset{}{\overset{O}{\underset{\|}{C}}}CH_3 + HOCH_2CH_2OH \xrightleftharpoons{\text{Acid catalyst}}$$

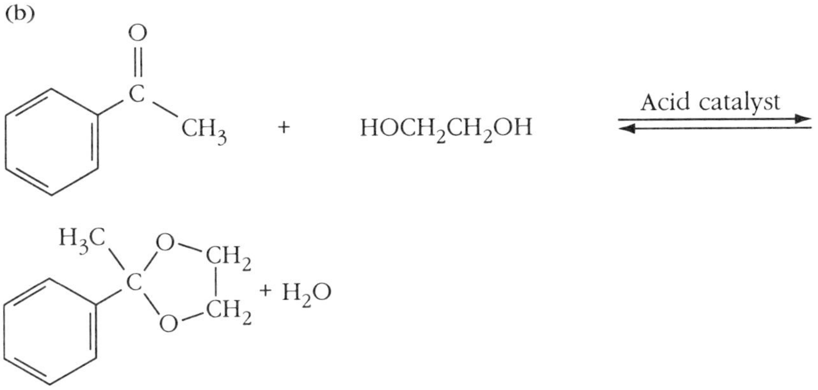

$$\underset{C_6H_5}{\overset{H_3C}{C}}\begin{matrix}O-CH_2\\ \\O-CH_2\end{matrix} + H_2O$$

(c)
$$CH_3CH_2CH_2\underset{H}{\overset{O}{\underset{\|}{C}}} + HOCH_2CH_2CH_2OH \xrightleftharpoons{\text{Acid catalyst}}$$

$$CH_3CH_2CH_2CH\begin{matrix}O-CH_2\\ \\ \ \ \ \ \ CH_2\\O-CH_2\end{matrix} + H_2O$$

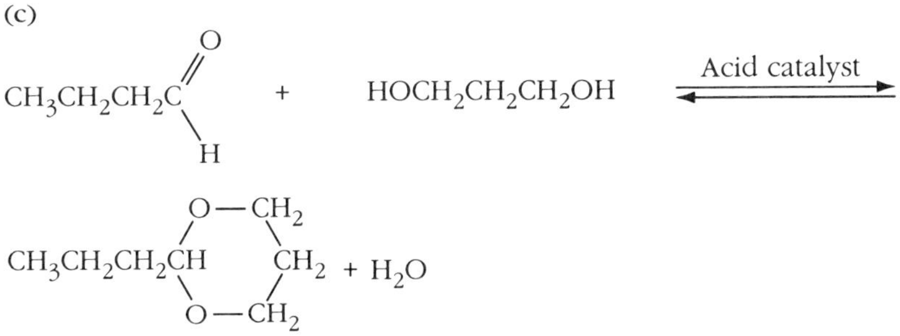

14.21

[Mechanism showing acid-catalyzed acetal formation from acetaldehyde and ethylene glycol, with successive protonation, nucleophilic addition, proton transfers, and loss of water steps.]

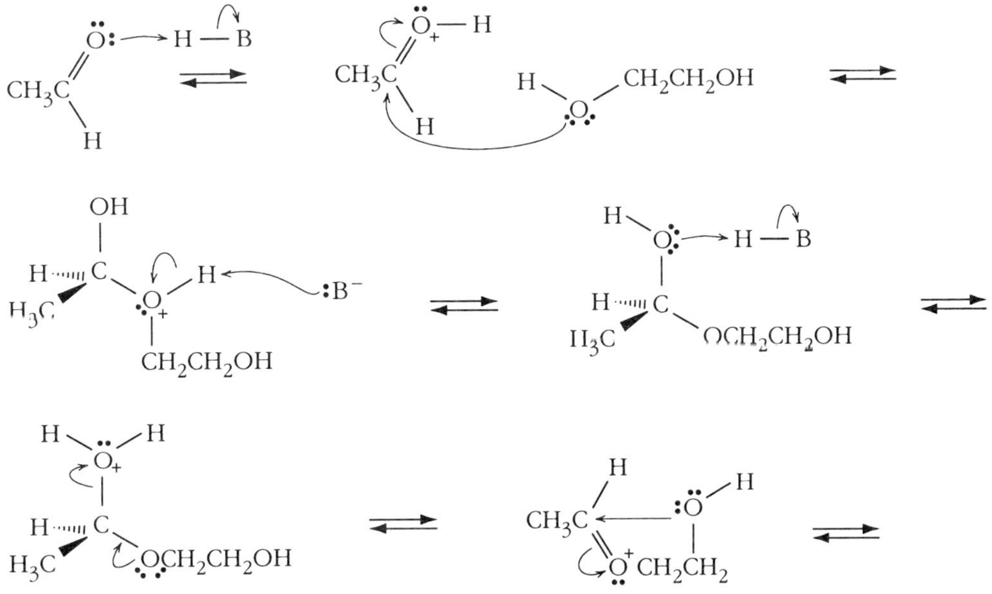

14.22 The target molecule, 1,6-heptanediol, has a longer carbon chain by two carbon atoms than the starting material, 5-bromo-2-pentanone. The most convenient way to add two carbon atoms and to create a terminal hydroxyl group is to react a Grignard reagent from 5-bromo-2-pentanone with oxirane. However, any Grignard reagent formed from 5-bromo-2-pentanone would react immediately with the carbonyl group of another molecule of 5-bromo-2-pentanone. In fact, the Grignard reagent self-destructs during its preparation.

A way to avoid this problem is to protect the carbonyl group of 5-bromo-2-pentanone by conversion to an acetal. Treatment of the protected 5-bromo-2-pentanone with magnesium in diethyl ether followed by addition of oxirane gives a magnesium alkoxide. Final treatment of the magnesium alkoxide accomplishes two things. First, protonation of the alkoxide gives the desired alcohol and second, hydrolysis of the cyclic acetal regenerates the carbonyl group. Now we can reduce the carbonyl group to the hydroxyl group.

Note: We should not try to carry out the reduction of the carbonyl group first. By doing so, we would form a hydroxyl group that would destroy any Grignard reagent formed later.

14.23

[Mechanism scheme showing acetaldehyde reacting with methanol through a tetrahedral intermediate with proton transfer steps to form the hemiacetal CH₃CH(OH)(OCH₃).]

[Second mechanism scheme showing acetaldehyde reacting with methylamine through a tetrahedral intermediate with proton transfer steps to form the carbinolamine CH₃CH(OH)(NHCH₃).]

The addition of the nucleophile (Step 1) in both mechanisms is identical. The structures of the first intermediates are similar because both are tetrahedral intermediates. The proton transfer occurs in several steps in both mechanisms, and the sequence of protonation-deprotonation may be different. Alcohols are weak acids and methanol may protonate the negatively charged oxygen atom first. Methylamine, on the other hand, would be more likely to remove a proton from —$\overset{+}{N}H_2CH_3$ rather than to donate its proton. Thus, the proton transfer process is different.

14.24

(a)

$$PhCOCH_3 + CH_3NH_2 \xrightarrow{\text{Buffer} \atop \text{pH 4-5}} PhC(=NCH_3)CH_3 + H_2O$$

(b)

$$CH_3(CH_2)_5\underset{H}{\overset{O}{C}} + H_2\ddot{N}-\text{cyclohexyl} \xrightarrow[\text{pH 4-5}]{\text{Buffer}} CH_3(CH_2)_5\underset{H}{\overset{}{C}}=\ddot{N}-\text{cyclohexyl} + H_2O$$

(c)

$$\text{PhCH}_2\ddot{N}H_2 + \underset{H}{\overset{O}{C}}-\text{cyclopentyl} \xrightarrow[\text{pH 4-5}]{\text{Buffer}}$$

$$\text{PhCH}_2\ddot{N}=\underset{H}{\overset{}{C}}-\text{cyclopentyl} + H_2O$$

14.25 Primary aliphatic amines (RNH$_2$) and primary aromatic amines (ArNH$_2$) react with the carbonyl group of aldehydes and ketones in the presence of an acid catalyst to give a product that contains a carbon-nitrogen double bond.

(a)

$$\text{CH}_3\text{CH}_2\underset{H}{\overset{O}{C}} + H_2\ddot{N}CH_3 \xrightarrow[\text{pH 4-5}]{\text{Buffer}} \text{CH}_3\text{CH}_2\underset{H}{\overset{}{C}}=\ddot{N}\text{CH}_3$$

(b)

$$(CH_3)_2C=O + H_2\ddot{N}-CH(CH_3)_2 \xrightarrow[\text{pH 4-5}]{\text{Buffer}} (CH_3)_2C=\ddot{N}-CH(CH_3)_2$$

(c)

$$\text{cyclohexyl-C(=O)CH}_3 + H_2\ddot{N}-\text{Ph} \xrightarrow[\text{pH 4-5}]{\text{Buffer}} \text{cyclohexyl-C(CH}_3\text{)}=\ddot{N}-\text{Ph}$$

14.26

$$\text{4-methylcyclohexanone} + H_2\ddot{N}-OH \xrightarrow{\text{Acidic buffer}} \text{4-methylcyclohexanone =}\ddot{N}-OH$$

Oxime

$$\text{4-methylcyclohexanone} + H_2N-NH_2 \xrightarrow{\text{Acidic buffer}} \text{4-methylcyclohexanone =}\ddot{N}-NH_2$$

Hydrazone

[Reaction 1: 4-methylcyclohexanone + 2,4-dinitrophenylhydrazine (H₂N—NH—Ar(NO₂)₂) → 2,4-Dinitrophenylhydrazone, acidic buffer]

[Reaction 2: 4-methylcyclohexanone + semicarbazide (H₂N—NH—C(=O)—NH₂) → Semicarbazone, acidic buffer]

14.27

Structure	¹H NMR	¹³C NMR
3-methylcyclohexanone	doublet at δ 1.0 (3H), multiplet at δ 1.5–2.5 (9H)	7 signals
4-methylcyclohexanone	doublet at δ 1.0 (3H) multiplet at δ 1.5–2.5 (9H)	5 signals

First, note that the symmetry in 4-methylcyclohexanone makes C2 equivalent to C6, and C3 equivalent to C5. As a result of that symmetry, both the ¹H and ¹³C NMR spectra of 4-methylcyclohexanone will be simpler. However, the ¹H NMR spectrum is still complex and is difficult to interpret. The ¹³C NMR spectra are much more useful in this case because the difference is more noticeable.

14.28 (a) 2-butanone + H₂N—OH → oxime (acidic buffer)

(b) benzaldehyde + 2,4-dinitrophenylhydrazine → benzaldehyde 2,4-dinitrophenylhydrazone (acidic buffer)

(c)

cyclopentanone + H$_2$NCH$_2$CH$_3$ $\xrightarrow{\text{Acidic buffer}}$ cyclopentanone N-ethylimine

14.29 (a)

$(C_6H_5)_3P \xrightarrow[\text{(2) CH}_3\text{CH}_2\text{CH}_2\text{CH}_2\text{Li}]{\text{(1) BrCH}_2\text{CH}_2\text{CH}_2\text{CH}_3} (C_6H_5)_3\overset{+}{P} - \overset{-}{C}HCH_2CH_2CH_3$

(b)

$(C_6H_5)_3P \xrightarrow[\text{(2) CH}_3\text{CH}_2\text{CH}_2\text{CH}_2\text{Li}]{\text{(1) BrCH}_2\text{C} \equiv \text{CCH}_2\text{CH}_3} (C_6H_5)_3\overset{+}{P} - \overset{-}{C}HC \equiv CCH_2CH_3$

(c)

$(C_6H_5)_3P \xrightarrow[\text{(2) CH}_3\text{CH}_2\text{CH}_2\text{CH}_2\text{Li}]{\text{(1) BrCH}(C_6H_5)_2} (C_6H_5)_3\overset{+}{P} - \overset{-}{C}(C_6H_5)_2$

(d)

$(C_6H_5)_3P \xrightarrow[\text{(2) CH}_3\text{CH}_2\text{CH}_2\text{CH}_2\text{Li}]{\text{(1) BrCH}_2\text{CH}_2-\text{cyclohexyl}} (C_6H_5)_3\overset{+}{P} - \overset{-}{C}HCH_2-\text{cyclohexyl}$

14.30

(a)

PhC(=O)CH$_3$ + $(C_6H_5)_3\overset{+}{P} - \overset{-}{C}H_2$ → PhC(=CH$_2$)CH$_3$

(b)

CH$_3$CH$_2$CH$_2$CHO + $(C_6H_5)_3\overset{+}{P} - \overset{-}{C}(CH_3)_2$ → CH$_3$CH$_2$CH$_2$CH=C(CH$_3$)$_2$

(c)

cyclopentanone + $(C_6H_5)_3\overset{+}{P} - \overset{-}{C}HC(=O)OCH_3$ → cyclopentylidene=CHC(=O)OCH$_3$

(d)

$(CH_3)_2C=O$ + $(C_6H_5)_3\overset{+}{P} - \overset{-}{C}HCH=CH_2$ → $(CH_3)_2C=CHCH=CH_2$

14.31 In order to identify the carbonyl compound and the ylide required to produce a given alkene, mentally disconnect the double bond so that one of its carbons is derived from a carbonyl group and the other is derived from an ylide. Typically there will be two Wittig routes to an unsymmetrical alkene, and any choice between them is made on the basis of availability of the particular starting materials. We should remember that phosphorus ylides are prepared from haloalkanes by a nucleophilic substitution reaction that occurs by an S_N2 mechanism. Thus, primary haloalkanes would be better substrates for the preparation of the Wittig reagent than secondary haloalkanes.

(a) Retrosynthesis:

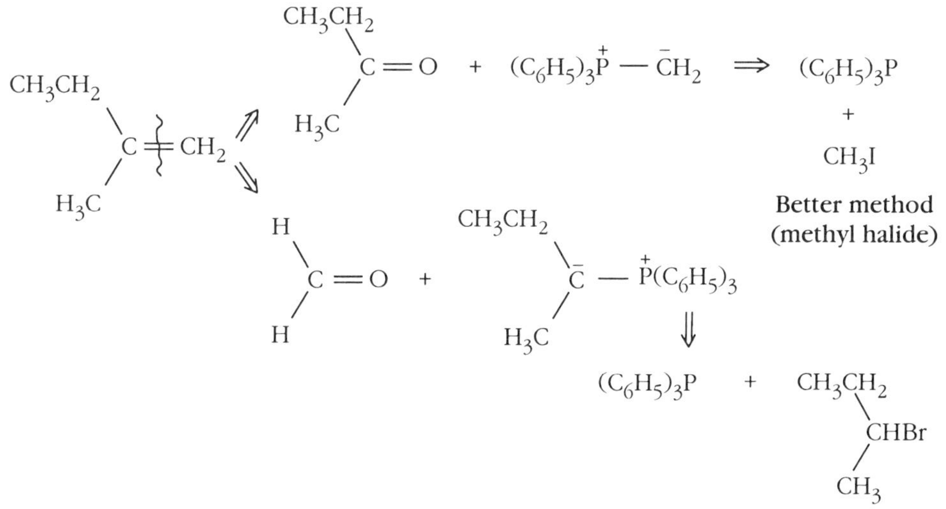

$$\underset{H_3C}{\overset{CH_3CH_2}{\diagdown}}C=CH_2 \Longrightarrow \underset{H_3C}{\overset{CH_3CH_2}{\diagdown}}C=O + (C_6H_5)_3\overset{+}{P}-\overset{-}{C}H_2 \Longrightarrow (C_6H_5)_3P + CH_3I$$

Better method (methyl halide)

$$\underset{H}{\overset{H}{\diagdown}}C=O + \underset{H_3C}{\overset{CH_3CH_2}{\diagdown}}\overset{-}{C}-\overset{+}{P}(C_6H_5)_3 \Downarrow$$

$$(C_6H_5)_3P + \underset{CH_3}{\overset{CH_3CH_2}{\diagdown}}CHBr$$

Synthesis:

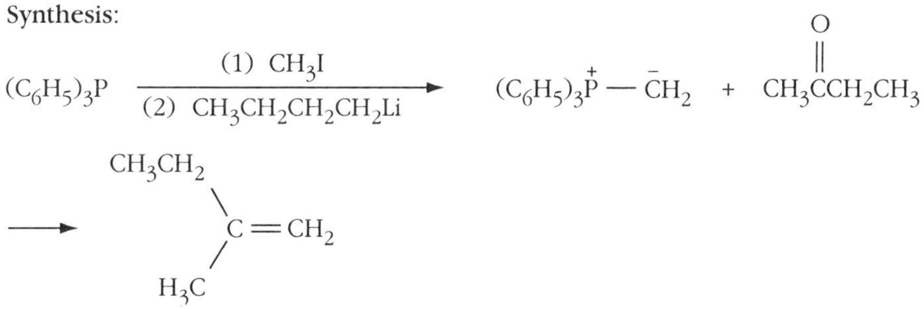

$$(C_6H_5)_3P \xrightarrow[\text{(2) } CH_3CH_2CH_2CH_2Li]{\text{(1) } CH_3I} (C_6H_5)_3\overset{+}{P}-\overset{-}{C}H_2 + CH_3\overset{O}{\overset{\|}{C}}CH_2CH_3$$

$$\longrightarrow \underset{H_3C}{\overset{CH_3CH_2}{\diagdown}}C=CH_2$$

(b) Retrosynthesis:

[cyclohexyl]=CHCH$_3$ $\Longrightarrow$ $CH_3\overset{O}{\overset{\|}{C}}\diagdown H$ + $(C_6H_5)_3\overset{+}{P}-\overset{-}{C}$(cyclohexyl) $\Longrightarrow$ $(C_6H_5)_3P$ + Br—(cyclohexyl)

cyclohexanone + $(C_6H_5)\overset{+}{P}-\overset{-}{C}HCH_3$ $\Longrightarrow$ $(C_6H_5)_3P$ + $BrCH_2CH_3$

Much better method (1° haloalkane)

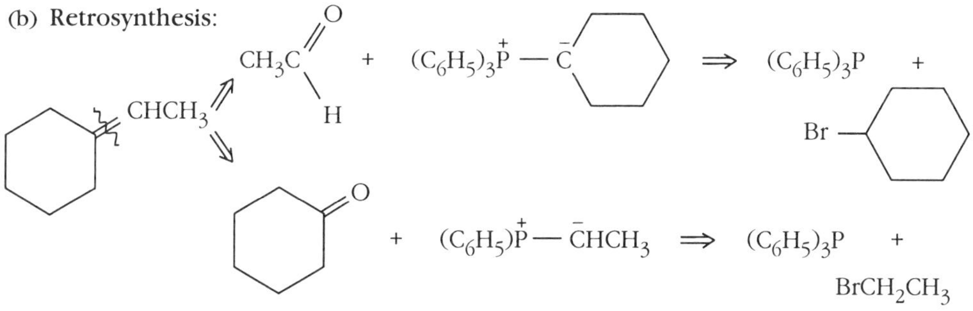

Synthesis:

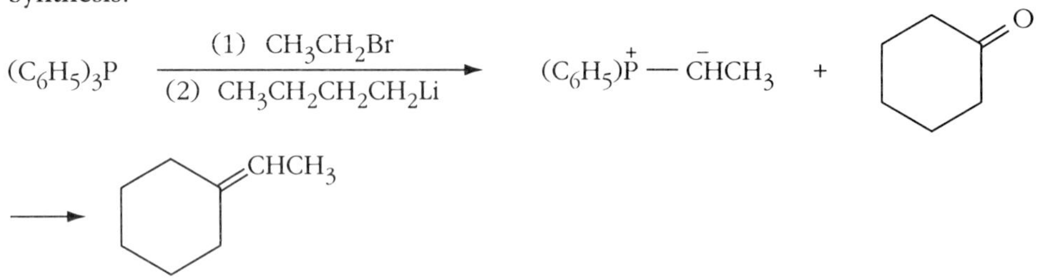

$(C_6H_5)_3P \xrightarrow[(2)\ CH_3CH_2CH_2CH_2Li]{(1)\ CH_3CH_2Br} (C_6H_5)_3\overset{+}{P}-\overset{-}{C}HCH_3\ +\ $ cyclohexanone

$\longrightarrow$ cyclohexylidene=CHCH$_3$

(c) Retrosynthesis:

$CH_3CH_2CH \overset{CH_3}{\underset{CH_3}{\cancel{=}C}} \begin{array}{l} \nearrow\ CH_3CH_2C\overset{O}{\underset{H}{\diagdown}}\ +\ (C_6H_5)_3\overset{+}{P}-\overset{-}{C}\overset{CH_3}{\underset{CH_3}{\diagdown}} \Rightarrow (C_6H_5)_3P\ +\ BrCH\overset{CH_3}{\underset{CH_3}{\diagdown}} \\ \searrow\ O=C\overset{CH_3}{\underset{CH_3}{\diagdown}}\ +\ (C_6H_5)_3\overset{+}{P}-\overset{-}{C}HCH_2CH_3 \end{array}$

$\Downarrow$

$(C_6H_5)_3P\ +\ BrCH_2CH_2CH_3$

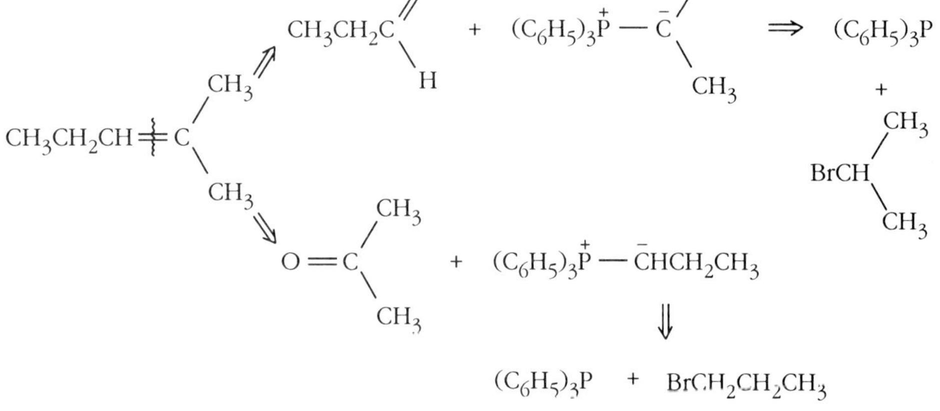

Better method
(1° haloalkane)

Synthesis:

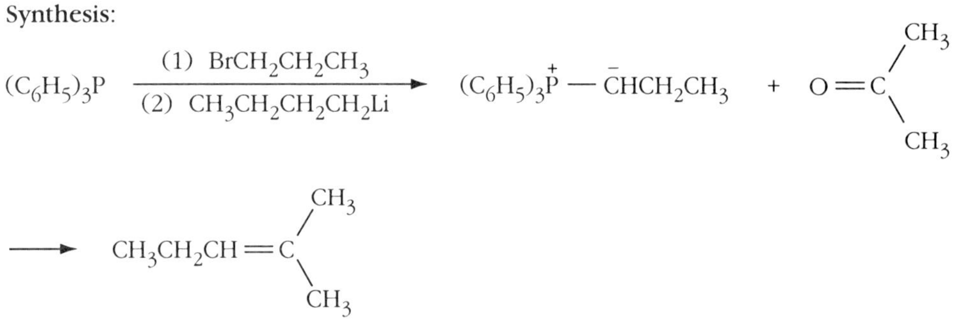

$(C_6H_5)_3P \xrightarrow[(2)\ CH_3CH_2CH_2CH_2Li]{(1)\ BrCH_2CH_2CH_3} (C_6H_5)_3\overset{+}{P}-\overset{-}{C}HCH_2CH_3\ +\ O=C\overset{CH_3}{\underset{CH_3}{\diagdown}}$

$\longrightarrow CH_3CH_2CH=C\overset{CH_3}{\underset{CH_3}{\diagdown}}$

14.32 (a) A **hemiacetal** is a compound that contains a carbon atom bearing an —OH and either —OR or —OAr groups. Hemiacetals are formed by addition of one molecule of alcohol to the carbonyl group of an aldehyde or ketone.

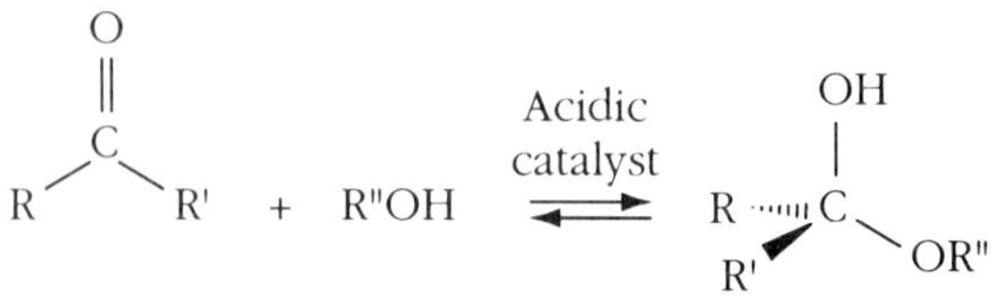

$$R-\underset{R'}{\overset{O}{\overset{\|}{C}}}-\ +\ R''OH\ \underset{}{\overset{\text{Acidic catalyst}}{\rightleftharpoons}}\ R\cdots\underset{R'}{\overset{OH}{\underset{}{\overset{|}{C}}}}-OR''$$

Hemiacetal

(b) **A nucleophilic addition reaction of a ketone** takes place when a nucleophile adds to the carbonyl carbon atom of a ketone to form a tetrahedral intermediate. Nucleophiles that can add to a carbonyl carbon atom include Grignard reagents, organolithium reagents, anions of terminal alkynes, water, ammonia, amines, hydrazine, and its derivatives.

(c) An **imine** (or a Schiff base) is a molecule containing a carbon-nitrogen double bond. Imines are formed when ammonia or a primary amine (aliphatic or aromatic) reacts with the carbonyl group of aldehydes or ketones in the presence of an acid catalyst.

(d) An **acetal** is a compound that contains a carbon atom bearing a combination of two —OR or —OAr groups. Acetals are stable to nucleophilic reagents and aqueous base but undergo hydrolysis in aqueous acid. Acetals are often used as protecting groups for the carbonyl groups of aldehydes and ketones or the hydroxyl groups of alcohols.

(e) A **semicarbazone** is an imine-type product formed from aldehydes or ketones and semicarbazide. This functional group consists of a carbon-nitrogen double bond between the carbonyl carbon and the hydrazine terminal nitrogen of semicarbazide.

$$\underset{\underset{\text{or ketone}}{\text{Aldehyde}}}{\overset{R}{\underset{R'}{>}}C=O} + \underset{\text{Semicarbazide}}{H_2N-NH\overset{}{\underset{H_2N}{\diagdown}}C=O} \xrightarrow{\text{Acid catalyst}} \underset{\text{Semicarbazone}}{\overset{R}{\underset{R'}{>}}C=N\overset{}{\underset{}{\diagdown}}NH\overset{}{\underset{H_2N}{\diagdown}}C=O}$$

(f) A **tetrahedral intermediate** is the intermediate formed in the nucleophilic addition mechanisms of aldehydes and ketones. Such an intermediate contains an sp^3 hybridized carbon atom that was the original carbonyl carbon atom of the aldehyde or ketone.

$$\underset{R \quad R'}{C=O} + :Nu-H \rightleftharpoons \underset{\underset{\text{Tetrahedral intermediate}}{R' \quad Nu-H}}{R\cdots C-O^-} \rightleftharpoons \underset{R' \quad Nu}{R\cdots C-OH}$$

(g) An **oxime** is a compound that contains a nitrogen atom bearing an —OH group and bonded to a carbon atom by a double bond. Oximes are formed when aldehydes or ketones react with hydroxylamine.

$$\overset{R}{\underset{R'}{>}}C=O + H_2N-OH \xrightarrow{\text{Acidic catalyst}} \overset{R}{\underset{R'}{>}}C=N\diagdown OH$$

(h) **Cyanohydrin** is a molecule containing —OH and —CN groups bonded to the same carbon atom. Cyanohydrins are formed when hydrogen cyanide adds to an aldehyde or a ketone.

$$\underset{R \quad R'}{\overset{O}{\|}}C \xrightarrow[H_3O^+]{NaCN} \underset{R'}{R\cdots C}\overset{OH}{\underset{C\equiv N}{|}}$$

(i) **2,4-Dinitrophenylhydrazone** is a solid product of the reaction of an aldehyde or a ketone with 2,4-dinitrophenylhydrazine. Before the use of IR, ^{1}H NMR, and ^{13}C NMR spectroscopy to identify compounds, these derivatives were used to identify aldehydes and ketones.

14.33 (a) *trans*-3-Chlorocyclopentanecarbaldehyde or (1*R*,3*R*)-3-chlorocyclopentanecarbaldehyde
(b) 2,3-Dimethylbutanedial
(c) Heptanedial
(d) Hexafluoroacetone or hexafluoropropanone
(e) 5-Methyl-1,3-cyclohexanedione
(f) (*E*)-3-Chloro-2-methylpropenal

14.34 (a) (b) (c) (d) (e) (f)

14.35 (a)

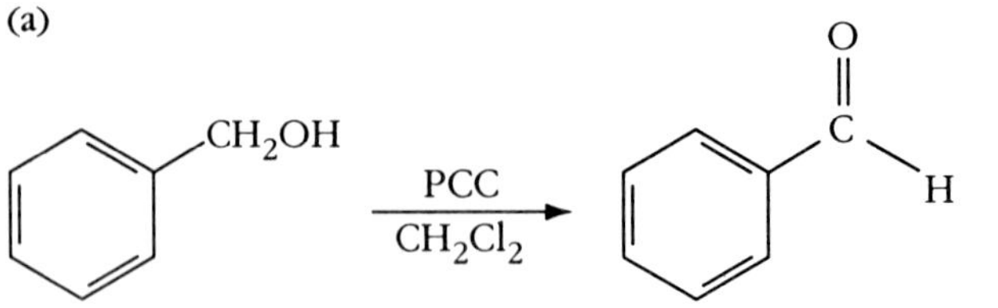

PhCH$_2$OH $\xrightarrow[\text{CH}_2\text{Cl}_2]{\text{PCC}}$ PhCHO

(b)

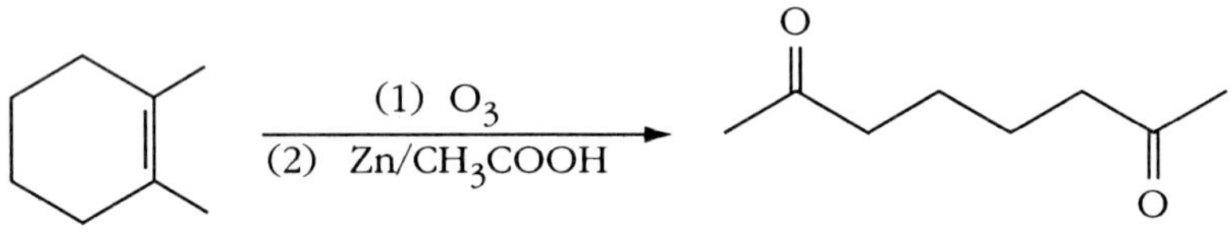

1,2-dimethylcyclohexene $\xrightarrow[\text{(2) Zn/CH}_3\text{COOH}]{\text{(1) O}_3}$ CH$_3$CO(CH$_2$)$_3$COCH$_3$ (heptane-2,6-dione)

(c)

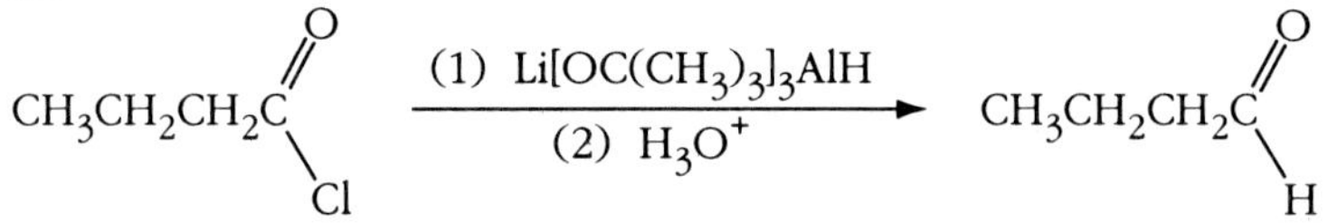

CH$_3$CH$_2$CH$_2$COCl $\xrightarrow[\text{(2) H}_3\text{O}^+]{\text{(1) Li[OC(CH}_3)_3]_3\text{AlH}}$ CH$_3$CH$_2$CH$_2$CHO

(d)

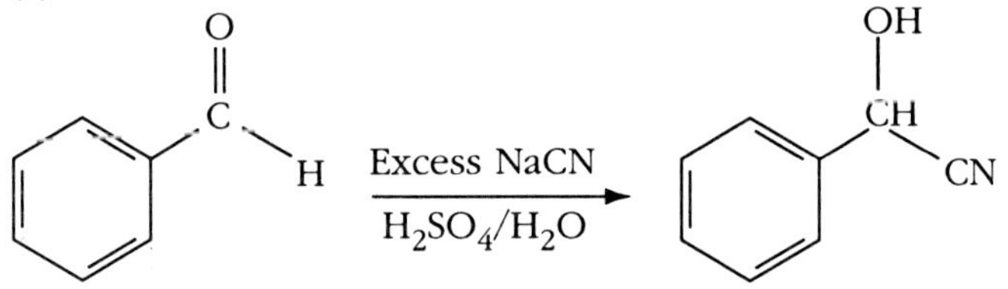

PhCHO $\xrightarrow[\text{H}_2\text{SO}_4/\text{H}_2\text{O}]{\text{Excess NaCN}}$ PhCH(OH)CN

(e)

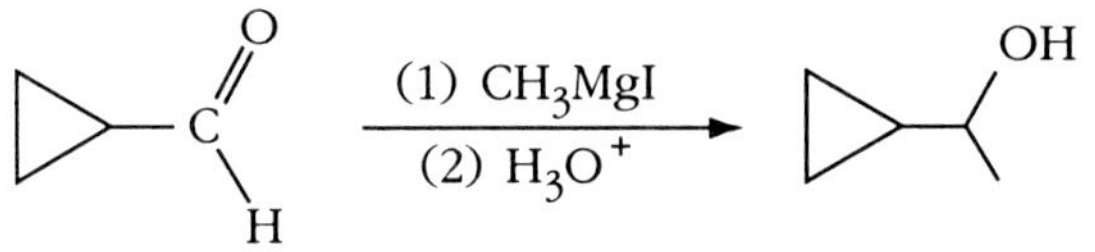

cyclopropyl-CHO $\xrightarrow[\text{(2) H}_3\text{O}^+]{\text{(1) CH}_3\text{MgI}}$ cyclopropyl-CH(OH)CH$_3$

(f)

cyclohexanone $\xrightarrow{\text{[H]}}$ cyclohexanol

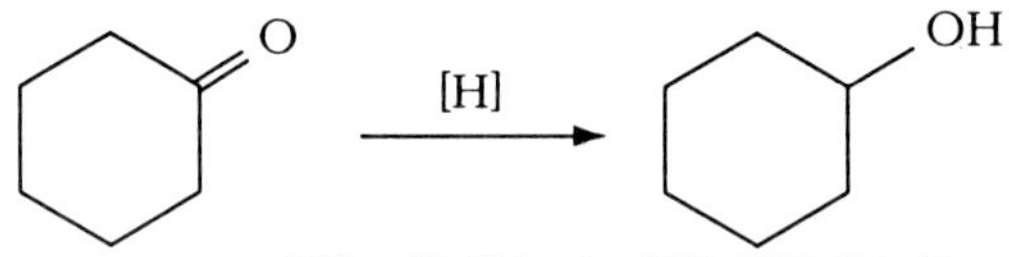

[H] = NaBH$_4$ in CH$_3$OH, H$_2$/Pt, or LiAlH$_4$ in diethyl ether

(g)

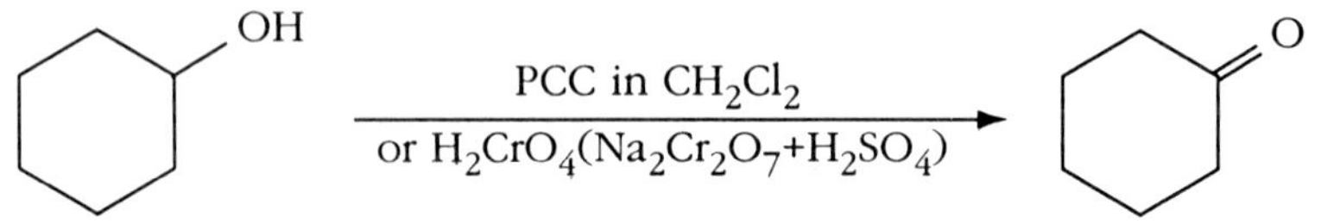

cyclohexanol $\xrightarrow[\text{or H}_2\text{CrO}_4\text{(Na}_2\text{Cr}_2\text{O}_7+\text{H}_2\text{SO}_4)]{\text{PCC in CH}_2\text{Cl}_2}$ cyclohexanone

(h)

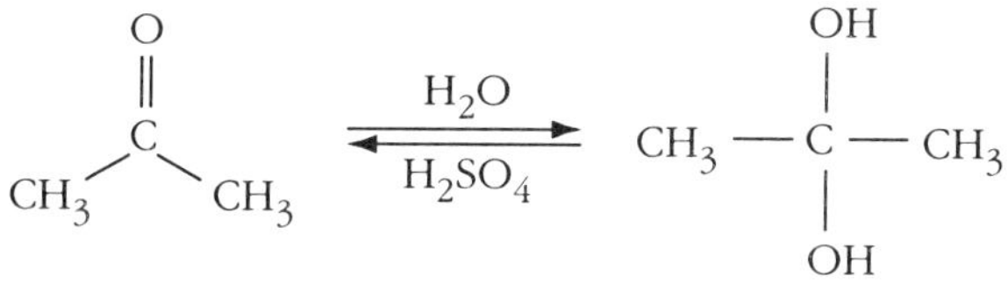

$$CH_3-\underset{CH_3}{\overset{O}{\underset{\|}{C}}}-CH_3 \;\underset{H_2SO_4}{\overset{H_2O}{\rightleftharpoons}}\; CH_3-\underset{OH}{\overset{OH}{\underset{|}{\overset{|}{C}}}}-CH_3$$

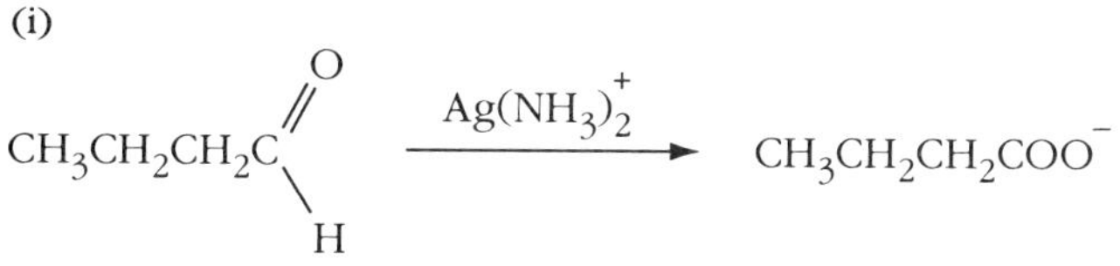

(i)

$$CH_3CH_2CH_2\overset{O}{\underset{H}{\overset{\|}{C}}} \;\xrightarrow{Ag(NH_3)_2^+}\; CH_3CH_2CH_2COO^-$$

(j)

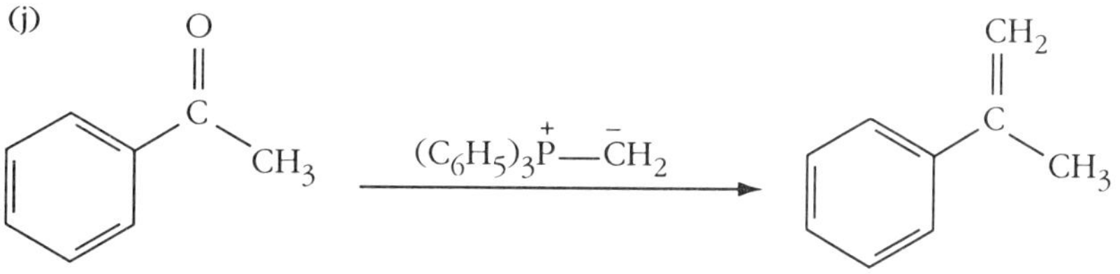

Acetophenone + $(C_6H_5)_3\overset{+}{P}-\overset{-}{CH_2}$ → α-methylstyrene

14.36

(a)

pentanal + $Ag(NH_3)_2^+ \xrightarrow[H_2O]{NH_3}$ pentanoate $O^- \; \overset{+}{NH_4}$ + Ag

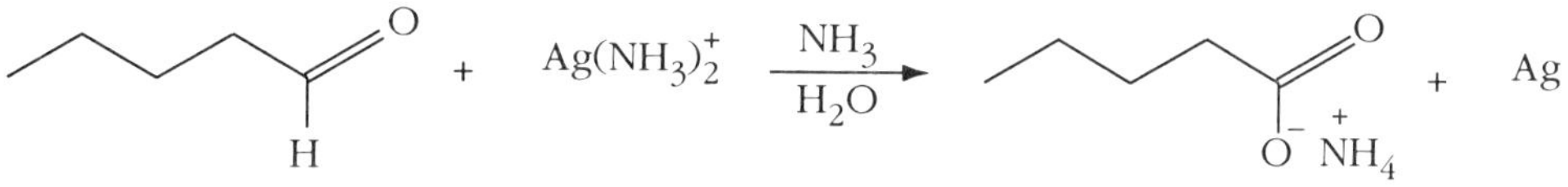

(b)

pentanal + EtOH $\xrightarrow{\text{Dry HCl}}$ hemiacetal

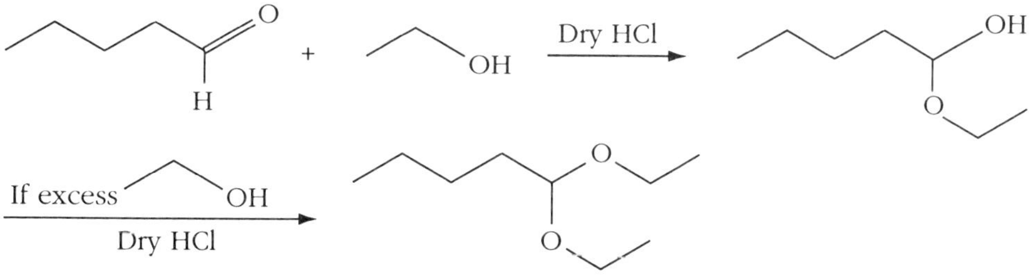

$\xrightarrow[\text{Dry HCl}]{\text{If excess EtOH}}$ acetal (diethyl acetal)

(c)

pentanal + NaCN $\xrightarrow{\text{aq. HCl}}$ cyanohydrin (OH, CN)

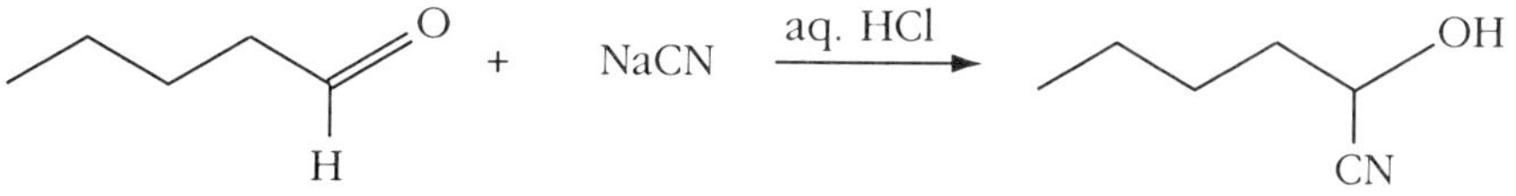

(d)

$$\text{CH}_3\text{CH}_2\text{CH}_2\text{CH}_2\text{CHO} + \text{H}_2\text{N}-\text{OH} \xrightarrow[\text{pH 4–5}]{\text{Buffer}} \text{CH}_3\text{CH}_2\text{CH}_2\text{CH}_2\text{CH}=\text{N}-\text{OH}$$

(e)

$$\text{CH}_3\text{CH}_2\text{CH}_2\text{CH}_2\text{CHO} \xrightarrow[(2)\ \text{H}_3\text{O}^+]{(1)\ \text{LiAlH}_4} \text{CH}_3\text{CH}_2\text{CH}_2\text{CH}_2\text{CH}_2\text{OH}$$

(f)

$$\text{CH}_3\text{CH}_2\text{CH}_2\text{CH}_2\text{CHO} \xrightarrow[(2)\ \text{H}_3\text{O}^+]{(1)\ \text{CH}_3\text{MgI}} \text{CH}_3\text{CH}_2\text{CH}_2\text{CH}_2\text{CH(OH)CH}_3$$

(g)

$$\text{CH}_3\text{CH}_2\text{CH}_2\text{CH}_2\text{CHO} \xrightarrow[\text{H}_2\text{SO}_4/\text{H}_2\text{O}]{\text{CrO}_3} \text{CH}_3\text{CH}_2\text{CH}_2\text{CH}_2\text{COOH}$$

(h)

$$\text{CH}_3\text{CH}_2\text{CH}_2\text{CH}_2\text{CHO} \xrightarrow[(2)\ \text{H}_3\text{O}^+]{(1)\ \text{HC}\equiv\text{C}^-\text{Na}^+} \text{CH}_3\text{CH}_2\text{CH}_2\text{CH}_2\text{CH(OH)C}\equiv\text{CH}$$

(i)

pentanal + 2,4-dinitrophenylhydrazine $\xrightarrow[\text{pH 4–5}]{\text{Buffer}}$ pentanal 2,4-dinitrophenylhydrazone

(j)

$$\text{CH}_3\text{CH}_2\text{CH}_2\text{CH}_2\text{CHO} + (\text{C}_6\text{H}_5)_3\overset{+}{\text{P}}-\overset{-}{\text{C}}\text{HCH}_3 \longrightarrow \text{CH}_3\text{CH}_2\text{CH}_2\text{CH}_2\text{CH}=\text{CHCH}_3$$

14.37 (a)

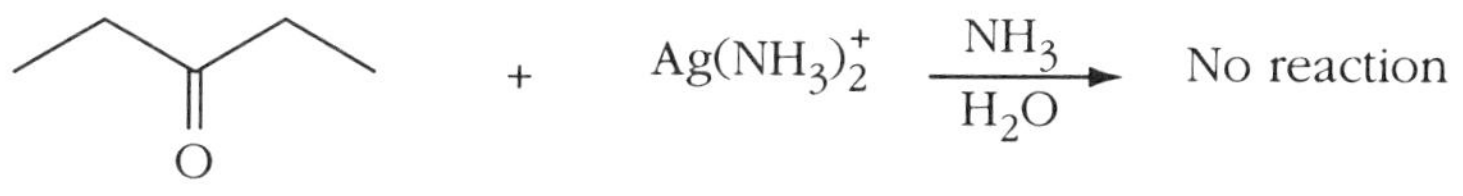

CH₃CH₂COCH₂CH₃ + Ag(NH₃)₂⁺ →(NH₃/H₂O)→ No reaction

(b)

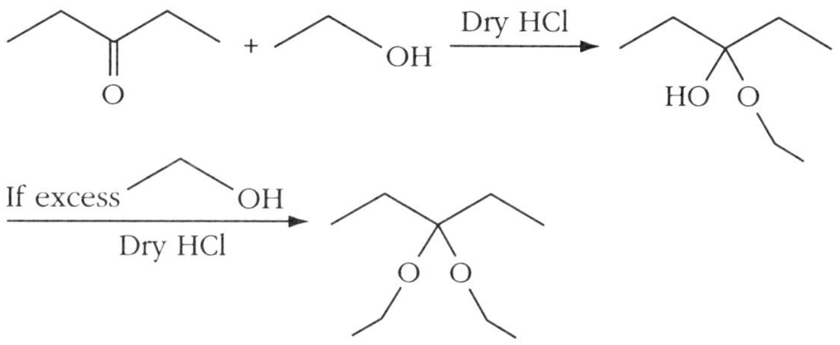

CH₃CH₂COCH₂CH₃ + CH₃CH₂OH →(Dry HCl)→ 3-pentanol hemiketal (HO, OEt on C3)

If excess CH₃CH₂OH, Dry HCl → 3,3-diethoxypentane (ketal)

(c)

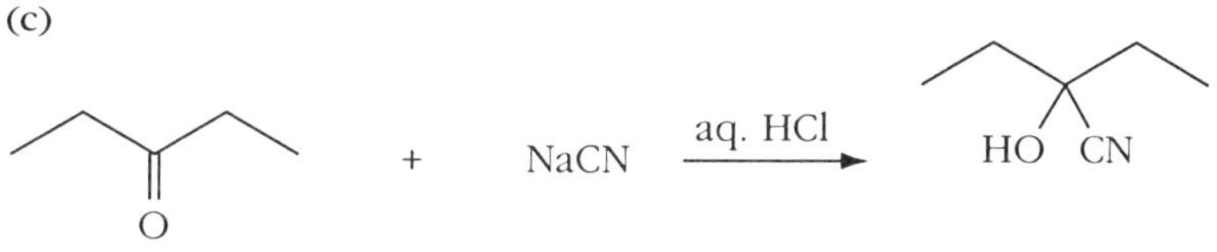

CH₃CH₂COCH₂CH₃ + NaCN →(aq. HCl)→ 3-hydroxy-3-cyanopentane (HO, CN on C3)

(d)

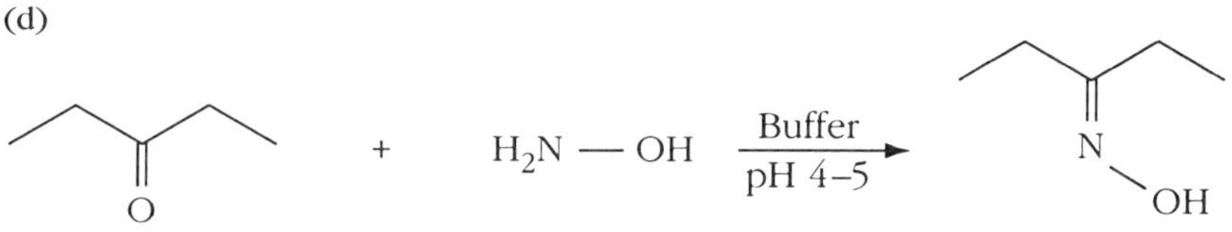

CH₃CH₂COCH₂CH₃ + H₂N—OH →(Buffer, pH 4–5)→ 3-pentanone oxime (=N–OH)

(e)

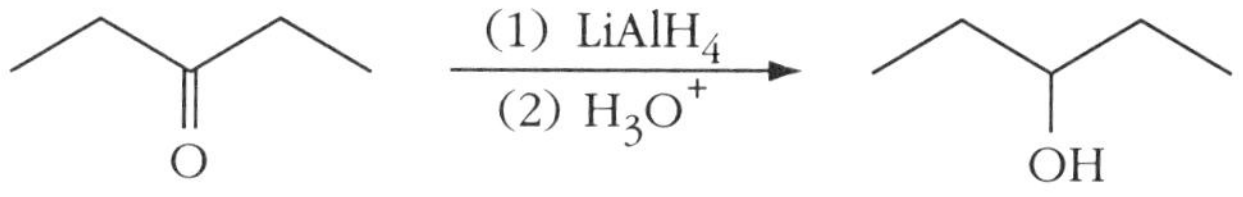

CH₃CH₂COCH₂CH₃ →(1) LiAlH₄ / (2) H₃O⁺→ 3-pentanol

(f)

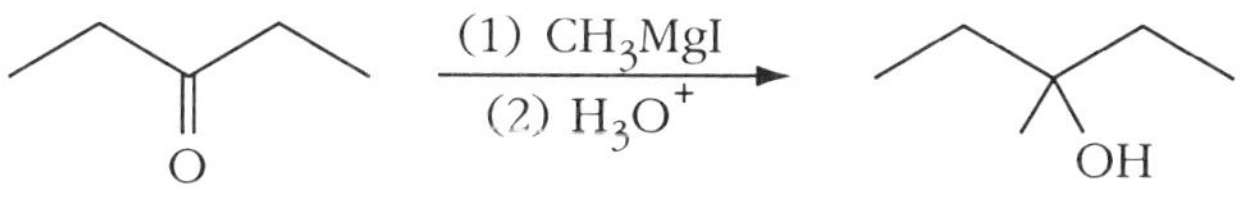

CH₃CH₂COCH₂CH₃ →(1) CH₃MgI / (2) H₃O⁺→ 3-methyl-3-pentanol

(g)

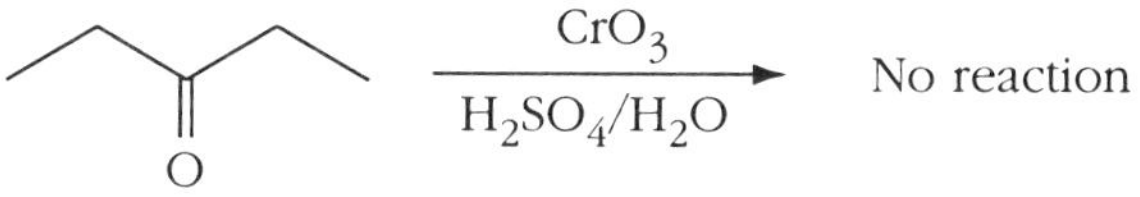

CH₃CH₂COCH₂CH₃ →(CrO₃, H₂SO₄/H₂O)→ No reaction

(h)

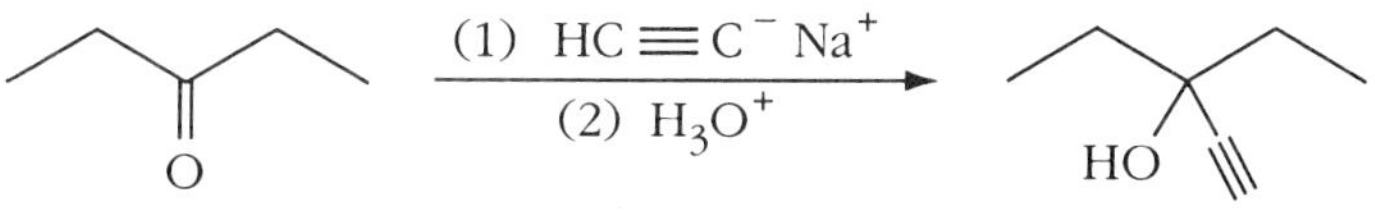

CH₃CH₂COCH₂CH₃ →(1) HC≡C⁻Na⁺ / (2) H₃O⁺→ 3-ethynyl-3-pentanol

(i)

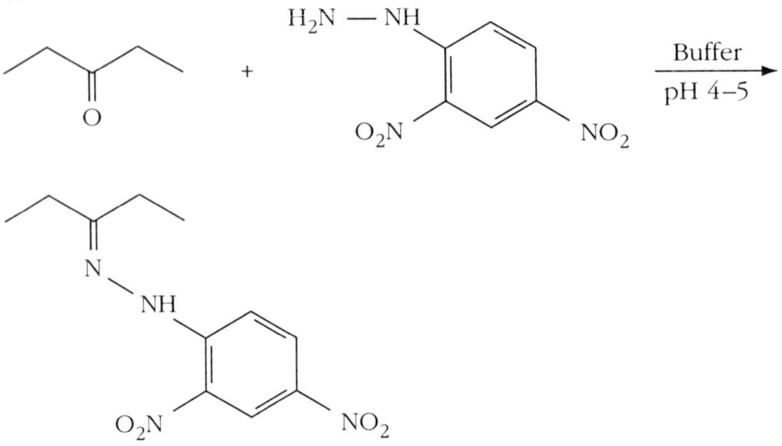

CH₃CH₂-C(=O)-CH₂CH₃ + H₂N—NH—C₆H₃(NO₂)₂ (2,4-dinitrophenylhydrazine) →(Buffer, pH 4–5)

CH₃CH₂-C(=N-NH-C₆H₃(NO₂)₂)-CH₂CH₃

(j)

CH₃CH₂-C(=O)-CH₂CH₃ + (C₆H₅)₃P⁺—⁻CHCH₃ ⟶ 3-ethyl-2-pentene (CH₃CH₂)(CH₃CH₂)C=CHCH₃

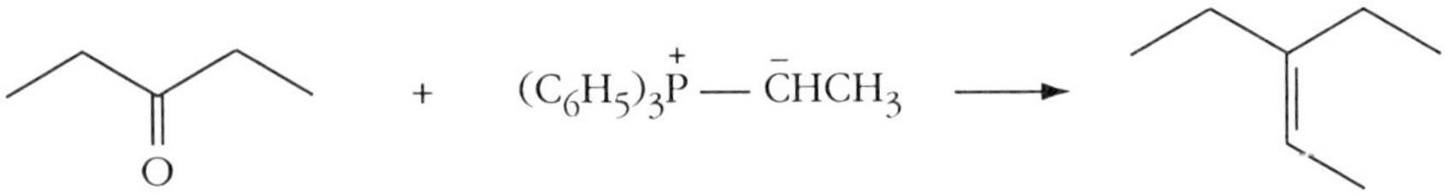

14.38 A = CH₃C≡C⁻Na⁺
B = Lindlar's catalyst/H₂
C = PCC/CH₂Cl₂

14.39 (a)

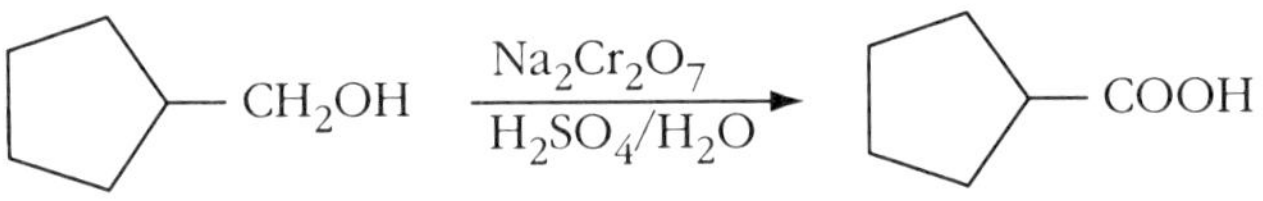

cyclopentyl-CH₂OH $\xrightarrow{\text{Na}_2\text{Cr}_2\text{O}_7 / \text{H}_2\text{SO}_4/\text{H}_2\text{O}}$ cyclopentyl-COOH

(b)

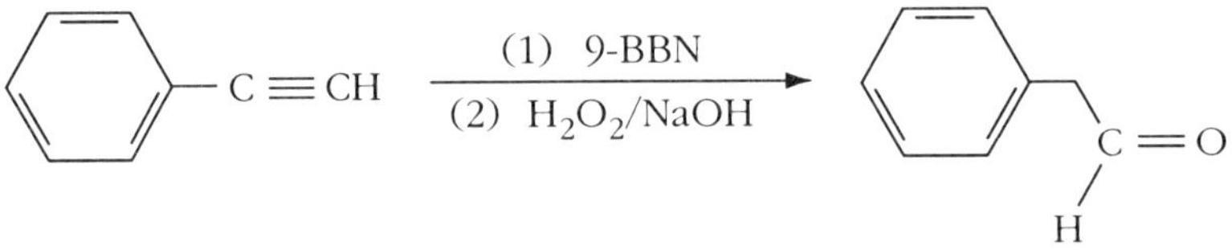

C₆H₅—C≡CH $\xrightarrow[\text{(2) H}_2\text{O}_2/\text{NaOH}]{\text{(1) 9-BBN}}$ C₆H₅—CH₂—CHO

(c)

cyclopentyl-CH₂OH $\xrightarrow{\text{PCC} / \text{CH}_2\text{Cl}_2}$ cyclopentyl-CHO

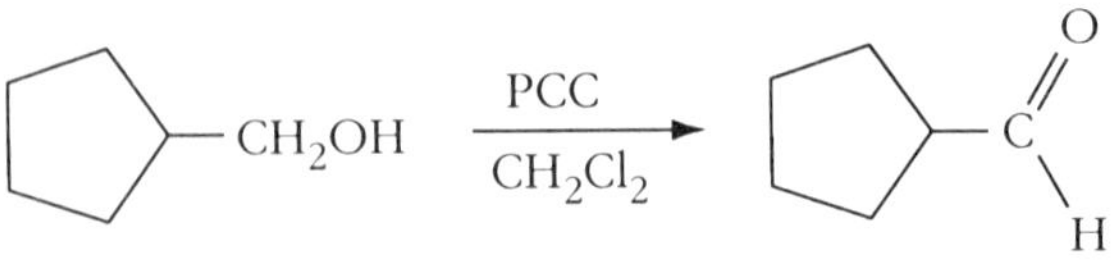

(d)

octahydronaphthalene (with C=C) $\xrightarrow[\text{(2) H}_2\text{O}_2/\text{NaOH}]{\text{(1) 9-BBN}}$ decahydronaphthalen-2-ol

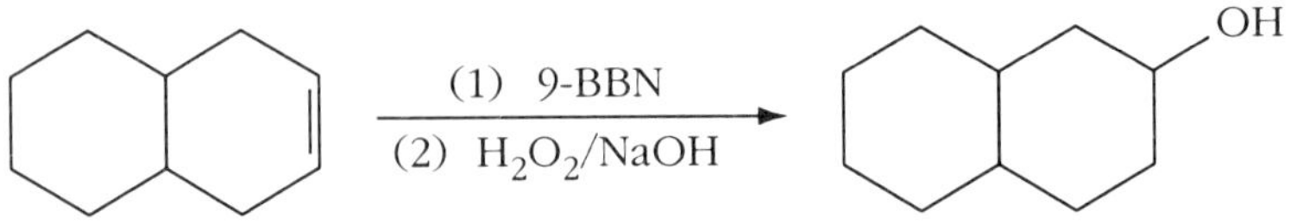

(e)

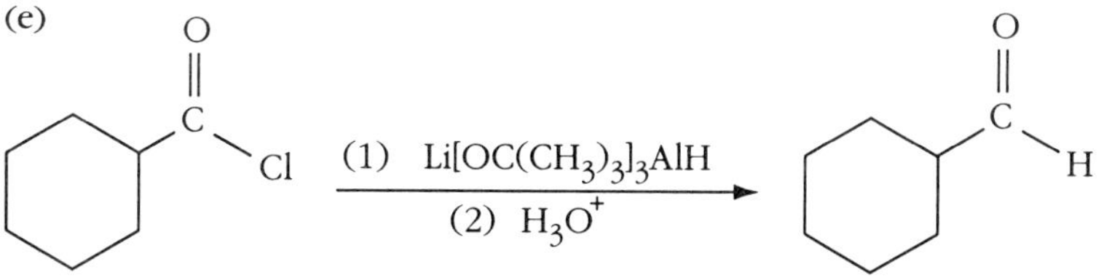

Cyclohexanecarbonyl chloride $\xrightarrow{\text{(1) Li[OC(CH}_3)_3]_3\text{AlH}}_{\text{(2) H}_3\text{O}^+}$ cyclohexanecarbaldehyde

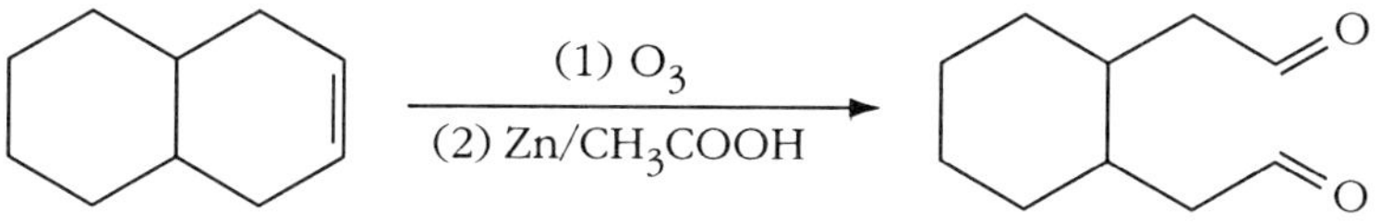

(f)

Octahydronaphthalene (with one C=C) $\xrightarrow{\text{(1) O}_3}_{\text{(2) Zn/CH}_3\text{COOH}}$ cyclohexane-1,2-diyl-bis(acetaldehyde)

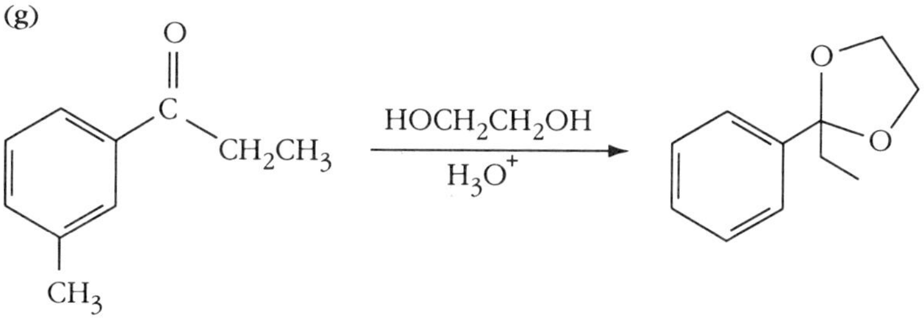

(g)

1-(3-methylphenyl)propan-1-one $\xrightarrow{\text{HOCH}_2\text{CH}_2\text{OH}}_{\text{H}_3\text{O}^+}$ 2-ethyl-2-phenyl-1,3-dioxolane

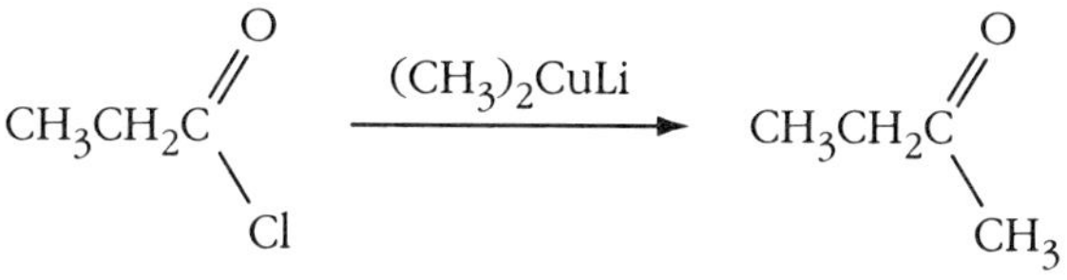

(h)

$CH_3CH_2\overset{O}{\underset{Cl}{C}} \xrightarrow{(CH_3)_2CuLi} CH_3CH_2\overset{O}{\underset{CH_3}{C}}$

(i)

3,4-dihydroxybutanal derivative $\xrightarrow{\text{NaCN}}_{\text{H}_2\text{SO}_4}$ cyanohydrin with added OH/CN

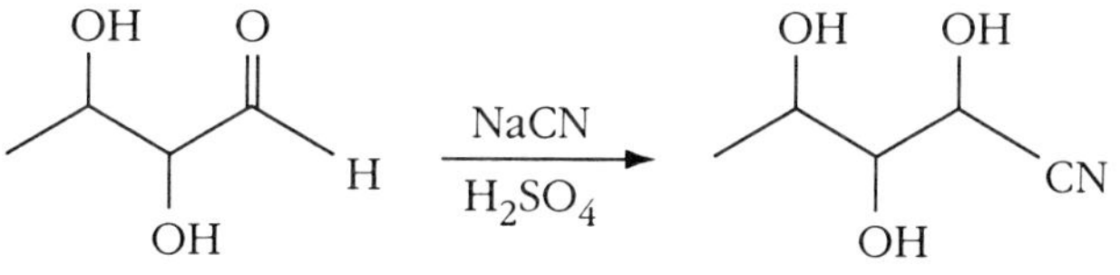

(j)

Cyclopentanone $\xrightarrow{(C_6H_5)_3\overset{+}{P}-\overset{-}{C}HCOOCH_3}$ methyl 2-cyclopentylideneacetate

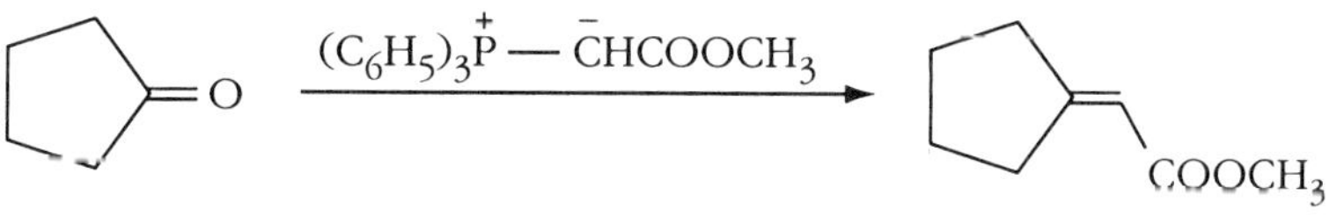

14.40

A = 4-methylcyclohex-3-ene-1-carbaldehyde

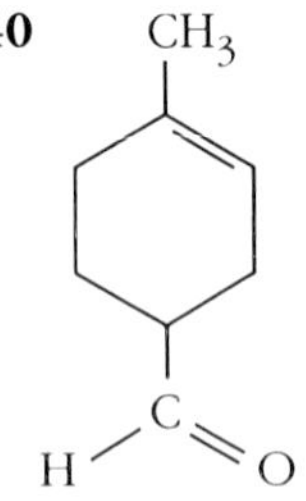

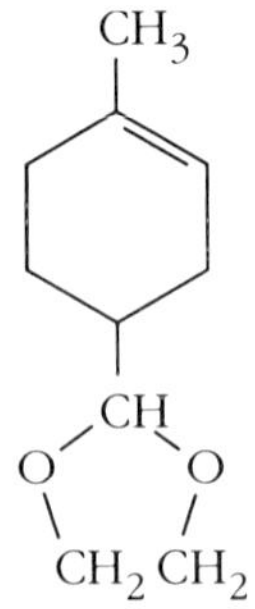

B = 4-methylcyclohex-3-ene-1-carbaldehyde ethylene acetal

C = keto-aldehyde with 1,3-dioxolane protecting group

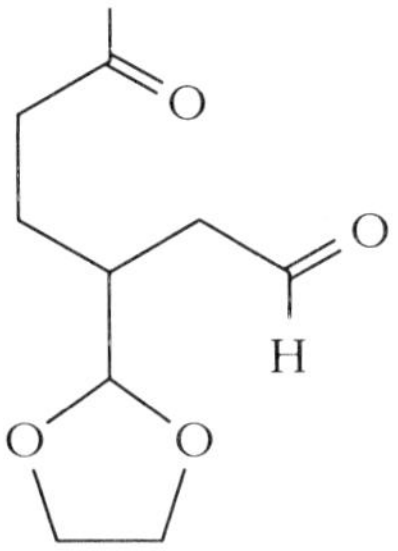

Acetic acid is not strong enough to catalyze hydrolysis of the acetal in B.

14.41

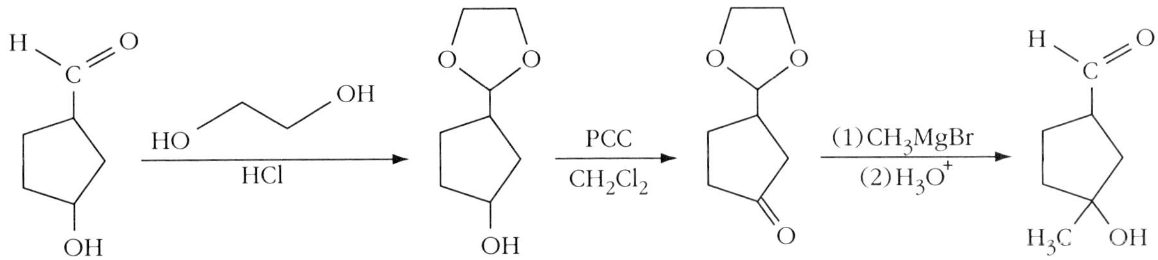

cyclopentane-1,3-diol-carbaldehyde →(HOCH$_2$CH$_2$OH, HCl)→ dioxolane-protected alcohol →(PCC, CH$_2$Cl$_2$)→ dioxolane-protected ketone →(1) CH$_3$MgBr (2) H$_3$O$^+$→ aldehyde with tertiary alcohol (H$_3$C, OH)

14.42

(a) Retrosynthesis:

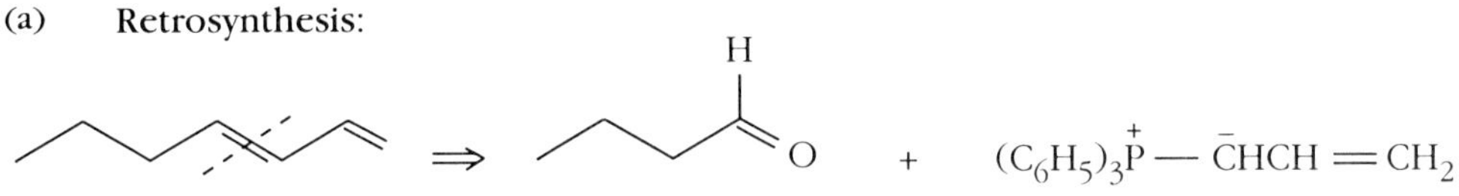

hept-2,6-diene ⇒ butanal + $(C_6H_5)_3\overset{+}{P}—\overset{-}{C}HCH=CH_2$

Synthesis:

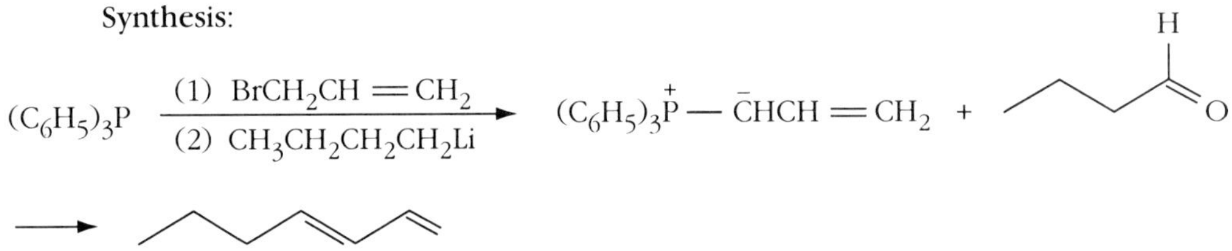

$(C_6H_5)_3P \xrightarrow{\text{(1) BrCH}_2\text{CH}=\text{CH}_2\ \text{(2) CH}_3\text{CH}_2\text{CH}_2\text{CH}_2\text{Li}} (C_6H_5)_3\overset{+}{P}—\overset{-}{C}HCH=CH_2 +$ butanal

⟶ hepta-2,6-diene

(b) Retrosynthesis:

hexan-3-ol ⇒ butanal + CH$_3$CH$_2$MgBr

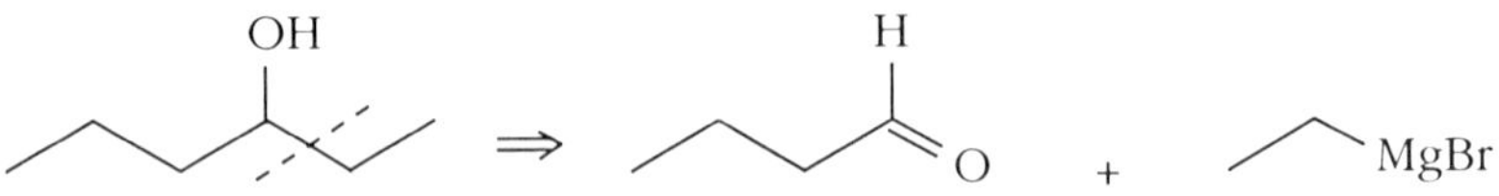

Synthesis:

butanal →(1) CH$_3$CH$_2$MgBr (2) H$_3$O$^+$→ hexan-3-ol

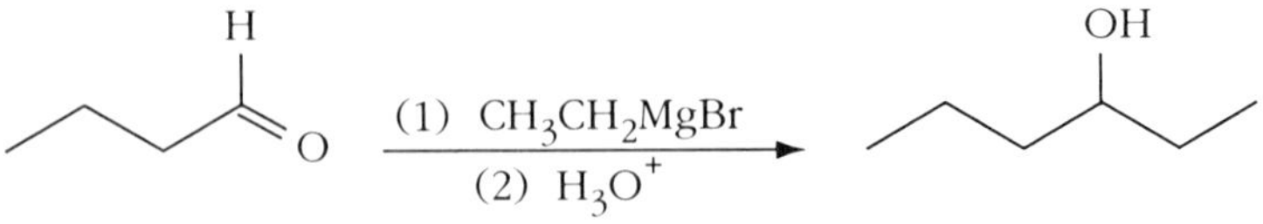

(c) Retrosynthesis:

$$\text{CH}_3\text{CH}_2\text{CH}_2\text{CH}_2\text{CH}_2\text{OH} \Rightarrow \text{CH}_3\text{CH}_2\text{CH}_2\text{CH}=\text{CH}_2 \Rightarrow \text{CH}_3\text{CH}_2\text{CH}_2\text{CHO} \;+\; (C_6H_5)_3\overset{+}{P}-\overset{-}{C}H_2$$

Synthesis:

$$(C_6H_5)_3P \xrightarrow[\text{(2) BuLi}]{\text{(1) CH}_3\text{I}} (C_6H_5)_3\overset{+}{P}-\overset{-}{C}H_2 \;+\; \text{CH}_3\text{CH}_2\text{CH}_2\text{CHO} \longrightarrow$$

$$\text{CH}_3\text{CH}_2\text{CH}_2\text{CH}=\text{CH}_2 \xrightarrow[\text{(2) H}_2\text{O}_2/\text{NaOH}]{\text{(1) 9-BBN}} \text{CH}_3\text{CH}_2\text{CH}_2\text{CH}_2\text{CH}_2\text{OH}$$

(d) Retrosynthesis:

$$\text{CH}_3\text{CH}_2\text{CH}_2\text{COOH} \Rightarrow \text{CH}_3\text{CH}_2\text{CH}_2\text{CHO}$$

Synthesis:

$$\text{CH}_3\text{CH}_2\text{CH}_2\text{CHO} \xrightarrow[(\text{Na}_2\text{Cr}_2\text{O}_7+\text{H}_2\text{SO}_4)]{\text{H}_2\text{CrO}_4} \text{CH}_3\text{CH}_2\text{CH}_2\text{COOH}$$

(Or one of many other oxidizing agents)

(e) Retrosynthesis:

$$\text{CH}_3\text{CH}_2\text{CH}_2\text{CH}=\text{CHCOOCH}_2\text{CH}_3 \Rightarrow \text{CH}_3\text{CH}_2\text{CH}_2\text{CHO} \;+\;$$

$$(C_6H_5)_3\overset{+}{P}-\overset{-}{C}H\text{C}(=O)\text{OCH}_2\text{CH}_3$$

Synthesis:

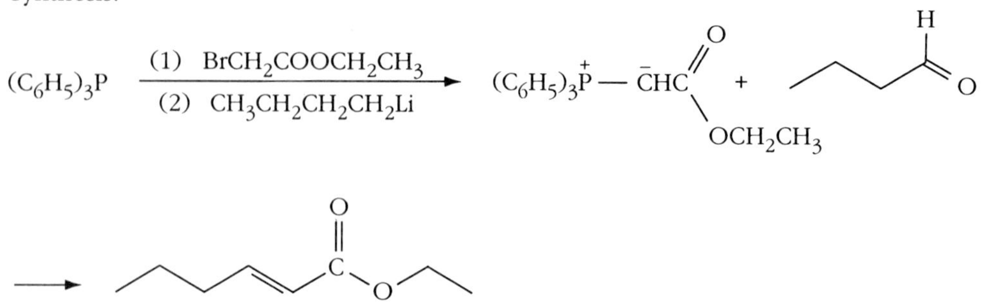

$(C_6H_5)_3P \xrightarrow{\text{(1) } BrCH_2COOCH_2CH_3}_{\text{(2) } CH_3CH_2CH_2CH_2Li} (C_6H_5)_3\overset{+}{P}-\bar{C}HC\underset{OCH_2CH_3}{\overset{O}{\parallel}} + \text{CH}_3\text{CH}_2\text{CH}_2\text{CHO}$

$\longrightarrow$ CH$_3$CH$_2$CH=CH—CH=CH—COOCH$_2$CH$_3$ (ethyl (E)-2-hexenoate)

(f) Retrosynthesis:

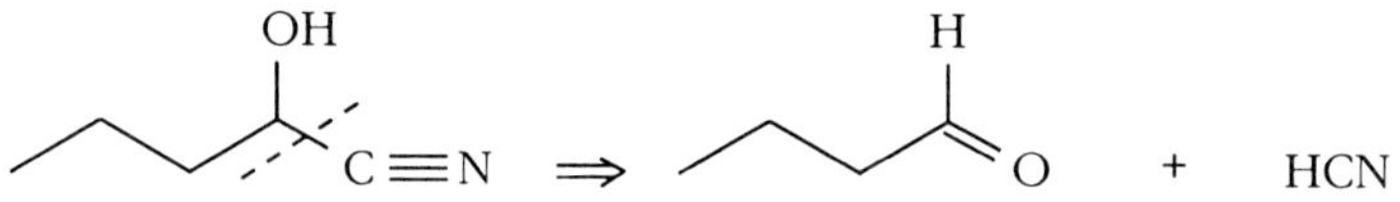

[structure: 2-hydroxy-2-cyanopentane (from butanal + HCN)] ⇒ butanal + HCN

Synthesis:

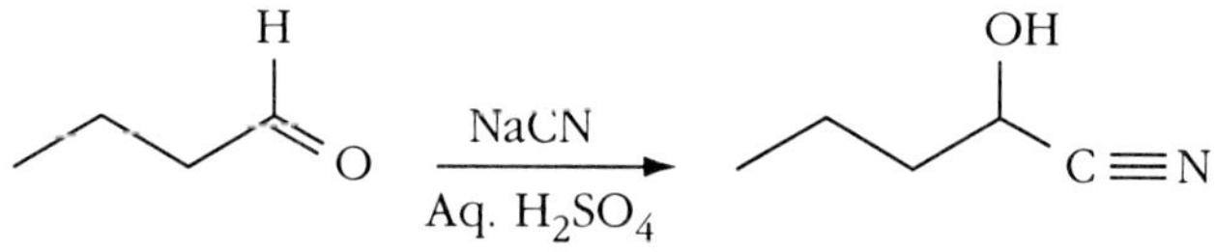

butanal $\xrightarrow[\text{Aq. H}_2\text{SO}_4]{\text{NaCN}}$ 2-hydroxypentanenitrile

(g) Retrosynthesis:

2-propyl-1,3-dioxane ⇒ butanal + HOCH$_2$CH$_2$CH$_2$OH

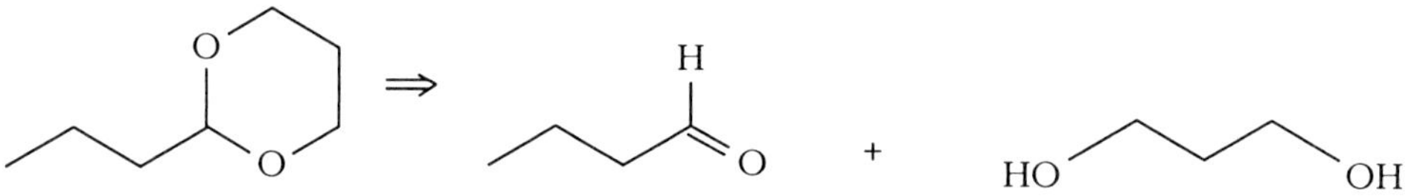

Synthesis:

butanal + HO—CH$_2$CH$_2$CH$_2$—OH $\xrightarrow{\text{Dry HCl}}$ 2-propyl-1,3-dioxane

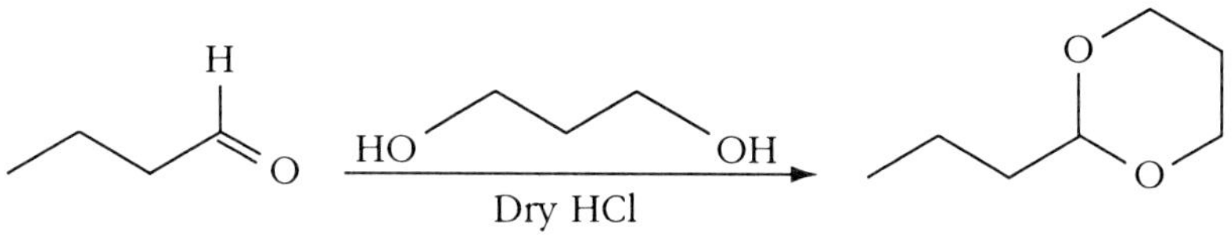

14.43

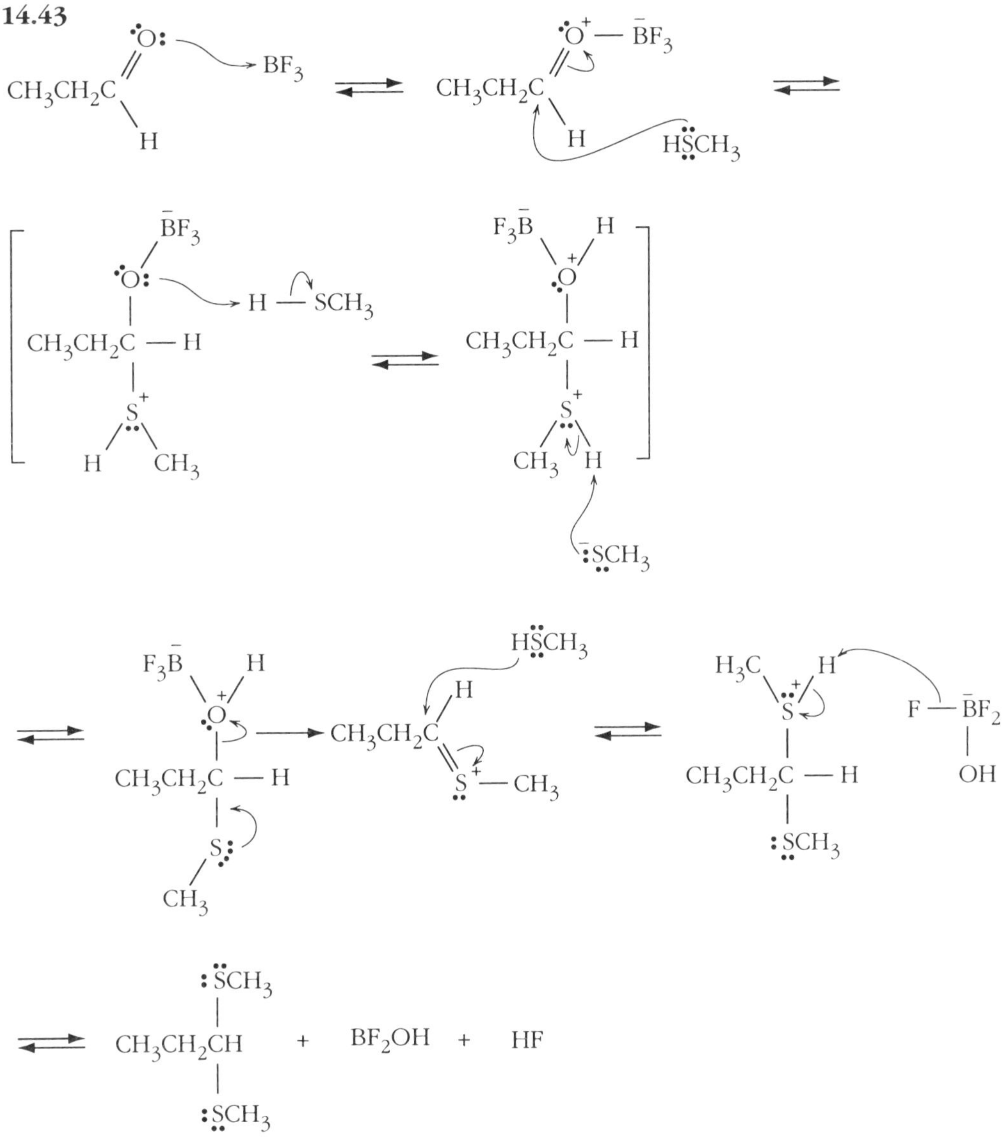

[Mechanism showing protonation of propanal by BF₃, followed by addition of HSCH₃, proton transfers, loss of water (as BF₂OH), addition of second HSCH₃, and deprotonation to give the dithioacetal]

$$CH_3CH_2CHO + BF_3 \rightleftharpoons CH_3CH_2C(H)=O^+-\bar{B}F_3$$

Step 1: BF₃ coordinates to the carbonyl oxygen of propanal; methanethiol attacks the activated carbonyl carbon.

Step 2: Proton transfer from S to a second molecule of methanethiol, giving a neutral hemithioacetal with BF₃ still bound to O.

Step 3: Protonation of the O-BF₃ by CH₃SH₂⁺; departure of BF₂OH generates a thiocarbocation (stabilized as sulfonium ion with C=S⁺).

Step 4: Attack of a second CH₃SH on the sulfonium ion.

Step 5: Deprotonation gives the dithioacetal.

$$\rightleftharpoons CH_3CH_2CH(SCH_3)_2 + BF_2OH + HF$$

Note: BF₂OH reacts with two more moles of propanal before it is converted to boric acid, H₃BO₃.

14.44

	IR	¹H NMR	¹³C NMR
(a) 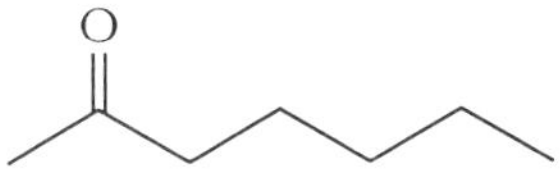 2-heptanone (CH₃COCH₂CH₂CH₂CH₂CH₃)	1710 cm⁻¹	singlet (δ 2.1, 3H), triplet (δ 2, 2H), multiplet (δ 1.1, 6H), triplet (δ 0.9, 3H)	7 signals
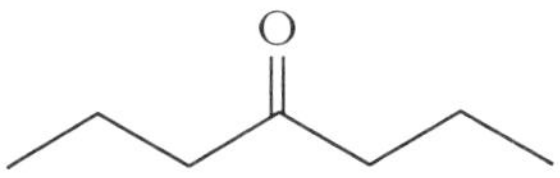 4-heptanone (CH₃CH₂CH₂COCH₂CH₂CH₃)	1710 cm⁻¹	triplet (δ 2.1, 2H), sextet (δ 1.2, 2H), triplet (δ 0.9, 3H)	4 signals

The infrared spectra are not useful because both compounds show a strong carbonyl stretch at 1710 cm^{-1}.

The ^{1}H NMR spectra are more useful. The seven carbons in 2-heptanone are unique and give rise to a complex aliphatic proton spectrum. In 4-heptanone, the two alkyl groups attached to the carbonyl carbon are equivalent. This compound therefore exhibits three types of hydrogen atoms (typical pattern of a propyl group).

The ^{13}C NMR is even more informative. There are seven signals representing the seven nonequivalent carbon atoms in the spectrum of 2-heptanone and only four signals in the spectrum of 4-heptanone.

(b)

	IR	^{1}H NMR	^{13}C NMR
cyclohexanone	1715 cm^{-1}	triplet (δ 2.1, 4H), multiplet (δ 1.1, 6H)	4 signals
3-cyclohexenol	3350 cm^{-1}	multiplet (δ 5–6, 2H) triplet (δ 3.5, 1H) multiplet (δ 1.5–2, 6H)	6 signals

The infrared spectra are distinctly different. Cyclohexanone shows the carbonyl absorption at 1715 cm^{-1}, which 3-cyclohexenol lacks. Conversely, 3-cyclohexenol exhibits a broad band for the O—H stretch at 3350 cm^{-1}, which is missing from the IR spectrum of cyclohexanone.

The ^{1}H NMR spectra are also useful. The six carbons in 3-cyclohexenol are unique and give rise to a more complex spectrum. Especially characteristic are the signals for the vinyl hydrogens of 3-cyclohexenol appearing in the range δ 5 to 6. Cyclohexanone shows two broad sets of peaks at δ 2 and δ 1.1.

The ^{13}C NMR spectra are also useful. The six carbons in 3-cyclohexenol are unique and give rise to six signals while cyclohexanone shows only four signals, including that of the carbonyl carbon at δ 211.

	IR	^{1}H NMR	^{13}C NMR
(c) CH$_3$CH$_2$CH$_2$CH$_2$COOH	3200–2500 cm^{-1} 1710 cm^{-1} (C=O)	singlet (δ 10–13, 1H), triplet (δ 2.1, 2H), multiplet (δ 1.1, 4H), triplet (δ 0.9, 3H)	5 signals
CH$_3$CH(OH)CH$_2$CH$_2$CHO	3500–3200 cm^{-1} 1725 cm^{-1} (C=O) 2700 cm^{-1}	singlet (δ 9.5, 1H), sextet (δ 3.6, 1H), singlet (δ 1.5–4, 1H), multiplet (δ 0.9–1.2, 7H)	5 signals

The infrared spectra differ in the position of the O—H stretch. Carboxylic acids show a wide band in the 3200 to 2500 cm^{-1} region, whereas

4-hydroxypentanal exhibits a strong absorption in the 3500 to 3200 cm^{-1} region.

The ^{1}H and ^{13}C NMR spectra are not useful in this case. One may look for a difference in the chemical shift of the carboxylic proton (δ 10-13) versus the aldehyde proton (δ 9.5) or the position of the carbonyl carbon of the carboxylic acid (δ 180) versus the carbonyl carbon of the aldehyde (δ 200) but, in general, the NMR spectra are less useful than the IR spectra for distinguishing between pentanoic acid and 4-hydroxypentanal.

14.45 Let's represent the acid and its conjugate base by HB and B:$^-$, respectively.

Formation of nucleophile:

$$HO\overset{+}{N}H_3Cl + B:^- \rightleftharpoons HB + H_2\ddot{N}-OH$$

Hydroxylamine hydrochloride   Hydroxylamine

Addition of nucleophile:

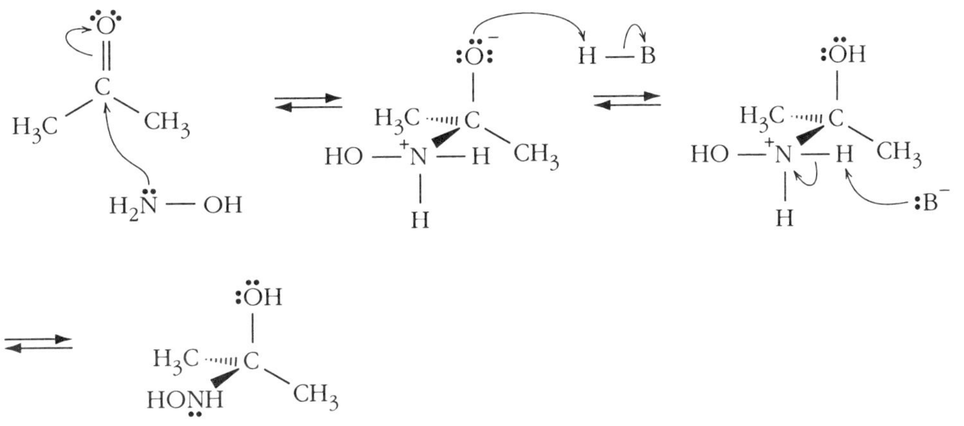

[Mechanism showing acetone reacting with hydroxylamine through tetrahedral intermediates to give the carbinolamine $CH_3-C(OH)(CH_3)-NHOH$]

Elimination of water:

[Mechanism showing deprotonation of OH, protonation of OH to form OH_2^+, loss of water to form iminium ion, then deprotonation to give the oxime $(CH_3)_2C=N-OH$]

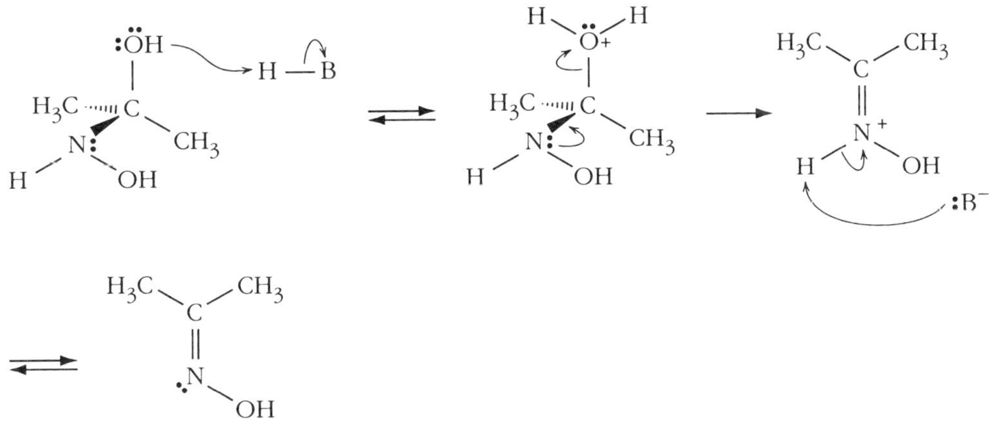

14.46 Formation of hemiacetal:

[mechanism showing acetaldehyde + ethanol forming protonated hemiacetal intermediate, then hemiacetal]

Formation of acetal:

[mechanism showing protonation of hemiacetal OH by ethanol is unfavorable, marked with ✗]

Ethanol is a weak acid (pK_a ≈16.9). It is too weak to protonate the —OH group of the hemiacetal, so the reaction stops at this point.

14.47 Step 1: Addition of hydrazine to the carbonyl group

[mechanism showing acetone + hydrazine addition, proton transfers with base B]

Step 2: Loss of water

Step 3: Addition of hydrazine to a second carbonyl group

Step 4: Loss of water

14.48 Step 1: Addition of water

Step 2: Loss of ethylamine

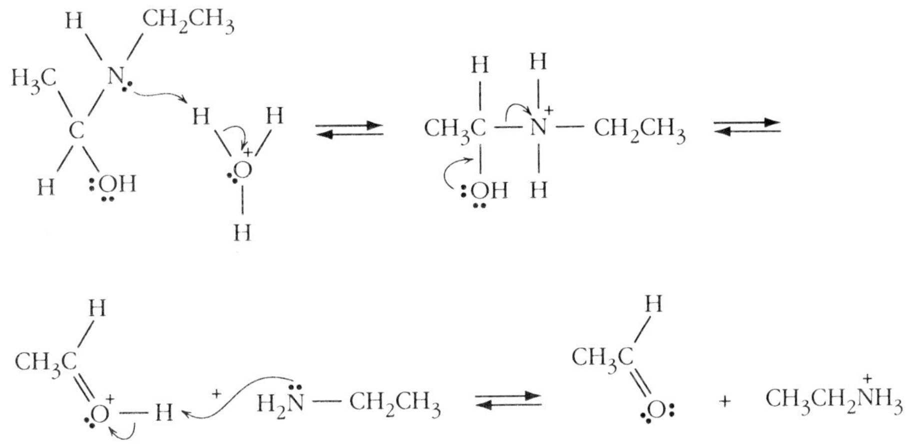

14.49 (a) Neutral solution.
Addition of water:

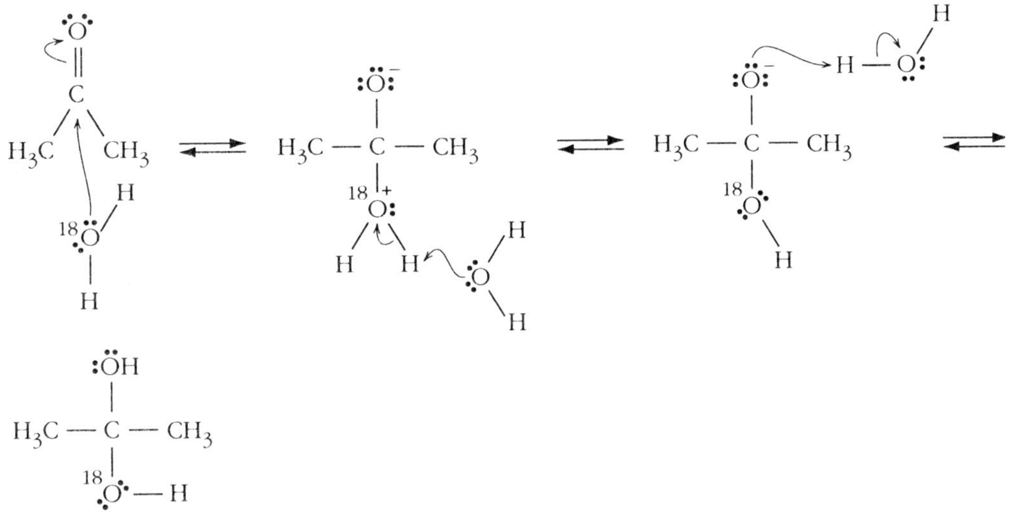

2,2-Propanediol

Loss of water:

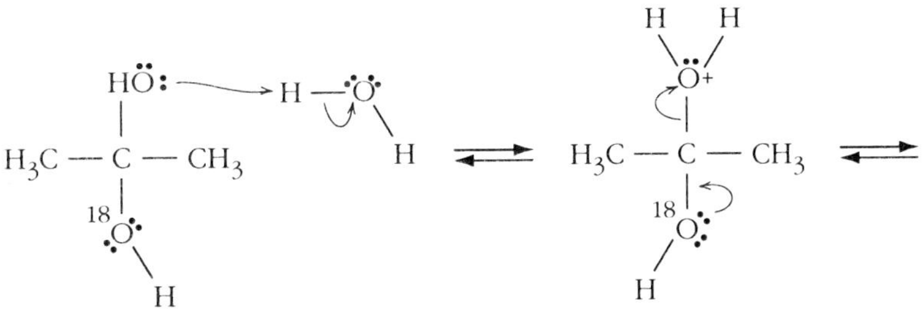

(b) Acidic solution.
Addition of water:

Loss of water:

(c) Basic solution.
Addition of $^{18}OH^-$:

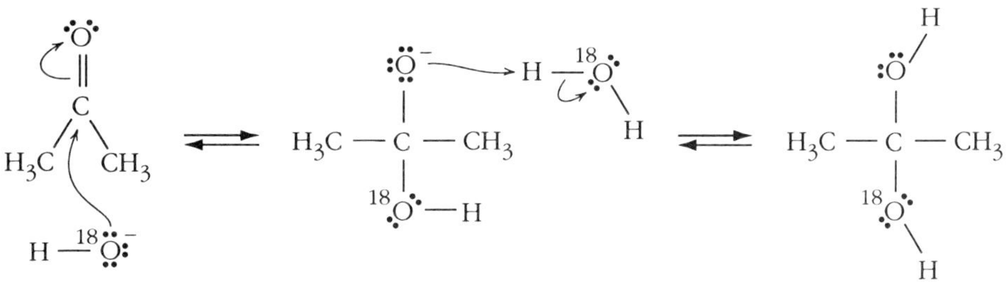

[Mechanism showing acetone + $H^{18}O^-$ → tetrahedral intermediate with $^{18}O^-$ → proton transfer with water → neutral tetrahedral intermediate with OH and ^{18}OH]

Loss of HO⁻:

[Mechanism showing neutral intermediate → deprotonation of ^{18}OH → collapse to give acetone with ^{18}O + HO⁻]

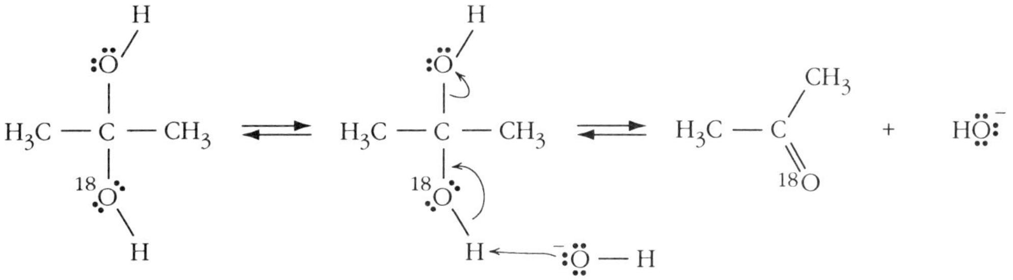

14.50 Formation of hemiacetal:

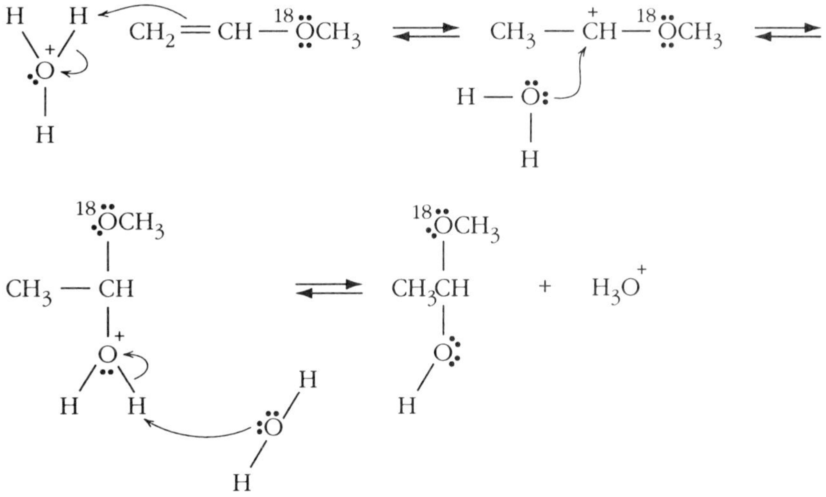

[Mechanism showing H_3O^+ protonating $CH_2=CH-^{18}OCH_3$ → $CH_3-\overset{+}{C}H-^{18}OCH_3$ carbocation, then water attacks]

[Second line: oxocarbenium with water adduct → deprotonation by water → CH_3CH with $^{18}OCH_3$ and OH (hemiacetal) + H_3O^+]

Hydrolysis of hemiacetal:

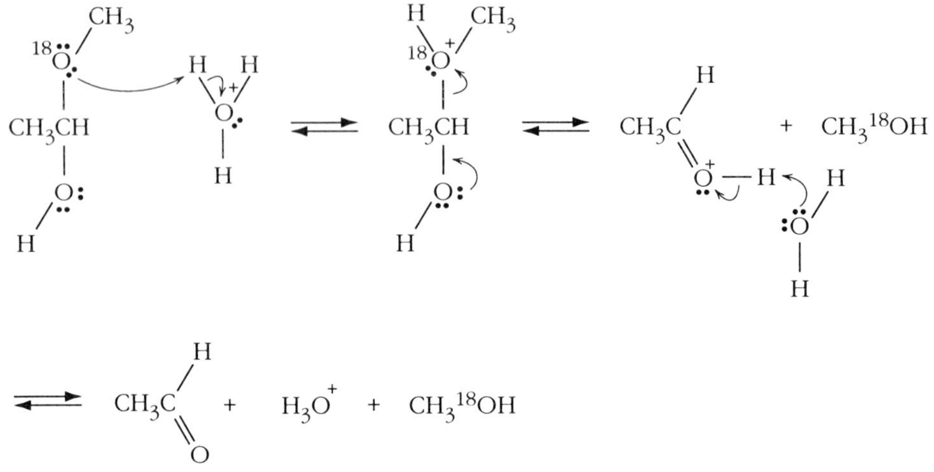

$$CH_3CH(O^{18}CH_3)(OH) + H_3O^+ \rightleftharpoons CH_3CH(O^{18}H^+CH_3)(OH) \rightleftharpoons CH_3C^+H(OH) + CH_3{}^{18}OH$$

$$\rightleftharpoons CH_3CHO + H_3O^+ + CH_3{}^{18}OH$$

Note: Remember that a compound that contains an electron deficient carbon atom adjacent to an oxygen atom exists as the resonance hybrid:

$$[R-\overset{+}{C}H-\ddot{O}CH_3 \longleftrightarrow R-CH=\overset{+}{O}CH_3]$$

14.51 Compound **H** is prepared by a Wittig reaction so it must be an alkene. The molecular formula indicates five degrees of unsaturation, one of which is a carbon-carbon double bond. The ^{1}H NMR spectrum of compound **H** shows the following signals:

δ 7.2-7.6 (multiplet, 5H)—monosubstituted benzene ring (accounts for the other four degrees of unsaturation),
δ 5.1 (s, 1H)—vinyl proton,
δ 5.4 (s, 1H)—vinyl proton,
δ 2.15 (s, 3H)—methyl group bonded to a carbon-carbon double bond.

We identify the following fragments:

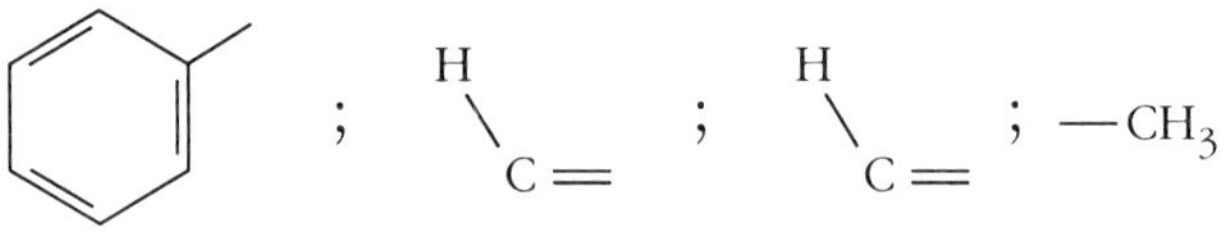

C$_6$H$_5$— ; H\C= ; H\C= ; —CH$_3$

These fragments can be joined in two ways:

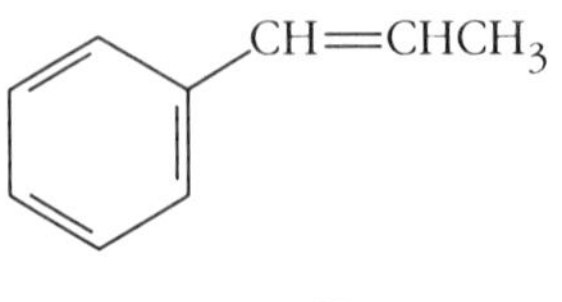

C$_6$H$_5$—CH=CHCH$_3$ C$_6$H$_5$—C(CH$_3$)=CH$_2$

H$_1$ 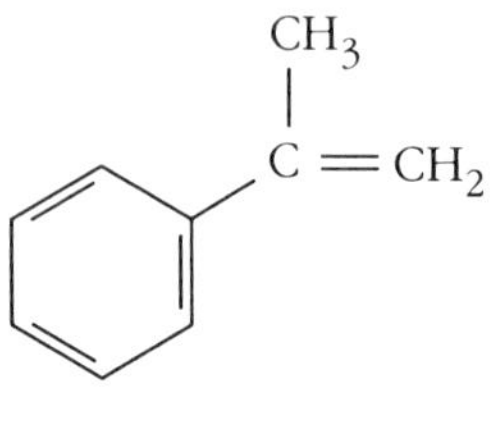   **H$_2$**

Structure **H$_1$** is incorrect because the —CH$_3$ signal would be a doublet. Structure **H$_2$** is consistent with the spectral data. Compound **H** is

As with any nonsymmetrical alkene, compound **H** can be prepared from two different sets of reagents. Retrosynthesis:

Of these two Wittig reactions, the best one is

$(C_6H_5)_3P \;+\; CH_3I \longrightarrow (C_6H_5)_3\overset{+}{P}CH_3 \xrightarrow{\text{BuLi}} (C_6H_5)_3\overset{+}{P}-\overset{-}{C}H_2 \;+$

14.52 (a) The formula, C_4H_7ClO, shows one unit of unsaturation (a ring, C=O, or C=C).

The IR spectrum exhibits an absorption band at 1722 cm^{-1}, which indicates the presence of a carbonyl group. The ^{13}C NMR spectrum shows four peaks, which indicates four nonequivalent carbon atoms, including the one of the carbonyl group at δ 203.2 (ketone). The ^1H NMR spectrum shows three signals:

δ 1.58 (d, 3H)—a methyl group bonded to a carbon with one hydrogen,
δ 2.30 (s, 3H)—a methyl group bonded to a carbonyl group,
δ 4.28 (quartet, 1H)—a hydrogen attached to a carbon; the hydrogen is split by three neighboring hydrogens (a methyl group).

Thus, the unknown is **3-chloro-2-butanone**.

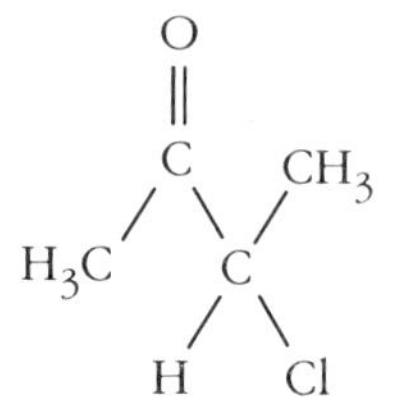

$$\underset{H_3C}{}\overset{O}{\underset{\|}{C}}-\underset{\underset{Cl}{|}}{\overset{\overset{CH_3}{|}}{C}}-H$$

(b) The formula, $C_5H_8O_2$, shows two units of unsaturation (a combination of a ring, C=O, or C=C). The IR spectrum exhibits an absorption band at 1715 cm^{-1}, which indicates the presence of a carbonyl group. The ^{13}C NMR spectrum shows five peaks, which indicates five nonequivalent carbon atoms. Two signals are in the carbonyl carbon region (δ 199.8 and δ 197.5), indicating that the compound has two carbonyl groups. The ^{1}H NMR spectrum shows three signals:

δ 1.10 (t, 3H)—a methyl group bonded to a carbon with two hydrogens,
δ 2.23 (s, 3H)—a methyl group bonded to a carbonyl group,
δ 2.72 (quartet, 2H)—a methylene group bonded to a methyl group and a carbonyl group; the carbonyl group deshields the methylene group so the chemical shift of the methylene group is farther downfield than usual.

The methylene group and one methyl group (δ 1.10) resemble the characteristic feature of an ethyl group, so the unknown is **2,3-pentanedione**:

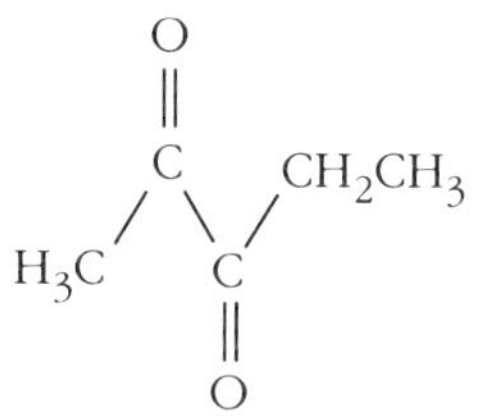

$$H_3C-\overset{O}{\underset{\|}{C}}-\overset{O}{\underset{\|}{C}}-CH_2CH_3$$

(c) The formula, $C_3H_6O_2$, shows one unit of unsaturation (a ring, C=O, or C=C). The IR spectrum has an absorption band at 1720 cm^{-1}, which indicates the presence of a carbonyl group. The absorption band at 2980 cm^{-1} is characteristic of a formic hydrogen (bonded to the carboxyl group). However, frequently this absorption band overlaps with that of alkane C—H bands. The ^{13}C NMR spectrum shows three peaks, which indicates three nonequivalent carbon atoms, including the one of the carboxyl group (δ 161). The ^{1}H NMR spectrum shows three signals:

δ 1.30 (t, 3H)—a methyl group bonded to a carbon with two hydrogens,
δ 4.22 (quartet, 2H)—a methylene group bonded to a methyl group and an oxygen (deshielding),
δ 8.04 (s, 1H)—a hydrogen bonded to a carboxyl group (formic hydrogen).

The triplet-quartet (3:2 ratio) is a characteristic feature of an ethyl group. The unknown is **ethyl formate:**

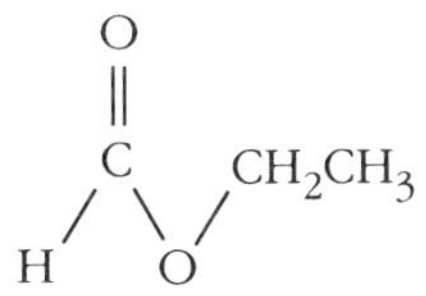

(d) The formula, $C_6H_{12}O_2$, shows one unit of unsaturation (a ring, C=O, or C=C). The IR spectrum exhibits an absorption band at 1710 cm^{-1}, which indicates the presence of a carbonyl group. The absorption band at 3250 cm^{-1} is characteristic of a hydroxyl group. The ^{13}C NMR spectrum shows five signals, which indicates some symmetry in the molecule (two equivalent carbon atoms). The spectrum also shows a signal for the carbonyl group carbon at δ 210.7. The ^{1}H NMR spectrum shows three signals:

δ 1.25 (s, 6H)—two methyl groups,
δ 2.18 (s, 3H)—a methyl group attached to a carbonyl group,
δ 2.60 (s, 2H)—a methylene group attached to a carbonyl group,

Note that the hydrogen atom of the OH group does not appear in the ^{1}H NMR spectrum.
The unknown is **4-hydroxy-4-methyl-2-pentanone:**

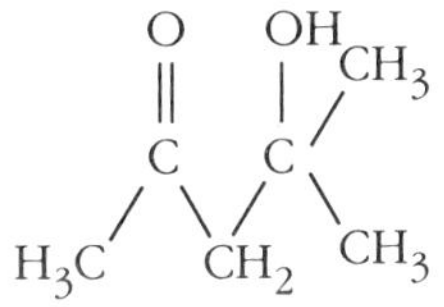

14.53 (a)

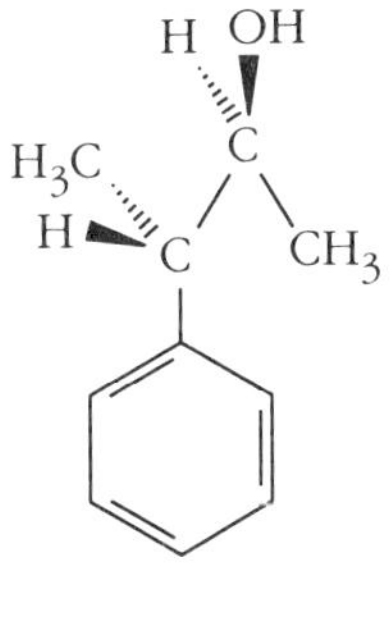

M (2*S*,3*S*)-3-Phenyl-2-butanol 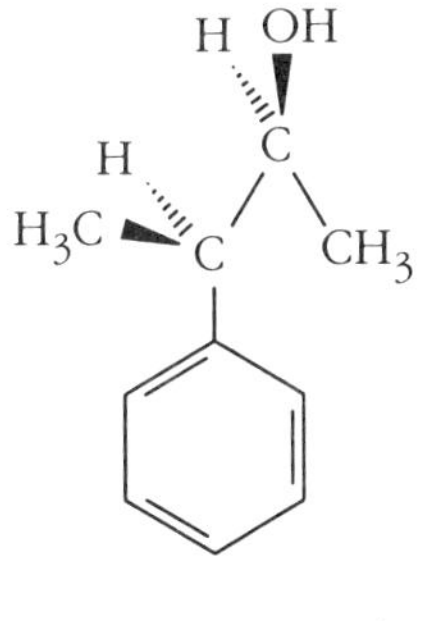   **N** (2*S*,3*R*)-3-Phenyl-2-butanol

(b)

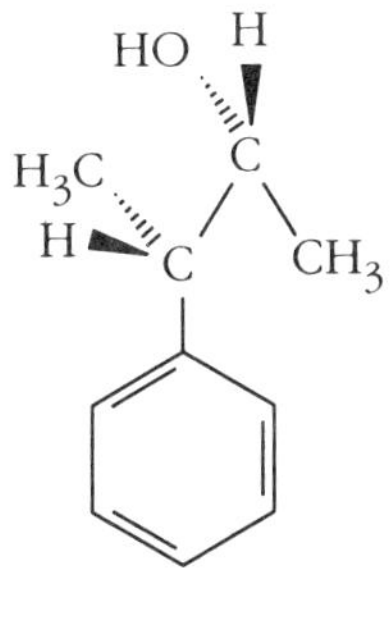

O (2*R*,3*S*)-3-Phenyl-2-butanol 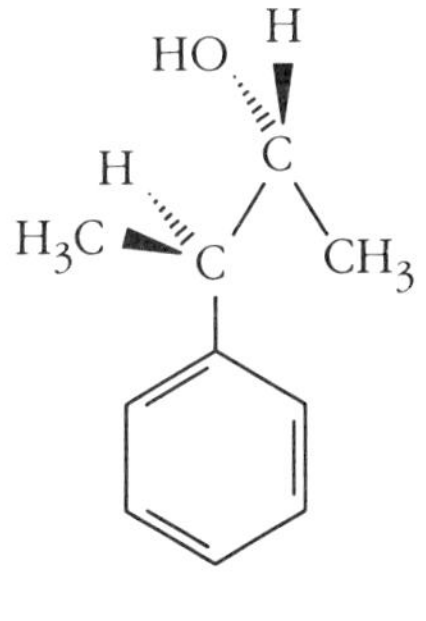   **P** (2*R*,3*R*)-3-Phenyl-2-butanol

In order to estimate relative amounts of different reduction products, it is best to examine the molecule of the carbonyl starting material, 3-phenyl-2-butanone:

(S)-3-Phenyl-2-butanone   (R)-3-Phenyl-2-butanone

The attack of the hydride ion may take place from either Re or Si side of the carbonyl group. It is easier to visualize these reactions with Newman projections along the C2-C3 bond (with C2 in front):

(S)-3-Phenyl-2-butanone   (R)-3-Phenyl-2-butanone

The two faces of the carbonyl substrates are not identical; one side has a methyl group and the other a hydrogen atom located very close to the carbonyl carbon. From the point of view of steric hindrance, the approach from the Si side to (S)-3-phenyl-2-butanone is identical to the one from the Re side of (R)-3-phenyl-2-butanone. As a result, the amounts of product formed by those two approaches *must* always be the same. The same is true when we compare the Re face of (S)-3-phenyl-2-butanone and the Si face of (R)-3-phenyl-2-butanone.

What products are formed as a result of reaction taking place at each side of each compound?

Substrate	Side	Product
(S)-3-Phenyl-2-butanone	Si	**O** (2R,3S)-3-phenyl-2-butanol
(S)-3-Phenyl-2-butanone	Re	**M** (2S,3S)-3-phenyl-2-butanol
(R)-3-Phenyl-2-butanone	Si	**P** (2R,3R)-3-phenyl-2-butanol
(R)-3-Phenyl-2-butanone	Re	**N** (2S,3R)-3-phenyl-2-butanol

Compounds **O** and **N** are enantiomers; compounds **M** and **P** are enantiomers, too. The remaining pairs are diastereomers. Now we can answer questions (c) and (d).

(c) Steric hindrance on the Si side of (S)-3-phenyl-2-butanone is identical to the steric hindrance on the Re side of (R)-3-phenyl-2-butanone. As a result, compound **O** will be always formed in the same amount as compound **N**. Compounds **N** and **O** are enantiomers.

Compounds **M** and **P** will be always formed in the same amounts for the same reason. Compounds **M** and **P** are also enantiomers.

(d) The steric hindrance on the Si side and on the Re side of each carbonyl compound is not identical. As a result, the amounts of products formed may not be the same. That is the amount of product **O** and product **M** may not be the same. The same is true with the amounts of product **N** and product **P**. The relationship between compounds in those pairs: they are diastereomers.

14.54 The molecular formula, $C_6H_{12}O$, indicates one degree of unsaturation. Compound **I** shows the following signals in the ^1H NMR spectrum:

δ 0.93 (t, 3H)—a methyl group bonded to a methylene group,
δ 1.1 (t, 3H)—a methyl group bonded to a methylene group,
δ 1.6 (sextet, 2H)—a methylene group with five neighboring hydrogen atoms,
δ 2.35-2.5 (multiplet, 4H)—it looks like two overlapping signals; chemical shift and integration data suggest two methylene groups bonded to a carbonyl group.

We can group the above signals and identify two alkyl groups (ethyl and propyl) bonded to a carbonyl group. Thus, compound **I** is **3-hexanone**:

$$\underset{\underset{\text{CH}_3\text{CH}_2\text{CCH}_2\text{CH}_2\text{CH}_3}{\|}}{O}$$

Compound **J** shows the following signals in the ^1H NMR spectrum:

δ 1.15 (s, 9H)—3 methyl groups bonded to a carbon with no hydrogens,
δ 2.1 (s, 3H)—a methyl group bonded to a carbonyl group.

Thus, compound **J** is **3,3-dimethyl-2-butanone**:

$$\begin{array}{c} O \quad\quad CH_3 \\ \| \quad\quad | \\ CH_3C-C-CH_3 \\ | \\ CH_3 \end{array}$$

Compound **K** shows the following signals in the ^1H NMR spectrum:

δ 0.9 (t, 3H)—a methyl groups bonded to a methylene group,
δ 1.3 (sextet, 2H)—a methylene group with five neighboring hydrogen atoms,
δ 1.55 (quintet, 2H)—a methylene group with four neighboring hydrogen atoms,
δ 2.15 (s, 3H)—a methyl group bonded to a carbonyl group,
δ 2.43 (t, 2H)—a methylene group bonded to another methylene group and to a carbonyl group; the carbonyl group deshields the methylene group.

One part of the molecule is CH_3C-; the other is a butyl group. Thus, compound **K** is **2-hexanone**:

$$\underset{\underset{\text{CH}_3\text{CCH}_2\text{CH}_2\text{CH}_2\text{CH}_3}{\|}}{O}$$

Compound **L** shows the following signals in the ^{1}H NMR spectrum:

δ 1.05 (t, 3H)—a methyl group bonded to a methylene group,
δ 1.1 (d, 6H)—two methyl groups bonded to a carbon with one hydrogen,
δ 2.49 (quartet, 2H)—a methylene group bonded to a methyl group and a carbonyl group,
δ 2.6 (heptet, 1H)—a hydrogen attached to a carbon bonded to a carbonyl group; the hydrogen is split by six neighboring hydrogens.

We can group the above signals and identify two alkyl groups (ethyl and isopropyl) bonded to a carbonyl group. Thus, compound **L** is **2-methyl-3-pentanone:**

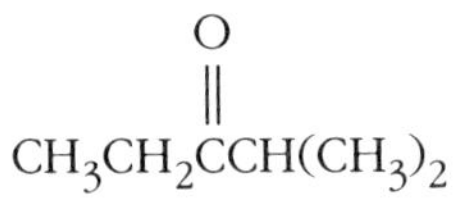

$$\underset{\underset{\text{CH}_3\text{CH}_2\text{CCH(CH}_3)_2}{\|}}{\text{O}}$$

14.55 (a) The formula, C_4H_8O, shows one unit of unsaturation (a ring, C=O, or C=C). The IR spectrum exhibits an absorption band at 1710 cm^{-1}, which indicates the presence of a carbonyl group. The ^{1}H NMR spectrum shows three signals:

δ 1.15 (d, 6H)—two methyl groups bonded to a carbon with one hydrogen,
δ 2.45 (septet, 1H)—a hydrogen attached to a carbon bonded to a carbonyl group; the hydrogen is split by six neighboring hydrogens,
δ 9.65 (1H)—aldehyde hydrogen.

The two signals in the alkane region exhibit a pattern characteristic of an isopropyl group [—CH(CH$_3$)$_2$]. Thus, the unknown is **2-methylpropanal:**

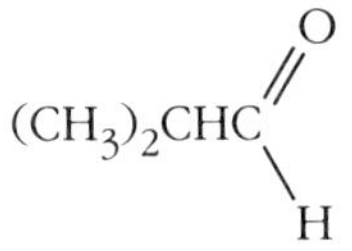

$$(\text{CH}_3)_2\text{CHC}\underset{\text{H}}{\overset{\overset{\text{O}}{\|}}{\diagdown}}$$

Notice the coupling between the α-hydrogen atom (δ 2.45) and the hydrogen atom bonded to the carbonyl group (δ 9.65). The peak at δ 9.65 is a doublet while each peak of the septet is split into a doublet.

(b) The formula, $C_5H_8O_2$, shows two units of unsaturation (a combination of ring, C=O, or C=C). The ^{13}C NMR spectrum exhibits five signals for five nonequivalent carbon atoms. The signal at δ 196 indicates the presence of a carbonyl carbon (ketone). The signal at δ 162 is an alkene carbon atom bonded to an oxygen atom. The signal at δ 107 is in the alkene region and the two signals at δ 27 and δ 56 are in the alkane region.
The ^{1}H NMR spectrum shows three signals:

δ 2.05 (s, 3H)—a methyl group bonded to a carbonyl group,
δ 3.4 (s, 3H)—a methyl group bonded to oxygen (CH$_3$O—),
δ 5.8 and δ 7.3 (a pair of doublets, 2H)—two vinyl hydrogens.

Thus, putting all these pieces together, the unknown is **4-methoxy-3-buten-2-one**:

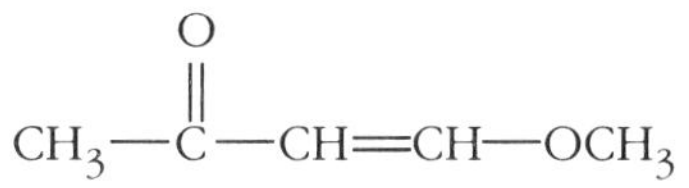

$$CH_3-\overset{\overset{O}{\|}}{C}-CH=CH-OCH_3$$

(c) The formula, $C_4H_{10}O_2$, shows zero units of unsaturation (saturated compound).

The ^{13}C NMR spectrum shows only three signals so there must be some symmetry in the compound (two equivalent carbon atoms). The ^1H NMR spectrum shows three signals:

δ 1.2 (d, 3H)—a methyl group bonded to a carbon with one hydrogen atom,
δ 3.35 (s, 6H)—two methyl groups each bonded to oxygen atoms (two methoxy groups),
δ 4.55 (quartet, 1H)—a hydrogen bonded to a carbon attached to a methyl group and oxygen atom or two oxygen atoms (chemical shift !).

Thus, the unknown is **1,1-dimethoxyethane** (acetaldehyde dimethyl acetal):

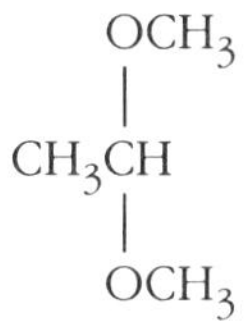

$$CH_3CH \begin{matrix} OCH_3 \\ | \\ \\ | \\ OCH_3 \end{matrix}$$

EXAMINATION 2

To get the maximum benefit from the following two-hour exam, choose a quiet location and spend exactly two hours answering the questions. Check your answers against those at the end of the Study Guide to evaluate your knowledge of the material.

Write all answers in the space provided.

Chapters 1–14

Time allowed: 2 hours, Maximum: 200 points

1. (*22 points*) Draw structures and give the IUPAC names of all alkynes of molecular formula C_6H_{10}.

    (a) (*5 points*) Indicate which of the above isomers will react with $NaNH_2$ in liquid ammonia.

    (b) (*5 points*) Indicate which of the isomers exist as a pair of enantiomers. Draw these enantiomers and give their IUPAC names.

2. (*20 points*) Draw structures and give the IUPAC names of all dienes of molecular formula C_6H_{10}.

3. (*30 points*) Write the missing reagents, organic products, or substrates in the following transformations. Indicate the stereochemistry where applicable.

(a) [norbornane with vinyl and methyl substituents] $\xrightarrow{\text{(1) BH}_3 \quad \text{(2) H}_2\text{O}_2, \text{HO}^-}$

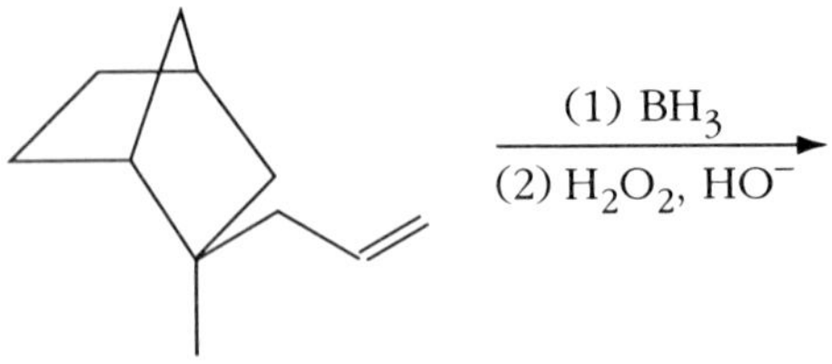

(b) HC≡C–C(CH₃)₂–H ⟶ HC≡C–C(CH₃)₂–CH₃

(c) [cyclopentane with CH₃ (wedge) and Br (dash)] $\xrightarrow{\text{CH}_3\text{CH}_2\text{O}^-\text{Na}^+}{\text{CH}_3\text{CH}_2\text{OH, 50 °C}}$

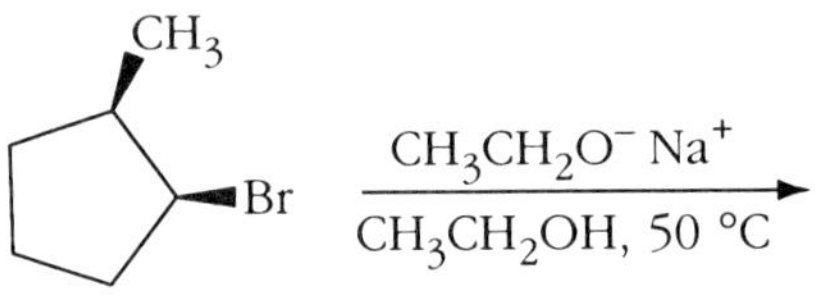

(d) H₂C=CHBr $\xrightarrow{\text{CH}_3\text{CH}_2\text{O}^-\text{Na}^+}{\text{CH}_3\text{CH}_2\text{OH, 30 °C}}$

(e) [cyclopentene] ⟶ [trans-cyclopentane-1,2-diol]

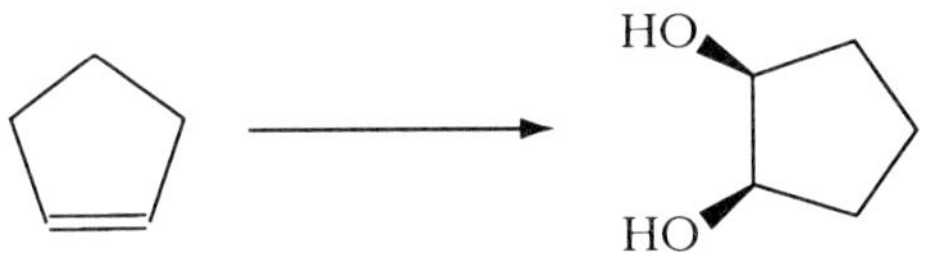

(f) ? + ? $\xrightarrow{p\text{-Toluenesulfonic acid}}$ [2-phenyl-1,3-dioxolane]

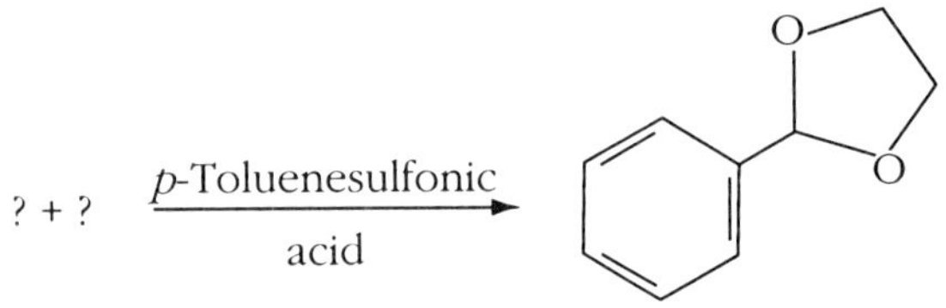

4. (*20 points*) Starting with acetylene (ethyne), any haloalkane, any inorganic reagents, and any solvent, propose a synthesis of each of the following compounds. Use the retrosynthesis first.

(a) [pentanal] ⟹

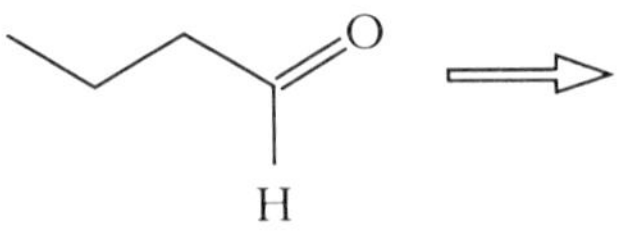

(b) [hexan-3-one] ⟹

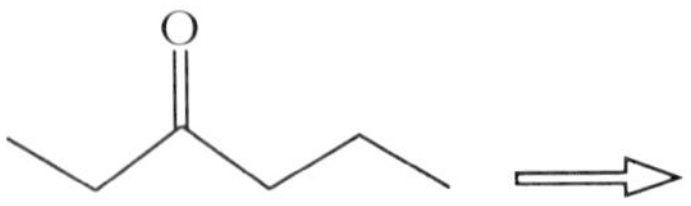

(c) [5-methylhexan-3-one (or similar)] ⟹

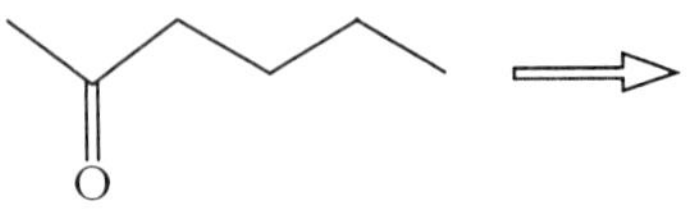

(d) 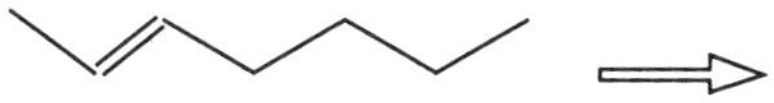 [structure] ⟹

5. (*20 points*) Consider the following structures (**A–E**). Indicate which of the structures

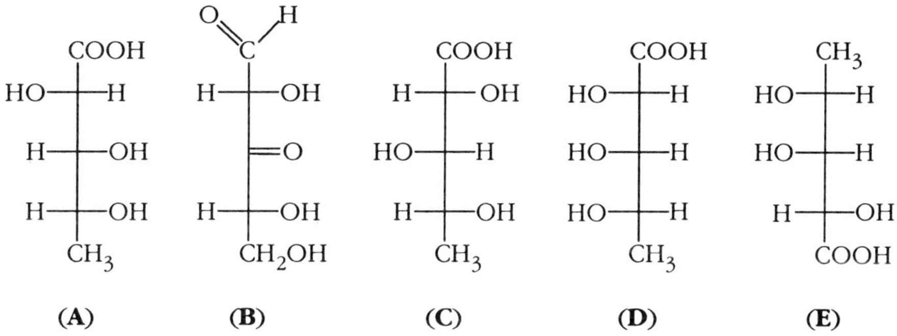

(A)	(B)	(C)	(D)	(E)
COOH	O=C–H	COOH	COOH	CH₃
HO—⊢—H	H—⊢—OH	H—⊢—OH	HO—⊢—H	HO—⊢—H
H—⊢—OH	=O	HO—⊢—H	HO—⊢—H	HO—⊢—H
H—⊢—OH	H—⊢—OH	H—⊢—OH	HO—⊢—H	H—⊢—OH
CH₃	CH₂OH	CH₃	CH₃	COOH

(a) is the *meso* compound

(b) are a pair of enantiomers

(c) are a pair of diastereomers

(d) will rotate plane-polarized light

(e) are two representations of the same compound

(f) are a pair of constitutional isomers

6. (*30 points*) Using curved arrows to show the electron flow, write detailed reaction mechanisms for the following reactions. Please make sure to write all intermediates formed and indicate all resonance stabilizations and rearrangements, if applicable.

(a) 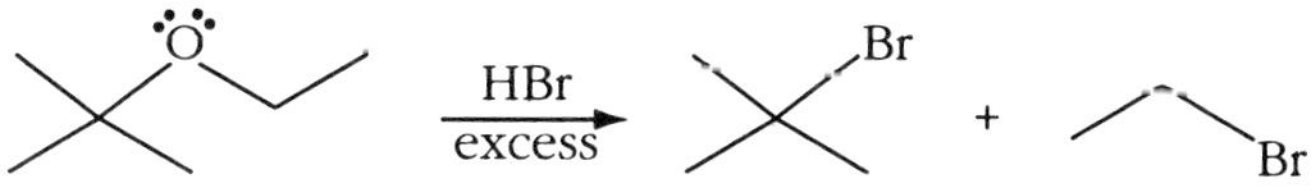 [(CH₃)₃C–O–CH₂CH₃] $\xrightarrow{\text{HBr, excess}}$ [(CH₃)₃C–Br] + [CH₃CH₂–Br]

(b) 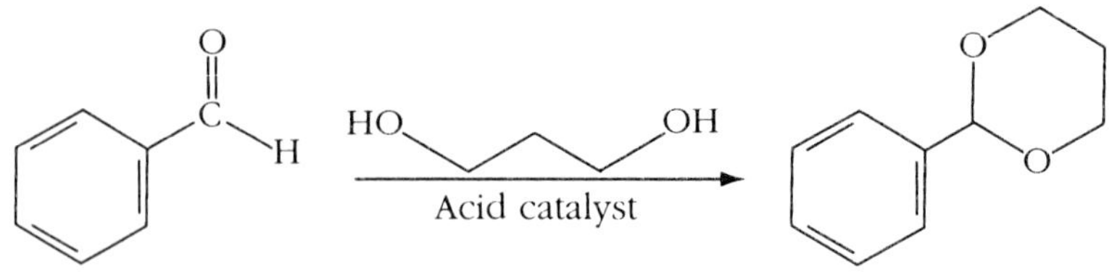 [Benzaldehyde + HO-CH₂CH₂CH₂-OH with acid catalyst → 2-phenyl-1,3-dioxane]

7. (**20 points**) Propose structures for compounds **A** to **D** that are consistent with the following spectra and molecular formulas:

   A C_4H_8O IR: 2820–2750 cm^{-1}; 1725 cm^{-1}. ^{1}H NMR: δ 1.16 d (6H); δ 2.45 septet (1H); δ 9.6 s (1H).

   B C_5H_{10} IR: 3080 cm^{-1}; 1660 cm^{-1}. ^{1}H NMR: δ 0.9 t (3H); δ 1.45 sextet (2H); δ 2.05 q (2H); δ 4.8–5.1 multiplet (2H); δ 5.5–6.1 multiplet (1H).

   C $C_5H_{10}O_2$ IR: 1735 cm^{-1}. ^{1}H NMR: δ 0.92 t (3H); δ 1.65 sextet (2H); δ 2.05 s (3H); δ 4.0 t (2H).

   D C_9H_{12} IR: 3030 cm^{-1}; 1500 cm^{-1}; 1460 cm^{-1}; 770–735 cm^{-1}. ^{1}H NMR: δ 1.2 t (3H); δ 2.25 s (3H); δ 2.55 q (2H); δ 7.05 s (4H).

8. (*28 points*) Propose structures for compounds **E** to **K** that are consistent with the following reactions:

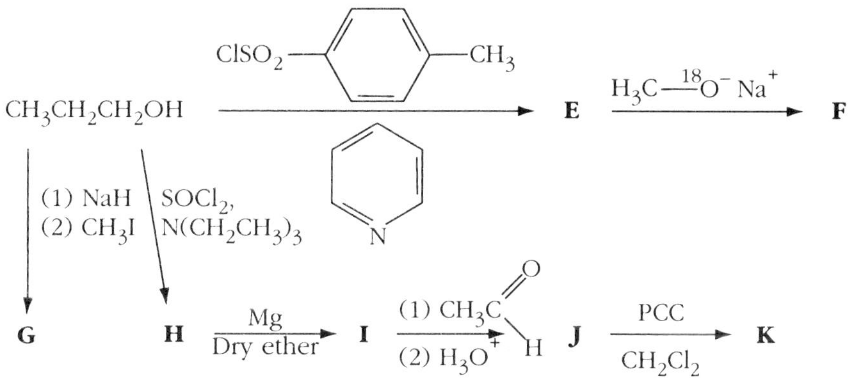

E =

F =

G =

H =

I =

J =

K =

CHAPTER 15

CARBOXYLIC ACIDS
Concepts

A **carboxylic acid** is an organic compound containing the **carboxyl group, —COOH.** The carboxyl group contains a hydroxyl group that is directly bonded to the carbon atom of the carbonyl group. The interactions of these two groups lead to a chemical reactivity that is unique to carboxylic acids.

Lone-pair donation from the hydroxyl group stabilizes the carbonyl group and makes it less electrophilic than that of an aldehyde or a ketone. Electron density is increased at the carbonyl oxygen. The hydrogen atom of the hydroxyl group is much more easily lost as a proton than is the hydrogen atom of the hydroxyl group of an alcohol.

Nomenclature. According to the IUPAC system, a carboxylic acid is named by dropping the letter -*e* from the name of the longest chain that contains the carboxyl group, and replacing it with the suffix -*oic acid*. A carboxylic acid containing a carboxyl group attached to a cycloalkane ring is named by giving the name of the ring and adding the suffix -*carboxylic acid*.

Ethanoic acid   3-Hydroxybutanoic acid   2-Chlorobenzoic acid   2-Cyclohexene-
Acetic acid   *o*-Chlorobenzoic acid   carboxylic acid

The carboxyl group carbon is always number 1 and the carboxyl group takes precedence over most other functional groups. The decreasing order of precedence for the major functional groups is given in the following Table.

Functional Group	Suffix if Highest Precedence	Prefix if Lower Precedence
—COOH	-oic acid	—
—CHO	-al	oxo-
\C=O/	-one	oxo-
—OH	-ol	hydroxy-
—NH$_2$	-amine	amino-
—SH	-thiol	mercapto-
—X, —R, —Ar, —NO$_2$	—	halo-, alkyl-, aryl-, nitro-

Solutions to the Exercises

15.1 (a) 2,2-Dimethylpropanoic acid
(b) 2-Methylpentanedioic acid
(c) 5-Bromo-2,2-dichloro-5-methylhexanoic acid
(d) Cyclobutanecarboxylic acid
(e) 3-Methyl-2-butenoic acid
(f) 2-Butynoic acid

15.2

(a), (b), 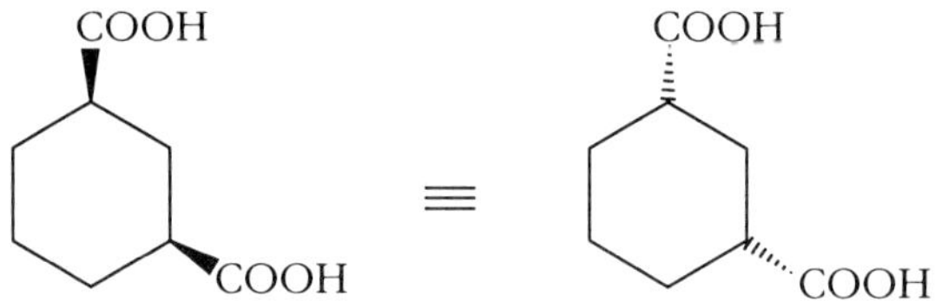 (c), 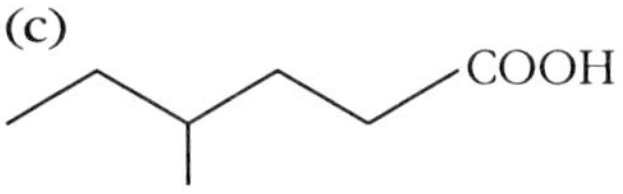 (d), 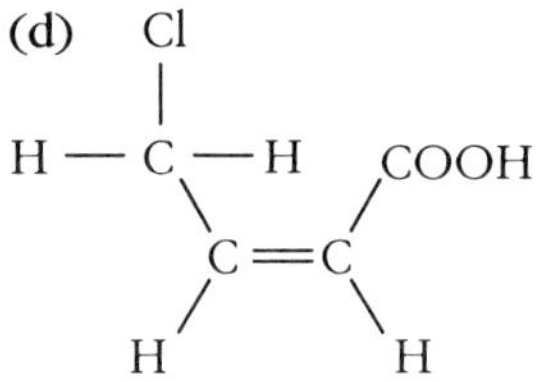 (e) I$_2$CHCOOH

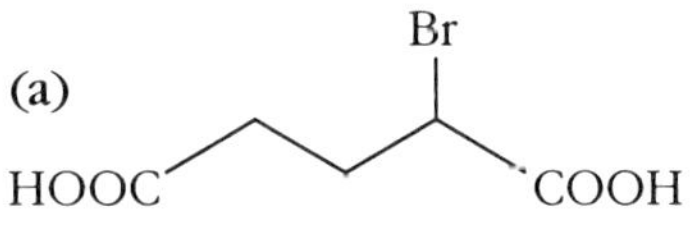

15.3 (a) Isopropyl butanoate or 1-methylethyl butanoate
(b) Hexyl 2-chloro-3-methylbutanoate
(c) Propyl cyclopentanecarboxylate
(d) Sodium 2-methylpropanoate
(e) Diisopropyl succinate
(f) Potassium propenoate

15.4

(a), (b), 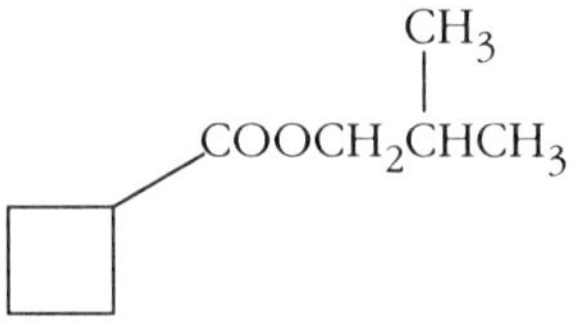 (c), 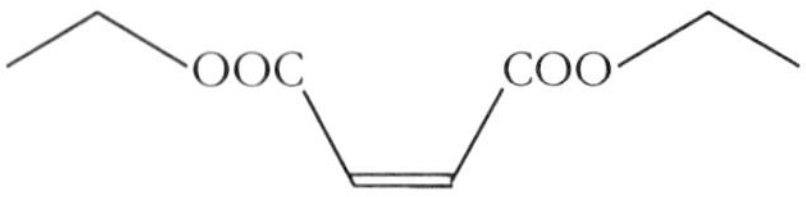 (d)

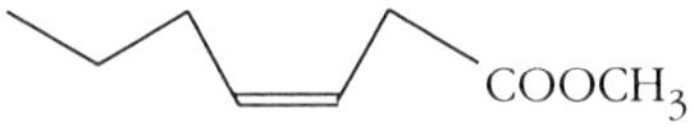

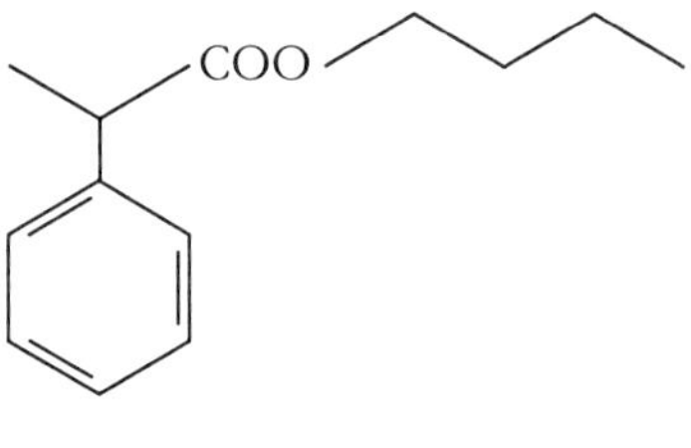

(e) (CH$_3$CH$_2$COO$^-$)$_2$Ca^{2+}

15.5

```
                    sp²-sp³
                    σ-bond
        sp²-sp²    ·Ö·  H      sp³-1s
        σ-bond      \  /        σ-bond
                     C—Ö:
        sp²-1s      / \ ··
        σ-bond    H    \     sp³ hybrid
                        \    oxygen atom
                    sp² hybrid
                    carbon atom
```

15.6 First, analyze the difference in structure of each pair of compounds, and then choose a spectroscopic method that will reveal that difference.

(a)
	IR	¹H NMR	¹³C NMR
$CH_3\overset{O}{\underset{\|}{C}}CH_2CH_3$	1710 cm⁻¹ (C=O)	δ 2.1 (singlet, 3H), δ 2.42 (quartet, 2H), δ 1.05 (triplet, 3H).	4 signals
$CH_3\overset{O}{\underset{\|}{C}}CH_3$	1715 cm⁻¹ (C=O)	δ 2.2 (singlet)	2 signals

IR cannot be used because there are no significant differences between the IR spectra of the two ketones. The extra methylene group in 2-butanone can be easily detected by either ¹H NMR or ¹³C NMR spectroscopy, so these are the methods of choice.

(b)
	IR	¹H NMR	¹³C NMR
$CH_3\overset{O}{\underset{\|}{C}}CH_3$	1715 cm⁻¹ (C=O)	δ 2.2 (singlet)	2 signals
$CH_3C\overset{\diagup O}{\underset{\diagdown H}{}}$	1725 cm⁻¹ (C=O) 2720 cm⁻¹ (C—H)	δ 2.18 (singlet, 3H), δ 9.8 (quartet, 1H)	2 signals

Either IR or ¹H NMR can be used to distinguish between these two compounds.

IR: The absorption band of the carbon-aldehydic hydrogen bond is observed in the spectrum of acetaldehyde at 2720 cm⁻¹ and this band is absent in the spectrum of acetone.

¹H NMR: Particularly significant is the quartet at δ 9.8 in acetaldehyde due to the aldehydic hydrogen, which is absent in the spectrum of acetone. The ¹H NMR spectrum of acetaldehyde shows a singlet at δ 2.2 (acetyl group) and a quartet at δ 9.8 (aldehydic hydrogen), while that of acetone shows *only* one singlet at about δ 2.0 for the methyl group.

The ¹³C NMR is the least helpful. There are two signals in the spectrum of each compound and the chemical shifts are very similar too (e.g., δ 201 for the acetaldehyde acetyl group and δ 207 for the acetone acetyl group).

(c) | | IR | ^{1}H NMR | ^{13}C NMR

CH$_3$CCH$_3$ (with C=O)
1715 cm^{-1} (C=O) | δ 2.2 (singlet) | 2 signals

CH$_3$COOH
1714 cm^{-1} (C=O)
3300-2500 cm^{-1} (COOH) | δ 2.1 (singlet, 3H)
δ 12.6 (singlet, 1H) | 2 signals

Infrared spectroscopy easily distinguishes between acetic acid and acetone. The O—H bond of the carboxyl group gives rise to a very broad absorption over the range 3300 to 2500 cm^{-1}. Such absorption is absent from the IR spectrum of acetone.

^{1}H NMR spectroscopy is less useful. Acetone shows just one singlet at about δ 2.0 while acetic acid shows two singlets that integrate in the ratio 3:1 (at δ 2.1 and at δ 10-13, respectively). The signal at δ 10 to 13, however, is sometimes not visible.

^{13}C NMR spectroscopy is also useful because there is a difference in the chemical shift of the two carbonyl carbon atoms: δ 178 for acetic acid and δ 210 for acetone.

(d) | | IR | ^{1}H NMR | ^{13}C NMR

benzaldehyde
1700 cm^{-1} (C=O) | δ 7-8 (multiplet, 5H)
δ 9.5 (singlet, 1H) | 5 signals

benzoic acid
1710 cm^{-1} (C=O)
3300-2500 cm^{-1} (COOH) | δ 7-8 (multiplet, 5H)
δ 11 (singlet, 1H) | 5 signals

Infrared spectroscopy easily distinguishes between benzoic acid and benzaldehyde. The O—H bond of the carboxyl group gives rise to a very broad absorption over the range 3300 to 2500 cm^{-1}. Such absorption is absent from the IR spectrum of benzaldehyde.

^{1}H NMR spectroscopy is also useful. The aromatic protons of both compounds show similar multiplets at δ 7 to 8 but the aldehydic proton of benzaldehyde that appears at δ 9.5 clearly identifies benzaldehyde.

^{13}C NMR spectroscopy is less useful because each compound shows five signals. One can note a difference in the chemical shift of the carbonyl carbon, however: δ 173 for benzoic acid and δ 192 for benzaldehyde.

(e)

	IR	¹H NMR	¹³C NMR
CH₃C(=O)OCH₂CH₃	1735 cm⁻¹ (C=O)	δ 2.0 (singlet, 3H), δ 4.2 (quartet, 2H), δ 1.3 (triplet, 3H)	4 signals
CH₃CH₂C(=O)H	1725 cm⁻¹ (C=O) 2720 cm⁻¹ (C—H)	δ 9.5 (singlet, 1H), δ 2.0 (quartet, 2H), δ 1.1 (triplet, 3H)	3 signals

IR, ¹H NMR, and ¹³C NMR all can distinguish between the two compounds.

IR: The characteristic C—H bond stretching of the aldehyde functional group around 2720 cm⁻¹ serves to distinguish propanal from ethyl acetate.

The ¹H NMR spectra of the two compounds are different. The deshielded aldehydic proton of propanal gives rise to a singlet at δ 9.5 and integrates to 1H. This peak is absent in the ¹H NMR spectrum of ethyl acetate. The singlet in the ¹H NMR spectrum of ethyl acetate appears at δ 2.0 (3H). This peak does not appear in the spectrum of propanal. The position of the quartet is also different (δ 4.2 in ethyl acetate and δ 2.0 in propanal).

The ¹³C NMR spectrum of ethyl acetate shows four signals while that of propanal only three signals.

(f)

	IR	¹H NMR	¹³C NMR
CH₃C(=O)OCH₃	1735 cm⁻¹ (C=O)	δ 4.0 (singlet, 3H) δ 2.0 (singlet, 3H)	3 signals
CH₃COOH	1760-1710 cm⁻¹ (C=O) 3300-2500 cm⁻¹ (COOH)	δ 2.1 (singlet, 3H) δ 12.6 (singlet, 1H)	2 signals

All three spectroscopic methods can be used to distinguish methyl acetate from acetic acid.

IR: The O—H bond of the carboxyl group gives rise to a very broad absorption over the range 3300 to 2500 cm⁻¹ in the IR spectrum of acetic acid. This kind of absorption is absent from the IR spectrum of methyl acetate.

¹H NMR spectroscopy is also useful. The two singlets of methyl acetate integrate in a ratio of 1:1 whereas the singlets of acetic acid integrate in a ratio of 1:3. The carboxyl proton gives rise to a signal at between δ 10 and δ 13, often as a broad peak (sometimes missing). The methyl groups of methyl acetate appear at δ 2.0 (acetic CH₃) and at δ 4.0 (methyl group bonded to the oxygen).

The ¹³C NMR spectrum of methyl acetate shows three signals, whereas that of acetic acid shows only two signals. One can also note a small difference in the chemical shift of the carbonyl carbon: δ 177 for acetic acid and δ 171 for methyl acetate.

15.7 The *expected* signals of the two compounds in their ^{1}H NMR spectra are the following:

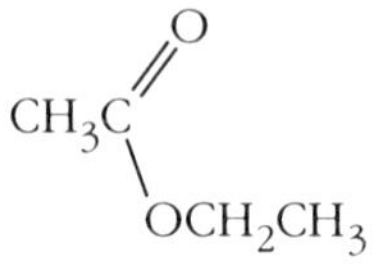

CH$_3$C(=O)OCH$_2$CH$_3$

δ 1.2 (triplet, 3H); —CH$_3$ of the ethyl group,
δ 4.1 (quartet, 2H); —CH$_2$— of the ethyl group,
δ 2.1 (singlet, 3H); —CH$_3$ of the acyl group.

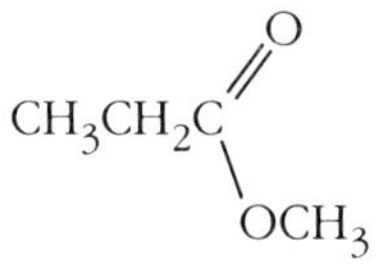

CH$_3$CH$_2$C(=O)OCH$_3$

δ 1.1 (triplet, 3H); —CH$_3$ of the ethyl group,
δ 2.1 (quartet, 2H); —CH$_2$— of the ethyl group,
δ 3.8 (singlet, 3H); —CH$_3$ bonded to the oxygen.

The signals of both compounds show the same splitting pattern, but the positions of those signals are different. The most characteristic is the position of the hydrogen atoms of the methyl group: δ 2.1 in the spectrum of ethyl acetate and δ 3.8 in the spectrum of methyl propanoate. The presence of a quartet at δ 2.3 in the *actual* spectrum (Figure 15.4) indicates that the spectrum is that of methyl propanoate.

15.8

(a) (isobutyl alcohol) $\xrightarrow{Na_2Cr_2O_7, H_2SO_4}$ (isobutyric acid, COOH)

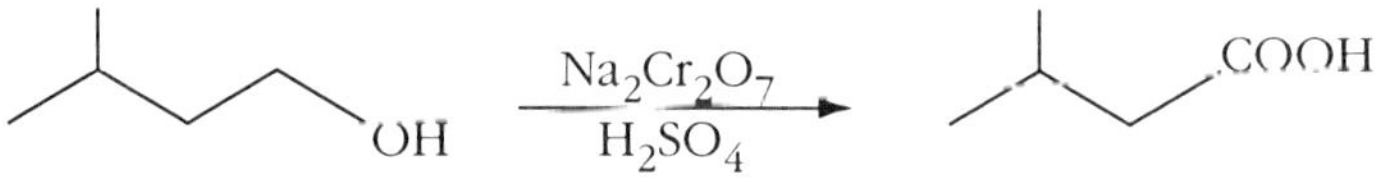

(b) (alkene) $\xrightarrow{CrO_3, H_2SO_4}$ 2 (isobutyric acid, COOH)

or (alkene) $\xrightarrow{CrO_3, H_2SO_4}$ (isobutyric acid, COOH) + CO$_2$

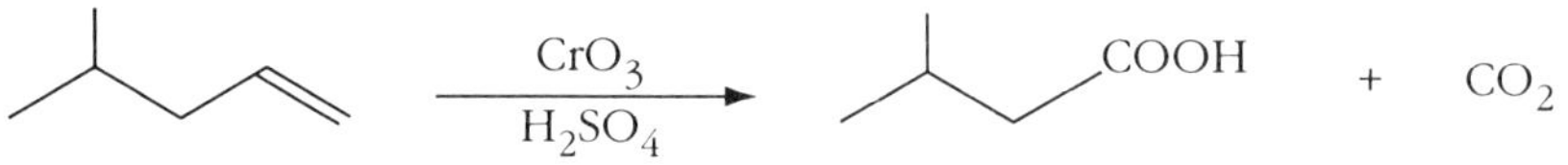

(c) cyclopentene $\xrightarrow{KMnO_4, H_2SO_4/H_2O}$ HOOC-(CH$_2$)$_3$-COOH

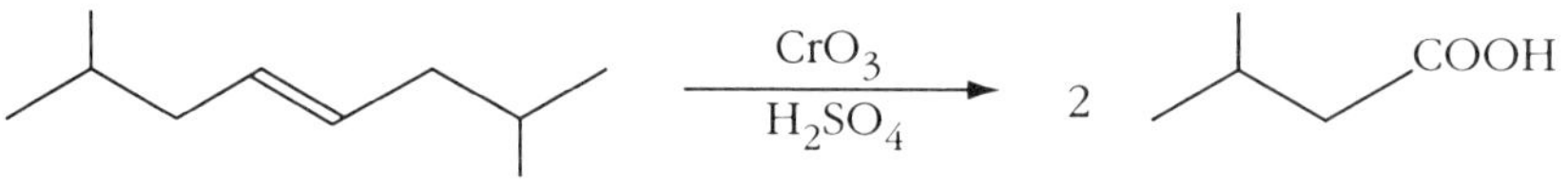

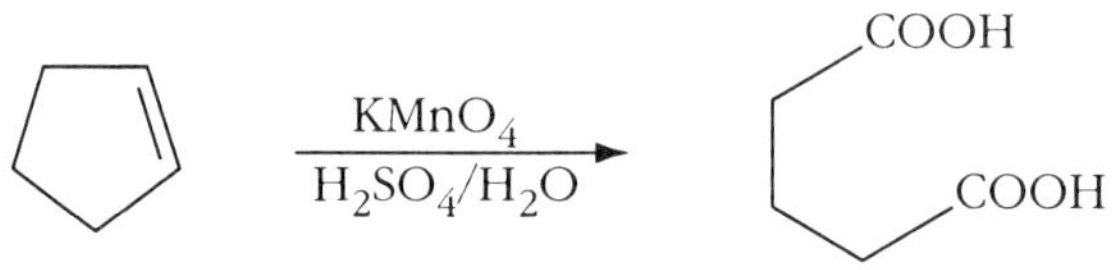

(d) 1-ethyl-3-chloro-5-methoxybenzene $\xrightarrow{KMnO_4, H_2SO_4}$ 3-chloro-5-methoxybenzoic acid

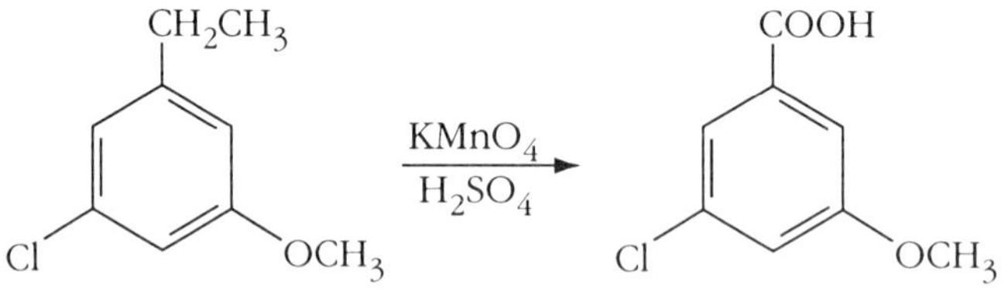

Note: You can oxidize any alkyl group with at least one hydrogen in the benzylic position.

15.9 Carboxylation of a Grignard reagent is an excellent method of converting a haloalkane to a carboxylic acid that contains one more carbon atom.

(a) [cyclopentyl-Br] $\xrightarrow{\text{Mg, Dry ether}}$ [cyclopentyl-MgBr] $\xrightarrow{(1)\ CO_2\ (2)\ H_3O^+}$ [cyclopentyl-COOH]

(b) Retrosynthesis:

[1-ethylcyclohexyl-COOH] ⇒ [1-ethylcyclohexyl-MgCl] ⇒ [1-ethylcyclohexyl-Cl] ⇒

[1-ethylcyclohexyl-OH] ⇒ [cyclohexanone]

Synthesis:

[cyclohexanone] $\xrightarrow{(1)\ CH_3CH_2MgBr\ (2)\ H_3O^+}$ [1-ethylcyclohexanol] $\xrightarrow{HCl}$

[1-ethylcyclohexyl-Cl] $\xrightarrow{\text{Mg, Dry ether}}$ [1-ethylcyclohexyl-MgCl] $\xrightarrow{(1)\ CO_2\ (2)\ H_3O^+}$ [1-ethylcyclohexyl-COOH]

(c) There are two more carbon atoms in the product than in the starting material. Reacting a Grignard reagent with oxirane is a good method of increasing a carbon chain by two carbon atoms.

[butyl-Br] $\xrightarrow{\text{Mg, Dry ether}}$ [butyl-MgBr] $\xrightarrow{(1)\ \text{oxirane}\ (2)\ H_3O^+}$

[hexyl-OH] $\xrightarrow{(1)\ KMnO_4,\ HO^-,\ \text{heat}\ (2)\ H_3O^+}$ [pentyl-COOH]

(d) Retrosynthesis:

[pentyl-COOH] ⇒ [pentyl-OH] ⇒

[butyl-CH(H)-Li + H_2C=O] ⇒ [butyl-Br] ⇒ [propene]

Synthesis:

[propene] $\xrightarrow{\text{HBr, ROOR, heat}}$ [butyl-Br] $\xrightarrow{\text{Li, Dry ether}}$ [butyl-Li]

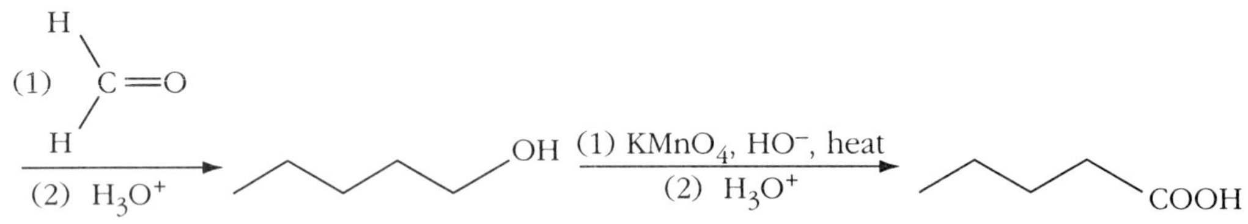

[Reaction scheme: H₂C=O (1), then (2) H₃O⁺ → HOCH₂CH₂CH₂CH₂OH → (1) KMnO₄, HO⁻, heat; (2) H₃O⁺ → HOOC-CH₂CH₂CH₂-COOH type product shown as CH₃CH₂CH₂CH₂COOH]

15.10 (a) The Grignard reagent (HOCH₂CH₂CH₂MgBr) cannot be obtained in the first step. The Grignard reagent is a very strong base and it will react with the hydroxyl group. Thus, any Grignard reagent formed will decompose according to the following equation:

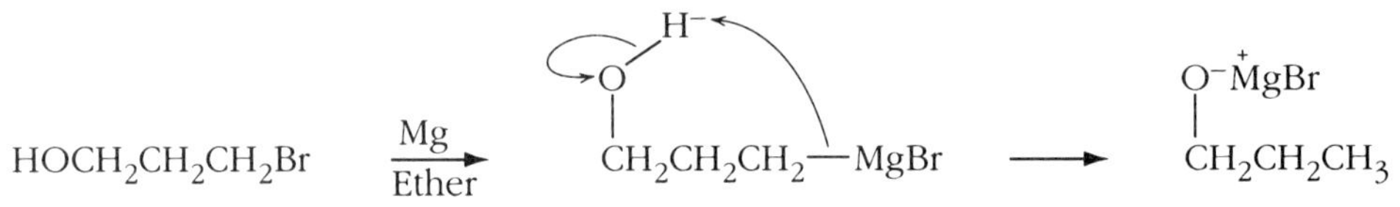

HOCH₂CH₂CH₂Br $\xrightarrow{\text{Mg, Ether}}$ CH₂CH₂CH₂—MgBr $\longrightarrow$ O⁻⁺MgBr | CH₂CH₂CH₃

(b) The preparation of nitrile (first step) follows an S_N2 mechanism. However, the starting material, 2-chloro-2-methylpropane (*tert*-butyl chloride), is a tertiary haloalkane and it will not react according to the S_N2 mechanism.

15.11 The Henderson-Hasselbalch equation provides a direct way to calculate the ratio of conjugate base to weak acid at any given pH:

$$pH = pK_a + \log \frac{[\text{conjugate base}]}{[\text{weak acid}]}$$

In general, the conjugate base predominates at pH values greater than pK_a by more than one unit. Within one unit of the pK_a, appreciable concentrations of both the acid and its conjugate base are present in solution. At any pH lower than pK_a by more than one unit, the acid form predominates.

The following are calculations for the given pH values to demonstrate these general guidelines:

(a) $\log \dfrac{[RCOO^-]}{[RCOOH]} = pH - pK_a = 7.0 - 8.54 = -1.54$

$\dfrac{[RCOO^-]}{[RCOOH]} = \dfrac{1}{34.7}$

$RCOOH = \dfrac{[RCOOH] \times 100\%}{[RCOO^-] + [RCOOH]} = \dfrac{34.7 \times 100}{34.7 + 1} = 97\%$

(b) $\log \dfrac{[RCOO^-]}{[RCOOH]} = pH - pK_a = 9.0 - 8.54 = 0.46$

$\dfrac{[RCOO^-]}{[RCOOH]} = \dfrac{2.88}{1}$

$RCOO^- = \dfrac{[RCOO^-] \times 100\%}{[RCOO^-] + [RCOOH]} = \dfrac{2.88 \times 100}{2.88 + 1} = 74\%$

(c) $\log \dfrac{[RCOO^-]}{[RCOOH]} = pH - pK_a = 11.0 - 8.54 = 2.46$

$\dfrac{[RCOO^-]}{[RCOOH]} = \dfrac{288}{1}$

$RCOO^- = \dfrac{[RCOO^-] \times 100\%}{[RCOO^-] + [RCOOH]} = \dfrac{288 \times 100}{288 + 1} = 99.6\%$

(d) $\log \dfrac{[RCOO^-]}{[RCOOH]} = pH - pK_a = 3.0 - 8.54 = -5.54$

$\dfrac{[RCOO^-]}{[RCOOH]} = \dfrac{1}{346\,737}$

$RCOOH = \dfrac{[RCOOH] \times 100\%}{[RCOO^-] + [RCOOH]} = \dfrac{346\,737 \times 100}{346\,737 + 1} = 99.9997\%$

15.12 An electron-withdrawing group bonded to the α-carbon atom of a carboxylic acid increases its acidity (pK_a less positive) while an electron-donating group bonded to the α-carbon atom has the opposite effect.
   (a) The electron-withdrawing inductive effect of the oxygen atom of the —OCH_3 group makes methoxyacetic acid the stronger acid.
   CH_3COOH CH_3OCH_2COOH
   pK_a 4.75 pK_a 3.6
   (b) FCH_2COOH F_3CCOOH
   pK_a 2.6 pK_a 0.23
   (c) O_2NCH_2COOH CH_3CH_2COOH
   pK_a 1.7 pK_a 4.9
   (d) The methyl group, —CH_3, is electron-donating compared to a hydrogen atom so methanoic acid (formic acid) is the stronger acid.
   HCOOH CH_3COOH
   pK_a 3.7 pK_a 4.75

15.13

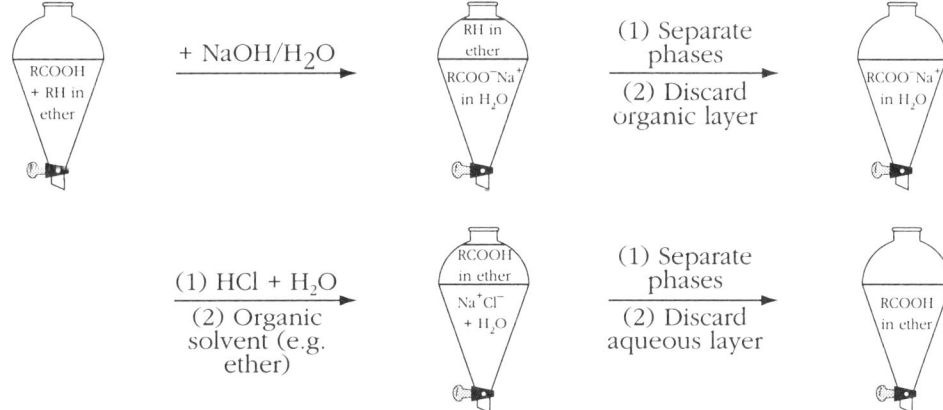

15.14 Step 1: Addition of water

Step 2: Elimination of methanol

15.15 To answer this question, we must write the mechanism, carefully noting the position of the ^{18}O.

[mechanism scheme showing nucleophilic addition of HO^- to ester $R-C(=O)-^{18}OR'$ forming tetrahedral intermediate $R-C(O^-)(^{18}OR')(OH)$, then collapse to $R-COOH$ + $R'-^{18}O^-$, followed by proton transfer to give $R-COO^-$ + H_2O and $R'-^{18}OH$ + HO^-]

$R' = CH_3CH_2-$

$R = CH_3CH_2-$

The ^{18}O will be found in $CH_3CH_2{}^{18}OH$.

15.16 Hydrolysis of a triacylglycerol gives one equivalent of glycerol and three equivalents of sodium palmitate. This means that the triacylglycerol has three identical fatty acids bonded to glycerol, and that this fatty acid is palmitic acid. The structure of the triacylglycerol and its saponification reaction are as follows.

$$\begin{array}{c} CH_2O-CO-(CH_2)_{14}CH_3 \\ | \\ CHO-CO-(CH_2)_{14}CH_3 \\ | \\ CH_2O-CO-(CH_2)_{14}CH_3 \end{array} \xrightarrow{3\ NaOH} \begin{array}{c} CH_2OH \\ | \\ CHOH \\ | \\ CH_2OH \end{array} + 3\ CH_3(CH_2)_{14}COO^-Na^+$$

15.17 (a)

$$CH_3(CH_2)_8COOH \xrightarrow[(2)\ H_3O^+]{(1)\ LiAlH_4} CH_3(CH_2)_8CH_2OH \xrightarrow{PBr_3} CH_3(CH_2)_8CH_2Br$$

(b) The product contains one more carbon atom than the starting material. There are two ways of carrying out this synthesis:

$$CH_3(CH_2)_8CH_2Br \text{ From (a)} \xrightarrow{NaCN} CH_3(CH_2)_8CH_2C\equiv N \xrightarrow[\text{Heat}]{H_3O^+} CH_3(CH_2)_8CH_2COOH$$

$$CH_3(CH_2)_8CH_2Br \xrightarrow[\text{Dry ether}]{Mg} CH_3(CH_2)_8CH_2MgBr \xrightarrow[(2)\ H_3O^+]{(1)\ CO_2} CH_3(CH_2)_8CH_2COOH$$

(c) The product contains two more carbon atoms than the starting material. A good method of extending a carbon chain by two units is a reaction of a Grignard reagent (or an organolithium reagent) with oxirane. The resulting hydroxyl group can be converted to a carboxylic group by oxidation:

$$CH_3(CH_2)_8CH_2Br \xrightarrow[\text{Dry ether}]{\text{Mg}} CH_3(CH_2)_8CH_2MgBr \xrightarrow[\text{(2) } H_3O^+]{\text{(1) oxirane}} CH_3(CH_2)_{10}CH_2OH$$

From (a)

$$\xrightarrow[\text{(2) } H_3O^+]{\text{(1) KMnO}_4, \text{HO}^-, \text{heat}} CH_3(CH_2)_{10}COOH$$

(d) $CH_3(CH_2)_8CH_2Br \xrightarrow[\text{Dry ether}]{\text{Mg}} CH_3(CH_2)_8CH_2MgBr \xrightarrow{H_2O} CH_3(CH_2)_8CH_3$

From (a)

(e) $CH_3(CH_2)_7CH_2CH_2Br \xrightarrow[\text{Heat}]{(CH_3)_3CO^-K^+} CH_3(CH_2)_7CH=CH_2$

From (a)

(f) $CH_3(CH_2)_7CH=CH_2 \xrightarrow[\text{CCl}_4]{Br_2} CH_3(CH_2)_7\overset{\overset{\text{Br}}{|}}{CH}-CH_2Br$

From (e)

$\xrightarrow[\text{(2) } H_2O]{\text{(1) 3 NaNH}_2} CH_3(CH_2)_7C\equiv CH$

(g) $CH_3(CH_2)_8CH_2Br \xrightarrow{HC\equiv C^-Na^+} CH_3(CH_2)_8CH_2C\equiv CH$

From (a)

$\xrightarrow[H_2SO_4]{HgSO_4, H_2O} CH_3(CH_2)_8CH_2\overset{\overset{O}{\|}}{C}CH_3$

(h) $CH_3(CH_2)_9C\equiv CH \xrightarrow[\text{Liq. NH}_3]{NaNH_2} CH_3(CH_2)_9C\equiv C^-Na^+$

From (g)

$\xrightarrow{CH_3CH_2Br} CH_3(CH_2)_9C\equiv CCH_2CH_3$

15.18 (a) A **carboxylic acid** is an organic compound that contains the —COOH functional group. The *carboxyl* group contains a *carb*onyl group and hydr*oxyl* group attached to the carbonyl carbon. The hydrogen atom of the carboxyl group is acidic; the pK_a of an unsubstituted carboxylic acid is in the range 4 to 6.

(b) **Carboxylate ion** is the conjugate base of a carboxylic acid.

$$R-\underset{\underset{H}{O}}{\overset{O}{\overset{\|}{C}}} + {}^-OH \longrightarrow \left[R-\underset{\underset{O^-}{}}{\overset{O}{\overset{\|}{C}}} \longleftrightarrow R-\underset{\underset{O}{}}{\overset{O^-}{C}} \right] + H-OH$$

Carboxylate ion

(c) A **fatty acid** is a long, straight-chain carboxylic acid containing an even number of carbon atoms (usually 12 or more carbon atoms). Such acids are found in natural fat and that is why they are called *fatty acids*; the most abundant fatty acids in nature are $C_{15}H_{31}COOH$ (palmitic acid) and $C_{17}H_{35}COOH$ (stearic acid).

(d) An **unsaturated fat** is a triester of glycerol and unsaturated fatty acid (a long, straight-chain carboxylic acid containing at least one carbon-carbon double bond along the chain). Unsaturated fats have lower melting points than their saturated counterparts and that is why they are found in oils. In most unsaturated fatty acids of fats, oils, and biological membranes, the *Z* isomer (*cis*) predominates; the *E* isomer (*trans*) is rare. The most abundant unsaturated fatty acid in nature is oleic acid [$CH_3(CH_2)_7CH=CH(CH_2)_7COOH$].

(e) An **ester** is a functional derivative of a carboxylic acid in which the hydrogen atom of the carboxyl group has been replaced by a group containing a carbon atom, RCOOR' or RCOOAr.

(f) **Saponification** is the hydrolysis of an ester with aqueous hydroxide to give a carboxylate anion and an alcohol. The term is derived from use of this chemical reaction in the manufacture of soaps (Latin: *saponem*, soap).

$$R-\underset{OR'}{\overset{O}{\overset{\|}{C}}} + HO^- \longrightarrow R-\underset{O^-}{\overset{O}{\overset{\|}{C}}} + R'OH$$

An ester   Hydroxide ion   Carboxylate ion   Alcohol

(g) **Acid-catalyzed esterification** is a reaction of carboxylic acids with alcohols in the presence of a catalytic amount of a strong acid, such as dry hydrogen chloride, concentrated sulfuric acid, or *p*-toluenesulfonic acid, to give esters and water. The reaction is reversible and generally considerable amounts of starting materials and products are present at equilibrium. The yield of ester can be increased either by removing water as it is formed or by increasing concentration of one of the reactants.

$$R-\underset{OH}{\overset{O}{\overset{\|}{C}}} + R'OH \underset{}{\overset{\text{Acid catalyst}}{\rightleftharpoons}} R-\underset{OR'}{\overset{O}{\overset{\|}{C}}} + H_2O$$

(h) The **salt of a carboxylic acid** (carboxylate salt) is the product of the reaction of a carboxylic acid with base.

$$R-\underset{OH}{\underset{|}{C}}=O \quad + \quad KOH \quad \longrightarrow \quad R-\underset{O^-K^+}{\underset{|}{C}}=O \quad + \quad H_2O$$

An acid · A base · A salt

(i) **Soap** is usually the sodium or potassium salt of a long-chain carboxylic acid. Soaps and detergents are examples of molecules with two structural parts that interact with water in opposing ways: (i) they have a polar head group, which is solvated by water (carboxylate ion), (ii) they have a hydrocarbon tail, which is not solvated by water (hydrocarbon chain).

(j) **Fat** is an ester of glycerol [$CH_2(OH)CH(OH)CH_2OH$] and fatty acid(s). Glycerol has three hydroxyl groups and can form esters with one (monoglyceride), two (diglyceride), or three molecules of fatty acid (triglyceride).

A monoglyceride · A diglyceride · A triglyceride

(k) **Detergent** is a term for any molecule with a long hydrocarbon chain, preferably 12 to 20 carbon atoms, and a polar group at one end of the molecule (hydrogen sulfate, sulfate, or sulfonate group).

Sodium stearate (octadecanoate)
Soap

Sodium dodecyl sulfate
Synthetic detergent

Soaps and detergents are examples of molecules with two structural parts that interact with water in opposing ways: (i) they have a polar head group, which is solvated by water (carboxylate ion), (ii) they have a hydrocarbon tail, which is not solvated by water (hydrocarbon chain).

(l) A **saturated fat** is an ester of glycerol and fatty acids. Butter, margarine, and lard are examples of saturated fats.
(m) **Fischer esterification** is a conversion of a carboxylic acid and alcohol to an ester in the presence of an acid catalyst (see *Acid-catalyzed esterification*).

15.19 (a) *cis*-1,2-Cyclopentanedicarboxylic acid
(b) (*R*)-1-Methylpropyl acetate or (*R*)-2-butyl acetate
(c) (*Z*)-3-Cyclopentylpropenoic acid or *cis*-3-cyclopentylpropenoic acid
(d) Potassium (*Z*)-2-butenoate or potassium *cis*-2-butenoate
(e) (*E*)-3-Methyl-2-pentenoic acid (Note: you must *not* use the term *trans* because *cis* and *trans* can be used only to name disubstituted alkenes)
(f) (*S*)-2-Bromopropanoic acid

15.20

(a) [PhCH₂COOH structure]   (b) Mg(OOCH)₂   (c) CH₃CH₂CH(CH₃)COOH

(d) [PhC(=O)O⁻Na⁺ structure]   (e) [CH₂=CH–C*H(Cl)–COOH with H, Cl stereochemistry]   (f) [CH₃CH=CHCH₂COOH, *trans*]

(g) CH₃C(=O)OCH₂CH₂CH₂CH₂CH₂CH₃   (h) [cyclohexane with NO₂ and COOH — two stereoisomers shown "or"]

(i) [isobutyl group–CH₂–C(=O)–O–isopropyl ester structure]

(j) [cyclopropane with H₃C and C(=O)O-*t*-butyl, two stereoisomers shown "or"]

15.21

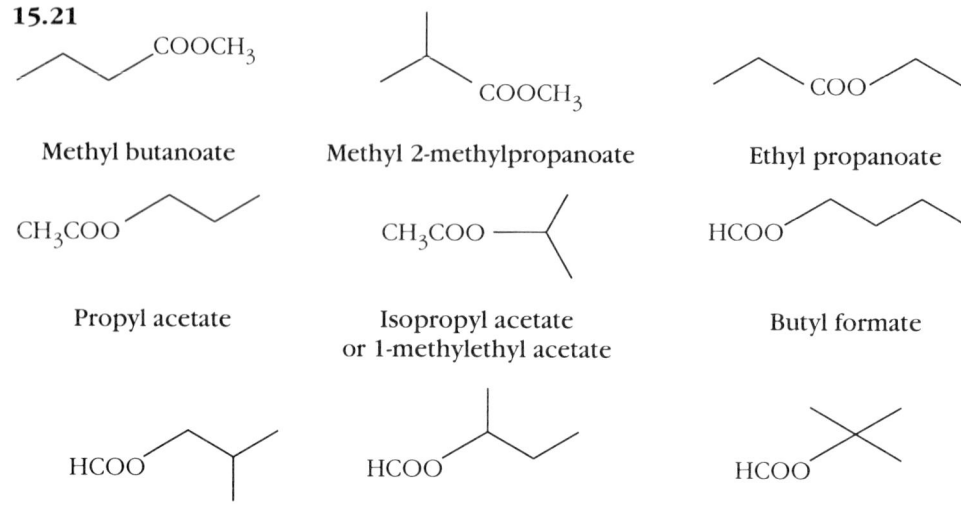

Methyl butanoate

Methyl 2-methylpropanoate

Ethyl propanoate

Propyl acetate

Isopropyl acetate
or 1-methylethyl acetate

Butyl formate

Isobutyl formate
or 2-methylpropyl formate

1-Methylpropyl formate
(racemic mixture)

1,1-Dimethylethyl formate
Common name: *tert*-butyl formate

15.22 Increasing acid strength →

(a) Ethanol   Acetic acid   Formic acid
    pK_a 15.9   pK_a 4.75   pK_a 3.7
(b) Butanoic acid   3-Chlorobutanoic acid 2-Chlorobutanoic acid
    pK_a 4.82   pK_a 3.98   pK_a 2.83
(c) 1-Butyne   Butanoic acid   Sulfuric acid
    pK_a 26   pK_a 4.82   pK_a ~ −9
(d) 1-Butyne   1-Butanol   Water
    pK_a 26   pK_a 17   pK_a 15.7

15.23 Strong acids have weak conjugate bases, while weak acids have strong conjugate bases. Keeping that in mind, we have to identify the weaker conjugate acid.

(a) CH$_3$COONa   Conjugate acid: CH$_3$COOH
    Stronger base   pK_a 4.75
                                       Weaker acid
    CH$_2$ClCOONa   Conjugate acid: CH$_2$ClCOOH
                                       pK_a 2.86

Acetic acid, CH$_3$COOH, is a weaker acid (pK_a 4.75) than chloroacetic acid (pK_a 2.86), so sodium acetate is a stronger base.

(b) CH$_3$COONa   Conjugate acid: CH$_3$COOH
                                       pK_a 4.75
    HC≡CNa   Conjugate acid: HC≡CH
    Stronger base   pK_a 25
                                       Weaker acid

Acetylene, HC≡CH, is a weaker acid (pK_a ≈ 25) than acetic acid (pK_a 4.75), so sodium acetylide is a stronger base.

(c) CH$_3$COONa   Conjugate acid: CH$_3$COOH
                          pK_a 4.75

CH$_3$CH$_2$ONa   Conjugate acid: CH$_3$CH$_2$OH
Stronger base   pK_a 17
                          Weaker acid

Ethanol, CH$_3$CH$_2$OH, is a weaker acid (pK_a ≈ 17) than acetic acid (pK_a 4.75), so sodium ethoxide is a stronger base.

(d) CH$_3$COONa   Conjugate acid: CH$_3$COOH
                          pK_a 4.75

NaOH   Conjugate acid: H$_2$O
Stronger base   pK_a 15.7
                          Weaker acid

Water, H$_2$O, is a weaker acid (pK_a 15.7) than acetic acid (pK_a 4.75), so sodium hydroxide is a stronger base.

(e) CH$_3$COONa   Conjugate acid: CH$_3$COOH
Stronger base   pK_a 4.75
                          Weaker acid

NaCl   Conjugate acid: HCl
                          pK_a ≈ −7

Acetic acid, CH$_3$COOH, is a weaker acid (pK_a 4.75) than hydrochloric acid (pK_a ≈ −7), so sodium acetate is a stronger base.

15.24 (a) CH$_3$CH$_2$CH$_2$COOH + NH$_3$ ⇌ CH$_3$CH$_2$CH$_2$COO$^-$ $\overset{+}{\text{N}}$H$_4$
           pK_a 4.82   pK_a 9.4

Ammonium ion, NH$_4^+$, is the weaker acid so the equilibrium is shifted to the right.

(b) CH$_3$COOH + CCl$_3$COONa ⇌ CH$_3$COONa + CCl$_3$COOH
    pK_a 4.75   pK_a 0.64

Acetic acid is the weaker acid so the equilibrium is shifted to the left.

(c) C$_6$H$_5$COOH + CH$_3$CH$_2$ONa ⇌ C$_6$H$_5$COONa + CH$_3$CH$_2$OH

pK_a 4.2   pK_a 16.9

Ethanol is the weaker acid so the equilibrium is shifted to the right.

15.25 CH$_3$(CH$_2$)$_{10}$COOH + HO$^-$ ⇌ CH$_3$(CH$_2$)$_{10}$COO$^-$ + H$_2$O
pK_a 4.89   pK_a 15.7

Water is the weaker acid so the position of this equilibrium lies to the right.

$$K_{eq} = \frac{K_a(\text{RCOOH})}{K_a(\text{H}_2\text{O})} = \frac{1.29 \times 10^{-5}}{2 \times 10^{-16}} = 6.45 \times 10^{10}$$

15.26 Ethyl acetate   1-Chlorobutane   2-Pentanone   Butanoic acid
b.p. 77 °C   b.p. 78 °C   b.p. 101 °C   b.p. 163 °C

All of the above compounds have comparable molecular weights (between 86 and 92). Thus, the difference in boiling points will reflect intermolecular interactions between molecules in the liquid state.

Carboxylic acids are associated by hydrogen bonding in the liquid state. Hydrogen bonding is the strongest of the intermolecular interactions so butanoic acid has the highest boiling point of all these compounds.

Because of the polarity of the carbonyl group, ketones are polar compounds and interact by dipole-dipole interactions. Ketones cannot associate by hydrogen bonding, so their boiling points are lower than those of alcohols and carboxylic acids. The boiling points of ketones, however, are higher than those of the less polar esters and chloroalkanes of comparable molecular weight.

15.27 (a) The mixture of hexanoic acid and ethyl hexanoate is first dissolved in diethyl ether. When the ether solution is shaken with aqueous sodium hydroxide, hexanoic acid is converted to its water-soluble salt. Then the ether and aqueous phases are separated. The ether solution is distilled, yielding first ether (b.p. 35 °C) and then ethyl hexanoate (b.p. 168 °C). The aqueous solution is acidified with HCl (or some other mineral acid), and hexanoic acid comes out as a water-insoluble liquid that can be easily extracted by ether and then separated by distillation (b.p. of hexanoic acid is 202 to 203 °C; b.p. of ether is 35 °C).

(b) The mixture of propanoic acid and dipropyl ether is first dissolved in diethyl ether. When the ether solution is shaken with aqueous NaOH or some other base, propanoic acid is converted to its water-soluble salt. Next, the ether and aqueous phases are separated. The aqueous solution is acidified with HCl (or some other mineral acid), and propanoic acid can be easily extracted by ether and then separated by distillation (b.p. of propanoic acid is 141 °C; b.p. of ether is 35 °C).

(c) The mixture of hexanoic acid and 1-hexanol is first dissolved in diethyl ether. When the ether solution is shaken with NaOH or some other base, hexanoic acid is converted to its water-soluble salt. Next, the ether and aqueous phases are separated. The ether solution is distilled, yielding first ether (b.p. 35 °C) and then 1-hexanol (b.p. 157 °C). The aqueous solution is acidified with HCl (or some other mineral acid), and hexanoic acid comes out as a water-insoluble liquid that can be easily extracted by ether and then separated by distillation (b.p. of hexanoic acid is 202 to 203 °C; b.p. of ether is 35 °C).

15.28

(a) $CH_3CH_2COOH + KOH \longrightarrow CH_3CH_2COOK + H_2O$

(b) $CH_3CH_2COOH \xrightarrow[(2)\ H_3O^+]{(1)\ LiAlH_4} CH_3CH_2CH_2OH$

(c) 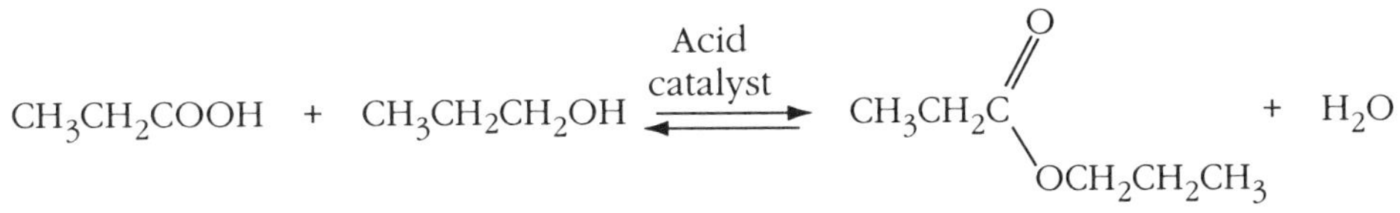 $CH_3CH_2COOH + CH_3CH_2CH_2OH \underset{}{\overset{\text{Acid catalyst}}{\rightleftharpoons}} CH_3CH_2C(=O)OCH_2CH_2CH_3 + H_2O$

(d) $CH_3CH_2COOH \xrightarrow{NaBH_4} H_2 + CH_3CH_2COO^-Na^+ \xrightarrow{H_3O^+} CH_3CH_2COOH$

(e) $CH_3CH_2COOH \xrightarrow[H_3O^+]{KMnO_4} CH_3CH_2COOH$ (No reaction)

15.29

(a) 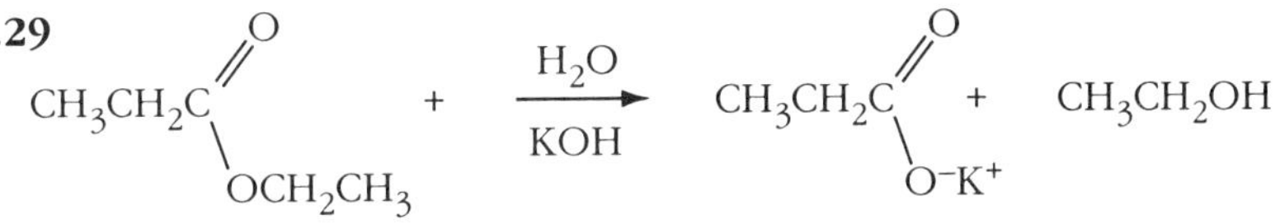 $CH_3CH_2C(=O)OCH_2CH_3 \xrightarrow[KOH]{H_2O} CH_3CH_2C(=O)O^-K^+ + CH_3CH_2OH$

(b) 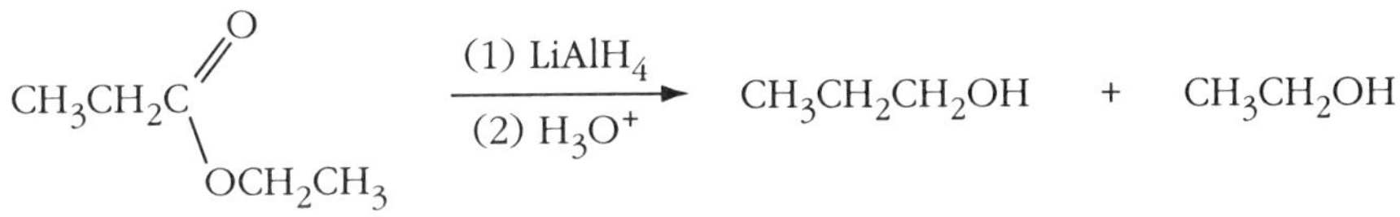 $CH_3CH_2C(=O)OCH_2CH_3 \xrightarrow[(2)\ H_3O^+]{(1)\ LiAlH_4} CH_3CH_2CH_2OH + CH_3CH_2OH$

(c) 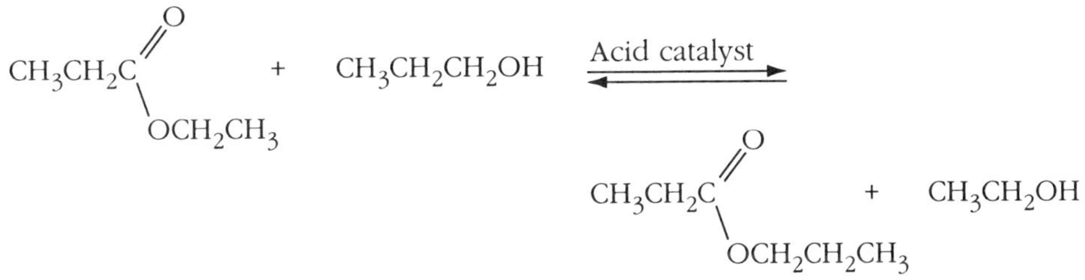 $CH_3CH_2C(=O)OCH_2CH_3 + CH_3CH_2CH_2OH \overset{\text{Acid catalyst}}{\rightleftharpoons}$

$CH_3CH_2C(=O)OCH_2CH_2CH_3 + CH_3CH_2OH$

(d) 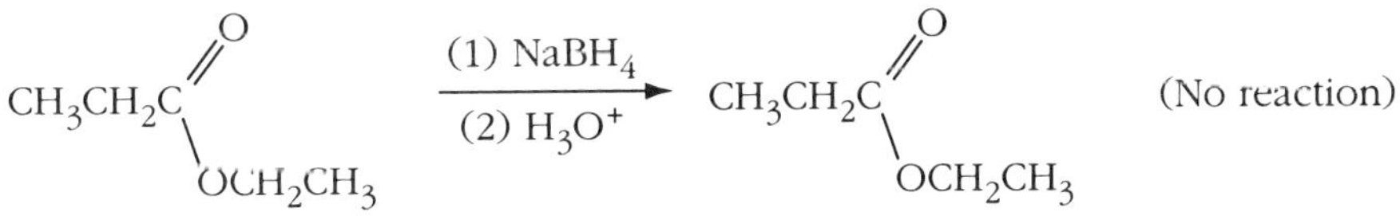 $CH_3CH_2C(=O)OCH_2CH_3 \xrightarrow[(2)\ H_3O^+]{(1)\ NaBH_4} CH_3CH_2C(=O)OCH_2CH_3$ (No reaction)

(e) Oxidation with potassium permanganate is usually carried out in water. Under the acidic conditions of the reaction, ethyl propanoate will hydrolyse and the ethanol formed by hydrolysis will be oxidized to acetic acid.

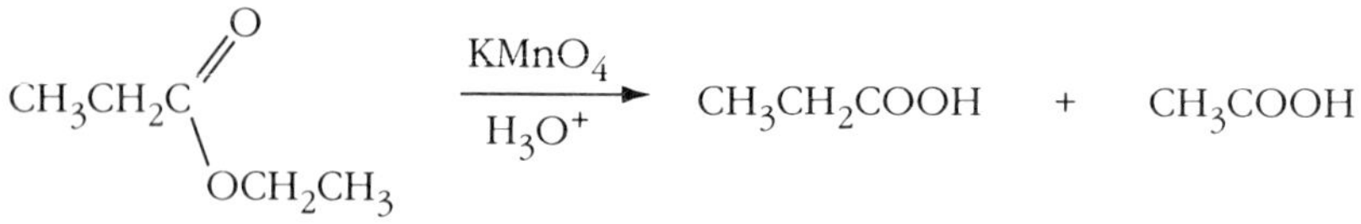

$$CH_3CH_2\overset{\displaystyle O}{\underset{\displaystyle OCH_2CH_3}{C}} \xrightarrow[H_3O^+]{KMnO_4} CH_3CH_2COOH \;+\; CH_3COOH$$

15.30

(a) $CH_3CH_2CH_2COO^-K^+ \xrightarrow[H_2O]{HCl} CH_3CH_2CH_2COOH \;+\; KCl$

(b) Cyclohexyl–MgBr $\xrightarrow[\text{(2) }H_3O^+]{\text{(1) }CO_2}$ Cyclohexyl–COOH

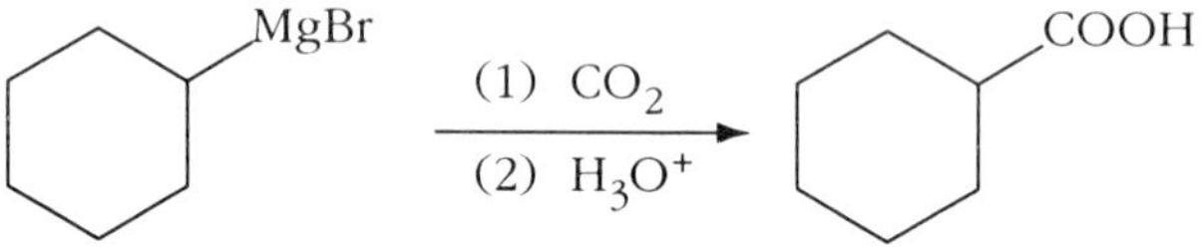

(c) CH₃CH₂CH₂CH₂CH₂Br $\xrightarrow{KCN}$ CH₃CH₂CH₂CH₂CH₂C≡N

(d) CH₃CH₂CH₂CH₂CH₂C≡N $\xrightarrow[\text{Heat}]{H_2SO_4/H_2O}$ CH₃CH₂CH₂CH₂CH₂COOH

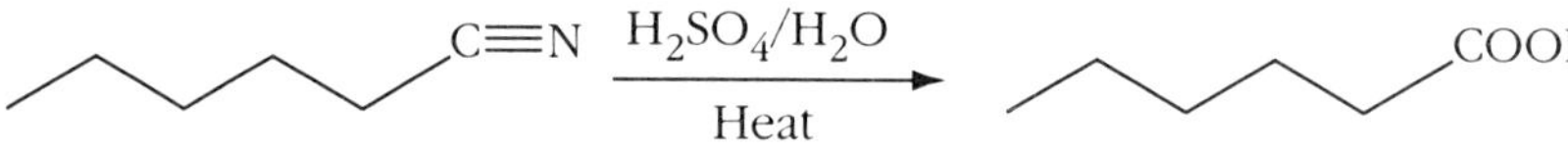

(e) $\underset{H}{\overset{CH_3(CH_2)_7}{>}}C=C\underset{H}{\overset{(CH_2)_7COOH}{<}} \xrightarrow[H_2SO_4/\text{Heat}]{KMnO_4} CH_3(CH_2)_7COOH \;+\; HOOC(CH_2)_7COOH$

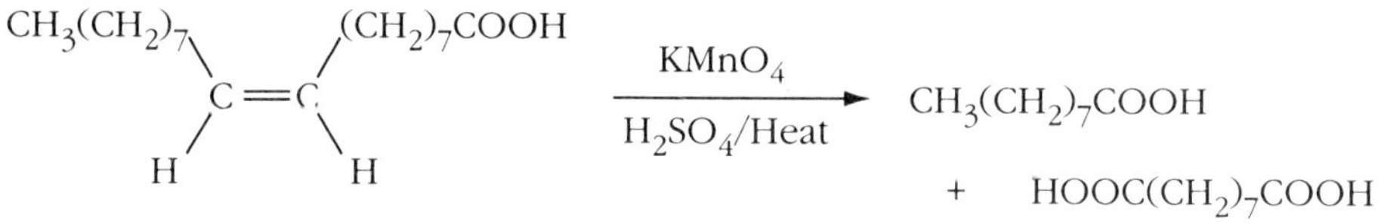

(f) $(CH_3)_2CHCOOH \xrightarrow[H_2O]{KOH} (CH_3)_2CHCOOK$

(g) $\underset{H}{\overset{CH_3(CH_2)_7}{>}}C=C\underset{H}{\overset{(CH_2)_7COOH}{<}} \xrightarrow[CCl_4]{I_2}$

$$CH_3(CH_2)_7-\underset{\displaystyle I}{CH}-\underset{\displaystyle I}{CH}-(CH_2)_7COOH$$

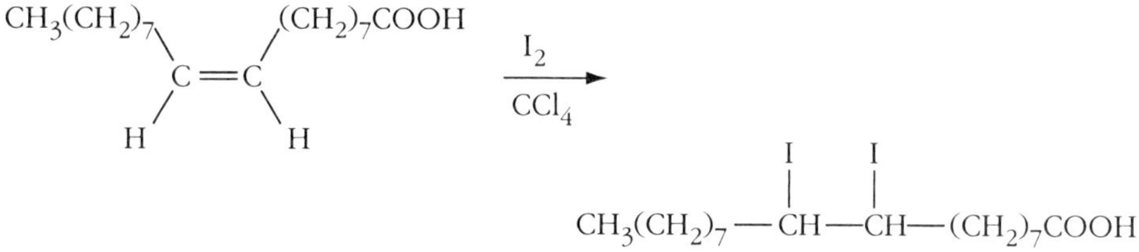

This is an equilibrium reaction in which the unreacted alkene is the major product at equilibrium.

(h) CH₃(CH₂)₃CH=CHCH₂CH=CHCH₂CH=CH(CH₂)₇COOH

$\downarrow 3H_2\;\;Ni$

CH₃(CH₂)₁₆COOH

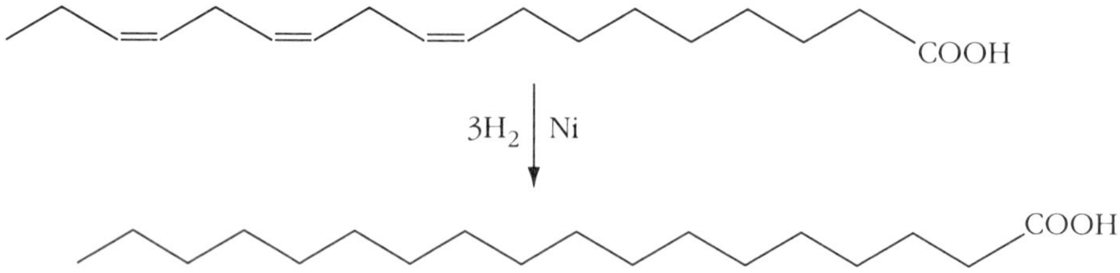

(i)

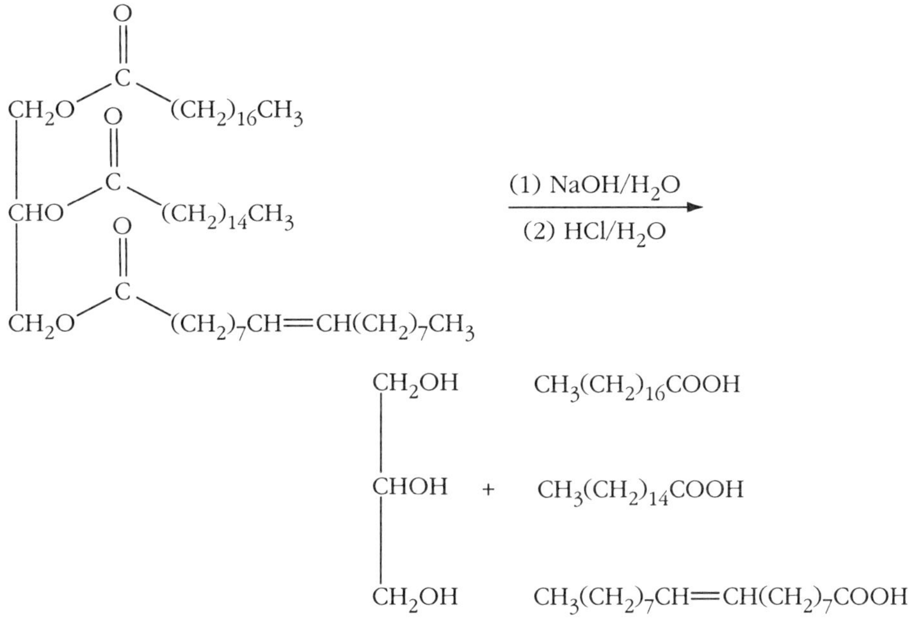

$$\begin{array}{c}\text{CH}_2\text{O}-\overset{\overset{\displaystyle O}{\|}}{\text{C}}-(\text{CH}_2)_{16}\text{CH}_3\\ \text{CHO}-\overset{\overset{\displaystyle O}{\|}}{\text{C}}-(\text{CH}_2)_{14}\text{CH}_3\\ \text{CH}_2\text{O}-\overset{\overset{\displaystyle O}{\|}}{\text{C}}-(\text{CH}_2)_7\text{CH}=\text{CH}(\text{CH}_2)_7\text{CH}_3\end{array} \xrightarrow[(2)\ \text{HCl/H}_2\text{O}]{(1)\ \text{NaOH/H}_2\text{O}}$$

$$\begin{array}{c}\text{CH}_2\text{OH}\\ \text{CHOH}\\ \text{CH}_2\text{OH}\end{array} + \begin{array}{l}\text{CH}_3(\text{CH}_2)_{16}\text{COOH}\\ \text{CH}_3(\text{CH}_2)_{14}\text{COOH}\\ \text{CH}_3(\text{CH}_2)_7\text{CH}=\text{CH}(\text{CH}_2)_7\text{COOH}\end{array}$$

(j)

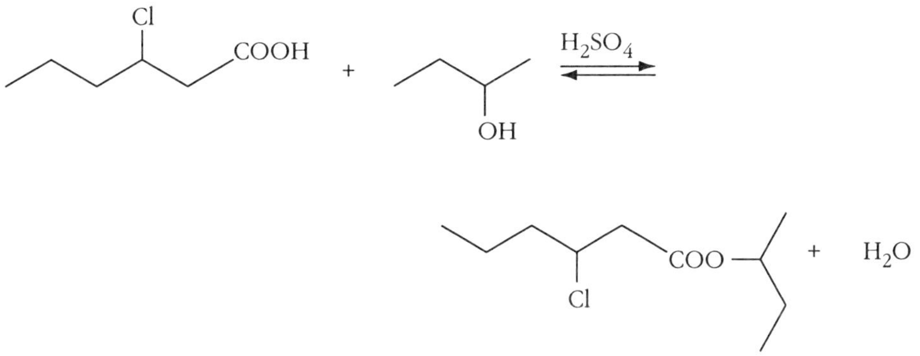

3-chloropentanoic acid + sec-butanol $\underset{}{\overset{\text{H}_2\text{SO}_4}{\rightleftharpoons}}$

sec-butyl 3-chloropentanoate + H_2O

15.31

(a)

$(\text{CH}_3)_3\text{CO}-\overset{\overset{\displaystyle O}{\|}}{\text{C}}-\text{CH}_2\text{CH}_2\text{CHOH}-\text{[4-ethoxycyclohexyl]}\ \xrightarrow[\text{H}_2\text{O}]{\text{KOH}}\ \text{HOCHCH}_2\text{CH}_2-\overset{\overset{\displaystyle O}{\|}}{\text{C}}-\text{[4-ethoxycyclohexyl]}-\text{O}^-\text{K}^+\ +\ (\text{CH}_3)_3\text{COH}$

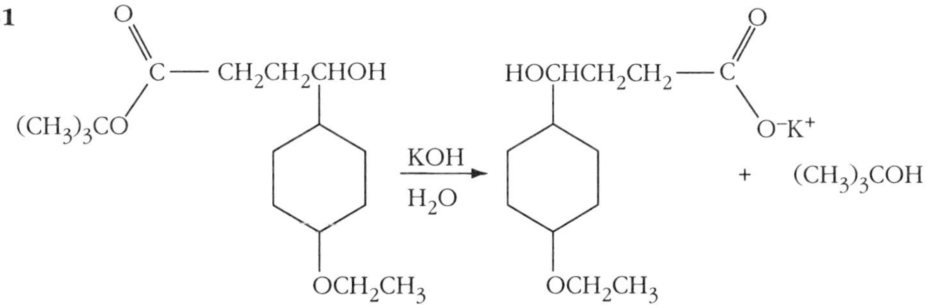

(b)

[structure: linoleic acid] → (1) LiAlH$_4$ / (2) H$_3$O$^+$ → [structure: linoleyl alcohol, CH$_2$OH]

(c) The oxidation of a primary alcohol gives a carboxylic acid. The carboxylic acid may react with the alcohol in the presence of an acidic catalyst to yield an ester. In most cases, however, the ester can hydrolyse back to the alcohol and acid. If the ester is not soluble and precipitates, however, it cannot hydrolyse. As a result, alcohols are more efficiently oxidized in base medium because the reaction is driven to completion without this side reaction.

[octahydronaphthalene-CH$_2$OH] → Na$_2$Cr$_2$O$_7$ / H$_2$SO$_4$/H$_2$O → [octahydronaphthalene-COOH]

+ [octahydronaphthalene-COOCH$_2$—] [octahydronaphthalene]

(d) The carbonyl group of an aldehyde or ketone is reduced to an alcohol group by hydrogen in the presence of a transition metal catalyst (most commonly finely divided Pd, Pt, Ni, Ru, or Cu-Cr complex.) The carbon-carbon double bond undergoes catalytic hydrogenation under the same reaction conditions, but it is a faster reaction. Therefore, it is usually possible to selectively reduce an alkene double bond in the presence of a carbonyl group by catalytic hydrogenation:

[diketone with cyclohexene] → H$_2$ (1 equivalent) / PtO$_2$ → [diketone with cyclohexane]

15.32

(a) Retrosynthesis:

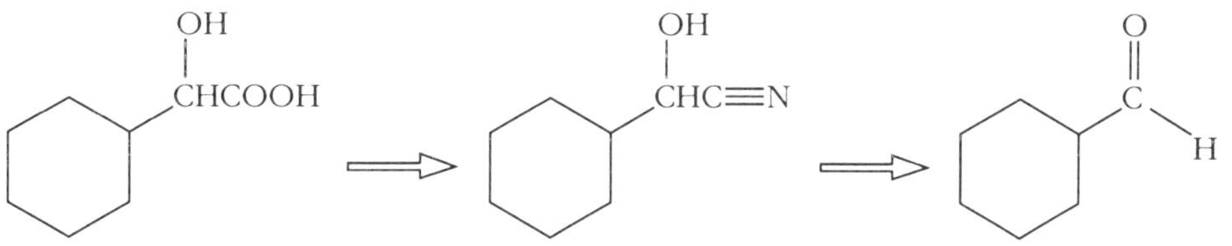

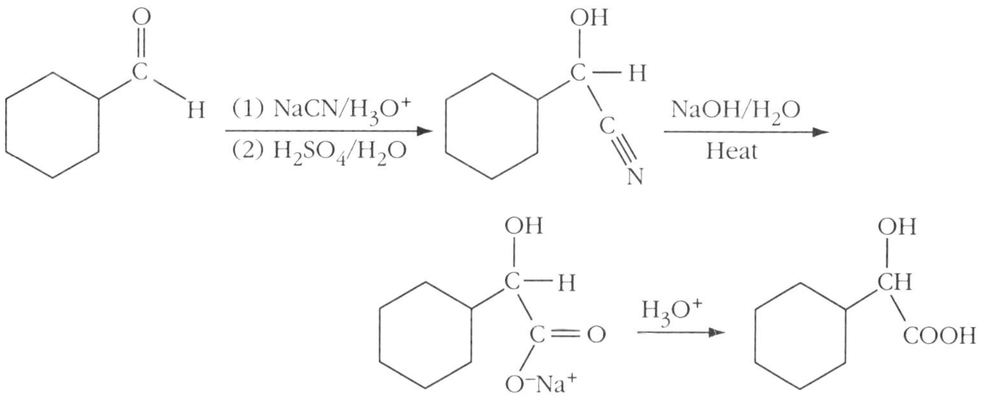

Cyclohexyl-CH(OH)COOH ⟹ Cyclohexyl-CH(OH)C≡N ⟹ Cyclohexyl-CHO

Cyclohexyl-CHO $\xrightarrow[\text{(2) } H_2SO_4/H_2O]{\text{(1) NaCN/}H_3O^+}$ Cyclohexyl-CH(OH)-C≡N $\xrightarrow[\text{Heat}]{\text{NaOH/}H_2O}$

Cyclohexyl-CH(OH)-C(=O)O⁻Na⁺ $\xrightarrow{H_3O^+}$ Cyclohexyl-CH(OH)-COOH

Note: Acid-catalyzed hydrolysis of the cyanohydrin would be accompanied by dehydration.

(b) Retrosynthesis:

$(CH_3)_2CHCOOH \Longrightarrow (CH_3)_2CHMgBr \Longrightarrow (CH_3)_2CHBr$

Synthesis:

$(CH_3)_2CHBr \xrightarrow[\text{Dry ether}]{Mg} (CH_3)_2CHMgBr \xrightarrow[\text{(2) } H_3O^+]{\text{(1) }CO_2} (CH_3)_2CHCOOH$

(c) Retrosynthesis:

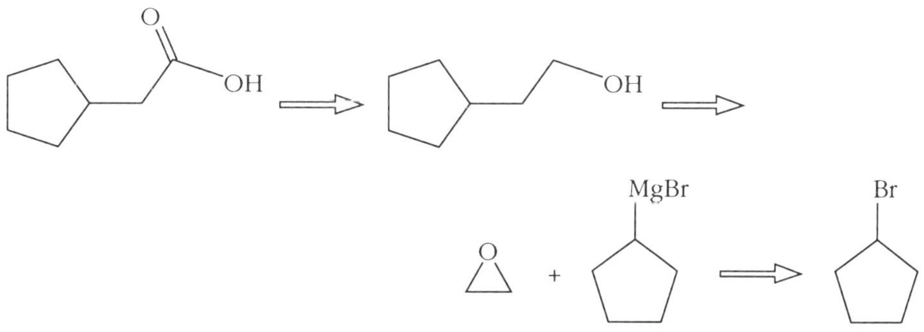

Cyclopentyl-CH₂COOH ⟹ Cyclopentyl-CH₂CH₂OH ⟹

ethylene oxide + Cyclopentyl-MgBr ⟹ Cyclopentyl-Br

Synthesis:

$$\text{cyclopentyl-Br} \xrightarrow[\text{Dry ether}]{\text{Mg}} \text{cyclopentyl-MgBr} \xrightarrow[\text{(2) H}_3\text{O}^+]{\text{(1) ethylene oxide}} \text{cyclopentyl-CH}_2\text{CH}_2\text{OH}$$

$$\xrightarrow[\text{(2) H}_2\text{SO}_4/\text{H}_2\text{O}]{\text{(1) KMnO}_4,\ \text{NaOH, heat}} \text{cyclopentyl-CH}_2\text{COOH}$$

(d) Retrosynthesis:

$$CH_3(CH_2)_6COOCH_3 \Longrightarrow CH_3(CH_2)_6COOH \Longrightarrow CH_3(CH_2)_6CH_2OH$$

$$CH_3(CH_2)_3C\equiv C^- \ + \ \triangle\hspace{-1.2em}O \ \Longleftarrow\ CH_3(CH_2)_3C\equiv CCH_2CH_2OH$$

$$\Downarrow$$

$$CH_3(CH_2)_3C\equiv CH$$

Synthesis:

$$CH_3(CH_2)_3C\equiv CH \xrightarrow[\text{Liq. NH}_3]{\text{NaNH}_2} CH_3(CH_2)_3C\equiv C^-Na^+ \xrightarrow[\text{(2) H}_3\text{O}^+]{\text{(1) ethylene oxide}}$$

$$CH_3(CH_2)_6CH_2OH \xleftarrow{\text{H}_2/\text{Pd}} CH_3(CH_2)_3C\equiv CCH_2CH_2OH$$

$$\xrightarrow[\text{(2) H}_3\text{O}^+]{\text{(1) KMnO}_4/\text{NaOH}}$$

$$CH_3(CH_2)_6COOH \xrightarrow[\text{H}_2\text{SO}_4]{\text{CH}_3\text{OH}} CH_3(CH_2)_6COOCH_3$$

(e) Retrosynthesis:

$$\underset{\text{CH}_3\text{CH}_2\text{CHCH}_2\text{Br}}{\overset{\text{CH}_3}{|}} \Longrightarrow \underset{\text{CH}_3\text{CH}_2\text{C}=\text{CH}_2}{\overset{\text{CH}_3}{|}}$$

$$\underset{\text{CH}_3\text{CH}_2\text{CHOH}}{\overset{\text{CH}_3}{|}} \Longleftarrow \underset{\text{CH}_3\text{CH}_2\text{C}=\text{O}}{\overset{\text{CH}_3}{|}} \ + \ (C_6H_5)_3\overset{+}{P}-\bar{C}H_2$$

$$\Downarrow$$

$$CH_3CH_2CH=CH_2$$

Synthesis:

$$CH_3CH_2CH=CH_2 \xrightarrow[\text{(2) NaBH}_4,\text{ HO}^-]{\text{(1) Hg(OAc)}_2,\text{ H}_2\text{O}} CH_3CH_2\underset{\underset{\text{OH}}{|}}{CH}CH_3$$

$$\xrightarrow{\text{PCC } | \text{ CH}_2\text{Cl}_2}$$

$$CH_3CH_2\underset{\underset{\text{CH}_2}{\|}}{C}CH_3 \xleftarrow{(C_6H_5)_3\overset{+}{P}-\overset{-}{C}H_2} CH_3CH_2\underset{\underset{\text{O}}{\|}}{C}CH_3$$

$$\xrightarrow[\text{RO—OR, heat}]{\text{HBr}}$$

$$CH_3CH_2\underset{\underset{\text{CH}_2\text{Br}}{|}}{CH}CH_3$$

15.33 (a) Retrosynthesis:

[Retrosynthesis scheme: Br-CH2-CH(Br)-CH2-COOH ⟹ Br-CH2-CH(Br)-CH2-C≡N ⟹ CH2=CH-CH2-C≡N ⟸ CH2=CH-CH2-Br]

Synthesis:

[Allyl bromide →(NaCN)→ CH2=CH-CH2-C≡N →(Br2/CCl4)→ Br-CH2-CH(Br)-CH2-C≡N →(H2SO4/H2O, Heat)→ Br-CH2-CH(Br)-CH2-COOH]

(b) Retrosynthesis:

[CH3(CH2)4COOH ⟹ CH3(CH2)4OH ⟹ ethylene oxide + CH3CH2CH2MgBr ⟹ CH3CH2CH2Br]

Synthesis:

[CH3CH2CH2Br →(Mg, Dry ether)→ CH3CH2CH2MgBr →(1) ethylene oxide (2) H3O+→ CH3(CH2)4OH →(1) KMnO4/HO⁻ (2) H3O+→ CH3(CH2)4COOH]

(c) Retrosynthesis:

$$\text{CH}_3\text{CH}=\text{CHCOOCH}_3 \implies \text{HC}\equiv\text{C-COOCH}_3 \implies \text{HC}\equiv\text{C-Li} \implies \text{HC}\equiv\text{CH}$$

Synthesis:

$$\text{HC}\equiv\text{CH} \xrightarrow{\text{CH}_3-\text{Li}} \text{HC}\equiv\text{C-Li} + \text{CH}_4 \xrightarrow{\underset{\text{OCH}_3}{\overset{\text{Cl}}{\text{C=O}}}} \text{HC}\equiv\text{C-C(=O)OCH}_3$$

$$\xrightarrow[\text{Lindlar's catalyst}]{\text{H}_2} \text{CH}_3\text{CH}=\text{CH-C(=O)-OCH}_3$$

(d) Retrosynthesis:

$$\text{CH}_3(\text{CH}_2)_4\text{Cl} \implies \text{CH}_3(\text{CH}_2)_4\text{OH} \implies \text{CH}_3(\text{CH}_2)_3\text{COOH}$$

Synthesis:

$$\text{CH}_3(\text{CH}_2)_3\text{COOH} \xrightarrow[\text{(2) H}_3\text{O}^+]{\text{(1) LiAlH}_4} \text{CH}_3(\text{CH}_2)_4\text{OH} \xrightarrow[\text{or SOCl}_2/\text{N(CH}_2\text{CH}_3)_3]{\text{PCl}_5} \text{CH}_3(\text{CH}_2)_4\text{Cl}$$

(e) Retrosynthesis:

$$\text{HOOC-(CH}_2)_3\text{-COOH} \implies \text{N}\equiv\text{C-(CH}_2)_3\text{-C}\equiv\text{N} \implies$$

$$\text{Br-(CH}_2)_3\text{-Br} \implies \text{HO-(CH}_2)_3\text{-OH}$$

Synthesis:

$$HO\text{-}CH_2CH_2CH_2CH_2\text{-}OH \xrightarrow{PBr_3} Br\text{-}CH_2CH_2CH_2CH_2\text{-}Br \xrightarrow{NaCN}$$

$$N\equiv C\text{-}CH_2CH_2CH_2CH_2\text{-}C\equiv N \xrightarrow[\text{Heat}]{H_2SO_4,\ H_2O} HOOC\text{-}CH_2CH_2CH_2CH_2\text{-}COOH$$

15.34

(a) $Na\ +\ H_2{}^{18}O \longrightarrow Na^{18}OH \xrightarrow{CH_3Br} CH_3{}^{18}OH$

$$\xrightarrow[\text{or}\ CH_3C(O)\text{-}O\text{-}C(O)CH_3]{CH_3COCl,\ \text{pyridine}} CH_3C(O)\text{-}{}^{18}OCH_3$$

(b) $CH_3C\equiv N \xrightarrow[(2)\ H_3O^+]{(1)\ H_2{}^{18}O,\ Na^{18}OH,\ \text{heat}} CH_3C({}^{18}O)\text{-}{}^{18}OH\ +\ CH_3OH \xrightleftharpoons{H_2SO_4} CH_3C({}^{18}O)\text{-}OCH_3$

(c) $CH_3C\equiv N \xrightarrow[\text{Heat}]{Na^{18}OH,\ H_2{}^{18}O} CH_3C({}^{18}O)\text{-}{}^{18}O^-Na^+\ +\ CH_3Br \longrightarrow CH_3C({}^{18}O)\text{-}{}^{18}OCH_3$

(d) $CH_3Br \xrightarrow[\text{Dry ether}]{Mg} CH_3MgBr \xrightarrow[(2)\ H_3O^+]{(1)\ {}^{13}CO_2} CH_3{}^{13}C(O)\text{-}OH \xrightarrow[H_2SO_4]{CH_3OH} CH_3{}^{13}C(O)\text{-}OCH_3$

(e) $^{13}CH_3OH \xrightarrow{PBr_3} {}^{13}CH_3Br \xrightarrow[\text{Dry ether}]{Mg} {}^{13}CH_3MgBr \xrightarrow[\text{(2) } H_3O^+]{\text{(1) } CO_2} {}^{13}CH_3COOH$

$\xrightarrow[H_2SO_4]{CH_3OH} {}^{13}CH_3C(=O)OCH_3$

15.35 First, analyze the differences in structure of each pair of compounds, and then devise a spectroscopic method that will reveal that difference.

(a) Ethyl acetate has one more carbon atom than methyl acetate.

	IR	^{1}H NMR	^{13}C NMR
$CH_3C(=O)OCH_2CH_3$	1735 cm^{-1} (C=O)	δ 2.0 (singlet, 3H), δ 4.1 (quartet, 2H), δ 1.3 (triplet, 3H).	4 signals
$CH_3C(=O)OCH_3$	1735 cm^{-1} (C=O)	δ 2.0 (singlet, 3H), δ 3.7 (singlet, 3H)	3 signals

Infrared spectroscopy is incapable of distinguishing between ethyl acetate and methyl acetate because both spectra show absorption bands at 1735 cm^{-1}.

^{1}H NMR and ^{13}C NMR spectroscopy will easily distinguish one ester from another because the additional carbon atom in ethyl acetate makes a difference in both spectra, namely a different splitting pattern in the ^{1}H NMR spectrum and one more signal in the ^{13}C NMR spectrum.

(b) 2-Chloropropanoic acid has one more carbon atom than chloroacetic acid.

	IR	^{1}H NMR	^{13}C NMR
ClCH$_2$COOH	2500–3200 cm^{-1} (COOH) 1715 cm^{-1} (C=O)	δ 10.8 (singlet, 1H), δ 4.1 (singlet, 3H)	2 signals
CH$_3$CHClCOOH	2500–3200 cm^{-1} (COOH) 1715 cm^{-1} (C=O)	δ 12.7 (singlet, 1H), δ 4.45 (quartet, 1H), δ 1.75 (doublet, 3H)	3 signals

Infrared spectroscopy is incapable of distinguishing between chloroacetic acid and 2-chloropropanoic acid because both compounds have the same functional groups and will show very similar spectra.

^{1}H NMR and ^{13}C NMR spectroscopy will easily distinguish one acid from another because the additional carbon atom in 2-chloropropanoic acid makes a difference in both spectra, namely a different splitting pattern in the ^{1}H NMR spectrum and one more signal in the ^{13}C NMR spectrum.

(c) The carbon-carbon double bond in propenoic acid (trivial name: acrylic acid) gives it an extra functional group as compared with propanoic acid. This difference will be seen in each of the spectral methods.

	IR	^{1}H NMR	^{13}C NMR
CH$_2$=CHCOOH	3200–2500 cm^{-1} (COOH) 1715 cm^{-1} (C=O) 1650 cm^{-1} (C=C)	δ 11.9 (singlet, 1H), δ 5.9–6.8 (multiplet, 3H).	3 signals
CH$_3$CH$_2$COOH	3200–2500 cm^{-1} (COOH) 1715 cm^{-1} (C=O)	δ 12.2 (singlet, 1H), δ 2.4 (quartet, 2H), δ 1.2 (triplet, 3H).	3 signals

IR spectroscopy: C=C stretching at 1650 cm^{-1} is present in the IR spectrum of acrylic acid and absent in that of propanoic acid.

^{13}C NMR: Chemical shifts of *alkene* carbon atoms in acrylic acid are expected to range from δ 116 to δ 140 while the *alkane* carbons atoms give rise to signals at δ 34 and δ 20. The signal of carboxylic carbon appears at δ 181 in the spectrum of propanoic acid and at δ 173 in the spectrum of acrylic acid.

The ^{1}H NMR spectra will show the most profound difference and they will be easiest to use. Propanoic acid shows a typical splitting pattern of the ethyl group (a quartet and a triplet), which is absent from the spectrum of acrylic acid; acrylic acid shows a multiplet of vinyl hydrogen atoms in the range δ 5.9 to 6.8, which is absent from the spectrum of propanoic acid.

(d) IR spectroscopy can distinguish between these two compounds. The IR spectra of both compounds show a strong absorption band in the region 1720 to 1680 cm^{-1} (C=O). The IR spectrum of the hydroxyketone also shows a strong band in the region 3400 to 3200 cm^{-1} (—OH), which is sharper than the broad band at 2400 to 3200 cm^{-1} (COOH) present in the IR spectrum of the carboxylic acid.

(e) The ester can be distinguished from the alkene by IR spectroscopy. The ester contains a strong band near 1735 cm^{-1} (C=O) in its IR spectrum, which is absent in the IR spectrum of the alkene. The absorption band due to the carbon-carbon double bond of the alkene at about 1645 cm^{-1} is weak or absent in the IR spectrum of this symmetrically substituted alkene.

15.36

15.37 Two possible triglycerides will give these results: oleic acid can be bonded to C1 and C2 *or* C1 and C3 of glycerol.

15.38 Addition of water:

Loss of Water:

15.39

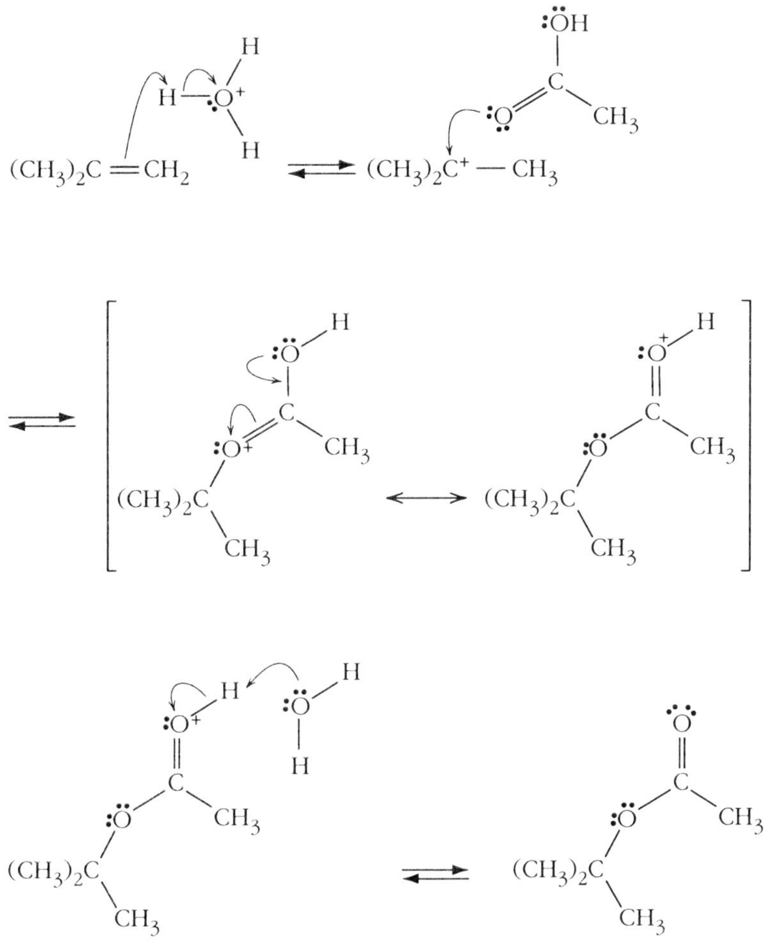

15.40

F = 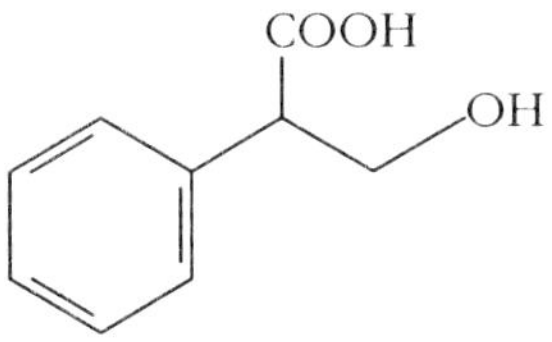 or 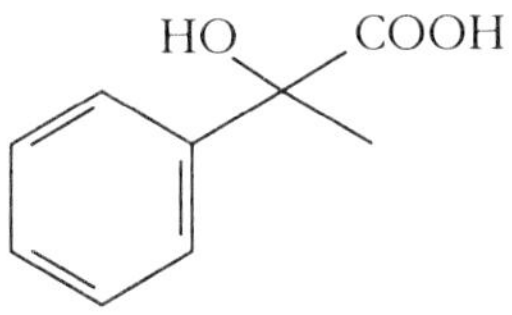

Can be either. No data are given to differentiate between the two.

G = 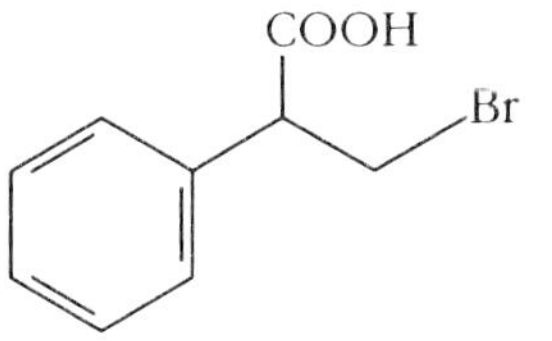 or 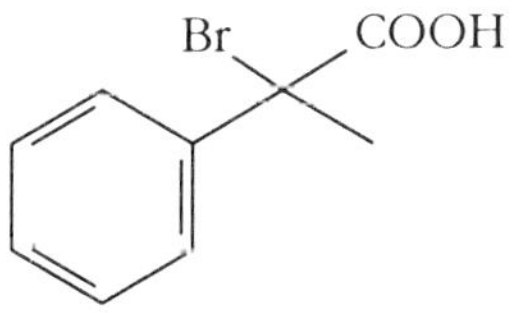

H = 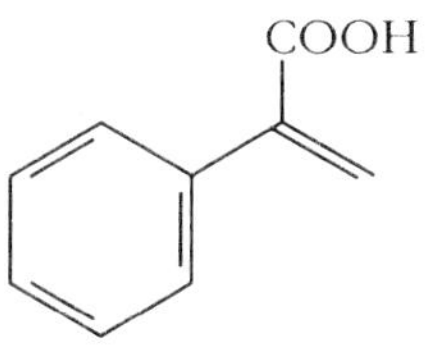

I = 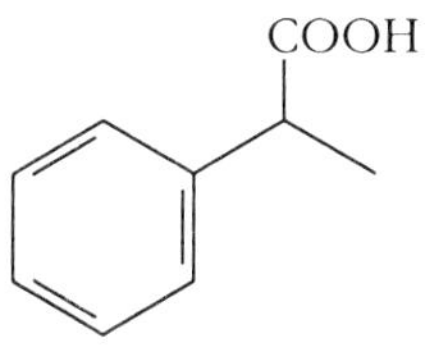

J = [1-chloroethyl benzene structure]   K = [1-hydroxyethyl benzene structure]

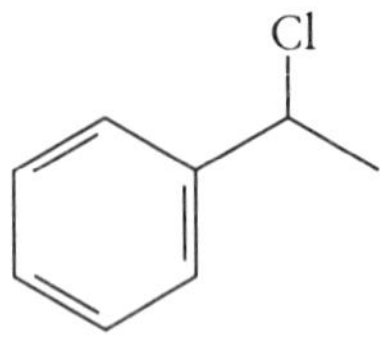

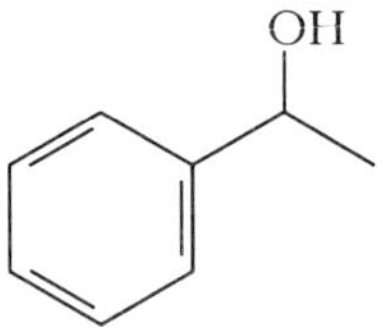

15.41 (a) The molecular formula, $C_5H_9ClO_2$, shows one unit of unsaturation. The IR spectrum indicates the presence of a carboxyl group. The signal for the acidic hydrogen is not observed in the ^1H NMR spectrum but the signal of the carboxylic carbon appears in the ^{13}C NMR spectrum at δ 182.1. The ^1H NMR spectrum shows two signals:

δ 1.33 (s, 3H)—a methyl group,
δ 3.62 (s, 1H)—a hydrogen atom bonded to a carbon bonded to an electron-withdrawing atom (chlorine).

When we add the integration data we get four hydrogen atoms (plus one of the carboxyl group which is not observed). The molecular formula shows nine hydrogen atoms, which means we have to multiply by two.

δ 1.33 (s, 6H)—two methyl groups,
δ 3.62 (s, 2H)—a methylene group bonded to an electron-withdrawing atom (chlorine).

Now we have the following fragments that account for $C_4H_9ClO_2$:

—COOH, —CH$_3$, —CH$_3$, —CH$_2$—, —Cl

What is missing is a carbon atom having no hydrogen atoms and bonded to four of the above fragments. Joining all the fragments, we can construct two structures:

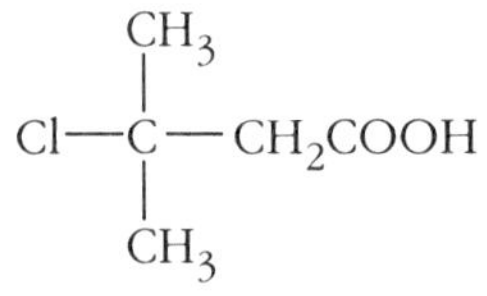

$$\begin{array}{c} CH_3 \\ | \\ Cl-C-CH_2COOH \\ | \\ CH_3 \end{array} \qquad \begin{array}{c} CH_3 \\ | \\ Cl-CH_2-C-COOH \\ | \\ CH_3 \end{array}$$

3-Chloro-3-methylbutanoic acid 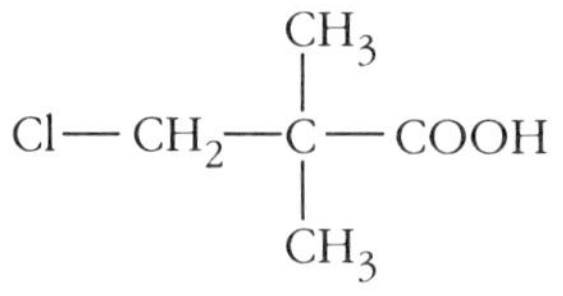   3-Chloro-2,2-dimethylpropanoic acid

The splitting patterns in the spectra of these two compounds are identical. The only difference is the chemical shift of the singlet for the methylene group. In the spectrum of 3-chloro-3-methylbutanoic acid that signal appears at δ ≈ 2.7 while in that of 3-chloro-2,2-dimethylpropanoic acid the signal appears at δ 3.6 because the electron-withdrawing chlorine is directly bonded to the —CH$_2$—group. Thus, the unknown is **3-chloro-2,2-dimethylpropanoic acid.**

(b) The molecular formula, $C_5H_9ClO_2$, shows one unit of unsaturation. The IR spectrum indicates the presence of an ester group. The signal of the carboxylic carbon appears in the ^{13}C NMR spectrum at δ 170.9. The ^1H NMR spectrum shows four signals:

δ 2.07 (s, 3H)—a methyl group bonded to a carbonyl group,
δ 2.10 (quintet, 2H)—a methylene group bonded to two other methylene groups,

δ 3.62 (t, 2H)—a methylene group bonded to a chlorine atom,
δ 4.22 (t, 2H)—a methylene group bonded to oxygen.

The two big fragments are CH$_3$C=O and —O—CH$_2$CH$_2$CH$_2$Cl and so the unknown is **3-chloropropyl acetate:**

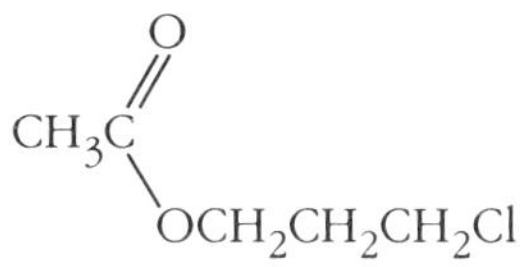

$$\text{CH}_3\text{C}(\text{=O})\text{OCH}_2\text{CH}_2\text{CH}_2\text{Cl}$$

(c) The molecular formula, C$_4$H$_8$O$_3$, shows one unit of unsaturation. The IR spectrum indicates the presence of a carboxyl group. The signal of the acidic hydrogen is not observed in the ^{1}H NMR spectrum but the signal of the carboxylic carbon appears in the ^{13}C NMR spectrum at δ 175.7. The ^{1}H NMR spectrum shows three signals:

δ 1.26 (t, 3H)—a methyl group bonded to a methylene group,
δ 3.63 (quartet, 2H)—a methylene group bonded to a methyl group and to an electron-withdrawing element (oxygen is the only electron-withdrawing element given in the molecular formula),
δ 4.14 (s, 2H)—a methylene group bonded to an electron-withdrawing element (oxygen).

We get the following fragments: —COOH, —OCH$_2$CH$_3$, —CH$_2$—. Those fragments account for C$_4$H$_8$O$_3$ (the whole molecule). There is only one way these fragments can be joined and that is to give the structure of **ethoxyacetic acid:**

$$\text{CH}_3\text{CH}_2\text{OCH}_2\text{COOH}$$

(d) The molecular formula, C$_5$H$_8$O$_4$, shows two units of unsaturation. The IR spectrum indicates the presence of a carboxyl group. The signal of the acidic hydrogen is not observed in the ^{1}H NMR spectrum but the signal of the carboxylic carbon appears in the ^{13}C NMR spectrum at δ 171. The ^{1}H NMR spectrum shows three signals:

δ 0.94 (t, 3H)—a terminal methyl group in an alkyl chain bonded to a methylene group,
δ 1.80 (quintet, 2H)—a methylene group having four neighboring hydrogen atoms,
δ 3.10 (t, 1H)—a hydrogen atom on a carbon atom bonded to a methylene group.

These three sets of signals are characteristic of a 1,1-disubstituted propyl group: CH$_3$CH$_2$CH$\diagdown^{\diagup}$. What is unaccounted for is C$_2$H$_2$O$_4$, which is two carboxyl groups (—COOH).

The structure of the unknown is **2-ethyl-1,3-propanedioic acid**.

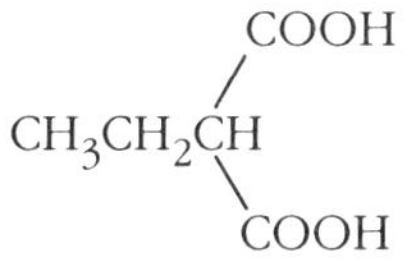

$$\text{CH}_3\text{CH}_2\text{CH}(\text{COOH})_2$$

15.42 The molecular formula, $C_3H_5ClO_2$, indicates one unit of unsaturation. The IR spectrum indicates the presence of a carboxyl group. The signal of the acidic hydrogen is not observed in the ^1H NMR spectra and the ratio of relative areas under the signals in Spectrum **A** is 1:3 and in Spectrum **B** is 1:1 (which represents 2 and 2 hydrogen atoms).

Spectrum **A**:

δ 4.45 (quartet, 1H)—a —CH group bonded to an electron-withdrawing element (molecular formula suggests chlorine and having three neighboring hydrogen atoms (a methyl group?),

δ 1.72 (doublet, 3H)—a methyl group bonded to a —CH group.

The two sets of signals indicate a 1,1-disubstituted ethyl group: CH₃CH. When we add —Cl and —COOH all the atoms will be accounted for and the structure of the compound **A** is **2-chloro-propanoic acid**:

$$\begin{array}{c} Cl \\ | \\ CH_3CHCOOH \end{array}$$

Spectrum **B**:

δ 2.87 (triplet, 2H)—a methylene group bonded to another methylene group,
δ 3.75 (triplet, 2H)—a methylene group bonded to another methylene group and an electron-withdrawing element (chlorine).

The two sets of signals indicate a 1,2-disubstituted ethyl group: Cl—CH₂CH₂—. When we add the carboxyl group the unknown becomes **3-chloropropanoic acid**:

ClCH₂CH₂COOH

15.43 (a) 2-Methylpropanoic acid will show the following signals:
—two equivalent methyl groups will show a doublet at δ ≈ 1.2 (6H),
—the —CH group will show a septet at δ ≈ 2.5 (1H),
—the carboxylic hydrogen will show a singlet at δ 12 to 13 or may not be seen at all.
The spectrum will look as follows:

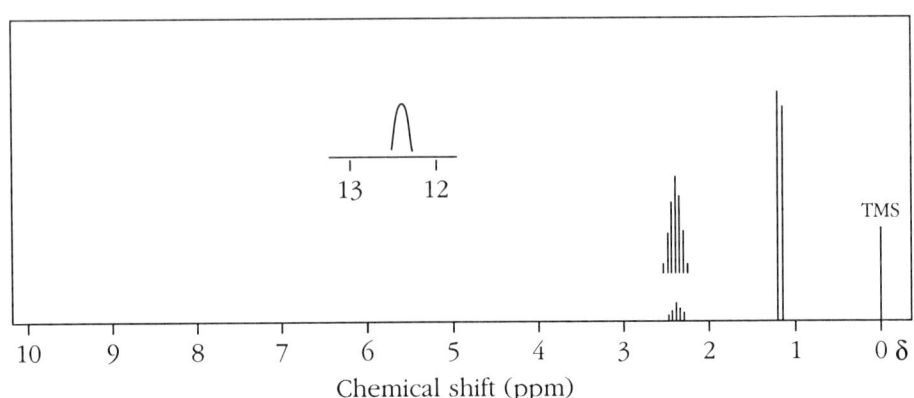

(b) Ethyl bromoacetate will show the following signals:
—the methylene group bonded to —Br and the carbonyl group will show a singlet at δ ≈ 3.8 (2H),
—the methylene group bonded to oxygen and the methyl group will be a quartet at δ 4.2 (2H),
—the methyl group bonded to the methylene group will show a triplet at δ 1.3 (3H).

The spectrum will look as follows:

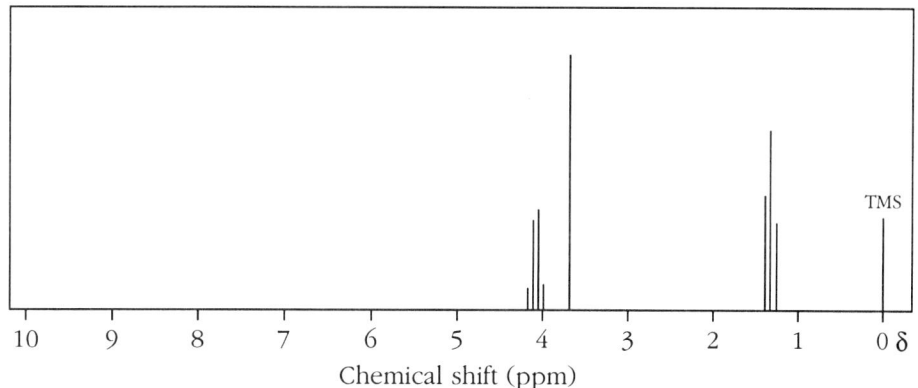

15.44 (a) Molecular formula, $C_5H_{10}O_2$, shows one unit of unsaturation. The IR spectrum shows a strong absorption band at 1730 cm^{-1}, which indicates the presence of a carbonyl group. Two oxygen atoms in the molecular formula and the lack of —OH and —COOH absorption bands in the IR spectrum indicate an ester.

The ^{1}H NMR spectrum shows the following signals:

δ 3.6 (s, 3H)—a methyl group bonded to oxygen,
δ 2.5 (septet, 1H)—a —CH group bonded to two methyl groups and a carbonyl group,
δ 1.11 (doublet, 6H)—two methyl groups bonded to the —CH group.

The unknown is **methyl 2-methylpropionate:**

$$\text{CH}_3\text{CH}-\underset{\underset{\text{CH}_3}{|}}{\text{C}}-\overset{\overset{\text{O}}{\|}}{\text{C}}-\text{OCH}_3$$

(b) The molecular formula, $C_5H_8O_2$, shows two units of unsaturation. The IR spectrum shows a strong absorption band at 3200 to 2500 cm^{-1} and a band at 1690 cm^{-1} indicating the presence of a carboxyl group. The absorption band at 1620 cm^{-1} indicates a carbon-carbon double bond.

The ^{1}H NMR spectrum shows the following signals:

δ 5.7 (multiplet, 1H)—one vinyl hydrogen,
δ 2.05 (a pair of doublets, 6H)—two methyl groups attached to C=C carbon atoms and coupled with one hydrogen.

From these data we get the following fragments:

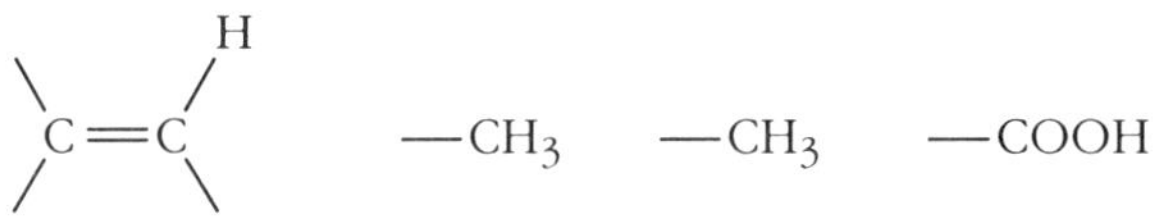

$$\ce{>C=C<^{H}} \quad -CH_3 \quad -CH_3 \quad -COOH$$

The fragments can be joined in three ways:

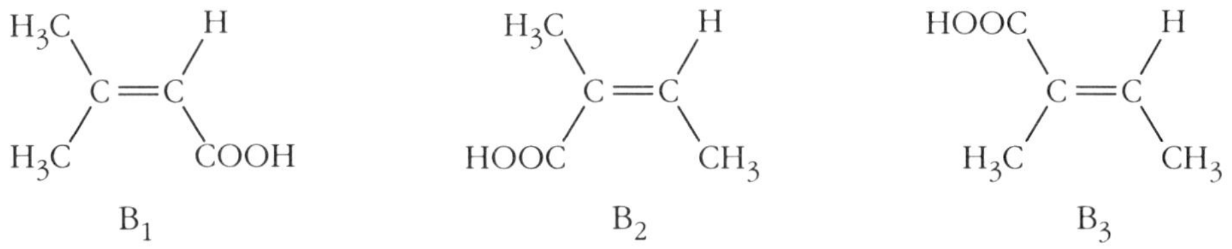

$$\underset{B_1}{\ce{(H_3C)(H_3C)C=C(H)(COOH)}} \quad \underset{B_2}{\ce{(H_3C)(HOOC)C=C(H)(CH_3)}} \quad \underset{B_3}{\ce{(HOOC)(H_3C)C=C(H)(CH_3)}}$$

All the coupling constants are very small, which suggests that the coupling is transmitted through the carbon-carbon double bond. The vinyl hydrogen has direct neighboring hydrogen atoms only in structure B_1, so the unknown is **3-methyl-2-butenoic acid** (B_1).

(c) The molecular formula, $C_4H_6O_2$, shows two units of unsaturation. The IR spectrum shows a strong absorption band at 1750 cm^{-1}, which indicates the presence of a carbonyl group. The absorption band at 1640 cm^{-1} indicates a carbon-carbon double bond. Two oxygen atoms in the molecular formula and the lack of —OH and —COOH absorption bands in the IR spectrum indicate an ester.

The ^{1}H NMR spectrum shows the following signals:

δ 7.21 (a pair of doublets, 1H)—one strongly deshielded vinyl hydrogen,
δ 4.81 (a pair of doublets, 1H)—one vinyl hydrogen,
δ 4.5 (a pair of doublets, 1H)—one vinyl hydrogen,
δ 2.0 (singlet, 3H)—a methyl group bonded to a carbonyl group.

These data give us two fragments: $CH_3C=O$ and $CH_2=CH-$. What is unaccounted for is one oxygen atom. We need one oxygen atom to have an ester group, however, so the unknown is **vinyl acetate:**

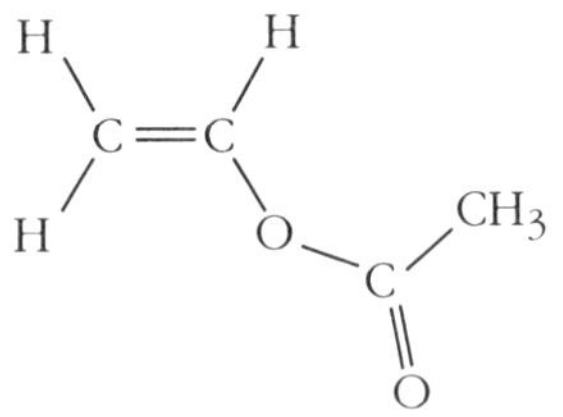

$$\ce{CH2=CH-O-C(=O)-CH3}$$

CHAPTER 16

ACYL TRANSFER REACTIONS
Concepts

Acid chlorides, acid anhydrides, esters, amides, and nitriles are derivatives of carboxylic acids that undergo a common reaction, **nucleophilic acyl substitution** (also called an **acyl transfer reaction**).

The mechanism of nucleophilic acyl substitution reactions involves two steps. Step 1 is *addition* of a nucleophile to the carbon atom of the carbonyl group, with formation of a tetrahedral intermediate. Step 2 is *elimination* to form a leaving group, with formation of the carbonyl group.

$$\underset{\text{Tetrahedral intermediate}}{\left[\begin{array}{c}\text{R}-\overset{\text{O}}{\underset{\text{LG}}{\text{C}}}-\text{Nu}\end{array}\right]}$$

The nature of the leaving group determines the fate of the tetrahedral intermediate. The reactions of acid derivatives differ from those of aldehydes and ketones, which do not have leaving groups bonded to the carbonyl group. The first step of the reaction with nucleophiles is the same for acid derivatives as it is for aldehydes and ketones. Unlike aldehydes and ketones, acid derivatives undergo nucleophilic substitution rather than nucleophilic addition.

The characteristic feature of **acid halides** is a halogen atom, usually chlorine, attached to the carbonyl group. Acid halides are named by identifying first the acyl group and then the halide. The acyl group name is derived from the carboxylic acid name by replacing the ending *-ic acid* with *-yl,* or the *-carboxylic acid* ending with *-carbonyl.*

The characteristic feature of **acid anhydrides** is two acyl groups attached to the same atom of oxygen. Symmetrical anhydrides are named by replacing the word *acid* with *anhydride.* If the anhydride is derived from two different acids, we list alphabetically the names of the acids and add the third word *anhydride.*

The characteristic feature of **esters** is an alkoxy group, —OR, bonded to the carbon atom of the carbonyl group. Systematic names for esters are derived by first giving the name of the alkyl (or aryl) group attached to oxygen and then identifying the carboxylic acid and replacing its *-ic acid* ending by *-ate.*

Thioesters are compounds in which the oxygen atom of the alkoxy group of an ester has been replaced by a sulfur atom. A thioester is more reactive than the corresponding oxyester. The names of thioesters consist of two words. The first word identifies the alkyl or aryl group bonded to the carboxyl sulfur atom. The second word identifies the acyl group. The name is formed by adding *thioate* to the name of the parent hydrocarbon containing the acyl group.

The characteristic feature of **amides** is a trivalent nitrogen atom bonded to the carbon atom of the carbonyl group. Amides with an unsubstituted —NH_2 group are named by replacing the *-oic acid* or *-ic acid* ending with *-amide* or by replacing the *-carboxylic acid* ending with *-carboxamide.* If the nitrogen atom is further substituted, the compound is named by first identifying the substituent group and then citing the parent amide name. The substituents are preceded by the letter N to identify them as being attached to nitrogen.

The characteristic feature of **nitriles** is a carbon-nitrogen triple bond. Nitriles are named by adding *-nitrile* as a suffix to the name of the alkane containing the same number of carbon atoms, counting the nitrile carbon, which is numbered C1.

The carboxylic acid derivatives can be interconverted to one another according to the following scheme:

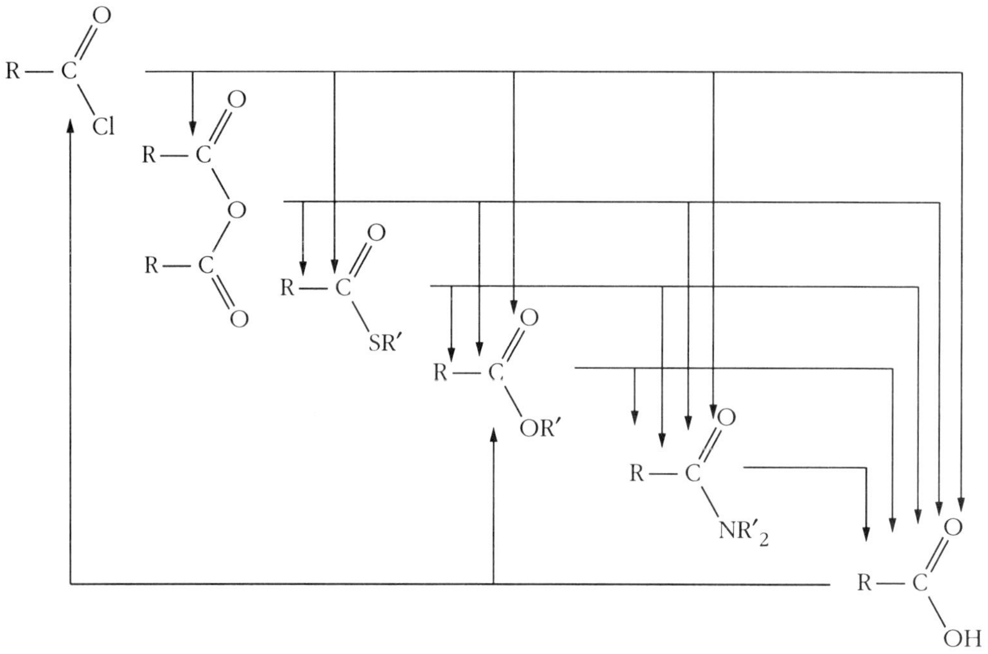

$$\begin{array}{c} R-C(=O)Cl \\ \downarrow \\ R-C(=O)-O-C(=O)-R \\ \downarrow \\ R-C(=O)SR' \\ \downarrow \\ R-C(=O)OR' \\ \downarrow \\ R-C(=O)NR'_2 \\ \downarrow \\ R-C(=O)OH \end{array}$$

Carboxylic acids are converted to acid chlorides by treatment with thionyl chloride (SOCl$_2$) or phosphorus trichloride (PCl$_3$) or phosphorus oxychloride (POCl$_3$).

The relative reactivities of the acid derivatives toward nucleophilic substitution decrease in the same order as they are shown in the diagram. That is, the most reactive acid chlorides are at the top and the least reactive carboxylic acids are at the bottom. An important conclusion of the observed reactivity order is that it is usually possible to transform a *more* reactive acid derivative into a *less* reactive one. Remembering the reactivity order is therefore a way to keep track of a large number of reactions.

Solutions to the Exercises

16.1 (a) 3-Methylpentanenitrile
(b) Succinic anhydride (butanedioic anhydride)
(c) Hexanoyl chloride
(d) 4-Methylpentamide
(e) *N*-Methylpropamide
(f) *N*-Cyclopentyl-*N*-Ethylbenzamide

16.2

(a) butanamide: CH₃CH₂CH₂C(=O)NH₂

(b) N-propyl acrylamide: CH₂=CHC(=O)NHCH₂CH₂CH₃

(c) 3,3-dichloropentanenitrile: CH₃CH₂C(Cl)₂CH₂C≡N

(d) propanoyl chloride: CH₃CH₂C(=O)Cl

(e) S-isopropyl butanethioate: CH₃CH₂CH₂C(=O)SCH(CH₃)₂

(f) octanoic anhydride (symmetric): [CH₃(CH₂)₆C(=O)]₂O

(g) 2-cyclohexylacetamide: C₆H₁₁CH₂C(=O)NH₂

(h) S-methyl 2-methyl-2-butenethioate: (CH₃)CH=C(CH₃)C(=O)SCH₃

(i) cyclopentyl 2-methylpropanoate: (CH₃)₂CHC(=O)OC₅H₉

(j) N-ethyl-N-isopropyl-4-pentenamide: CH₂=CHCH₂CH₂C(=O)N(CH₂CH₃)CH(CH₃)₂

16.3 The better the hydrogen bonding capability, the higher the boiling point. If neither compound is capable of forming hydrogen bonds, the more polar the molecule, the higher the boiling point.

In order of decreasing boiling point the compounds are:

(a) Pentanoic acid (strong hydrogen bond donor as well as acceptor): b.p. 186.4 °C; 1-Chloropentane (no hydrogen bonding): b.p. 108 °C.

(b) Pentanamide (hydrogen bond donor as well as acceptor) has the higher boiling point. In fact, it's a solid at room temperature (m.p. 106 °C). Ethyl butanoate (only hydrogen bond acceptor): b.p. 122 °C.

(c) Hexanenitrile: b.p. 164 °C;
3-Hexanone: b.p. 125 °C.

(d) 1-Hexanol (strong hydrogen bond donor as well as acceptor): b.p. 158 °C; Butanoyl chloride (only a hydrogen bond acceptor): b.p. 102 °C.

16.4

	IR	¹H NMR	¹³C NMR
(a) CH₃C(=O)NH₂	1683 cm⁻¹ (C=O) 3325 cm⁻¹ (N—H) } (two peaks) 3163 cm⁻¹ (N—H)	δ 1.9 (s, 3H) δ 6.5-7.0 (broad doublet, 2H)	two signals
CH₃C(=O)Cl	1807 cm⁻¹ (C=O)	δ 2.65	two signals

IR is the best spectral method to differentiate between acetamide and acetyl chloride. There is a difference in the position of the carbonyl group

(1683 and 1807 cm^{-1}, respectively). Moreover, the N—H absorption is observed in the spectrum of acetamide at 3300 to 3150 cm^{-1} and this absorption is absent from the spectrum of acetyl chloride.

^{1}H NMR can also distinguish between the two compounds. Acetamide shows a signal of the —NH$_2$ group at δ 6.5 to 7.0 and this signal is missing from the spectrum of acetyl chloride.

^{13}C NMR spectroscopy is not useful because the spectra of both compounds contain two signals at approximately the same chemical shifts.

(b)

CH$_3$C(=O)—O—C(=O)CH$_3$

| | 1827 cm^{-1} (C=O), 1755 cm^{-1} (C=O) (2 peaks) | δ 2.2 (s) | 2 signals |

CH$_3$COOH

| | 1715 cm^{-1} (C=O), 3400-2500 cm^{-1} (O—H) | δ 2.1 (s, 3H), δ 10-13 depending on concentration of the acid (singlet- often broad, 1H) | 2 signals |

IR is the best spectral method to differentiate between acetic anhydride and acetic acid. The absorption band of the carbonyl group of acetic acid appears at 1715 cm^{-1} while that of acetic anhydride appears as two bands at 1755 and 1827 cm^{-1}. Moreover, the —COOH absorption is observed in the spectrum of acetic acid at 3400 to 2500 cm^{-1} and this absorption is absent from the spectrum of acetic anhydride.

^{1}H NMR can also distinguish between the two compounds. Acetic acid shows a signal of the carboxylic hydrogen at about δ 12 and this signal is missing from the spectrum of acetic anhydride. Sometimes, however, the carboxylic hydrogen is not observed and then ^{1}H NMR cannot be used to differentiate between the two compounds.

^{13}C NMR spectroscopy is not useful because the spectra of both compounds contain two signals at approximately the same chemical shifts.

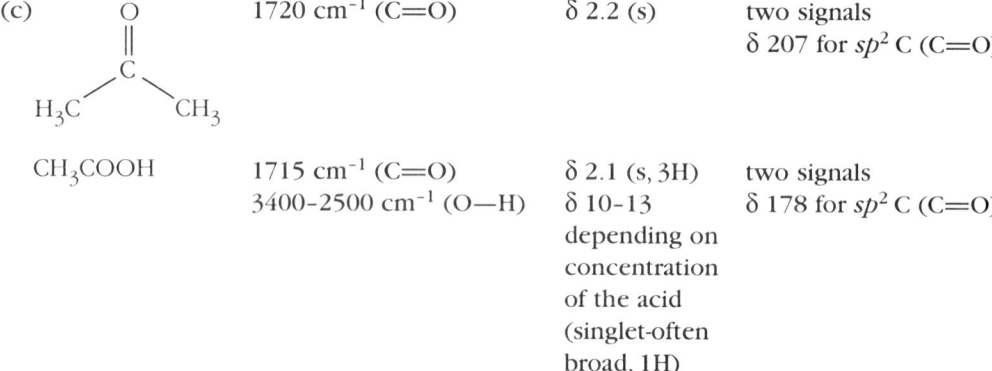

(c) H$_3$C—C(=O)—CH$_3$

| | 1720 cm^{-1} (C=O) | δ 2.2 (s) | two signals, δ 207 for sp^2 C (C=O) |

CH$_3$COOH

| | 1715 cm^{-1} (C=O), 3400-2500 cm^{-1} (O—H) | δ 2.1 (s, 3H), δ 10-13 depending on concentration of the acid (singlet-often broad, 1H) | two signals, δ 178 for sp^2 C (C=O) |

IR is the best spectral method to differentiate between acetone and acetic acid. The absorption band of the carbonyl group of acetic acid and that of acetone appear at the same frequency but the —COOH absorption is observed in the spectrum of acetic acid at 3400 to 2500 cm^{-1} and this absorption is absent from the spectrum of acetone.

^{1}H NMR can also distinguish between the two compounds. Acetic acid shows a signal of the carboxylic hydrogen at about δ 12 and this signal is missing in the spectrum of acetone. Sometimes, however, the carboxylic hydrogen is not observed and then the ^{1}H NMR cannot be used to differentiate between the two compounds.

^{13}C NMR spectroscopy may be useful although the spectra of both compounds contain two signals. The chemical shift of the carbonyl carbon is δ 207 for acetone and δ 178 for acetic acid.

(d) CH$_3$C(=O)SCH$_3$

1690 cm^{-1} (C=O)

δ 2.0 (s, 3H)
δ 2.4 (s, 3H)

three signals
δ 195 ppm for sp^2 C (C=O)

CH$_3$C(=O)OCH$_3$

1735 cm^{-1} (C=O)

δ 2.1 (s, 3H)
δ 3.7 (s, 3H)

three signals
δ 171 ppm for sp^2 C (C=O)

All three spectral techniques show small differences between the spectrum of methyl ethanethioate and methyl ethanoate (acetate).

IR: The absorptions of the carbonyl groups are observed at slightly different frequencies.

^{1}H NMR: The signal of the —S—CH$_3$ group appears at δ 2.4 and that of —O—CH$_3$ at δ 3.7.

^{13}C NMR: The chemical shifts of the carbonyl carbon atoms are noticeably different.

(e) CH$_3$C(=O)N(CH$_3$)$_2$

1676 cm^{-1} (C=O)

δ 2.1 (s, 3H)
δ 2.95 (s, 3H)
δ 3.05 (s, 3H)

three signals

Note: Rotation about the C—N bond is restricted (see page 703 of the text) so the two N-methyl groups are in different magnetic environments. As a result, they have different chemical shifts.

CH$_3$C(=O)NH$_2$

1683 cm^{-1} (C=O)
3325 cm^{-1} (N—H)
3163 cm^{-1} (N H)
(2 peaks)

δ 1.9 (s, 3H)
δ 6.5-7.0
(broad doublet, 2H)

two signals

All three spectral techniques show differences between the spectrum of acetamide and N,N-dimethylacetamide.

IR: The N—H absorption is observed in the spectrum of acetamide at 3300 to 3150 cm^{-1} and this absorption is absent from the spectrum of N,N-dimethylacetamide.

^{1}H NMR: Acetamide shows a signal of the —NH$_2$ group at δ 6.5 to 7.0 and this signal is missing in the spectrum of N,N-dimethylacetamide.

^{13}C NMR: *N,N*-Dimethylacetamide shows three signals while acetamide shows only two signals.

16.5

(a) 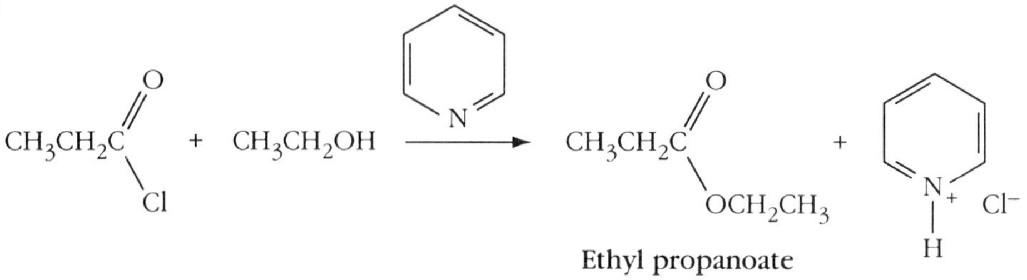 $CH_3CH_2COCl + CH_3CH_2OH \xrightarrow{\text{pyridine}} CH_3CH_2COOCH_2CH_3 + \text{pyridinium}^+ Cl^-$

Ethyl propanoate

(b) 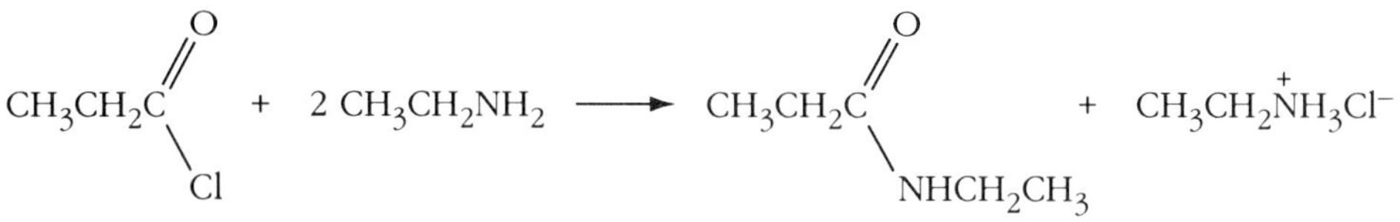 $CH_3CH_2COCl + 2\ CH_3CH_2NH_2 \longrightarrow CH_3CH_2CONHCH_2CH_3 + CH_3CH_2\overset{+}{N}H_3Cl^-$

N-Ethylpropamide

(c) 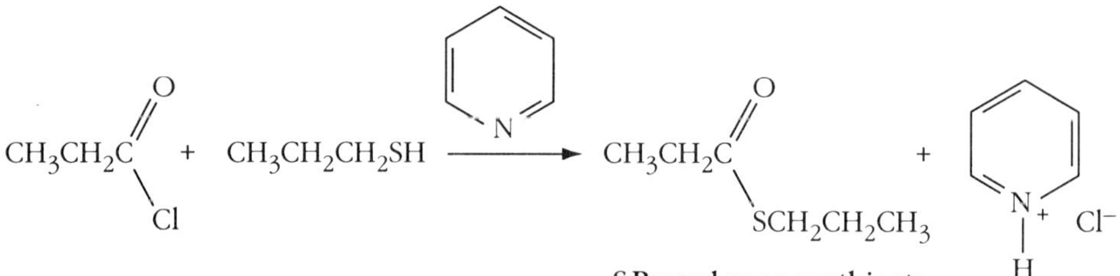 $CH_3CH_2COCl + CH_3CH_2CH_2SH \xrightarrow{\text{pyridine}} CH_3CH_2COSCH_2CH_2CH_3 + \text{pyridinium}^+ Cl^-$

S-Propyl propanethioate

(d) 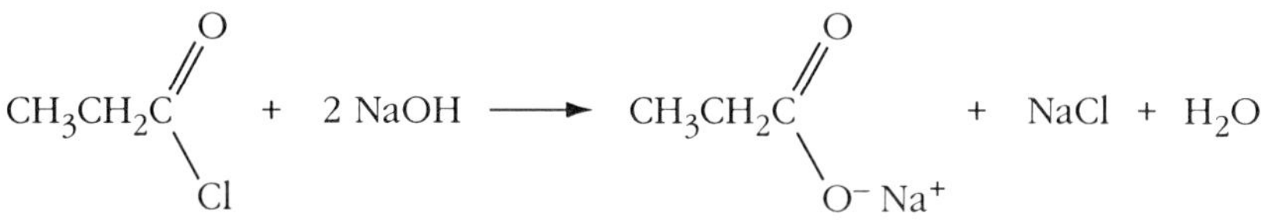 $CH_3CH_2COCl + 2\ NaOH \longrightarrow CH_3CH_2COO^-Na^+ + NaCl + H_2O$

Sodium propanoate

(e) 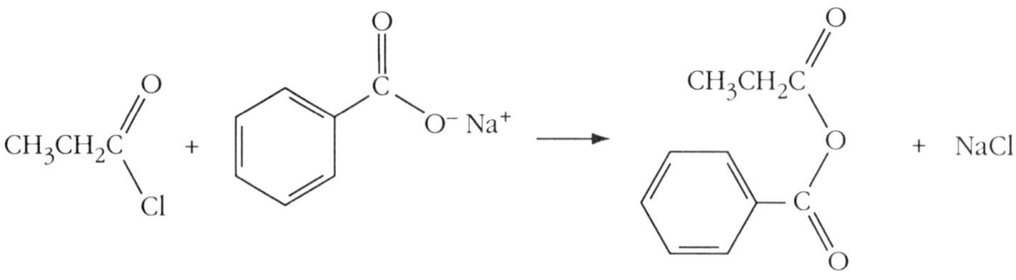 $CH_3CH_2COCl + C_6H_5COO^-Na^+ \longrightarrow CH_3CH_2CO\text{-}O\text{-}COC_6H_5 + NaCl$

Benzoic propanoic anhydride

(f) 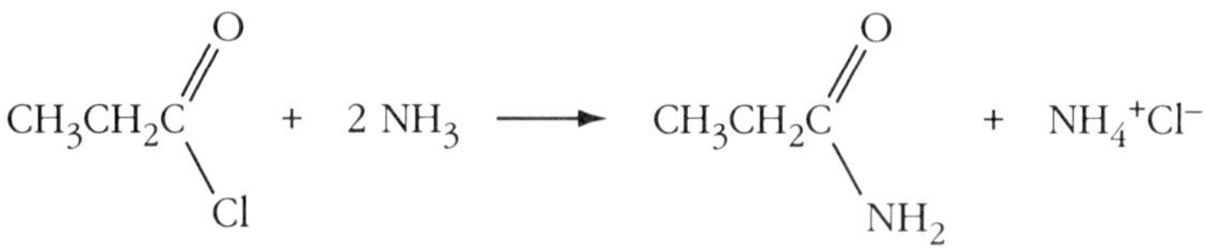 $CH_3CH_2COCl + 2\ NH_3 \longrightarrow CH_3CH_2CONH_2 + NH_4^+Cl^-$

Propamide

(g) 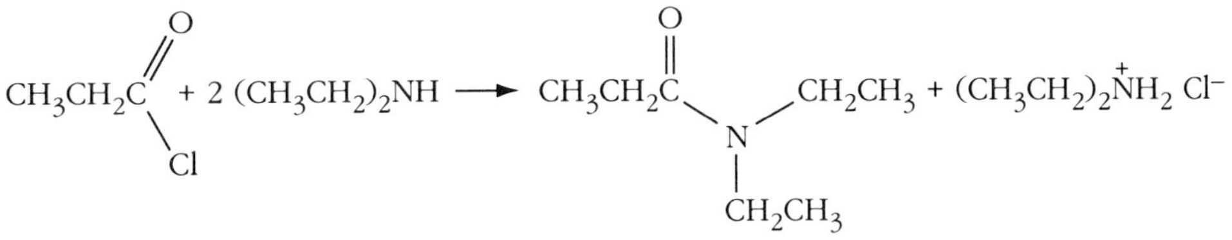

N,N-Diethylpropamide

(h) 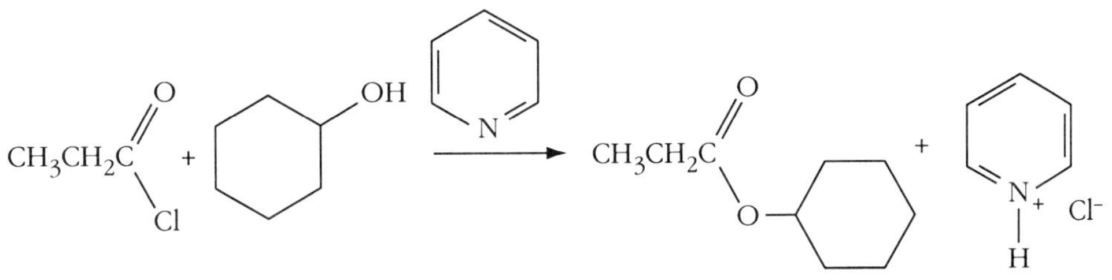

Cyclohexyl propanoate

16.6

(a) 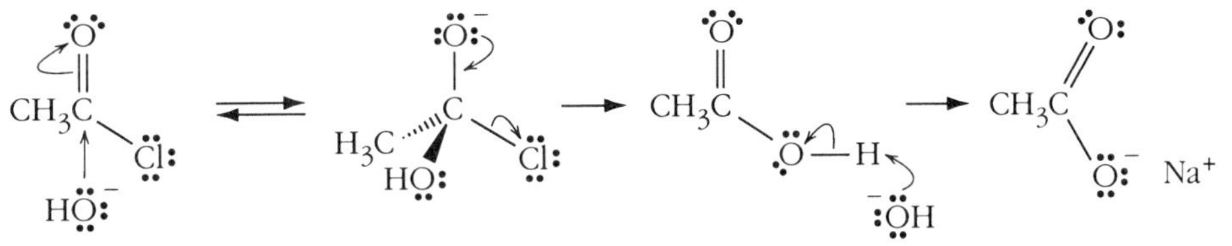

(b) 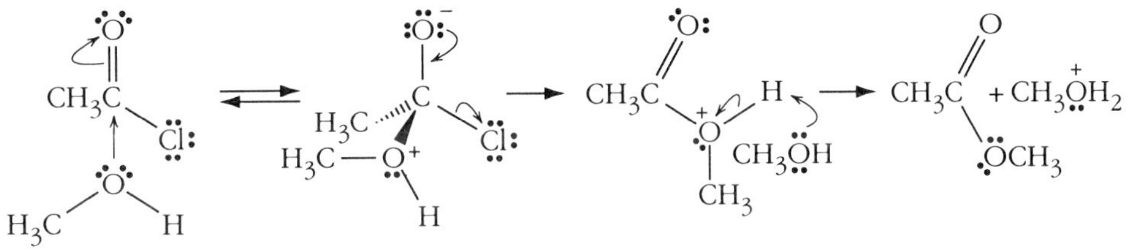

(c) 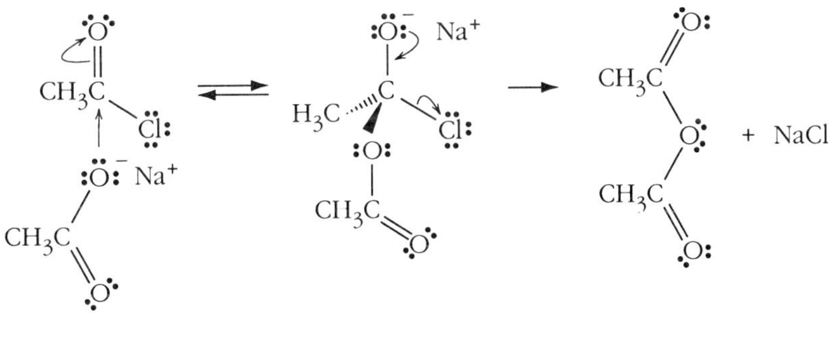

(d) 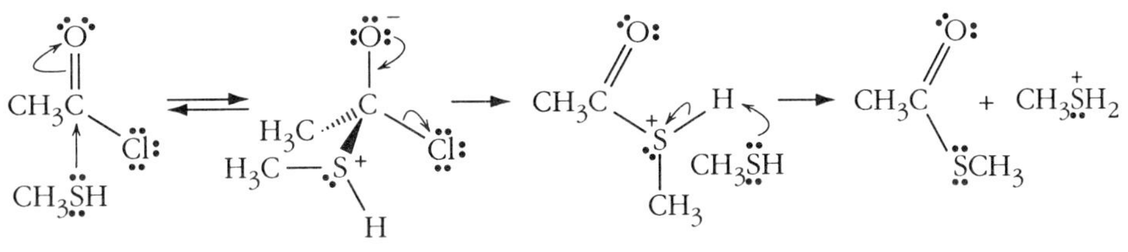

16.7 Use a retrosynthetic analysis in the following way: (1) Draw the structure of the target molecule to be synthesized, (2) Disconnect one of the bonds leading to the carbonyl group, (3) Add —Cl to the carbonyl group, (4) Complete the structure of the other reactant, (5) Write the synthesis.

(a) Retrosynthesis:

(b) Retrosynthesis:

Synthesis:

(c) Retrosynthesis:

Synthesis:

(d) Synthesis:

$$CH_3CH_2CH_2COCl + CH_3CH_2SH \xrightarrow{\text{pyridine}} CH_3CH_2CH_2C(O)SCH_2CH_3$$

16.8 Recall from Chapter 12 that benzenesulfonate is a very good leaving group and is readily displaced by nucleophilic substitution reactions that occur by an S_N2 mechanism.

$$H-{}^{18}\ddot{\text{O}}{:}^- \curvearrowright CH_3 - O - S(O)(O^-)(C_6H_5) \longrightarrow CH_3{}^{18}OH + C_6H_5SO_3^-$$

The hydrolysis of an ester occurs by an addition-elimination mechanism, as discussed in Chapter 15:

$$CH_3C(=O)OCH_3 + {}^{18}OH^- \rightleftharpoons CH_3C(O^-)({}^{18}OH)(OCH_3) \longrightarrow CH_3C(=O){}^{18}OH + {}^-OCH_3 \longrightarrow CH_3C(=O){}^{18}O^- + CH_3OH$$

Thus although in both cases we observe the same product (methanol), the reaction mechanisms are different. Methanol obtained by the S_N2 substitution of sulfonyl ester will contain ^{18}O from $H^{18}O^-$. The hydrolysis of esters, on the other hand, will give an alcohol with the same oxygen atom that was part of the alkoxy group of the ester.

16.9 The reactivity of acid derivatives toward nucleophilic substitution reactions at the carbonyl group follows the following order:

$$R-C(=O)Cl > R-C(=O)-O-C(=O)-R > R-C(=O)SR' > R-C(=O)OR' > R-C(=O)NH_2 > R-C(=O)OH$$

Thus,

(a) $CH_3-C(=O)Cl$ will react faster than $CH_3-C(=O)OCH_2CH_3$

(b) $CH_3CH_2-C(=O)SCH_3$ will react faster than $CH_3CH_2CH_2-C(=O)OCH_3$

(c) $CH_3-C(=O)-O-C(=O)-CH_3$ will react faster than $CH_3-C(=O)NH_2$

(d) ethyl propanoate will react faster than N,N-diethyl propanamide

16.10 Step 1: Protonation of the amide

[mechanism showing protonation of CH₃C(=O)NH(CH₂CH₃) by H₃O⁺ to give resonance structures of the protonated amide + H₂O]

Step 2: Nucleophilic addition of water

Step 3: Proton transfer (intermolecular)

Step 4: Elimination of amine

Step 5: Proton transfer to the amine or any other base present

16.11

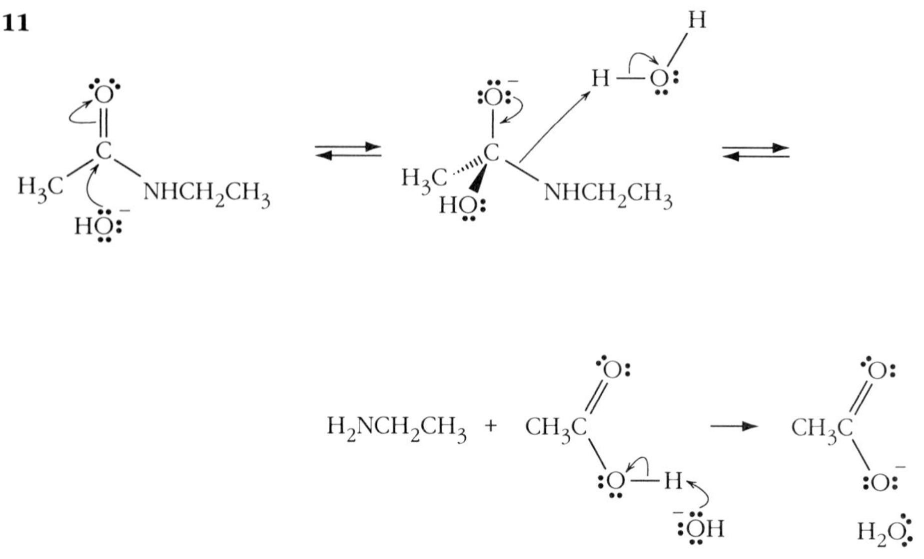

[Mechanism showing basic hydrolysis of N-ethylacetamide: hydroxide attacks the carbonyl carbon of CH₃C(=O)NHCH₂CH₃ to form a tetrahedral intermediate, which is then deprotonated by water, followed by C–N bond cleavage with the amide anion abstracting a proton from water to give ethylamine (H₂NCH₂CH₃) and acetate (CH₃CO₂⁻).]

For the carbon-nitrogen bond to break in the basic hydrolysis of N-ethylacetamide, the leaving group must be the conjugate base of the amine (the amide anion). In basic solution, it is unlikely that protonation of the nitrogen in the tetrahedral intermediate will take place before the carbon-nitrogen bond begins to break. However, the amide anion is such a strong base that it is improbable that it has any real existence in water. The mechanism accounts for this by showing the amide anion taking a proton from water as the tetrahedral intermediate breaks up.

16.12 The general mechanism for the reaction of methanol with different carboxylic acid derivatives is the following:

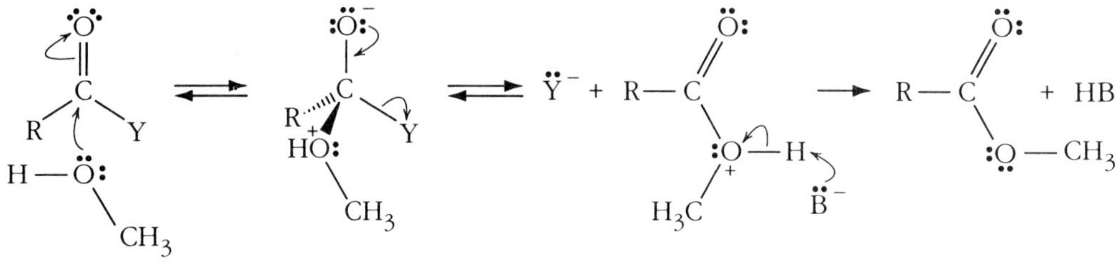

[General mechanism: methanol adds to R–C(=O)–Y to form a tetrahedral intermediate, proton transfer, loss of Y⁻, and deprotonation by base B to give the methyl ester R–C(=O)–OCH₃ + HB.]

(a)

$$\text{CH}_3\text{CH}_2\text{C(=O)Cl} + \text{CH}_3\text{OH} \xrightarrow{\text{pyridine}} \text{CH}_3\text{CH}_2\text{C(=O)OCH}_3 + \text{pyridinium Cl}^-$$

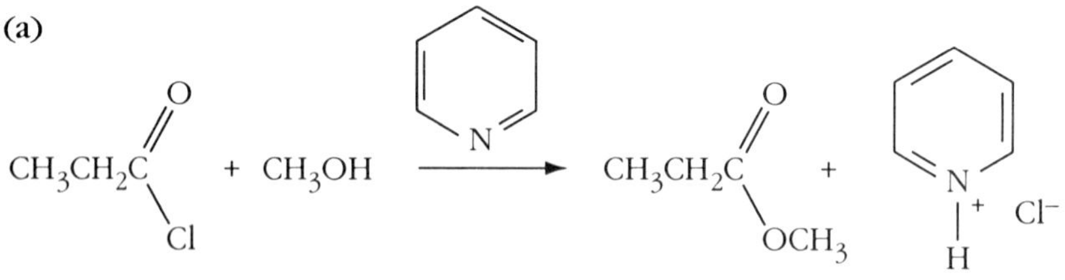

The reactions of acid chlorides with alcohols are usually carried out in the presence of pyridine or another base (Na₂CO₃, triethylamine, or NaOH) to react with the HCl formed and prevent it from causing side reactions (such as reaction with the alcohol to form an alkyl chloride).

(b)

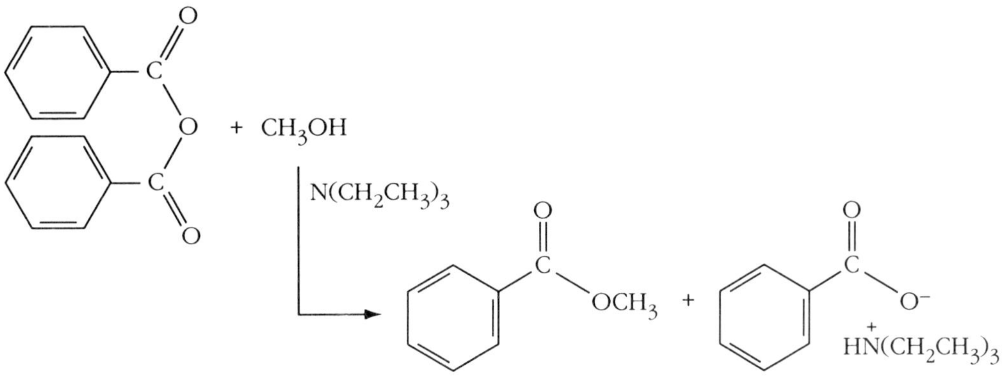

[Benzoic anhydride] + CH₃OH —N(CH₂CH₃)₃→ [methyl benzoate: PhC(=O)OCH₃] + [benzoate: PhC(=O)O⁻] + ⁺HN(CH₂CH₃)₃

(c)

CH₃CH₂CH₂C(=O)SCH₃ + CH₃OH —Base→ CH₃CH₂CH₂C(=O)OCH₃ + CH₃SH

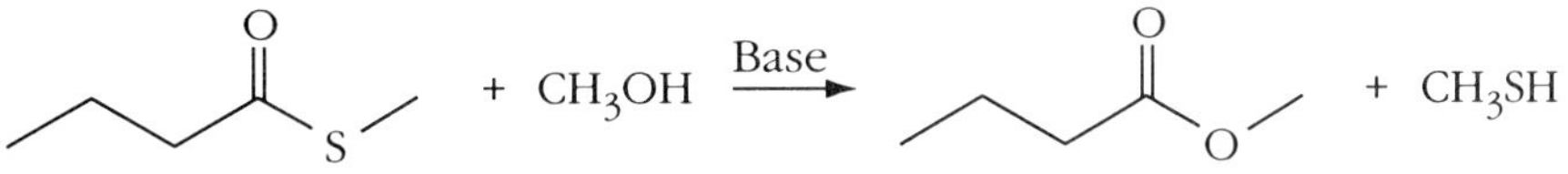

(d) This is a transesterification reaction:

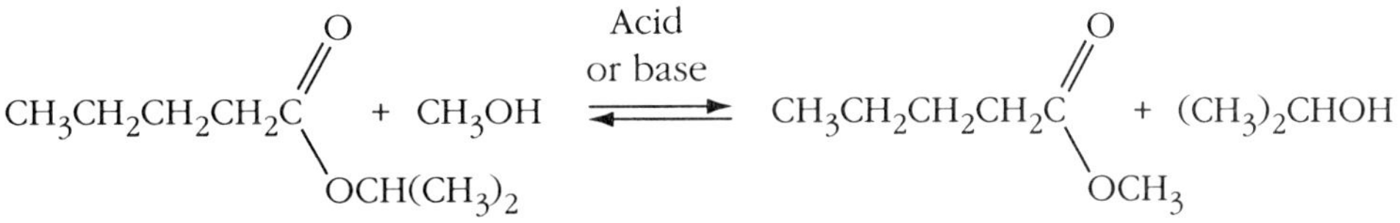

$$CH_3CH_2CH_2CH_2C(=O)OCH(CH_3)_2 + CH_3OH \underset{\text{or base}}{\overset{\text{Acid}}{\rightleftharpoons}} CH_3CH_2CH_2CH_2C(=O)OCH_3 + (CH_3)_2CHOH$$

(e) Only a very small amount (if any) of methyl butanoate will be formed in neutral solution because the amide is more stable than the ester and the equilibrium lies to the left:

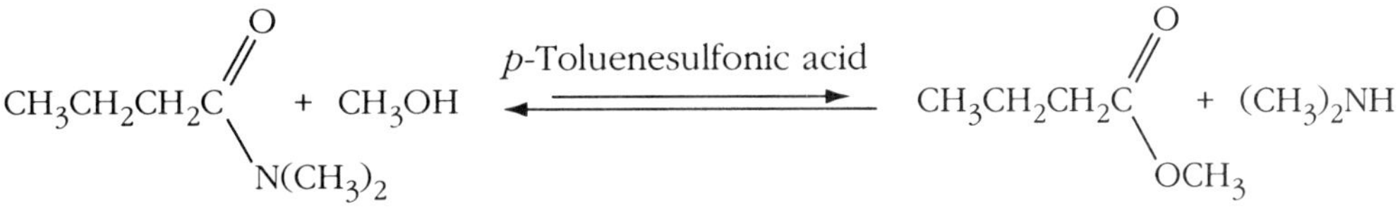

$$CH_3CH_2CH_2C(=O)N(CH_3)_2 + CH_3OH \overset{p\text{-Toluenesulfonic acid}}{\rightleftharpoons} CH_3CH_2CH_2C(=O)OCH_3 + (CH_3)_2NH$$

The ester is the major product when an excess of acid is used. The excess acid protonates the amine and drives the reaction to completion:

$$(CH_3)_2NH + HB \longrightarrow (CH_3)_2\overset{+}{N}H_2 + B^-$$

16.13 The general mechanism for the reaction of ammonia with different carboxylic acid derivatives is the following:

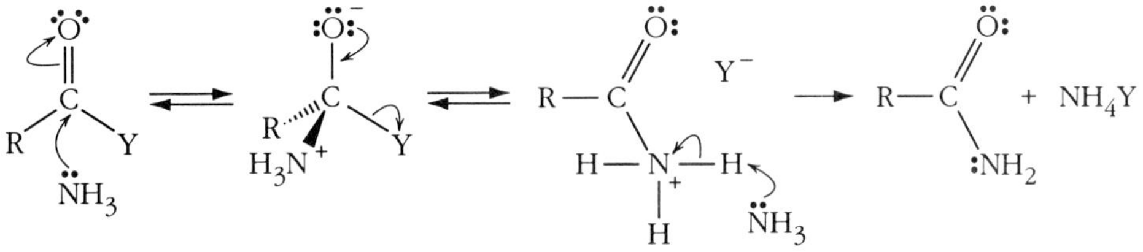

[Mechanism: R–C(=O)–Y with :NH₃ attacking ⇌ tetrahedral intermediate R–C(O⁻)(Y)(NH₃⁺) ⇌ R–C(=O)–NH₂ (via proton transfer involving NH₃, loss of Y⁻) → R–C(=O)–NH₂ + NH₄Y]

(a) $CH_3CH_2CH_2CH_2C(=O)Cl + 2\,NH_3 \longrightarrow CH_3CH_2CH_2CH_2C(=O)NH_2 + NH_4Cl$

(b) $(CH_3C(=O))_2O + 2\,NH_3 \longrightarrow CH_3C(=O)NH_2 + CH_3C(=O)O^- \; {}^+NH_4$

(c) $CH_3CH_2CH_2C(=O)SCH_3 + NH_3 \longrightarrow CH_3CH_2CH_2C(=O)NH_2 + CH_3SH$

(d) $(CH_3)_2CHCH_2CH_2C(=O)OCH(CH_3)_2 + NH_3 \longrightarrow (CH_3)_2CHCH_2CH_2C(=O)NH_2 + (CH_3)_2CHOH$

16.14 The reaction mechanism for the reaction of methylamine (CH_3NH_2) with different carboxylic acid derivatives is identical to that shown in Exercise 16.13 for ammonia. The only difference is that we obtain *N*-methyl amides. Thus,

(a) $CH_3CH_2CH_2CH_2C(=O)NHCH_3 + CH_3\overset{+}{N}H_3Cl^-$

(b) $CH_3C(=O)NHCH_3 + CH_3C(=O)O^- \; {}^+NH_3CH_3$

(c) $CH_3CH_2CH_2C(=O)NHCH_3 + CH_3SH$

(d) $(CH_3)_2CHCH_2CH_2C(=O)NHCH_3 + (CH_3)_2CHOH$

16.15 The reaction mechanism for the reaction of dimethylamine [$(CH_3)_2NH$] with different carboxylic acid derivatives is identical to that shown in Exercise 16.13 for ammonia. The only difference is that we obtain *N,N*-dimethyl amide derivatives.

Thus,

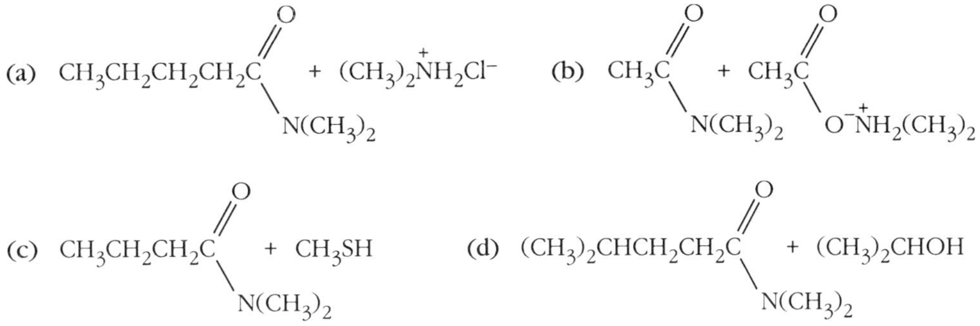
(a) CH3CH2CH2CH2C(=O)N(CH3)2 + (CH3)2NH2Cl−
(b) CH3C(=O)N(CH3)2 + CH3C(=O)O− NH2(CH3)2
(c) CH3CH2CH2C(=O)N(CH3)2 + CH3SH
(d) (CH3)2CHCH2CH2C(=O)N(CH3)2 + (CH3)2CHOH

16.16 Addition of ethanol:

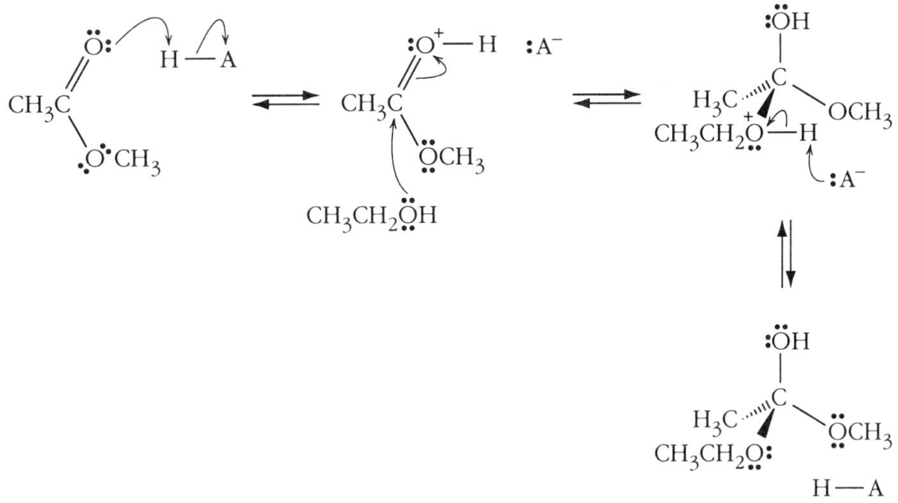
H—A
CH3C
OCH3
:A−
CH3CH2OH
OH
H3C
OCH3
CH3CH2O—H
OH
H3C
CH3CH2O
OCH3
H—A

Elimination of methanol:

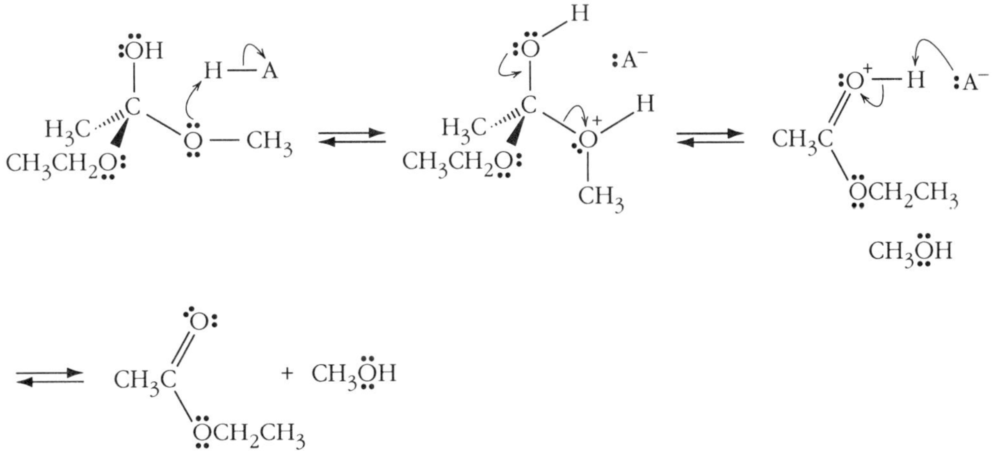
OH
H—A
H3C
CH3CH2O
O—CH3
:A−
H
CH3
CH3C
OCH2CH3
CH3OH
+ CH3OH

16.17 Addition of Enzyme—OH to carbonyl carbon atom:

Elimination of $HOCH_2CH_2\overset{+}{N}(CH_3)_3$:

16.18

(a) (CH₃)₂CHCH₂C(O)Cl $\xrightarrow{(1)\ LiAlH_4,\ (2)\ H_3O^+}$ (CH₃)₂CHCH₂CH₂OH

(b) cyclohexyl-CHO $\xrightarrow{(1)\ LiAlH_4,\ (2)\ H_3O^+}$ cyclohexyl-CH₂OH

(c) CH₂=CHC(O)NH₂ $\xrightarrow{(1)\ LiAlH_4,\ (2)\ H_3O^+}$ CH₂=CHCH₂NH₃⁺ $\xrightarrow{NaOH}$ CH₂=CHCH₂NH₂

(d) CH₃CH₂OCH₂CH₂C(O)OCH(CH₃)₂ $\xrightarrow{(1)\ LiAlH_4,\ (2)\ H_3O^+}$ CH₃CH₂OCH₂CH₂CH₂OH + HOCH(CH₃)₂

(e) CH₃CH₂CH₂CH₂C(O)N(CH₃)₂ $\xrightarrow{(1)\ LiAlH_4,\ (2)\ H_3O^+}$ CH₃CH₂CH₂CH₂CH₂N⁺H(CH₃)₂ $\xrightarrow{NaOH}$ CH₃CH₂CH₂CH₂CH₂N(CH₃)₂

(f) PhC(O)NHCH₃ $\xrightarrow{(1)\ LiAlH_4,\ (2)\ H_3O^+}$ PhCH₂N⁺H₂CH₃ $\xrightarrow{NaOH}$ PhCH₂NHCH₃

(g) (CH₃CH₂C(O))₂O $\xrightarrow{(1)\ LiAlH_4,\ (2)\ H_3O^+}$ 2 CH₃CH₂CH₂OH

(h) CH₃CH₂C(O)SCH₂CH₃ $\xrightarrow{(1)\ LiAlH_4,\ (2)\ H_3O^+}$ CH₃CH₂CH₂OH + CH₃CH₂SH

16.19

(a) (CH₃)₂CHCH₂C(O)Cl + CH₃OH → (CH₃)₂CHCH₂C(O)OCH₃ —NaBH₄→ No reaction

(b) cyclohexanecarbaldehyde —(1) NaBH₄/CH₃OH; (2) H₃O⁺→ cyclohexyl-CH₂OH

(c) CH₂=CH–C(O)NH₂ —NaBH₄/CH₃OH→ No reaction

(d) CH₃CH₂–O–CH₂CH₂–C(O)–O–CH(CH₃)₂ —NaBH₄/CH₃OH (Very slow reaction)→ CH₃CH₂–O–CH₂CH₂CH₂–OH + (CH₃)₂CHOH (Very low yield)

(e) CH₃CH₂CH₂CH₂–C(O)–N(CH₃)₂ —NaBH₄/CH₃OH→ No reaction

(f) C₆H₅–C(O)–NH–CH₃ —NaBH₄/CH₃OH→ No reaction

(g) (CH₃CH₂C(O))₂O —NaBH₄/CH₃OH→ CH₃CH₂C(O)OCH₃ + CH₃CH₂C(O)O⁻ Na⁺

(h) CH₃CH₂–C(O)–S–CH₂CH₃ —NaBH₄/CH₃OH (Slow reaction)→ CH₃CH₂CH₂OH + CH₃CH₂SH (Low yield)

16.20 Acid halides, acid anhydrides, thioesters, and esters react with two equivalents of Grignard reagent to form, after acidic workup, 3° alcohols in which two of the three groups bonded to the carbon atom with an —OH group come from the Grignard reagent. Carboxylic acids and amides have acidic protons and compounds with acidic protons decompose the Grignard reagents.

16.21

(a) (CH₃)₂CHCOCl + H₂O → (CH₃)₂CHCOOH

(b) (CH₃)₂CHCOCl + 2 HN(CH₂CH₃)₂ → (CH₃)₂CHCON(CH₂CH₃)₂ + (CH₃CH₂)₂NH₂⁺Cl⁻

(c) (CH₃)₂CHCOCl + (1) 2 CH₃MgI (2) H₃O⁺ → (CH₃)₂CHC(CH₃)₂OH

(d) (CH₃)₂CHCOCl + CH₃CH₂CH₂SH + pyridine → (CH₃)₂CHC(O)SCH₂CH₂CH₃

(e) 2 (CH₃)₂CHCOCl + (CH₃CH₂)₂CuLi → 2 (CH₃)₂CHC(O)CH₂CH₃   or

(CH₃)₂CHCOCl (1) LiAl[OC(CH₃)₃]₃H in THF, −78 °C (2) H₃O⁺ → (CH₃)₂CHCHO

(1) CH₃CH₂MgI / (2) H₃O⁺ → (CH₃)₂CHCH(OH)CH₂CH₃ →[PCC, CH₂Cl₂] (CH₃)₂CHC(O)CH₂CH₃

(f) (CH₃)₂CHCOCl (1) LiAl[OC(CH₃)₃]₃H in THF, −78°C (2) H₃O⁺ → (CH₃)₂CHCHO

(g) (CH₃)₂CHCOCl (1) LiAlH₄ / dry ether (2) H₃O⁺ → (CH₃)₂CHCH₂OH

(h) Shows isobutyryl chloride + sodium propanoate → mixed anhydride (isobutyric propionic anhydride)

(i) Shows isobutyryl chloride + isopropanol, with pyridine → isopropyl isobutyrate

16.22

(a) (CH₃)₂CHC≡N $\xrightarrow[\text{(2) H}_3\text{O}^+]{\text{(1) CH}_3\text{CH}_2\text{MgBr in dry ether}}$ 2-methyl-3-pentanone (ethyl isopropyl ketone)

(b) CH₃CH=CHCH₂C≡N $\xrightarrow[\substack{\text{(2) H}_3\text{O}^+ \\ \text{(3) NaOH}}]{\text{(1) LiAlH}_4}$ CH₃CH=CHCH₂CH₂NH₂

(c) (CH₃)₂CHCH₂CH₂CH₂C≡N $\xrightarrow[\text{(2) H}_3\text{O}^+]{\text{(1) NaOH / H}_2\text{O / heat}}$ (CH₃)₂CHCH₂CH₂CH₂COOH

(d) CH₃CH₂CH₂C≡N $\xrightarrow[\text{(2) H}_3\text{O}^+]{\text{(1) LiAl[OC(CH}_3)_3]_3\text{H}}$ CH₃CH₂CH₂CHO

(e) (CH₃)₂CHC≡N $\xrightarrow[\text{(2) H}_3\text{O}^+]{\text{(1) CH}_3\text{CH}_2\text{CH}_2\text{MgBr in dry ether}}$ 2-methyl-3-hexanone

or CH₃CH₂CH₂C≡N $\xrightarrow[\text{(2) H}_3\text{O}^+]{\text{(1) (CH}_3)_2\text{CHMgBr in dry ether}}$ 2-methyl-3-hexanone

(f) C₆H₅C≡N $\xrightarrow[\text{Heat}]{\text{NaOH / H}_2\text{O}}$ C₆H₅COO⁻Na⁺

16.23 (a) An **acyl chloride** is a compound that contains a chlorine atom attached to the carbon atom of a carbonyl group.

$$R-\overset{\overset{\displaystyle O}{\|}}{C}-Cl$$

(b) An **acyl group** is a carbonyl group with an alkyl or aryl group attached:

$$R-\overset{\overset{\displaystyle O}{\|}}{C}- \quad \text{or} \quad Ar-\overset{\overset{\displaystyle O}{\|}}{C}-$$

(c) An **amide** is a compound containing a trivalent nitrogen atom attached to the carbon atom of a carbonyl group. An ***N*-substituted amide** has an alkyl or aryl group attached to the nitrogen atom.

$$R-\overset{\overset{\displaystyle O}{\|}}{C}-NH_2 \qquad R-\overset{\overset{\displaystyle O}{\|}}{C}-NHR' \qquad R-\overset{\overset{\displaystyle O}{\|}}{C}-NR'_2$$

An amide   *N*-Substituted amide   *N,N*-Disubstituted amide

(d) The **addition-elimination mechanism** in an acyl transfer reaction involves the addition of a nucleophile to the carbonyl carbon atom to form a tetrahedral intermediate, which eliminates a group to form the product.

$$Nu:^- \quad R-\overset{\overset{\displaystyle O:}{\|}}{\underset{Y}{C}} \quad \xrightleftharpoons{\text{Addition}} \quad \left[\begin{array}{c} :\ddot{O}:^- \\ | \\ R\cdots C\cdots Y \\ Nu \end{array} \right] \quad \xrightleftharpoons{\text{Elimination}} \quad R-\overset{\overset{\displaystyle \ddot{O}:}{\|}}{\underset{Nu}{C}} \quad + \quad Y^-$$

Nucleophilic attack   Tetrahedral intermediate

(e) An **acid anhydride** (or carboxylic acid anhydride) is a carboxylic acid derivative formed from two acid molecules with loss of water.

$$2 \ R-\overset{\overset{\displaystyle O}{\|}}{C}-OH \quad \rightleftharpoons \quad \begin{array}{c} R-\overset{\overset{\displaystyle O}{\|}}{C} \\ \diagdown \\ O \\ \diagup \\ R-\underset{\underset{\displaystyle O}{\|}}{C} \end{array} \quad + \quad H_2O$$

A mixed anhydride is an anhydride derived from two different acid molecules.

(f) A **nitrile** is an organic compound containing the cyano group, —C≡N.

$$R-C\equiv N$$

(g) A **thioester,** previously known as a thiol ester, is a sulfur analogue of an ester formed from a carboxylic acid and a thiol:

$$R-\underset{OH}{\overset{O}{C}} + R'-SH \rightleftharpoons R-\underset{SR'}{\overset{O}{C}} + H_2O$$

Thioesters are more reactive toward nucleophilic acyl substitution than normal esters.

(h) An **enzyme inhibitor** is a structural analog of the substrate (S) that prevents an enzyme from carrying out its catalytic function. The enzyme inhibitor can form a *reversible* or *irreversible* bond to the enzyme (E). There are two types of reversible inhibitors: (i) a competitive inhibitor that competes with the substrate to form the enzyme-substrate (E-S) complex and (ii) a noncompetitive inhibitor that interferes only with the catalytic action of the enzyme in decomposing the E-S complex to the products. The noncompetitive inhibitor does not affect the binding of substrate to the enzyme to form the E-S complex.

An irreversible inhibitor forms such a strong union with the enzyme that it may not be removed readily. This type of inhibitor may form covalent bonds with one of the amino acid residues at the enzyme's active site, which effectively prevents any catalytic action on the part of the enzyme.

(i) An **acyl transfer reaction,** also known as nucleophilic acyl substitution, is a reaction in which a nucleophile substitutes for a leaving group on a carbonyl carbon atom. It is called an acyl transfer reaction because it transfers the acyl group from the leaving group to the attacking nucleophile. This reaction usually takes place through the addition-elimination mechanism described in Exercise 16.23(d).

16.24 (a) (*R*)-2-Ethyl-2-methylpentanenitrile
(b) (*E*)-3-Chloro-2-methyl-2-pentenoyl chloride
(c) Methyl *trans*-2-nitrocyclohexanecarboxylate or methyl (1*R*,2*R*)-2-nitrocyclohexanecarboxylate
(d) *N*-[(*S*)-2-Bromo-2-methylbutyl]-2-methylpropanamide
or *N*-[(*S*)-2-bromo-2-methylbutyl]-isobutanamide
(e) 2-Methylpropyl (*S*)-2-fluoropropanethiolate
or isobutyl (*S*)-2-fluoropropanethiolate

16.25

(a) 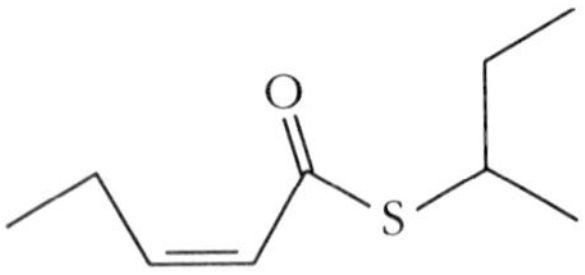 S-sec-butyl (Z)-pent-2-enethioate

(b) 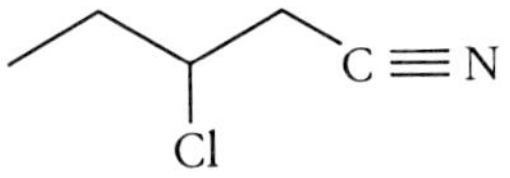 3-chloropentanenitrile

(c) 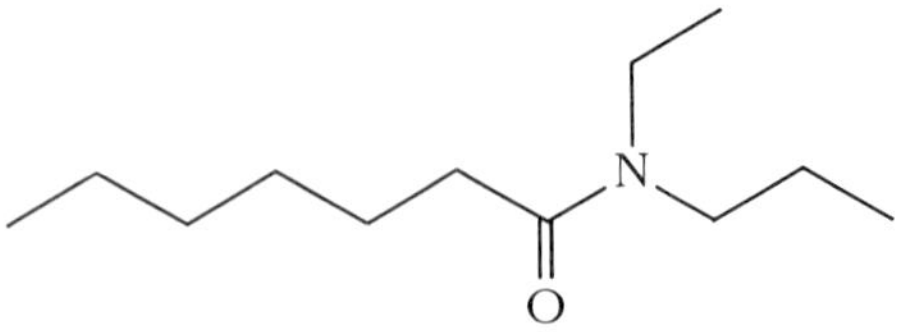 N-ethyl-N-propylheptanamide

(d) 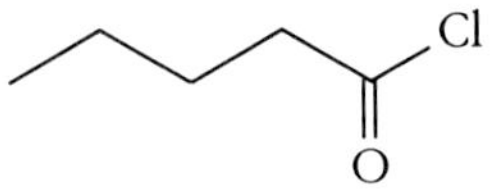 pentanoyl chloride

(e) 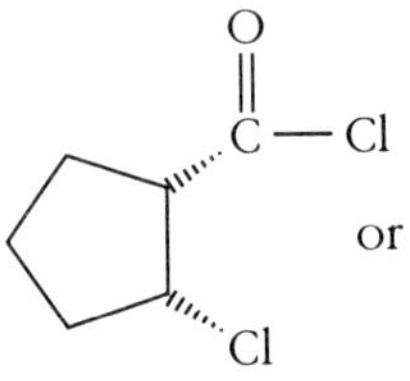 trans-2-chlorocyclopentanecarbonyl chloride (the structure on the right)

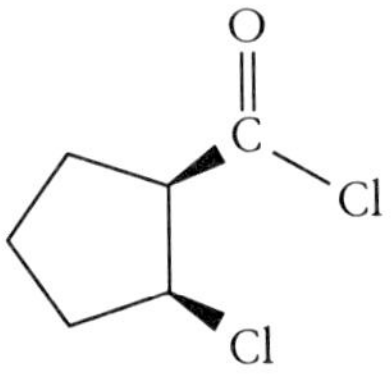

(f) 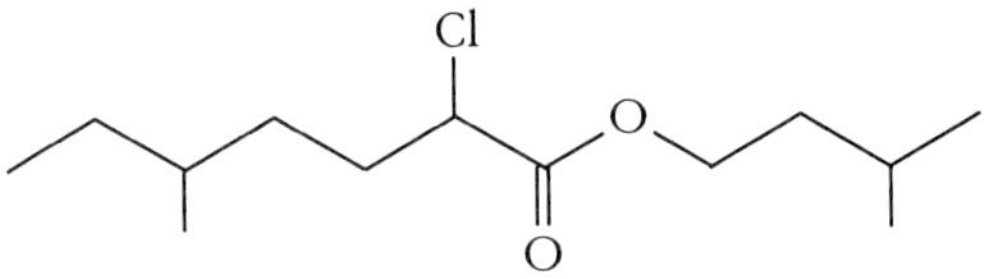 3-methylbutyl 2-chloro-5-methylheptanoate

16.26

(a) 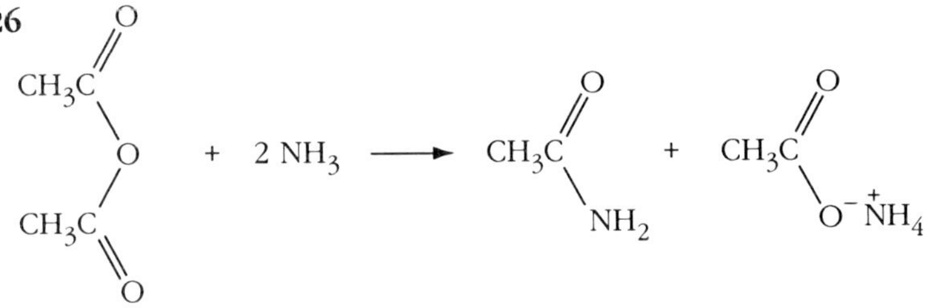 (CH₃CO)₂O + 2 NH₃ ⟶ CH₃CONH₂ + CH₃COO⁻ NH₄⁺

(b) 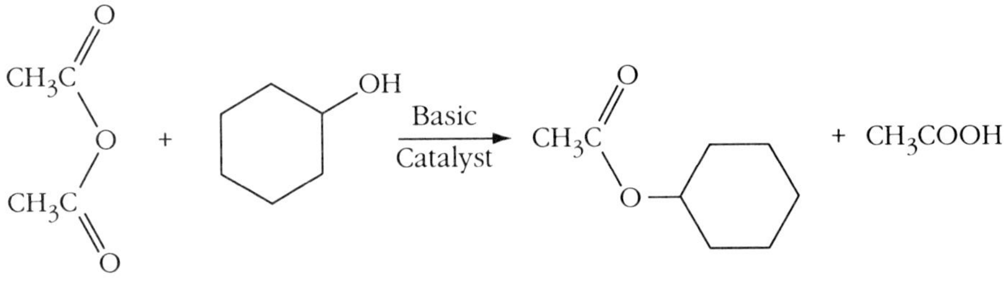 (CH₃CO)₂O + cyclohexanol →[Basic Catalyst] CH₃COO-cyclohexyl + CH₃COOH

(c) 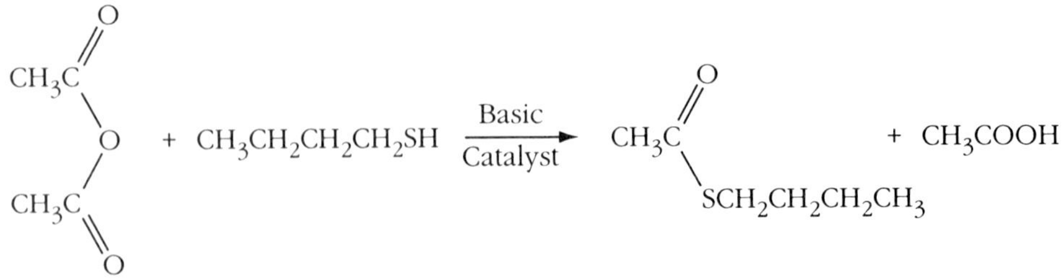 (CH₃CO)₂O + CH₃CH₂CH₂CH₂SH →[Basic Catalyst] CH₃C(O)SCH₂CH₂CH₂CH₃ + CH₃COOH

(d) $\text{CH}_3\text{C}(=\text{O})-\text{O}-\text{C}(=\text{O})\text{CH}_3$ + 3 CH$_3$MgI $\longrightarrow$ (CH$_3$)$_3$C–O$^-$$^+$MgI + CH$_3$C(=O)–O$^-$$^+$MgI

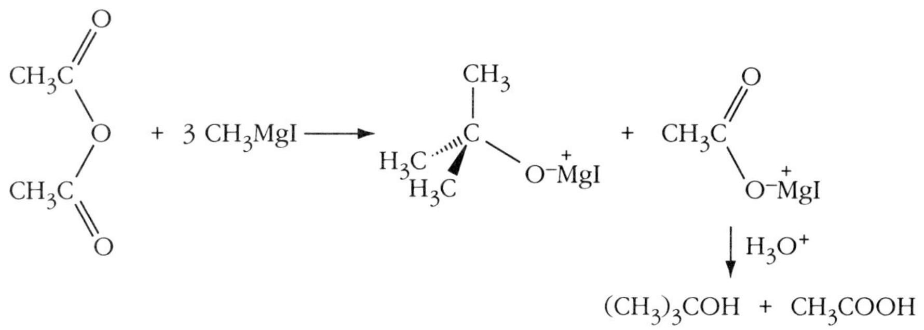

$\downarrow$ H$_3$O$^+$

(CH$_3$)$_3$COH + CH$_3$COOH

(e) $\text{CH}_3\text{C}(=\text{O})-\text{O}-\text{C}(=\text{O})\text{CH}_3$ $\xrightarrow{\text{(1) Excess LiAlH}_4}{\text{(2) H}_3\text{O}^+}$ 2 CH$_3$CH$_2$OH

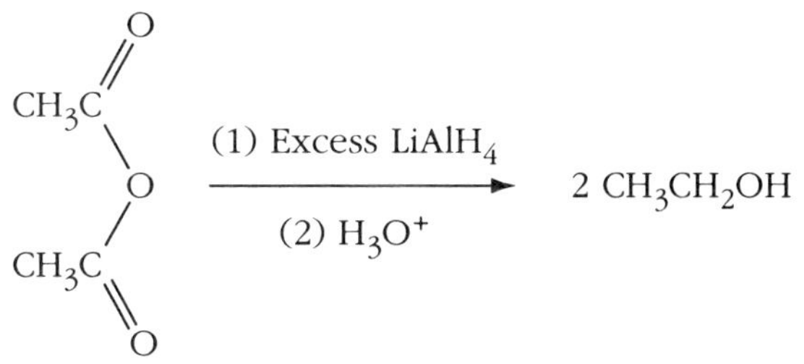

(f) $\text{CH}_3\text{C}(=\text{O})-\text{O}-\text{C}(=\text{O})\text{CH}_3$ $\xrightarrow{\text{NaOH}}{\text{H}_2\text{O}}$ 2 CH$_3$C(=O)O$^-$ Na$^+$

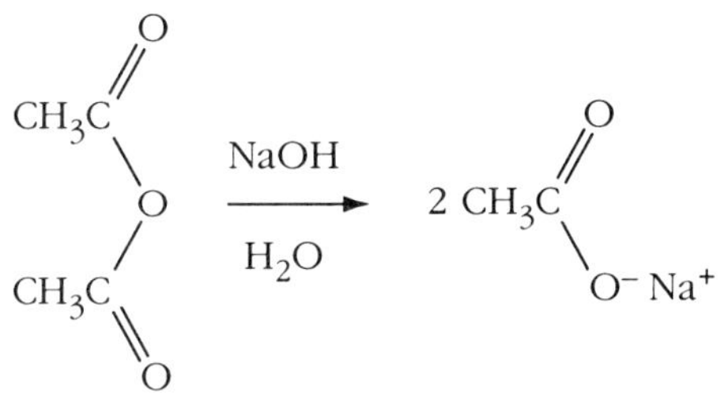

(g) $\text{CH}_3\text{C}(=\text{O})-\text{O}-\text{C}(=\text{O})\text{CH}_3$ $\xrightarrow{\text{NaBH}_4 / \text{CH}_3\text{OH}}$ CH$_3$C(=O)OCH$_3$ + CH$_3$C(=O)O$^-$ Na$^+$

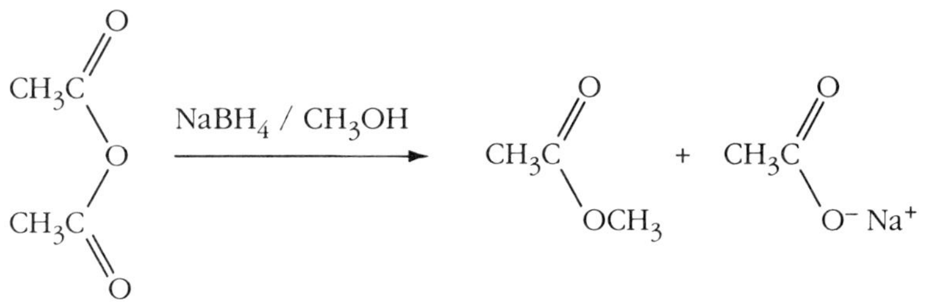

16.27

(a) 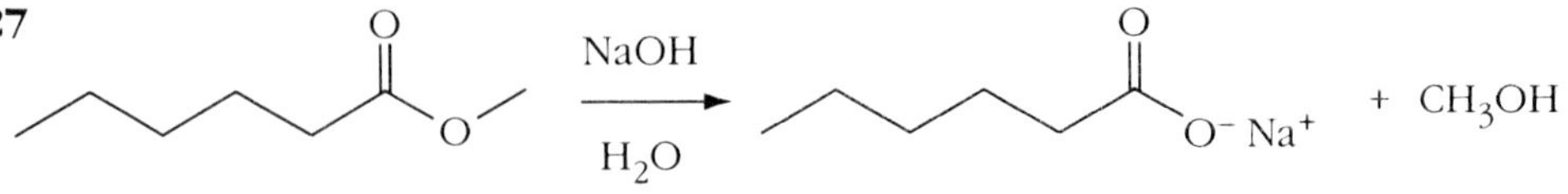 CH₃(CH₂)₄C(=O)OCH₃ + NaOH / H₂O → CH₃(CH₂)₄C(=O)O⁻ Na⁺ + CH₃OH

(b) 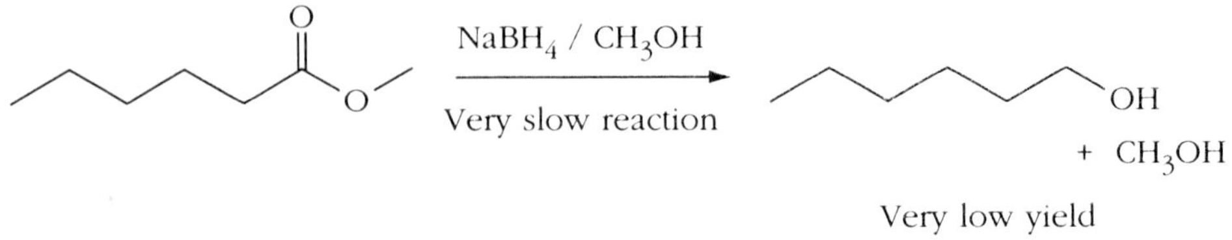 CH₃(CH₂)₄C(=O)OCH₃ + NaBH₄ / CH₃OH (Very slow reaction) → CH₃(CH₂)₅OH + CH₃OH (Very low yield)

(c) 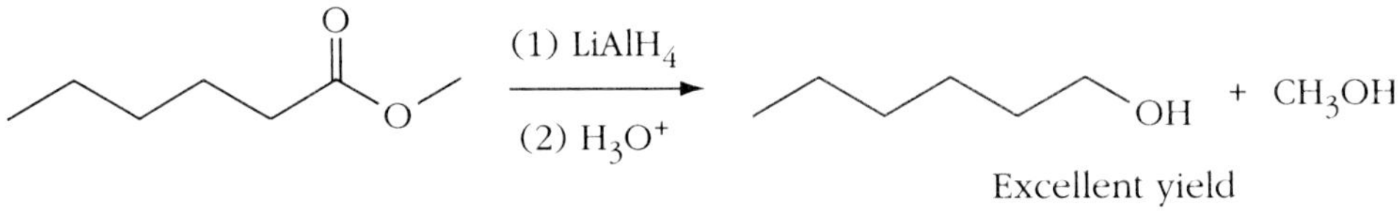 CH₃(CH₂)₄C(=O)OCH₃ (1) LiAlH₄ (2) H₃O⁺ → CH₃(CH₂)₅OH + CH₃OH (Excellent yield)

(d) 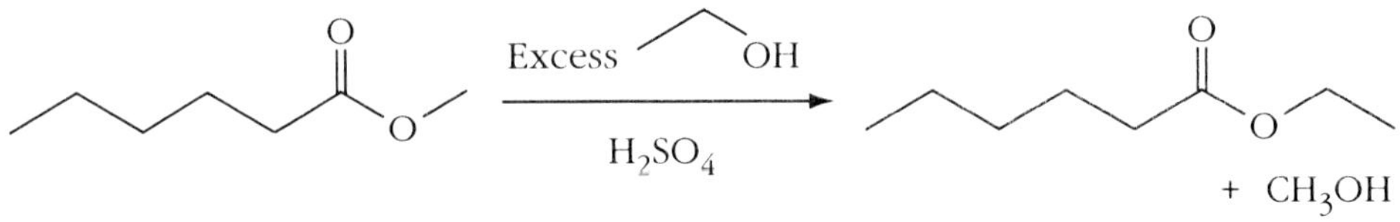 CH₃(CH₂)₄C(=O)OCH₃ + Excess CH₃CH₂OH, H₂SO₄ → CH₃(CH₂)₄C(=O)OCH₂CH₃ + CH₃OH

(e) 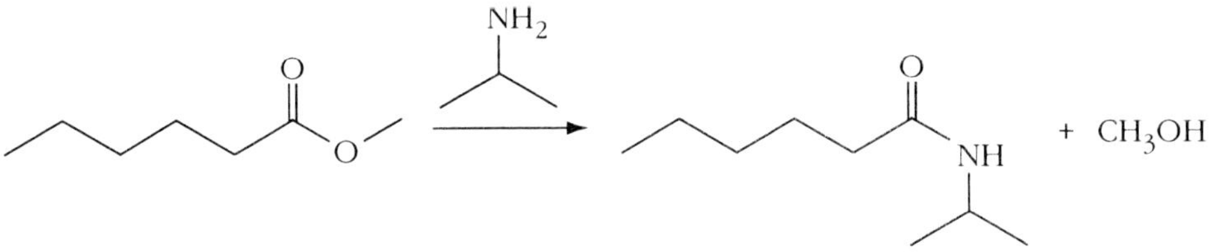 CH₃(CH₂)₄C(=O)OCH₃ + (CH₃)₂CHNH₂ → CH₃(CH₂)₄C(=O)NHCH(CH₃)₂ + CH₃OH

16.28

(a) N≡CCH₂CH₂C≡N $\xrightarrow[\text{Heat}]{\text{H}_2\text{O/HCl}}$ HOOCCH₂CH₂COOH

(b) 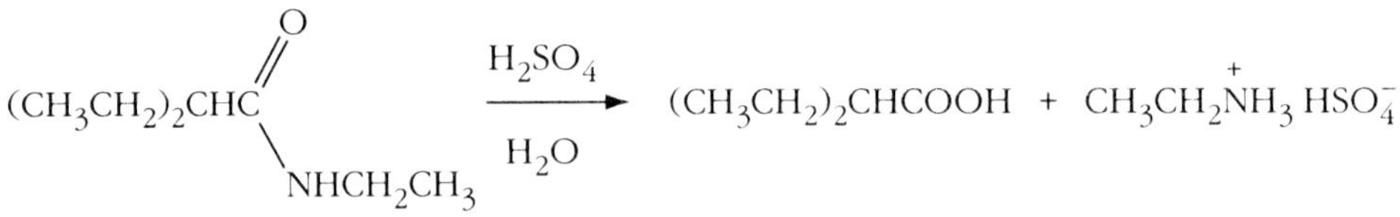 (CH₃CH₂)₂CHC(=O)NHCH₂CH₃ $\xrightarrow[\text{H}_2\text{O}]{\text{H}_2\text{SO}_4}$ (CH₃CH₂)₂CHCOOH + CH₃CH₂NH₃⁺ HSO₄⁻

(c) 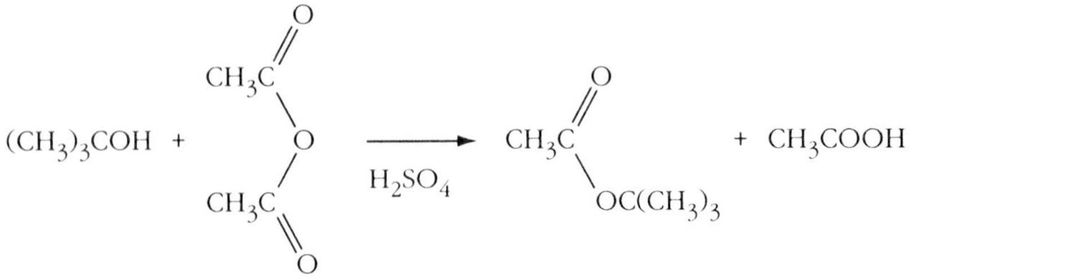 (CH₃)₃COH + (CH₃C(=O))₂O $\xrightarrow{\text{H}_2\text{SO}_4}$ CH₃C(=O)OC(CH₃)₃ + CH₃COOH

(d) The only reaction that takes place is that of decomposition of the Grignard reagent:

C₆H₅C(=O)NH₂ $\xrightarrow{\text{CH}_3\text{MgI, Dry ether}}$ C₆H₅C(=O)NH⁻ MgI⁺ + CH₄

$\xrightarrow{\text{H}_3\text{O}^+}$ C₆H₅C(=O)NH₂

(e) $CH_3CH_2C(=O)NH_2$ + CH_3OH $\xrightarrow[\text{Heat}]{\text{Excess BF}_3}$ $CH_3CH_2C(=O)OCH_3$ + F_3BNH_3

(f) $Cl_3CC(=O)NH_2$ + P_4O_{10} $\xrightarrow{\text{Heat}}$ $Cl_3CC\equiv N$

(g) $CH_3C(=O)SCH_2CH_3$ $\xrightarrow[\text{(2) H}_3\text{O}^+]{\text{(1) Excess CH}_3\text{MgI/dry ether}}$ $(CH_3)_3COH + CH_3CH_2SH$

(h) C₆H₅C(=O)SCH₂CH₃ + CH₃CH₂NH₂ ⟶ C₆H₅C(=O)NHCH₂CH₃ + CH₃CH₂SH

(i) (C₆H₅CO)₂O + CH₃CH₂SH $\xrightarrow{\text{Basic Catalyst}}$ C₆H₅C(=O)SCH₂CH₃ + C₆H₅COOH

(j)

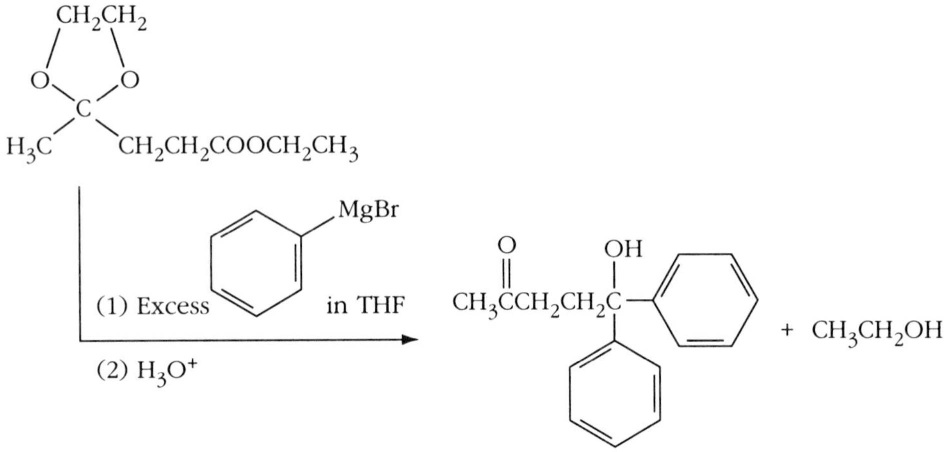

Structure: central C with OCH₂CH₂O (cyclic ketal), CH₃, and CH₂CH₂COOCH₂CH₃ groups

$$\xrightarrow[\text{(2) H}_3\text{O}^+]{\text{(1) Excess PhMgBr in THF}}$$

CH₃CCH₂CH₂C(OH)(Ph)₂ + CH₃CH₂OH

(with C=O ketone regenerated)

(k) (CH₃CH₂O)₂C=O

$$\xrightarrow[\text{(2) H}_3\text{O}^+]{\text{(1) Excess CH}_3\text{MgI/dry ether}}$$

(CH₃)₃COH + 2 CH₃CH₂OH

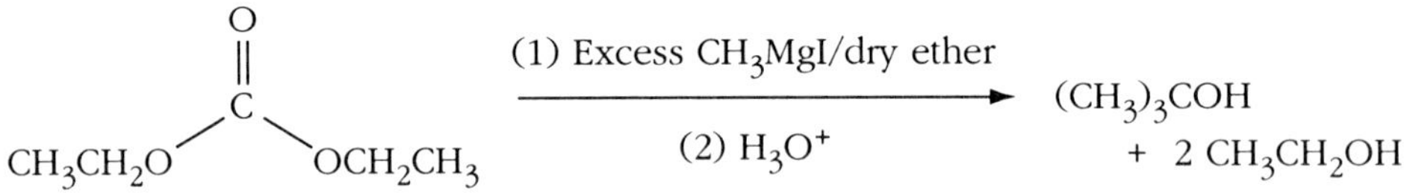

16.29

(a) $CH_3COOH + (CH_3)_2CHCH_2OH \underset{}{\overset{H_2SO_4}{\rightleftharpoons}} CH_3COOCH_2CH(CH_3)_2 + H_2O$

(b) PhCOCl + 2 NH₃ ⟶ PhCONH₂ + NH₄Cl

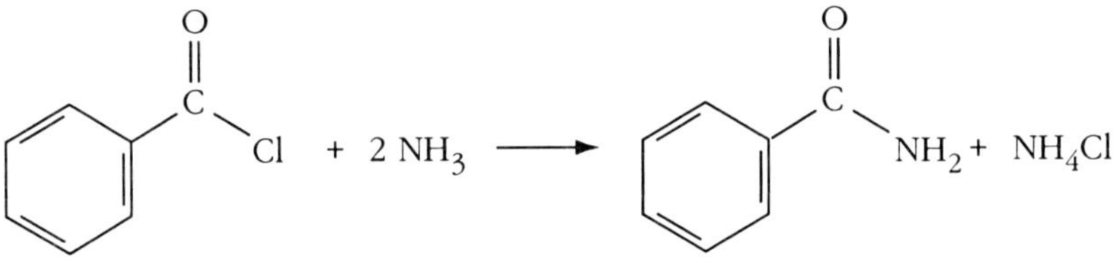

(c) cyclopentanone + H₂NNH-C₆H₃(NO₂)₂

$$\xrightarrow{\text{Acidic buffer}}$$

cyclopentanone 2,4-dinitrophenylhydrazone + H₂O

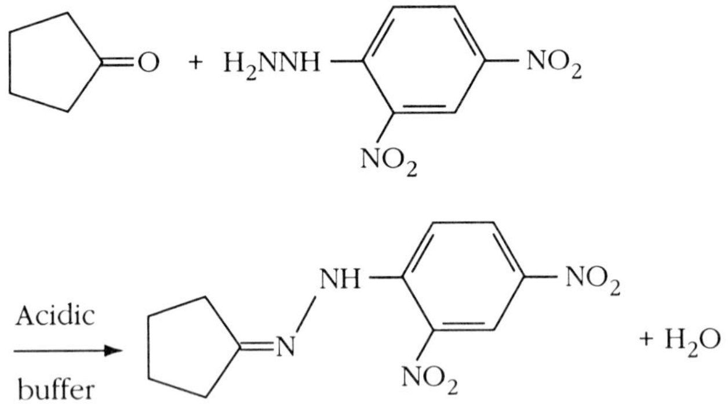

(d) C₆H₅MgBr + H—CHO →(1) Dry ether (2) H₃O⁺→ C₆H₅CH₂OH

(e) C₆H₅COO—CH₂C₆H₅ →(1) LiAlH₄/THF (2) H₃O⁺→ 2 C₆H₅CH₂OH

(f) CH₃CH₂COOCH₂CH₂CH₃ →HN(CH₂CH₃)₂→ CH₃CH₂C(=O)N(CH₂CH₃)₂ + CH₃CH₂CH₂OH

(g) HO—C(CH₃)(CH₂CH₃)(H) + (CH₃CO)₂O →H₂SO₄→ CH₃C(=O)O—C(CH₃)(CH₂CH₃)(H) + CH₃COOH

16.30

(a) A = (CH₃)₃CCH₂CH(CH₃)O⁻Li⁺

B = (CH₃)₃CCH₂CH(CH₃)O—C(=O)CH₃ + CH₃C(=O)O⁻Li⁺

(b) C = cyclopentane with CH₂OH and COOCH₃ substituents

D = cyclopentane with CH₂Cl and COOCH₃ substituents

E = cyclopentane with CH₂C≡N and COOCH₃ substituents

F = cyclopentane with CH₂CH₂NH₂ and CH₂OH substituents

(c) G = [1,4-dioxaspiro[4.4]nonane with COOCH₃ substituent]   H = [3-(hydroxymethyl)cyclopentanone]

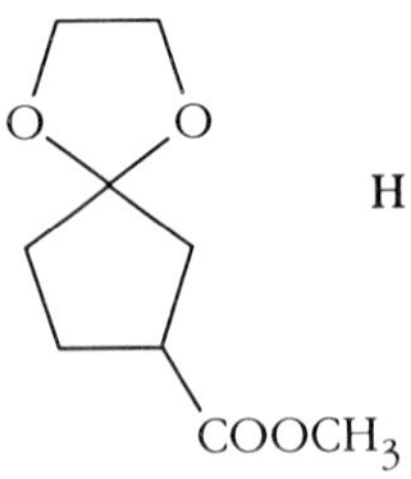

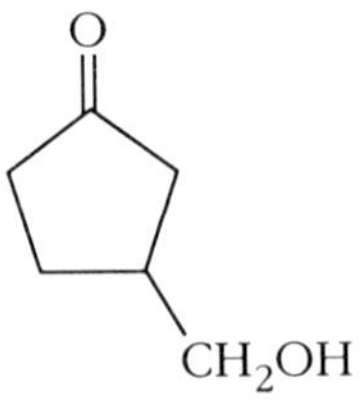

(d) I = CH₃C≡C⁻Na⁺   J = CH₃C≡CCH₂CH₂CH₂Cl

K = CH₃C≡CCH₂CH₂CH₂COOH   L = (E)-CH₃CH=CHCH₂CH₂CH₂COO⁻Li⁺ (with H₃C and H on one carbon, H and CH₂CH₂CH₂COO⁻Li⁺ on the other)

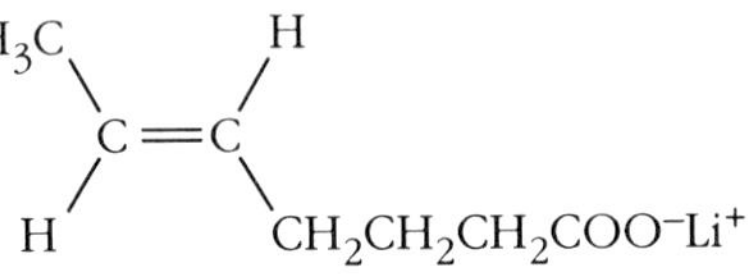

M = (E)-CH₃CH=CHCH₂CH₂CH₂C(=O)Cl

N = (E)-CH₃CH=CHCH₂CH₂CH₂CHO

16.31

(a) CH₃CH₂C(=O)Cl $\xrightarrow[\text{(2) H}_3\text{O}^+]{\text{(1) LiAl[OC(CH}_3)_3]_3\text{H in THF, }-78\,°\text{C}}$ CH₃CH₂C(=O)H

(b) CH₃CH₂C(=O)Cl + 2 NH₃ ⟶ CH₃CH₂C(=O)NH₂ + NH₄Cl

(c) CH₃CH₂C(=O)Cl + CH₃CH₂OH $\xrightarrow{\text{pyridine}}$ CH₃CH₂C(=O)OCH₂CH₃

(d) CH₃CH₂C(=O)Cl + (CH₃)₂CuLi ⟶ CH₃CH₂CCH₃ (with C=O)

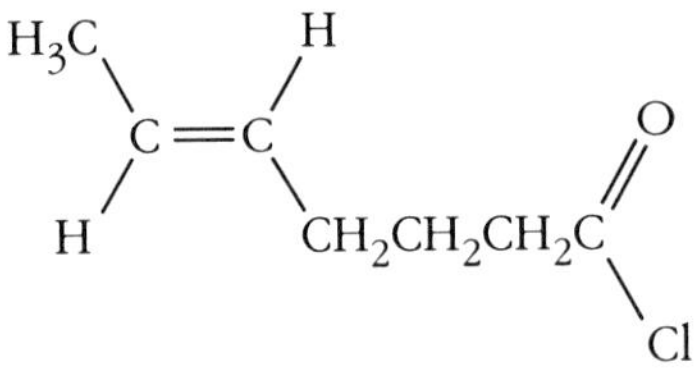

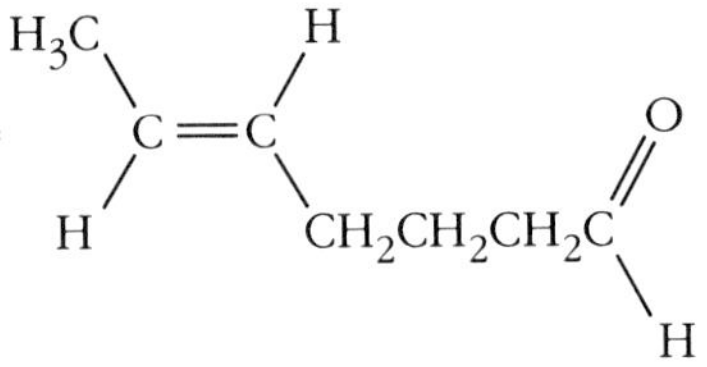

(e) CH₃CH₂C(=O)Cl $\xrightarrow[\text{(2) H}_3\text{O}^+]{\text{(1) LiAlH}_4/\text{dry ether}}$ CH₃CH₂CH₂OH

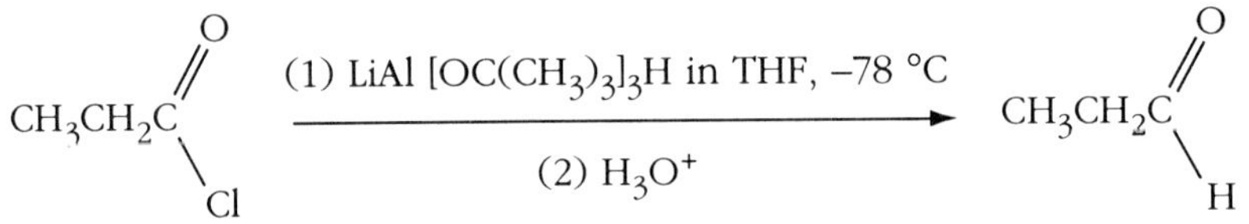

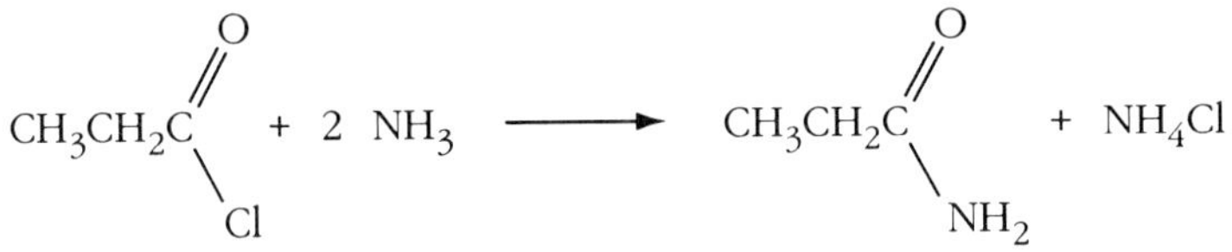

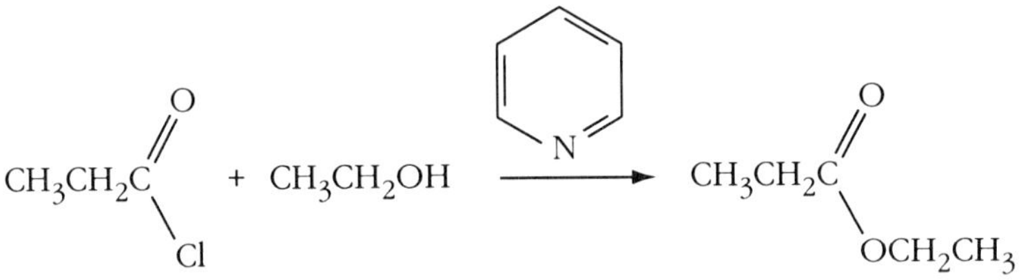

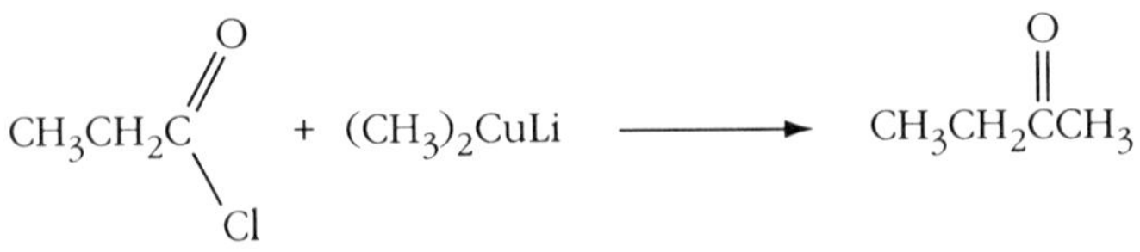

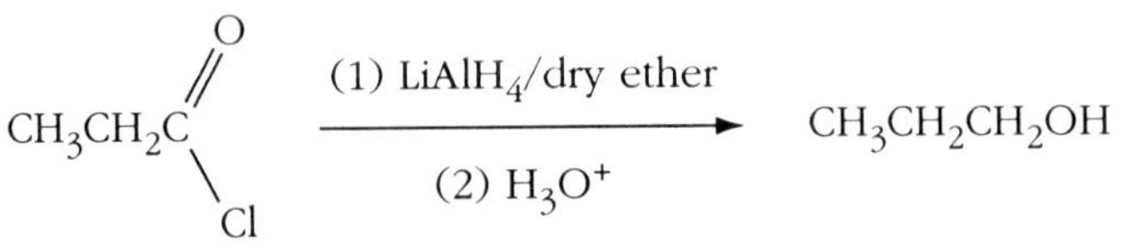

(f) $CH_3CH_2\overset{O}{\underset{Cl}{C}}$ $\xrightarrow[\text{pyridine}]{CH_3CH_2CH_2SH}$ $CH_3CH_2\overset{O}{\underset{SCH_2CH_2CH_3}{C}}$

(g) $CH_3CH_2\overset{O}{\underset{Cl}{C}}$ + $C_6H_5\overset{O}{C}-O^-\;Na^+$ $\longrightarrow$ $CH_3CH_2\overset{O}{C}-O-\overset{O}{C}-C_6H_5$

16.32

(a) [δ-valerolactone] $\xrightarrow[H_2O]{NaOH}$ $HO-CH_2CH_2CH_2CH_2-\overset{O}{C}-O^-\;Na^+$

(b) [δ-valerolactam] $\xrightarrow[\text{Heat}]{HCl/H_2O}$ $H_3\overset{+}{N}-CH_2CH_2CH_2CH_2-\overset{O}{C}-OH$ Cl^-

(c) [glycolide, cyclic diester] $\xrightarrow[H_2O]{NaOH}$ $HO-CH_2CH_2-OH$ + $\overset{O}{\underset{O^-\;Na^+}{C}}-\overset{O}{\underset{O^-\;Na^+}{C}}$

(d) $H-\overset{O}{C}-O-C_6H_5$ $\xrightarrow[H_2O]{NaOH}$ $H-\overset{O}{C}-O^-\;Na^+$ + $Na^+\;^-O-C_6H_5$

The pK_a of phenol (C_6H_5OH) is 10, so in an aqueous NaOH solution it exists as an anion (see Section 23.7).

(e) [cholesterol] + [acetic anhydride] →(H₂SO₄) [cholesteryl acetate] + CH₃COOH

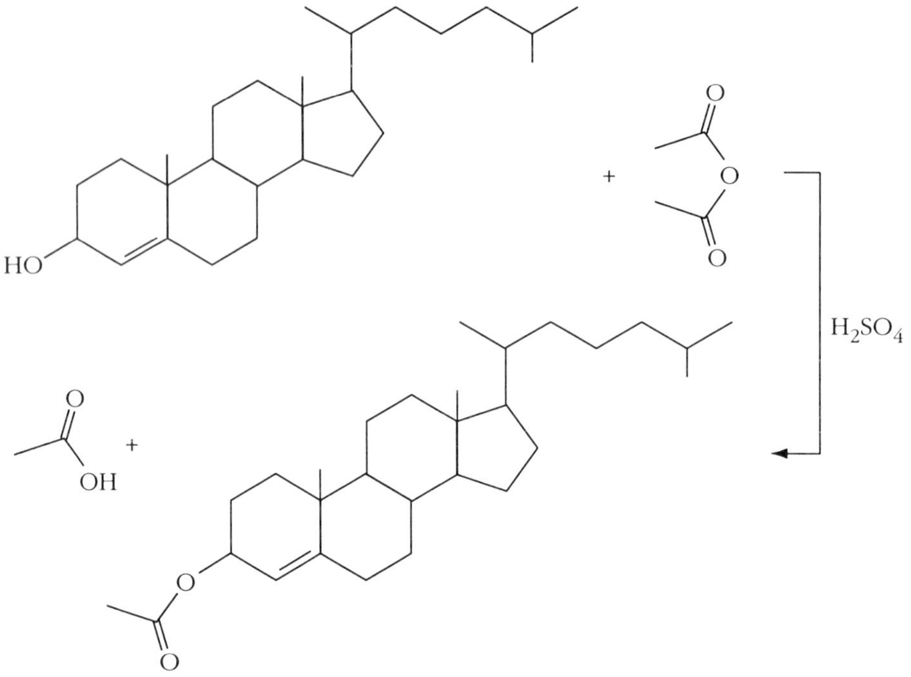

16.33

(a) $CH_3CH_2OH \xrightarrow[(2)\ H_3O^+]{(1)\ KMnO_4,\ HO^-} CH_3\underset{OH}{\overset{O}{C}}$

(b) $CH_3\underset{OH}{\overset{O}{C}} + CH_3CH_2OH \xrightleftharpoons{H_2SO_4} CH_3\underset{OCH_2CH_3}{\overset{O}{C}} + H_2O$

From (a)

(c) $CH_3\underset{OCH_2CH_3}{\overset{O}{C}} \xrightarrow[(2)\ H_3O^+]{(1)\ Excess\ CH_3CH_2MgBr} CH_3CH_2-\underset{\underset{CH_3}{|}}{\overset{\overset{OH}{|}}{C}}-CH_2CH_3$

From (b)

(d) $CH_3CH_2OH \xrightarrow[Heat]{KMnO_4,\ NaOH} CH_3\underset{O^-M^+}{\overset{O}{C}}$

M⁺ = Na⁺ or K⁺

(e) $CH_3CH_2OH \xrightarrow{PBr_3} CH_3CH_2Br \xrightarrow[\text{Ethanol}]{NaCN} CH_3CH_2C\equiv N$

16.34 Addition of the first mole of the Grignard reagent:

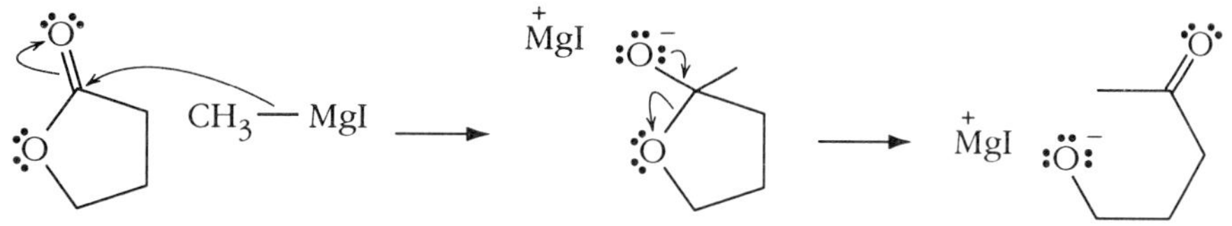

Addition of a second mole of the Grignard reagent:

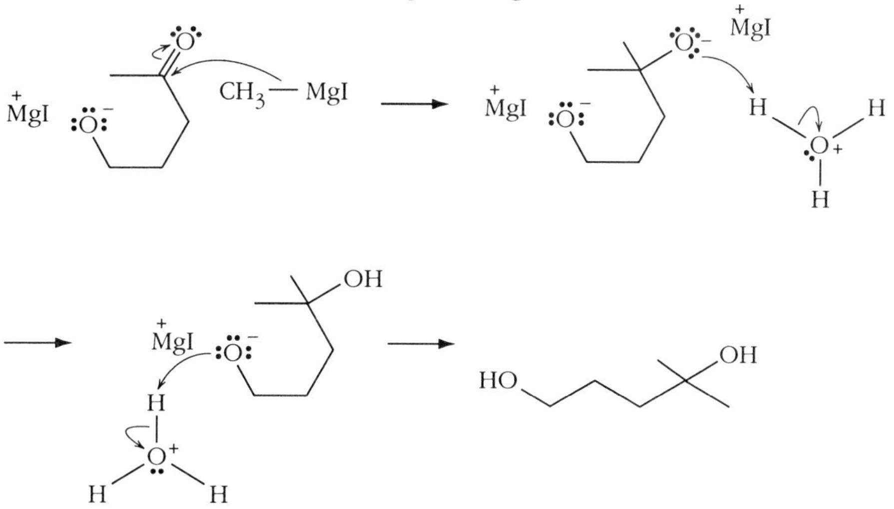

16.35

(a) $CH_3CH_2COOH \xrightarrow{SOCl_2} CH_3CH_2\overset{O}{\underset{Cl}{C}} \xrightarrow[\text{(2) } H_3O^+]{\text{(1) Excess } CH_3MgI} CH_3CH_2\underset{CH_3}{\overset{OH}{\underset{|}{\overset{|}{C}}}}CH_3$

(b) $CH_3CH_2COOH \xrightarrow[\text{(2) } H_3O^+]{\text{(1) LiAlH}_4} CH_3CH_2CH_2OH \xrightarrow[160\,°C]{95\%\ H_2SO_4} CH_3CH=CH_2$

$CH_3CH_2CH_2OH \xrightarrow{SOCl_2} CH_3CH_2CH_2Cl \xrightarrow[\text{Heat*}]{CH_3CH_2O^-Na^+} CH_3CH=CH_2$

*Heating the reaction mixture increases the amount of elimination product ($CH_3CH=CH_2$).

(c) $CH_3CH_2COOH \xrightarrow[\text{(2) } H_3O^+]{\text{(1) LiAlH}_4/\text{Dry ether}} CH_3CH_2CH_2OH$

(d) $CH_3CH_2CH_2OH$ $\xrightarrow{SOCl_2}$ $CH_3CH_2CH_2Cl$ $\xrightarrow[\text{Ethanol}]{NaCN}$ $CH_3CH_2CH_2C\equiv N$
From (c)

(e) 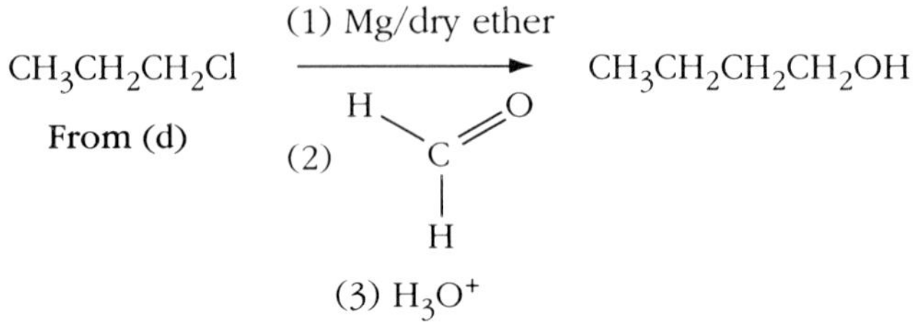 $CH_3CH_2CH_2Cl$ $\xrightarrow{\text{(1) Mg/dry ether}}$ $CH_3CH_2CH_2CH_2OH$
From (d)

(2) HCHO (formaldehyde)

(3) H_3O^+

(f) CH_3CH_2COOH $\xrightarrow{SOCl_2}$ $CH_3CH_2C(=O)Cl$ + 2 NH_3 $\longrightarrow$ $CH_3CH_2C(=O)NH_2$

(g) $CH_3CH_2C(=O)NH_2$ $\xrightarrow[\text{(2) }H_3O^+]{\text{(1) LiAlH}_4}$ $CH_3CH_2CH_2NH_2$

(3) NaOH/H_2O

From (f)

(h) $CH_3CH_2CH_2Cl$ $\xrightarrow[\text{Dry Ether}]{Mg}$ $CH_3CH_2CH_2MgCl$ $\xrightarrow[\text{(2) }H_3O^+]{\text{(1) }CH_2\text{—}CH_2\text{ (epoxide)}}$ $CH_3CH_2CH_2CH_2CH_2OH$

From (d)

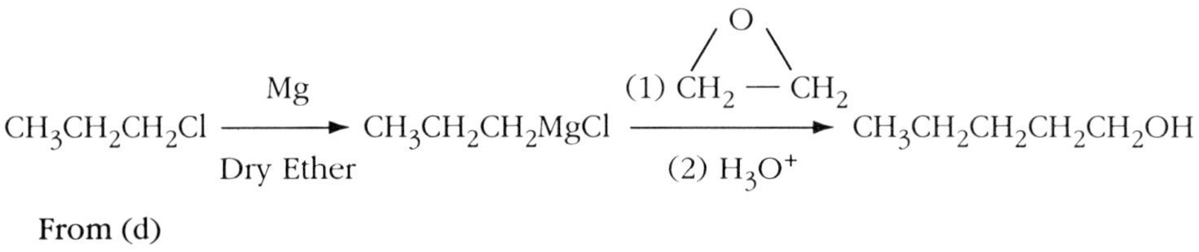

16.36 Addition step to form tetrahedral intermediate:

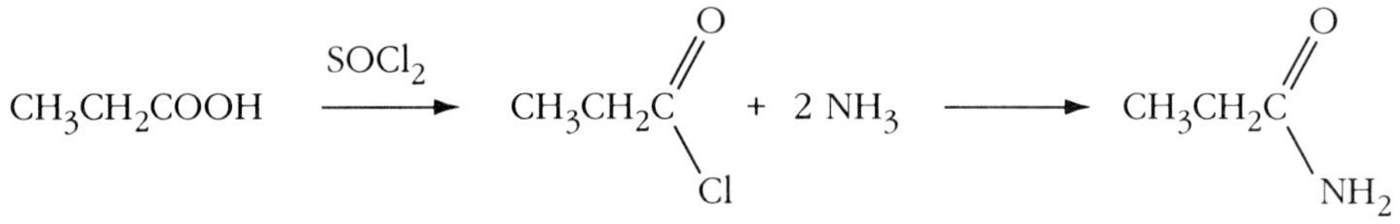

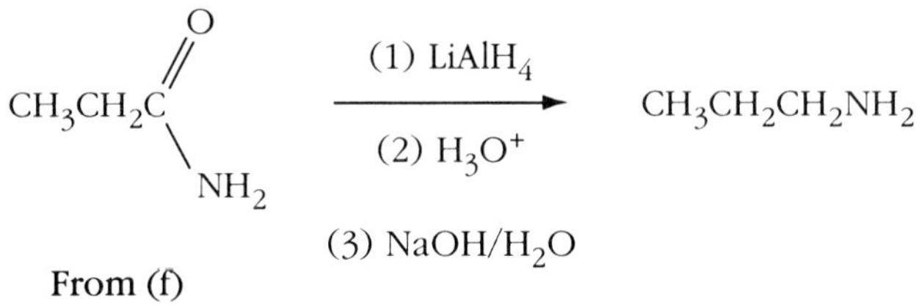

[Mechanism showing acetic anhydride being protonated by $H-OSO_3H$, giving HSO_4^-, then attack by $HOCH_2CH(CH_3)_2$ on the carbonyl carbon, forming the tetrahedral intermediate, followed by proton transfer to HSO_4^- to give the neutral tetrahedral intermediate with $:OCH_2CH(CH_3)_2$ group]

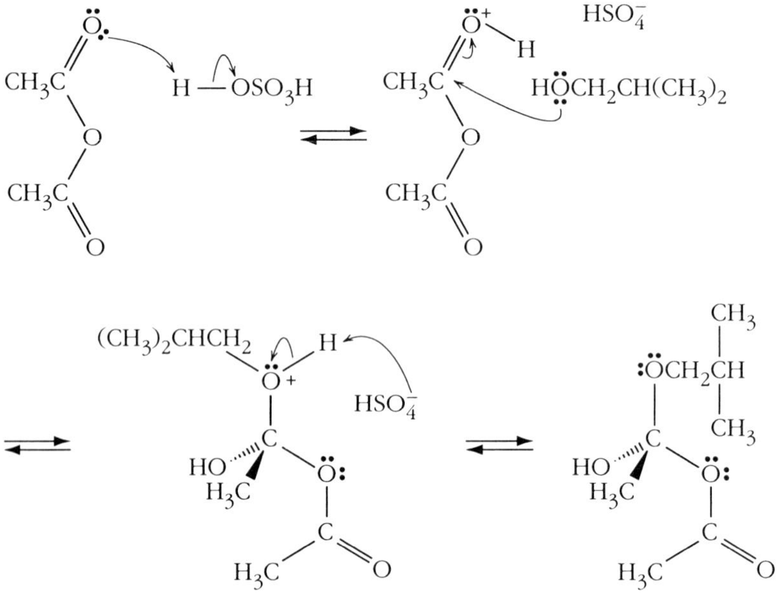

CHAPTER 16 — 467

Elimination step:

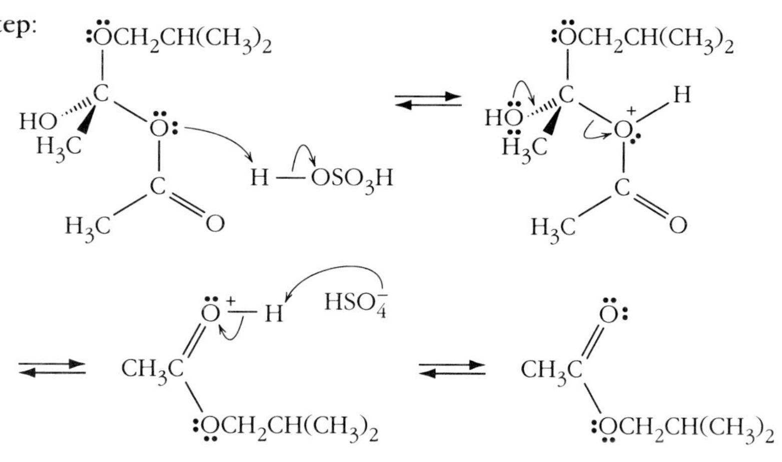

[Structures showing elimination step with :ÖCH₂CH(CH₃)₂, HO, H₃C, and H—OSO₃H, progressing through equilibria to CH₃C(=O)OCH₂CH(CH₃)₂]

16.37 See Figure 16.8.

$$\underset{NH_2}{CH_3C(=O)} < \underset{OCH_3}{CH_3C(=O)} < \underset{SCH_3}{CH_3C(=O)} < \underset{CH_3C(=O)-O-C(=O)CH_3}{} < \underset{Cl}{CH_3C(=O)}$$

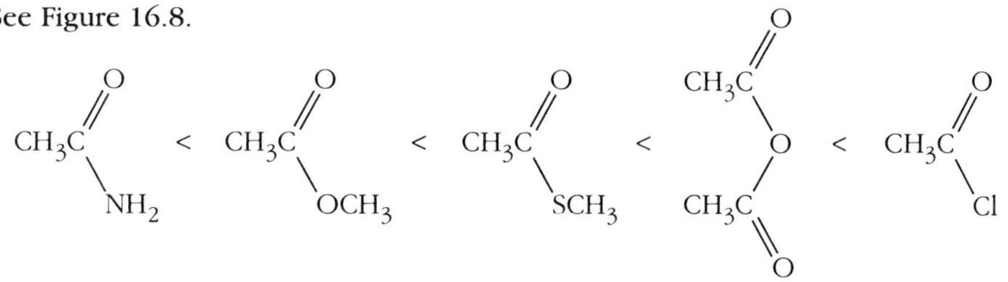

16.38 Step 1: Addition of hydrazine

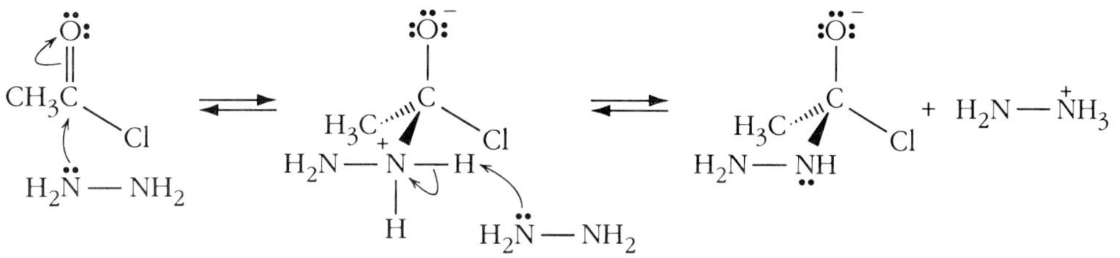

[Mechanism showing CH₃C(=O)Cl + H₂N—NH₂ forming tetrahedral intermediate, then deprotonation by H₂N—NH₂ to give H₃C–C(O⁻)(Cl)(NH–NH₂) + H₂N–N⁺H₃]

Step 2: Elimination of chloride ion

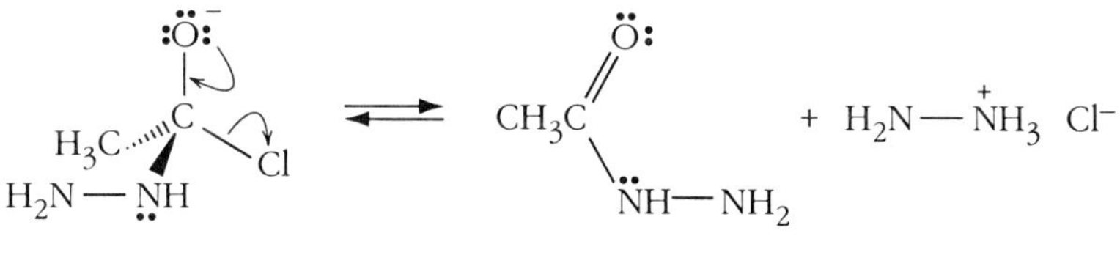

[Tetrahedral intermediate eliminates Cl⁻ to give CH₃C(=O)NH–NH₂ + H₂N–N⁺H₃ Cl⁻]

16.39

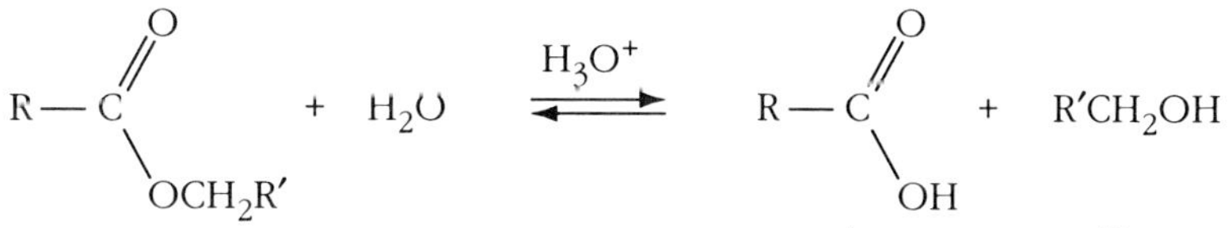

$$R-C(=O)-OCH_2R' + H_2O \underset{}{\overset{H_3O^+}{\rightleftharpoons}} \underset{A}{R-C(=O)-OH} + \underset{B}{R'CH_2OH}$$

Because

$$\underset{B}{R'CH_2OH} \xrightarrow[H_2SO_4, H_2O]{KMnO_4} \underset{A}{R'-C(=O)-OH}$$

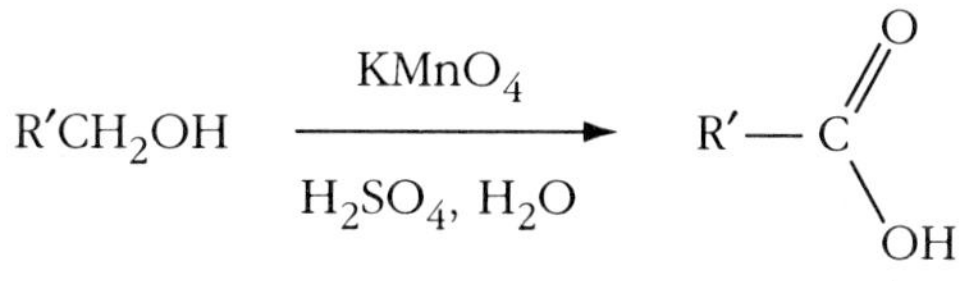

then R' = R. This means that the original ester must be a symmetrical compound: RCOOCH$_2$R. The formula of the alkyl group R can be calculated by subtracting C$_2$H$_2$O$_2$ from C$_8$H$_{16}$O$_2$ (that gives C$_6$H$_{14}$) and dividing the result by two (we have two R groups in the formula). Thus, the formula for R is C$_3$H$_7$. We can have two alkyl groups of formula C$_3$H$_7$, a propyl group and an isopropyl group. Thus, we can write two structures of the original ester:

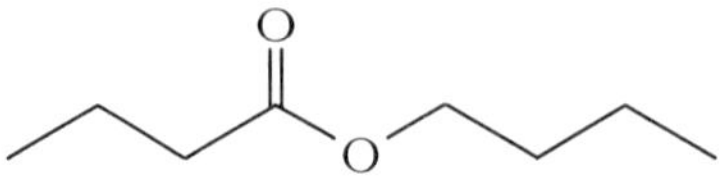

Butyl butanoate

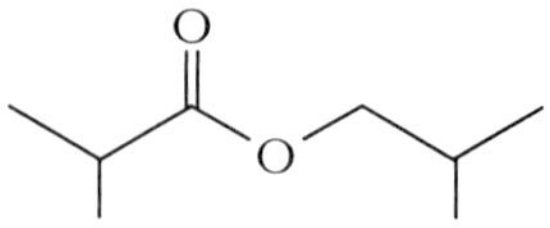

Isobutyl isobutanoate
(or 2-methylpropyl 2-methylpropanoate)

16.40

(a) 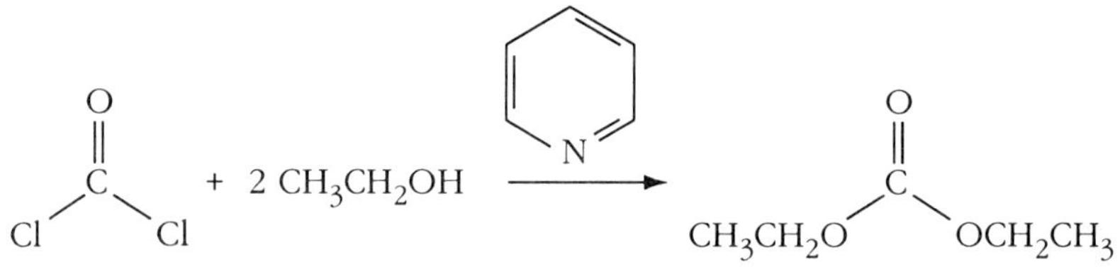 Cl–CO–Cl + 2 CH$_3$CH$_2$OH → (pyridine) → CH$_3$CH$_2$O–CO–OCH$_2$CH$_3$

(b)  Cl–CO–Cl + 2 H$_2$O → HO–CO–OH + 2 HCl

(c) 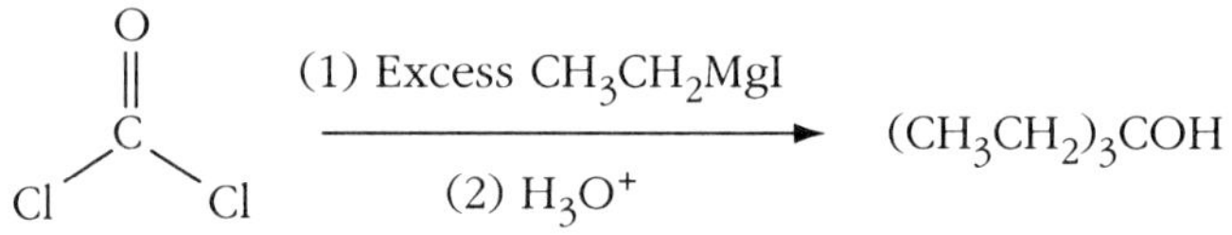 Cl–CO–Cl (1) Excess CH$_3$CH$_2$MgI (2) H$_3$O$^+$ → (CH$_3$CH$_2$)$_3$COH

(d) 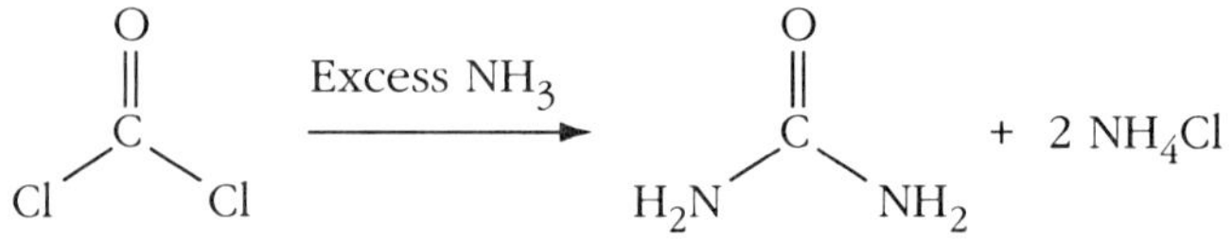 Cl–CO–Cl Excess NH$_3$ → H$_2$N–CO–NH$_2$ + 2 NH$_4$Cl

16.41 First addition-elimination sequence of steps:

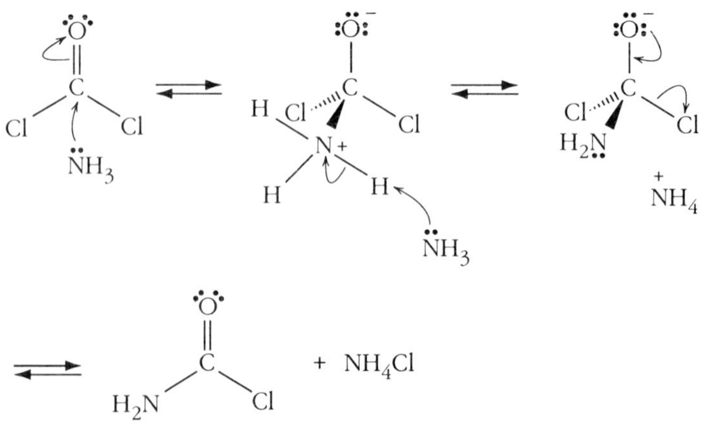

⇌ ⇌ ⇌ H$_2$N–CO–Cl + NH$_4$Cl

Second addition-elimination sequence of steps:

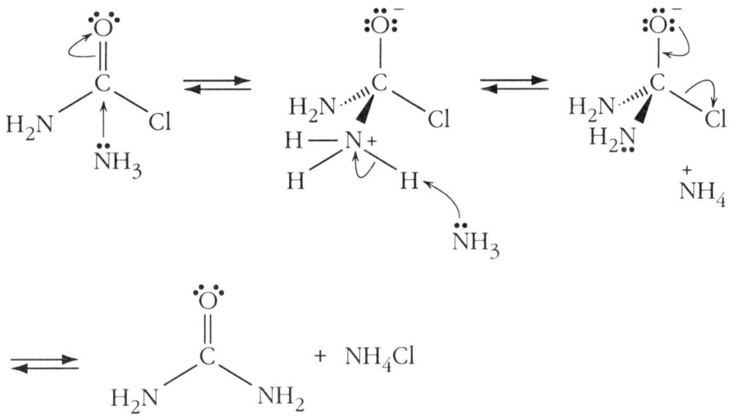

[Mechanism showing addition of NH$_3$ to carbamoyl chloride intermediate, proton transfer, and elimination of Cl to give urea + NH$_4$Cl]

$$\rightleftharpoons \quad H_2N-\underset{\underset{NH_2}{|}}{\overset{\overset{\ddot{O}:}{\|}}{C}} \quad + \; NH_4Cl$$

16.42 (a) Retrosynthesis:

[Retrosynthetic scheme: target 8-oxooctanoic acid type aldehyde-acid ⇒ dioxolane-protected aldehyde carboxylic acid ⇒ protected-aldehyde alcohol ⇒ protected-aldehyde Grignard reagent (MgBr) ⇒ protected-aldehyde alkyl bromide]

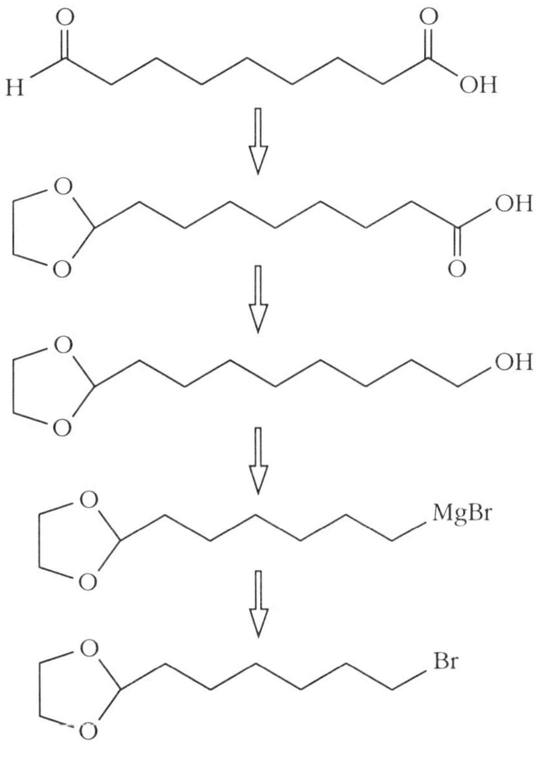

Synthesis: The starting material has a halide and protected carbonyl group. The product contains two more carbon atoms, a carboxyl group, and a free carbonyl group. The synthesis strategy should involve preparation of a Grignard reagent while the carbonyl group is still protected, reaction of the Grignard reagent with oxirane to elongate the carbon chain, oxidation of the hydroxyl group in *basic* conditions (in acidic medium the protecting group would be removed and a dioic acid would form), and a work-up in acidic conditions during which the carbonyl group will be finally unprotected.

[Reaction scheme showing synthesis starting from a bromoalkyl-dioxolane]

Mg, Dry ether → Grignard reagent (MgBr-alkyl-dioxolane)

(1) ethylene oxide
(2) H₃O⁺
→ alcohol intermediate (dioxolane-alkyl-OH)

KMnO₄ | KOH
→ carboxylate (dioxolane-alkyl-CO₂⁻ K⁺)

H₃O⁺
→ aldehyde-acid (OHC-(CH₂)ₙ-COOH)

(b) Retrosynthesis:

$$CH_3C{\equiv}CC(=O)H \Longrightarrow CH_3C{\equiv}CCH_2OH \Longrightarrow CH_3C{\equiv}CC(=O)OCH_3$$

Synthesis: The starting material contains a carbon-carbon triple bond and an ester group, while the product contains a carbon-carbon triple bond and an aldehyde group. The position of the triple bond in the molecule remains unchanged and the ester group is replaced by the aldehyde group. Thus the sequence of operations should involve the reduction of the ester with lithium aluminum hydride and the selective oxidation of the resulting alcohol with PCC, PDC, or Collin's reagent.

$$CH_3C{\equiv}CC(=O)OCH_3 \xrightarrow[(2)\ H_3O^+]{(1)\ LiAlH_4} CH_3C{\equiv}CCH_2OH \xrightarrow[CH_2Cl_2]{PCC} CH_3C{\equiv}CC(=O)H$$

(c) Retrosynthesis:

[Scheme: hexyl-CH(OH)-C≡C-C(=O)OCH₃ ⟹ Na⁺ ⁻C≡C-C(=O)OCH₃ ⟹ HC≡C-C(=O)OCH₃ ⟹ C≡C-C(=O)OCH₃]

Synthesis: The starting material contains a methyl ester group and a terminal carbon-carbon triple bond. The target molecule contains the same two functional groups *plus* a hydroxyl group and a seven-carbon alkane chain. The obvious extension of the carbon chain can be achieved through the formation of an acetylide ion from the terminal alkyne group and its reaction with heptanal.

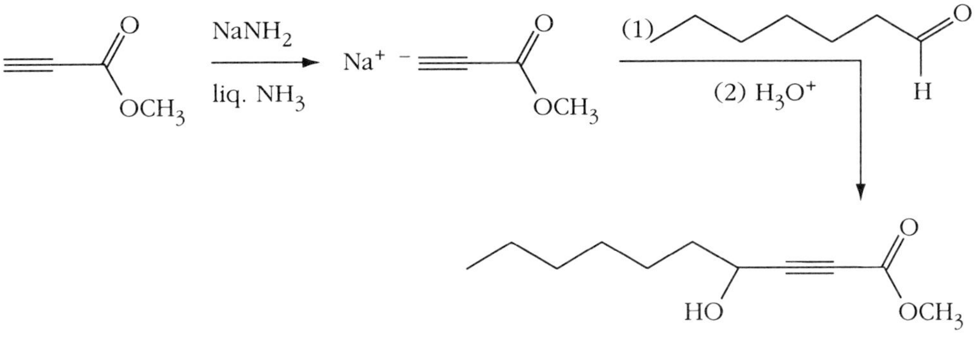

(d) The starting material contains a carbon-carbon double bond and an ester group while the product contains the same carbon-carbon double bond in the same position as well as a —CD$_2$OH group in the place of the ester. It suggests a reduction of the ester with a deuterated lithium aluminum hydride.

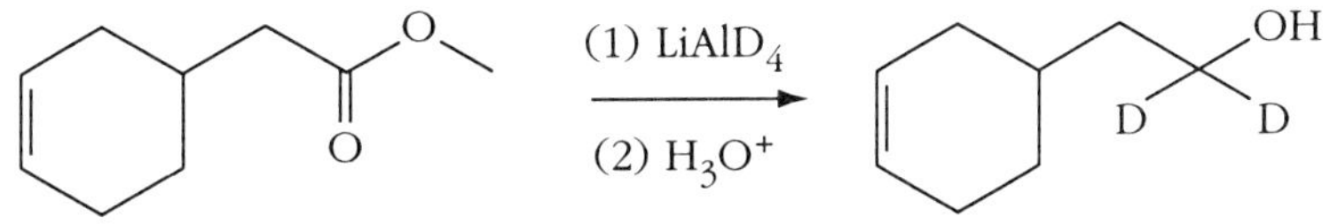

(e) Retrosynthesis:

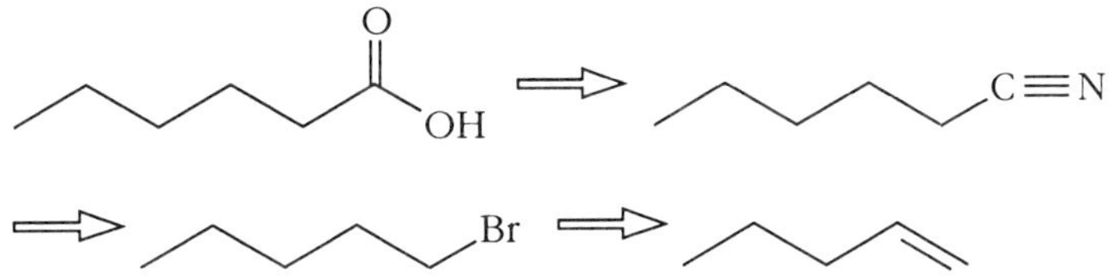

Synthesis: The starting material is a terminal alkene and the product is a carboxylic acid with one more carbon atom. A simple synthetic route would involve preparation of a terminal haloalkane by free-radical addition of HBr followed by substitution of the terminal halogen with a nitrile group and hydrolysis of the nitrile.

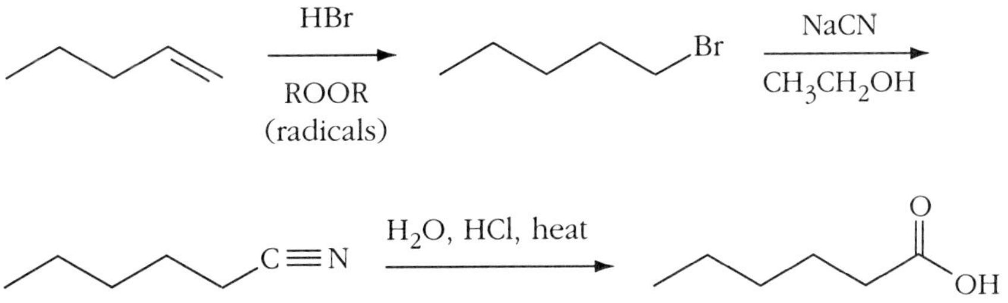

(f) The starting material contains an aldehyde group, a carbon-carbon double bond, and an ester group. The product contains the carbon-carbon double bond (its position is unchanged), a carboxyl group in place of the aldehyde, and a hydroxyl group and two methyl groups in place of the ester group. It is clear that the aldehyde group can be easily oxidized to the carboxyl group and an excess of methylmagnesium halide would react with the ester group to give the expected part of the molecule. However, the Grignard reagent cannot be used in the presence of both an aldehyde group and a carboxyl group. As a result, the carbonyl group must be protected first, then the Grignard reagent used, and finally the aldehyde group must be oxidized. Note: Dehydration of the tertiary alcohol will take place when a strong acid or base is used during the oxidation reaction.

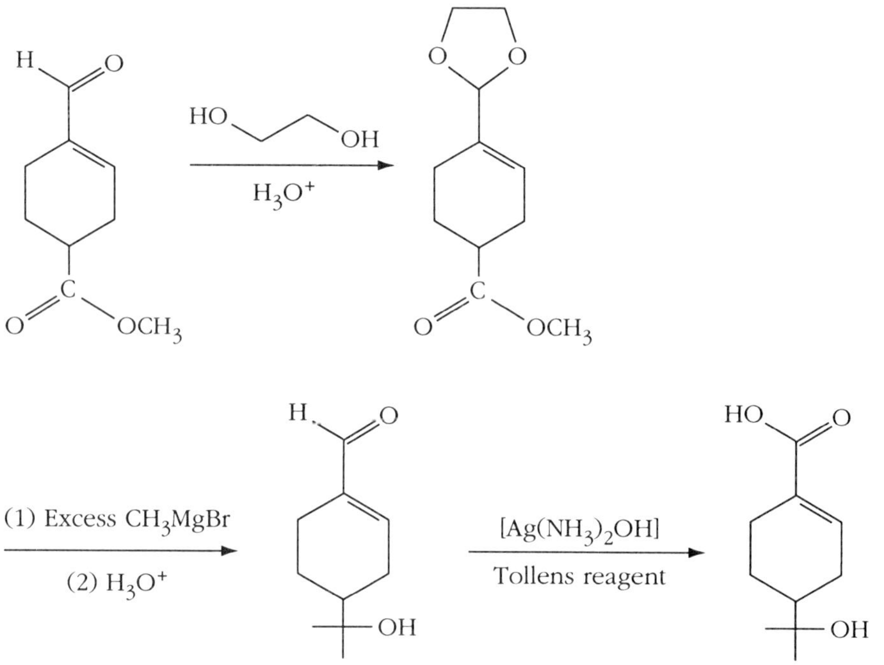

(g) Retrosynthesis:

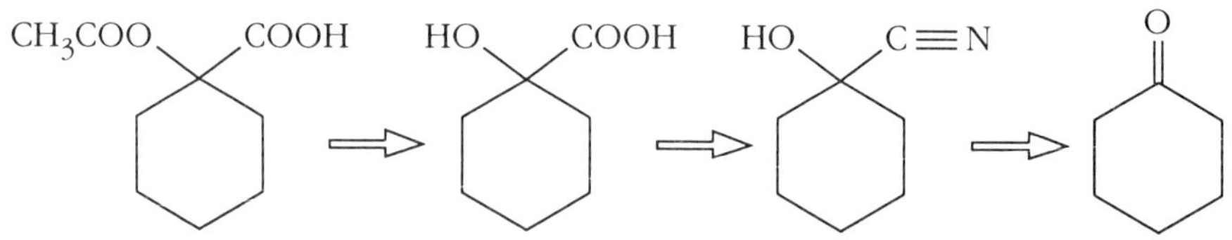

Synthesis: The starting material is cyclohexanone and the product contains a carboxyl group and an ester group in the place of the carbonyl group. The reaction sequence involves formation of the cyanohydrin, hydrolysis of the nitrile group to the carboxyl group, and esterification of the hydroxyl group. Note: (1) Usually hydrolysis of nitriles is carried out in acidic solution. In this case, however, acidification and an elevated temperature would dehydrate the tertiary alcohol, so basic hydrolysis must be used instead. (2) The final step, esterification, requires some care because the ester is formed from a compound that contains both the carboxyl group and a hydroxyl group. Fisher esterification (catalyzed by a mineral acid) would give a mixture of products, including the product resulting from an intermolecular condensation. This problem is overcome by the use of a reactive carboxyl derivative that reacts selectively with the hydroxyl group.

16.43 Methyl acetate hydrolysis by a nucleophilic acyl substitution mechanism:

The hydrolysis of *tert*-butyl acetate occurs according to the following mechanism, which involves formation of a *tert*-butyl cation:

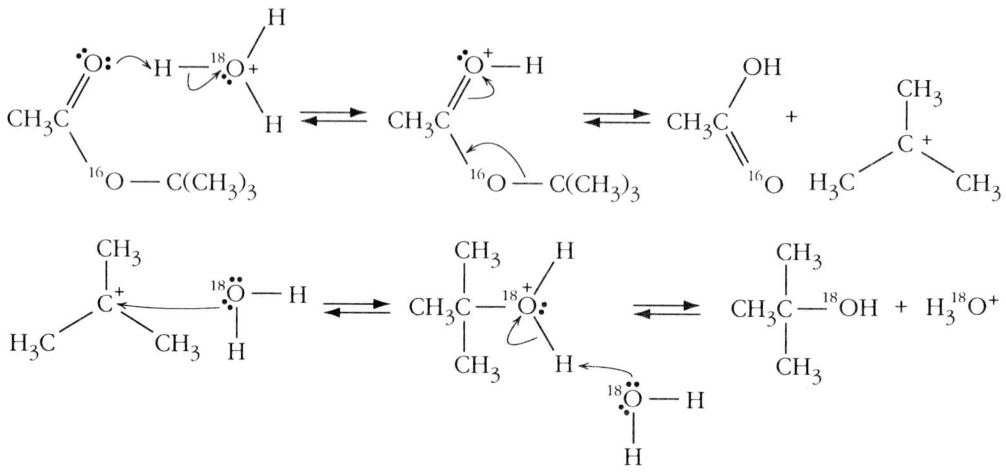

[Mechanism scheme showing protonation of tert-butyl acetate by $H_3{}^{18}O^+$, formation of tert-butyl cation and acetic acid, then attack of $H_2{}^{18}O$ on the cation and deprotonation to give tert-butanol with ^{18}O label and $H_3{}^{18}O^+$]

16.44 (a) The formula, $C_6H_{10}O_3$, indicates two units of unsaturation. The IR spectrum shows two absorption bands at 1820 cm^{-1} and 1750 cm^{-1}. The fact that there are two absorptions in the carbonyl region is unique for an anhydride functional group.

The ^{1}H NMR spectrum shows only two signals:

δ 1.19 (t, 3H)—a methyl group coupled with two hydrogens,
δ 2.49 (quartet, 2H)—a methylene group bonded to a carbonyl group and a methyl group.

The triplet-quartet pattern is a characteristic feature of an ethyl group. There must be two ethyl group to account for ten hydrogen atoms in the absence of other signals. Thus, the unknown is **propanoic anhydride:**

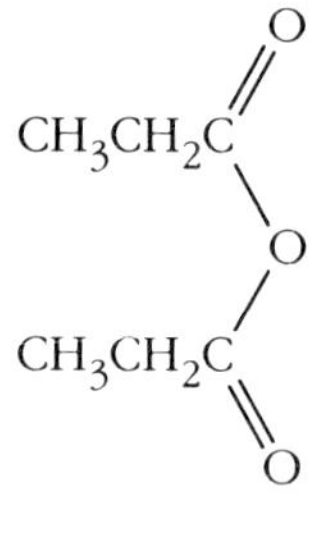

$$\text{CH}_3\text{CH}_2\text{C}(=O)\text{—O—C}(=O)\text{CH}_2\text{CH}_3$$

(b) The formula, C_3H_7NO, indicates one unit of unsaturation. The IR spectrum shows the presence of a carbonyl group. The position of the absorption, 1680 cm^{-1}, suggests an amide. The hypothesis of an amide group is supported by the ^{13}C NMR spectrum. The signal at δ 162.4 is characteristic of a carboxyl group derivative.

The ^{1}H NMR spectrum is the most informative. It shows the following three signals:

δ 2.88 (s, 3H)—a methyl group bonded to a nitrogen,
δ 2.98 (s, 3H)—a methyl group bonded to a nitrogen,
δ 8.0 (s, 1H)—a hydrogen bonded to a carboxylic carbon.

The unknown is *N,N*-dimethylformamide:

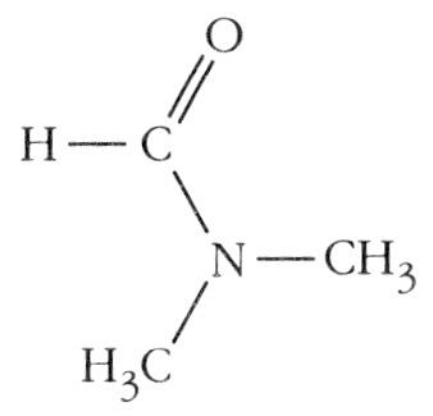

$$\text{H–C(=O)–N(CH}_3\text{)}_2$$

Please note that we observe two separate signals for the two methyl groups. This is due to the fact that there is a partial double bond character between the carbon atom of the carbonyl group and the nitrogen atom, which prevents free rotation. Because there is not free rotation, the two methyl groups are in different environments and thus have different chemical shifts.

(c) The formula, $C_5H_6Cl_2O_2$, indicates two units of unsaturation. The IR spectrum shows an absorption at 1795 cm^{-1}, which indicates a carbonyl group (most likely an acid chloride). The ^{13}C NMR shows only three signals, which demonstrates symmetry in the molecule. The signal at δ 173.1 is characteristic of a carboxyl group derivative.

The ^{1}H NMR spectrum shows only two signals:

δ 2.08 (quintet, 2H)—a methylene group seeing four neighboring hydrogen atoms,
δ 3.03 (triplet, 4H)—two equivalent methylene groups bonded to an acyl chloride group and the internal methylene group.

The unknown is **1,5-pentanedioyl chloride**:

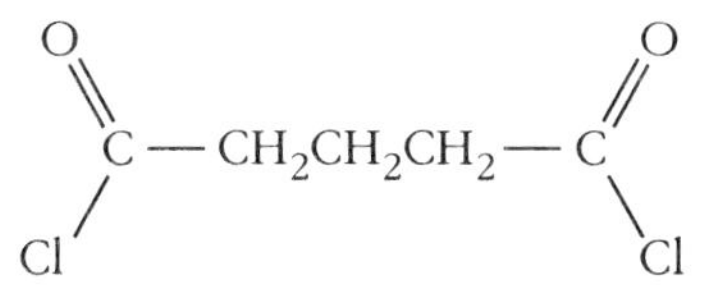

$$\text{Cl–C(=O)–CH}_2\text{CH}_2\text{CH}_2\text{–C(=O)–Cl}$$

(d) The formula, $C_5H_9ClO_2$, indicates one unit of unsaturation. The IR spectrum shows the presence of a carbonyl group. The position of the absorption, 1735 cm^{-1}, suggests an ester.

The ^{1}H NMR spectrum shows the following four signals:

δ 1.21 (t, 3H)—a methyl group bonded to a methylene group,
δ 2.82 (t, 2H)—a methylene group bonded to a carbonyl carbon,
δ 3.80 (t, 2H)—a methylene group bonded to a halide (chlorine),
δ 4.20 (q, 2H)—a methylene group bonded to an oxygen atom and coupled with a methyl group.

We can group some of the fragments. For instance, the triplet-quartet pattern is characteristic of an ethyl group attached to the oxygen of the ester group. The two remaining triplets indicate two nonequivalent methylene groups bonded together, with one attached to the carbonyl carbon and the other attached to the chlorine atom.

The unknown is **ethyl 3-chloropropanoate**:

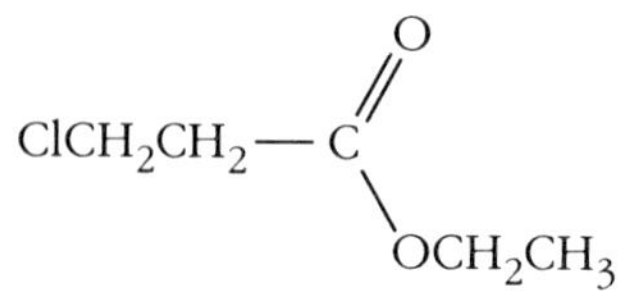

$$\text{ClCH}_2\text{CH}_2-\overset{\overset{\displaystyle O}{\|}}{\text{C}}-\text{OCH}_2\text{CH}_3$$

16.45 (a) The formula, $C_5H_{10}O_2$, indicates one unit of unsaturation. The IR spectrum shows the presence of a carbonyl group. The position of the absorption, around 1730 cm^{-1}, suggests an ester.

The ^{1}H NMR spectrum shows the following three signals:

δ 1.15 (d, 6H)—two methyl groups bonded to a carbon with one hydrogen,
δ 1.95 (s, 3H)—a methyl group bonded to a carbonyl carbon,
δ 4.9 (septet, 1H)—a hydrogen attached to a carbon bonded to an oxygen and two methyl groups.

The septet-doublet pattern is very characteristic of an isopropyl group (methylethyl group). Thus, the unknown is **isopropyl acetate** (IUPAC name: 1-methylethyl ethanoate):

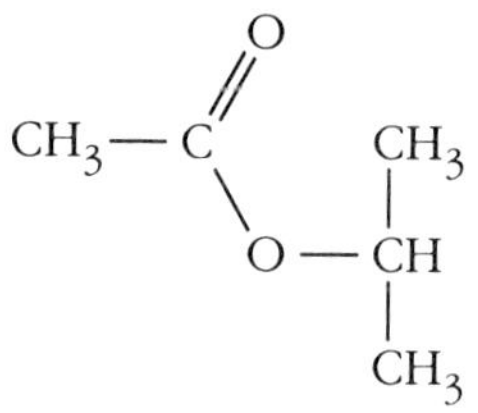

$$\text{CH}_3-\overset{\overset{\displaystyle O}{\|}}{\text{C}}-\text{O}-\underset{\underset{\displaystyle \text{CH}_3}{|}}{\overset{\overset{\displaystyle \text{CH}_3}{|}}{\text{CH}}}$$

(b) The formula, $C_6H_{13}NO$, indicates one unit of unsaturation. The IR spectrum shows the presence of a carbonyl group. The position of the absorption, around 1650 cm^{-1}, suggests an amide.

The ^{1}H NMR spectrum shows the following signals:

δ 0.98 (t, 3H)—a methyl group bonded to a methylene group,
δ 1.05 (t, 3H)—a methyl group bonded to a methylene group,
δ 2.0 (s, 3H)—a methyl group bonded to a carbonyl carbon,
δ 3.18 (q, 2H)—a methylene group bonded to nitrogen and a methyl group,
δ 3.23 (q, 2H)—a methylene group bonded to nitrogen and a methyl group.

After assigning the two quartet-triplet coupling patterns to two ethyl groups, we can identify the unknown as ***N,N*-diethylacetamide**:

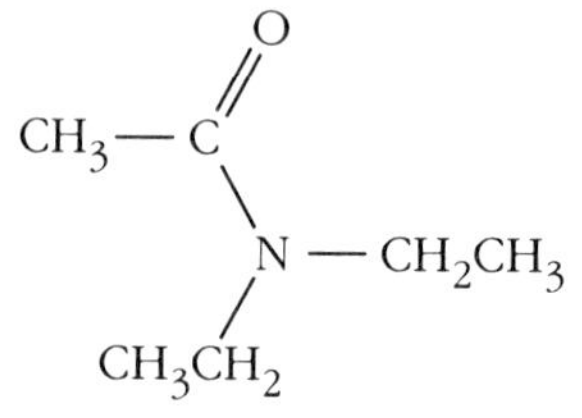

$$\text{CH}_3-\overset{\overset{\displaystyle O}{\|}}{\text{C}}-\underset{\underset{\displaystyle \text{CH}_3\text{CH}_2}{/}}{\text{N}}-\text{CH}_2\text{CH}_3$$

Please note that we observe two separate signals for the two methyl groups. This is due to the fact that there is partial double bond character between the carbon atom of the carbonyl group and the nitrogen atom, which prevents free rotation. Because there is not free rotation, the two methyl groups are in different environments and thus have two different chemical shifts.

(c) The formula, C_4H_6BrN, indicates two units of unsaturation (a combination of a ring, carbon-carbon double bond, a carbon-nitrogen double bond, or a carbon-nitrogen triple bond). The IR spectrum shows a sharp absorption band at 2220 cm^{-1} that is characteristic of a carbon-nitrogen triple bond.

The ^{1}H NMR spectrum shows three peaks:

δ 2.22 (quintet, 2H)—a methylene group bonded to two other methylene groups,
δ 2.60 (triplet, 2H)—a methylene group bonded to a nitrile group and a methylene group,
δ 3.53 (triplet, 2H)—a methylene group bonded to a bromine atom and a methylene group.

Thus, the unknown is **4-bromobutanenitrile:**

$$BrCH_2CH_2CH_2C\equiv N$$

(d) The formula, $C_4H_6O_2$, indicates two units of unsaturation. The IR spectrum shows a strong absorption band at 1760 cm^{-1}, which indicates the presence of a carbonyl group (an ester?).

The ^{1}H NMR spectrum shows three peaks:

δ 2.33 (quintet, 2H)—a methylene group bonded to two other methylene groups,
δ 2.55 (triplet, 2H)—a methylene group bonded to another methylene group and a carbonyl group,
δ 4.4 (triplet, 2H)—a methylene group bonded to a methylene group and oxygen.

We get the following fragment:

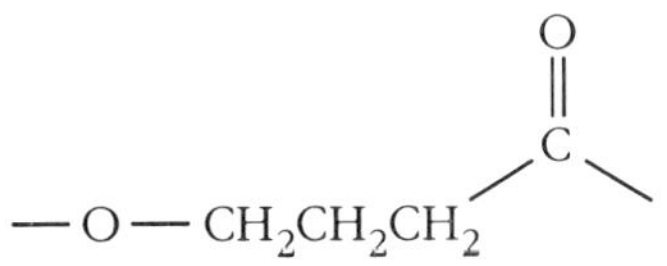

$$-O-CH_2CH_2CH_2-C(=O)-$$

This fragment accounts for all atoms so we join the ends to get the structure of the unknown:

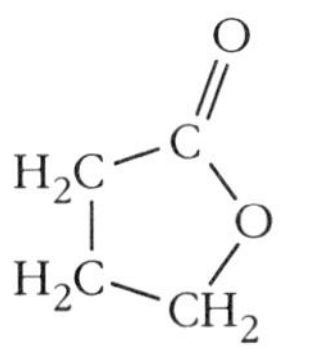

Lactone of 4-hydroxybutanoic acid

CHAPTER 17

ENOLS AND ENOLATE ANIONS
Concepts

Carbonyl compounds that have hydrogen atoms bonded to their α carbon atoms rapidly interconvert with their corresponding **enols** (*ene* + *ol*, unsaturated alcohol). This rapid interconversion between the chemically distinct species is a special kind of isomerism known as **tautomerism**.

The process of converting a carbonyl compound into its enol form is called **enolization**. There are two mechanisms, namely **base-catalyzed enolization** and **acid-catalyzed enolization**.

(a) Base-catalyzed enolization:

Keto form   Enolate anion   Enol form

(b) Acid-catalyzed enolization:

Keto form   Enol form

Most carbonyl compounds exist almost exclusively in the keto form at equilibrium. Equilibrium constants for enolate formation from simple aldehydes and ketones are in the 10^{-16} to 10^{-20} range. When alkali metal hydroxides or alkoxides are used as bases, the position of equilibrium still favors reactants rather than products.

$$\text{H}_3\text{C}-\text{CO}-\text{CH}_3 + \text{NaOH} \rightleftharpoons \text{H}_3\text{C}-\text{C}(\text{O}^-\text{Na}^+)=\text{CH}_2 + \text{H}_2\text{O} \qquad K_{eq} = 4 \times 10^{-5}$$

pK_a 19 (weaker acid)   pK_a 15.7

$$\underset{\substack{\text{p}K_a\ 17 \\ \text{(weaker acid)}}}{\overset{\overset{\displaystyle O}{\|}}{H-C-CH_3}} + CH_3O^- Na^+ \rightleftharpoons \underset{\text{p}K_a\ 15.5}{\overset{\overset{\displaystyle O^- Na^+}{|}}{H-C=CH_2}} + CH_3OH \quad K_{eq} = 10^{-3}$$

A very strong base, such as **lithium diisopropylamide (LDA),** is needed to drive the equilibrium of formation of enolate anions to the right.

$$\underset{\substack{\text{p}K_a\ 22}}{\overset{\overset{\displaystyle O}{\|}}{H_3C-C-OCH_3}} + [(CH_3)_2CH]_2N^-Li^+ \longrightarrow \underset{\substack{\text{p}K_a\ 38 \\ \text{(weaker acid)}}}{\overset{\overset{\displaystyle O^-Li^+}{|}}{H_2C=C-OCH_3}} + [(CH_3)_2CH]_2NH \quad K_{eq} = 10^{14}$$

The formation of enolate anions is very highly **regioselective** for a carbonyl compound with two sets of α-hydrogens. Under the condition of **kinetic control** (low temperature, –78 °C, and a slight excess of base), the less substituted lithium enolate is formed quantitatively and there is no equilibrium established.

(2-methylcyclohexanone) + [(CH$_3$)$_2$CH]$_2$N$^-$Li$^+$ (LDA) ⟶ (enolate toward substituted side) 1% + (enolate toward less substituted side) 99% + [(CH$_3$)$_2$CH]$_2$NH

Under conditions of **thermodynamic control** (room temperature and a slight excess of the carbonyl compound), the more stable (more highly substituted) enolate predominates.

(2-methylcyclohexanone) + [(CH$_3$)$_2$CH]$_2$N$^-$Li$^+$ (LDA) ⟶ (more substituted enolate) 72% + (less substituted enolate) 28% + [(CH$_3$)$_2$CH]$_2$NH

Enamines are compounds in which a trivalent nitrogen atom is bonded to an sp^2 hybridized carbon atom of the carbon-carbon double bond. Enamines are nitrogen analogs of enols and they are synthesized from a secondary amine and an aldehyde or a ketone.

$$\underset{}{\overset{\overset{\displaystyle O}{\|}}{H_3C-C-H}} + \text{(pyrrolidine, N-H)} \xrightarrow{\text{Acid catalyst}} \text{(pyrrolidine-N-CH=CH}_2\text{)}$$

Solutions to the Exercises

17.1

Keto tautomer		Enol tautomer
CH_3CHO	⇌	$CH_2=CHOH$
$(CH_3)_2CHCHO$	⇌	$(CH_3)_2C=CHOH$
CH_3COCH_3	⇌	$H_3C-C(OH)=CH_2$
cyclohexanone	⇌	cyclohexenol

17.2 When a carbonyl compound (aldehyde and ketone) reacts with one equivalent of halogen in acidic solution usually only one α-hydrogen atom is substituted by a halogen atom:

(a) $CH_3COCH_3 + Br_2 \xrightarrow{CH_3COOH} CH_3COCH_2Br$

(b) cyclopentanone $+ Br_2 \xrightarrow{CH_3COOH}$ 2-bromocyclopentanone

(c) $CH_3CH_2COCH_2CH_3 + Br_2 \xrightarrow{CH_3COOH} CH_3CH_2COCHBrCH_3$

(d) $CH_3CHO + Br_2 \xrightarrow{CH_3COOH} BrCH_2CHO$

(e)

$(CH_3)_2CHC(=O)H + Br_2 \xrightarrow{CH_3COOH} (CH_3)_2CBrC(=O)H$

17.3 The halogenation of ketones occurs according to a three-step mechanism.
Step 1 is an acid-catalyzed enolization of the ketone;
Step 2 is an electrophilic addition of the halide to enol; and
Step 3 is deprotonation of the cationic intermediate.

Step 1 is the rate-determining step and the halogen does not participate in the reaction until after the enolization of the ketone. Thus the overall reaction rate does *not* depend on the type of halogen and that is why the rates of chlorination, bromination, and iodination of (R)-(+)-sec-butyl phenyl ketone are the same within experimental error.

17.4 The enolization of (S)-2,2,4-trimethyl-3-heptanone forms an achiral enol, which can form either the (R) or (S) enantiomer. As a result, racemization occurs.

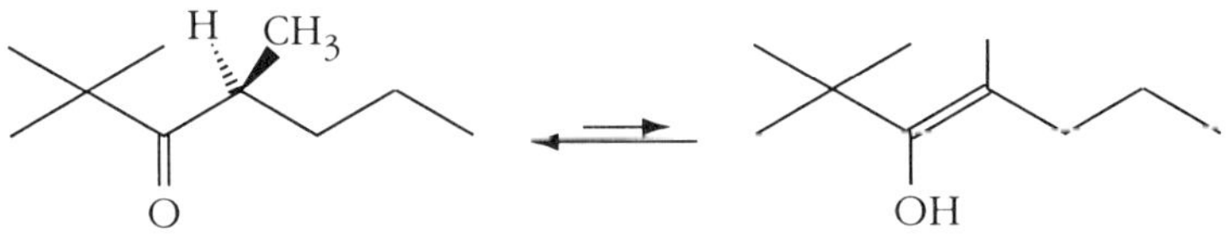

(S)-2,2,4-Trimethyl-3-heptanone ⇌ Enol

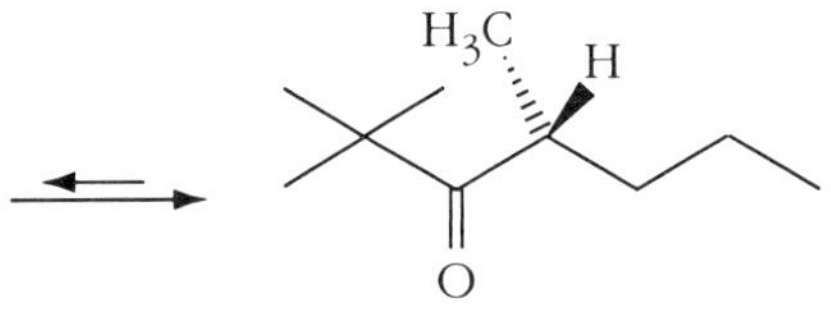

⇌ (R)-2,2,4-Trimethyl-3-heptanone

The enolization of (S)-2,2,5-trimethyl-3-heptanone does not affect the stereocenter in the molecule so enolization does *not* affect the optical activity of the compound.

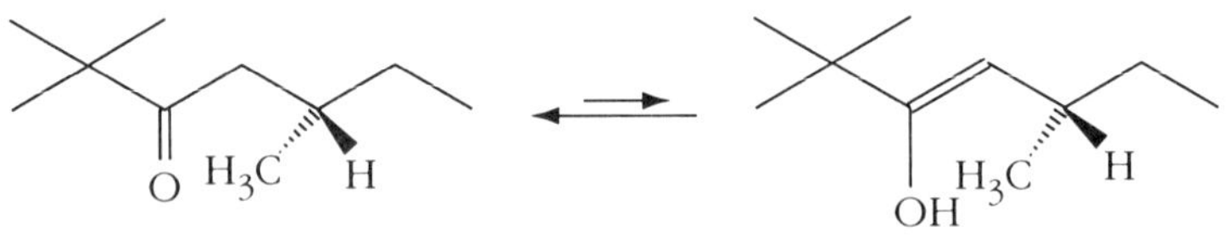

(S)-2,2,5-Trimethyl-3-heptanone ⇌ Optically active enol

17.5 (a)

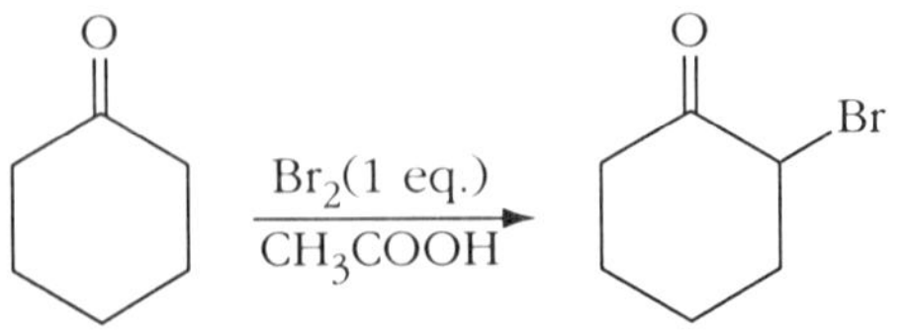

cyclohexanone $\xrightarrow{Br_2 \text{ (1 eq.)}}{CH_3COOH}$ 2-bromocyclohexanone

CHAPTER 17   483

(b)

[Structure: 3-pentanone] $\xrightarrow{\text{Cl}_2 (2 \text{ eq.})}_{\text{HCl/H}_2\text{O}}$ [Structure: 2,4-dichloro-3-pentanone]

(c)

[Structure: 2-methylbutanoic acid] $\xrightarrow{\text{Br}_2 (1 \text{ eq.})}_{\text{PBr}_3}$ [Structure: 2-bromo-2-methylbutanoic acid]

17.6 (a)

$$\text{ClCH}_2\text{COOH} \; + \; \text{NaOH/H}_2\text{O} \; \longrightarrow \; \text{HOCH}_2\text{COO}^-\text{Na}^+$$

(b)

[Structure: phenylacetic acid] $\xrightarrow{\text{Br}_2 (1 \text{ eq.})}_{\text{PBr}_3}$ [Structure: α-bromophenylacetic acid]

(c)

[Structure: 2-chlorocyclopentanone] $+ \; \text{S(CH}_3)_2 \; \longrightarrow$ [Structure: 2-(dimethylsulfonio)cyclopentanone chloride]

17.7 (a)

[Structure: acetone] $+ \; \text{NaNH}_2 \; \underset{\text{NH}_3}{\rightleftharpoons}$ [Structure: sodium enolate of acetone] $+ \; \text{NH}_3$

$\text{p}K_a \; 19.3 \hspace{4cm} \text{p}K_a \approx 38$

$$K = \frac{K_{a \text{ (acetone)}}}{K_{a \text{ (ammonia)}}} = \frac{5 \times 10^{-20}}{1 \times 10^{-38}} \approx 5 \times 10^{18}$$

(b)

[Structure: acetone] $+ \; \text{NaH} \; \underset{\text{THF}}{\rightleftharpoons}$ [Structure: sodium enolate of acetone] $+ \; \text{H}_2$

$\text{p}K_a \; 19.3 \hspace{4cm} \text{p}K_a \approx 35$

$$K = \frac{K_{a \text{ (acetone)}}}{K_{a \text{ (hydrogen)}}} = \frac{5 \times 10^{-20}}{1 \times 10^{-35}} \approx 5 \times 10^{15}$$

(c)

$$\underset{\text{p}K_a\ 19.3}{\underset{H_3C\diagdown\diagup CH_3}{C=O}} + NH_3 \underset{}{\overset{H_2O}{\rightleftarrows}} \underset{\text{p}K_a\ 9.2}{\underset{H_3C\diagdown\diagup CH_2}{C=O^-}} + \overset{+}{N}H_4$$

$$K = \frac{K_{a\ (\text{acetone})}}{K_{a\ (\text{ammonium ion})}} = \frac{5 \times 10^{-20}}{6.3 \times 10^{-10}} \approx 7.9 \times 10^{-11}$$

(d)

$$\underset{\text{p}K_a\ 19.3}{\underset{H_3C\diagdown\diagup CH_3}{C=O}} + CH_3O^-Na^+ \underset{}{\overset{CH_3OH}{\rightleftarrows}} \underset{\text{p}K_a\ 15.2}{\underset{H_3C\diagdown\diagup CH_2}{C=O^-Na^+}} + CH_3OH$$

$$K = \frac{K_{a\ (\text{acetone})}}{K_{a\ (\text{methanol})}} = \frac{5 \times 10^{-20}}{6.3 \times 10^{-16}} \approx 7.9 \times 10^{-5}$$

17.8 (a)

$$^-\ddot{C}H_2-\underset{OCH_3}{\overset{\overset{\curvearrowleft \ddot{O}:}{\|}}{C}} \longleftrightarrow CH_2=\underset{OCH_3}{\overset{\ddot{O}:^-}{C}}$$

(b)

$$^-\ddot{C}H_2-\underset{H}{\overset{\overset{\curvearrowleft \ddot{O}:}{\|}}{C}} \longleftrightarrow CH_2=\underset{H}{\overset{\ddot{O}:^-}{C}}$$

(c)

$$CH_2=\underset{H}{\overset{\overset{\curvearrowleft \ddot{O}:^-}{}}{C}} \longleftrightarrow {}^-\ddot{C}H_2-\underset{H}{\overset{\ddot{O}:}{C}}$$

(d)

$$^-\ddot{C}H_2-CH=CH_2 \longleftrightarrow CH_2=CH-\ddot{C}H_2^-$$

(e)

$$^-\ddot{C}H_2-C\equiv N \longleftrightarrow CH_2=C=\ddot{N}:^-$$

17.9 Aldol condensation reactions are reactions in which the enolate ion of one carbonyl compound reacts with the carbonyl group of another one to form a larger molecule from two smaller ones:

(a)

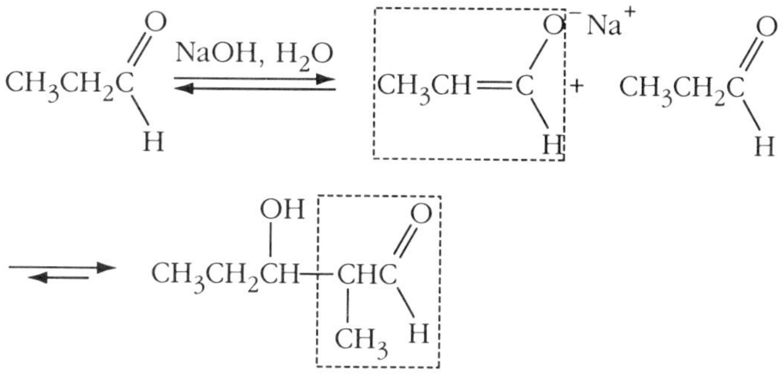

$$CH_3CH_2\overset{O}{\underset{H}{C}} \xrightleftharpoons[]{NaOH, H_2O} \left[CH_3CH=\overset{O^-Na^+}{\underset{H}{C}}\right] + CH_3CH_2\overset{O}{\underset{H}{C}}$$

$$\rightleftharpoons CH_3CH_2\underset{}{\overset{OH}{\underset{}{CH}}}\!\!-\!\!\left[\underset{CH_3}{\overset{}{CH}}\overset{O}{\underset{H}{C}}\right]$$

(b) In condensation reactions involving aromatic compounds, the intermediate loses water with great ease because the product of that dehydration has a double bond in conjugation not only with the carbonyl group but with the aromatic ring as well.

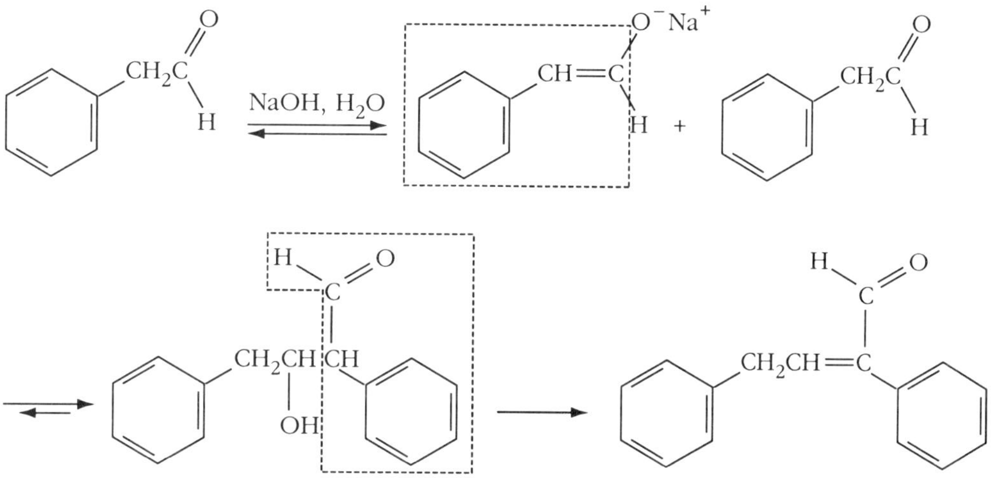

(c)

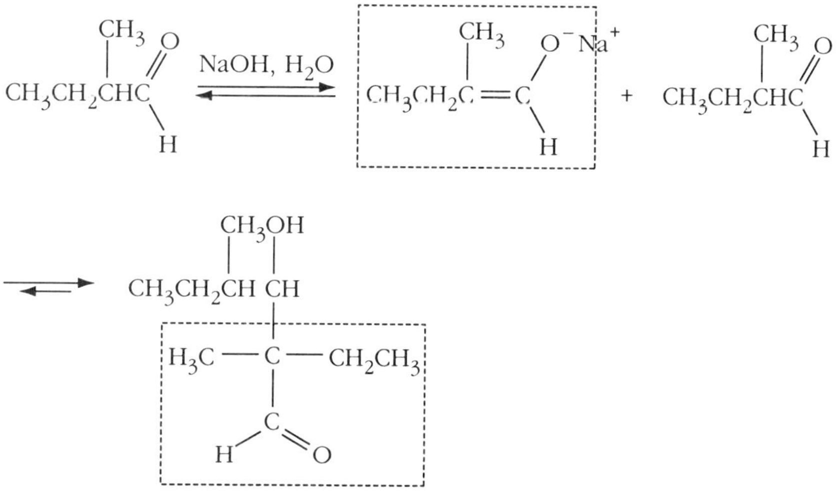

17.10

(a)

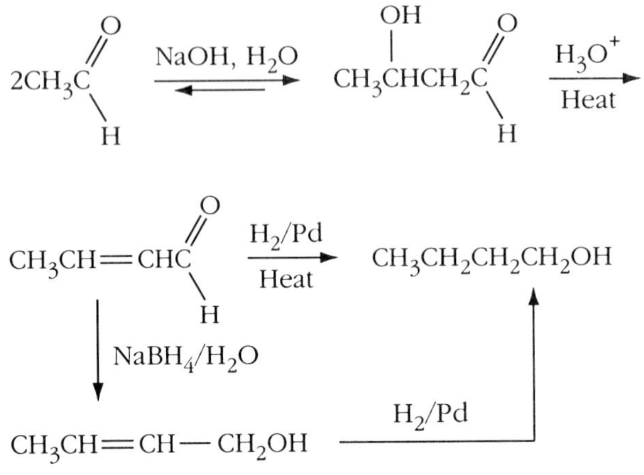

$$2CH_3C(=O)H \xrightleftharpoons[]{NaOH, H_2O} CH_3CH(OH)CH_2C(=O)H \xrightarrow[Heat]{H_3O^+}$$

$$CH_3CH=CHC(=O)H \xrightarrow[Heat]{H_2/Pd} CH_3CH_2CH_2CH_2OH$$

$$\downarrow NaBH_4/H_2O$$

$$CH_3CH=CH-CH_2OH \xrightarrow{H_2/Pd}$$

(b)

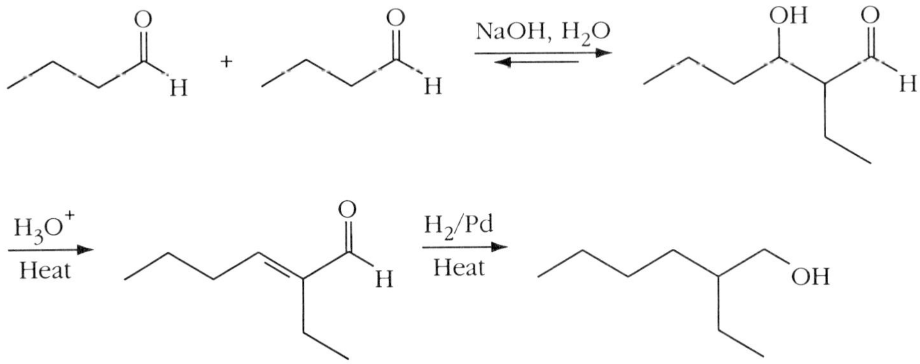

Butanal + butanal $\xrightleftharpoons[]{NaOH, H_2O}$ 2-ethyl-3-hydroxyhexanal

$\xrightarrow[Heat]{H_3O^+}$ 2-ethyl-2-hexenal $\xrightarrow[Heat]{H_2/Pd}$ 2-ethyl-1-hexanol

(c)

Acetophenone + acetophenone $\xrightleftharpoons[]{NaOH, H_2O}$ β-hydroxy ketone

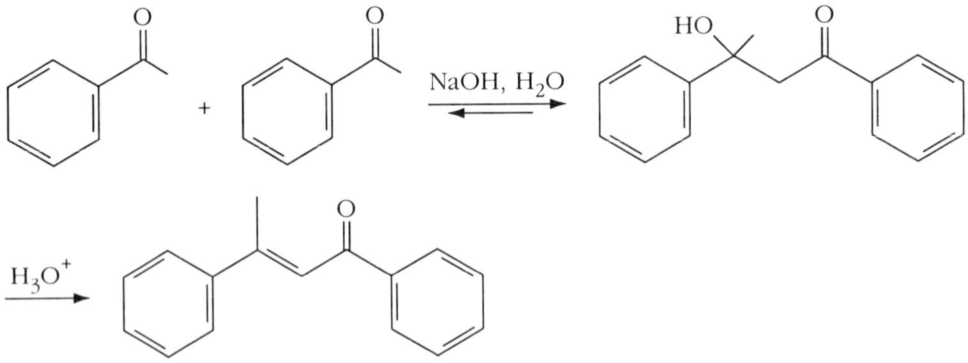

$\xrightarrow{H_3O^+}$ α,β-unsaturated ketone

17.11 (a)

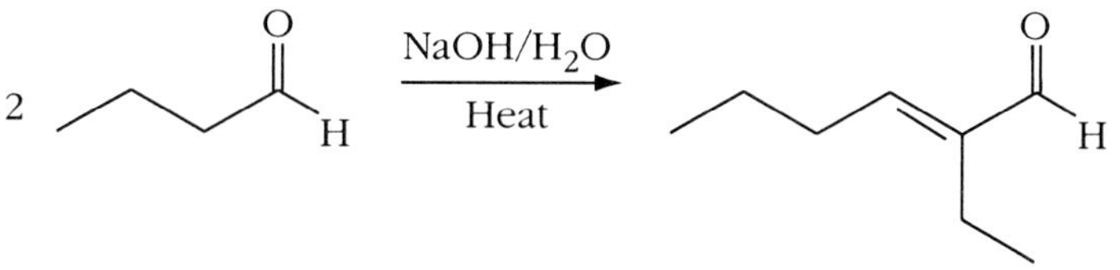

$$2 \; CH_3CH_2CH_2CHO \xrightarrow[\text{Heat}]{\text{NaOH/H}_2\text{O}} CH_3CH_2CH_2CH=C(CH_2CH_3)CHO$$

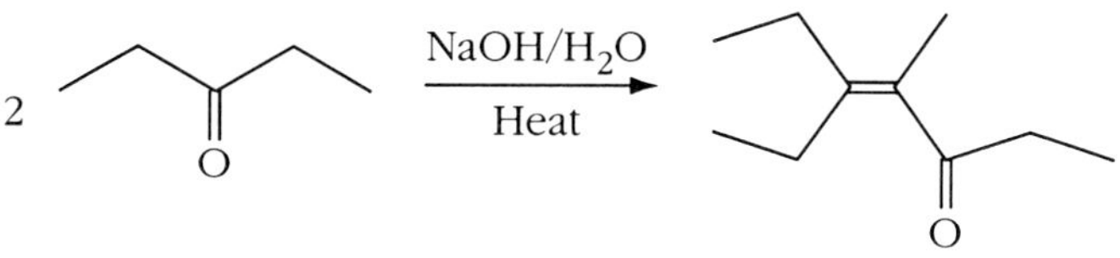

(b)

$$2 \; CH_3CH_2COCH_2CH_3 \xrightarrow[\text{Heat}]{\text{NaOH/H}_2\text{O}} \text{(ethyl-substituted enone product)}$$

(c)

$$2 \; C_6H_5COCH_3 \xrightarrow[\text{Heat}]{\text{NaOH/H}_2\text{O}} C_6H_5C(CH_3)=CHCOC_6H_5$$

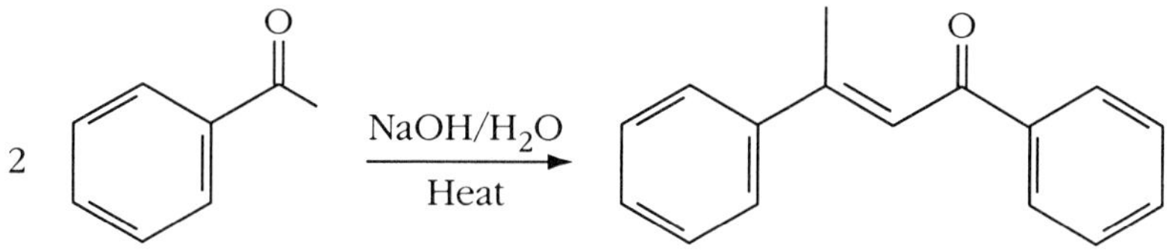

(d)

$$2 \; \text{cyclohexanone} \xrightarrow[\text{Heat}]{\text{NaOH/H}_2\text{O}} \text{2-cyclohexylidenecyclohexanone}$$

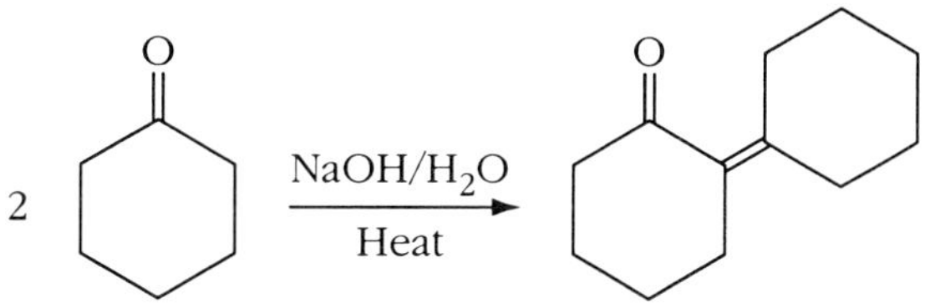

17.12 When the enolate anion from an ester reacts with the carbonyl group of another ester, the reaction is known as the **Claisen condensation.** The alkoxy group on the ester serves as a leaving group from the tetrahedral intermediate formed when the carbonyl group is attacked by an enolate ion (see Exercise 17.13).

(a)

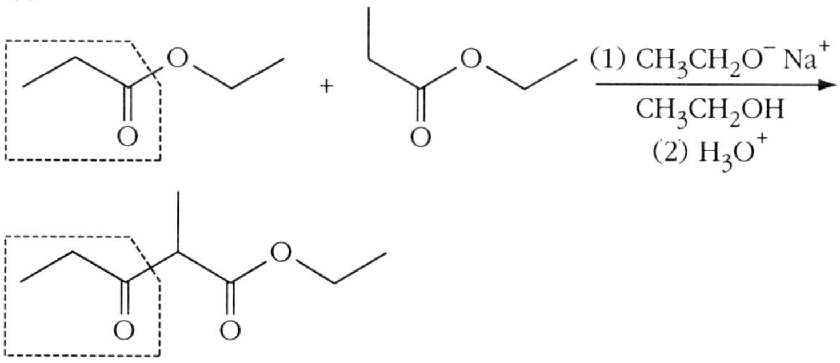

$$CH_3CH_2CO_2CH_2CH_3 + CH_3CH_2CO_2CH_2CH_3 \xrightarrow[\text{CH}_3\text{CH}_2\text{OH}]{\text{(1) CH}_3\text{CH}_2\text{O}^-\text{Na}^+} \xrightarrow{\text{(2) H}_3\text{O}^+} CH_3CH_2COCH(CH_3)CO_2CH_2CH_3$$

(b)

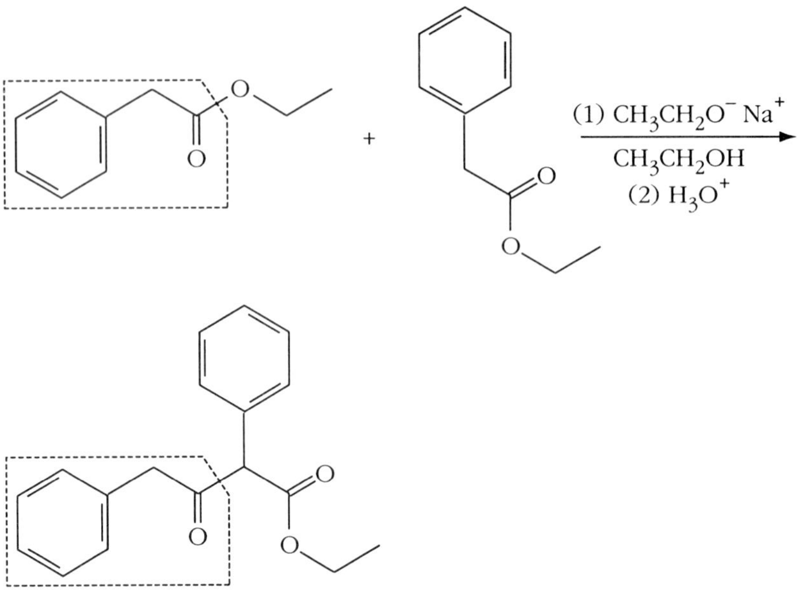

[Reaction: ethyl phenylacetate + ethyl phenylacetate, (1) CH₃CH₂O⁻ Na⁺ / CH₃CH₂OH, (2) H₃O⁺ → product]

(c)

[Reaction: ethyl pentanoate + ethyl pentanoate, (1) CH₃CH₂O⁻ Na⁺ / CH₃CH₂OH, (2) H₃O⁺ → product]

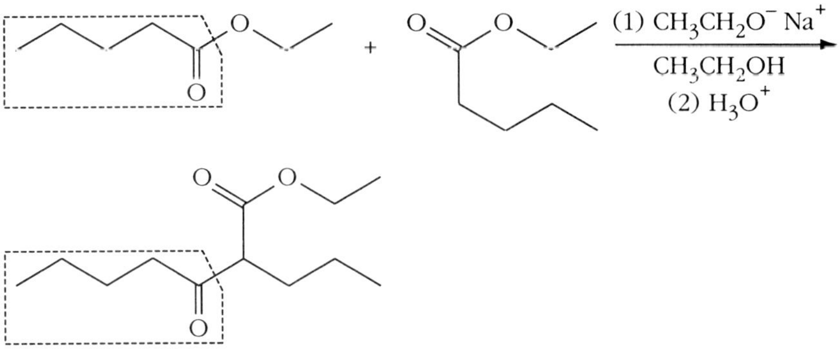

17.13 Step 1: Ethoxide base abstracts an acidic α-hydrogen atom from an ester molecule, yielding an ester enolate ion

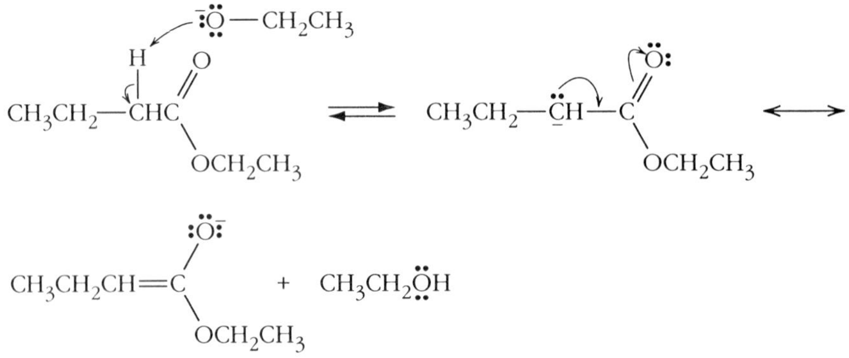

$$CH_3CH_2-CHC(=O)OCH_2CH_3 + {}^-\!:\!\ddot{O}\!-\!CH_2CH_3 \rightleftharpoons CH_3CH_2-\ddot{C}H-C(\ddot{O}:)OCH_2CH_3 \leftrightarrow$$

$$CH_3CH_2CH=C(\ddot{O}:^-)OCH_2CH_3 \;+\; CH_3CH_2\ddot{O}H$$

Step 2: Nucleophilic addition of the ester enolate ion to a second ester molecule forms a tetrahedral intermediate

$$CH_3CH_2\ddot{C}H-C(\ddot{O})(OCH_2CH_3) + CH_3CH_2CH_2-C(\ddot{O})(OCH_2CH_3) \rightleftharpoons CH_3CH_2CH_2-\underset{\underset{OCH_2CH_3}{|}}{\overset{\overset{:\ddot{O}:^-}{|}}{C}}-\underset{CH_3CH_2\overset{\|}{C}HC\overset{O}{\diagdown}}{}$$

Tetrahedral intermediate

Step 3: The tetrahedral intermediate eliminates ethoxide ion to yield the new carbonyl compound, ethyl 2-ethyl-3-oxohexanoate

$$CH_3CH_2H_2C\cdots C(:\ddot{O}:^-)(\ddot{O}CH_2CH_3)(CH_3CH_2\overset{\|}{C}HC-OCH_2CH_3) \rightleftharpoons CH_3CH_2CH_2\overset{O}{\overset{\|}{C}}-\underset{CH_3CH_2}{\overset{|}{C}H}C\overset{O}{\overset{\|}{-}}OCH_2CH_3 + CH_3CH_2\ddot{O}:^-$$

Step 4: The β-keto ester product still contains an acidic hydrogen atom, which is removed by ethoxide ion. The ionization of the β-keto ester shifts the equilibrium and drives the reaction to completion

$$CH_3CH_2CH_2\overset{O}{\overset{\|}{C}}\underset{CH_3CH_2}{\overset{H}{\underset{|}{C}}}\overset{O}{\overset{\|}{C}}-OCH_2CH_3 \quad :\ddot{O}CH_2CH_3 \longrightarrow CH_3CH_2CH_2\overset{O}{\overset{\|}{C}}\underset{CH_2CH_3}{\overset{-}{\underset{|}{C}}}\overset{O}{\overset{\|}{C}}-OCH_2CH_3$$

Step 5: Protonation by addition of an acid yields the condensation product

$$CH_3CH_2CH_2\overset{O}{\overset{\|}{C}}\underset{CH_2CH_3}{\overset{-}{\underset{|}{C}}}\overset{O}{\overset{\|}{C}}-OCH_2CH_3 \quad H\overset{H}{\overset{|}{\underset{|}{\overset{+}{O}}}}H \longrightarrow CH_3CH_2CH_2\overset{O}{\overset{\|}{C}}\underset{CH_2CH_3}{\overset{H}{\underset{|}{C}}}\overset{O}{\overset{\|}{C}}-OCH_2CH_3$$

17.14 (a) Ethyl benzoate lacks an α-hydrogen atom that must be removed in Step 1 of the Claisen condensation reaction (see Exercise 17.13).
(b) Ethyl 2-methylbutanoate is an ester containing one α-hydrogen atom. As a result, Step 1, Step 2, and Step 3 can take place but Step 4, which shifts the equilibrium to products, cannot occur. The equilibrium constants of Steps 1, 2, and 3 favor the reactants so no condensation reaction occurs.

Conclusion: If the starting ester has more than one acidic α-hydrogen, the product β-keto ester has a highly acidic, doubly activated hydrogen atom that can be abstracted by a base. The deprotonation serves to drive the Claisen equilibrium completely to the product side so that high yields are obtained.

17.15

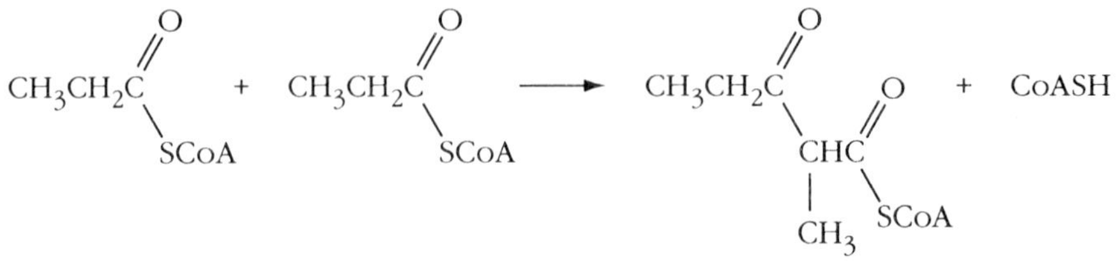

$$CH_3CH_2C(=O)SCoA + CH_3CH_2C(=O)SCoA \longrightarrow CH_3CH_2C(=O)-CH(CH_3)-C(=O)SCoA + CoASH$$

17.16 (a) Benzaldehyde has no α-hydrogen atoms so it cannot form an enolate ion. It can, however, act as the electrophilic acceptor component in a mixed aldol condensation reaction:

PhCHO + PhC(=O)CH$_3$ $\xrightarrow{\text{CH}_3\text{CH}_2\text{O}^-\text{Na}^+ / \text{CH}_3\text{CH}_2\text{OH}}$ PhC(=O)-CH=CH-Ph

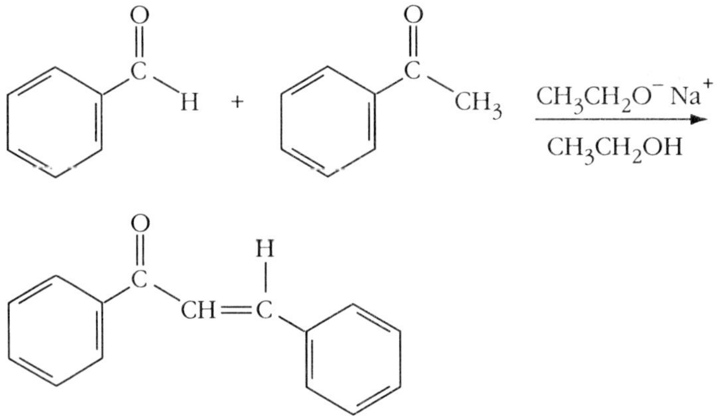

(b) Methyl benzoate has no α-hydrogen atoms so it cannot form an enolate ion. It can, however, act as the electrophilic acceptor component in a mixed Claisen condensation reaction:

PhC(=O)OCH$_3$ + CH$_3$CH$_2$C(=O)OCH$_3$ $\xrightarrow[\text{(2) H}_3\text{O}^+]{\text{(1) CH}_3\text{O}^-\text{Na}^+ \text{(1 eq.)}}$ PhC(=O)-CH(CH$_3$)-C(=O)OCH$_3$

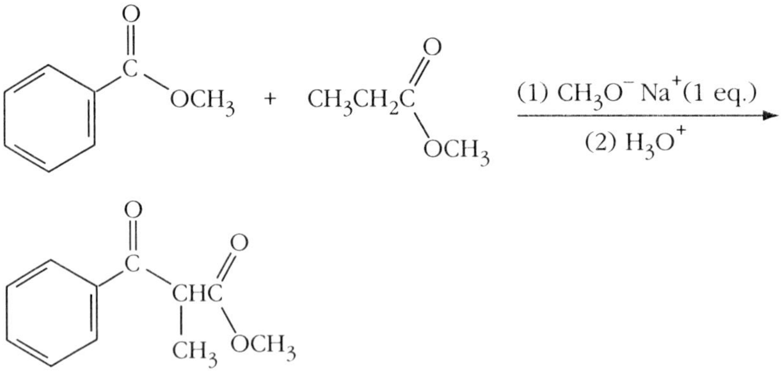

(c) Ethyl formate has no α-hydrogen atoms so it cannot form an enolate ion. It can, however, act as the electrophilic acceptor component in a mixed Claisen condensation reaction:

$$\underset{H}{\overset{O}{\underset{\|}{C}}}-OCH_2CH_3 \;+\; C_6H_5CH_2\underset{}{\overset{O}{\underset{\|}{C}}}-OCH_2CH_3 \;\xrightarrow[(2)\;H_3O^+]{(1)\;CH_3CH_2O^-Na^+\,(1\;eq.)}$$

$$\underset{H}{\overset{O}{\underset{\|}{C}}}-\underset{\underset{C_6H_5}{|}}{CH}-\underset{\underset{O}{\|}}{C}-OCH_2CH_3$$

(d) Ethyl formate has no α-hydrogen atoms so it cannot form an enolate ion. It can, however, act as the electrophilic acceptor component in a mixed Claisen condensation reaction:

$$\begin{array}{c}CH_2COOCH_2CH_3\\|\\CH_2COOCH_2CH_3\end{array} \;+\; \underset{H}{\overset{O}{\underset{\|}{C}}}-OCH_2CH_3 \;\xrightarrow[(2)\;H_3O^+]{(1)\;CH_3CH_2O^-Na^+\,(1\;eq.)}$$

$$\underset{H}{\overset{O}{\underset{\|}{C}}}-\underset{\underset{CH_2COOCH_2CH_3}{|}}{CHCOOCH_2CH_3}$$

17.17 A retro-condensation reaction is a reverse process to aldol and Claisen condensations. The mechanism is simply the reverse of the forward condensation reaction.

(a) Step 1: A base removes a proton from the hydroxyl group of 3-hydroxy-2-methylpentanal

$$CH_3CH_2-\underset{\underset{CH_3CHC}{|}\atop\underset{H}{\diagdown}\mkern-6mu\overset{O}{\diagup}}{CH}-\overset{\overset{:\ddot{O}-H\;\curvearrowleft\;:\ddot{O}H\;(or\;:\ddot{O}CH_3)}{}}{} \;\rightleftharpoons\; CH_3CH_2-\underset{\underset{CH_3CHC}{|}\atop\underset{H}{\diagdown}\mkern-6mu\overset{O}{\diagup}}{\overset{:\ddot{O}:^-}{\underset{|}{CH}}} \;+\; H_2\ddot{O}:$$

Step 2: A carbon-carbon bond breaks between the α and β carbon atoms to form propanal and the propanal enolate ion

[Step 2 continued - retro-aldol mechanism showing propanal enolate formation]

$CH_3CH_2\overset{\overset{\displaystyle :\overset{-}{O}:}{|}}{\underset{\underset{\displaystyle CH_3}{|}}{C}}\cdots H\!-\!CHC\overset{O}{\diagdown}_H \rightleftharpoons CH_3CH_2\overset{O}{\underset{H}{\diagdown C}} + CH_3\overset{(-)}{CH}C\overset{O}{\underset{H}{\diagdown}} \longleftrightarrow CH_3CH=C\overset{:\overset{-}{O}:}{\underset{H}{\diagup}}$

Step 3: The enolate ion is protonated to form the second molecule of propanal as the product of retro-aldol condensation

$CH_3CH=C\overset{:\overset{-}{O}:}{\underset{H}{\diagup}} \;+\; H_3O^+ \rightleftharpoons CH_3CH_2\overset{O}{\underset{H}{\diagdown C}} + H_2\ddot{O}:$

(b) [cyclohexanone with C(OH)(CH₃)₂ at α-position] $\xrightarrow[\;]{HO^-,\,H_2O}$ cyclohexanone + acetone (CH₃COCH₃)

(c) PhCH(OH)CH₂C(O)Ph $\xrightarrow[\;]{HO^-,\,H_2O}$ PhCHO + PhCOCH₃

17.18 The enolate anions of aldehydes, ketones, and esters are nucleophilic and they react with alkyl halides. As a result, alkylation at the α-carbon atom of a carbonyl compound takes place.

The most convenient base used to convert a carbonyl compound to its enolate ion completely and irreversibly is LDA (lithium diisopropylamide) because it is a very strong base and it cannot add to the carbonyl group because it is sterically hindered.

(a) cyclohexanone $\xrightarrow[-78\,°C]{LDA/THF}$ cyclohexenyl-O⁻Li⁺ + $CH_3CH_2Br \longrightarrow$

2-ethylcyclohexanone + LiBr

(b)

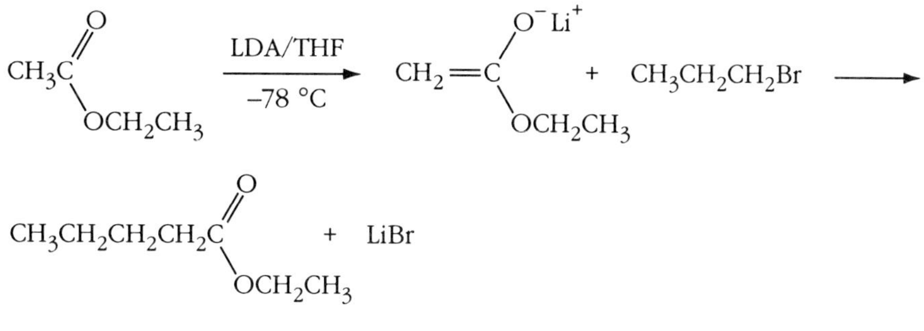

$$CH_3C(=O)OCH_2CH_3 \xrightarrow[-78\,°C]{LDA/THF} CH_2=C(O^-Li^+)(OCH_2CH_3) + CH_3CH_2CH_2Br \longrightarrow$$

$$CH_3CH_2CH_2CH_2C(=O)OCH_2CH_3 + LiBr$$

(c)

$$CH_3C\equiv N \xrightarrow[-78\,°C]{LDA/THF} CH_2=C=N^- \; Li^+ + PhCH_2Br \longrightarrow$$

$$PhCH_2CH_2C\equiv N + LiBr$$

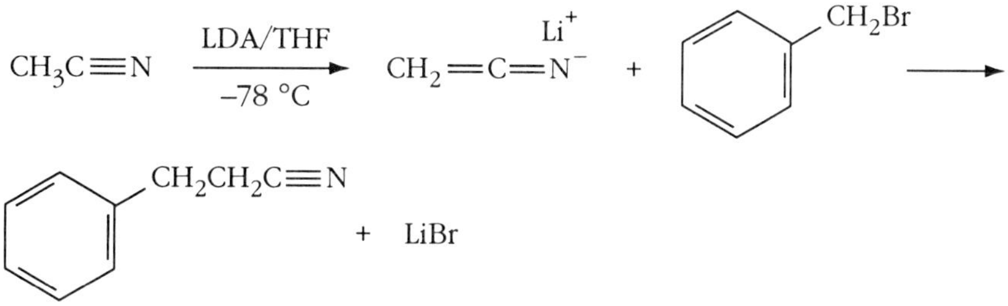

17.19 LDA quantitatively converts ethyl acetate to its enolate ion. The enolate ion, which is nucleophilic, reacts with benzyl bromide in a nucleophilic substitution reaction that occurs by an S_N2 mechanism.

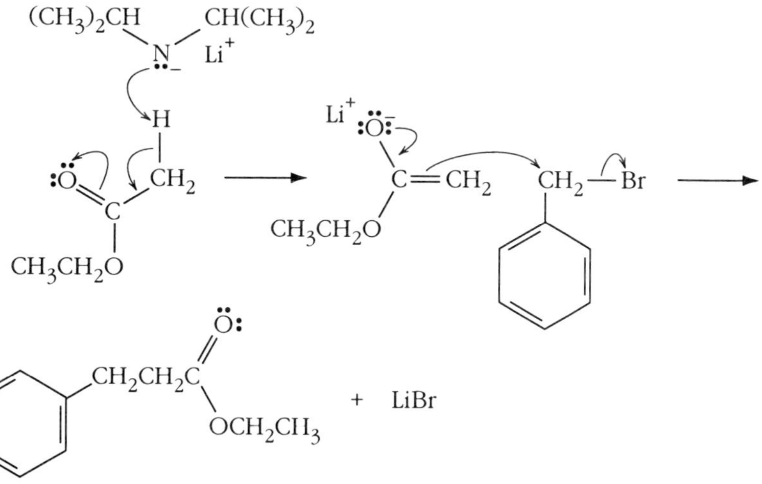

[Mechanism diagram showing LDA deprotonation of ethyl acetate to form the lithium enolate, followed by S_N2 attack on benzyl bromide]

$$PhCH_2CH_2C(=O)OCH_2CH_3 + LiBr$$

17.20 LDA quantitatively converts esters, carboxylic acids, amides, and nitriles to the corresponding enolate ions, which can attack the electrophilic carbon atom of an aldehyde or a ketone.

(a)

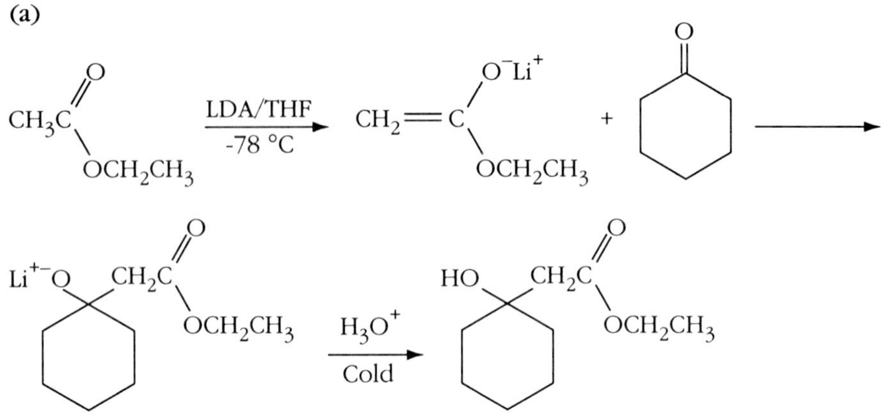

$$CH_3C(=O)OCH_2CH_3 \xrightarrow[-78\ °C]{LDA/THF} CH_2=C(O^-Li^+)(OCH_2CH_3) + \text{cyclohexanone} \longrightarrow$$

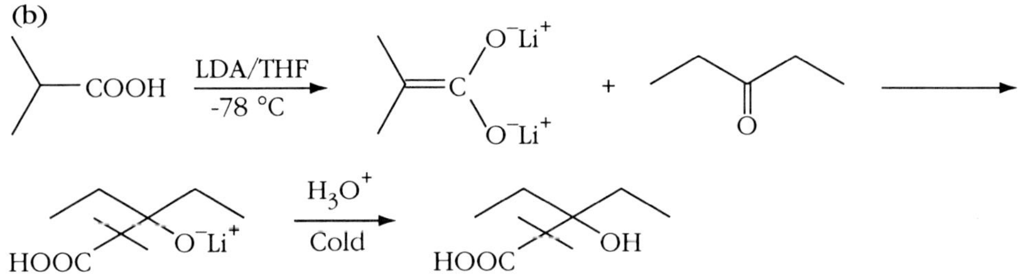

cyclohexane-C(O$^-$Li$^+$)–CH$_2$C(=O)OCH$_2$CH$_3$ $\xrightarrow[\text{Cold}]{H_3O^+}$ cyclohexane-C(OH)–CH$_2$C(=O)OCH$_2$CH$_3$

(b)

$$(CH_3)_2CH-COOH \xrightarrow[-78\ °C]{LDA/THF} (CH_3)_2C=C(O^-Li^+)_2 + \text{pentan-3-one} \longrightarrow$$

HOOC–C(CH$_3$)$_2$–C(Et)(Et)(O$^-$Li$^+$) $\xrightarrow[\text{Cold}]{H_3O^+}$ HOOC–C(CH$_3$)$_2$–C(Et)(Et)(OH)

(c)

$$CH_3C\equiv N \xrightarrow[-78\ °C]{LDA/THF} CH_2=C=N\overset{..}{:}Li^+ + \text{pentan-3-one} \longrightarrow$$

Li^{+-}O–C(Et)(Et)–CH$_2$C≡N $\xrightarrow[\text{Cold}]{H_3O^+}$ HO–C(Et)(Et)–CH$_2$C≡N

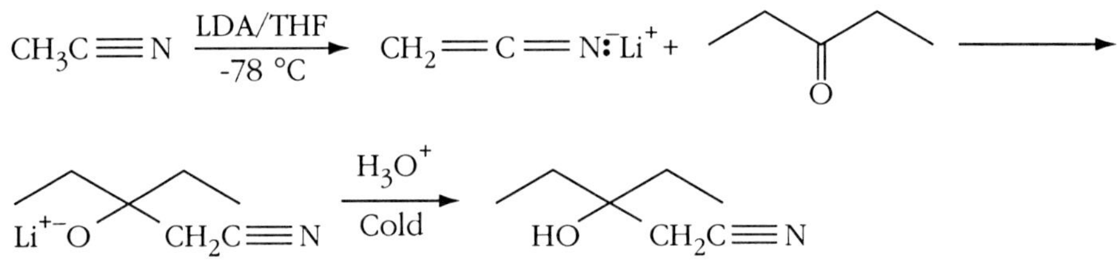

(d)

$$CH_3C(=O)N(CH_3)_2 \xrightarrow[-78\ °C]{LDA/THF} CH_2=C(O^-Li^+)N(CH_3)_2 + \text{cyclohexanone} \longrightarrow$$

cyclohexane-C(O$^-$Li$^+$)–CH$_2$C(=O)N(CH$_3$)$_2$ $\xrightarrow[\text{Cold}]{H_3O^+}$ cyclohexane-C(OH)–CH$_2$C(=O)N(CH$_3$)$_2$

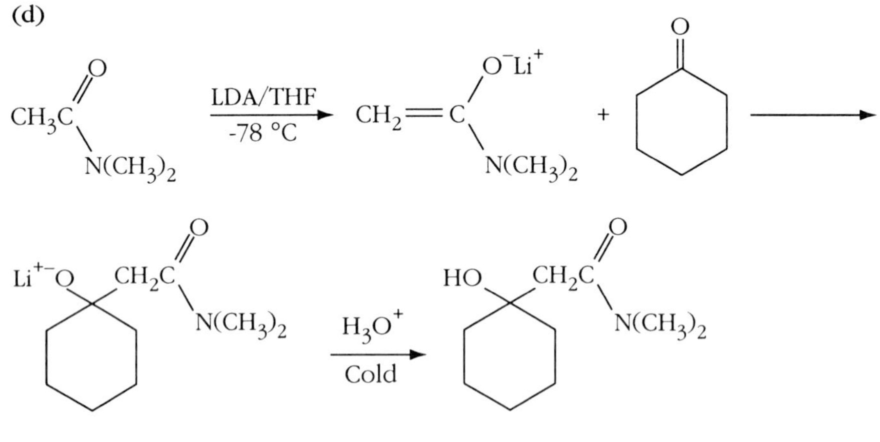

Note: Tertiary alcohols, formed as products, can easily dehydrate in the presence of an acid catalyst at an elevated temperature.

17.21 LDA quantitatively converts ethyl acetate to its enolate ion. The enolate ion, which is nucleophilic, adds to the electrophilic carbon atom of 2,2-dimethylpropanal.

17.22 By choosing proper experimental conditions we can prepare one or the other enolate ion of unsymmetrical ketones. Alkylation of the major enolate ion leads to the preparation of selectively alkylated ketones.

(a) Retrosynthesis:

Synthesis:

(b) Retrosynthesis:

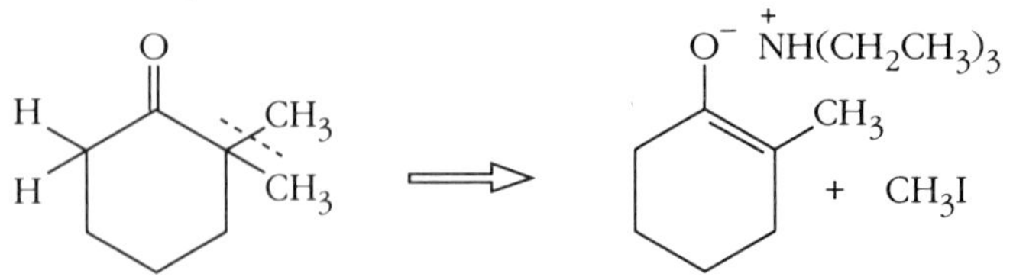

[Structure: 2,2-dimethylcyclohexanone] ⟹ [Structure: enolate with O⁻ NH(CH₂CH₃)₃⁺, CH₃ substituent] + CH₃I

Synthesis:

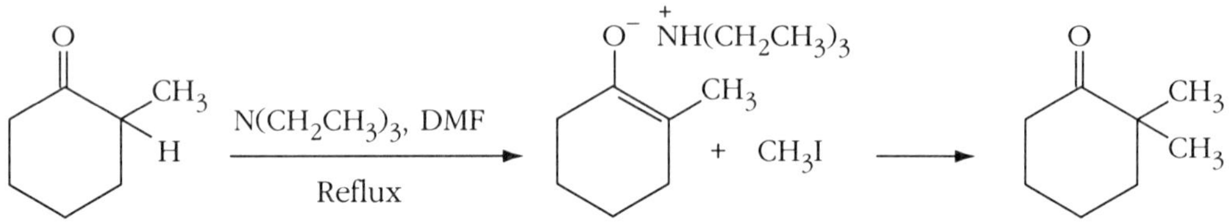

[2-methylcyclohexanone] →(N(CH₂CH₃)₃, DMF, Reflux)→ [enolate O⁻ NH(CH₂CH₃)₃⁺, CH₃] + CH₃I → [2,2-dimethylcyclohexanone]

(c) Retrosynthesis:

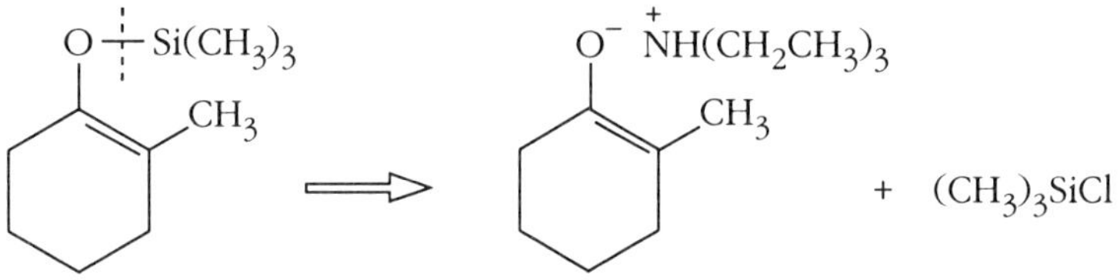

[Silyl enol ether: O–Si(CH₃)₃, CH₃] ⟹ [enolate O⁻ NH(CH₂CH₃)₃⁺, CH₃] + (CH₃)₃SiCl

Synthesis:

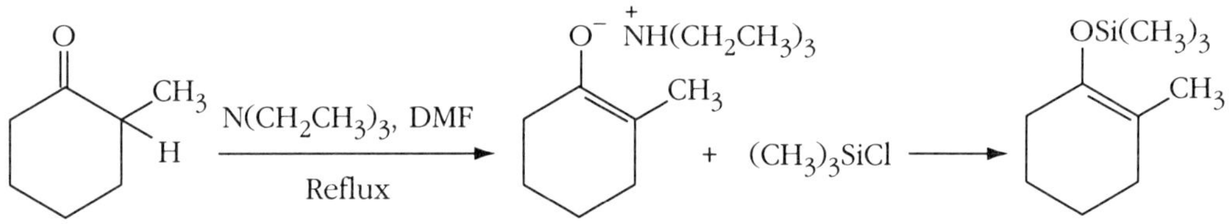

[2-methylcyclohexanone] →(N(CH₂CH₃)₃, DMF, Reflux)→ [enolate O⁻ NH(CH₂CH₃)₃⁺, CH₃] + (CH₃)₃SiCl → [OSi(CH₃)₃, CH₃ silyl enol ether]

17.23 Cyclohexanone reacts with a secondary amine (like pyrrolidine) to form enamines. These enamines react rapidly with haloalkanes in a nucleophilic substitution reaction leading to imminium salts that give alkylated cyclohexanone upon hydrolysis.

(a)

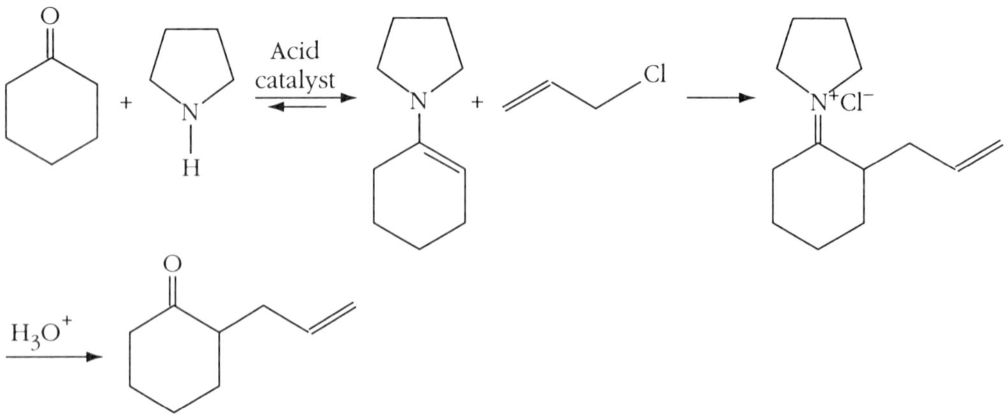

Cyclohexanone + pyrrolidine →(Acid catalyst)⇌ enamine + allyl chloride → iminium salt (N⁺Cl⁻) →(H₃O⁺)→ 2-allylcyclohexanone

(b)

[Reaction scheme: Cyclohexanone + pyrrolidine → (acid catalyst) → enamine + benzyl bromide (CH₂Br) → iminium bromide intermediate → (H₃O⁺) → 2-benzylcyclohexanone]

17.24 (a) Retrosynthesis:

[Retrosynthesis scheme: 2-butylcyclohexanone ⇒ butyl bromide + enamine ⇒ cyclohexanone]

Synthesis:

[Synthesis scheme: cyclohexanone + pyrrolidine → (acid catalyst) → enamine + butyl bromide → iminium bromide intermediate → (H₃O⁺) → 2-butylcyclohexanone]

(b) Retrosynthesis:

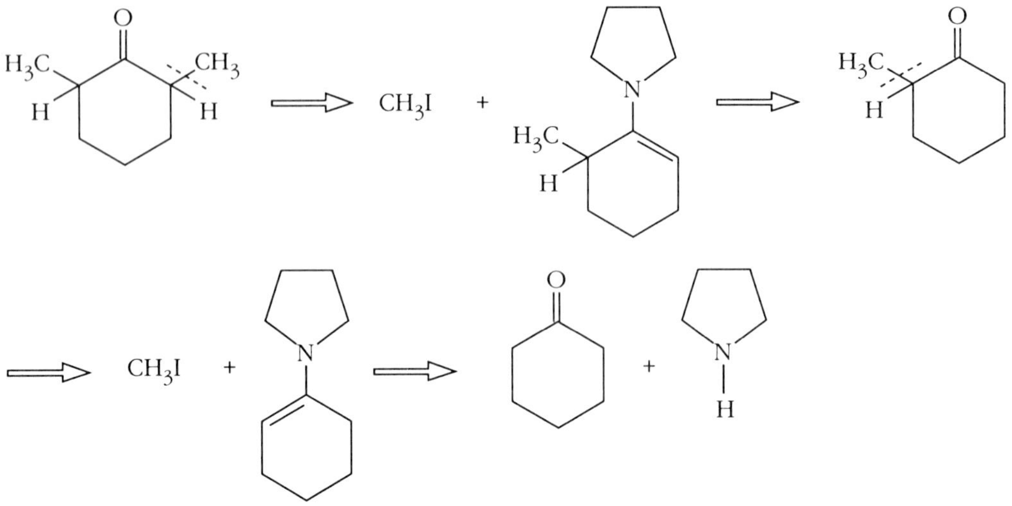

2,6-dimethylcyclohexanone ⟹ CH₃I + 1-(6-methylcyclohex-1-en-1-yl)pyrrolidine ⟹ 2-methylcyclohexanone

⟹ CH₃I + 1-(cyclohex-1-en-1-yl)pyrrolidine ⟹ cyclohexanone + pyrrolidine

Synthesis:

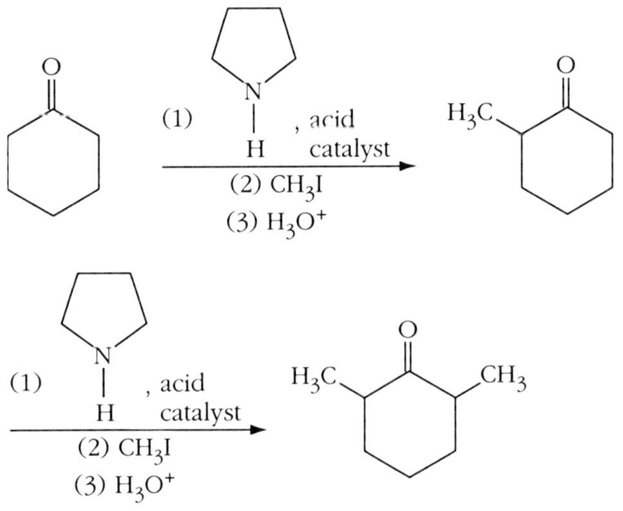

cyclohexanone $\xrightarrow[\substack{(2)\ CH_3I \\ (3)\ H_3O^+}]{(1)\ \text{pyrrolidine, acid catalyst}}$ 2-methylcyclohexanone

$\xrightarrow[\substack{(2)\ CH_3I \\ (3)\ H_3O^+}]{(1)\ \text{pyrrolidine, acid catalyst}}$ 2,6-dimethylcyclohexanone

(c) Retrosynthesis:

3-methylpentan-3... (3-methylpentan-2-one precursor shown: pentan-3-one with methyl) ⟹ CH₃I + enamine ⟹ pentan-3-one

+ pyrrolidine

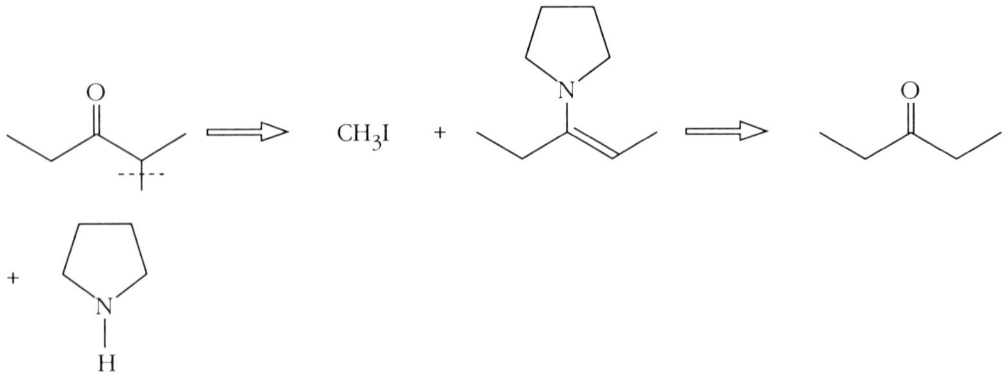

Synthesis:

[Reaction scheme: 3-pentanone + pyrrolidine (acid catalyst) → enamine; then (1) CH₃I, (2) H₃O⁺ → 2-methyl-3-pentanone]

An alternate synthesis starting with an unsymmetrical ketone, such as 3-methyl-2-butanone, is unsatisfactory because a mixture of products may result.

17.25 (a) An **enol tautomer** is a vinyl alcohol that exists in equilibrium with its keto form.

[Structure: Enol tautomer (C=C with OH) ⇌ (Acid or base catalyst) Keto tautomer (C–C with H and C=O)]

(b) **Claisen condensation** is a base-catalyzed conversion of two esters into a β-keto ester.

[Reaction: 2 CH₃—C(=O)—OCH₂CH₃ → (1) CH₃CH₂O⁻ Na⁺, (2) H₃O⁺ → CH₃—C(=O)—CH₂—C(=O)—OCH₂CH₃]

Claisen condensation occurs when an ester undergoes nucleophilic acyl substitution by an ester enolate. The intermediate has an alkoxy (—OR) group that acts as a leaving group.

(c) A **keto tautomer** is a carbonyl compound with an enolizable hydrogen atom on its α-carbon atom. In the presence of a strong base or acid, an equilibrium is set up with its enol tautomer, which is a vinyl alcohol.

[Structure: Keto tautomer ⇌ (Acid or base catalyst) Enol tautomer]

(d) **Tautomerism** is an isomerism involving movement of a proton and the corresponding double bond. An example is the keto-enol tautomerism of a ketone or aldehyde with its enol form. The keto-enol tautomerism is catalyzed by either base or acid. In the presence of strong bases, a proton on the α-carbon atom (the carbon adjacent to the carbonyl group) is abstracted to form a resonance-stabilized enolate ion. Reprotonation can occur either on the α-carbon atom (returning to the keto form) or on the oxygen atom, giving a vinyl alcohol (enol form):

[Keto form] + HO⁻ ⇌ [Enolate ion] + H₂O

⇌ [Enol form with C=C, OH] + HO⁻

Keto-enol tautomerism is also catalyzed by acid. In acid, a proton is moved from the α-carbon atom to oxygen by first protonating oxygen and then removing a proton from carbon.

[Keto form] ⇌ [Protonated carbonyl] + H₂O

⇌ [Enol form] + H_3O^+

(e) The **Hell-Volhard-Zelinsky reaction** is a reaction of a carboxylic acid with Br_2 and PBr_3 to give an α-bromo acyl bromide, which is often hydrolyzed to an α-bromo acid.

$$R-CH_2-C(=O)OH \xrightarrow{Br_2/PBr_3} R-CH(Br)-C(=O)OH + HBr$$

(f) **Aldol condensation** is a base-catalyzed conversion of two ketones or aldehydes to a β-hydroxy ketone or aldehyde. The aldol condensation involves the nucleophilic addition of an enolate ion to another carbonyl group.

$$\text{R—CH}_2\text{—C(=O)} \;+\; \text{R—CH}_2\text{—C(=O)R'R'} \xrightleftharpoons{\text{HO}^-} \underset{\text{Aldol product}}{\text{R—CH}_2\text{—C(OH)(R')—CH(R)—C(=O)R'}} \xrightarrow[\text{Heat}]{\text{H}_3\text{O}^+ + \text{HO}^-}$$

Aldehyde or ketone

$$\underset{\alpha,\beta\text{-Unsaturated aldehyde or ketone}}{\text{RCH}_2\text{—C(R')=C(R)—C(R')(=O)}} \;+\; \text{H}_2\text{O}$$

Aldol condensations often take place with subsequent dehydration to give α,β-unsaturated aldehydes and ketones.

(g) **A mixed Claisen condensation,** also known as a *crossed Claisen condensation,* is a condensation of two *different* esters. Mixed Claisen reactions are generally successful only when one of the two ester components has *no* α-hydrogens and thus cannot form an enolate ion.

$$\underset{\text{Acceptor}}{\text{C}_6\text{H}_5\text{C(=O)OCH}_3} \;+\; \underset{\text{Donor}}{\text{CH}_3\text{CH}_2\text{C(=O)OCH}_3} \xrightarrow[\text{(2) H}_3\text{O}^+]{\text{(1) NaH/THF}}$$

$$\text{C}_6\text{H}_5\text{C(=O)CH(CH}_3\text{)C(=O)OCH}_3 \;+\; \text{CH}_3\text{OH}$$

For example, methyl benzoate cannot form enolate ions and thus cannot serve as a donor. It can, however, act as the electrophilic acceptor of an enolate ion donor formed from methyl propanoate.

(h) An **enamine** (vinyl amine) is the nitrogen analog of an enol and it is usually generated by the acid-catalyzed reaction of a secondary amine with an aldehyde or a ketone.

(i) **Enolization,** also called keto-enol tautomerization, is an interconversion of the keto-enol tautomers. The enolization can be catalyzed by either acids or bases (see Exercise 17.25 d).

(j) A **mixed aldol condensation,** also known as a *crossed aldol condensation,* is a condensation between two *different* aldehydes or ketones. The mixed aldol condensation is likely to be successful when one of the carbonyl components contains *no* α-hydrogens (and thus cannot form an enolate ion to become a donor), but does contain a reactive carbonyl group that is a good acceptor of nucleophiles.

17.26 Tautomers are constitutional isomers in equilibrium with each other that differ in location of a hydrogen atom and a double bond. In order to find a missing tautomer, we need to move a double bond toward an atom with a hydrogen, and move the hydrogen to the atom previously bonded by the double bond.

(c) 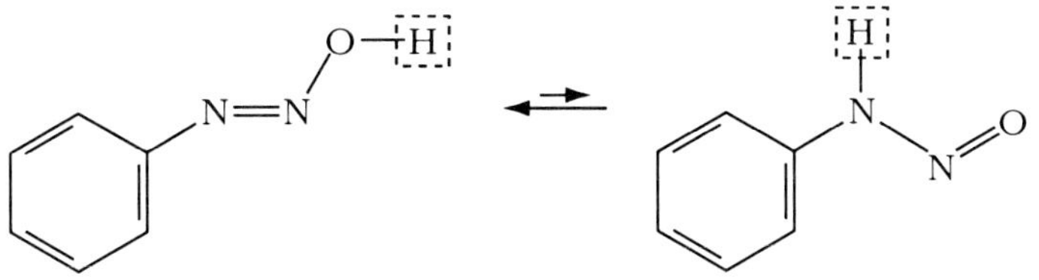
O
H
N=N
H
N
N
O

17.27 (a) 2-Pentanone is a ketone. α-Hydrogen atoms in ketones are acidic ($pK_a \approx 19 - 20$).

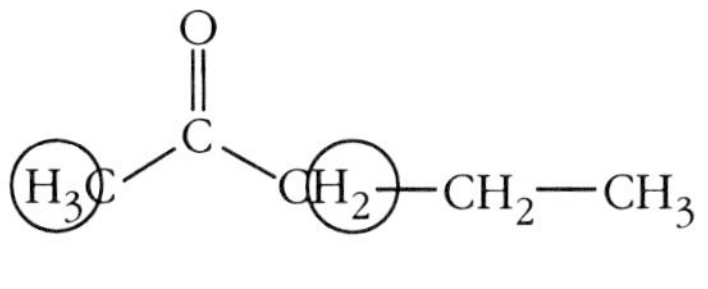
O
C
H3C
CH2
CH2
CH3

(b) Propanoic anhydride.

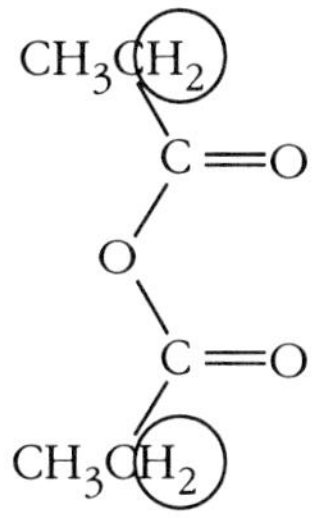
CH3CH2
C=O
O
C=O
CH3CH2

(c) Butanoic acid is a carboxylic acid ($pK_a = 4.82$).

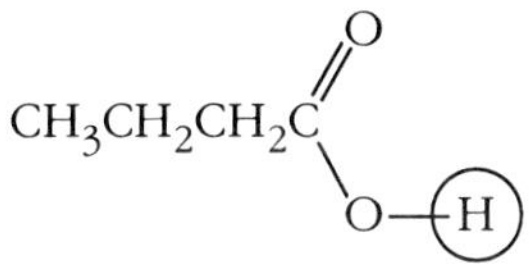
O
CH3CH2CH2C
O
H

(d) 3-Methyl-2-butenal

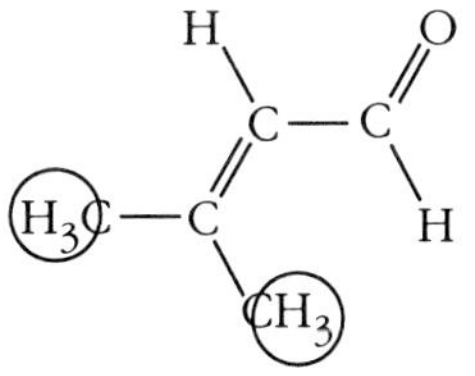
H
O
C
C
H3C
C
H
CH3

(e) Methyl 2-hydroxypropanoate

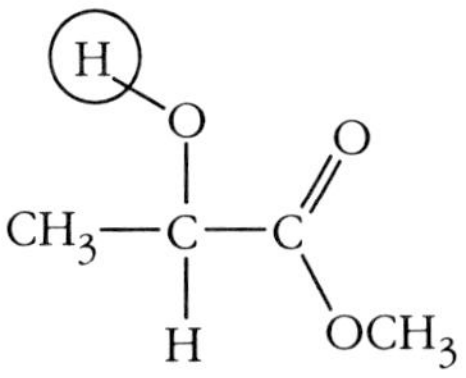
H
O
O
CH3
C
C
H
OCH3

17.28

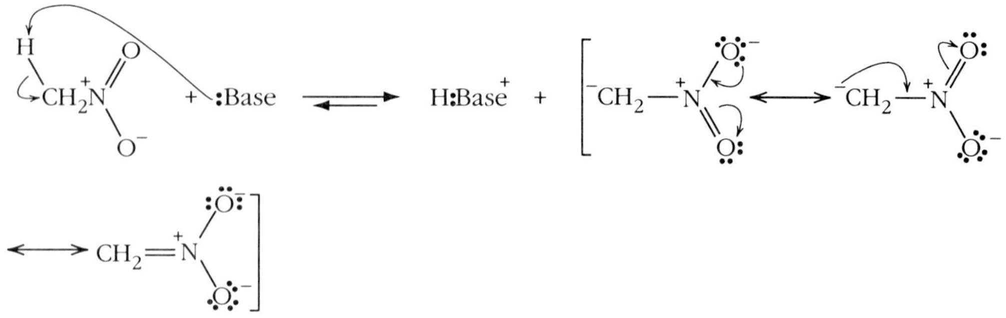

[Mechanism showing deprotonation of nitromethane (CH₃NO₂) by a base, producing H:Base⁺ and the resonance-stabilized nitronate anion with three resonance structures: ⁻CH₂—N⁺(=O)(O⁻) ↔ ⁻CH₂—N⁺(O⁻)(=O) ↔ CH₂=N⁺(O⁻)₂]

17.29

Increasing Acidity →

CH₃CH₂C(H)=CH₂ < CH₃CH₂C(H)₂C(=O)OCH₂CH₃ ≈ CH₃CH₂C≡C(H) < CH₃C(=O)C(H)₂CH₃ < CH₃CH₂CH₂C(=O)O(H)

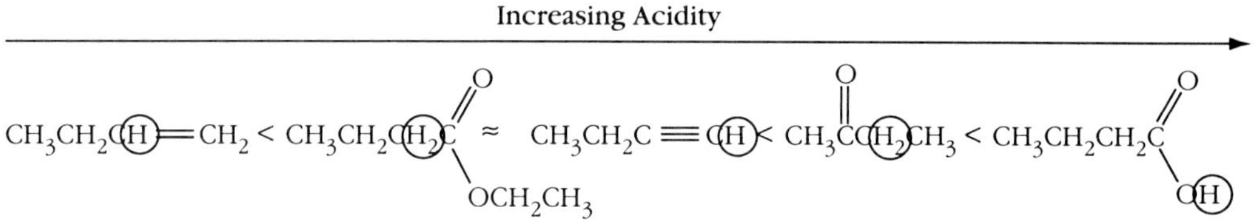

1-Butene	Ethyl butyrate	1-Butyne	2-Butanone	Butanoic acid
pK_a ≈ 44	pK_a ≈ 25	pK_a ≈ 25	pK_a ≈ 19	pK_a ≈ 4.8

17.30 Acid-catalyzed keto–enol tautomerization of methylpropanal:

[Mechanism: (CH₃)₂CH—CHO + H₃O⁺ → protonated carbonyl (resonance structures shown) → enol (CH₃)₂C=CHOH + H₃O⁺]

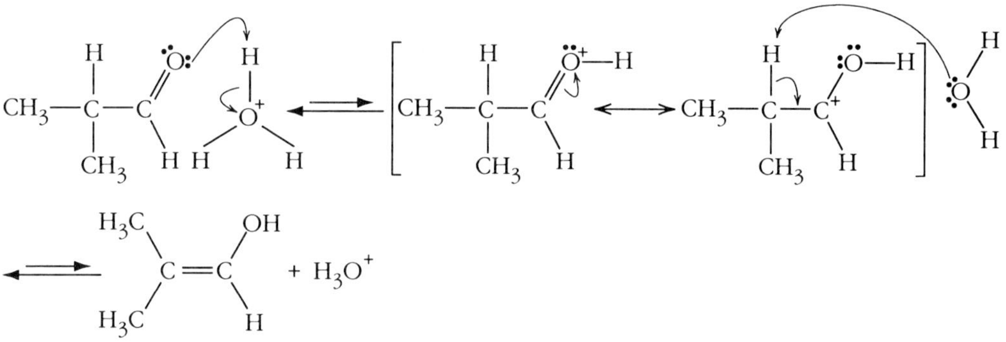

Base-catalyzed keto–enol tautomerization of methylpropanal:

[Mechanism: (CH₃)₂CH—CHO + HO⁻ → enolate (resonance structures shown) → enol (CH₃)₂C=CHOH + HO⁻]

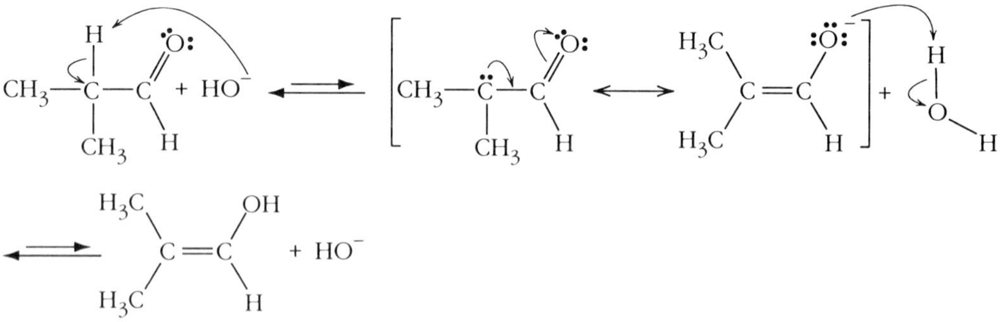

17.31 Acid-catalyzed enolization of acetone:

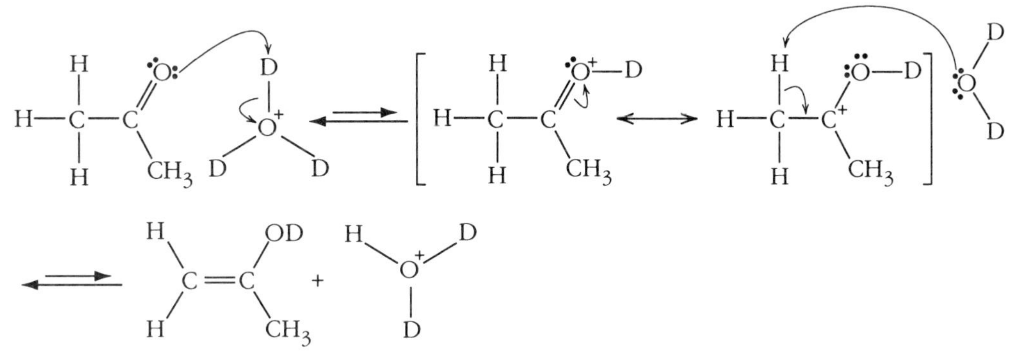

[Mechanism showing acetone being protonated by D_3O^+, forming the protonated carbonyl with resonance structures, then losing a proton to give the enol with OD group plus D_3O^+.]

Reaction of the acetone enol with D_3O^+ to form deuteroacetone:

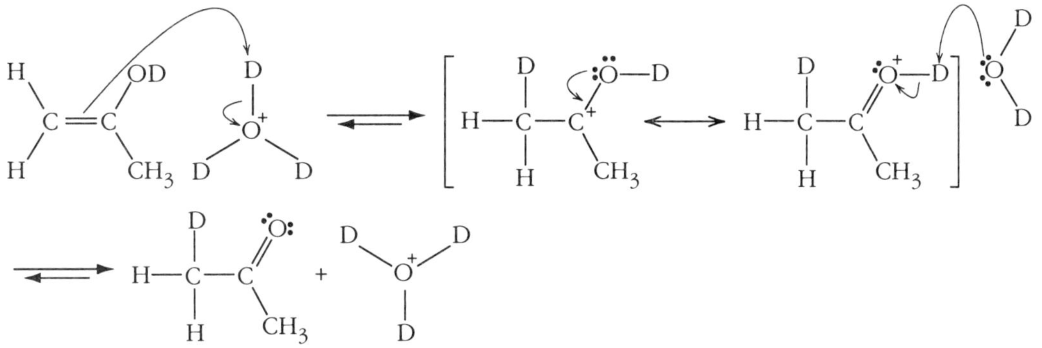

[Mechanism showing the enol reacting with D_3O^+ at the alpha carbon to give a protonated carbonyl intermediate (shown with resonance structures), which then loses D from oxygen to form deuteroacetone plus D_3O^+.]

Reaction of the acetone enol with bromine:

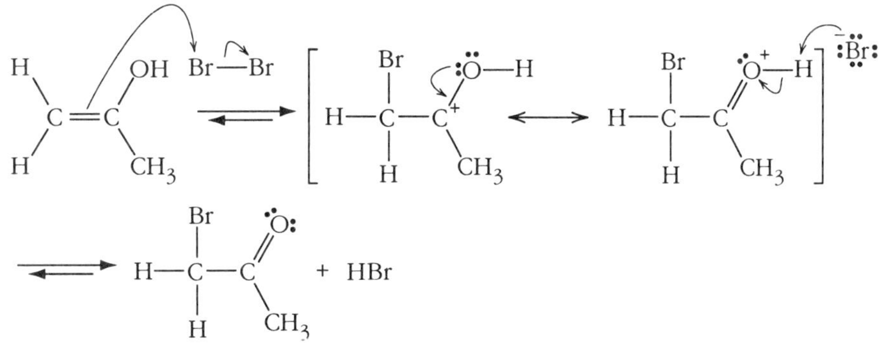

[Mechanism showing the enol attacking Br_2 at the alpha carbon, giving a protonated carbonyl intermediate (with resonance structures) and bromide, which then loses a proton to give α-bromoacetone + HBr.]

The acid-catalyzed enolization of acetone is the slow, rate-determining step of both bromination and hydrogen-deuterium exchange. The mechanism of the reaction of the acetone enol is also identical in both reactions. As a result, the rate constant of bromination and hydrogen-deuterium exchange are the same.

17.32 (a) Acid-catalyzed bromination of a ketone with one equivalent of Br_2 gives an α-monobrominated ketone:

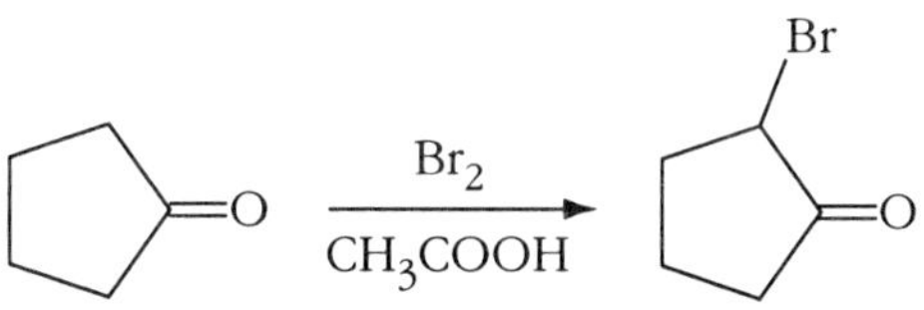

[Cyclopentanone + Br_2 / CH_3COOH → 2-bromocyclopentanone]

(b) Base-catalyzed hydrogen-deuterium exchange will replace all α-hydrogen atoms by deuterium atoms:

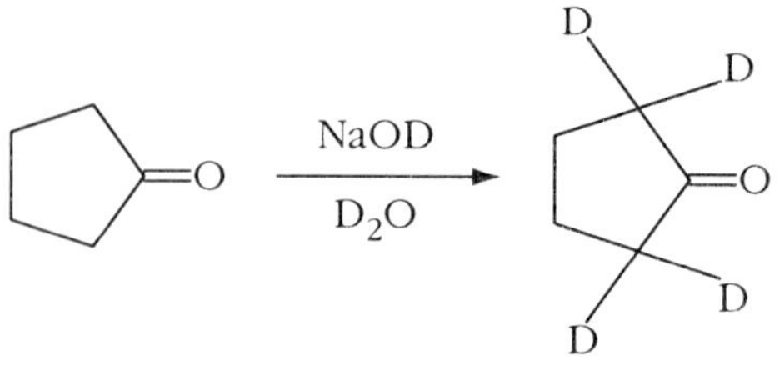

(c) The Wittig reaction involves the reaction of an aldehyde or a ketone with an ylide to give an alkene:

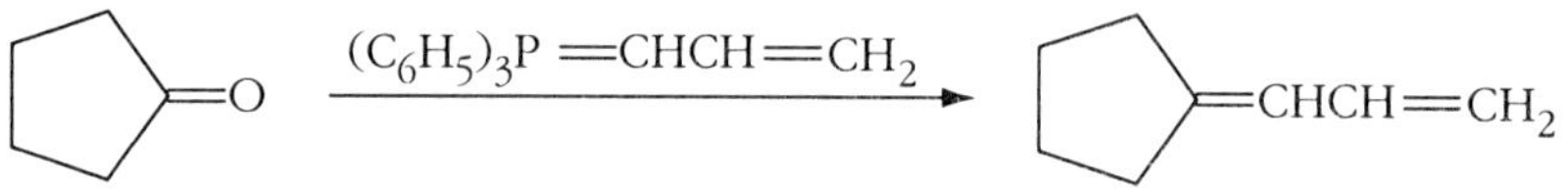

(d) LDA converts a carbonyl compound into its enolate ion, which undergoes alkylation with an alkyl halide:

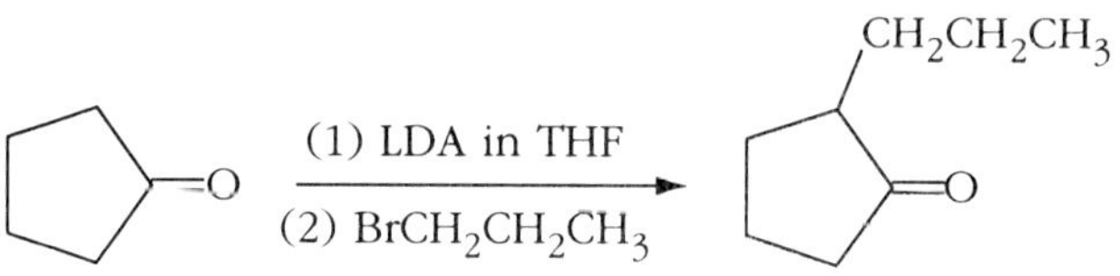

17.33 Compound **A** enolizes in basic aqueous solution to form enol **E**. Enolization is reversible, so the keto tautomer is formed again. However, in this case we can have two tautomers formed because hydrogen can be added to either side of the carbon-carbon double bond in enol **E** to form Compound **A** or Compound **B**. Compound **B** is more stable because the two cyclohexane rings are *trans* as compared with *cis* in Compound **A** (compare stability of *trans*-decalin and *cis*-decalin, Exercise 4.14 d). Therefore at equilibrium compound **B** is favored.

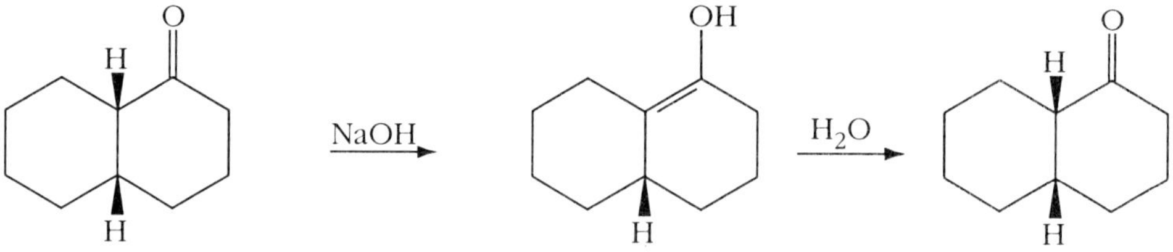

Compound A   Compound E   Compound A

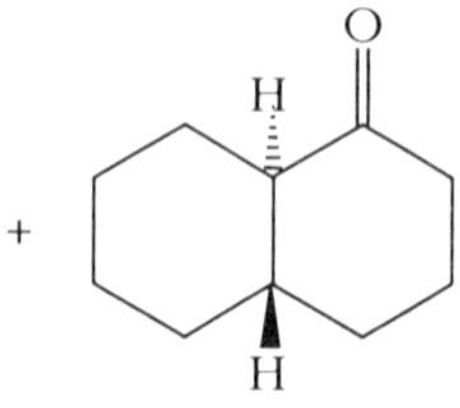

+

Compound B

Compound **C** does not have a hydrogen atom on the bridgehead carbon atom and cannot form an enol with a carbon-carbon double bond in the same position as that in Compound **E**. The only enol it can form is Compound **F**, which can tautomerize back to Compound **C** only.

Compound C →(NaOH) Compound F →(H_2O) Compound C

Compound **D**, therefore, cannot be formed by isomerization of Compound **C**.

17.34 (a) 2,2,6,6-Tetradeuterocyclohexanone can be easily obtained from cyclohexanone by hydrogen-deuterium exchange in either basic or acidic solution of D_2O.

cyclohexanone →(D_2O, D_3O^+ or D_2O, DO^-)→ 2,2,6,6-tetradeuterocyclohexanone

(b) Retrosynthetic analysis:

3,3-dideutero-1-deutero-cyclohexene ⇒ deuterated cyclohexanol ⇒ 2,2,6,6-tetradeuterocyclohexanone ⇒ cyclohexanone

Synthesis:

cyclohexanone →(D_2O, D_3O^+ or D_2O, DO^-)→ 2,2,6,6-tetradeuterocyclohexanone →($NaBD_4$, CH_3CH_2OD)→ deuterated cyclohexanol

→(D_2SO_4, Heat)→ deuterated cyclohexene

17.35 (a) One substrate is furfural and the other is acetone. Furfural has no α-hydrogen atom so it cannot enolize. Acetone can form an enol in a basic aqueous solution. The equilibrium for the aldol condensation of acetone favors the reactants so only a mixed aldol condensation reaction occurs:

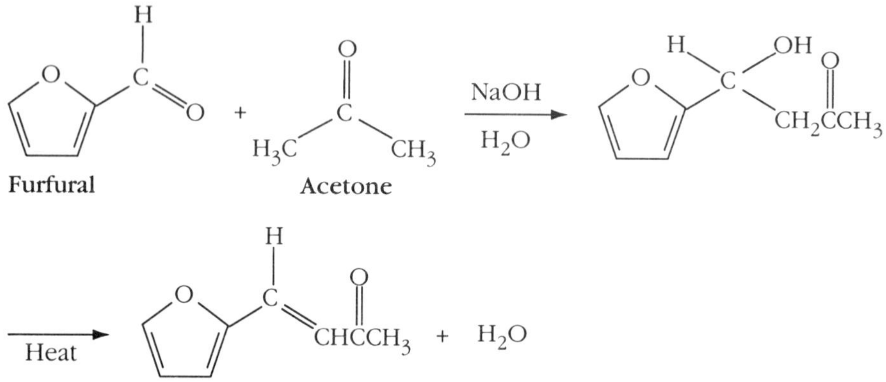

If there is an excess of furfural, a second molecule of furfural can react with the product shown above.

(b) The benzaldehyde derivative cannot form an enolate ion because it does not have an α-hydrogen atom. Acetonitrile, on the other hand, *can* form a resonance-stabilized ion that reacts in the same manner as an enolate ion. The workup in acidic aqueous solution will remove the protecting acetal group and it may also cause dehydration.

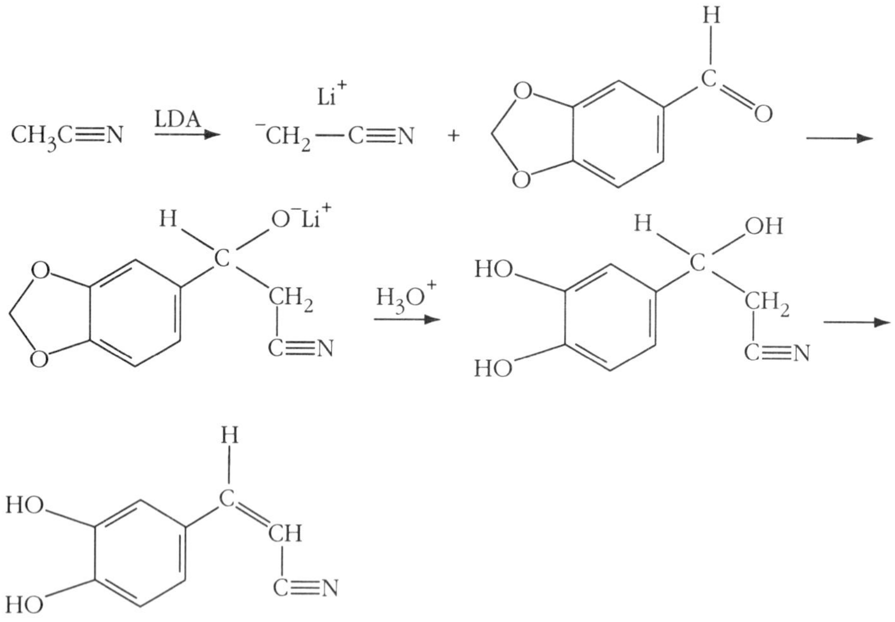

(c) Ph–CO–CH₃ + CH₃CH₂O–CO–OCH₂CH₃ →(1) NaOH/H₂O (2) H₃O⁺→ Ph–CO–CH₂–CO–OCH₂CH₃

(d) CH₃C(=O)N(CH₃)₂ →LDA/THF, −78 °C→ CH₂=C(O⁻Li⁺)N(CH₃)₂ →(1) CH₃COCl (2) H₃O⁺→ CH₃C(=O)CH₂C(=O)N(CH₃)₂

(e) CH₂(CO₂CH₂CH₃)₂ →CH₃CH₂O⁻Na⁺ / CH₃CH₂OH→ Na⁺⁻O–C(OCH₂CH₃)=CH–C(=O)OCH₂CH₃ →(1) Ph₂C=O (2) H₃O⁺→ Ph₂C=C(CO₂CH₂CH₃)(CH₂CO₂CH₂CH₃)

(f)

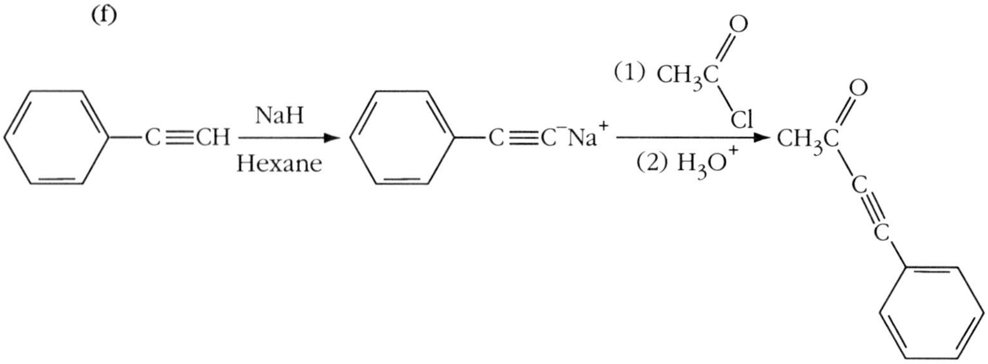

PhC≡CH →[NaH, Hexane] PhC≡C⁻Na⁺ →[(1) CH₃COCl] [(2) H₃O⁺] CH₃C(O)C≡CPh

(g)

(CH₃)₂CHC(O)OCH₂CH₃ →[LDA, THF] (CH₃)₂C=C(O⁻Li⁺)(OCH₂CH₃) →[(1) PhCOCl] [(2) H₃O⁺] PhC(O)C(CH₃)₂C(O)OCH₂CH₃

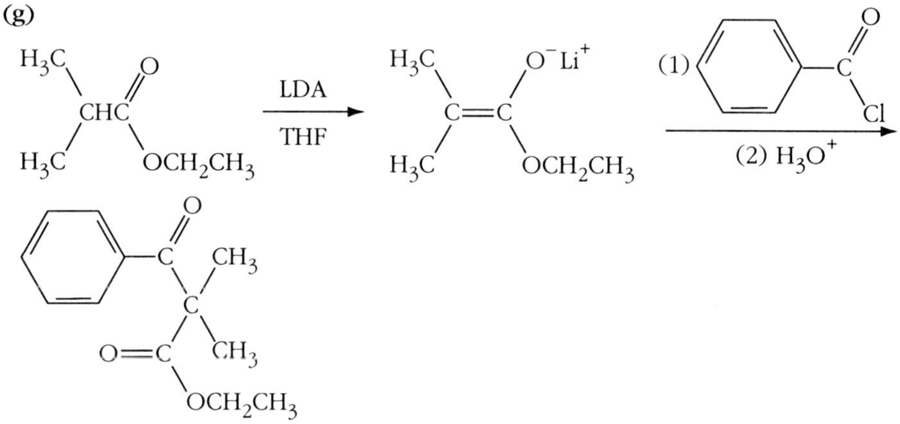

(h)

CH₃C(O)CH₃ →[LDA, DME] CH₃C(O⁻Li⁺)=CH₂ →[(1) CH₃CH₂C(O)OCH₂CH₃] [(2) H₃O⁺] CH₃C(O)CH₂C(O)CH₂CH₃

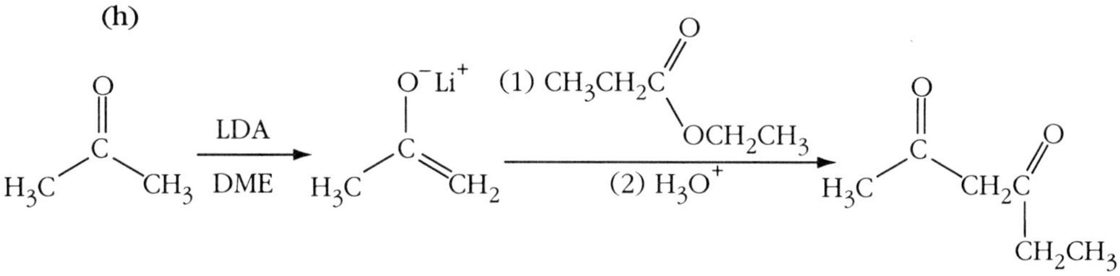

(i)

(decalone) →[LDA/THF] (lithium enolate) →[(1) CH₃I] [(2) H₃O⁺] (2-methyldecalone)

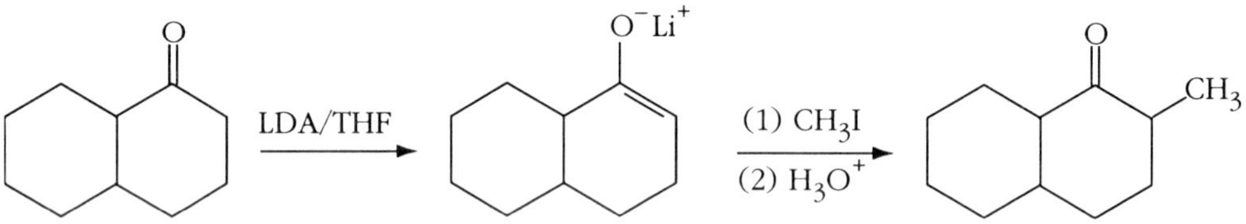

(j)

17.36 First we have an acid-catalyzed enolization of an aldehyde; then quick tautomerization of an enol to a more stable keto form. This last step is a driving force of the overall sequence of equilibrium reactions.

17.37

(a) Cyclohexanone + pyrrolidine —Acid catalyst→ 1-(cyclohex-1-en-1-yl)pyrrolidine (enamine) —(1) BrCH$_2$C(=O)OCH$_2$CH$_3$; (2) H$_3$O$^+$→ 2-(2-oxocyclohexyl)acetic acid ethyl ester (ethyl (2-oxocyclohexyl)acetate)

(b) 7-ethyl-1-tetralone —(1) LDA/THF/−78 °C; (2) CH$_3$CHBrC(=O)OCH$_2$CH$_3$; (3) H$_3$O$^+$→ 2-[1-(ethoxycarbonyl)ethyl]-7-ethyl-1-tetralone

(c) Methyl cyclohexanecarboxylate —(1) LDA/THF/−78 °C; (2) CH$_3$(CH$_2$)$_5$CH$_2$I→ methyl 1-heptylcyclohexanecarboxylate

(d) (CH$_3$)$_2$CHCOOH —2 LDA, THF, −78 °C→ (CH$_3$)$_2$C=C(O$^-$Li$^+$)(O$^-$Li$^+$) —CH$_3$CH$_2$CH$_2$Br→ CH$_3$CH$_2$CH$_2$C(CH$_3$)$_2$COOH

(e) bicyclic lactone —LDA/THF, −78 °C→ lithium enolate —(1) CH$_3$I; (2) H$_3$O$^+$→ α-methyl bicyclic lactone

17.38 (a)

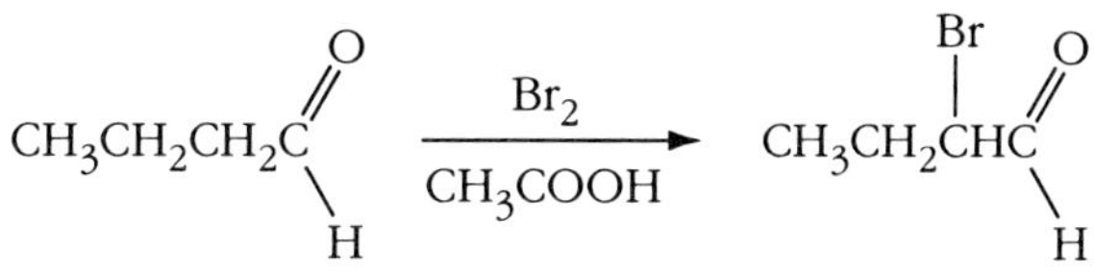

$$CH_3CH_2CH_2\underset{H}{\overset{O}{\overset{\|}{C}}} \xrightarrow[CH_3COOH]{Br_2} CH_3CH_2\underset{H}{\overset{Br}{\underset{|}{C}H}}\overset{O}{\overset{\|}{C}}H$$

(b)

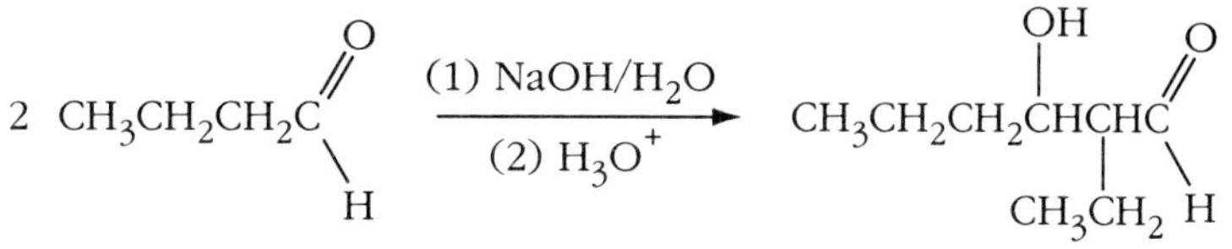

$$2\ CH_3CH_2CH_2\underset{H}{\overset{O}{\overset{\|}{C}}} \xrightarrow[(2)\ H_3O^+]{(1)\ NaOH/H_2O} CH_3CH_2CH_2\underset{}{\overset{OH}{\underset{|}{C}H}}\underset{CH_2CH_3}{\overset{}{\underset{|}{C}H}}\overset{O}{\overset{\|}{C}}H$$

(c)

$$CH_3CH_2CH_2\underset{H}{\overset{O}{\overset{\|}{C}}} \xrightarrow[H_2SO_4/H_2O]{CrO_3} CH_3CH_2CH_2\underset{OH}{\overset{O}{\overset{\|}{C}}}$$

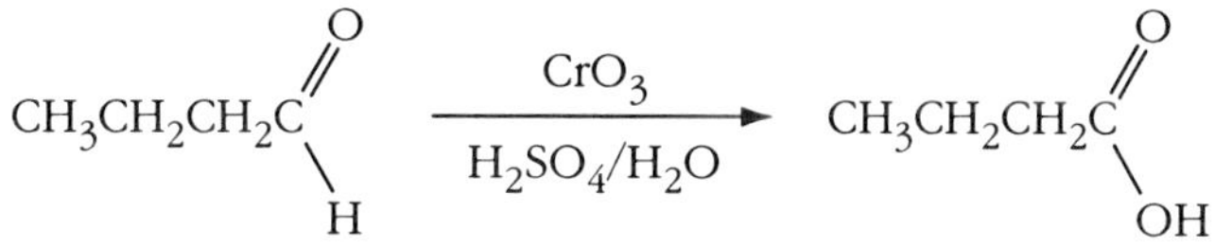

(d) Retrosynthesis:

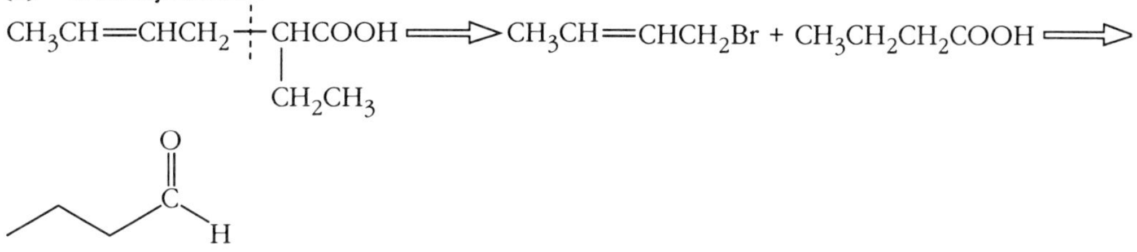

$$CH_3CH=CHCH_2 \!\mid\! \underset{CH_2CH_3}{\overset{}{\underset{|}{C}H}}COOH \Longrightarrow CH_3CH=CHCH_2Br + CH_3CH_2CH_2COOH \Longrightarrow$$

butanal (structure shown)

Synthesis:

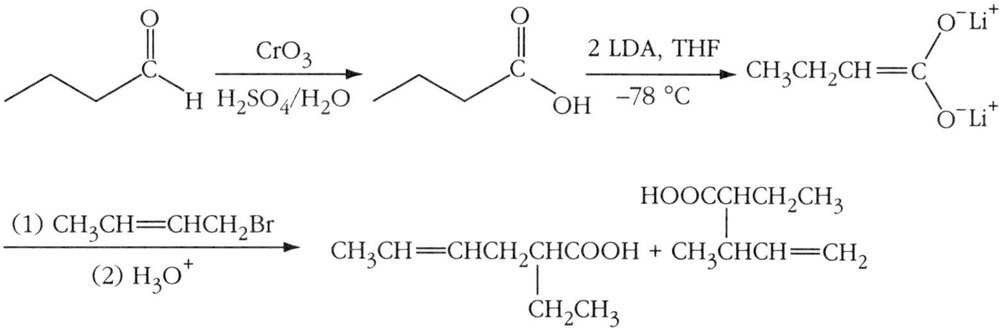

butanal $\xrightarrow[H_2SO_4/H_2O]{CrO_3}$ butanoic acid $\xrightarrow[-78\ °C]{2\ LDA,\ THF}$ $CH_3CH_2CH=C\underset{O^-Li^+}{\overset{O^-Li^+}{\underset{|}{}}}$

$\xrightarrow[(2)\ H_3O^+]{(1)\ CH_3CH=CHCH_2Br}$ $CH_3CH=CHCH_2\underset{CH_2CH_3}{\overset{}{\underset{|}{C}H}}COOH + CH_3\underset{CH_2CH_3}{\overset{HOOCCHCH_2CH_3}{\underset{|}{C}H}}CH=CH_2$

(e) Retrosynthesis:

$$CH_3CH=CHC\overset{O}{\overset{\|}{\underset{H}{}}} \Longrightarrow \underset{Br}{\overset{}{\underset{|}{}}}\overset{O}{\overset{\|}{C}}-H \Longrightarrow \overset{O}{\overset{\|}{C}}H + Br_2/CH_3COOH$$

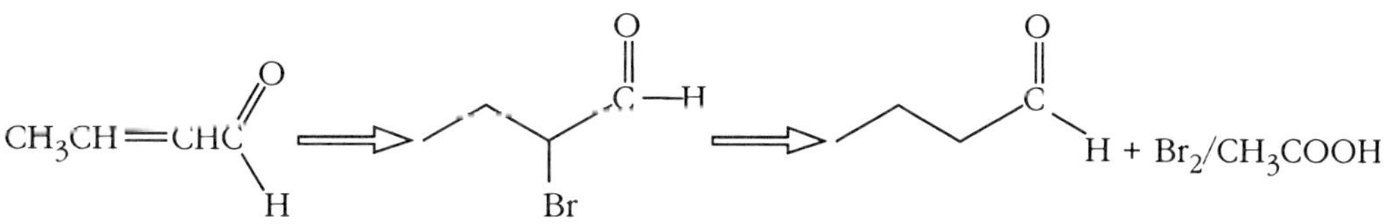

Synthesis:

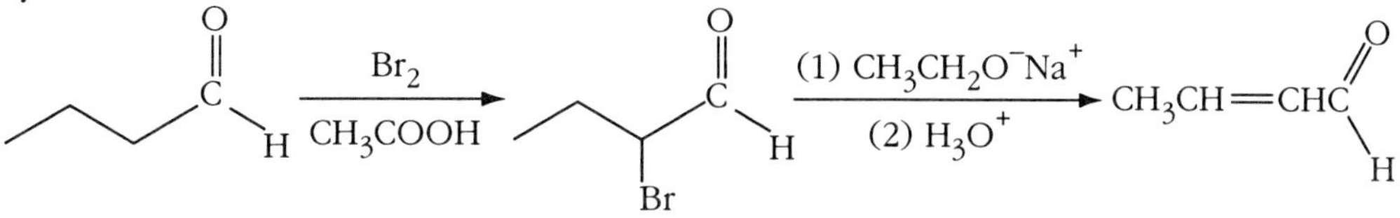

butanal $\xrightarrow[CH_3COOH]{Br_2}$ 2-bromobutanal $\xrightarrow[(2)\ H_3O^+]{(1)\ CH_3CH_2O^-Na^+}$ $CH_3CH=CHC\overset{O}{\overset{\|}{\underset{H}{}}}$

(f) Retrosynthesis:

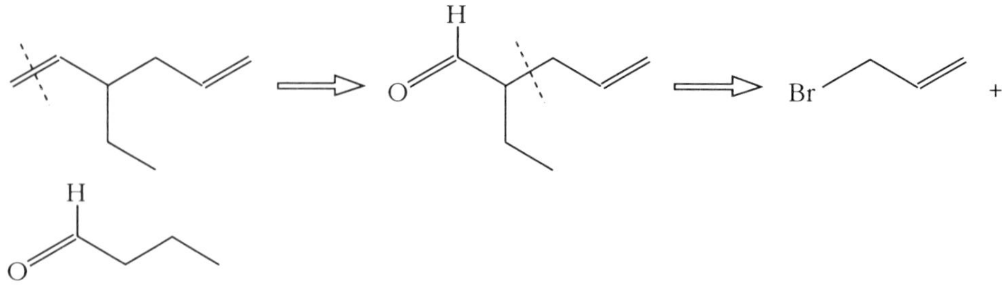
H
O
Br
+
H
O

Synthesis:

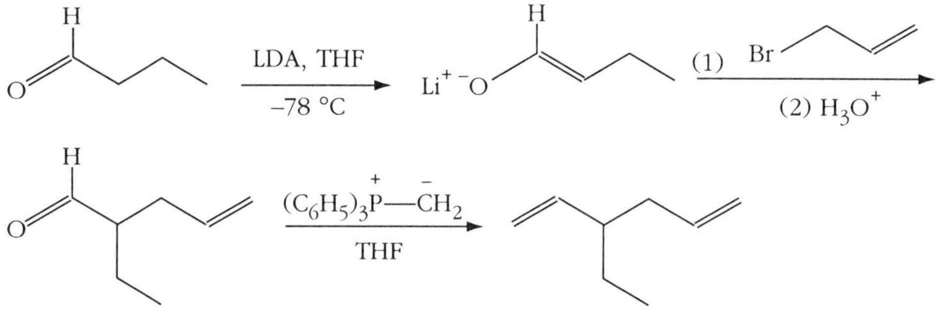
H
O
LDA, THF
−78 °C
H
Li⁺ ⁻O
(1) Br
(2) H₃O⁺
H
O
(C₆H₅)₃P—CH₂
THF

17.39 The mechanism of the Perkin condensation reaction:

Step 1: Acetate ion abstracts an α-proton from acetic anhydride to form the enolate anion

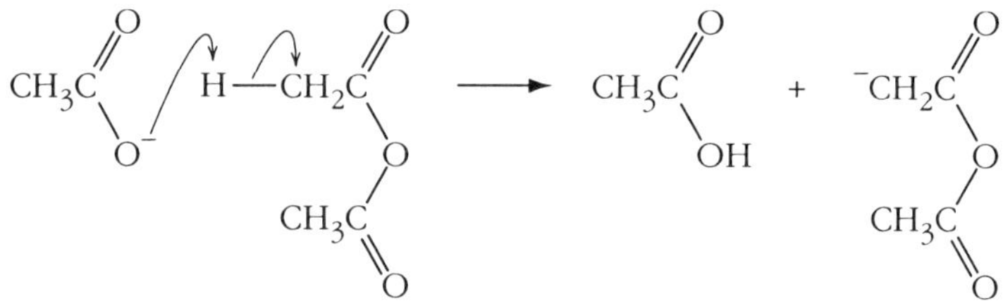
CH₃C
O
O⁻
H—CH₂C
O
O
CH₃C
O
CH₃C
O
OH
+
⁻CH₂C
O
O
CH₃C
O

Step 2: The enolate anion adds to the electrophilic carbonyl carbon atom of benzaldehyde

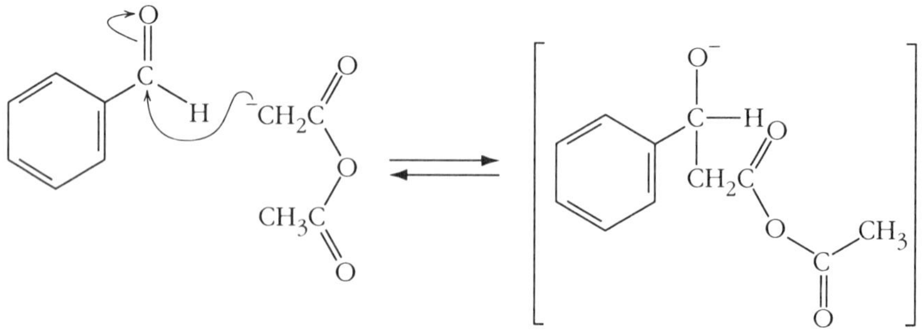
O
C
H
⁻CH₂C
O
O
CH₃C
O
O⁻
C—H
O
CH₂C
O
CH₃
C
O

Step 3: Protonation of the adduct, followed by heat, forms an unsaturated intermediate

$$\downarrow \text{Heat}$$

Step 4: Treatment of the unsaturated intermediate with water cleaves the anhydride and yields cinnamic acid

Cinnamic acid

17.40 (a) Retrosynthetic analysis:

2-ethyl-1-butene ⟹ pentan-3-one

Synthesis:

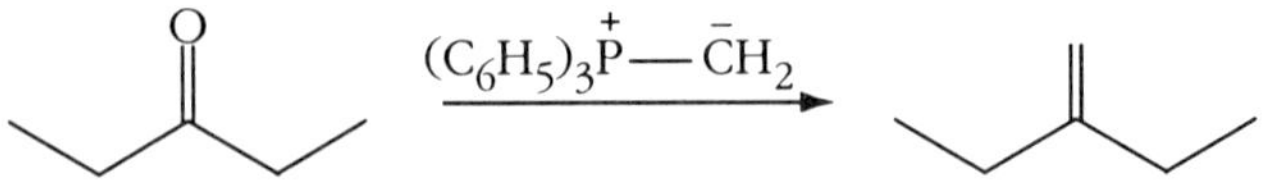

pentan-3-one + $(C_6H_5)_3\overset{+}{P}-\overset{-}{C}H_2$ → 2-ethyl-1-butene

(b) Retrosynthesis:

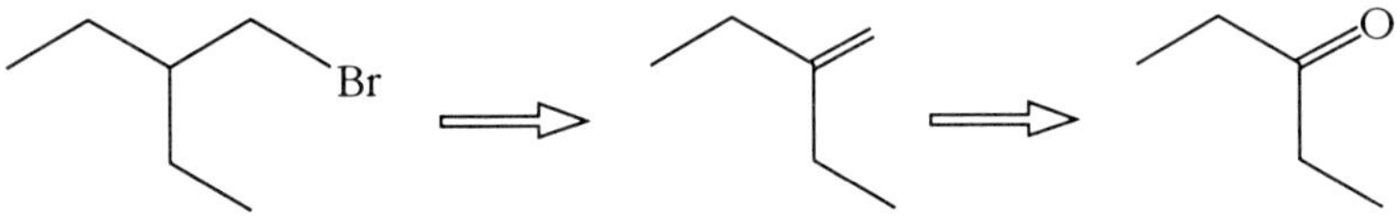

1-bromo-2-ethylbutane ⟹ 2-ethyl-1-butene ⟹ pentan-3-one

Synthesis:

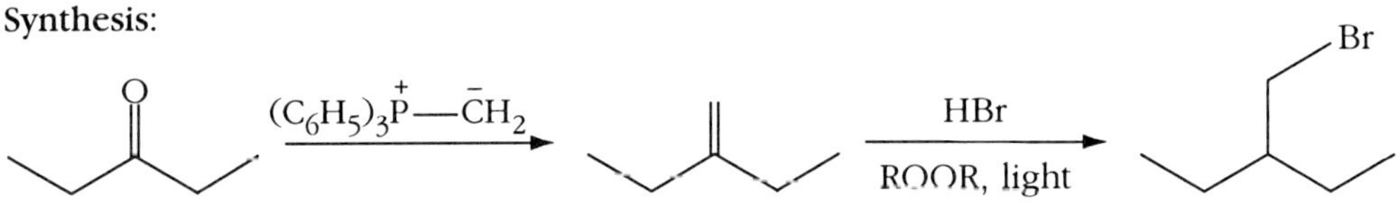

pentan-3-one $\xrightarrow{(C_6H_5)_3\overset{+}{P}-\overset{-}{C}H_2}$ 2-ethyl-1-butene $\xrightarrow[\text{ROOR, light}]{\text{HBr}}$ 1-bromo-2-ethylbutane

(c) Retrosynthesis:

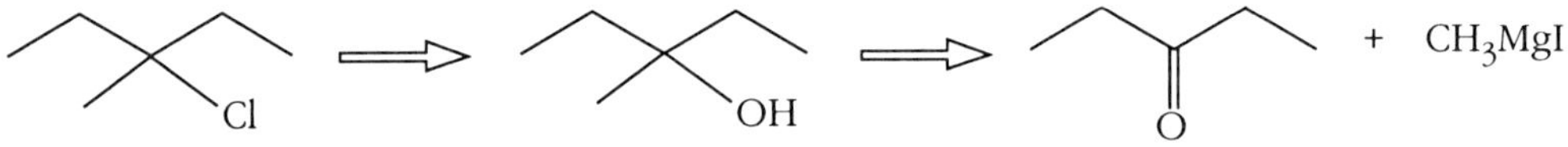

3-chloro-3-methylpentane ⟹ 3-methylpentan-3-ol ⟹ butan-2-one + CH_3MgI

Synthesis:

butan-2-one $\xrightarrow{CH_3MgI}$ alkoxide ($O^-\overset{+}{M}gI$) $\xrightarrow[\text{Ether}]{\text{Aq. HCl gas}}$ 3-chloro-3-methylpentane

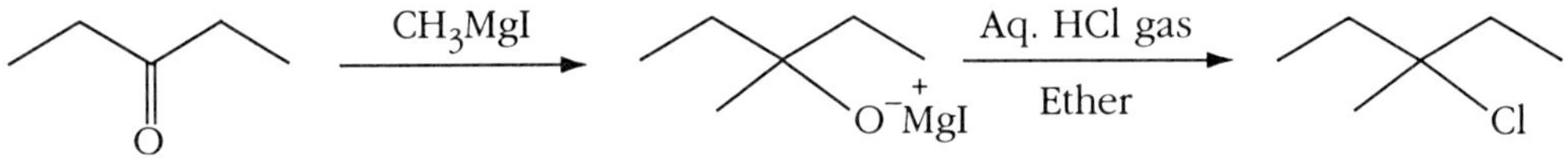

Both steps of the synthesis are carried out in the same reaction vessel so the alcohol does not need to be isolated.

(d) Retrosynthesis:

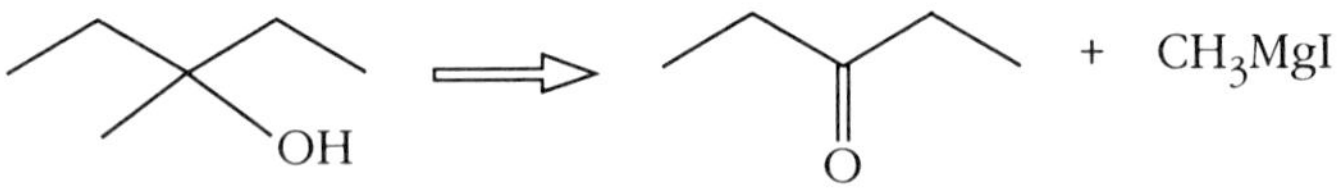

3-methylpentan-3-ol ⟹ pentan-3-one + CH_3MgI

Synthesis:

pentan-3-one $\xrightarrow[\text{Dry ether}]{CH_3MgI}$ alkoxide ($O^-\overset{+}{M}gI$) $\xrightarrow[\text{Cold}]{H_3O^+}$ 3-methylpentan-3-ol

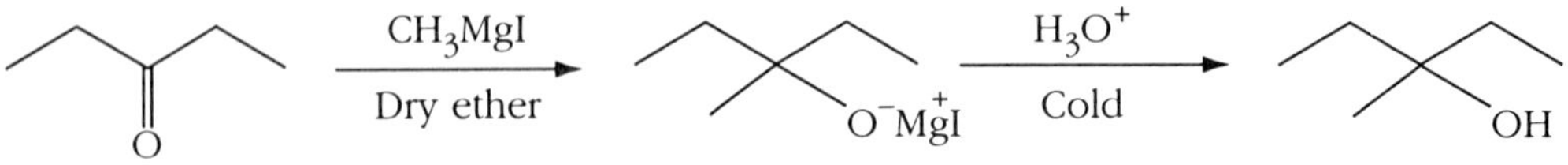

(e) Retrosynthesis:

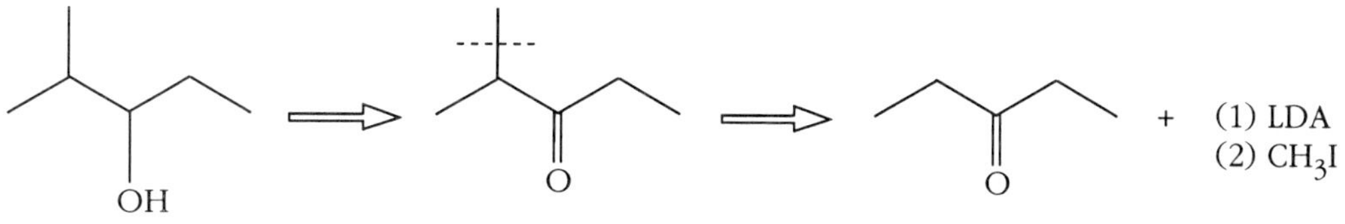

[structures showing retrosynthesis from alcohol to ketone to smaller ketone + (1) LDA (2) CH₃I]

Synthesis:

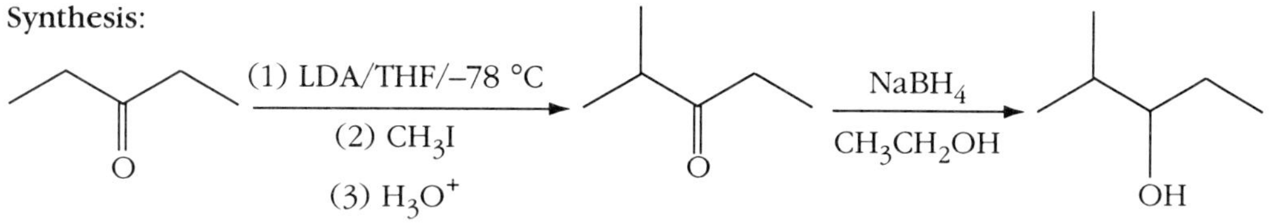

[ketone] $\xrightarrow{\text{(1) LDA/THF/−78 °C}}_{\text{(2) CH}_3\text{I} \quad \text{(3) H}_3\text{O}^+}$ [methylated ketone] $\xrightarrow{\text{NaBH}_4}_{\text{CH}_3\text{CH}_2\text{OH}}$ [alcohol]

(f) Retrosynthesis:

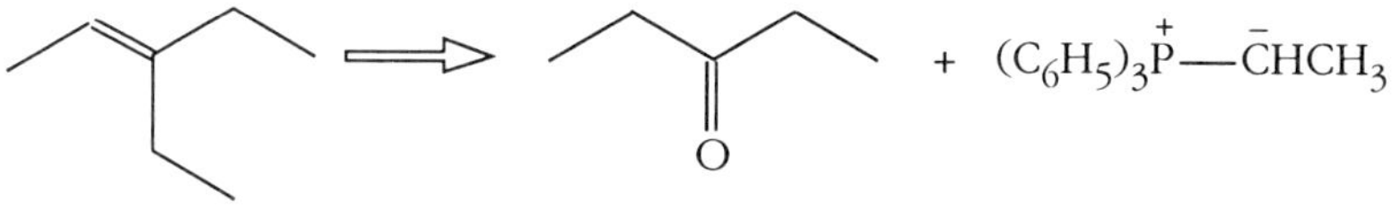

[alkene] ⟹ [ketone] + $(C_6H_5)_3\overset{+}{P}\text{—}\overset{-}{C}HCH_3$

Synthesis:

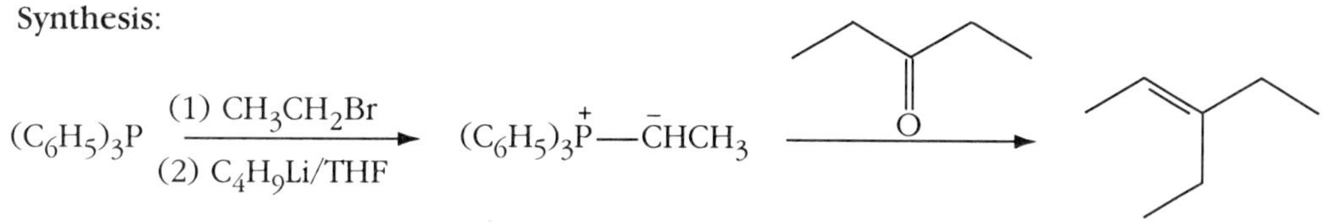

$(C_6H_5)_3P \xrightarrow{\text{(1) CH}_3\text{CH}_2\text{Br}}_{\text{(2) C}_4\text{H}_9\text{Li/THF}} (C_6H_5)_3\overset{+}{P}\text{—}\overset{-}{C}HCH_3 \xrightarrow{\text{[ketone]}}$ [alkene]

17.41 The acylation of *N*-(1-cyclohexenyl)-pyrrolidine.

Steps 1 and 2: Addition of the enamine to the carbon atom and elimination of the chloride ion

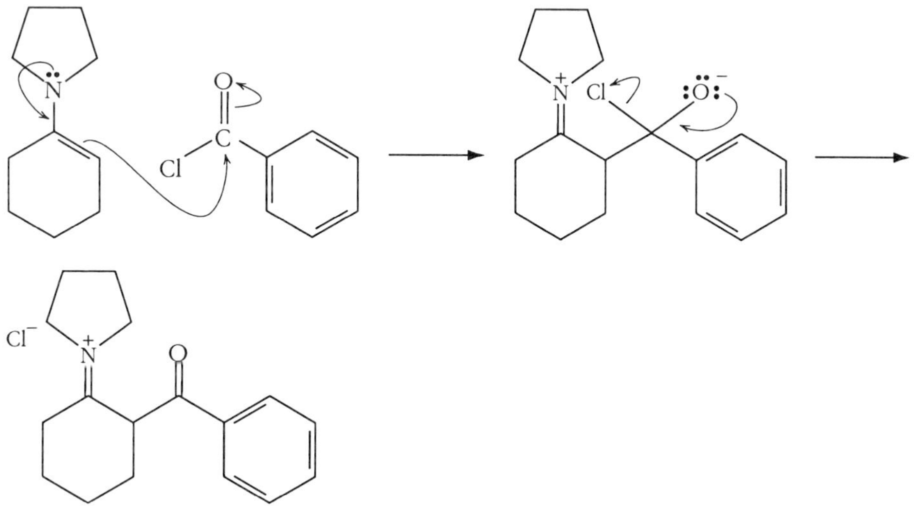

[mechanism diagram showing enamine attacking benzoyl chloride, forming iminium intermediate with chloride leaving, then resulting iminium-ketone product with Cl⁻]

Step 3: Proton transfer to another molecule of the enamine

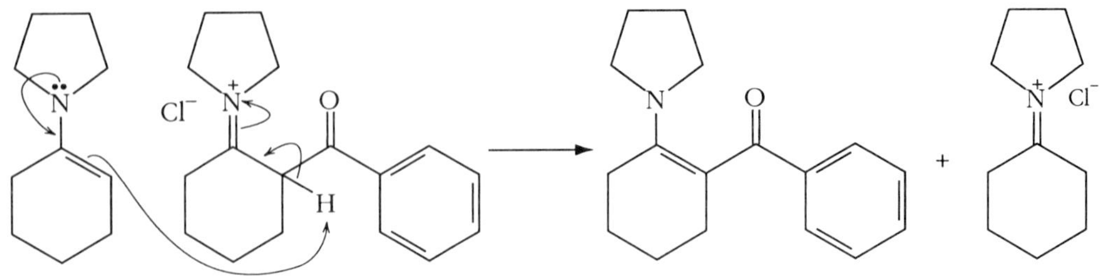

The mechanism of the reaction of benzoyl chloride with ammonia.
Steps 1 and 2: Addition of ammonia to the carbon atom and elimination of the chloride ion

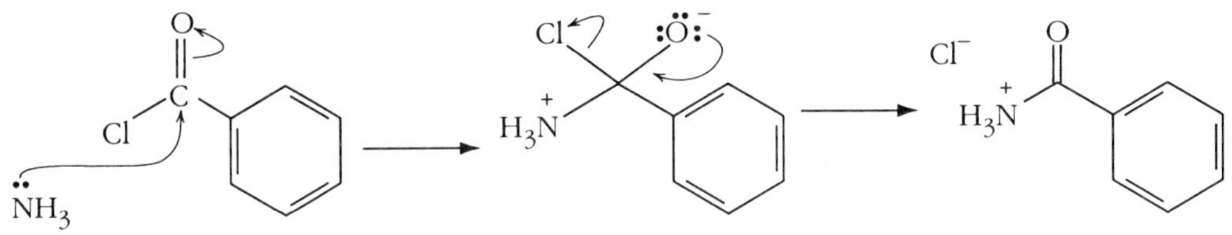

Step 3: Proton transfer to another molecule of ammonia

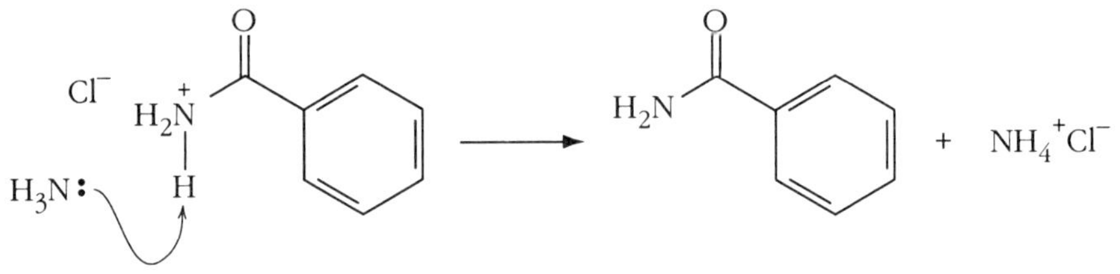

Both reactions occur by an addition-elimination mechanism. The major difference is that a carbon-carbon bond is formed in the acylation of an enamine and a carbon-nitrogen bond is formed in the reaction of ammonia and an acyl chloride.

17.42 (a) The molecular formula, $C_5H_9BrO_2$, indicates one unit of unsaturation. A carbon-carbon double bond is unlikely because there are no signals of vinyl hydrogen atoms in the region of δ 5 to 6. It may be a carbon-oxygen double bond or a ring.

The ^1H NMR spectrum shows four signals:

δ 1.2 (triplet, 3H)—a methyl group bonded to a methylene group,
δ 2.85 (triplet, 2H)—a methylene group bonded to another methylene group,
δ 3.52 (triplet, 2H)—a methylene group bonded to another methylene group and to an electron-withdrawing group or atom (most likely —Br because bromoalkanes show signals at around δ 3.5), δ 4.12 (quartet, 2H)—a methylene group bonded to a methyl group and an electronegative atom (most likely oxygen).

Thus we have the following fragments:

$BrCH_2CH_2-$ and CH_3CH_2O-

These fragments account for C₄H₉BrO and what remains unaccounted for is CO. Thus a carbonyl group is a third fragment. We can join the three fragments in two ways:

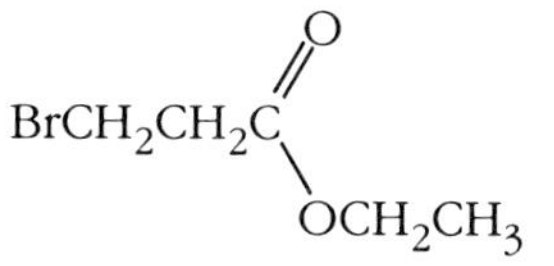

$$\text{BrCH}_2\text{CH}_2\text{C}(=O)\text{OCH}_2\text{CH}_3 \qquad \text{CH}_3\text{CH}_2\text{C}(=O)\text{OCH}_2\text{CH}_2\text{Br}$$

  Ethyl 3-bromopropanoate 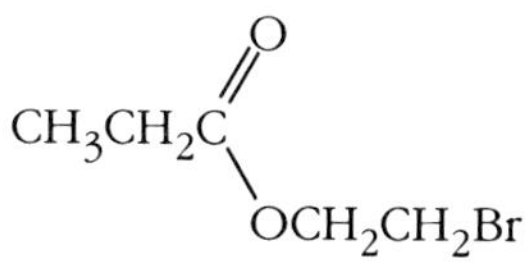   2-Bromoethyl propanoate

The signals of the ethyl group in ethyl 3-bromopropanoate should appear as a quartet at δ 4.1 and a triplet at δ 1.2 (which is what we observe in the spectrum shown). The signals of an ethyl group in 2-bromoethyl propanoate would appear as a quartet at δ 2.4 and a triplet at δ 1.0 (which is not observed in the spectrum shown).

The signals of the two methylene groups in ethyl 3-bromopropanoate appear as two triplets at δ 3.5 and δ 2.8 (which is what we observe). On the other hand, these signals in 2-bromoethyl propanoate would appear at δ 4.4 and 3.8 (we do not observe such signals).

We conclude that the unknown is **ethyl 3-bromopropanoate:**

$$\text{BrCH}_2\text{CH}_2\text{C}(=O)\text{OCH}_2\text{CH}_3$$

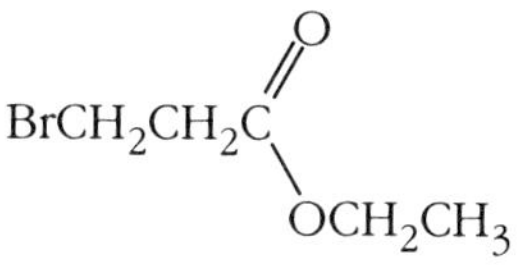

(b) The molecular formula, C₄H₈O₂, indicates one unit of unsaturation. A carbon-carbon double bond is unlikely because there are no signals of vinyl hydrogen atoms in the region of δ 5 to δ 6. It may be a carbon-oxygen double bond or a ring.

The ¹H NMR spectrum shows four signals:

δ 0.98 (triplet, 3H)—a methyl group bonded to a methylene group,
δ 1.58 (sextet, 2H)—a methylene group bonded to another methylene group and to a methyl group,
δ 2.32 (triplet, 2H)—a methylene group bonded to another methylene group and to an electron-withdrawing group (like a carbonyl group),
δ 11.1 (singlet, 1H)—a signal characteristic of a carboxylic hydrogen.

Thus, we have two fragments, namely a propyl group and a carboxylic group. The unknown is **butanoic acid:**

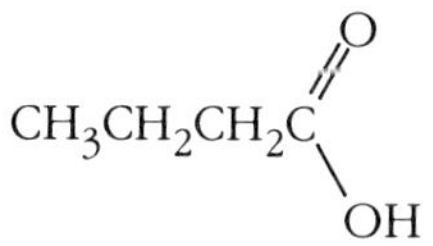

$$\text{CH}_3\text{CH}_2\text{CH}_2\text{C}(=O)\text{OH}$$

CHAPTER 18

FREE RADICAL REACTIONS
Concepts

Radicals or **free radicals** are species that have an odd number of electrons and are neutral. The chlorine radical is an example of a free radical.

$$Cl:Cl \xrightarrow{\text{Heat or light}} 2\ Cl^{\bullet}$$

The chlorine-chlorine bond was broken in a manner that each initially bonded atom retains one of the electrons in the bond. Such a cleavage of a bond is called a **homolytic cleavage**. "**Fish-hook**" **arrows** are used to show the change in position of single electrons. **Heterolytic cleavage**, on the other hand, occurs when one atom retains both the electrons from the bond that is broken. The dissociation of *tert*-butyl chloride to form a tertiary carbocation and chloride anion is an example of heterolytic cleavage.

$$H_3C-\underset{\underset{CH_3}{|}}{\overset{\overset{CH_3}{|}}{C}}-Cl \rightleftharpoons \underset{H_3C}{\overset{CH_3}{C^+}}{\diagdown}CH_3 + Cl^-$$

Like carbocations, free radicals are sp^2-hybridized, planar, and electron-deficient species. Also like carbocations, free radicals are stabilized by electron-donating alkyl substituents, making more highly substituted radicals more stable. Like carbocations, radicals can be stabilized by resonance. Overlap with the p orbitals of a π bond allows the odd electron to be delocalized over two carbon atoms. Benzyl and allyl radicals are even more stable than tertiary radicals.

The structure of a radical; the p orbital contains a single electron

Decreasing stability of radicals

$$\underset{R}{\overset{R}{\diagup}}C^{\bullet}\diagdown R > \underset{R}{\overset{R}{\diagup}}C^{\bullet}\diagdown H > \underset{H}{\overset{H}{\diagup}}C^{\bullet}\diagdown R > \underset{H}{\overset{H}{\diagup}}C^{\bullet}\diagdown H$$

Resonance stabilization

Halogenation of alkanes is one way to synthesize haloalkanes. There are many limitations of such syntheses: (i) mixtures of products are frequently obtained and (ii) only chlorination and bromination are practical (fluorine undergoes explosive reactions with hydrocarbons, iodine is nonreactive toward alkanes).

The mechanism of free-radical halogenation is best thought of as a series of stepwise reactions:

1. **Initiation:** $Cl:Cl \xrightarrow{\text{Heat or light}} 2\ Cl^\bullet$

2. **Propagation:**
$$Cl^\bullet + H:CH_3 \longrightarrow H:Cl + H_3C^\bullet$$
$$H_3C^\bullet + Cl:Cl \longrightarrow H_3C:Cl + Cl^\bullet$$
n times

3. **Termination:**
$$Cl^\bullet + Cl^\bullet \longrightarrow Cl_2$$
$$H_3C^\bullet + Cl^\bullet \longrightarrow CH_3Cl$$
$$H_3C^\bullet + H_3C^\bullet \longrightarrow CH_3CH_3$$

Radical hydrobromination of alkenes takes place in the presence of radical initiators such as peroxides. As a result, HBr adds to alkenes via a radical chain reaction mechanism. The reaction does not follow Markownikoff's rule because the bromine ends up on the less substituted carbon atom of the alkene.

$$(CH_3)_2C=CH_2 \xrightarrow[\text{Peroxides}]{HBr} (CH_3)_2CH-CH_2Br$$

Chain-growth polymers (also known as **addition polymers**) are made by adding monomers to the growing end of a chain. The monomers used most commonly are ethylene and substituted ethylenes. Chain-growth polymerization usually occurs by a radical mechanism or by use of Ziegler-Natta catalysts.

Radical polymerization has three distinct steps: (a) an **initiation** step (the initiator breaks homolytically into radicals), (b) **propagation** steps (a radical adds to an alkene monomer and converts it into a radical; the radical thus formed reacts with another monomer, adding a new unit), and (c) a **termination** step (two radicals combine).

The stereochemistry of the polymer depends on the mechanism by which polymerization occurs. Karl Ziegler and Giulio Natta found that the stereochemistry of a polymer can be controlled if the growing end of the chain and the incoming monomer are coordinated with an aluminum-titanium initiator.

$$(CH_3CH_2)_3Al + TiCl_3\ (\text{or } TiCl_4) \longrightarrow \text{Ziegler-Natta catalyst}$$

These initiators are called **Ziegler-Natta catalysts.** These catalysts allow the synthesis of stronger and stiffer polymers that have more resistance to cracking and heat (e.g., high-density polyethylene).

Mass spectrometry differs from the other types of spectroscopy discussed so far in that a mass spectrum is not a record of the energy absorbed by a molecule in going from one energy level to another. A mass spectrum is a record of the exact masses of a series of ions that are formed by fragmentation of a molecular species that is created by the collision of a molecule with a high-energy particle (usually an electron). The collision knocks an electron out of the molecule, giving rise to a radical-cation (a species with a positive charge and an unpaired electron) called the *molecular ion*.

$$\text{H:}\overset{\overset{\text{H}}{..}}{\underset{\underset{\text{H}}{..}}{\text{C}}}\text{:H} + e^- \longrightarrow 2\,e^- + \text{H:}\overset{\overset{\text{H}}{..}}{\underset{\underset{\text{H}}{..}}{\overset{+}{\text{C}}\cdot}}\text{H} \quad \text{abbreviated as} \quad CH_4^{+\cdot}$$

Methane   Methane radical-cation

Following its formation, the methane radical-cation decomposes in a series of reactions called **fragmentation reactions.** In one such reaction, it loses a hydrogen atom to generate the methyl cation.

$$\text{H:}\overset{\overset{\text{H}}{..}}{\underset{\underset{\text{H}}{}}{\overset{+}{\text{C}}\cdot}}\text{H} \longrightarrow \text{H}^\bullet + {}^+CH_3$$

Methane radical-cation   Methyl cation
$m/z = 16$ $m/z = 15$

The ion derived from electron ejection before any fragmentation takes place is known as the **molecular ion,** and is abbreviated M$^{+\cdot}$. The molecular ion occurs at an m/z value equal to the molecular mass of the sample molecule. In the mass spectrum of methane, for example, the molecular ion occurs at $m/z = 16$.

The **base peak** is the ion of greatest relative abundance. The base peak is arbitrarily assigned a relative abundance of 100%, and the other peaks in the mass spectrum are scaled relative to it.

Associated with each peak in a mass spectrum are other peaks at higher or lower mass that arise from the presence of isotopes in their natural abundance. Such isotopic peaks are particularly useful for diagnosing the presence of elements that consist of more than one isotope with high natural abundance, such as chlorine or bromine.

Solutions to the Exercises

18.1

(a)  $Cl \overset{\frown\frown}{\text{—}} Cl \longrightarrow 2\ Cl^\bullet$

(b) 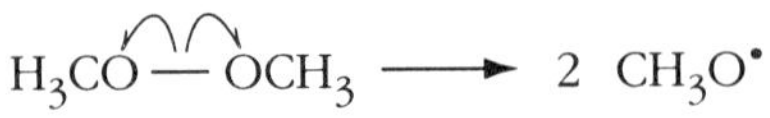 $H_3CO \overset{\frown\frown}{\text{—}} OCH_3 \longrightarrow 2\ CH_3O^\bullet$

(c) 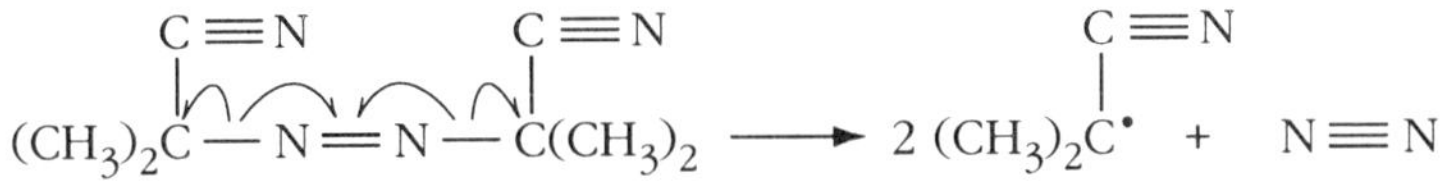 $(CH_3)_2\underset{\underset{C\equiv N}{|}}{C} \overset{\frown\frown}{\text{—}} N=N \overset{\frown}{\text{—}} \underset{\underset{C\equiv N}{|}}{C}(CH_3)_2 \longrightarrow 2\ (CH_3)_2\underset{\underset{C\equiv N}{|}}{C}^\bullet\ +\ N\equiv N$

18.2 Homolytic cleavage of a bond occurs in any reaction that involves splitting of one or more electron-pair bonds into unpaired electrons, or any reaction that involves the joining of two unpaired electrons to give an electron-pair bond. In contrast, any reaction involving the movement of only electron-pair bonds and/or unshared electron-pairs is heterolytic.

(a) Homolytic cleavage (each atom retains one electron),

$$H \overset{\frown\frown}{\text{—}} Br \longrightarrow H^\bullet\ +\ Br^\bullet$$

(b) Heterolytic cleavage (oxygen retains two electrons and forms the hydroxide ion),

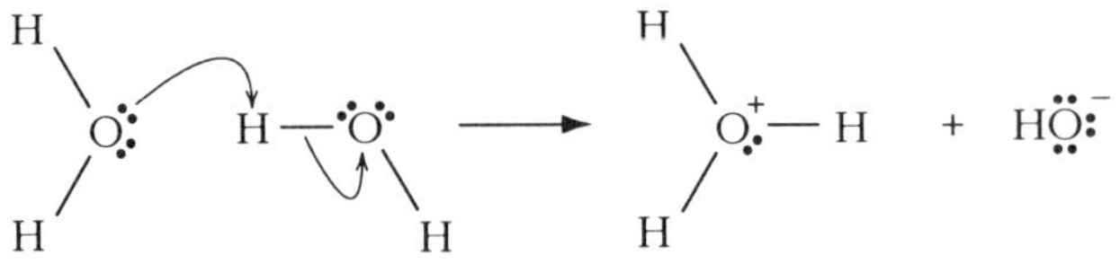

[diagram of water attacking H—O—H to give H$_3$O$^+$ + HO$^-$]

(c) Heterolytic cleavage (chlorine retains two electrons and forms chloride ion),

$CH_2{=}CH_2\quad H{\text{—}}\ddot{\underset{..}{Cl}}{:} \longrightarrow CH_3{-}\overset{+}{C}H_2\ +\ :\ddot{\underset{..}{Cl}}{:}^-$

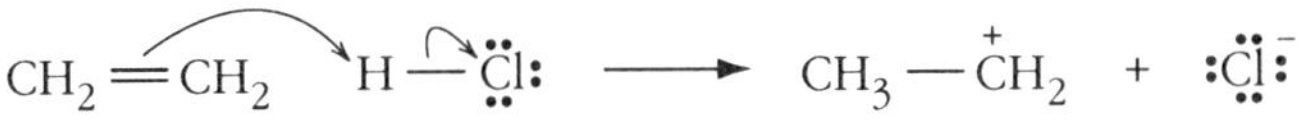

(d) Homolytic cleavage (each carbon atom retains one electron),

$\overset{\frown\frown}{CH_2}{=}CH_2\ \overset{\frown}{\ }^\bullet SCH_3 \longrightarrow\ ^\bullet CH_2{-}CH_2SCH_3$

(e) Heterolytic cleavage (chlorine retains two electrons and forms chloride ion),

$H_3N{:}\quad H{\text{—}}\ddot{\underset{..}{Cl}}{:} \longrightarrow H_4N^+\quad :\ddot{\underset{..}{Cl}}{:}^-$

(f) Homolytic cleavage (each carbon atom retains one electron).

$\overset{\frown\frown}{CH_2}{=}CH_2\ \overset{\frown}{\ }^\bullet CCl_3 \longrightarrow\ ^\bullet CH_2{-}CH_2CCl_3$

18.3 The regeneration of a free radical is characteristic of a propagation step of a chain reaction. In the first propagation step, a chlorine radical collides with a chloromethane molecule and removes a hydrogen atom from chloromethane.

$$Cl\cdot \quad H—CH_2Cl \longrightarrow HCl + \cdot CH_2Cl$$

In the second propagation step, the chloromethyl radical reacts with a molecule of chlorine to form dichloromethane. Thus, the second propagation step reproduces another chlorine radical, which can react with another molecule of chloromethane to continue the chain reaction.

$$Cl—Cl \quad \cdot CH_2Cl \longrightarrow Cl\cdot + CH_2Cl_2$$

18.4 In order to obtain principally chloromethane, we should use an excess of methane. In such a case, the chlorine radical formed in the initiation step will have a high probability of colliding with a methane molecule rather than one of the chlorination products. We will still have some dichlorination (or polychlorination) and some unreacted methane but the major product will be chloromethane.

In order to form tetrachloromethane, an excess of chlorine should be used. The overall reaction stoichiometry shows that four moles of chlorine are needed to replace four hydrogen atoms of methane. Thus, more than four moles would assure the completion of the reaction.

$$CH_4 + 4\,Cl_2 \longrightarrow CCl_4 + 4\,HCl$$

18.5 After examining the chlorination products of many hydrocarbons, it became apparent that, at room temperature, it is five times easier for a chlorine radical to abstract a hydrogen from a tertiary carbon than from a primary carbon and 3.8 times easier to abstract a hydrogen atom from a secondary carbon than from a primary carbon. Thus, the most easily replaced hydrogen atoms are those attached to the most highly branched carbon atoms.

(a) Only secondary and primary hydrogens are present in butane. The secondary hydrogens will be replaced more easily than the primary ones.

$$CH_3—(CH_2)(CH_2)—CH_3$$

(b) Primary, secondary, and one tertiary hydrogen are present in 2-methylbutane. The tertiary hydrogen bonded to C2 will be the most easily replaced by chlorine.

$$\begin{array}{c} CH_3 \\ | \\ CH_3—CH—CH_2—CH_3 \end{array}$$

(c) Primary, secondary, and one tertiary hydrogen are present in methylcyclohexane. The tertiary hydrogen bonded to C1 will be the most easily replaced by chlorine.

(d) Primary and secondary hydrogens are present in 2,2-dimethylbutane. The secondary hydrogens bonded to C3 will be the most easily replaced by chlorine.

(e) Only secondary and primary hydrogens are present in 3,3-diethylpentane. The secondary hydrogens will be replaced more easily than the primary ones.

18.6 The relative stability of free radicals can be determined by comparing their homolytic bond dissociation energies. By doing so, we obtain the following relative stability sequence:

Tertiary > Secondary > Primary > Methyl

Applying the above relative stability sequence to radicals given in this Exercise, we find the primary radical (a) to be the least stable, the secondary radical (c) to be in the middle, and the tertiary radical (b) to be the most stable.

$$\cdot CH_2CHCH_2CH_3 \ < \ CH_3\overset{\cdot}{C}HCHCH_3 \ < \ CH_3\overset{\cdot}{C}CH_2CH_3$$
(each with a CH_3 substituent)

18.7 The free radical mechanism for the chlorination of an alkane consists of three kinds of steps: an initiation step, propagation steps, and termination steps. For the chlorination of methylpropane, those steps are as follows:
Initiation step:

$$Cl-Cl \xrightarrow[\text{or light}]{\text{Heat}} 2\ Cl\cdot$$

Propagation steps: There are two types of hydrogen atoms in methylpropane, so two different radicals are formed in the first propagation step.

$$CH_3CH(CH_3)-CH_2-H \quad \cdot Cl \longrightarrow HCl + CH_3CH(CH_3)-\dot{C}H_2 \quad (1° \text{ radical})$$

$$CH_3C(CH_3)(CH_3)-H \quad \cdot Cl \longrightarrow HCl + CH_3\dot{C}(CH_3)(CH_3) \quad (3° \text{ radical})$$

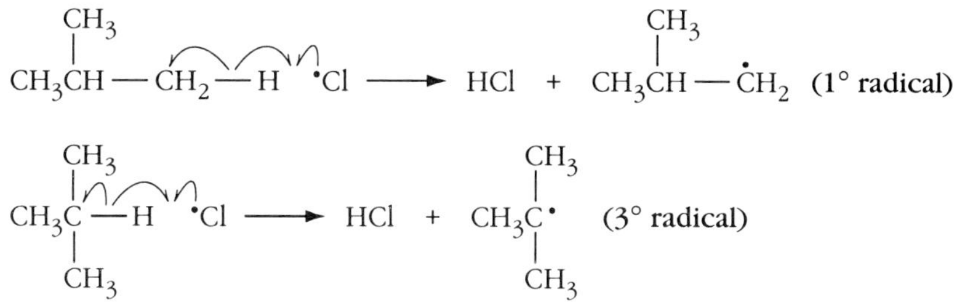

There are two radical intermediates formed in the first propagation step. Therefore, two different products are formed in the second propagation step.

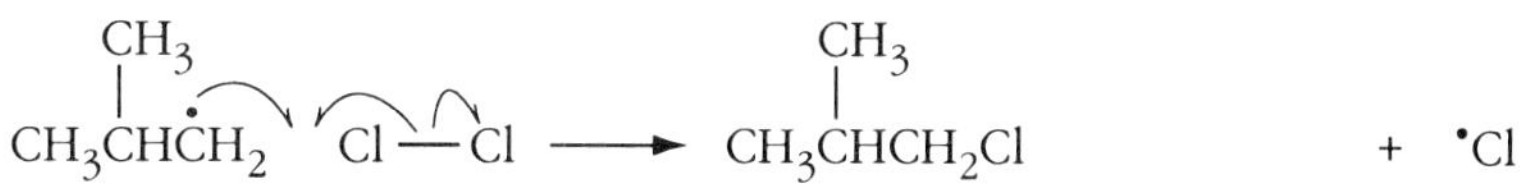

$$CH_3CH(CH_3)\dot{C}H_2 \quad Cl-Cl \longrightarrow CH_3CH(CH_3)CH_2Cl + \cdot Cl$$

1-Chloro-2-methylpropane

$$CH_3\dot{C}(CH_3)(CH_3) \quad Cl-Cl \longrightarrow CH_3C(CH_3)(CH_3)-Cl + \cdot Cl$$

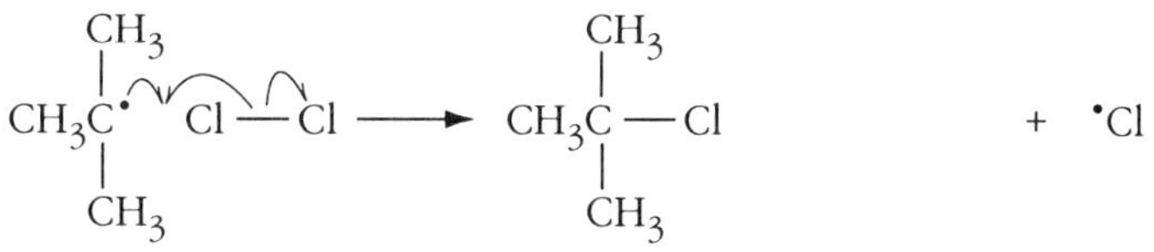

2-Chloro-2-methylpropane

Termination step: Any two of the three free-radicals collide to form a stable molecule.

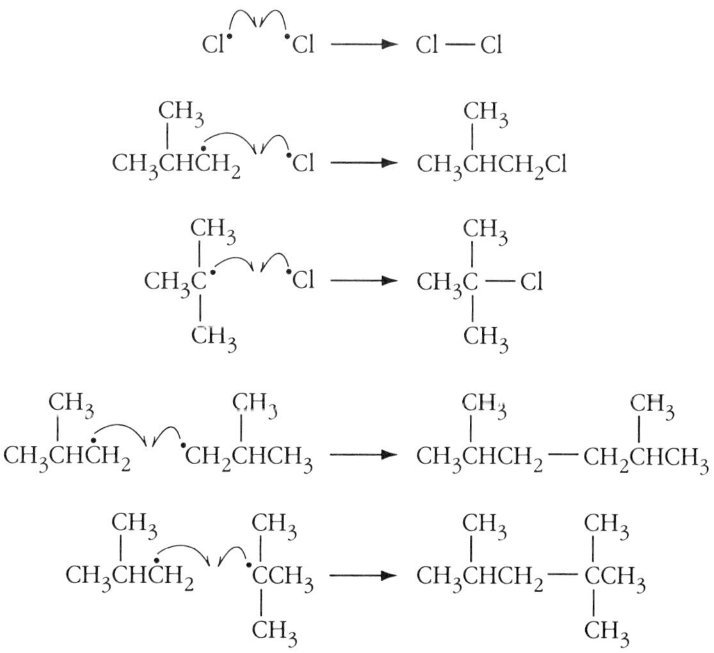

$$Cl\cdot \quad \cdot Cl \longrightarrow Cl-Cl$$

$$CH_3CH(CH_3)\dot{C}H_2 \quad \cdot Cl \longrightarrow CH_3CH(CH_3)CH_2Cl$$

$$CH_3\dot{C}(CH_3)(CH_3) \quad \cdot Cl \longrightarrow CH_3C(CH_3)(CH_3)-Cl$$

$$CH_3CH(CH_3)\dot{C}H_2 \quad \dot{C}H_2CH(CH_3)CH_3 \longrightarrow CH_3CH(CH_3)CH_2-CH_2CH(CH_3)CH_3$$

$$CH_3CH(CH_3)\dot{C}H_2 \quad \dot{C}(CH_3)(CH_3)CH_3 \longrightarrow CH_3CH(CH_3)CH_2-C(CH_3)(CH_3)CH_3$$

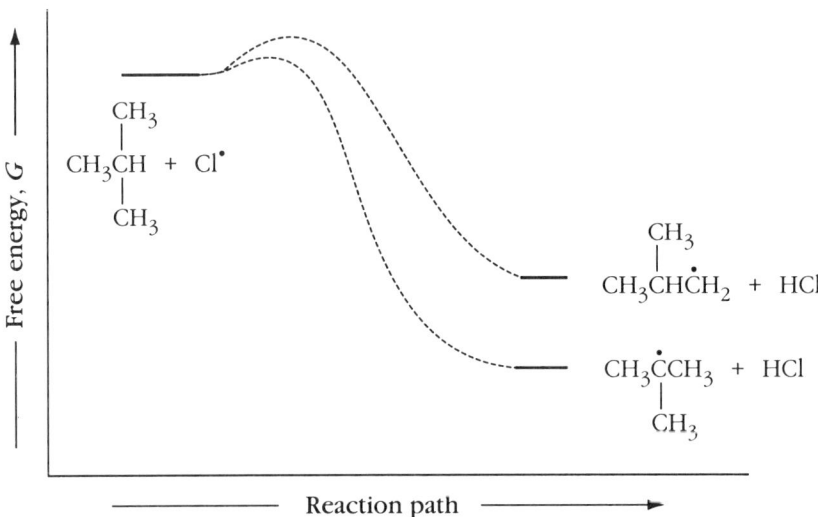

The free energy-reaction path diagram for the first reaction in the propagation cycle of the chlorination of methylpropane:

18.8

Initiation, Propagation steps showing formation of (R)-3-Chloro-3-methylhexane and (S)-3-Chloro-3-methylhexane.

18.9 (a) In the presence of a free radical initiator (In•), a free-radical mechanism operates and the addition of HBr to alkenes occurs such that the hydrogen is bound to the carbon of the double bond bearing the greater number of alkyl branches (anti-Markownikoff product).

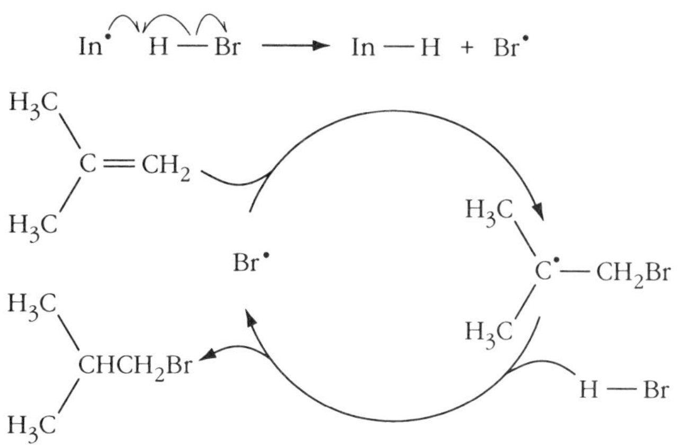

1-Bromo-2-methylpropane

(b) In the presence of a free radical inhibitor, an ionic mechanism operates and the "Markownikoff" product is formed.

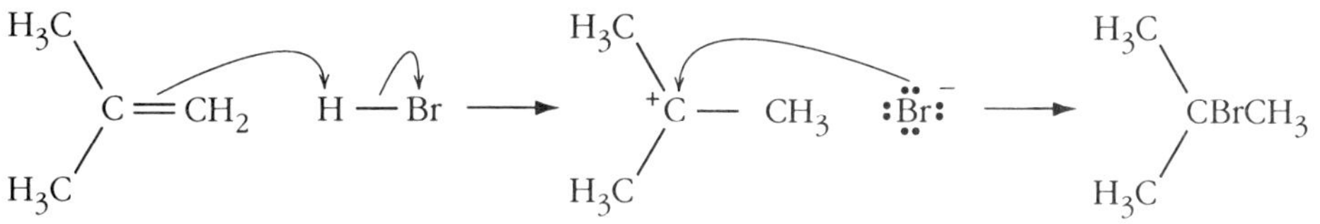

2-Bromo-2-methylpropane

18.10 Mechanism:

Initiation: $(CH_3)_3C-O-O-C(CH_3)_3 \xrightarrow{\text{Heat}} 2\,(CH_3)_3C-O^\bullet$

Propagation steps: $(CH_3)_3C-O^\bullet + H-Br \longrightarrow (CH_3)_3C-OH + Br^\bullet$

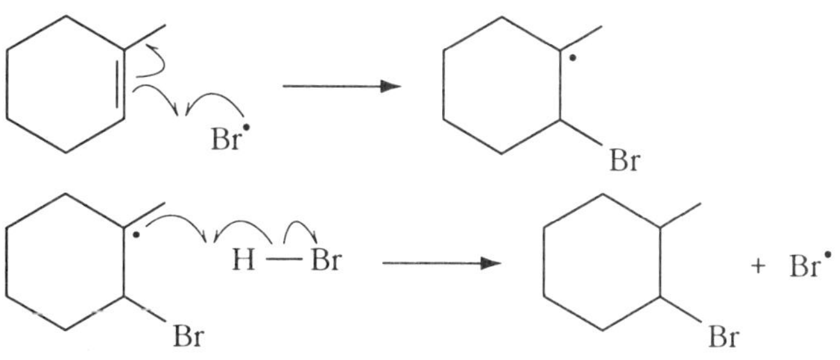

Overall equation:

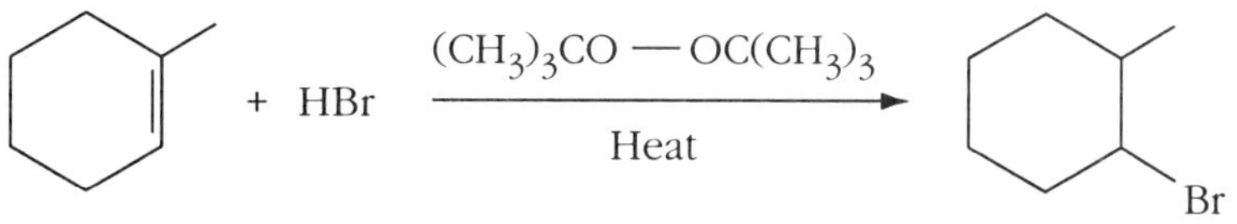

1-Methylcyclohexene + HBr $\xrightarrow[\text{Heat}]{(CH_3)_3CO-OC(CH_3)_3}$ 1-Bromo-2-methylcyclohexane

18.11 Vitamin E (α-tocopherol) reacts with the hydroxy radical, HO•, and forms the phenoxy radical. Phenoxy radicals are resonance stabilized and are much less reactive than hydroxy radicals (OH•), so they do not initiate any further propagation steps. In this manner, a reactive hydroxy radical is eliminated from the medium.

18.12

(a) Styrene polymerization to polystyrene.

(b) 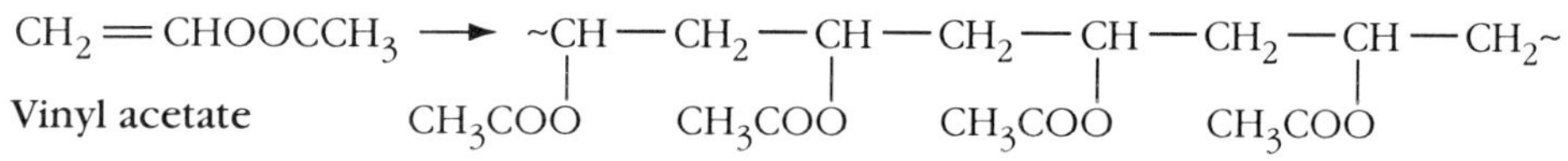 $4\ CH_2\!=\!CHOOCCH_3\ \longrightarrow$ ~CH—CH$_2$—CH—CH$_2$—CH—CH$_2$—CH—CH$_2$~
Vinyl acetate | | | |
              CH$_3$COO CH$_3$COO CH$_3$COO CH$_3$COO

(c) 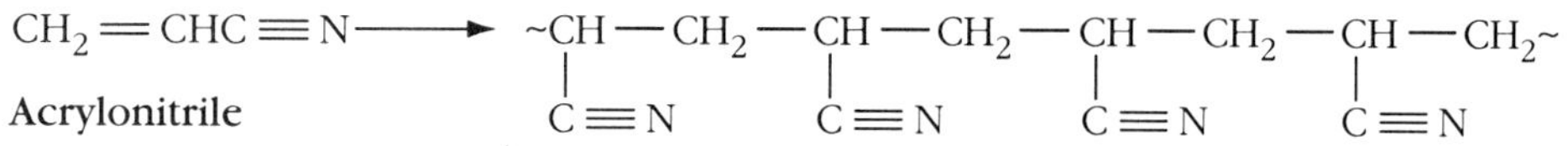 $4\ CH_2\!=\!CHC\!\equiv\!N \longrightarrow$ ~CH—CH$_2$—CH—CH$_2$—CH—CH$_2$—CH—CH$_2$~
Acrylonitrile | | | |
              C≡N   C≡N   C≡N   C≡N

(d) 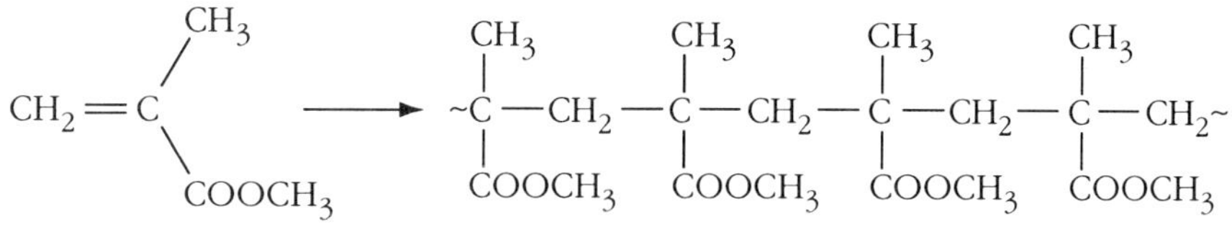 $4\ CH_2\!=\!C\!\begin{smallmatrix}CH_3\\ \\COOCH_3\end{smallmatrix}\longrightarrow$
          CH$_3$ CH$_3$ CH$_3$ CH$_3$
           | | | |
        ~C—CH$_2$—C—CH$_2$—C—CH$_2$—C—CH$_2$~
           | | | |
        COOCH$_3$ COOCH$_3$ COOCH$_3$ COOCH$_3$

Methyl methacrylate

18.13 Polymers are grouped into three classes according to their predominant stereochemistry:

(a) **Atactic** polystyrene has the side group (phenyl group) randomly placed on either side of the polymer backbone (main chain):

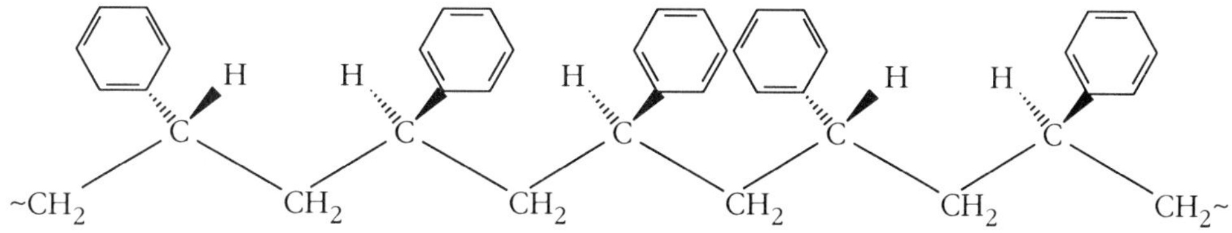

(b) **Isotactic** polystyrene has all the side groups (phenyl groups) generally on the same side of the polymer backbone (main carbon chain):

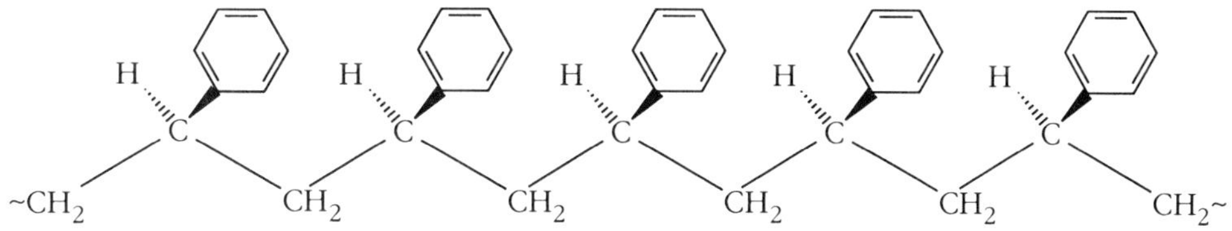

(c) **Syndiotactic** polystyrene has the side groups generally alternating from one side to the other:

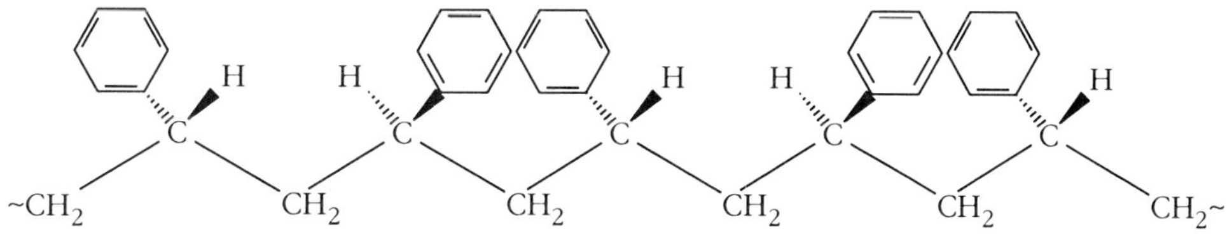

18.14 Each of the stereocenters formed by polymerization exists as a mixture of the two possible stereoisomers. As a result, the polymer is optically inactive.

18.15 It is impossible to form syndiotactic or isotactic polyvinylidene by the polymerization of vinylidene chloride because both side groups are identical (chlorine atoms). Each carbon atom in the polymer has two identical substituents (hydrogen atoms or chlorine atoms). That is why we do not have a stereocenter on any of the carbon atoms and we cannot talk about the stereochemistry of the polymer.

18.16 The molecular ion is a species with an unpaired electron formed when the electron beam hits a molecule and knocks an electron out of the compound. If a compound contains nonbonded electrons, electron bombardment dislodges a nonbonded electron because a molecule does not hold onto its nonbonded electrons as tightly as it holds onto its bonding electrons. Therefore the electron ejected from methanol is that from a lone pair of oxygen. The same is observed in the case of formaldehyde. In the case of methylamine, the first electron ejected is that from the lone pair of nitrogen.

(a) H:C̈:Ö:H (b) H₂C::Ö⁺ (c) H:C̈:N̈:H

18.17 The molecular ion is formed when an electron is ejected from the compound, so the m/z value of the molecular ion is the same as the molecular weight of the compound.

(a) Ethane has molecular mass = 30. The m/z of its molecular ion is also 30.
(b) 2-Chloropropane has molecular mass = 78 (for the ^{35}Cl isotope). The m/z of its molecular ion is also 78.
(c) Methanol has molecular mass = 32. The m/z of its molecular ion is also 32.

18.18 An (M + 1) peak is a peak that occurs at m/z one mass unit higher than the molecular ion (M). This ion occurs when the molecule contains at least one element that exists as a mixture of isotopes with significant abundance. For the elements present in molecules listed in Exercise 18.17 we have the following isotopic abundances of the main isotopes:

$$^{12}C = 98.90\% \quad ^{13}C = 1.10\%$$
$$^{35}Cl = 75.77\% \quad ^{37}Cl = 24.23\%$$
$$^{16}O = 99.76\% \quad ^{18}O = 0.2\%$$

(a) Ethane has a peak at (M+1)⁺ due to the ^{13}C isotope of carbon. The ratio of (M+1) to the M peak is 0.01.
(b) 2-Chloropropane has a peak at (M+1)⁺ due to the ^{13}C isotope of carbon. The ratio of (M+1) to the M peak is 0.01. 2-Chloropropane also has a peak at (M+2)⁺ due to the ^{37}Cl isotope of chlorine. The ratio of the (M+2) peak to the M peak is 0.25.
(c) Methanol has a peak at (M+1)⁺ due to the ^{13}C isotope of carbon. The ratio of (M+1) to the M peak is 0.01. Methanol also has a tiny (M+2) peak due to the presence of the ^{18}O isotope of oxygen. The ratio of the (M+2) peak to the M peak is 0.002.

18.19 Bromomethane contains nonbonded electrons on bromine, and electron bombardment dislodges an electron from a lone pair because nonbonded electrons are held less tightly than bonding electrons. The easiest bond to break in this compound is the carbon-bromine bond because it is the weakest. The bond breaks heterolytically, with the electrons going to the more electronegative of the atoms that formed the bond; consequently, the base peak in the mass spectrum of bromomethane is at $m/z = 15$ [M - 79 or (M + 2) - 81].

$$CH_3Br \xrightarrow{-e^-} [CH_3 - ^{79}Br]^{+\cdot} + [CH_3 - ^{81}Br]^{+\cdot} \longrightarrow {}^+CH_3 + Br^{\cdot}$$

$$m/z = 79 \qquad m/z = 81 \qquad\qquad m/z = 15$$

18.20 The isotopic distribution of chlorine is the following: ^{35}Cl = 75.77 % and ^{37}Cl = 24.13%. The molecular ion of chloromethane is at $m/z = 50$ and the (M + 2) peak is at $m/z = 52$. The peaks at $m/z = 50$ and $m/z = 52$ should have a 75.77 : 24.23 (or 3:1) ratio.

18.21 (a) Fragmentation of alcohols is characterized by the loss of water and α-cleavage:

$$CH_3CH_2CH_2OH \xrightarrow{-e^-} \begin{cases} [CH_3\overset{\frown}{C}H_2 - CH_2 - \overset{\cdot}{O}H]^+ \longrightarrow CH_3\overset{\cdot}{C}H_2 + \overset{H}{\underset{H}{\diagdown}}C=\overset{+}{O}H \\ m/z = 60 \qquad\qquad\qquad\qquad\qquad\qquad m/z = 31 \\ \\ [CH_3\overset{H}{\underset{|}{C}}H - \overset{\overset{\cdot}{O}H}{\underset{|}{C}}H_2]^+ \longrightarrow [CH_3CH=CH_2]^+ + H_2O \\ m/z = 60 \qquad\qquad m/z = 42 \end{cases}$$

(b) α-Cleavage predominates and is responsible for a base peak at $m/z = 45$:

$$\underset{H_3C}{\overset{OH}{\underset{|}{CH}}}\diagdown_{CH_3} \xrightarrow{-e^-} [\underset{H_3C}{\overset{\overset{\cdot}{O}H}{\underset{|}{CH}}}\diagdown_{CH_3}]^+ \longrightarrow {}^{\cdot}CH_3 + CH_3\overset{+OH}{\underset{H}{C}} \longrightarrow \overset{H}{\underset{H}{\diagdown}}C=\overset{+}{\underset{H}{C}} + H_2O$$

$$m/z = 60 \qquad\qquad\qquad m/z = 45 \qquad\qquad m/z = 27$$

(c) Alkanes cleave to give the most stable carbocations:

$$CH_3CH_2CH_2CH_3 \xrightarrow{-e^-} \begin{cases} [CH_3CH_2 - CH_2 \cdot CH_3]^+ \longrightarrow {}^{\cdot}CH_3 + CH_3CH_2\overset{+}{C}H_2 \\ m/z = 58 \qquad\qquad\qquad\qquad\qquad m/z = 43 \\ \\ [CH_3CH_2 \cdot CH_2CH_3]^+ \\ m/z = 58 \end{cases}$$

(d) 2-Methylpropane has the same molecular formula as butane, and therefore it also has a molecular ion with $m/z = 58$. Its mass spectrum is similar to that of butane with one notable exception: the peak at $m/z = 43$ is much more abundant and becomes the base peak. The peak at $m/z = 43$ for 2-methylpropane is so abundant because loss of a methyl group from

2-methylpropane forms a secondary carbocation, while loss of the same group from butane forms a less stable primary carbocation.

$$\text{(CH}_3)_3\text{CH} \xrightarrow{-e^-} [(\text{CH}_3)_3\text{CH}]^{+\cdot} \longrightarrow {}^\cdot\text{CH}_3 + \text{CH}_3\overset{+}{\text{C}}\text{HCH}_3$$

$$m/z = 58 \qquad\qquad m/z = 43$$

18.22 (a) The mass spectrum of heptane shows several characteristics typical of straight-chain alkanes. The base peak ($m/z = 43$) corresponds to propyl cation. Note that the different fragments are not formed with the same relative abundances. Furthermore, some possible fragment peaks are weak or missing. Thus, the fragment for methyl cation ($m/z = 15$) is very weak and the peaks corresponding to fragments at m/z greater than 71 are very weak.

$$[\text{CH}_3\text{CH}_2-\text{CH}_2-\text{CH}_2-\text{CH}_2-\text{CH}_2-\text{CH}_3]^{+\cdot}$$

Molecular ion
$m/z = 100$

$\longrightarrow \text{CH}_3\text{CH}_2\text{CH}_2\overset{\cdot}{\text{C}}\text{H}_2 + \text{CH}_3\text{CH}_2\overset{+}{\text{C}}\text{H}_2 \longrightarrow \text{H}_3\text{C}\text{HC}=\overset{+}{\text{C}}\text{H} + \text{H}_2$

$m/z = 43 \qquad\qquad m/z = 41$

$\longrightarrow \text{CH}_3\text{CH}_2\text{CH}_2\text{CH}_2\overset{\cdot}{\text{C}}\text{H}_2 + \text{CH}_3\overset{+}{\text{C}}\text{H}_2 \longrightarrow \text{CH}_2=\overset{+}{\text{C}}\text{H} + \text{H}_2$

$m/z = 29 \qquad\qquad m/z = 27$

$\longrightarrow \text{CH}_3\text{CH}_2\overset{\cdot}{\text{C}}\text{H}_2 + \text{CH}_3\text{CH}_2\text{CH}_2\overset{+}{\text{C}}\text{H}_2$

$m/z = 57$

$\longrightarrow \text{CH}_3\overset{\cdot}{\text{C}}\text{H}_2 + \text{CH}_3\text{CH}_2\text{CH}_2\text{CH}_2\overset{+}{\text{C}}\text{H}_2$

$m/z = 71$

(b) Fragmentation of a branched alkane commonly occurs at a branch carbon atom to give the most highly substituted cation and radical. Thus we observe isopropyl cation ($m/z = 43$) and 2-butyl cation ($m/z = 57$) as the most abundant peaks.

$$[\text{CH}_3\text{CH}_2\underset{\underset{\text{CH}_3}{|}}{-\text{CH}}\underset{\underset{\text{CH}_3}{|}}{-\text{CH}}-\text{CH}_3]^{+\bullet}$$

Molecular ion
$m/z = 100$

$\longrightarrow \ ^{\bullet}\underset{\underset{\text{CH}_3}{|}}{\text{CHCH}_3} \ + \ \text{CH}_3\text{CH}_2\overset{+}{\underset{\underset{\text{CH}_3}{|}}{\text{CH}}}$

$m/z = 57$

$\longrightarrow \ \text{CH}_3\text{CH}_2\underset{\underset{\text{CH}_3}{|}}{\overset{\bullet}{\text{CH}}} \ + \ ^{+}\underset{\underset{\text{CH}_3}{|}}{\text{CHCH}_3} \ \longrightarrow \ \underset{\underset{+}{}}{\overset{\text{CH}_3}{\diagdown}}\text{C}=\text{CH}_2 \ + \ \text{H}_2$

$m/z = 43 \qquad\qquad m/z = 41$

$\longrightarrow \ ^{\bullet}\text{CH}_3 \ + \ \text{CH}_3\text{CH}_2\overset{+}{\text{CH}}\text{CH}(\text{CH}_3)_2$

$m/z = 85$

$\longrightarrow \ \text{CH}_3\overset{\bullet}{\text{CH}}_2 \ + \ \text{CH}_3\overset{+}{\text{CH}}\text{CH}(\text{CH}_3)_2$

$m/z = 71$

(c) The compound contains chlorine because the M+2 peak is one-third the height of the molecular ion peak. The base peak results from cleavage of the carbon-chlorine bond, which is the weakest bond in the compound.

$$\text{Cl}\overset{\bullet}{\text{C}}\text{H}_2 \ + \ ^{+}\text{CH}_2-^{35}\text{Cl} \ + \ ^{+}\text{CH}_2-^{37}\text{Cl}$$

$\qquad\qquad\qquad m/z = 49 \qquad m/z = 51$

$[\text{Cl}-\text{CH}_2-\text{CH}_2-\text{Cl}]^{+\bullet}$

Molecular ion
$m/z = 98$
$m/z = 100$

$\text{HCl} \ + \ \underset{H}{\overset{^{35}\text{Cl}}{\diagdown}}\text{C}=\overset{+}{\text{CH}}_2^{\bullet} \ + \ \underset{H}{\overset{^{37}\text{Cl}}{\diagdown}}\text{C}=\overset{+}{\text{CH}}_2^{\bullet}$

$\qquad\qquad m/z = 62 \qquad\qquad m/z = 64$

$\qquad\qquad\qquad \searrow -^{35}\text{Cl}^{\bullet} \qquad \swarrow -^{37}\text{Cl}^{\bullet}$

$\qquad\qquad\qquad\qquad \underset{H}{\overset{+}{\diagup}}\text{C}=\text{CH}_2$

$\qquad\qquad\qquad\qquad m/z = 27$

(d) The molecular ion for 2-methyl-1-butanol ($m/z = 88$) is not observed because it loses water so readily. Thus, the largest fragment observed is the M − 18 peak ($m/z = 70$). In addition to losing water, alcohols may fragment next to the carbinol carbon atom (α-cleavage). Consequently, the base peak at $m/z = 57$ corresponds to the 2-butyl carbocation.

$$CH_3CH_2CHCH_2OH \text{ (with } CH_3 \text{ substituent)} \xrightarrow[-H_2O^-]{-e^-} [CH_3CH_2\overset{+}{C}=CH_2]^{\cdot} \text{ } m/z = 70$$

branching to:
- $CH_3\dot{C}H_2 + [CH_3\overset{+}{C}=CH_2]$, $m/z = 41$
- $H_3C^{\cdot} + [CH_3CH_2\overset{+}{C}=CH_2]$, $m/z = 55$ (−H_2)
- $\overset{H}{\underset{H}{>}}C=\overset{+}{O}H$, $m/z = 31$ + $[CH_3CH_2\overset{+}{CH}]$ (with CH_3), $m/z = 57$
- $[\overset{+}{C}H_2-CH=CH_2] + CH_4$, $m/z = 31$

(e) The molecular ion ($m/z = 74$) is not observed. Tertiary alcohols, however, do not lose water first. Instead, they fragment next to the carbinol carbon atom to give a resonance-stabilized carbocation. Thus we observe a base peak at $m/z = 59$. The loss of a second methyl group gives the acylium ion at $m/z = 31$. Acylium ions are commonly observed during fragmentations of aldehydes and ketones.

$$H_3C-\underset{CH_3}{\overset{OH}{\underset{|}{C}}}-CH_3 \xrightarrow{-e^-} [H_3C-\underset{CH_3}{\overset{\cdot OH}{\underset{|}{C}}}-CH_3]^+ \longrightarrow {}^{\cdot}CH_3 + CH_3\overset{\overset{+}{OH}}{\underset{CH_3}{\overset{\|}{C}}} \longleftrightarrow CH_3\overset{OH}{\underset{CH_3}{\overset{|}{C^+}}}$$

$m/z = 59$

$$\downarrow$$

$CH_3C\equiv O^+ + CH_4$

$m/z = 43$

18.23 (a) A **free radical** (radical) is an atom or molecule in which one of the atoms has an odd number of electrons. Most commonly, a radical contains a carbon atom with three bonds and an odd number of valence electrons (seven), one of which is unpaired. The unpaired electron is always indicated by a dot:

$Cl^{\cdot}$ $Br^{\cdot}$ $H_3C^{\cdot}$

(b) **Homolytic bond breaking** (homolysis or radical cleavage) is breaking a bond with the result that each of the atoms gets one of the bonding electrons:

$$Br:Br \xrightarrow{\text{Heat or light}} 2\ Br^{\bullet}$$

(c) A **copolymer** is a polymeric material made by polymerizing two or more different monomers together to give products with desired properties.

(d) A **molecular ion** (parent ion) is the radical cation with the same mass as the molecular weight of the original compound. The molecular ion is abbreviated **M⁺̇**.

(e) An **addition polymer** is a polymer in which no atoms of the monomer unit have been lost as a result of the polymerization reaction. Addition polymers result from the addition of monomers one at a time to a growing polymer chain, usually with a reactive intermediate (cation, anion, or radical) at the growing end of the chain. Addition polymers are sometimes called chain-growth polymers because one molecule is added at a time to the end of the polymer chain. The monomers are usually alkenes, and polymerization involves successive additions across the double bonds.

(f) A **monomer** is a molecule (usually a small molecule) that reacts with other molecules to form a polymer. As a result, the monomer becomes a repeating unit in the polymer chain.

(g) **Free radical polymerization** is the mechanism of forming an addition polymer by chain-growth polymerization involving free radicals as intermediates.

(h) The **initiation step** is the preliminary step in a chain reaction. The reactive intermediate forms during the initiation step. The reactive intermediates created in the initiation step are most commonly radicals.

(i) An **atactic polymer** contains stereocenters whose configurations are random.

(j) The **termination step** is a reaction in which a reactive intermediate is consumed without another one being formed. In free radical reactions, two radicals combine to produce a molecule in which all the electrons are paired.

(k) An **isotactic polymer** contains stereocenters whose configurations are all (R) or all (S).

(l) **Heterolytic bond breaking** (heterolytic cleavage) is breaking a bond with the result that one of the atoms retains both bonding electrons. As a result, ions are formed:

$$(CH_3)_3C-\ddot{Br}: \longrightarrow (CH_3)_3C^+ + :\ddot{Br}:^-$$

(m) A **propagation step** is one of several (usually two) steps in a chain reaction that are repeated over and over to form the product. The reactive intermediate reacts with a stable molecule to form another reactive intermediate, allowing the chain to continue until the supply of reactants is exhausted or the reactive intermediate is destroyed. The sum of the propagation steps should give the net reaction.

(n) A **syndiotactic polymer** contains stereocenters whose configurations alternate (R),(S), (R), (S), and so on.

(o) A **free radical initiator** is any compound that can readily undergo homolysis (dissociate to form radicals). Peroxides are typical free radical initiators:

$$RO-OR \xrightarrow{\text{Light or heat}} 2\ RO^\bullet$$

(p) A **free radical chain mechanism** is a multistep process where a reactive radical formed in one step brings about a second step that generates the radical needed for the first step.

(q) A **free radical inhibitor** is a compound that traps radicals and prevents reactions that take place by mechanisms involving radicals. Hydroquinone is a radical inhibitor. Hydroquinone reacts with a reactive radical to form a semiquinone radical that is stabilized by resonance and is therefore unreactive compared to other radicals.

Hydroquinone + RO• (Reactive radical) ⟶ Semiquinone + ROH

Similar mechanisms of trapping unwanted radicals are demonstrated by such radical inhibitors as vitamin C, vitamin E, BHA, and BHT.

(r) A **mass spectrum** is the graph produced by a mass spectrometer showing the ratio m/z (where m = mass of the positively charged fragment, and z = number of charges) along the x axis and the relative number of ions of each ratio m/z on the y axis.

(s) A **Ziegler-Natta catalyst** is any one of a group of addition polymerization catalysts involving titanium-aluminum complexes. A typical Ziegler-Natta catalyst is formed by adding a solution of $TiCl_4$ (titanium tetrachloride) to a solution of $(CH_3CH_2)_3Al$ (triethyl aluminum). With a Ziegler-Natta catalyst, a high-density polyethylene (or linear polyethylene) can be produced with almost no chain branching and with much greater strength than common low-density polyethylene.

18.24 A chain reaction mechanism consists of three kinds of steps: initiation, propagation, and termination. We can very easily identify each step by the following characteristic features: (i) An initiation step is a reaction that generates free radicals. In short, the number of free radicals increases. (ii) A propagation step is a reaction in which a free radical reacts with a stable molecule to form another radical and a new stable molecule. In short, the number of free radicals remains unchanged. (iii) A termination step is a reaction in which a free radical is consumed without another one being formed. In short, the number of radicals decreases.

(a) An initiation step (radicals are generated)
(b) A termination step (two radicals react giving two stable products)
(c) A propagation step (one radical is used and one is generated)
(d) A termination step (two radicals react giving a stable product)
(e) A propagation step (one radical is used and one is generated)
(f) An initiation step (radicals are generated)

18.25 An alkane has a general molecular formula C_nH_{2n+2}. To calculate the molecular formula, we can use the following equation: $12 \times (n + 1) \times (2n + 2) = MW$. We can substitute 72 for MW (given) and calculate the number of carbon atoms, n:

$$12n + 2n + 2 = 72;\ 14n = 70;\ n = 5.$$

Hence, the molecular formula of the hydrocarbon is C_5H_{12} (MW = 72).

There are three possible constitutional isomers of C_5H_{12}:

$CH_3CH_2CH_2CH_2CH_3$ $CH_3CHCH_2CH_3$ (with CH_3 branch) $H_3C-C(CH_3)_2-CH_3$

Pentane   2,2-Methylbutane (Isopentane)   2,2-Dimethylpropane (Neopentane)

Pentane has three groups of equivalent hydrogen atoms and can give three monochlorinated products. 2-Methylbutane has four groups of equivalent hydrogen atoms and can give four monochlorinated products. 2,2-Dimethylpropane has only one group of equivalent hydrogen atoms and can give only a single monochlorinated product:

$H_3C-C(CH_3)_2-CH_3 + Cl_2 \xrightarrow{\text{Light or heat}} H_3C-C(CH_3)_2-CH_2Cl + HCl$

1-Chloro-2,2-dimethylpropane

18.26 (a) Free-radical addition of HBr gives the anti-Markownikoff product, in which hydrogen was added to the end of the double bond that started with fewer hydrogen atoms:

(methylcyclohexene) + HBr, ROOR → (cyclohexane with CH3 and CH2Br)

(b) Ionic addition of HBr gives products with Markownikoff orientation:

(methylcyclohexene) + HBr, CH_3COOH → (tertiary Br product) Major + (tertiary OOCCH3 product) Minor

(c) The addition of bromine is a stereospecific *anti* addition:

[Reaction: 1-methylcyclohexene + Br₂/CCl₄ → trans-1,2-dibromo-1-methylcyclohexane (two enantiomers shown)]

(d) Hydroboration-oxidation gives an alcohol with anti-Markownikoff orientation. The simultaneous addition of boron and hydrogen to the double bond in the first step leads to a *syn* addition and oxidation in the second step replaces boron with a hydroxyl group in the same stereochemical position:

[Reaction: 1-methylcyclohexene (1) 9-BBN (2) H₂O₂/NaOH → trans-2-methylcyclohexanol (two enantiomers)]

(e) Acid-catalyzed hydration of an alkene gives an alcohol with Markownikoff orientation:

[Reaction: 1-methylcyclohexene + H₂SO₄/H₂O → 1-methylcyclohexanol]

18.27 Halogenation of an alkane or a cycloalkane is a free radical reaction and it involves the following three distinct steps:

1. Initiation: The free radicals that take part in subsequent steps are formed from a molecule that readily undergoes homolysis.

$$Cl:Cl \xrightarrow{Light} 2\ Cl^\bullet$$

2. Propagation: The free radicals formed in the initiation step react with a molecule of an alkane to form a free radical by-product that serves as the starting material for the next step. The propagation steps continue in a chainlike fashion until the reactants are consumed. As a result, products are formed with no *net* consumption or destruction of free radicals.

[Cl• + cyclopentane-H → HCl + cyclopentyl radical]

[cyclopentyl• + Cl:Cl → cyclopentyl-Cl + Cl•]

3. Termination: Free radicals are destroyed by reacting with one another. Such a reaction of two free radicals coming together to form a covalent bond is the reverse of a homolysis reaction and it is known as a *recombination reaction*.

$$Cl\cdot + \cdot Cl \longrightarrow Cl_2$$

[cyclopentyl radical] + $\cdot Cl \longrightarrow$ [cyclopentyl–Cl]

[cyclopentyl radical] + [cyclopentyl radical] $\longrightarrow$ [bicyclopentyl]

18.28 First, we must analyze the structure of the polymer and try to identify the smallest repeating unit in the molecule of the polymer. Chain-growth polymers have carbon-carbon covalent bonds joining the molecules of a monomer while condensation polymers have other atoms involved in the chain formation (usually ester or amide bonding). After the repeating unit is identified, we disconnect the bonds joining it with other molecules and restore the original bonds in the monomer.

(a) The smallest repeating unit is $-CF_2-$. However, we know that there are no one-carbon monomers, so we take the two-carbon unit, $-CF_2CF_2-$. We know that monomers for chain-growth polymers are usually alkenes, so the structure of the monomer is

$$\underset{F}{\overset{F}{\diagdown}}C=C\underset{F}{\overset{F}{\diagup}}$$

(b) The structure of the monomer is

$$\underset{H}{\overset{H}{\diagdown}}C=C\underset{CN}{\overset{H}{\diagup}}$$

(c) The structure of the monomer is

$$\underset{H}{\overset{H}{\diagdown}}C=C\underset{NO_2}{\overset{H}{\diagup}}$$

(d) The structure of the monomer is

$$\text{H}_2\text{C}=\text{CH}-\text{OOCCH}_3$$

(e) The structure of the monomer is

$$\text{H}_2\text{C}=\text{CHCl}$$

(f) Here we have two two-carbon repeating units:

$$-\text{CH}_2\text{CH}_2- \quad \text{and} \quad -\text{CH}_2\overset{\overset{\displaystyle \text{OOCCH}_3}{|}}{\text{CH}}-.$$

Thus, the structures of the two monomers are

$$\text{CH}_2=\text{CH}_2 \quad \text{and} \quad \text{CH}_2=\overset{\overset{\displaystyle \text{OOCCH}_3}{|}}{\text{CH}}$$

(g) The repeating unit here is *cis*-$\text{CH}_2\text{CH}=\text{CH}-\text{CH}_2-$. The monomer is 1,3-butadiene:

$$\text{CH}_2=\text{CH}-\text{CH}=\text{CH}_2$$

The stereochemistry of the polymer is achieved by use of an appropriate catalyst.

(h) The structure of the monomer is

$$\underset{\text{H}}{\overset{\text{H}}{}}\text{C}=\text{C}\underset{\text{CH}_3}{\overset{\text{CH}_3}{}}$$

18.29 The presence of dialkyl peroxide, a well-known free radical initiator, indicates a free radical mechanism that consists of the three typical steps:

1. Initiation:

$$\text{RO}:\text{OR} \xrightarrow{\text{Light or heat}} 2\ \text{RO}^\bullet$$

2. Propagation (the two last steps form a typical cycle of the chain reaction):

RO• + H—CCl₃ ⟶ ROH + Cl₃C•

Cl₃C• + CH₂=CHCH₃ ⟶ Cl₃C—CH₂—ĊHCH₃

Cl₃C—CH₂—ĊHCH₃ + H—CCl₃ ⟶ Cl₃C—CH₂CH₂CH₃ + Cl₃C•

3. Termination:

Cl₃C• + •CCl₃ ⟶ Cl₃C—CCl₃

Cl₃C—CH₂—ĊHCH₃ + •CHCH₃(CH₂CCl₃) ⟶ Cl₃CCH₂CH(CH₃)—CH(CH₃)CH₂CCl₃

Only two of many possible termination steps are shown.

18.30 Chlorination of (S)-2-chlorobutane yields five products:

(structures A, B, C, D, E shown)

A (R)-1,2-Dichlorobutane is optically active,
B 2,2-Dichlorobutane is *not* optically active (C2 lost its chirality because it has two identical substituents),
C (S)-1,3-Dichlorobutane is optically active,
D (2S,3S)-2,3-Dichlorobutane is optically active,
E (2R,3S)-2,3-Dichlorobutane in *not* optically active (a *meso* compound).

18.31 (a) Atactic polyacrylonitrile is a polymer with the side groups on random sides of the polymer backbone. The side group here is —C≡N:

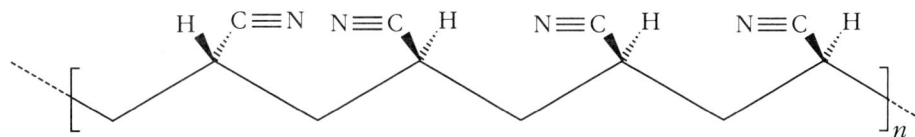

(b) Isotactic polyvinyl acetate is a polymer with all the side groups on the same side of the polymer backbone. The side group here is —OOCCH$_3$:

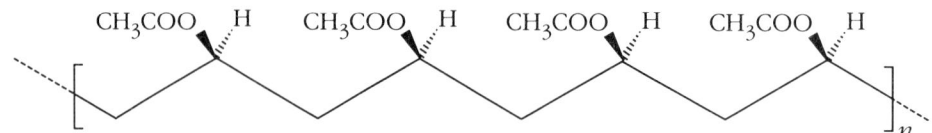

(c) Syndiotactic polystyrene is a polymer with the side groups on alternating sides of the polymer backbone. The side group here is a benzene ring, —C$_6$H$_5$:

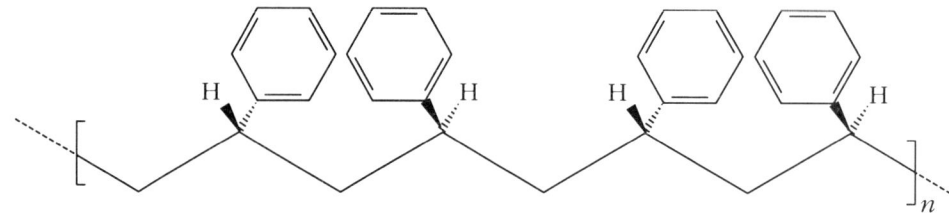

18.32 (1) The fragmentation of a molecular ion leads to formation of neutral molecules that are not observed in the mass spectrum. The following are some typical small molecules:

(a) H$_2$O
(b) N$_2$
(c) CH$_2$=CH$_2$
(d) HCl (confirmed by the presence of a (M+2)$^+_\bullet$ peak before fragmentation and the absence of such a peak after the loss of HCl); or two water molecules (possible but very rare)
(e) CO$_2$

(2) We do not observe neutral molecules in the mass spectrum. However, we observe peaks corresponding to loss of small, stable molecules. Loss of a small molecule is usually indicated by a peak with an even mass number, corresponding to loss of an even mass molecule. For example, we assume the formation of a water molecule by observing an M-18 peak. Sometimes (for instance in the case of many alcohols) the molecular ion (M)$^+_\bullet$ is not observed because the molecule loses water so readily. In such case, the (M - 18)$^+_\bullet$ peak is the highest peak observed in the spectrum.

Similar observations apply to other small molecules.

18.33 Fragmentation of alcohols is characterized by α-cleavage which, in the case of ethanol, is responsible for a base peak at $m/z = 31$.

$$CH_3-CH_2-OH \xrightarrow{-e^-} [CH_3 \frown CH_2 \frown \overset{\bullet}{O}H]^+ \longrightarrow {}^\bullet CH_3 + \underset{H \quad H}{\overset{\overset{+OH}{\|}}{C}}$$

$$m/z = 46 \qquad\qquad m/z = 31 \text{ (base peak)}$$

(a) Fragmentation of ethanol indicates that the ^{13}C isotope will be found in the base peak observed at $m/z = 32$ and the molecular ion will be observed at $m/z = 47$:

$$CH_3-{}^{13}CH_2-OH \xrightarrow{-e^-} [CH_3-{}^{13}CH_2-\overset{\cdot}{O}H]^+ \longrightarrow {}^\cdot CH_3 + \underset{HH}{{}^{13}C\overset{{}^+OH}{\underset{\|}{}}}$$

$m/z = 47 \qquad\qquad m/z = 32$ (base peak)

(b) Fragmentation of ethanol indicates that the ^{18}O isotope will be found in the base peak observed at $m/z = 33$ and the molecular ion will be observed at $m/z = 48$:

$$CH_3-CH_2-{}^{18}OH \xrightarrow{-e^-} [CH_3-CH_2-{}^{18}\overset{\cdot}{O}H]^+ \longrightarrow {}^\cdot CH_3 + \underset{HH}{C\overset{{}^{18}\overset{+}{O}H}{\underset{\|}{}}}$$

$m/z = 48 \qquad\qquad m/z = 33$ (base peak)

(c) Fragmentation of ethanol indicates that the ^{13}C isotope will be found in the methyl radical (not observed in the spectrum). The base peak will be observed at $m/z = 31$ and the molecular ion will be observed at $m/z = 47$:

$$^{13}CH_3-CH_2-OH \xrightarrow{-e^-} [{}^{13}CH_3-CH_2-\overset{\cdot}{O}H]^+ \longrightarrow {}^{13}\overset{\cdot}{C}H_3 + \underset{HH}{C\overset{{}^+OH}{\underset{\|}{}}}$$

$m/z = 47 \qquad\qquad m/z = 31$ (base peak)

As the above analysis shows, mass spectrometry is very useful for differentiating between ethanol and the three isotopically labelled samples.

18.34 Benzoyl peroxide is a well-known free radical initiator, so polymerization will occur according to a free radical mechanism. Free radical polymerization typically involves the following three steps:

1. Initiation: In the initiation steps, the free radicals that take part in subsequent steps of the reaction are formed from a molecule of initiator that readily undergoes homolysis.

PhC(O)-O-O-C(O)Ph ⟶ 2 PhC(O)-O• ⟶ 2 Ph• + 2 CO₂

2. Propagation: In the propagation steps, the free radical formed in the initiation step reacts with molecules of the monomer to form the product. The free radical product of one propagation step serves as the starting material for another.

[Propagation reaction scheme: phenyl radical adds to CH₂=C(COOCH₃)(CH₃) to give PhCH₂—C•(COOCH₃)(CH₃); this radical adds another methyl methacrylate monomer to give PhCH₂—C(COOCH₃)(CH₃)—CH₂—C•(COOCH₃)(CH₃); which then adds a third monomer unit to give PhCH₂—C(COOCH₃)(CH₃)—CH₂—C(COOCH₃)(CH₃)—CH₂—C•(COOCH₃)(CH₃).]

Note: The initiator becomes bonded to one end of the polymer.
3. Termination: In the termination step(s), two free radicals come together to form a covalent bond. The Exercise asked only for the first three units of the polymer but the reaction does not stop after the three units are added and termination takes place after many units of monomer are added.

18.35 Bromine can react with propene in two different ways, namely free radical bromination in the allylic position and an addition to the carbon-carbon double bond. Thus the free radical bromination will form 3-bromopropene:

1. Initiation:

$$Br:Br \xrightarrow{\text{Light}} 2\ Br^\bullet$$

2. Propagation:

$$Br^\bullet + H-CH_2-CH=CH_2 \rightarrow HBr + {}^\bullet CH_2-CH=CH_2 \leftrightarrow CH_2=CH-{}^\bullet CH_2$$

$$CH_2=CH-{}^\bullet CH_2 + Br-Br \rightarrow CH_2=CH-CH_2Br + Br^\bullet$$

Electrophilic addition of bromine to propene can form 1,2-dibromopropane:

$$Br-Br\ \ CH_2=CH-CH_3 \rightarrow \underset{Br^-}{\overset{Br^+}{CH_2-CH-CH_3}} \rightarrow BrCH_2-\underset{Br}{CH}-CH_3$$

Electrophilic addition of bromine to 3-bromopropene can form 1,2,3-tribromopropane:

$$Br-Br\ \ CH_2=CH-CH_2Br \rightarrow \underset{Br^-}{\overset{Br^+}{CH_2-CH-CH_2Br}} \rightarrow BrCH_2-\underset{Br}{CH}-CH_2Br$$

18.36 The mass spectrum of a ketone usually has a molecular ion. Ketones fragment homolytically at the carbon-carbon bond adjacent to the carbon-oxygen double bond to give acylium ions:

(a) An α-cleavage can take place on either side of the carbonyl group in 2-butanone:

$$\underset{CH_3CH_2}{\overset{\overset{\displaystyle \ddot{O}\!\bullet}{\|}}{C}}\underset{CH_3}{} \xrightarrow{-e^-} \left[\underset{CH_3CH_2}{\overset{\overset{\displaystyle \ddot{O}\!\overset{+}{\bullet}}{\|}}{C}}\underset{CH_3}{}\right]$$

$$m/z = 72$$

$$\nearrow CH_3\dot{C}H_2 + [CH_3C\equiv \overset{+}{O}]\quad m/z = 43$$

$$\searrow H_3C^\bullet + [CH_3CH_2C\equiv \overset{+}{O}]\quad m/z = 57$$

We observe the molecular ion at $m/z = 72$ and the base peak at $m/z = 43$.

(b) An α-cleavage gives the acyl ion observed at m/z = 29, cleavage of a methyl group gives an ion observed at m/z = 57, and the McLafferty rearrangement gives the base peak at m/z = 44:

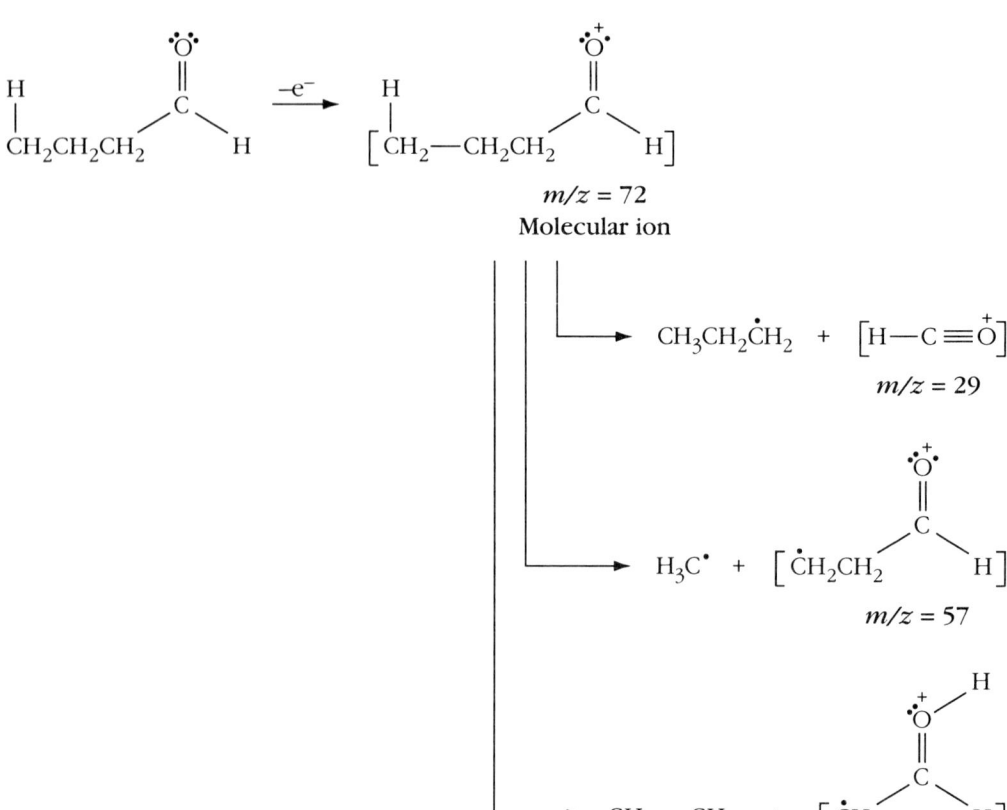

(c) An α-cleavage can take place as follows:

$$\underset{\underset{CH_3}{|}}{H_3C-\overset{H}{\underset{|}{C}}CH_2CH_2}-\overset{\overset{\cdot\overset{..}{O}\cdot}{\|}}{C}-CH_3 \xrightarrow{-e^-} \left[\underset{\underset{CH_3}{|}}{H_3C-\overset{H}{\underset{|}{C}}CH_2CH_2}-\overset{\overset{\cdot\overset{..}{O}\cdot^+}{\|}}{C}-CH_3\right]$$

$m/z = 114$

$(CH_3)_2CHCH_2\dot{C}H_2 \;+\; [H_3C-C\equiv\overset{+}{O}]$

$m/z = 43$

$CH_3\dot{C}HCH_3 \;+\; \left[\dot{C}H_2CH_2-\overset{\overset{\cdot\overset{..}{O}\cdot^+}{\|}}{C}-CH_3\right]$

$m/z = 71$

$(CH_3)_2CH=CH_2 \;+\; \left[{}^\cdot CH_2-\overset{\overset{\overset{+}{\overset{..}{O}}\diagup H}{\|}}{C}-CH_3\right]$

$m/z = 58$

The molecular ion at $m/z = 114$ has very low intensity and is barely seen. By far the most abundant peak is observed at $m/z = 43$ (base peak). The second most abundant peak is the product of the McLafferty rearrangement ($m/z = 58$).

(d) α-Cleavage predominates in the fragmentation of secondary alcohols and the α-cleavage is responsible for the base peak at $m/z = 59$ and another major peak at $m/z = 73$:

$$\underset{\underset{CH_3}{|}}{H_2C}\overset{OH}{\underset{|}{CH}}\underset{\underset{CH_2CH_3}{|}}{CH_2} \xrightarrow{-e^-} \left[\underset{\underset{CH_3}{|}}{H_2C}\overset{\overset{\cdot OH}{\underset{(a)\swarrow\;\searrow(b)}{CH}}}{}\underset{\underset{CH_2CH_3}{|}}{CH_2}\right]^+$$

(a) → $CH_3\dot{C}H_2 \;+\; CH_3CH_2CH_2\overset{\overset{+OH}{\diagup}}{\underset{\diagdown H}{C}}$

$m/z = 73$

(b) → $CH_3CH_2\dot{C}H_2 \;+\; CH_3CH_2\overset{\overset{+OH}{\diagup}}{\underset{\diagdown H}{C}}$

$m/z = 59$

18.37 (a) 2-Octanone has a base peak at $m/z = 43$ and a second most abundant peak at $m/z = 58$. Both peaks are formed as a result of the McLafferty rearrangement. The base peak at $m/z = 43$ may also be formed by an α-cleavage (acylium ion, $CH_3C\equiv O^+$).

[structure of 2-octanone radical cation with McLafferty H-transfer] → [H,H C=C CH$_2$CH$_2$CH$_3$,H propene fragment] + [$^+$:O(H)–C(CH$_3$)=C(H)(H)] $m/z = 58$

$^\bullet CH_3$ + [$^+$:O(H)–C(H)=C(H)$_2$ ...] $m/z = 43$

(b) The McLafferty rearrangement of 3-methylpentanal would lead to a formation of a peak at $m/z = 44$. This peak is observed but it is not very abundant, which indicates that the rearrangement takes place but it is not a major fragmentation pathway. 3-Methylpentanal shows a base peak at $m/z = 56$, which is formed in other fragmentations.

[structure of 3-methylpentanal radical cation with McLafferty H-transfer] → [enol radical cation H,O(H) C=C H,H] + $CH_3CH=CHCH_3$

$m/z = 44$

(c) The McLafferty rearrangement of 5-nonanone leads to formation of a peak at $m/z = 100$. This ion undergoes subsequent fragmentation to give a peak at $m/z = 85$ (second most abundant peak) and $m/z = 57$ (the base peak). Thus the McLafferty rearrangement is a major fragmentation pathway of 5-nonanone.

18.38 (a) Butanal and 2-butanone are constitutional isomers (molecular formula, C_4H_8O), so they both show their molecular ions at $m/z = 72$. Butanal shows the base peak at $m/z = 44$ (see Exercise 18.36b) while 2-butanone shows the base peak at $m/z = 43$ (see Exercise 18.36a).

(b) 2-Methylbutanal and 3-methylbutanal are constitutional isomers (molecular formula, $C_5H_{10}O$), so they both show their molecular ions at $m/z = 86$. 2-Methylbutanal shows the following fragmentation pattern:

$$\text{CH}_3\text{CH}_2\overset{\underset{|}{\text{CH}_3}}{\text{CH}}\overset{\ddot{\ddot{\text{O}}}\cdot}{\text{C}}\text{H} \xrightarrow{-e^-} \left[\overset{\text{H}}{\underset{\underset{\text{CH}_3}{|}}{\text{CH}_2-\text{CH}_2-\text{CH}}}\overset{\ddot{\ddot{\text{O}}}^+}{\text{C}}\text{H}\right] \longrightarrow \text{CH}_3\text{CH}_2\overset{+}{\text{CH}}\text{CH}_3$$

$$m/z = 86 \qquad\qquad m/z = 57$$

$$\text{CH}_3\text{CH}_2\overset{\underset{|}{\text{CH}_3}}{\overset{\bullet}{\text{CH}}} + [\text{H}-\text{C}\equiv\overset{+}{\text{O}}]$$

$$m/z = 29$$

$$\text{CH}_3\overset{\bullet}{\text{CH}}_2 + \left[\text{CH}_3-\overset{\bullet}{\text{CH}}\overset{\overset{\ddot{\text{O}}\cdot^+}{\parallel}}{\text{C}}\underset{\text{H}}{}\right]$$

$$m/z = 71$$

$$\text{CH}_2=\text{CH}_2 + \left[\text{CH}_3\overset{\bullet}{\text{CH}}\overset{\overset{\ddot{\text{O}}\cdot^+-\text{H}}{\parallel}}{\text{C}}\underset{\text{H}}{}\right]$$

$$m/z = 58$$

3-Methylbutanal will fragment according to the following fragmentation pattern:

$$\text{CH}_3\text{CHCH}_2\text{-CHO (with CH}_3\text{ branch)} \xrightarrow{-e^-} [\text{CH}_2\text{—CH(CH}_3\text{)—CH}_2\text{—CHO}]^{+\cdot}$$

$m/z = 86$

$$\text{CH}_3\text{CHCH}_2^\cdot \text{ (with CH}_3\text{ branch)} + [\text{H—C}\equiv\overset{+}{\text{O}}]$$

$m/z = 29$

$$\text{H}_3\text{C}^\cdot + [\,^\cdot\text{CH(CH}_3\text{)CH}_2\text{CHO}\,]^{+}$$

$m/z = 57$

$$\text{CH}_2\text{=CHCH}_3 + [\,^\cdot\text{CH}_2\text{—CH(OH)}\,]^{+}$$

$m/z = 44$

Thus both 2-methylbutanal and 3-methylbutanal will show the same molecular ion at $m/z = 86$. The base peak for 2-methylbutanal appears at $m/z = 57$ and is due to α-cleavage because α-cleavage forms a secondary radical-cation. The α-cleavage in 3-methylbutanal would produce a primary radical-cation so α-cleavage is not a major fragmentation route. In the case of 3-methylbutanal, the base peak appears at $m/z = 44$ and it is due to the McLafferty rearrangement.

(c) 2-Pentanone and 3-pentanone are constitutional isomers (molecular formula $C_5H_{10}O$), but they can be very easily distinguished from one another because each ketone shows different fragmentation patterns. There are two main mechanisms of fragmentations of ketones, namely α-cleavage and McLafferty rearrangement. The α-cleavage of 2-pentanone may result in a loss of the methyl group or the propyl group. Because a methyl radical is less stable than a propyl radical, the peak corresponding to loss of the methyl group ($m/z = 71$) is much weaker than the base peak ($m/z = 43$) from loss of the propyl group.

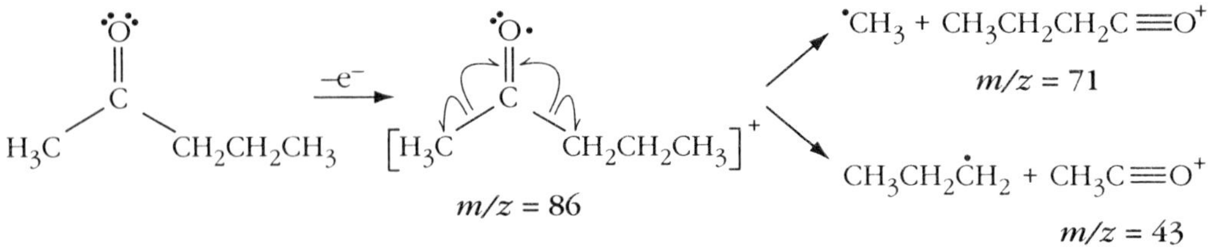

There is also a weak peak at $m/z = 58$ corresponding to loss of ethene (McLafferty rearrangement).

3-Pentanone has two equal alkyl (ethyl) groups attached to the carbonyl group. As a result, the base peak appears at $m/z = 57$ and corresponds to loss of one of the ethyl groups.

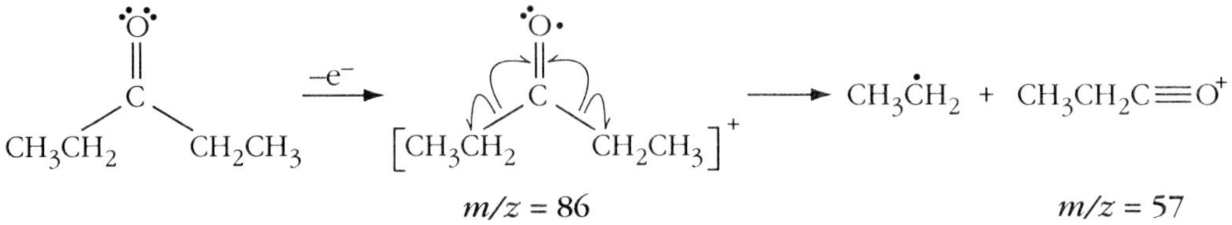

18.39 (a) CH_2=CHC≡N (acrylonitrile or propenenitrile) shows three sets of signals in the alkene region of its ^1H NMR spectrum:

$\delta_a \approx 6.1$
$\delta_b \approx 5.7$
$\delta_c \approx 6.0$

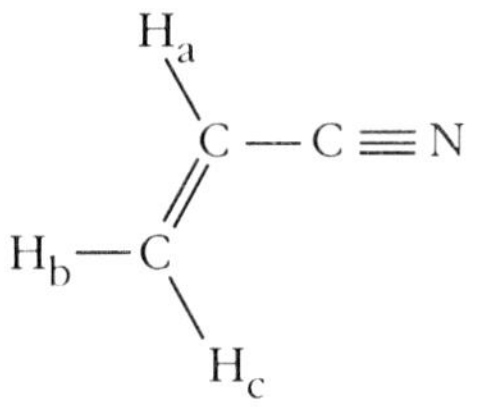

Spectrum 3 could be the ^1H NMR spectrum of acrylonitrile.

(b) CH_2=$CHBr$ (vinyl bromide or bromoethene) shows three sets of signals in the alkene region of the ^1H NMR spectrum:

$\delta_a \approx 6.5$
$\delta_b \approx 5.9$
$\delta_c \approx 6.0$

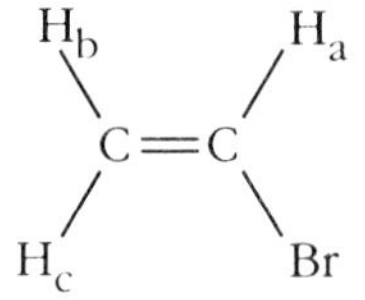

Spectrum 3 could also be the ^1H NMR spectrum of vinyl bromide. The chemical shift match is better for vinyl bromide than for acrylonitrile, but a definite assignment can be made only with the additional information provided by IR spectroscopy.

(c) Isoprene (2-methyl-1,3-butadiene) is expected to show the following signals:

$\delta = 1.9$ (singlet, 3H)—methyl group on C2,
$\delta = 5.05$ (singlet, 2H)—two hydrogen atoms on C1,
$\delta = 5.3$ (doublet, 2H)—two hydrogen atoms on C4,
$\delta = 6.6$ (a pair of doublets, 1H)—a hydrogen on C3.
None of the spectra matches this expected pattern.

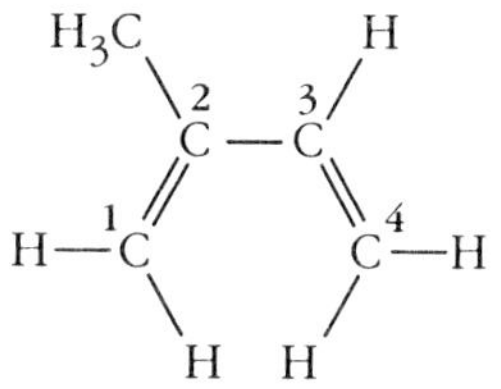

(d) Methyl methacrylate (methyl 2-methylpropenoate) shows four signals:

$\delta = 1.95$ (singlet, 3H)—methyl group on C2,
$\delta = 3.8$ (singlet, 3H)—methyl group bonded to oxygen,
$\delta = 5.5$ (doublet, 1H)—H_a,
$\delta = 6.1$ (doublet, 1H)—H_b.

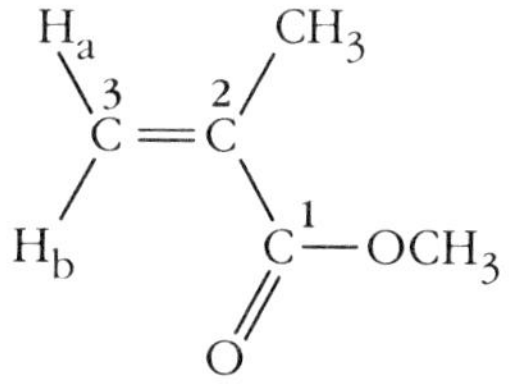

Spectrum 2 matches the expected ^{1}H NMR spectrum of methyl methacrylate. Please note that Spectrum 2 shows signals of hydrogen atoms H_a and H_b as singlets. If the signals were enlarged, they would show splitting with a very small coupling constant.

(e) Ispropenyl acetate shows three signals:

$\delta = 1.9$ (singlet, 3H)—methyl group bonded to the carbonyl group,
$\delta = 2.1$ (singlet, 3H)—methyl group of the isopropenyl group,
$\delta = 4.65$ (singlet, 2H)—vinyl hydrogen atoms of the isopropenyl group.

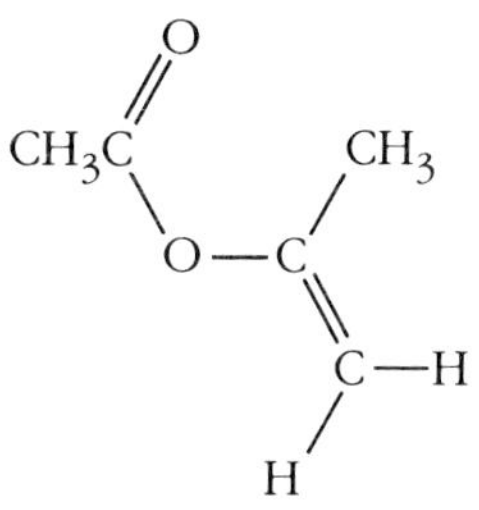

Spectrum 1 matches the expected ^{1}H NMR spectrum of isopropenyl acetate.

CHAPTER 19

π ELECTRON DELOCALIZATION IN ACYCLIC COMPOUNDS AND INTERMEDIATES

Concepts

Hydrocarbons that contain two double bonds are called **dienes**, those with three double bonds are called **trienes,** and so forth. **Polyenes** are hydrocarbons that contain several double bonds.

Carbon-carbon double bonds separated by two or more carbon-carbon single bonds are called **isolated double bonds**. Isolated double bonds react independently of each other. Carbon-carbon double bonds separated by only one carbon-carbon single bond are called **conjugated double bonds**. Conjugated double bonds react as a single functional group. **Cumulenes** contain at least one carbon atom that is part of two carbon-carbon double bonds.

Dienes and polyenes undergo addition reactions just like alkenes. A carbocation intermediate formed by such addition reactions to dienes is stabilized by π electron delocalization. As a result, more than one product is formed. The kinetically controlled product distribution is determined by the relative rates of formation of the products while the thermodynamically controlled product distribution is determined by the relative stabilities of the products.

Dienes and polyenes undergo reversible thermal or photochemical interconversions to form cyclic compounds in a process called an **electrocyclic reaction**. The stereochemistry of the products formed by an electrocyclic reaction depends on the number of π electrons in the polyene and whether the reaction occurs thermally or photochemically.

Dienes undergo a concerted **Diels-Alder reaction** with **dieneophiles** to form cyclohexene or cyclohexadiene derivatives.

Dienes are named by the IUPAC rules by designating the longest continuous chain containing both double bonds as the parent chain. The *-ane* ending of the name of this parent alkane is replaced by *-adiene.* The chain is numbered in the direction that gives the first double bond the lowest possible number.

2-Methyl-1,3-butadiene   5-Chloro-1,3-cyclohexadiene

Solutions to the Exercises

19.1 Allylic hydrogen atoms are hydrogen atoms bonded to a carbon atom adjacent to a double bond. The allylic hydrogen atoms are circled in each of the following structures.

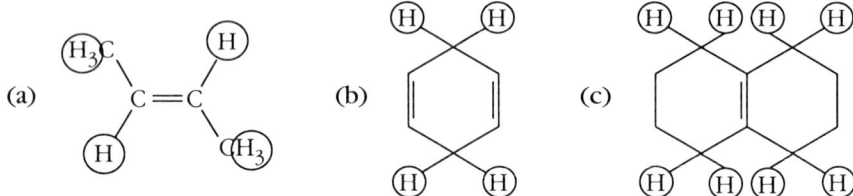

19.2 The π electron delocalization that stabilizes allylic carbocations can be represented as a resonance hybrid of the following resonance structures:

(a) [cyclopentenyl cation resonance structures]

(b) $H_2\overset{+}{C}-CH=C\begin{smallmatrix}CH_3\\CH_3\end{smallmatrix} \longleftrightarrow H_2C=CH-\overset{+}{C}\begin{smallmatrix}CH_3\\CH_3\end{smallmatrix}$

(c) [methylenecyclohexyl cation resonance structures]

19.3 The two isomeric products formed by the reaction of 2-buten-1-ol and HBr are 1-bromo-2-butene and 3-bromo-1-butene. The reaction occurs by the following mechanism involving a resonance stabilized allyllic cation:

Step 1: Formation of the allylic cation

$CH_3CH=CHCH_2\ddot{O}H \rightleftharpoons CH_3CH=CHCH_2-\overset{+}{O}\begin{smallmatrix}H\\H\end{smallmatrix} \longrightarrow$

$\left[\begin{array}{c}CH_3CH=CH-\overset{+}{C}H_2\\ \updownarrow\\ CH_3\overset{+}{C}H-CH=CH_2\end{array}\right]$

Resonance stabilized allylic cation

$+$

H_2O

Step 2: Reaction of allylic cation with bromide ion

$$\left[CH_3CH=CH\overset{+}{C}H_2 \quad \longleftrightarrow \quad CH_3\overset{+}{C}H-CH=CH_2 \right]$$

$$\downarrow Br^-$$

$$CH_3CH=CHCH_2Br \quad + \quad CH_3\underset{Br}{CH}CH=CH_2$$

19.4 (a) The products formed on hydrolysis are

$$\underset{OH}{CH_2\overset{|}{C}HCH=CH_2} \text{ and } CH_3CH=CHCH_2OH$$

The hydrolysis of $CH_3CH=CHCH_2Br$ forms the same two products. The resonance structures that contribute to the resonance hybrid of the intermediate allylic cation are the following:

$$\left[CH_3\overset{+}{C}H-CH=CH_2 \quad \longleftrightarrow \quad CH_3CH=CH-\overset{+}{C}H_2 \right]$$

(b) Two products of hydrolysis are

[cyclohexene ring with OH and CH₃ on same carbon] and [cyclohexene ring with OH on one carbon and CH₃ on another]

Another compound that forms these two products on hydrolysis is:

[cyclohexene ring with Cl and CH₃]

Contributing resonance structures are:

$$\left[\text{[cyclohexenyl cation with CH}_3\text{]} \quad \longleftrightarrow \quad \text{[cyclohexenyl cation with CH}_3\text{]} \right]$$

19.5 A trace of HCl catalyzes the reversible formation of an achiral allylic cation so all optical activity of the starting material is lost. The acid-catalyzed racemization of the optically active alcohol occurs according to the following mechanism:

19.6 An allyl radical has three electrons in its π molecular orbitals so its π electron structure is as follows:

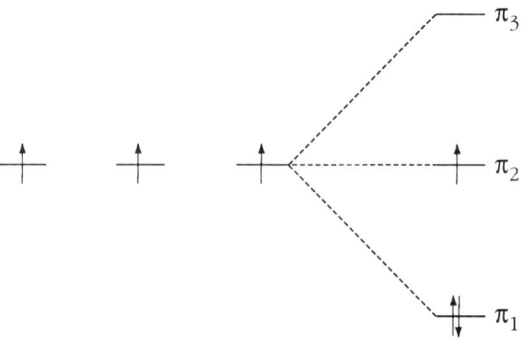

19.7 NBS reacts to replace an allylic hydrogen atom of an alkene with a bromine atom, so the products are as follows:

(a) [3-bromocyclopentene structure]

(b) H₃C\C=C/CH₂Br with H₃C and CH₃   +   Br\CH₃C—C/CH₂ with CH₃ and CH₃

(c) [1-bromo-2-methylenecyclohexane] + [1-(bromomethyl)cyclohexene]

19.8 Each of the target molecules has a carbon-carbon double bond and another functional group that can be conveniently introduced by a nucleophilic substitution reaction of 3-bromopropene. 3-Bromopropene is prepared by the reaction of NBS and propene.

$$CH_2=CHCH_2Z \Longrightarrow CH_2=CHCH_2Br \Longrightarrow CH_2=CHCH_3$$

(a) $CH_2=CHCH_3 \xrightarrow[CCl_4]{NBS} CH_2=CHCH_2Br \xrightarrow[CH_3CH_2OH]{NaCN} CH_2=CHCH_2CN$

(b) $CH_2=CHCH_2Br \xrightarrow[CH_3CH_2OH]{KI} CH_2=CHCH_2I$

(c) $CH_2=CHCH_2Br \xrightarrow{(H_2N)_2C=S} CH_2=CHCH_2\overset{+}{S}=C(NH_2)_2$

$CH_2=CHCH_2\overset{+}{S}=C(NH_2)_2 \xrightarrow[(2)\ H_3O^+]{(1)\ NaOH/H_2O} CH_2=CHCH_2SH$

(d) $CH_2=CHCH_2Br \xrightarrow[CH_3COOH]{CH_3C(=O)ONa} CH_2=CHCH_2-O-\overset{O}{\underset{\|}{C}}CH_3$

19.9 The double bonds that are separated by one single bond are conjugated so only the compounds in (b) and (c) are conjugated hydrocarbons.

19.10 (a) (1Z,5Z)-1,5-Cyclooctadiene (b) (1Z, 3Z, 5Z)-1,3,5-Cyclooctatriene
(c) 1,3-Cyclopentadiene (d) (2E,7E)-2,7-Nonadiene or 2-*trans*-7-*trans*-nonadiene

19.11 If the double bonds of 2-methyl-1,3-butadiene do not interact, the sum of the heats of hydrogenation of 3-methyl-1-butene and 2-methyl-1-butene should be equal to the heat of hydrogenation of 2-methyl-1,3-butadiene. If the double bonds interact, the heat of hydrogenation will be lower by the delocalization energy.

The typical value of the heat of hydrogenation of a monosubstituted alkene such as 3-methyl-1-butene is –29.8 kcal/mol (–125 kJ/mol) so we calculate the heat of hydrogenation of 2-methyl-1,3-butadiene as follows:

$$CH_2=\underset{\underset{CH_3}{|}}{C}-CH_2CH_3 + H_2 \longrightarrow CH_3\underset{\underset{CH_3}{|}}{CH}CH_2CH_3 \qquad \Delta H = -26.9 \text{ kcal/mol } (-113 \text{ kJ/mol})$$

$$CH_3-\underset{\underset{H}{|}}{\overset{\overset{CH_3}{|}}{C}}-CH=CH_2 + H_2 \longrightarrow CH_3\underset{\underset{CH_3}{|}}{CH}CH_2CH_3 \qquad \underline{\Delta H = -29.8 \text{ kcal/mol } (-125 \text{ kJ/mol})}$$

Calculated heat = –56.7 kcal/mol (–238 kJ/mol) of hydrogenation of

$$CH_2=\underset{\underset{CH_3}{|}}{C}-CH=CH_2$$

Observed heat = –53.4 kcal/mol (–223 kJ/mol) of hydrogenation of

$$CH_2=\underset{\underset{CH_3}{|}}{C}-CH=CH_2$$

Delocalization energy of ≈ 3.3 kcal/mol (≈ 15 kJ/mol).

$$CH_2=\underset{\underset{CH_3}{|}}{C}-CH=CH_2$$

The difference between the observed heat of hydrogenation and the calculated heat of hydrogenation, 3.3 kcal/mol (15 kJ/mol) is the delocalization energy of 2-methyl-1,3-butadiene.

19.12 The conditions of the addition of HBr to 1,3,5-hexatriene are not specified so the reaction may be either an ionic or a free radical mechanism.

Ionic mechanism:

Free radical mechanism:

19.13 Thermal ring opening of 3,4-dimethylcyclobutene occurs in a conrotatory fashion; that is, both methyl groups rotate in the same direction. In this case the rotation is counterclockwise.

19.14 The thermal ring opening reaction of a conjugated diene occurs by conrotatory rotation whereas conjugated trienes undergo disrotatory thermal ring openings.

19.15

(a) [structure: disrotatory ring opening of dimethyl-substituted cyclobutene to diene]

(b) Conrotatory ring closure: [structure of cyclooctatriene with CH₃, H, H, CH₃ substituents]

(c) [cyclooctatriene structure with CH₃, H, CH₃, H substituents]

(d) [open-chain tetraene structure with CH₃, H, CH₃, H substituents]

19.16 Good dienophiles in the Diels-Alder reaction must have an electron-withdrawing group or groups attached to the carbon atom of the double bond. Only the compound in (c) meets this criterion; that is, it is the only compound that has an electron-withdrawing group (—C(=O)Cl) bonded to the carbon atom of the double bond.

19.17 A diene must be conjugated and capable of assuming an *s-cis* conformation to undergo a Diels-Alder reaction. Compounds in (b) and (d) have the required conformation while (c) can rotate into an *s-cis* conformation.

(a) This compound is not a suitable diene for a Diels-Alder reaction because it is not conjugated.

(b) This compound is a suitable diene because it is conjugated and it has a permanent *s-cis* conformation.

(c) This compound is a suitable diene because it is conjugated and it can rotate into the *s-cis* conformation.

[structures showing rotation about single bond connecting cyclohexene to vinyl group, giving s-cis form]

(d) This compound is a suitable diene because it is conjugated and it has a permanent s-cis conformation.

19.18

(a) [butadiene] + [methyl acrylate] →(Heat) [methyl cyclohex-3-ene-1-carboxylate]

(b) [2,3-dimethylbutadiene] + HC≡CC(=O)OCH$_3$ →(Heat) [4,5-dimethylcyclohexa-1,4-diene-1-carboxylic acid methyl ester]

(c) cis-1,2-dicyano cyclohex-4-ene

(d) trans-1,2-dicyano cyclohex-4-ene

(e) 1,1-dicyano cyclohex-3-ene

19.19 A six-membered ring is formed by the diene and a dienophile in the Diels-Alder reaction. To determine the structures of the diene and the dienophile, we split the ring into two parts. One part, which comes from the diene, consists of four carbon atoms and contains a double bond between carbon atoms 2 and 3. Converting this four-carbon part into a butadiene gives the structure of the diene. The other two carbon atoms of the ring come from the dienophile. The structure of the dienophile is obtained by adding an additional bond between these two carbon atoms.

(a) The part of molecule obtained from the diene

[structure: bicyclic anhydride with diene portion circled] + [maleic anhydride] → [product]

The part of molecule obtained from the dienophile

(b) This part of molecule is obtained from the diene

This part of molecule is obtained from the dienophile

(c) This part of molecule is obtained from the diene

This part of molecule is obtained from the dienophile

Must use Z not E isomer

19.20 Bonding occurs between the terminal carbon atoms of the diene and the dienophile because the signs of the orbitals permit overlap that results in bond formation.

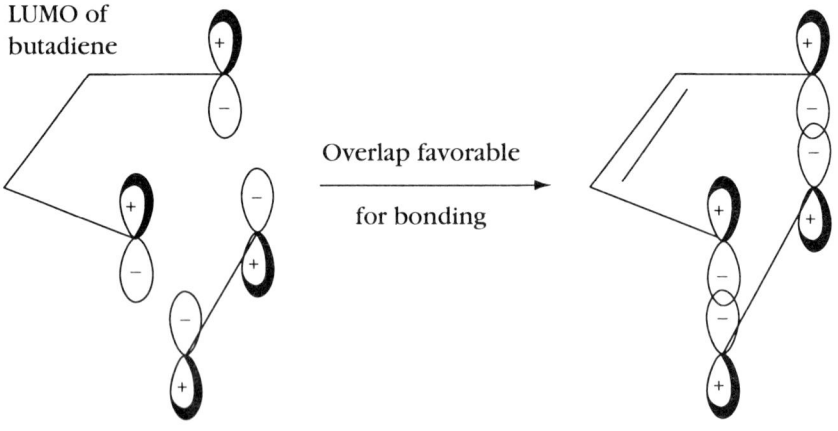

19.21 An E1 mechanism that involves an allylic cation can be proposed for this reaction.

19.22 A mechanism similar to the one given as the answer to Exercise 19.21 can be proposed for this reaction:

Myrcene

Ocimene

19.23 The extinction coefficient, ε, is calculated from the formula $\varepsilon = A/lc$ where A is the absorbance, l is the length of the light path in cm, and c is the concentration of the sample in moles/liter.

$$\text{Moles of compound} = \frac{0.01\text{g}}{240\text{g/mole}} = 4.2 \times 10^{-5} \text{ moles}$$

$$c = \frac{4.2 \times 10^{-5} \text{mol}}{0.010 \text{ L}} = 4.2 \times 10^{-3} \text{ mol/L}$$

$$l = 1 \text{ cm}$$

$$A = 0.65$$

$$\varepsilon = \frac{0.65}{(1)(4.2 \times 10^{-3})} = 155$$

19.24 (a) **Conrotatory rotation:** A term that describes the way in which substituents at the ends of a π system rotate in the same direction, either both clockwise or both counterclockwise, during an electrocyclic ring opening or ring closing.

(b) **Conjugated diene:** A molecule that contains two carbon-carbon double bonds that are separated by one carbon-carbon single bond.

(c) **Dienophile:** The alkene or alkyne that reacts with the diene in a Diels-Alder reaction.

(d) **Diels-Alder reaction:** The reaction of a conjugated diene with certain alkenes to form a product that contains a six-membered ring. The diene must be in an *s-cis* conformation and the reaction rate is enhanced by the presence of electron-withdrawing groups on the dienophile.

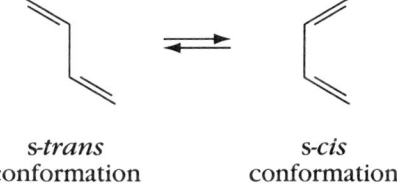

Z = Electron-withdrawing group

(e) **1,2-Addition:** The product formed by addition of the two parts of an electrophile to the adjacent carbon atoms of a conjugated π system.
(f) **Nonbonding MO:** A nonbonding molecular orbital has the same energy as the atomic orbitals combined to form it.
(g) **Concerted reaction:** A reaction that occurs in a single step without any intermediates. In a concerted reaction all bond making and bond breaking occur in the same single step.
(h) **HOMO** refers to the highest occupied molecular orbital.
(i) **Allyl cation:** The carbocation in which the electron-deficient carbon atom is adjacent to a carbon-carbon double bond: $CH_2=CHCH_2^+$
(j) **λ_{max}:** The wavelength at the maximum of the absorption peak of a compound.
(k) **Allyl radical:** The radical in which the electron-deficient carbon atom is adjacent to a carbon-carbon double bond: $CH_2=CH\dot{C}H_2$.
(l) **Products of kinetic control:** The product composition of a reaction mixture that is determined by the relative rates of formation of the products.
(m) **s-cis Conformation:** The conformation in which the two double bonds of a diene are in the same plane and on the same side of a single bond:

s-*trans*   s-*cis*
conformation   conformation

(n) **Products of thermodynamic control:** The product composition of a reaction that is determined by the relative stabilities of the products.
(o) **1,4-Addition:** The product formed by addition of an electrophile to carbon atoms 1 and 4 of a conjugated π system.
(p) **Disrotatory rotation:** A term that describes the way in which substituents at the ends of a π system rotate in opposite directions during an electrocyclic ring opening or ring closing.
(q) **A conjugated triene:** A compound that contains three carbon-carbon double bonds separated by two carbon-carbon single bonds:

(r) **Cumulative double bonds:** Double bonds in which one carbon atom is part of two carbon-carbon double bonds.

19.25 (a) 1,3-Cyclopentadiene
     (b) 1,2,8-Nonatriene
     (c) 4-Methyl-1,2-pentadiene
     (d) 4-*tert*-Butyl-1-chloro-2-methyl-1,3,5-cyclooctatriene
     (e) 3-Methylene-1,4-cyclohexadiene

(f) (Z)-1,1-Dichloro-6-methyl-1,3,5-heptatriene
(g) 1,4,8-Cycloundecatriene
(h) (Z)-3,4,5-Trimethyl-1,3-hexadiene

19.26

(a) $CH_3CH=C=CH(CH_2)_4CH_3$

(b)
```
    CH₃      H
      \    /
       C=C
      /    \
     H     CH₂   H
             \  /
              C=C
             /    \
            H    CH₂CH₂CH₃
```

(c) $CH_2=CHCH_2$—[cyclohexadiene ring]

(d)
```
              Cl
              |
         CH=C
        /     \
   CH₂=C     CH₃
        \
         Cl
```

(e) [cyclohexane ring with H, CH=CH₂, CH=CH₂, H substituents]

(f)
```
   CH₃OCH₂      H
         \    /
          C=C
         /    \
        H    CH₂OCH₃
```

19.27

(a)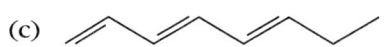
Isolated double bonds

(b)
Conjugated double bonds

(c) [structure] Conjugated double bonds

(d) [structure] Cumulative double bonds

(e)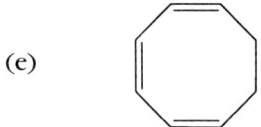
Conjugated double bonds

19.28

(a) [resonance structures of cyclopentadienyl anion]

(b) [resonance structures of naphthalene radical]

(c) $\left[\begin{array}{c} \text{CH}_2=\text{C}(\text{CH}_2^+)\text{CH}_2\text{CH}_3 \end{array} \leftrightarrow \begin{array}{c} \text{CH}_2=\text{C}-\text{CH}=\text{CH}-\text{CH}_3 \\ {}^+ \end{array} \right]$

(d) [cyclohexadienyl cation resonance structures]

19.29

(a) cyclopentadiene $\xrightarrow{\text{Cl}_2}$ 1,2-addition product (3,4-dichlorocyclopent-1-ene) + 1,4-addition product (3,5-dichlorocyclopent-1-ene)

                                   1,2 Addition   1,4 Addition
                                      product   product

(b) cyclopentadiene $\xrightarrow[\text{CCl}_4]{\text{NBS}}$ 5-bromocyclopenta-1,3-diene

(c) cyclopentadiene $\xrightarrow{\text{H}_2/\text{Pt}}$ cyclopentene $\xrightarrow{\text{H}_2/\text{Pt}}$ cyclopentane

(d) cyclopentadiene $\xrightarrow[(2)\ \text{Zn}/\text{H}_3\text{O}^+]{(1)\ \text{O}_3}$ OHC-CH$_2$-CHO + OHC-CHO

(e) cyclopentadiene $\xrightarrow{\text{HCl}}$ 3-chlorocyclopent-1-ene

19.30

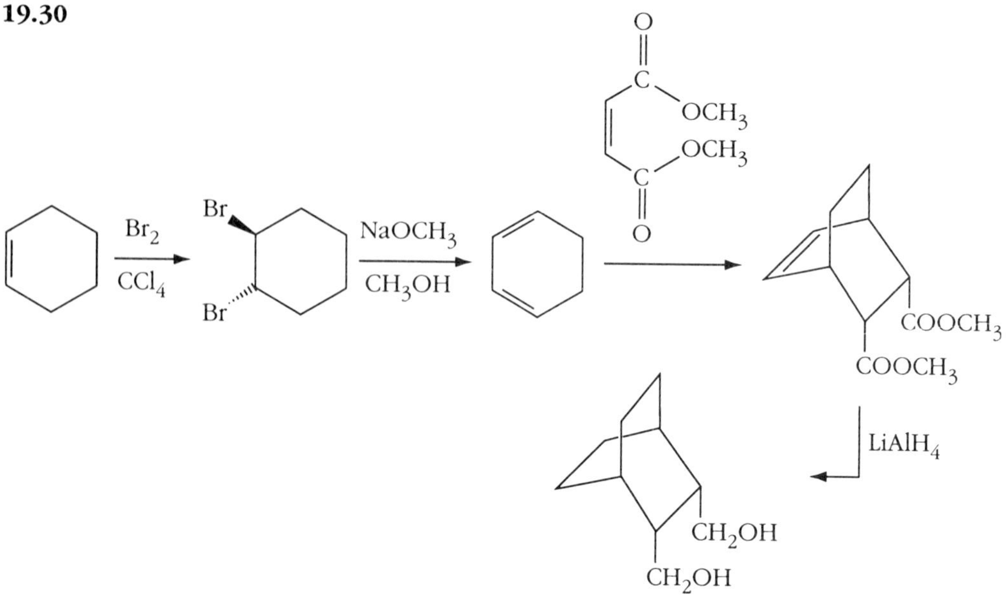

[Reaction scheme: cyclohexene → Br₂/CCl₄ → trans-1,2-dibromocyclohexane → NaOCH₃/CH₃OH → 1,3-cyclohexadiene → + dimethyl maleate → Diels-Alder adduct (bicyclic diester with two COOCH₃ groups) → LiAlH₄ → bicyclic diol with two CH₂OH groups]

19.31

(a) [3-bromocyclopentene structure: cyclopentene ring with Br substituent at allylic position]

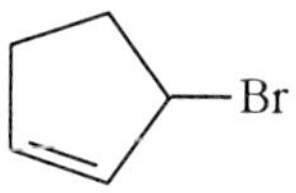

(b) Nucleophilic substitution reaction that occurs by an S_N1 mechanism to form an allylic cation in Step 1. The allylic cation reacts with methanol to form a mixture of products in Step 2.

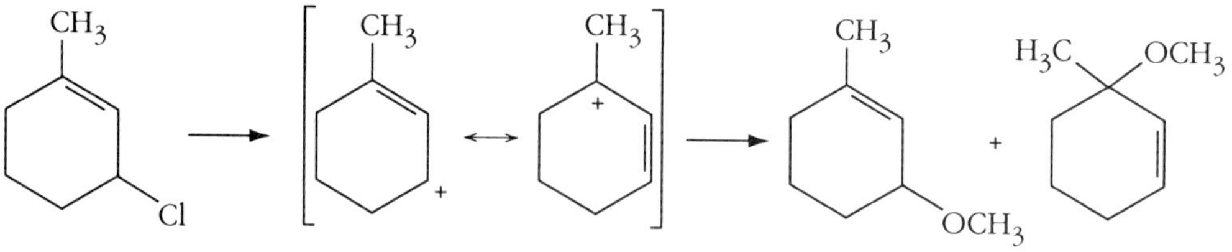

[Mechanism: 3-chloro-1-methylcyclohexene → resonance-stabilized allylic cation (two resonance structures shown in brackets) → two methyl ether products: 3-methoxy-1-methylcyclohexene and 1-methoxy-1-methyl-cyclohex-3-ene]

(c) $4n\ \pi$ electrons and a photochemical reaction so disrotatory ring closure occurs to form the product:

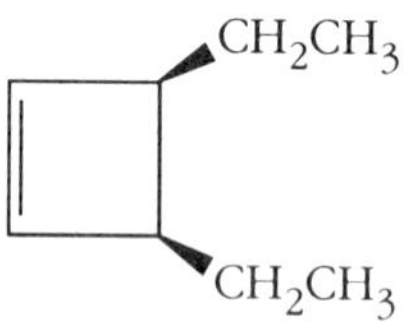

[cis-3,4-diethylcyclobutene structure with both CH₂CH₃ groups on the same face]

(d) A Diels-Alder reaction to form

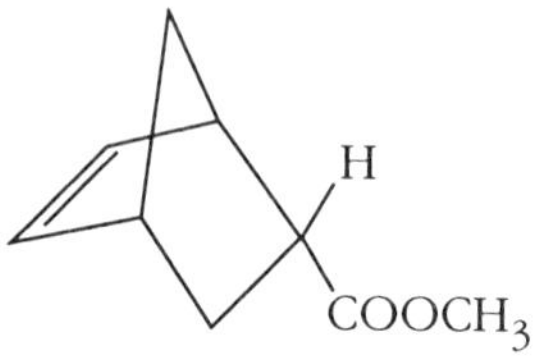

[norbornene-type bicyclic product with COOCH₃ group and H shown]

(e) A Diels-Alder reaction in which cyclopentadiene acts as both the diene and the dienophile to form a dimer.

The cyclopentadiene that reacted as a diene

The cyclopentadiene that reacted as a dienophile

(f) $4n + 2$ π electrons and a thermal reaction so disrotatory ring closure occurs to form the product.

(g) Diels-Alder reaction

19.32 The value of λ_{max} increases as the number of conjugated double bonds in the molecules increase so

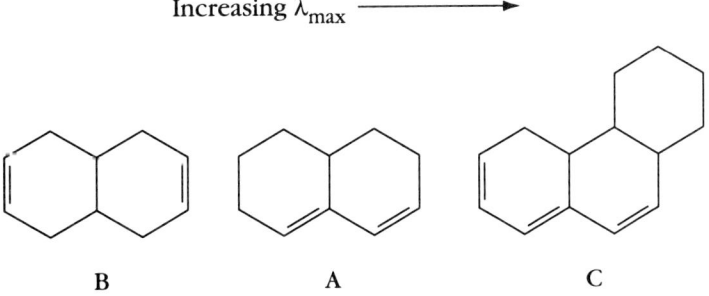

Increasing λ_{max} →

B   A   C

19.33 Step 1: Protonation of the alcohol

$$\underset{CH_2=CHCHCH_3}{\overset{:\ddot{O}H}{|}} + H\text{—}Cl \rightleftharpoons \underset{CH_2=CHCHCH_3}{\overset{\overset{H}{|}}{\overset{+}{\ddot{O}H}}}$$

Step 2: Loss of water molecule to form an allylic cation

$$\underset{CH_2=CHCHCH_3}{\overset{\overset{H}{|}}{\overset{+}{\ddot{O}H}}} \longrightarrow \left[CH_2=CH\overset{+}{C}HCH_3 \longleftrightarrow \overset{+}{C}H_2CH=CHCH_3 \right] + H_2O$$

Step 3: Reaction of allylic cation and chloride ion

$$\left[CH_2=CH\overset{+}{C}HCH_3 \longleftrightarrow \overset{+}{C}H_2CH=CHCH_3 \right] \xrightarrow{Cl^-} \begin{array}{c} \underset{CH_2=CHCHCH_3}{\overset{\overset{Cl}{|}}{}} \\ + \\ ClCH_2CH=CHCH_3 \end{array}$$

19.34 To solve this Exercise, we must determine the number of π electrons (4n or 4n +2) involved in the reaction and whether the ring opening or closing is conrotatory or disrotatory.

(a) 4n π electrons and conrotatory ring opening so it is a thermal reaction.
(b) 4n π electrons and conrotatory ring opening so it is a thermal reaction.
(c) 4n +2 π electrons and disrotatory ring closure so it is a thermal reaction.
(d) 4n π electrons and disrotatory ring closure so it is a photochemical reaction.
(e) 4n +2 π electrons and disrotatory ring opening so it is a thermal reaction.
(f) 4n π electrons and disrotatory ring closure so it is a photochemical reaction.
(g) 4n +2 π electrons and disrotatory ring closure so it is a thermal reaction.

19.35

sp^2 hybrid carbon atom — sp hybrid carbon atom

These two stereoisomers are enantiomers because they are nonsuperposable mirror images.

19.36 At −15 °C, 3-chloro-3-methyl-1-butene is formed in the greatest amount so this isomer is the product of kinetic control. At room temperature, 1-chloro-3-methyl-2-butene is formed in the greatest amount so it is the product of thermodynamic control.

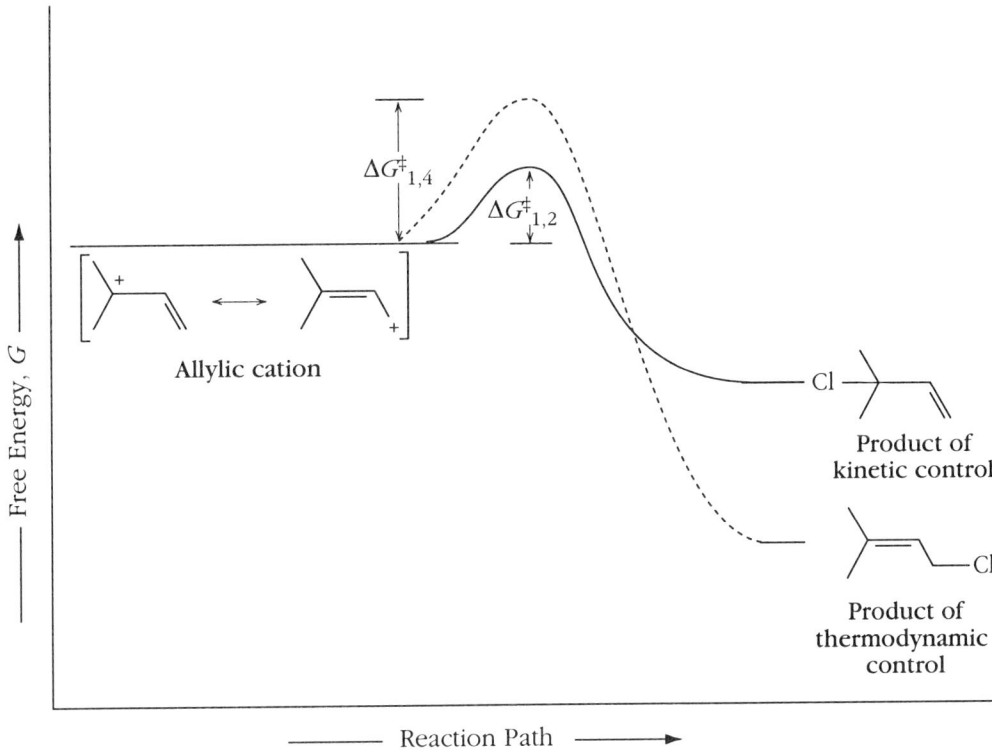

19.37 Clockwise conrotatory ring opening forms (2Z,4Z)-2,4-hexadiene while counterclockwise conrotatory ring opening forms (2E,4E)-2,4-hexadiene.

Steric hindrance between the two methyl groups of (2Z,4Z)-2,4-hexadiene makes it much less stable than (2E,4E)-2,4-hexadiene. The transition state leading to the formation of (2Z,4Z)-2,4-hexadiene, therefore, is much higher in energy than the transition state for the formation of (2E,4E)-2,4-hexadiene so only (2E,4E)-2,4-hexadiene is formed as product.

19.38

(a) 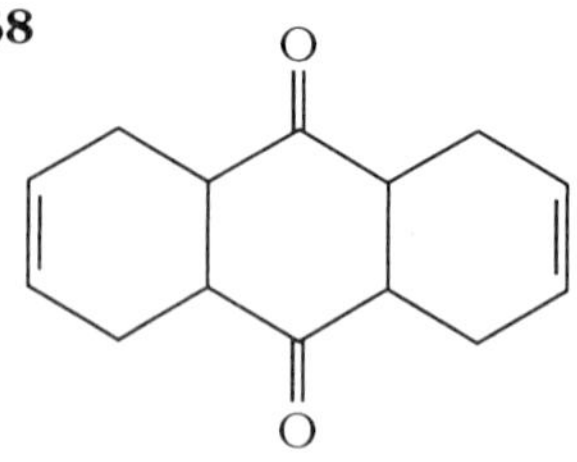 [structure: decahydroanthracene-9,10-dione with two double bonds]

(b) 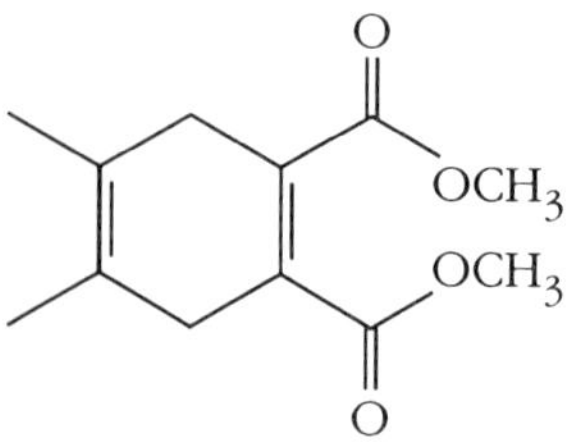 [structure: dimethyl 4,5-dimethylcyclohexa-1,4-diene-1,2-dicarboxylate]

(c) 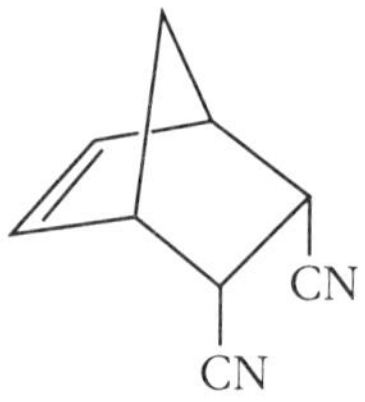 [structure: norbornene with two CN groups]

(d) 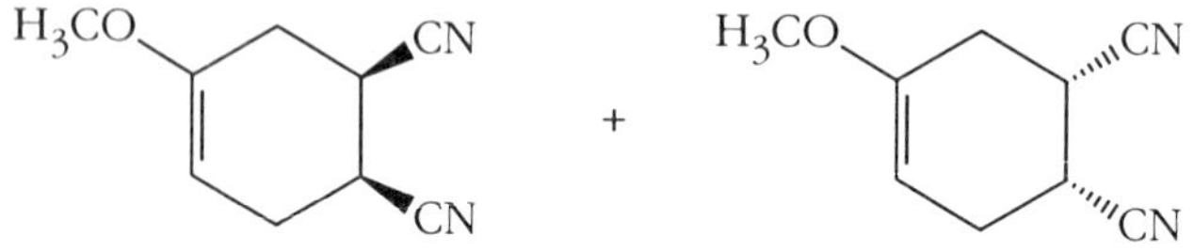 [structure: H₃CO-substituted cyclohexene with two CN groups (cis)] + [structure: H₃CO-substituted cyclohexene with two CN groups (trans)]

19.39 The solutions to this Exercise are obtained by the method explained in Exercise 19.19.

(a) 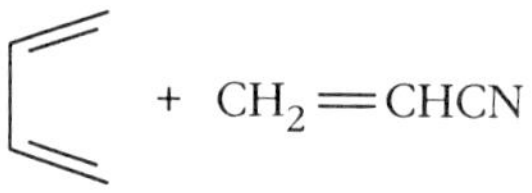 [butadiene] + CH$_2$=CHCN

(b) 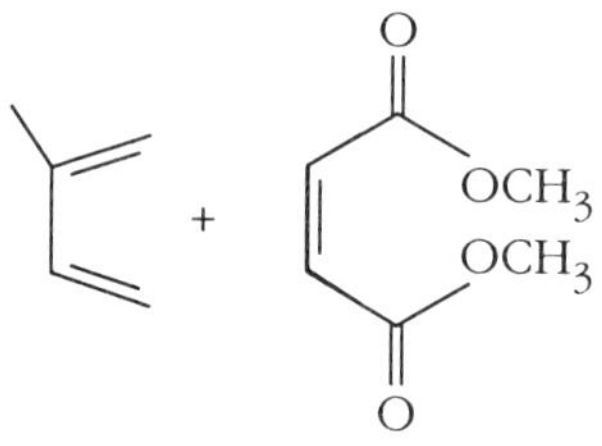 [isoprene] + [dimethyl maleate]

(c) 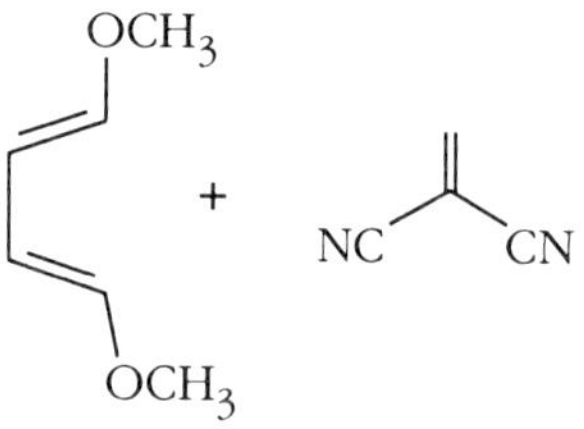 [1,4-dimethoxybutadiene] + NC–C(=)–CN (dicyanoethylene... methylenemalononitrile)

(d) 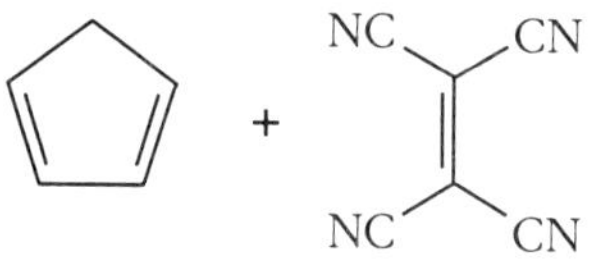 [cyclopentadiene] + [tetracyanoethylene, (NC)$_2$C=C(CN)$_2$]

(e) 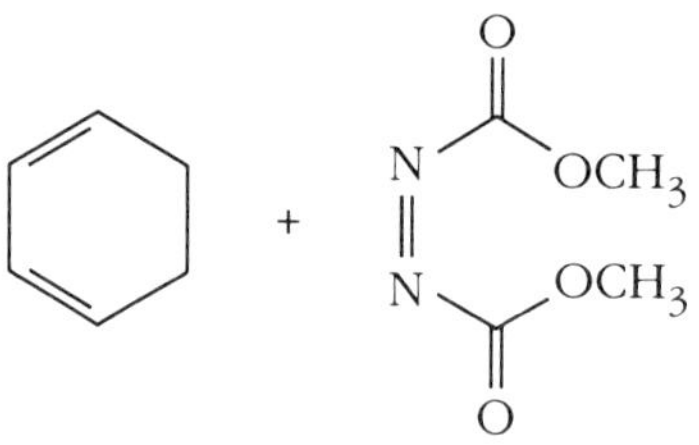 [1,3-cyclohexadiene] + [dimethyl azodicarboxylate, CH$_3$O-C(=O)-N=N-C(=O)-OCH$_3$]

19.40 The reaction of chlorine and 3,4-dimethylcyclobutene at room temperature forms the addition product of chlorine to 3,4-dimethylcyclobutene (compound I). When the solution is heated before adding chlorine, ring opening of the cyclobutene ring occurs to form 2,4-hexadiene. Chlorine reacts with 2,4-hexadiene by 1,2-and 1,4-addition to form compounds II and III.

19.41 An E1 mechanism similar to the one proposed as the answer to Exercise 19.21 can be proposed for this reaction. Notice that the allylic cation must undergo isomerization before it can cyclize.

19.42 **(a)** The molecular formula, $C_8H_{16}O$, indicates one degree of unsaturation, which is due to a carbonyl group because the IR spectrum shows an absorption band at 1710 cm^{-1} characteristic of a carbonyl group. The ^{1}H NMR spectrum contains a singlet at δ 1.2 (9H), a triplet at δ 0.9 (3H), a triplet at δ 2.46 (2H), and a sextet at δ 1.58 (2H). A singlet of relative area 9 is characteristic of a *tert*-butyl group so we can now write two partial structures:

$$C=O \text{ and } (CH_3)_3C$$

The triplet at δ 0.9 (3H) suggests a methyl group adjacent to a methylene group. The chemical shift of the triplet at δ 2.46 (2H) is consistent with a methylene group bonded to a carbonyl carbon atom and its splitting pattern shows that it is also bonded to a methylene group. The sextet at δ 1.58 (2H) is consistent with a methylene group bonded to both another methylene group and a methyl group. The structure of the compound is

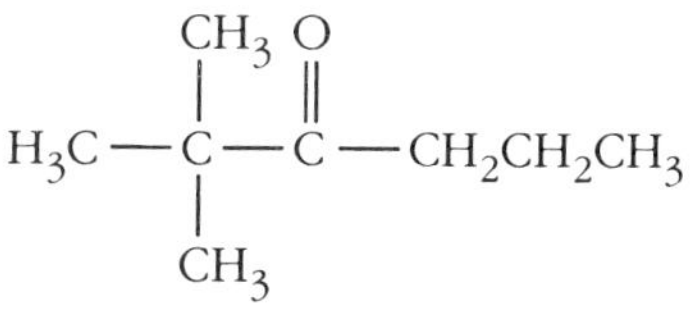

$$\text{H}_3\text{C}-\underset{\underset{\text{CH}_3}{|}}{\overset{\overset{\text{CH}_3}{|}}{\text{C}}}-\overset{\overset{\text{O}}{\|}}{\text{C}}-\text{CH}_2\text{CH}_2\text{CH}_3$$

2,2-Dimethyl-3-hexanone

(b) The molecular formula, $C_8H_{12}O_4$, indicates three degrees of unsaturation. The IR spectrum contains an absorption band at 1720 cm^{-1} indicating the presence of a carbonyl group. The ^{1}H NMR contains a triplet at δ 1.26 (3H), a quartet at δ 4.2 (2H) and a singlet at δ 6.1 (1H). The relative areas account for only six hydrogen atoms so the molecule must have symmetry. The splitting patterns suggest one or more ethyl groups and the chemical shift of the methylene group suggests that it is bonded to an oxygen atom. The singlet at δ 6.1 (1H) is characteristic of a vinyl hydrogen atom. The partial structures we have so far are

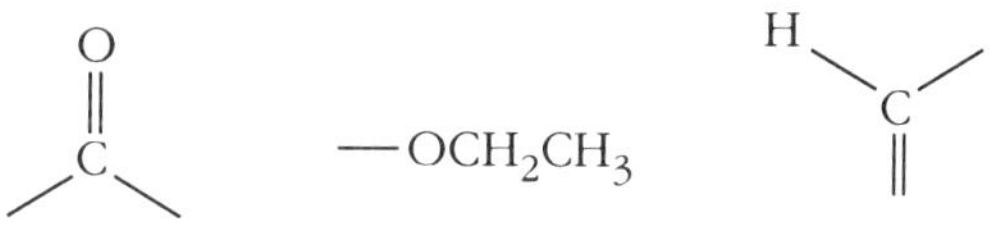

$$\overset{\overset{\text{O}}{\|}}{\underset{}{\text{C}}} \qquad -\text{OCH}_2\text{CH}_3 \qquad \overset{\text{H}}{\underset{\|}{\text{C}}}$$

Joining these part structures, we obtain the structure of the compound:

$$\underset{\text{CH}_3\text{CH}_2\text{O}}{\overset{\overset{\text{O}}{\|}}{\text{C}}}-\text{CH}=\text{CH}-\underset{\text{OCH}_2\text{CH}_3}{\overset{\overset{\text{O}}{\|}}{\text{C}}}$$

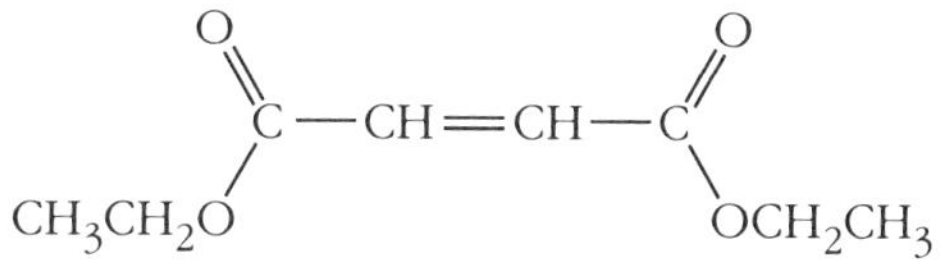

(c) The molecular formula, $C_3H_5BrO_2$, indicates one degree of unsaturation, which is due to a carbonyl group because the IR spectrum shows an absorption band at 1700 cm^{-1} characteristic of a carbonyl group. The carbonyl group is part of a carboxylic acid group because of the broad absorption band between 3400 to 2400 cm^{-1} in the IR spectrum. The ^{1}H NMR spectrum shows two triplets, one at δ 3.0 (2H) and the other at about δ 3.58 (2H). A broad peak near δ 8 (1H) is due to the hydrogen atom of the carboxyl group. The triplets at δ 3.0 and at about δ 3.58 suggest two methylene groups. Their coupling constants are identical and the triplets lean towards each other, which suggests that the methylene groups are joined together. The chemical shifts of the two methylene groups are downfield, which indicates that each one is bonded to an electron-withdrawing atom or group. Joining these part structures gives following structure of the compound:

BrCH$_2$CH$_2$COOH
3-Bromopropanoic acid

(d) The molecular formula, $C_6H_{12}O_2$, indicates one degree of unsaturation, which must be due to a carbonyl group because the IR spectrum shows an absorption band at 1700 cm^{-1} characteristic of a carbonyl group. The carbonyl group is part of a carboxylic acid group because of the broad absorption band between 3400 to 2400 cm^{-1} in the IR spectrum. The

^{1}H NMR spectrum shows a doublet at δ 0.94 (6H), a multiplet at δ 1.5 to 1.7 (3H), a triplet at δ 2.38 (2H), and a singlet at δ 12. The signal at δ 12 is due to the hydrogen atom of the hydroxyl group. The doublet at δ 0.94 (6H) indicates a methyl group split by a single neighboring hydrogen atom (probably the methyl groups of an isopropyl group). The hydrogen of the isopropyl group is part of the complex multiplet at δ 1.5 to 1.7 (3H). This accounts for four of the six carbon atoms of the compound so the partial structures are

$$—COOH \text{ and } —CH(CH_3)_2$$

The triplet at δ 2.38 (2H) is consistent with a methylene group bonded to the —COOH group and another methylene group. Adding the two adjacent methylene groups to the others of the molecule gives the following structure of the compound:

$$(CH_3)_2CHCH_2CH_2COOH$$
4-Methylpentanoic acid

CHAPTER 20

AROMATICITY
Concepts

Hückel's rule of aromaticity defines a compound that contains p atomic orbitals on all adjacent atoms of a planar ring as aromatic if it contains $4n + 2$ π electrons. ^{1}H NMR spectroscopy provides one of the best criteria of aromatic compounds. The hydrogen atoms outside an aromatic ring are deshielded so they are found downfield in the δ 6.5 to 8.5 region. Hydrogen atoms inside the ring are shielded so they are found upfield, often upfield of TMS. Benzene and its derivatives are the most common examples of aromatic compounds.

A site adjacent to an aromatic ring is more reactive that it would be in the absence of the ring. Alkyl benzenes, for example, undergo free radical bromination exclusively at the benzylic position, and their vigorous oxidation converts them into aromatic carboxylic acids.

Solutions to the Exercises

20.1

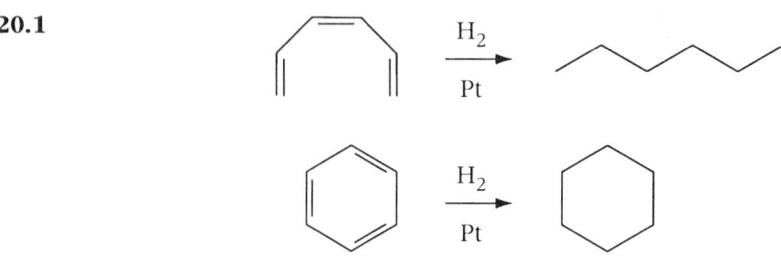

Heat of hydrogenation of (Z)-1,3,5-hexatriene = 80.5 kcal/mol (−337 kJ/mol)
Heat of hydrogenation of benzene = −49.3 kcal/mol (−206 kJ/mol)
Estimated resonance energy of benzene = 31 kcal/mol (131 kJ/mol)

   Notice that this value of about 31 kcal/mol (131 kJ/mol) for the resonance energy of benzene is different than the value of 36.5 kcal/mol (152 kJ/mol) given in Figure 20.1. The reason for this difference is that (Z)-1,3,5-hexatriene is used rather than cyclohexene as the compound from which we try to calculate the heat of hydrogenation of the imaginary benzene molecule without π electron delocalization. The difference in the estimated values of the resonance energy of benzene, which range from 30 to 80 kcal/mol (125 to 335 kJ/mol), are due to the use of different compounds to calculate the heat of hydrogenation of this imaginary compound.

20.2 The carbon atoms of benzene are relatively unaffected by the induced magnetic field because they are located between shielding and nonshielding regions of the induced magnetic field.

20.3 (a) This compound contains 12 π electrons so it is a $4n$ system and therefore it is nonaromatic. Remember that only two of the four π electrons of the carbon-carbon triple bond form part of the delocalized system.
   (b) This compound contains 10 π electrons so it is a $4n + 2$ system and therefore it is aromatic.
   (c) This compound contains 14 π electrons so it is a $4n + 2$ system and therefore it is aromatic.

20.4

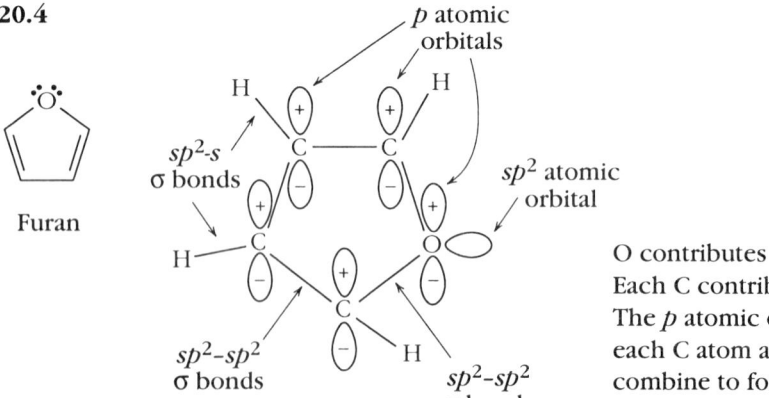

O contributes 2 π electrons. Each C contributes 1 π electron. The p atomic orbitals on each C atom and the O atom combine to form the molecular orbitals.

20.5 The ions in (b), (c), and (d) all contain $4n + 2$ π electrons.

(a) 2 π electrons

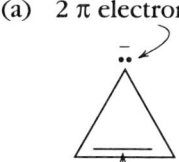

2 π electrons
Total of 4 π electrons
$4n$ π electrons ($n = 1$)
so *not* aromatic

(b) 2 π electrons   2 π electrons

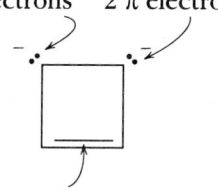

2 π electrons
Total of 6 π electrons
$4n + 2$ π electrons ($n = 1$)
so aromatic

(c)

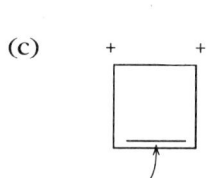

2 π electrons
Total of 2 π electrons
$4n + 2$ π electrons ($n = 0$)
so aromatic

(d) 2 π electrons

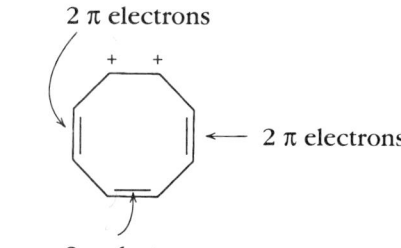

← 2 π electrons

2 π electrons
Total of 6 π electrons
$4n + 2$ π electrons ($n = 1$)
so aromatic

20.6 The molecular orbital diagram in Figure 20.6 indicates that a cyclic molecule containing nine sp^2 hybrid carbon atoms contains $4n + 1$ π electrons. A $4n + 2$ aromatic system can be obtained by the addition of one electron to the HOMO to form a planar cyclic molecule with 10 π electrons ($4n+2, n = 2$). An aromatic system can also be achieved by removing three π electrons to form a carbocation with a charge of +3 and 6 π electrons ($4n + 2; n = 1$).

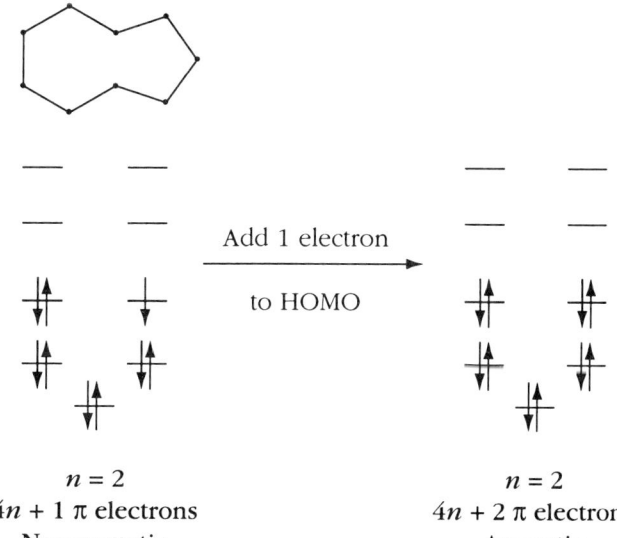

20.7

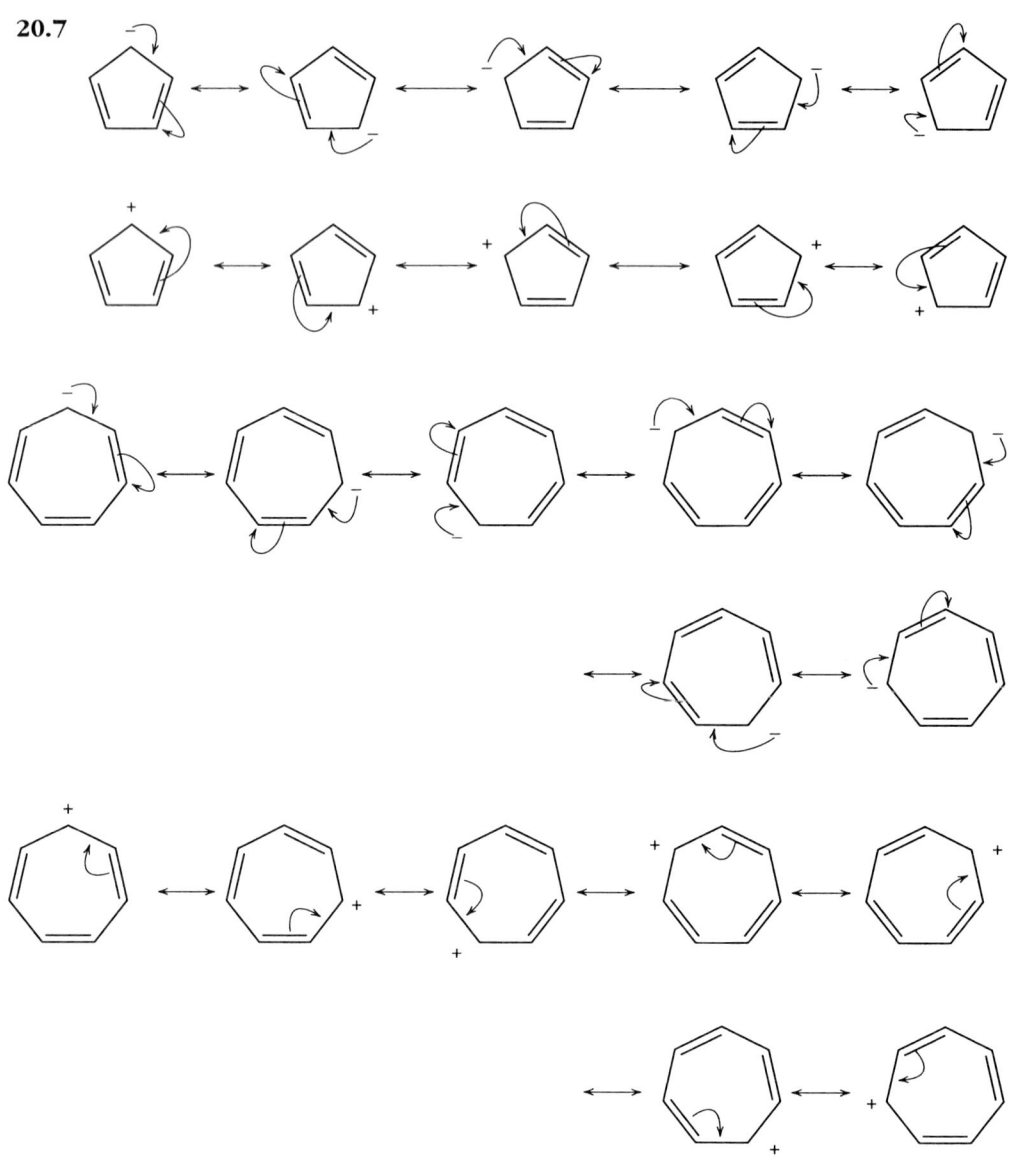

20.8 All of the carbon atoms of the cyclopentadienyl anion are equivalent so its ^{13}C NMR spectrum consists of one signal. Similarly all of the hydrogen atoms of the cyclopentadienyl anion are equivalent so its ^{1}H NMR spectrum also consists of a singlet. For the same reason, both the ^{1}H NMR spectrum and the ^{13}C NMR spectrum of the cycloheptatrienyl cation will have only one signal.

20.9 The singlet at δ 9.08 indicates aromatic character in the three-membered ring while a high dipole moment indicates a separation of charge in the molecule. These two observations are consistent with a structure of cyclopropenone in which the π electrons of the carbonyl group are highly polarized so that the oxygen atom bears most of the negative charge and the carbon atom bears most of the positive charge. In such a structure, the three-membered ring is a $4n + 2$ system ($n = 0$).

20.10 Alkyl groups bonded to a benzene ring that have at least one benzylic hydrogen atom are all oxidized to carboxylic acid groups. Alkyl groups that lack a benzylic hydrogen atom are not oxidized. In Exercise 20.10(c), for example, the *tert*-butyl group bonded to a benzene ring does not have a benzylic hydrogen atom so it is not oxidized to a carboxylic acid group.

20.11

20.12 Nucleophilic substitution reactions of benzyl chloride occur by an S_N2 mechanism to form the following products:

(a) $C_6H_5CH_2I$ (b) $C_6H_5CH_2CN$ (c) $C_6H_5CH_2OC(CH_3)_3$ (d) $C_6H_5CH_2OCH_3$

20.13 The nucleophilic substitution reactions of 2-chloro-2-phenylpropane with good nucleophiles occur by an S_N1 mechanism whereas strong bases favor elimination reactions.

(a) 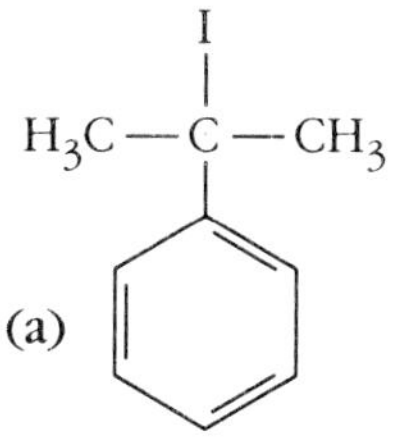 2-iodo-2-phenylpropane (b) 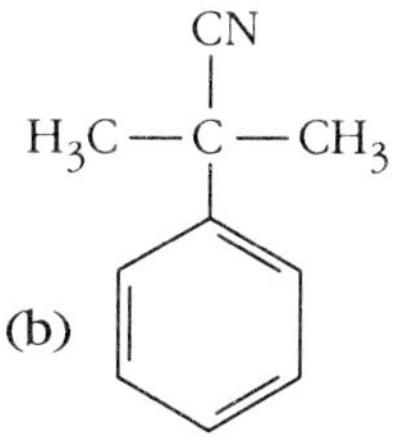 2-cyano-2-phenylpropane (c) 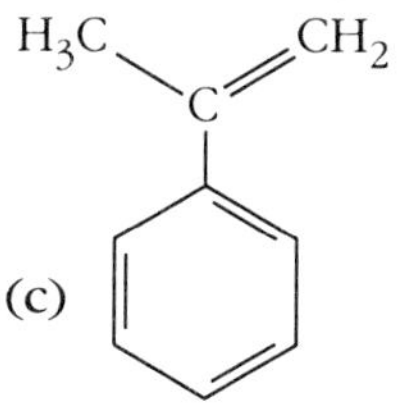 2-phenylpropene (d) 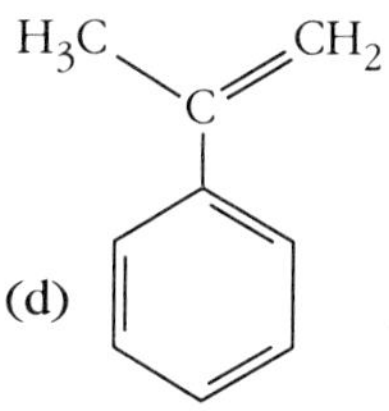 2-phenylpropene

20.14 Step 1: Addition of a proton to the double bond

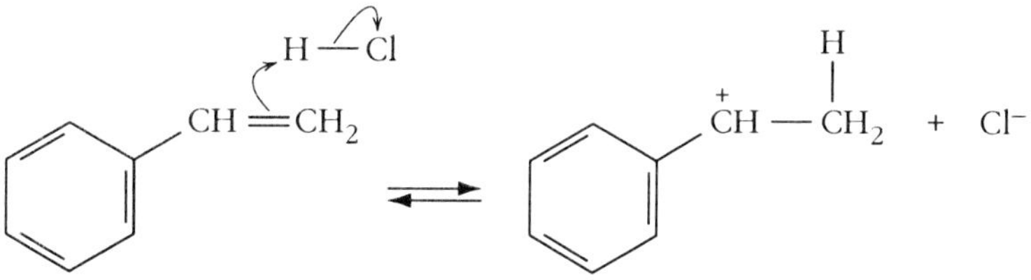

Step 2: Reaction of chloride ion and the carbocation to form the product

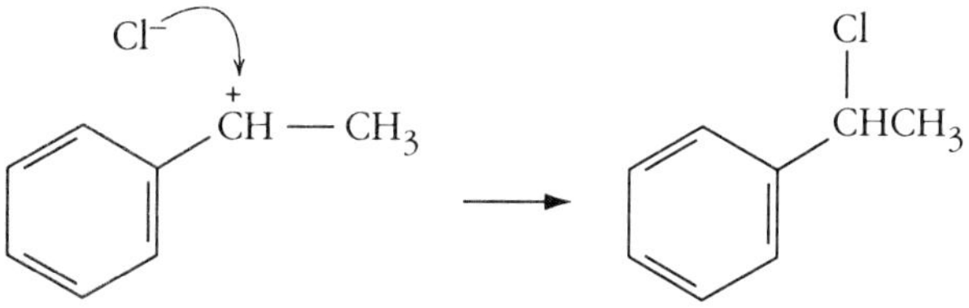

(Racemic mixture)

20.15

Initiation: $RO{-}OR \rightleftharpoons 2\,RO^\bullet$

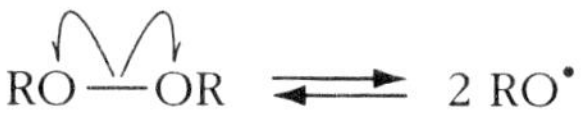

$RO^\bullet + H{-}Br \longrightarrow ROH + Br^\bullet$

Propagation: (as shown in scheme)

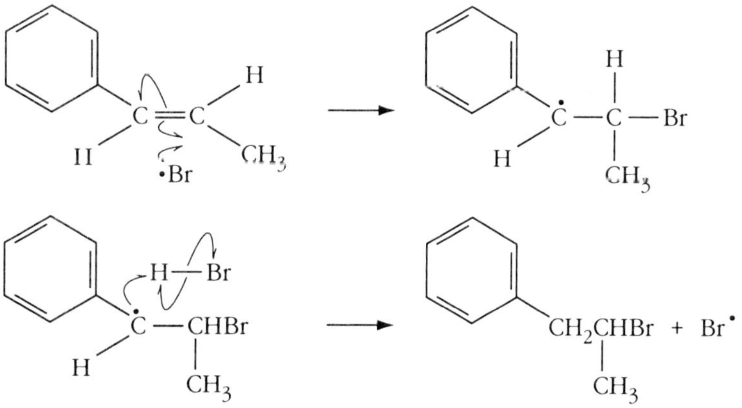

Termination: Br• + Br• ⇌ Br₂

[Ph-CH(•)-C(Br)(CH₃)-H structure] + Br• → [Ph-CH(Br)-CH(Br)(CH₃)]

20.16 The double bond in the five-membered ring of indene undergoes addition reactions that are typical of alkenes. The regioselectivity of electrophilic addition reactions is governed by the ability of the aromatic ring to stabilize the adjacent carbocation.

(a) 1-chloroindane
(b) cis-1,2-dihydroxyindane or trans-1,2-dihydroxyindane
(c) 1-hydroxyindane
(d) trans-1-hydroxy-2-bromoindane or cis-1-hydroxy-2-bromoindane
(e) 1-hydroxyindane
(f) 2-hydroxyindane
(g) indane
(h) phthalic acid (benzene-1,2-dicarboxylic acid)
(i) 1-bromoindane
(j) indene epoxide (exo) or indene epoxide (endo)

20.17 (a) **Resonance energy** is the difference in energy between a molecule or ion in which there is electron delocalization and a hypothetical model in which the electrons are localized. Resonance energy, therefore, is the energy by which an ion or molecule is stabilized by electron delocalization.

(b) **Hückel's rule of aromaticity** states that any compound that contains p atomic orbitals on adjacent atoms in a planar ring is aromatic if it contains $4n + 2$ π electrons.

(c) An **annulene** is a completely conjugated monocyclic polyene. Annulenes may or may not be aromatic. Benzene ([6]annulene) and [18]annulene, for example, are aromatic while cyclobutadiene ([4]annulene), and cyclooctatetraene ([8]annulene) are not aromatic.

(d) A **polycyclic aromatic hydrocarbon** is a molecule that contains two or more benzene-like rings fused or joined at adjacent carbon atoms in such a way that each ring shares two carbon atoms. The simplest polycyclic hydrocarbon is naphthalene:

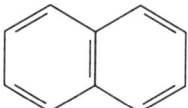

Naphthalene

(e) **Ring current** is the electric current generated by the circulation of π electrons of an aromatic ring when it is placed in a magnetic field. The ring current induces a local magnetic field that reinforces the applied field outside the ring, which accounts for the pronounced downfield shifts of aromatic hydrogen atoms in ^{1}H NMR spectroscopy.

(f) A **heterocyclic aromatic compound** is a planar, cyclic compound that contains one or more ring heteroatoms (atoms other than carbon) in a planar ring that contains $4n + 2$ π electrons.

(g) A **charged aromatic compound** is a positively or negatively charged cyclic and planar ion that obeys Hückel's rule of aromaticity. Examples include the cyclopropenyl cation and the cyclopentadienyl anion.

Cyclopropenyl cation   Cyclopentadienyl anion

20.18 (a) 8 π electron-containing compound so it is *not* aromatic.
    (b) 6 π electron-containing compound so it is aromatic. The lone pairs of electrons on the two nitrogen atoms are not part of the π system.
    (c) 10 π electron-containing compound so it is aromatic even though it contains two rings.
    (d) *Not* aromatic even though the ring contains 6 π electrons because the ring contains one sp^3 hybrid carbon atom.

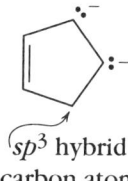

sp^3 hybrid carbon atom

    (e) 6 π electron-containing compound so it is aromatic. The lone pair of electrons on the oxygen atom are not part of the π system because they are located in a *p* orbital that is perpendicular to the plane of the ring.

(f) 22 π electron-containing compound so it is aromatic.
(g) 4 π electron-containing compound that also has an *sp*³ hybrid carbon atom in the ring so it is *not* aromatic.
(h) 6 π electron-containing compound so it is aromatic.
(i) Aromatic compound.
(j) *Not* aromatic.
(k) *Not* aromatic.

20.19 These three hydrocarbons gain additional stability by arranging their π electrons so that two parts of the molecule become aromatic. As a result, their centers of positive and negative charge do not coincide, so they have dipole moments.

(a) The electrons of the double bonds can be shared unequally so that the five-membered ring gains six π electrons and, consequently, a negative charge (so this part of the molecule resembles a cyclopentadienyl anion aromatic system). Two of these electrons are also shared with the fused seven-membered ring so it too has six electrons and a positive charge (so this part of the molecule resembles a cycloheptatrienyl cation). This separation of charge results in a dipole moment.

6 π electrons   6 π electrons

(b)

2 π electrons   6 π electrons

(c)

6 π electrons   6 π electrons

20.20 In each reaction, an intermediate or the product itself is aromatic so the reactions occur readily.

(a) The addition of AgNO₃ to the reaction mixture facilitates the breaking of the C—I bond. This ensures that the reaction occurs by an S_N1 mechanism to form the aromatic cycloheptatrienyl cation as an intermediate.

[Mechanism scheme showing cycloheptatrienyl iodide + Ag⁺ → tropylium cation + AgI, then + :OH₂ → protonated alcohol intermediate, then deprotonation by :B → cycloheptatrienyl alcohol + BH]

(b) Reaction of NaNH₂ with cyclopentadiene occurs readily to form the aromatic cyclopentadienyl anion, which is alkylated in the second step.

[Scheme: cyclopentadiene →(NaNH₂) cyclopentadienyl anion →(CH₃I) methylcyclopentadiene]

(c) Potassium metal donates two electrons to cyclooctatetraene (reduces cyclooctatetraene) to form an aromatic eight-membered ring that contains 10 electrons.

[Scheme: cyclooctatetraene →(2K) cyclooctatetraene dianion]

(d) HCl easily cleaves the ether to form a cyclopropenyl cation and ethanol by an S_N1 mechanism. Ethanol then reacts with excess HCl to form 1-chloroethane.

[Mechanism scheme: cyclopropenyl ethyl ether + HCl → protonated ether → cyclopropenyl cation + CH₃CH₂OH + Cl⁻; CH₃CH₂OH →(HCl) CH₃CH₂Cl]

20.21 In each of these reactions the intermediate or the transition state is not stabilized by aromaticity because each is a $4n$ π electron system. In fact the difficulty in carrying out these reactions suggests that an intermediate or transition state with $4n$ π electrons is destabilized compared to the saturated analog. As a result, $4n$ Hückel systems are known as *antiaromatic systems*.

(a) This reaction can occur by either an S_N1 or an S_N2 mechanism. In either mechanism the rate determining transition state receives no stabilization because it resembles a cyclopentadienyl cation (a $4n\,\pi$ system). The desired product is

[cyclopentadiene with CN substituent]

(b) The reaction of cycloheptatriene with $NaNH_2$ forms a cycloheptatrienyl anion, which is a $4n\,\pi$ system, so the reaction occurs slowly to form the following product:

[cycloheptatriene with CH_3 substituent]

(c) This reaction occurs by an E2 mechanism and its transition state resembles the product, which has $8\,\pi$ electrons. In fact, the elimination product (pentalene) is not formed by this reaction.

[Pentalene structure]

Pentalene

Instead, a substitution reaction forms the following product:

[bicyclic structure with $-O-t-C_4H_9$]

20.22 We learned in Chapter 15 that there are two factors that contribute to the acidity of a compound compared to water. One factor is the electron-withdrawing or electron-donating ability of the group bonded to the OH group compared to a hydrogen atom. Electron-withdrawing groups make the compound more acidic than water. In tropolone, the group bonded to the OH group is electron withdrawing compared to a hydrogen atom because by acquiring a positive charge the seven-membered ring becomes aromatic. This is shown by writing the structure of tropolone as a hybrid of the following resonance structures.

[Five resonance structures of tropolone]

The second factor that contributes to the acidity of a compound is the relative stability of its conjugate base compared to hydroxide ion. The conjugate base of tropolone is stabilized by resonance as shown by the following resonance hybrid.

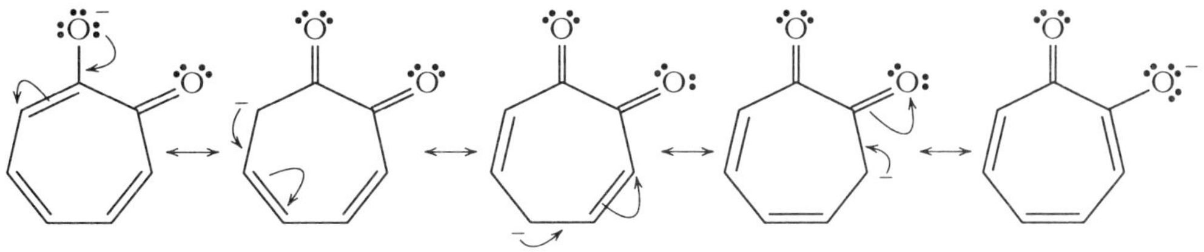

Contributions from both of these effects makes tropolone a stronger acid than most alcohols.

20.23 (a) The acid-base reaction of the two compounds is:

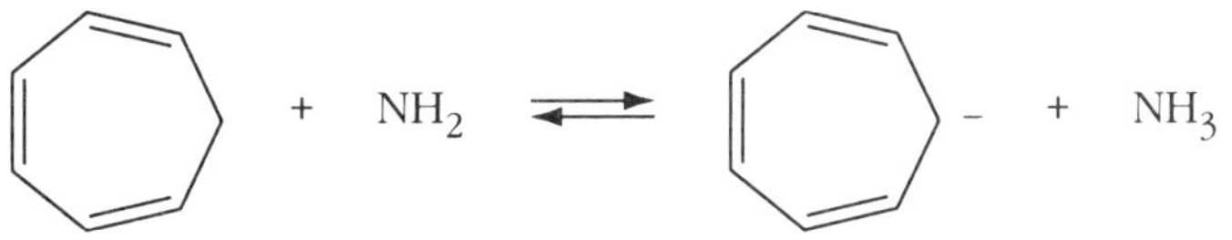

8 π electrons

6 π electrons

Loss of a proton from cyclopentadiene forms the aromatic cyclopentadienyl anion whereas loss of a proton from cycloheptatriene forms a cycloheptatrienyl anion, which gains no stability because it is not aromatic. Cyclopentadiene, therefore, is the stronger acid.

(b) We learned in Chapter 9 (Section 9.10) that the more s character in a hybrid atomic orbital that contains a pair of electrons, the less basic is that pair of electrons. The hybridization of the two nitrogen atoms of the two amines is as follows:

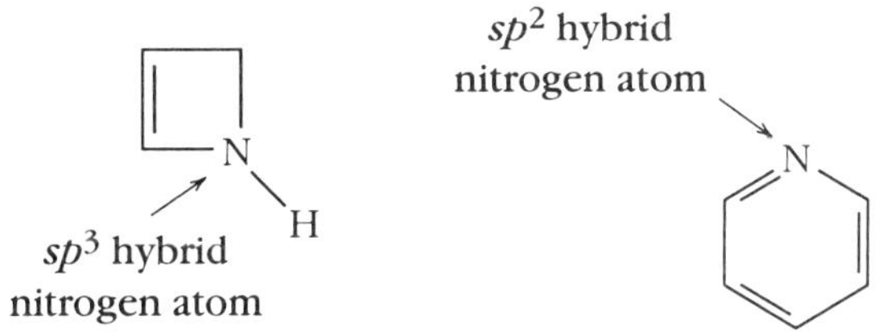

sp^3 hybrid nitrogen atom

sp^2 hybrid nitrogen atom

The electron pair in the sp^3 hybrid orbital is the more basic so its conjugate acid is the weaker acid. Thus, [azetidinium structure] is a weaker acid than [pyridinium structure]

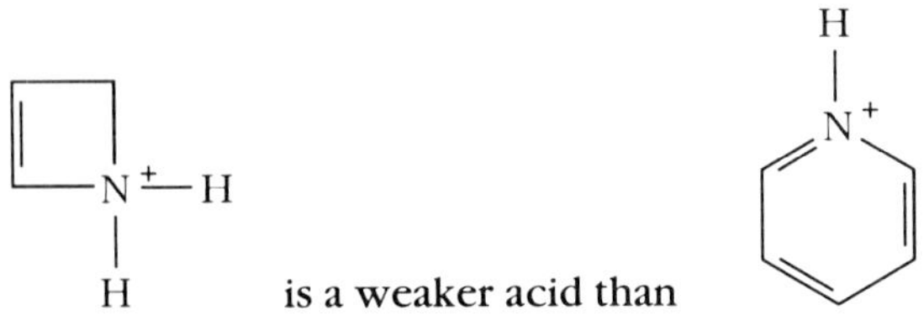

20.24 The planar ring of 4-pyrone is aromatic because it contains six π electrons (4n + 2). This aromaticity can be represented by the following resonance structures:

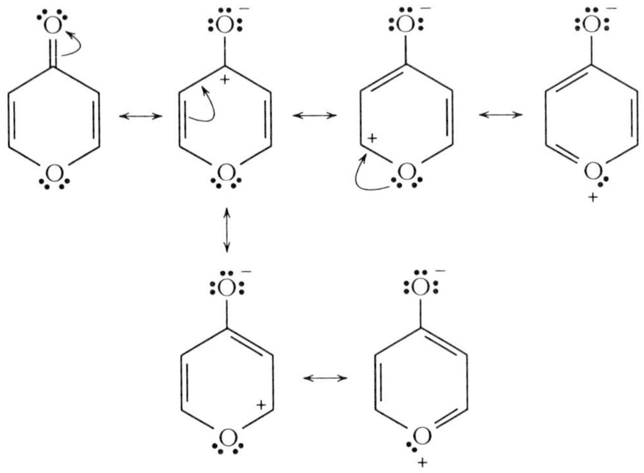

Protonation on the carbonyl oxygen atom of 4-pyrone forms a positively charged ring that is aromatic because it still contains six π electrons (4n + 2). The aromaticity of 4-pyrone is not lost on protonation of its carbonyl oxygen. The π electron delocalization of the carbonyl oxygen protonated 4-pyrone can be represented by the following resonance structures:

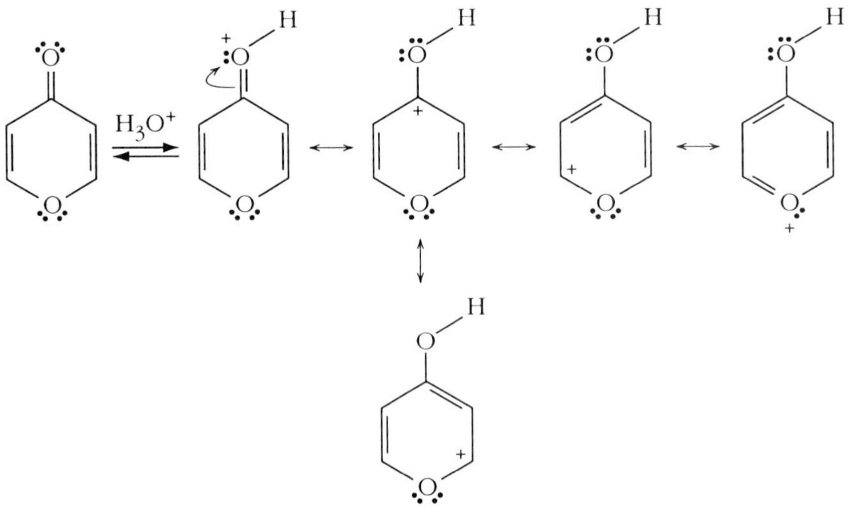

The positively charged ion formed by protonation of the ring oxygen of 4-pyrone is no longer aromatic. Loss of aromaticity, therefore, makes protonation of the ether oxygen atom energetically unfavorable compared to protonation of the carbonyl oxygen atom.

20.25 The polarization of the carbonyl group results in an aromatic system with six π electrons in the ring of cycloheptatrienone, whereas it results in a nonaromatic system with four π electrons in the ring of cyclopentadienone. As a result, introduction of a carbonyl group into a cycloheptatriene ring is energetically favorable while introduction of a carbonyl group into a cyclopentadiene ring is energetically unfavorable.

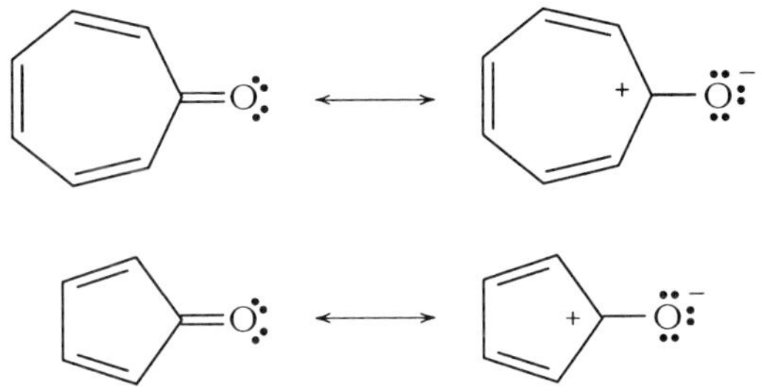

20.26

(a) 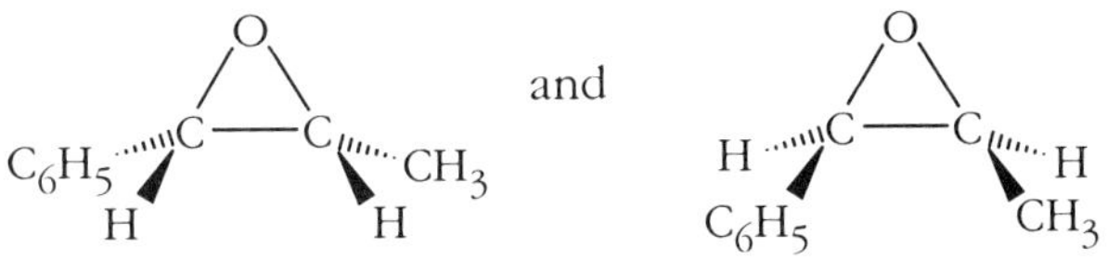 two epoxide stereoisomers: C_6H_5 and H,CH$_3$ trans arrangement and the enantiomer

(b) $C_6H_5CH_2O-C_6H_5$ (benzyl phenyl ether)

(c) $C_6H_5-CH=CH_2$ (styrene)

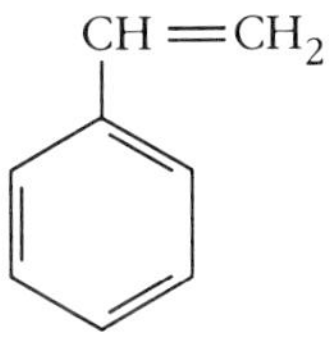

(d) 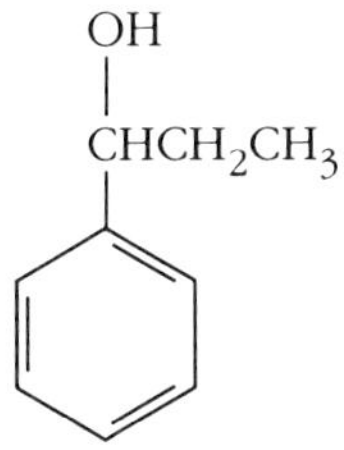 $C_6H_5-CH(OH)CH_2CH_3$ — Racemic mixture

(e) 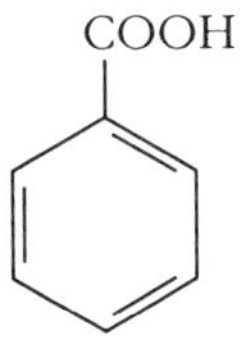 C_6H_5COOH (benzoic acid)

(f) 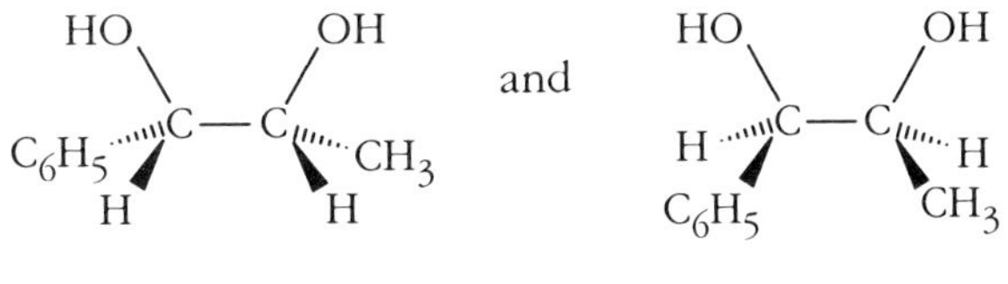 diol stereoisomers with HO, OH on adjacent carbons bearing C_6H_5 and CH_3 — and enantiomer

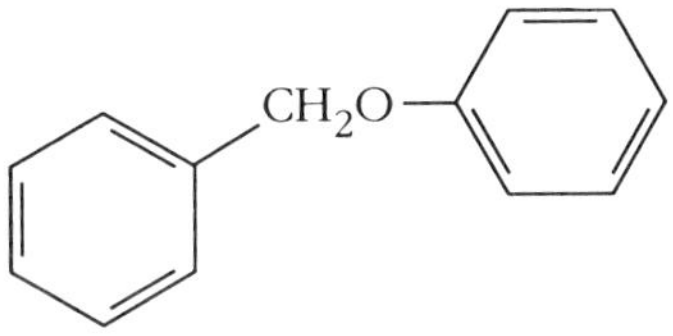

20.27

(a) 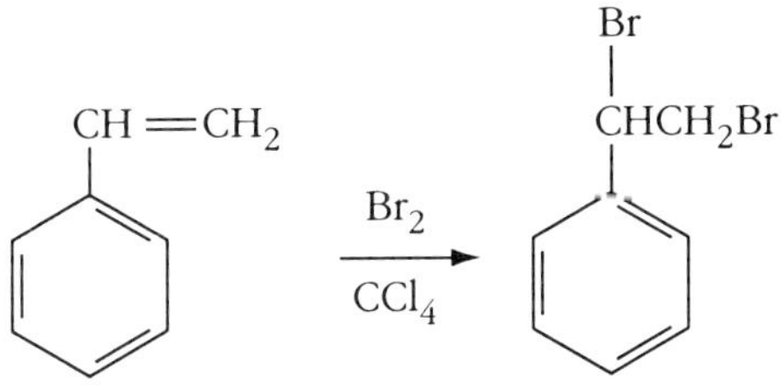 $C_6H_5-CH=CH_2$ $\xrightarrow{Br_2/CCl_4}$ $C_6H_5-CHBrCH_2Br$

(b) Ph-CH=CH₂ $\xrightarrow[\text{Free-radical inhibitor to ensure ionic addition}]{\text{HBr}}$ Ph-CHBrCH₃

(c) Ph-CHBrCH₂Br $\xrightarrow[\text{CH}_3\text{CH}_2\text{OH}]{2\text{ NaOCH}_2\text{CH}_3}$ Ph-C≡CH

From (a)

(d) Ph-C≡CH $\xrightarrow[\text{(2) CH}_3\text{CH}_2\text{Br}]{\text{(1) NaNH}_2/\text{NH}_3\text{(l)}}$ Ph-C≡CCH₂CH₃ $\xrightarrow[\text{Pt}]{\text{H}_2}$

From (c)

Ph-CH₂CH₂CH₂CH₃

(e) Ph-CH=CH₂ $\xrightarrow[\text{ROOR}]{\text{HBr}}$ Ph-CH₂CH₂Br $\xrightarrow{\text{NaC≡CH}}$ Ph-CH₂CH₂C≡CH

(f) Ph-CH₂CH₂C≡CH $\xrightarrow[\text{Lindlar's catalyst}]{\text{H}_2}$ Ph-CH₂CH₂CH=CH₂ $\xrightarrow[\text{H}_2\text{O}]{\text{Br}_2}$ Ph-CH₂CH₂CH(OH)CH₂Br

From (e)

20.28 (a) The molecular formula, C_9H_{12}, indicates four degrees of unsaturation. Such a high degree of unsaturation is usually due to a benzene ring. This is confirmed by the broad multiplet at δ 7.26 (5H) in the ¹H NMR spectrum and the four peaks at δ 125.7, δ 126.3, δ 128.2, and δ 148.7 in the ¹³C NMR spectrum. Furthermore, the relative area (5H) of the aromatic peak in the ¹H NMR spectrum and the presence of four different signals in the ¹³C NMR spectrum indicate that the benzene ring contains one substituent. The doublet at δ 1.25 (6H) and the septet at δ 2.89 (1H) are characteristic of an isopropyl group so the compound is

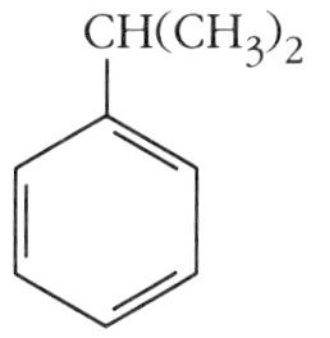

Isopropylbenzene

(b) The molecule has three degrees of unsaturation, two different kinds of hydrogen atoms and two different kinds of carbon atoms. The chemical shifts of the signals in both its ^{1}H NMR and ^{13}C NMR spectra indicate aromatic carbon and hydrogen atoms. We can write the following partial structures:

$$-CH=CH-, -CH=CH-, \text{ and } -O-$$

Joining these together we obtain the structure of furan:

(c) A compound with four degrees of unsaturation and a singlet in its ^{1}H NMR and ^{13}C NMR spectra must have symmetry. The chemical shifts are consistent with an aromatic compound. Furthermore the presence of two nitrogen atoms indicates a heteroaromatic compound. The following structure is consistent with all the data.

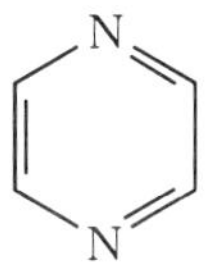

(d) This compound is a constitutional isomer of the one in Exercise 20.28 (c). This compound contains two nonequivalent aromatic hydrogen atoms and two nonequivalent carbon atoms. The following structure is consistent with all the data.

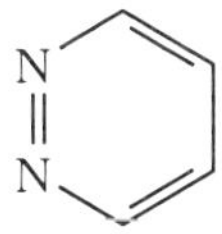

20.29

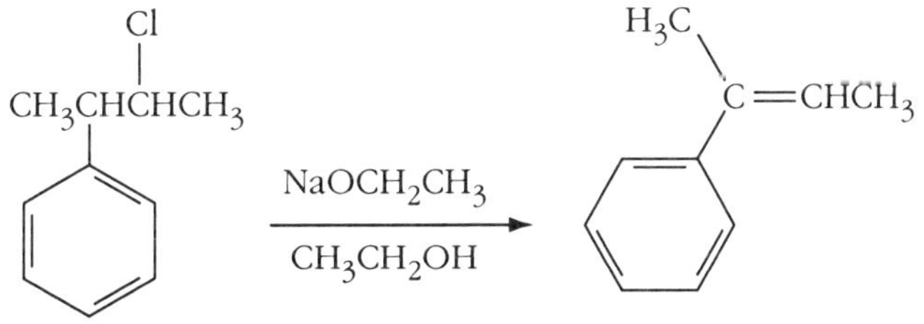

The four stereoisomers and their products formed by an elimination reaction that occurs by an E2 mechanism are as follows:

(Z)-2-Phenyl-2-butene   (E)-2-Phenyl-2-butene

20.30 The product of the reaction of chlorine with 2-pentene is 2,3-dichloropentane, which exists as four stereoisomers. This addition of chlorine is a stereospecific reaction because addition to the (Z)-isomer forms a racemic mixture of two stereoisomers while addition to the (E)-isomer forms a racemic mixture of the other two stereoisomers. This observation is explained by a mechanism that involves the formation of a chloronium ion in Step 1. Reaction of the chloronium ion with chloride ion in Step 2 can only form a racemic mixture of two of the four stereoisomers:

Notice that II is identical to III and I is identical to IV.

Addition of chlorine to (Z)-1-phenylpropene is not a stereospecific reaction because the product mixture contains all four stereoisomers. The mechanism of this reaction, therefore, cannot involve a chloronium ion. The intermediate formed must be capable of forming all four stereoisomers so an α-chloro carbocation is proposed as the intermediate. The following is the accepted mechanism of the chlorination of (Z)-1-phenylpropene:

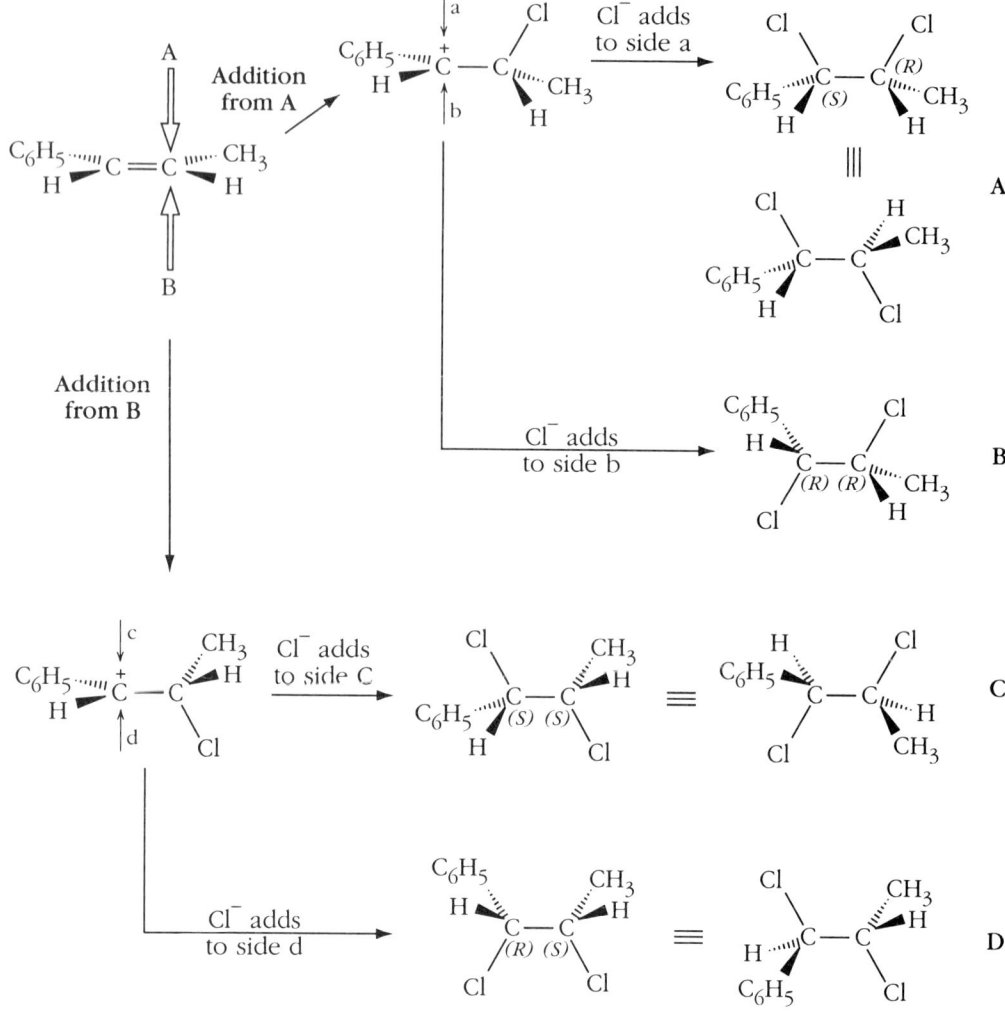

A, B, C, and D are the four stereoisomers. A and D are enantiomers; B and C are also enantiomers.

CHAPTER 21

CHEMISTRY OF BENZENE AND ITS DERIVATIVES
Concepts

The principal reactions of electrophiles and benzene and its derivatives are **electrophilic aromatic substitution** reactions in which a hydrogen atom on the aromatic ring is substituted by another atom or group. In this chapter, we discuss **halogenation, nitration, sulfonation,** and **Friedel-Crafts** reactions. All of these reactions occur by a similar three-step mechanism. Step 1 is formation of the electrophile, which is usually different for each electrophile. The electrophile adds to the aromatic ring in the rate-determining Step 2 to form a resonance-stabilized carbocation intermediate.

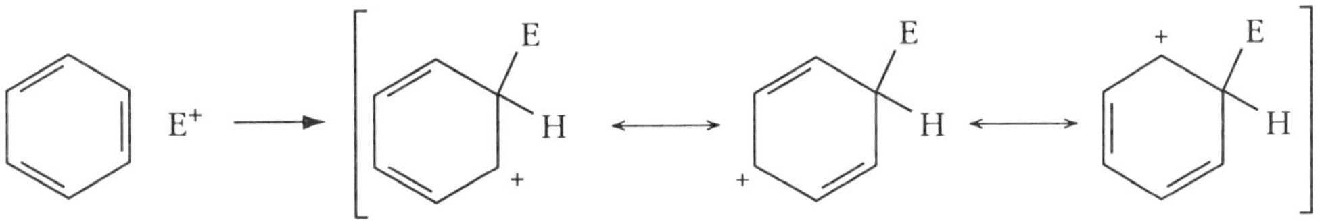

Resonance-stabilized carbocation intermediate

A base removes a proton from the carbocation intermediate in Step 3 to regenerate the aromatic ring.

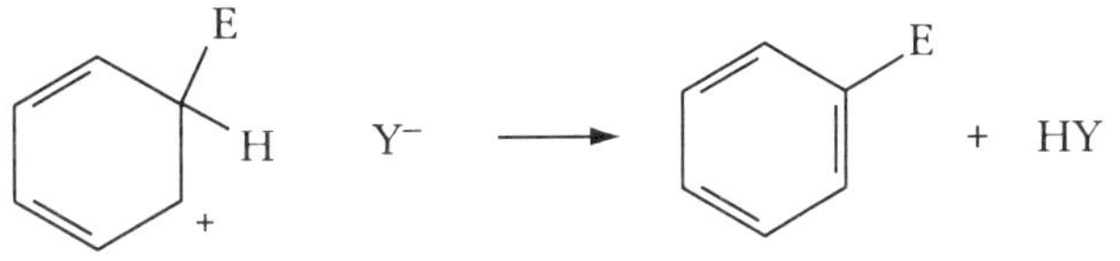

Substituents on the ring affect both the reactivity of the ring to further substitution and the position of substitution. All substituents fall into one of the following three categories:

Activating and *ortho, para*-directing. These substituents are electron-donating groups such as alkoxide groups, amino and substituted amino groups, hydroxyl groups, and alkyl groups.

Deactivating and *ortho, para*-directing. The halogens are the only example of such substituents.

Deactivating and *meta*-directing. These substituents are electron-withdrawing groups such as nitro groups, carbonyl-containing groups, and sulfonic acid.

Solutions to the Exercises

21.1 (a) Fluorobenzene
(b) *tert*-Butylbenzene
(c) 3-Isopropylphenol or 3-(1-methylethyl) phenol
(d) 4-Isobutylbenzoic acid or 4-(2-methylpropyl) benzoic acid
(e) 3,4-Dibromoaniline
(f) 4-Chloro-3-ethylbenzaldehyde
(g) 2,4,6-Trinitrotoluene
(h) 4-Bromo-5-chloro-2-iodoacetophenone

21.2

(a) Iodobenzene

(b) 2-fluorotoluene (CH$_3$ and F on adjacent positions)

(c) 4-propylbenzoic acid (COOH and CH$_2$CH$_2$CH$_3$ para)

(d) 3-nitroaniline (NH$_2$ and NO$_2$ meta)

(e) 1-tert-butyl-3-nitrobenzene (NO$_2$ and C(CH$_3$)$_3$ meta)

(f) 4-fluorophenol (OH and F para)

(g) 2-fluoro-1,3,5-trinitrobenzene (F with O$_2$N, NO$_2$, NO$_2$ substituents)

(h) 1-methoxy-2,4-dinitrobenzene (OCH$_3$ with two NO$_2$ groups)

21.3 (a) The molecular formula, C$_8$H$_{10}$, indicates four degrees of unsaturation. The ^{1}H NMR spectrum shows two singlets with relative areas of 2:3. The molecule actually contains ten hydrogen atoms according to the molecular formula, so the actual ratio of hydrogen atoms is 4:6. The singlet at δ 7.1 (4H) indicates that this unsaturation is due to a disubstituted benzene ring. Another singlet at δ 2.3 (6H) is consistent with a methyl group bonded to a benzene ring. The only way to place two methyl groups on a benzene ring such that all aromatic hydrogen atoms are equivalent in its ^{1}H NMR spectrum is by placing them in the 1 and 4 positions. The compound, therefore, is *p*-xylene.

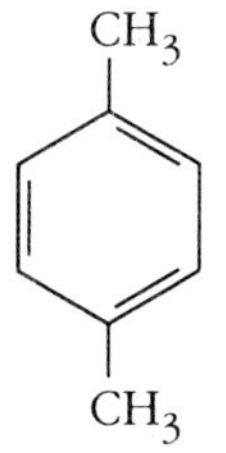

p-Xylene

(b) This compound has five degrees of unsaturation. The two doublets at δ 7.0 (2H) and δ 7.8 (2H) in its ^{1}H NMR spectrum indicate the presence of a benzene ring that is substituted in the 1,4-positions. The singlet at δ 9.8 (1H) is due to a hydrogen atom bonded to a carbonyl carbon atom. The singlet at δ 3.8 (3H) is shifted downfield from the usual range of a methyl group because it is bonded to an electronegative oxygen atom. This gives us the following partial structures;

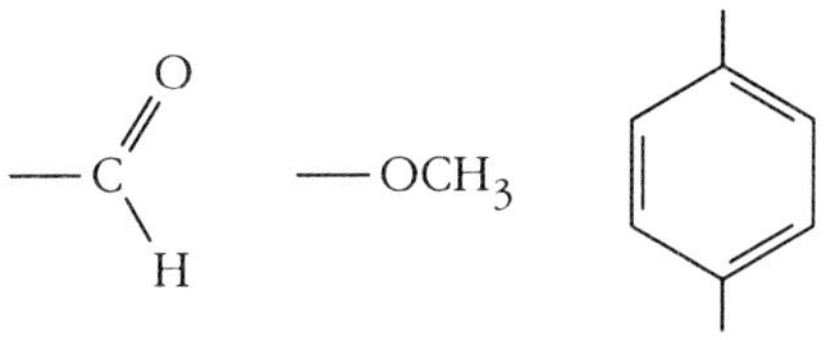

Joining the parts gives the structure of the compound:

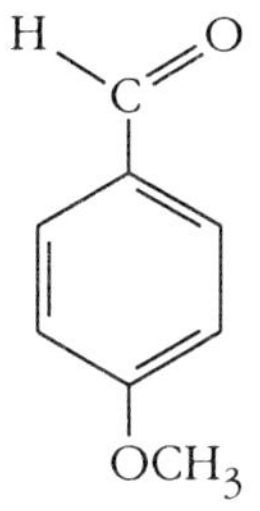

21.4 (a) The ^{13}C NMR spectrum will show four lines; one due to the methyl carbon atoms; one due to ring carbon atoms 1 and 2; one due to carbon atoms 3 and 6; and one due to carbon atoms 4 and 5.

(b) Two lines will appear in the ^{13}C NMR spectrum of this compound. One will be due to carbon atoms 1 and 4 and a second will be due to carbon atoms 2, 3, 5 and 6.

(c) 4
(d) 6
(e) 6

21.5 The three steps in this mechanism are formation of the electrophile, addition of the electrophile to the benzene ring, and loss of a proton to a base to regenerate the aromatic ring.

Step 1:

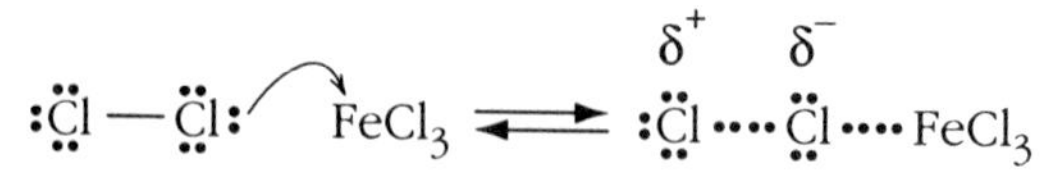

[Cl—Cl + FeCl₃ ⇌ Cl⋯Cl⋯FeCl₃ with δ+ on first Cl and δ− on second Cl]

Step 2:

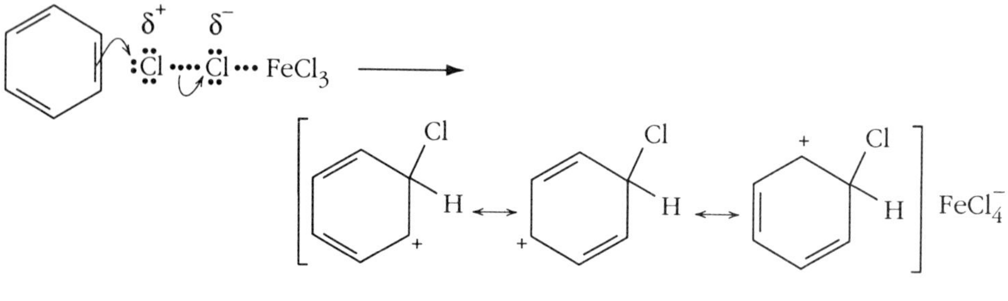

[Benzene attacks Cl(δ+)⋯Cl(δ−)⋯FeCl₃, giving three resonance structures of the arenium ion with Cl and H on sp³ carbon, and FeCl₄⁻]

Step 3:

[Arenium ion deprotonated by [Cl—FeCl₃]⁻ to give chlorobenzene + HCl + FeCl₃]

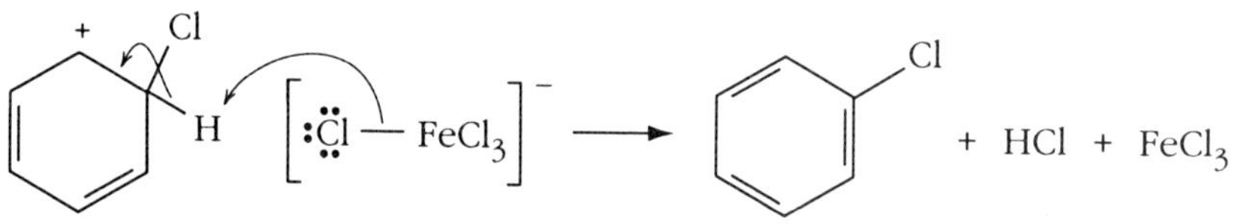

21.6 The equations for the protonation of nitric acid by sulfuric acid and the protonation of sulfuric acid by nitric acid are the following:

Protonation of nitric acid:
$$H_2SO_4 + HNO_3 \rightleftharpoons H_2NO_3^+ + HSO_4^-$$

Protonation of sulfuric acid:
$$H_2SO_4 + HNO_3 \rightleftharpoons H_3SO_4^+ + NO_3^-$$

Sulfuric acid is a stronger acid ($pK_a \approx -9$) than nitric acid ($pK_a \approx -1.4$) so HSO_4^- is a weaker base than NO_3^-. An acid-base reaction occurs to form the weaker base so sulfuric acid acts as an acid to protonate nitric acid to form HSO_4^-, the weaker base.

21.7

(a) [Two H₂SO₄ molecules react: one acts as base (O⁻) attacking the O—H of the other, giving HSO₄⁻ + H₃SO₄⁺]

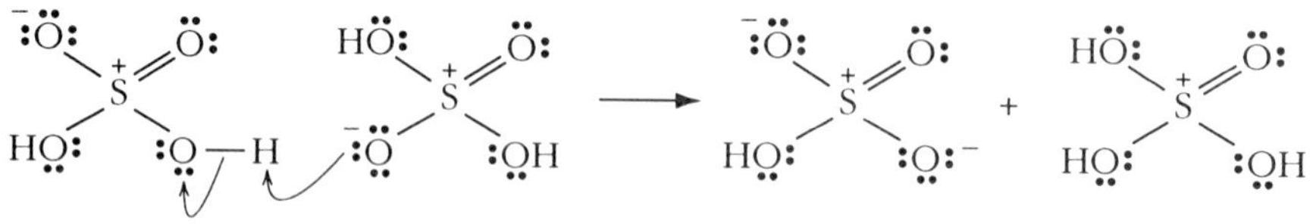

    H₂SO₄   H₂SO₄   HSO₄⁻   H₃SO₄⁺

(b) $2\ H_2SO_4 \rightleftharpoons SO_3 + H_3O^+ + HSO_4^-$

(c) $3\ H_2SO_4 \rightleftharpoons HSO_3^+ + H_3O^+ + 2HSO_4^-$

(d) $3\ H_2SO_4 \rightleftharpoons H_2S_2O_7 + H_3O^+ + HSO_4^-$

21.8

$$\text{CH}_3\text{CHCH}_2\ddot{\text{Cl}}: \;\curvearrowright\; \text{AlCl}_3 \;\rightleftharpoons\; \overset{\text{CH}_3}{\text{CH}_3\text{CHCH}_2}^+ \text{AlCl}_4^-$$

(with CH$_3$ on the middle carbon)

$$\underset{\underset{H}{|}}{\text{CH}_3\overset{\text{CH}_3}{\overset{|}{\text{C}}}}\!\!-\!\text{CH}_2^+\,\text{AlCl}_4^- \;\longrightarrow\; \text{CH}_3\overset{\text{CH}_3}{\underset{+}{\overset{|}{\text{C}}}}\!-\!\text{CH}_3 \quad \text{AlCl}_4^-$$

$(\text{CH}_3)_3\text{C}^+ \;\;\text{AlCl}_4^- \;+\; \text{C}_6\text{H}_6 \;\longrightarrow$

$$\left[\;(\text{CH}_3)_3\text{C}\text{–C}_6\text{H}_6^+ \text{ resonance structures}\;\right]\text{AlCl}_4^-$$

$$\left[\text{Cl}_3\text{–Al}\!-\!\ddot{\text{Cl}}:\right]^{-} \;+\; (\text{CH}_3)_3\text{C–C}_6\text{H}_6^+ \;\longrightarrow\; (\text{CH}_3)_3\text{C–C}_6\text{H}_5 \;+\; \text{AlCl}_3 \;+\; \text{HCl}$$

$\text{HF} + \text{BF}_3 \longrightarrow \text{HBF}_4$

21.9

$$\underset{\underset{H-BF_4}{\curvearrowright}}{\text{CH}_3\!-\!\overset{\text{CH}_3}{\overset{|}{\text{C}}}\!=\!\text{CH}_2} \;\rightleftharpoons\; \text{CH}_3\!-\!\overset{\text{CH}_3}{\underset{+}{\overset{|}{\text{C}}}}\!-\!\text{CH}_3 \quad \text{BF}_4^-$$

$$\text{CH}_3\!-\!\overset{\text{CH}_3}{\underset{\text{CH}_3}{\overset{|}{\text{C}}}}\!\!{}^+ \;+\; \text{C}_6\text{H}_6 \;\xrightarrow{\text{BF}_4^-}$$

$$\left[\;(\text{CH}_3)_3\text{C}\text{–C}_6\text{H}_6^+ \text{ resonance structures}\;\right]\text{BF}_4^-$$

$$\left[\text{F}_3\text{B}\!-\!\text{F}\right]^{-} \;+\; (\text{CH}_3)_3\text{C–C}_6\text{H}_6^+ \;\longrightarrow\; (\text{CH}_3)_3\text{C–C}_6\text{H}_5 \;+\; \text{HF} \;+\; \text{BF}_3$$

21.10 By breaking the bond between the carbonyl group and the benzene ring in the target molecule, we can determine the structure of the acyl chloride needed to react with benzene to form the desired product.

21.12

(a) PhCOCH₃ →(HCl, Zn(Hg))→ PhCH₂CH₃

(b) PhCOCH₃ →(1) LiAlH₄ (2) H₃O⁺→ PhCH(OH)CH₃

(c) PhCOCH₃ →(1) CH₃CH₂MgI (2) H₃O⁺→ PhC(OH)(CH₃)CH₂CH₃

(d) PhCH₂CH₃ →(NBS, CCl₄)→ PhCHBrCH₃ ;
from (a)

PhCH(OH)CH₃ →(HBr)→ PhCHBrCH₃
from (b)

(e) 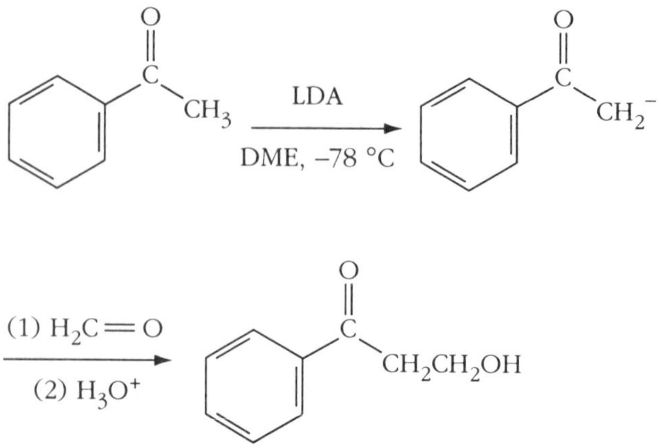 [PhCOCH₃] →[LDA, DME, −78 °C] [PhCOCH₂⁻] →[(1) H₂C=O; (2) H₃O⁺] PhCOCH₂CH₂OH

21.13

(a) Reaction occurs faster than with benzene

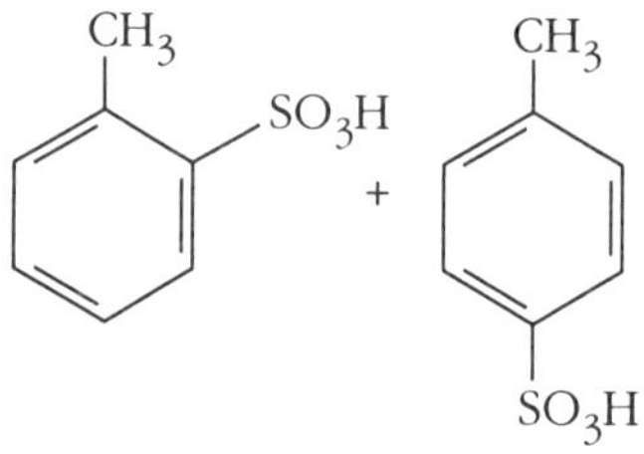

 2-methylbenzenesulfonic acid + 4-methylbenzenesulfonic acid

(b) Slower than benzene

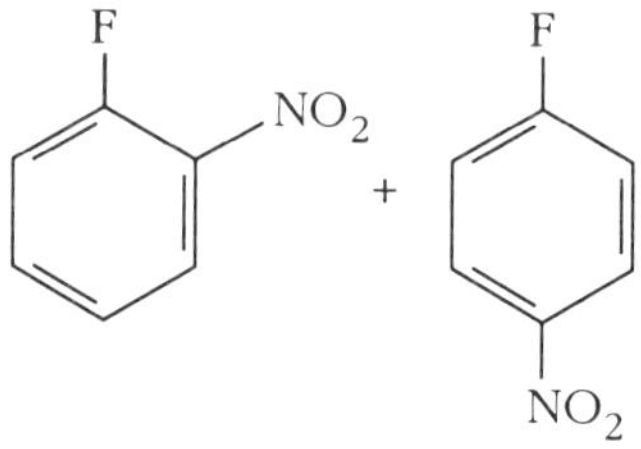

 2-fluoronitrobenzene + 4-fluoronitrobenzene

(c) No reaction

(d) Slower

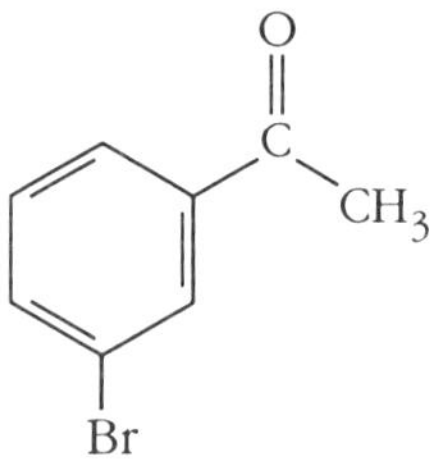

 3-bromoacetophenone

(e) Slower

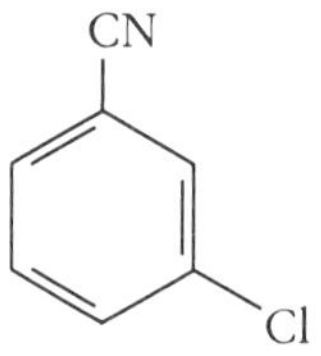

 3-chlorobenzonitrile

(f) Slower

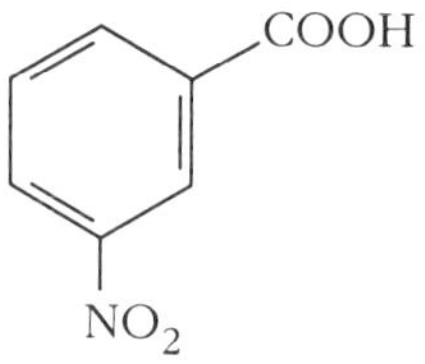

 3-nitrobenzoic acid

(g) Slower

 ethyl 3-chlorobenzoate

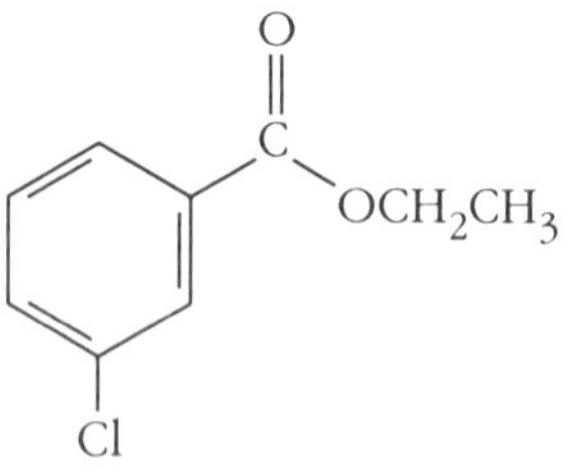

(h) Faster

 4-methoxyacetophenone + 2-methoxyacetophenone

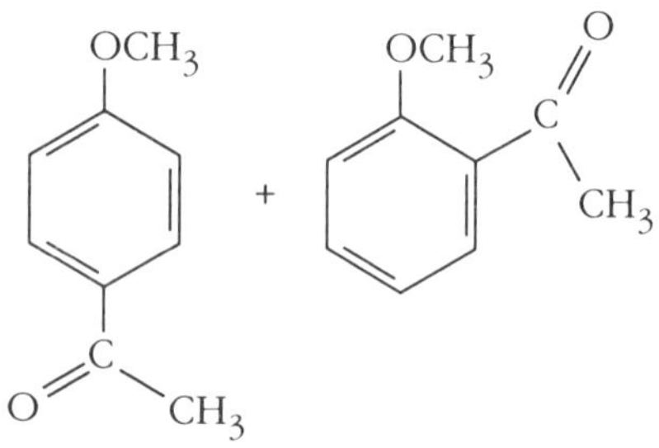

21.14 The electron structure of the carbocation intermediates formed by addition of an electrophile to the *ortho* and *para* positions of biphenyl can be represented as a resonance hybrid of the following resonance structures:

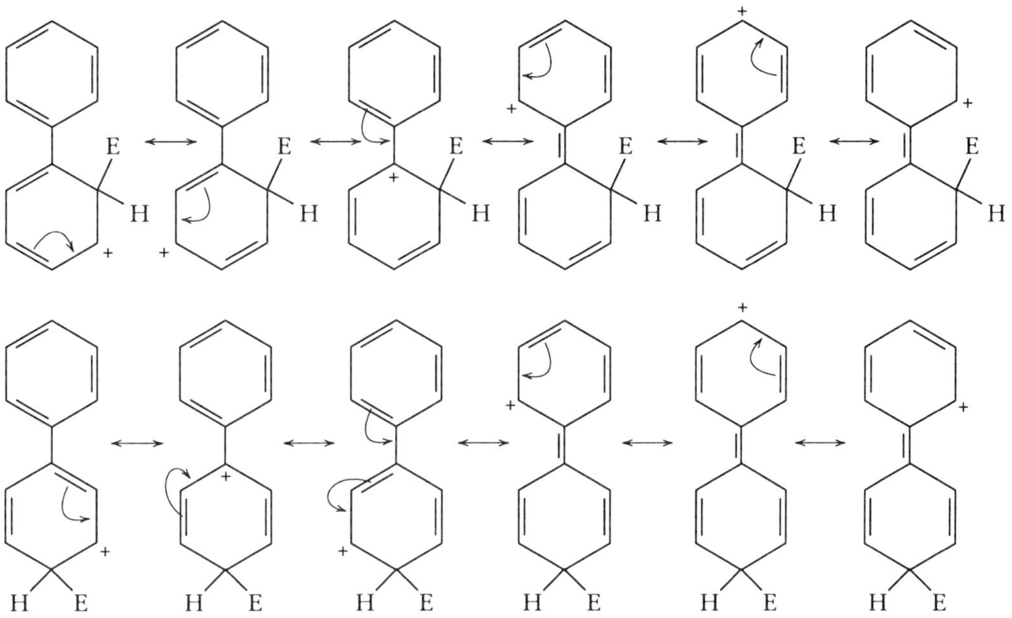

The electron structure of the carbocation intermediates formed by addition of an electrophile to the *meta* position of biphenyl can be represented as a resonance hybrid of the following resonance structures:

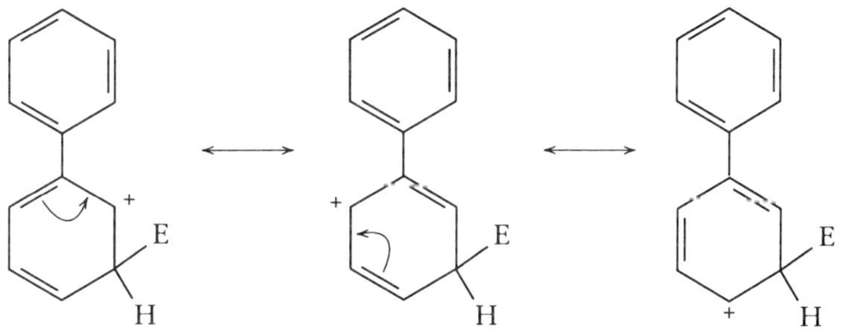

In the carbocation intermediates formed by addition to the *ortho* or *para* positions of biphenyl, the phenyl group is bonded to one of the electron-deficient carbon atoms. In the carbocation intermediate formed by addition to the *meta* position, in contrast, the phenyl group is not bonded to one of the electron-deficient carbon atoms. The phenyl group stabilizes an adjacent electron-deficient carbon atom by π-electron donation. No such electron donation is possible in the intermediate carbocation formed by addition to the *meta* position. As a result the intermediate carbocations for addition *ortho* and *para* and the transition states leading to their formation are more stable than the corresponding intermediate and transition state leading to *meta* substitution. Therefore the *ortho* and *para* positions of biphenyl are more reactive to electrophilic substitution.

21.15

benzene $\xrightarrow{\text{HNO}_3/\text{H}_2\text{SO}_4}$ nitrobenzene $\xrightarrow{\text{Br}_2/\text{FeBr}_3}$ m-bromonitrobenzene

21.16 (a) The two nitro groups are *meta* to each other. Reacting benzene with an excess of nitric acid in sulfuric acid will form the desired product because the first nitro group introduced into the ring is a *meta* director.

benzene $\xrightarrow{\text{HNO}_3/\text{H}_2\text{SO}_4}$ 1,3-dinitrobenzene

(b) The two substituents are located *meta*. The nitro group can be introduced by nitration and the ethyl group must be introduced by a Friedel-Crafts reaction. Even though the nitro group is a *meta* director, we cannot introduce it first because nitrobenzene does not undergo Friedel-Crafts reactions. The ethyl group, therefore, must be added first. An ethyl group is an *ortho, para* director, however, so we cannot introduce the ethyl group directly by a Friedel-Crafts alkylation reaction. The best method is to carry out a Friedel-Crafts acylation reaction to form a ketone (in this case acetophenone). The acyl group is a *meta* director so nitration forms *m*-nitroacetophenone, which forms the desired product upon reduction.

benzene $\xrightarrow[\text{AlCl}_3]{\text{CH}_3\text{COCl}}$ acetophenone $\xrightarrow[\text{H}_2\text{SO}_4]{\text{HNO}_3}$ m-nitroacetophenone

$\xrightarrow[\text{(2) Raney Ni/C}_2\text{H}_5\text{OH}]{\text{(1) HSCH}_2\text{CH}_2\text{SH/BF}_3}$ m-nitroethylbenzene

Clemmensen reduction is *not* a viable alternative to reduce the nitroacetophenone because the nitro group and the carbonyl group will be reduced.

(c) The ethyl and nitro groups are located *para*. The ethyl group is an *ortho, para* director so introducing it first by Friedel-Crafts alkylation (using an excess of benzene to prevent polyalkylation) followed by nitration forms a mixture of *o*- and *p*-nitroethylbenzene, which must be separated to obtain pure *p*-nitroethylbenzene.

(d) [benzene] $\xrightarrow[\text{H}_2\text{SO}_4]{\text{SO}_3}$ [benzenesulfonic acid] $\xrightarrow[\text{FeCl}_3]{\text{Cl}_2}$ [3-chlorobenzenesulfonic acid]

(e) [benzene] $\xrightarrow[\text{FeBr}_3]{\text{Br}_2}$ [bromobenzene] $\xrightarrow{\substack{(1)\ \text{Mg} \\ (2)\ \text{CO}_2 \\ (3)\ \text{H}_3\text{O}^+}}$ [benzoic acid] $\xrightarrow[\text{H}_2\text{SO}_4]{\text{HNO}_3}$ [3-nitrobenzoic acid]

(f) [benzene] $\xrightarrow[\text{AlCl}_3]{\text{CH}_3\text{COCl}}$ [acetophenone] $\xrightarrow[\text{FeBr}_3]{\text{Br}_2}$ [3-bromoacetophenone] $\xrightarrow{\substack{(1)\ \text{LiAlH}_4 \\ (2)\ \text{H}_3\text{O}^+}}$ [1-(3-bromophenyl)ethanol]

21.17 (a) Both the nitro and the carboxy group are *meta*-directing so the nitronium ion will add *meta* to each to form 3,5-dinitrobenzoic acid.

[Structure: 3,5-dinitrobenzoic acid with COOH at top, O_2N and NO_2 at positions 3 and 5]

(b) A nitro group is *meta*-directing while chlorine is *o, p*-directing, so electrophilic substitution can occur at two positions, ring carbon atoms 4 and 6:

[Structure: benzene ring with NO_2 and Cl on adjacent carbons, with arrows pointing to two positions]

The major product is 2,4-dinitrochlorobenzene because steric hindrance makes approach of the nitronium ion to carbon atom 6 energetically unfavorable.

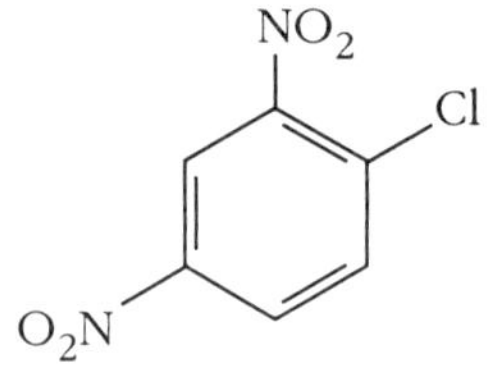

(c), 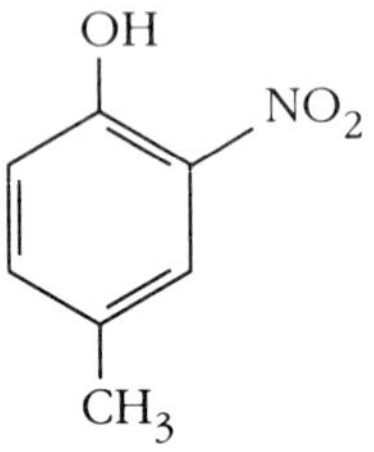 (d), 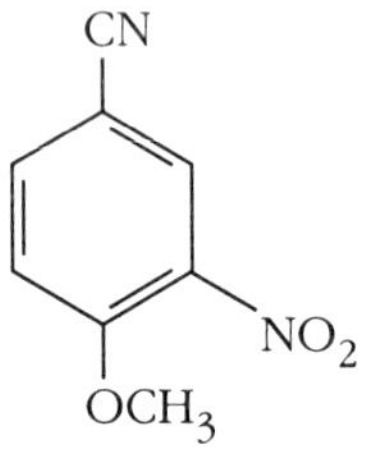 (e) 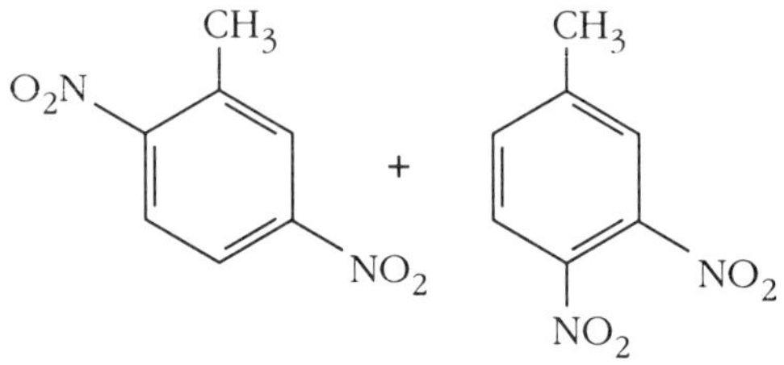 [structures shown]

The OH group is a more powerful *ortho-*, *para*-director than the methyl group.

21.18 (a) An ***ortho, para*-directing substituent** is an atom or group bonded to a benzene ring that directs an incoming electrophile to a position *ortho* or *para* to itself. For example:

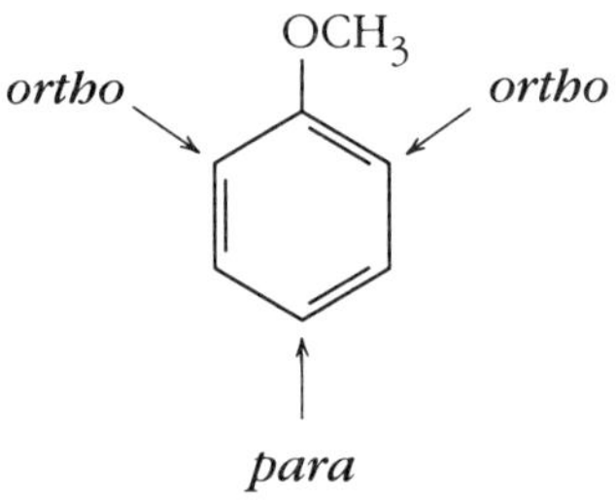

OCH$_3$ is an *ortho*, *para*-directing substituent because it directs incoming electrophiles to a position *ortho* or *para* to itself

(b) A **Friedel-Crafts alkylation reaction** is a reaction between a haloalkane (but not haloaromatic or vinyl halides) and benzene, or a substituted benzene derivative in which the substituent is electron donating, to form alkyl benzenes. Lewis acids such as FeBr$_3$ or AlCl$_3$ are needed as catalysts.

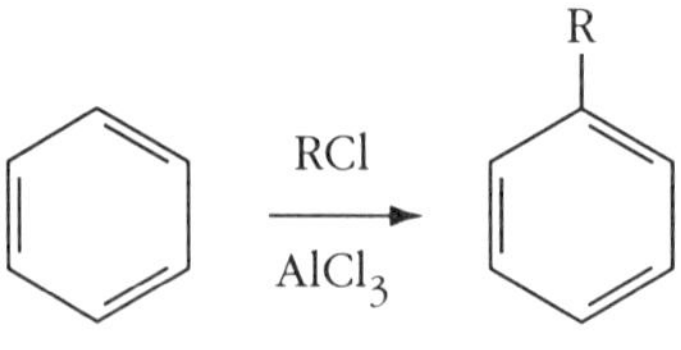

Friedel-Crafts alkylation reactions have two limitations. Polyalkylated benzenes are often formed as products and the carbon skeleton of the electrophile undergoes rearrangement if possible.

(c) An **activating group** is a substituent on a benzene ring that increases the rate of electrophilic aromatic substitution reactions of the compound compared to benzene. Anisole, for example, reacts faster than benzene in electrophilic aromatic substitution reactions so the methoxy group is an activating group.

(d) **Clemmenson reduction** reduces a carbonyl group of a ketone to a methylene group. The reducing agent is concentrated hydrochloric acid and a zinc amalgam, Zn(Hg). Acetophenone, for example, undergoes Clemmenson reduction to form ethylbenzene.

$$\text{Acetophenone} \xrightarrow[\text{Zn(Hg)}]{\text{HCl}} \text{Ethylbenzene}$$

(e) A **deactivating group** is a substituent on a benzene ring that decreases the rate of electrophilic aromatic substitution reactions of the compound compared to benzene. Nitrobenzene, for example, reacts slower than benzene in electrophilic aromatic substitution reactions so the nitro group is a deactivating group.

(f) A **Friedel-Crafts acylation reaction** is a reaction between an acyl chloride and benzene, or a substituted benzene derivative in which the substituent is electron donating, to form aryl ketones. Lewis acids such as $FeBr_3$ or $AlCl_3$ are needed as catalysts.

R = alkyl or aryl

In Friedel-Crafts acylation, unlike Friedel-Crafts alkylation, neither rearrangement of the electrophile nor polysubstitution occurs.

(g) A ***meta*-directing substituent** is an atom or group bonded to a benzene ring that directs an incoming electrophile to a position *meta* to itself. For example:

NO_2 is a *meta*-directing substituent because it directs incoming electrophiles to a position *meta* to itself

(h) **Wolff-Kishner reduction** reduces the carbonyl group of an aldehyde or ketone to a methylene group. The reducing agent is hydrazine, and the reaction is carried out in basic solution using potassium hydroxide. High temperatures are needed so the reaction is carried out in a high boiling solvent such as diethylene glycol ($HOCH_2CH_2OCH_2CH_2OH$). A recent modification of the procedure is to use dimethyl sulfoxide [DMSO, $(CH_3)_2SO$] as the solvent and potassium *tert*-butoxide as the base. Under these conditions the reaction occurs at a much lower temperature.

(i) A **thioacetal** is an acetal in which the oxygen atoms are replaced by sulfur.

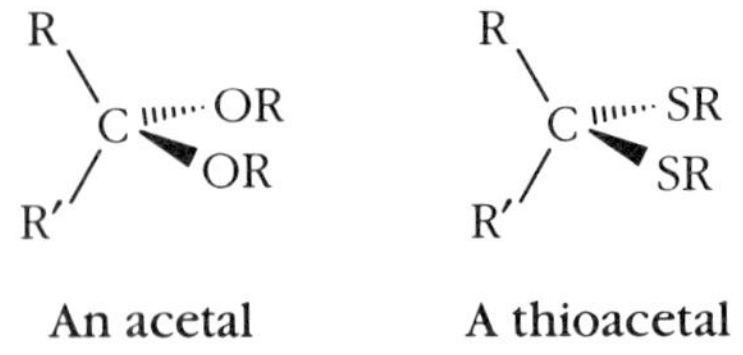

$$\underset{R'}{\overset{R}{>}}C\underset{OR}{\overset{OR}{<}} \qquad \underset{R'}{\overset{R}{>}}C\underset{SR}{\overset{SR}{<}}$$

An acetal   A thioacetal

Thioacetals are formed by the reaction of thiols, the sulfur analogs of alcohols, with aldehydes and ketones.

21.19

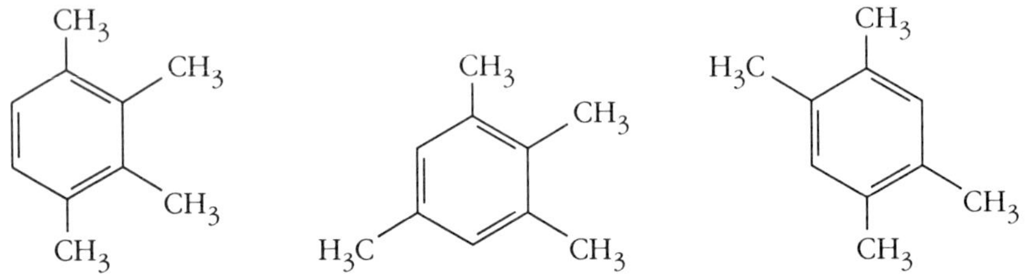

1,2,3,4-Tetramethyl-benzene   1,2,3,5-Tetramethyl-benzene   1,2,4,5-Tetramethyl-benzene

21.20

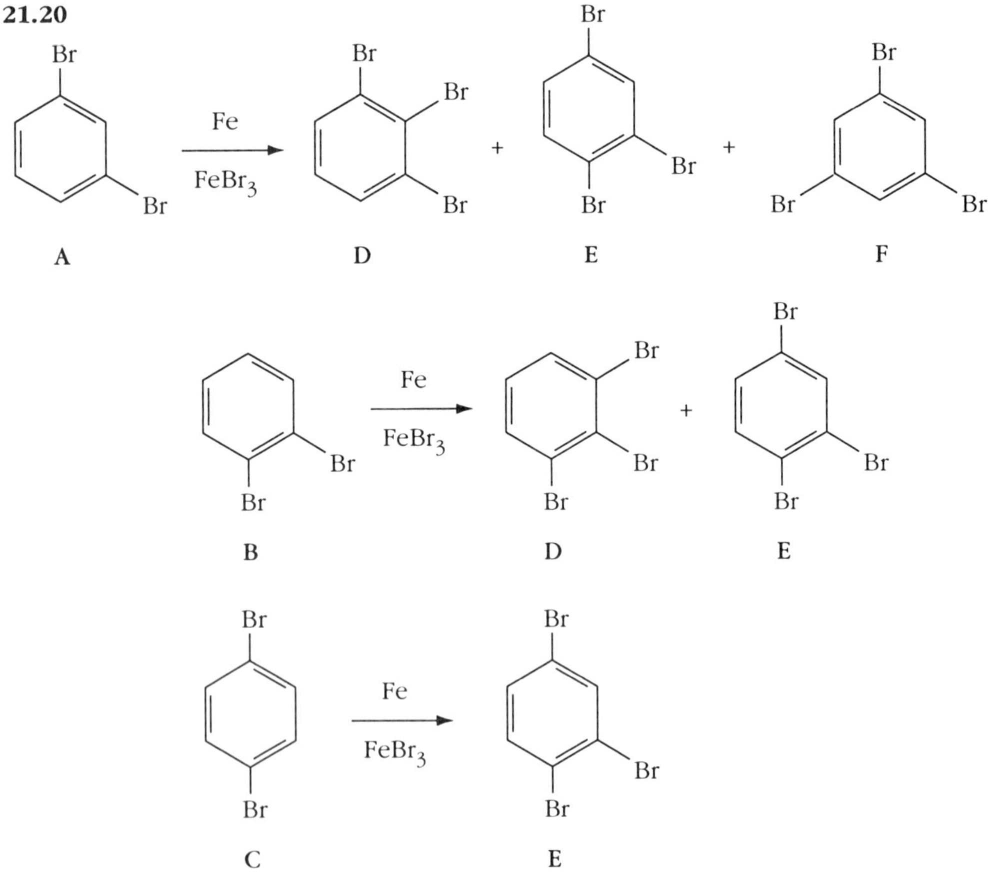

21.21

(a) [2-bromotoluene] + [4-bromotoluene] ; [2-nitrotoluene] + [4-nitrotoluene] ; [2-methylbenzenesulfonic acid] + [4-methylbenzenesulfonic acid]

Toluene reacts faster than benzene.

(b) [3-bromonitrobenzene] ; [1,3-dinitrobenzene] ; [3-nitrobenzenesulfonic acid]

Nitrobenzene reacts slower than benzene.

(c) [2-bromoanisole] + [4-bromoanisole] ; [2-nitroanisole] + [4-nitroanisole] ; [2-methoxybenzenesulfonic acid] + [4-methoxybenzenesulfonic acid]

Anisole reacts faster than benzene.

(d) [2-bromochlorobenzene] + [4-bromochlorobenzene] ; [2-nitrochlorobenzene] + [4-nitrochlorobenzene] ; [2-chlorobenzenesulfonic acid] + [4-chlorobenzenesulfonic acid]

Chlorobenzene reacts slower than benzene.

(e) [3-bromo(trifluoromethyl)benzene] ; [3-nitro(trifluoromethyl)benzene] ; [3-(trifluoromethyl)benzenesulfonic acid]

Trifluoromethylbenzene reacts slower than benzene.

A trifluoromethyl group is an electron-withdrawing and a deactivating group because of the strong inductive electronegative effect of the three fluorine atoms bonded to the methyl carbon atom.

21.22 The nitroso group can donate a pair of electrons to an adjacent electron-deficient carbon atom so it stabilizes the carbocation intermediates formed by addition of an electrophile to the *ortho* and *para* positions. Such electron donation by the nitroso group is not possible in the carbocation intermediate formed by addition of the electrophile to the *meta* position. This is shown by the following electron structures of the carbocation intermediates formed by addition of an electrophile to the *ortho*, *para*, and *meta* positions of nitrosobenzene:

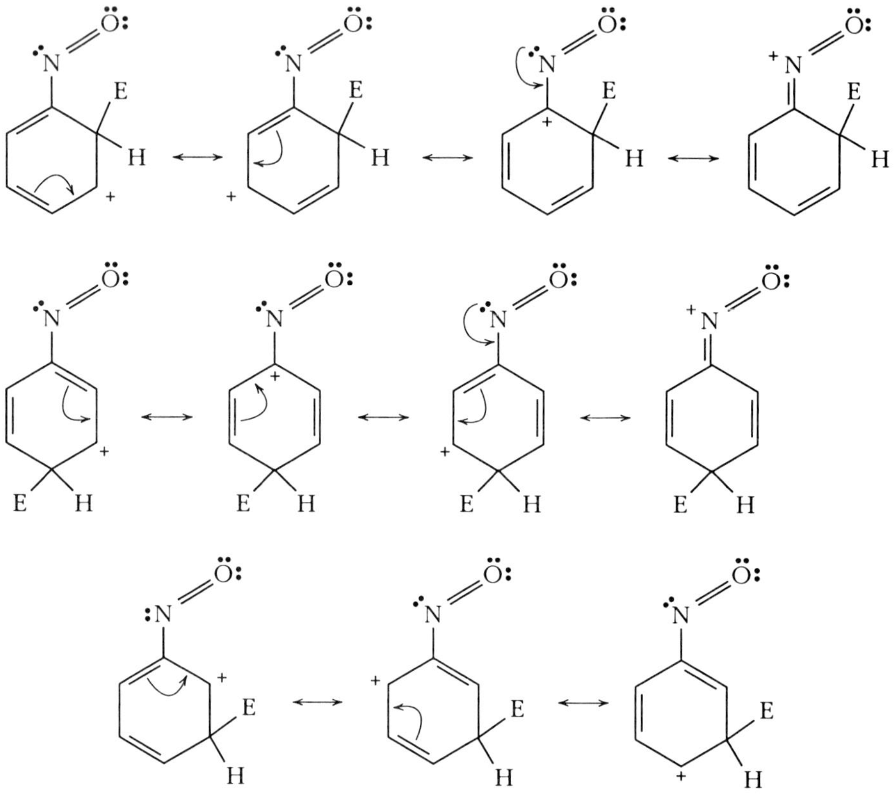

Therefore, the *ortho* and *para* positions of nitrosobenzene are more reactive to electrophilic substitution than the *meta* position.

21.23

(a) 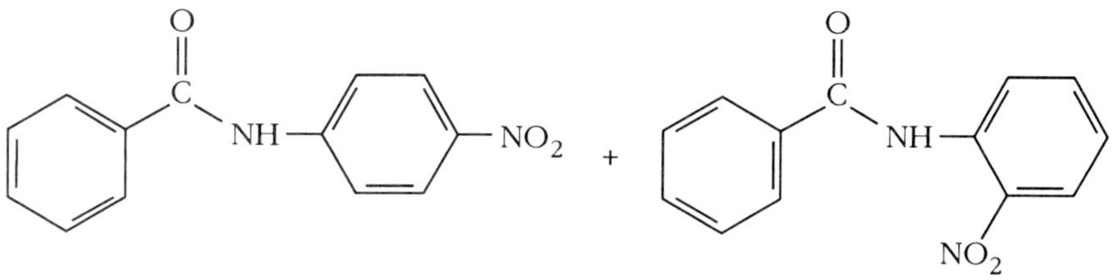 Ph-C(=O)-NH-C6H4-NO2 (para) + Ph-C(=O)-NH-C6H4-NO2 (ortho)

(b) 3-(trifluoromethyl)benzenesulfonic acid (CF3 and SO3H meta on benzene)

(c) tert-butylbenzene (PhC(CH3)3)

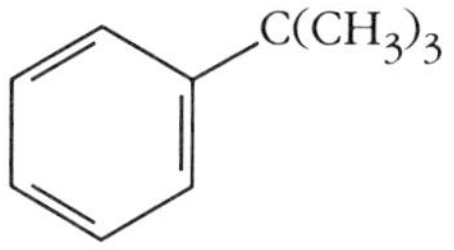

(d) 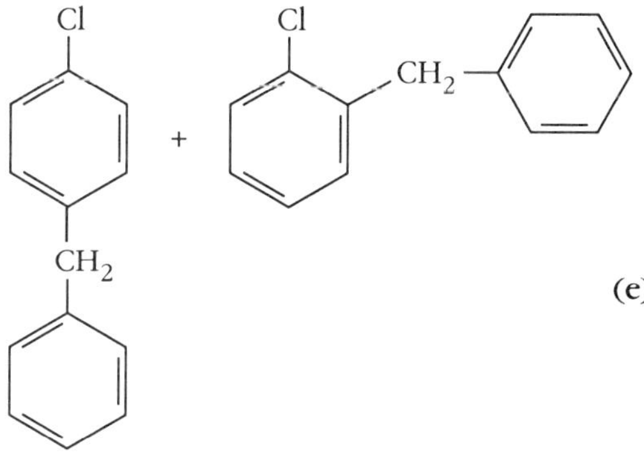 4-chloro(benzyl)benzene (p-Cl-C6H4-CH2-Ph) + 2-chloro(benzyl)benzene (o-Cl-C6H4-CH2-Ph)

(e) 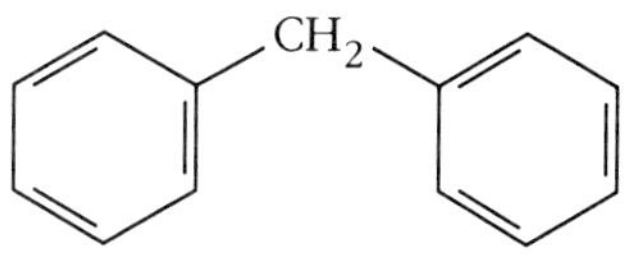 diphenylmethane (Ph-CH2-Ph)

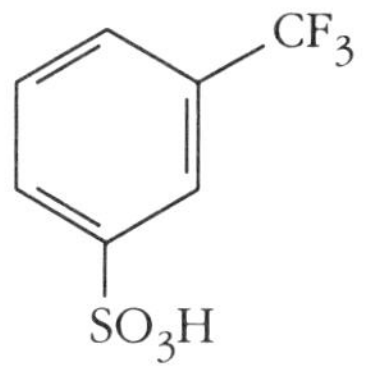

(f) 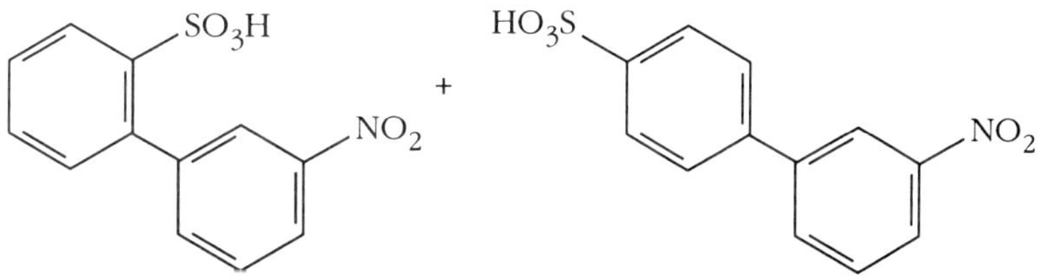 2'-sulfo-3-nitrobiphenyl (SO3H ortho on one ring, NO2 meta on other) + 4'-sulfo-3-nitrobiphenyl (SO3H para on one ring, NO2 meta on other)

(g) 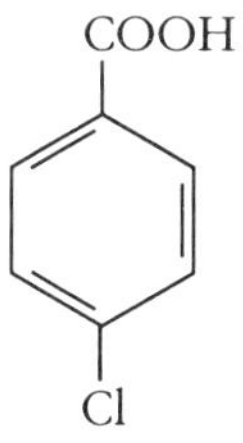 4-chlorobenzoic acid (COOH and Cl para)

(h) 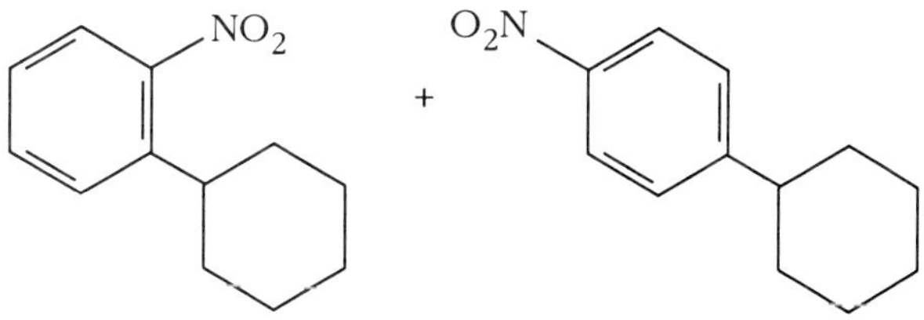 2-nitro(cyclohexyl)benzene (o-NO2-C6H4-cyclohexyl) + 4-nitro(cyclohexyl)benzene (p-NO2-C6H4-cyclohexyl)

(i) 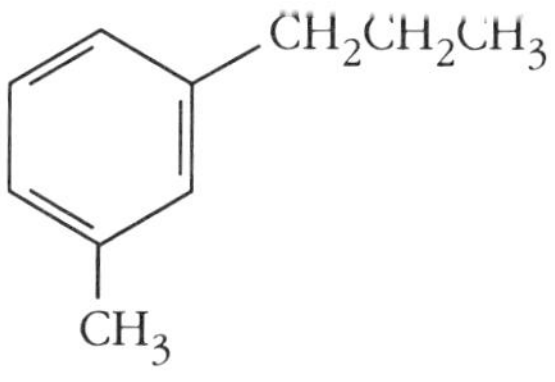 1-methyl-3-propylbenzene (CH3 and CH2CH2CH3 meta)

(j) 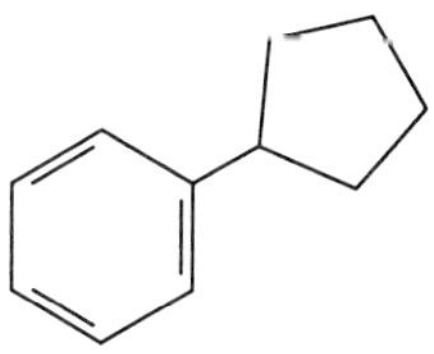 cyclopentylbenzene

21.24

(a) PhCH$_2$CH$_3$ ⟹ PhCOCH$_3$ ⟹ PhH + CH$_3$COCl

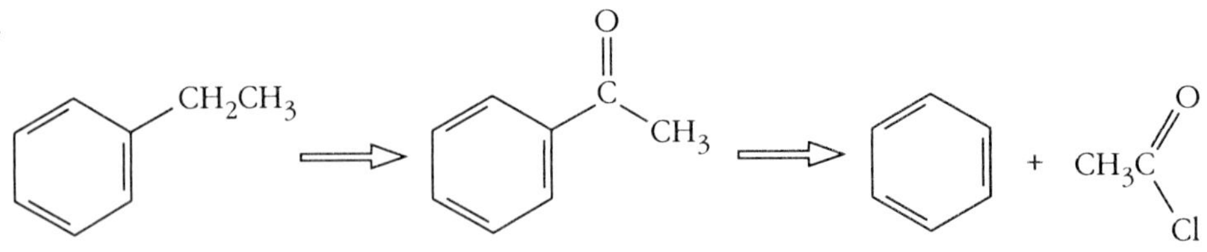

PhH $\xrightarrow[\text{AlCl}_3]{\text{CH}_3\text{COCl}}$ PhCOCH$_3$ $\xrightarrow[\text{Zn(Hg)}]{\text{HCl}}$ PhCH$_2$CH$_3$

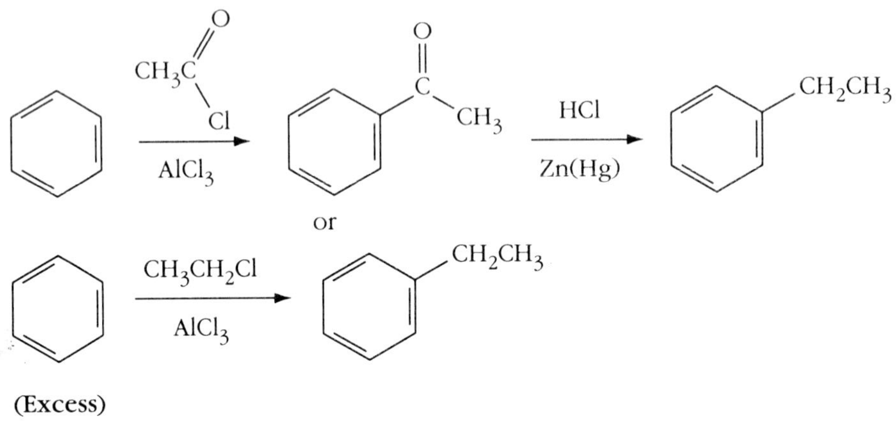

or

PhH (Excess) $\xrightarrow[\text{AlCl}_3]{\text{CH}_3\text{CH}_2\text{Cl}}$ PhCH$_2$CH$_3$

(b) 4-Br-C$_6$H$_4$-CH$_3$ ⟹ PhCH$_3$ ⟹ PhH + CH$_3$Cl

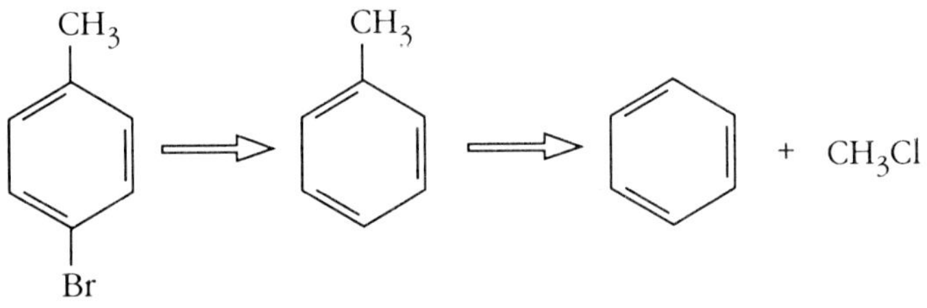

PhH (Excess) $\xrightarrow[\text{AlCl}_3]{\text{CH}_3\text{Cl}}$ PhCH$_3$ $\xrightarrow[\text{FeBr}_3]{\text{Br}_2}$ 4-Br-C$_6$H$_4$-CH$_3$

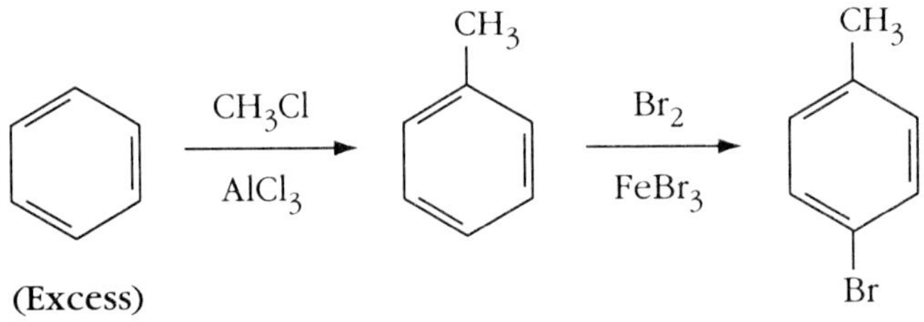

(c) PhCH=CH₂ ⇒ PhCH(OH)CH₃ ⇒ PhCOCH₃ ⇒ PhH + CH₃COCl

PhH + CH₃COCl →(AlCl₃) PhCOCH₃ →((1) LiAlH₄, (2) H₃O⁺) PhCH(OH)CH₃ →(H₂SO₄) PhCH=CH₂

(d) PhCN ⇒ PhCONH₂ ⇒ PhCOOH ⇒ PhBr ⇒ PhH

PhH →(Br₂, FeBr₃) PhBr →((1) Mg, (2) CO₂, (3) H₃O⁺) PhCOOH →((1) SOCl₂, (2) NH₃) PhCONH₂ →(SOCl₂, Heat) PhCN

(e) (CH₃)₂C(CH₃)(CH₂Cl)(C₆H₅) ⇒ (CH₃)₃C—C₆H₅ ⇒ C₆H₆ + (CH₃)₃CCl

$$C_6H_6 \xrightarrow[\text{AlCl}_3]{(CH_3)_3CCl} (CH_3)_3C\text{-}C_6H_5 \xrightarrow[h\nu]{Cl} (CH_3)_2C(CH_2Cl)\text{-}C_6H_5$$

(f) (C₆H₅)₂CHCH₃ ⇒ (C₆H₅)₂C=CH₂ ⇒ (C₆H₅)₂C=O ⇒ C₆H₆ + C₆H₅COCl

$$C_6H_6 + C_6H_5COCl \xrightarrow{\text{AlCl}_3} (C_6H_5)_2C=O \xrightarrow{(C_6H_5)_3\overset{+}{P}-\bar{C}H_2} (C_6H_5)_2C=CH_2 \xrightarrow[\text{Pt}]{H_2} (C_6H_5)_2CHCH_3$$

(g) Retrosynthetic analysis:

4-chloroacetophenone ⇒ 1-(4-chlorophenyl)ethanol ⇒ 1-bromo-1-(4-chlorophenyl)ethane ⇒ 4-chloroethylbenzene ⇒ ethylbenzene

Synthesis:

$$\text{benzene (excess)} \xrightarrow[\text{AlCl}_3]{\text{C}_2\text{H}_5\text{Cl}} \text{ethylbenzene} \xrightarrow[\text{FeCl}_3]{\text{Cl}_2} \text{4-chloroethylbenzene} \xrightarrow[\text{h}\nu]{\text{Br}_2} \text{1-bromo-1-(4-chlorophenyl)ethane} \xrightarrow{\text{H}_2\text{O}} \text{1-(4-chlorophenyl)ethanol} \xrightarrow[\text{HCCl}_3]{\text{PCC}} \text{4-chloroacetophenone}$$

(h) $$\text{benzene} \xrightarrow[\text{FeCl}_3]{\text{Cl}_2} \text{chlorobenzene} \xrightarrow[\text{H}_2\text{SO}_4]{2\ \text{HNO}_3} \text{1-chloro-2,4-dinitrobenzene}$$

21.25 The data in Table 21.1 (p. 942) of the text can be used to determine the relative reactivities of the compounds in each set.

—— Increasing reactivity ——▶

(a) Nitrobenzene, chlorobenzene, benzene
(b) *t*-Butylbenzene, toluene, phenol
(c) Benzenesulfonic acid, styrene, anisole
(d) Benzonitrile, benzoic acid, acetophenone
(e) Benzamide, ethyl benzoate, phenyl acetate

21.26

Step 1:

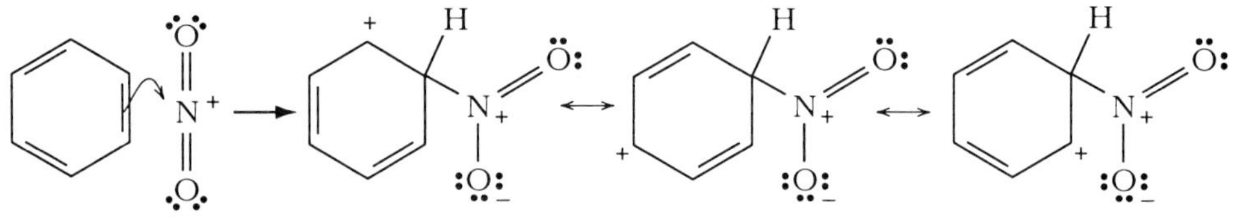

Step 2:

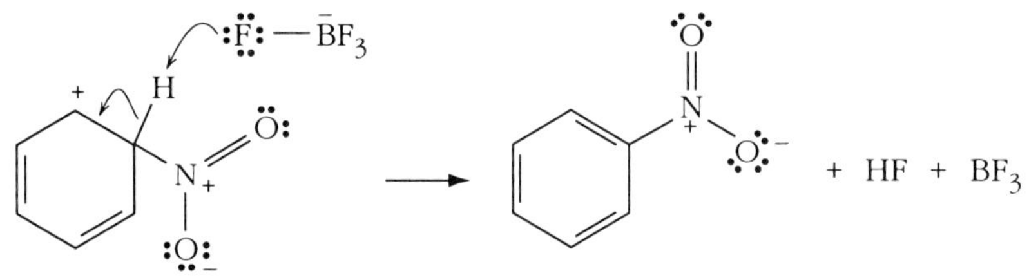

21.27

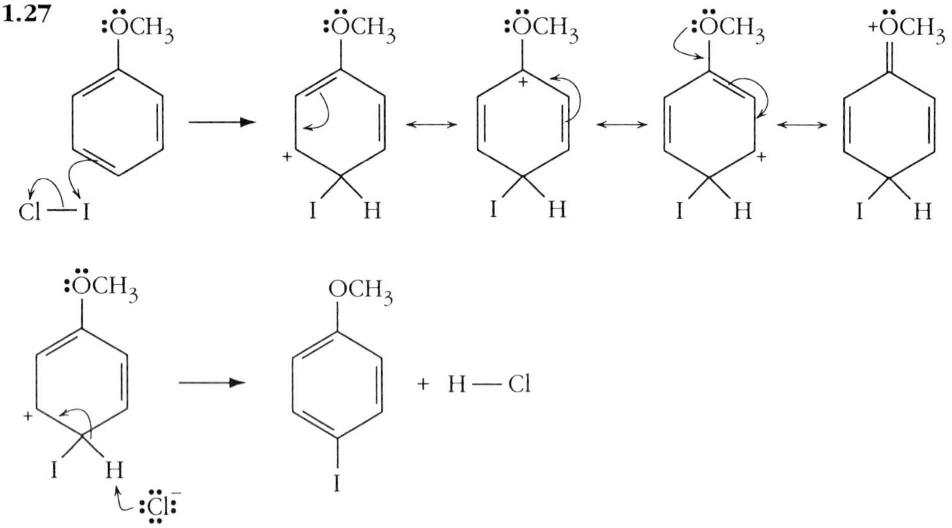

21.28 Step 1, formation of the carbocation, is the rate-determining step in the S_N1 mechanism. The free energies of activation of Step 1 determine the relative rates of nucleophilic substitution reactions of the two compounds. If we assume that the free energies of the two compounds are about the same, then their free energies of activation will be determined by the relative energies of the carbocation intermediates and the transition states leading to their formation. The compound that has the more stable carbocation intermediate, therefore, should react more quickly in a nucleophilic substitution reaction that occurs by an S_N1 mechanism. We can determine the relative stabilities of the carbocation intermediates by describing their electron structure in terms of the following resonance hybrid structures.

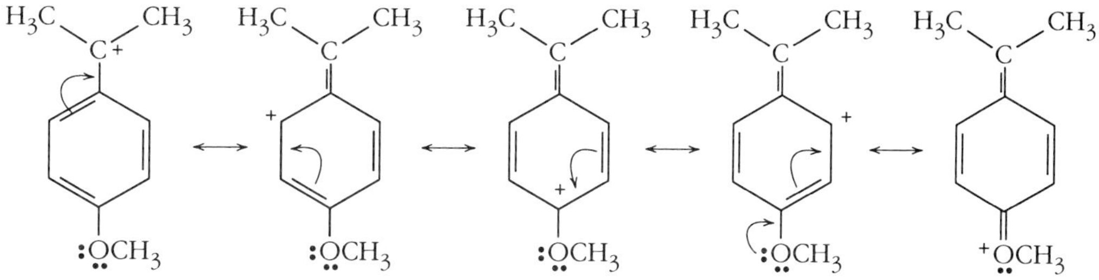

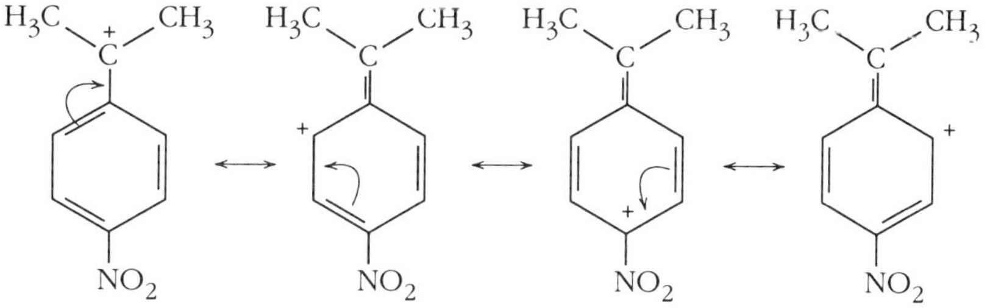

The electron-donating methoxy group is correctly positioned to stabilize the carbocation intermediate by electron delocalization while the electron-withdrawing nitro group destabilizes the carbocation intermediate. As a result, 2-chloro-2-(4-methoxyphenyl) propane will undergo nucleophilic substitution reactions much faster than 2-chloro-2-(4-nitrophenyl) propane.

21.29 The ^{1}H NMR spectrum of isomer A has two sets of doublets in the aromatic region, which is the pattern characteristic of a 1,4-disubstituted benzene derivative. Isomer A, therefore, is *p*-nitrotoluene and isomer B is *o*-nitrotoluene.

21.30 (a)

A is: phenyl–C(=O)–CH(CH₃)₂

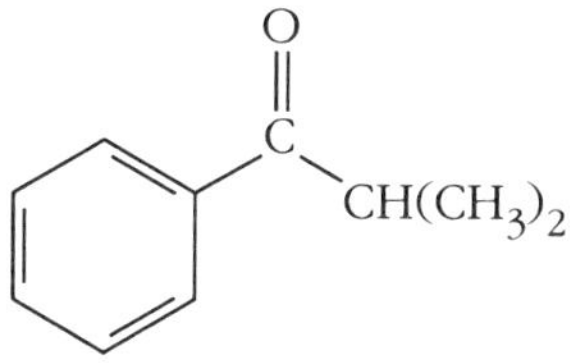

B is: 3-nitrophenyl–C(=O)–CH(CH₃)₂

C is: phenyl–CH=C(CH₃)₂

D is: phenyl–CH₂CH(CH₃)₂

E is: phenyl–CH(Br)CH(CH₃)₂

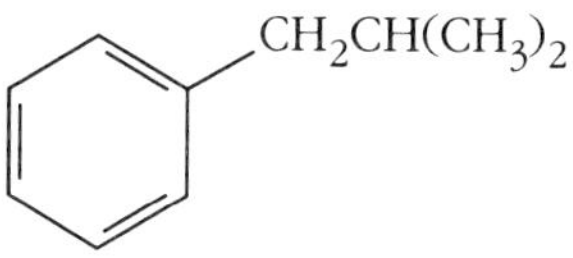

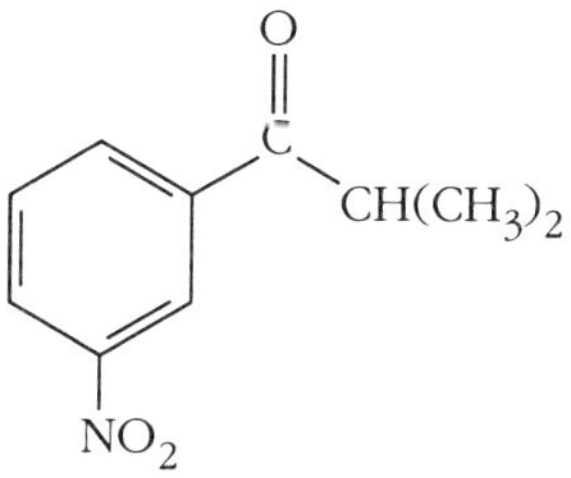

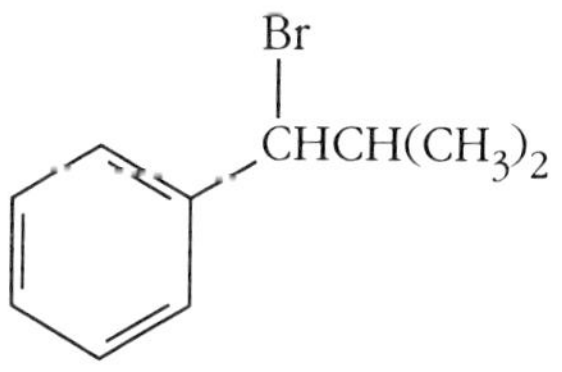

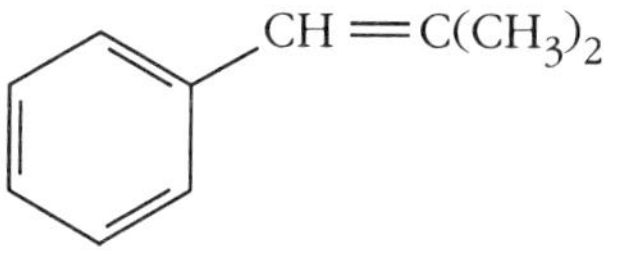

(b) F is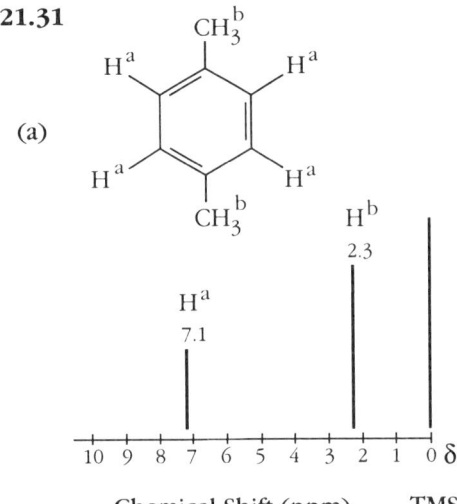

G is

H is

I is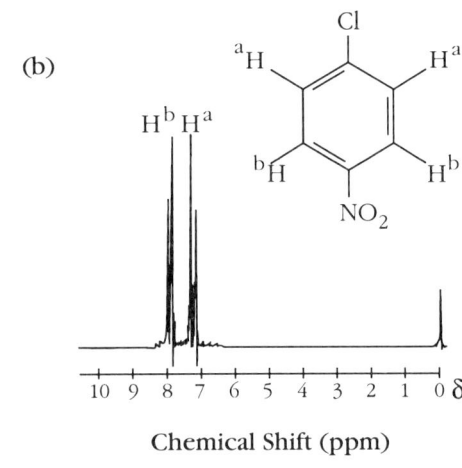

J is

21.31

(a)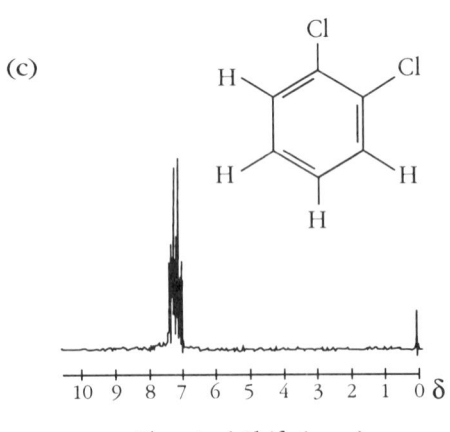

(b)

(c)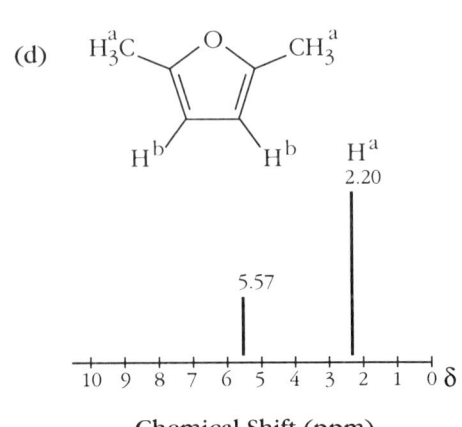

(d)

21.32 The desired compounds are all aryl ketones, which can be prepared by a Friedel-Crafts acylation reaction.

(a) [3,5-dimethylbenzoyl chloride] + [benzene] $\xrightarrow{AlCl_3}$ [3,5-dimethylphenyl phenyl ketone]

Notice that the other combination does not form the desired product:

[benzoyl chloride] + [1,3-dimethylbenzene] $\xrightarrow{AlCl_3}$ [2,4-dimethylphenyl phenyl ketone]

(b) [4-nitrobenzoyl chloride] + [benzene] $\xrightarrow{AlCl_3}$ [4-nitrophenyl phenyl ketone]

Remember that nitrobenzene does not undergo Friedel-Crafts acylation reactions.

(c) [3-chlorobenzoyl chloride] + [anisole] $\xrightarrow{AlCl_3}$ [3-chlorophenyl 4-methoxyphenyl ketone]

21.33

formaldehyde + HCl $\rightleftharpoons$ protonated formaldehyde

21.34

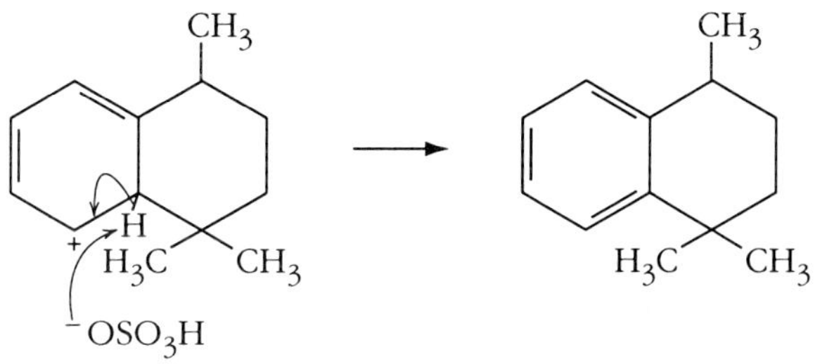

21.35 Step 1 of the mechanism of the addition of HCl to an alkene is proton transfer to form a carbocation. Formation of the more stable carbocation is favored and it determines the product. The two carbocations formed in step 1 are:

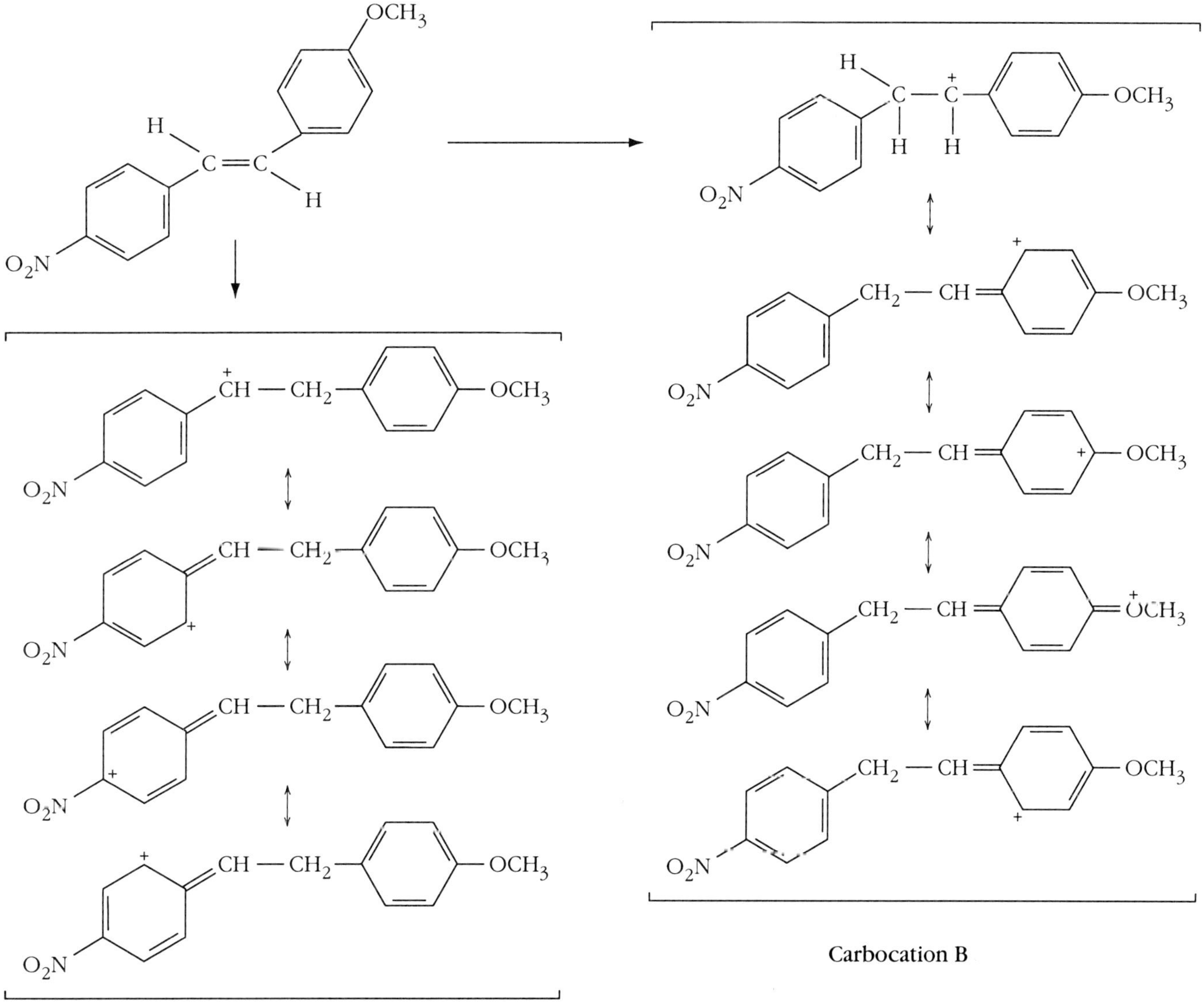

Carbocation A

Carbocation B

NO$_2$ is an electron-withdrawing group so it does not stabilize carbocation A; CH$_3$O is an electron-donating group so it stabilizes the carbocation B.

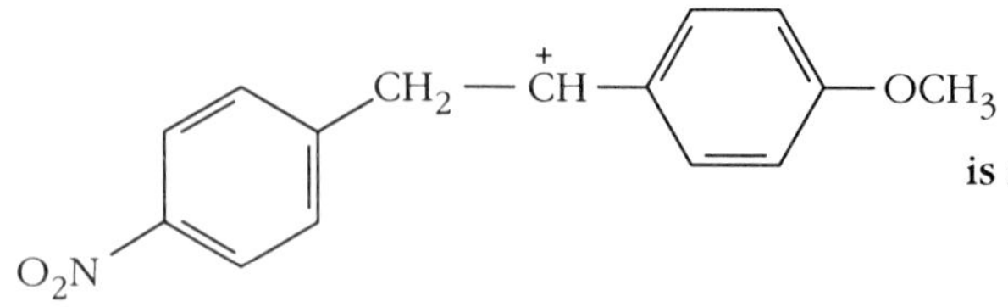

O$_2$N—C$_6$H$_4$—CH$_2$—$\overset{+}{\text{CH}}$—C$_6$H$_4$—OCH$_3$ is more stable than

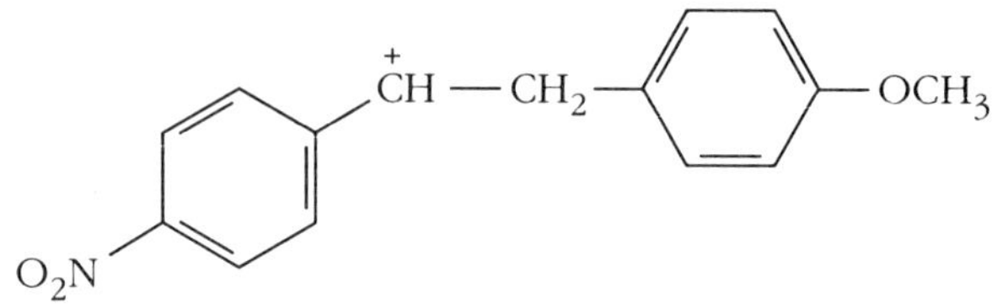

O$_2$N—C$_6$H$_4$—$\overset{+}{\text{CH}}$—CH$_2$—C$_6$H$_4$—OCH$_3$

So

(O$_2$N—C$_6$H$_4$)(H)C=C(H)(C$_6$H$_4$—OCH$_3$) + HCl ⟶

O$_2$N—C$_6$H$_4$—CH$_2$CH(Cl)—C$_6$H$_4$—OCH$_3$

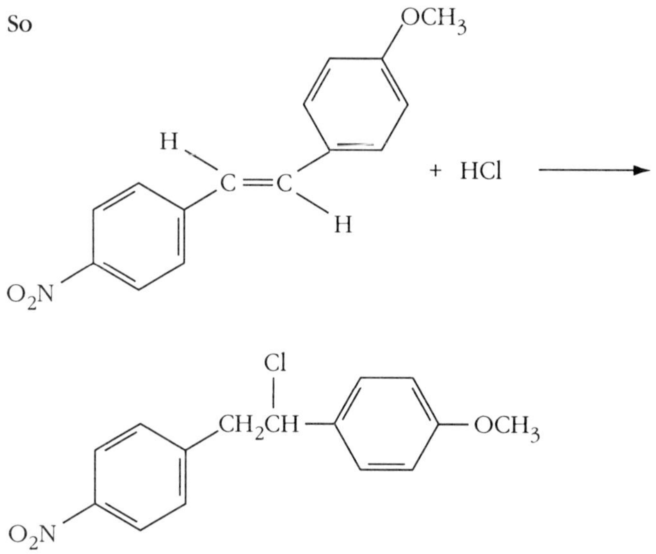

21.36 (a) The molecular formula indicates four degrees of unsaturation. The multiplet in the ^{1}H NMR spectrum at δ 7.0 to δ 7.5 (5H) is due to a monosubstituted benzene ring. Confirming this are the four signals in the aromatic region of the ^{13}C NMR spectrum. The triplets downfield at δ 3.65 (2H) and δ 3.08 (2H) are due to two -CH$_2$- groups. The structure is

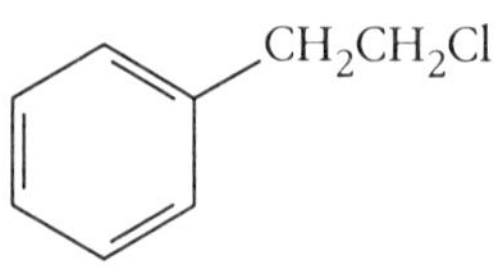

C$_6$H$_5$—CH$_2$CH$_2$Cl

(b) The carbonyl absorption band in the IR spectrum and the two multiplets in the ^{1}H NMR spectrum at δ 7.3 to δ 8 (5H) account for the five degrees of unsaturation. The doublet at δ 1.21 (6H) and the septet at δ 3.55 (1H) are characteristic of an isopropyl group. The benzene ring, the carbonyl carbon atom, and the isopropyl group account for all the hydrogen and carbon atoms in the molecule so the structure is

$$\text{C}_6\text{H}_5-\text{C}(=\text{O})-\text{CH}(\text{CH}_3)_2$$

(c) The carbonyl absorption in the IR spectrum and the signal at δ 9.9 (1H), which is due to a hydrogen atom bonded to a carbonyl carbon indicates that the compound contains an aldehyde group. The two sets of doublets in the aromatic region (4H) indicate a 1,4-disubstituted benzene ring. The singlet at δ 2.4 (3H) is due to a methyl group bonded to a benzene ring. The structure is

$$\text{4-CH}_3-\text{C}_6\text{H}_4-\text{CHO}$$

(d) The carbonyl absorption in the IR spectrum and the two sets of doublets at δ 9.7 (1H), which are due to a hydrogen atom bonded to a carbonyl carbon indicate that the compound contains an aldehyde group. The multiplet at δ 7.3 to δ 7.7 (5H) indicates a monosubstituted benzene ring. The benzene ring and the aldehyde group account for six hydrogen atoms, seven carbon atoms, and five of the six degrees of unsaturation. The sixth degree of unsaturation indicates that the remaining C_2H_2 is a carbon-carbon double bond. The structure is

$$\text{C}_6\text{H}_5-\text{CH}=\text{CH}-\text{CHO}$$

Notice that each of the signals for the two vinyl hydrogen atoms and the hydrogen bonded to the carbonyl carbon atom are split into a set of doublets, which indicates that the three hydrogen atoms are coupled to each other.

(e) The ^{13}C NMR spectrum has six signals in the aromatic region, which indicates that the ring has no symmetry. The downfield singlet at δ 4.5 (2H) is due to a methylene group bonded to the bromine atom. The ratio of the area of aromatic hydrogen atoms to the singlet is 2:1, which indicates that the aromatic ring is disubstituted. The two substituents are —CH$_2$Br and NO$_2$. The substituents are not in the 1,4-positions because the aromatic hydrogen atoms do not appear as a set of doublets. Often a decision as to whether the substituents are 1,2 or 1,3, can be made by carefully examining the splitting of the four aromatic hydrogen atoms. Generally the hydrogen atoms of a 1,2-disubstituted benzene appear as two triplets and two doublets, while the hydrogen atoms of a 1,3-disubstituted benzene appear as a triplet, two doublets, and a singlet:

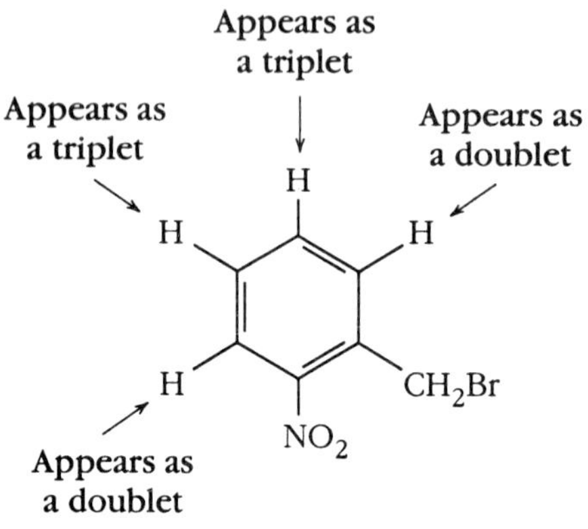

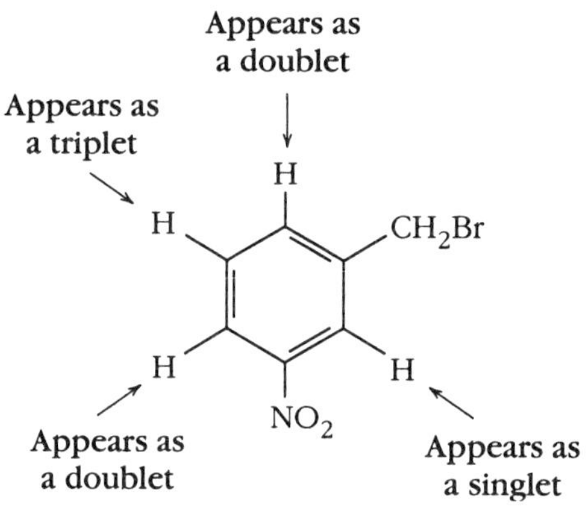

We observe a singlet at δ 8.25, a doublet at δ 8.15, another doublet at δ 7.72, and a triplet at δ 7.52, so the compound is 3-nitrobenzylbromide:

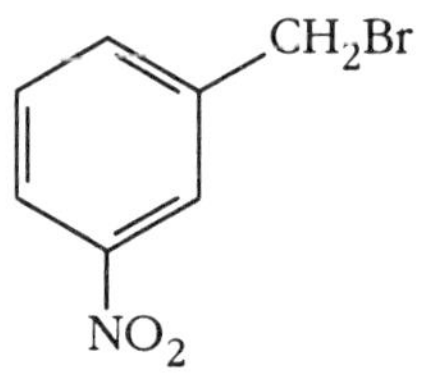

(f) The set of doublets in the aromatic region with relative area of four indicates that the compound is a 1,4-disubstituted benzene. The singlet at δ 2.3 (3H) is due to a methyl group bonded to a benzene ring. This accounts for all the hydrogen and carbon atoms. Only the bromine atom remains so the structure of the compound is

(g) The carbonyl absorption band in the IR spectrum and the three multiplets in the ^{1}H NMR spectrum at δ 6.8 to δ 7.7 (4H) account for the five degrees of unsaturation. The three complex multiplets in the aromatic region of relative area 1:1:2 indicate that the compound is a 1,2-disubstituted benzene derivative. The coupling of the aromatic hydrogen atoms supports this assignment. The aromatic hydrogen atoms appear as a doublet at δ 7.65, a triplet at δ 7.55, and an overlapping triplet and doublet at δ 6.85. The singlet at δ 2.5 is due to a methyl ketone while the other singlet at δ 3.8 is due to a methyl group bonded to an oxygen. The structure of the compound is *o*-methoxyacetophenone:

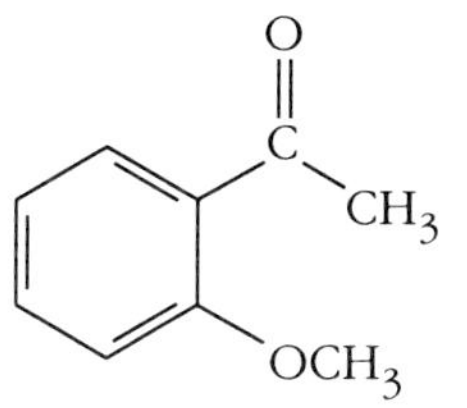

(h) The carbonyl absorption band in the IR spectrum and the multiplet in the ^{1}H NMR spectrum at δ 7.3 to δ 8.0 (5H) account for the five degrees of unsaturation. The two triplets and the quintet all have a relative area of two. This splitting pattern and the similarities in their coupling constants are consistent with the partial structure —$CH_2CH_2CH_2$—. The other partial structures are a monosubstituted benzene ring, a carbonyl group, and a bromine atom. These partial structures can be joined in two ways:

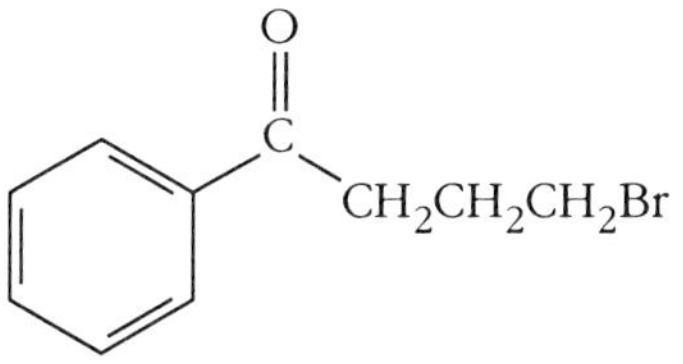

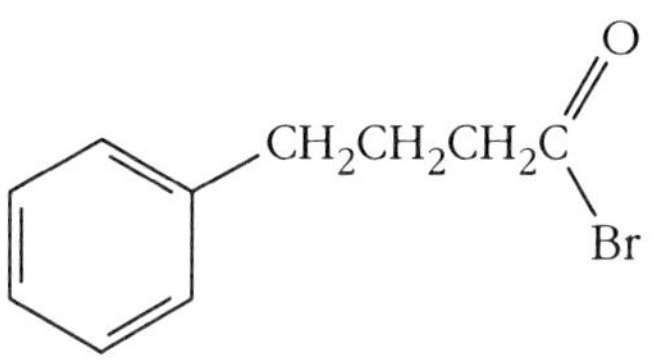

The chemical shift of the triplet due to the methylene group bonded to a benzene ring should appear at about δ 2.4 while the chemical shift of the triplet due to the methylene group bonded to a bromine atom should appear at about δ 3.5. There is no triplet in the spectrum at about δ 2.4 but there is one at δ 3.65 so the structure of the compound is

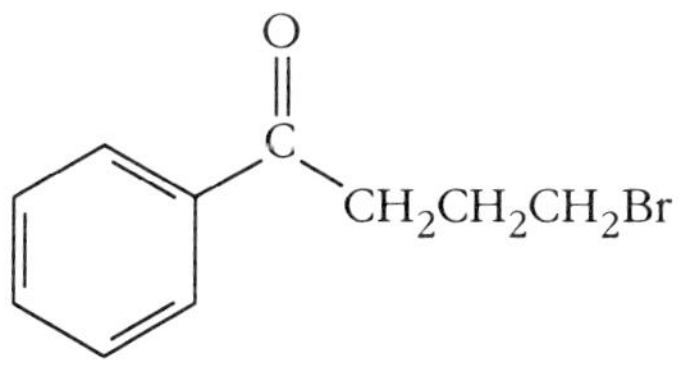

CHAPTER 22

AMINES

Concepts

Amines are derivatives of ammonia in which one or more hydrogen atoms are replaced by alkyl or aryl groups. **Primary amines** have one, **secondary amines** have two, and **tertiary amines** have three organic groups bonded to nitrogen.

The nonbonding pair of electrons on the nitrogen atom makes amines both bases and nucleophiles. The basicity of amines is usually expressed as the pK_a of the conjugate acids. The pK_a values of alkylammonium ions are in the range 9 to 11 while anilinium ions are more acidic with pK_a values in the range 4 to 6. Electron-withdrawing groups attached to the nitrogen atom of amines decrease the basicities of amines. Carboxylic acid amides ($pK_a = 0$) are an example of this effect. Electron-donating groups increase the basicities of amines.

Amines are nucleophiles that react with haloalkanes and carbonyl-containing compounds. Nucleophilic substitution reactions of amines and haloalkanes are difficult to control because a mixture of secondary amines, tertiary amines, and quaternary ammonium salts is often formed. Ammonia, primary amines, and secondary amines react with acyl halides to form amides, *N*-substituted amides, and *N,N*-disubstituted amides, respectively. Ammonia and primary amines react with aldehydes and ketones to form imines while secondary amines react with ketones to form enamines.

Primary amines are named by replacing the final *e* of the IUPAC name of the parent compound with *amine*. Unsymmetrically substituted secondary or tertiary amines are named as *N*-substituted primary amines. The largest of the alkyl substituent is chosen as the parent chain. For example:

$$\text{CH}_3\text{CH}_2\overset{\overset{\displaystyle \text{NH}_2}{|}}{\text{CH}}\text{CH}_3 \qquad \text{cyclohexyl-NHCH}_3 \qquad \text{CH}_3\text{CH}_2\overset{\overset{\displaystyle \text{N(CH}_2\text{CH}_3)_2}{|}}{\text{CH}}\text{CH}_2\text{CH}_3$$

2-Butanamine   *N*-Methylcyclohexanamine   *N,N*-Diethyl-3-pentanamine

Solutions to the Exercises

22.1 (a) There are two alkyl groups bonded to the nitrogen atom of this amine so it is a secondary amine.
(b) There are three groups, two methyl groups and one phenyl group, bonded to the nitrogen atom of this amine so it is a tertiary amine.
(c) Primary amine
(d) Secondary amine
(e) Tertiary amine

22.2 (a) Only one of the four hydrogen atoms of an ammonium ion has been replaced by an alkyl group so this is the salt of a primary amine.
(b) All four of the hydrogen atoms of an ammonium ion have been replaced by alkyl groups so this is a quaternary ammonium salt.
(c) Salt of a secondary amine
(d) Salt of a secondary amine
(e) Quaternary ammonium salt

22.3
(a) $CH_3CH_2CH_2CH_2NH_2$

(b) $CH_3CH_2\underset{\underset{NHCH_2CH_3}{|}}{CH}CH_2CH_2CH_2CH_3$

(c) $[(CH_3)_2CHCH_2]_2NH$

(d) $CH_3CH_2\underset{\underset{C_6H_5}{|}}{N}CH_2CH_3$ (N-phenyl)

(e) cyclopentane with NH_2 (wedge) and NH_2 (dash) at 1,3 positions

(f) $CH_3CH_2CH_2CH_2\overset{+}{N}H_3\overset{-}{Cl}$

(g) $CH_3(CH_2)_8CH_2\overset{+}{N}(CH_3)_3\overset{-}{O}OCCH_3$

(h) 3-chlorocyclopentan-1-amine (Cl and NH_2 on cyclopentane)

22.4 (a) *N,N*-Dimethyl-2-propanamine
(b) *N*-Methylcyclobutanamine
(c) *N,N,N*-Trimethylcyclohexylammonium iodide
(d) 1,3-Diaminobenzene
(e) *N*-Ethyl-*N*-methylcyclohexanamine
(f) Di-*tert*-butylamine

22.5 The nitrogen atom of amines rapidly inverts by a pyramidal inversion so amines whose chirality is due to the substituents on the nitrogen atom cannot be resolved. The barrier to pyramidal inversion is sufficiently high in nitrogen-containing three-membered rings, however, that the enantiomers do *not* interconvert. The enantiomers of quaternary ammonium ions do not interconvert.

(a)
$$CH_3\text{-}\underset{C_6H_5}{\overset{H}{N}}: \underset{\text{inversion}}{\overset{\text{Rapid}}{\rightleftharpoons}} :\underset{C_6H_5}{\overset{CH_3}{N}}\text{-}H$$

(b)
$$(CH_3)_2CH\text{-}\underset{\underset{CH_3}{|}}{\overset{H}{\underset{+}{N}}}\text{-}CH_2CH_3 \quad \rlap{/}{\rightleftharpoons} \quad CH_3CH_2\text{-}\underset{\underset{CH_3}{|}}{\overset{H}{\underset{+}{N}}}\text{-}CH(CH_3)_2$$

(c) [aziridine inversion structures — does not invert]

22.6 The triplet at δ 0.92 (3H) is due to a methyl group adjacent to a methylene group. The sextet at δ 1.47 (2H) is due to a methylene group with five neighboring hydrogen atoms whose coupling constants are about the same. The triplet at δ 2.65 (2H) is due to a methylene group adjacent to a methylene group and an electronegative atom. The broad signal at δ 1.70 (2H) suggests that the electronegative atom is the nitrogen atom of a primary amine. Thus the structure of the amine is **1-propanamine:**

$$CH_3CH_2CH_2NH_2$$

The two peaks at $m/z = 59$ and $m/z = 30$ are due to fragmentation by α-cleavage:

$$CH_3CH_2CH_2\ddot{N}H_2 \xrightarrow{e^-} [H_3C-CH_2-CH_2-\dot{N}H_2]^+ \longrightarrow CH_3CH_2\cdot + CH_2=\overset{+}{N}H_2$$

$$M^+ (m/z = 59) \qquad (m/z = 30)$$

22.7 $CH_3CH_2CH_2CH_2\overset{+}{N}H_3 + H_2O \rightleftharpoons CH_3CH_2CH_2CH_2NH_2 + H_3O^+$

$$pK_a = 10.6$$

$$pH = pK_a + \log\frac{[\text{Conjugate Base}]}{[\text{Acid}]} = pK_a + \log\frac{[CH_3CH_2CH_2CH_2NH_2]}{[CH_3CH_2CH_2CH_2\overset{+}{N}H_3]}$$

$$\log\frac{[CH_3CH_2CH_2CH_2NH_2]}{[CH_3CH_2CH_2CH_2\overset{+}{N}H_3]} = pH - pK_a$$

(a) pH = 5.0 Let R = $CH_3CH_2CH_2CH_2-$

$$\log\frac{[RNH_2]}{[RNH_3^+]} = 5.0 - 10.6 = -5.6$$

$$\frac{[RNH_2]}{[RNH_3^+]} = 10^{-5.6} = 10^{0.4} + 10^{-6} = 2.5 \times 10^{-6}$$

$$\frac{[RNH_2]}{[RNH_3^+]} = \frac{1}{400,000}$$

$$[RNH_3^+] = \frac{400,000}{400,001} \times 100\% = 99.99\%$$

(b) pH = 7.4

$$\frac{[RNH_2]}{[RNH_3^+]} = 10^{-3.2} = 6.31 \times 10^{-4}$$

$$\frac{[RNH_2]}{[RNH_3^+]} = \frac{1}{1580}$$

$$[RNH_3^+] = \frac{1580}{1580 + 1} \times 100\% = 99.94\%$$

(c) pH = 10.5

$$\frac{[RNH_2]}{[RNH_3^+]} = 0.776 = \frac{1}{1.29}$$

$$[RNH_3^+] = \frac{1.29}{1.29 + 1} \times 100\% = 56.3\%$$

(d) pH = 12.0

$$\frac{[RNH_2]}{[RNH_3^+]} = 24.5$$

$$[RNH_3^+] = \frac{1}{24.5 + 1} \times 100\% = 0.039\%$$

22.8 A value of the equilibrium constant (K_{eq}) for each of the following acid-base reactions can be obtained from the expression of K_{eq} in terms of the K_a of the two acids in solution. You may wish to review Section 3.5.

(a) $$CH_3\overset{+}{N}H_3 + H_2O \rightleftharpoons CH_3NH_2 + H_3O^+$$

$$pK_a(CH_3\overset{+}{N}H_3) = 10.63 \qquad K_a(CH_3\overset{+}{N}H_3) = 2.3 \times 10^{-11}$$

$$CH_3\overset{+}{O}H_2 + H_2O \rightleftharpoons CH_3OH + H_3O^+$$

$$pK_a(CH_3\overset{+}{O}H_2) \approx -2 \qquad K_a(CH_3\overset{+}{O}H_2) = 10^2$$

$$CH_3\overset{+}{N}H_3 + CH_3OH \overset{K_{eq}}{\rightleftharpoons} CH_3\overset{+}{O}H_2 + CH_3NH_2$$

$$K_{eq} = \frac{K_a(CH_3\overset{+}{N}H_3)}{K_a(CH_3\overset{+}{O}H_2)} = \frac{2.3 \times 10^{-11}}{10^2} = 2.3 \times 10^{-13}$$

(b) $$CH_3CH_2NH_2 + H_2O \overset{K_{eq}}{\rightleftharpoons} CH_3CH_2\overset{+}{N}H_3 + OH^-$$

$$K_{eq} = \frac{K_a(H_2O)}{K_a(CH_3CH_2\overset{+}{N}H_3)} = \frac{2 \times 10^{-16}}{1.7 \times 10^{-11}} = 1.1 \times 10^{-5}$$

(c) $$(CH_3CH_2)_2NH + CH_3COOH \overset{K_{eq}}{\rightleftharpoons} (CH_3CH_2)_2\overset{+}{N}H_2 + CH_3COO^-$$

$$K_{eq} = \frac{K_a(CH_3COOH)}{K_a((CH_3CH_2)_2\overset{+}{N}H_2)} = \frac{1.76 \times 10^{-5}}{1.14 \times 10^{-11}} = 1.54 \times 10^6$$

22.9 $$\log\frac{[C_6H_5NH_2]}{[C_6H_5\overset{+}{N}H_3]} = pH - pK_a$$

pK_a (Aniline) = 4.60

(a) $\dfrac{[C_6H_5NH_2]}{[C_6H_5NH_3^+]} = 3.98 \%$ $[C_6H_5NH_2] = 79.9\%$

(b) $\dfrac{[C_6H_5NH_2]}{[C_6H_5NH_3^+]} = 630 \%$ $[C_6H_5NH_2] = 99.8\%$

(c) $\dfrac{[C_6H_5NH_2]}{[C_6H_5NH_3^+]} = 7.9 \times 10^5$

(d) $\dfrac{[C_6H_5NH_2]}{[C_6H_5NH_3^+]} = 7.9 \times 10^7$

22.10 When Y is an atom or group that is electron-withdrawing, it stabilizes YNH_2 compared to NH_3 but destabilizes YNH_3^+ compared to NH_4^+.

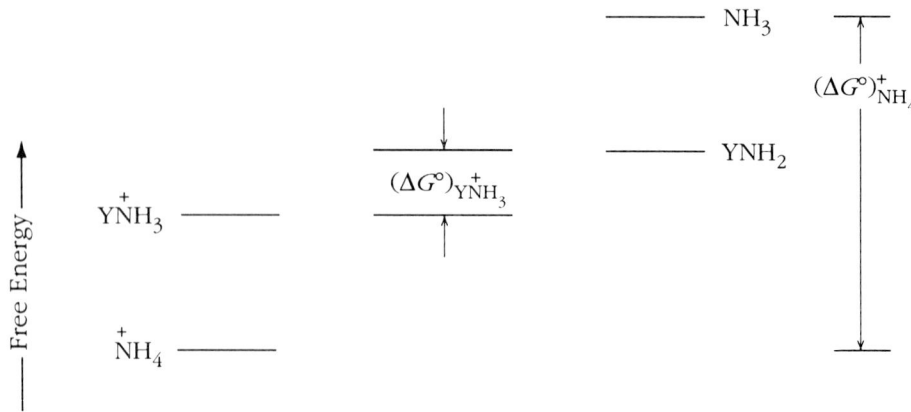

22.11

22.12

(a) $CH_3C(=O)NH_2 + HCl \xrightleftharpoons{K_{eq}} CH_3C(=O)\overset{+}{N}H_3Cl^-$

$$K_{eq} = \frac{K_a(HCl)}{K_a(CH_3C(=O)\overset{+}{N}H_3)} = \frac{1.5 \times 10^2}{1} = 158$$

$$\% \; CH_3C(=O)\overset{+}{N}H_3 = \frac{0.0063}{1.0 + 0.0063} \times 100\% = 0.63\%$$

(b) $CH_3CH_2NH_2 + HCl \xrightleftharpoons{K_{eq}} CH_3CH_2\overset{+}{N}H_3Cl^-$

$$K_{eq} = \frac{K_a(HCl)}{K_a(CH_3CH_2\overset{+}{N}H_3)} = \frac{1.58 \times 10^2}{1.77 \times 10^{-11}} = 8.92 \times 10^{12}$$

22.13 Base strengths of compounds are a measure of their ability to accept a proton from a hydronium ion according to the following equation, where B represents the compound that acts as a base.

$$B + H_3O^+ \rightleftharpoons BH^+ + H_2O$$

We can estimate the strengths of bases from the acid dissociation constants, pK_a, of their conjugate acids. The stronger an acid, the weaker is its conjugate base.

(a)

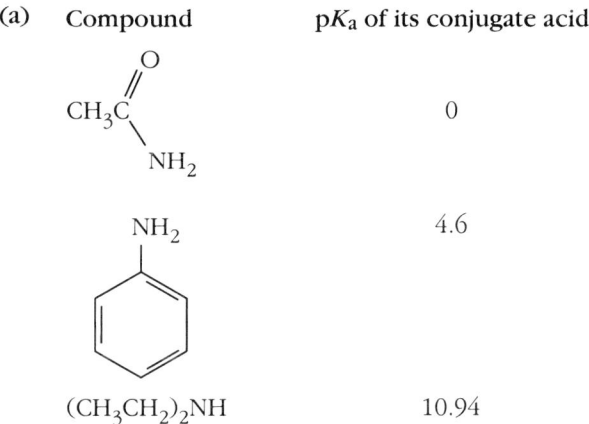

Compound	pK_a of its conjugate acid
$CH_3C(=O)NH_2$	0
$C_6H_5NH_2$ (aniline)	4.6
$(CH_3CH_2)_2NH$	10.94

The conjugate acid of diethylamine is the weakest acid so diethylamine is the strongest of the three bases. The conjugate acid of acetamide is the strongest acid so acetamide is the weakest base of the three compounds. The order from least basic to most basic is:

Increasing basicity →

Acetamide < aniline < diethylamine

(b) Acetic acid donates a proton to water so it is the weakest base of the three compounds. The nitro group of nitroaniline makes its conjugate acid a stronger acid than anilinium ion so nitroaniline is a weaker base than aniline. The order from least basic to most basic is:

Increasing basicity →

Acetic acid < nitroaniline < aniline

(c)

Increasing basicity →

Acetic acid < aniline < ethylamine

22.14 $K_{eq} = \dfrac{K_a(CH_3NH_3^+)}{K_a(NH_4^+)} = \dfrac{2.34 \times 10^{-11}}{5.62 \times 10^{-10}} = 0.042$

22.15

(a) $CH_3(CH_2)_6CH_2Br$ + [potassium phthalimide] ⟶ [N-octyl phthalimide: $NCH_2(CH_2)_6CH_3$]

[N-octyl phthalimide] $\xrightarrow{\text{(1) } H_2SO_4/H_2O/\text{heat} \quad \text{(2) } NaOH/H_2O}$ $H_2NCH_2(CH_2)_6CH_3$

(b) $(CH_3)_2CHCH_2Br$ $\xrightarrow{\text{(1) potassium phthalimide} \quad \text{(2) } H_2SO_4/H_2O/\text{heat} \quad \text{(3) } NaOH/H_2O}$ $(CH_3)_2CHCH_2NH_2$

(c) $BrCH_2CO_2^-$ $\xrightarrow{\text{(1) potassium phthalimide} \quad \text{(2) } H_2SO_4/H_2O/\text{heat} \quad \text{(3) } NaOH/H_2O}$ $H_2NCH_2CO_2^-Na^+$

22.16 *t*-Butyl chloride does not undergo nucleophilic substitution reactions by an S_N2 mechanism. Potassium phthalimide acts as a base with *t*-butyl chloride to eliminate HCl and form phthalimide, KCl, and methylpropene.

22.17 (a) Nitration of benzoic acid forms *m*-nitrobenzoic acid, which can be reduced to *m*-aminobenzoic acid.

(b) Nitration of anisole (methoxybenzene) forms a mixture of *o*- and *p*-nitroanisole. Reduction of these compounds forms a mixture of *o*- and *p*-methoxyaniline not *m*-methoxyaniline.

(c) Nitration of chlorobenzene forms a mixture of *o*- and *p*-nitrochlorobenzene. Reduction of these compounds forms a mixture of *o*- and *p*-chloroaniline. These can be separated to yield pure *p*-chloroaniline.

22.18

(a) $CH_3CH_2CH_2CH_2CH_2CHO \xrightarrow[\text{Weakly acidic buffer}]{NH_3} [CH_3(CH_2)_4CH=NH] \xrightarrow[\text{CH}_3\text{OH}]{\text{LiBH}_3\text{CN}} CH_3(CH_2)_4CH_2NH_2$

(b) CH_3CCH_3 (with =O) + aniline ($C_6H_5NH_2$) $\xrightarrow{\text{Weakly acidic buffer}}$ [$(CH_3)_2C=N-C_6H_5$] $\xrightarrow{\text{LiBH}_3\text{CN}/CH_3OH}$ $(CH_3)_2CH-NH-C_6H_5$

(c) C_6H_5CHO + $(CH_3CH_2)_2NH$ $\xrightarrow{\text{Weakly acidic catalyst}}$ [$C_6H_5CH=N^+(CH_2CH_3)_2$ H] $\xrightarrow{\text{LiBH}_3\text{CN}/CH_3OH}$ $C_6H_5CH_2N(CH_2CH_3)_2$

(d) cyclohexanone + piperidine $\xrightarrow[\text{(2) LiBH}_3\text{CN/CH}_3\text{OH}]{\text{(1) Weakly acidic catalyst}}$ N-cyclohexylpiperidine

22.19

(a) $CH_3(CH_2)_5CH=CH_2$ $\xrightarrow{\text{BH}_3/\text{THF}}$ $[CH_3(CH_2)_5CH_2CH_2]_3B$ $\xrightarrow[\text{(2) NaOH/H}_2\text{O}]{\text{(1) NH}_2\text{OSO}_3\text{H}}$

$3CH_3(CH_2)_5CH_2CH_2NH_2$

(b) $CH_2=C(CH_3)(CH_2CH_2CH_3)$ $\xrightarrow[\text{(2) NH}_2\text{OSO}_3\text{H} \text{ (3) NaOH/H}_2\text{O}]{\text{(1) BH}_3/\text{THF}}$ $H_2NCH_2-CH(CH_3)CH_2CH_2CH_3$

(c)

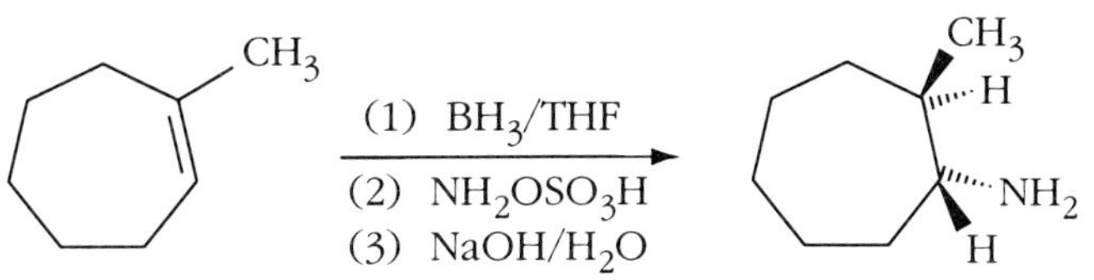

$$\text{1-methylcycloheptene} \xrightarrow[\substack{(2)\ NH_2OSO_3H \\ (3)\ NaOH/H_2O}]{(1)\ BH_3/THF} \text{trans-2-methylcycloheptylamine}$$

22.20 The following reactions occur in several steps and are catalyzed by the enzyme glutamate dehydrogenase. They represent examples of the biosynthesis and biodegradation of amino acids.

(a) 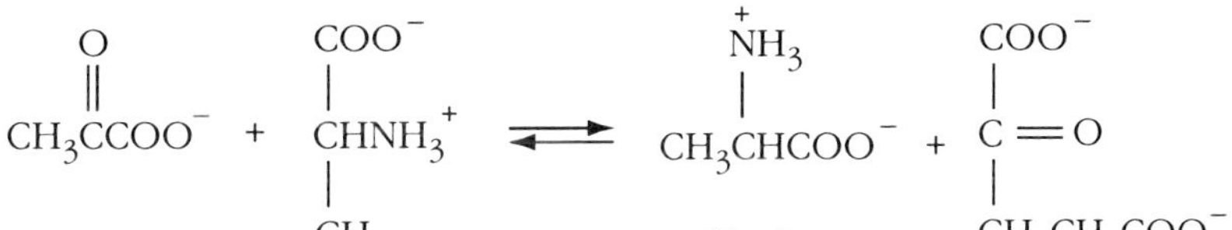 $CH_3CCOO^- + \begin{array}{c} COO^- \\ | \\ CHNH_3^+ \\ | \\ CH_2 \\ | \\ CH_2COO^- \end{array} \rightleftharpoons \begin{array}{c} \overset{+}{NH_3} \\ | \\ CH_3CHCOO^- \\ \\ \text{Alanine} \end{array} + \begin{array}{c} COO^- \\ | \\ C=O \\ | \\ CH_2CH_2COO^- \\ \\ \alpha\text{-Ketoglutarate} \end{array}$

(b) 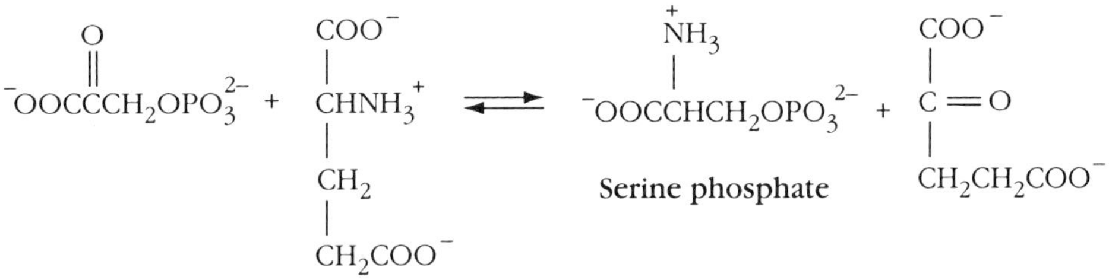 $^-OOCCCH_2OPO_3^{2-} + \begin{array}{c} COO^- \\ | \\ CHNH_3^+ \\ | \\ CH_2 \\ | \\ CH_2COO^- \end{array} \rightleftharpoons \begin{array}{c} \overset{+}{NH_3} \\ | \\ ^-OOCCHCH_2OPO_3^{2-} \\ \\ \text{Serine phosphate} \end{array} + \begin{array}{c} COO^- \\ | \\ C=O \\ | \\ CH_2CH_2COO^- \end{array}$

22.21

(a) 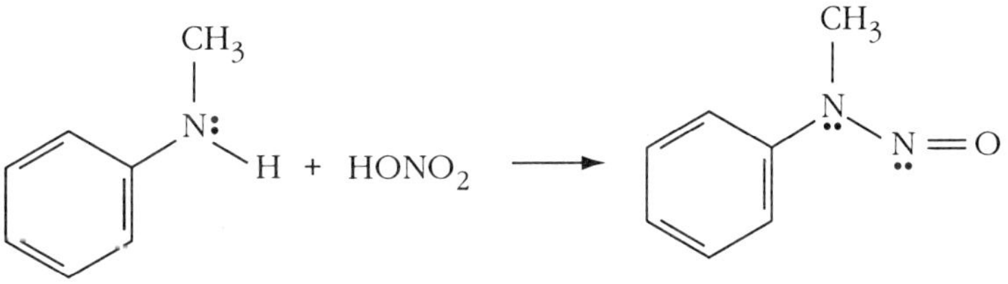 N-methylaniline + HONO₂ → N-Nitroso-N-methylaniline

(b)  $[(CH_3)_2CH]_2\ddot{N}H + HONO_2 \longrightarrow [(CH_3)_2CH]_2\ddot{N}-\ddot{N}=O$

N-Nitrosodiisopropylamine

(c) 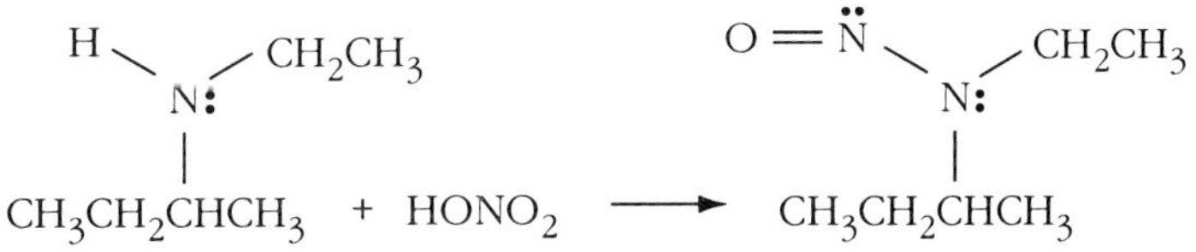 $\underset{CH_3CH_2CHCH_3}{\overset{H\diagdown\ \ \diagup CH_2CH_3}{\underset{|}{N:}}} + HONO_2 \longrightarrow \underset{CH_3CH_2CHCH_3}{\overset{O=\ddot{N}\diagdown\ \ \diagup CH_2CH_3}{\underset{|}{N:}}}$

N-Nitroso-N-ethyl-2-butanamine

22.22 Primary amines react with nitrous acid, formed by the reaction of sodium nitrite and aqueous acid, to form diazonium ions. Secondary amines react with nitrous acid to form N-nitrosoamines.

(a) $CH_3CH_2CHCH_3$ with $H-\underset{|}{\overset{..}{N}}-CH_3$ $\xrightarrow{\text{NaNO}_2}{\text{H}_3\text{O}^+}$ $CH_3CH_2CHCH_3$ with $O=N-\underset{|}{\overset{..}{N}}-CH_3$

N-Nitroso-*N*-methyl-2-butanamine

(b) 2,6-dichloro-4-chloroaniline $\xrightarrow{\text{NaNO}_2}{\text{H}_3\text{O}^+}$ 2,4,6-trichlorobenzenediazonium chloride

2,4,6-Trichlorobenzenediazonium chloride

(c) diphenylamine $\xrightarrow{\text{NaNO}_2}{\text{H}_3\text{O}^+}$ *N*-nitrosodiphenylamine

N-Nitrosodiphenylamine

22.23 (a) Bromine is added to the benzene ring first because subsequent substitution occurs *ortho-* and *para-* to it.

1-bromo-2,4-dinitrobenzene ⟹ bromobenzene ⟹ benzene

benzene $\xrightarrow[\text{FeBr}_3]{\text{Br}_2}$ bromobenzene $\xrightarrow[\text{H}_2\text{SO}_4]{2\text{HNO}_3}$ 1-bromo-2,4-dinitrobenzene

(b)

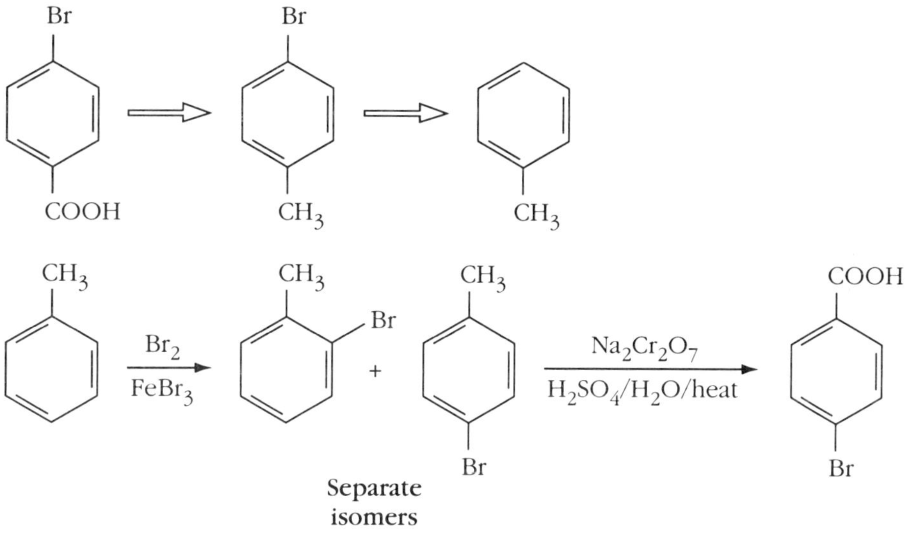

Retrosynthesis: 4-bromobenzoic acid ⟹ 4-bromotoluene ⟹ toluene

Toluene $\xrightarrow{\text{Br}_2/\text{FeBr}_3}$ 2-bromotoluene + 4-bromotoluene (Separate isomers) $\xrightarrow{\text{Na}_2\text{Cr}_2\text{O}_7/\text{H}_2\text{SO}_4/\text{H}_2\text{O}/\text{heat}}$ 4-bromobenzoic acid

(c)

Retrosynthesis: 4-methylbenzoic acid ⟹ 4-bromotoluene ⟹ toluene

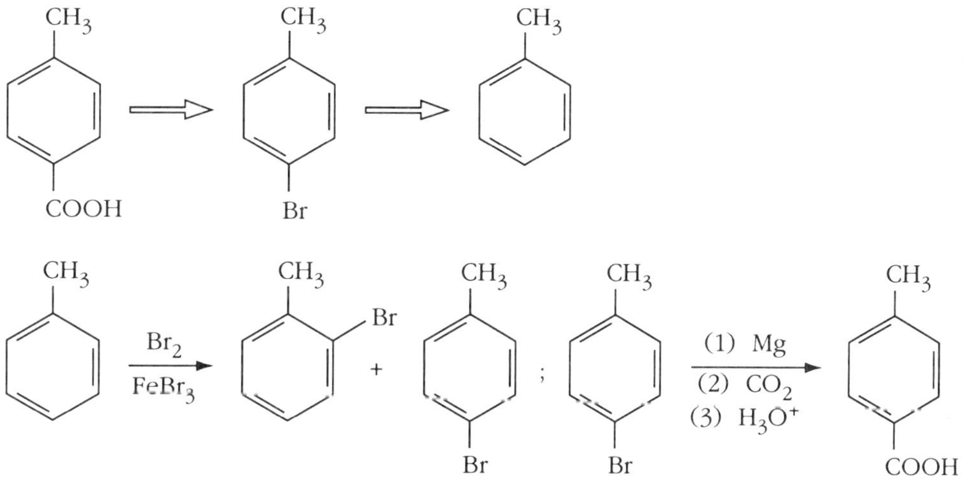

Toluene $\xrightarrow{\text{Br}_2/\text{FeBr}_3}$ 2-bromotoluene + 4-bromotoluene ; 4-bromotoluene $\xrightarrow[\text{(3) H}_3\text{O}^+]{\text{(1) Mg; (2) CO}_2}$ 4-methylbenzoic acid

(d)

Retrosynthesis: 4-hydroxyacetophenone ⟹ 4-diazoniumacetophenone ⟹ 4-aminoacetophenone ⟹ 4-acetamidoacetophenone ⟹

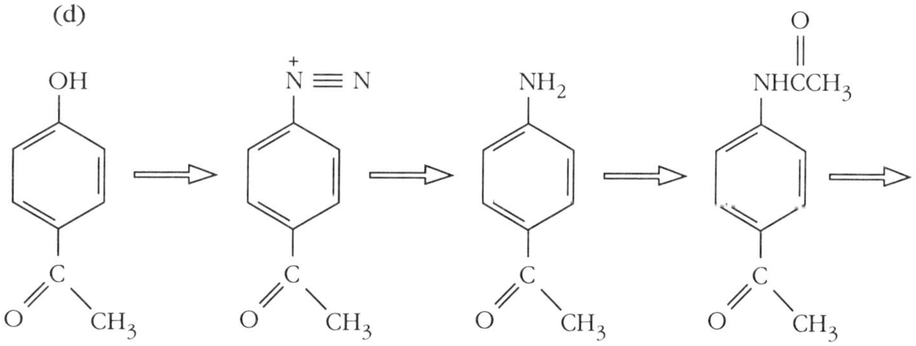

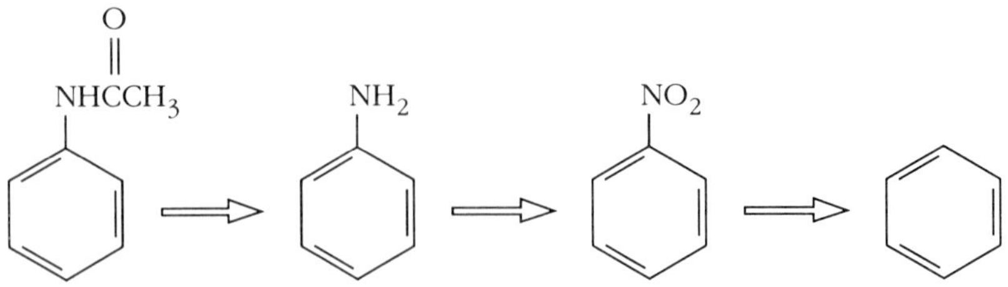

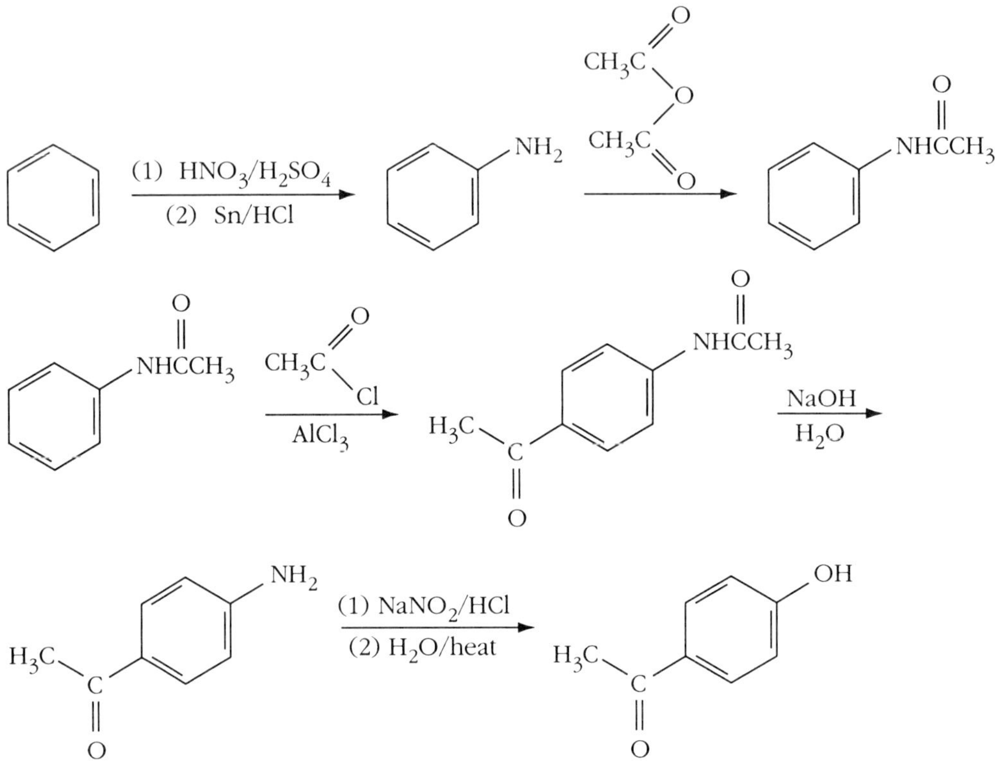

Retrosynthesis:

Ph-NHCOCH₃ ⟹ Ph-NH₂ ⟹ Ph-NO₂ ⟹ Ph-H

Forward synthesis:

Benzene —(1) HNO₃/H₂SO₄; (2) Sn/HCl→ aniline (PhNH₂) —(CH₃CO)₂O→ acetanilide (PhNHCOCH₃)

Acetanilide —CH₃COCl / AlCl₃→ 4′-acetamidoacetophenone (H₃C–CO–C₆H₄–NHCOCH₃) —NaOH / H₂O→ 4′-aminoacetophenone (H₃C–CO–C₆H₄–NH₂) —(1) NaNO₂/HCl; (2) H₂O/heat→ 4′-hydroxyacetophenone (H₃C–CO–C₆H₄–OH)

22.24 The azo group is usually introduced into a molecule by a diazo coupling reaction. In this reaction, benzenediazonium salt reacts with *N,N*-dimethylaniline, which is a highly activated benzene derivative:

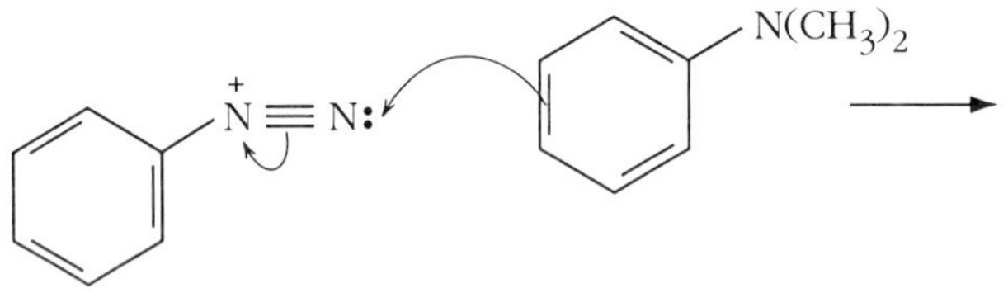

Ph–N⁺≡N: + C₆H₅–N(CH₃)₂ ⟶

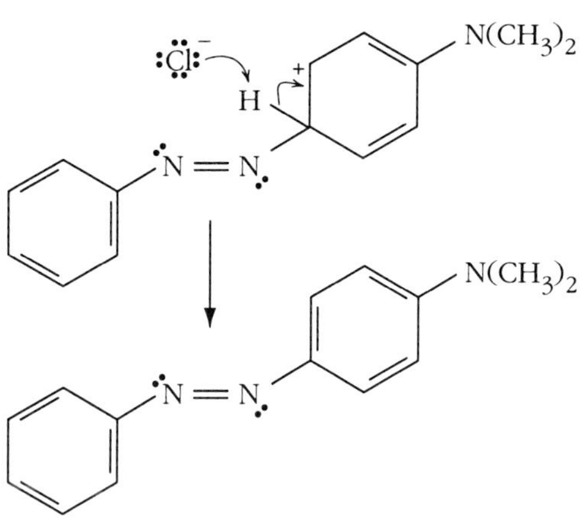

22.25 The synthesis of all three sulfa drugs involves a common intermediate, *p*-acetoamidobenzenesulfonyl chloride, which is formed by acetylation of the amino group of aniline to form acetanilide followed by reaction of acetanilide with chlorosulfonic acid.

Preparation of *p*-acetoamidobenzenesulfonyl chloride:

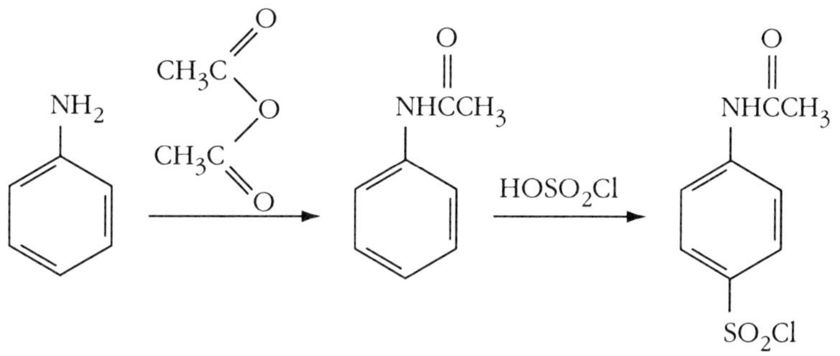

p-Acetoamido-benzenesulfonyl chloride

Preparation of sulfathiazole by the reaction of 2-aminothiazole with *p*-acetoamidobenzenesulfonyl chloride:

There are two amine groups in 2-aminothiazole. The ring nitrogen atom cannot form a stable product by reacting with a sulfonyl chloride (or an acyl chloride for that matter) because it is a tertiary amine (See Section 16.5).

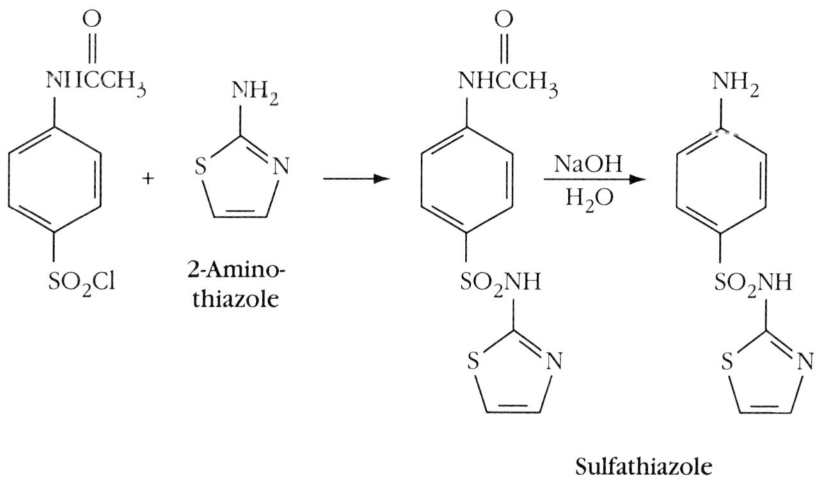

Sulfathiazole

Preparation of sulfapyridine by the reaction of 2-aminopyridine with *p*-acetoamidobenzenesulfonyl chloride:

There are two amine groups in 2-aminopyridine but the ring nitrogen atom cannot form a stable product by reacting with a sulfonyl chloride because it is a tertiary amine.

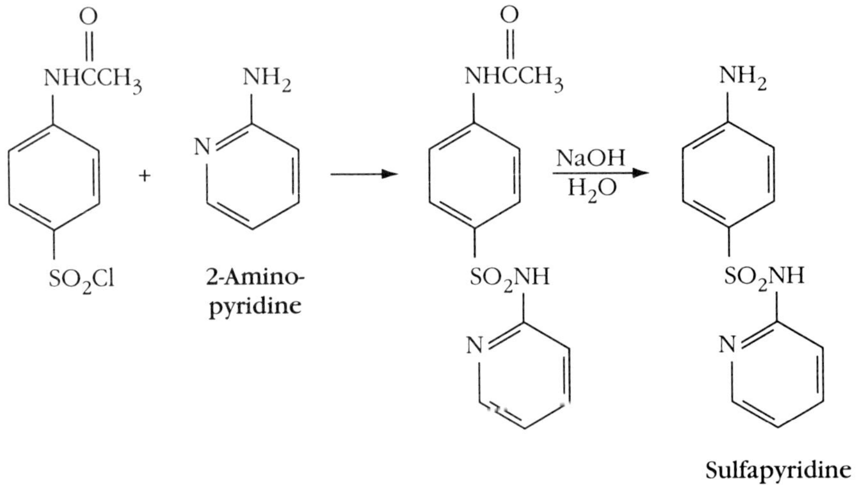

Preparation of sulfamethazine by the reaction of 3,5-dimethylaniline with *p*-acetoamidobenzenesulfonyl chloride:

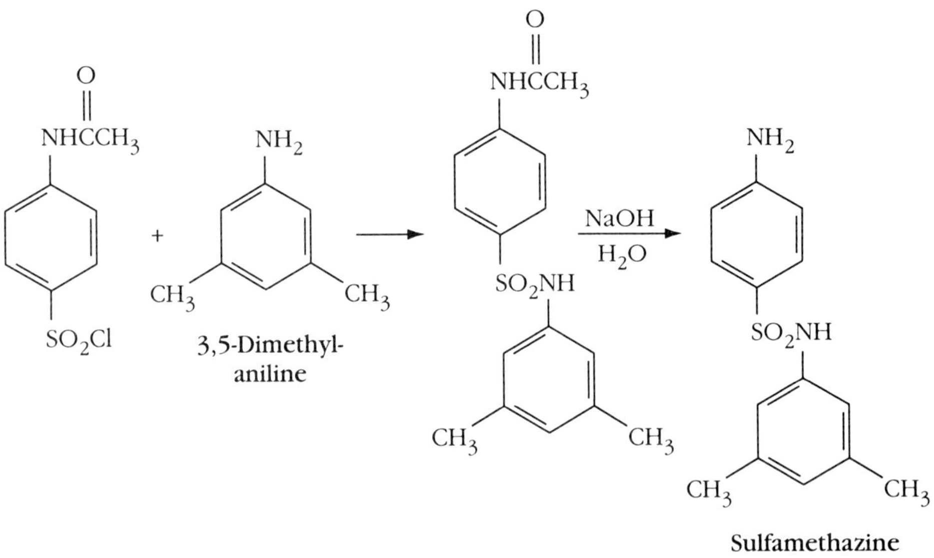

22.26 (a) **Gabriel amine synthesis** is a method of preparing primary amines. The nucleophilic substitution reaction of potassium phthalimide with a primary or secondary haloalkane forms an alkylphthalimide that forms a primary amine upon hydrolysis:

[Reaction scheme: Potassium phthalimide + RCH$_2$Br →(S_N2) N-alkylated phthalimide → (1) H$_3$O$^+$ (2) NaOH/H$_2$O → disodium phthalate + H$_2$NCH$_2$R]

(b) **Sulfa drugs** are compounds used to combat infectious diseases. Sulfa drugs are structurally similar to *p*-aminobenzoic acid, which bacteria need to form folic acid. The bacteria use the sulfa drug instead of *p*-aminobenzoic acid to form an altered folic acid, which is ineffective as a coenzyme in the bacteria so the infectious organism dies. Sulfa drugs have the following general formula:

[Structure: benzene ring with NH$_2$ at top and SO$_2$NHR at bottom (para)] R=H or aromatic ring

(c) A **1° amine** is an amine that has only one alkyl or aryl group bonded to its nitrogen atom.

$$RNH_2 \text{ or } ArNH_2$$

(d) **Nitrosation of amines** is the reaction of nitrous acid and amines. The product of the reaction depends on whether the amine is 1° or 2°. Nitrosation of 2° alkyl or arylamines forms a nitrosamine. Nitrosation of a 1° alkyl amine forms an alkane diazonium ion, which rapidly loses nitrogen to form a carbocation. This carbocation undergoes elimination to form an alkene, reacts with a nucleophile, and if possible, rearranges. Nitrosation of 1° arylamines forms arenediazonium ions, which are useful synthetic intermediates.

(e) **Arenediazonium ions** are formed by the reaction of nitrous acid and 1° arylamines. Arenediazonium ions are useful synthetically because the azo group (—NN+) can be replaced by a number of other groups or atoms such as a hydroxyl group, a cyano group, or a fluorine atom.

$$\text{PhNH}_2 \xrightarrow{\text{NaNO}_2, \text{HCl}/\text{H}_2\text{O}} \text{Ph-N}^+\equiv\text{N} \ \text{Cl}^-$$

(f) **N-Nitrosamine** is the product of the reaction of a 2° alkyl or aryl amine with nitrous acid:

$$\underset{R}{\overset{R}{\text{N-H}}} \xrightarrow{\text{NaNO}_2, \text{HCl}/\text{H}_2\text{O}} \underset{R}{\overset{R}{\text{N-N=O}}}$$

R = Alkyl or aryl groups

(g) **Diazotization** is the reaction of a 1° arylamine with nitrous acid. Diazotization converts a 1° arylamine into an arenediazonium ion (the reaction is shown in Exercise 22.26e).

(h) The **Sandmeyer reaction** is the reaction of an arenediazonium ion with Cu(I)Cl, Cu(I)Br, or Cu(I)CN to replace the diazo group with a Cl atom, Br atom, or CN group, respectively.

$$\text{Ph-N}^+\equiv\text{N} \ \text{Cl}^- \xrightarrow{\text{Cu(I)X}} \text{Ph-X}$$

(i) **Quaternary ammonium salts** are compounds in which all four hydrogen atoms of an ammonium ion are replaced by any combination of alkyl or aryl groups.

$$\text{R}-\overset{R}{\underset{R}{\overset{|}{\text{N}^+}}}-\text{R} \quad \text{X}^- \qquad \text{R = Alkyl or aryl group}$$

(j) **Reductive amination** is the reduction of imines, formed by the reaction of ammonia or 1° amines and aldehydes and ketones, to form primary or secondary amines. Iminium ions, formed in the reaction of 2° amines and aldehydes and ketones, can also be reduced to form 3° amines by reductive amination.

$$\underset{R}{\overset{R}{\text{C=O}}} \xrightarrow[\text{(2) NaBH}_3\text{C}\equiv\text{N}]{\text{(1) R'}_2\text{NH}} \underset{R}{\overset{R}{\text{CH-N}}}\overset{R'}{\underset{R'}{}}$$

(k) A **3° amine** is an amine in which three alkyl or aryl groups are bonded to its nitrogen atom.

$$R_3N \text{ or } Ar_3N$$

(l) **Arenesulfonamide** is a class of compounds in which an arenesulfonyl group replaces one hydrogen atom of ammonia, a 1° amine, or a 2° amine:

R = H, alkyl or aryl group

22.27

(a) $(CH_3)_2CHCH_2NH_2$

(b) C₆H₅CH₂NH—cyclopropyl

(c) $(CH_2=CHCH_2)_4\overset{+}{N} \; \bar{O}OCCH_3$

(d) 4-Cl-C₆H₄-N(CH₃)₂

(c) $CH_3CHCH_2CHCH_3$ with Br on C2 and NH_2 on C4

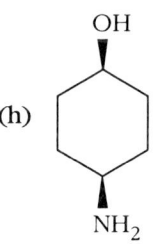

(f) $CH_3(CH_2)_6CH_2N(CH_2CH_3)(CH_2C_6H_5)$

(g) $BrCH_2CH_2CHCH_2CH_2NH_2$ with NH_2 on C3

(h) trans-4-aminocyclohexanol

22.28 (a) *N*-Ethyl-3,5-dinitroaniline
  (b) 2,5-Hexanediamine
  (c) *cis*-*N*-Methyl-4-methylcyclohexanamine
  (d) *N,N,N*-Trimethylbenzylammonium bromide
  (e) *N*-Cyclohexylacetamide
  (f) Triphenylamine
  (g) 3-Buten-1-amine

22.29 (a) 2° amine
  (b) 1° amine (diamine)
  (c) 2° amine
  (d) Quaternary ammonium ion
  (e) An *N*-substituted amide
  (f) 3° amine
  (g) 1° amine

22.30 (a) This compound has a plane of symmetry so it is achiral.
(b) This compound has a plane of symmetry so it is achiral.
(c) Rapid inversion occurs about the nitrogen atom but not about the carbon stereocenter so N-methyl-1-phenylethylamine exists as enantiomers.

(d) Two enantiomers of this compound exist and they are not interconvertible so they can be resolved.

22.31 Recall that aniline is a weaker base than ammonia and that alkylamines are slightly stronger bases than ammonia. Electron-withdrawing groups in the aromatic ring of aniline *decrease* their basicity while electron-donating groups *increase* their basicity relative to aniline.

Increasing basicity →

(a) Aniline < ammonia < methylamine
(b) 4-Nitroaniline < aniline < methylamine
(c) Acetamide < 4-chloroaniline < ammonia
(d) Acetic acid < aniline < methylamine

22.32 (a) $CH_3COOH + CH_3CH_2NH_2 \rightleftharpoons CH_3COO^- + CH_3CH_2NH_3^+$

Ethylammonium ion (pK_a = 10.75) is a weaker acid than acetic acid (pK_a = 4.74) so the products are favored at equilibrium.

(b) C₆H₅NH₂ + $CH_3COOH \rightleftharpoons CH_3COO^-$ + C₆H₅NH₃⁺

Anilinium ion (pK_a = 4.60) is a slightly stronger acid than acetic acid (pK_a = 4.74) so at equilibrium there are appreciable amounts of all four species in solution.

(c) [p-nitroaniline] + [anilinium ion] ⇌ [p-nitroanilinium ion] + [aniline]

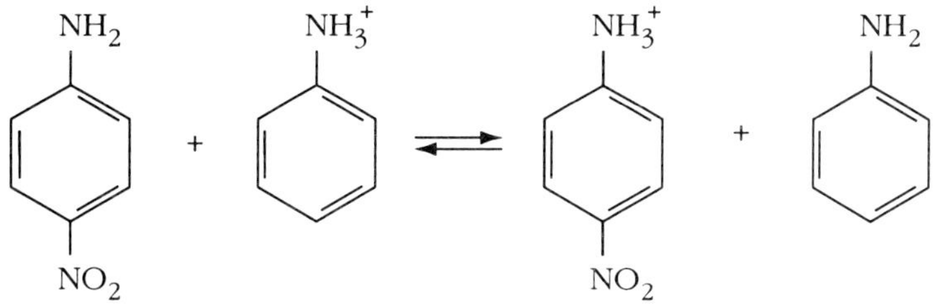

Anilinium ion (pK_a = 4.60) is a weaker acid than p-nitroanilinium ion (pK_a = 1.00) so the reactants are favored at equilibrium.

22.33 The nitro group of 4-nitroaniline is electron-withdrawing so it increases the delocalization of the unshared electron pair on nitrogen in 4-nitroaniline compared to aniline. As a result, 4-nitroaniline is stabilized compared to aniline.

The electron-withdrawing nitro group increases the electron deficiency in the benzene ring of 4-nitroanilinium compared to the anilinium ion with the result that the 4-nitroanilinium ion is less stable than the anilinium ion.

These effects are illustrated on the following free energy diagram:

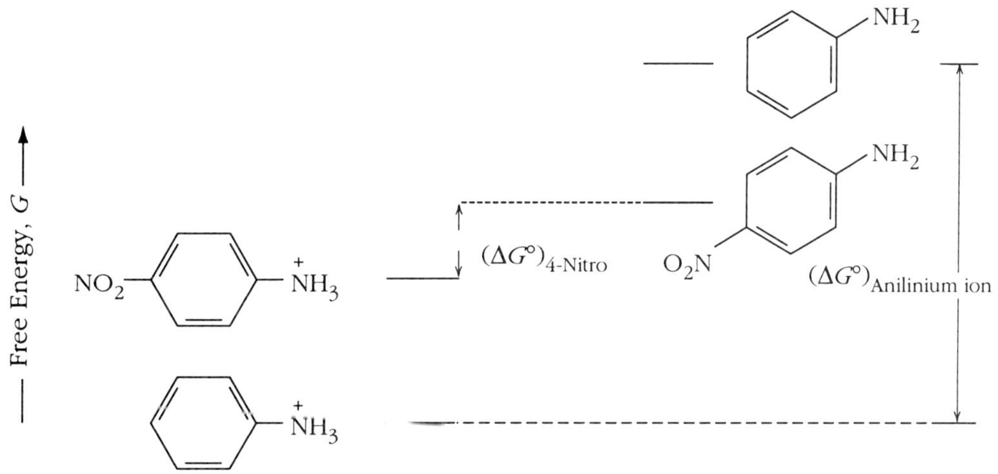

The overall effect of the nitro group, therefore, is to make $(\Delta G°_{\text{4-Nitroanilinium ion}}) < (\Delta G°_{\text{Anilinium ion}})$ so 4-nitroanilinium ion is a stronger acid than the anilinium ion. 4-Nitroaniline, therefore, is a weaker base than aniline.

22.34

(a) 2 PhNH$_2$ + (CH$_3$CO)$_2$O ⟶ PhNHCCH$_3$ (with C=O) + Ph$\overset{+}{N}$H$_3$ $\overset{-}{O}$OCCH$_3$

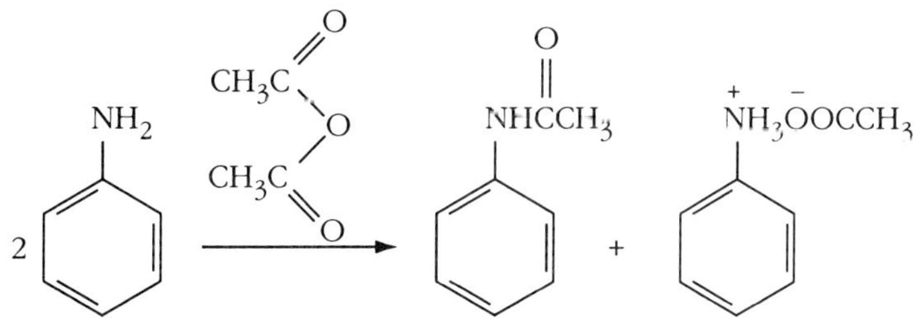

(b) [aniline] →(1) NaNO$_2$/HCl/H$_2$O/0°C (2) CuI→ [iodobenzene]

(c) [aniline] →(1) NaNO$_2$/HCl (2) H$_2$O/heat→ [phenol]

(d) [aniline] →CH$_3$C(=O)-O-C(=O)CH$_3$→ [acetanilide] →HNO$_3$/H$_2$SO$_4$→ [p-nitroacetanilide] →NaOH/H$_2$O→ [p-nitroaniline]

Major product

If aniline is not acetylated before nitration, the acidic conditions of nitration form the anilinium ion. The —$\overset{+}{N}H_3$ group is *meta*-directing and strongly deactivating.

[aniline] →H$_2$SO$_4$→ [anilinium ion] →HNO$_3$/H$_2$SO$_4$, Slow→ [m-nitroanilinium]

(e) [aniline] →(1) NaNO$_2$/H$_3$O$^+$/0°C (2) CuCN→ [benzonitrile] →H$_3$O$^+$/Heat→ [benzoic acid]

(f) [aniline] + excess CH$_3$I → [PhN$^+$(CH$_3$)$_3$ I$^-$]

(g) [aniline] →(1) NaNO₂/H₃O⁺/0°C (2) H₃PO₂ → [benzene]

22.35

(a) 2 [piperidine N–H] →(CH₃CO)₂O→ [N-acetylpiperidine] + [piperidinium acetate]

(b) CH_3CCH_3 (ketone) →NH₃, NaBH₃CN, Acidic buffer→ CH_3CHCH_3 with NH_2

(c) $CH_3C(=O)NHCH_2CH_3$ →(1) LiAlH₄ (2) NaOH/H₂O→ $CH_3CH_2NHCH_2CH_3$

(d) [nitrobenzene] →(1) Sn/HCl (2) NaOH/H₂O→ [aniline]

(e) $CH_3CH_2CH_2CN$ →(1) LiAlH₄ (2) NaOH/H₂O→ $CH_3CH_2CH_2CH_2NH_2$

(f) [1-methylcyclohexene] →(1) 9-BBN (2) H₂NCl→ [trans-2-methylcyclohexylamine]

(g) [aniline] →NaNO₂, HCl→ [benzenediazonium chloride, $\overset{+}{N}\equiv N\ Cl^-$]

(h) Ph-N₂⁺ + phenol → 4-(phenyldiazenyl)phenol

(i) 4-diazonium-benzene radical + CuC≡N → benzonitrile

(j) PhN₂⁺ $\xrightarrow[(2)\ \text{Heat}]{(1)\ \text{HBF}_4}$ PhF

22.36 (a) $CH_3CH_2CH_2CH_2CH_2OH \xrightarrow{PBr_3} CH_3CH_2CH_2CH_2CH_2Br$

$\xrightarrow[\substack{(2)\ H_3O^+ \\ (3)\ NaOH/H_2O}]{(1)\ \text{K}^{+-}\text{N(phthalimide)}} CH_3CH_2CH_2CH_2CH_2NH_2$

(b) $CH_3(CH_2)_3CH_2NH_2 \xrightarrow{\text{Excess } CH_3I} CH_3(CH_2)_3CH_2N(CH_3)_2$
From (a)

(c) $CH_3(CH_2)_3CH_2Br \xrightarrow{NaCN} CH_3(CH_2)_3CH_2CN$
From (a)

(d) $CH_3(CH_2)_3CH_2OH \xrightarrow[CH_2Cl_2]{PCC} CH_3(CH_2)_3C(=O)H$

(e) $CH_3(CH_2)_3CH_2OH \xrightarrow[H_2SO_4]{CrO_3} CH_3(CH_2)_3C(=O)OH$

(f) $2\,CH_3(CH_2)_3CH_2NH_2$ + $CH_3C(=O)Cl \longrightarrow CH_3(CH_2)_3CH_2NHCCH_3$ + $CH_3(CH_2)_3CH_2\overset{+}{N}H_3Cl^-$
From (a)

22.37

(a) $CH_3(CH_2)_3CH_2Br \xrightarrow[\text{(2) }H_3O^+\text{; (3) NaOH/}H_2O]{\text{(1) potassium phthalimide}} CH_3(CH_2)_3CH_2NH_2$

(b) $CH_3(CH_2)_2CH_2OH \xrightarrow{PBr_3} CH_3(CH_2)_2CH_2Br \xrightarrow[CH_3CH_2OH]{NaCN}$

$CH_3(CH_2)_2CH_2CN \xrightarrow[\text{(2) NaOH/}H_2O]{\text{(1) LiAlH}_4} CH_3(CH_2)_3CH_2NH_2$

(c) $CH_3(CH_2)_3C(=O)NH_2 \xrightarrow[\text{(2) NaOH/}H_2O]{\text{(1) LiAlH}_4} CH_3(CH_2)_3CH_2NH_2$

(d) $CH_3CH_2CH_2CH=CH_2 \xrightarrow[\text{(2) }H_2NCl]{\text{(1) 9-BBN}} CH_3(CH_2)_3CH_2NH_2$

(e) $CH_3(CH_2)_3COOH \xrightarrow[\text{(2) NH}_3]{\text{(1) SOCl}_2} CH_3(CH_2)_3C(=O)NH_2$

$\xrightarrow[\text{(2) NaOH/}H_2O]{\text{(1) LiAlH}_4} CH_3(CH_2)_3CH_2NH_2$

(f) $CH_3(CH_2)_3C(=O)H \xrightarrow[NaBH_3CN/CH_3OH]{NH_3/\text{weakly acidic buffer}} CH_3(CH_2)_3CH_2NH_2$

22.38

(a) CH₃CH₂NHC(=O)CH₃

(b) 3-methylaniline (NH₂ on benzene with CH₃ meta)

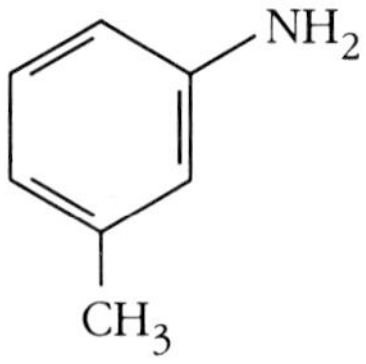

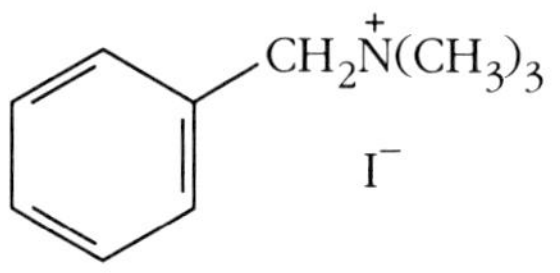

(c) benzyl-N⁺(CH₃)₃ I⁻

(d) C₆H₅CH₂NHCH₃

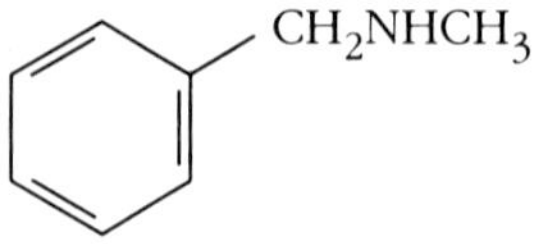

(e) (CH₃)₂CHCH₂NH₂

(f) (CH₃)₂CHCH₂NH₂

(g) cyclohexyl-N=C(CH₃)₂

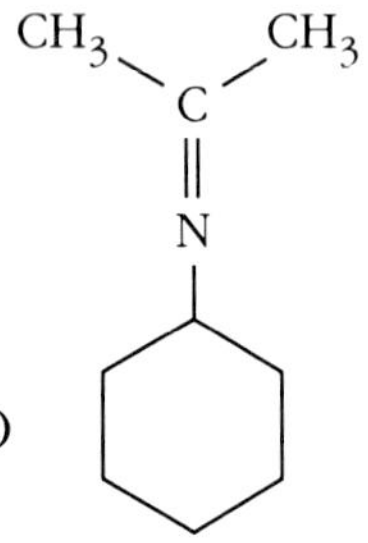

(h) 1-chloro-4-fluorobenzene

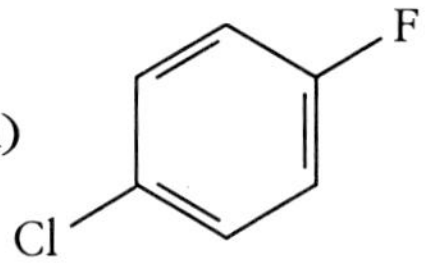

(i) N,N-diethylcyclohexylamine

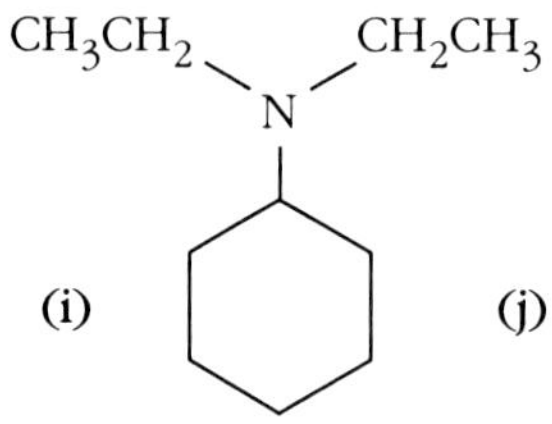

(j) CH₃(CH₂)₆CH₂NH₂

(k) CH₃CH₂CH(C≡N)CH₂CH₂CH₃

(l) C₆H₅N(CH₃)C(=O)CH₃

(m) 3-bromo-5-nitrophenylacetic acid —COOH is an electron withdrawing group

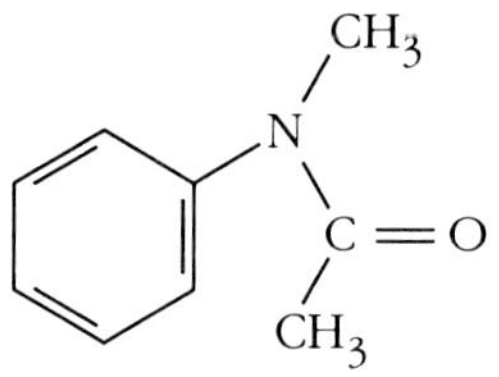

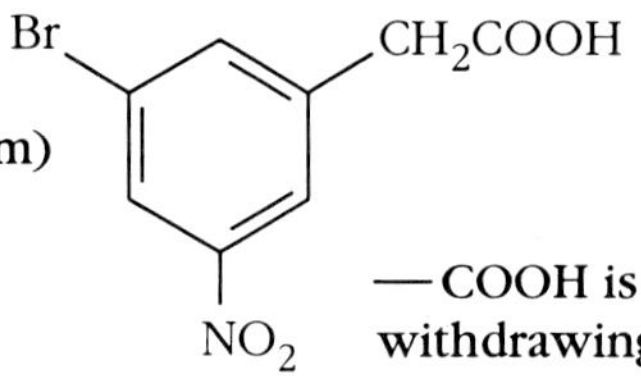

(n) 1-(2-aminoethyl)cyclohexene

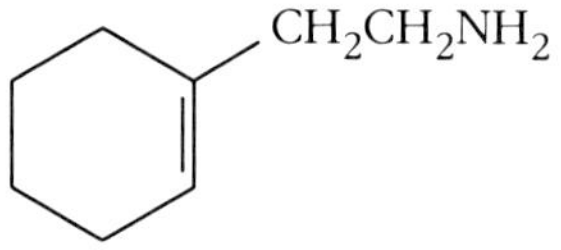

(o) 4-nitro-N-acetylaniline (O₂N-C₆H₄-NHC(=O)CH₃)

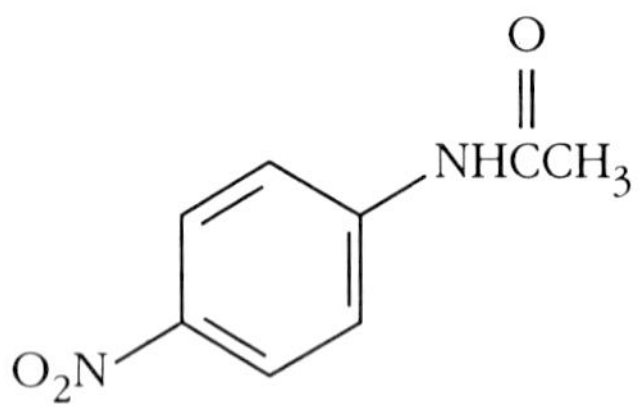

22.39

(a) $CH_3CH_2C(=O)Y + 2(CH_3)_2CHNH_2 \longrightarrow CH_3CH_2C(=O)NHCH(CH_3)_2 + (CH_3)_2CH\overset{+}{N}H_3Cl^-$

Y = Cl, $OC(=O)CH_2CH_3$, OR or SR

(b) 3-methylaniline $\xrightarrow[(2)\ CuI]{(1)\ NaNO_2/HCl}$ 3-iodotoluene

(c) $CH_3(CH_2)_4CHO \xrightarrow[NaBH_3CN/CH_3OH]{NH_3/\text{weak acid catalyst}} CH_3(CH_2)_5NH_2$

(d) $C_6H_5CH_2NH_2 \xrightarrow{\text{Excess } CH_3CH_2I} C_6H_5CH_2\overset{+}{N}(CH_2CH_3)_3\ I^-$

(e) 4-chloronitrobenzene $\xrightarrow[(2)\ NaOH/H_2O]{(1)\ Zn/HCl}$ 4-chloroaniline

(f) 3,4-methylenedioxy-β-nitrostyrene $\xrightarrow{H_2 / Pd}$ 3,4-methylenedioxyphenethylamine

(g) PhNHCH₃ →(NaNO₂/HCl) Ph-N(CH₃)-N=O

(h) cyclohexyl-NH₂ + PhSO₂Cl → PhSO₂NH-cyclohexyl + cyclohexyl-NH₃⁺

(i) Ph-N⁺≡N Cl⁻ + PhOH → Ph-N=N-C₆H₄-OH (para)

22.40

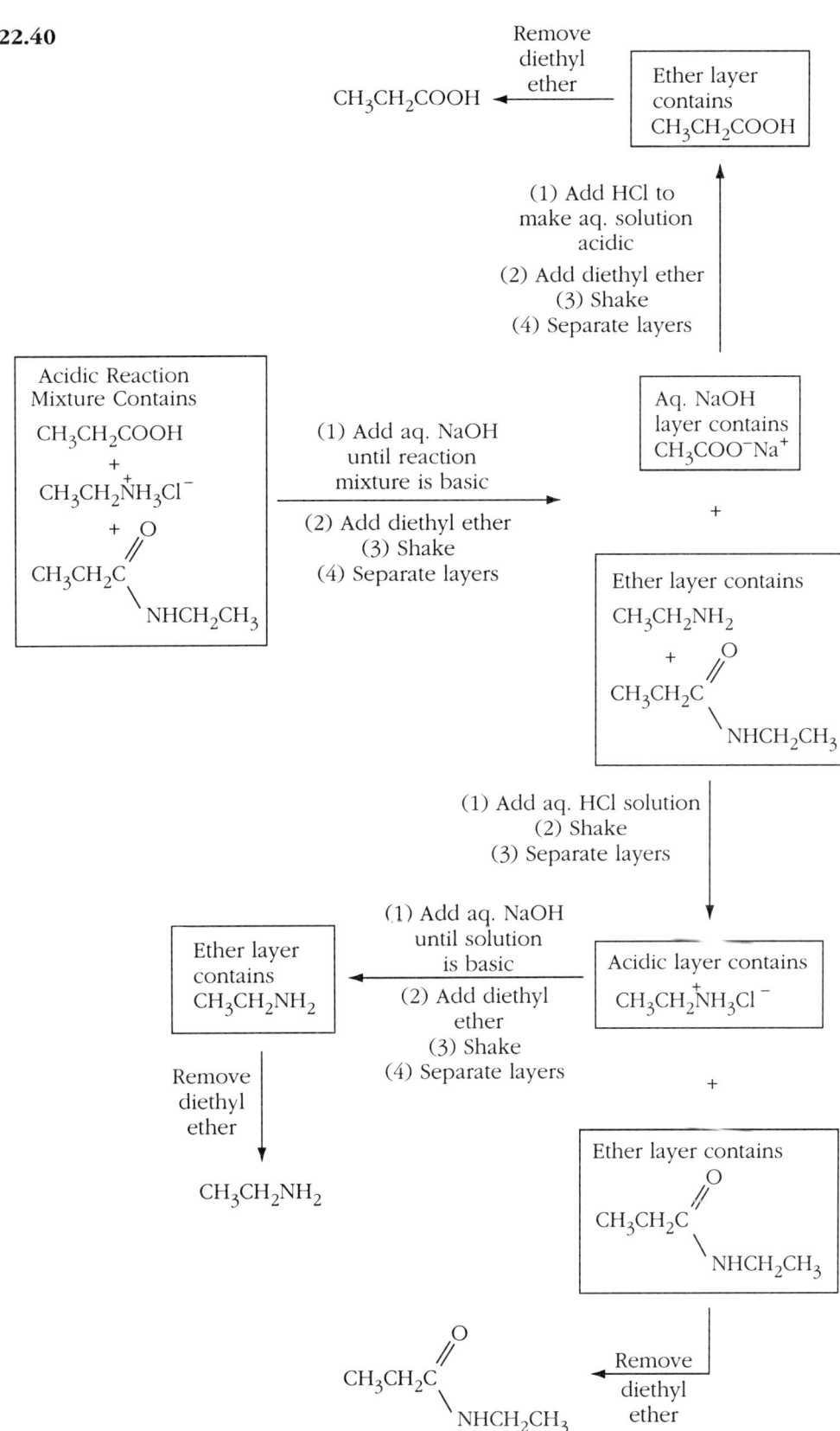

22.41

$$CH_3CH_2CH_2CH_2NH_2 \xrightarrow[H_2O]{NaNO_2/HCl} CH_3CH_2CH-CH_2-\overset{+}{N}\equiv N: \longrightarrow$$
$$\underset{H}{|}$$

$$CH_3CH_2\overset{+}{C}HCH_3 \;+\; :N\equiv N:$$

Product-Forming Steps

(1) $CH_3CH_2CH_2CH_2-\overset{+}{N}\equiv N: \;+\; H_2\ddot{O} \longrightarrow CH_3CH_2CH_2CH_2-\overset{H}{\underset{+}{\ddot{O}}}\cdots H-\ddot{O}-H$

$$\rightleftharpoons CH_3CH_2CH_2CH_2-\ddot{O}\overset{H}{\underset{\cdot\cdot}{}}$$

(2) $CH_3CH_2CH_2CH_2-\overset{+}{N}\equiv N: \longrightarrow CH_3CH_2CH_2CH_2Cl \;+\; :N\equiv N:$
    Cl^-

(3) $CH_3CH_2CH-CH_2-\overset{+}{N}\equiv N: \longrightarrow CH_3CH_2CH=CH_2$
    $\underset{H}{|}$
    $B:\qquad B: = Cl^-, H_2O, NO_2^-$

(4) $CH_3CH_2\overset{+}{C}HCH_3 \;+\; H_2\ddot{O} \longrightarrow CH_3CH_2CHCH_3 \rightleftharpoons CH_3CH_2CHCH_3$
    $\qquad\qquad\qquad\qquad\qquad\qquad\overset{+}{O} \cdots H-\ddot{O}-H \qquad\qquad |$
    $\qquad\qquad\qquad\qquad\qquad H\;\;\;H\qquad\qquad\qquad\qquad OH$

(5) $CH_3CH_2\overset{+}{C}HCH_3 \longrightarrow CH_3CH_2CHCH_3$
    $Cl^-\qquad\qquad\qquad\qquad\qquad |$
    $\qquad\qquad\qquad\qquad\qquad Cl$

(6) $Cl^- \; H \; \overset{+}{C}-CH_3 \longrightarrow \underset{CH_3}{\overset{H}{}}C=C\underset{CH_3}{\overset{H}{}}$
    $\quad H \; \; CH_3$

(7) $Cl^- \; H \; \overset{+}{C}-CH_3 \longrightarrow \underset{H}{\overset{CH_3}{}}C=C\underset{CH_3}{\overset{H}{}}$
    $CH_3 \; H$

22.42

(a) Retrosynthesis: 2-methyl-N,N-diethylcyclohexylamine ⟹ 2-methylcyclohexanone ⟹ cyclohexanone

Cyclohexanone + pyrrolidine (H on N) —[Acid catalyst]→ 1-pyrrolidinyl-cyclohexene —[CH₃I]→ iminium salt (with CH₃) —[H₂O]→ 2-methylcyclohexanone

2-methylcyclohexanone —[(CH₃CH₂)₂NH/acid catalyst; NaBH₃CN/CH₃OH]→ 2-methyl-N,N-diethylcyclohexylamine

An acceptable alternative synthesis is:

Cyclohexanone —[(1) LDA, (2) CH₃I]→ 2-methylcyclohexanone —[(CH₃CH₂)₂NH/acid catalyst; NaBH₃CN/CH₃OH]→ 2-methyl-N,N-diethylcyclohexylamine

(b) Retrosynthesis: p-NHCOCH₃-toluene ⟹ p-NH₂-toluene ⟹ p-NO₂-toluene ⟹ toluene

Toluene —[HNO₃, H₂SO₄]→ p-nitrotoluene —[Sn/HCl]→ p-toluidine —[(CH₃CO)₂O]→ p-acetamidotoluene

(c) $H_2NCH_2CH_2CH_2CH_2NH_2 \implies BrCH_2CH_2CH_2CH_2Br \implies$ tetrahydrofuran

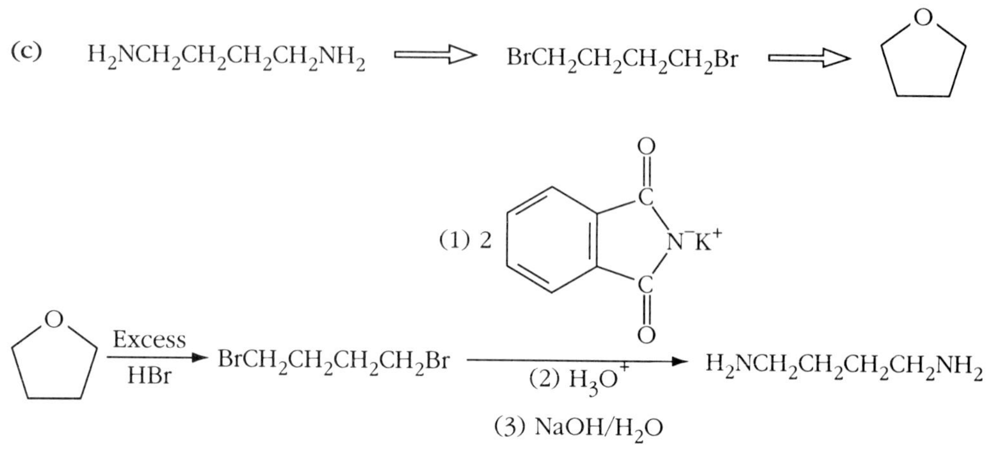

tetrahydrofuran $\xrightarrow{\text{Excess HBr}}$ $BrCH_2CH_2CH_2CH_2Br$ $\xrightarrow[\text{(2) } H_3O^+]{\text{(1) 2 potassium phthalimide}}$ $H_2NCH_2CH_2CH_2CH_2NH_2$
(3) NaOH/H$_2$O

(d) $CH_3CH_2CH_2CHNH_2$ (with CH_2CH_3) $\implies CH_3CH_2CH_2\overset{O}{\overset{\|}{C}}CH_2CH_3 \implies$

$CH_3CH_2CH_2\overset{O}{\overset{\|}{C}}Cl \implies CH_3CH_2CH_2COOH$

$\Downarrow$

$CH_3CH_2CH_2CH_2OH$

$\Downarrow$

$CH_3CH_2CH_2CH_2Br$

$CH_3(CH_2)_2CH_2Br \xrightarrow{\text{NaOH}/H_2O} CH_3(CH_2)_2CH_2OH \xrightarrow{CrO_3/H_2SO_4} CH_3(CH_2)_2COOH \xrightarrow{SOCl_2}$

$CH_3(CH_2)_2\overset{O}{\overset{\|}{C}}Cl \xrightarrow{(CH_3CH_2)_2CuLi} CH_3(CH_2)_2\overset{O}{\overset{\|}{C}}CH_2CH_3 \xrightarrow[\text{NaBH}_3\text{CN}]{NH_3}$

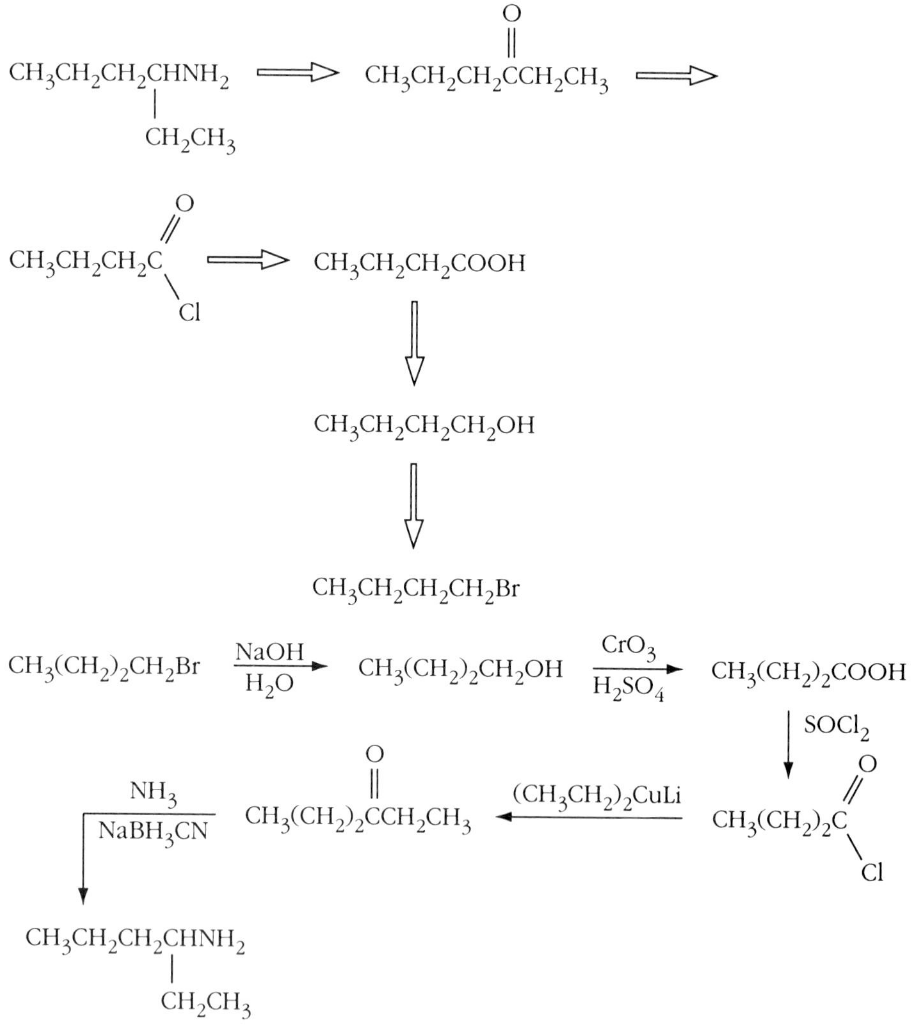

$CH_3CH_2CH_2CHNH_2$
    |
    CH_2CH_3

22.43

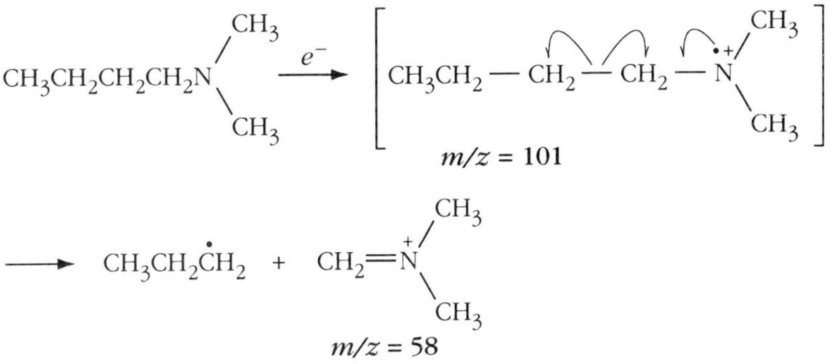

$$CH_3CH_2CH_2CH_2N(CH_3)_2 \xrightarrow{e^-} \left[CH_3CH_2-CH_2-CH_2-\overset{\bullet+}{N}(CH_3)_2 \right]$$

$m/z = 101$

$$\longrightarrow CH_3CH_2\overset{\bullet}{C}H_2 + CH_2=\overset{+}{N}(CH_3)_2$$

$m/z = 58$

22.44

(a) 3-chloroaniline $\xrightarrow{\text{NaNO}_2, \text{HCl/H}_2\text{O}}$ 3-chlorobenzenediazonium $\xrightarrow{\text{CuBr}}$ 1-bromo-3-chlorobenzene

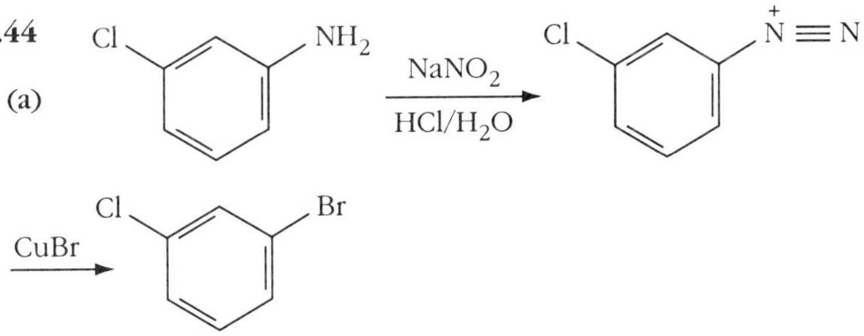

(b) 3-chloroaniline $\xrightarrow[\text{(2) CuCN}]{\text{(1) NaNO}_2/\text{HCl/H}_2\text{O}}$ 3-chlorobenzonitrile

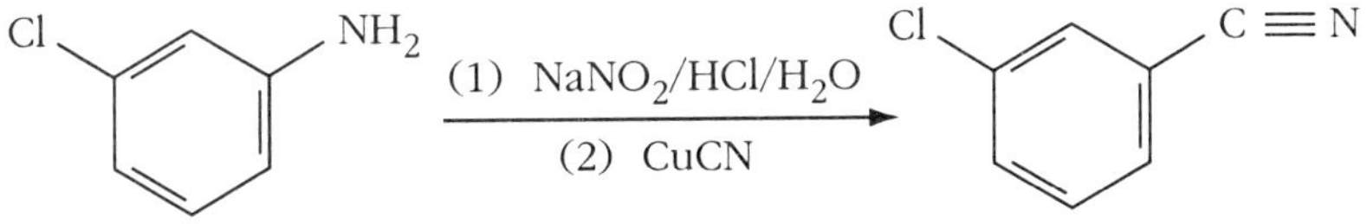

(c) 3-chloroaniline $\xrightarrow{\text{NaNO}_2/\text{HCl/H}_2\text{O}}$ 3-chlorophenol

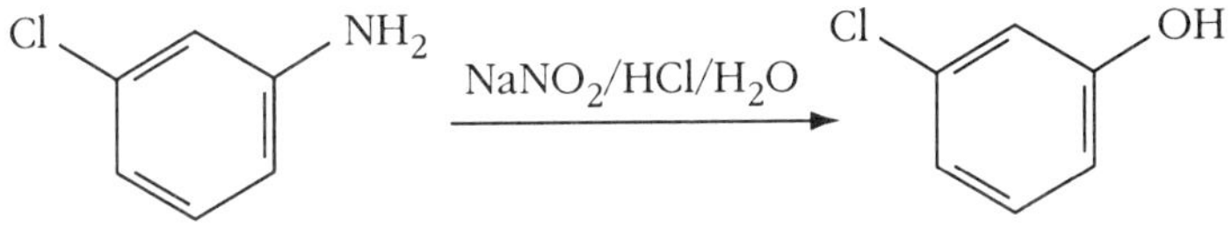

(d) 3-chlorobenzonitrile (Obtained as in 22.44 b) $\xrightarrow{\text{H}_3\text{O}^+}$ 3-chlorobenzoic acid $\xrightarrow[\text{Acid catalyst}]{\text{CH}_3\text{OH}}$ methyl 3-chlorobenzoate

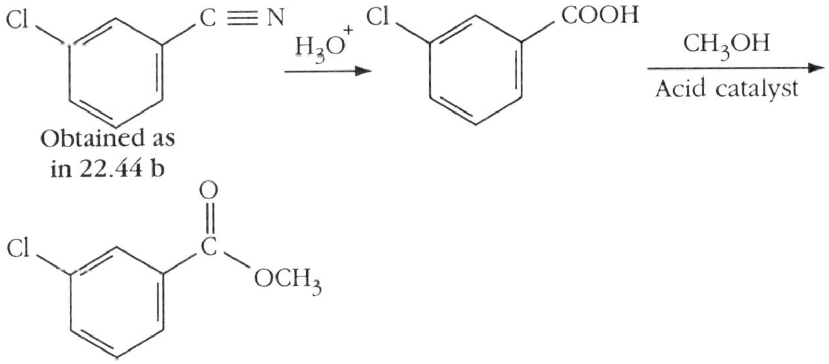

(e) 3-chlorobenzonitrile $\xrightarrow[\text{(3) NaOH/H}_2\text{O}]{\text{(1) LiAlH}_4, \text{(2) H}_3\text{O}^+}$ 3-chlorobenzylamine

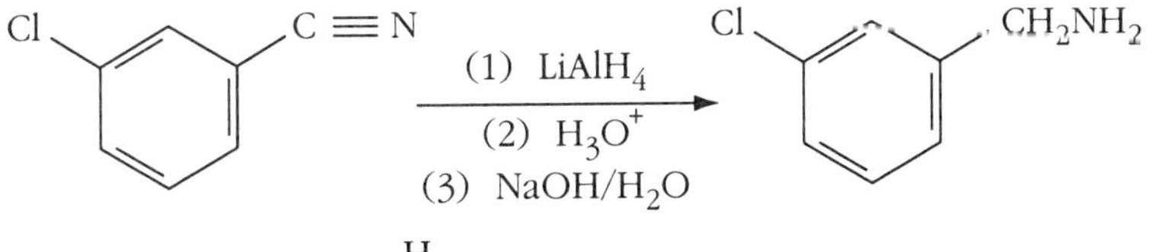

(f) 3-chloroaniline $\xrightarrow[\text{NaBH}_3\text{CN/CH}_3\text{OH}]{\text{HCHO/acid catalyst}}$ N-methyl-3-chloroaniline

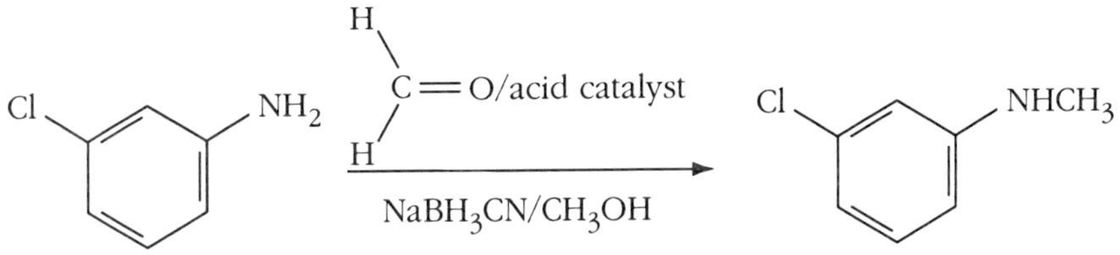

(g) 3-chloroaniline + (CH$_3$CO)$_2$O → 3'-chloroacetanilide → HNO$_3$/H$_2$SO$_4$ → 3-chloro-4-nitroacetanilide → NaOH/H$_2$O → 3-chloro-4-nitroaniline

(h) 3'-chloroacetanilide + HOSO$_2$Cl → 3-chloro-4-(chlorosulfonyl)acetanilide

Obtained as in 22.44 g

3,5-dimethylaniline → [chloro-acetanilide-sulfonamide intermediate] → NaOH/H$_2$O → [aminated sulfonamide product]

22.45

(a) cyclopentenyl-CH$_2$NH$_2$ ⟹ cyclopentenyl-CN ⟹ cyclopentenyl-Br ⟹

cyclopentene ⟹ bromocyclopentane

bromocyclopentane — NaOC$_2$H$_5$ / CH$_3$CH$_2$OH → cyclopentene — NBS / CCl$_4$/heat → 3-bromocyclopentene — NaCN / CH$_3$CH$_2$OH →

3-cyanocyclopentene — (1) LiAlH$_4$ (2) H$_3$O$^+$ (3) NaOH/H$_2$O → 3-(aminomethyl)cyclopentene

(b) Retrosynthesis:

4'-fluoropropiophenone ⇒ 4-propionyl benzenediazonium ⇒ 4'-aminopropiophenone ⇒ 4'-(propanamido)propiophenone ⇒

aniline ⇒ nitrobenzene ⇒ benzene

Forward synthesis:

benzene —HNO$_3$/H$_2$SO$_4$→ nitrobenzene —Sn/HCl→ aniline —Excess CH$_3$CH$_2$C(O)Cl / AlCl$_3$→

4'-(propanamido)propiophenone —H$_2$O / NaOH→ 4'-aminopropiophenone —(1) NaNO$_2$/HCl/H$_2$O (2) HBF$_4$/heat→ 4'-fluoropropiophenone

(c) Retrosynthesis:

2-bromotoluene ⇒ 2-bromo-4-diazonium toluene ⇒ 2-bromo-4-aminotoluene ⇒ 2-bromo-4-nitrotoluene ⇒ 4-nitrotoluene

Synthesis:

4-nitrotoluene $\xrightarrow{\text{Br}_2/\text{FeBr}_3}$ 2-bromo-4-nitrotoluene $\xrightarrow[\text{HCl}]{\text{Sn}}$ 2-bromo-4-aminotoluene $\xrightarrow[(2)\ \text{H}_3\text{PO}_2]{(1)\ \text{NaNO}_2/\text{HCl}}$ 2-bromotoluene

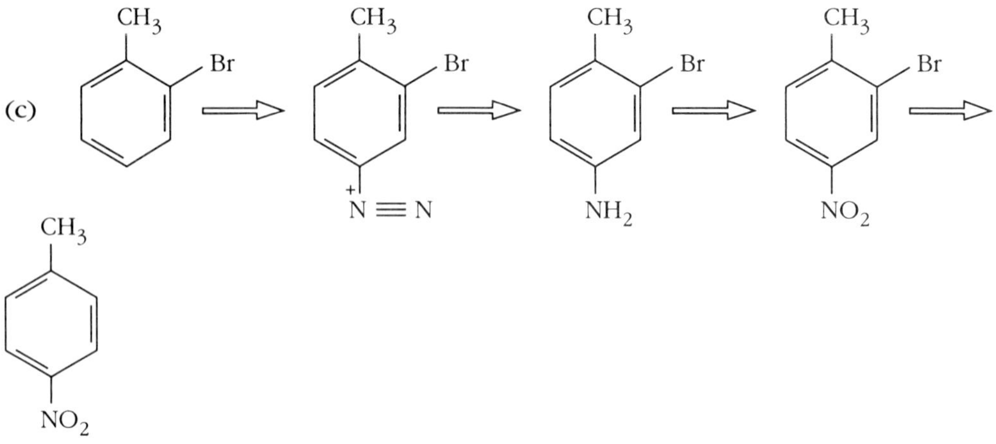

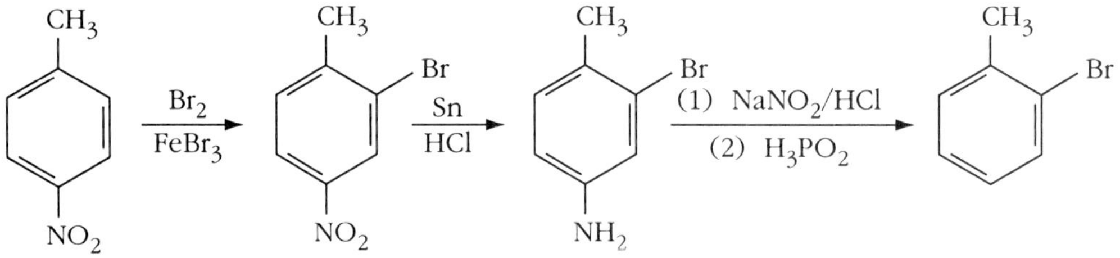

(d) Retrosynthesis:

3,4-dihydroxyphenylpropanoic acid ⇒ 3,4-dihydroxycinnamic acid ⇒ methyl 3,4-methylenedioxycinnamate ⇒ 3,4-methylenedioxybenzaldehyde + CH$_3$C(O)OCH$_3$

Synthesis:

3,4-methylenedioxybenzaldehyde + CH$_3$C(O)OCH$_3$ $\xrightarrow[(2)\ \text{H}_3\text{O}^+]{(1)\ \text{NaOCH}_3}$ 3,4-dihydroxycinnamic acid $\xrightarrow{\text{H}_2/\text{Pd}}$ 3,4-dihydroxyphenylpropanoic acid

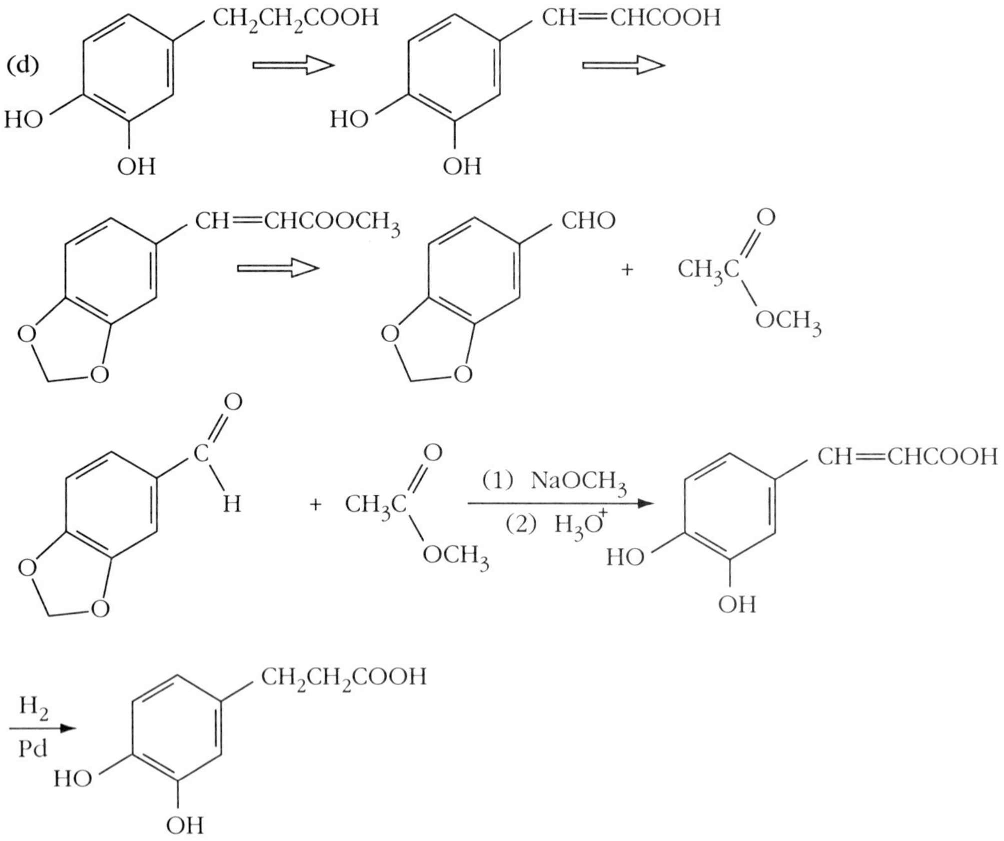

H$_2$/Pd reduces CH=CH but not COOH

(e) Retrosynthesis:

1,3-dichloro-5-nitrobenzene ⇒ 2,6-dichloro-4-nitrobenzenediazonium ⇒ 2,6-dichloro-4-nitroaniline ⇒ 4-nitroaniline

Forward synthesis:

4-nitroaniline →[Cl$_2$]→ 2,6-dichloro-4-nitroaniline →[(1) NaNO$_2$/HCl/H$_2$O; (2) H$_3$PO$_2$]→ 1,3-dichloro-5-nitrobenzene

(f) PhNH$_2$ + CH$_2$—CHCH$_3$ (epoxide, propylene oxide) ⟶ PhNHCH$_2$CH(CH$_3$)OH

(g) Retrosynthesis:

PhCH$_2$NHCH$_2$CH(CH$_3$)$_2$ ⇒ PhCH$_2$NH$_2$ + (CH$_3$)$_2$CHCHO ⇒ PhCN ⇒ PhNH$_2$ ⇒ PhNO$_2$ ⇒ PhH

Forward synthesis:

PhH →[HNO$_3$/H$_2$SO$_4$]→ PhNO$_2$ →[Sn/HCl]→ PhNH$_2$ →[(1) NaNO$_2$/HCl; (2) CuCN]→ PhCN →[(1) LiAlH$_4$; (2) H$_3$O$^+$; (3) NaOH/H$_2$O]→ PhCH$_2$NH$_2$ →[(CH$_3$)$_2$CHCHO, Acid catalyst]→ PhCH$_2$N=CHCH(CH$_3$)$_2$ →[NaBH$_3$CN]→ PhCH$_2$NHCH$_2$CH(CH$_3$)$_2$

22.46

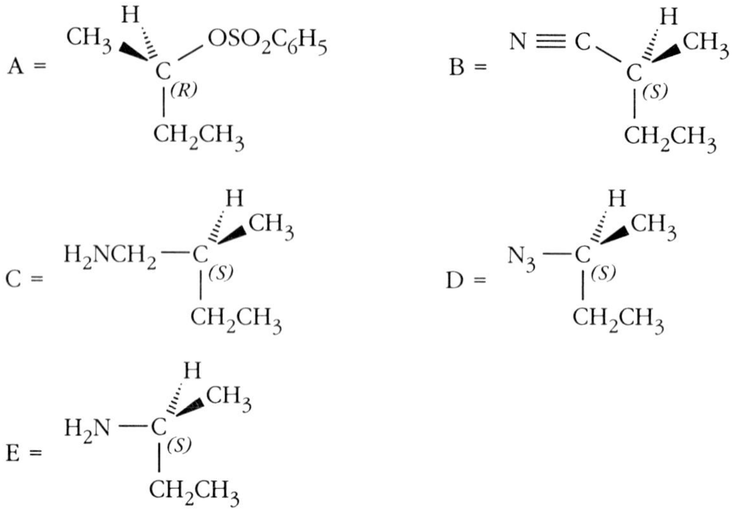

A = CH₃(H)C(R)(OSO₂C₆H₅)(CH₂CH₃)

B = N≡C–C(S)(H)(CH₃)(CH₂CH₃)

C = H₂NCH₂–C(S)(H)(CH₃)(CH₂CH₃)

D = N₃–C(S)(H)(CH₃)(CH₂CH₃)

E = H₂N–C(S)(H)(CH₃)(CH₂CH₃)

22.47

$C_6H_5NH_2 + H_2C=O$ (excess) $\xrightarrow{\text{NaBH}_3\text{CN/CH}_3\text{OH, Mild acid catalyst}}$ $C_6H_5N(CH_3)_2$

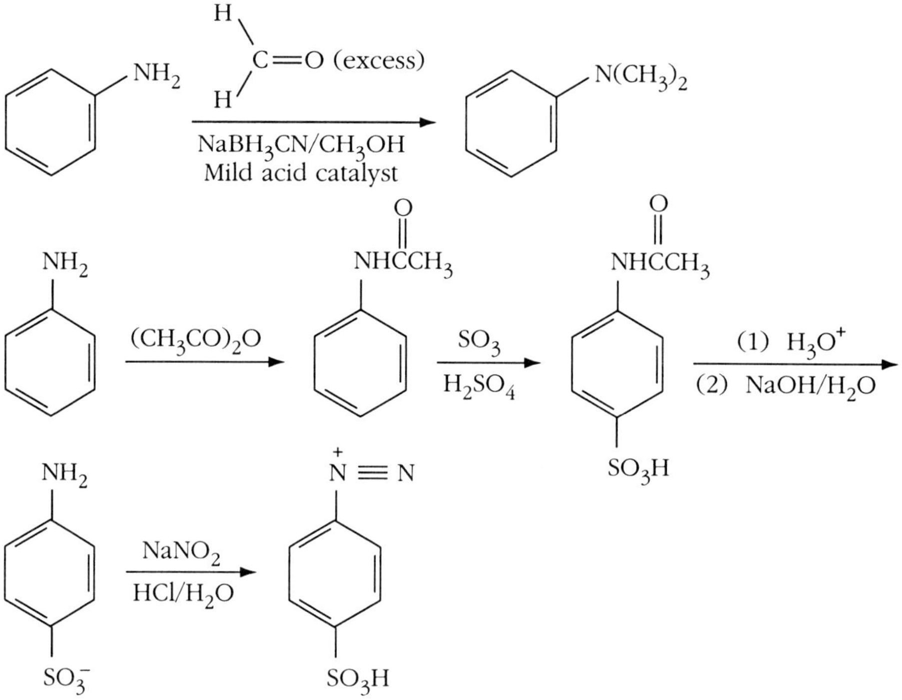

$C_6H_5NH_2 \xrightarrow{(CH_3CO)_2O} C_6H_5NHCOCH_3 \xrightarrow{SO_3 / H_2SO_4}$ 4-CH₃CONH–C₆H₄–SO₃H $\xrightarrow{(1)\ H_3O^+ \ (2)\ NaOH/H_2O}$

4-H₂N–C₆H₄–SO₃⁻ $\xrightarrow{\text{NaNO}_2 / \text{HCl}/H_2O}$ 4-(N₂⁺)–C₆H₄–SO₃H

22.48 1st Reductive Amination:

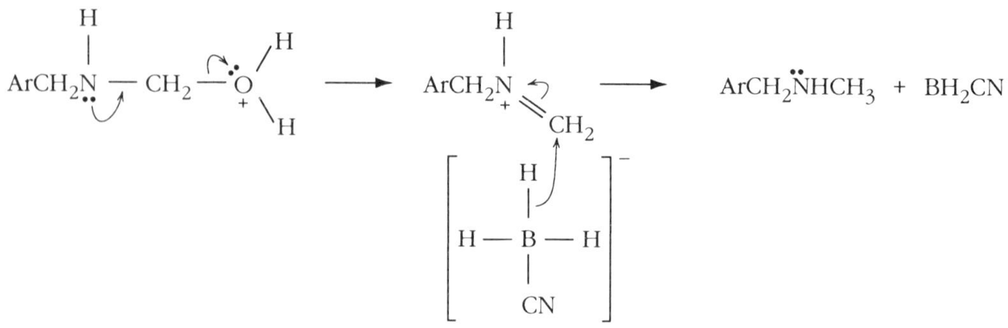

2nd Reductive Amination:

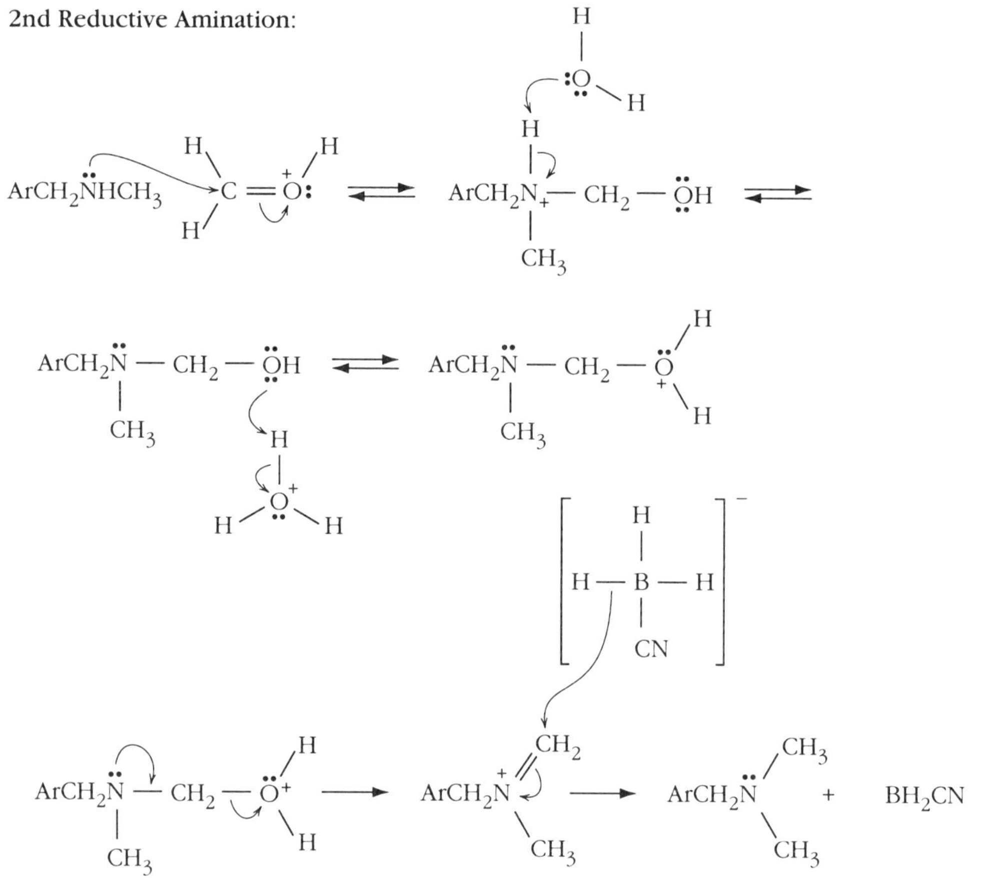

22.49 (a) The molecular formula, $C_4H_{11}N$, indicates that this is a saturated compound. The doublet at δ 1.05 (6H) and the septet at δ 2.68 (1H) are characteristic of an isopropyl group. The chemical shift of the methine hydrogen atom indicates that the isopropyl group is bonded to a nitrogen atom. The singlet at δ 2.38 (3H) indicates a methyl group that is also bonded to a nitrogen atom. The signal at δ 2.68 (1H) can be assigned to a hydrogen atom bonded to a nitrogen atom. The presence of a secondary amine is confirmed by a single absorption band at about 3300 cm^{-1} in the IR spectrum. The structure is the following:

$(CH_3)_2CHNHCH_3$
N-Methyl-2-propanamine

(b) The molecular formula indicates five degrees of unsaturation. The carbonyl absorption band in the IR spectrum and the two sets of doublets in the aromatic region (4H), which indicate a 1,4-disubstituted benzene ring, account for all five degrees of unsaturation. The two sharp absorption bands in the 3300 to 3500 cm^{-1} region of the IR spectrum indicate a primary amine, which is confirmed by the signal at about δ 4.0 (2H) in the ^{1}H NMR spectrum. The singlet at δ 3.8 indicates a methyl group bonded to an oxygen atom. The structure is the following:

Methyl 4-aminobenzoate

(c) The molecular formula indicates five degrees of unsaturation. The carbonyl absorption band in the IR spectrum and the multiplet in the aromatic region (5H), which indicates a monosubstituted benzene ring, account for all five degrees of unsaturation. The singlet at δ 8.5 (1H) indicates a hydrogen atom bonded to the carbonyl carbon atom of a formamide while the singlet at δ 3.30 (3H) indicates a methyl group isolated from any hydrogen atom to which it could be coupled. The lack of absorption bands in the 3300 to 3500 cm^{-1} region of the IR spectrum indicates a tertiary amine. The structure is the following:

N-Formyl-N-methylaniline

(d) The molecular formula indicates five degrees of unsaturation. The carbonyl absorption band in the IR spectrum and the two sets of doublets in the aromatic region (4H), which indicate a 1,4-disubstituted benzene ring, account for all five degrees of unsaturation. The chemical shift of the singlet at δ 2.15 (3H) is typical of a methyl group bonded to a carbonyl group while the chemical shift of the other singlet at δ 2.33 (3H) is typical of a methyl group bonded to a benzene ring. This gives the following partial structures:

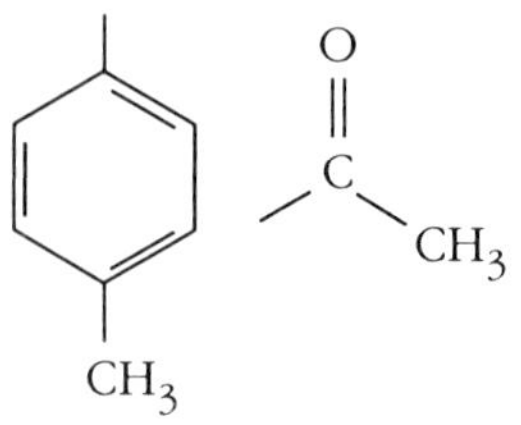

[Structure: 1,4-disubstituted benzene with CH₃ group and C(=O)CH₃ group]

These partial structures account for all of the atoms except one hydrogen atom and one nitrogen atom. Adding these two atoms to the partial structures gives the following structure:

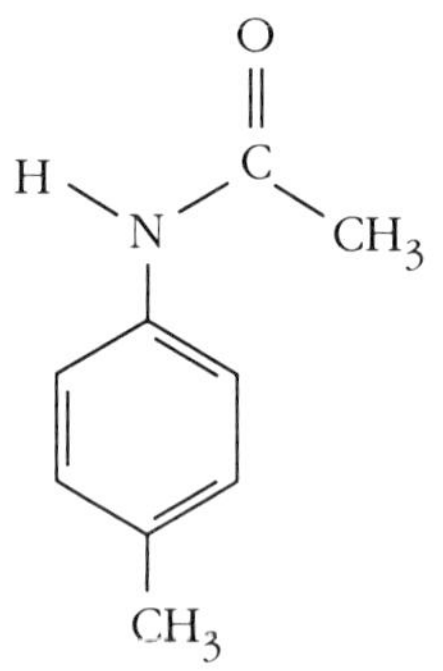

[Structure of N-Acetyl-4-methylaniline]

N-Acetyl-4-methylaniline

22.50 (a) The molecular formula indicates that this is a saturated compound. The single absorption band at 3300 cm^{-1} indicates a secondary amine. The ^{1}H NMR spectrum shows a pattern typical of an ethyl group and a singlet due to a methyl group. Thus the structure of the compound is the following:

$$CH_3NHCH_2CH_3$$
Ethylmethylamine

(b) The molecular formula indicates that this is a saturated compound. The integrated area of the signals in the ^{1}H NMR spectrum are less than the total number of hydrogen atoms in the molecular formula so the actual ratio must be a multiple of the integrated area. The ^{1}H NMR data therefore is δ 2.24 s (12H) and δ 2.39 s (4H). This molecule contains two types of hydrogen atoms. The singlet at δ 2.24 (12H) indicates four methyl groups bonded to nitrogen atoms while the singlet at δ 2.39 (4H) indicates two methylene groups bonded to nitrogen atoms. Combining these partial structures gives the following structure:

$$(CH_3)_2NCH_2CH_2N(CH_3)_2$$
N,N,N,N-Tetramethyl-1,2-ethanediamine

(c) The molecular formula indicates that this compound contains four degrees of unsaturation. The two sets of doublets at δ 6.58 (2H) and δ 6.96 (2H), which indicate a 1,4-disubstituted benzene ring, account for all four degrees of unsaturation. The two absorption bands at 3320 and 3410 cm^{-1} in the IR spectrum indicate a primary amine. The ^{1}H NMR spectrum also shows a typical ethyl group splitting pattern. Combining these partial structures gives the final structure as the following:

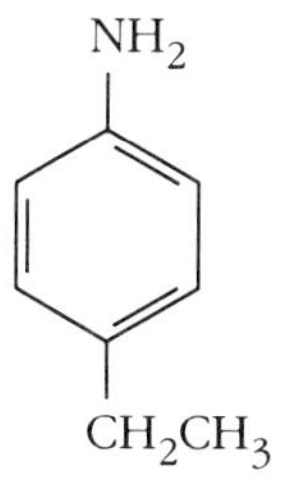

$$\underset{\text{4-Ethylaniline}}{\text{4-H}_2\text{N-C}_6\text{H}_4\text{-CH}_2\text{CH}_3}$$

(d) The molecular formula indicates that this compound has one degree of unsaturation. The absorption band at 1640 cm^{-1} in the IR spectrum and the signal at δ 179.1 in the ^{13}C NMR spectrum indicate the presence of a carbonyl group. The septet at δ 2.40 and the doublet at δ 1.06 are typical of an isopropyl group. The two absorption bands at 3330 and 3180 cm^{-1} in the IR spectrum are characteristic of two N-H bonds. This information provides us with the following partial structures:

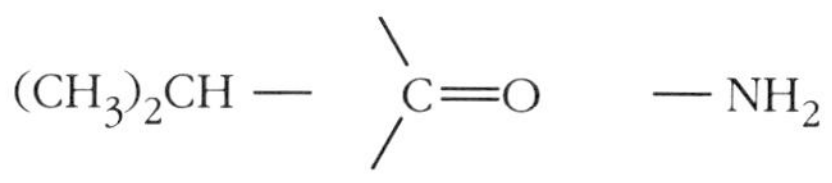

$$(CH_3)_2CH— \quad \text{C}=\text{O} \quad —NH_2$$

These are combined to give the following structure of the compound:

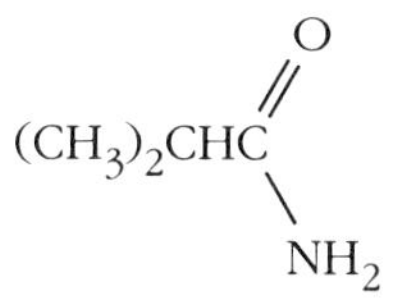

$$(CH_3)_2CHC(=O)NH_2$$

2-Methylpropanamide

EXAMINATION 3

To get the maximum benefit from the following one-hour exam, choose a quiet location and spend exactly one hour answering the questions. Check your answers against those at the end of the Study Guide to evaluate your knowledge of the material.

Chapters 15–22

Time allowed: 1 hour, Maximum: 100 points

1. *(16 points)* Name each of the following compounds according to the IUPAC rules, including stereochemistry when shown:

   (a)

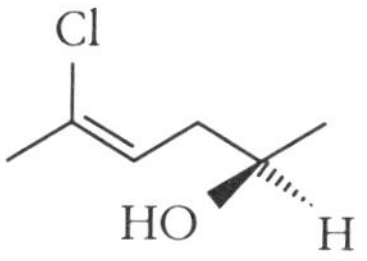

   (b)

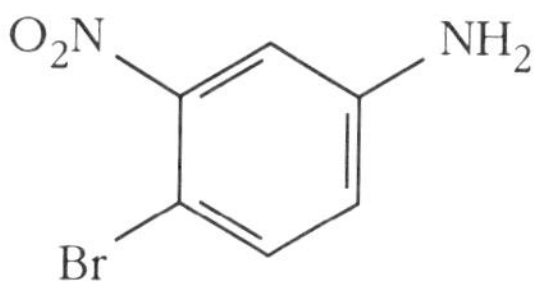

   (c)

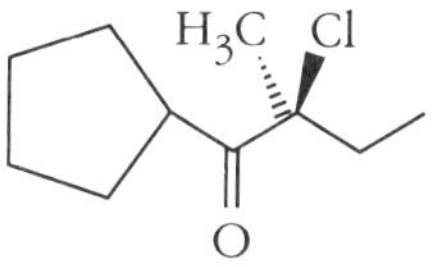

   (d)

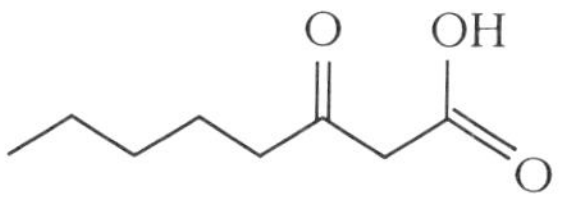

2. *(10 points)* Arrange the following compounds in order of increasing acidity (weakest acid first):

   (a) $CH_3C \equiv CH$ (b) CCl_3COOH (c)

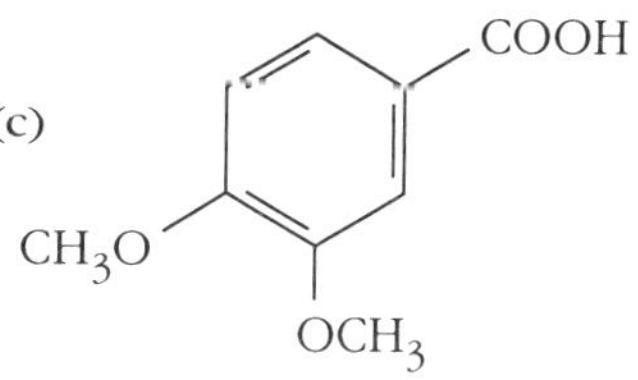

   (d) 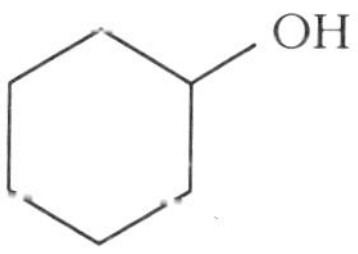   (e)

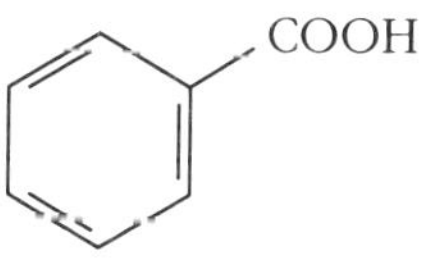

3. **(24 points)** Write structures of products, substrates, or necessary organic and inorganic reagents for the following transformations. Indicate if more than one step is needed.

(a) Cyclohexanecarbaldehyde + HOCH$_2$CH$_2$OH, Acid catalyst ⇌

(b) Furan-2-carbaldehyde + H—C≡N / NaCN →

(c) Acetone → phenylhydrazone (CH$_3$C(=N-NH-C$_6$H$_5$)CH$_2$CH$_3$ type structure) — (propiophenone phenylhydrazone)

(d) _____ $\xrightarrow{\text{(1) LiAlH}_4\text{, dry ether} \atop \text{(2) H}_2\text{O}}$ C$_6$H$_5$—CH$_2$OH

(e) Anisole → o-methoxyacetophenone + p-methoxyacetophenone

(f) Cyclohexanone $\xrightarrow[\text{(2) CH}_3\text{Br (2 moles)} \atop \text{(3) H}_3\text{O}^+]{\text{(1) pyrrolidine N—H, Acid catalyst}}$

4. **(10 points)** Using curved arrows to show the electron flow, write a detailed reaction mechanism for the following conversion. Please make sure to write all intermediates formed and indicate all reagents needed to isolate the final product.

$$CH_3CHO \xrightarrow[\text{Acidic buffer}]{CH_3NH_2} CH_3CH=NCH_3$$

5. *(30 points)* Plan a synthesis of each of the following compounds from the substrates shown and any organic reagents containing three or fewer carbon atoms, and any inorganic reagents you need. The maximum number of marks will be given for the shortest route with the highest yield.

(a)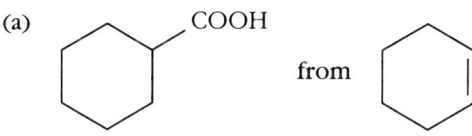

(b)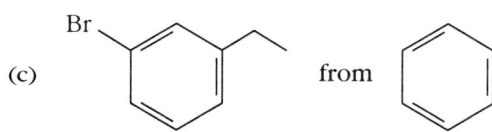

(c)

Br
  \
   ⌬—CH₂—   from   ⌬

6. *(10 points)* Write the structural formula of each of the following compounds based on its molecular formula and spectral data.

(a) $C_8H_{10}O$. IR: 3300 cm^{-1}. ^{1}H NMR: δ 1.42 d (3H); δ 2.70 broad singlet that disappears on shaking the sample with D_2O (1H); δ 4.80 q (1H); δ 7.25 broad singlet (5H).

(b) $C_5H_{13}NO_2$ IR: 3300 cm^{-1}. ^{13}C NMR: δ 43.2; δ 59.2; δ 61.0. ^{1}H NMR: δ 2.32 s (3H); δ 2.54 t (4H); δ 3.62 t (4H); δ 4.40 singlet that disappears on shaking the sample with D_2O (2H).

CHAPTER 23

HALOBENZENES, PHENOLS, AND QUINONES
Concepts

Halobenzenes are compounds that have a halogen atom bonded directly to a benzene ring. Halobenzenes, like haloalkanes, undergo nucleophilic substitution reactions, but the mechanisms of the two classes of compounds are entirely different. Nucleophilic substitution reactions of haloalkanes occur by either an S_N1 or an S_N2 mechanism. Halobenzenes that contain strongly electron-withdrawing substituents in positions *ortho-* or *para-* to the position of the leaving group easily undergo substitution by an **addition-elimination mechanism** that involves addition of the nucleophile to the halogen-bearing carbon atom to form an anionic intermediate, which then eliminates a halide ion to form the product.

Step 1: Addition of the nucleophile

Step 2: Elimination of the leaving group

Nonactivated halobenzenes react by an **elimination-addition mechanism** that involves first elimination of HX to form a **benzyne** intermediate, which then adds the nucleophile to form the product.

Step 1: Elimination to form the benzyne intermediate

Benzyne

Step 2: Addition of nucleophile

Phenols are compounds that contain a hydroxyl group bonded directly to an aromatic ring. Phenols are stronger acids than alcohols. Electron-withdrawing substituents in the ring increase the acidity of phenols. Phenols undergo many of the same reactions as alcohols. Phenols, for example, form esters and ethers. Oxidation of phenols, dihydroxybenzenes, and dihydroxypolycyclic benzenoid compounds forms **quinones.** Quinones and hydroquinones can be interconverted easily and rapidly via a reversible oxidation and reduction reaction.

Solutions to the Exercises

23.1

(a)

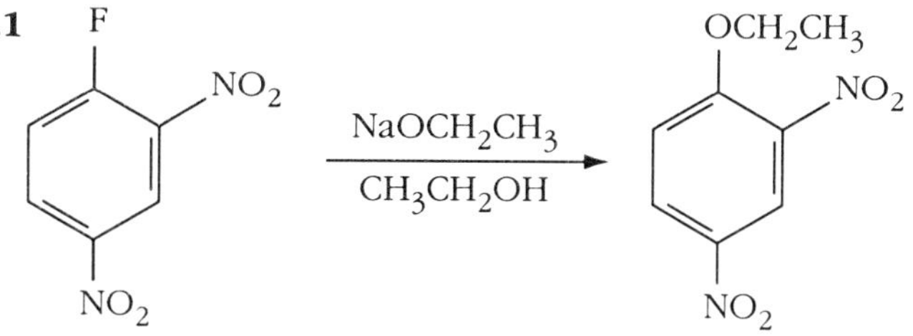
F
NO2
NaOCH2CH3
CH3CH2OH
NO2
OCH2CH3
NO2
NO2

(b)

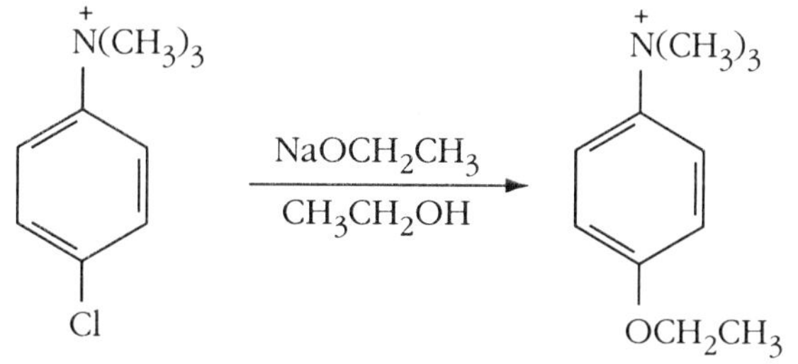
N(CH3)3
Cl
NaOCH2CH3
CH3CH2OH
N(CH3)3
OCH2CH3

(c)

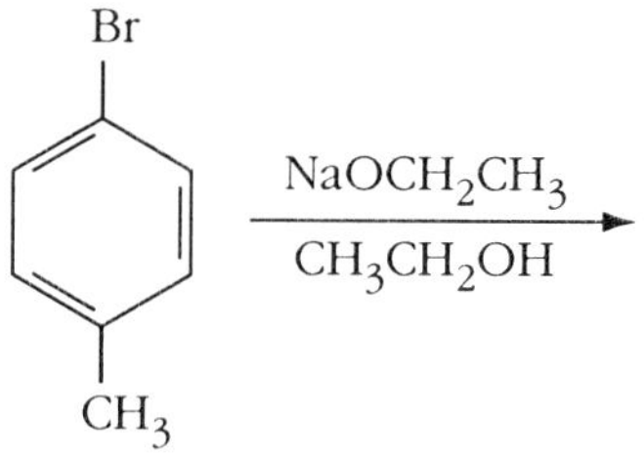
Br
CH3
NaOCH2CH3
CH3CH2OH

No reaction because molecule lacks electron-withdrawing groups to activate the benzene ring to nucleophilic aromatic substitution reactions.

(d)

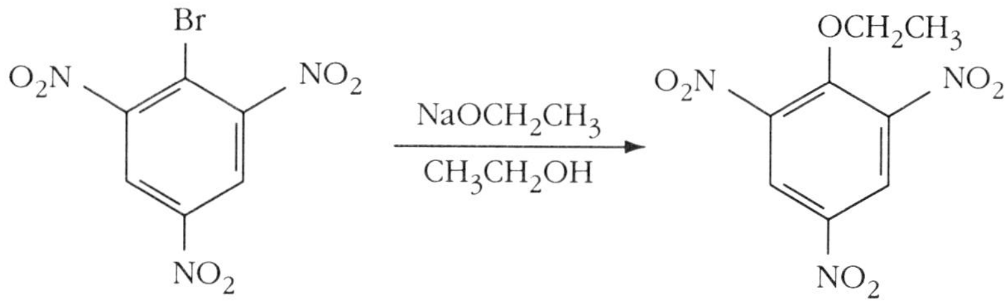
Br
O2N
NO2
NO2
NaOCH2CH3
CH3CH2OH
OCH2CH3
O2N
NO2
NO2

23.2

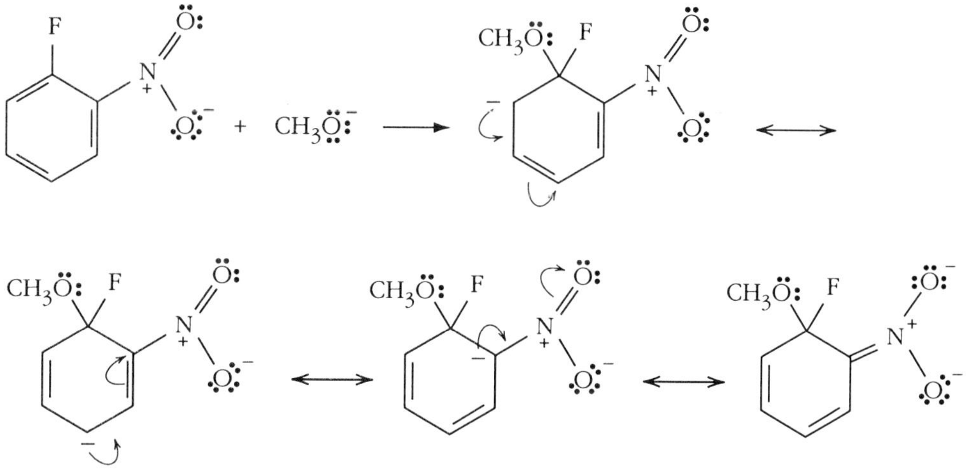

23.3

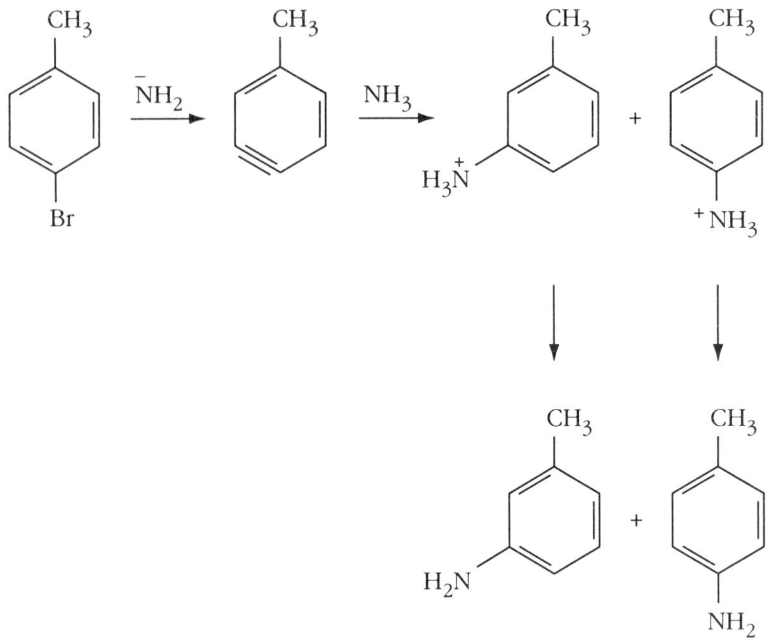

4-bromotoluene $\xrightarrow{^-NH_2}$ benzyne (with CH₃) $\xrightarrow{NH_3}$ 3-methylanilinium + 4-methylanilinium

↓ ↓

3-methylaniline (m-toluidine) + 4-methylaniline (p-toluidine)

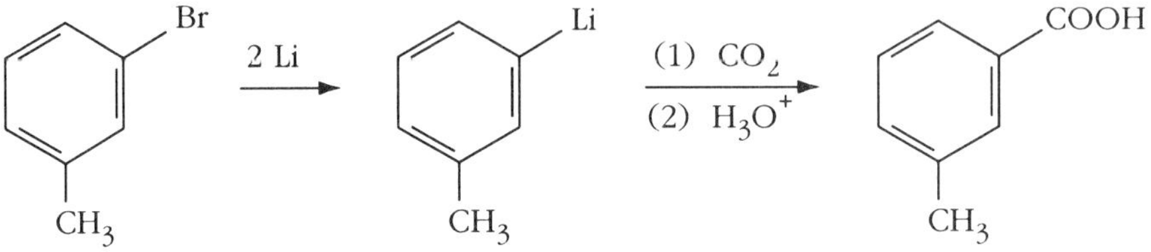

23.4

(a) 3-bromotoluene $\xrightarrow{2\ Li}$ 3-methylphenyllithium $\xrightarrow[(2)\ H_3O^+]{(1)\ CO_2}$ 3-methylbenzoic acid

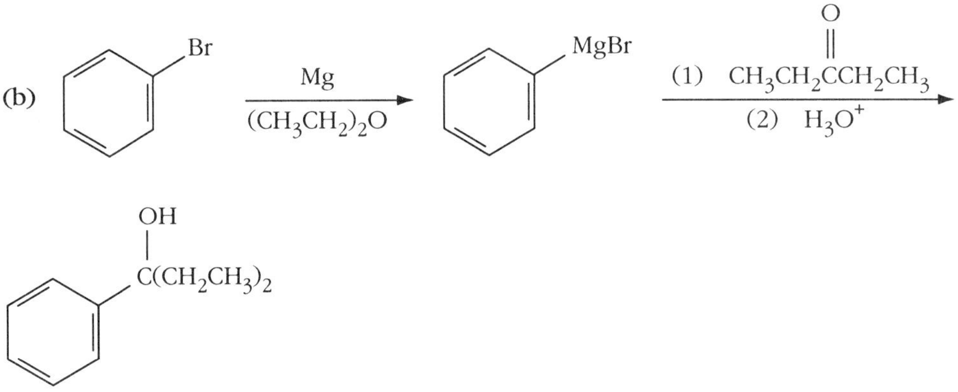

(b) bromobenzene $\xrightarrow[(CH_3CH_2)_2O]{Mg}$ PhMgBr $\xrightarrow[(2)\ H_3O^+]{(1)\ CH_3CH_2\overset{O}{\underset{\|}{C}}CH_2CH_3}$

Ph–C(OH)(CH₂CH₃)₂

(c) [PhBr] →(2 Li)→ [PhLi] →(1) CH₂—CHCH₃ epoxide; (2) H₃O⁺→ PhCH₂CH(OH)CH₃

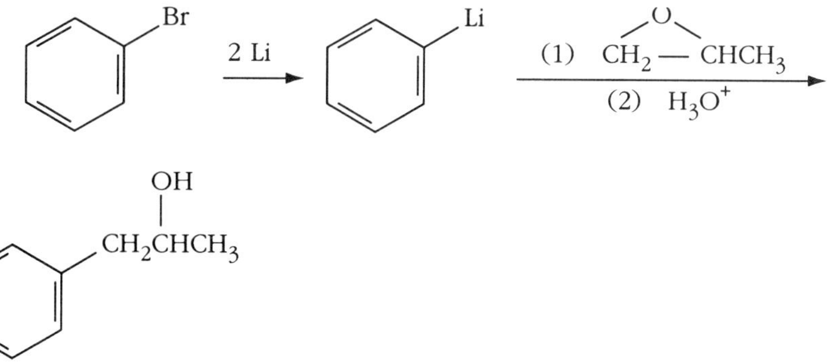

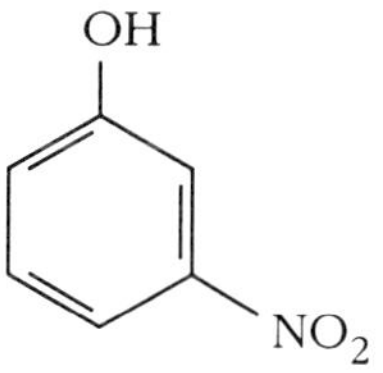

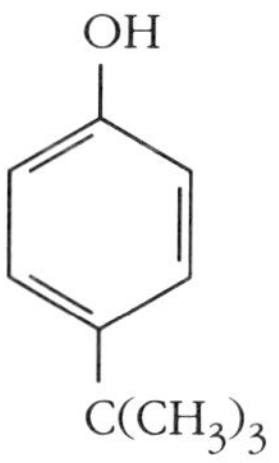

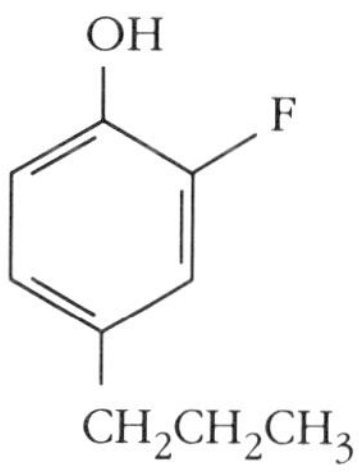

23.5
(a) 3-nitrophenol
(b) 4-tert-butylphenol
(c) 2-fluoro-4-propylphenol
(d) 3-isopropoxyphenol
(e) (1-ethylpropoxy)benzene

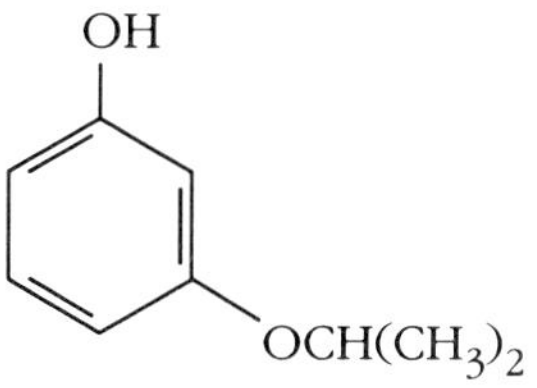

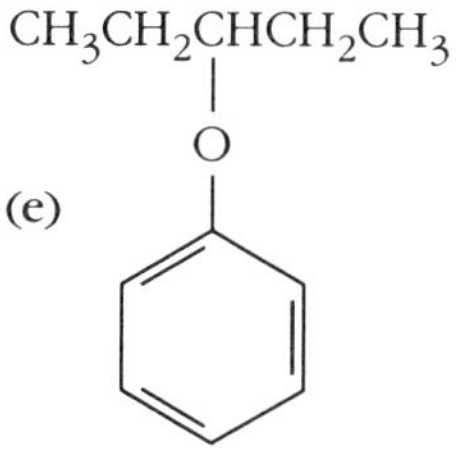

23.6
(a) 3-Ethylphenol
(b) 2-Hydroxybenzaldehyde
(c) 3-Chloro-5-hydroxyacetophenone
(d) Propoxybenzene

23.7 The biosynthetic analog of the Claisen condensation occurs by way of a carbanion-like mechanism even though a carbanion is not formed in the mechanism.

23.8 Claisen-like condensation:

Aldol-like condensation:

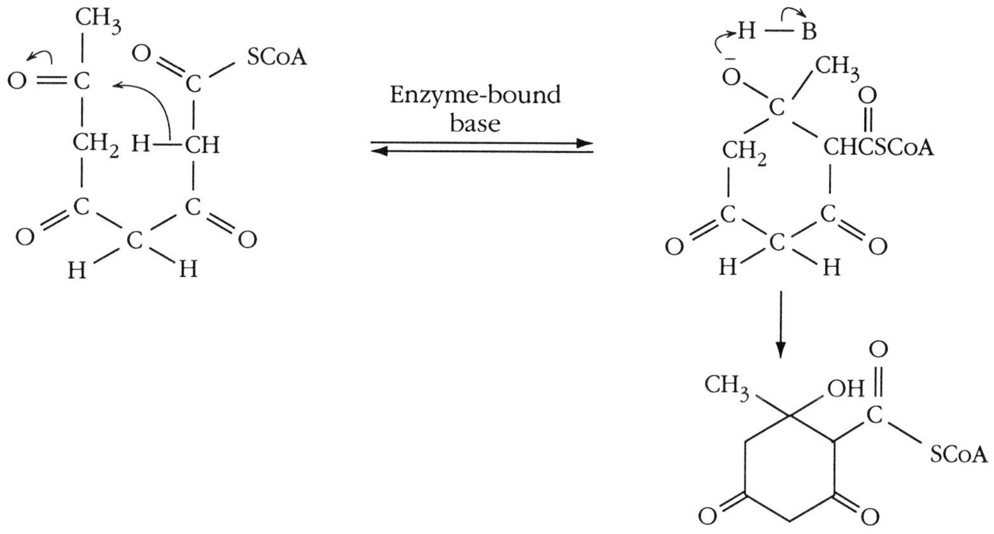

23.9

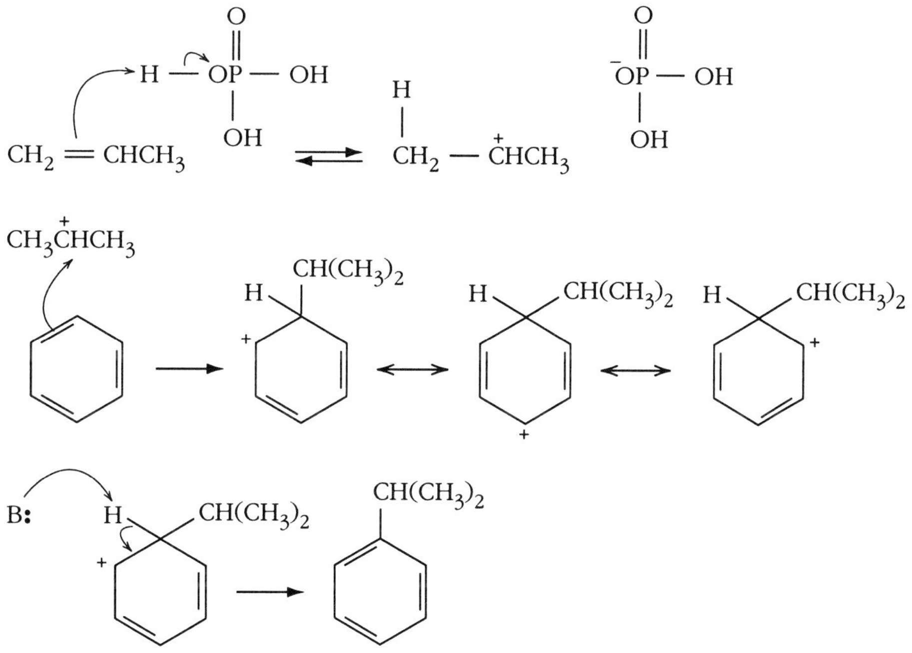

23.10

(a) PhOH + HCO$_3^-$ ⇌ PhO$^-$ + H$_2$CO$_3$ (H$_2$O + CO$_2$) K_{eq}

$$K_{eq} = \frac{K_a \text{(phenol)}}{K_a \text{(H}_2\text{CO}_3\text{)}} = \frac{1.0 \times 10^{-10}}{4.5 \times 10^{-7}} = 2.3 \times 10^{-4}$$

(b) $K_{eq} = \dfrac{5.5 \times 10^{-5}}{4.5 \times 10^{-7}} = 122$

(c) $K_{eq} = \dfrac{1.0 \times 10^{-10}}{4.7 \times 10^{-11}} = 2.1$

(d) $K_{eq} = \dfrac{1.77}{4.5 \times 10^{-7}} = 3.9 \times 10^6$

23.11

⎯⎯ Decreasing acidity ⎯⎯→

(a) Acetic acid   phenol   ethanol
    pK_a = 4.74   pK_a = 10   pK_a ≈ 16

(b) Electron-withdrawing substituents increase acidity of phenol

⎯⎯ Decreasing acidity ⎯⎯→

  4-nitrophenol   phenol   4-methoxyphenol
  pK_a = 7.15   pK_a = 10   pK_a = 10.21

(c) ⎯⎯ Decreasing acidity ⎯⎯→

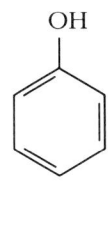

4-nitrophenol   phenol   HCO$_3^-$

(d)

◄——— Decreasing acidity ———►

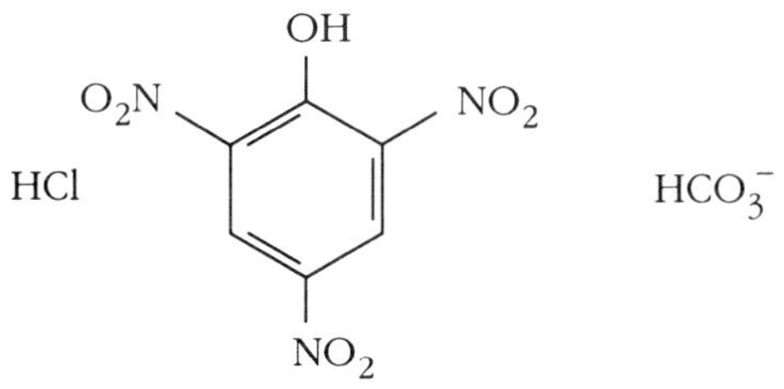

HCl   2,4,6-trinitrophenol   HCO₃⁻

23.12 Step 1: Protonation of the anhydride

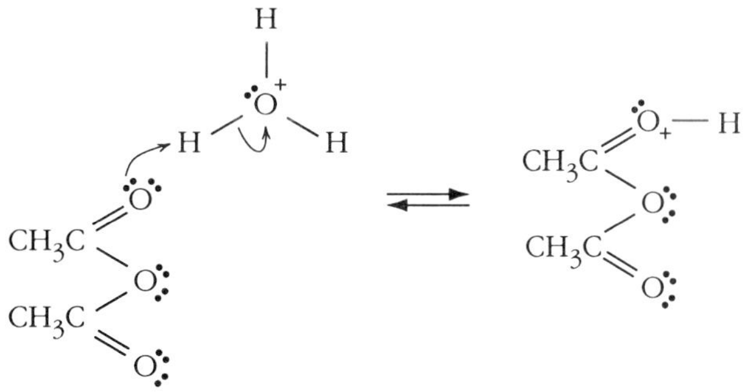

Step 2: Nucleophilic addition to the carbonyl carbon atom

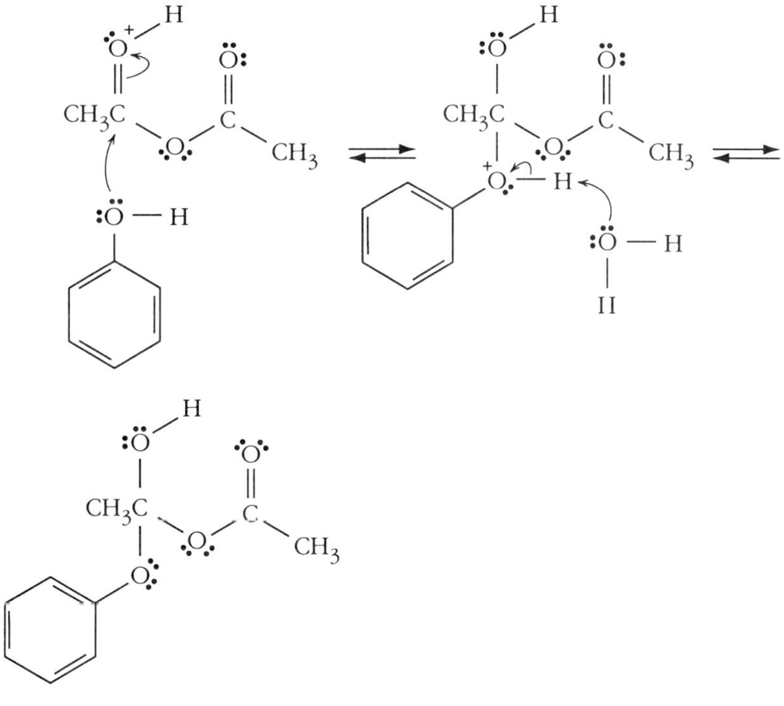

Step 3: Elimination of acetate ion

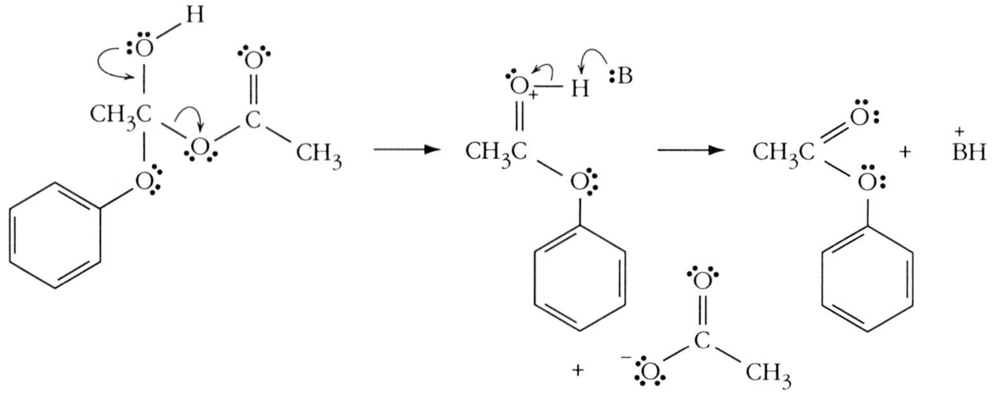

23.13 Only the two —OH groups can be acetylated because the ether oxygen atom and the tertiary amine do not react with anhydrides.

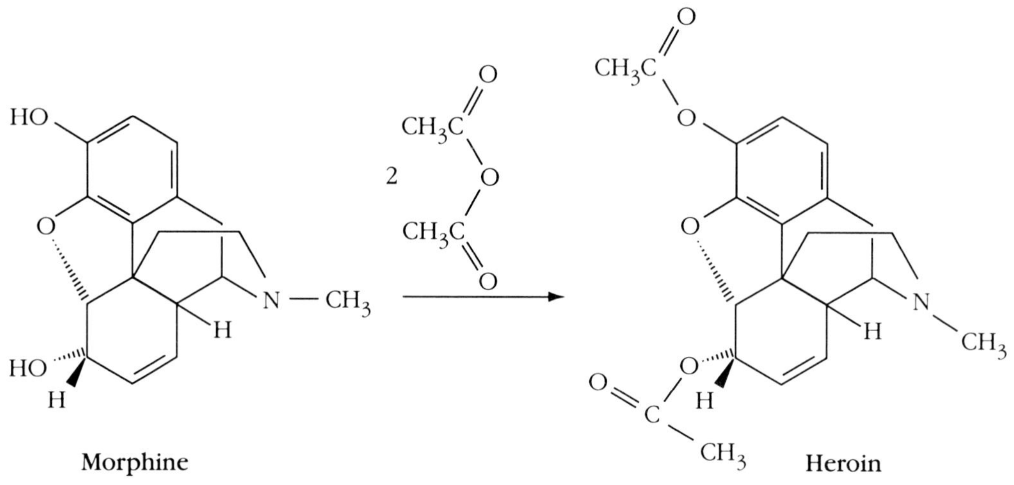

Morphine → Heroin

23.14 Anisole reacts with HI by the following mechanism:

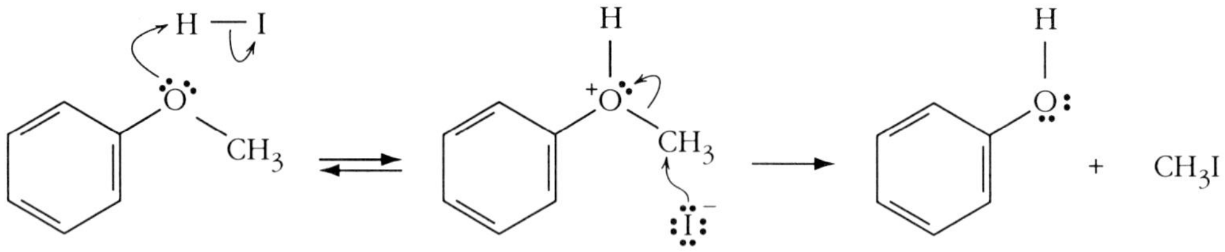

Diphenyl ether is protonated by HI but nucleophilic aromatic substitution at an unactivated benzene carbon atom is extremely difficult (Section 23.3) so the ether cleavage does not occur.

23.15 PCl$_5$ reacts with the —OH group, which is a poor leaving group, and converts it into OPCl$_4$, which is a good leaving group. The OPCl$_4$ group of the phosphate ester is replaced by chloride ion in a subsequent addition-elimination mechanism.

23.16

(b) C₆H₅OH + H₂SO₄ ⇌ C₆H₅OH₂⁺ + HSO₄⁻ → No further reaction

(c) C₆H₅OH + Cl₂ —H₂O→ 2,4,6-trichlorophenol

(d) C₆H₅OH + Cl₂ —CH₃COOH→ 4-chlorophenol

(e) C₆H₅OH + C₆H₅SO₂Cl —NH₂R→ C₆H₅—O—SO₂—C₆H₅

23.17

$$H_2SO_4 + NaNO_2 \rightleftharpoons HNO_2 + NaHSO_4$$

$$HNO_2 + H_2SO_4 \rightleftharpoons H_2\overset{+}{N}O_2 + HSO_4^-$$

$$H_2\overset{+}{N}O_2 \rightleftharpoons \left[:\overset{+}{N}=\overset{..}{\underset{..}{O}} \longleftrightarrow :N\equiv\overset{+}{\underset{..}{O}} \right] + H_2O$$

See page 987 for the detailed mechanism.

B⁻ = HSO₄⁻ or H₂O, depending on the concentration of H₂SO₄

23.18

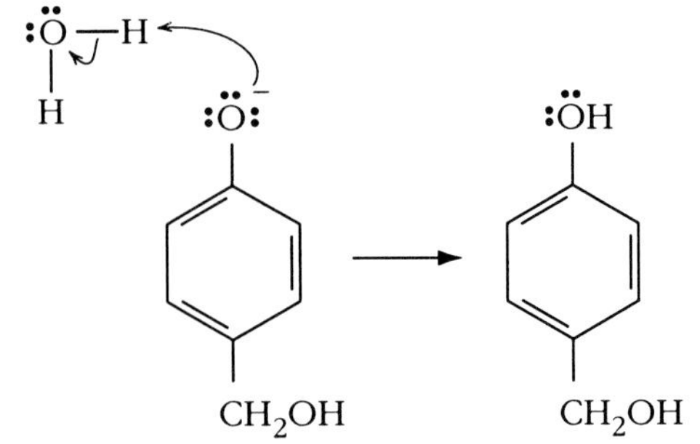

23.19

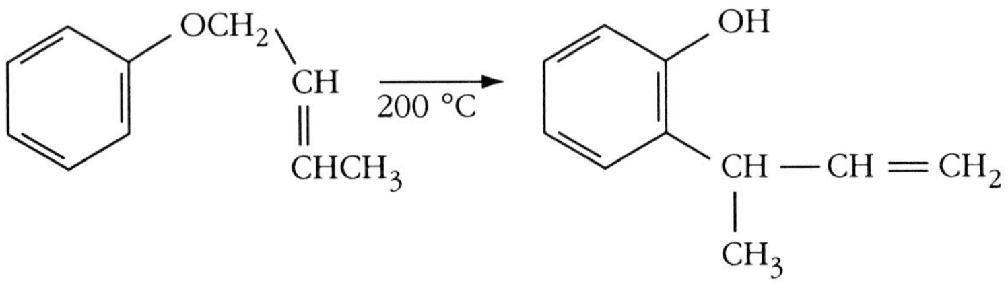

23.20

(a)

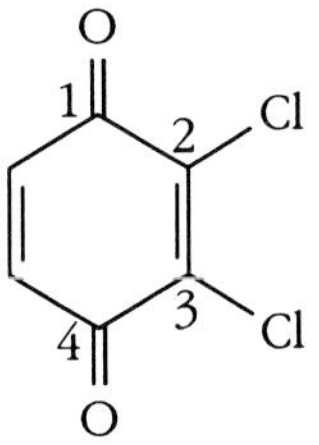

(b)

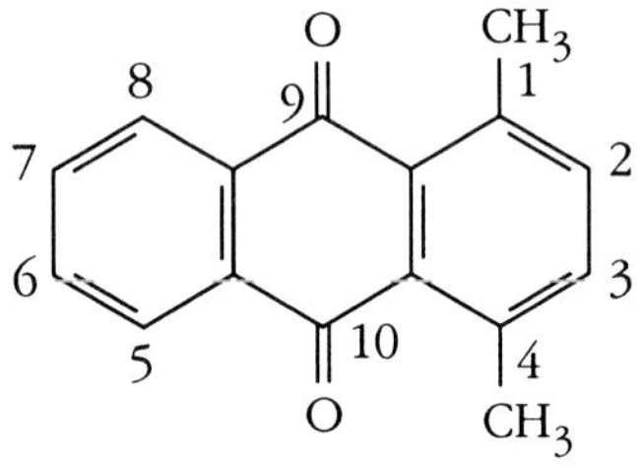

(c)

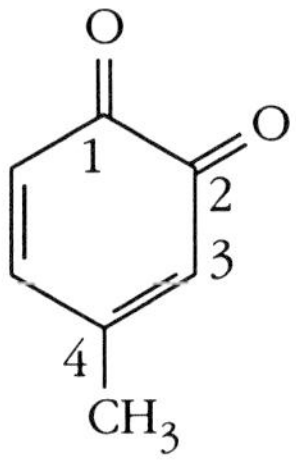

23.21

(a)

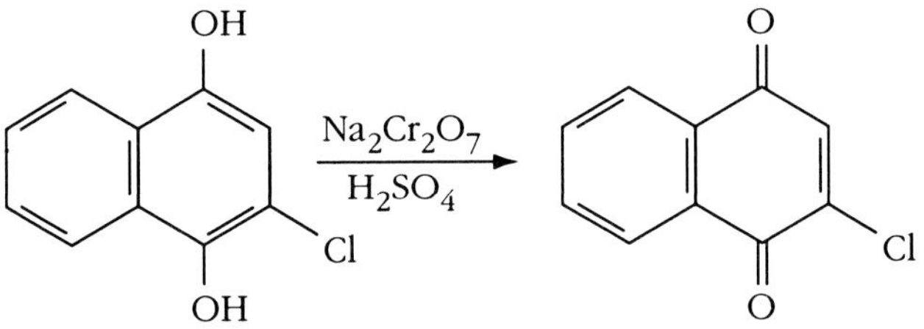

(b)

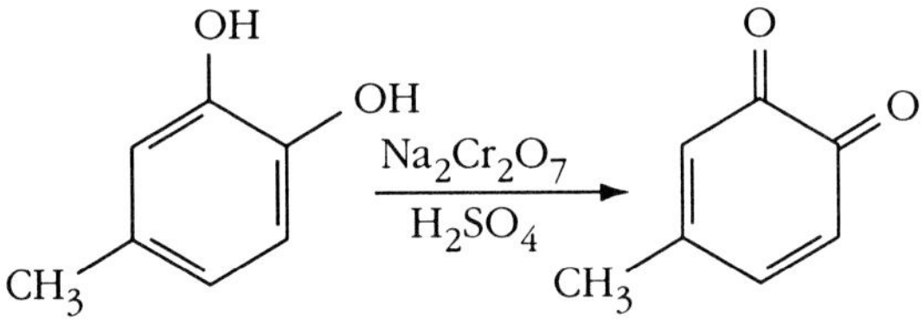

(c)

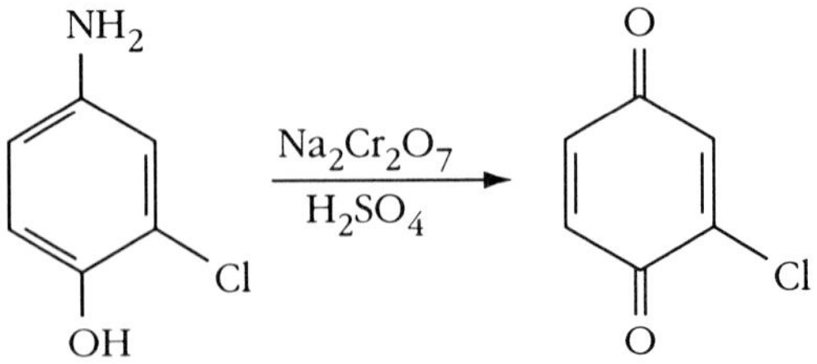

23.22

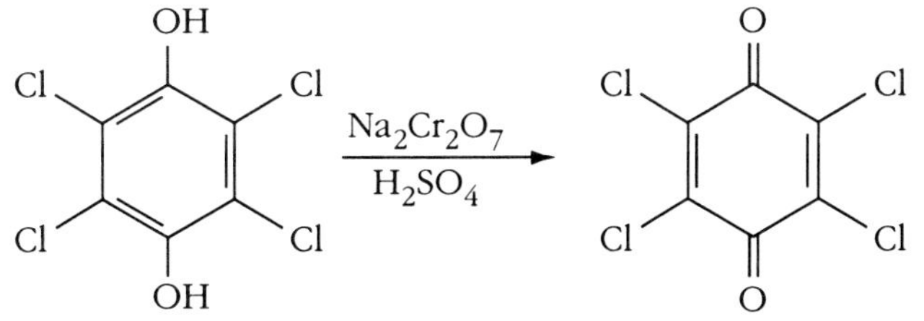

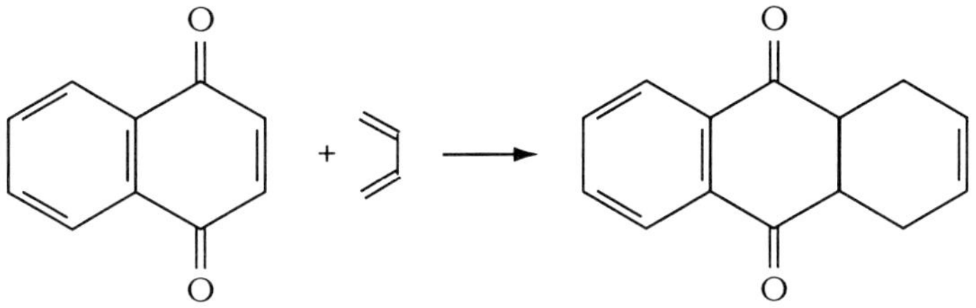

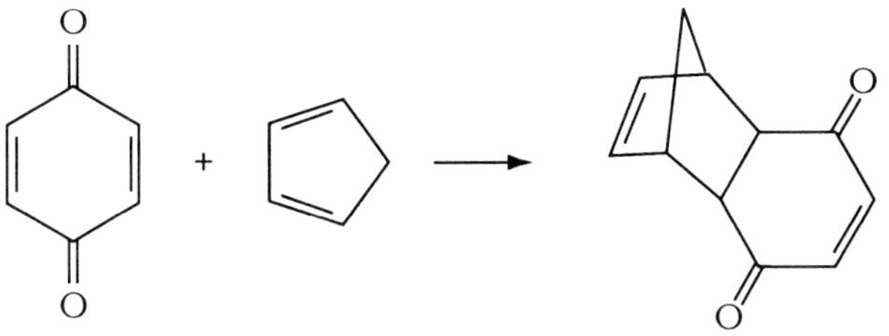

23.23

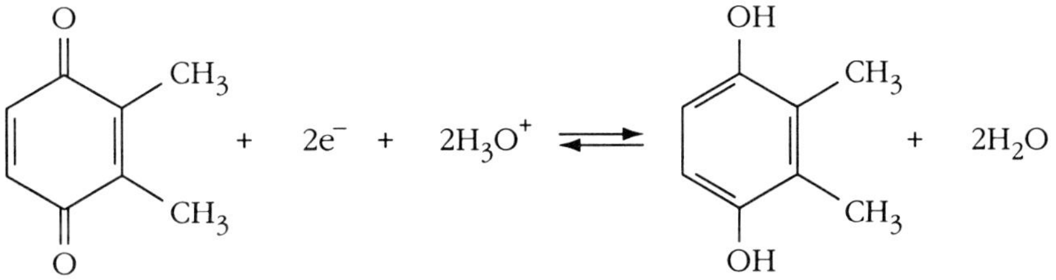

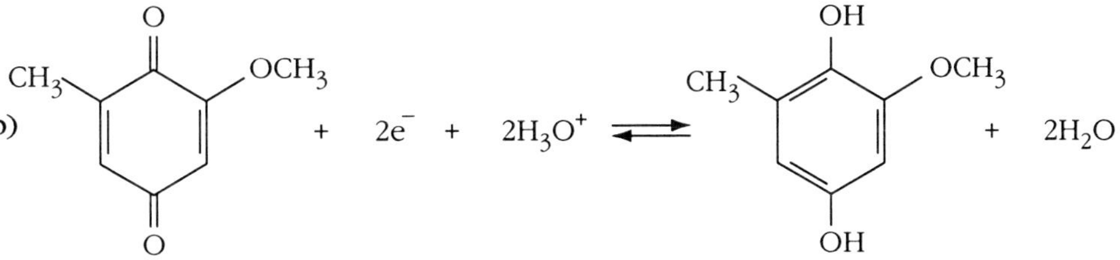

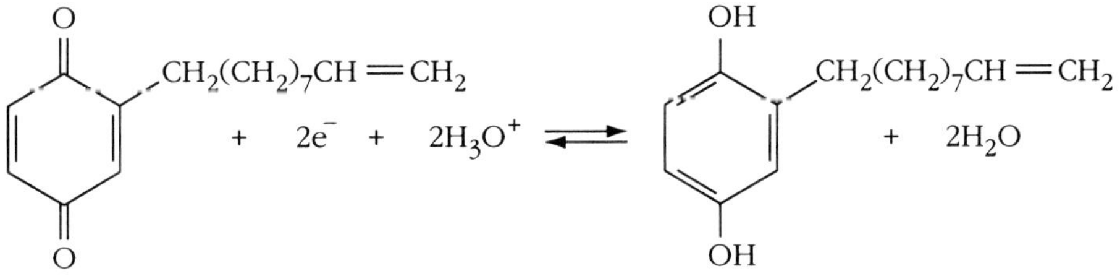

23.24 (a) A **phenol** is a compound that contains a hydroxy group bonded directly to a carbon atom of a benzene ring or fused benzene rings. The word *phenol* refers to both a specific compound as well as a class of compounds.

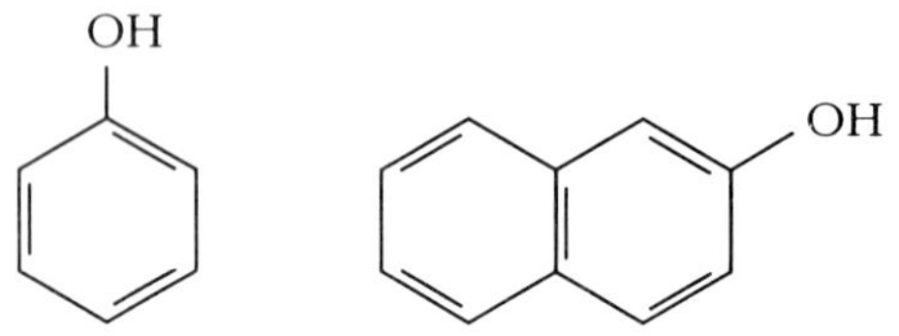

Phenol

β-Naphthol
(A phenol)

(b) The **nucleophilic aromatic substitution reaction** is a reaction in which a nucleophile replaces a group or atom bonded to one of the carbon atoms of a benzene ring. Electron-withdrawing groups activate the benzene ring to nucleophilic aromatic substitution reactions.

$$\text{2,4-dinitrochlorobenzene} \xrightarrow{\text{NaOCH}_3, \text{CH}_3\text{OH}} \text{2,4-dinitroanisole}$$

Groups that activate the benzene ring to nucleophilic aromatic substitution reactions

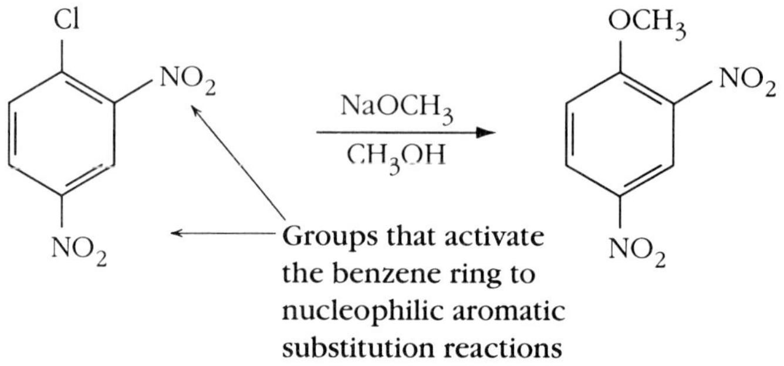

(c) A **hydroquinone** is a compound that contains two hydroxyl groups bonded to the carbon atoms of a benzene ring. Hydroquinones are oxidized to quinones by a variety of oxidizing reagents. Examples of hydroquinones include 1,4-dihydroxybenzene and 1,4-dihydroxynaphthalene:

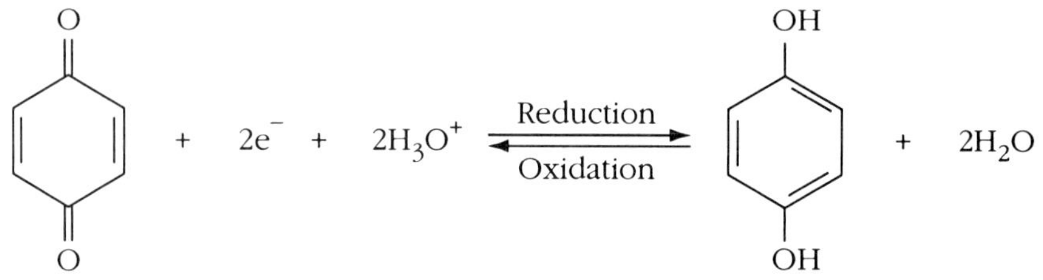

$$\text{1,4-Benzoquinone} + 2e^- + 2H_3O^+ \underset{\text{Oxidation}}{\overset{\text{Reduction}}{\rightleftharpoons}} \text{1,4-Dihydroxybenzene} + 2H_2O$$

1,4-Benzoquinone
(Quinone)

1,4-Dihydroxybenzene
(Hydroquinone)

1,4-Naphthoquinone + 2e⁻ + 2H₃O⁺ ⇌ 1,4-Dihydroxynaphthalene + 2H₂O (Reduction / Oxidation)

1,4-Naphthoquinone
(A quinone)

1,4-Dihydroxynaphthalene
(A hydroquinone)

(d) The **Kolbe reaction** is a method of forming sodium salicylate by the reaction of sodium phenoxide and carbon dioxide at 125 °C and 100 atm pressure.

Sodium phenoxide + CO_2 ⇌ (125 °C, 100 atm) Sodium salicylate

(e) A **halobenzene** is a compound that contains a halogen atom (F, Cl, Br, or I) bonded to the carbon atom of a benzene ring. The mechanism of nucleophilic substitution reactions of halobenzenes is different that those of haloalkanes. When the aromatic ring of a halobenzene is activated by electron-withdrawing groups, the reaction occurs by an addition-elimination mechanism. Unactivated halobenzenes undergo nucleophilic aromatic substitution reactions by an elimination-addition mechanism that involves a benzyne intermediate.

(f) **Claisen rearrangement** is the migration of the allyl group of an allyl aryl ether from the oxygen atom to the ring carbon atom *ortho* to it when heated.

Phenyl allyl ether →(200 °C) 2-allylphenol

(g) The **addition-elimination mechanism** is a two-step mechanism for nucleophilic aromatic substitution reactions. In Step 1, the addition step, a nucleophile adds to the carbon atom of the aromatic ring that is bonded to the leaving group to form an anionic intermediate. In Step 2, the elimination step, the leaving group is expelled from the intermediate.

Step 1: Addition

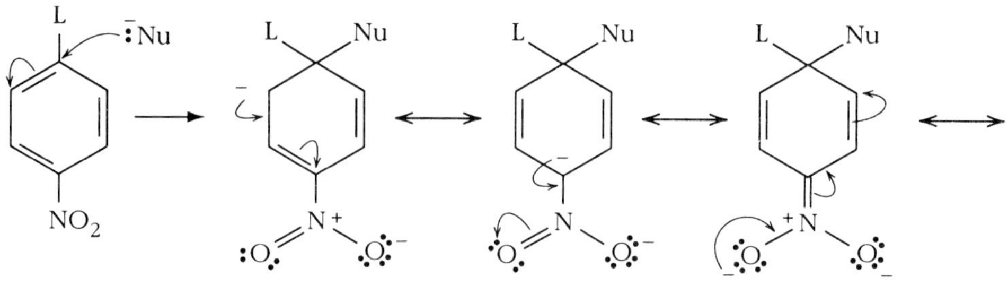

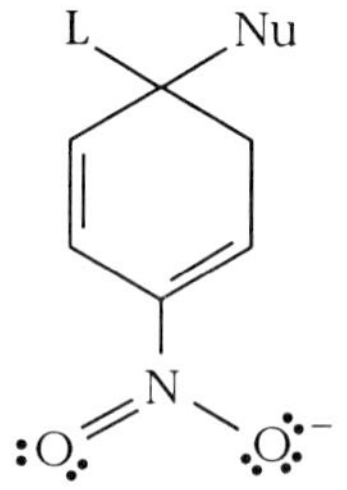

L = Leaving group
Nu:⁻ = Nucleophile

Step 2: Elimination

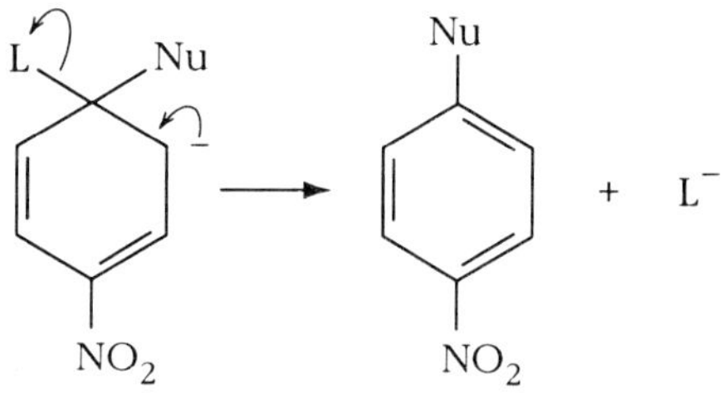

(h) **Benzyne** is the reactive intermediate formed in the first step of the elimination-addition mechanism of nucleophilic aromatic substitution reactions.

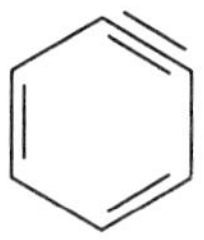

Benzyne

(i) The **elimination-addition mechanism** involves the elimination of two atoms or groups from adjacent carbon atoms of an unactivated benzene derivative in Step 1 to form a benzyne intermediate. Step 2 is the addition of a nucleophile to benzyne. Additional steps involving proton transfer may be needed to form the final product. The reaction of a halobenzene and an amide ion occurs by an elimination-addition mechanism:

Step 1: Elimination

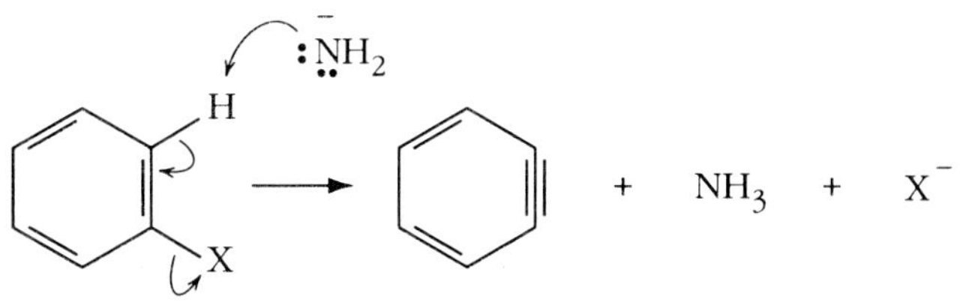

[Mechanism showing ortho-H on haloarene being removed by :NH₂⁻, with loss of X⁻, forming benzyne + NH₃ + X⁻]

Step 2: Addition

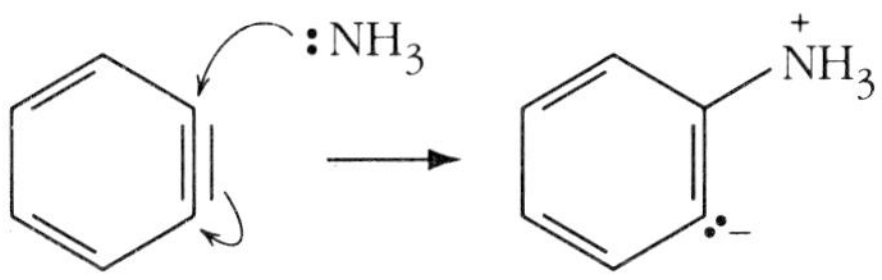

[Benzyne reacts with :NH₃ to give phenyl carbanion with +NH₃ substituent]

Proton transfer steps:

[Carbanion with +NH₃ group abstracts H from H—NH₂, giving arene with +NH₃ and ortho-H, plus :NH₂⁻]

[Anilinium (+NH₂—H) transfers proton to :NH₂⁻, giving aniline (NH₂) + :NH₃]

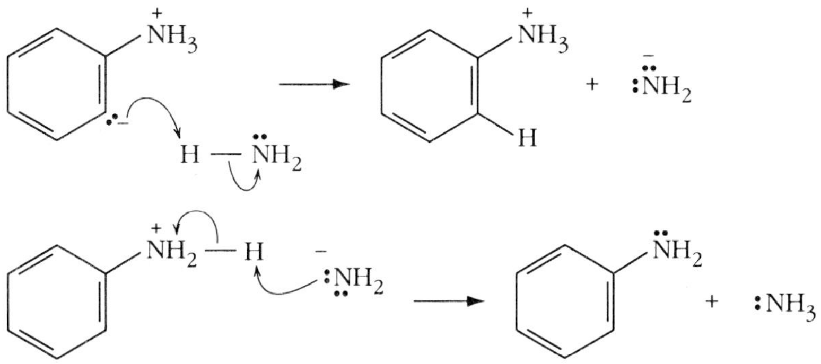

23.25
(a) 3-Bromoethylbenzene
(b) 3-Bromo-5-chlorophenol
(c) 6-Methoxy 1,4 naphthoquinone
(d) Methyl 5-bromo-2-hydroxybenzoate
(e) 3-Methylphenyl acetate
(f) 2,4-Dichlorobenzoic acid
(g) 2,5-Dichloro-3,6-dimethoxy-1,4-benzoquinone

23.26

(a) 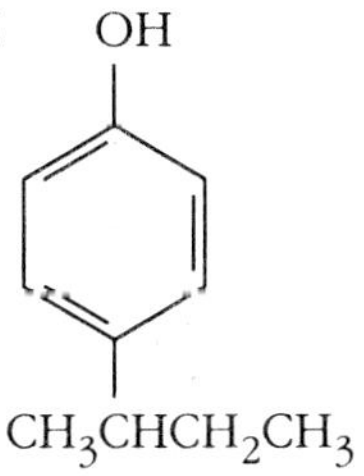 4-(CH₃CHCH₂CH₃)-phenol (para sec-butyl phenol with OH)

(b) 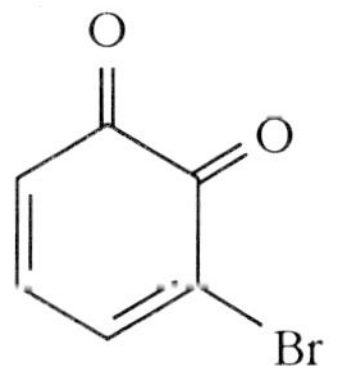 3-Bromo-1,2-benzoquinone

(c) 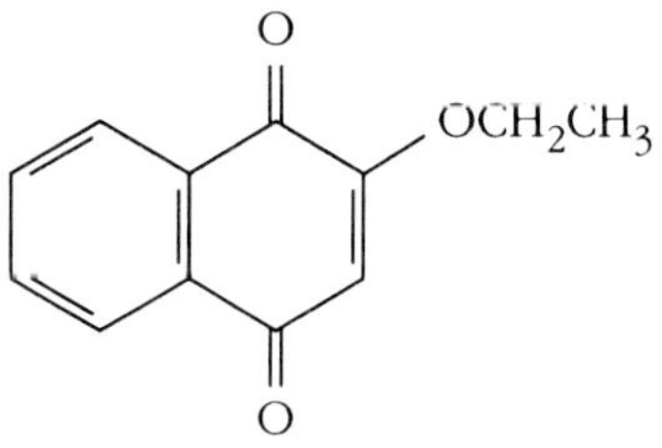 2-Ethoxy-1,4-naphthoquinone (OCH₂CH₃)

(d) 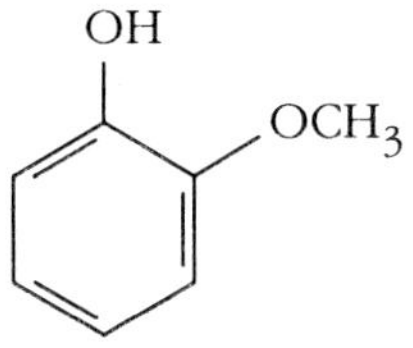 2-Methoxyphenol (OH, OCH₃)

(e) 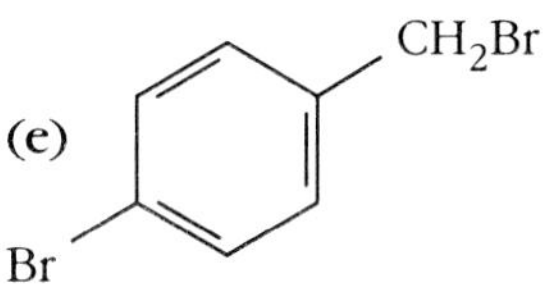 4-Bromobenzyl bromide (CH₂Br, Br para)

(f) 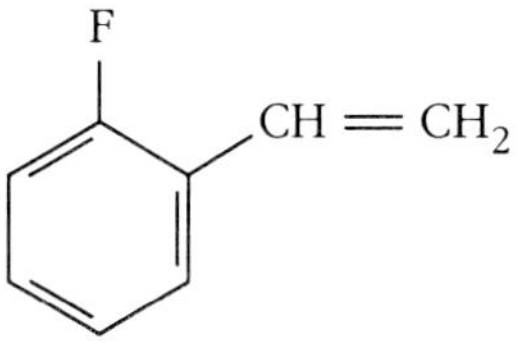 2-Fluorostyrene (F, CH=CH₂)

(g) benzene-1,3,5-triol (HO, OH, OH on benzene)

(h) 4-methylbenzene-1,2-diol (CH₃ with two OH)

(i) aspirin: 2-(acetyloxy)benzoic acid

23.27

(a) 4-methylphenol $\xrightarrow{\text{NaOH}}$ 4-methylphenoxide $\xrightarrow{\text{CH}_3\text{I}}$ 4-methylanisole (OCH₃ with para CH₃)

(b) 4-methylphenol $\xrightarrow[\text{CH}_3\text{COOH}]{\text{Br}_2}$ 2-bromo-4-methylphenol

(c) 4-methylphenol $\xrightarrow[\text{CH}_3\text{COOH}]{(\text{CH}_3\text{CO})_2\text{O}}$ 4-methylphenyl acetate

(d) 4-methylphenol $\xrightarrow[\text{H}_2\text{O}]{\text{Br}_2}$ 2,6-dibromo-4-methylphenol

(e)

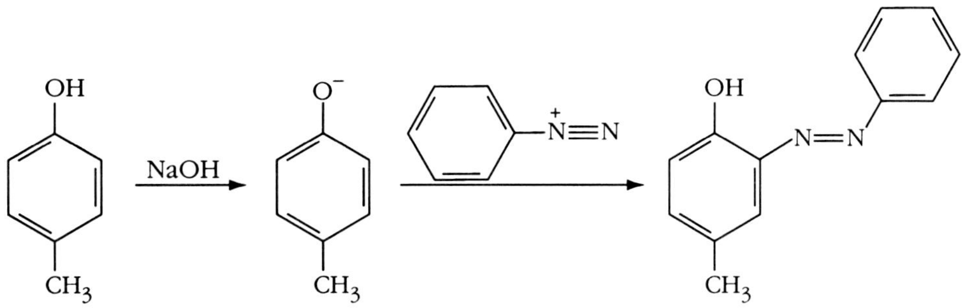
OH
NaOH
O⁻
N≡N
OH
N=N
CH3
CH3
CH3

(f)

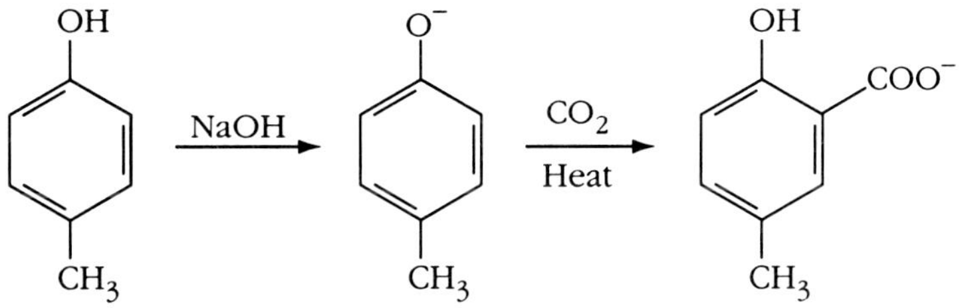
OH
NaOH
O⁻
CO2
Heat
OH
COO⁻
CH3
CH3
CH3

(g)

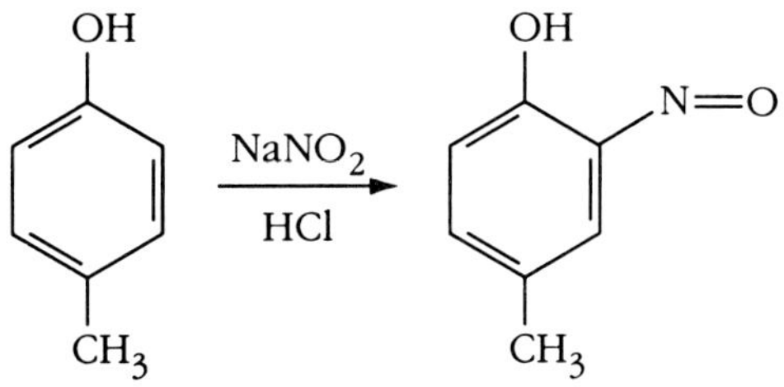
OH
NaNO2
HCl
OH
N=O
CH3
CH3

(h)

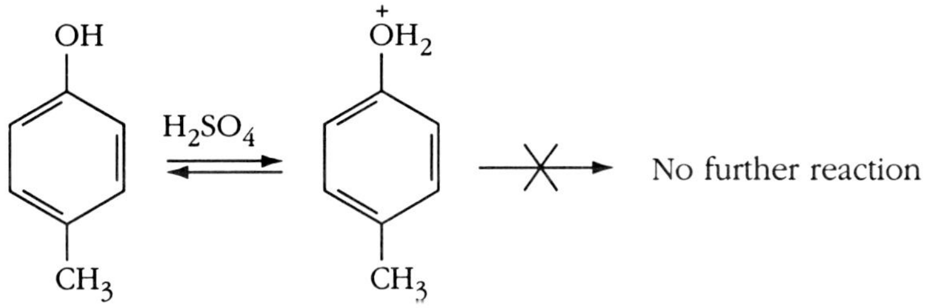
OH
H2SO4
OH2
No further reaction
CH3
CH3

23.28

(a)

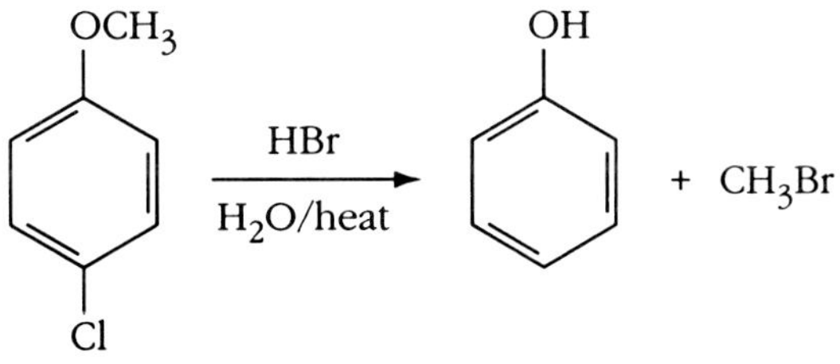
OCH3
HBr
H2O/heat
OH
+ CH3Br
Cl

(b)

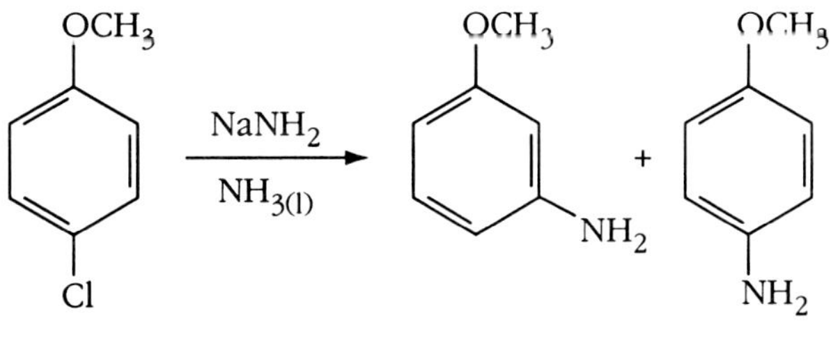
OCH3
NaNH2
NH3(l)
OCH3
+
OCH3
NH2
Cl
NH2

(c) [p-chloroanisole] + HNO₃/H₂SO₄ → 4-chloro-2-nitroanisole

(d) [p-chloroanisole] + Mg, Dry ether → [p-methoxyphenyl-MgCl] + (1) CH₃COCH₃ (2) H₃O⁺ → 4-methoxyphenyl-C(CH₃)₂OH

(e) [p-chloroanisole] + NaOH/H₂O, Heat/pressure → 4-methoxyphenol + 3-methoxyphenol

(f) [p-chloroanisole] + Br₂/FeBr₃ → 2-bromo-4-chloroanisole

23.29 Most of the hydrogen bonds in 4-hydroxyacetophenone are intermolecular while in 2-hydroxyacetophenone they are intramolecular. More energy is needed, therefore, to break the intermolecular hydrogen bonds of 4-hydroxyacetophenone, which are not present in 2-hydroxyacetophenone so 2-hydroxyacetophenone is the more volatile. In fact, 2-hydroxyacetophenone is a liquid at room temperature (melting point = 4–6 °C) while 4-hydroxyacetophenone is a solid (melting point = 109–110 °C).

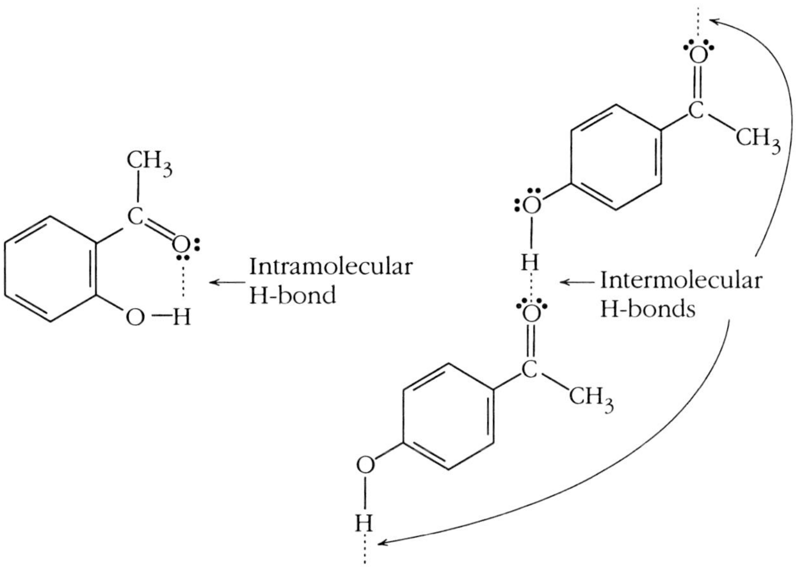

Intramolecular H-bond

Intermolecular H-bonds

23.30

(a) Chlorobenzene $\xrightarrow{\text{HONO}_2 / \text{H}_2\text{SO}_4 / \text{heat}}$ 4-nitrochlorobenzene $\xrightarrow[\text{Slow}]{\text{HONO}_2 / \text{H}_2\text{SO}_4 / \text{heat}}$ 2,4-dinitrochlorobenzene $\xrightarrow[\text{Slow}]{\text{HONO}_2 / \text{H}_2\text{SO}_4 / \text{heat}}$ 2,4,6-trinitrochlorobenzene

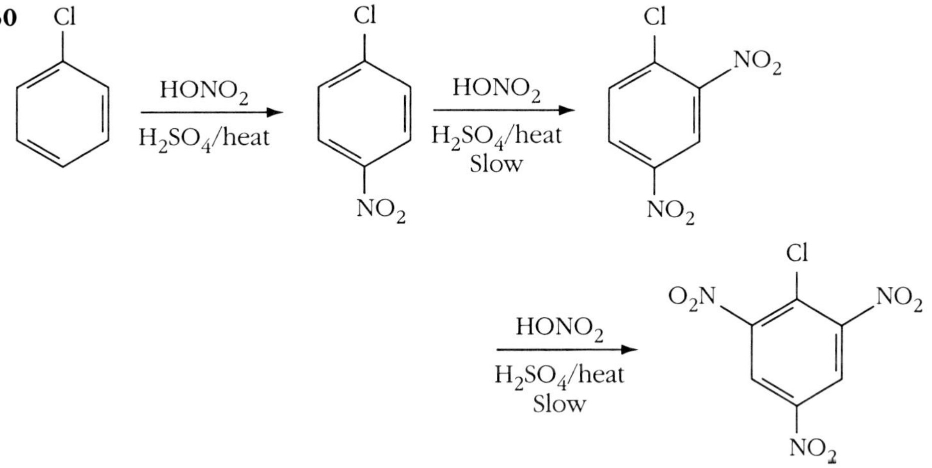

Difficult to form because of three electron-withdrawing groups of 2,4-Dinitrochlorobenzene

(b) 2,4-dinitrochlorobenzene $\xrightarrow{\text{NH}_3}$ 2,4-dinitroaniline

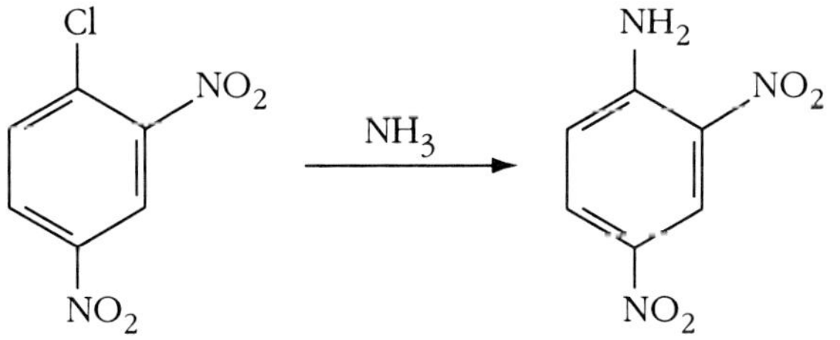

(c) 1,2-dimethoxybenzene $\xrightarrow[\text{Heat}]{\text{Excess HBr}}$ catechol (1,2-dihydroxybenzene)

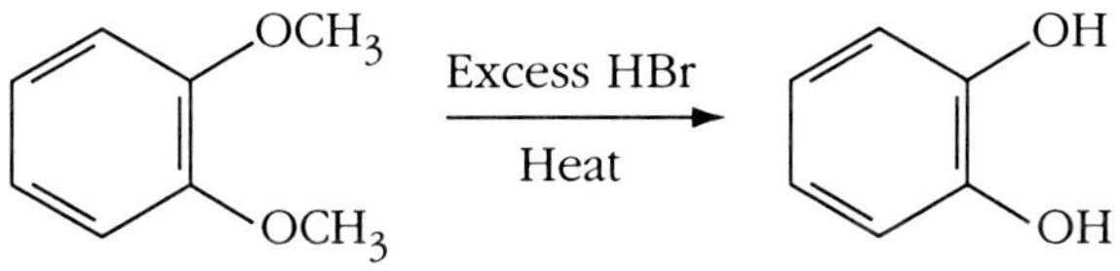

(d) Catechol + Na$_2$Cr$_2$O$_7$ / H$_2$SO$_4$ → ortho-benzoquinone

(e) Salicylic acid + NaHCO$_3$ → sodium salicylate

(f) p-Cresol + (CH$_3$CO)$_2$O / OH$^-$ → p-tolyl acetate

(g) 3-Bromotoluene + NaNH$_2$ / NH$_3$(l) / -33 °C → benzyne intermediates → p-toluidine + m-toluidine + o-toluidine

(h) 4-Sulfobenzenediazonium chloride + phenol → 4-hydroxy-4′-sulfo azobenzene

(i) 4-methylphenol + HNO₃/CH₃COOH → 4-methyl-2-nitrophenol

(j) PhCH₂-C₆H₄-OH + Br₂/CH₃COOH → PhCH₂-C₆H₃(Br)-OH
(The more reactive ring because of OH)

(k) PhOH + NaOH/H₂O → PhO⁻ → (1) epoxide/H₂O, (2) H₃O⁺ → PhOCH₂CH(OH)CH₃

(l) 2-methylphenol + K₂CO₃ → 2-methylphenoxide + CH₂=CHCH₂Br → 2-methylphenyl allyl ether → Heat → 2-allyl-6-methylphenol (Claisen rearrangement product: 2,6-disubstituted phenol with CH₂CH=CH₂ and CH₃)

23.31

$$CH_3CH=CHCH_3 \rightleftharpoons CH_3\overset{+}{C}HCH_2CH_3$$

with H₃O⁺ (hydronium ion shown donating H⁺)

PhOH (4-methylphenol, :ÖH with CH₃ para) + ⁺CHCH₂CH₃(CH₃) →

arenium ion intermediate: cyclohexadienyl cation with :ÖH, CH(CH₃)CH₂CH₃, H at sp³ carbon, CH₃ at para position, positive charge

↔ resonance structure with Ö⁺H, CH(CH₃)CH₂CH₃, CH₃

23.32

23.33 Step 1: Isomerization of one carbonyl group

Step 2: Isomerization of the second carbonyl group

23.34

23.35

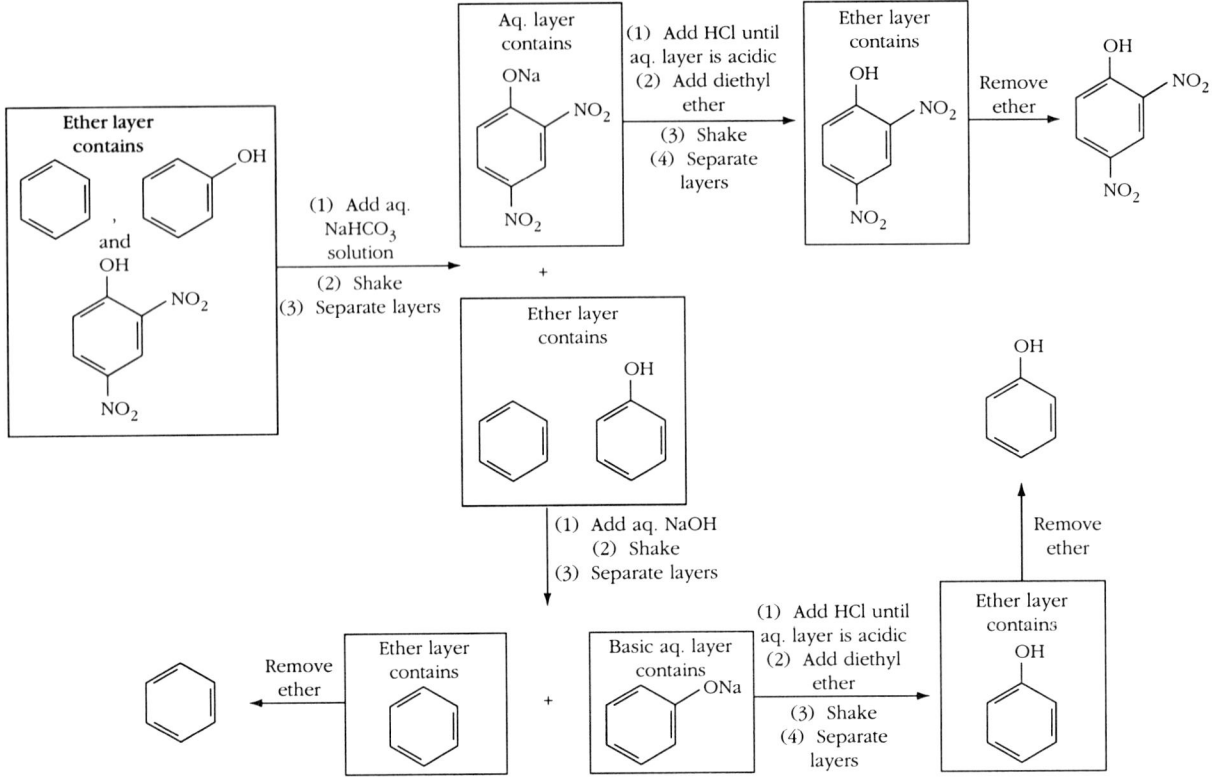

23.36 Step 1: Formation of electrophile

$$HCl + ZnCl_2 \rightleftharpoons \overset{+}{H}ZnCl_3^- \quad \text{(a Lewis acid-base complex)}$$

$$RCH_2C \equiv N: \overset{\curvearrowleft}{\ } \overset{+}{H}ZnCl_3^- \rightleftharpoons RCH_2C \equiv \overset{+}{N}\overset{H}{\diagup} \longleftrightarrow RCH_2\overset{+}{C}=\overset{..}{N}-H$$

Step 2: Addition of electrophile to aromatic ring

Step 3: Regeneration of aromatic ring

Compound A

Step 4: Hydrolysis of the imine; let Ar =

[2,4-dihydroxyphenyl group with OH at top and OH at bottom]

[Mechanism scheme showing protonation of imine nitrogen, water attack on carbon, proton transfers, and loss of ammonia to give ketone]

$$RCH_2\!-\!\underset{Ar}{\underset{|}{C}}\!=\!O \;\equiv\; \text{[2,4-dihydroxyphenyl ketone with } CH_2R]$$

23.37 The peaks that disappear upon shaking with D_2O are due to the hydrogen atom of an —OH group. The hydrogen atoms in the aromatic region in both spectra have an area of four relative to the —OH hydrogen atom so the aromatic ring must be disubstituted. One substituent is an —OH group and the other must be a chlorine atom. There are three chlorophenols:

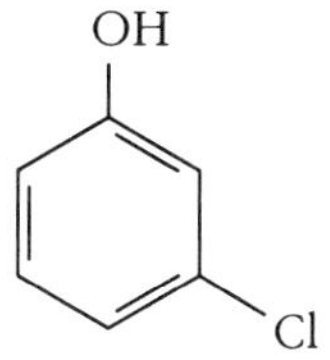

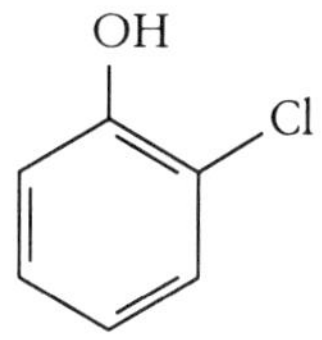

4-Chlorophenol   3-Chlorophenol   2-Chlorophenol

Neither spectrum A nor spectrum B shows a pair of doublets in the aromatic region so the two substituents are *not* located 1,4 to each other.

3-Chlorophenol is expected to show the following coupling pattern for the aromatic hydrogen atoms: a singlet for C2, a doublet for C4, a triplet for C5, and a doublet for C6. Spectrum A shows this coupling pattern so it is the spectrum of 3-chlorophenol.

2-Chlorophenol is expected to show the following coupling pattern for the aromatic hydrogen atoms: a doublet for C3, a triplet for C4, a triplet for C5, and a doublet for C6. Spectrum B shows this pattern so it is the spectrum of 2-chlorophenol.

Moreover, 2-chlorophenol is more acidic than 3-chlorophenol because the electron-withdrawing chlorine atom is closer to the —OH group in the aromatic ring in 2-chlorophenol than in 3-chlorophenol. As a result, the more deshielded —OH hydrogen atom of 2-chlorophenol would be expected to appear at a lower field than the —OH hydrogen atom of 3-chlorophenol. This expectation is borne out by the observation that the —OH hydrogen atom appears at δ 5.7 in the spectrum of 2-chlorophenol and at δ 5.2 in the spectrum of 3-chlorophenol.

23.38 The singlet at δ 7.37 (5H) in spectrum C indicates a monosubstituted benzene ring. The broad peak that disappears upon shaking with D_2O is due to the hydrogen atom of an —OH group. The remaining singlet at δ 4.68 (2H) suggests a methylene group. Spectrum C is the spectrum of benzyl alcohol:

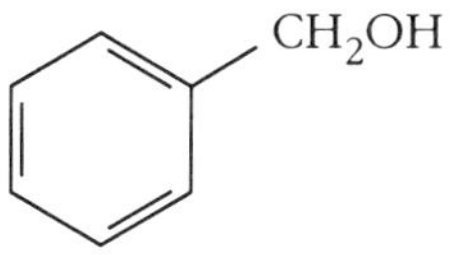

The singlet at δ 2.30 (3H) indicates a methyl group bonded to a benzene ring. The other substituent is a hydroxyl group. The coupling pattern of the aromatic hydrogen atoms appears as a triplet, a doublet, a singlet, and another doublet as we move upfield. This is typical of a 1,3-disubstituted benzene ring (compare with Exercise 23.37). Spectrum D, therefore, is the spectrum of 3-methylphenol:

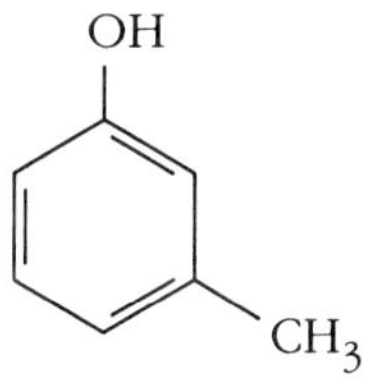

CHAPTER 24

CHEMISTRY OF DIFUNCTIONAL COMPOUNDS
Concepts

The reactions of compounds that contain more than one functional group depend on the relative locations of their functional groups. If the functional groups are sufficiently far apart, each functional group undergoes its normal reactions unaffected by the other functional groups.

If two functional groups that can react with each other are present in the same molecule, reaction may occur intermolecularly or intramolecularly. When the two functional groups are placed so that a five- or six-membered ring can be formed, intramolecular ring formation is favored. If the two reacting functional groups are located so that ring formation is difficult or impossible, intermolecular reaction occurs to form a dimer, a trimer, and finally a polymer.

If two functional groups are bonded to the same carbon atom, one may affect the rate or equilibrium constant of the other. Two carbonyl-containing functional groups bonded to the same carbon atoms is an example. Such compounds are called **1,3-dicarbonyl compounds.** The second carbonyl group enhances the enol content of the compound and increases the acidity of the methylene hydrogen atoms.

1,3-Dicarbonyl compounds are widely used as the starting material for the synthesis of ketones and carboxylic acids because of the ease of forming enolate anions, which can react with haloalkanes and most carbonyl-containing compounds to form a new carbon-carbon bond. The **malonic ester synthesis** and the **ethyl acetoacetate synthesis** are two examples of such syntheses.

Malonic Ester Synthesis:

$$\underset{\text{Diethyl malonate}}{H_2C(COOCH_2CH_3)_2} \xrightarrow[\substack{(2)\ RCH_2X \\ (3)\ H_3O^+/\text{heat}}]{\substack{(1)\ Na^+\ {}^-OCH_2CH_3 \\ CH_3CH_2OH}} \underset{\text{Carboxylic acid}}{RCH_2CH_2COOH}$$

Ethyl Acetoacetate Synthesis:

$$\underset{\text{Ethyl acetoacetate}}{H_2C\begin{pmatrix}COCH_3 \\ COOCH_2CH_3\end{pmatrix}} \xrightarrow[\substack{(2)\ RCH_2X \\ (3)\ H_3O^+/\text{heat}}]{\substack{(1)\ Na^+\ {}^-OCH_2CH_3 \\ CH_3CH_2OH}} \underset{\text{Methyl ketone}}{RCH_2CH_2COCH_3}$$

Solutions to the Exercises

24.1 (a) Methyl 3-oxopentanoate
(b) 5-Methyl-2,4-hexanedione
(c) 4-Chloro-3-oxopentanal
(d) 4-Phenyl-4-oxobutanoic acid

24.2

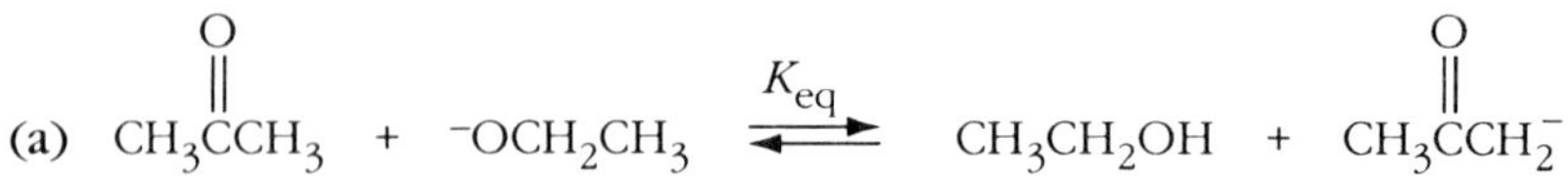

(a) $CH_3\overset{O}{\overset{\|}{C}}CH_3$ + $^-OCH_2CH_3$ $\underset{}{\overset{K_{eq}}{\rightleftharpoons}}$ CH_3CH_2OH + $CH_3\overset{O}{\overset{\|}{C}}CH_2^-$

$$K_{eq} = \frac{K_a(\text{Acetone})}{K_a(\text{Ethanol})} = \frac{5.0 \times 10^{-20}}{1.0 \times 10^{-16}} = 5.0 \times 10^{-4}$$

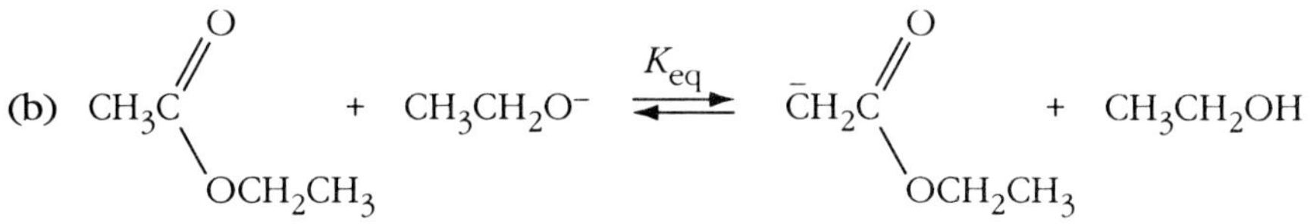

(b) $CH_3C(\!\!=\!\!O)OCH_2CH_3$ + $CH_3CH_2O^-$ $\underset{}{\overset{K_{eq}}{\rightleftharpoons}}$ $^-CH_2C(\!\!=\!\!O)OCH_2CH_3$ + CH_3CH_2OH

$$K_{eq} = \frac{K_a(\text{Ethyl acetate})}{K_a(\text{Ethanol})} = \frac{3.2 \times 10^{-25}}{1.0 \times 10^{-16}} = 3.2 \times 10^{-9}$$

24.3 Acetone (pK_a = 19.3) is a much stronger acid than ethyl acetate (pK_a = 24.5) so it will react with all the acetone and leave ethyl acetate untouched.

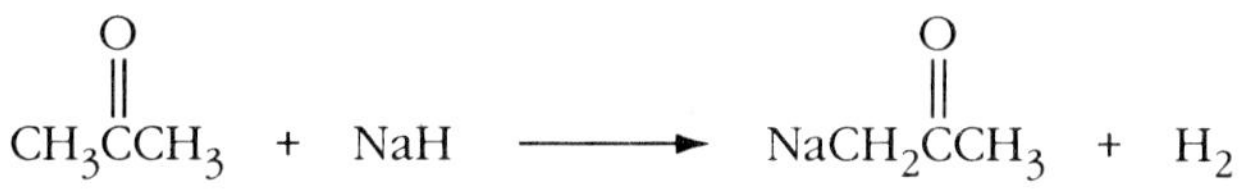

$$CH_3\overset{O}{\overset{\|}{C}}CH_3 + NaH \longrightarrow NaCH_2\overset{O}{\overset{\|}{C}}CH_3 + H_2$$

Therefore, the reaction mixture contains only ethyl acetate and the enolate anion of acetone so that the only reaction that occurs in solution is the following:

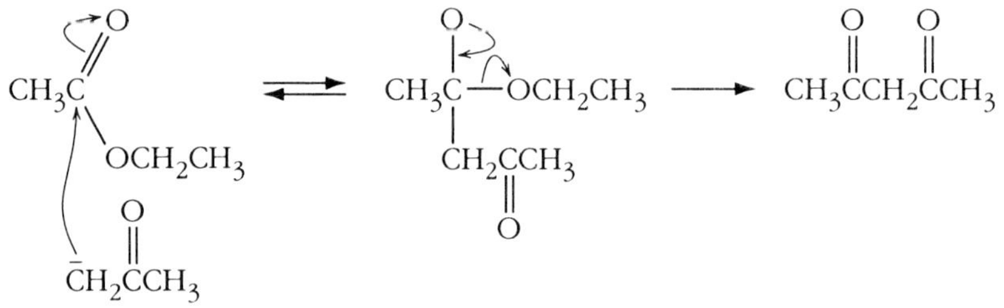

$CH_3C(\!\!=\!\!O)OCH_2CH_3$ + $^-CH_2CCH_3$ $\rightleftharpoons$ $CH_3C(O^-)(OCH_2CH_3)(CH_2CCH_3{=}O)$ $\longrightarrow$ $CH_3\overset{O}{\overset{\|}{C}}CH_2\overset{O}{\overset{\|}{C}}CH_3$

Excess base removes a proton from 2,4-pentanedione to form an enolate anion:

$CH_3\overset{O}{\overset{\|}{C}}CH_2\overset{O}{\overset{\|}{C}}CH_3$ + $B{:}^-$ $\longrightarrow$ $CH_3\overset{O}{\overset{\|}{C}}\overset{-}{C}H\overset{O}{\overset{\|}{C}}CH_3$ + BH

Acid work-up is needed to obtain 2,4-pentanedione.

24.4

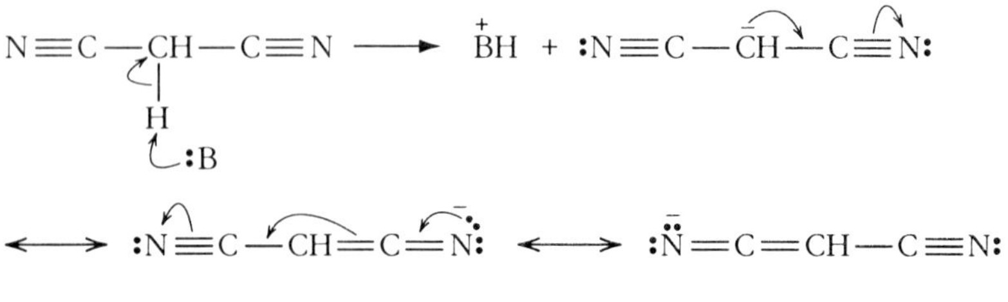

$N\equiv C-CH(H)-C\equiv N$ + :B $\longrightarrow$ $\overset{+}{B}H$ + $:N\equiv C-\overset{-}{C}H-C\equiv N:$

$\longleftrightarrow$ $:N\equiv C-CH=C=\overset{..}{N}:^-$ $\longleftrightarrow$ $:\overset{..}{\overset{-}{N}}=C=CH-C\equiv N:$

24.5

$$\overset{..}{\underset{..}{O}}=C=\overset{..}{\underset{..}{O}}$$
$$+$$

[Mechanism: malonic acid structure with curved arrows] →Heat→ [Enol intermediate: $H\overset{..}{\underset{..}{O}}-C(=CH_2)-\overset{..}{\underset{..}{O}}-H$] → $H\overset{..}{\underset{..}{O}}-C(CH_3)=\overset{..}{\underset{..}{O}}$ Acetic acid

24.6

(a) Ph–CH₂–|CH₂COOH| ⟹ Ph–CH₂Br + CH₂(COOCH₂CH₃)₂

Part of carboxylic acid obtained from diethyl malonate

Part of carboxylic acid obtained from haloalkane

Diethyl malonate $\xrightarrow[\text{(2) PhCH}_2\text{Br}]{\text{(1) NaOCH}_2\text{CH}_3/\text{CH}_3\text{CH}_2\text{OH}}$ Ph–CH(COOCH₂CH₃)₂ (benzylated diethyl malonate)

$\xrightarrow{H_3O^+/150\,°C}$ Ph–CH₂CH₂COOH

(b) Diethyl malonate $\xrightarrow[\substack{\text{(2) CH}_3(\text{CH}_2)_4\text{CH}_2\text{Br} \\ \text{(3) H}_3\text{O}^+/150\,°C}]{\text{(1) NaOCH}_2\text{CH}_3/\text{CH}_3\text{CH}_2\text{OH}}$ CH₃(CH₂)₄CH₂CH₂COOH

(c)
$$\text{CH}_2\text{COOCH}_2\text{CH}_3 \atop |\atop \text{COOCH}_2\text{CH}_3 \quad \xrightarrow[\text{(2) CH}_3(\text{CH}_2)_2\text{CH}_2\text{Br}]{\text{(1) NaOCH}_2\text{CH}_3/\text{CH}_3\text{CH}_2\text{OH}}} \quad \text{CH}_3(\text{CH}_2)_2\text{CH}_2\text{CHCOOCH}_2\text{CH}_3 \atop |\atop \text{COOCH}_2\text{CH}_3$$

$$\downarrow \text{(1) NaOCH}_2\text{CH}_3/\text{CH}_3\text{CH}_2\text{OH, (2) CH}_3\text{Br, (3) H}_3\text{O}^+/150\ °\text{C}$$

$$\text{CH}_3(\text{CH}_2)_2\text{CH}_2\text{CHCOOH} \atop | \atop \text{CH}_3$$

24.7 The key step in a malonic ester synthesis is the alkylation reaction of diethylmalonate (a nucleophilic substitution reaction), which occurs by an S_N2 mechanism. The syntheses of 2-phenylacetic acid and 2,2-dimethylpropanoic acid by the malonic ester synthesis require the reaction of diethylmalonate with two bromine-containing compounds that cannot undergo nucleophilic substitution reactions by an S_N2 mechanism, namely bromobenzene and *tert*-butyl bromide. Consequently, the malonic ester synthesis cannot be used to synthesize these two compounds.

(a) $\bar{\text{C}}\text{HCOOCH}_2\text{CH}_3 \atop | \atop \text{COOCH}_2\text{CH}_3$ + C$_6$H$_5$Br $\longrightarrow$ No reaction

(b) $\bar{\text{C}}\text{HCOOCH}_2\text{CH}_3 \atop | \atop \text{COOCH}_2\text{CH}_3$ + (CH$_3$)$_3$C—Br $\longrightarrow$ CH$_2$=C(CH$_3$)$_2$ + $\text{CH}_2\text{COOCH}_2\text{CH}_3 \atop | \atop \text{COOCH}_2\text{CH}_3$

24.8

(a) $\underbrace{\text{CH}_3\text{CCH}_2}_{\text{Obtained from ethyl acetoacetate}} \boxed{\text{CH}_2\text{CHCH}_3 \atop | \atop \text{CH}_3}_{\text{Obtained from haloalkane}} \Longrightarrow \text{CH}_3\overset{\text{O}}{\overset{\|}{\text{C}}}\text{CH}_2\overset{\text{O}}{\overset{\diagup}{\text{C}}}_{\text{OCH}_2\text{CH}_3} + \text{Br}-\boxed{\text{CH}_2\text{CHCH}_3 \atop | \atop \text{CH}_3}$

$\text{CH}_3\overset{\text{O}}{\overset{\|}{\text{C}}}\text{CH}_2\overset{\text{O}}{\overset{\diagup}{\text{C}}}_{\text{OCH}_2\text{CH}_3} \xrightarrow[\text{(2) (CH}_3)_2\text{CHCH}_2\text{Br}]{\text{(1) NaOCH}_2\text{CH}_3/\text{CH}_3\text{CH}_2\text{OH}} \text{CH}_3\overset{\text{O}}{\overset{\|}{\text{C}}}\text{CHCOOCH}_2\text{CH}_3 \atop | \atop \text{CH}_2\text{CH(CH}_3)_2$

$\text{CH}_3\overset{\text{O}}{\overset{\|}{\text{C}}}\text{CH}_2\text{CH}_2\text{CH(CH}_3)_2 \xleftarrow[\text{(2) H}_3\text{O}^+/\text{heat}]{\text{(1) NaOH/H}_2\text{O}}$

(b) $\underset{\parallel}{\text{CH}_3\overset{O}{\text{C}}\text{CH}_2}\vdots\text{CH}_2-\text{C}_6\text{H}_5 \Longrightarrow \text{CH}_3\overset{O}{\underset{\parallel}{\text{C}}}\text{CH}_2\text{COOCH}_2\text{CH}_3 + \text{C}_6\text{H}_5\text{CH}_2\text{Br}$

$\text{CH}_3\overset{O}{\underset{\parallel}{\text{C}}}\text{CH}_2\text{COOCH}_2\text{CH}_3 \xrightarrow[\substack{(2)\ \text{C}_6\text{H}_5\text{CH}_2\text{Br} \\ (3)\ \text{H}_3\text{O}^+/\text{heat}}]{(1)\ \text{NaOCH}_2\text{CH}_3/\text{CH}_3\text{CH}_2\text{OH}} \text{CH}_3\overset{O}{\underset{\parallel}{\text{C}}}\text{CH}_2\text{CH}_2-\text{C}_6\text{H}_5$

(c) $\text{CH}_3\overset{O}{\underset{\parallel}{\text{C}}}\text{CH}_2\vdots\text{CH}_2\text{CH}=\text{CH}_2 \Longrightarrow \text{CH}_3\overset{O}{\underset{\parallel}{\text{C}}}\text{CH}_2\text{COOCH}_2\text{CH}_3 + \text{BrCH}_2\text{CH}=\text{CH}_2$

$\text{CH}_3\overset{O}{\underset{\parallel}{\text{C}}}\text{CH}_2\text{COOCH}_2\text{CH}_3 \xrightarrow[\substack{(2)\ \text{BrCH}_2\text{CH}=\text{CH}_2 \\ (3)\ \text{NaOH}/\text{H}_2\text{O} \\ (4)\ \text{H}_3\text{O}^+}]{(1)\ \text{NaOCH}_2\text{CH}_3/\text{CH}_3\text{CH}_2\text{OH}} \text{CH}_3\overset{O}{\underset{\parallel}{\text{C}}}\text{CH}_2\text{CH}_2\text{CH}=\text{CH}_2$

24.9

(a) $\underbrace{\text{CH}_3\text{CH}_2\text{CH}_2\text{CH}}_{\substack{\text{Obtained} \\ \text{from} \\ \text{butanal}}}=\underbrace{\text{C}\begin{array}{c}\text{COCH}_3 \\ \diagdown \\ \diagup \\ \text{COCH}_2\text{CH}_3 \\ \parallel \\ \text{O}\end{array}}_{\substack{\text{Obtained} \\ \text{from ethyl} \\ \text{acetoacetate}}} \Longrightarrow \text{CH}_3\text{CH}_2\text{CH}_2\text{CHO} + \text{CH}_3\overset{O}{\underset{\parallel}{\text{C}}}\text{CH}_2\text{COOCH}_2\text{CH}_3$

$\text{CH}_3\text{CH}_2\text{CH}_2\text{CHO} + \text{CH}_3\overset{O}{\underset{\parallel}{\text{C}}}\text{CH}_2\overset{O}{\underset{\parallel}{\text{C}}}\text{OCH}_2\text{CH}_3 \xrightarrow[\text{Benzene/heat}]{\overset{+}{\text{R}}\text{NH}_3\overset{-}{\text{O}}\text{OCCH}_3}$

$\text{CH}_3\text{CH}_2\text{CH}_2\text{CH}=\text{C}\begin{array}{c}\overset{O}{\underset{\parallel}{\text{C}}}\text{CH}_3 \\ \diagdown \\ \diagup \\ \overset{}{\underset{\parallel}{\text{C}}}\text{OCH}_2\text{CH}_3 \\ \text{O}\end{array}$

(b) $CH_3CH=CHCOOH \implies CH_3CHO + \underset{\underset{COOCH_2CH_3}{|}}{CH_2COOCH_2CH_3}$

↑ Obtained from diethyl malonate

↑ Obtained from CH_3CHO

$$CH_3CHO + \underset{\underset{COOCH_2CH_3}{|}}{CH_2COOCH_2CH_3} \xrightarrow[\text{Benzene/heat}]{RNH_3^+ \, ^-OOCCH_3} CH_3CH=C\begin{matrix} COCH_2CH_3 \\ \parallel \\ O \\ COCH_2CH_3 \\ \parallel \\ O \end{matrix}$$

$$CH_3CH=CHCOOH \xleftarrow[\text{(2) } H_3O^+/\text{heat}]{\text{(1) NaOH/}H_2O}$$

(c) [3-methylphenyl]-CH=CHCOOH $\implies$ [3-methylphenyl]-CHO + $\underset{\underset{COOCH_2CH_3}{|}}{CH_2COOCH_2CH_3}$

[3-methylphenyl]-CHO + $CH_2\begin{matrix}COOCH_2CH_3 \\ COOCH_2CH_3\end{matrix}$ $\xrightarrow[\begin{array}{l}\text{(2) NaOH/}H_2O \\ \text{(3) } H_3O^+/\text{heat}\end{array}]{\text{(1) } RNH_3^+ \, ^-OOCCH_3/\text{benzene/heat}}$ [3-methylphenyl]-CH=CHCOOH

(d) 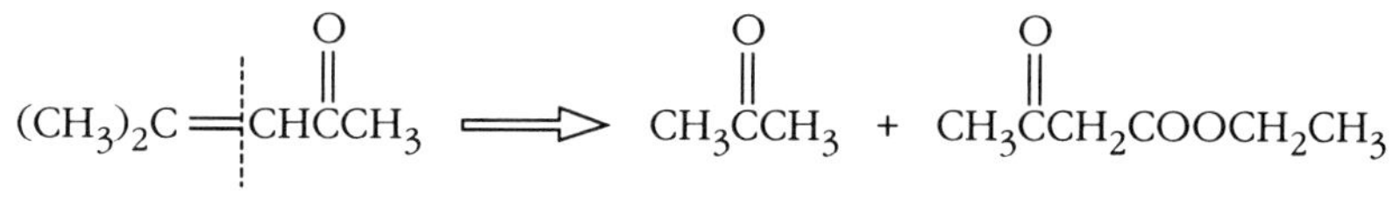 $(CH_3)_2C\!=\!\!\!=\!CHCCH_3 \implies CH_3\overset{O}{\overset{\|}{C}}CH_3 + CH_3\overset{O}{\overset{\|}{C}}CH_2COOCH_2CH_3$

$(CH_3)_2C\!=\!O + \underset{\underset{O}{\|}{\underset{C-OCH_2CH_3}{}}}{\overset{\overset{O}{\|}}{\overset{C}{\underset{|}{CH_2}}}}\!\!\!CH_3 \xrightarrow[\text{Benzene/heat}]{R\overset{+}{N}H_3\overset{-}{O}OCCH_3}$ [product shown]

$\xleftarrow[\text{(2) }H_3O^+]{\text{(1) NaOH/H}_2\text{O}}$ $\underset{H_3C}{\overset{H_3C}{}}\!\!\!C\!=\!CH\overset{O}{\overset{\|}{C}}CH_3$

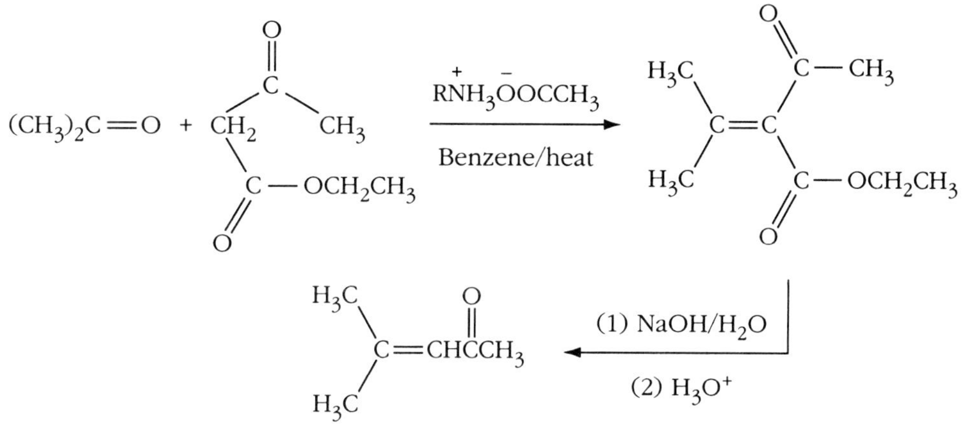

24.10 Step 1: **Addition** of enolate anion to carbonyl carbon atom to form a tetrahedral intermediate

Tetrahedral intermediate

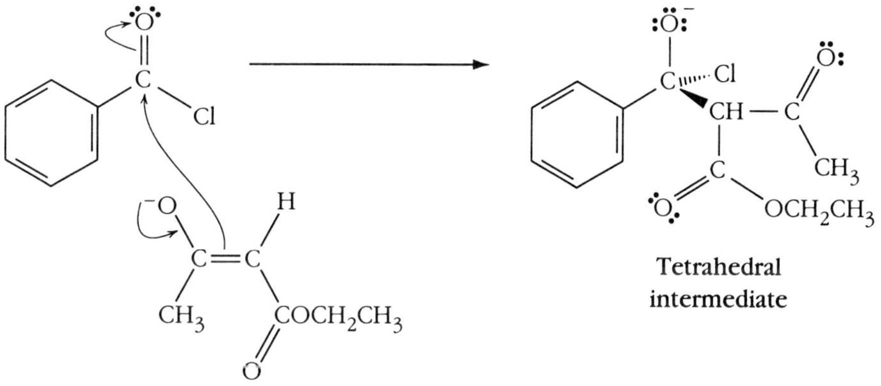

Step 2: **Elimination** of chloride ion from the tetrahedral intermediate

$+ \ Cl^-$

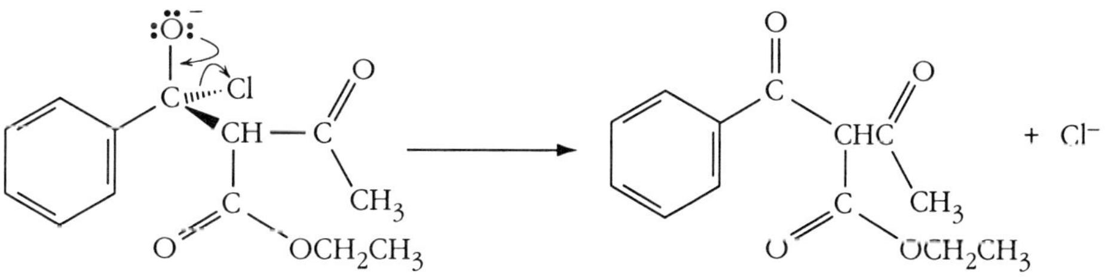

24.11

(a) PhCH=CHCOOCH₂CH₃ ⟹ PhCH=CHCOOH ⟹ PhCHO + CH₂(COOCH₂CH₃)₂

PhCHO + CH₂(COOCH₂CH₃)₂ $\xrightarrow{\text{(1) RNH}_3^+ \text{}^-\text{OOCCH}_3/\text{benzene/heat} \quad \text{(2) NaOH/H}_2\text{O} \quad \text{(3) H}_3\text{O}^+/\text{heat}}$ PhCH=CHCOOH $\xrightarrow[\text{Acid catalyst}]{\text{CH}_3\text{CH}_2\text{OH}}$ PhCH=CHCOOCH₂CH₃

(b) CH₃COCH=CHPh ⟹ CH₃COCH₃ + PhCHO

CH₃COCH₃ + PhCHO $\xrightarrow{\text{(1) NaOH/H}_2\text{O} \quad \text{(2) H}_3\text{O}^+/\text{heat}}$ CH₃COCH=CH−Ph

(c) CH₂=CHCOOH ⟹ CH₃CHBrCOOH ⟹ CH₃CH₂COOH ⟹ CH₃CH₂Br ⟹ CH₃CH₂OH

CH₃CH₂OH $\xrightarrow{\text{PBr}_3}$ CH₃CH₂Br $\xrightarrow{\text{(1) Mg} \quad \text{(2) CO}_2 \quad \text{(3) H}_3\text{O}^+}$ CH₃CH₂COOH $\xrightarrow[\text{PBr}_3]{\text{Br}_2}$ CH₃CHBrCOOH $\xrightarrow{\text{(1) NaOH/H}_2\text{O/heat} \quad \text{(2) H}_3\text{O}^+}$ CH₂=CHCOOH

(d) [PhCH=CHCOOH, obtained from benzaldehyde (PhCH portion) and from acetic anhydride (=CHCOOH portion)] ⟹ PhCHO + (CH₃CO)₂O

PhCHO + (CH₃CO)₂O →[CH₃COO⁻ / CH₃COOH] PhCH(OH)CH₂C(O)OH (with OH on acid) →[H₃O⁺, Heat] PhCH=CHCOOH

24.12 Strongly basic nucleophiles such as CH₃Li and LiAlH₄ react predominantly or exclusively by 1,2-addition while weaker bases such as ⁻CN or NH₃ usually react by 1,4-addition.

(a) $CH_3\overset{O}{\overset{\|}{C}}CH=CHCH_3$ →[NaCN / H₂O] $CH_3\overset{O}{\overset{\|}{C}}CH_2CH(C\equiv N)CH_3$

(b) $CH_3\overset{O}{\overset{\|}{C}}CH=CHCH_3$ →[(1) LiAlH₄ (2) H₃O⁺] $CH_3CH(OH)CH=CHCH_3$

(c) $CH_3\overset{O}{\overset{\|}{C}}CH=CHCH_3$ →[(1) CH₃Li (2) H₃O⁺] $CH_3C(OH)(CH_3)CH=CHCH_3$

(d) $CH_3\overset{O}{\overset{\|}{C}}CH=CHCH_3$ →[NH₃ / H₂O] $CH_3\overset{O}{\overset{\|}{C}}CH_2CH(NH_2)CH_3$

24.13 (a) α,β-Unsaturated carbonyl compounds react with enolate anions to form a 1,4-addition product. This product is then hydrolyzed and decarboxylated by heating.

$$CH_2=CHCHO + \bar{C}H(COOCH_2CH_3)COCH_3 \longrightarrow CH_2(CH(CHO)H)(CH(CH_3)(COOCH_2CH_3)CO) \xrightarrow[\text{Heat}]{H_3O^+} CH_2CH_2CHO\text{-}CH_2COCH_3$$

(b) $CH_3CH=CHCOOCH_2CH_3 \xrightarrow{\bar{C}H(COOCH_2CH_3)COCH_3} CH_3CH(CH(CH_3)(COOCH_2CH_3)CO)\bar{C}HCOOCH_2CH_3 \xrightarrow[\text{Heat}]{H_3O^+} CH_3CHCH_2COOH\text{ with }CH_2COCH_3$

(c) $CH_2=CH-C\equiv N \xrightarrow[(2)\ H_3O^+/\text{heat}]{(1)\ \bar{C}H(COOCH_2CH_3)COCH_3} CH_2CH_2COOH\text{ with }CH_2COCH_3$

(d) cyclohexenone + $\bar{C}H(COOCH_2CH_3)COCH_3 \longrightarrow$ 3-substituted cyclohexanone with $\bar{C}H(COOCH_2CH_3)COCH_3$ group $\xrightarrow[\text{Heat}]{H_3O^+}$ 3-(2-oxopropyl)cyclohexanone

24.14

(a) $\underset{\text{O}}{\text{CH}_3\overset{\|}{\text{C}}\text{CH}=\text{CHCH}_3}$ + $^-\text{CH(COOCH}_2\text{CH}_3)_2$ ⟶ $\text{CH}_3\overset{\overset{\text{O}}{\|}}{\text{C}}\text{CHCHCH}_3$
　　　　　　　　　　　　　　　　　　　　　　　　　　　　　　　　　　　　　　|
　　　　　　　　　　　　　　　　　　　　　　　　　　　　　　　　　　　　　CHCOOCH$_2$CH$_3$
　　　　　　　　　　　　　　　　　　　　　　　　　　　　　　　　　　　　　　|
　　　　　　　　　　　　　　　　　　　　　　　　　　　　　　　　　　　　　COOCH$_2$CH$_3$

$\xrightarrow[\text{Heat}]{\text{H}_3\text{O}^+}$ $\text{CH}_3\overset{\overset{\text{O}}{\|}}{\text{C}}\text{CH}_2\text{CHCH}_3$
　　　　　　　　　　　　　　|
　　　　　　　　　　CH$_2$COOH

(b) $\text{CH}_3\overset{\overset{\text{O}}{\|}}{\text{C}}\text{CH}=\text{CHCH}_3$ + $\text{CH}_3\text{CH}_2\overset{\overset{\text{O}}{\|}}{\text{C}}\underset{\underset{\underset{\text{O}}{\|}}{\text{CH}_3\text{CH}_2\text{C}}}{\overset{|}{\text{CH}}}$ ⟶ $\text{CH}_3\overset{\overset{\text{O}}{\|}}{\text{C}}-\overset{|}{\text{C}}\text{HCHCH}_3$
　　　O=C　CH　C=O
　　　|　　　　　|
　　CH$_2$CH$_3$　CH$_2$CH$_3$

$\xrightarrow{\text{H}_3\text{O}^+}$ $\text{CH}_3\overset{\overset{\text{O}}{\|}}{\text{C}}\text{CH}_2\text{CHCH}_3$
　　　　　　　　　　　　　　　|
　　　　　　　CH$_3$CH$_2$CCHCCH$_2$CH$_3$
　　　　　　　　　　　　‖　　‖
　　　　　　　　　　　 O　 O

(c) $\text{CH}_3\overset{\overset{\text{O}}{\|}}{\text{C}}\text{CH}=\text{CHCH}_3$ + $^-\overset{\overset{\text{O}}{\|}}{\text{C}}\text{HC}\underset{\text{COOCH}_2\text{CH}_3}{\overset{\text{CH}_3}{{<}}}$ ⟶ $\text{CH}_3\overset{\overset{\text{O}}{\|}}{\text{C}}\text{CHCHCH}_3$
　　|
　　　　　　　　　　　　　　　　　　　　　　　　　　　　　　　　　　　　　　CHCCH$_3$
　　　　　　　　　　　　　　　　　　　　　　　　　　　　　　　　CH$_3$CH$_2$OOC　‖
　　O

$\xrightarrow{\text{H}_3\text{O}^+}$ $\text{CH}_3\overset{\overset{\text{O}}{\|}}{\text{C}}\text{CH}_2\text{CHCH}_3$
　　　　　　　　　　　　　　|
　　　　　　　　　CH$_2$CCH$_3$
　　　　　　　　　　　　‖
　　　　　　　　　　　 O

(d) $\text{CH}_3\overset{\overset{\text{O}}{\|}}{\text{C}}\text{CH}=\text{CHCH}_3$ + $\text{CH}_3\overset{\overset{\text{NO}_2}{|}}{\underset{}{\text{C}}}\text{CH}_3$ ⟶ $\text{CH}_3\overset{\overset{\text{O}}{\|}}{\text{C}}\text{CHCHCH}_3$ $\xrightarrow{\text{H}_3\text{O}^+}$ $\text{CH}_3\overset{\overset{\text{O}}{\|}}{\text{C}}\text{CH}_2\text{CHCH}_3$
　　|　　　　　　　　　　　　　|
　　　　　　　　　　　　　　　　　　　　　　　　　　　　　　　　　　　　CH$_3$—C—CH$_3$　　　　CH$_3$—C—CH$_3$
　　　　　　　　　　　　　　　　　　　　　　　　　　　　　　　　　　　　　　　|　　　　　　　　　　　　　　|
　　　　　　　　　　　　　　　　　　　　　　　　　　　　　　　　　　　　　　NO$_2$　　　　　　　　　　　NO$_2$

24.15

$$\text{CH}_3\text{COCH}_2\text{COOCH}_2\text{CH}_3 \xrightarrow[\text{CH}_3\text{CH}_2\text{OH}]{\text{NaOCH}_2\text{CH}_3} \text{CH}_3\text{CO}\overset{-}{\text{CH}}\text{COOCH}_2\text{CH}_3$$

$$\text{CH}_3\text{COCH}=\text{CH}_2 + \text{CH}_3\text{CO}\overset{-}{\text{CH}}\text{COOCH}_2\text{CH}_3 \longrightarrow \text{CH}_3\text{CO}\overset{-}{\text{C}}(\text{COCH}_3)(\text{COOCH}_2\text{CH}_3)\text{CH}_2\text{CH}_2\text{—see structure}$$

Intermediate: CH$_3$COCHCH$_2$CH(COCH$_3$)(COOCH$_2$CH$_3$)

$$\text{CH}_3\text{COCHCH}_2\text{CH}(\text{COCH}_3)(\text{COOCH}_2\text{CH}_3) \xrightarrow[\text{Heat}]{\text{H}_3\text{O}^+} \text{CH}_3\text{COCH}_2\text{CH}_2\text{CH}_2\text{COCH}_3 + \text{CO}_2 + \text{CH}_3\text{CH}_2\text{OH}$$

24.16

(a) $\text{CH}_2(\text{COOCH}_2\text{CH}_3)_2 \xrightarrow[\text{CH}_3\text{CH}_2\text{OH}]{\text{NaOCH}_2\text{CH}_3} {}^-\text{CH}(\text{COOCH}_2\text{CH}_3)_2 \xrightarrow{\text{CH}_3\text{CH}=\text{CHCOOCH}_2\text{CH}_3}$

$$\text{CH}_3\text{CH}(\text{CH}(\text{COOCH}_2\text{CH}_3)_2)(\text{CH}_2\text{COOCH}_2\text{CH}_3) \xrightarrow[\text{H}_3\text{O}^+]{\text{Heat}} \text{CH}_3\text{CH}(\text{CH}_2\text{COOH})_2$$

(b) $\text{H}_2\text{C}(\text{COOCH}_2\text{CH}_3)(\text{C}\equiv\text{N}) \xrightarrow[\text{CH}_3\text{CH}_2\text{OH}]{\text{NaOCH}_2\text{CH}_3} \text{HC}^-(\text{COOCH}_2\text{CH}_3)(\text{C}\equiv\text{N}) \xrightarrow{\text{CH}_2=\text{CHCHO}}$

$$\text{H—C}(\text{CH}_2\text{CH}_2\text{CHO})(\text{COOCH}_2\text{CH}_3)(\text{C}\equiv\text{N}) \xrightarrow[\text{Heat}]{\text{H}_3\text{O}^+} \text{CH}(\text{CH}_2\text{CH}_2\text{CHO})(\text{CH}_2\text{COOH})$$

(c) 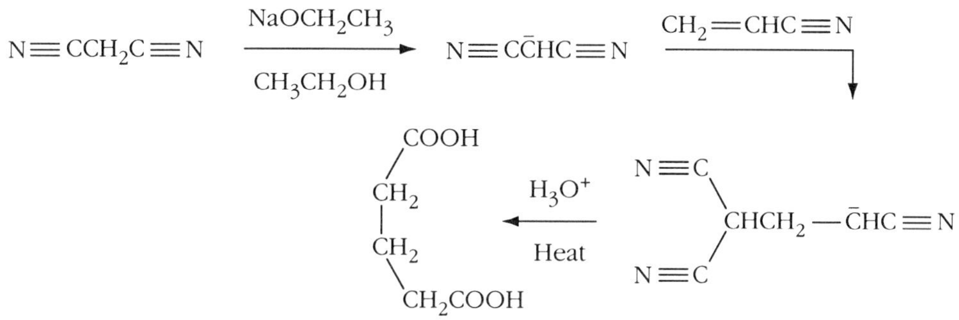 $N\equiv CCH_2C\equiv N$ $\xrightarrow[CH_3CH_2OH]{NaOCH_2CH_3}$ $N\equiv C\bar{C}HC\equiv N$ $\xrightarrow{CH_2=CHC\equiv N}$

$\underset{\text{Heat}}{\xleftarrow{H_3O^+}}$

Left structure: COOH–CH$_2$–CH$_2$–CH(COOH)–CH$_2$COOH chain (glutaric-type: HOOC–CH$_2$–CH$_2$–CH(—)–CH$_2$COOH)

Right structure: (N≡C)$_2$CH–CH$_2$–CH(C≡N)–CH$_2$–... — $(N\equiv C)_2CHCH_2-\bar{C}HC\equiv N$

24.17

(a) 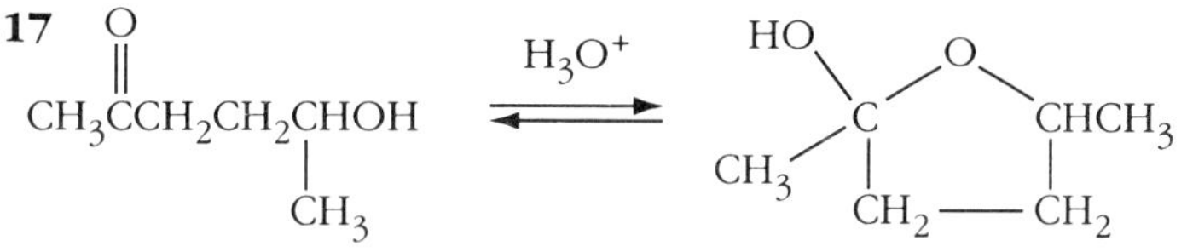 $CH_3\overset{O}{\underset{\|}{C}}CH_2CH_2\overset{}{\underset{CH_3}{C}}HOH$ $\xrightleftharpoons{H_3O^+}$ (cyclic hemiacetal with HO, CH$_3$ on one carbon; O ring; CHCH$_3$; CH$_2$—CH$_2$)

(A cyclic hemiacetal)

(b) 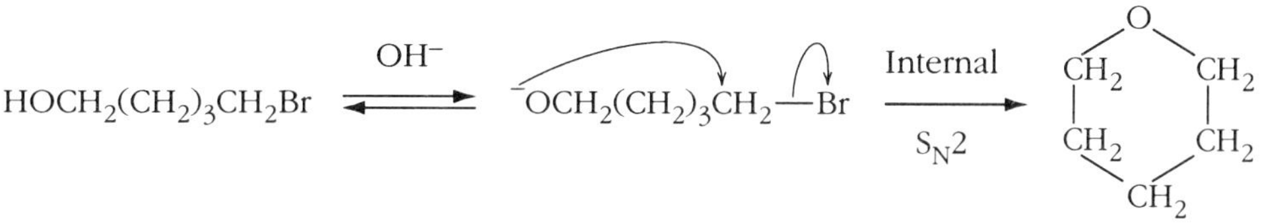 $HOCH_2(CH_2)_3CH_2Br$ $\xrightleftharpoons{OH^-}$ $\bar{O}CH_2(CH_2)_3CH_2\!-\!Br$ $\xrightarrow[S_N2]{\text{Internal}}$ (cyclic ether: oxane ring)

(c) 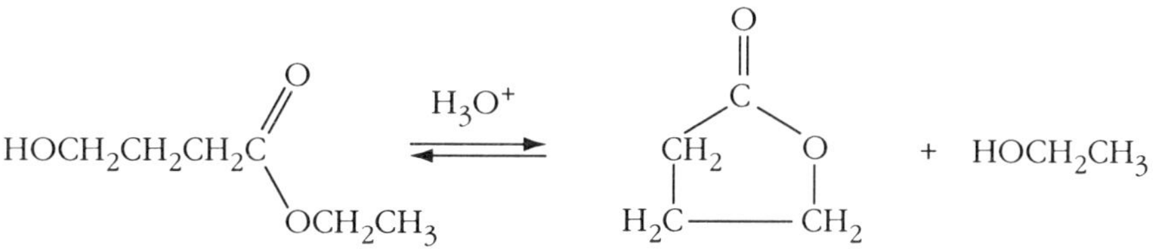 $HOCH_2CH_2CH_2\overset{O}{\underset{OCH_2CH_3}{C}}$ $\xrightleftharpoons{H_3O^+}$ (γ-butyrolactone ring) $+\ HOCH_2CH_3$

Lactone

24.18

(a) 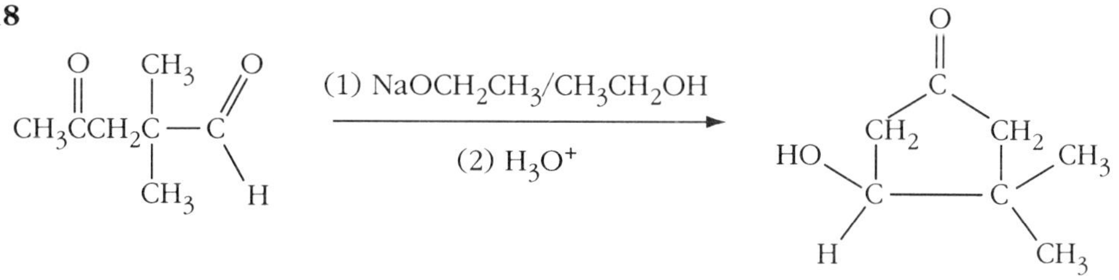 $CH_3\overset{O}{\underset{}{C}}CH_2\underset{CH_3}{\overset{CH_3}{C}}\overset{O}{\underset{H}{C}}$ $\xrightarrow[(2)\ H_3O^+]{(1)\ NaOCH_2CH_3/CH_3CH_2OH}$ (cyclopentanone with HO, H on one α-carbon; CH$_3$, CH$_3$ on the other α-carbon)

(b) $NCCH_2CH_2CH_2CH_2CN \underset{CH_3CH_2OH}{\overset{NaOCH_2CH_3}{\rightleftharpoons}} NCCH_2CH_2CH_2\bar{C}HCN$

[mechanism showing cyclization to iminyl anion intermediate, then H_3O^+/Heat to give 2-oxocyclopentanecarboxylic acid, which loses CO_2 to give cyclopentanone + CO_2]

(c) [2-(3-oxobutyl)cyclohexanone] $\underset{H_2O}{\overset{KOH}{\rightleftharpoons}}$ [enolate intermediate] → [bicyclic alkoxide intermediate] $\underset{Heat}{\overset{H_3O^+}{\longrightarrow}}$ [octahydronaphthalen-2(1H)-one with C=C]

24.19 Start by making the enolate anion of ethyl acetoacetate, then

[mechanism showing Michael addition of ethyl acetoacetate enolate to an α,β-unsaturated ketone, with proton transfer from $HOCH_2CH_3$]

24.20

24.21

24.22 A diol and a dicarboxylic acid form a polyester, which is a condensation polymer.

24.23 (a) A **1,3-dicarbonyl compound** has two carbonyl-containing functional groups bonded to the same carbon atom. Diethyl malonate and ethyl acetoacetate are two examples of 1,3-dicarbonyl compounds.

Diethyl malonate   Ethyl acetoacetate

(b) A **Michael addition reaction** is the 1,4-addition of enolate anions of 1,3-dicarbonyl compounds to α,β-unsaturated carbonyl compounds. For example:

$$CH_3C(=O)-CH=CH_2 \quad {}^-CH(COOCH_2CH_3)_2$$

$$\downarrow$$

$$CH_3C(\bar{O})=CH-CH_2-CH(COOCH_2CH_3)_2 \quad H-B$$

$$\downarrow$$

$$CH_3CCH_2CH_2CH(COOCH_2CH_3)_2$$
$$\quad \|$$
$$\quad O$$

(c) An α,β-**unsaturated carbonyl compound** contains a carbon-carbon double bond in conjugation with a carbonyl group. The double bond in an α,β-unsaturated carbonyl compound is polarized by the adjacent carbonyl group so that nucleophiles add to the β-position.

β-Carbon atom: the site of addition of nucleophiles

Propenal
(An α,β-unsaturated carbonyl compound)

(d) A **Michael acceptor** in the Michael addition reaction is the α,β-unsaturated carbonyl compound that adds a nucleophilic enolate anion to its β-carbon atom.

(e) A **condensation polymer** is formed by the reaction of two compounds, each having a different functional group. The other product of the reaction is a small molecule such as an alcohol or water. The reaction of terephthalic acid and 1,2-ethanediol forms a condensation polymer and water:

$$n \text{ HOOC-C}_6\text{H}_4\text{-COOH} + n \text{ HOCH}_2\text{CH}_2\text{OH} \xrightarrow{H_3O^+} \left[\text{-OC-C}_6\text{H}_4\text{-CO-OCH}_2\text{CH}_2\text{O-} \right]_n + n \text{ H}_2\text{O}$$

Terephthalic acid → Polyester

(f) A **Michael donor** is the nucleophilic enolate anion that adds to the β-carbon atom of an α,β-unsaturated carbonyl compound in a Michael addition reaction.

(g) The **Robinson annulation reaction** is a sequence of two reactions used to form a new six-membered ring. The first reaction is a Michael addition reaction followed by an aldol condensation reaction.

(h) The **Knoevenagel reaction** is the reaction of aldehydes or unhindered ketones with a compound containing an active methylene group to form a condensation product that contains a carbon-carbon double bond. The reaction is catalyzed by amines or amine salts.

$$\text{RCHO} + \text{H}_2\text{C}(\text{COR'})(\text{COR''}) \xrightarrow[\text{Benzene/heat}]{\text{RNH}_3^+ \text{}^-\text{OOCCH}_3} \text{RCH=C}(\text{COR'})(\text{COR''})$$

(i) An **intramolecular reaction** is a reaction that occurs within a molecule, which often results in the formation of a ring.

(j) The **malonic ester synthesis** is a sequence of reactions in which diethyl malonate is converted into a carboxylic acid. The first reaction is the quantitative formation of the enolate anion of diethyl malonate. Alkylation of the enolate anion forms a diethyl alkylmalonate as product. Diethylalkylmalonate still has an α-hydrogen atom so the alkylation step can be repeated. Hydrolysis and decarboxylation of either the monoalkyl or dialkyl malonic ester forms a carboxylic acid.

(k) The **acetoacetic ester synthesis** is a sequence of reactions in which ethyl acetoacetate is converted into a methyl ketone. The first reaction is the quantitative formation of the enolate anion of ethyl acetoacetate. Alkylation of the enolate anion forms an ethyl alkylacetoacetate as a product. Ethyl alkylacetoacetate still has an α-hydrogen atom so the alkylation step can be repeated. Hydrolysis and decarboxylation of either the monoalkyl or dialkylalkylacetoacetate forms a methyl ketone.

(l) An **intermolecular reaction** is a reaction that occurs between two separate molecules.

24.24 (a) 2-Oxocyclohexanecarbaldehyde
(b) 3-Isopropyl-2-cyclohexenone or 3-(1-methylethyl)-2-cyclohexenone
(c) 5,5-Dimethyl-1,3-cyclohexanedione
(d) Methyl 4-oxo-2-pentenoate
(e) 2-Cyclohexene-1,4-dione
(f) Phenyl 3-oxo-4-phenylpentanoate

24.25

(a) $CH_3\overset{O}{\overset{\|}{C}}CH=C(CH_3)_2$

(b) [structure: 2,2-dimethyl-1,3-cyclohexanedione]

(c) [structure: PhC(O)CH$_2$CH(Ph)C≡N]

(d) [structure: cyclopentanone with -C(O)OCH$_2$CH$_3$ substituent]

(e) $CH_3CH_2O\overset{O}{\overset{\|}{C}}-\underset{CH_3}{\overset{|}{CH}}-\underset{\overset{\|}{O}}{\overset{CH_2CH_3}{\overset{|}{C}}}-CH(CH_2)_3CH_3$

24.26

(a) [cyclopentane with two COOH groups] $\xrightarrow{150\ °C}$ [cyclopentane-COOH] + CO_2

(b) $CH_2(COOCH_2CH_3)_2 \xrightarrow[\substack{(2)\ BrCH_2(CH_2)_6CH_3 \\ (3)\ H_3O^+/heat}]{(1)\ NaOCH_2CH_3/CH_3CH_2OH} CH_3(CH_2)_6CH_2CH_2COOH$

(c) $(CH_3)_2C=CHCCH_3 \xrightarrow[(2)\ H_3O^+]{(1)\ CH_3MgI} (CH_3)_2C=CHC(OH)(CH_3)CH_3$

(with C=O on left and C(OH)(CH_3) on right)

(d) HO–(CH$_2$)$_3$–CHO $\xrightarrow{H_3O^+}$ 2-hydroxytetrahydropyran

(e) $(CH_3)_2C=CHCHO \xrightarrow[CH_3OH]{NaCN} (CH_3)_2C(CN)-CH_2CHO$

(f) PhCOCH$_3$ + PhCHO $\xrightarrow[(2)\ H_3O^+/heat]{(1)\ NaOCH_2CH_3/CH_3CH_2OH}$ PhCO–CH=CH–Ph

(g) CH$_3$COCH$_2$CH$_2$COCH$_3$ $\xrightarrow[(2)\ H_3O^+/heat]{(1)\ NaOCH_2CH_3/CH_3CH_2OH}$ 3-methylcyclopent-2-enone

(h) $HOCH_2(CH_2)_2CH_2Br \xrightarrow[\text{Hexane}]{\text{NaH}}$ tetrahydrofuran ring (O with H_2C—CH_2—CH_2—CH_2)

(i) $CH_3\overset{O}{\underset{SCoA}{C}} + CO_2 \xrightarrow{\text{Acetyl CoA carboxylase}} \,^-O\overset{O}{\underset{}{C}}CH_2\overset{O}{\underset{SCoA}{C}}$

(j) $(CH_3)_2C=CH\overset{O}{\underset{}{C}}CH_3 \xrightarrow{CH_3NH_2} (CH_3)_2\underset{NHCH_3}{C}-CH_2\overset{O}{\underset{}{C}}CH_3$

(k) $Ph-CH=CHC\overset{O}{\underset{H}{}} + H_2\underset{\underset{O}{\overset{\|}{C}-OCH_2CH_3}}{\overset{\overset{O}{\|}{CCH_3}}{C}} \xrightarrow{\text{piperidine}} Ph-CH=CHCH=\underset{\underset{O}{\overset{\|}{C}-OCH_2CH_3}}{\overset{\overset{O}{\|}{CCH_3}}{C}}$

(l) 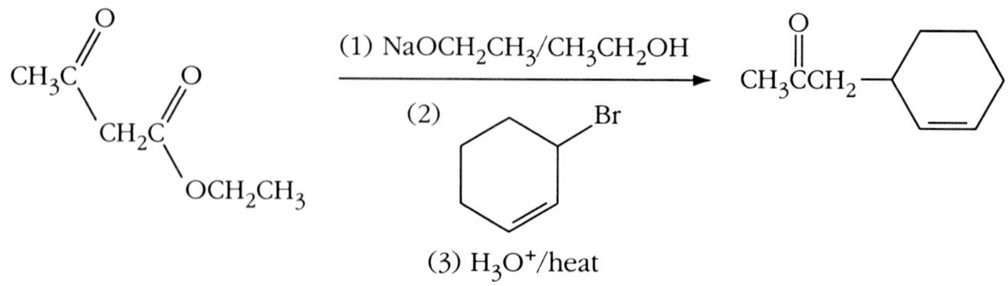 $CH_3C(O)CH_2C(O)OCH_2CH_3$

(1) NaOCH$_2$CH$_3$/CH$_3$CH$_2$OH
(2) [3-bromocyclohexene]
(3) H$_3$O$^+$/heat

→ CH_3CCH_2—[cyclohexenyl]

24.27 A = NaCN or KCN in an alcohol (CH$_3$OH or CH$_3$CH$_2$OH)
B = (1) NaOCH$_2$CH$_3$/CH$_3$CH$_2$OH (2) H$_2$SO$_4$/H$_2$O/heat
C = Excess CH$_3$OH with an acid catalyst (e.g., H$_2$SO$_4$)
D = (1) NaOCH$_2$CH$_3$/CH$_3$CH$_2$OH (2) BrCH$_2$CH=CH$_2$
E = H$_3$O$^+$/heat

24.28

(a) $CH_3CH(CH_3)CH_2COOH \implies CH_2(COOCH_2CH_3)_2 + (CH_3)_2CHBr$

$CH_2(COOCH_2CH_3)_2$
(1) NaOCH$_2$CH$_3$/CH$_3$CH$_2$OH
(2) (CH$_3$)$_2$CHBr
(3) H$_3$O$^+$/heat
→ $CH_3CH(CH_3)CH_2COOH$

(b) 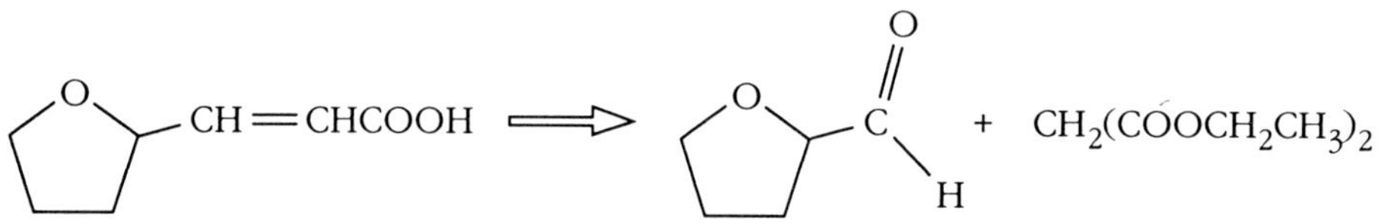 [furan-2-yl]—CH=CHCOOH $\implies$ [furan-2-yl]—C(=O)H + $CH_2(COOCH_2CH_3)_2$

[furan-2-yl]—C(=O)H + $CH_2(COOCH_2CH_3)_2$
(1) RNH$_3^+$ $^-$OOCCH$_3$
(2) NaOH/H$_2$O
(3) H$_3$O$^+$/heat
→ [furan-2-yl]—CH=CHCOOH

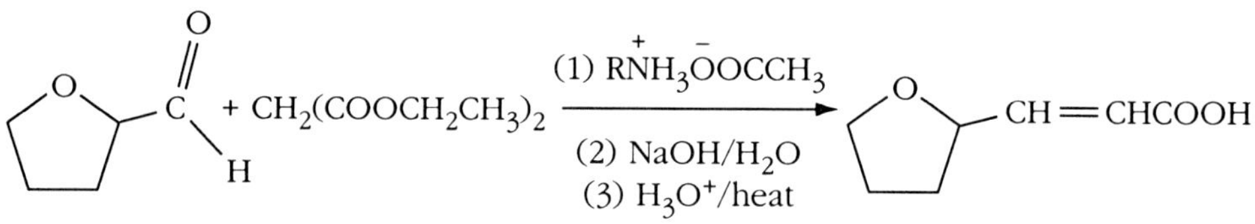

(c) [cyclopentane-COOH] ⟹ BrCH$_2$CH$_2$CH$_2$CH$_2$Br + CH$_2$(COOCH$_2$CH$_3$)$_2$

CH$_2$(COOCH$_2$CH$_3$)$_2$ $\xrightarrow{\text{(1) NaOCH}_2\text{CH}_3/\text{CH}_3\text{CH}_2\text{OH}}_{\text{(2) Br-CH}_2\text{CH}_2\text{CH}_2\text{CH}_2\text{-Br}}$ Br-CH$_2$CH$_2$CH$_2$CH$_2$-CH(COOCH$_2$CH$_3$)$_2$

$\xrightarrow{\text{NaOCH}_2\text{CH}_3/\text{CH}_3\text{CH}_2\text{OH}}$ [cyclopentane with C(COOCH$_2$CH$_3$)$_2$] $\xrightarrow{\text{H}_3\text{O}^+/\text{heat}}$ [cyclopentane-COOH]

(d) [PhC(O)-CH(COOCH$_2$CH$_3$)$_2$] ⟹ PhC(O)Cl + CH$_2$(COOCH$_2$CH$_3$)$_2$

CH$_2$(COOCH$_2$CH$_3$)$_2$ $\xrightarrow[\text{(2) PhC(O)Cl}]{\text{(1) Na/benzene}}$ PhC(O)-CH(COOCH$_2$CH$_3$)$_2$

24.29

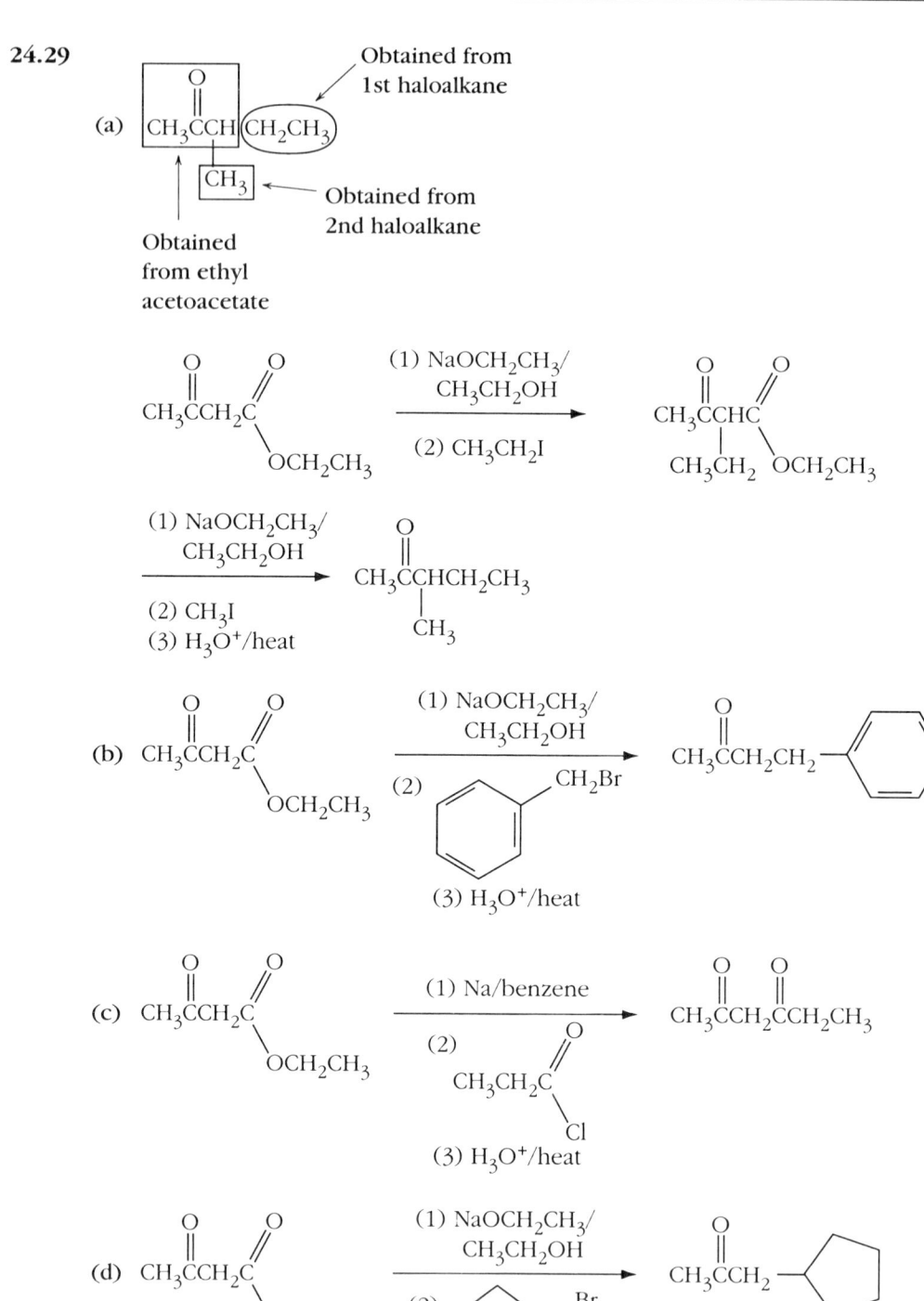

24.30

[Mechanism showing Dieckmann cyclization: methyl ester of a diester with α-hydrogen is deprotonated by methoxide, the enolate attacks the other ester carbonyl intramolecularly to form a tetrahedral intermediate, methoxide is eliminated to give a cyclic β-ketoester, which is deprotonated by methoxide, and finally protonated by H₃O⁺ to yield 2-(methoxycarbonyl)cyclopentanone + H₂O.]

24.31

(a) $CH_3\overset{O}{\overset{\|}{C}}CH_2CH_2CH_2COOH$ contains both a carboxylic acid group and a methyl ketone so either ethyl acetoacetate or diethyl malonate can be used as starting material.

$CH_3\overset{O}{\overset{\|}{C}}CH_2CH_2CH_2COOH \implies CH_3\overset{O}{\overset{\|}{C}}CH_2\overset{O}{\overset{\|}{C}}OCH_2CH_3 \;+\; BrCH_2CH_2\overset{O}{\overset{\|}{C}}OCH_2CH_3$

$\Downarrow$

$CH_2(COOCH_2CH_3)_2 \;+\; CH_3\overset{O}{\overset{\|}{C}}CH_2CH_2Br$

$CH_3\overset{O}{\overset{\|}{C}}CH_2COOCH_2CH_3 \quad\xrightarrow[\text{(2) BrCH}_2\text{CH}_2\text{COOCH}_2\text{CH}_3]{\text{(1) NaOCH}_2\text{CH}_3/\text{CH}_3\text{CH}_2\text{OH}}\quad CH_3\overset{O}{\overset{\|}{C}}CH_2CH_2CH_2COOH$
(3) H₃O⁺/heat

(b) $CH_3\overset{O}{\overset{\|}{C}}CH_2\overset{O}{\overset{\|}{C}}OCH_2CH_3 \quad\xrightarrow[\substack{\text{(2) CH}_3(\text{CH}_2)_2\text{CHO}\\\text{(3) H}_3\text{O}^+/\text{heat}}]{\substack{\text{(1) R}\overset{+}{\text{NH}}_3\overset{-}{\text{O}}\text{OCCH}_3\\\text{benzene/heat}}}\quad CH_3CH_2CH_2CH=CHCCH_3$

(c) $CH_2(COOCH_2CH_3)_2 \quad\xrightarrow[\substack{\text{(2) cyclopentenyl-Br}\\\text{(3) H}_3\text{O}^+/\text{heat}}]{\text{(1) NaOCH}_2\text{CH}_3/\text{CH}_3\text{CH}_2\text{OH}}\quad$ (cyclopentenyl)–CH₂COOH

(d) $CH_2(COOCH_2CH_3)_2$ $\xrightarrow{(2)\ CH_3CH_2I}$ $CH_3CH_2CH(COOCH_2CH_3)_2$

$\downarrow$ (1) $NaOCH_2CH_3$/ CH_3CH_2OH
(2) CH_3I
(3) H_3O^+/heat

$CH_3CH_2\underset{\underset{CH_3}{|}}{C}H\underset{\underset{OCH_2CH_3}{}}{\overset{\overset{O}{\|}}{C}}$ $\xleftarrow[\text{Acid catalyst}]{\text{Excess } CH_3CH_2OH}$ $CH_3CH_2\underset{\underset{CH_3}{|}}{C}HCOOH$

24.32

(a) PhCHO + $CH_2(COOCH_2CH_3)_2$ $\xrightarrow[(2)\ H_3O^+/\text{heat}]{(1)\ NaOCH_2CH_3/\ CH_3CH_2OH}$ Ph−CH=CHCOOH

(b) $HOCH_2CH_2CH_2COOH$ $\xrightarrow{H_3O^+}$ γ-butyrolactone + H_2O

(c) ethyl 2-benzoylacetate $\xrightarrow[(3)\ H_3O^+/\text{heat}]{(1)\ NaOCH_2CH_3/CH_3CH_2OH\ (2)\ \text{5-methylhexyl bromide}}$ 1-phenyl-6-methyloctan-1-one (PhCO-CH₂CH₂CH₂CH₂CH(CH₃)CH₂CH₃)

(d) 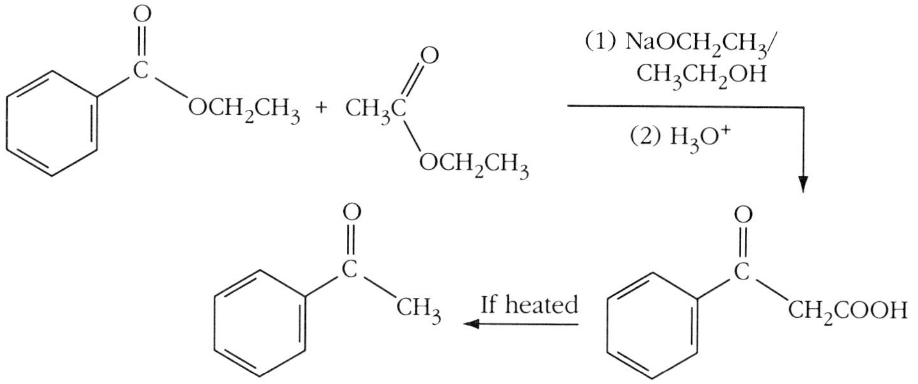 Ph-C(=O)-OCH₂CH₃ + CH₃C(=O)-OCH₂CH₃
(1) NaOCH₂CH₃/CH₃CH₂OH
(2) H₃O⁺
→ Ph-C(=O)-CH₂COOH —If heated→ Ph-C(=O)-CH₃

(e) 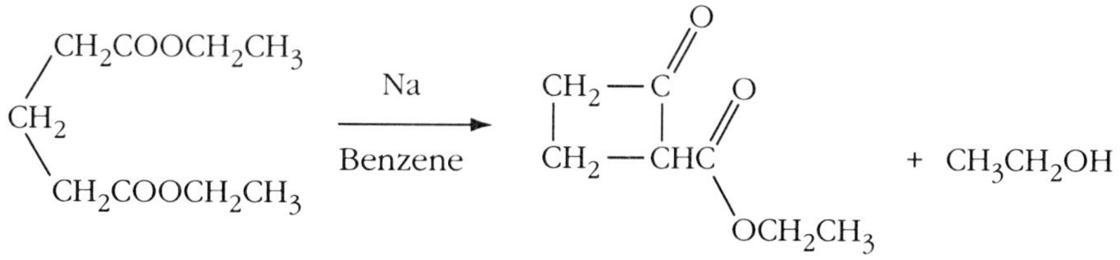 CH₂(CH₂COOCH₂CH₃)₂ —Na/Benzene→ cyclic β-ketoester + CH₃CH₂OH

(f) 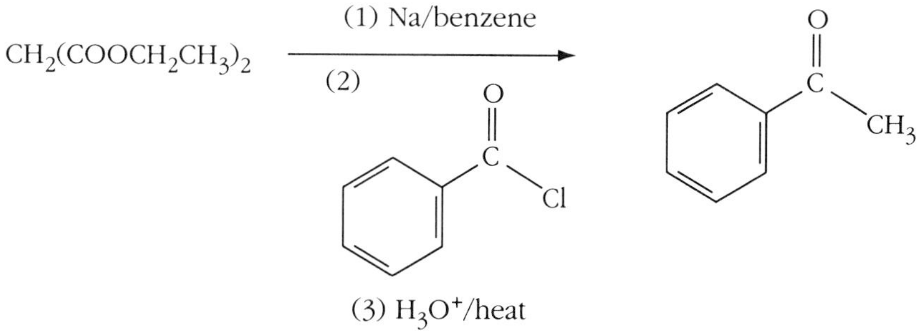 CH₂(COOCH₂CH₃)₂
(1) Na/benzene
(2) PhC(=O)Cl
(3) H₃O⁺/heat
→ Ph-C(=O)-CH₃

24.33

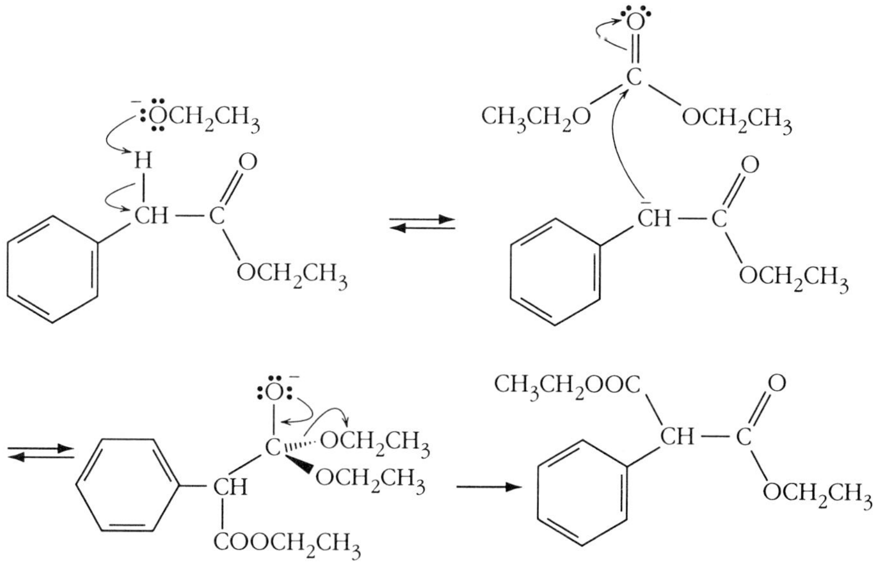

[Mechanism showing ethoxide deprotonation of ethyl phenylacetate, the resulting carbanion attacking diethyl carbonate, tetrahedral intermediate collapse, yielding diethyl phenylmalonate]

Diethyl phenylmalonate

Diethyl ethylphenylmalonate

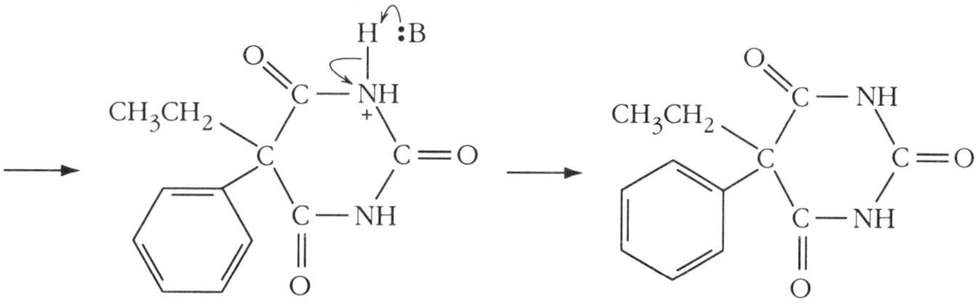

24.34

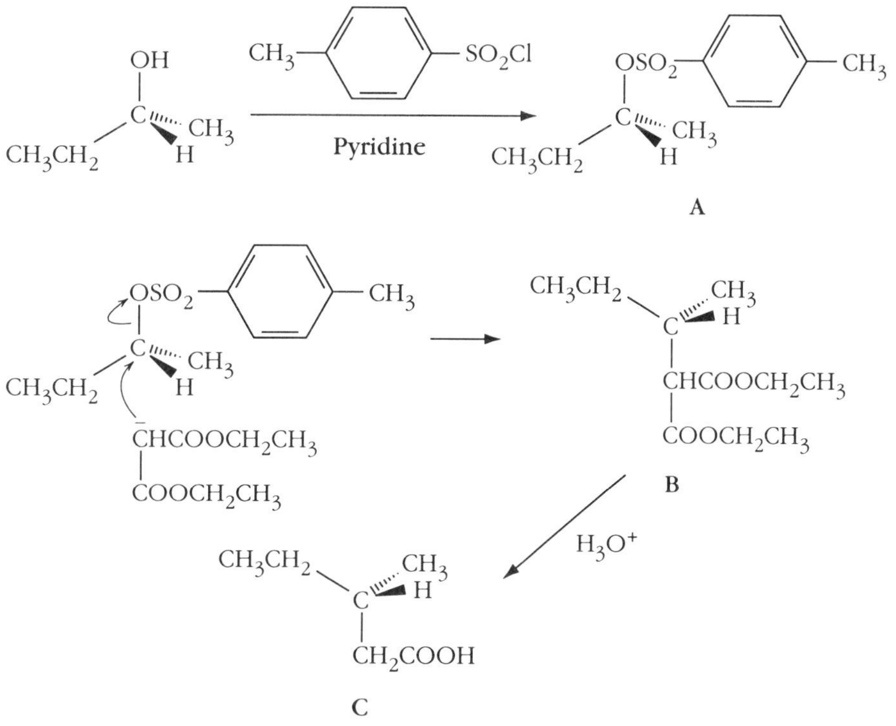

24.35

(a)

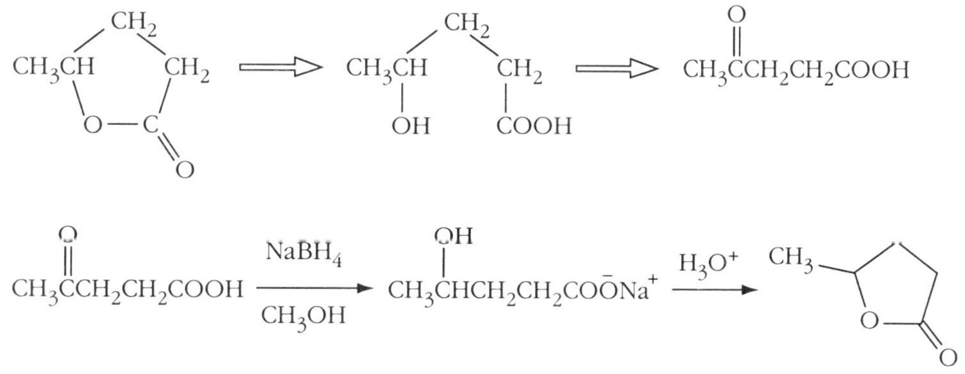

(b) [reaction scheme: cyclohexanol with ester side chain + NaH/Benzene → alkoxide attacks ester → tetrahedral intermediate → bicyclic lactone]

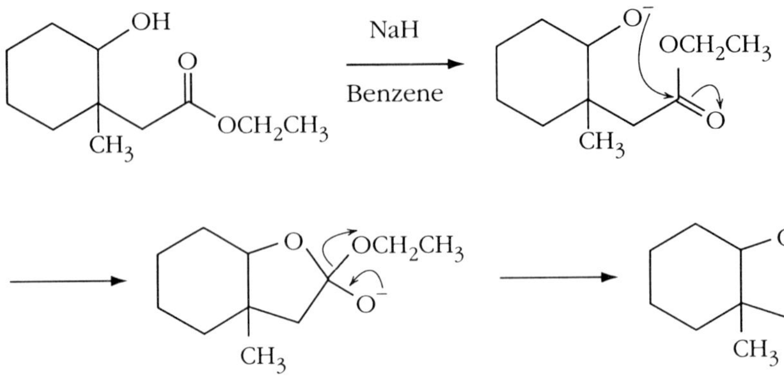

(c) $CH_3\overset{O}{\underset{||}{C}}CHCH_2$—Ph with CH_3 branch ⟹ $CH_3\overset{O}{\underset{||}{C}}CH_2COOCH_2CH_3$

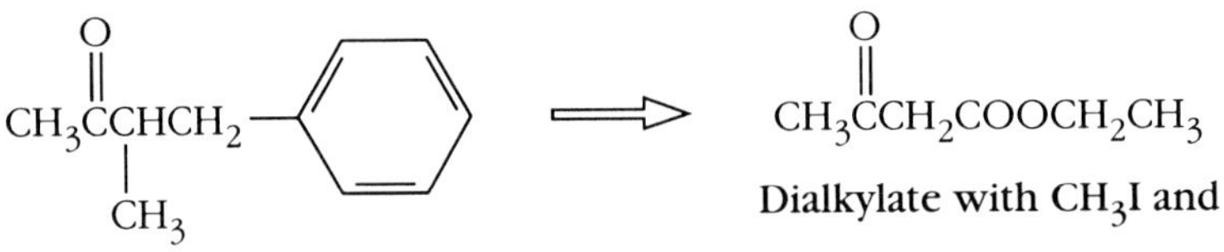

Dialkylate with CH_3I and

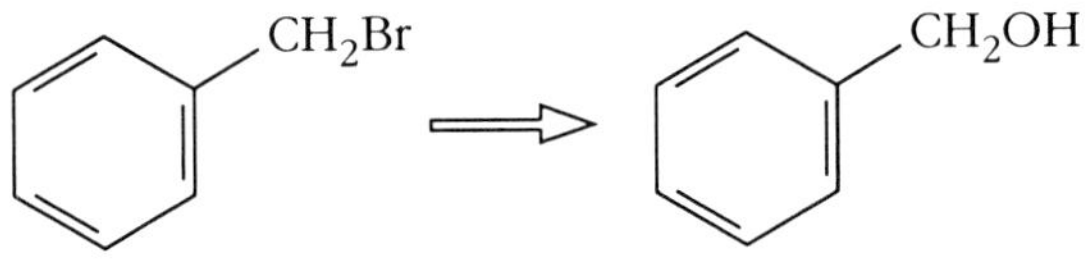

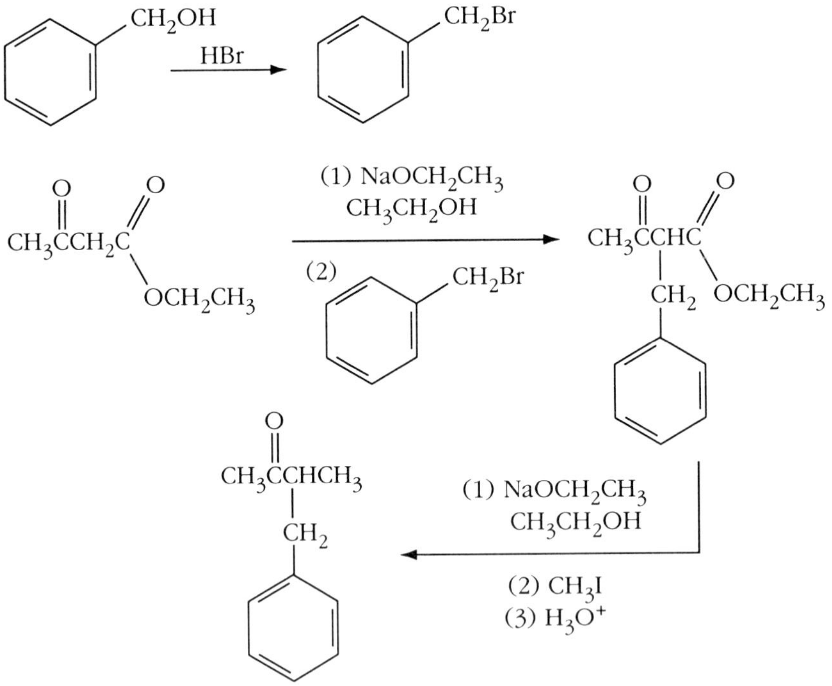

Ph–CH_2Br ⟹ Ph–CH_2OH

Ph–CH_2OH →(HBr)→ Ph–CH_2Br

$CH_3\overset{O}{\underset{||}{C}}CH_2\overset{O}{\underset{||}{C}}OCH_2CH_3$ →(1) $NaOCH_2CH_3$, CH_3CH_2OH (2) Ph–CH_2Br→ $CH_3\overset{O}{\underset{||}{C}}CH\overset{O}{\underset{||}{C}}OCH_2CH_3$ with CH_2–Ph

→(1) $NaOCH_2CH_3$, CH_3CH_2OH (2) CH_3I (3) H_3O^+→ $CH_3\overset{O}{\underset{||}{C}}CHCH_3$ with CH_2–Ph

(d)

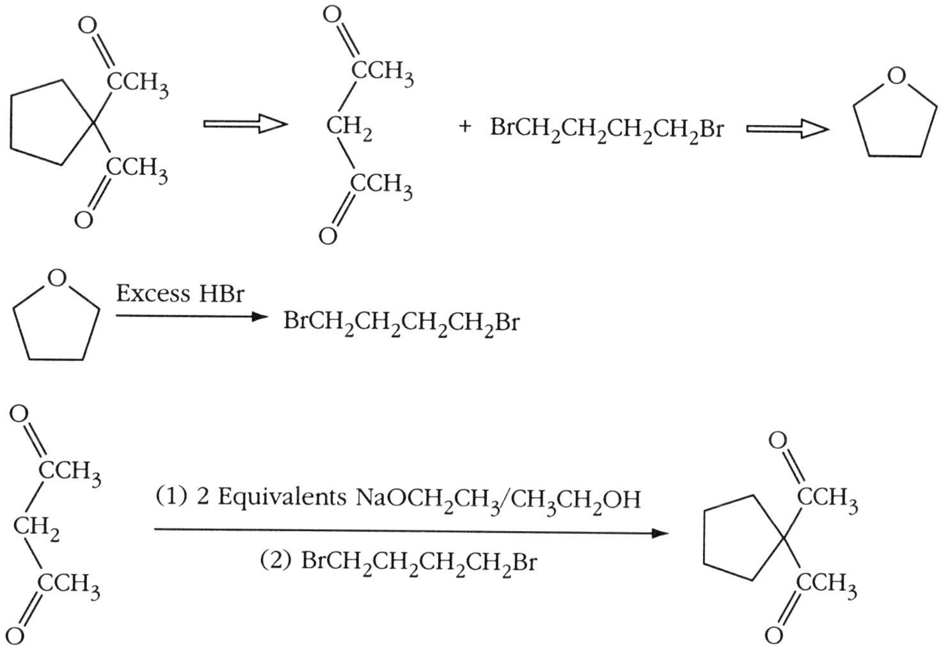

[Retrosynthesis: 1,1-diacetylcyclopentane ⇒ pentane-2,4-dione (acetylacetone) + BrCH$_2$CH$_2$CH$_2$CH$_2$Br ⇒ tetrahydrofuran]

Tetrahydrofuran $\xrightarrow{\text{Excess HBr}}$ BrCH$_2$CH$_2$CH$_2$CH$_2$Br

Pentane-2,4-dione $\xrightarrow[\text{(2) BrCH}_2\text{CH}_2\text{CH}_2\text{CH}_2\text{Br}]{\text{(1) 2 Equivalents NaOCH}_2\text{CH}_3/\text{CH}_3\text{CH}_2\text{OH}}$ 1,1-diacetylcyclopentane

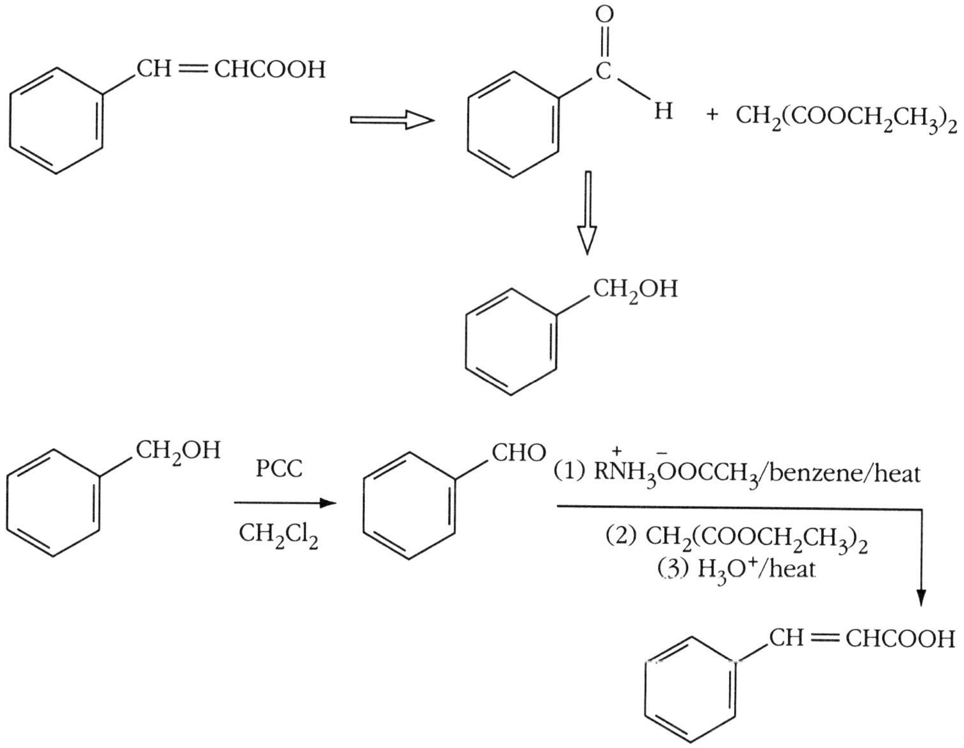

(e)

PhCH=CHCOOH ⇒ PhCHO + CH$_2$(COOCH$_2$CH$_3$)$_2$

PhCHO ⇒ PhCH$_2$OH

PhCH$_2$OH $\xrightarrow[\text{CH}_2\text{Cl}_2]{\text{PCC}}$ PhCHO $\xrightarrow[\substack{\text{(2) CH}_2\text{(COOCH}_2\text{CH}_3)_2 \\ \text{(3) H}_3\text{O}^+/\text{heat}}]{\text{(1) R}\overset{+}{\text{N}}\text{H}_3\overset{-}{\text{O}}\text{OCCH}_3/\text{benzene/heat}}$ PhCH=CHCOOH

24.36

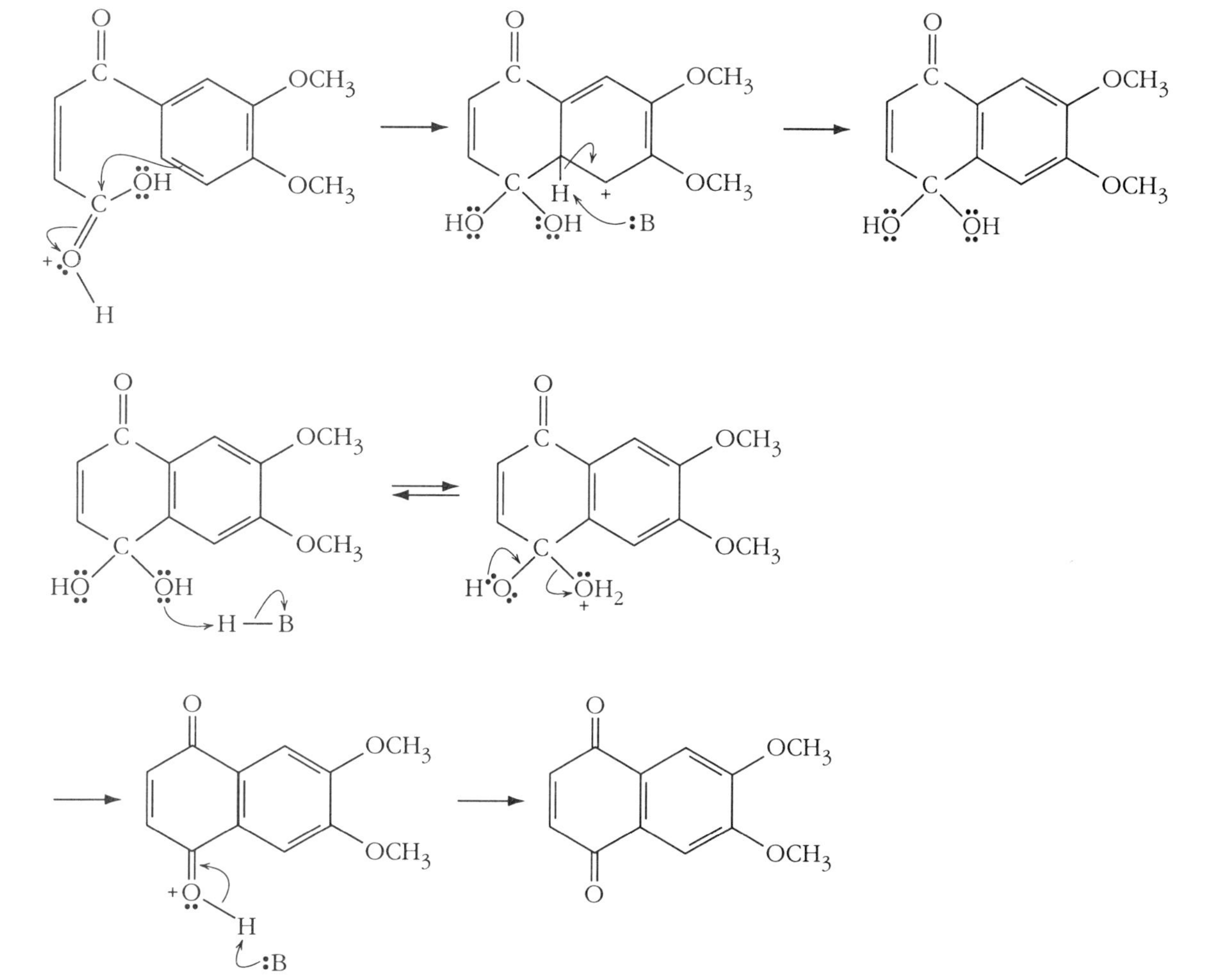

24.37

(a)

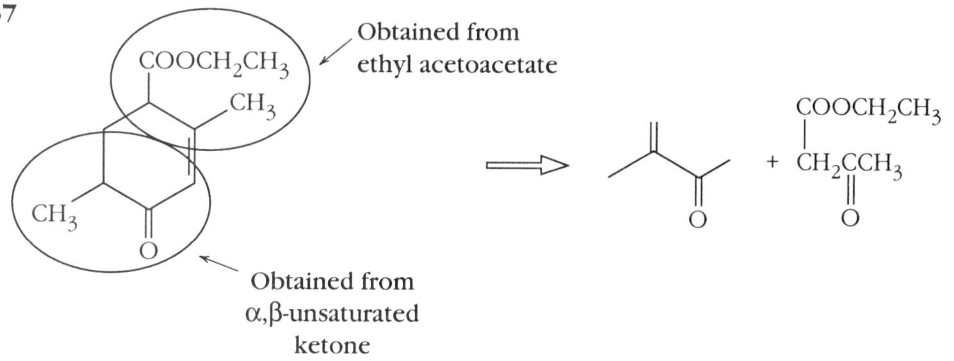
COOCH2CH3
CH3
CH3
O
Obtained from
ethyl acetoacetate
Obtained from
α,β-unsaturated
ketone
O
+
COOCH2CH3
CH2CCH3
O

(c)

Obtained from α,β-unsaturated ketone (CH₃-substituted portion with C=O)
Obtained from cyclohexanone

⟹ cyclohexanone + ethyl vinyl ketone (CH₂=CH–CO–CH₂CH₃)

cyclohexanone + CH₂=CH–C(=O)–CH₂CH₃ $\xrightarrow[\text{CH}_3\text{CH}_2\text{OH}]{\text{Michael addition} \atop \text{NaOCH}_2\text{CH}_3}$ (cyclohexanone α-substituted with –CH₂–CH(–)–C(=O)–CH₂CH₃, with CH₃ group)

$\xrightarrow{\text{Mixed Aldol}}$ 1-methyl-bicyclic enone (octahydronaphthalenone with CH₃ on the α,β-unsaturated ketone)

24.38

$$\text{CH}_3\text{C}(=\overset{..}{\overset{\displaystyle ..}{\text{O}}})\text{CH}=\text{CH}_2 \;+\; \text{H–Cl}$$

$$\Updownarrow$$

$$\left[\; \underset{\text{CH}_3\overset{\|}{\text{C}}-\text{CH}=\text{CH}_2}{\overset{\overset{+}{\text{O}}-\text{H}}{}} \;\longleftrightarrow\; \underset{\text{CH}_3\overset{+}{\text{C}}-\text{CH}=\text{CH}_2}{\overset{:\!\overset{..}{\text{O}}\text{H}}{}} \;\longleftrightarrow\; \underset{\text{CH}_3\text{C}=\text{CH}-\overset{+}{\text{C}}\text{H}_2}{\overset{:\!\overset{..}{\text{O}}\text{H}}{}} \;\right]$$

$\downarrow$ Cl⁻

$$\underset{\text{CH}_3\text{C}=\text{CH}-\text{CH}_2\text{Cl}}{\overset{:\!\overset{..}{\text{O}}\text{H}}{}} \;\dashrightarrow\; \text{CH}_3\overset{\text{O}}{\overset{\|}{\text{C}}}\text{CH}_2\text{CH}_2\text{Cl}$$

The $\text{CH}_3\text{C}(\!\!\begin{array}{c}\diagup\!\!\!\text{O}\\ \diagdown\end{array}\!\!)$ group is electron withdrawing so electrophiles add to the double bond of 3-buten-2-one more slowly than to the double bond of simple alkenes.

24.39 Formation of enolate anion of diethyl malonate:

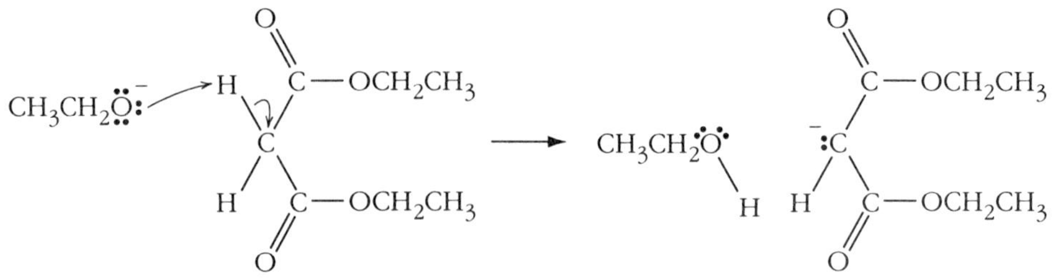

$$CH_3CH_2\ddot{\underset{..}{O}}{:}^- + \underset{\underset{O}{\overset{H}{\underset{|}{C}}}}{\overset{\overset{O}{\|}}{\underset{|}{C}}}\overset{H}{\underset{H}{\overset{\|}{C}-OCH_2CH_3}} \longrightarrow CH_3CH_2\ddot{\underset{..}{O}}-H + \underset{\underset{O}{\overset{H}{\underset{|}{C}}}}{\overset{\overset{O}{\|}}{\underset{|}{C}}}\overset{:C^-}{\underset{H}{\overset{\|}{C}-OCH_2CH_3}}$$

Reaction with first equivalent of formaldehyde:

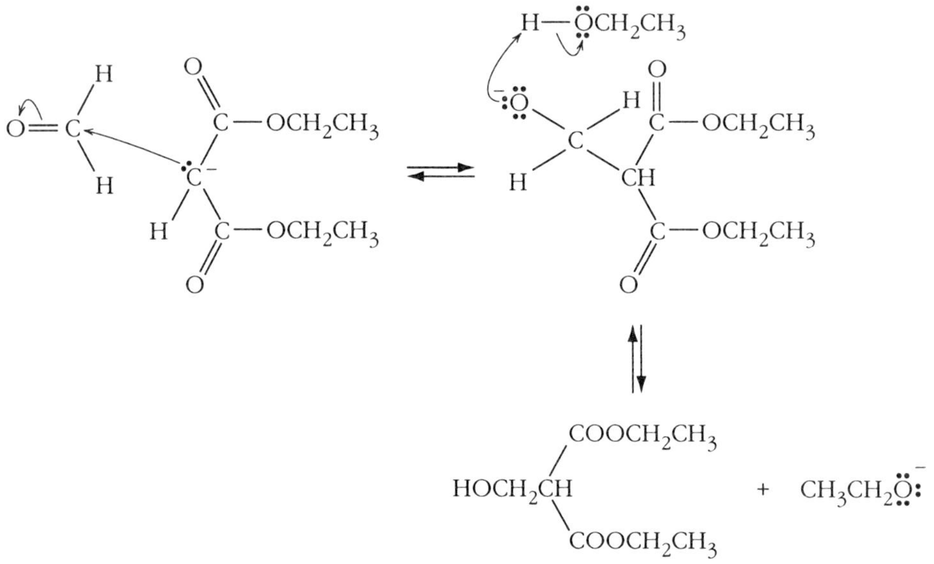

[Mechanism showing nucleophilic addition of malonate enolate to formaldehyde, followed by proton transfer from ethanol, yielding the alkoxide intermediate, which is protonated to give]

$$HOCH_2CH\underset{COOCH_2CH_3}{\overset{COOCH_2CH_3}{\diagup}} + CH_3CH_2\ddot{\underset{..}{O}}{:}^-$$

Reaction with second equivalent of formaldehyde:

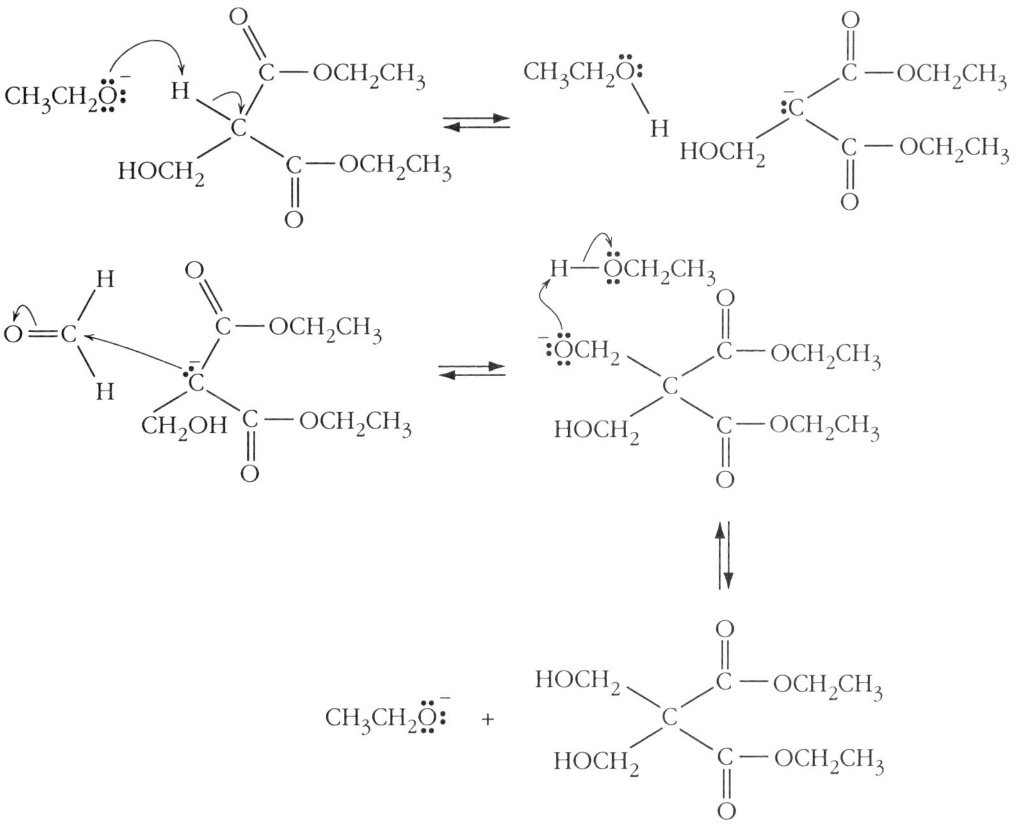

24.40 (a) The molecular formula indicates that this compound has two degrees of unsaturation. The absorption band at 1740 cm^{-1} in the IR spectrum indicates the presence of a carbonyl group of an ester. The carbonyl absorption is actually two bands close together, which suggests that the compound may be the diester of a dicarboxylic acid (see Section 16.4). The triplet δ 1.20 (3H) and the quartet at δ 4.15 (2H) are typical of an ethyl group attached to an oxygen atom or an ethoxy group bonded to a carbonyl carbon atom. There is another singlet at δ 3.22 (1H). The integrated areas, however, add up to only six hydrogen atoms instead of the 12 of the molecular formula. This means that the actual ratio is 6:4:2. Therefore, there are two equivalent ethyl groups in the molecule, one methylene group, and two ester groups. Joining these partial structures gives the structure of diethyl malonate:

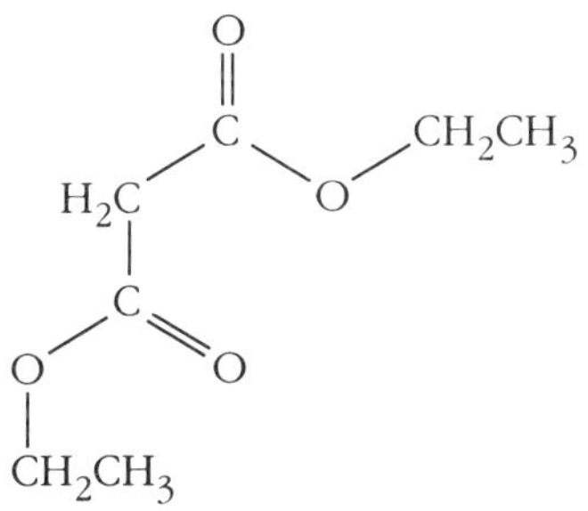

Diethyl malonate

(b) The molecular formula indicates that this compound has two degrees of unsaturation. The weak absorption band at about 2100 cm^{-1} in the IR spectrum indicates the presence of a carbon-carbon triple bond. The strong and sharp absorption band at about 3300 cm^{-1} indicates that the compound is a terminal alkyne. The strong and broad absorption band at about 3400 cm^{-1} indicates the presence of an alcohol functional group. All of the signals in the ^{1}H NMR spectrum are singlets so there is no coupling between the hydrogen atoms of the molecule. The singlet at δ 0.98 (9H) is due to the three equivalent methyl groups of a *tert*-butyl group, the singlet at δ 1.40 (3H) is due to a methyl group, and the sharp singlet at δ 2.40 (1H) is due to the hydrogen atom bonded to the carbon atom of the terminal triple bond. The broad peak at δ 2.3 (1H) is due to the hydrogen atom of the —OH group. We have identified four groups in the molecule, namely a *tert*-butyl group, a methyl group, a H—C≡C— group, and an —OH group that account for all but one carbon atom of the molecule. Attaching these four groups to the remaining carbon atom gives the following structure for the compound:

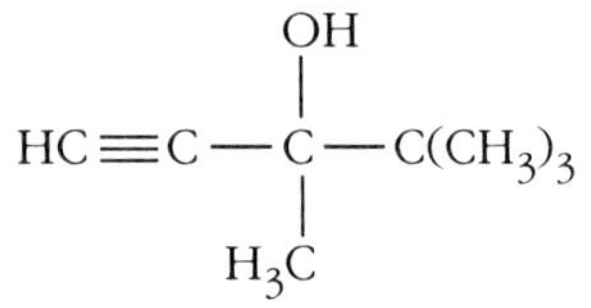

$$HC\equiv C-\underset{\underset{H_3C}{|}}{\overset{\overset{OH}{|}}{C}}-C(CH_3)_3$$

3,4,4-Trimethyl-1-pentyne-3-ol

(c) The ^{1}H NMR spectrum shows the typical coupling pattern of an ethyl group. Also, there are two singlets of relative area 3:2, which are due to a methyl and methylene group, respectively. An ethyl group, a methyl group, and a methylene group account for all the hydrogen atoms of the molecular formula and four of the six carbon atoms. The IR spectrum shows two absorption bands in the carbonyl region. The one at about 1740 cm^{-1} is due to an ester that is confirmed by the strong absorption band at about 1210 cm^{-1} due to a C—O bond stretching. The ester group accounts for a carbon atom and two oxygen atoms. The remaining carbon and oxygen atoms make up a carbonyl group that is responsible for the second absorption band in the carbonyl region of the IR spectrum. These various pieces can be combined in the two following ways:

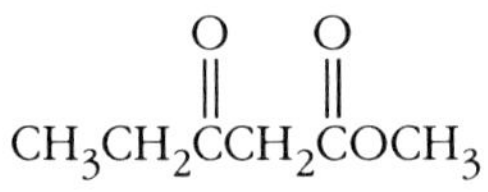

$$CH_3CH_2\overset{\overset{O}{\|}}{C}CH_2\overset{\overset{O}{\|}}{C}OCH_3 \quad \text{or} \quad CH_3\overset{\overset{O}{\|}}{C}CH_2\overset{\overset{O}{\|}}{C}OCH_2CH_3$$

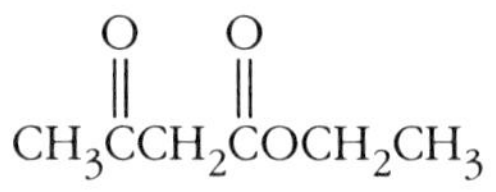

The chemical shift of the methylene quartet for the compound that has the ethyl group bonded to the carbonyl carbon atom would be expected at δ 5.2. The chemical shift of the methylene group of the compound that has the ethyl group bonded to an oxygen atom would be expected at δ 4.2. The quartet in the ^{1}H NMR spectrum appears at about δ 4.2 so the compound is ethyl acetoacetate:

$$\underset{\text{CH}_3\text{CCH}_2\text{COCH}_2\text{CH}_3}{\overset{\overset{\text{O}}{\|}\quad\overset{\text{O}}{\|}}{}}$$

(d) The ^{1}H NMR spectrum shows the typical coupling pattern of an ethyl group. A complicated multiplet in the aromatic region of relative area 5 indicates that the compound contains a monosubstituted benzene ring. The two singlets at δ 6.38 (1H) and δ 6.47 (1H) are due to vinyl hydrogen atoms. This is confirmed by the presence of an absorption band at about 1650 cm^{-1} in the IR spectrum. The absorption band at about 1730 cm^{-1} is due to the carbonyl group of an ester. This is confirmed by the presence of a strong absorption band at 1250 cm^{-1}. A monosubstituted benzene ring, a disubstituted carbon-carbon double bond, and an ester group and an ethyl group account for all the atoms in the molecule. These various parts can be combined in the two following ways:

Ph–C(=CH$_2$)–COOCH$_2$CH$_3$ or   Ph–CH=CHCOOCH$_2$CH$_3$

The chemical shifts indicate that the two vinyl hydrogen atoms are bonded to different carbon atoms so the compound is ethyl 3-phenylpropenoate (ethyl cinnamate):

Ph–CH=CHCOOCH$_2$CH$_3$

(e) There are five sets of signals in the ^{1}H NMR spectrum of this compound, namely a broad triplet at δ 0.82 (3H), a triplet overlapping a multiplet at around δ 1.2 (10H), a quartet at δ 1.82 (2H), a triplet at δ 3.23 (1H), and a quartet at δ 4.15 (4H). The quartet at δ 4.15 (4H) and the triplet at δ 1.22 are due to the two ethyl groups of a diethyl ester (compare their chemical shifts with the chemical shifts of the methylene and methyl hydrogen atoms of the two ethyl groups in the ^{1}H NMR spectrum of diethyl malonate, Exercise 24.40a). The triplet at δ 3.23 (1H) is due to a —CH group bonded to a methylene group and to the two ester functional groups. We can write the following partial structure of the compound:

$$\text{—CH}_2\text{—}\underset{\underset{\underset{\text{OCH}_2\text{CH}_3}{\text{O}}}{\overset{\|}{\text{C}}}}{\overset{\overset{\overset{\text{O}}{\|}}{\text{C}}\diagup \text{OCH}_2\text{CH}_3}{\text{C}}}\text{—H}$$

This structure accounts for eight of the 11 carbon atoms, all of the oxygen atoms, and 13 of the 20 hydrogen atoms. The partial structure and the remaining atoms (C_3H_7) indicate that the methylene group is bonded to either a propyl or an isopropyl group. The ^{1}H NMR spectrum is in accord with a propyl group. The triplet at δ 0.82 (3H) is due to the methyl group bonded to a methylene group. The complex splitting pattern of the hydrogens of the two remaining methyl groups overlap the triplet at about δ 1.2. Therefore, the compound is diethyl butylmalonate:

$$\text{CH}_3\text{CH}_2\text{CH}_2\text{CH}_2\text{—}\underset{\underset{\underset{\text{OCH}_2\text{CH}_3}{\text{O}}}{\overset{\|}{\text{C}}}}{\overset{\overset{\overset{\text{O}}{\|}}{\text{C}}\diagup \text{OCH}_2\text{CH}_3}{\text{C}}}\text{—H}$$

CHAPTER 25

CARBOHYDRATES
Concepts

Carbohydrates are polyhydroxy aldehydes and ketones or substances that yield such compounds on acid-catalyzed hydrolysis. **Monosaccharides** or simple sugars are carbohydrates that cannot be hydrolyzed into smaller compounds. **Disaccharides** can be hydrolyzed into two monosaccharides and hydrolysis of **polysaccharides** yields many (up to several thousand) monosaccharides.

Monosaccharides are classified according to the number of carbon atoms (trioses, tetroses, pentoses, etc.) and the kind of carbonyl group they contain (aldoses or ketoses). A four-carbon monosaccharide that contains an aldehyde group is classified as an aldotetrose.

Monosaccharides undergo the usual reactions of the functional groups they contain (aldehydes or ketones and alcohols). Monosaccharides that contain five or six carbon atoms exist primarily as cyclic hemiacetals rather than open chain aldehydes or ketones. A new stereocenter called an **anomeric center** is formed on formation of a cyclic hemiacetal. As a result, carbohydrates can exist as two diastereomeric hemiacetals, which are called the alpha (α-) anomer and the beta (β-) anomer. Equilibration of the two anomers occurs in aqueous solution through the open chain form as an intermediate by a process called **mutarotation.**

The stereochemical relationship among open chain monosaccharides can be shown by Fischer projection formulas. Five-membered ring cyclic hemiacetals are usually portrayed by Haworth projections while the chair form of the six-membered ring cyclic hemiacetal or acetal most accurately represents their stereochemistry.

All monosaccharides belong to either the *D-* or *L-*family. The Fischer projection formulas of members of the *D-*family have the —OH group on the stereocenter farthest from the carbonyl group pointing to the right.

$$
\begin{array}{c}
\text{CHO} \\
\text{H}\!\!-\!\!\!\!-\!\!\text{OH} \\
\text{H}\!\!-\!\!\!\!-\!\!\boxed{\text{OH}} \\
\text{CH}_2\text{OH}
\end{array}
$$

D-Erythrose

Monosaccharides are joined by glycoside bonds to form disaccharides and polysaccharides. The most common glycoside bonds are 1,4 bonds. The glycoside bond can be either alpha (α-) or beta (β-). Cellulose, whose partial structure is shown on p. 1121 of the text, is an example of a polysaccharide composed of *D*-glucopyranose molecules joined by β-1,4 bonds. Starch, whose partial structure is shown on p. 1122 of the text, is an example of a polysaccharide composed of *D*-glucopyranose molecules joined by α-1,4 bonds.

Solutions to the Exercises

25.1 A Fischer projection formula of a monosaccharide is written by convention so that the —CHO group is at the top of the vertical line and the —CH$_2$OH group is at the bottom. Groups located at the ends of the vertical line in a Fischer projection are located behind the plane of the page. To transform a three-dimensional representation of an aldotriose into a Fischer projection formula, therefore, the three carbon atoms of the three-dimensional representation must be aligned vertically with the —CHO group at the top and the —CH$_2$OH group at the bottom. Make sure that the molecule is oriented so that both the —CHO and —CH$_2$OH groups point away from the viewer. This establishes the location of the groups attached to the horizontal lines, which point toward the viewer.

(a)
$$\text{HO}\overset{\text{CH}_2\text{OH}}{\underset{\text{CHO}}{\overset{|}{\underset{|}{\text{C}}}}}\text{H} \equiv \text{H}\blacktriangleright\overset{\text{CHO}}{\underset{\text{CH}_2\text{OH}}{\text{C}}}\blacktriangleleft\text{OH} \equiv \begin{array}{c}\text{CHO}\\\text{H}\!-\!\!\!-\!\text{OH}\\\text{CH}_2\text{OH}\end{array}$$

(b)
$$\text{HO}\blacktriangleright\overset{\text{H}}{\underset{\text{CH}_2\text{OH}}{\text{C}}}\blacktriangleleft\text{CHO} \equiv \text{H}\blacktriangleright\overset{\text{CHO}}{\underset{\text{CH}_2\text{OH}}{\text{C}}}\blacktriangleleft\text{OH} \equiv \begin{array}{c}\text{CHO}\\\text{H}\!-\!\!\!-\!\text{OH}\\\text{CH}_2\text{OH}\end{array}$$

(c)
$$\overset{\text{HO}}{\underset{\text{H}}{\text{C}}}\overset{\text{CH}_2\text{OH}}{\underset{\text{CHO}}{\phantom{\text{C}}}} \equiv \text{HO}\blacktriangleright\overset{\text{CHO}}{\underset{\text{CH}_2\text{OH}}{\text{C}}}\blacktriangleleft\text{H} \equiv \begin{array}{c}\text{CHO}\\\text{HO}\!-\!\!\!-\!\text{H}\\\text{CH}_2\text{OH}\end{array}$$

(d)
$$\text{H}\blacktriangleright\overset{\text{CHO}}{\underset{\text{OH}}{\text{C}}}\blacktriangleleft\text{CH}_2\text{OH} \equiv \text{HO}\blacktriangleright\overset{\text{CHO}}{\underset{\text{CH}_2\text{OH}}{\text{C}}}\blacktriangleleft\text{H} \equiv \begin{array}{c}\text{CHO}\\\text{HO}\!-\!\!\!-\!\text{H}\\\text{CH}_2\text{OH}\end{array}$$

25.2 (a) Tollen's reagent oxidizes the aldehyde group of a monosaccharide to a carboxylic acid group.

$$\begin{array}{c}\text{CHO}\\\text{HO}\!-\!\!\!-\!\text{H}\\\text{H}\!-\!\!\!-\!\text{OH}\\\text{CH}_2\text{OH}\end{array} \xrightarrow[\text{NH}_3]{\text{Ag(NH}_3)_2^+} \begin{array}{c}\text{COO}^-\text{NH}_4^+\\\text{HO}\!-\!\!\!-\!\text{H}\\\text{H}\!-\!\!\!-\!\text{OH}\\\text{CH}_2\text{OH}\end{array}$$

(b) Aqueous sodium hydroxide interconverts aldoses and ketoses and causes epimerization.

$$\begin{array}{c} \text{CHO} \\ \text{HO}-\text{H} \\ \text{H}-\text{OH} \\ \text{CH}_2\text{OH} \end{array} \xrightarrow{\text{NaOH} \atop \text{H}_2\text{O}} \begin{array}{c} \text{CH}_2\text{OH} \\ =\text{O} \\ \text{H}-\text{OH} \\ \text{CH}_2\text{OH} \end{array} + \begin{array}{c} \text{CHO} \\ \text{H}-\text{OH} \\ \text{H}-\text{OH} \\ \text{CH}_2\text{OH} \end{array} + \begin{array}{c} \text{CHO} \\ \text{HO}-\text{H} \\ \text{H}-\text{OH} \\ \text{CH}_2\text{OH} \end{array}$$

(c) NaBH$_4$ reduces the aldehyde group to a primary alcohol.

$$\begin{array}{c} \text{CHO} \\ \text{HO}-\text{H} \\ \text{H}-\text{OH} \\ \text{CH}_2\text{OH} \end{array} \xrightarrow{\text{NaBH}_4} \begin{array}{c} \text{CH}_2\text{OH} \\ \text{HO}-\text{H} \\ \text{H}-\text{OH} \\ \text{CH}_2\text{OH} \end{array}$$

(d) Bromine water oxidizes the aldehyde group to a carboxylic acid group.

$$\begin{array}{c} \text{CHO} \\ \text{HO}-\text{H} \\ \text{H}-\text{OH} \\ \text{CH}_2\text{OH} \end{array} \xrightarrow{\text{Br}_2 \atop \text{H}_2\text{O}} \begin{array}{c} \text{COOH} \\ \text{HO}-\text{H} \\ \text{H}-\text{OH} \\ \text{CH}_2\text{OH} \end{array}$$

(e) Nitric acid oxidizes both the aldehyde group and the —CH$_2$OH group to carboxylic acid groups.

$$\begin{array}{c} \text{CHO} \\ \text{HO}-\text{H} \\ \text{H}-\text{OH} \\ \text{CH}_2\text{OH} \end{array} \xrightarrow{\text{HNO}_3} \begin{array}{c} \text{COOH} \\ \text{HO}-\text{H} \\ \text{H}-\text{OH} \\ \text{COOH} \end{array}$$

25.3

25.4

$$\underset{D\text{-}(-)\text{-Ribose}}{\begin{array}{c}\text{CHO}\\ \text{H}\!-\!\!\!-\!\text{OH}\\ \text{H}\!-\!\!\!-\!\text{OH}\\ \text{H}\!-\!\!\!-\!\text{OH}\\ \text{CH}_2\text{OH}\end{array}} \xrightarrow{\underset{\text{H}_2\text{O}}{\text{Br}_2}} \begin{array}{c}\text{COOH}\\ \text{H}\!-\!\!\!-\!\text{OH}\\ \text{H}\!-\!\!\!-\!\text{OH}\\ \text{H}\!-\!\!\!-\!\text{OH}\\ \text{CH}_2\text{OH}\end{array} \xrightarrow{\text{Ca(OH)}_2} \left[\begin{array}{c}\text{COO}^-\\ \text{H}\!-\!\!\!-\!\text{OH}\\ \text{H}\!-\!\!\!-\!\text{OH}\\ \text{H}\!-\!\!\!-\!\text{OH}\\ \text{CH}_2\text{OH}\end{array}\right]_2 \text{Ca}^{2+} \xrightarrow{\underset{\text{Fe(III)}}{\text{H}_2\text{O}_2}} \underset{D\text{-}(-)\text{-Erythrose}}{\begin{array}{c}\text{CHO}\\ \text{H}\!-\!\!\!-\!\text{OH}\\ \text{H}\!-\!\!\!-\!\text{OH}\\ \text{CH}_2\text{OH}\end{array}}$$

25.5 Xylose is an aldopentose so it contains five carbon atoms. Ruff degradation forms (−)-threose so the configurations of C3 and C4 of (+)-xylose are the same as the configurations of C2 and C3 of (−)-threose. On the basis of this information, we can write the following partial structure of (+)-xylose:

$$\begin{array}{c}\text{CHO}\\ |\\ \text{CHOH}\\ \text{HO}\!-\!\!\!-\!\text{H}\\ \text{H}\!-\!\!\!-\!\text{OH}\\ \text{CH}_2\text{OH}\end{array}$$

Nitric acid oxidation of (+)-xylose forms an optically inactive acid so the —OH group on C2 of (+)-xylose must point to the right. If the —OH group pointed to the left, the dicarboxylic acid would be optically active.

$$\begin{array}{c}\text{COOH}\\ \text{H}\!-\!\!\!-\!\text{OH}\\ \text{HO}\!-\!\!\!-\!\text{H} \quad \text{Plane of symmetry}\\ \text{H}\!-\!\!\!-\!\text{OH}\\ \text{COOH}\end{array} \qquad \begin{array}{c}\text{COOH}\\ \text{HO}\!-\!\!\!-\!\text{H}\\ \text{HO}\!-\!\!\!-\!\text{H}\\ \text{H}\!-\!\!\!-\!\text{OH}\\ \text{COOH}\end{array}$$

Optically inactive *(meso)*   Optically active

The structure of (+)-xylose is

$$\begin{array}{c}\text{CHO}\\ \text{H}\!-\!\!\!-\!\text{OH}\\ \text{HO}\!-\!\!\!-\!\text{H}\\ \text{H}\!-\!\!\!-\!\text{OH}\\ \text{CH}_2\text{OH}\end{array}$$

25.6 Members of the *D*-family of monosaccharides have the —OH group pointing to the right on the stereocenter farthest from the carbonyl group.

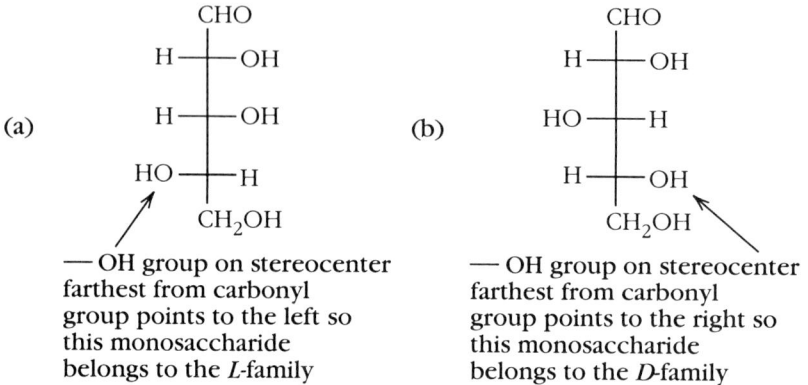

(a) —OH group on stereocenter farthest from carbonyl group points to the left so this monosaccharide belongs to the *L*-family

(b) —OH group on stereocenter farthest from carbonyl group points to the right so this monosaccharide belongs to the *D*-family

(c) *D*-Monosaccharide
(d) *L*-Monosaccharide

25.7 Members of the *L*-family are the enantiomers of the corresponding members of the *D*-family. *L*-Erythrose, therefore, is the mirror image of *D*-erythrose.

```
         CHO              CHO              CHO              CHO
(a) HO─┼─H      (b) HO─┼─H      (c) H─┼─OH     (d) HO─┼─H
    HO─┼─H          HO─┼─H          HO─┼─H          H─┼─OH
       CH₂OH          HO─┼─H          HO─┼─H          H─┼─OH
                       CH₂OH            CH₂OH       HO─┼─H
                                                       CH₂OH
  L-Erythrose       L-Ribose       L-Arabinose       L-Galactose
```

25.8 A hydroxyl group on a stereocenter that points to the left in a Fischer projection appears above the plane of the ring in a Haworth formula. A hydroxyl group that points to the right appears below the plane of the ring. In the α-anomer, the —OH group on C1 points down in the Haworth formula while the —OH group points up in the β-anomer.

(a) Open-chain form of *D*-Erythose → β-*D*-Erythrofuranose

(b) Open-chain form of D-Xylose α-D-Xylofuranose   (c) Open-chain form of D-Lyxose β-D-Lyxofuranose

25.9 The chair conformation of a particular monosaccharide is easily drawn by recognizing the differences between the configuration about C2, C3, and C4 of the monosaccharide and glucose. First let's draw the cyclic chair conformation of D-glucose.

(a) In D-glucose, the —OH groups on C2, C3, and C4 all occupy equatorial positions. The α-anomer has the —OH group on C1 *trans* to the —CH$_2$OH group while in the β-anomer the two groups are *cis*.

Open-chain form of D-Glucose   α-D-Glucopyranose

(b) D-Mannose is the C2 epimer of glucose so the cyclic chair form of D-mannose has the hydroxyl group on C2 in an axial position instead of the equatorial position as in glucose.

Open-chain form of D-Glucose   Open-chain form of D-Mannose   β-D-Mannopyranose

(c) *D*-Galactose is the C4 epimer of glucose so the cyclic chair form of *D*-galactose has the hydroxyl group on C4 in an axial position instead of the equatorial position as in glucose.

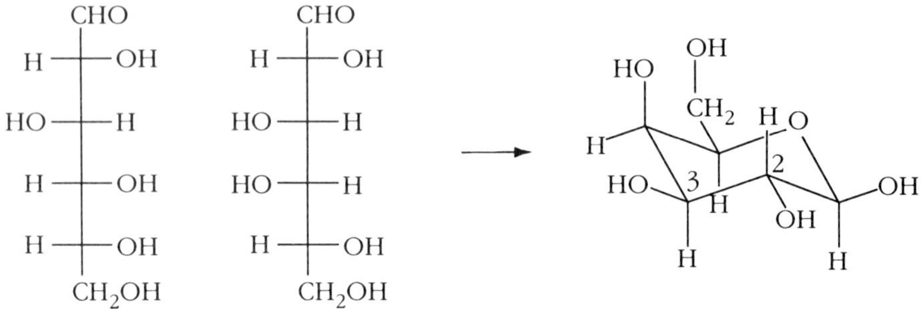

Open-chain form of *D*-Glucose

Open-chain form of *D*-Galactose

β-*D*-Galactopyranose

25.10 The carbon atom of a hemiacetal group is bonded to a hydroxyl group and an alkoxy group.

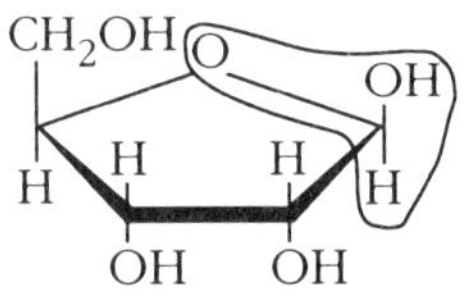

(a) β-Form

(b) α-Form

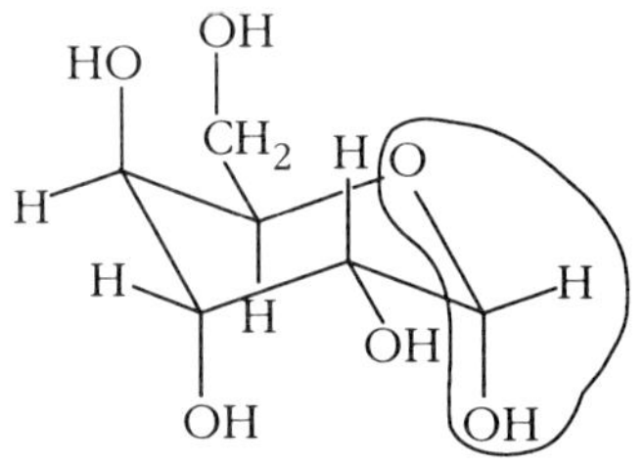

(c) α-Form

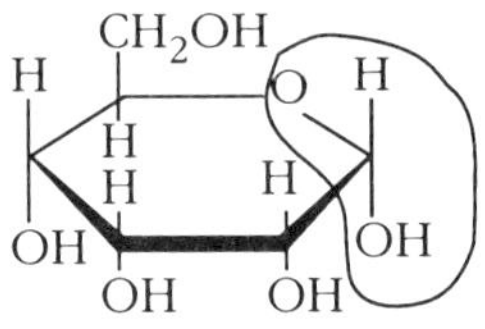

(d) β-Form

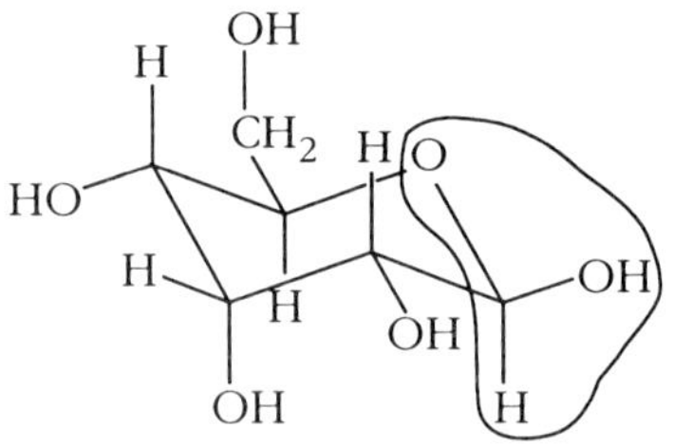

25.11 The Fischer projection formula of a deoxy sugar is obtained by writing the Fischer projection of the sugar and replacing the hydroxyl group at the specified carbon atom by a hydrogen atom.

(a)

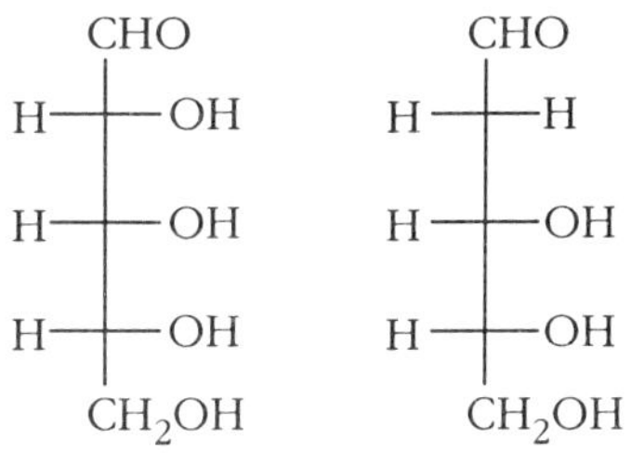

D-Ribose 2-Deoxy-*D*-Ribose

(b)

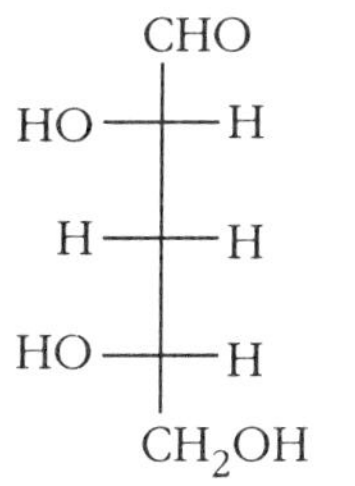

3-Deoxy-*L*-Ribose

(c)

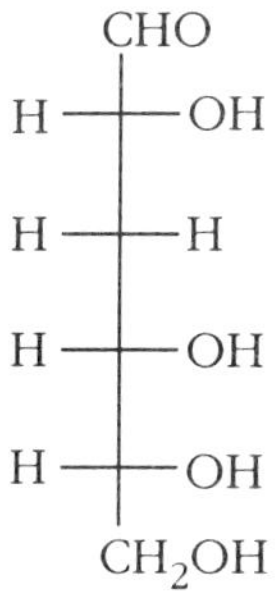

3-Deoxy-*D*-Glucose

25.12

(a)

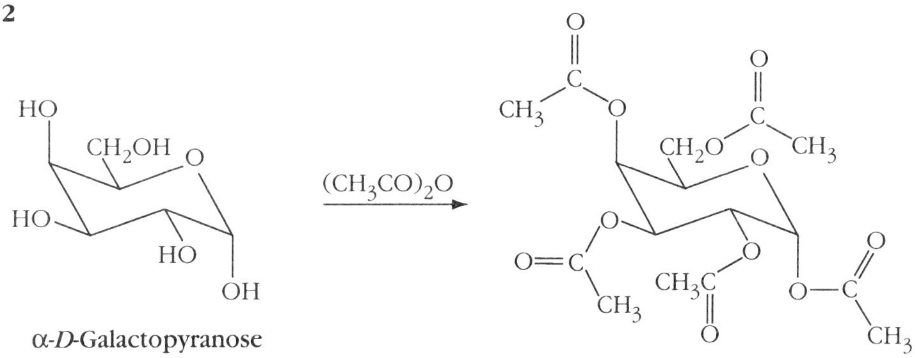

α-*D*-Galactopyranose

(b)

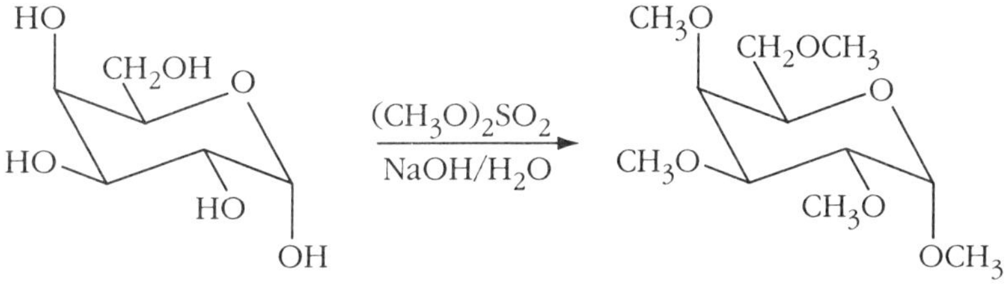

(c)

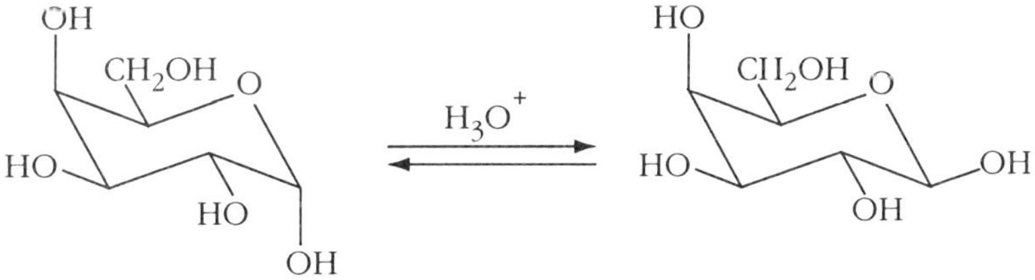

α-*D*-Galactopyranose β-*D*-Galactopyranose

(d)

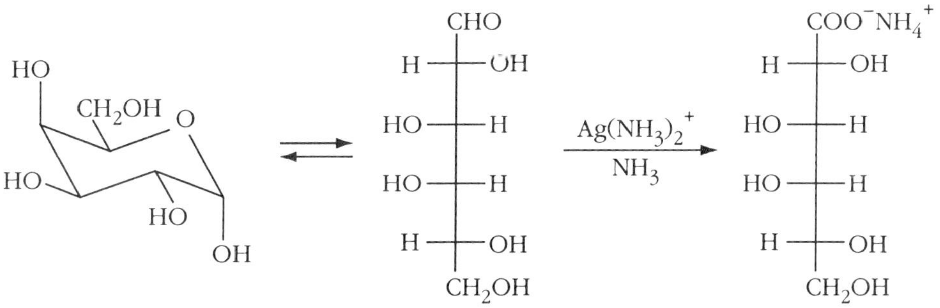

(e) 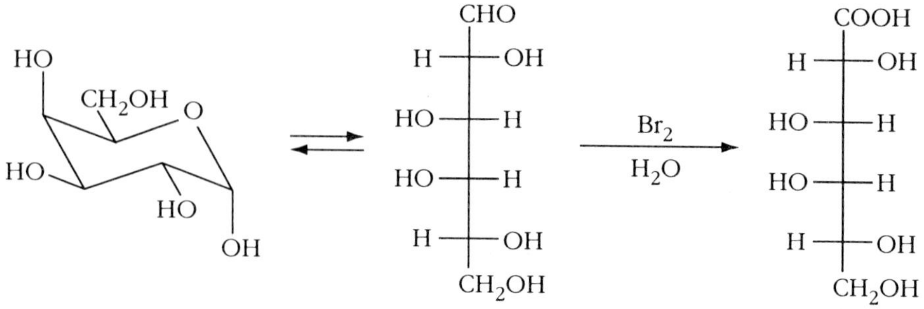 [structure of sugar in pyranose form] ⇌ [open chain aldose Fischer projection: CHO, H-OH, HO-H, HO-H, H-OH, CH₂OH] →(Br₂/H₂O)→ [COOH, H-OH, HO-H, HO-H, H-OH, CH₂OH]

25.13

(a) 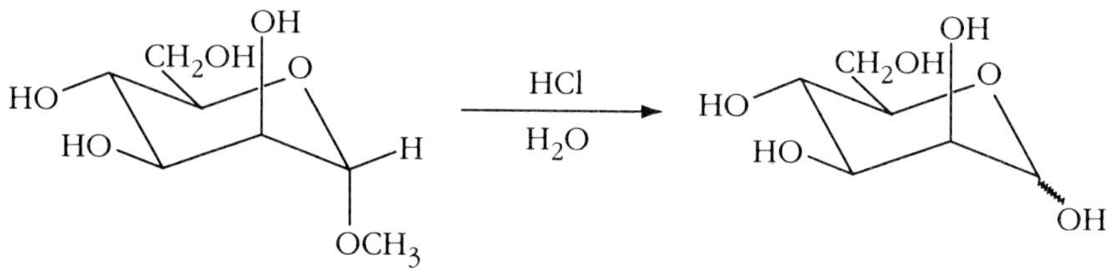 Methyl α-D-mannopyranoside →(HCl, H₂O)→ Mixture of α- and β-anomers

(b) No reaction because an acetal group is not oxidized by Br₂/H₂O
(c) No reaction because Fehling's solution does not oxidize acetals

(d) 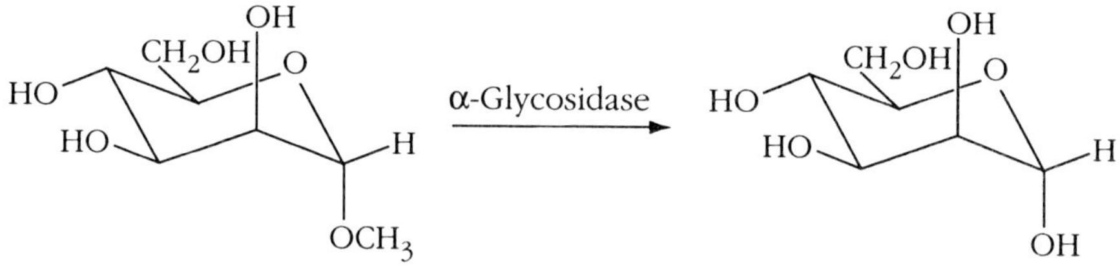 [methyl α-D-mannopyranoside] →(α-Glycosidase)→ [D-mannopyranose]

(e) No reaction because an enzyme that catalyzes the hydrolysis of only β-glycosides does not hydrolyze α-glycosides

25.14

(a) 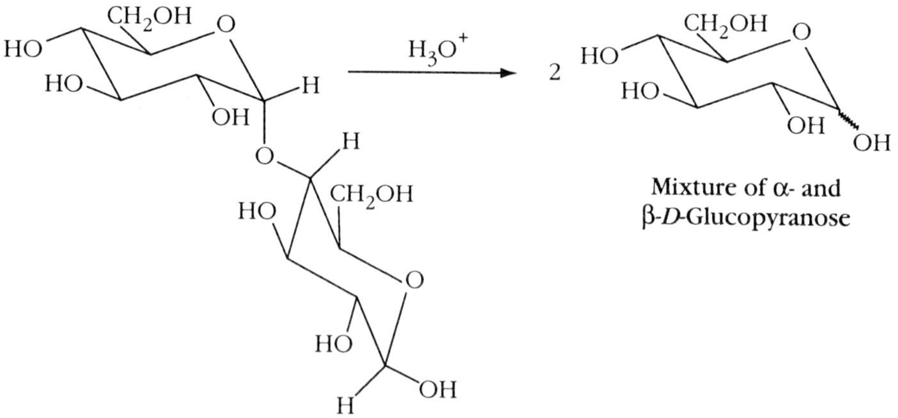 [disaccharide structure] →(H₃O⁺)→ 2 [glucopyranose]   Mixture of α- and β-D-Glucopyranose

(b) and (c)   One ring of maltose contains a carbonyl group that is in a hemiacetal form so it can be oxidized by either Tollen's reagent or Br_2/H_2O.

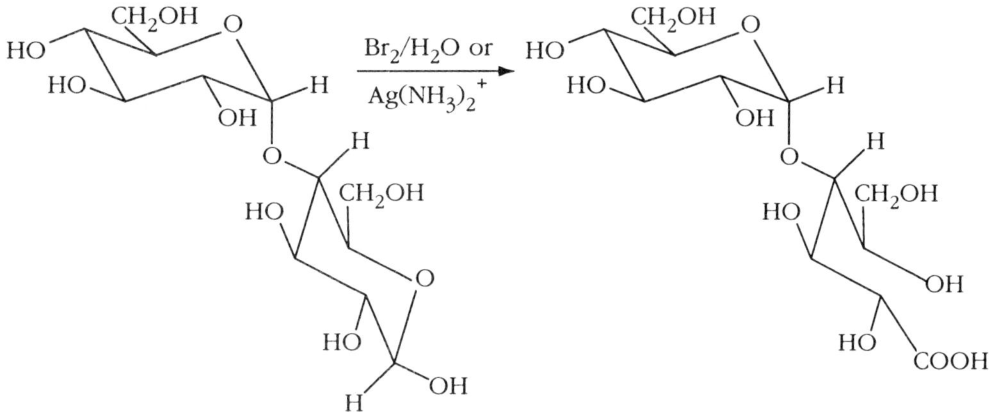

(d)

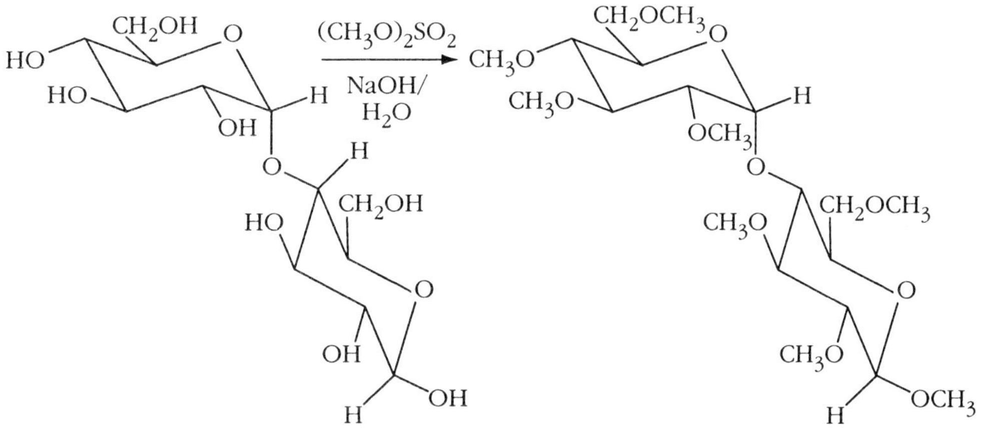

25.15

(a) α-Cellobiose

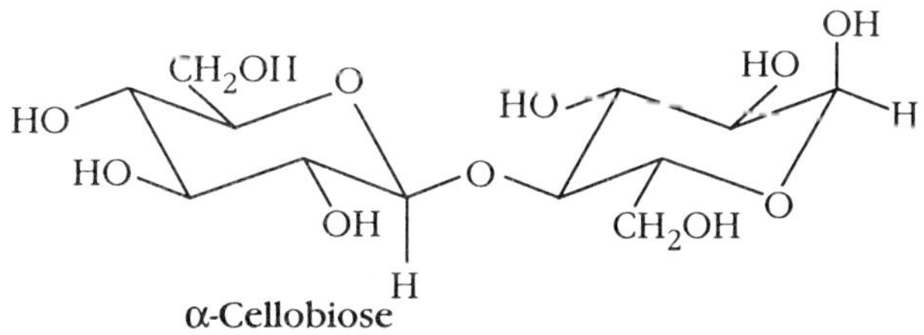

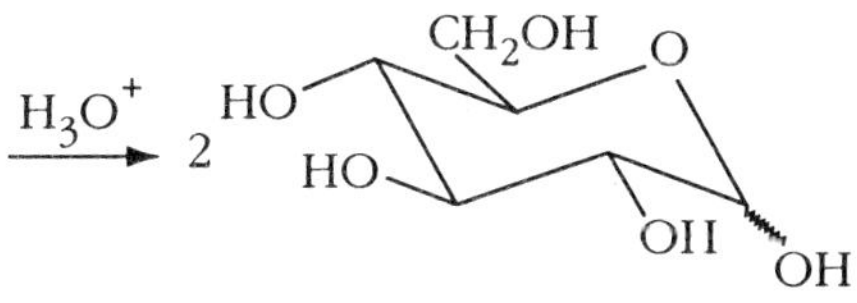

$\xrightarrow{H_3O^+}$ 2

(b) [structure: disaccharide with CH₂OH, HO, HO, OH, H groups and second ring with HO, HO, OH, H, OH, CH₂OH]

$\xrightarrow{\text{Br}_2 / \text{H}_2\text{O}}$ [structure: disaccharide oxidized to carboxylic acid at right end]

(c) [structure: disaccharide similar to (b)]

$\xrightarrow{(\text{CH}_3\text{CO})_2\text{O} / \text{Na}^+\bar{\text{O}}\text{OCCH}_3}$ [structure: fully acetylated disaccharide (octaacetate)]

(d) [structure: disaccharide with open-chain reduced right end showing CH₂OH, OH, HO, OH, CH₂OH]

(e) No reaction

(f) [structure: disaccharide with open-chain right end terminating in COOH]

25.16

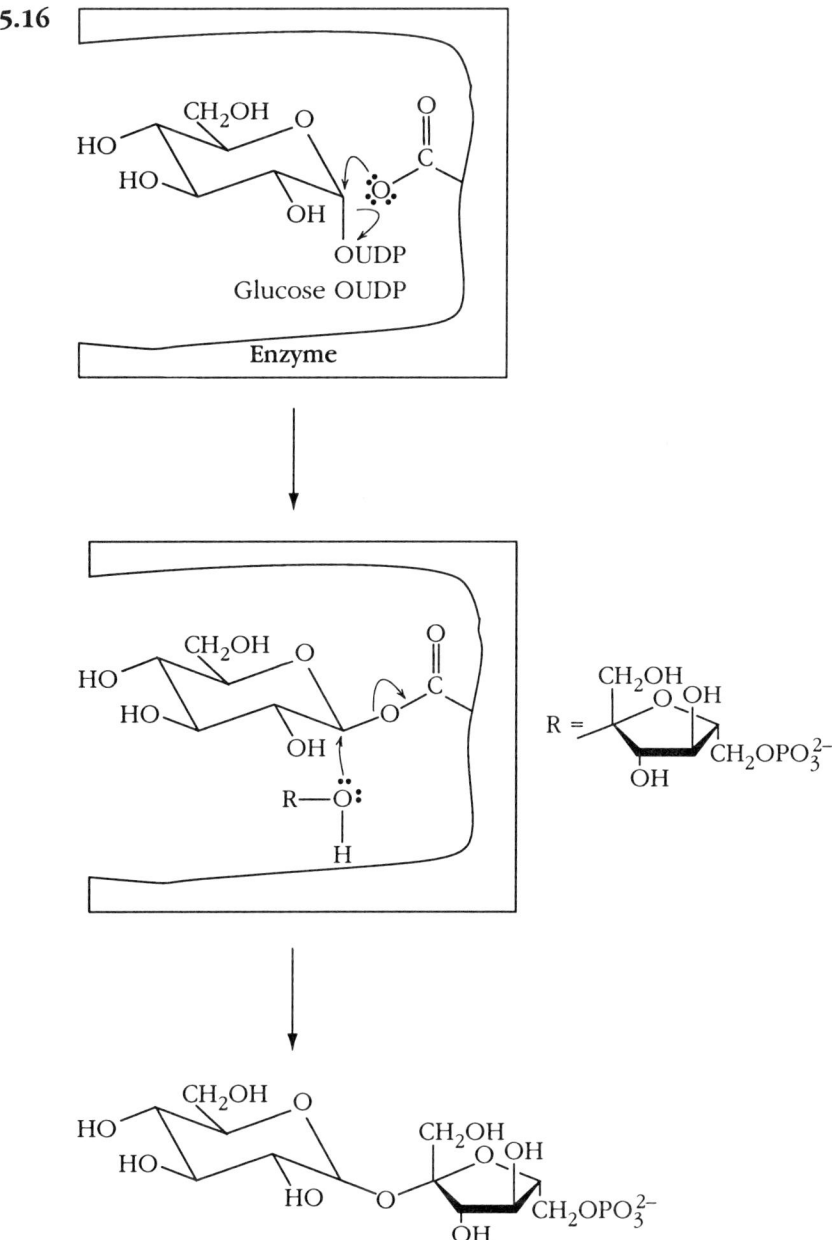

Sucrose phosphate

25.17 (a) **Monosaccharides** or simple sugars are carbohydrates that cannot be hydrolyzed into smaller compounds. See p. 1108 of the text for some examples of monosaccharides.

(b) **D-(+)-Glucose** is the aldohexose that has the following configuration represented as a Fischer projection, a Haworth formula, and pyranose ring in its chair conformation:

Notice that in the chair conformation of *D*-glucose, the hydroxyl groups on C2, C3, and C4 are all equatorial.

(c) **Polysaccharides** are carbohydrates that yield many (up to several thousand) monosaccharides on hydrolysis.

(d) **Epimers** are diastereomers that differ from each other in the configuration at only one of their stereocenters. *D*-Glucose and *D*-mannose are an example of epimers:

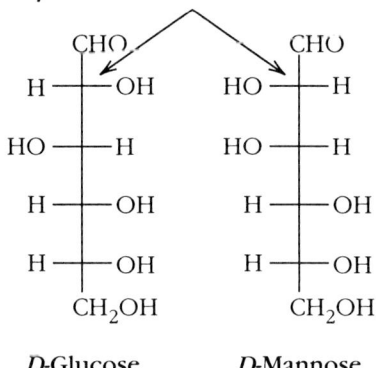

(e) An **aldonic acid** is the product of the mild oxidation of an aldose in which only the aldehyde group is oxidized.

Mild oxidizing agents include Br$_2$/H$_2$O, Fehling's or Tollen's reagents

(f) **Mutarotation** is the spontaneous change in the optical rotation observed when a pure anomer is dissolved in water. A freshly prepared solution of α-D-(+)-glucopyranose, for example, has a specific rotation of +112°, which gradually changes with time until a constant value of +53° is reached.

(g) The **Killiani-Fischer synthesis** is a sequence of reactions that lengthens an aldose chain by one carbon atom. An aldopentose, for example, is converted to an aldohexose by this method (See Section 25.3E).

(h) **Ruff degradation** is a method of shortening the chain of aldoses by one carbon atom (See Section 25.3F).

(i) A **furanose ring** is a saturated five-membered oxygen-containing ring formed by intramolecular nucleophilic addition of the hydroxy groups on C4 of an aldose to the carbonyl carbon atom.

```
      CHO
   H——OH                H    OH                     H    O    OH
                           \ /    CHO                  \ / \ /
   H——OH                    X                           X   X
                           / \  /                      / \ / \
      CH₂OH              H    H                       H   H   H
                              OH  OH                      OH  OH
  Open-chain                                         β-D-Erythrofuranose
   form of
  D-Erythose
```

(j) **Anomers** are epimers that differ in stereochemistry only at the stereocenter created at the hemiacetal carbon atom by ring formation. For example, α-glucopyranose and β-glucopyranose are anomers.

(k) A **pyranose ring** is a saturated six-membered oxygen-containing ring formed by intramolecular nucleophilic addition of the hydroxy group on C5 of an aldose to the carbonyl carbon atom.

(l) A **deoxy sugar** is a saccharide in which one hydroxy group is replaced by a hydrogen atom. 2-Deoxy-D-ribose is an example of a deoxy sugar.

```
       CHO                CHO
   H——OH              H——H
   H——OH              H——OH
   H——OH              H——OH
      CH₂OH              CH₂OH
     D-Ribose       2-Deoxy-D-Ribose
```

(m) An **amino sugar** is a monosaccharide in which one hydroxyl group is replaced by an amino group. 2-Acetamido-2-deoxy-β-D-glucopyranose is an example of a derivative of an amino sugar.

2-Acetamido-2-deoxy-β-*D*-glucopyranose

(n) A **glycoside** is a carbohydrate acetal.
(o) **Glycogen** is a polysaccharide consisting of coiled chains of *D*-glucopyranose molecules joined by α-1,4-glycoside bonds. Glycogen is used to store carbohydrates in mammals.
(p) **Disaccharides** are dimers made up of two monosaccharide molecules joined by a glycoside linkage between the anomeric carbon atom of one monosaccharide and an —OH group of the other. The monosaccharides may be the same or different.
(q) **Glycolysis** is a series of ten enzyme-catalyzed reactions that converts glucose into pyruvate and ATP.
(r) An **enediol intermediate** is the intermediate that is involved in the base-catalyzed interconversion of aldoses and ketoses. The enediol intermediate contains one carbon-carbon double bond (ene) whose two carbon atoms are bonded to two —OH groups, one to each carbon atom of the double bond (diol):

Enediol intermediate

25.18 (a) The compound contains four carbon atoms (tetrose) and an aldehyde group (aldose) so it is an aldotetrose.
(b) The compound contains six carbon atoms (hexose) and a ketone group (ketose) so it is a ketohexose.
(c) Ketopentose
(d) Aldohexose

25.19 The maximum number of stereoisomers of a compound with n stereocenters is 2^n.

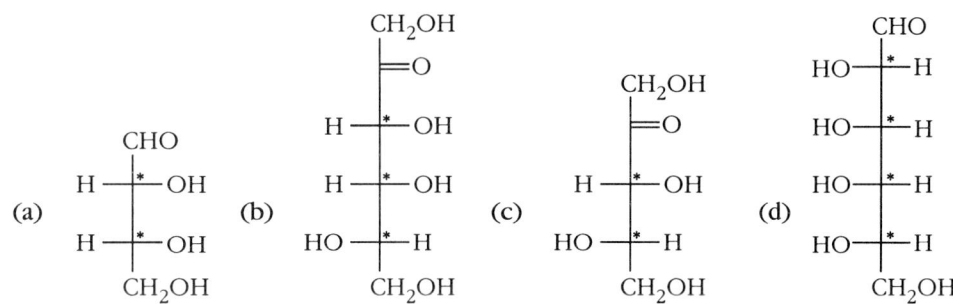

2 Stereocenters   3 Stereocenters   2 Stereocenters   4 Stereocenters
$2^2 = 4$ stereoisomers $2^3 = 6$ stereoisomers $2^2 = 4$ stereoisomers $2^4 = 16$ stereoisomers

25.20

(a) The —OH group on the stereocenter the farthest from the carbonyl group points to the right so this monosaccharide belongs to the D-family

(b) L-Family
(c) L-Family
(d) L-Family

25.21

25.22 Any configuration of the stereocenters is acceptable. We have arbitrarily chosen the configurations as shown in the following Fischer projections.

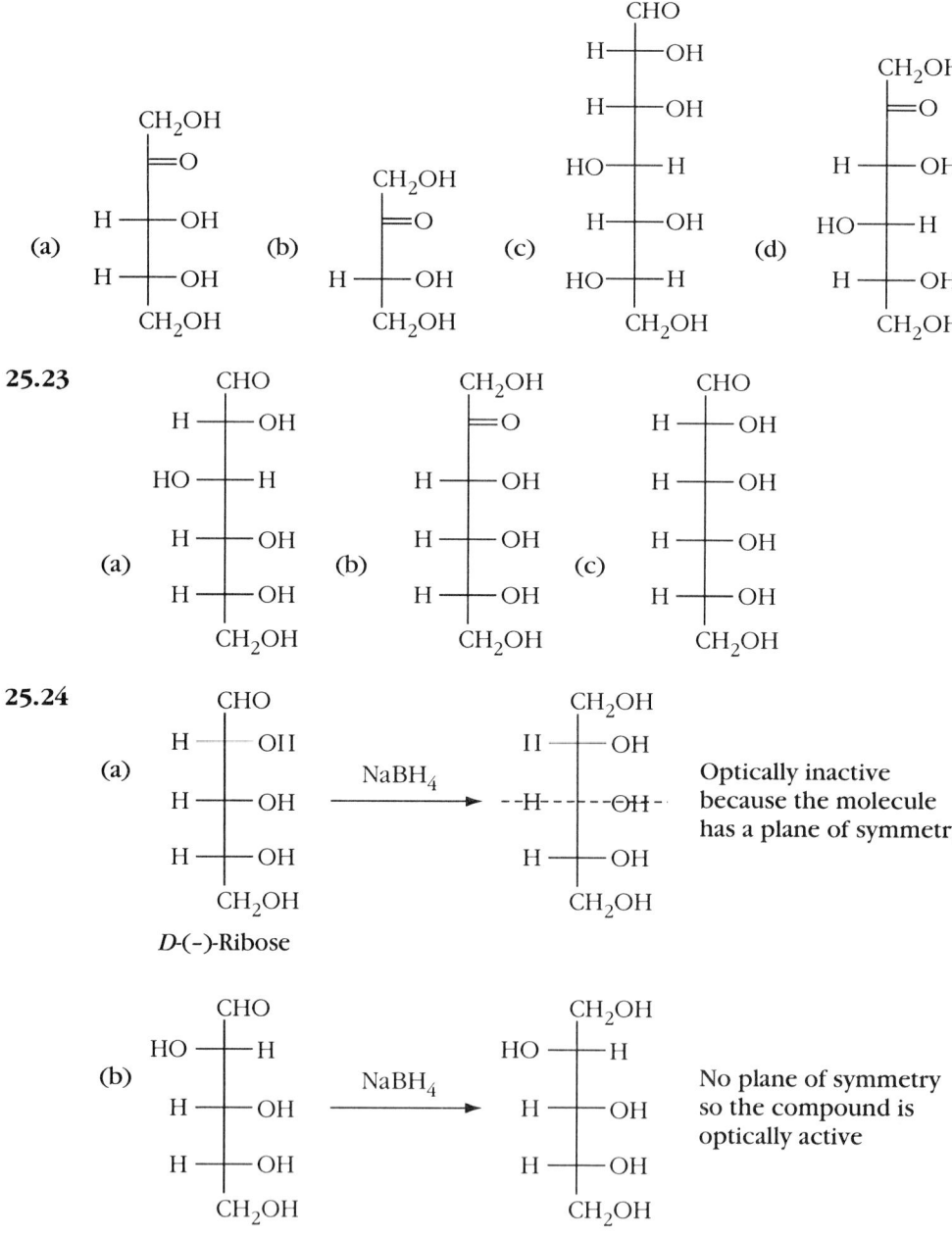

(c)

```
        CHO                              CH₂OH
  HO ——|—— H                       HO ——|—— H
   H ——|—— OH      NaBH₄          --H---|--OH--      Optically inactive
  HO ——|—— H       ——————→         HO ——|—— H        because the molecule
        CH₂OH                            CH₂OH        has a plane of symmetry
    L-(+)-Xylose
```

(d)

```
        CHO                              CH₂OH
  HO ——|—— H                       HO ——|—— H
  HO ——|—— H       NaBH₄          HO ——|—— H         No plane of symmetry
   H ——|—— OH      ——————→         H ——|—— OH         so the compound is
        CH₂OH                            CH₂OH        optically active
    D-(−)-Lyxose
```

25.25

(a)

```
        CHO                              COOH
   H ——|—— OH                      H ——|—— OH
   H ——|—— OH       HNO₃          --H---|--OH--      Optically inactive
   H ——|—— OH       ——————→        H ——|—— OH        because the molecule
        CH₂OH                            COOH         has a plane of symmetry
    D-(−)-Ribose
```

(b)

```
        CHO                              COOH
  HO ——|—— H                       HO ——|—— H
   H ——|—— OH       HNO₃           H ——|—— OH        No plane of symmetry
   H ——|—— OH       ——————→        H ——|—— OH        so the compound is
        CH₂OH                            COOH         optically active
    D-(−)-Arabinose
```

(c)

```
        CHO                              COOH
  HO ——|—— H                       HO ——|—— H
   H ——|—— OH       HNO₃          --H---|--OH--      Optically inactive
  HO ——|—— H        ——————→        HO ——|—— H        because the molecule
        CH₂OH                            COOH         has a plane of symmetry
    L-(+)-Xylose
```

(d) D-(−)-Lyxose structure →[HNO₃] oxidized product — No plane of symmetry so the compound is optically active

25.26

(a) Sugar + (CH₃CO)₂O / CH₃COO⁻Na⁺ → pentaacetate

(b) Sugar + (CH₃O)₂SO₂ / NaOH/H₂O → permethylated sugar

(c) Sugar ⇌ [HCl / H₂O] Mixture of α and β anomers

(d) Sugar ⇌ open-chain aldose →[Cu²⁺, Tartarate buffer] aldonic acid

(e) 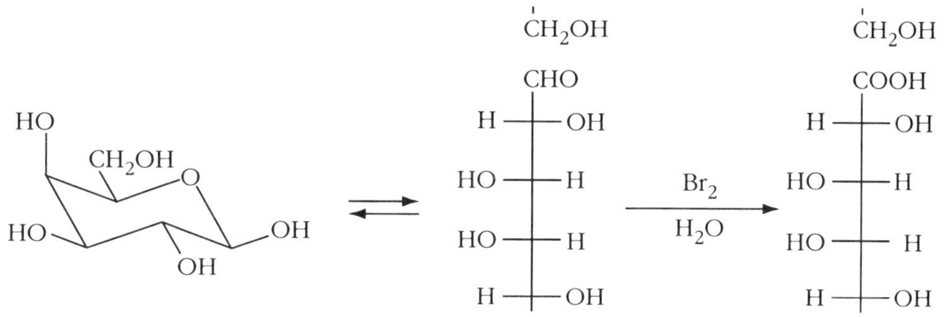 [Structure: pyranose ring with HO, CH₂OH, HO, OH, OH groups] ⇌

$$\begin{array}{c} CHO \\ H-\!\!\!-OH \\ HO-\!\!\!-H \\ HO-\!\!\!-H \\ H-\!\!\!-OH \\ CH_2OH \end{array} \xrightarrow{Br_2, H_2O} \begin{array}{c} COOH \\ H-\!\!\!-OH \\ HO-\!\!\!-H \\ HO-\!\!\!-H \\ H-\!\!\!-OH \\ CH_2OH \end{array}$$

25.27 (a) Methyl β-glucopyranoside is an acetal, which is not oxidized by an aqueous bromine solution.

(b) A ketohexose is not oxidized by an aqueous bromine solution.

(c) An acetal does not react with an aqueous solution of bromine.

25.28 The alderic acid formed by the nitric acid oxidation of D-(-)-arabinose has the following structure:

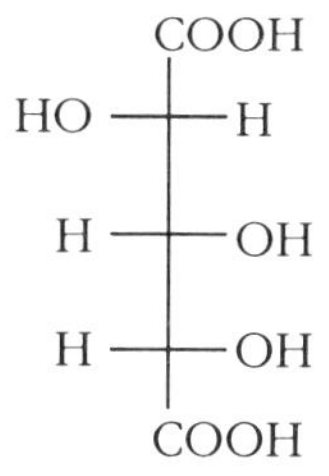

$$\begin{array}{c} COOH \\ HO-\!\!\!-H \\ H-\!\!\!-OH \\ H-\!\!\!-OH \\ COOH \end{array}$$

An aldopentose with the same relative configuration at C2, C3, and C4 as D-(-)-arabinose will form the same aldaric acid on oxidation. D-(-)-Lyxose has the same relative configuration at C2, C3, and C4 as D-(-)-arabinose so it will be oxidized by nitric acid to the same aldaric acid.

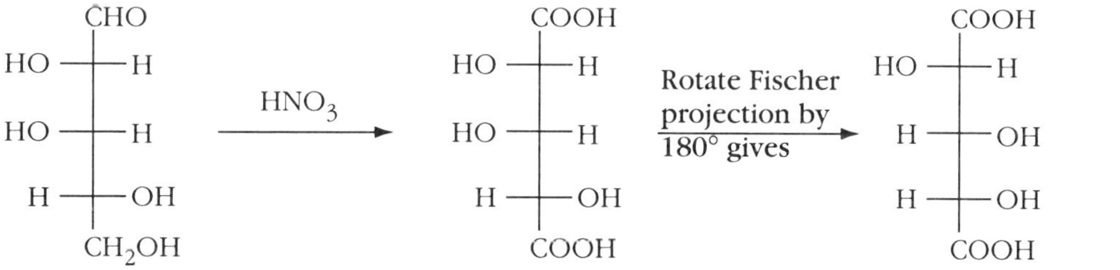

$$\begin{array}{c} CHO \\ HO-\!\!\!-H \\ HO-\!\!\!-H \\ H-\!\!\!-OH \\ CH_2OH \end{array} \xrightarrow{HNO_3} \begin{array}{c} COOH \\ HO-\!\!\!-H \\ HO-\!\!\!-H \\ H-\!\!\!-OH \\ COOH \end{array} \xrightarrow{\text{Rotate Fischer projection by } 180° \text{ gives}} \begin{array}{c} COOH \\ HO-\!\!\!-H \\ H-\!\!\!-OH \\ H-\!\!\!-OH \\ COOH \end{array}$$

D-(-)-Lyxose

The aldaric acid obtained by oxidation of D-(-)-lyxose and D-(-)-arabinose

25.29 An aldonic acid is the product of mild oxidation of an aldose in which only the —CHO group is oxidized while both the —CH₂OH and the —CHO group of an aldose have been oxidized in an aldaric acid.

(a)

$$\begin{array}{c} \text{CHO} \\ \text{H}\!-\!\!-\!\text{OH} \\ \text{HO}\!-\!\!-\!\text{H} \\ \text{CH}_2\text{OH} \end{array} \longrightarrow \begin{array}{c} \text{COOH} \\ \text{H}\!-\!\!-\!\text{OH} \\ \text{HO}\!-\!\!-\!\text{H} \\ \text{CH}_2\text{OH} \end{array} \longrightarrow \begin{array}{c} \text{COOH} \\ \text{H}\!-\!\!-\!\text{OH} \\ \text{HO}\!-\!\!-\!\text{H} \\ \text{COOH} \end{array}$$

Aldonic acid   Aldaric acid

(b)

$$\begin{array}{c} \text{COOH} \\ \text{H}\!-\!\!-\!\text{OH} \\ \text{HO}\!-\!\!-\!\text{H} \\ \text{H}\!-\!\!-\!\text{OH} \\ \text{CH}_2\text{OH} \end{array} \qquad \begin{array}{c} \text{COOH} \\ \text{H}\!-\!\!-\!\text{OH} \\ \text{HO}\!-\!\!-\!\text{H} \\ \text{H}\!-\!\!-\!\text{OH} \\ \text{COOH} \end{array}$$

Aldonic acid   Aldaric acid

(c)

[furanose ring structure] ⇌

$$\begin{array}{c} \text{CHO} \\ \text{H}\!-\!\!-\!\text{OH} \\ \text{H}\!-\!\!-\!\text{H} \\ \text{H}\!-\!\!-\!\text{OH} \\ \text{HO}\!-\!\!-\!\text{H} \\ \text{CH}_2\text{OH} \end{array} \longrightarrow \begin{array}{c} \text{COOH} \\ \text{H}\!-\!\!-\!\text{OH} \\ \text{H}\!-\!\!-\!\text{H} \\ \text{H}\!-\!\!-\!\text{OH} \\ \text{HO}\!-\!\!-\!\text{H} \\ \text{CH}_2\text{OH} \end{array}$$

Aldonic acid

$$\longrightarrow \begin{array}{c} \text{COOH} \\ \text{H}\!-\!\!-\!\text{OH} \\ \text{H}\!-\!\!-\!\text{H} \\ \text{H}\!-\!\!-\!\text{OH} \\ \text{HO}\!-\!\!-\!\text{H} \\ \text{COOH} \end{array}$$

Aldaric acid

(d) [structure] ⇌ [Fischer with CHO, 4× H-OH, CH₂OH] →

[Aldonic acid: COOH, 4× H-OH, CH₂OH] → [Aldaric acid: COOH, 4× H-OH, COOH]

Aldonic acid   Aldaric acid

25.30 To determine the family to which ascorbic acid belongs, we must write the Fischer projection formula of the open chain form of ascorbic acid. The Fischer projection formula is obtained by breaking the bond between the carbonyl carbon atom and the oxygen atom of the ester bond. Stretching out the open chain form allows us to draw the Fischer projection formula of ascorbic acid. The hydroxyl group the farthest from the carbonyl group points to the left so ascorbic acid belongs to the *L*-family.

25.31

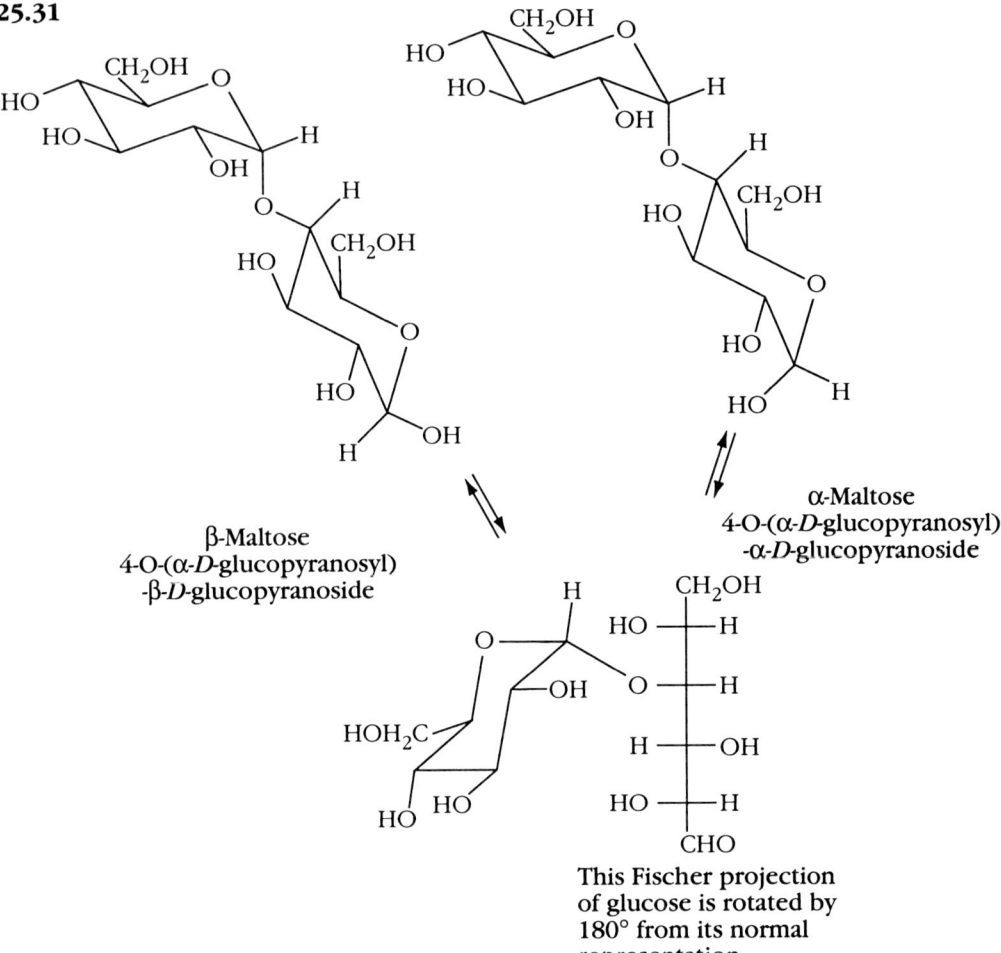

β-Maltose
4-O-(α-D-glucopyranosyl)
-β-D-glucopyranoside

α-Maltose
4-O-(α-D-glucopyranosyl)
-α-D-glucopyranoside

This Fischer projection of glucose is rotated by 180° from its normal representation

25.32 Step 1: Protonation of acetone

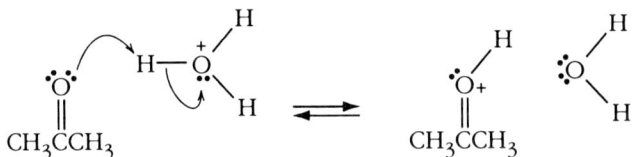

Step 2: Nucleophilic addition of —OH group of monosaccharide to carbonyl C atom to form a hemiacetal

Let R =

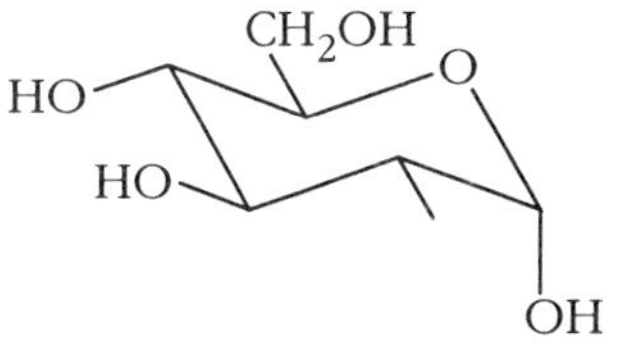

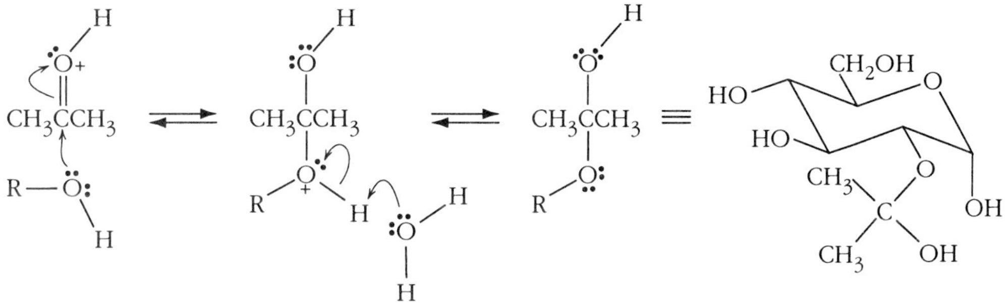

Step 3: Acetal formation

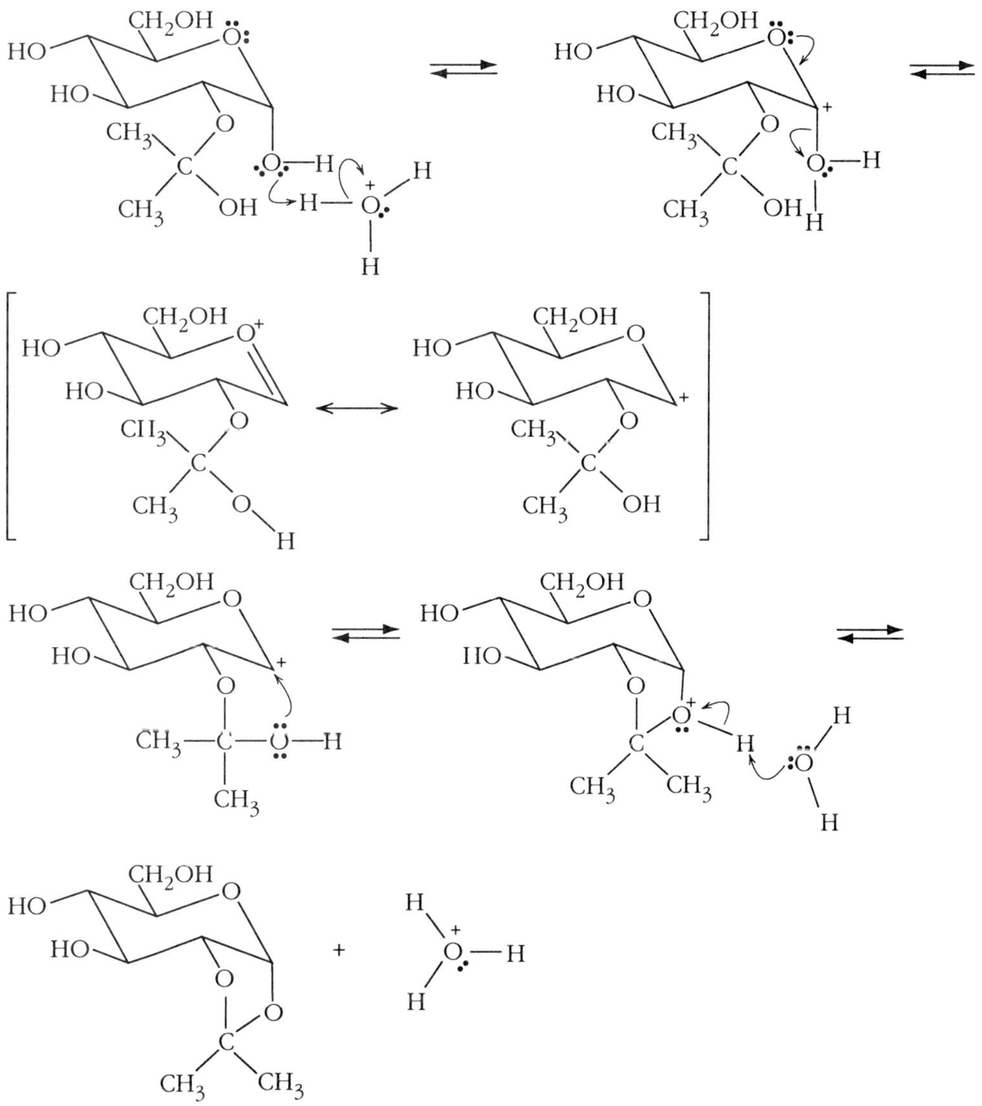

25.33 (a) (1), (3)
(b) (1), (2)
(c) (1), (2)
(d) (1), (2), (5)
(e) (2), (6)

25.34 The same resonance-stabilized carbocation is formed from either methyl α-D-glucopyranoside or methyl β-D-glucopyranoside and it can react with water from either face to form both the α- and the β-anomers.

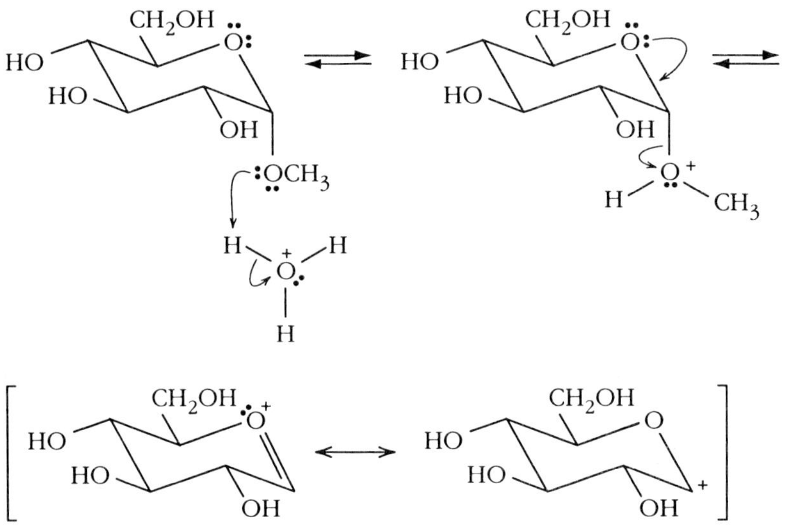

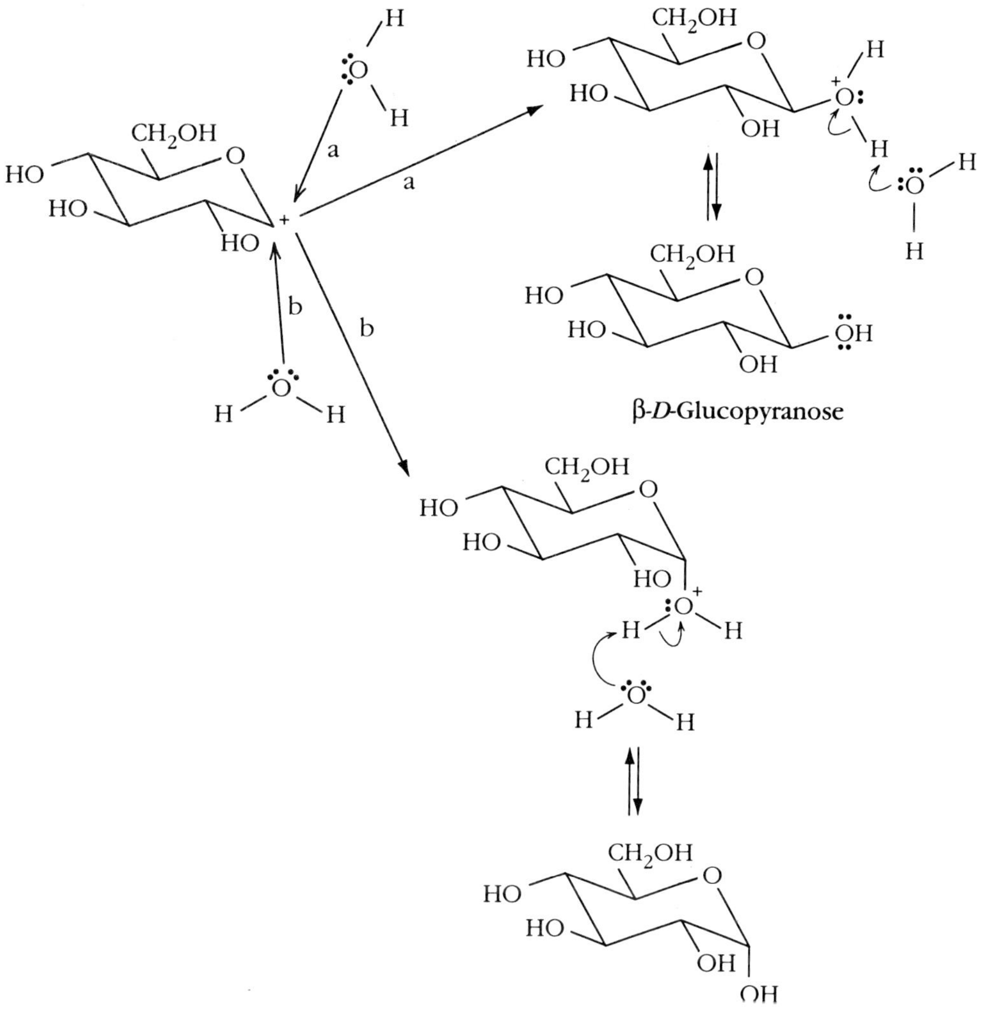

β-*D*-Glucopyranose

α-*D*-Glucopyranose

25.35

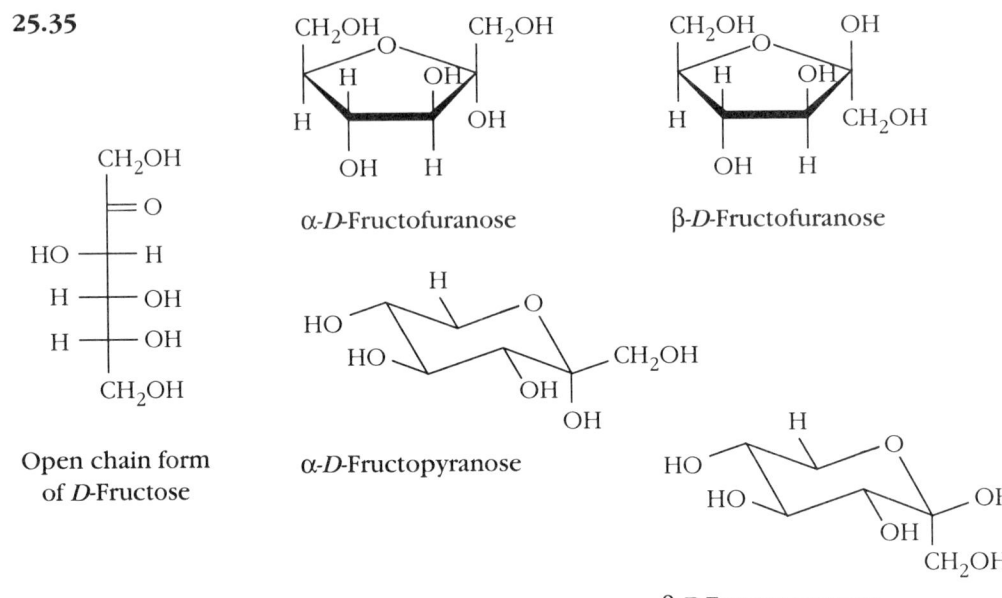

25.36 (a) The ring contains six atoms so the structure shown is the pyranose form of talose.
(b) The —OH group on C1 is *cis* to the —CH$_2$OH group on C5 so this is the β-anomer.
(c) This structure differs from the structure of *D*-glucose only in the configuration of C2 and C4 so this is the structure of *D*-talose. This can be confirmed by converting the cyclic structure of talose into its Fischer projection in which the —OH group farthest from the carbonyl group points to the right.
(d) This cyclic form of talose is in equilibrium with its open chain form (which is an aldohexose), so talose is a reducing sugar.
(e) The configurations of C2 and C4 are opposite to those of *D*-glucose so if we start with the Fischer projection formula of glucose, we need only change the configurations about C2 and C4 to obtain the Fischer projection of talose:

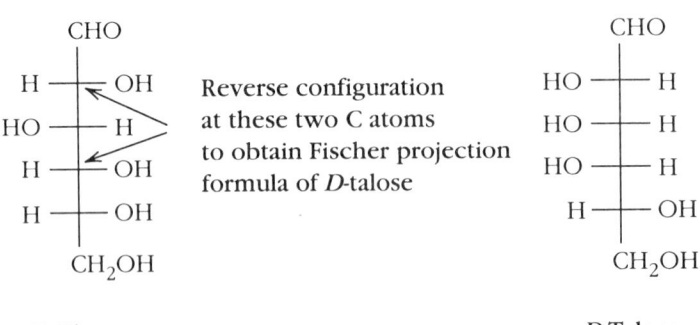

25.37 Step 1: First cyclization

Step 2: Second cyclization

25.38 (a) This information tells us that trehalose is a disaccharide.
(b) Trehalose does not reduce Tollen's reagent or Fehling's solution so trehalose lacks a hemiacetal group. This means that the disaccharide is joined by a 1,1-glucoside bond.
(c) Trehalose contains two glucose units joined together.
(d) This information tells us that the two glucose monomers are in their pyranose forms and that the two units are joined at C1 of each glucose unit.
(e) This information tells us that an α-linkage joins the two glucose monomers.

Trehalose

25.39

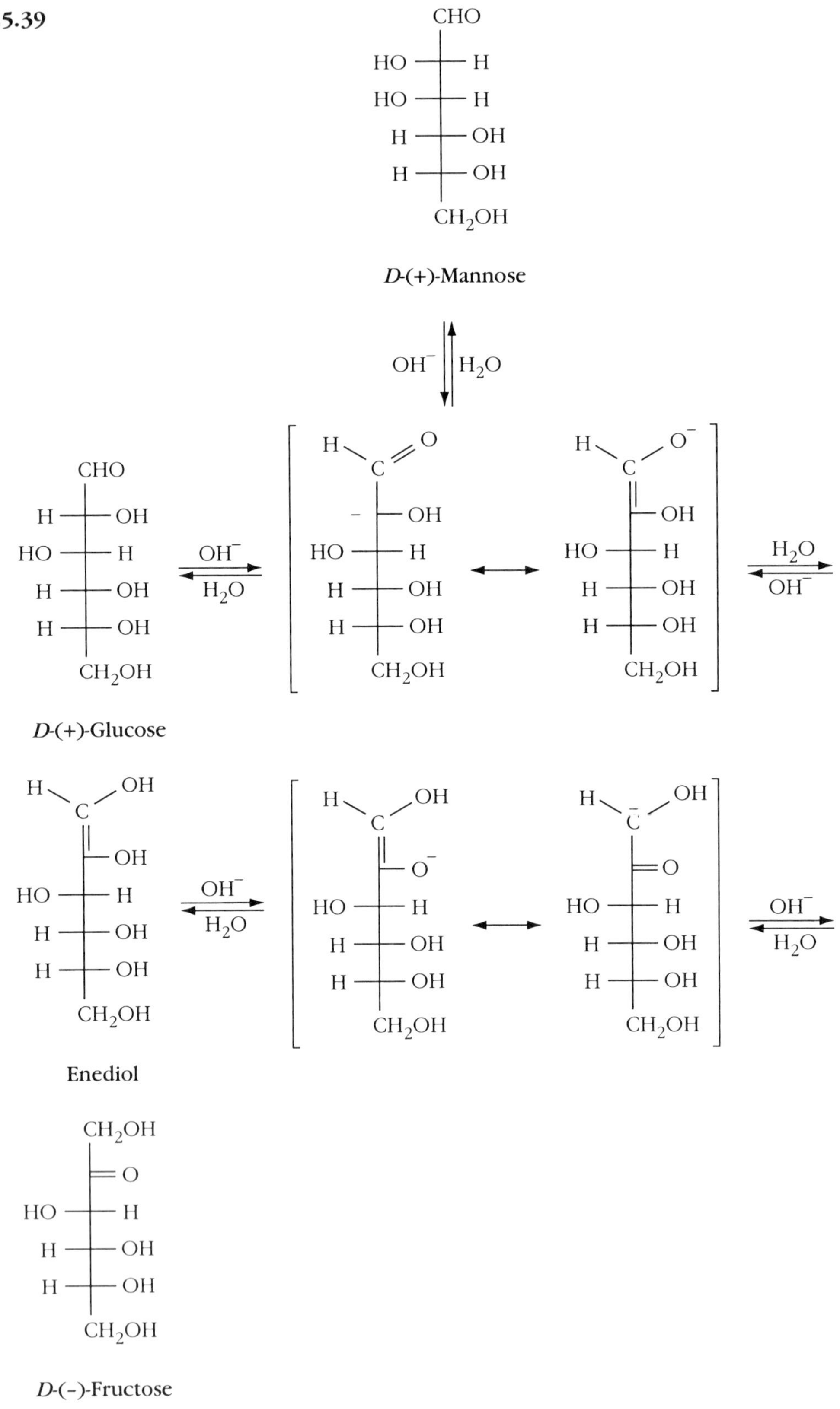

```
        CHO
   HO ──┼── H
   HO ──┼── H
    H ──┼── OH
    H ──┼── OH
        CH₂OH
```
D-(+)-Mannose

⇅ OH⁻ / H₂O

```
        CHO
    H ──┼── OH
   HO ──┼── H
    H ──┼── OH
    H ──┼── OH
        CH₂OH
```
D-(+)-Glucose

$\underset{H_2O}{\overset{OH^-}{\rightleftharpoons}}$

$\left[\begin{array}{c} H-C=O \\ -OH \\ HO-H \\ H-OH \\ H-OH \\ CH_2OH \end{array} \longleftrightarrow \begin{array}{c} H-C-O^- \\ =OH \\ HO-H \\ H-OH \\ H-OH \\ CH_2OH \end{array}\right]$

$\underset{OH^-}{\overset{H_2O}{\rightleftharpoons}}$

```
    H     OH
     \   /
      C
      ‖
      ──── OH
   HO ──┼── H
    H ──┼── OH
    H ──┼── OH
        CH₂OH
```
Enediol

$\underset{H_2O}{\overset{OH^-}{\rightleftharpoons}}$

$\left[\begin{array}{c} H \quad OH \\ C \\ \| \\ -O^- \\ HO-H \\ H-OH \\ H-OH \\ CH_2OH \end{array} \longleftrightarrow \begin{array}{c} H \quad OH \\ \bar{C} \\ =O \\ HO-H \\ H-OH \\ H-OH \\ CH_2OH \end{array}\right]$

$\underset{H_2O}{\overset{OH^-}{\rightleftharpoons}}$

```
        CH₂OH
        =O
   HO ──┼── H
    H ──┼── OH
    H ──┼── OH
        CH₂OH
```
D-(−)-Fructose

25.40

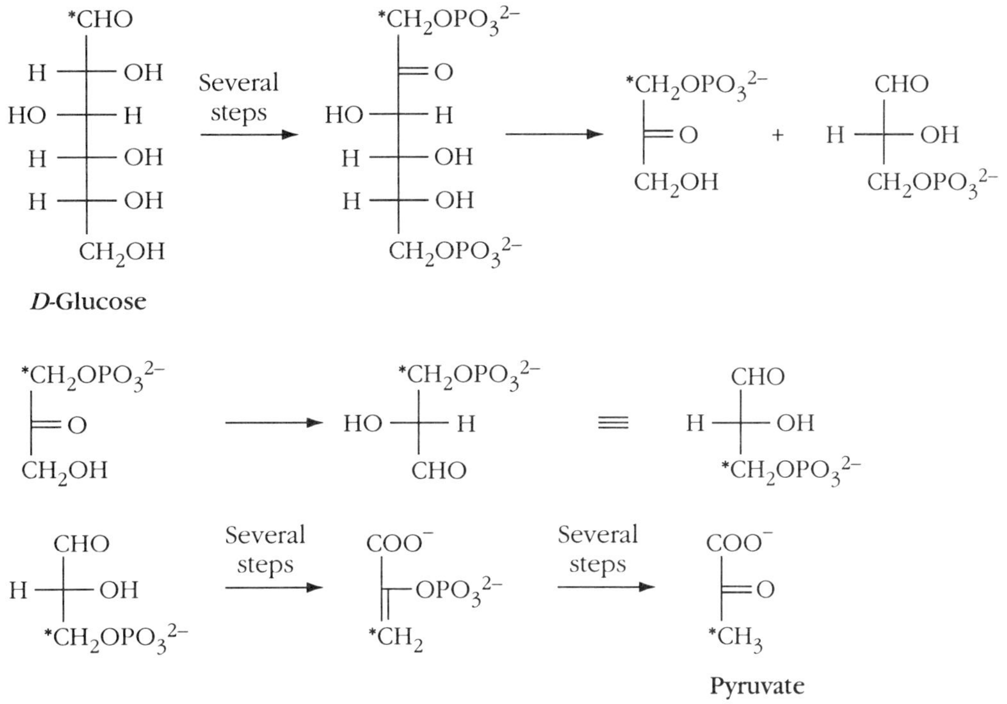

*CHO / H—OH / HO—H / H—OH / H—OH / CH₂OH (D-Glucose) →(Several steps)→ *CH₂OPO₃²⁻ / =O / HO—H / H—OH / H—OH / CH₂OPO₃²⁻ → *CH₂OPO₃²⁻ / =O / CH₂OH + CHO / H—OH / CH₂OPO₃²⁻

*CH₂OPO₃²⁻ / =O / CH₂OH → *CH₂OPO₃²⁻ / HO—H / CHO ≡ CHO / H—OH / *CH₂OPO₃²⁻

CHO / H—OH / *CH₂OPO₃²⁻ →(Several steps)→ COO⁻ / —OPO₃²⁻ / *CH₂ →(Several steps)→ COO⁻ / =O / *CH₃ (Pyruvate)

* = Location of ¹⁴C

25.41 Amylose is a biopolymer of glucopyranose molecules joined by 1,4-α-glycoside bonds.

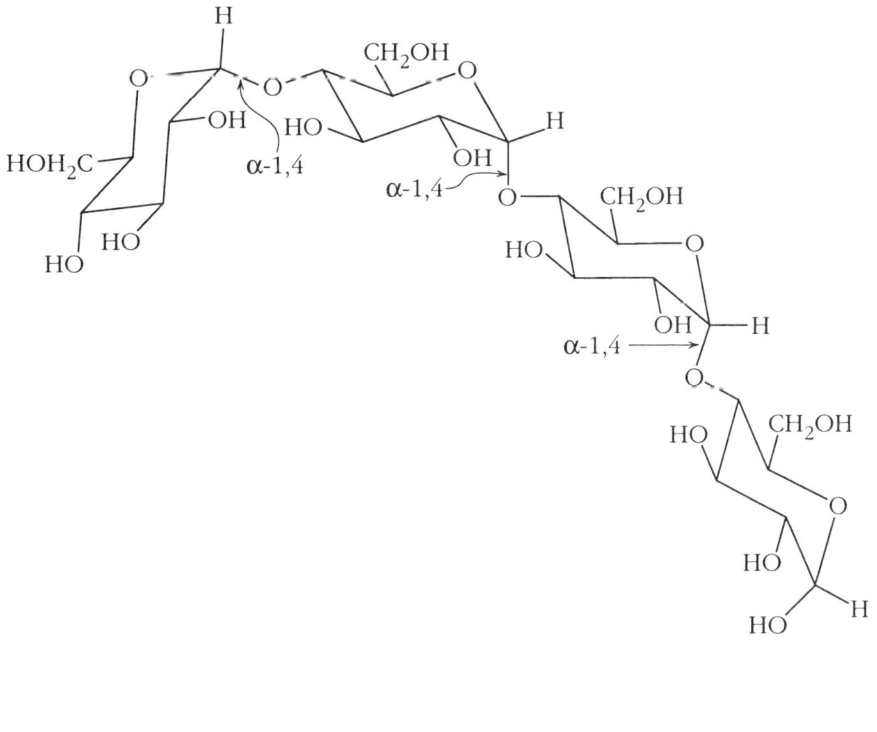

25.42 Let R = 2 Glucose units joined by an α-1,4-linkage.

Step 1: Cleavage of terminal glucose unit of tetrasaccharide

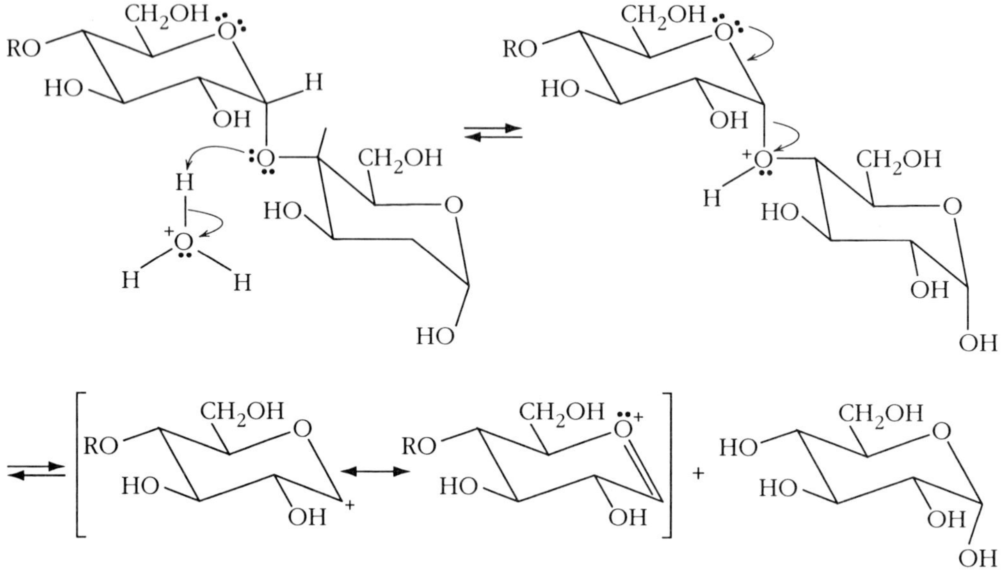

Step 2: Formation of α- and β-anomer of trisaccharide

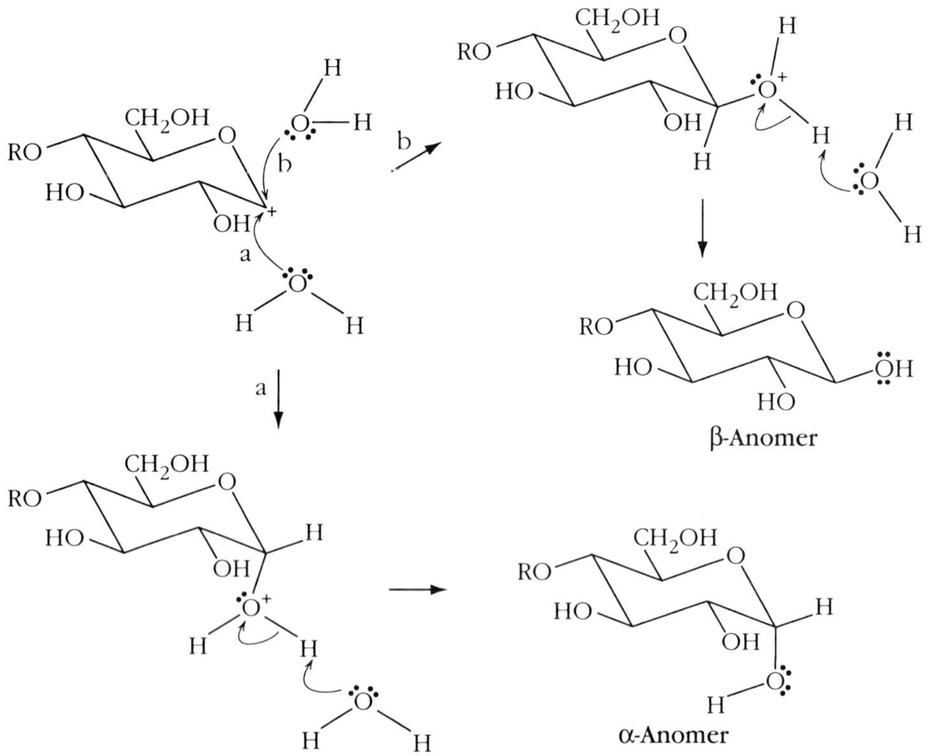

β-Anomer

α-Anomer

Step 3: Repeat Steps 1 and 2 twice more to hydrolyze the four glucose units of amylose to four *D*-glucopyranose molecules

CHAPTER 26

AMINO ACIDS, POLYPEPTIDES, PROTEINS, AND ENZYMES
Concepts

Polypeptides are linear biopolymers that consist of α-amino acid residues joined together by amide or peptide bonds. **Proteins** are molecules that consist of one or more polypeptide chains. Twenty α-amino acids are commonly found in proteins and, except for glycine, they all contain a stereocenter that has the *S*-configuration. The stereocenter has been related to *L*-glyceraldehyde so the α-amino acids in proteins belong to the *L*-family.

The **primary structure of a polypeptide** is its sequence of α-amino acids in a polypeptide. One method of determining this sequence is by the **Edman procedure,** which removes the amino acid residue, one at a time from the N-terminal end of the polypeptide. Chemical synthesis of polypeptides can be accomplished by the **Merrifield solid-phase method**. This method adds amino acids one at a time to the lengthening polypeptide chain, which is anchored to an insoluble copolymer support.

The conformations of a polypeptide that are repeated regularly along its main chain are its **secondary structure**. The **tertiary structure** is the folding of the structures of proteins together with the spatial arrangements of the side chains of its amino acid residues. The **quaternary structure** of a protein is the assembly of two or more individual polypeptide chains.

The information needed to fold a polypeptide chain is contained in its primary structure. In general the polypeptide chain folds spontaneously so that the hydrophobic side chains are placed in the interior away from water and the polar and charged side chains are placed on the surface in contact with aqueous solutions.

Solutions to the Exercises

26.1

(a) $\overset{+}{H_3N}-\overset{\displaystyle COO^-}{\underset{\displaystyle CH_3}{\overset{\displaystyle |}{C}}}-H$
(b) $\overset{+}{H_3N}-\overset{\displaystyle COO^-}{\underset{\displaystyle CH_2C_6H_5}{\overset{\displaystyle |}{C}}}-H$
(c) $\overset{+}{H_3N}-\overset{\displaystyle COO^-}{\underset{\displaystyle CH_2}{\overset{\displaystyle |}{C}}}-H$ with imidazole ring
(d) $\overset{+}{H_3N}-\overset{\displaystyle COO^-}{\underset{\displaystyle H-\overset{\displaystyle |}{C}-CH_3,\; OH}{\overset{\displaystyle |}{C}}}-H$

26.2 We want the ratio $\dfrac{[HA]}{[A^-]}$ so the Henderson-Hasselbach equation must be rearranged as follows:

$$pH = pK_a + \log\dfrac{[A^-]}{[HA]} = pK_a - \log\dfrac{[HA]}{[A^-]}$$

$$\log\dfrac{[HA]}{[A^-]} = pK_a - pH$$

(a) $\log\dfrac{[H_3\overset{+}{N}CH(CH_3)COOH]}{[H_3\overset{+}{N}CH(CH_3)COO^-]} = p(K_a)_1 - pH = 2.3 - 1.5 = 0.8$

$\dfrac{[H_3\overset{+}{N}CH(CH_3)COOH]}{[H_3\overset{+}{N}CH(CH_3)COO^-]} = 6.3$

(b) $\dfrac{[H_3\overset{+}{N}CH(CH_3)COOH]}{[H_3\overset{+}{N}CH(CH_3)COO^-]} = 0.020$

(c) $\log\dfrac{[H_3\overset{+}{N}CH(CH_3)COO^-]}{[H_2NCH(CH_3)COO^-]} = p(K_a)_2 - pH = 9.9 - 8.2 = 1.7$

$\dfrac{[H_3\overset{+}{N}CH(CH_3)COO^-]}{[H_2NCH(CH_3)COO^-]} = 50$

(d) $\dfrac{[H_3\overset{+}{N}CH(CH_3)COO^-]}{[H_2NCH(CH_3)COO^-]} = 7.9 \times 10^{-3}$

26.3

(a) $H_3\overset{+}{N}\overset{\displaystyle |}{C}HCOO^-$ with $CH(CH_3)_2$

(b) $H_3\overset{+}{N}\overset{\displaystyle |}{C}HCOO^-$ with $CH_2CH_2COO^-$

(c) $H_3\overset{+}{N}\overset{\displaystyle |}{C}HCOO^-$ with $CH_2CH_2CH_2CH_2\overset{+}{N}H_3$

(d) $H_3\overset{+}{N}CH_2COO^-$

26.4 (a) H$_3$⁺NCHCOOH
          |
          CH$_2$CH$_2$CH$_2$CH$_2$NH$_3$⁺

(b) H$_2$NCHCOO⁻
     |
     CH$_2$CH$_2$CH$_2$CH$_2$NH$_3$⁺

(c) H$_2$NCHCOO⁻
     |
     CH$_2$CH$_2$CH$_2$CH$_2$NH$_2$

(d) H$_3$⁺NCHCOO⁻
     |
     CH$_2$CH$_2$CH$_2$CH$_2$NH$_3$⁺

26.5

These steric interactions are absent in the *trans*-conformation

26.6 By convention, peptides are written with the N-terminal amino acid to the left and the C-terminal amino acid to the right.

26.7

(a)
```
        Gln
   Ile /
    |  Asn
   Tyr  \
    |   Cys — Pro — Leu — GlyNH₂
    |  S
    | /
    |S
   Cys
```
↓ 2HSCH₂CH₂OH

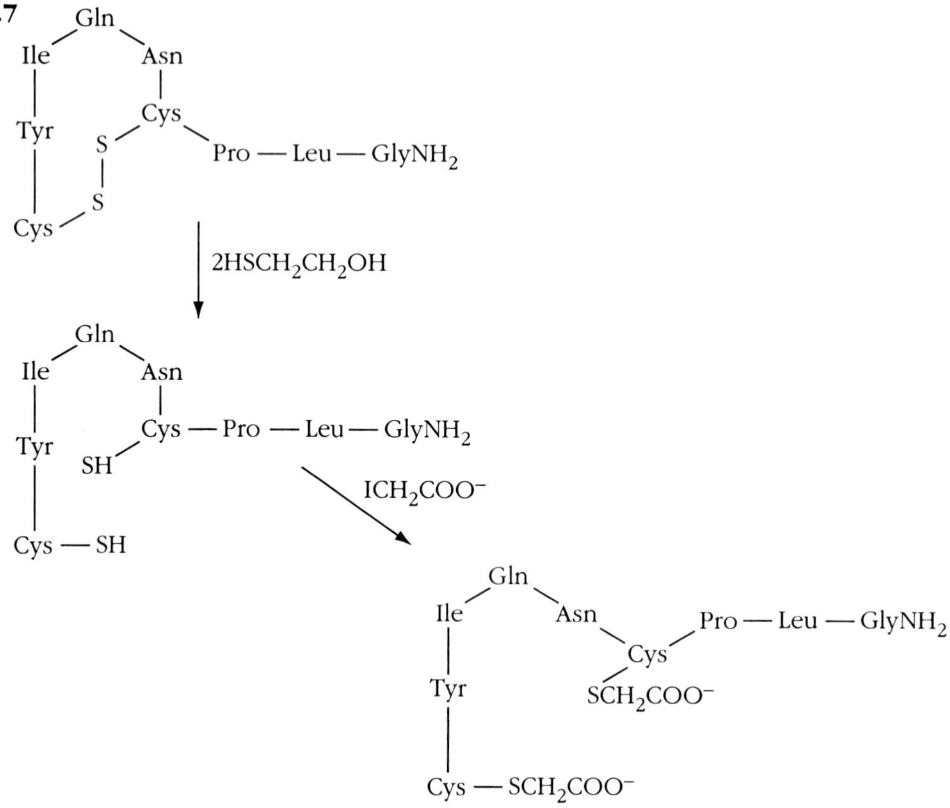

```
        Gln
   Ile /
    |  Asn
   Tyr  |
    |   Cys — Pro — Leu — GlyNH₂
    SH
    
   Cys — SH
```
→ ICH₂COO⁻

```
           Gln
      Ile /
       |   Asn        Pro — Leu — GlyNH₂
      Tyr   \        /
       |    Cys
       |    SCH₂COO⁻
       |
      Cys — SCH₂COO⁻
```

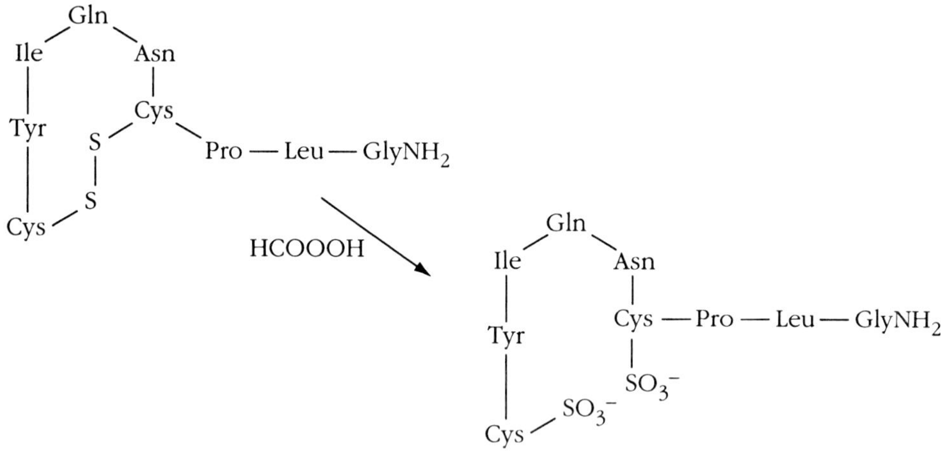

(b)
```
        Gln
   Ile /
    |  Asn
   Tyr  |
    |   Cys — Pro — Leu — GlyNH₂
    |  S
    | /
    |S
   Cys
```
→ HCOOOH

```
           Gln
      Ile /
       |   Asn
      Tyr   |
       |    Cys — Pro — Leu — GlyNH₂
       |    SO₃⁻
       
      Cys — SO₃⁻
```

26.8

$$\underset{\substack{|\\ CH_2\\ |\\ CH(CH_3)_2}}{\overset{+}{H_3N}CH-\underset{\substack{\|\\ O^-}}{\overset{O}{C}}} \xrightarrow{OH^-} \underset{\substack{|\\ CH_2\\ |\\ CH(CH_3)_2}}{H_2NCH-COO^-} \longrightarrow$$

[ninhydrin-type structure with 2]

↓

[purple Ruhemann's anion structure]

+

$(CH_3)_2CHCH_2\underset{\substack{\|\\ H}}{\overset{O}{C}}$ + CO_2

26.9 N≡CBr cleaves the C-terminal end of methionine (Met).

Ala—Val—Lys—Met—|Ile—Pro—Tyr—Thr—Arg—Ser—Met|—Leu—His—Gln
                      N≡CBr   N≡CBr
                 cleaves here   cleaves here

The enzyme trypsin cleaves the C-terminal end of arginine (Arg) and lysine (Lys).

Ala—Val—Lys|—Met—Ile—Pro—Tyr—Thr—Arg|—Ser—Met—Leu—His—Gln
              Trypsin   Trypsin
          cleaves here   cleaves here

The enzyme chymotrypsin cleaves the C-terminal end of phenylalanine (Phe), tyrosine (Tyr), and tryptophan (Trp)

Ala—Val—Lys—Met   Ile—Pro—Tyr|—Thr—Arg—Ser—Met—Leu—His—Gln
                                Chymotrypsin
                                cleaves here

26.10 (a) Match the overlapping regions of the various polypeptides
                             Asp-Arg-Leu
                                  Leu-Tyr-Ile
                                      Tyr-Ile-Phe
Complete sequence: Asp-Arg-Leu-Tyr-Ile-Phe

(b)
```
                    Ile-Gln-Asn-Cys
                    Gln-Asn-Cys-Pro
                    Cys-Pro-Leu-Gly
                        Pro-Leu-Gly
            Cys-Tyr-Ile-Gln-Asn
Complete sequence:  Cys-Tyr-Ile-Gln-Asn-Cys-Pro-Leu-Gly
```

26.11

Ala—Ala

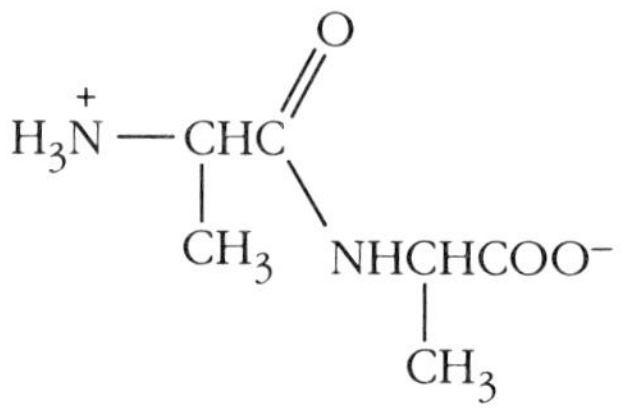

$$\overset{+}{H_3N}-CHC(=O)-NHCHCOO^-$$
with CH$_3$ on each α-carbon

Val—Val

$$\overset{+}{H_3N}-CH-C(=O)-NHCHCOO^-$$
with CH(CH$_3$)$_2$ on each α-carbon

Ala—Val

$$\overset{+}{H_3N}-CHC(=O)-NHCHCOO^-$$
with CH$_3$ (Ala) and CH(CH$_3$)$_2$ (Val)

Val—Ala

$$\overset{+}{H_3N}-CH-C(=O)-NHCHCOO^-$$
with CH(CH$_3$)$_2$ (Val) and CH$_3$ (Ala)

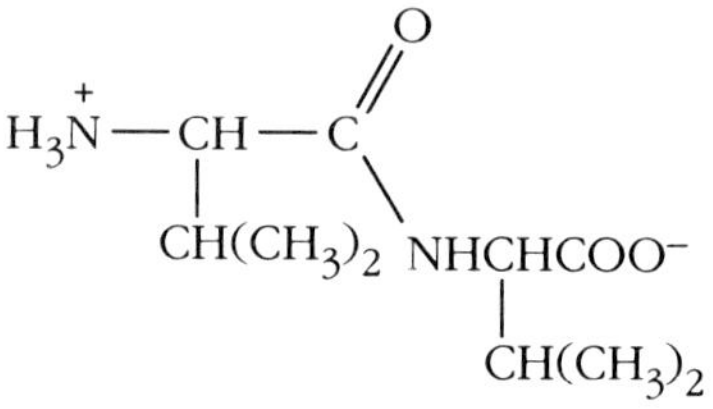

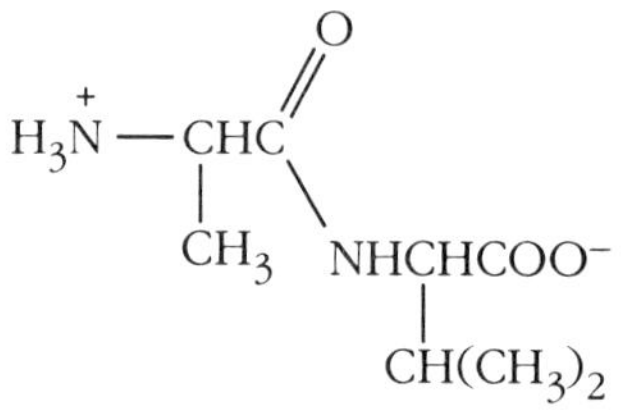

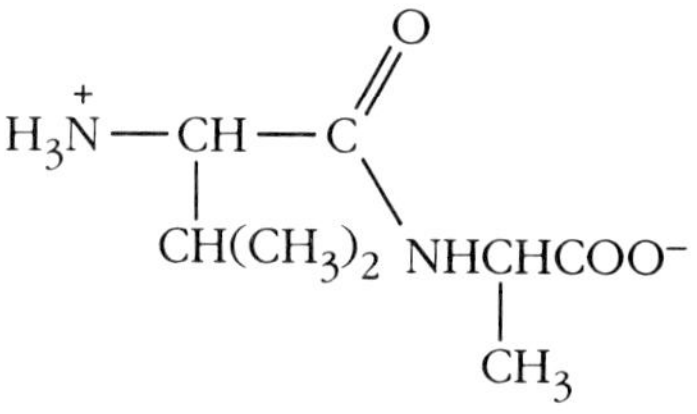

26.12 Step 1: Nucleophilic addition to carbonyl carbon atom of di-*t*-butyl dicarbonate

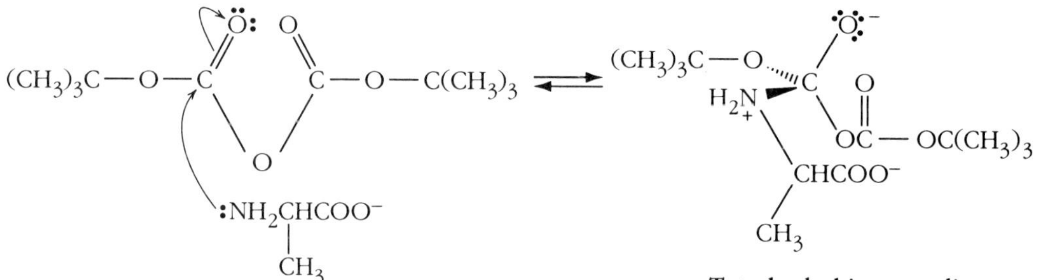

[Mechanism: attack of :NH$_2$CH(CH$_3$)COO$^-$ on (CH$_3$)$_3$C—O—C(=O)—O—C(=O)—O—C(CH$_3$)$_3$ gives the Tetrahedral intermediate]

Step 2: Elimination of $^-$OOCOC(CH$_3$)$_3$

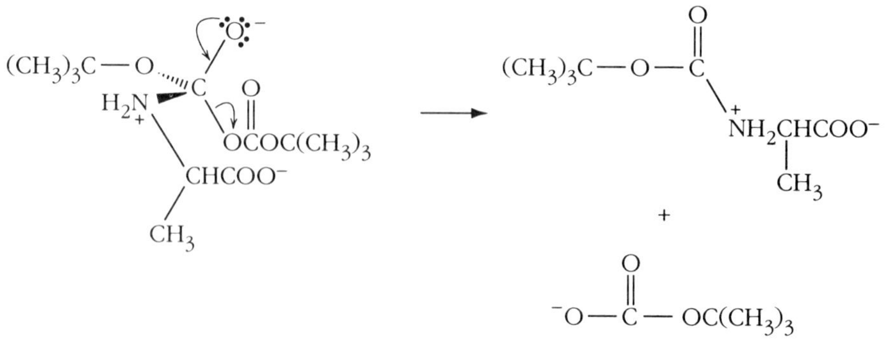

[The tetrahedral intermediate collapses to give (CH$_3$)$_3$C—O—C(=O)—$\overset{+}{N}$H$_2$CHCOO$^-$ (with CH$_3$) plus $^-$O—C(=O)—OC(CH$_3$)$_3$]

Step 3: Proton transfer

$$(CH_3)_3C-O-\underset{\underset{H}{\overset{+}{N}HCHCOO^-}}{\overset{O}{\underset{|}{C}}} \rightleftharpoons (CH_3)_3C-O-\underset{\underset{CH_3}{NHCHCOO^-}}{\overset{O}{\underset{|}{C}}} + H\overset{+}{B}$$

B:

t-Boc alanine

Step 4: Elimination of CO_2

$$^-O-\overset{O}{\underset{\|}{C}}-OC(CH_3)_3 \longrightarrow \overset{O}{\underset{\|}{C}}{\underset{\|}{\overset{}{O}}} + {}^-OC(CH_3)_3 \xrightarrow{BH} HOC(CH_3)_3 + :B$$

26.13 Preparation of Leu-Ala:

Leu + $(CH_3)_3C-O-CO-O-CO-OC(CH_3)_3$ → $(CH_3)_3COC(O)NHCHCOOH$ with side chain $CH_2CH(CH_3)_2$

Ala $\xrightarrow{(1)\ CH_3OH/HCl}{(2)\ NaOH/H_2O}$ $H_2NCHC(O)OCH_3$ with CH_3 side chain

$(CH_3)_3COC(O)NHCH(CH_2CH(CH_3)_2)COOH$ + $H_2NCH(CH_3)C(O)OCH_3$ $\xrightarrow{DCC}$

Preparation of Leu-Ala-Val:

Leu — Ala — Val

26.14 Reaction 1: Prepare the *t*-Boc protected form of the three amino acids

$$\underset{\text{Alanine}}{\underset{|}{\text{H}_2\text{NCHCOOH}}\atop\text{CH}_3} \xrightarrow{[(\text{CH}_3)_3\text{COCO}]_2\text{O}} \underset{\text{t-Boc-Ala}}{(\text{CH}_3)_3\text{COC}(=\text{O})\text{NHCH}(\text{CH}_3)\text{COOH}}$$

$$\underset{\text{Valine}}{\underset{|}{\text{H}_2\text{NCHCOOH}}\atop(\text{CH}_3)_2\text{CH}} \xrightarrow{[(\text{CH}_3)_3\text{COCO}]_2\text{O}} \underset{\text{t-Boc-Val}}{(\text{CH}_3)_3\text{COC}(=\text{O})\text{NHCH}(\text{CH}(\text{CH}_3)_2)\text{COOH}}$$

$$\underset{\text{Phenylalanine}}{\underset{|}{\text{H}_2\text{NCHCOOH}}\atop\text{C}_6\text{H}_5\text{CH}_2} \xrightarrow{[(\text{CH}_3)_3\text{COCO}]_2\text{O}} \underset{\text{t-Boc-Phe}}{(\text{CH}_3)_3\text{COC}(=\text{O})\text{NHCH}(\text{CH}_2\text{C}_6\text{H}_5)\text{COOH}}$$

Reaction 2: Phe is the C-terminal amino acid of the tripeptide so it is added first to the polymer

$$t\text{-Boc}—\underset{\underset{\text{CH}_2\text{C}_6\text{H}_5}{|}}{\text{NHCHCOO}^-} \quad \text{ClCH}_2—\text{Copolymer}$$

$$\downarrow$$

$$t\text{-Boc}—\underset{\underset{\text{CH}_2\text{C}_6\text{H}_5}{|}}{\text{NHCH}}—\overset{\overset{\text{O}}{\|}}{\text{C}}\diagdown_{\text{OCH}_2—\text{Copolymer}}$$

Reaction 3:   (1) Wash
Remove *t*-Boc   (2) CF$_3$COOH
group   (3) NaOH/H$_2$O wash

$$\downarrow$$

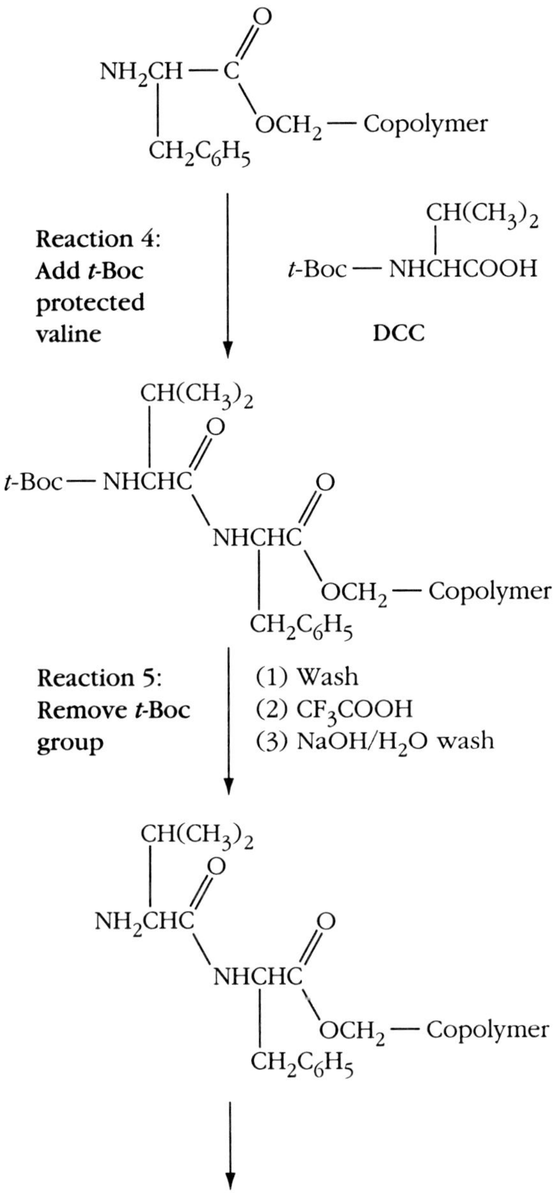

Reaction 4:
Add *t*-Boc protected valine

Reagents: *t*-Boc—NHCHCOOH with CH(CH$_3$)$_2$ side chain; DCC

Reaction 5:
Remove *t*-Boc group

(1) Wash
(2) CF$_3$COOH
(3) NaOH/H$_2$O wash

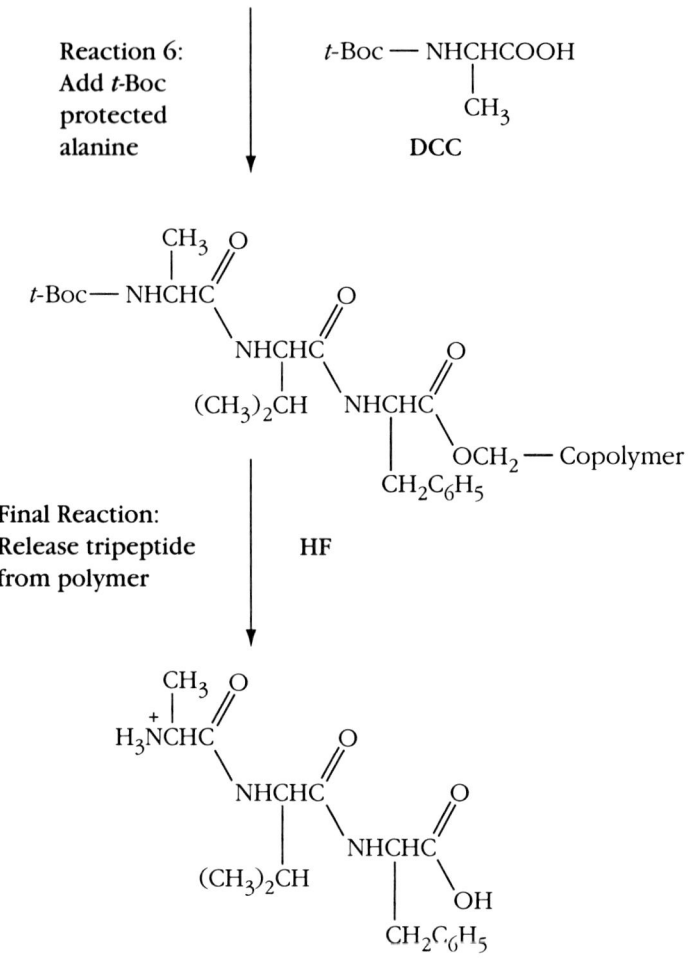

26.15 Step 1: Proton transfer from acid catalyst to glycoside oxygen atom

Step 2: Cleavage of the glycoside bond forms an oxonium ion

Oxonium ion

Step 3: Reaction of oxonium ion with water

Step 4: Proton transfer regenerates acid catalyst

Similarities:
1. Proton transfer from a catalyst in Step 1 of both mechanisms.
2. Cleavage of the glycoside bond to form an oxonium ion in Step 2 of both mechanisms.
3. Reaction of oxonium ion with water in Step 3 of both mechanisms.
4. Proton transfer to regenerate the catalyst in Step 4.

Differences:
1. The source of the proton in Step 1 is different. In an aqueous acid solution, the hydronium ion is the source of a proton. In the lysozyme-catalyzed reaction, the proton comes from the —COOH group of Glu 35.
2. Addition of water to the oxonium ion intermediate occurs from both sides to form a mixture of the α- and β-anomers in the acid-catalyzed hydrolysis. Only the β-anomer is formed in the enzyme-catalyzed hydrolysis because Asp 52 shields one side of the oxonium ion so that addition of water occurs only to the other side.

26.16 (a) A **zwitterion** is a neutral dipolar molecule in which the two unlike charges are not adjacent to each other. The zwitterion form of α-amino acids, for example, contains a protonated amino group and the carboxyl group exists as a carboxylate ion.

Zwitterion form
of α-amino acids

(b) α-**Amino acids** are compounds that contain an amino group and another substituent, called the *amino acid side chain*, on the α-carbon atom of acetic acid. The twenty α-amino acids differ in the structure of the amino acid side chain. They are classified as amino acids with (1) nonpolar side chains, (2) uncharged polar side chains, and (3) charged polar side chains.

(c) A **disulfide bond** is a bond between two sulfur atoms formed by the mild oxidation of two thio groups. The most common disulfide bond in nature joins two cysteine molecules to form cystine:

$$\text{Cysteine} \xrightarrow{O_2 \text{ (Air oxidation)}} \text{Cystine}$$

(d) The **primary structure of a polypeptide** is the sequence of its amino acids. The information for the three-dimensional structure of a polypeptide is contained within the primary structure.

(e) A **peptide bond** is another name for an amide bond between two amino acids:

Peptide bond (amide bond)

(f) **Essential amino acids** are amino acids that must be obtained from food because they cannot be synthesized by the human body. The following amino acids are essential: Ile, Leu, Met, Phe, Thr, Trp, Val, Arg, His, and Lys.

(g) **Secondary structure of a polypeptide** is the conformations of a polypeptide along its main chain. Three types of secondary structures are the α-helix, the β-pleated sheet, and the β-bend.

(h) **Solid-phase peptide synthesis** is a technique that forms a growing polypeptide chain on a solid support that consists of beads of a copolymer prepared by copolymerizing styrene with a few percent 4-(chloromethyl) styrene. The polypeptide synthesis starts by covalently bonding the C-terminal amino acid of the polypeptide to be synthesized to the copolymer. The *t*-butoxycarbonyl protected α-amino acids are then joined in the desired sequence one at a time. Finally, the desired polypeptide is cleaved from the copolymer and isolated.

(i) An **amphoteric compound** contains acidic and basic functional groups so it is capable of accepting *and* donating a proton. Amino acids are amphoteric compounds because they contain an acidic functional group (the ammonium ion, NH_3^+) and a basic functional group (the carboxylate ion —COO^-).

(j) The **isoelectric point** is the pH at which the number of positive charges and the number of negative charges on a protein or an amino acid are equal.

26.17

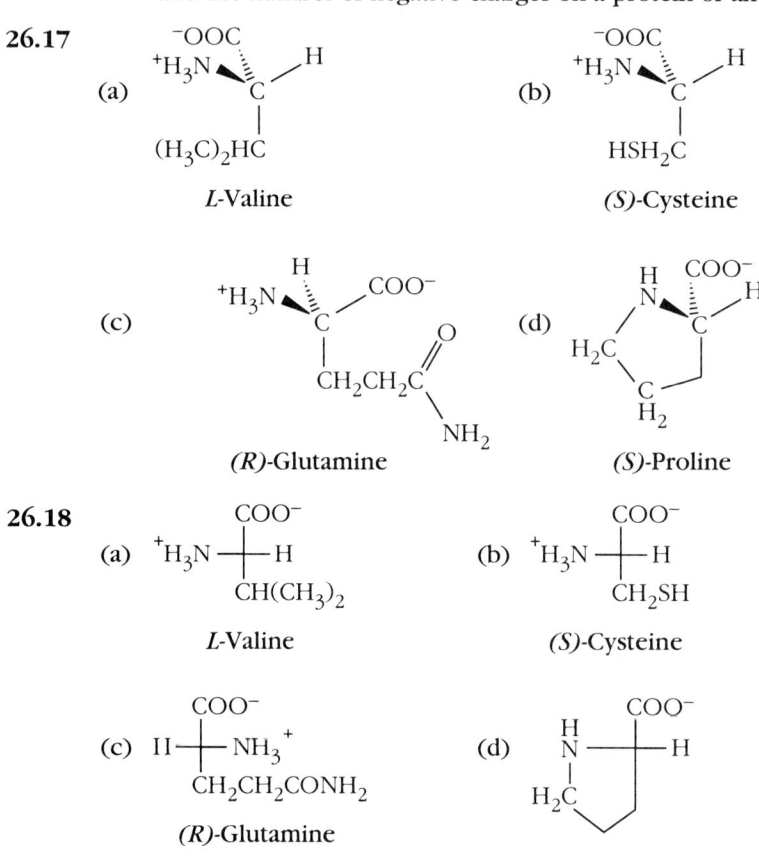

26.18

26.19

(d) Structure: +H₃N-CH(H)-C(=O)-NH-CH(CH(CH₃)₂)-C(=O)-NH-CH(CH₂CH₂COO⁻)-COO⁻

(e) Structure: +H₃N-CH(CH₂SH)-C(=O)-NH-CH((CH₂)₄NH₃⁺)-C(=O)-NH-CH(CH₂-imidazole)-C(=O)... with Arg and Asp residues

26.20 pH 3:

(a) ^+H_3N—C(H)(CH₂CH(CH₃)₂)—COOH
Leu

(b) ^+H_3N—C(H)(CH₂CH₂SCH₃)—COOH
Met

(c) ^+H_3N—C(H)(CH₂COOH)—COOH
Asp

(d) ^+H_3N—C(H)((CH₂)₄NH₃⁺)—COOH
Lys

pH 7:

(a) ^+H_3N—C(H)(CH₂CH(CH₃)₂)—COO⁻

(b) ^+H_3N—C(H)(CH₂CH₂SCH₃)—COO⁻

(c) ^+H_3N—C(H)(CH₂COO⁻)—COO⁻

(d) ^+H_3N—C(H)((CH₂)₄NH₃⁺)—COO⁻

pH 11:

(a) H_2N—C(H)(CH₂CH(CH₃)₂)—COO⁻

(b) H_2N—C(H)(CH₂CH₂SCH₃)—COO⁻

(c) H_2N—C(H)(CH₂COO⁻)—COO⁻

(d) H_2N—C(H)((CH₂)₄NH₂)—COO⁻

26.21 (a) It takes three equivalents of base to titrate all the acid or basic groups of this amino acid so we know that it has a carboxylic acid group, an ammonium ion, and another acidic group on the side chain. The first proton to be removed from an amino acid when base is added is the one from the carboxylic acid because it is the most acidic. When 0.5 equivalent of base is added, therefore, one half the carboxylic acid group is neutralized so at that pH, pH = p$(K_a)_1$. Therefore the value of p$(K_a)_1$ is 2.2. This value is within the range of p$(K_a)_1$ for all the amino acids (1.8 to 3.0).

We know that (p$K_a)_2$ for amino acids is in the range 9 to 10. The pH when 3.0 equivalents of base is added is equal to p$(K_a)_2$ so the value of p$(K_a)_2$ is 9.9.

The K_a of the remaining acidic functional group is determined by the pH when 1.5 equivalents of base has been added because at that pH, pH = p(K_R). The value of p(K_R) is 4.2.

(b) The amino acid is glutamic acid.

(c)

Point A

$$\begin{array}{c} \text{COOH} \\ {}^+\text{H}_3\text{N}\!\!-\!\!\text{C}\!\!-\!\!\text{H} \\ \text{CH}_2\text{CH}_2\text{COOH} \end{array}$$

Point B

$$\begin{array}{c} \text{COO}^- \\ {}^+\text{H}_3\text{N}\!\!-\!\!\text{C}\!\!-\!\!\text{H} \\ \text{CH}_2\text{CH}_2\text{COOH} \end{array}$$

Point C

$$\begin{array}{c} \text{COO}^- \\ {}^+\text{H}_3\text{N}\!\!-\!\!\text{C}\!\!-\!\!\text{H} \\ \text{CH}_2\text{CH}_2\text{COO}^- \end{array}$$

Point D

$$\begin{array}{c} \text{COO}^- \\ \text{H}_2\text{N}\!\!-\!\!\text{C}\!\!-\!\!\text{H} \\ \text{CH}_2\text{CH}_2\text{COO}^- \end{array}$$

26.22

$$\begin{array}{c} \text{COOH} \\ {}^+\text{H}_3\text{N}\!\!-\!\!\text{C}\!\!-\!\!\text{H} \\ (\text{CH}_2)_4\text{NH}_3^+ \end{array} + \text{H}_2\text{O} \xrightleftharpoons{\text{p}(K_a)_1} \begin{array}{c} \text{COO}^- \\ {}^+\text{H}_3\text{N}\!\!-\!\!\text{C}\!\!-\!\!\text{H} \\ (\text{CH}_2)_4\text{NH}_3^+ \end{array} + \text{H}_3\text{O}^+$$

$$\begin{array}{c} \text{COO}^- \\ {}^+\text{H}_3\text{N}\!\!-\!\!\text{C}\!\!-\!\!\text{H} \\ (\text{CH}_2)_4\text{NH}_3^+ \end{array} + \text{H}_2\text{O} \xrightleftharpoons{\text{p}(K_a)_2} \begin{array}{c} \text{COO}^- \\ \text{H}_2\text{N}\!\!-\!\!\text{C}\!\!-\!\!\text{H} \\ (\text{CH}_2)_4\text{NH}_3^+ \end{array} + \text{H}_3\text{O}^+$$

$$\begin{array}{c} \text{COO}^- \\ \text{H}_2\text{N}\!\!-\!\!\text{C}\!\!-\!\!\text{H} \\ (\text{CH}_2)_4\text{NH}_3^+ \end{array} + \text{H}_2\text{O} \xrightleftharpoons{\text{p}K_R} \begin{array}{c} \text{COO}^- \\ \text{H}_2\text{N}\!\!-\!\!\text{C}\!\!-\!\!\text{H} \\ (\text{CH}_2)_4\text{NH}_2 \end{array} + \text{H}_3\text{O}^+$$

26.23 We learned in Section 15.9 that replacing an α-hydrogen atom of a carboxylic acid by an electron-withdrawing group increases the acidity of the carboxylic acid. Replacing an α-hydrogen of acetic acid (pK_a = 4.74) by an electron-withdrawing NH_3^+ group increases acidity. Alanine (pK_a = 3) is a stronger acid than acetic acid.

26.24 The pK_R values of arginine (12.5), lysine (10.8), tyrosine (10.1), and cysteine (8.3) which can be taken from data in Table 26.3 tell us that at pH 7 they do not transfer a proton from their side chain acidic groups to water. The pK_R values of aspartic acid (3.9), and glutamic acid (4.1), tell us that they have already transferred a proton from their side chain acidic groups to water. The side chains of arginine and lysine, which each contain a protonated nitrogen atom, and aspartic acid and glutamic acid, which each contain a carboxylate ion, are charged at pH 7. The pK_R value of histidine (6.0) indicates that an appreciable concentration of the charged side chain form exists in solution of pH 7.

26.25

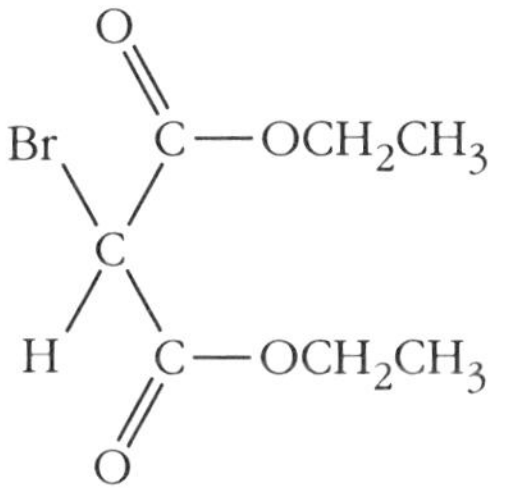

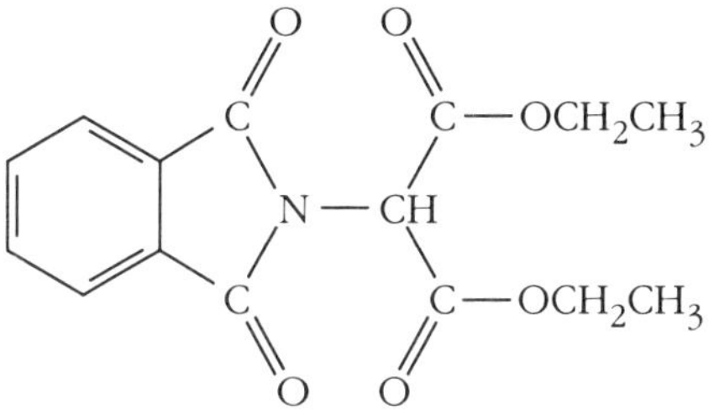

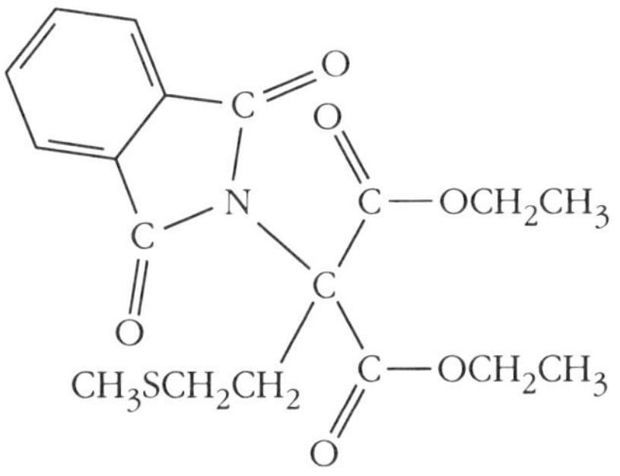

26.26

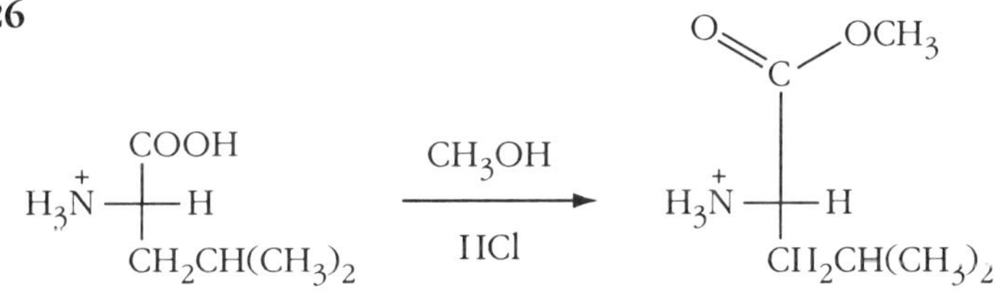

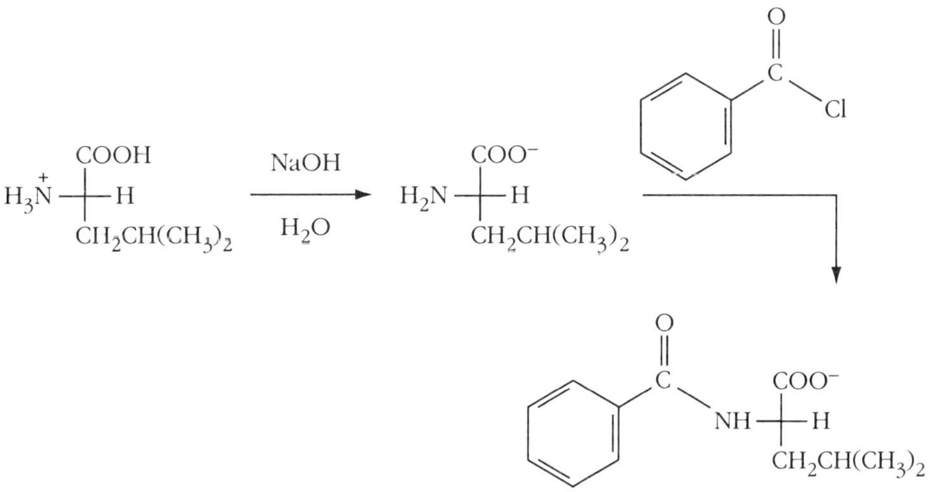

(c)
$$\underset{\underset{CH_2CH(CH_3)_2}{|}}{\overset{\overset{COOH}{|}}{H_2N-\!\!\!-\!\!\!-H}} \quad + \quad \underset{(CH_3)_3C-O-\overset{\overset{O}{\|}}{C}}{\overset{(CH_3)_3C-O-\overset{\overset{O}{\|}}{C}}{\underset{\underset{O}{\|}}{\overset{}{\diagdown}\!\!O\!\!\diagup}}} \quad \longrightarrow$$

$$(CH_3)_3C-O-\overset{\overset{O}{\|}}{C}-\underset{\underset{CH_2CH(CH_3)_2}{|}}{\overset{\overset{COOH}{|}}{NH-\!\!\!-\!\!\!-H}}$$

(d)
$$\underset{\underset{CH_2CH(CH_3)_2}{|}}{\overset{\overset{COOH}{|}}{H_2N-\!\!\!-\!\!\!-H}} \quad + \quad \underset{\underset{CH_3\overset{O}{\overset{\|}{C}}}{}}{\overset{CH_3\overset{O}{\overset{\|}{C}}}{\diagdown\!\!O\!\!\diagup}} \quad \longrightarrow \quad CH_3\overset{\overset{O}{\|}}{C}-\underset{\underset{CH_2CH(CH_3)_2}{|}}{\overset{\overset{COOH}{|}}{NH-\!\!\!-\!\!\!-H}}$$

26.27

(a)
$$\underset{\underset{\underset{NH_2}{\diagdown}}{\underset{CH_2\overset{\overset{O}{\|}}{C}}{|}}}{\overset{\overset{COO^-}{|}}{H_2N-\!\!\!-\!\!\!-H}} \quad \xrightarrow[\text{Heat}]{NaOH/H_2O} \quad \underset{\underset{CH_2COO^-}{|}}{\overset{\overset{COO^-}{|}}{H_2N-\!\!\!-\!\!\!-H}}$$

Asn

(b) Val — Gln — Lys $\xrightarrow[H_2O]{HCl/H_2O}$ $\underset{\underset{CH(CH_3)_2}{|}}{\overset{\overset{COOH}{|}}{^+H_3N-\!\!\!-\!\!\!-H}}$ +

$$\underset{\underset{\underset{NH_2}{\diagdown}}{\underset{CH_2CH_2\overset{\overset{O}{\|}}{C}}{|}}}{\overset{\overset{COOH}{|}}{^+H_3N-\!\!\!-\!\!\!-H}} \quad + \quad \underset{\underset{CH_2CH_2CH_2CH_2NH_3^+}{|}}{\overset{\overset{COOH}{|}}{^+H_3N-\!\!\!-\!\!\!-H}}$$

(c)
$(CH_3)_3COC(O)NH-CH(CH_3)-COOH$ + $H_2N-CH(CH(CH_3)_2)-COOCH_3$ $\xrightarrow{DCC}$

$(CH_3)_3COC(O)-NHCH(CH_3)C(O)-NHCH(CH(CH_3)_2)C(O)OCH_3$

(d) $H_3\overset{+}{N}-CH(CH_2COOH)-COOH$ $\xrightarrow[CH_3OH]{HCl}$ $H_3\overset{+}{N}-CH(CH_2C(O)OCH_3)-C(O)OCH_3$

(e) $H_2N-CH(CH_2C_6H_5)-COO^-$ + 2 (ninhydrin: indane-1,3-dione-2,2-diol) $\xrightarrow{^-OH}$ (Ruhemann's purple: bis-indanedione imine anion) + $C_6H_5CH_2CHO$ + CO_2

26.28

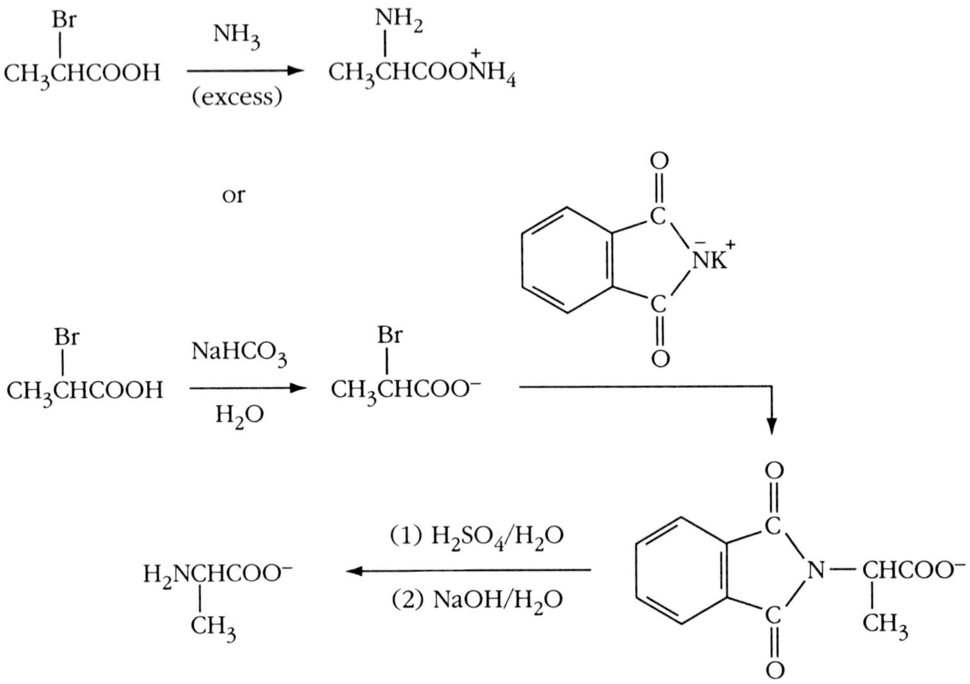

$$\underset{\text{CH}_3\text{CHCOOH}}{\overset{\text{Br}}{|}} \xrightarrow[\text{(excess)}]{\text{NH}_3} \underset{\text{CH}_3\text{CHCOO}^-\overset{+}{\text{NH}_4}}{\overset{\text{NH}_2}{|}}$$

or

$$\underset{\text{CH}_3\text{CHCOOH}}{\overset{\text{Br}}{|}} \xrightarrow[\text{H}_2\text{O}]{\text{NaHCO}_3} \underset{\text{CH}_3\text{CHCOO}^-}{\overset{\text{Br}}{|}} \xrightarrow{\text{phthalimide-K}^+}$$

$$\underset{\text{CH}_3}{\overset{\text{H}_2\text{NCHCOO}^-}{|}} \xleftarrow[\text{(2) NaOH/H}_2\text{O}]{\text{(1) H}_2\text{SO}_4/\text{H}_2\text{O}} \text{phthalimide-CHCH}_3\text{COO}^-$$

All of these reactions are carried out in a single reaction vessel.

26.29

(a) Gly—Met—Gly—Phe—Ala—Val—Arg—|Met—Leu—Tyr—Lys|—Gly—Ala

                                              Trypsin   Trypsin
                                        cleaves here   cleaves here

Gly—Met—Gly—Phe|—Ala—Val—Arg—Met—Leu—Tyr—|Lys—Gly—Ala

                       Chymotrypsin   Chymotrypsin
                       cleaves here   cleaves here

Gly—Met—|Gly—Phe—Ala—Val—Arg—Met—|Leu—Tyr—Lys—Gly—Ala

         N≡CBr   N≡CBr
    cleaves here   cleaves here

(b) Phe — Ile — Met — Lys ⫯ Tyr — Asp — Gly — Arg ⫯ Ala — Val — Leu — Pro — Cys
                              Trypsin   Trypsin
                           cleaves here   cleaves here

    Phe ⫯ Ile — Met — Lys — Tyr ⫯ Asp — Gly — Arg — Ala — Val — Leu — Pro — Cys
     Chymotrypsin   Chymotrypsin
     cleaves here   cleaves here

    Phe — Ile — Met ⫯ Lys — Tyr — Asp — Gly — Arg — Ala — Val — Leu — Pro — Cys
               N≡CBr
            cleaves here

(c) Val — Cys — Gly — Glu — Met — Pro — Leu — Arg ⫯ Ala — Ile — Tyr — Gly — Ala
                                                    Trypsin
                                                 cleaves here

    Val — Cys — Gly — Glu — Met — Pro — Leu — Arg — Ala — Ile — Tyr ⫯ Gly — Ala
                                                              Chymotrypsin
                                                              cleaves here

    Val — Cys — Gly — Glu — Met ⫯ Pro — Leu — Arg — Ala — Ile — Tyr — Gly — Ala
                          N≡CBr
                       cleaves here

26.30 The Edman degradation begins with the reaction of phenyl isothiocyanate with the free amino group of an N-terminal amino acid.

(a) Reaction 1:

Reaction 2:

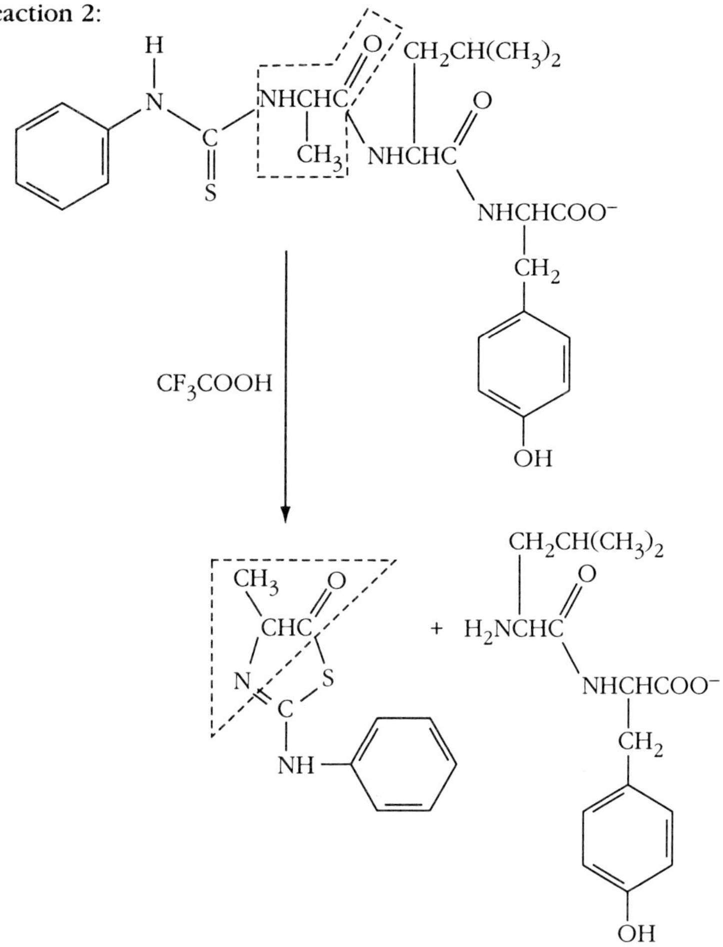
H
N
C
S
O
NHCHC
CH3
CH2CH(CH3)2
O
NHCHC
NHCHCOO−
CH2
OH
CF3COOH
CH3
O
CHC
N
S
C
NH
+ H2NCHC
CH2CH(CH3)2
O
NHCHCOO−
CH2
OH

Reaction 3:

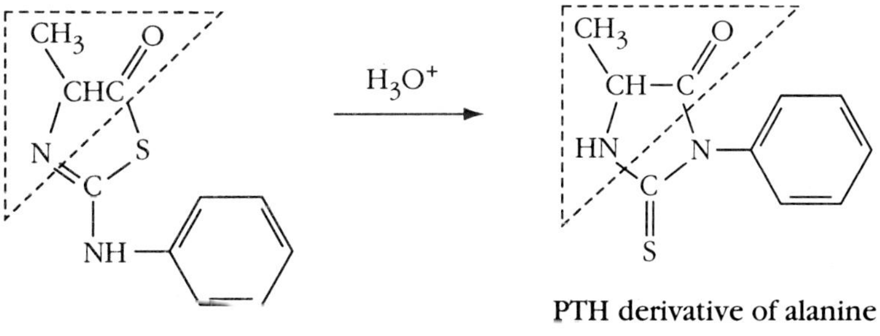
CH3
O
CHC
N
S
C
NH
H3O+
CH3
O
CH—C
HN
N
C
S

PTH derivative of alanine

Repeat reactions 1–3:

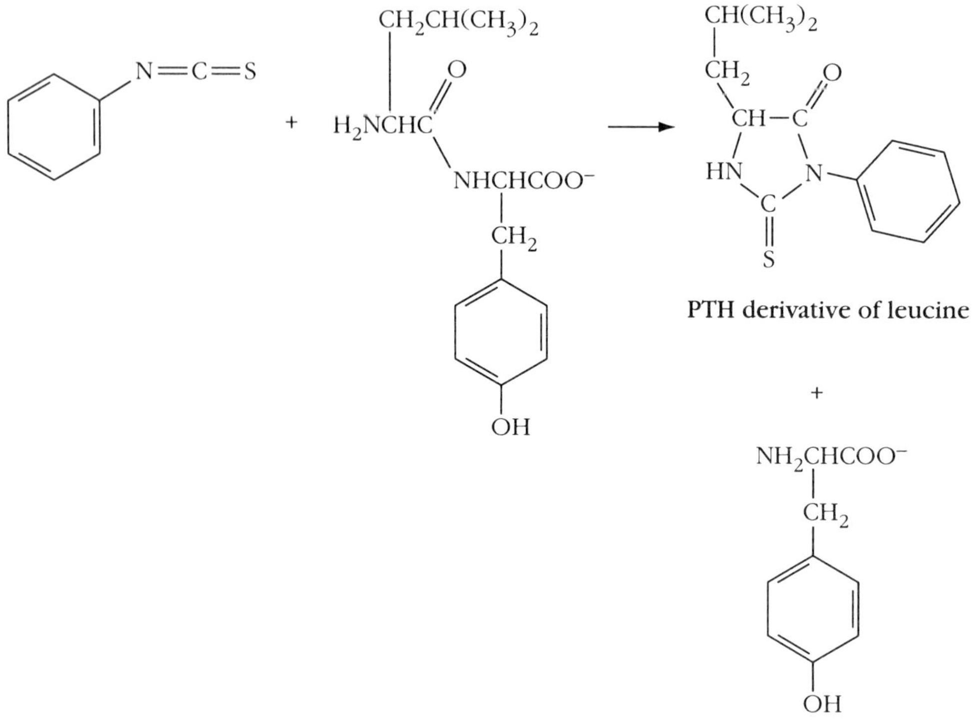

Round 1

Phenyl isothiocyanate + dipeptide (Leu-Tyr) → PTH derivative of leucine + Tyrosine

(b)

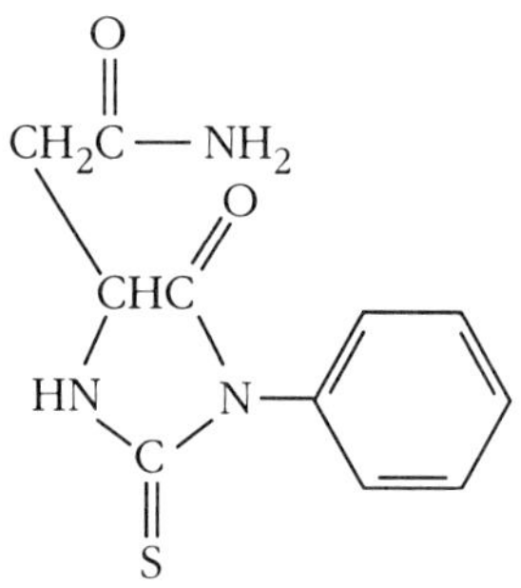

Round 2: PTH derivative with side chain $CH_2C(=O)NH_2$ (asparagine)

Round 3: PTH derivative with side chain $CH_2CH_2COO^-$ (glutamate)

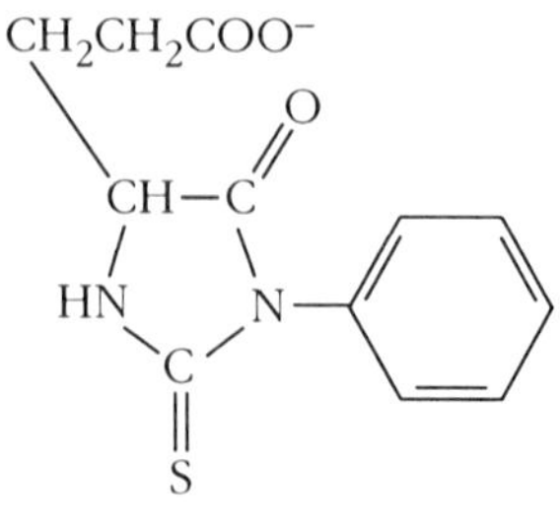

Round 4: PTH derivative with side chain $CH(OH)CH_3$ (threonine)

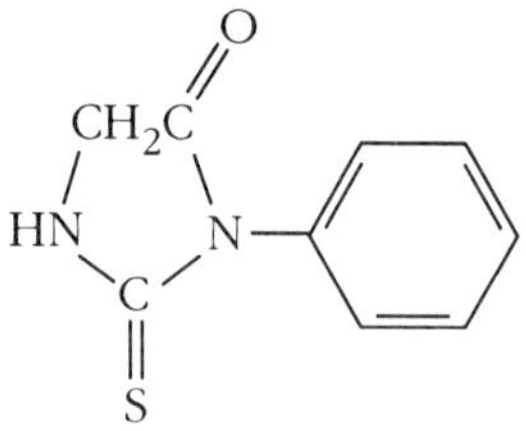

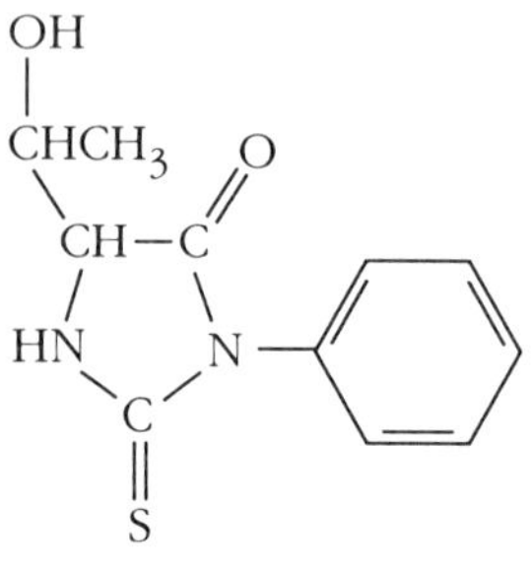

Round 1

(c) [structure: PTH derivative with CH(CH₃)₂ side chain — Val]

Round 2

[structure: PTH derivative with –CH₂CH₂CH₂– forming ring to N — Pro]

Round 3

[structure: PTH derivative with CH₃CHCH₂CH₃ side chain — Ile]

Round 4

[structure: PTH derivative with CH₂-imidazole side chain — His]

26.31 Subject the dipeptide to the Edman degradation procedure. If the sequence is Asp-Phe, the structure of the PTH derivative formed would be

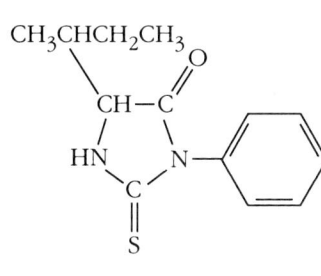

PTH derivative of Asp

If the sequence is Phe-Asp, the structure of the PTH derivative formed would be

[structure: PTH derivative with CH₂-phenyl side chain]

PTH derivative of Phe

26.32 Reaction 1 of the Edman degradation procedure involves nucleophilic addition to the carbon atom of the C=S bond of phenyl isothiocyanate in which the N-terminal amino group is the nucleophile.

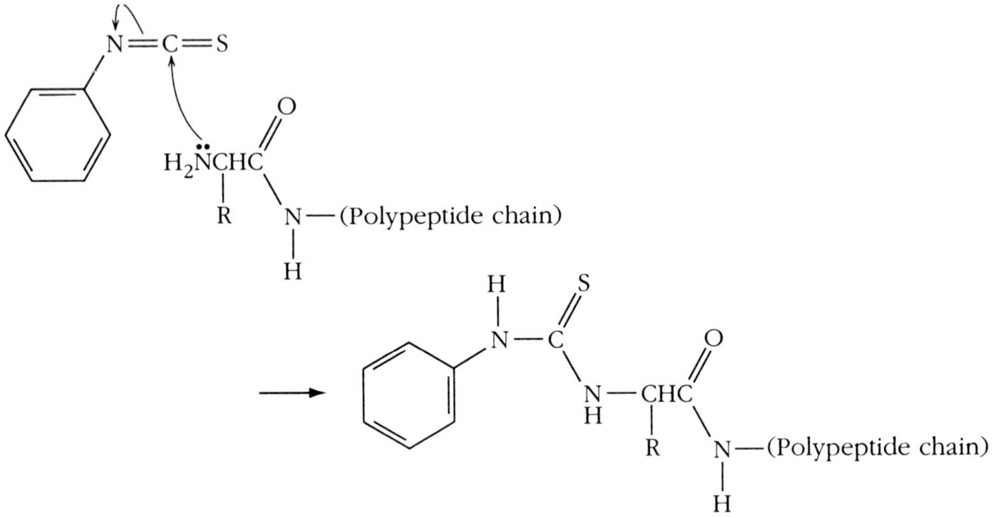

At pH 3, the terminal amino group is protonated to form its conjugate acid which is *not* a nucleophile so the reaction will not occur.

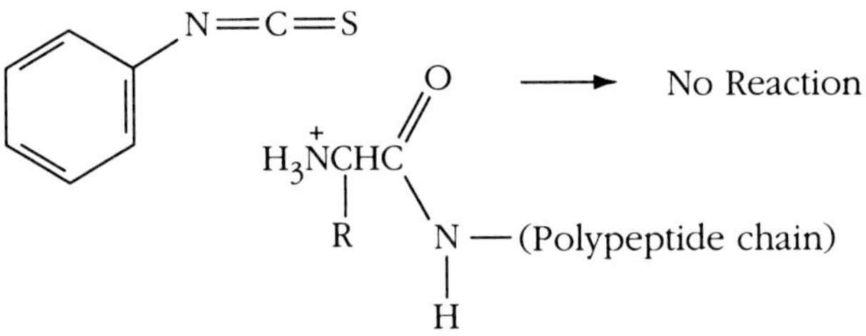

26.33 The part of the PTH derivative obtained from the amino acid is the following:

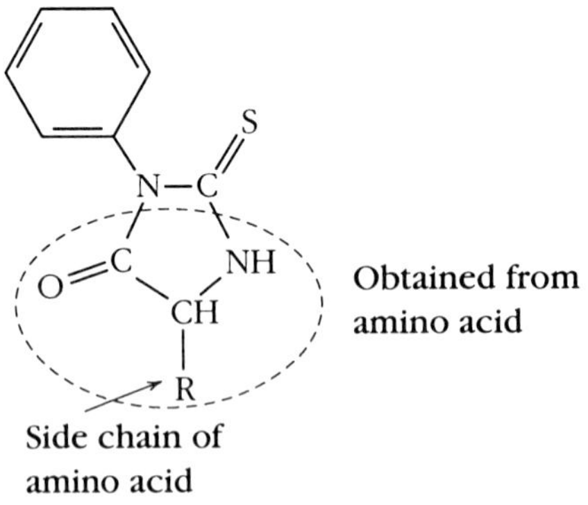

(a) The composition is

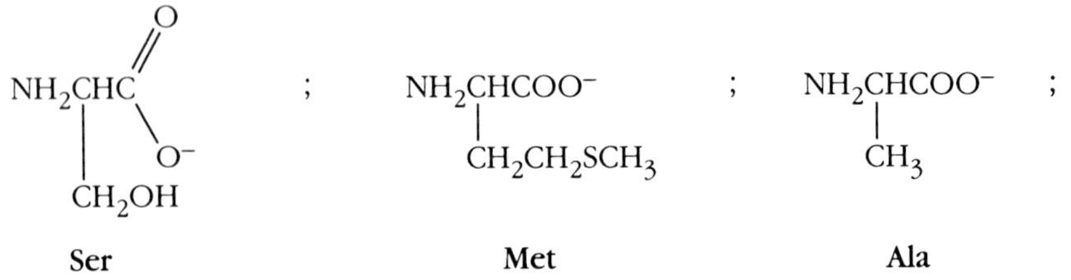

$NH_2CHC(=O)O^-$; NH_2CHCOO^- ; NH_2CHCOO^- ;
| | |
CH_2OH $CH_2CH_2SCH_3$ CH_3

Ser　　　　　　　　　　Met　　　　　　　　　　Ala

NH₂CHCOO⁻ ; NH₂CHCOO⁻ ; NH₂CHCOO⁻
| | |
CH₂C=O   CH₃   CH₂
| Ala |
NH₂   [phenyl ring]
Asn   Phe

(b) The N-terminal amino acid is removed in round 1. The next amino acid is removed in round 2 and so on. The sequence is therefore Ser-Met-Ala-Gln-Ala-Phe.

26.34

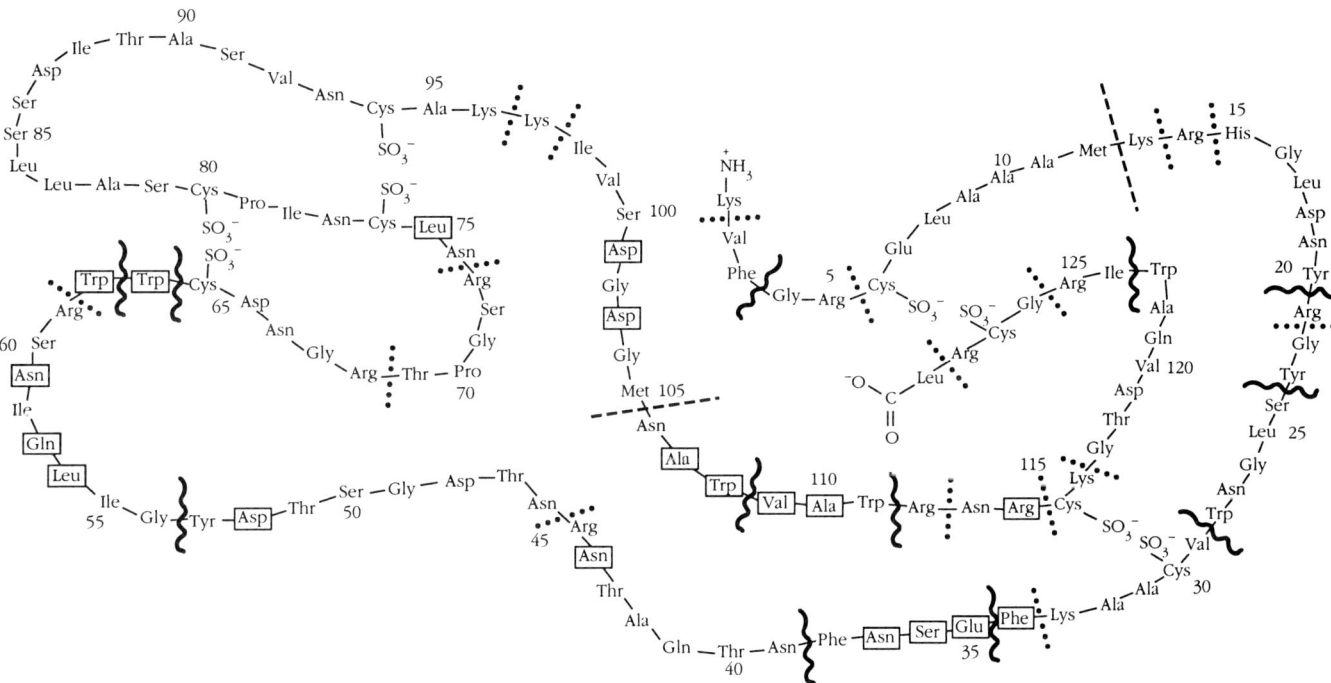

(a) N≡CBr cleaves at positions indicated by --------
(b) Trypsin cleaves at positions indicated by ·······
(c) Chymotrypsin cleaves at positions indicated by ∿∿

26.35 The reaction of 2,4-dinitrofluorobenzene with the N-terminal amino acid of a polypeptide chain is a nucleophilic aromatic substitution reaction. Electron-withdrawing nitro groups *o*- and *p*- to the fluorine atom, which is substituted by the amino groups, suggests an addition-elimination mechanism (Section 23.2).

Step 1: Addition of nucleophile

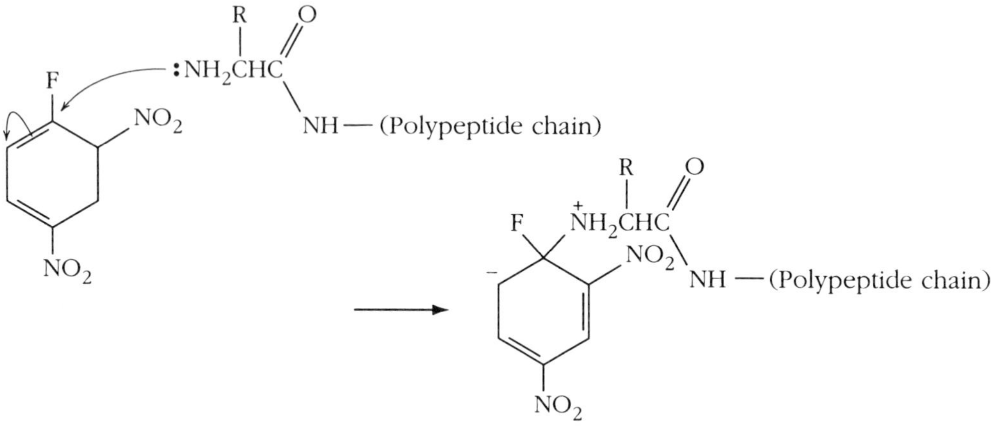

Step 2: Elimination of fluoride ion

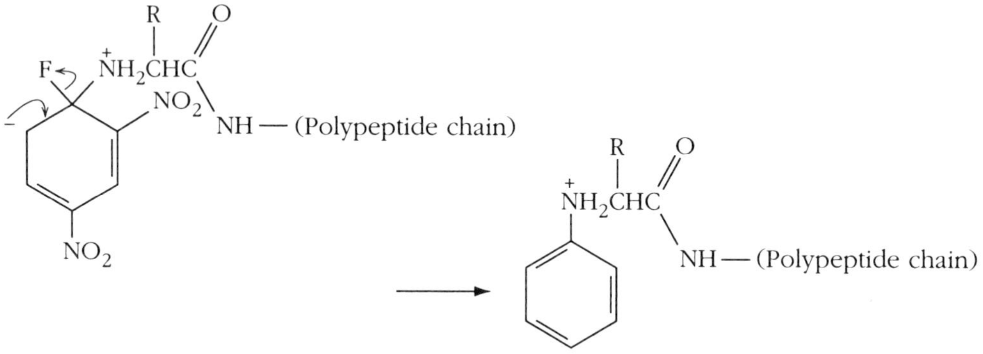

Step 3: Proton transfer

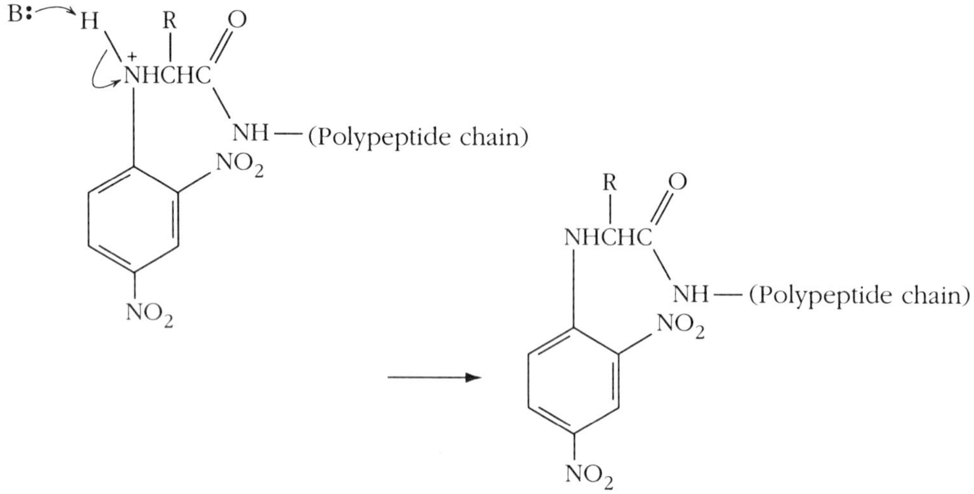

26.36   Arg-Gly-Pro
Pro-Leu-Phe-Ile-Val
Gly-Pro-Leu
The primary structure of heptapeptide is Arg-Gly-Pro-Leu-Phe-Ile-Val

26.37 Val-Trp-Met-Gly-Lys
Met-Gly-Lys-Ala-Ile-Pro-Met-Asp-Arg-Tyr
Val-Trp
Ala-Ile-Pro-Met-Asp-Arg
Tyr-Ala-Gly-Cys
Ala-Gly-Cys

Primary structure: Val-Trp-Met-Gly-Lys-Ala-Ile-Pro-Met-Asp-Arg-Tyr-Ala-Gly-Cys
CHECK:
There are 15 amino acids in the polypeptide in agreement with the complete acid hydrolysis.

The composition also agrees   2 of Met, Gly, and Ala
1 each of Val, Trp, Lys, Ile, Pro, Asp, Arg, Tyr, Cys

26.38

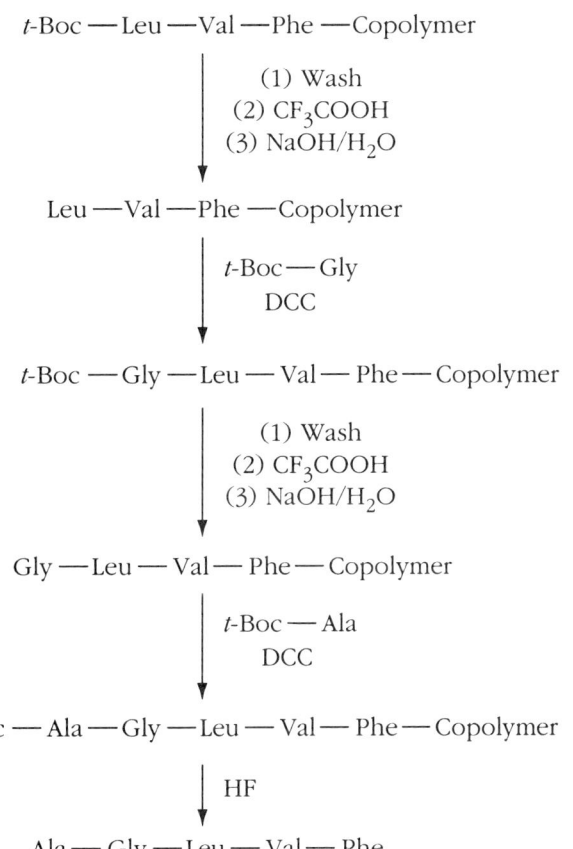

26.39

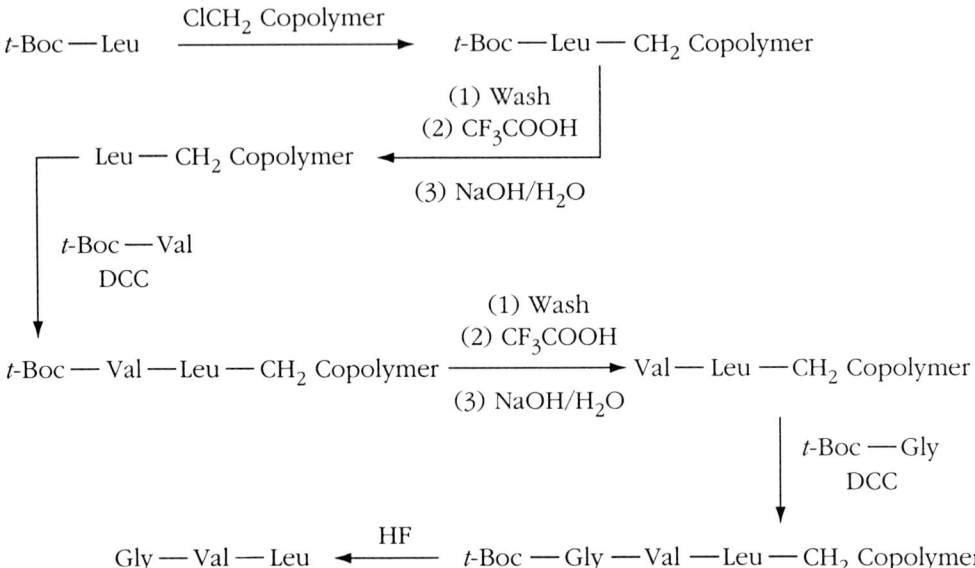

CHAPTER 27

NUCLEIC ACIDS
Concepts

Deoxyribonucleic acids (DNA) are the master repository of genetic information in cells while **ribonucleic acids (RNA)** implement the genetic information in cells. Both DNA and RNA are biopolymers consisting of repeating nucleotide units. A **nucleotide** contains a phosphate group, a pentose, and a heterocyclic base. In DNA, the pentose is 2-deoxy-β-D-ribose and the heterocyclic bases are adenine (A), guanine (G), thymine (T) and cytosine (C). In RNA, the pentose is β-D-ribose and uracil (U) replaces thymine (T) as one of the four heterocyclic bases.

Molecules of DNA consist of two nucleic acid chains that form a double helix. The two chains are held together by hydrogen bonds between heterocyclic bases on the different chains. Adenine forms hydrogen bonds with thymine; cytosine forms hydrogen bonds with guanine. The sequence of bases in DNA contains the genetic information of a particular species. DNA forms exact copies of itself by a process called **replication**. DNA is also the template for the synthesis of the three kinds of RNA (mRNA, tRNA, and rRNA). Instructions for protein synthesis are transcribed from DNA to mRNA. Ribosome-bound rRNA translates these instructions to tRNA, which results in protein synthesis.

Solution to the Exercises

27.1

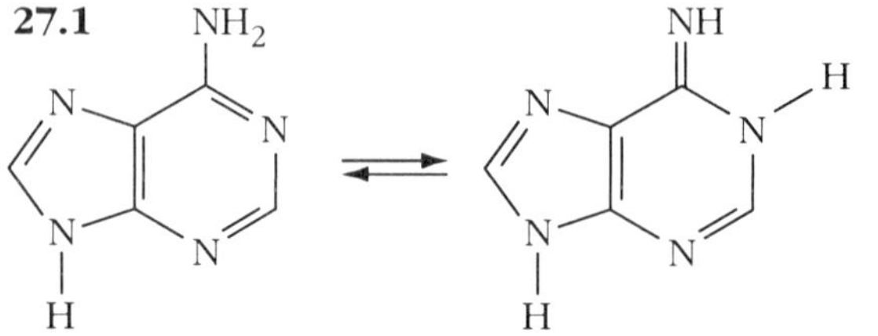

Adenine

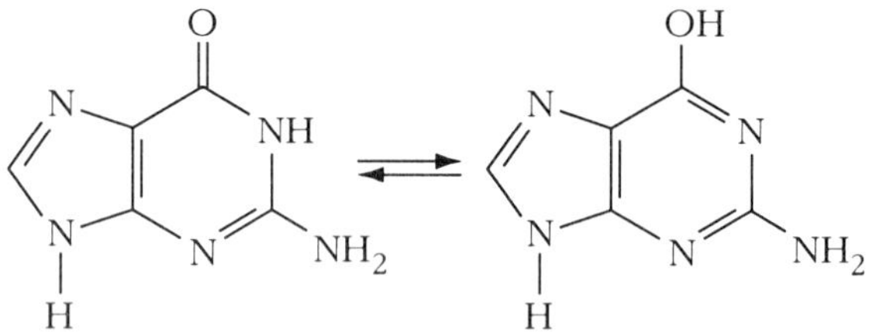

Guanine

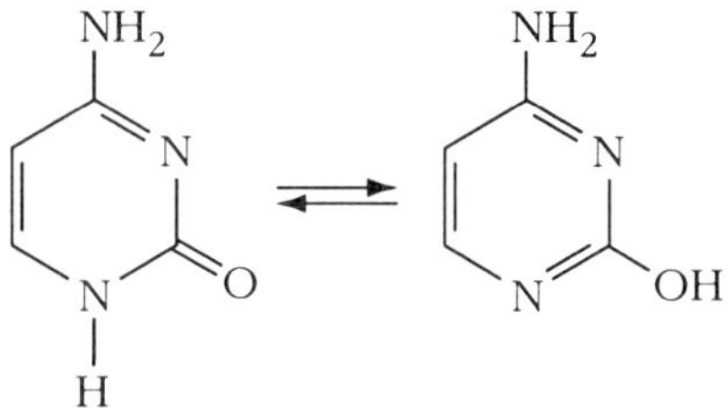

Cytosine

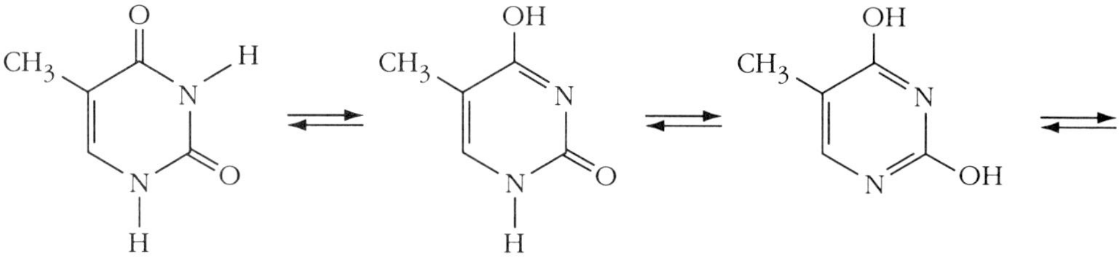

Thymine

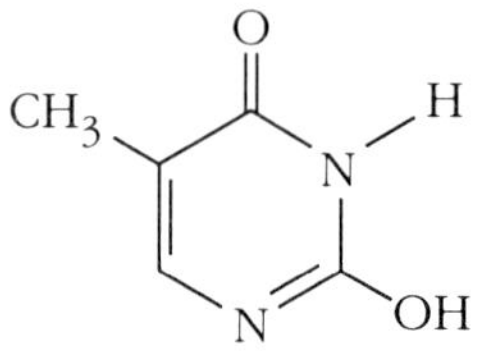

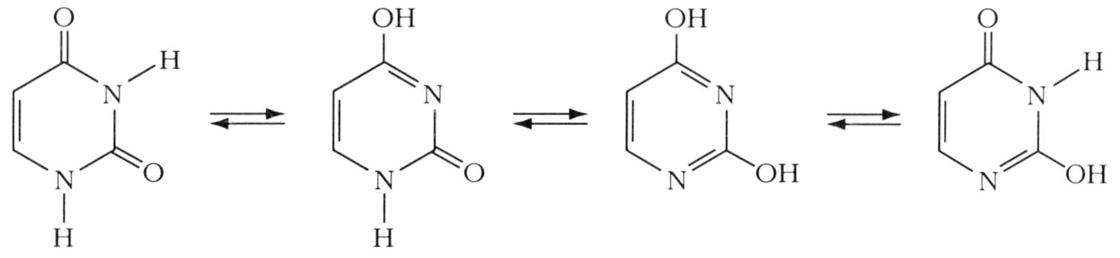

Uracil

27.2

(a) 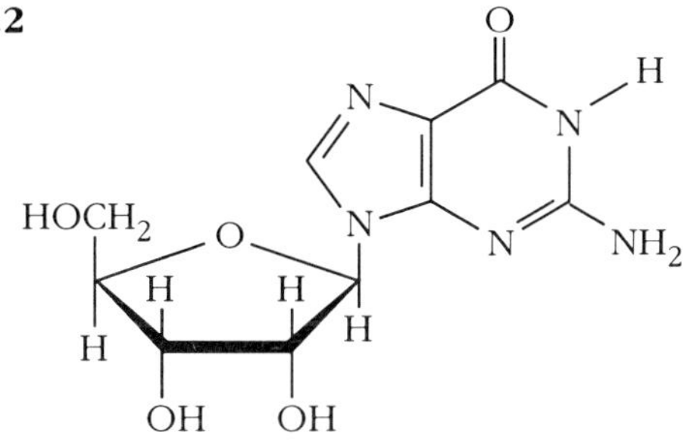

(b) 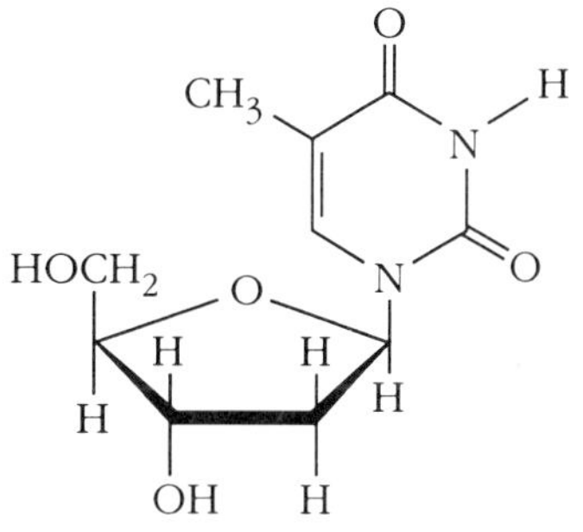

(c) 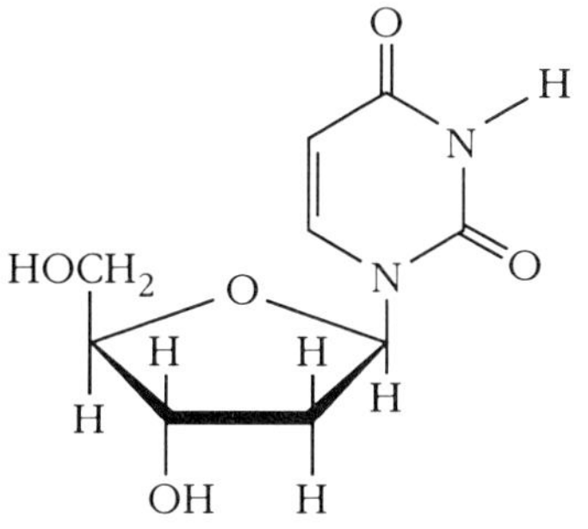

27.3

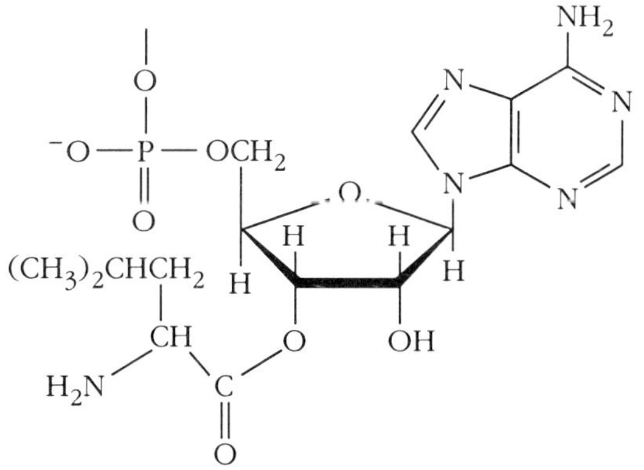

27.4 CGU ; CGC ; CGA ; CGG ; AGA ; AGG

27.5 (a) AUU = Ile
(b) GAU = Asp
(c) GCU = Ala
(d) GGG = Gly
(e) CGA = Arg
(f) CAU - His

27.6 UCA | UUG | ACC | GAG
Ser   Leu   Thr   Glu

Ser — Leu — Thr — Glu

27.7

(a) ACA | UCA | UUU | GCU | CAU
Thr   Ser   Phe   Ala   His

Thr — Ser — Phe — Ala — His

(b) AUU | GGU | GUG | UAC | CAG | GGG | UAG
Ile   Gly   Val   Tyr   Gln   Gly   End

Ile — Gly — Val — Tyr — Gln — Gly

27.8

Original DNA

AGT ACG CGA ACT GCA TGC

↓ Transcription

UCA UGC GCU UGA CGU ACG

↓ Translation

Ser — Cys — Ala — End

Ser — Cys — Ala

Mutant DNA

AGA CGC GAA CTG CAT

↓ Transcription

UCU GCG CUU GAC GUA

↓ Translation

Ser — Ala — Leu — Asp — Val

27.9 (a) **DNA replication** is the process by which DNA molecules reproduce themselves in the nuclei of cells. This enzyme-catalyzed process begins by partially unwinding the DNA double helix and the separated portions act as templates on which enzyme-catalyzed synthesis of a new complementary chain occurs.

(b) A **codon** is a set of three nucleotide bases on a molecule of mRNA that directs the incorporation of a specific amino acid into a growing polypeptide chain.

(c) **Transcription** is the synthesis of RNA by an RNA polymerase–catalyzed process that starts by the unwinding of a section of the double helix of DNA. The sequence of nucleotide bases on one of the two exposed chains is the code for the synthesis of RNA.

(d) A **mutagen** is anything that interferes with the correct transmission of genetic information on DNA by altering the sequence of nucleotide bases in DNA molecules.

(e) **Translation** is the process by which the genetic information transcribed from DNA to mRNA is read by tRNA to direct protein biosynthesis.

(f) **tRNA (transfer RNA)** transports specific amino acids to ribosomes where they are joined together to form proteins. Each of the 20 amino acids has at least one tRNA molecule that carries it to the ribosome.

(g) A **nucleotide** is a constituent of a nucleic acid that consists of three parts: a pentose, which is β-D-ribose in RNA and 2'deoxy-β-D-ribose in DNA, any one of four heterocyclic bases, which are adenine, guanine, cytosine, and thymine in DNA and adenine, guanine, cytosine, and uracil in RNA, and a phosphate group bonded to the C5' position of the 2'deoxyribose unit in DNA and the C5' position of the ribose unit in RNA. The general structure of a nucleotide found in DNA and RNA is

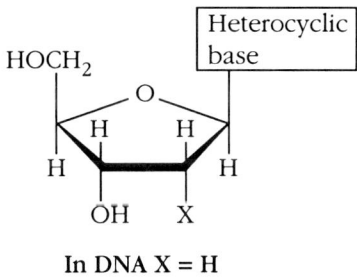

In DNA X = H
In RNA X = OH

(h) **Clones** are collections of identical organisms that are derived from a single ancestor.
(i) **Anticodon** is the sequence of three bases located at the bottom of a tRNA molecule that is a specific code for the amino acid carried by the tRNA.
(j) **mRNA (messenger RNA)** carries the genetic information from DNA to the ribosomes where it directs protein synthesis.
(k) A **nucleoside** is a constituent of a nucleic acid that consists of a sugar residue bonded to a heterocyclic purine or pyrimidine base. The general structure of a nucleoside is

In DNA X = H
In RNA X = OH

27.10 Nucleotides are the repeating units in nucleic acids:

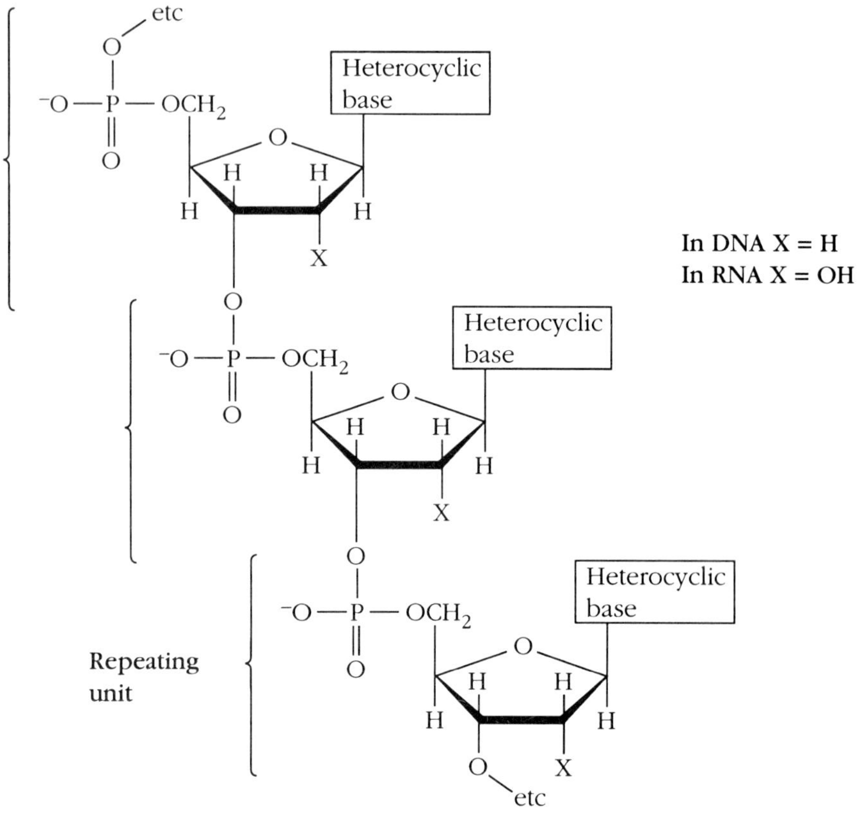

In DNA X = H
In RNA X = OH

α-Amino acids are the repeating units in proteins:

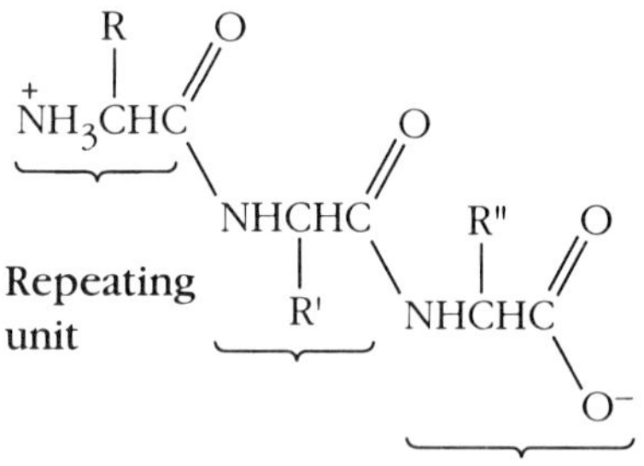

D-Glucopyranose is the repeating unit in carbohydrates. D-Glucopyranose molecules are joined by either α-1,4 or β-1,4 bonds. The structure of D-glucopyranose molecules joined together by β-1,4 bonds is

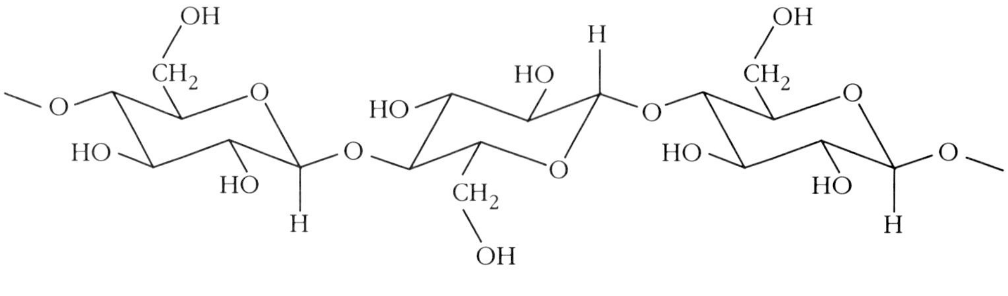

27.11 The pentose in DNA is 2-deoxy-β-D-ribose:

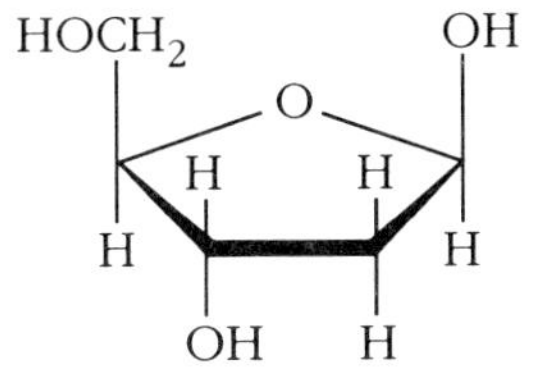

2-Deoxy-β-D-ribose

The pentose in RNA is β-D-ribose:

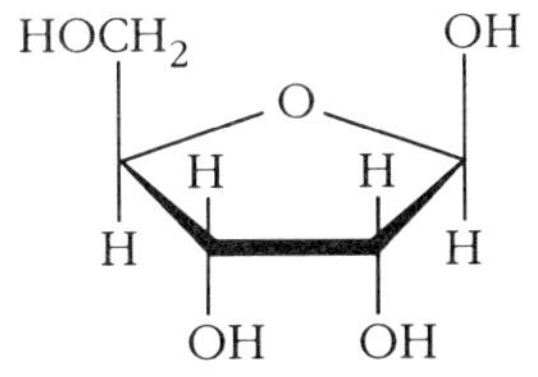

β-D-ribose

27.12 DNA and RNA differ in two respects. First the pentose in DNA is 2-deoxy-β-D-ribose while the pentose in RNA is β-D-ribose. Second the heterocyclic bases in the nucleotides of DNA are adenine, guanine, cytosine, and thymine while the heterocyclic bases in the nucleotides of RNA are adenine, guanine, cytosine, and uracil.

27.13 DNA is the master repository of genetic information in cells while RNA implements this genetic information in the synthesis of proteins.

27.14

(a) Adenine or Guanine

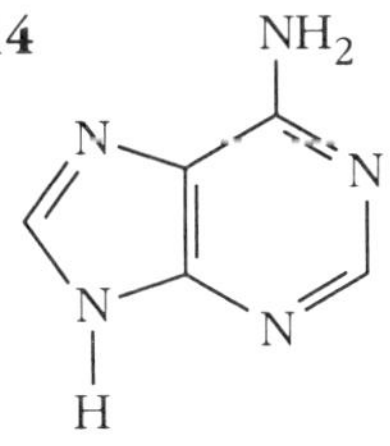

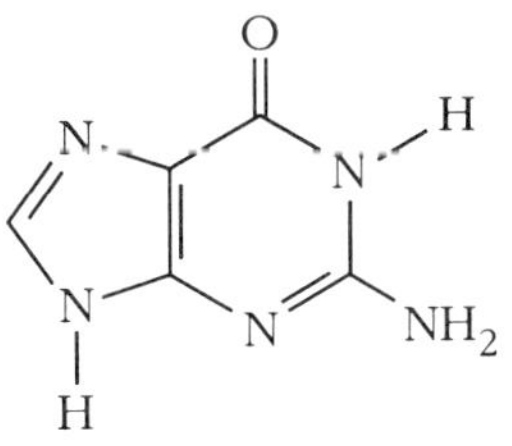

(b) Cytosine or Uracil

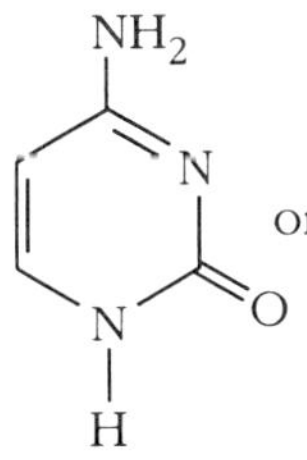

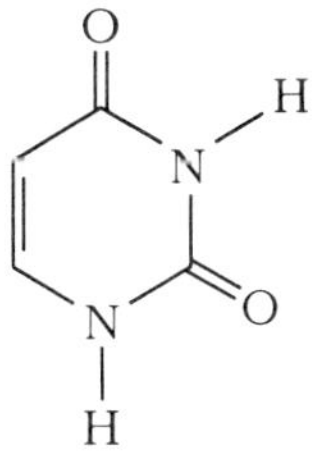

(c) Uracil

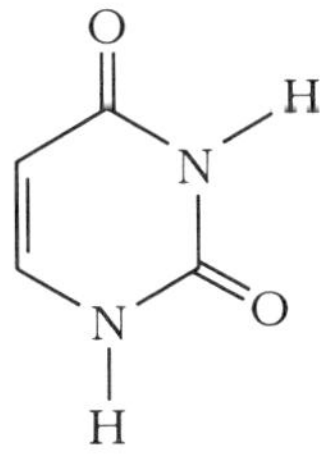

27.15 (a) DNA

Adenine — Thymine: The AT base pair

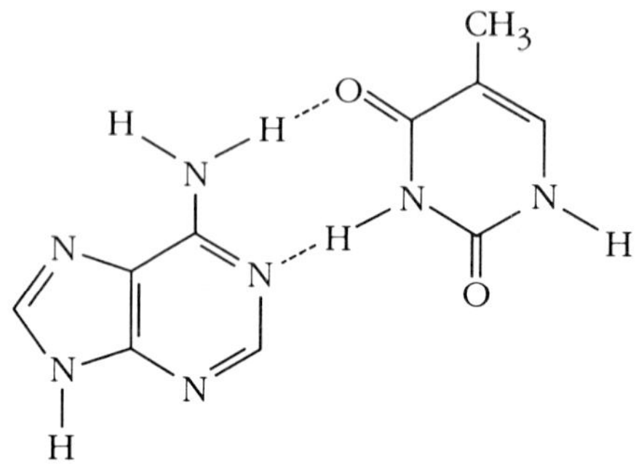

Guanine — Cytosine: The GC base pair

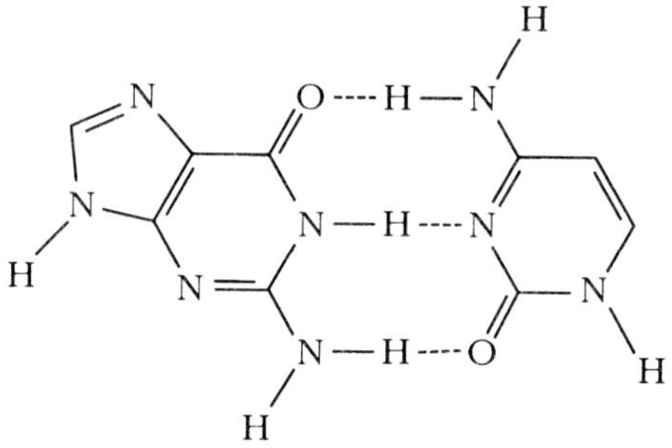

(b) Adenine — Uracil

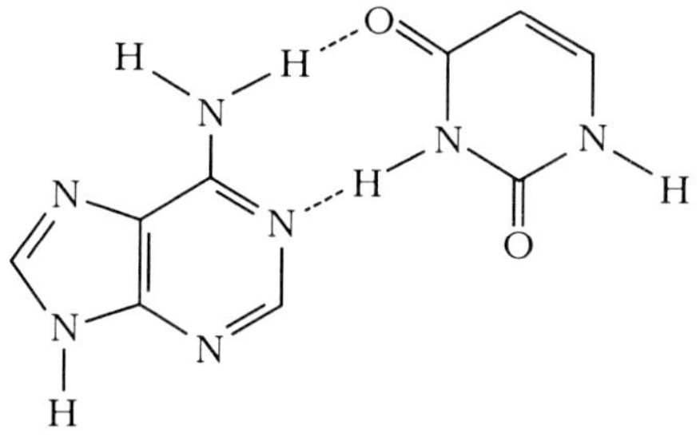

Guanine — Cytosine

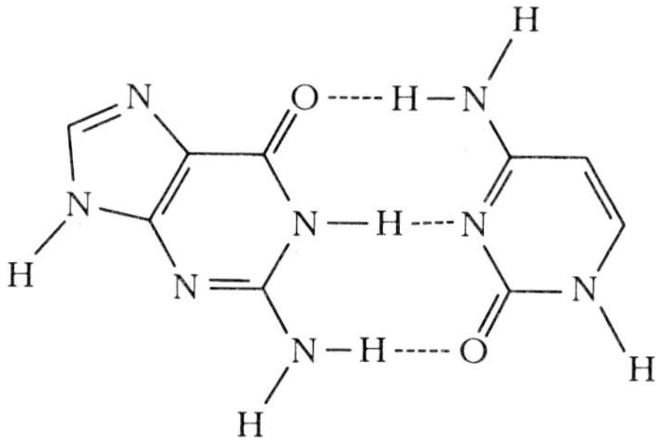

27.16

(a) Adenosine

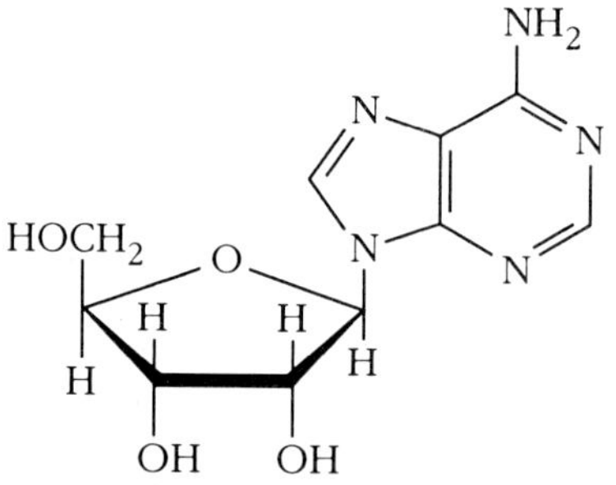

(b) Uridine

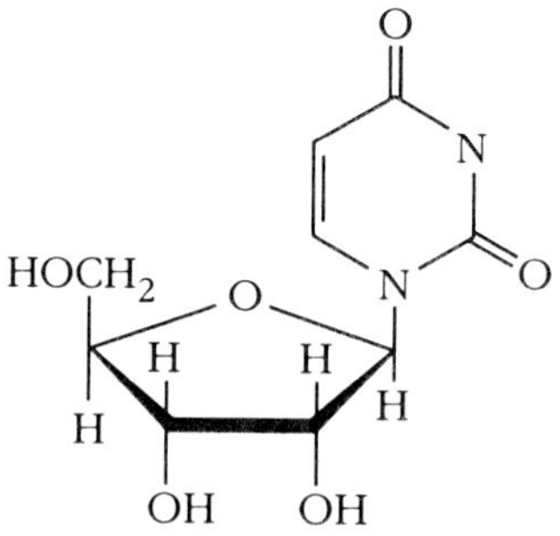

(c) 2'-Deoxycytidine

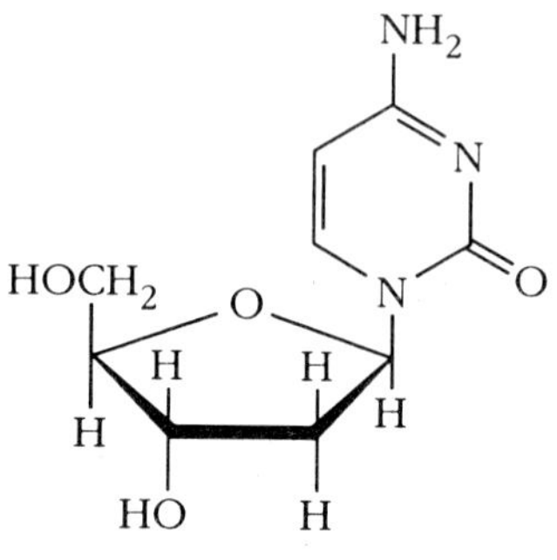

27.17

(a) Cytidine 5'-monophosphate

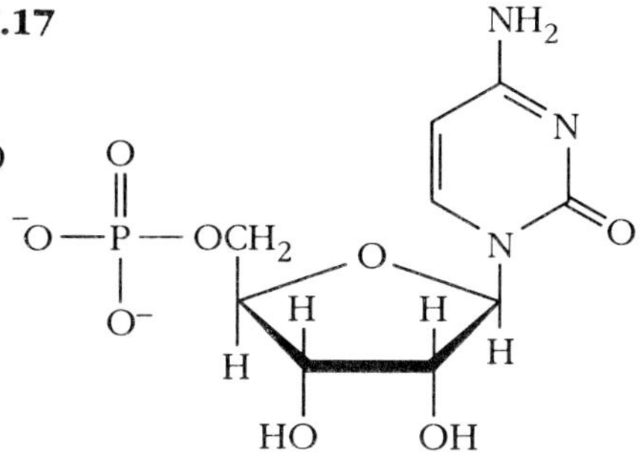

(b) 2'-Deoxyadenosine 5'-monophosphate

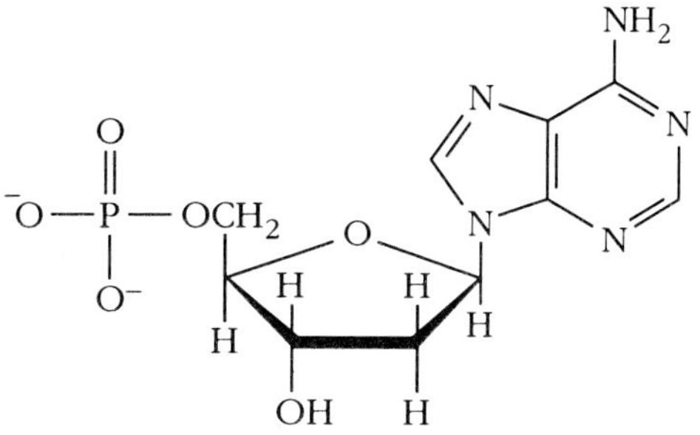

(c) Thymidine 5'-monophosphate

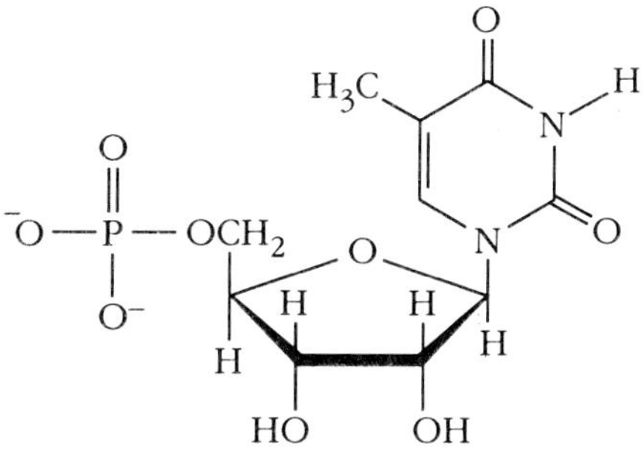

27.18

The structure of adenosine 5'-monophosphate is

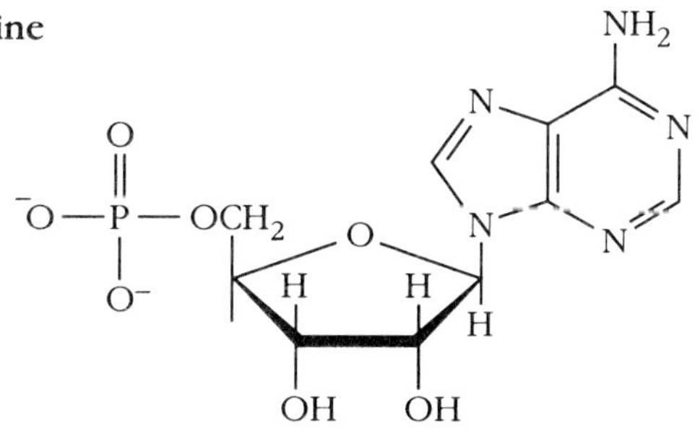

so the structure of adenosine triphosphate (ATP) is

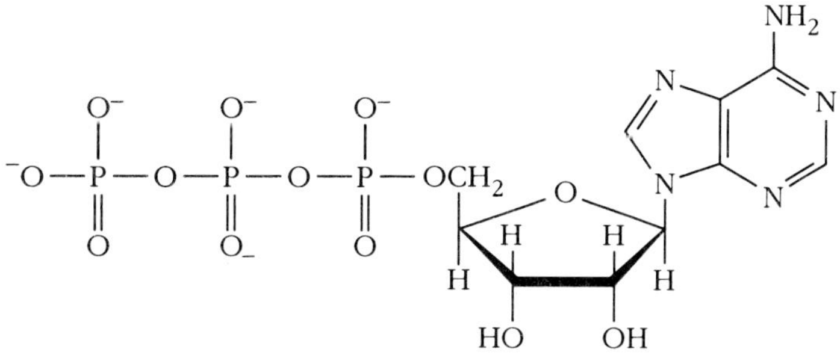

ATP

27.19 The complementary chain of DNA contains a G opposite each C and vice versa and an A opposite each T and vice versa of the other chain:

5'-end AGGCTATTCGT-3' end
3'-end TCCGATAAGCA-5' end

27.20 Most molecules of RNA consist of a single nucleotide chain rather than a double helix so RNA does not have a complementary chain. It is the double helix and its complementary base arrangement that accounts for the ratios G/C = 1; A/T = 1 in DNA. RNA does not have such a complementary base arrangement so G/C ≠ 1 and A/T ≠ 1.

27.21

	(a) Leu	(b) Val	(c) Glu	(d) Ile	(e) Pro	(f) His
	UUA	GUU	GAA	AUU	CCC	CAU
	UUG	GUC	GAG	AUC	CCA	CAC
	CUU	GUA		AUA	CCG	
	CUC	GUG			CCU	
	CUA					
	CUG					

27.22 (a)
5'-end   3'-end
-CGC   UUU   UUA   ACU   AUG   UCU
Arg   Phe   Leu   Thr   Met   Ser

This is not the only correct answer because all of the amino acids, except Met, have other codons that are signals to incorporate them into a growing polypeptide chain.

(b) 5'-end   3'-end
-GGU   CCG   AUU   UGC   GAC   UAC   CAG   GAG
Gly   Pro   Ile   Cys   Asp   Tyr   Gln   Glu

27.23 (a) A-A-T
(b) G-T-C
(c) T-A-T
(d) C-C-C
(e) T-T-G

27.24 Codons and anticodons are complementary. The anticodons in tRNA recognize and combine with the complementary bases of the codons of mRNA. An A base, therefore, combines with U and a G combines with C.

27.25 (a) One of the codons for Val is GUU. The anticodons are complementary so the sequence is CAA.
(b) Codon GCU; anticodon CGA
(c) Codon UUA; anticodon AAU
(d) Codon CAU; anticodon GUA
(e) Codon AUG; anticodon UAC
(f) Codon AUU; anticodon UAA.

27.26 (a) The codons for Arg are CGU, CGC, CGA, CGG; the codons for Glu are GAA and GAG

5'-end   3'-end
CGU GAA   CGC GAA   CGA GAA   CGG GAA
CGU GAG   CGC GAG   CGA GAG   CGG GAG

(b) Val: GUU, GUC, GUA, GUG; Ser: AGU and AGC.

```
5'-end      3'-end
   GUU AGU     GUC AGU     GUA AGU     GUG AGU
   GUU AGC     GUC AGC     GUA AGC     GUG AGC
```

(c) Ala: GCU, GCC, GCA, GCG; Tyr: UAU, UAC.

```
5'-end      3'-end
   GCU UAU     GCC UAU     GCA UAU     GCG UAU
   GCU UAC     GCC UAC     GCA UAC     GCG UAC.
```

(d) Glu: GAA, GAG; Arg: CGU, CGC, CGA, CGG.

```
5'-end      3'-end
   GAA CGU     GAG CGU
   GAA CGC     GAG CGC
   GAA CGA     GAG CGA
   GAA CGG     GAG CGG
```

27.27 (a) Leu-Ile-Val
(b) Phe-Thr-Ala-Ile-Leu
(c) Pro-Ile

27.28 (a) ACC
(b) ACC

27.29 (a) UUCGGCAUGCUC
(b) Phe-Gly-Met-Leu

27.30 The mRNA chain becomes UUCGGCAUCCUC so the polypeptide chain is Phe-Gly-Ile-Leu.

27.31 Codons that direct synthesis of the original polypeptide:

Cys — Ser — His — Ala — Ser — Gly
UGU   UCA   CAC   GCG   UCU   GGU

Deletion of C from serine codon and insertion of A between Ser and Gly codons leads to the following codons:

Codons that direct mutant polypeptide:

UGU   UAC   ACG   CGU   UAG
Cys — Tyr — Thr — Arg — End

Sequence of bases on original DNA: ACA   AGT   GTG   CGC   AGA   CCA
Sequence of bases on mutant DNA: ACA   ATG   TGC   GCA   ATC

The mutant polypeptide is shorter because after four amino acids have been combined a codon is encountered that signals the end of polypeptide synthesis.

EXAMINATION 4

To get the maximum benefit from the following one-hour exam, choose a quiet location and spend exactly one hour answering the questions. Check your answers against those at the end of the Study Guide to evaluate your knowledge of the material.

Chapters 23–27

Time Allowed: 1 Hour Maximum 100 points

1. *(10 points)* Determine the configuration of the aldopentose based on the following experimental results.

$$\text{Aldopentose} \xrightarrow{\text{Ruff degradation}} \begin{array}{c} \text{CHO} \\ \text{H}\!\!-\!\!\!\!-\!\!\text{OH} \\ \text{H}\!\!-\!\!\!\!-\!\!\text{OH} \\ \text{CH}_2\text{OH} \end{array}$$

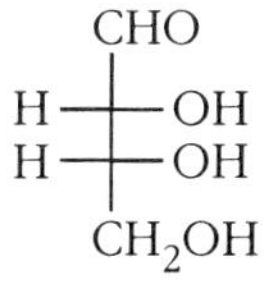

$$\text{Aldopentose} \xrightarrow{\text{HNO}_3} \text{Optically active aldaric acid}$$

2. *(20 points)* Write the reactions to prepare each of the following compounds starting with either diethyl malonate or ethyl acetoacetate, and any other organic or inorganic reagents needed.

   (a) 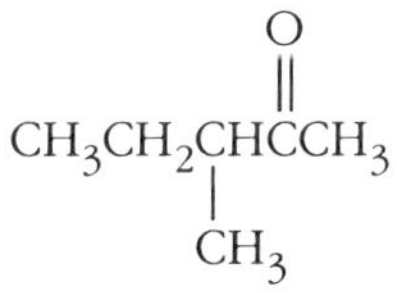 $CH_3CH_2\underset{\underset{CH_3}{|}}{CH}\overset{\overset{O}{\|}}{C}CH_3$

   (b) 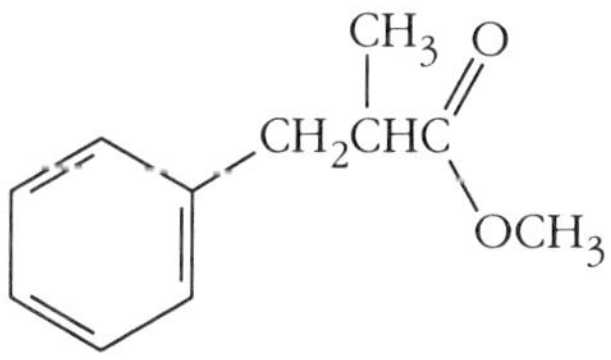 Ph–$CH_2\underset{\underset{CH_3}{|}}{CH}\overset{\overset{O}{\|}}{C}OCH_3$

3. *(40 points)* Write the structure of the product or products of each of the following reactions.

   (a) 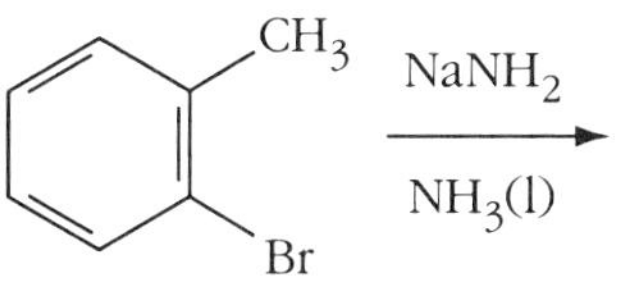 2-bromotoluene $\xrightarrow[\text{NH}_3(l)]{\text{NaNH}_2}$

   (b) 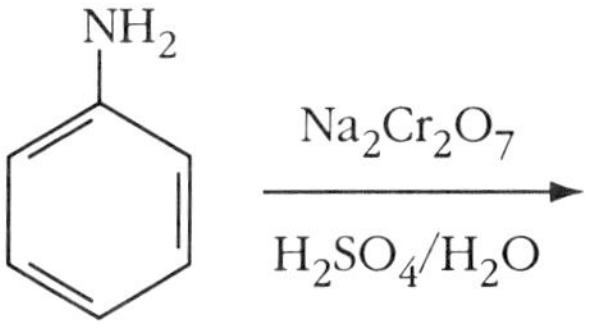 aniline $\xrightarrow[\text{H}_2\text{SO}_4/\text{H}_2\text{O}]{\text{Na}_2\text{Cr}_2\text{O}_7}$

(c) $\text{H}_3\overset{+}{\text{N}}\text{CHC}(=\text{O})\text{NHCHC}(=\text{O})\text{NHCH}_2\text{COO}^-$ with CH$_3$ on first α-carbon and CH$_2$C$_6$H$_5$ on second α-carbon $\xrightarrow[\text{Heat}]{\text{HCl}}$

(d) $\text{CH}_3\text{CH(OH)(CH}_2)_3\text{C(=O)OH}$ $\xrightarrow{\text{H}_3\text{O}^+}$

(e) β-D-glucopyranose (HO, HO, CH$_2$OH, OH, OH) $\xrightarrow{\text{Excess (CH}_3\text{O)}_2\text{SO}_2}$

(f) benzyl β-D-glucopyranoside (HO, HO, CH$_2$OH, OH, OCH$_2$C$_6$H$_5$) $\xrightarrow[\text{H}_2\text{O}]{\text{HCl}}$

(g) C$_6$H$_5$—OCH$_2$C(CH$_3$)=CH$_2$ $\xrightarrow{\text{Heat}}$

(h) cyclohexanone + pyrrolidine $\xrightarrow{\text{Acid catalyst}}$

(i) 4-methylbenzenesulfonyl chloride + aniline $\longrightarrow$

(j) 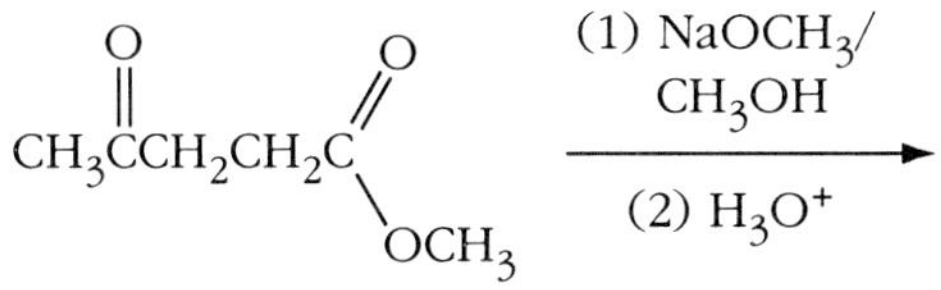 $CH_3\overset{O}{\overset{\|}{C}}CH_2CH_2\overset{O}{\overset{\|}{C}}\!-\!OCH_3$ $\xrightarrow{\text{(1) NaOCH}_3/\text{CH}_3\text{OH}}_{\text{(2) H}_3\text{O}^+}$

4. *(20 points)* Write the mechanism for each of the following reactions.

(a) 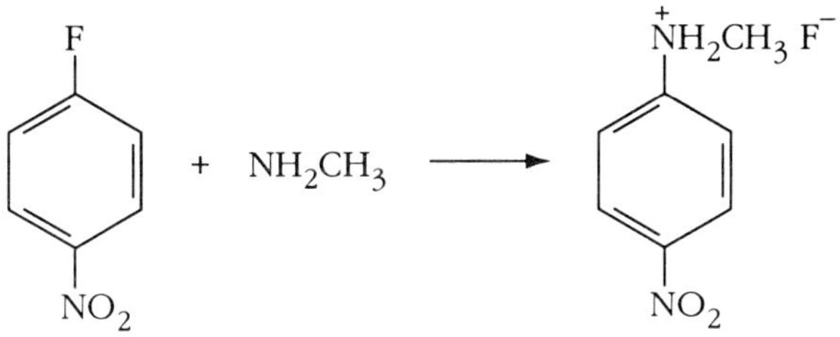 4-fluoronitrobenzene + NH_2CH_3 ⟶ 4-nitro-N-methylanilinium fluoride

(b) 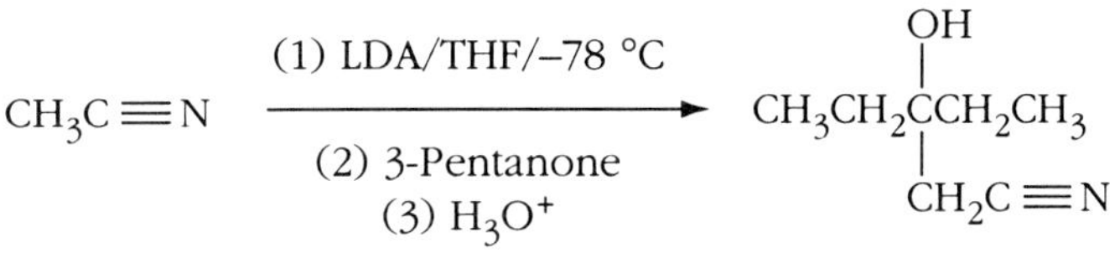 $CH_3C\equiv N$ $\xrightarrow{\text{(1) LDA/THF/}-78\,°C}_{\substack{\text{(2) 3-Pentanone}\\\text{(3) H}_3\text{O}^+}}$ $CH_3CH_2\underset{\underset{CH_2C\equiv N}{|}}{\overset{\overset{OH}{|}}{C}}CH_2CH_3$

5. *(10 points)* Determine the structure of each of the following compounds from the data provided.

(a) $C_3H_7NO_2$ ^{1}H NMR: δ 1.00 t (3H); δ 2.02 sextet (2H); δ 4.32 t (2H)
(b) $C_5H_7NO_2$ IR: 2270 cm^{-1}; 1745 cm^{-1}. ^{1}H NMR: δ 1.37 t (3H); δ 3.50 s (2H); δ 4.25 q (2H)

APPENDIX A

SYNTHESIS OF IMPORTANT FUNCTIONAL GROUPS

This section summarizes the preparation of important functional groups introduced in the text. The functional groups are listed alphabetically, followed by reference to the appropriate section in the text, an example of the method, and any special features or limitations of the synthetic method.

Acetals (Section 14.11)

Acetals are formed by the reaction of aldehydes and ketones with at least two equivalents of alcohol in the presence of an acid catalyst such as H_2SO_4.

$$CH_3CHO + 2CH_3CH_2OH \xrightleftharpoons{H_2SO_4} CH_3CH(OCH_2CH_3)_2$$

The equilibrium constant for the reaction usually favors reactants so to obtain good yields of acetals, water must be removed as it is formed.

Acid Anhydrides (Section 16.5)

Acid anhydrides, including mixed anhydrides, are formed by the reaction of a carboxylate ion with acyl chlorides.

$$CH_3COCl + CH_3CH_2COO^- \longrightarrow CH_3CO\text{-}O\text{-}COCH_2CH_3$$

Acyl Chlorides (Section 16.5).

Acyl chlorides are formed by the reaction of a carboxylic acid with thionyl chloride ($SOCl_2$), phosphorus trichloride (PCl_3) or phosphorus pentachloride (PCl_5).

$$C_6H_5COOH \xrightarrow{SOCl_2} C_6H_5COCl + HCl + SO_2$$

Alcohols

There are many ways to synthesize alcohols.

1. Nucleophilic substitution reactions (Section 12.18). Water is a sufficiently strong nucleophile to convert tertiary haloalkanes into alcohols while hydroxide ion, a stronger nucleophile, is needed to convert primary and secondary haloalkanes into alcohols. Alkenes, formed by an elimination reaction, are frequently side products in this reaction.

$$\underset{\substack{H_3C \\ H \\ CH_2CH_3}}{C-Br} + HO^- \longrightarrow \underset{\substack{CH_3 \\ H \\ CH_2CH_3}}{HO-C}$$

$$\text{(cyclohexyl-Cl)} \xrightarrow{H_2O} \text{(cyclohexyl-OH)}$$

2. Grignard and organolithium reactions.
   (a) Grignard reagents and organolithium reagents react with formaldehyde to form primary alcohols (Section 11.7).

   $$\text{Cy}-MgCl \xrightarrow[\text{(2) } H_3O^+]{\text{(1) } H_2C=O} \text{Cy}-CH_2-OH$$

   (b) Grignard reagents and organolithium reagents react with any aldehyde except formaldehyde to form secondary alcohols (Section 11.7).

   $$\text{Cy}-MgCl \xrightarrow[\text{(2) } H_3O^+]{\text{(1) } CH_3C(=O)H} \text{Cy}-C(CH_3)(H)-OH$$

   (c) Grignard reagents and organolithium reagents react with ketones to form tertiary alcohols (Section 11.7).

   $$CH_3-MgCl \xrightarrow[\text{(2) } H_3O^+]{\text{(1) cyclopentanone}} \text{1-methylcyclopentanol}$$

(d) Grignard reagents and organolithium reagents react with formate esters to form secondary alcohols (Section 16.13).

$$\text{H}-\underset{\text{OR}'}{\overset{\text{O}}{\text{C}}} \xrightarrow[\text{(2) H}_3\text{O}^+]{\text{(1) CH}_3\text{CH}_2\text{MgBr (2 moles)}} \text{CH}_3\text{CH}_2\underset{\text{OH}}{\text{CHCH}_2\text{CH}_3}$$

(e) Grignard reagents and organolithium reagents react with any ester except formate esters to form tertiary alcohols in which two identical groups, obtained from the Grignard or organolithium reagent, and a hydroxyl group are bonded to a carbon atom (Section 16.13).

$$\text{H}_3\text{C}-\underset{\text{OR}'}{\overset{\text{O}}{\text{C}}} \xrightarrow[\text{(2) H}_3\text{O}^+]{\text{(1) CH}_3\text{CH}_2\text{MgBr (2 moles)}} \text{CH}_3\text{CH}_2-\underset{\underset{\text{CH}_3}{|}}{\overset{\overset{\text{OH}}{|}}{\text{C}}}-\text{CH}_2\text{CH}_3$$

(f) Grignard reagents and organolithium reagents react with ethylene oxide (oxirane) to form primary alcohols (Section 13.12).

3. Reduction of carbonyl compounds.
   (a) Reduction of aldehydes forms primary alcohols (Section 11.10).

   (b) Reduction of ketones forms secondary alcohols (Section 11.10).

   (c) Reduction of carboxylic acids forms primary alcohols (Section 15.17). $NaBH_4$ does not reduce carboxylic acids but $LiAlH_4$ and catalytic hydrogenation do.

   $$\text{CH}_3(\text{CH}_2)_6\overset{\text{O}}{\underset{\text{OH}}{\text{C}}} \xrightarrow[\text{(2) H}_3\text{O}^+]{\text{(1) LiAlH}_4/(\text{CH}_3\text{CH}_2)_2\text{O}} \text{CH}_3(\text{CH}_2)_6\text{CH}_2\text{OH}$$

(d) Reduction of acyl chlorides forms primary alcohols (Section 16.12).

$$CH_3CH_2CH_2C(=O)Cl \xrightarrow[\text{(2) } H_3O^+]{\text{(1) LiAlH}_4/(CH_3CH_2)_2O} CH_3CH_2CH_2CH_2OH$$

(e) Reduction of esters forms two alcohols as product. One is a primary alcohol obtained from the acyl part of the ester. The other is obtained from the alkoxy part (—OR′) of the ester (Section 16.12).

PhC(=O)OCH$_2$CH$_2$CH$_3$ $\xrightarrow[\text{(2) } H_3O^+]{\text{(1) LiAlH}_4/(CH_3CH_2)_2O}$ PhCH$_2$OH + CH$_3$CH$_2$CH$_2$OH

4. Hydration of alkenes.
   (a) Acid-catalyzed hydration of alkenes forms the product of Markownikoff addition of water (Section 7.14). The reaction generally occurs in low yield and rearrangement takes place whenever possible.

   (methylcyclohexene) $\xrightarrow{H_3O^+}$ (1-methylcyclohexanol)

   (b) Oxymercuration-demercuration of alkenes forms the product of Markownikoff addition of water (Section 8.10). The reaction generally occurs in high yield with no rearrangement.

   $\xrightarrow[\text{(2) NaBH}_4]{\text{(1) Hg(OAc)}_2, H_2O}$

   (c) Hydroboration-oxidation of alkenes occurs by syn addition to form products of anti-Markownikoff addition of water (Sections 8.1–8.4). The reaction generally occurs in high yield with no rearrangement.

   $\xrightarrow[\text{(2) } H_2O_2, \text{NaOH}]{\text{(1) BH}_3, \text{THF}}$

Aldehydes

1. Oxidation of primary alcohols (Section 11.18). Oxidation of primary alcohols can be stopped at the aldehyde stage only by using special reagents like pyridinium chlorochromate (PCC).

$$R{-}CH_2OH \xrightarrow{PCC} R{-}C(=O)H$$

2. Ozonolysis of alkenes followed by reductive workup (Section 14.5). Synthetically useful only with terminal alkenes, symmetrically substituted alkenes, and cyclic alkenes.

cyclohexene $\xrightarrow[\text{(2) Zn/CH}_3\text{COOH}]{\text{(1) O}_3}$ OHC-CH$_2$CH$_2$CH$_2$CH$_2$-CHO

3. Reduction of acyl chlorides (Section 16.14A).

$$R-\underset{Cl}{\overset{O}{\|}}C \xrightarrow[\text{(2) H}_3\text{O}^+]{\text{(1) LiAlH[OC(CH}_3)_3]} R-\underset{H}{\overset{O}{\|}}C$$

4. Hydroboration-oxidation of terminal alkynes (Section 9.7B).

$$CH_3CH_2CH_2C\equiv CH \xrightarrow[\text{(2) H}_2\text{O}_2/\text{NaOH}]{\text{(1) R}_2\text{BH}} CH_3CH_2CH_2CH_2\underset{H}{\overset{O}{\|}}C$$

Alkanes

1. Catalytic hydrogenation of alkenes (Section 8.11). Catalytic hydrogenation occurs by syn addition of hydrogen gas.

1,2-dimethylcyclohexene $\xrightarrow[\text{Pt}]{\text{H}_2}$ cis-1,2-dimethylcyclohexane

2. Catalytic hydrogenation of alkynes (Section 9.8).

$$CH_3CH_2C\equiv CCH_3 \xrightarrow[\text{Pt}]{2H_2} CH_3CH_2CH_2CH_2CH_3$$

3. Reaction of Grignard and organolithium reagents with water (Section 11.8).

$$(CH_3)_2CHCH_2CH_2Br \xrightarrow{2Li} (CH_3)_2CHCH_2CH_2Li \xrightarrow{H_2O} (CH_3)_2CHCH_2CH_3$$

4. Reduction of carbonyl-containing compounds (Section 21.10).
   (a) Wolff-Kishner reduction of aldehydes or ketones.

   PhCOCH₃ + NH₂NH₂ →[KOH, 200 °C / Triethylene glycol] PhCH₂CH₃

   (b) Clemmensen reduction of aldehydes or ketones.

   PhCOCH₃ →[Zn(Hg) / Concentrated HCl] PhCH₂CH₃

   (c) Reduction of a thioacetal.

   CH₃CH₂C(=O)Ph + HSCH₂CH₂SH →[BF₃] (thioketal with CH₃CH₂ and Ph)

   →[Raney Ni / CH₃CH₂OH] CH₃CH₂CH₂Ph

Alkenes

1. β-Elimination reactions of haloalkanes (dehydrohalogenation) or alkylsulfonates (Section 12.12). The more substituted alkene is generally formed.

   $CH_3CH_2C(CH_3)(Br)CH_3$ →[$CH_3CH_2O^-Na^+$ / CH_3CH_2OH] $CH_3CH=C(CH_3)_2$ (70%) + $CH_3CH_2C(CH_3)=CH_2$ (30%)

2. Acid-catalyzed dehydration of an alcohol (Section 7.18 and 12.16). Best results are obtained with secondary and tertiary alcohol to form the more substituted alkene.

   cyclohexanol →[H_2SO_4] cyclohexene

3. Hydrogenation of alkynes (Section 9.8).
   (a) Lindlar's catalyst partially hydrogenates the triple bond to form the (Z)-isomer of the alkene.

   $$CH_3CH_2C{\equiv}CCH_2CH_3 \xrightarrow[\text{Lindlar's catalyst}]{H_2} \begin{array}{c} CH_3CH_2 \\ \diagdown \\ H \end{array} C{=}C \begin{array}{c} CH_2CH_3 \\ \diagup \\ H \end{array}$$

   (b) Lithium or sodium metal in liquid ammonia partially hydrogenates the triple bond to form the (E)-isomer of the alkene.

   $$CH_3CH_2C{\equiv}CCH_2CH_3 \xrightarrow[\text{NH}_{3(l)}]{Li} \begin{array}{c} CH_3CH_2 \\ \diagdown \\ H \end{array} C{=}C \begin{array}{c} H \\ \diagup \\ CH_2CH_3 \end{array}$$

   (c) Hydroboration-protonolysis converts an internal alkyne into the (Z)-isomer of an alkene.

   $$CH_3C{\equiv}CCH_3 \xrightarrow{H-BR_2} \begin{array}{c} CH_3 \\ \diagdown \\ H \end{array} C{=}C \begin{array}{c} CH_3 \\ \diagup \\ BR_2 \end{array} \xrightarrow{CH_3CH_2COOH} \begin{array}{c} CH_3 \\ \diagdown \\ H \end{array} C{=}C \begin{array}{c} CH_3 \\ \diagup \\ H \end{array}$$

4. The Wittig reaction (Section 14.16).

   cyclohexanone + $(C_6H_5)_3\overset{+}{P}{-}\overset{-}{C}H_2 \longrightarrow$ methylenecyclohexane + $(C_6H_5)_3\overset{+}{P}{-}O^-$

Alkynes

1. Alkylation of acetylide ions (Section 9.11). Best results are obtained by reaction of a methyl or primary haloalkane with an acetylide ion.

   $$CH_3C{\equiv}C^-Na^+ + (CH_3)_2CHCH_2Br \longrightarrow CH_3C{\equiv}CCH_2CH(CH_3)_2$$

2. Double dehydrohalogenation of dihalides, either geminal or vicinal (Section 12.12).

   $$C_6H_5{-}CHBr{-}CHBr{-}C_6H_5 \xrightarrow[CH_3CH_2OH]{2\ KOH} C_6H_5{-}C{\equiv}C{-}C_6H_5$$

Amides

1. Reaction of acyl chlorides and ammonia or an amine (Section 16.5). Acyl chlorides react with ammonia to form unsubstituted amides, with primary amines to form *N*-substituted amides, and with secondary amines to form *N*,*N*-disubstituted amides.

$$CH_3COCl + 2NH_3 \longrightarrow CH_3CONH_2 + \overset{+}{N}H_4Cl^-$$

$$CH_3COCl + 2CH_3NH_2 \longrightarrow CH_3CONHCH_3 + CH_3\overset{+}{N}H_3Cl^-$$

$$CH_3COCl + 2(CH_3)_2NH \longrightarrow CH_3CON(CH_3)_2 + (CH_3)_2\overset{+}{N}H_2Cl^-$$

2. Reaction of carboxylic acid anhydrides with ammonia or amines (Section 16.9).

$$(CH_3CO)_2O + 2NH_3 \longrightarrow CH_3CONH_2 + CH_3COO^-\overset{+}{N}H_4$$

$$(CH_3CO)_2O + 2CH_3NH_2 \longrightarrow CH_3CONHCH_3 + CH_3COO^-\overset{+}{N}H_3CH_3$$

3. Reaction of esters with ammonia or amines (Section 16.9).

$$CH_3COOCH_2CH_3 + NH_3 \longrightarrow CH_3CONH_2 + CH_3CH_2OH$$

4. By heating a carboxylic acid with ammonia (Section 24.17).

$$CH_3CH_2COOH \xrightarrow{NH_3} CH_3CH_2COO^-\overset{+}{N}H_4 \xrightarrow{\text{Heat}} CH_3CH_2CONH_2 + H_2O$$

5. The reaction of a carboxylic acid and an amine with dicyclohexylcarbodiimide (DCC) (Section 26.7).

$$CH_3CH_2C(=O)OH \xrightarrow[DCC]{CH_3CH_2NH_2} CH_3CH_2C(=O)NHCH_2CH_3$$

Amines

1. Reduction of amides by LiAlH$_4$ (Section 16.12). Unsubstituted amides are reduced to primary amines, *N*-substituted amides are reduced to secondary amines, and *N,N*-disubstituted amides are reduced to tertiary amines.

$$CH_3C(=O)NH_2 \xrightarrow[\substack{(2)\ H_3O^+ \\ (3)\ NaOH/H_2O}]{(1)\ LiAlH_4} CH_3CH_2NH_2$$

$$CH_3C(=O)NHCH_3 \xrightarrow[\substack{(2)\ H_3O^+ \\ (3)\ NaOH/H_2O}]{(1)\ LiAlH_4} CH_3CH_2NHCH_3$$

$$CH_3C(=O)N(CH_3)_2 \xrightarrow[\substack{(2)\ H_3O^+ \\ (3)\ NaOH/H_2O}]{(1)\ LiAlH_4} CH_3CH_2N(CH_3)_2$$

2. Reduction of nitriles forms primary amines (Section 16.15D). The reducing agent can be either hydrogen gas and a metal catalyst or LiAlH$_4$.

$$CH_3CH_2CH_2C \equiv N \xrightarrow[\substack{(2)\ H_3O^+ \\ (3)\ NaOH/H_2O}]{(1)\ LiAlH_4} CH_3CH_2CH_2CH_2NH_2$$

3. Reaction of a haloalkane and potassium phthalimide (Gabriel synthesis) (Section 22.9). The reaction occurs by an S$_N$2 mechanism so best results are obtained by using a primary haloalkane.

phthalimide-N$^-$K$^+$ + BrCH$_2$CH$_2$CH$_3$ ⟶ phthalimide-N—CH$_2$CH$_2$CH$_3$

$$\xrightarrow[(2)\ NaOH/H_2O]{(1)\ H_3O^+} H_2NCH_2CH_2CH_3$$

4. Nucleophilic substitution reaction of a primary haloalkane and excess ammonia (Section 22.8). Secondary and tertiary amines and quaternary ammonium ions are often formed in appreciable quantities despite using an excess of ammonia.

$$CH_3CH_2Br + NH_3 \text{ (Excess)} \longrightarrow CH_3CH_2NH_2 + \overset{+}{N}H_4\bar{B}r$$

5. Reaction of α-bromocarboxylic acids with ammonia or amines (Section 17.4).

$$CH_3\underset{Br}{CHCOO^-} + NH_3 \longrightarrow CH_3\underset{NH_2}{CHCOO^-} + \overset{+}{N}H_4\bar{B}r$$

6. Reactions of epoxides with ammonia or amines (Section 13.12A). Ammonia and amines react at the least substituted carbon atom of the epoxide ring.

$$CH_3NH_2 + CH_2\overset{O}{-\!\!\!-\!\!\!-}CHCH_3 \longrightarrow CH_3NHCH_2\underset{OH}{CHCH_3}$$

7. Reaction of aldehydes or ketones with ammonia or amines and NaBH₃CN (reductive amination) (Section 22.10). Reaction of an aldehyde or ketone with ammonia forms primary amines; reaction with primary amines forms secondary amines; reaction with secondary amines forms tertiary amines.

cyclohexanone $\xrightarrow[\text{(2) NaBH}_3\text{CN/methanol}]{\text{(1) (CH}_3)_2\text{NH/acidic buffer}}$ N,N-dimethylcyclohexylamine

8. Hydroboration-amination of an alkene (Section 22.11). Syn anti-Markownikoff addition of NH_3 to a carbon-carbon double bond.

1-methylcyclopentene $\xrightarrow[\text{(2) H}_2\text{NCl}]{\text{(1) BH}_3 \cdot \text{THF}}$ trans-2-methylcyclopentylamine

9. Reduction of azides (Section 22.10).

$$CH_3(CH_2)_5CH_2N_3 \xrightarrow[\text{(2) NaOH/H}_2\text{O}]{\text{(1) LiAlH}_4} CH_3(CH_2)_5CH_2NH_2$$

10. Reduction of nitro groups (Section 22.10). The method of choice for preparing arylamines.

$$\text{C}_6\text{H}_5\text{NO}_2 \xrightarrow[\text{(2) NaOH/H}_2\text{O}]{\text{(1) Sn/HCl}} \text{C}_6\text{H}_5\text{NH}_2$$

Arenediazonium Ions

Arenediazonium ions are prepared by the reaction of primary arylamines and nitrous acid (Section 22.15).

$$\text{C}_6\text{H}_5\text{NH}_2 \xrightarrow[\text{H}_2\text{O/0 °C}]{\text{NaNO}_2/\text{HCl}} \text{C}_6\text{H}_5\overset{+}{\text{N}}\equiv\text{N} \; \text{Cl}^-$$

Arenesulfonic Acids

Arenesulfonic acids are prepared from arenes by electrophilic aromatic substitution reaction with SO_3/H_2SO_4 (Section 21.7).

$$\text{C}_6\text{H}_6 \xrightarrow[\text{H}_2\text{SO}_4]{\text{SO}_3} \text{C}_6\text{H}_5\text{SO}_3\text{H}$$

Azides

Alkyl azides are prepared by the reaction of azide ion and a haloalkane. This nucleophilic substitution reaction occurs by an S_N2 mechanism so best results are obtained by methyl, primary, and secondary haloalkanes (Section 12.18).

$$\text{CH}_3(\text{CH}_2)_5\text{CH}_2\text{Br} + \text{N}_3^- \longrightarrow \text{CH}_3(\text{CH}_2)_5\text{CH}_2\text{N}_3 + \text{Br}^-$$

Carboxylic Acids

1. Oxidation of primary alcohols (Section 11.18).

$$\text{CH}_3\text{CH}_2\text{CH}_2\text{CH}_2\text{OH} \xrightarrow[\text{H}_2\text{SO}_4/\text{H}_2\text{O}]{\text{CrO}_3} \text{CH}_3\text{CH}_2\text{CH}_2\text{COOH}$$

2. Oxidation of aldehydes (Section 11.18).

$$\text{CH}_3\text{CH}_2\text{CHO} \xrightarrow[\text{or Fehling's solution}]{\text{Tollen's reagent}} \text{CH}_3\text{CH}_2\text{COOH}$$

3. Oxidation of alkylbenzenes (Section 20.9B). Vigorous oxidation occurs at benzylic carbon atoms containing at least one benzylic hydrogen atom to form

benzoic acid or one of its derivatives. Notice that the benzene ring is not oxidized by most common oxidizing agents.

$$\underset{C(CH_3)_3}{\underset{|}{C_6H_4}}-CH_3 \xrightarrow[H_2SO_4/H_2O]{CrO_3} \underset{C(CH_3)_3}{\underset{|}{C_6H_4}}-COOH$$

4. Oxidation of alkenes (Section 15.5). Synthetically useful only with terminal alkenes, symmetrically disubstituted alkenes, or cycloalkenes.

$$CH_3CH_2CH_2CH=CH_2 \xrightarrow[H_2SO_4/H_2O]{K_2Cr_2O_7} CH_3CH_2CH_2COOH + CO_2$$

5. Reaction of Grignard or organolithium reagents with CO_2 (Section 15.6).

$$CH_3CH_2I + Mg \longrightarrow CH_3CH_2MgI \xrightarrow[(2)\ H_3O^+]{(1)\ CO_2} CH_3CH_2COOH$$

$$C_6H_5Br + Li \longrightarrow C_6H_5Li \xrightarrow[(2)\ H_3O^+]{(1)\ CO_2} C_6H_5COOH$$

6. Hydrolysis of nitriles (Section 15.7).

$$CH_3CH_2CH_2C\equiv N \xrightarrow[\text{Heat}]{H_3O^+} CH_3CH_2CH_2COOH$$

7. Hydrolysis of acyl chlorides, acid anhydrides, thioesters, esters, and amides (Section 16.8).

$$CH_3CH_2\overset{O}{\underset{\|}{C}}-Y \xrightarrow{H_3O^+} CH_3CH_2\overset{O}{\underset{\|}{C}}-OH$$

$$Y = Cl,\ O\overset{O}{\underset{\|}{C}}CH_2CH_3,\ SCH_3,\ OCH_3,\ NH_2,\ NHCH_3,\ \text{or}\ N(CH_3)_2$$

Carboxylic Acid Esters

1. Reaction of carboxylic acids and alcohols in the presence of an acid catalyst (Fischer esterification, Section 15.11). To obtain a good yield of ester, the water must be removed as it forms or an excess of one of the reactants must be used.

$$CH_3CH_2COOH + CH_3OH \xrightleftharpoons[]{H_2SO_4} CH_3CH_2COOCH_3 + H_2O$$

2. Nucleophilic substitution reaction of a carboxylate ion and a haloalkane, a reaction that occurs by an S_N2 mechanism (Section 12.18).

$$CH_3COO^-Na^+ + CH_3CH_2I \longrightarrow CH_3COOCH_2CH_3$$

3. Reactions of acyl chlorides, acid anhydrides, and thioesters with an alcohol (Section 16.9).

$$CH_3COY + CH_3CH_2OH \longrightarrow CH_3COOCH_2CH_3$$

$$Y = Cl,\ OCCH_3,\ SCH_3$$

(with the middle Y being an acetate group $-O-\overset{O}{\underset{\|}{C}}-CH_3$)

4. Alkylation of enolate ions of esters (Section 17.11).

cyclohexyl-C(=O)-OCH$_2$CH$_3$ $\xrightarrow[\text{(2) CH}_3\text{I}]{\text{(1) LDA/THF/0 °C}}$ 1-methylcyclohexyl-C(=O)-OCH$_2$CH$_3$

5. Esters of phenol are prepared by the reaction of phenoxide ion with acyl chlorides or acid anhydrides (Section 23.9).

PhOH + NaOH ⟶ PhO$^-$Na$^+$

$$\text{C}_6\text{H}_5\text{O}^-\text{Na}^+ + \text{CH}_3\text{COCl} \longrightarrow \text{C}_6\text{H}_5\text{OCCH}_3\text{ (O)}$$

$$\text{C}_6\text{H}_5\text{O}^-\text{Na}^+ + (\text{CH}_3\text{CO})_2\text{O} \longrightarrow \text{C}_6\text{H}_5\text{OCCH}_3\text{ (O)}$$

Cyanohydrins

Cyanohydrins are prepared by the reaction of HCN (obtained by the reaction of NaCN and acid) with aldehydes or ketones (Section 14.9).

$$\text{CH}_3\text{CHO} \xrightarrow[\text{H}_2\text{SO}_4]{\text{NaCN}} \text{CH}_3\text{CH(OH)C}\equiv\text{N}$$

Cyclic Compounds

1. The Diels-Alder reaction forms six-membered carbon-containing rings (Section 19.12).

$$\text{butadiene} + \text{dimethyl maleate} \longrightarrow \text{cyclohexene-cis-dicarboxylate}$$

2. Cyclopropane rings are formed by the additions of carbenes or carbenoids to carbon-carbon double bonds (Section 8.7).

$$(\text{CH}_3)_2\text{C}=\text{CH}_2 \xrightarrow[\substack{\text{K}^+\bar{\text{O}}\text{C}(\text{CH}_3)_3 \\ (\text{CH}_3)_3\text{COH}}]{\text{CHCl}_3} \text{1,1-dichloro-2,2-dimethylcyclopropane}$$

$$(\text{CH}_3)_2\text{C}=\text{CH}_2 \xrightarrow[\text{Zn(Cu)}]{\text{CH}_2\text{I}_2} \text{1,1-dimethylcyclopropane}$$

3. Most reactions that can form new bonds between functional groups in separate molecules can also form rings if the reactive functional groups are both present in the same molecule. Intramolecular reactions to form three, five, and six-membered rings are favored (Section 24.14).

$$\text{CH}_3\text{CH(OH)(CH}_2)_2\text{C(=O)OH} \xrightarrow{\text{H}_3\text{O}^+} \text{γ-methyl-γ-butyrolactone}$$

$$\text{BrCH}_2(\text{CH}_2)_4\text{NH}_2 \longrightarrow \text{piperidinium} + \text{Br}^-$$

Disulfides

Disulfides are formed by oxidation of thiols (Section 11.19).

$$\text{CH}_3\text{CH}_2\text{SH} + \text{HSCH}_2\text{CH}_3 \xrightarrow{\text{O}_2} \text{CH}_3\text{CH}_2\text{S}-\text{SCH}_2\text{CH}_3$$

Enamines

Enamines are formed by the reaction of an aldehyde or ketone with a secondary amine in the presence of an acid catalyst (Section 17.14).

cyclohexanone + pyrrolidine $\xrightarrow{\text{Acid catalyst}}$ 1-(cyclohex-1-en-1-yl)pyrrolidine + H_2O

Epoxides

1. Reaction of alkenes with peroxycarboxylic acids (Sections 8.6 and 13.10). Syn addition of an oxygen atom to the double bond maintains the stereochemistry of the alkene in the product epoxide.

cis-2-butene + $\text{CH}_3\text{C(=O)O-OH}$ ⟶ cis-2,3-dimethyloxirane

2. Reaction of halohydrins with base (Section 13.10). The —OH group and the halogen atom must be anti to each other.

Ethers

1. Reaction of an alkoxide ion with a haloalkane (Williamson ether synthesis, Section 13.8C). A nucleophilic substitution reaction that occurs by an S_N2 mechanism.

$$(CH_3)_3C\text{—}O^-K^+ + CH_3I \longrightarrow (CH_3)_3C\text{—}O\text{—}CH_3$$

2. Alkoxymercuration-demercuration of an alkene (Sections 8.10 and 13.8B). Markownikoff addition of an alcohol to the carbon-carbon double bond without rearrangement.

$$CH_3(CH_2)_5CH=CH_2 \xrightarrow[\text{(2) NaBH}_4/\text{NaOH}/\text{H}_2\text{O}]{\text{(1) Hg(OAc)}_2/\text{CH}_3\text{OH}} CH_3(CH_2)_5\underset{\underset{OCH_3}{|}}{CH}CH_3$$

3. Reaction of halobenzenes with alkoxide ions (Sections 23.2 and 23.3). Reaction occurs much more easily when electron-withdrawing groups are o- and p- to the halogen atom on the aromatic ring.

4. Phenyl ethers are prepared by the reaction of phenoxide ion with haloalkanes (Section 23.9).

Haloalkanes

1. Electrophilic addition of HX to alkenes (Section 7.7) and alkynes (Section 9.6A). Addition occurs to form products of Markownikoff addition. Mechanism of addition involves a carbocation intermediate, which rearranges if possible.

$$(CH_3)_2C=CH_2 \xrightarrow{HCl} CH_3-\underset{\underset{Cl}{|}}{\overset{\overset{CH_3}{|}}{C}}-CH_3$$

2. Electrophilic addition of chlorine and bromine to alkenes (Section 8.8) and alkynes (Section 9.6B). Halogenation of aliphatic alkenes and cycloalkenes occurs by anti addition.

[cyclopentene + Br$_2$ → trans-1,2-dibromocyclopentane]

$$CH_3C\equiv CCH_3 \xrightarrow{Cl_2} \underset{Cl}{\overset{CH_3}{\underset{|}{C}}}=\underset{CH_3}{\overset{Cl}{\underset{|}{C}}} + \underset{Cl}{\overset{CH_3}{\underset{|}{C}}}=\underset{Cl}{\overset{CH_3}{\underset{|}{C}}}$$

3. Electrophilic addition of an aqueous solution of bromine or chlorine to an alkene (Section 8.9). The anti addition of OH and X (X = Cl or Br) to a double bond follows Markownikoff rule; the —OH group bonds to the more substituted carbon atom of the double bond.

[1-methylcyclopentene + Br$_2$/H$_2$O → 2-bromo-1-methylcyclopentanol]

4. Free radical addition of HBr to alkenes (Section 18.5). Addition occurs to form products of anti-Markownikoff orientation.

$$(CH_3)_2C=CH_2 \xrightarrow[\text{(peroxide)}]{\text{HBr, ROOR}} CH_3-\underset{\underset{H}{|}}{\overset{\overset{CH_3}{|}}{C}}-CH_2Br$$

5. Free radical chlorination and bromination of alkanes (Section 18.1).

$$CH_3-\underset{\underset{CH_3}{|}}{\overset{\overset{H}{|}}{C}}CH_2CH_3 \xrightarrow[h\nu]{Br_2} CH_3\underset{\underset{CH_3}{|}}{\overset{\overset{Br}{|}}{C}}CH_2CH_3$$

6. Allylic bromination of alkenes with *N*-bromosuccinimide (NBS) (Section 19.3).

cyclohexene $\xrightarrow[CCl_4]{NBS}$ 3-bromocyclohexene

7. Benzylic bromination of alkylbenzenes with *N*-bromosuccinimide (Section 20.9A).

$PhCH_2CH_3 \xrightarrow[CCl_4]{NBS} PhCHBrCH_3$

8. Cleavage of ethers with excess HCl, HBr, or HI (Section 13.9).

$CH_3CH_2OCH_2CH_3 \xrightarrow{HBr} CH_3CH_2Br + CH_3CH_2OH \xrightarrow{HBr} CH_3CH_2Br$

$PhOCH_2CH_3 \xrightarrow{HBr} CH_3CH_2Br + PhOH \xrightarrow{HBr}$ No reaction

9. Reactions of alcohols (Section 11.17).
   (a) With $SOCl_2$.

   $CH_3CH_2CH_2OH + SOCl_2 \longrightarrow CH_3CH_2CH_2Cl + SO_2 + HCl$

   (b) Reaction with PBr_3.

   $CH_3CH_2CH_2OH \xrightarrow{PBr_3} CH_3CH_2CH_2Br$

   (c) Reaction with HCl, HBr, or HI.

   $CH_3CH_2CH_2CH_2OH \xrightarrow[H_2SO_4]{NaBr} CH_3CH_2CH_2CH_2Br$

10. Nucleophilic substitution reaction of haloalkanes or alkyl tosylates with halide ions (Section 12.18).

    $PhSO_3CH_2CH_3 \xrightarrow[CH_3CH_2OH]{Na^+ I^-} CH_3CH_2I + PhSO_3^-$

Halobenzenes

1. Aromatic substitution reaction of chlorine or bromine with arenes in the presence of a Lewis acid catalyst (Section 21.5).

$$C_6H_6 \xrightarrow[FeCl_3]{Cl_2} C_6H_5Cl$$

2. Reaction of arenediazonium ion with cuprous halides (Sandmeyer reaction, Section 22.16).

$$\text{4-CH}_3\text{-C}_6\text{H}_4\text{-N}{\equiv}\text{N Cl}^- \xrightarrow{CuI} \text{4-CH}_3\text{-C}_6\text{H}_4\text{-I}$$

Halohydrins

1. Reaction of aqueous solutions of bromine or chlorine with alkenes (Section 8.9). The product is formed by anti addition and according to the Markownikoff rule: the —OH group adds to the more substituted carbon atom of the double bond.

(1-methylcyclopentene) $\xrightarrow[H_2O]{Cl_2}$ (trans-1-methyl-2-chloro-1-hydroxycyclopentane)

2. Ring opening of an epoxide by hydrogen halides (Section 13.12).

$$CH_3CH\overset{O}{-\!\!\!-\!\!\!-}CH_2 \xrightarrow{HCl} CH_3CHCH_2OH$$
$$\quad\quad\quad\quad\quad\quad\quad\quad\quad\quad\quad\quad\quad |$$
$$\quad\quad\quad\quad\quad\quad\quad\quad\quad\quad\quad\quad\quad Cl$$

Hemiacetal

Hemiacetals are formed by the reaction of one equivalent of alcohol with an aldehyde or ketone (Section 14.11).

$$CH_3CHO + CH_3OH \rightleftharpoons CH_3CH(OH)(OCH_3)$$

Imines

Imines are prepared by the reaction of aldehydes or ketones with primary amines (Section 14.13).

$$\underset{\text{CH}_3\text{CCH}_3}{\overset{\text{O}}{\|}} + \text{CH}_3\text{CH}_2\text{CH}_2\text{NH}_2 \longrightarrow \underset{\underset{\text{CH}_3 \quad \text{CH}_3}{|}}{\overset{\text{N}-\text{CH}_2\text{CH}_2\text{CH}_3}{\|}}{\text{C}}$$

Ketones

1. Oxidation of secondary alcohols (Section 11.18).

$$\underset{\underset{\text{CH}_3\text{CHCH}_2\text{CH}_3}{|}}{\text{OH}} \xrightarrow[\text{H}_2\text{SO}_4/\text{H}_2\text{O}]{\text{CrO}_3} \underset{\text{CH}_3\text{CCH}_2\text{CH}_3}{\overset{\text{O}}{\|}}$$

2. Ozonolysis of alkenes (Section 14.6).

$$\text{alkene} \xrightarrow[\text{(2) Zn/CH}_3\text{COOH}]{\text{(1) O}_3} 2 \; \text{ketone}$$

3. Friedel-Crafts acylation (Section 21.9).

$$\text{C}_6\text{H}_6 + \text{CH}_3\text{C}(\text{O})\text{Cl} \xrightarrow{\text{AlCl}_3} \text{C}_6\text{H}_5\text{CCH}_3$$

4. Acid-catalyzed hydration of alkynes (Section 9.7A). Synthetically useful only for terminal alkynes and symmetrical internal alkynes.

$$\text{C}_6\text{H}_{11}-\text{C}\equiv\text{CH} \xrightarrow[\text{HgSO}_4]{\text{H}_2\text{SO}_4/\text{H}_2\text{O}} \text{C}_6\text{H}_{11}\text{CCH}_3$$

5. Reaction of organocopper reagent and acyl chlorides (Section 16.14B).

$$\underset{\text{CH}_3(\text{CH}_2)_3\text{C}(\text{O})\text{Cl}}{} + (\text{CH}_3\text{CH}_2)_2\text{CuLi} \longrightarrow \underset{\text{CH}_3(\text{CH}_2)_3\text{CCH}_2\text{CH}_3}{\overset{\text{O}}{\|}}$$

6. Hydroboration-oxidation of symmetrically substituted and terminal alkynes (Section 9.7B). Synthetically useful only for terminal alkynes and symmetrical internal alkynes.

$$CH_3CH_2C{\equiv}CCH_2CH_3 \xrightarrow[\text{(2) } H_2O_2/NaOH/H_2O]{\text{(1) } H-BR_2} CH_3CH_2\overset{O}{\underset{\|}{C}}CH_2CH_2CH_3$$

7. Reaction of nitriles with Grignard reagents (Section 16.15E).

$$\text{C}_6\text{H}_{11}\text{-C}{\equiv}\text{N} \xrightarrow[\text{(2) } H_3O^+]{\text{(1) } CH_3MgI} \text{C}_6\text{H}_{11}\text{-C(=O)-CH}_3$$

Nitriles

1. Nucleophilic substitution reaction of haloalkanes with cyanide ion (Section 12.18).

$$(CH_3)_2CHCH_2Br \xrightarrow[CH_3CH_2OH]{NaCN} (CH_3)_2CHCH_2C{\equiv}N$$

2. Dehydration of unsubstituted amides by P_4O_{10} or $SOCl_2$ (Section 16.15B).

$$(CH_3)_3CC(=O)NH_2 \xrightarrow[\text{Heat}]{P_4O_{10}} (CH_3)_3CC{\equiv}N$$

3. Reaction of arenediazonium ions with CuCN (Section 22.16).

$$\text{Ph-}\overset{+}{N}{\equiv}N\ Cl^- \xrightarrow{CuCN} \text{Ph-C}{\equiv}N$$

4. Alkylation of the conjugate base of nitriles (Section 17.11).

$$CH_3CH_2C{\equiv}N \xrightarrow[\text{(2) } CH_3I]{\text{(1) } LDA/THF/0\ °C} CH_3\underset{\underset{CH_3}{|}}{CH}C{\equiv}N$$

Nitrobenzenes

Nitrobenzenes are formed by electrophilic aromatic substitution reaction of benzene or its derivatives with HNO_3/H_2SO_4 (Section 21.6).

Organometallics

1. Grignard reagents are formed by the reaction of haloalkanes or halobenzenes with magnesium metal (Section 11.7).

$$CH_3CH_2CH_2I \xrightarrow[(CH_3CH_2)_2O]{Mg} CH_3CH_2CH_2MgI$$

2. Organolithium compounds are formed by the reaction of haloalkanes or halobenzenes with lithium metal (Section 11.7).

$$CH_3CH_2CH_2CH_2Br + 2Li \longrightarrow CH_3CH_2CH_2CH_2Li + LiBr$$

3. Lithium dialkylcuprates are formed by the reaction of two equivalents of organolithium reagents with one equivalent of CuCl (Section 16.14B).

$$2CH_3CH_2Li + CuCl \longrightarrow (CH_3CH_2)_2CuLi + LiCl$$

Phenols

1. Heating arene diazonium ions in aqueous acid solution (Section 23.6A).

2. Nucleophilic aromatic substitution reaction of halobenzenes with hydroxide ion (Section 23.2). The reaction is facilitated by the presence of electron-withdrawing groups o- and p- to the halogen atom.

Quinones

Oxidation of dihydroxybenzenes or arylamines forms quinones (Section 23.14).

$$\text{hydroquinone} \xrightarrow[\text{H}_2\text{SO}_4/\text{H}_2\text{O}]{\text{Na}_2\text{Cr}_2\text{O}_7} \text{1,4-benzoquinone}$$

$$\text{aniline} \xrightarrow[\text{H}_2\text{SO}_4/\text{H}_2\text{O}]{\text{Na}_2\text{Cr}_2\text{O}_7} \text{1,4-benzoquinone}$$

Sulfides

Sulfides are prepared by the nucleophilic substitution reaction of primary or secondary haloalkanes with a thiolate ion (RS$^-$) (Section 13.15).

$$CH_3CH_2I \; + \; CH_3S^-Na^+ \longrightarrow CH_3CH_2SCH_3$$

Sulfones

Sulfones are prepared by the peroxycarboxylic acid oxidation of sulfides (Section 13.15).

$$CH_3SCH_3 \xrightarrow{CH_3C(=O)O-OH} \underset{O^-}{\overset{O^-}{CH_3\overset{|}{\underset{|}{S^{2+}}}CH_3}}$$

Sulfoxides

Sulfoxides are prepared by the oxidation of sulfides with one equivalent of hydrogen peroxide (Section 13.15).

$$CH_3SCH_3 \xrightarrow{\text{1 Equivalent H}_2\text{O}_2} CH_3\overset{O^-}{\underset{+}{\overset{|}{S}}}CH_3$$

Thioacetals

Thioacetals are the sulfur analog of acetals and are prepared by the acid-catalyzed reaction of aldehydes or ketones and thiols (Section 21.10).

$$C_6H_5COCH_2CH_3 + HSCH_2CH_2SH \rightleftharpoons \text{(cyclic dithiolane)} + H_2O$$

Thiols

1. Nucleophilic substitution reaction of haloalkanes with excess hydrosulfide ion (HS⁻) (Section 11.19). Yields are often poor because the product thiol can react further with the haloalkane to form a symmetrical disulfide.

$$CH_3CH_2CH_2I + \overset{-}{S}H \longrightarrow CH_3CH_2CH_2SH + I^-$$

2. Reaction of thiourea and haloalkanes followed by hydrolysis (Section 11.19).

$$CH_3CH_2CH_2Br \xrightarrow{NH_2-C(=S)-NH_2} \underset{H_2N}{\overset{H_2N}{>}}C=\overset{+}{S}-CH_2CH_2CH_3$$

$$\xrightarrow[\text{(2) } H_3O^+]{\text{(1) NaOH/H}_2\text{O}} HSCH_2CH_2CH_3$$

APPENDIX B

REACTIONS OF IMPORTANT FUNCTIONAL GROUPS

This section summarizes the reactions of important functional groups introduced in the text. The functional groups are listed alphabetically, followed by reference to the appropriate section in the text, an example of the reaction, and any special features or limitations of the reaction.

Acetals (Section 14.11)

Acetals react with excess water (hydrolysis) in the presence of an acid catalyst to form an alcohol and an aldehyde or ketone.

$$\underset{\underset{OCH_2CH_3}{|}}{\overset{\overset{OCH_2CH_3}{|}}{CH_3CH}} \xrightarrow{\text{Excess } H_2O} CH_3C(=O)H + 2CH_3CH_2OH$$

Acid Anhydrides (Section 16.9)

Acid anhydrides react with the following nucleophiles:

(a) Water to form two equivalents of carboxylic acid.

$$(CH_3CO)_2O \xrightarrow{H_2O} 2CH_3C(=O)OH$$

(b) Alcohol to form an ester and a carboxylic acid.

$$(CH_3CO)_2O \xrightarrow{CH_3CH_2OH} CH_3C(=O)OCH_2CH_3 + CH_3C(=O)OH$$

(c) Ammonia to form an unsubstituted amide and a carboxylate ion.

$$\underset{\underset{\underset{O}{\|}}{CH_3C}}{\overset{\overset{O}{\|}}{CH_3C}}\!\!\!>\!\!O \xrightarrow{NH_3} CH_3\underset{NH_2}{\overset{O}{\overset{\|}{C}}} + CH_3\underset{O^-}{\overset{O}{\overset{\|}{C}}}$$

(d) Primary amines to form an *N*-substituted amide and a carboxylate ion.

$$\underset{\underset{\underset{O}{\|}}{CH_3C}}{\overset{\overset{O}{\|}}{CH_3C}}\!\!\!>\!\!O \xrightarrow{CH_3NH_2} CH_3\underset{NHCH_3}{\overset{O}{\overset{\|}{C}}} + CH_3\underset{O^-}{\overset{O}{\overset{\|}{C}}}$$

(e) Secondary amines to form *N,N*-disubstituted amides and a carboxylate ion.

$$\underset{\underset{\underset{O}{\|}}{CH_3C}}{\overset{\overset{O}{\|}}{CH_3C}}\!\!\!>\!\!O \xrightarrow{(CH_3)_2NH} CH_3\underset{N(CH_3)_2}{\overset{O}{\overset{\|}{C}}} + CH_3\underset{O^-}{\overset{O}{\overset{\|}{C}}}$$

Acyl Chlorides

1. Reaction with water forms carboxylic acids (Section 16.5).

$$CH_3\underset{Cl}{\overset{O}{\overset{\|}{C}}} \xrightarrow{H_2O} CH_3\underset{OH}{\overset{O}{\overset{\|}{C}}}$$

2. Reaction with alcohols forms esters (Section 16.5).

$$CH_3\underset{Cl}{\overset{O}{\overset{\|}{C}}} \xrightarrow{CH_3CH_2OH} CH_3\underset{OCH_2CH_3}{\overset{O}{\overset{\|}{C}}}$$

3. Reaction with thiols forms thioesters (Section 16.5).

PhC(=O)Cl + HSCH₃ ⟶ PhC(=O)SCH₃

4. Reaction with ammonia forms unsubstituted amides; reaction with primary amines forms *N*-substituted amides; reaction with secondary amines forms *N,N*-disubstituted amides (Section 16.5).

$$CH_3CH_2C(=O)Cl \xrightarrow{2CH_3NH_2} CH_3CH_2C(=O)NHCH_3 + CH_3\overset{+}{N}H_3\overset{-}{Cl}$$

5. Reaction with carboxylate ion forms acid anhydrides (Section 16.5).

$$CH_3C(=O)Cl + CH_3CH_2C(=O)O^- \longrightarrow CH_3C(=O)OC(=O)CH_2CH_3$$

6. Reaction with Grignard or organolithium reagents forms tertiary alcohols (Section 16.13).

$$CH_3C(=O)Cl \xrightarrow[(2)\ H_3O^+]{(1)\ 2CH_3CH_2MgI} CH_3C(OH)(CH_2CH_3)CH_2CH_3$$

7. Reduction with LiAlH₄ forms primary alcohols (Section 16.12).

$$PhC(=O)Cl \xrightarrow[(2)\ H_3O^+]{(1)\ LiAlH_4} PhCH_2OH$$

8. Partial reduction forms aldehydes (Section 16.14).

$$CH_3CH_2CH_2C(=O)Cl \xrightarrow[(2)\ H_3O^+]{(1)\ LiAlH[OC(CH_3)_3]_3} CH_3CH_2CH_2CHO$$

9. Reaction with lithium dialkylcuprates forms ketones (Section 16.14).

$$CH_3(CH_2)_3\overset{O}{\underset{Cl}{C}} \xrightarrow{(CH_3CH_2)_2CuLi} CH_3(CH_2)_3\overset{O}{\underset{CH_2CH_3}{C}}$$

Alcohols

1. Oxidation (Section 11.18).

    (a) Aldehydes from primary alcohols.

    $$CH_3CH_2CH_2CH_2OH \xrightarrow[CH_2Cl_2]{PCC} CH_3CH_2CH_2\overset{O}{\underset{H}{C}}$$

    (b) Carboxylic acids from primary alcohols.

    $$(CH_3)_2CHCH_2OH \xrightarrow[H_2SO_4/H_2O]{CrO_3} (CH_3)_2CH\overset{O}{\underset{OH}{C}}$$

    (c) Ketones from secondary alcohols.

    $$\underset{CH_3CHCH_2CH_3}{\overset{OH}{|}} \xrightarrow[H_2SO_4/H_2O]{CrO_3} CH_3\overset{O}{\underset{}{\overset{||}{C}}}CH_2CH_3$$

2. Alkoxide formation (Section 11.5).

    (a) Reactions with bases whose conjugate acids have $pK_a > 25$.

    $$CH_3CH_2OH \xrightarrow{NaH} CH_3CH_2O^- Na^+ + H_2$$

    (b) Reaction with alkali metals.

    $$CH_3CH_2OH + Na \longrightarrow CH_3CH_2O^- Na^+ + 1/2\ H_2$$

3. Reaction with carboxylic acids forms esters (Fischer esterification, Section 15.11).

    $$PhC(=O)OH + CH_3CH_2OH \underset{H_2SO_4}{\rightleftharpoons} PhC(=O)OCH_2CH_3 + H_2O$$

4. Reaction with acyl chlorides forms esters (Section 16.5).

$$CH_3C(=O)Cl \xrightarrow[\text{Pyridine}]{CH_3CH_2OH} CH_3C(=O)OCH_2CH_3$$

5. Dehydration forms alkenes (Sections 7.18 and 12.16).

$$\underset{\underset{CH_3}{|}}{\overset{\overset{OH}{|}}{CH_3CCH_2CH_3}} \xrightarrow{\text{Concentrated } H_2SO_4} (CH_3)_2C=CHCH_3$$

6. Conversion into haloalkanes (Section 11.17).

   (a) Reaction with $SOCl_2$.

   $$CH_3CH_2CH_2CH_2OH \xrightarrow{SOCl_2} CH_3CH_2CH_2CH_2Cl$$

   (b) Reaction with PBr_3.

   $$\underset{}{\overset{\overset{OH}{|}}{CH_3CH_2CHCH_3}} \xrightarrow{PBr_3} \underset{}{\overset{\overset{Br}{|}}{CH_3CH_2CHCH_3}}$$

   (c) Reaction with HBr or HCl.

   $$CH_3CH_2CH_2CH_2OH \xrightarrow{HBr} CH_3CH_2CH_2CH_2Br$$

7. Ethers from alcohols (Section 13.8).

   (a) Williamson ether synthesis (Section 13.8C). The alcohol is first converted into an alkoxide ion which then reacts with a primary or secondary haloalkane.

   $$(CH_3)_3C-OH \xrightarrow{K} (CH_3)_3C-O^-K^+ \xrightarrow{CH_3I} (CH_3)_3C-OCH_3$$

   (b) Alkoxymercuration-demercuration of alkenes (Section 13.8B).

   $$CH_3CH=CH_2 \xrightarrow[Hg(OAc)_2]{CH_3OH} \underset{\underset{OCH_3}{|}}{CH_3CHCH_2HgOAc} \xrightarrow[NaOH/H_2O]{NaBH_4} \underset{\underset{OCH_3}{|}}{CH_3CHCH_3}$$

8. Ester formation (Section 11.16).

   (a) Arenesulfonate esters.

   $C_6H_5SO_2Cl + CH_3CH_2OH \longrightarrow C_6H_5SO_3CH_2CH_3$

   (b) Sulfate esters.

   $CH_3OH + HOSO_2OH \rightleftharpoons CH_3OSO_2OH \xrightarrow{CH_3OH} CH_3OSO_2OCH_3$

Aldehydes

1. Oxidation of aldehydes to carboxylic acids (Section 14.18).

   $(CH_3)_2CHCHO \xrightarrow{Ag(NH_3)_2^+} (CH_3)_2CHCOOH$

2. Reduction of aldehydes to primary alcohols (Section 11.10).

   $(CH_3)_2CHCHO \xrightarrow[(2)\ H_3O^+]{(1)\ LiAlH_4} (CH_3)_2CHCH_2OH$

3. Nucleophilic addition reactions.

   (a) Reaction of aldehydes, except formaldehyde, with Grignard and organolithium reagents forms secondary alcohols (Section 11.7).

   $CH_3CH_2CHO \xrightarrow[(2)\ H_3O^+]{(1)\ CH_3CH_2CH_2MgBr} CH_3CH_2CH(OH)CH_2CH_2CH_3$

(b) Reaction of formaldehyde with Grignard and organolithium reagents forms primary alcohols (Section 11.7).

$$CH_3CH_2CH_2Li \xrightarrow[(2)\ H_3O^+]{(1)\ H_2C=O} CH_3CH_2CH_2CH_2OH$$

(c) Reaction with HCN forms cyanohydrins (Section 14.9).

$$C_6H_5-CHO + HCN \rightleftharpoons C_6H_5-CH(OH)-C\equiv N$$

(d) Acid- or base- catalyzed reaction with water (Section 14.10).

$$CH_3CHO + H_2O \underset{H_2SO_4}{\rightleftharpoons} CH_3CH(OH)_2$$

(e) Reaction with alcohols forms hemiacetals as the first product, which then react with more alcohol in acid solutions to form acetals (Section 14.11).

$$CH_3CHO \underset{HCl}{\overset{CH_3OH}{\rightleftharpoons}} \underset{\text{Hemiacetal}}{CH_3CH(OH)OCH_3} \underset{HCl}{\overset{CH_3OH}{\rightleftharpoons}} CH_3CH(OCH_3)_2 + H_2O$$

(f) Reaction with thiols forms thioacetals (Section 21.10).

$$CH_3CHO + CH_3SH \rightleftharpoons CH_3CH(OH)SCH_3 \underset{\text{catalyst}}{\overset{CH_3SH}{\underset{\text{Acid}}{\rightleftharpoons}}} CH_3CH(SCH_3)_2$$

(g) Reaction with primary amines forms imines (Section 14.13).

$$CH_3(CH_2)_3CHO + CH_3CH_2NH_2 \longrightarrow CH_3(CH_2)_3CH=NCH_2CH_3 + H_2O$$

(h) Reaction with Wittig reagents forms alkenes (Section 14.16).

$(C_6H_5)_3\overset{+}{P}-\overset{-}{C}H_2$ + [cyclohexyl-CHO] ⟶ [cyclohexyl-C(=CH$_2$)H] + $(C_6H_5)_3\overset{+}{P}-\overset{-}{O}$

(i) Aldol condensation reaction to form β-hydroxyaldehydes (Section 17.6).

CH_3CHO + CH_3CHO $\underset{H_2O}{\overset{NaOH}{\rightleftharpoons}}$ $CH_3CH(OH)CH_2CHO$

Alkanes

Radical reaction of chlorine and bromine with alkanes forms haloalkanes (Section 18.2).

$(CH_3)_3C-H \xrightarrow{\underset{h\nu}{Br_2}} (CH_3)_3C-Br + HBr$

Alkenes

1. Electrophilic addition of hydrogen halides (Section 7.8). Additions occur to form products of Markownikoff regiochemistry; hydrogen always adds to the less substituted carbon atom of the double bond.

$(CH_3)_2C=CH_2 \xrightarrow{HCl} (CH_3)_2C(Cl)-CH_3$

2. Hydroxylation forms 1,2-diols (Section 8.5). Products are formed by syn addition.

[cyclohexene] $\xrightarrow{\underset{H_2O_2}{OsO_4}}$ [cis-cyclohexane-1,2-diol]

3. Epoxidation forms epoxides (oxiranes, Section 8.6). The stereochemistry of the alkene is maintained in the epoxide.

4. Addition of carbenes forms cyclopropanes (Section 8.7).

5. Addition of carbenoids forms cyclopropanes (Section 8.7).

6. Electrophilic addition of bromine and chlorine forms 1,2-dihalides (Section 8.8). Bromination and chlorination of aliphatic alkenes and cycloalkenes occurs by antiaddition.

7. Addition of aqueous solutions of bromine or chlorine form halohydrins (Section 8.9). Product is formed by antiaddition and according to Markownikoff Rule; the —OH bonds to the more substituted carbon atom of the double bond.

$$(CH_3)_2C=CH_2 \xrightarrow[H_2O]{Br_2} (CH_3)_2C(OH)-CH_2Br$$

8. Catalytic hydrogenation forms alkanes (Section 8.11).

9. Acid-catalyzed hydration forms alcohols (Section 7.14). Product is formed according to Markownikoff Rule. Carbocation formed as an intermediate will rearrange if possible.

$$(CH_3)_2C=CH_2 \xrightarrow{H_3O^+} (CH_3)_2C(OH)-CH_3$$

10. Hydroboration-oxidation forms alcohols with anti-Markownikoff orientation (Sections 8.1-8.4). Syn addition of water is observed.

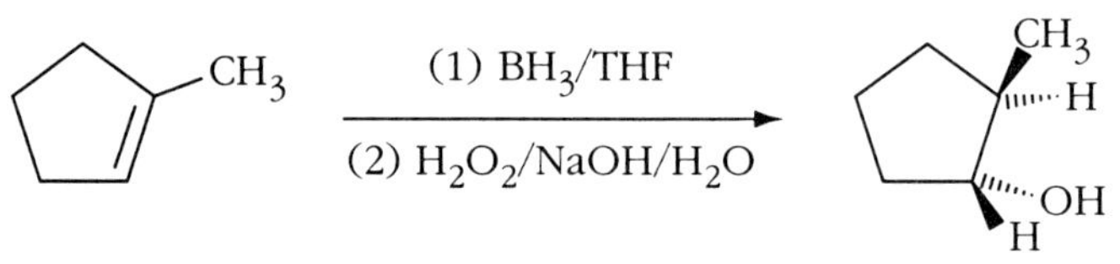

1-methylcyclopentene $\xrightarrow[(2) H_2O_2/NaOH/H_2O]{(1) BH_3/THF}$ trans-2-methylcyclopentanol

11. Oxymercuration-demercuration forms alcohols with Markownikoff orientation (Section 8.10).

$$CH_3CH_2CH=CH_2 \xrightarrow[(2) NaBH_4/H_2O/NaOH]{(1) Hg(OAc)_2 THF/H_2O} CH_3CH_2CH(OH)CH_3$$

12. Oxidative cleavage forms carbonyl compounds (Sections 14.5 and 14.6).

(a) Aldehydes

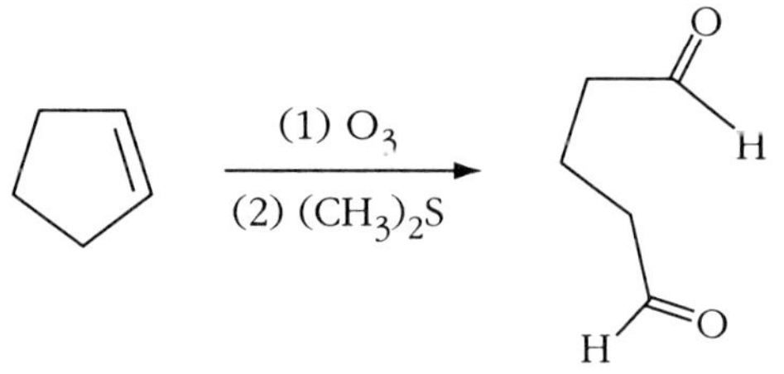

cyclopentene $\xrightarrow[(2) (CH_3)_2S]{(1) O_3}$ pentanedial

(b) Carboxylic acids

cyclopentene $\xrightarrow[H_2SO_4]{CrO_3}$ glutaric acid

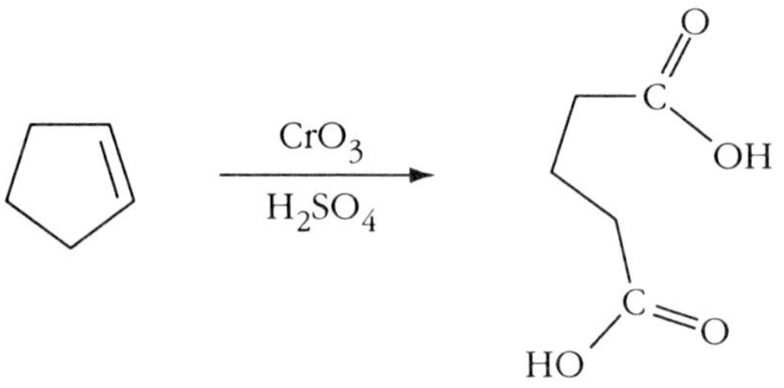

(c) Ketones $(CH_3)_2C=C(CH_3)_2 \xrightarrow[(2) Zn/CH_3COOH]{(1) O_3} (CH_3)_2C=O + O=C(CH_3)_2$

13. Radical addition of HBr forms haloalkanes (Section 18.5). Product is formed by anti-Markownikoff addition.

$$(CH_3)_2C=CH_2 \xrightarrow[\text{Peroxide}]{\text{HBr}} (CH_3)_2\overset{H}{\underset{|}{C}}-CH_2Br$$

Alkynes

1. Addition of Brønsted-Lowry acids (Section 9.6A). Addition can be stopped after the addition of one equivalent of acid.

$$CH_3C\equiv CCH_3 \xrightarrow{\text{HCl}} \underset{\text{Mixture of }(E)\text{- and }(Z)\text{-isomers}}{CH_3\overset{H}{\underset{\underset{Cl}{|}}{C}}=CCH_3} \xrightarrow{\text{HCl}} CH_3\overset{H}{\underset{\underset{H}{|}}{\overset{|}{C}}}-\overset{Cl}{\underset{\underset{Cl}{|}}{\overset{|}{C}}}CH_3$$

2. Addition of chlorine and bromine (Section 9.6B). Addition of one equivalent of halogen forms a mixture of (E)- and (Z)-isomers while addition of two equivalents forms a tetrahaloalkane.

$$CH_3C\equiv CCH_3 \xrightarrow{Cl_2} \underset{\text{Mixture of }(E)\text{- and }(Z)\text{-isomers}}{CH_3\overset{Cl}{\underset{\underset{Cl}{|}}{C}}=CCH_3} \xrightarrow{Cl_2} CH_3\overset{Cl}{\underset{\underset{Cl}{|}}{\overset{|}{C}}}-\overset{Cl}{\underset{\underset{Cl}{|}}{\overset{|}{C}}}CH_3$$

3. Acid-catalyzed hydration (Section 9.7A).

$$CH_3CH_2C\equiv CCH_2CH_3 \xrightarrow[H_2SO_4]{H_2O} \left[\underset{\text{Enol}}{CH_3CH_2\overset{OH}{\underset{|}{C}}=CHCH_2CH_3}\right]$$

$$\longrightarrow CH_3CH_2\overset{O}{\overset{\|}{C}}CH_2CH_2CH_3$$

$$CH_3CH_2CH_2C\equiv CH \xrightarrow[HgSO_4/H_2SO_4]{H_2O} \left[CH_3CH_2CH_2\overset{OH}{\underset{|}{C}}=CH_2\right]$$

$$\longrightarrow CH_3CH_2CH_2\overset{O}{\overset{\|}{C}}CH_3$$

4. Hydroboration-oxidation forms aldehydes and ketones (Section 9.7B). Terminal alkynes form aldehydes while internal symmetrical alkynes form ketones.

$$CH_3CH_2CH_2C\equiv CH \xrightarrow[\text{(2) } H_2O_2/\text{NaOH/}H_2O]{\text{(1) 9-BBN/THF}} \left[CH_3CH_2CH_2CH=CH(OH) \right] \longrightarrow CH_3CH_2CH_2CH_2CHO$$

Enol

$$CH_3CH_2C\equiv CCH_2CH_3 \xrightarrow[\text{(2) } H_2O_2/\text{NaOH/}H_2O]{\text{(1) } BH_3/\text{THF}} \left[CH_3CH_2C(OH)=CHCH_2CH_3 \right] \longrightarrow CH_3CH_2COCH_2CH_2CH_3$$

5. Catalytic hydrogenation forms alkanes (Section 9.8).

$$CH_3C\equiv CCH_3 \xrightarrow[\text{Pt}]{2H_2} CH_3CH_2CH_2CH_3$$

6. Partial hydrogenation (Section 9.8).

   (a) Lindlar's catalyst reduces internal alkynes to the *(Z)*-isomer of an alkene.

   $$CH_3C\equiv CCH_3 \xrightarrow[\text{Lindlar's catalyst}]{H_2} \begin{array}{c} CH_3 \\ \diagdown \\ H \end{array} C=C \begin{array}{c} CH_3 \\ \diagup \\ H \end{array}$$

   (b) Hydroboration-protonolysis reduces internal alkynes to the *(Z)*-isomer of an alkene.

   $$CH_3C\equiv CCH_3 \xrightarrow[\text{(2) } CH_3CH_2COOH]{\text{(1) 9-BBN}} \begin{array}{c} CH_3 \\ \diagdown \\ H \end{array} C=C \begin{array}{c} CH_3 \\ \diagup \\ H \end{array}$$

   (c) Lithium or sodium metal in liquid ammonia reduces internal alkynes to the *(E)*-isomer of an alkene.

   $$CH_3C\equiv CCH_3 \xrightarrow[NH_{3(l)}]{Na} \begin{array}{c} CH_3 \\ \diagdown \\ H \end{array} C=C \begin{array}{c} H \\ \diagup \\ CH_3 \end{array}$$

7. Reaction of terminal alkynes with any base whose conjugate acid has a $pK_a > 25$ removes the terminal hydrogen atom to form an acetylide ion (Section 9.10).

$$CH_3CH_2C \equiv CH \xrightarrow[NH_{3(l)}]{NaNH_2} CH_3CH_2C \equiv C^-Na^+$$

8. Alkylation of acetylide ions forms internal alkynes (Section 9.11).

$$CH_3CH_2C \equiv C^-Na^+ \xrightarrow{CH_3CH_2I} CH_3CH_2C \equiv CCH_2CH_3$$

Amides

1. Hydrolysis under acidic or basic conditions forms carboxylic acids (Section 16.8).

$$C_6H_5-C(=O)-NH_2 \xrightarrow[\text{Heat}]{H_3O^+} C_6H_5-C(=O)-OH$$

2. Reduction to amines (Section 16.12). Unsubstituted amides are reduced to primary amines; N-substituted amides are reduced to secondary amines; and N,N-disubstituted amides are reduced to tertiary amines.

$$CH_3C(=O)NH_2 \xrightarrow[\substack{(2)\ H_3O^+ \\ (3)\ NaOH}]{(1)\ LiAlH_4} CH_3CH_2NH_2$$

3. Dehydration of amides forms nitriles (Section 16.15B).

$$C_6H_5-C(=O)-NH_2 \xrightarrow{P_4O_{10}} C_6H_5-C \equiv N$$

Amines

1. Reaction with acids (Section 22.5).

$$CH_3CH_2CH_2NH_2 + HCl \rightleftharpoons CH_3CH_2CH_2\overset{+}{N}H_3\ Cl^-$$

2. Reaction with arenesulfonyl chlorides forms arenesulfonamides (Section 22.18).

$$C_6H_5-SO_2Cl + 2\ NH_2CH_3 \longrightarrow C_6H_5-SO_2NHCH_3 + CH_3\overset{+}{N}H_3Cl^-$$

3. Nitrosation (Section 22.14).

   (a) Secondary amines form *N*-nitrosamines.

   $$CH_3NHCH(CH_3)_2 \xrightarrow[HCl]{NaNO_2} \begin{array}{c} CH_3 \\ (CH_3)_2CH \end{array}\!\!\!N\!-\!N\!=\!O$$

   (b) Primary amines form diazonium ions. Only arenediazonium ions are synthetically useful.

   $$C_6H_5NH_2 \xrightarrow[HCl]{NaNO_2} C_6H_5\overset{+}{N}\!\equiv\!N\ Cl^-$$

4. Reaction with aldehydes and ketones.

   (a) Ammonia and primary amines react with aldehydes and ketones to form imines (Section 14.13).

   cyclohexanone + $CH_3CH_2NH_2$ $\underset{\text{Acid catalyst}}{\rightleftharpoons}$ N-ethylimine of cyclohexanone + H_2O

   (b) Secondary amines react with aldehydes and ketones to form enamines (Section 17.14).

   cyclohexanone + pyrrolidine $\underset{\text{Acid catalyst}}{\rightleftharpoons}$ 1-pyrrolidinylcyclohexene + H_2O

5. Reaction of ammonia and primary or secondary amine with carboxylic acid derivatives (Sections 16.5 and 16.9). Ammonia forms unsubstituted amides; primary amines form *N*-substituted amides; and secondary amines form *N,N*-disubstituted amides.

   $$CH_3C(=O)Y + NH_2CH_3 \longrightarrow CH_3C(=O)NHCH_3$$

   $$Y = Cl,\ OC(=O)\!-\!R,\ SR,\ OR$$

6. Nucleophilic substitution reactions (Section 12.18). Ammonia reacts with a primary alkyl derivative to form a primary amine; a primary amine reacts to form a secondary amine; a secondary amine reacts to form a tertiary amine; and a tertiary amine reacts to form a quaternary ammonium salt.

$$CH_3I \xrightarrow{NH_3} CH_3\overset{+}{N}H_3 \xrightarrow[H_2O]{NaOH} CH_3NH_2 \xrightarrow{CH_3I} (CH_3)_2\overset{+}{N}H_2 \longrightarrow \text{etc.}$$

Arenes

1. Electrophilic aromatic substitution reactions.

    (a) Halogenation (Section 21.5).

    $$\text{C}_6\text{H}_6 \xrightarrow[FeX_3]{X_2} \text{C}_6\text{H}_5\text{X} + HX$$

    X = Cl or Br

    (b) Nitration (Section 21.6).

    $$\text{C}_6\text{H}_6 \xrightarrow[H_2SO_4]{HNO_3} \text{C}_6\text{H}_5\text{NO}_2$$

    (c) Sulfonation (Section 21.7).

    $$\text{C}_6\text{H}_6 \xrightarrow[H_2SO_4]{SO_3} \text{C}_6\text{H}_5\text{SO}_3\text{H}$$

    (d) Friedel-Crafts reaction.

    i. Alkylation (Section 21.8). Polyalkylation products are often formed. Rearrangement of the haloalkane during the reaction can occur.

    $$\text{C}_6\text{H}_6 \xrightarrow[AlCl_3]{CH_3I} \text{C}_6\text{H}_5\text{CH}_3 \xrightarrow[AlCl_3]{CH_3I} (CH_3)_2\text{C}_6\text{H}_4 \longrightarrow$$

    ii. Acylation (Section 21.9). Reaction does not occur when electron-withdrawing groups are on the aromatic ring.

    $$\text{C}_6\text{H}_6 \xrightarrow[AlCl_3]{CH_3COCl} \text{C}_6\text{H}_5\text{COCH}_3$$

2. Side chain reactions.
   (a) Oxidation occurs at benzylic carbon atoms containing at least one benzylic hydrogen atom to form benzoic acid or its derivatives (Section 20.9B).

   $$4\text{-}(CH_3)_3C\text{-}C_6H_4\text{-}CH_3 \xrightarrow[H_2SO_4]{Na_2Cr_2O_7} 4\text{-}(CH_3)_3C\text{-}C_6H_4\text{-}COOH$$

   (b) Radical bromination occurs exclusively at the benzylic carbon atom (Section 20.9A).

   $$C_6H_5\text{-}CH_2CH_2CH_3 \xrightarrow[h\nu]{Br_2} C_6H_5\text{-}CHBrCH_2CH_3$$

Arenediazonium Ions

1. Replacement of the diazonium group (Section 22.16).

   $C_6H_5\text{-}N\equiv N\ Cl^-$ reacts with:
   - H_2O → OH (phenol)
   - (1) HBF_4 (2) Heat → F (fluorobenzene)
   - CuCl → Cl (chlorobenzene)
   - CuBr → Br (bromobenzene)
   - CuI → I (iodobenzene)
   - CuCN → CN (benzonitrile)
   - H_3PO_2 → H (benzene)

2. Diazo coupling (Section 22.17).

$$\text{C}_6\text{H}_5\text{N}_2^+\text{Cl}^- + \text{C}_6\text{H}_5\text{OH} \longrightarrow \text{C}_6\text{H}_5\text{-N=N-C}_6\text{H}_4\text{-OH}$$

Azides

Azides form primary amines by reduction with LiAlH$_4$ (Section 22.10).

$$\text{CH}_3\text{CH}_2\text{CH}_2\text{CH}_2\text{N}_3 \xrightarrow[\text{(3) NaOH/H}_2\text{O}]{\text{(1) LiAlH}_4 \quad \text{(2) H}_3\text{O}^+} \text{CH}_3\text{CH}_2\text{CH}_2\text{CH}_2\text{NH}_2$$

Carboxylic Acids

1. Reaction with bases (Section 15.8). pK_a of unsubstituted aliphatic carboxylic acids is 4 to 5.

$$\text{CH}_3\text{COOH} + \text{NaOH} \rightleftharpoons \text{CH}_3\text{COO}^-\text{Na}^+ + \text{H}_2\text{O}$$

2. Acid-catalyzed reaction with alcohols forms esters (Fischer esterification, Section 15.11).

$$\text{CH}_3\text{COOH} + \text{CH}_3\text{CH}_2\text{OH} \underset{\text{H}_2\text{SO}_4}{\rightleftharpoons} \text{CH}_3\text{COOCH}_2\text{CH}_3 + \text{H}_2\text{O}$$

3. Reduction by LiAlH$_4$ or catalytic hydrogenation forms primary alcohols (Section 15.17)

$$\text{CH}_3\text{COOH} \xrightarrow[\text{(2) H}_3\text{O}^+]{\text{(1) LiAlH}_4} \text{CH}_3\text{CH}_2\text{OH}$$

4. Reaction with SOCl$_2$ forms acyl chlorides (Section 16.5).

$$\text{C}_6\text{H}_5\text{COOH} \xrightarrow{\text{SOCl}_2} \text{C}_6\text{H}_5\text{COCl}$$

5. Heating with ammonia forms an amide (Section 24.17).

$$CH_3CH_2COOH + NH_3 \xrightarrow{\text{Heat}} CH_3CH_2CONH_2 + H_2O$$

6. Reaction with an amine and dicyclohexylcarbodiimide (DCC) forms an amide (Section 26.7).

$$CH_3CH_2COOH + NH_2CH_2CH_3 \xrightarrow{\text{DCC}} CH_3CH_2CONHCH_2CH_3$$

Carboxylic Acid Esters

1. Acid-catalyzed hydrolysis forms a carboxylic acid and an alcohol (Section 15.11).

$$CH_3CH_2COOCH_2CH_3 + H_2O \xrightleftharpoons{H_2SO_4} CH_3CH_2COOH + CH_3CH_2OH$$

2. Saponification (hydrolysis in basic solution) forms a carboxylate ion and an alcohol (Section 15.12).

$$CH_3CH_2COOCH_2CH_3 \xrightarrow[H_2O]{\text{NaOH}} CH_3CH_2COO^- + CH_3CH_2OH$$

3. Reduction by LiAlH$_4$ forms two alcohols. One is a primary alcohol obtained from the acyl part of the ester. The other is obtained from the alkoxy (—OR') part of the ester (Section 16.12).

$$CH_3CH_2COOCH_2CH_3 \xrightarrow[(2)\ H_3O^+]{(1)\ LiAlH_4} CH_3CH_2CH_2OH + CH_3CH_2OH$$

4. Transesterification: The acid-catalyzed reaction of an ester with an alcohol to form another ester (Section 16.9).

$$CH_3COOCH_3 + CH_3CH_2CH_2OH \xrightleftharpoons{H_2SO_4} CH_3COOCH_2CH_2CH_3 + CH_3OH$$

5. Reaction with Grignard or organolithium reagents forms tertiary alcohols (Section 16.13).

$$CH_3C(=O)OCH_3 \xrightarrow[(2)\ H_3O^+]{(1)\ CH_3CH_2CH_2Li} CH_3C(OH)(CH_2CH_2CH_3)(CH_2CH_2CH_3)$$

6. Claisen condensation reaction forms β-keto esters (Section 17.7).

$$CH_3C(=O)OCH_3 + CH_3C(=O)OCH_3 \xrightarrow[CH_3OH]{Na^+\ ^-OCH_3} CH_3C(=O)CH_2C(=O)OCH_3 + CH_3OH$$

7. Reaction with amines forms amides (Section 16.9). Reaction with ammonia forms unsubstituted amides; reaction with primary amines forms *N*-substituted amides; and reaction with secondary amines forms *N,N*-disubstituted amides.

$$CH_3C(=O)OCH_3 + CH_3NH_2 \longrightarrow CH_3C(=O)NHCH_3 + CH_3OH$$

Epoxides

1. Reactions with nucleophiles under basic conditions cleaves the epoxide ring (Section 13.12). Nucleophiles react at the least sterically hindered carbon atom of the epoxide ring.

$$CH_3CH\overset{O}{-\!\!\!-}CH_2 + CH_3CH_2CH_2Li \longrightarrow CH_3CH(OH)CH_2CH_2CH_2CH_3$$

2. Acid-catalyzed epoxide ring opening (Section 13.12). Nucleophiles react with the most sterically hindered carbon atom of the conjugated acid of the epoxide ring.

$$CH_3CH\overset{O}{-\!\!\!-}CH_2 + HCl \longrightarrow CH_3CH(Cl)CH_2OH$$

Ethers

Heating ethers in aqueous solutions of HCl, HBr, or HI cleaves the C—O bond of the ether to form a haloalkane and an alcohol. When an excess of acid is used the initially formed alcohol reacts further to form a haloalkane (Section 13.9).

$$CH_3CH_2OCH_2CH_3 \xrightarrow[Heat]{HBr} CH_3CH_2Br + CH_3CH_2OH \xrightarrow{HBr} CH_3CH_2Br$$

Haloalkanes

1. Reaction with magnesium metal forms a Grignard reagent (Section 11.7).

$$CH_3CH_2CH_2I \xrightarrow{Mg} CH_3CH_2CH_2MgI$$

2. Nucleophilic substitution reactions (Sections 12.1-12.9). The reaction occurs by either an S_N1 or S_N2 mechanism.

$$CH_3CH_2CH_2I \xrightarrow[S_N2 \text{ mechanism}]{^-C\equiv N} CH_3CH_2CH_2C\equiv N$$

$$(CH_3)_3C-Cl \xrightarrow[S_N1 \text{ mechanism}]{CH_3OH} (CH_3)_3C-OCH_3$$

3. Dehydrohalogenation (Sections 12.10-12.16). The reaction occurs by either an E1 or E2 mechanism.

$$CH_3CH_2\underset{\underset{CH_3}{|}}{\overset{\overset{Br}{|}}{C}}CH_3 \xrightarrow[CH_3CH_2OH]{Na^+\ ^-OCH_2CH_3} CH_3CH=C\underset{CH_3}{\overset{CH_3}{<}}$$

Major product

Halohydrins

Reaction with base forms epoxides when the —OH group and the halogen atom are in an anti conformation (Section 13.10).

Hemiacetals

Hemiacetals are hydrolyzed to an alcohol and an aldehyde or ketone (Section 14.11).

$$CH_3\overset{\overset{OH}{|}}{C}HOCH_3 \xrightarrow[\text{Acid catalyst}]{\text{Excess } H_2O} CH_3\overset{O}{\overset{\|}{C}}H + CH_3OH$$

Imines

Imines are hydrolyzed to a primary amine and an aldehyde or ketone (Section 14.13).

$$\underset{CH_3\ \ \ CH_3}{\overset{NCH_2CH_3}{\overset{\|}{C}}} \xrightarrow[\text{Acid catalyst}]{\text{Excess } H_2O} \underset{CH_3\ \ \ CH_3}{\overset{O}{\overset{\|}{C}}} + H_2NCH_2CH_3$$

Ketones

1. **Reduction**

   (a) LiAlH$_4$ or catalytic hydrogenation form secondary alcohols (Section 11.10).

   $$\text{CH}_3\text{CCH}_2\text{CH}_3 \xrightarrow[\text{(2) H}_3\text{O}^+]{\text{(1) LiAlH}_4} \text{CH}_3\text{CHCH}_2\text{CH}_3$$
   (with C=O on left, OH on right)

   (b) Zinc amalgam (Clemmensen reduction) reacts to form alkanes (Section 21.10).

   $$\text{PhCOCH}_3 \xrightarrow[\text{HCl}]{\text{Zn(Hg)}} \text{PhCH}_2\text{CH}_3$$

   (c) Reaction with hydrazine (Wolff-Kishner reduction) forms an alkane (Section 21.10).

   $$\text{PhCOCH}_3 \xrightarrow{\text{NH}_2\text{NH}_2} \text{PhCH}_2\text{CH}_3$$

2. Reaction with Grignard or organolithium reagents forms tertiary alcohols (Section 11.7).

   $$\text{PhCOCH}_3 \xrightarrow[\text{(2) H}_3\text{O}^+]{\text{(1) CH}_3\text{CH}_2\text{MgBr}} \text{Ph-C(OH)(CH}_3\text{)(CH}_2\text{CH}_3\text{)}$$

3. Reaction with HCN forms cyanohydrins (Section 14.9).

   $$\text{CH}_3\text{CCH}_2\text{CH}_3 + \text{HCN} \rightleftharpoons \text{CH}_3\text{C(OH)(C}\equiv\text{N)CH}_2\text{CH}_3$$

4. Reaction with alcohols forms acetals (Section 14.11). The first formed product is a hemiacetal that reacts with alcohol in an acid solution to form an acetal.

$$CH_3CCH_2CH_3 \text{ (with =O)} \xrightleftharpoons{CH_3OH} \underset{\text{Hemiacetal}}{CH_3C(OH)(OCH_3)CH_2CH_3} \xrightleftharpoons[H_2SO_4]{CH_3OH} \underset{\text{Acetal}}{CH_3C(OCH_3)_2CH_2CH_3}$$

5. Reaction with thiols forms thioacetals (Section 21.10).

$$CH_3CCH_2CH_3 \text{ (with =O)} \xrightleftharpoons{CH_3SH} CH_3C(OH)(SCH_3)CH_2CH_3 \xrightleftharpoons[\text{Acid catalyst}]{CH_3SH} CH_3C(SCH_3)_2CH_2CH_3$$

6. The Wittig reaction converts ketones into alkenes (Section 14.16).

$$(C_6H_5)_3\overset{+}{P}-\overset{-}{C}HCH_3 \;+\; \text{cyclohexanone} \longrightarrow \text{cyclohexane=CHCH}_3 \;+\; (C_6H_5)_3\overset{+}{P}-\overset{-}{O}$$

7. Reaction with primary amines forms imines (Section 14.13).

$$\text{cyclohexanone} \;+\; H_2NCH_2CH_3 \xrightleftharpoons{\text{Acidic buffer}} \text{cyclohexane=NCH}_2CH_3 \;+\; H_2O$$

8. Aldol reaction forms β-hydroxy ketones (Section 17.6).

$$CH_3CCH_3 \;+\; CH_3CCH_3 \xrightleftharpoons[H_2O]{NaOH} CH_3C(OH)(CH_3)CH_2CCH_3$$

9. Reaction with bromine in acid solution forms α-bromo ketones (Section 17.4).

$$CH_3CH_2CCH_2CH_3 \xrightarrow[H_3O^+]{Br_2} CH_3CHBrCCHBrCH_3 \text{ (with C=O)}$$

Nitriles

1. Hydrolysis under acidic or basic conditions forms carboxylic acids (Section 16.15C).

$$CH_3CH_2CH_2C\equiv N \xrightarrow[\text{Heat}]{H_3O^+} CH_3CH_2CH_2COOH$$

$$CH_3CH_2CH_2C\equiv N \xrightarrow[\text{(2) } H_3O^+]{\text{(1) NaOH/H}_2\text{O/heat}} CH_3CH_2CH_2COOH$$

2. Reduction by LiAlH$_4$ forms primary amines (Section 16.15D).

$$(CH_3)_2CHC\equiv N \xrightarrow[\text{(3) NaOH/H}_2\text{O}]{\substack{\text{(1) LiAlH}_4 \\ \text{(2) H}_3\text{O}^+}} (CH_3)_2CHCH_2NH_2$$

3. Reaction with Grignard reagents forms ketones (Section 16.15E).

$$CH_3CH_2C\equiv N \xrightarrow[\text{(2) H}_3\text{O}^+]{\text{(1) CH}_3\text{CH}_2\text{CH}_2\text{MgBr}} CH_3CH_2\overset{\overset{\displaystyle O}{\|}}{C}CH_2CH_2CH_3$$

Nitrobenzenes

Reduction forms arylamines (Section 22.10).

$$\text{4-methylnitrobenzene} \xrightarrow[\text{HCl}]{\text{Sn}} \text{4-methylaniline}$$

Organometallics

1. Reaction of Grignard and organolithium reagents with carbonyl compounds.

    (a) Reaction with formaldehyde forms primary alcohols (Section 11.7).

$$\text{PhMgBr} \xrightarrow[\text{(2) H}_3\text{O}^+]{\text{(1) H}_2\text{C=O}} \text{PhCH}_2\text{OH}$$

(b) Reaction with any aldehyde, except formaldehyde, forms secondary alcohols (Section 11.7).

[Cyclohexyl-CH$_2$CHO] $\xrightarrow{\text{(1) CH}_3\text{CH}_2\text{CH}_2\text{CH}_2\text{Li} \quad \text{(2) H}_3\text{O}^+}$ [Cyclohexyl-CH$_2$CH(OH)CH$_2$CH$_2$CH$_2$CH$_3$]

(c) Reaction with ketones forms tertiary alcohols (Section 11.7).

[Phenyl-CH$_2$C(=O)CH$_2$CH$_3$] $\xrightarrow{\text{(1) CH}_3\text{MgI} \quad \text{(2) H}_3\text{O}^+}$ [Phenyl-CH$_2$C(OH)(CH$_3$)CH$_2$CH$_3$]

(d) Reaction of formate ester forms secondary alcohols (Section 16.13).

[HC(=O)OCH$_2$CH$_3$] $\xrightarrow{\text{(1) 2CH}_3\text{CH}_2\text{MgBr} \quad \text{(2) H}_3\text{O}^+}$ CH$_3$CH$_2$CH(OH)CH$_2$CH$_3$

(e) Reaction with any ester except formate esters forms tertiary alcohols (Section 16.13).

[CH$_3$C(=O)OCH$_2$CH$_3$] $\xrightarrow{\text{(1) 2CH}_3\text{CH}_2\text{MgBr} \quad \text{(2) H}_3\text{O}^+}$ CH$_3$C(OH)(CH$_2$CH$_3$)$_2$

(f) Reaction with acyl chlorides forms tertiary alcohols (Section 16.13).

[CH$_3$C(=O)Cl] $\xrightarrow{\text{(1) 2CH}_3\text{CH}_2\text{MgBr} \quad \text{(2) H}_3\text{O}^+}$ CH$_3$C(OH)(CH$_2$CH$_3$)$_2$

(g) Reaction with CO$_2$ forms carboxylic acids (Section 15.6).

[Cyclohexyl-CH$_2$MgBr] $\xrightarrow{\text{(1) CO}_2 \quad \text{(2) H}_3\text{O}^+}$ [Cyclohexyl-CH$_2$COOH]

(h) Reaction with nitriles forms ketones (Section 16.15E).

$$CH_3CH_2CH_2MgBr \xrightarrow[\text{(2) } H_3O^+]{\text{(1) } CH_3CH_2C\equiv N} CH_3CH_2\overset{\overset{\displaystyle O}{\|}}{C}CH_2CH_2CH_3$$

(i) Reaction with acidic hydrogen atoms forms alkanes (Section 11.8).

$$CH_3CH_2MgBr \xrightarrow{H_2O} CH_3CH_3 \; + \; Mg(OH)Br$$

$$CH_3CH_2Li \xrightarrow{CH_3OH} CH_3CH_3 \; + \; LiOCH_3$$

Phenols

1. Reaction with bases (Section 23.7).

PhOH + NaOH ⟶ PhO⁻Na⁺ + H₂O

2. Reaction with acyl chlorides and acid anhydrides forms esters (Section 23.9).

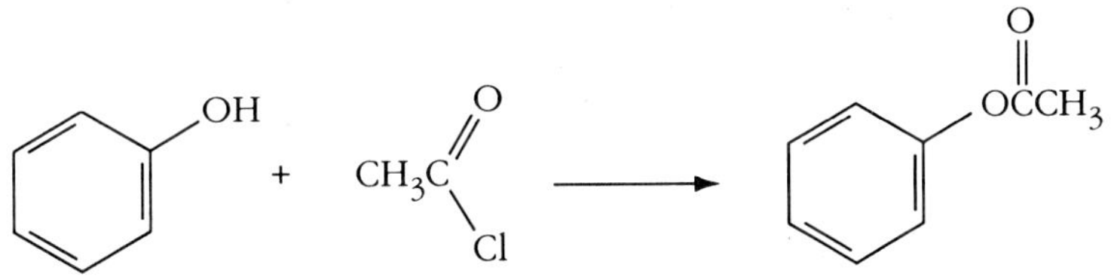

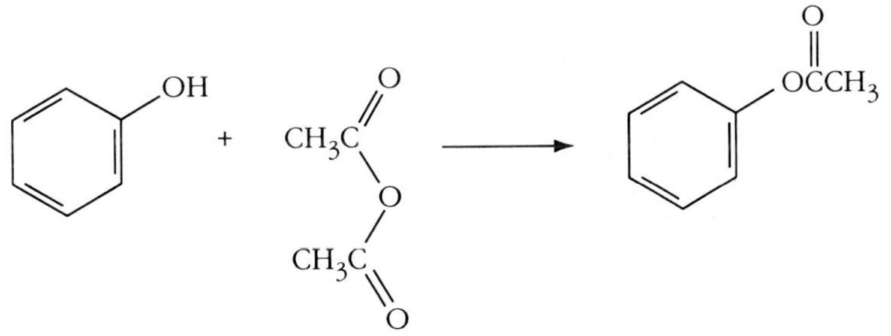

PhOH + CH₃COCl ⟶ PhOCOCH₃

PhOH + (CH₃CO)₂O ⟶ PhOCOCH₃

3. Reaction of a primary or secondary haloalkane with phenoxide ions forms ethers (Williamson ether synthesis, Section 23.9).

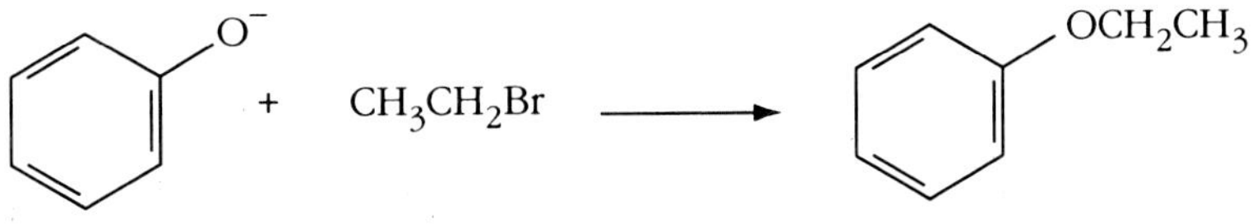

PhO⁻ + CH₃CH₂Br ⟶ PhOCH₂CH₃

4. Oxidation forms quinones (Section 23.14).

[phenol] $\xrightarrow{\text{Na}_2\text{Cr}_2\text{O}_7}_{\text{H}_2\text{SO}_4/\text{H}_2\text{O}}$ [1,4-benzoquinone]

Quinones

Reduction of quinones forms hydroquinones (Section 23.16).

[benzoquinone] + Fe^{3+} $\underset{\text{H}_3\text{O}^+}{\rightleftharpoons}$ [hydroquinone] + Fe^{2+}

Sulfides

1. Oxidation with one equivalent of H_2O_2 forms sulfoxides (Section 13.15).

$$CH_3SCH_3 \xrightarrow{H_2O_2} CH_3\overset{O^-}{\underset{+}{S}}CH_3$$

2. Oxidation with peroxycarboxylic acid forms a sulfone (Section 13.15).

$$CH_3SCH_3 \xrightarrow{CH_3C(O)O-OH} CH_3-\overset{O^-}{\underset{O^-}{\overset{|}{\underset{|}{S}}}}^{2+}-CH_3$$

Thiols

1. Reaction with aldehydes and ketones forms thioacetals (Section 21.10).

$$CH_3CH_2\underset{H}{\overset{O}{\overset{\|}{C}}} \xrightleftharpoons{CH_3SH} CH_3CH_2\underset{SCH_3}{\overset{OH}{\underset{|}{\overset{|}{CH}}}} \xrightleftharpoons[\text{Acid catalyst}]{CH_3SH} CH_3CH_2\underset{SCH_3}{\overset{SCH_3}{\underset{|}{\overset{|}{CH}}}} + H_2O$$

2. Oxidation forms disulfides (Section 13.15).

$$CH_3SH \;+\; CH_3SH \xrightarrow{O_2} CH_3S-SCH_3$$

3. The conjugate bases of thiols react with primary or secondary haloalkanes to form sulfides (Section 13.15).

$$CH_3S^- \;+\; CH_3CH_2CH_2Br \longrightarrow CH_3SCH_2CH_2CH_3$$

APPENDIX C

USES OF IMPORTANT REAGENTS

This section lists some of the more commonly used reagents in organic chemistry. They are listed in alphabetical order and reference is given to sections in the textbook where more information and examples of their use can be found.

Aluminum Chloride, $AlCl_3$ is a Lewis acid that is used as a catalyst in the chlorination reaction (Section 21.5) and the Friedel-Crafts alkylation (Section 21.8) and acylation (Section 21.9) of benzene derivatives.

Borane, BH_3, which exists in equilibrium with its dimer diborane B_2H_6, is used in the hydroboration reactions of alkenes (Section 8.1) and alkynes (Section 9.7B) to form organoboranes.

9-BBN, 9-borabicyclo[3.3.1]nonane, is used in place of borane in the hydroboration reactions of alkenes (Section 8.1) and alkynes (Section 9.7B) when a bulky hydroborating reagent is needed.

Bromine, Br_2, undergoes electrophilic addition reactions with alkenes (Section 8.8) and alkynes (Section 9.6B). An aqueous solution of bromine is also used to oxidize aldoses (Section 25.3C).

Chromium trioxide, CrO_3, is a general oxidizing agent that as an acidic aqueous solution reacts with many organic functional groups. It oxidizes primary alcohols to carboxylic acids (Section 11.18), secondary alcohols to ketones (Section 11.18), alkylbenzenes to derivatives of benzoic acid (Section 20.9B), and oxidizes the double bond of alkenes to carboxylic acids and ketones (Section 14.6). See also sodium dichromate and potassium permanganate.

Collins' reagent, $(C_6H_5N)_2CrO_3$, is one of the special oxidizing reagents that oxidizes alcohols to aldehydes but does not oxidize aldehydes to carboxylic acids (Section 11.18). See also PCC.

DCC, N, N'-**dicyclohexylcarbodiimide,** $C_6H_{11}N=C=NC_6H_{11}$, is used to form peptide bonds. An equimolar mixture of DCC, a carboxylate ion, and amine forms an amide bond (Section 26.7).

Fehling's solution is an alkaline solution of Cu(II) in a tartarate buffer that is used as a test for aldoses and ketoses (Sections 14.18 and 25.3). See also Tollen's reagent.

Grignard reagents, RMgX, are reagents that react with aldehydes (Section 11.7), ketones (Section 11.7), esters (Section 16.13), and acyl chlorides (Section 16.13) to form alcohols; with carbon dioxide to form carboxylic acids (Section 15.6); and with nitriles to form ketones (Section 16.15E). See also organolithium compounds.

Hydrogen gas and a metal catalyst, catalytic hydrogenation, are reagents that reduce alkenes (Section 8.11) and alkynes (Section 9.8) to alkanes. While catalytic hydrogenation also reduces carbonyl-containing compounds to alcohols (Section 15.17) and amides and nitriles to amines (Section 16.15D), it is more convenient to use lithium aluminum hydride for these purposes.

Hydrogen peroxide, H_2O_2 is an oxidizing agent that has many uses in organic chemistry. It converts organoboranes to alcohols (Section 8.4) and it is used in the hydroxylation of alkenes to regenerate the toxic and expensive osmium tetraoxide (Section 8.5).

Lindlar's catalyst catalyzes the hydrogenation of alkynes to alkenes but not the hydrogenation of alkenes to alkanes (Section 9.8). The Z-isomer is formed in cases where isomers are possible in the hydrogenation of alkynes.

Lithium aluminum hydride, $LiAlH_4$, is a reducing agent that reduces aldehydes and ketones to alcohols (Section 11.10), carboxylic acid derivatives to alcohols (Section 16.12), and amides and nitriles to amines (Section 16.15D). See also sodium borohydride.

Lithium aluminum tri(t-butoxy)hydride, $Li(t\text{-}OC_4H_9)_3AlH$, is a reagent that converts acyl chlorides into aldehydes (Section 16.14A).

Lithium dialkylcuprates, R_2CuLi, are reagents that convert acyl chlorides into ketones (Section 16.14B).

Lithium diisopropylamide, $LiN[CH(CH_3)_2]_2$, is a strong base that is used to convert ketones and nitriles quantitatively into their enolate ions (Section 17.11).

Mercury (II) acetate, $Hg(OAc)_2$, adds to alkenes (Section 8.10) to form organomercury compounds in a reaction called oxymercuration. Oxymercuration is the first of the two-reaction sequence called oxymercuration-demercuration in which water is added to an alkene according to Markownikoff's rule.

N-Bromosuccinimide reacts with alkenes by replacing an allylic hydrogen atom by a bromine atom (Section 19.3). It also reacts with alkylbenzenes by replacing the benzylic hydrogen atom by a bromine atom (Section 20.9A).

Nitric acid, HNO_3, in concentrated sulfuric acid is used to nitrate benzene derivatives (Section 21.6). Nitric acid is also used to oxidize aldoses to aldaric acids (Section 25.3C).

Nitrous acid, HNO_2, reacts with primary amines to form diazonium ions (Sections 22.15 and 22.16) and with secondary amines to form nitrosoamines (Section 22.14).

Organolithium reagents, RLi, are reagents that react with aldehydes (Section 11.7), ketones (Section 11.7), esters (Section 16.13), and acyl chlorides (Section 16.13) to form alcohols; and with carbon dioxide to form carboxylic acids (Section 15.6). See also Grignard reagents.

Osmium tetraoxide, OsO_4, converts alkenes into vicinal glycols in a reaction called hydroxylation (Section 8.5).

Ozone, O_3, is an oxidizing agent that cleaves carbon-carbon double bonds to form aldehydes and ketones (Sections 14.5 and 14.6).

Peroxycarboxylic acids, RCOOOH, are oxidizing agents. Their major use is the conversion of alkenes into epoxides (oxiranes) in a reaction called epoxidation (Section 8.6).

Potassium permanganate, $KMnO_4$, is a general oxidizing agent that as an acidic aqueous solution reacts with many organic functional groups. It oxidizes primary alcohols to carboxylic acids (Section 11.18), secondary alcohols to ketones (Section 11.18), alkylbenzenes to derivatives of benzoic acid (Section 20.9B), and oxidizes the double bond of alkenes to carboxylic acids and ketones (Section 14.6). See also sodium dichromate and chromium trioxide.

Pyridinium chlorochromate, $C_5H_5N^+$ $ClCrO_3^-$, is one of the special oxidizing reagents that oxidizes alcohols to aldehydes but does not oxidize aldehydes to carboxylic acids (Section 11.18). See also Collins' reagent.

Sodium amide, $NaNH_2$, a strong base used to form acetylide ions from terminal alkynes (Section 9.10) and alkoxide ions from alcohols (Section 11.15). See sodium hydride.

Sodium borohydride, $NaBH_4$, is a milder reducing agent than $LiAlH_4$. It reduces aldehydes and ketones to alcohols (Section 11.10), but it does not reduce carboxylic acid derivatives and nitriles. See also lithium aluminum hydride.

Sodium dichromate, $Na_2Cr_2O_7$, is a general oxidizing agent that as an acidic aqueous solution reacts with many organic functional groups. It oxidizes primary alcohols to carboxylic acids (Section 11.18), secondary alcohols to ketones (Section 11.18), alkylbenzenes to derivatives of benzoic acid (Section 20.9B), and oxidizes the double bond of alkenes to carboxylic acids and ketones (Section 14.6). See also potassium permanganate and chromium trioxide.

Sodium ethoxide, $NaOCH_2CH_3$, is a weaker base than sodium hydride or sodium amide and is widely used in elimination reactions (Section 12.12), Claisen condensation reactions (Section 17.7), the malonic ester synthesis (Section 24.7), and the acetoacetic ester synthesis (Section 24.8).

Sodium hydride, NaH, a strong base used to form acetylide ions from terminal alkynes (Section 9.10) and alkoxide ions from alcohols (Section 11.15). See sodium amide.

Sulfuric acid, H_2SO_4, is often used as a strong acid catalyst. See, for example, the acid-catalyzed hydration of alkenes (Section 7.14), alkynes (Section 9.7A), and as a catalyst in the Fischer esterification (Section 15.11). A solution of SO_3 in sulfuric acid is used to sulfonate benzene derivatives (Section 21.7).

Thionyl chloride, $SOCl_2$, reacts with carboxylic acids to form acyl chlorides (Section 16.5) and reacts with alcohols to form chloroalkanes (Section 11.17).

Tollen's reagent is used as a test for aldoses and ketoses (Sections 14.18 and 25.3). The oxidizing agent in Tollen's reagent is $Ag(NH_3)_2^+$. The silver (I) is reduced to silver metal, which is deposited as a silver mirror on the sides of the test tube. See also Fehling's solution.

Wittig reagent converts aldehydes and ketones into alkenes (Section 14.16).

Answers to Examination 1

1. (a)

   CH₃—C(H)(Cl)(CH₂CH₃)

   (b)

   H₃C\   /CH₂CH₂CH₃
       C=C
   H /   \CH₃

   (c)

   (cyclohexane with H₃C and CH(CH₃)₂ substituents)

   (d)

   Br, H, Br, H on C's with H₃C, CH₂, CH₃   or   H, Br, H, Br on C's with H₃C, CH₂, CH₃

2. (a) (*R*)-2-Bromo-3-methylbutane
   (b) *cis*-1,2-Dichlorocyclohexane
   (c) 2-Chloro-2,5-dimethylhexane

3. $CH_3CH_2CH_2CH_2CH_3$ < $CH_3CHClCH_2CH_3$ < $CH_3CH_2CH_2OH$

   bp 36.1 °C   bp 68.3 °C   bp 97.4 °C

4.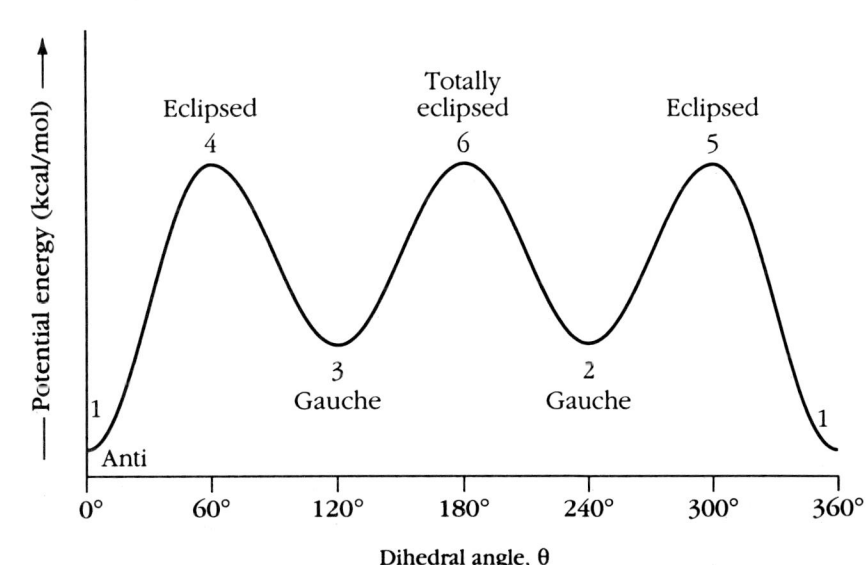

5. (a) $CH_3CH_2O^- Na^+/CH_3CH_2OH$, 50 °C (or any other alkoxide at an elevated temperature)
   (b) H_2/Pd, Pt, Ni or Rh

(c) [bicyclic diol structures: trans-diol + cis-diol]

(d) [chlorohydrin structures on hydrindane] + [stereoisomer]

(e) (1) Hg(OAc)$_2$, H$_2$O; (2) NaBH$_4$ or H$_3$O$^+$ (not as good)
(f) (1) BH$_3$, (2) H$_2$O$_2$/HO$^-$

(g) [trans-1,2-dibromocyclohexane, diaxial chair]

(h) [two enantiomeric epoxides of cyclohexene]

Answers to Examination 2

1.
HC≡CCH₂CH₂CH₂CH₃
1-Hexyne
A

HC≡CCH₂CH(CH₃)₂
4-Methyl-1-pentyne
B

HC≡CCH(CH₃)CH₂CH₃
3-Methyl-1-pentyne
C

HC≡CC(CH₃)₃
3,3-Dimethyl-1-butyne
D

CH₃C≡CCH₂CH₂CH₃
2-Hexyne
E

CH₃C≡CCH(CH₃)₂
4-Methyl-2-pentyne
F

CH₃CH₂C≡CCH₂CH₃
3-Hexyne
G

(a) A, B, C, and D
(b) C

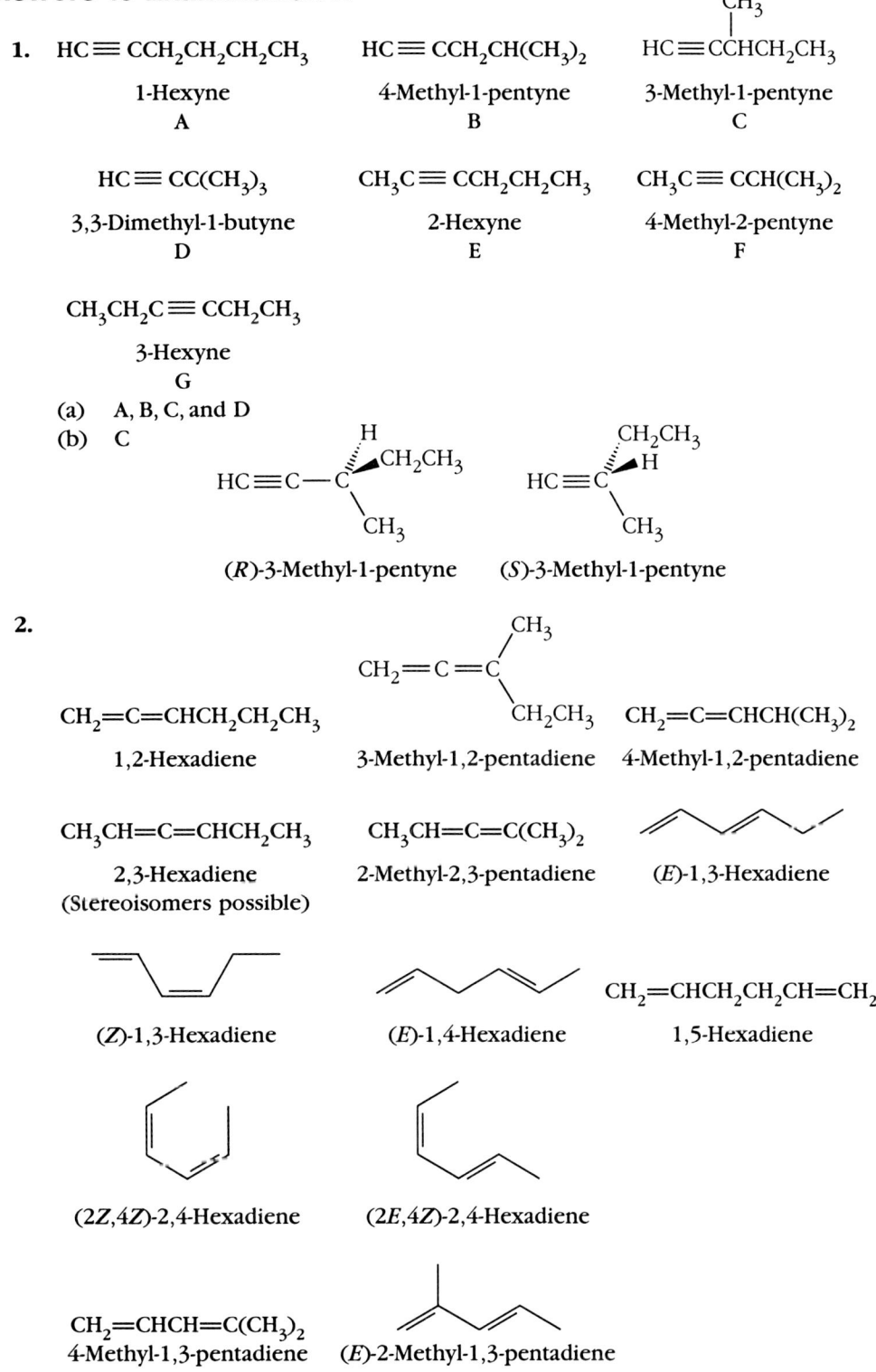

HC≡C—C(H)(CH₂CH₃)(CH₃)
(R)-3-Methyl-1-pentyne

HC≡C—C(CH₂CH₃)(H)(CH₃)
(S)-3-Methyl-1-pentyne

2.

CH₂=C=CHCH₂CH₂CH₃
1,2-Hexadiene

CH₂=C=C(CH₃)(CH₂CH₃)
3-Methyl-1,2-pentadiene

CH₂=C=CHCH(CH₃)₂
4-Methyl-1,2-pentadiene

CH₃CH=C=CHCH₂CH₃
2,3-Hexadiene
(Stereoisomers possible)

CH₃CH=C=C(CH₃)₂
2-Methyl-2,3-pentadiene

(E)-1,3-Hexadiene

(Z)-1,3-Hexadiene

(E)-1,4-Hexadiene

CH₂=CHCH₂CH₂CH=CH₂
1,5-Hexadiene

(2Z,4Z)-2,4-Hexadiene

(2E,4Z)-2,4-Hexadiene

CH₂=CHCH=C(CH₃)₂
4-Methyl-1,3-pentadiene

(E)-2-Methyl-1,3-pentadiene

(Z)-2-Methyl-1,3-pentadiene   (Z)-3-Methyl-1,3-pentadiene

(E)-3-Methyl-1,3-pentadiene   2-Methyl-1,4-pentadiene

$CH_2=CHCH(CH_3)CH=CH_2$
3-Methyl-1,4-pentadiene   2,3-Dimethyl-1,3-butadiene

3. (a)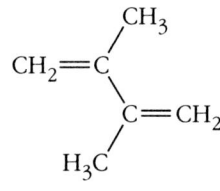

(b) $\xrightarrow{(1)\ NaNH_2/liq.\ NH_3\quad (2)\ CH_3I}$

(c) (cyclopentene)

(d) No reaction

(e) $\xrightarrow{Cold\ KMnO_4/H_2O/KOH}$ or $\xrightarrow{(1)\ OsO_4\quad (2)\ H_2O/KOH}$

(f) benzaldehyde + $HOCH_2CH_2OH$

4. **(a)** Retrosynthesis:

$$CH_3CH_2CH_2CHO \implies CH_3CH_2C\equiv CH \implies HC\equiv C^-Na^+ + CH_3CH_2Br$$

Synthesis:

$$HC\equiv CH \xrightarrow[\text{liq. NH}_3]{\text{NaNH}_2} HC\equiv C^-Na^+ \xrightarrow{CH_3CH_2Br} HC\equiv CCH_2CH_3$$

$$\xrightarrow[\text{(2) H}_2O_2/OH^-]{\text{(1) 9-BBN}} CH_3CH_2CH_2CHO$$

(b) Retrosynthesis:

$$CH_3CH_2COCH_2CH_2CH_3 \implies CH_3CH_2C\equiv CCH_2CH_3 \implies CH_3CH_2Br + HC\equiv CCH_2CH_3$$

$$\implies HC\equiv CH + CH_3CH_2Br$$

Synthesis:

$$HC\equiv CH \xrightarrow[\text{liq. NH}_3]{\text{NaNH}_2} HC\equiv C^-Na^+ \xrightarrow{CH_3CH_2Br} HC\equiv CCH_2CH_3$$

$$\xrightarrow[\text{liq. NH}_3]{\text{NaNH}_2} Na^+{}^-C\equiv CCH_2CH_3 \xrightarrow{CH_3CH_2Br} CH_3CH_2C\equiv CCH_2CH_3$$

$$\xrightarrow[\text{H}_2SO_4]{\text{HgSO}_4/\text{H}_2O} CH_3CH_2COCH_2CH_2CH_3$$

(c) Retrosynthesis:

$$CH_3COCH_2CH_2CH_2CH_3 \implies CH_3C\equiv CCH_2CH_2CH_3 \implies HC\equiv CH + CH_3CH_2CH_2CH_2Br$$

Synthesis:

$HC\equiv CH \xrightarrow[\text{liq. NH}_3]{\text{NaNH}_2} HC\equiv C^-Na^+ \xrightarrow{CH_3CH_2CH_2CH_2Br} HC\equiv CCH_2CH_2CH_2CH_3 \xrightarrow[\text{H}_2\text{SO}_4]{\text{HgSO}_4/\text{H}_2\text{O}} CH_3\overset{O}{\underset{\|}{C}}CH_2CH_2CH_2CH_3$

(d) Retrosynthesis:

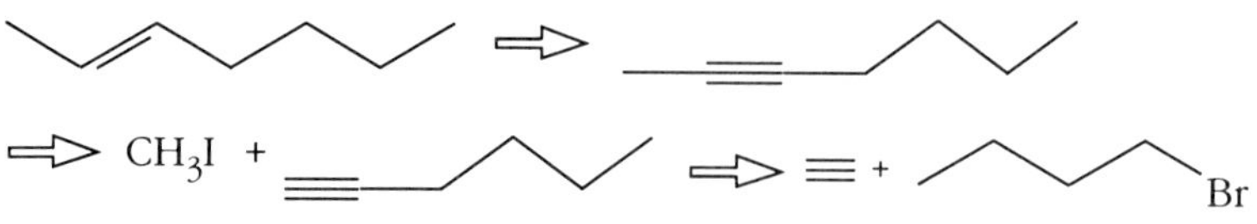

$\diagup\!\!\diagdown\!\!\diagup\!\!\diagdown\!\!\diagup \Rightarrow \diagup\!\!\equiv\!\!\diagdown\!\!\diagup \Rightarrow CH_3I + \equiv\!\!\diagdown\!\!\diagup\!\!\diagdown \Rightarrow \equiv + \diagdown\!\!\diagup\!\!\diagdown\!\!Br$

Synthesis:

$HC\equiv CH \xrightarrow[\text{liq. NH}_3]{\text{NaNH}_2} HC\equiv C^-Na^+ \xrightarrow{CH_3CH_2CH_2CH_2Br}$

$HC\equiv CCH_2CH_2CH_2CH_3 \xrightarrow[\text{liq. NH}_3]{\text{NaNH}_2} Na^{+\ -}C\equiv CCH_2CH_2CH_2CH_3$

$\xrightarrow{CH_3I} CH_3C\equiv CCH_2CH_2CH_2CH_3 \xrightarrow[\text{liq. NH}_3]{\text{Na or Li}} CH_3CH=CHCH_2CH_2CH_2CH_3$

5. (a) None
   (b) None
   (c) AC or AD or CD or CE or DE
   (d) A, B, C, D, and E
   (e) A and E
   (f) AB or BC or BD or BE

6. (a)

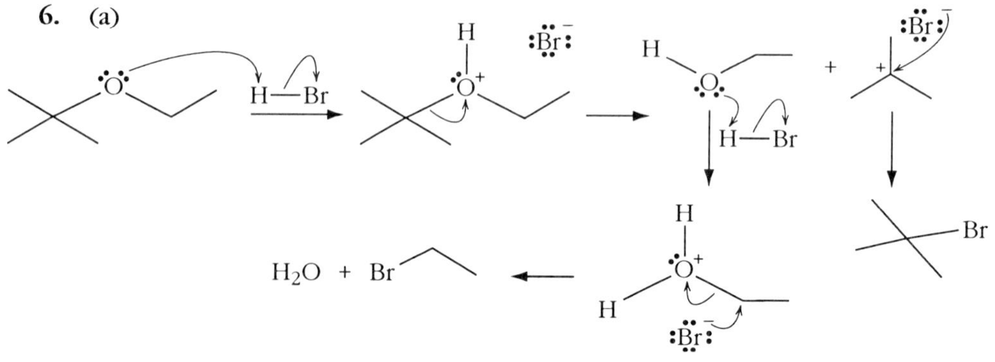

(b)

7.

$$A = (CH_3)_2CHC\overset{O}{\underset{H}{\diagdown}}$$

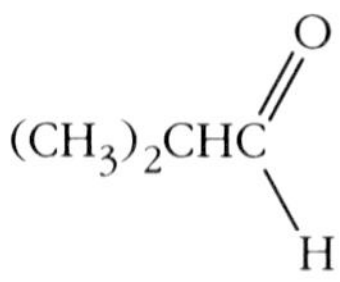

$B = CH_2=CHCH_2CH_2CH_3$

$$C = CH_3\overset{O}{\overset{\|}{C}}OCH_2CH_2CH_3$$

D =

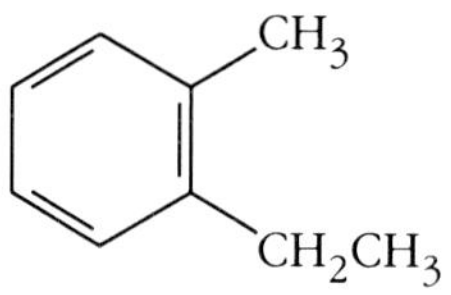

[1,2-disubstituted benzene with CH$_3$ and CH$_2$CH$_3$ groups]

8.

$E = CH_3CH_2CH_2OSO_2$—[p-tolyl]—CH_3

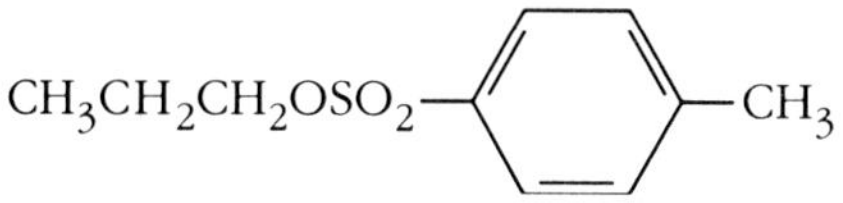

$F = CH_3CH_2CH_2$—^{18}O—CH_3
$G = CH_3CH_2CH_2OCH_3$
$H = CH_3CH_2CH_2Cl$
$I = CH_3CH_2CH_2MgCl$

$$J = CH_3CH_2CH_2\underset{OH}{CH}CH_3$$

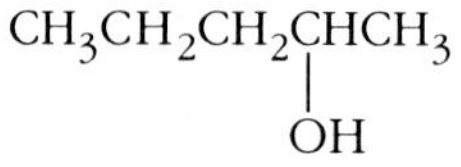

$$K = CH_3CH_2CH_2\overset{O}{\overset{\|}{C}}CH_3$$

Answers to Examination 3

1. (a) (R)-(Z)-5-Chloro-4-hexen-2-ol
   (b) 4-Bromo-3-nitroaniline
   (c) (S)-2-Chloro-1-cyclopentyl-2-methyl-1-butanone
   (d) 3-Oxooctanoic acid

2. $CH_3C\equiv CH$ < cyclohexanol < 3,4-dimethoxybenzoic acid < benzoic acid < CCl_3COOH

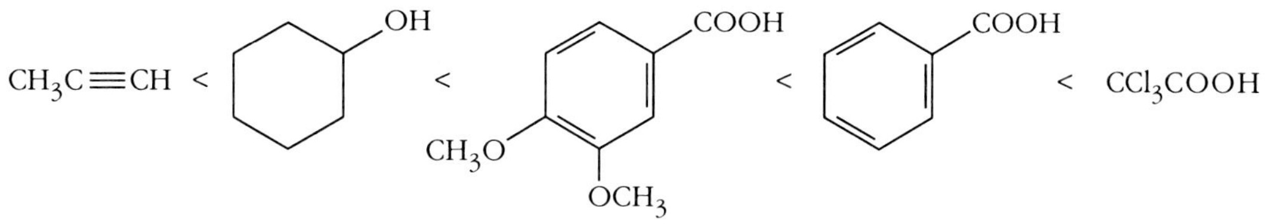

3.
(a) cyclohexyl-1,3-dioxolane

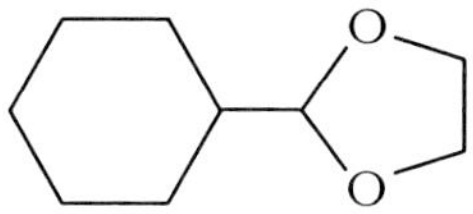

(b) 2-furyl-CH(OH)-C≡N

(c) H_2N-NH-phenyl, Acid catalyst

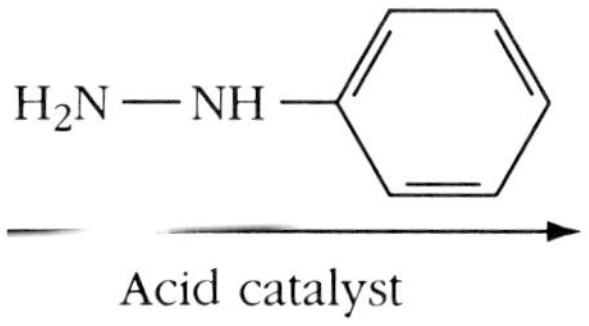

(d) PhCOOH or PhCOCl or PhCOOR' or PhCHO

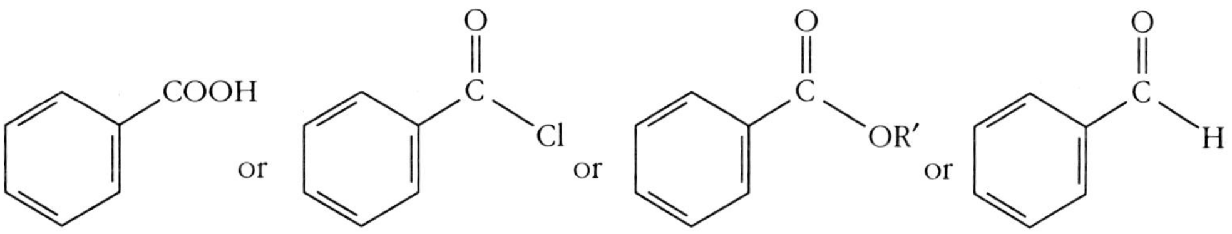

(e) CH_3COCl / $AlCl_3$

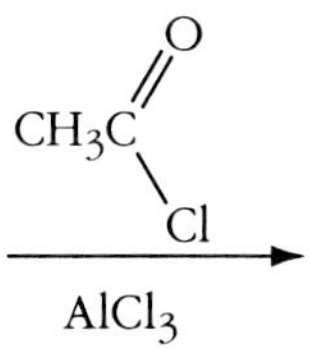

(f) 2-methylcyclohexanone

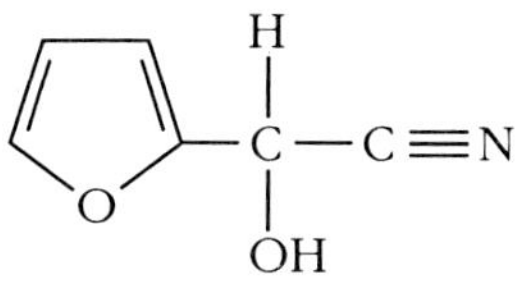

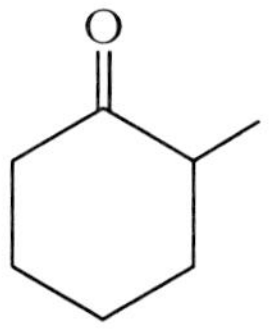

4.

$$CH_3\overset{\ddot{\ddot{O}}:}{\underset{H}{C}} + H_2\ddot{N}CH_3 \rightleftharpoons CH_3\overset{:\ddot{\ddot{O}}:^-}{\underset{\underset{H}{\overset{|}{N}^+}-CH_3}{C-H}} \overset{H-B}{\longrightarrow} \rightleftharpoons CH_3\overset{:\ddot{O}H}{\underset{\underset{H}{\overset{|}{N}}-CH_3}{C-H}} \overset{H-B}{\longrightarrow}$$

$$CH_3\overset{H}{\underset{\ddot{N}-CH_3}{C}} \rightleftharpoons CH_3\overset{H}{\underset{\underset{H}{\overset{|}{N}^+}-CH_3}{C}} \rightleftharpoons CH_3\overset{\overset{H}{\underset{}{\overset{+}{O}}}\overset{H}{}}{\underset{\underset{H}{\overset{|}{N}}-CH_3}{C-H}}$$

5. (a) cyclohexene $\xrightarrow{HBr}$ bromocyclohexane $\xrightarrow[\text{(2) CO}_2]{\text{(1) Mg/dry ether}}$ cyclohexanecarboxylic acid
    (3) H_3O^+

(b)

$CH_3C\equiv CH \xrightarrow[\text{liq. NH}_3]{NaNH_2} CH_3C\equiv C^-Na^+ \xrightarrow{CH_3I} CH_3C\equiv CCH_3 \xrightarrow[H_2SO_4]{H_2O/HgSO_4}$

$CH_3\overset{O}{\underset{\|}{C}}CH_2CH_3$

(c) benzene $\xrightarrow[\text{AlCl}_3]{CH_3COCl}$ acetophenone $\xrightarrow[\text{FeBr}_3]{Br_2}$ 3-bromoacetophenone $\xrightarrow[\text{HCl}]{Zn(Hg)}$ 3-bromotoluene

6. (a) C₆H₅−CH(OH)−CH₃

1-phenylethanol (structure: benzene ring with −CH(OH)CH₃ substituent)

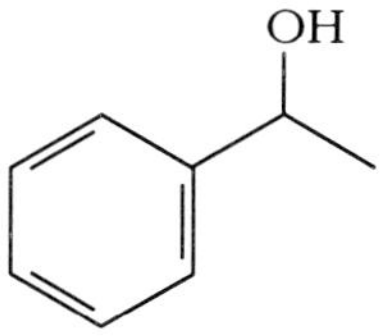

(b) H₃C−N(CH₂CH₂OH)₂

N-methyldiethanolamine

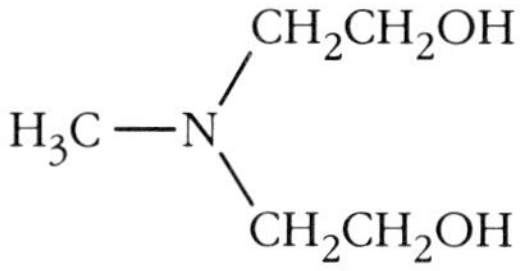

Answers to Examination 4

1.

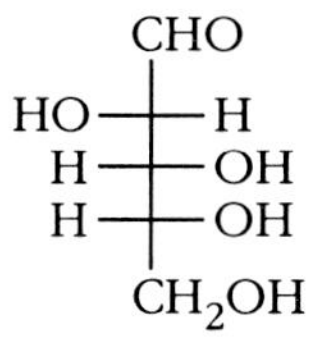

```
        CHO
   HO ──┼── H
    H ──┼── OH
    H ──┼── OH
       CH₂OH
```

Fischer projection:
- CHO (top)
- HO—H
- H—OH
- H—OH
- CH$_2$OH (bottom)

2. (a)

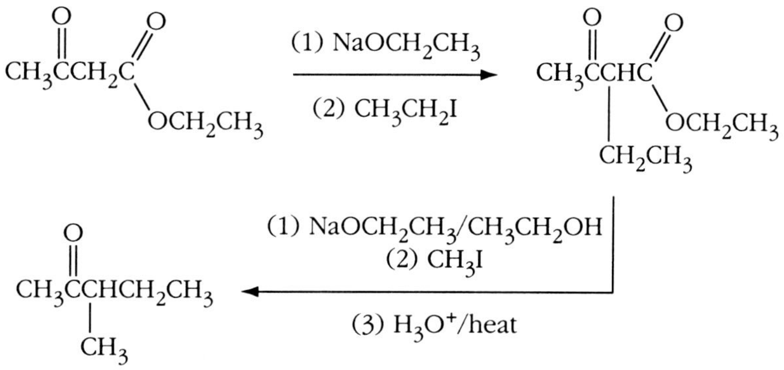

$$CH_3\overset{O}{\underset{\|}{C}}CH_2\overset{O}{\underset{\|}{C}}OCH_2CH_3 \xrightarrow[\text{(2) } CH_3CH_2I]{\text{(1) NaOCH}_2CH_3} CH_3\overset{O}{\underset{\|}{C}}CH\underset{CH_2CH_3}{\overset{O}{\underset{\|}{C}}}OCH_2CH_3$$

$$CH_3\overset{O}{\underset{\|}{C}}CH(CH_3)CH_2CH_3 \xleftarrow[\substack{\text{(2) } CH_3I \\ \text{(3) } H_3O^+/\text{heat}}]{\text{(1) NaOCH}_2CH_3/CH_3CH_2OH}$$

Product: $CH_3\overset{O}{\underset{\|}{C}}CHCH_2CH_3$ with CH_3 substituent

(b)

$$CH_2(COOCH_2CH_3)_2 \xrightarrow[\text{(2) } PhCH_2Cl]{\text{(1) NaOCH}_2CH_3/CH_3CH_2OH} PhCH_2CH(COOCH_2CH_3)_2$$

$$\xrightarrow[\substack{\text{(2) } CH_3I \\ \text{(3) } H_3O^+/\text{heat}}]{\text{(1) NaOCH}_2CH_3/CH_3CH_2OH} PhCH_2\underset{CH_3}{CH}COOH$$

$$\xrightarrow[H_2SO_4]{\text{Excess } CH_3OH} PhCH_2\underset{CH_3}{CH}\overset{O}{\underset{\|}{C}}OCH_3$$

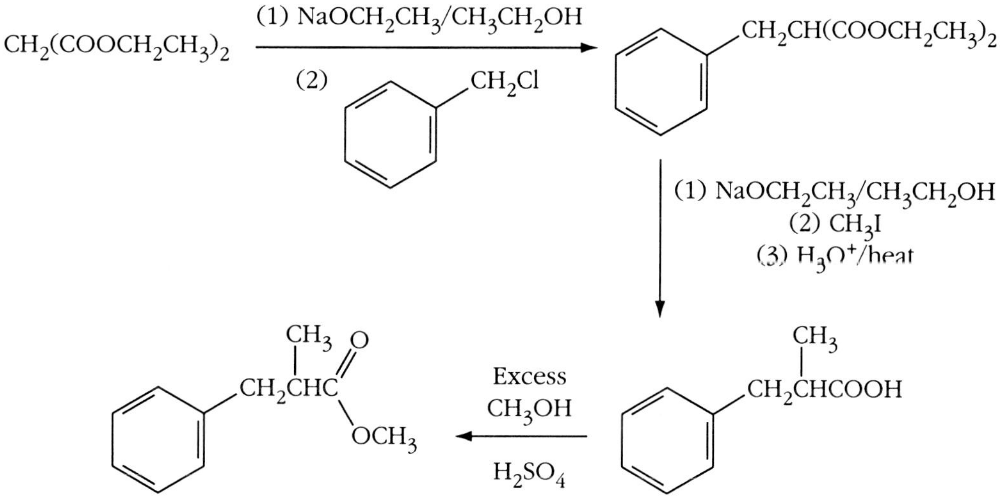

3.

(a) 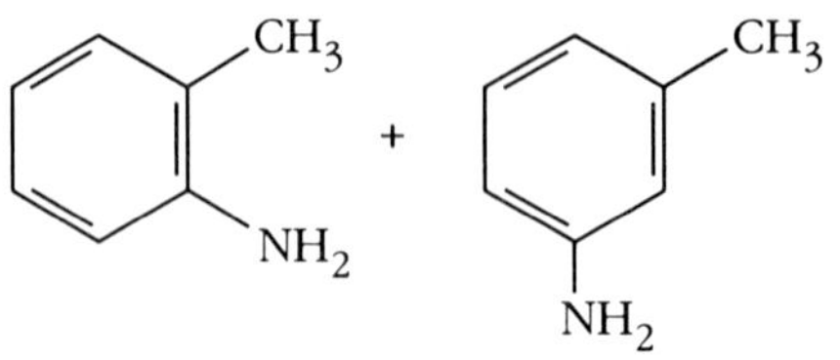 2-methylaniline (o-toluidine) + 3-methylaniline (m-toluidine)

(b) 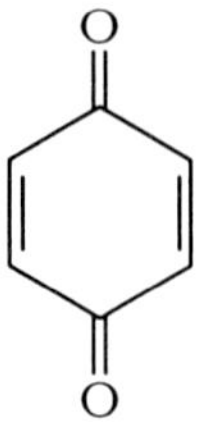 1,4-benzoquinone

(c) 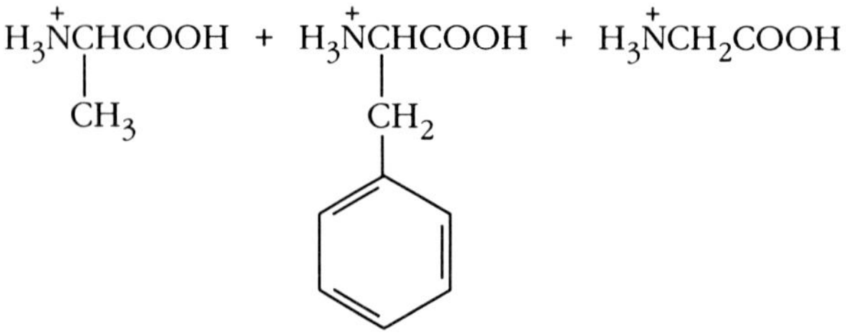 $H_3\overset{+}{N}CHCOOH$ + $H_3\overset{+}{N}CHCOOH$ + $H_3\overset{+}{N}CH_2COOH$
      | |
      CH_3 CH_2
                             |
                             C_6H_5

(d) δ-valerolactone (6-membered ring lactone: CH_2–C(=O)–O–CH_2CH_2–CH_2)

(e) methyl 2,3,4,6-tetra-O-methyl-glucopyranoside

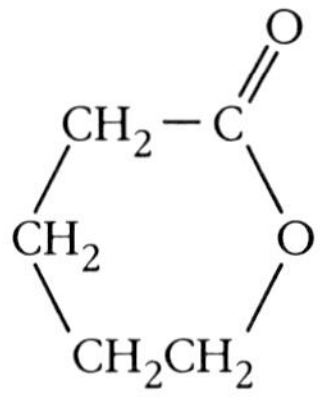

(f) glucopyranose (with CH$_2$OH, HO, HO, OH, OH substituents) + $C_6H_5CH_2OH$

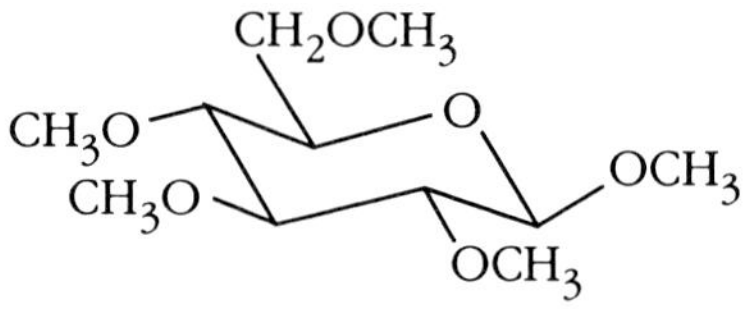

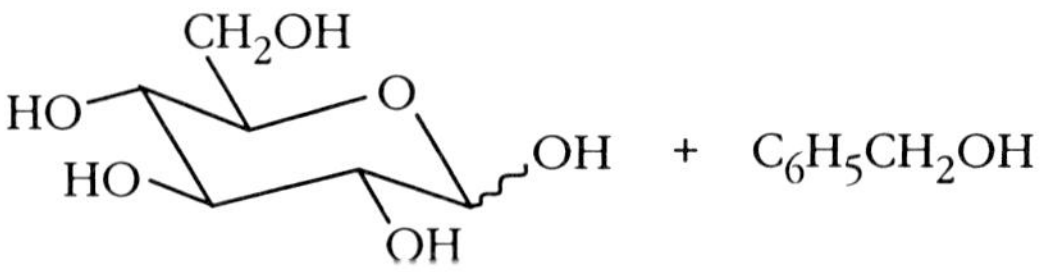

(g) 2-(2-methylallyl)phenol: ortho-OH phenyl–CH_2–C(=CH_2)–CH_3

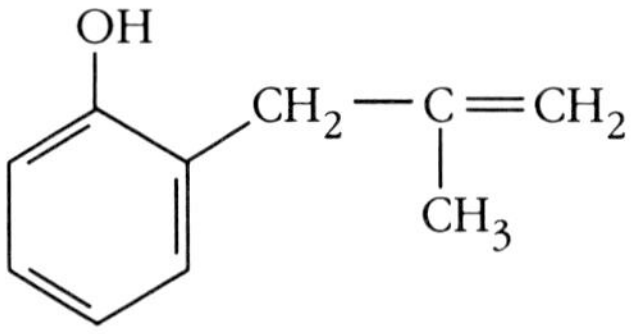

(h) 1-(1-pyrrolidinyl)cyclohexene

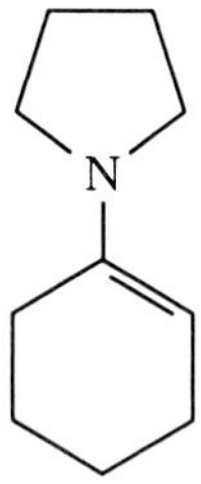

(i) 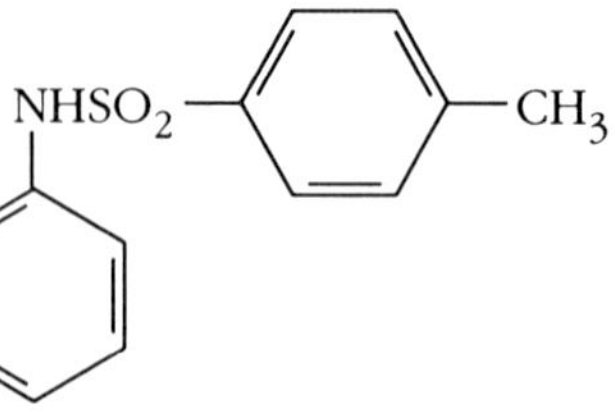 C_6H_5–NHSO$_2$–C$_6H_4$–CH_3 (para) [N-phenyl-p-toluenesulfonamide]

(j) 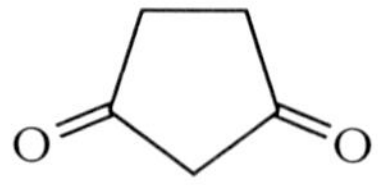 1,3-cyclopentanedione

4.

(a) [Mechanism showing nucleophilic aromatic substitution of 4-fluoronitrobenzene with methylamine, proceeding through Meisenheimer-like intermediate to give p-nitro-N-methylanilinium fluoride]

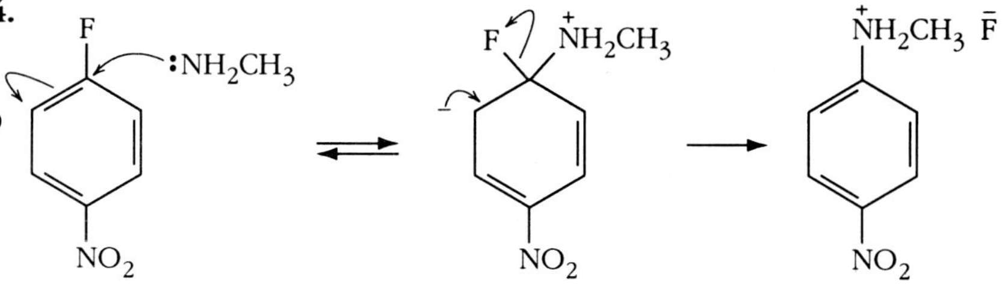

(b) $[(CH_3)_2CH]_2\ddot{N}^-\;\;\;\;[(CH_3)_2CH]_2NH$

$N\equiv CCH_2-H \;\;\rightleftharpoons\;\; N\equiv C-CH_2^- \;+\; (CH_3CH_2)_2C=O \;\longrightarrow$

$N\equiv CCH_2-C(CH_2CH_3)_2-O^- \;+\; H-O^+H_2 \;\longrightarrow\; N\equiv CCH_2-C(CH_2CH_3)_2-OH$

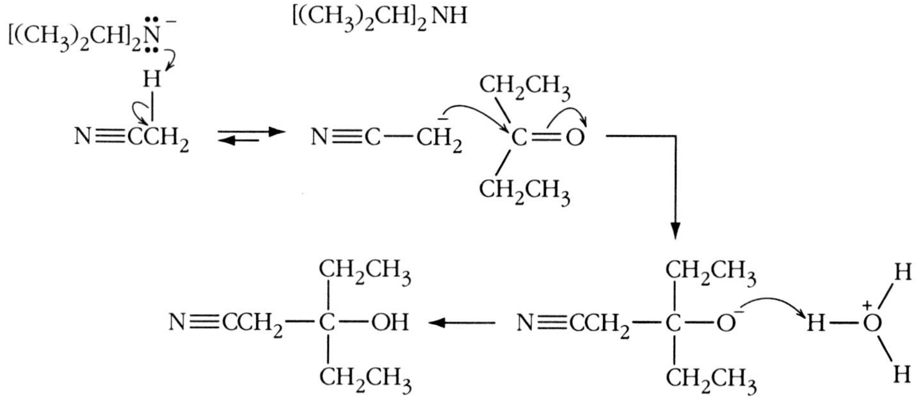

5.

(a) $CH_3CH_2CH_2NO_2$

(b) $N\equiv CCH_2C(=O)OCH_2CH_3$

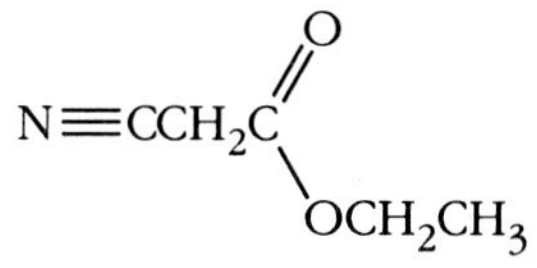